D1423476

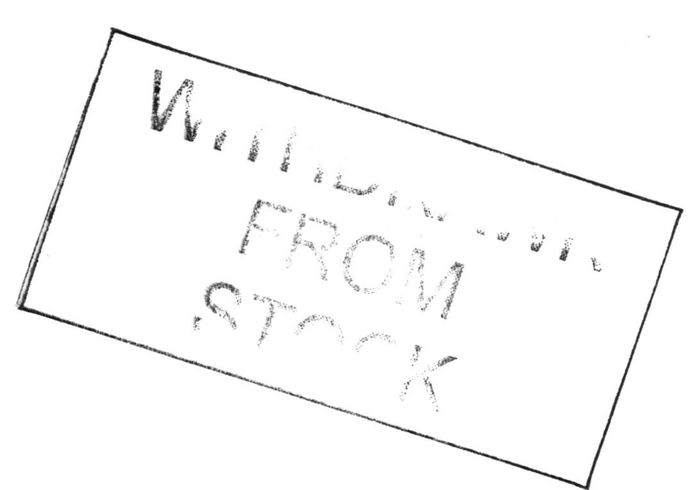
FROM

5 092 981 X

# The Regional Impacts of Climate Change

## An Assessment of Vulnerability

Edited by

**Robert T. Watson**
*The World Bank*

**Marufu C. Zinyowera**
*Zimbabwe Meteorological Services*

**Richard H. Moss**
*Battelle Pacific Northwest National Laboratory*

*Project Administrator*
**David J. Dokken**

A Special Report of IPCC Working Group II

*Published for the Intergovernmental Panel on Climate Change*

PUBLISHED BY THE PRESS SYNDICATE OF THE UNIVERSITY OF CAMBRIDGE
The Pitt Building, Trumpington Street, Cambridge CB2 1RP, United Kingdom

CAMBRIDGE UNIVERSITY PRESS
The Edinburgh Building, Cambridge CB2 2RU, UK http://www.cup.cam.ac.uk
40 West 20th Street, New York, NY 10011-4211, USA http://www.cup.org
10 Stamford Road, Oakleigh, Melbourne 3166, Australia



First published 1998

Printed in the United States of America

*A catalog record for this book is available from the British Library*

*Library of Congress Cataloging-in-Publication Data available*

ISBN 0-521- 632560 Hardback
ISBN 0-521- 634555 Paperback

*Photo Credits* – Cover imagery derived from the following sources: 'The World Bank' CD-ROMs (Aztech New Media Corporation); 'Photo Gallery' CD-ROMs (SoftKey International, Inc.); and the 'Earth, Air, Fire, and Water' CD-ROM (MediaRights, Inc.).
*Cover Art Designer* – Mark Sutton, Global Vision Works.

# Contents

# Foreword

The Intergovernmental Panel on Climate Change (IPCC) was jointly established by the World Meteorological Organization and the United Nations Environment Programme in 1988 to assess the scientific and technical literature on climate change, the potential impacts of changes in climate, and options for adaption to and mitigation of climate change. Since its inception, the IPCC has produced a series of Assessment Reports, Special Reports, Technical Papers, methodologies, and other products which have become standard works of reference, widely used by policymakers, scientists, and other experts.

This Special Report, which has been produced by Working Group II of the IPCC, builds on the Working Group's contribution to the Second Assessment Report (SAR), and incorporates more recent information made available since mid-1995. It has been prepared in response to a request from the Subsidiary Body for Scientific and Technological Advice (SBSTA) of the UN Framework Convention on Climate Change (UNFCCC). It addresses an important question posed by the Conference of the Parties (COP) to the UNFCCC, namely, the degree to which human conditions and the natural environment are vulnerable to the potential effects of climate change. The report establishes a common base of information regarding the potential costs and benefits of climatic change, including the evaluation of uncertainties, to help the COP determine what adaptation and mitigation measures might be justified. The report consists of vulnerability assessments for 10 regions that comprise the Earth's entire land surface and adjoining coastal seas: Africa, Arid Western Asia (including the Middle East), Australasia, Europe, Latin America, North America, the Polar Regions (The Arctic and the Antarctic), Small Island States, Temperate Asia, and Tropical Asia. It also includes several annexes that provide information about climate observations, climate projections, vegetation distribution projections, and socioeconomic trends.

As usual in the IPCC, success in producing this report has depended on the enthusiasm and cooperation of numerous scientists and other experts worldwide. These individuals have given generously of their time, often going beyond reasonable demands of duty. We applaud, admire, and are grateful for their commitment to the IPCC process. We are pleased to note the continuing efforts made by the IPCC to ensure participation of scientists and other experts from the developing countries and countries with economies in transition. Given the regional focus of this report, their participation was especially essential to its successful completion. We also express our thanks to the many governments, including those in the developing regions and regions with economies in transition, that supported these scientists and experts in their work.

We take this opportunity to express our gratitude to the following individuals for nurturing another IPCC report through to completion:

- Professor B. Bolin, the Chairman of the IPCC
- The Co-Chairs of Working Group II, Dr. R.T. Watson (USA) and Dr. M.C. Zinyowera (Zimbabwe)
- The Vice-Chairs of the Working Group, Dr. M. Beniston (Switzerland), Dr. O. Canziani (Argentina), Dr. J. Friaa (Tunisia), Ing. (Mrs.) M. Perdomo (Venezuela), Dr. S.K. Sharma (India), Mr. H. Tsukamoto (Japan), and Professor P. Vellinga (The Netherlands)
- Dr. R.H. Moss, Head of the Technical Support Unit (TSU) of Working Group II, Mr. D.J. Dokken, the Project Administrator, and the other members of the TSU, including Ms. S. MacCracken, Ms. L. Van Wie McGrory, and Ms. F. Ormond
- Dr. N. Sundararaman, the Secretary of the IPCC, and his staff, including Ms. R. Bourgeois, Ms. C. Ettori, and Ms. C. Tanikie.

**G.O.P. Obasi**

Secretary-General
World Meteorological Organization

**Ms. E. Dowdeswell**

Executive Director
United Nations Environment Programme

# Preface

The Intergovernmental Panel on Climate Change (IPCC) has produced a series of Assessment Reports, Special Reports, Technical Papers, and methodologies. As an intergovernmental body, the IPCC has procedures governing the production of each of these. This Special Report on the Regional Impacts of Climate Change was first requested by the Subsidiary Body for Scientific and Technological Advice (SBSTA) of the Conference of the Parties (COP) to the United Nations Framework Convention on Climate Change (UNFCCC) as a Technical Paper, which restricted the authors to using only materials already in IPCC Assessment Reports and Special Reports. In the course of drafting the paper, the authors felt that the inclusion of new literature that had become available since the completion of the IPCC Second Assessment Report (SAR), including work undertaken under the auspices of several "country studies programs," would make the paper more complete, up-to-date, and broadly representative of trends and vulnerabilities in the regions. Including these materials in the report would not have conformed to the IPCC procedures for Technical Papers; hence, the IPCC decided at its Twelfth Session (Mexico City, 11–13 September 1996) to rewrite the Technical Paper as a Special Report, and SBSTA was informed accordingly.

The Special Report explores the potential consequences of changes in climate for ten continental- or subcontinental-scale regions. Because of the uncertainties associated with regional projections of climate change, the report necessarily takes the approach of assessing sensitivities and vulnerabilities of each region, rather than attempting to provide quantitative predictions of the impacts of climate change. As in the SAR, "vulnerability" is the extent to which climate change may damage or harm a system; it is a function of both sensitivity to climate and the ability to adapt to new conditions.

This assessment confirms the findings of the SAR and underlines the potential for climate change to alter the ability of the Earth's physical and biological systems (land, atmosphere, and oceans) to provide goods and services essential for sustainable economic development.

The report represents an important step in the evolution of the impact assessment process for the IPCC. Previous impact assessments have examined the potential effects of climate change primarily at a global scale. This report analyzes impacts at a continental or subcontinental scale that is of more practical interest to decisionmakers. This regional approach reveals wide variation in the vulnerability of different populations and environmental systems. This variation stems from differences in local environmental conditions; economic, social, and political conditions; and degrees of dependence on climate-sensitive resources, among other factors. Because of its smaller scale of analysis, the report provides more information regarding the potential for the adaptation of systems, activities, and infrastructure to climate change than did the SAR. The chapters indicate, however, that far more research and analysis of adaptation options and adjustment processes are necessary if private sector and governmental entities are to make climate-sensitive sectors more resilient to today's climate variability, and to limit damage from—or take advantage of—potential long-term changes in climate.

The report is also an initial step in examining how projected changes in climate could interact with other environmental changes (e.g., biodiversity loss, land degradation, stratospheric ozone depletion, and degradation of water resources) and social trends (e.g., population growth, economic development, and technological progress). The assessment indicates that additional research into the interlinkages among environmental issues also is needed.

This report will provide a foundation for impacts assessment in the Third Assessment Report (TAR), which is expected to be completed in late 2000. An important early step in the process of preparing the IPCC TAR will be to review and refine the approach—and the regional groupings—used in this assessment. In doing so, advances in the ability to project climatic and environmental changes on finer scales will be an important consideration. The report provides a foundation for the TAR in another important respect, as it represents a substantial further step forward in increasing the level of participation of scientists and technical experts from developing countries and countries with economies in transition. The IPCC remains committed to building on this accomplishment, and will not relax its efforts to identify experts from these regions and secure their participation in future assessments.

*Acknowledgments*

We would like to acknowledge the contributions of numerous individuals and organizations to the successful completion of this report. First and foremost, we are grateful for the voluntary efforts of the members of the scientific and technical community who prepared and peer-reviewed the chapters and annexes of the report. These individuals served in several capacities, including Convening Lead Authors, Lead Authors, Contributors/Reviewers, Regional Coordinators, and Sector Contributors (authors of the SAR who extracted regional information from their sector-oriented chapters as starting points for the regional assessments). We also gratefully acknowledge the assistance provided by governments to a number of these lead authors.

All of these contributions would have come to nothing had it not been for the tireless and good-natured efforts of David Jon Dokken, Project Administrator, whose roles and responsibilities in preparation of this report are too numerous to mention, and without whom the report would not have been assembled in such a timely and efficient fashion. Other members of the Working Group II Technical Support Unit also provided significant help in preparation of the report, including Sandy MacCracken, Laura Van Wie McGrory, and Flo Ormond. The staff of the IPCC Secretariat, including Rudie Bourgeois, Chantal Ettori, and Cecilia Tanikie, provided essential support and welcome advice.

Others who contributed to the report in various analytical and organizational roles and to whom we wish to express our thanks include Tererei Abete, Isabel Alegre, Ron Benioff, Carroll Curtis, Paul Desanker, Robert Dixon and his colleagues at the U.S. Country Studies Program, Roland Fuchs, Christy Goodale, David Gray, Mike Hulme, Jennifer Jenkins, Richard Klein, S.C. Majumdar, Scott Ollinger, Erik Rodenberg, Robert Scholes, Joel Smith, Regina Tannon, David Theobald, and Hassan Virji.

**Bert Bolin**
**Robert Watson**
**Marufu Zinyowera**
**Narasimhan Sundararaman**
**Richard Moss**

# Summary for Policymakers

# The Regional Impacts of Climate Change: An Assessment of Vulnerability

*A Special Report of Working Group II of the Intergovernmental Panel on Climate Change*

# CONTENTS

## 1. Scope of the Assessment

This report has been prepared at the request of the Conference of the Parties to the United Nations Framework Convention on Climate Change (UNFCCC) and its subsidiary bodies (specifically, the Subsidiary Body for Scientific and Technological Advice—SBSTA). The special report provides, on a regional basis, a review of state-of-the-art information on the vulnerability to potential changes in climate of ecological systems, socioeconomic sectors (including agriculture, fisheries, water resources, and human settlements), and human health. The report reviews the sensitivity of these systems as well as options for adaptation. Though this report draws heavily upon the sectoral impact assessments of the Second Assessment Report (SAR), it also draws upon more recent peer-reviewed literature (*inter alia*, country studies programs).

## 2. Nature of the Issue

Human activities (primarily the burning of fossil fuels and changes in land use and land cover) are increasing the atmospheric concentrations of greenhouse gases, which alter radiative balances and tend to warm the atmosphere, and, in some regions, aerosols—which have an opposite effect on radiative balances and tend to cool the atmosphere. At present, in some locations primarily in the Northern Hemisphere, the cooling effects of aerosols can be large enough to more than offset the warming due to greenhouse gases. Since aerosols do not remain in the atmosphere for long periods and global emissions of their precursors are not projected to increase substantially, aerosols will not offset the global long-term effects of greenhouse gases, which are long-lived. Aerosols can have important consequences for continental-scale patterns of climate change.

These changes in greenhouse gases and aerosols, taken together, are projected to lead to regional and global changes in temperature, precipitation, and other climate variables—resulting in global changes in soil moisture, an increase in global mean sea level, and prospects for more severe extreme high-temperature events, floods, and droughts in some places. Based on the range of sensitivities of climate to changes in the atmospheric concentrations of greenhouse gases (IPCC 1996, WG I) and plausible changes in emissions of greenhouse gases and aerosols (IS92a-f, scenarios that assume no climate policies), climate models project that the mean annual global surface temperature will increase by 1–3.5°C by 2100, that global mean sea level will rise by 15–95 cm, and that changes in the spatial and temporal patterns of precipitation would occur. The average rate of warming probably would be greater than any seen in the past 10,000 years, although the actual annual to decadal rate would include considerable natural variability, and regional changes could differ substantially from the global mean value. These long-term, large-scale, human-induced changes will interact with natural variability on time scales of days to decades [e.g., the El Niño-Southern Oscillation (ENSO) phenomenon] and thus influence social and economic well-being. Possible local climate effects which are due to unexpected events like a climate change-induced change of flow pattern of marine water streams like the Gulf Stream have not been considered, because such changes cannot be predicted with confidence at present.

Scientific studies show that human health, ecological systems, and socioeconomic sectors (e.g., hydrology and water resources, food and fiber production, coastal systems, and human settlements), all of which are vital to sustainable development, are sensitive to changes in climate—including both the magnitude and rate of climate change—as well as to changes in climate variability. Whereas many regions are likely to experience adverse effects of climate change—some of which are potentially irreversible—some effects of climate change are likely to be beneficial. Climate change represents an important additional stress on those systems already affected by increasing resource demands, unsustainable management practices, and pollution, which in many cases may be equal to or greater than those of climate change. These stresses will interact in different ways across regions but can be expected to reduce the ability of some environmental systems to provide, on a sustained basis, key goods and services needed for successful economic and social development, including adequate food, clean air and water, energy, safe shelter, low levels of disease, and employment opportunities. Climate change also will take place in the context of economic development, which may make some groups or countries less vulnerable to climate change—for example, by increasing the resources available for adaptation; those that experience low rates of growth, rapid increases in population, and ecological degradation may become increasingly vulnerable to potential changes.

## 3. Approach of the Assessment

This report assesses the vulnerability of natural and social systems of major regions of the world to climate change. Vulnerability is defined as the extent to which a natural or social system is susceptible to sustaining damage from climate change. Vulnerability is a function of the sensitivity of a system to changes in climate (the degree to which a system will respond to a given change in climate, including both beneficial and harmful effects) and the ability to adapt the system to changes in climate (the degree to which adjustments in practices, processes, or structures can moderate or offset the potential for damage or take advantage of opportunities created, due to a given change in climate). Under this framework, a highly vulnerable system would be one that is highly sensitive to modest changes in climate, where the sensitivity includes the potential for substantial harmful effects, and one for which the ability to adapt is severely constrained.

Because the available studies have not employed a common set of climate scenarios and methods, and because of uncertainties regarding the sensitivities and adaptability of natural and social systems, the assessment of regional vulnerabilities is necessarily qualitative. However, the report provides substantial and indispensable information on what currently is known about vulnerability to climate change.

In a number of instances, quantitative estimates of impacts of climate change are cited in the report. Such estimates are dependent upon the specific assumptions employed regarding future changes in climate, as well as upon the particular methods and models applied in the analyses. To interpret these estimates, it is important to bear in mind that uncertainties regarding the character, magnitude, and rates of future climate change remain. These uncertainties impose limitations on the ability of scientists to project impacts of climate change, particularly at regional and smaller scales.

It is in part because of the uncertainties regarding how climate will change that this report takes the approach of assessing vulnerabilities rather than assessing quantitatively the expected impacts of climate change. The estimates are best interpreted as illustrative of the potential character and approximate magnitudes of impacts that may result from specific scenarios of climate change. They serve as indicators of sensitivities and possible vulnerabilities. Most commonly, the estimates are based upon changes in equilibrium climate that have been simulated to result from an equivalent doubling of carbon dioxide ($CO_2$) in the atmosphere. Usually the simulations have excluded the effects of aerosols. Increases in global mean temperatures corresponding to these scenarios mostly fall in the range of 2–5°C. To provide a temporal context for these scenarios, the range of projected global mean warming by 2100 is 1–3.5°C accompanied by a mean sea-level rise of 15–95 cm, according to the IPCC Second Assessment Report. General circulation model (GCM) results are used in this analysis to justify the order of magnitude of the changes used in the sensitivity analyses. They are not predictions that climate will change by specific magnitudes in particular countries or regions. The amount of literature available for assessment varies in quantity and quality among the regions.

## 4. Overview of Regional Vulnerabilities to Global Climate Change

Article 2 of the UNFCCC explicitly acknowledges the importance of natural ecosystems, food production, and sustainable economic development (see Box 1). This report's assessment of regional vulnerability to climate change focuses on ecosystems, hydrology and water resources, food and fiber production, coastal systems, human settlements, human health, and other sectors or systems (including the climate system) important to 10 regions that encompass the Earth's land surface. Wide variation in the vulnerability of similar sectors or systems is to be expected across regions, as a consequence of regional differences in local environmental conditions, preexisting stresses to ecosystems, current resource-use patterns, and the framework of factors affecting decisionmaking—including government policies, prices, preferences, and values. Nonetheless, some general observations, based on information contained in the SAR and synthesized from the regional analyses in this assessment, provide a global context for the assessment of each region's vulnerability.

**Box 1. Article 2 of the UNFCCC: Objective**

The ultimate objective of this Convention and any related legal instruments that the Conference of the Parties may adopt is to achieve, in accordance with the relevant provisions of the Convention, stabilization of greenhouse gas concentrations in the atmosphere at a level that would prevent dangerous anthropogenic interference with the climate system. Such a level should be achieved within a time-frame sufficient to allow ecosystems to adapt naturally to climate change, to ensure that food production is not threatened and to enable economic development to proceed in a sustainable manner.

### 4.1. *Ecosystems*

Ecosystems are of fundamental importance to environmental function and to sustainability, and they provide many goods and services critical to individuals and societies. These goods and services include: (i) providing food, fiber, fodder, shelter, medicines, and energy; (ii) processing and storing carbon and nutrients; (iii) assimilating wastes; (iv) purifying water, regulating water runoff, and moderating floods; (v) building soils and reducing soil degradation; (vi) providing opportunities for recreation and tourism; and (vii) housing the Earth's entire reservoir of genetic and species diversity. In addition, natural ecosystems have cultural, religious, aesthetic, and intrinsic existence values. Changes in climate have the potential to affect the geographic location of ecological systems, the mix of species that they contain, and their ability to provide the wide range of benefits on which societies rely for their continued existence. Ecological systems are intrinsically dynamic and are constantly influenced by climate variability. The primary influence of anthropogenic climate change on ecosystems is expected to be through the rate and magnitude of change in climate means and extremes—climate change is expected to occur at a rapid rate relative to the speed at which ecosystems can adapt and reestablish themselves—and through the direct effects of increased atmospheric $CO_2$ concentrations, which may increase the productivity and efficiency of water use in some plant species. Secondary effects of climate change involve changes in soil characteristics and disturbance regimes (e.g., fires, pests, and diseases), which would favor some species over others and thus change the species composition of ecosystems.

Based on model simulations of vegetation distribution, which use GCM-based climate scenarios, large shifts of vegetation boundaries into higher latitudes and elevations can be expected. The mix of species within a given vegetation class likely will change. Under equilibrium GCM climate scenarios, large regions show drought-induced declines in vegetation, even when the direct effects of $CO_2$ fertilization are included. By comparison, under transient climate scenarios—in which trace gases increase slowly over a period of years—the full effects of changes in temperature and precipitation lag the effects of a change in atmospheric composition by a number of decades; hence, the positive effects of $CO_2$ precede the full effects of changes in climate.

Climate change is projected to occur at a rapid rate relative to the speed at which forest species grow, reproduce, and reestablish themselves (past tree species' migration rates are believed to be on the order of 4–200 km per century). For mid-latitude regions, an average warming of 1–3.5°C over the next 100 years would be equivalent to a poleward shift of the present geographic bands of similar temperatures (or "isotherms") approximately 150–550 km, or an altitude shift of about 150–550 m. Therefore, the species composition of forests is likely to change; in some regions, entire forest types may disappear, while new assemblages of species and hence new ecosystems may be established. As a consequence of possible changes in temperature and water availability under doubled equivalent-$CO_2$ equilibrium conditions, a substantial fraction (a global average of one-third, varying by region from one-seventh to two-thirds) of the existing forested area of the world likely would undergo major changes in broad vegetation types—with the greatest changes occurring in high latitudes and the least in the tropics. In tropical rangelands, major alterations in productivity and species composition would occur due to altered rainfall amount and seasonality and increased evapotranspiration, although a mean temperature increase alone would not lead to such changes.

Inland aquatic ecosystems will be influenced by climate change through altered water temperatures, flow regimes, water levels, and thawing of permafrost at high latitudes. In lakes and streams, warming would have the greatest biological effects at high latitudes—where biological productivity would increase and lead to expansion of cool-water species' ranges—and at the low-latitude boundaries of cold- and cool-water species ranges, where extinctions would be greatest. Increases in flow variability, particularly the frequency and duration of large floods and droughts, would tend to reduce water quality, biological productivity, and habitat in streams. The geographical distribution of wetlands is likely to shift with changes in temperature and precipitation, with uncertain implications for net greenhouse gas emissions from non-tidal wetlands. Some coastal ecosystems (saltwater marshes, mangrove ecosystems, coastal wetlands, coral reefs, coral atolls, and river deltas) are particularly at risk from climate change and other stresses. Changes in these ecosystems would have major negative effects on freshwater supplies, fisheries, biodiversity, and tourism.

Adaptation options for ecosystems are limited, and their effectiveness is uncertain. Options include establishment of corridors to assist the "migration" of ecosystems, land-use management, plantings, and restoration of degraded areas. Because of the projected rapid rate of change relative to the rate at which species can reestablish themselves, the isolation and fragmentation of many ecosystems, the existence of multiple stresses (e.g., land-use change, pollution), and limited adaptation options, ecosystems (especially forested systems, montane systems, and coral reefs) are vulnerable to climate change.

## 4.2. *Hydrology and Water Resources*

Water availability is an essential component of welfare and productivity. Currently, 1.3 billion people do not have access to adequate supplies of safe water, and 2 billion people do not have access to adequate sanitation. Although these people are dispersed throughout the globe—reflecting sub-national variations in water availability and quality—some 19 countries (primarily in the Middle East and north and southern Africa) face such severe shortfalls that they are classified as either water-scarce or water-stressed; this number is expected to roughly double by 2025, in large part because of increases in demand resulting from economic and population growth. For example, most policy makers now recognize drought as a recurrent feature of Africa's climate. However, climate change will further exacerbate the frequency and magnitude of droughts in some places.

Changes in climate could exacerbate periodic and chronic shortfalls of water, particularly in arid and semi-arid areas of the world. Developing countries are highly vulnerable to climate change because many are located in arid and semi-arid regions, and most derive their water resources from single-point systems such as bore holes or isolated reservoirs. These systems, by their nature, are vulnerable because there is no redundancy in the system to provide resources, should the primary supply fail. Also, given the limited technical, financial, and management resources possessed by developing countries, adjusting to shortages and/or implementing adaptation measures will impose a heavy burden on their national economies. There is evidence that flooding is likely to become a larger problem in many temperate and humid regions, requiring adaptations not only to droughts and chronic water shortages but also to floods and associated damages, raising concerns about dam and levee failures.

The impacts of climate change will depend on the baseline condition of the water supply system and the ability of water resources managers to respond not only to climate change but also to population growth and changes in demands, technology, and economic, social, and legislative conditions.

Various approaches are available to reduce the potential vulnerability of water systems to climate change. Options include pricing systems, water efficiency initiatives, engineering and structural improvements to water supply infrastructure, agriculture policies, and urban planning/management. At the national/regional level, priorities include placing greater emphasis on integrated, cross-sectoral water resources management, using river basins as resource management units, and encouraging sound pricing and management practices. Given increasing demands, the prevalence and sensitivity of many simple water management systems to fluctuations in precipitation and runoff, and the considerable time and expense required to implement many adaptation measures, the water resources sector in many regions and countries is vulnerable to potential changes in climate.

## 4.3. *Food and Fiber Production*

Currently, 800 million people are malnourished; as the world's population increases and incomes in some countries rise, food consumption is expected to double over the next three to four decades. The most recent doubling in food production occurred over a 25-year period and was based on irrigation, chemical inputs, and high-yielding crop varieties. Whether the remarkable gains of the past 25 years will be repeated is uncertain: Problems associated with intensifying production on land already in use (e.g., chemical and biological runoff, waterlogging and salinization of soils, soil erosion and compaction) are becoming increasingly evident. Expanding the amount of land under cultivation (including reducing land deliberately taken out of production to reduce agricultural output) also is an option for increasing total crop production, but it could lead to increases in competition for land and pressure on natural ecosystems, increased agricultural emissions of greenhouse gases, a reduction in natural sinks of carbon, and expansion of agriculture to marginal lands—all of which could undermine the ability to sustainably support increased agricultural production.

Changes in climate will interact with stresses that result from actions to increase agricultural production, affecting crop yields and productivity in different ways, depending on the types of agricultural practices and systems in place. The main direct effects will be through changes in factors such as temperature, precipitation, length of growing season, and timing of extreme or critical threshold events relative to crop development, as well as through changes in atmospheric $CO_2$ concentration (which may have a beneficial effect on the growth of many crop types). Indirect effects will include potentially detrimental changes in diseases, pests, and weeds, the effects of which have not yet been quantified in most available studies. Evidence continues to support the findings of the IPCC SAR that "global agricultural production could be maintained relative to baseline production" for a growing population under $2xCO_2$ equilibrium climate conditions. In addition, the regional findings of this special report lend support to concerns over the "potential serious consequences" of increased risk of hunger in some regions, particularly the tropics and subtropics. Generally, middle to high latitudes may experience increases in productivity, depending on crop type, growing season, changes in temperature regimes, and the seasonality of precipitation. In the tropics and subtropics—where some crops are near their maximum temperature tolerance and where dryland, nonirrigated agriculture predominates—yields are likely to decrease. The livelihoods of subsistence farmers and pastoral peoples, who make up a large portion of rural populations in some regions, also could be negatively affected. In regions where there is a likelihood of decreased rainfall, agriculture could be significantly affected.

Fisheries and fish production are sensitive to changes in climate and currently are at risk from overfishing, diminishing nursery areas, and extensive inshore and coastal pollution. Globally, marine fisheries production is expected to remain about the same in response to changes in climate; high-latitude freshwater and aquaculture production is likely to increase, assuming that natural climate variability and the structure and strength of ocean currents remain about the same. The principal impacts will be felt at the national and local levels, as centers of production shift. The positive effects of climate change—such as longer growing seasons, lower natural winter mortality, and faster growth rates in higher latitudes—may be offset by negative factors such as changes in established reproductive patterns, migration routes, and ecosystem relationships.

Given the many forces bringing profound changes to the agricultural sector, adaptation options that enhance resilience to current natural climate variability and potential changes in means and extremes and address other concerns (e.g., soil erosion, salinization) offer no- or low-regret options. For example, linking agricultural management to seasonal climate predictions can assist in incremental adaptation, particularly in regions where climate is strongly affected by ENSO conditions. The suitability of these options for different regions varies, in part because of differences in the financial and institutional ability of the private sector and governments in different regions to implement them. Adaptation options include changes in crops and crop varieties, development of new crop varieties, changes in planting schedules and tillage practices, introduction of new biotechnologies, and improved water-management and irrigation systems, which have high capital costs and are limited by availability of water resources. Other options, such as minimum- and reduced-tillage technologies, do not require such extensive capitalization but do require high levels of agricultural training and support.

In regions where agriculture is well adapted to current climate variability and/or where market and institutional factors are in place to redistribute agricultural surpluses to make up for shortfalls, vulnerability to changes in climate means and extremes generally is low. However, in regions where agriculture is unable to cope with existing extremes, where markets and institutions to facilitate redistribution of deficits and surpluses are not in place, and/or where adaptation resources are limited, the vulnerability of the agricultural sector to climate change should be considered high. Other factors also will influence the vulnerability of agricultural production in a particular country or region to climate change—including the extent to which current temperatures or precipitation patterns are close to or exceed tolerance limits for important crops; per capita income; the percentage of economic activity based on agricultural production; and the preexisting condition of the agricultural land base.

## 4.4. *Coastal Systems*

Coastal zones are characterized by a rich diversity of ecosystems and a great number of socioeconomic activities. Coastal human populations in many countries have been growing at double the national rate of population growth. It is currently estimated that about half of the global population lives in coastal zones, although there is large variation among countries. Changes in climate will affect coastal systems through sea-level rise and an increase in storm-surge hazards and possible changes in the frequency and/or intensity of extreme events.

Coasts in many countries currently face severe sea-level rise problems as a consequence of tectonically and anthropogenically induced subsidence. An estimated 46 million people per year currently are at risk of flooding from storm surges. Climate change will exacerbate these problems, leading to potential impacts on ecosystems and human coastal infrastructure. Large numbers of people also are potentially affected by sea-level rise—for example, tens of millions of people in Bangladesh would be displaced by a 1-m increase (the top of the range of IPCC Working Group I estimates for 2100) in the absence of adaptation measures. A growing number of extremely large cities are located in coastal areas, which means that large amounts of infrastructure may be affected. Although annual protection costs for many nations are relatively modest—about 0.1% of gross domestic product (GDP)—the average annual costs to many small island states total several percent of GDP. For some island nations, the high cost of providing storm-surge protection would make it essentially infeasible, especially given the limited availability of capital for investment.

Beaches, dunes, estuaries, and coastal wetlands adapt naturally and dynamically to changes in prevailing winds and seas, as well as sea-level changes; in areas where infrastructure development is not extensive, planned retreat and accommodation to changes may be possible. It also may be possible to rebuild or relocate capital assets at the end of their design life. In other areas, however, accommodation and planned retreat are not viable options, and protection using hard structures (e.g., dikes, levees, floodwalls, and barriers) and soft structures (e.g., beach nourishment, dune restoration, and wetland creation) will be necessary. Factors that limit the implementation of these options include inadequate financial resources, limited institutional and technological capability, and shortages of trained personnel. In most regions, current coastal management and planning frameworks do not take account of the vulnerability of key systems to changes in climate and sea level or long lead times for implementation of many adaptation measures. Inappropriate policies encourage development in impact-prone areas. Given increasing population density in coastal zones, long lead times for implementation of many adaptation measures, and institutional, financial, and technological limitations (particularly in many developing countries), coastal systems should be considered vulnerable to changes in climate.

### *4.5. Human Health*

In much of the world, life expectancy is increasing; in addition, infant and child mortality in most developing countries is dropping. Against this positive backdrop, however, there appears to be a widespread increase in new and resurgent vector-borne and infectious diseases, such as dengue, malaria, hantavirus, and cholera. In addition, the percentage of the developing world's population living in cities is expected to increase from 25% (in 1960) to more than 50% by 2020, with percentages in some regions far exceeding these averages. These changes will bring benefits only if accompanied by increased access to services such as sanitation and potable water supplies; they also can lead to serious urban environmental problems, including air pollution (e.g., particulates, surface ozone, and lead), poor sanitation, and associated problems in water quality and potability, if access to services is not improved.

Climate change could affect human health through increases in heat-stress mortality, tropical vector-borne diseases, urban air pollution problems, and decreases in cold-related illnesses. Compared with the total burden of ill health, these problems are not likely to be large. In the aggregate, however, the direct and indirect impacts of climate change on human health do constitute a hazard to human population health, especially in developing countries in the tropics and subtropics; these impacts have considerable potential to cause significant loss of life, affect communities, and increase health-care costs and lost work days. Model projections (which entail necessary simplifying assumptions) indicate that the geographical zone of potential malaria transmission would expand in response to global mean temperature increases at the upper part of the IPCC-projected range (3–5°C by 2100), increasing the affected proportion of the world's population from approximately 45% to approximately 60% by the latter half of the next century. Areas where malaria is currently endemic could experience intensified transmission (on the order of 50–80 million additional annual cases, relative to an estimated global background total of 500 million cases). Some increases in non-vector-borne infectious diseases—such as salmonellosis, cholera, and giardiasis—also could occur as a result of elevated temperatures and increased flooding. However, quantifying the projected health impacts is difficult because the extent of climate-induced health disorders depends on other factors—such as migration, provision of clean urban environments, improved nutrition, increased availability of potable water, improvements in sanitation, the extent of disease vector-control measures, changes in resistance of vector organisms to insecticides, and more widespread availability of health care. Human health is vulnerable to changes in climate—particularly in urban areas, where access to space conditioning may be limited, as well as in areas where exposure to vector-borne and communicable diseases may increase and health-care delivery and basic services, such as sanitation, are poor.

## 5. Anticipatory Adaptation in the Context of Current Policies and Conditions

A key message of the regional assessments in this report is that many systems and policies are not well-adjusted even to today's climate and climate variability. Increasing costs, in terms of human life and capital, from floods, storms, and droughts demonstrate current vulnerability. This situation suggests that there are adaptation options that would make many sectors more resilient to today's conditions and thus would help in adapting to future changes in climate. These options—so-called "win-win" or "no-regrets" options—could have multiple benefits and most likely would prove to be beneficial even in the absence of climate change impacts.

In many countries, the economic policies and conditions (e.g., taxes, subsidies, and regulations) that shape private decision making, development strategies, and resource-use patterns (and hence environmental conditions) hinder implementation of adaptation measures. In many countries, for example, water is subsidized, encouraging over-use (which draws down existing sources) and discouraging conservation measures—which may well be elements of future adaptation strategies. Other examples are inappropriate land-use zoning and/or subsidized disaster insurance, which encourage infrastructure development in areas prone to flooding or other natural disasters—areas that could become even more vulnerable as a result of climate change. Adaptation and better incorporation of the long-term environmental consequences of resource use can be brought about through a range of approaches, including strengthening legal and institutional frameworks, removing preexisting market distortions (e.g., subsidies), correcting market failures (e.g., failure to reflect environmental damage or resource depletion in prices or inadequate economic valuation of biodiversity), and promoting public participation and education. These types of actions would adjust resource-use patterns to current environmental conditions and better prepare systems for potential future changes.

The challenge is to identify opportunities that facilitate sustainable development by making use of existing technologies and developing policies that make climate-sensitive sectors resilient to today's climate variability. This strategy will require many regions of the world to have more access to appropriate technologies, information, and adequate financing. In addition, the regional assessments suggest that adaptation will require anticipation and planning; failure to prepare systems for projected changes in climate means, variability, and extremes could lead to capital-intensive development of infrastructure or technologies that are ill-suited to future conditions, as well as missed opportunities to lower the costs of adaptation. Additional analysis of current vulnerability to today's climate fluctuations and existing coping mechanisms is needed and will offer lessons for the design of effective options for adapting to potential future changes in climate.

## 6. Regional Vulnerability to Global Climate Change

### 6.1. *Africa*

Several climate regimes characterize the African continent; the wet tropical, dry tropical, and alternating wet and dry climates are the most common. Many countries on the continent are prone to recurrent droughts; some drought episodes, particularly in southeast Africa, are associated with ENSO phenomena. Deterioration in terms of trade, inappropriate policies, high population growth rates, and lack of significant investment—coupled with a highly variable climate—have made it difficult for several countries to develop patterns of livelihood that would reduce pressure on the natural resource base. Under the assumption that access to adequate financing is not provided, Africa is the continent most vulnerable to the impacts of projected changes because widespread poverty limits adaptation capabilities.

*Ecosystems*: In Africa today, tropical forests and rangelands are under threat from population pressures and systems of land use. Generally apparent effects of these threats include loss of biodiversity, rapid deterioration in land cover, and depletion of water availability through destruction of catchments and aquifers. Changes in climate will interact with these underlying changes in the environment, adding further stresses to a deteriorating situation. A sustained increase in mean ambient temperatures beyond 1°C would cause significant changes in forest and rangeland cover; species distribution, composition, and migration patterns; and biome distribution. Many organisms in the deserts already are near their tolerance limits, and some may not be able to adapt further under hotter conditions. Arid to semi-arid subregions and the grassland areas of eastern and southern Africa, as well as areas currently under threat from land degradation and desertification, are particularly vulnerable. Were rainfall to increase as projected by some GCMs in the highlands of east Africa and equatorial central Africa, marginal lands would become more productive than they are now. These effects are likely to be negated, however, by population pressure on marginal forests and rangelands. Adaptive options include control of deforestation, improved rangeland management, expansion of protected areas, and sustainable management of forests.

*Hydrology and Water Resources*: Of the 19 countries around the world currently classified as water-stressed, more are in Africa than in any other region—and this number is likely to increase, independent of climate change, as a result of increases in demand resulting from population growth, degradation of watersheds caused by land-use change, and siltation of river basins. A reduction in precipitation projected by some GCMs for the Sahel and southern Africa—if accompanied by high inter-annual variability—could be detrimental to the hydrological balance of the continent and disrupt various water-dependent socioeconomic activities. Variable climatic conditions may render the management of water resources more difficult both within and between countries. A drop in water level in dams and rivers could adversely affect the quality of water by increasing the concentrations of sewage waste and industrial effluents, thereby increasing the potential for the outbreak of diseases and reducing the quality and quantity of fresh water available for domestic use. Adaptation options include water harvesting, management of water outflow from dams, and more efficient water usage.

*Agriculture and Food Security*: Except in the oil-exporting countries, agriculture is the economic mainstay in most African countries, contributing 20–30% of GDP in sub-Saharan Africa and 55% of the total value of African exports. In most African countries, farming depends entirely on the quality of the rainy season—a situation that makes Africa particularly vulnerable to climate change. Increased droughts could seriously impact the availability of food, as in the horn of Africa and southern Africa during the 1980s and 1990s. A rise in mean winter temperatures also would be detrimental to the production of winter wheat and fruits that need the winter chill. However, in subtropical Africa, warmer winters would reduce the incidence of damaging frosts, making it possible to grow horticultural produce susceptible to frosts at higher elevations than is possible at present. Productivity of freshwater fisheries may increase, although the mix of fish species could be altered. Changes in ocean dynamics could lead to changes in the migratory patterns of fish and possibly to reduced fish landings, especially in coastal artisinal fisheries.

*Coastal Systems*: Several African coastal zones—many of which already are under stress from population pressure and conflicting uses—would be adversely affected by sea-level rise associated with climate change. The coastal nations of west and central Africa (e.g., Senegal, Gambia, Sierra Leone, Nigeria, Cameroon, Gabon, Angola) have low-lying lagoonal coasts that are susceptible to erosion and hence are threatened by sea-level rise, particularly because most of the countries in this area have major and rapidly expanding cities on the coast. The west coast often is buffeted by storm surges and currently is at risk from erosion, inundation, and extreme storm events. The coastal zone of east Africa also will be affected, although this area experiences calm conditions through much of the year. However, sea-level rise and climatic variation may reduce the buffer effect of coral and patch reefs along the east coast, increasing the potential for erosion. A number of studies indicate that a sizable proportion of the northern part of the Nile delta will be lost through a combination of inundation and erosion, with consequent loss of agricultural land and urban areas. Adaptation measures in African coastal zones are available but would be very costly, as a percentage of GDP, for many countries. These measures could include erection of sea walls and relocation of vulnerable human settlements and other socioeconomic facilities.

*Human Settlement, Industry, and Transportation*: The main challenges likely to face African populations will emanate from extreme climate events such as floods (and resulting landslides in some areas), strong winds, droughts, and tidal waves. Individuals living in marginal areas may be forced to migrate to urban areas (where infrastructure already is approaching its limits as a result of population pressure) if the marginal lands become less productive under new climate conditions. Climate change could worsen current trends in depletion of biomass energy resources. Reduced stream flows would cause reductions in hydropower production, leading to negative effects on industrial productivity and costly relocation of some industrial plants. Management of pollution, sanitation, waste disposal, water supply, and public health, as well as provision of adequate infrastructure in urban areas, could become more difficult and costly under changed climate conditions.

*Human Health*: Africa is expected to be at risk primarily from increased incidences of vector-borne diseases and reduced nutritional status. A warmer environment could open up new areas for malaria; altered temperature and rainfall patterns also could increase the incidence of yellow fever, dengue fever, onchocerciasis, and trypanosomiasis. Increased morbidity and mortality in subregions where vector-borne diseases increase following climatic changes would have far-reaching economic consequences. In view of the poor economic status of most African nations, global efforts will be necessary to tackle the potential health effects.

*Tourism and Wildlife*: Tourism—one of Africa's fastest-growing industries—is based on wildlife, nature reserves, coastal resorts, and an abundant water supply for recreation. Projected droughts and/or reduction in precipitation in the Sahel and eastern and southern Africa would devastate wildlife and reduce the attractiveness of some nature reserves, thereby reducing income from current vast investments in tourism.

*Conclusions*: The African continent is particularly vulnerable to the impacts of climate change because of factors such as widespread poverty, recurrent droughts, inequitable land distribution, and overdependence on rain-fed agriculture. Although adaptation options, including traditional coping strategies, theoretically are available, in practice the human, infrastructural, and economic response capacity to effect timely response actions may well be beyond the economic means of some countries.

### 6.2. *Polar Regions: The Arctic and the Antarctic*

The polar regions include some very diverse landscapes, and the Arctic and the Antarctic are very different in character. The Arctic is defined here as the area within the Arctic Circle; the Antarctic here includes the area within the Antarctic Convergence, including the Antarctic continent, the Southern Ocean, and the sub-Antarctic islands. The Arctic can be described as a frozen ocean surrounded by land, and the Antarctic as a frozen continent surrounded by ocean. The projected warming in the polar regions is greater than for many other regions of the world. Where temperatures are close to freezing on average, global warming will reduce land ice and sea ice, the former contributing to sea-level rise. However, in the interiors of ice caps, increased temperature may not be sufficient to lead to melting of ice and snow, and will tend to have the effect of increasing snow accumulation.

*Ecosystems*: Major physical and ecological changes are expected in the Arctic. Frozen areas close to the freezing point will thaw and undergo substantial changes with warming. Substantial loss of sea ice is expected in the Arctic Ocean. As warming occurs, there will be considerable thawing of permafrost—leading to changes in drainage, increased slumping, and altered landscapes over large areas. Polar warming probably will increase biological production but may lead to different species composition on land and in the sea. On land, there will be a tendency for polar shifts in major biomes such as tundra and boreal forest and associated animals, with significant impacts on species such as bear and caribou. However, the Arctic Ocean geographically limits northward movement. Much smaller changes are likely for the Antarctic, but there may be species shifts. In the sea, marine ecosystems will move poleward. Animals dependent on ice may be disadvantaged in both polar areas.

*Hydrology and Water Resources*: Increasing temperature will thaw permafrost and melt more snow and ice. There will be more running and standing water. Drainage systems in the Arctic are likely to change at the local scale. River and lake ice will break up earlier and freeze later.

*Food and Fiber Production*: Agriculture is severely limited by the harsh climate. Many limitations will remain in the future, though some small northern extension of farming into the Arctic may be possible. In general, marine ecological productivity should rise. Warming should increase growth and development rates of nonmammals; ultraviolet-B (UV-B) radiation is still increasing, however, which may adversely affect primary productivity as well as fish productivity.

*Coastal Systems*: As warming occurs, the Arctic could experience a thinner and reduced ice cover. Coastal and river navigation will increase, with new opportunities for water transport, tourism, and trade. The Arctic Ocean could become a major global trade route. Reductions in ice will benefit offshore oil production. Increased erosion of Arctic shorelines is expected from a combination of rising sea level, permafrost thaw, and increased wave action as a result of increased open water. Further breakup of ice shelves in the Antarctic peninsula is likely. Elsewhere in Antarctica, little change is expected in coastlines and probably in its large ice shelves.

*Human Settlements*: Human communities in the Arctic will be substantially affected by the projected physical and ecological changes. The effects will be particularly important for indigenous peoples leading traditional lifestyles. There will be new opportunities for shipping, the oil industry, fishing, mining, tourism, and migration of people. Sea ice changes projected for the Arctic have major strategic implications for trade, especially between Asia and Europe.

*Conclusions*: The Antarctic peninsula and the Arctic are very vulnerable to projected climate change and its impacts. Although the number of people directly affected is relatively small, many native communities will face profound changes that impact on traditional lifestyles. Direct effects could include ecosystem shifts, sea- and river-ice loss, and permafrost thaw. Indirect effects could include feedbacks to the climate system such as further releases of greenhouse gases, changes in ocean circulation drivers, and increased temperature and higher precipitation with loss of ice, which could affect climate and sea level globally. The interior of Antarctica is less vulnerable to climate change, because the temperature changes envisaged over the next century are likely to have little impact and very few people are involved. However, there are considerable uncertainties about the mass balance of the Antarctic ice sheets and the future behavior of the West Antarctic ice sheet (low probability of disintegration over the next century). Changes in either could affect sea level and Southern Hemisphere climates.

### *6.3. Arid Western Asia (Middle East and Arid Asia)*

This region includes the predominantly arid and semi-arid areas of the Middle East and central Asia. The region extends from Turkey in the west to Kazakstan in the east, and from Yemen in the south to Kazakstan in the north. The eastern part of the region has a large area dominated by mountains.

*Ecosystems*: Vegetation models project little change in most arid or desert vegetation types under climate change projections—i.e., most lands that are deserts are expected to remain deserts. Greater changes in the composition and distribution of vegetation types of semi-arid areas—for example, grasslands, rangelands, and woodlands—are anticipated. Small increases in precipitation are projected, but these increases are likely to be countered by increased temperature and evaporation. Improved water-use efficiency by some plants under elevated $CO_2$ conditions may lead to some improvement in plant productivity and changes in ecosystem composition. Grasslands, livestock, and water resources are likely to be the most vulnerable to climate change in this region because they are located mostly in marginal areas. Appropriate land-use management, including urban planning, could reduce some of the pressures on land degradation. Management options, such as better stock management and more integrated agro-ecosystems, could improve land conditions and counteract pressures arising from climate change. The region is an important refuge for wild relatives of many important crop species; with appropriate conservation measures it may continue to provide a source of genetic material for future climatic conditions.

*Hydrology and Water Resources*: Water shortage, already a problem in many countries of this arid region, is unlikely to be reduced, and may be exacerbated, by climate change. Changes in cropping practices and improved irrigation practices could significantly improve the efficiency of water use in some countries. Glacial melt is projected to increase under climate change—leading to increased flows in some river systems for a few decades, followed by a reduction in flow as the glaciers disappear.

*Food and Fiber Production*: Land degradation problems and limited water supplies restrict present agricultural productivity and threaten the food security of some countries. There are few projections of the impacts of climate change on food and fiber production for the region. The adverse impacts that may result in the region are suggested by the results of studies that estimate that wheat production in Kazakstan and Pakistan would decline under selected scenarios of climate change. The studies, however, are too few to draw strong conclusions regarding agriculture across the entire region. Many of the options available to combat existing problems would contribute to reducing the anticipated impacts of climate change. Food and fiber production, concentrated on more intensively managed land, could lead to greater reliability in food production and reduce the detrimental impacts of extreme climatic events. Countries of the former Soviet Union are undergoing major economic changes, particularly in agricultural systems and management. This transition is likely to provide opportunities to change crop types and introduce more efficient irrigation—providing significant win-win options for conservation of resources to offset the projected impacts of climate change.

*Human Health*: Heat stress, affecting human comfort levels, and possible spread in vector-borne diseases are likely to result from changes in climate. Decreases in water availability and food production would lead to indirect impacts on human health.

*Conclusions*: Water is an important limiting factor for ecosystems, food and fiber production, human settlements, and human health in this arid region of the world. Climate change is anticipated to alter the hydrological cycle, and is unlikely to relieve the limitations placed by water scarcity upon the region. Climate change and human activities may further influence the levels of the Caspian and Aral Seas, which will affect associated ecosystems, agriculture, and human health in the surrounding areas. Win-win opportunities exist which offer the potential to reduce current pressures on resources and human welfare in the region and also offer the potential to reduce their vulnerability to adverse impacts from climate change.

## 6.4. Australasia

Australasia includes Australia, New Zealand, and their outlying islands. The region spans the tropics to mid-latitudes and has varied climates and ecosystems, ranging from interior deserts to mountain rainforests. The climate is strongly affected by the oceanic environment and the ENSO phenomenon.

*Ecosystems*: Some of the region's ecosystems appear to be very vulnerable to climate change, at least in the long term, because alterations to soils, plants, and ecosystems are very likely, and there may be increases in fire occurrence and insect outbreaks. Many species will be able to adapt, but in some instances, a reduction of species diversity is highly likely. Any changes will occur in a landscape already fragmented by agricultural and urban development; such changes will add to existing problems such as land degradation, weeds, and pest infestations. Impacts on aquatic ecosystems from changes in river flow, flood frequency, and nutrient and sediment inputs are likely to be greatest in the drier parts of the region. Coastal ecosystems are vulnerable to the impacts of sea-level rise and possible changes in local meteorology. Tropical coral reefs, including the Great Barrier Reef, may be able to keep pace with sea-level rise—but will be vulnerable to bleaching and death of corals induced by episodes of higher sea temperatures and other stresses. Measures to facilitate adaptation include better rangeland management; plantings along waterways; and research, monitoring, and prediction. Active manipulation of species generally will not be feasible in the region's extensive natural and lightly managed ecosystems.

*Hydrology and Water Resources*: Vulnerability appears to be potentially high. Any reduction of water availability, especially in Australia's extensive drought-prone areas, would sharpen competition among uses, including agriculture and wetland ecosystem needs. Freshwater supplies on low-lying islands are also vulnerable. More frequent high-rainfall events may enhance groundwater recharge and dam-filling events, but they also may increase the impacts of flooding, landslides, and erosion, with flood-prone urban areas being heavily exposed to financial loss. Reduced snowpack and a shorter snow season appear likely, and New Zealand's glaciers are likely to shrink further. Some adaptation options are available, but the cost involved would be high.

*Food and Fiber Production*: Vulnerability appears to be low, at least in the next few decades (potentially high sensitivity coupled with high adaptability). Agriculture in the region is adaptable, and production increases are likely in some cases. However, there may be a trend toward increased vulnerability in the longer term—especially in warmer and more water-limited parts of Australia, where initial gains for some crops are eroded later as the delayed full effects of climate change (e.g., changes in temperature and precipitation) tend to outweigh the more immediate benefits of increased atmospheric $CO_2$ concentrations. Impacts will vary widely from district to district and crop to crop. There will be changes in growth and quality of crops and pastures; shifts in the suitability of districts for particular crops; and possibly increased problems with weeds, pests, and diseases. Rangeland pastoralism and irrigated agriculture will be especially affected where rainfall changes occur. Changes in food production elsewhere in the world, which affect prices, would have major economic impacts on the region. With regard to forestry, the longer time to maturity results in a relatively large exposure to financial loss from extreme events, fire, or any locally rapid change in climate conditions.

*Coastal Systems*: Parts of the region's coasts and rapidly growing coastal settlements and infrastructure are very vulnerable to any increase in coastal flooding and erosion arising from sea-level rise and meteorological changes. Indigenous coastal and island communities in the Torres Strait and in New Zealand's Pacific island territories are especially vulnerable. Many adaptation options exist, although these measures are not easily implemented on low-lying islands. Moreover, climate change and sea-level rise generally are not well accommodated in current coastal management planning frameworks.

*Human Settlements*: In addition to hydrological and coastal risks, moderate vulnerability is present from a variety of impacts on air quality, drainage, waste disposal, mining, transport, insurance, and tourism. Overall, these effects are likely to be small relative to other economic influences, but they still may represent significant costs for large industries.

*Human Health*: Some degree of vulnerability is apparent. Indigenous communities and the economically disadvantaged may be more at risk. Increases are expected in heat-stress mortality, vector-borne diseases such as dengue, water and sewage-related diseases, and urban pollution-related respiratory problems. Though small compared with the total burden of ill health, these impacts have the potential to cause considerable community impact and cost.

*Conclusions*: Australia's relatively low latitude makes it particularly vulnerable to impacts on its scarce water resources and on crops growing near or above their optimum temperatures, whereas New Zealand's cooler, wetter, mid-latitude location may lead to some benefit through the ready availability of suitable crops and likely increases in agricultural production. In both countries, however, there is a wide range of situations where vulnerability is thought to be moderate to high—particularly in ecosystems, hydrology, coastal zones, human settlements, and human health.

## 6.5. Europe

Europe constitutes the western part of the Eurasian continent. Its eastern boundary is formed by the Ural Mountains, the Ural River, and part of the Caspian Sea. The proximity of the relatively warm Gulf Stream and typical atmospheric circulation contribute to the large spatial and temporal variability of the region's temperature and precipitation. South of the main Alpine divide, the climate is of the Mediterranean type.

*Ecosystems*: Natural ecosystems generally are fragmented, disturbed, and confined to poor soils. This situation makes them more sensitive to climate change. Mediterranean and boreal grasslands may shift in response to changes in the amount and the seasonal distribution of precipitation. The northern boundaries of forests in Fennoscandia and northern Russia would likely expand into tundra regions, reducing the extent of tundra, mires, and permafrost areas. Survival of some species and forest types may be endangered by the projected movement of climate zones at rates faster than migration speeds. High-elevation ecosystems and species are particularly vulnerable because they have nowhere to migrate. An increase in temperature, accompanied by decreases in soil moisture, would lead to a substantial reduction in peat formation in Fennoscandian and northern Russian peatlands. Thawing of the permafrost layer would lead to lowered water tables in some areas and would flood thaw lakes in others, altering current wetland ecosystem types. Although the diversity of freshwater species may increase in a warmer climate, particularly in middle and high latitudes, there may be an initial reduction in species diversity in cool temperate and boreal regions. Ecosystems in southern Europe would be threatened mainly by reduced precipitation and subsequent increases in water scarcity.

*Hydrology and Water Resources*: Most of Europe experienced temperature increases this century larger than the global average, and enhanced precipitation in the northern half and decreases in the southern half of the region. Projections of future climate, not taking into account the effect of aerosols, indicate that precipitation in high latitudes of Europe may increase, with mixed results for other parts of Europe. The current uncertainties about future precipitation are mainly exacerbated by the effects of aerosols.

Water supply may be affected by possible increases in floods in northern and northwest Europe and by droughts in southern portions of the continent. Many floodplains in western Europe already are overpopulated, which hampers effective additional flood protection. Pollution is a major problem for many rivers; a warmer climate could lead to reduced water quality, particularly if accompanied by reduced runoff. Warmer summers would lead to increased water demand, although increased demand for irrigation would be at least partly offset for many crops by increased water-use efficiency associated with $CO_2$ fertilization.

Expected changes in snow and ice will have profound impacts on European streams and rivers. Up to 95% of Alpine glacier mass could disappear by 2100, with subsequent consequences for the water flow regime—affecting, for example, summer water supply, shipping, and hydropower. Also, in some areas, winter tourism would be negatively affected.

Water management is partly determined by legislation and cooperation among government entities, within countries and internationally; altered water supply and demand would call for a reconsideration of existing legal and cooperative arrangements.

*Food and Fiber Production*: Risks of frost would be reduced in a warmer climate, allowing winter cereals and other winter crops to expand to areas such as southern Fennoscandia and western Russia. Potential yields of winter crops are expected to increase, especially in central and southern Europe, assuming that neither precipitation nor irrigation are limiting and that water-use efficiency increases with the ambient atmospheric concentration of $CO_2$. Increasing spring temperatures would extend suitable zones for most summer crops. Summer crop yield increases are possible in central and eastern Europe, though decreases are possible in western Europe. Decreases in precipitation in southern Europe would reduce crop yields and make irrigation an even larger competitor to domestic and industrial water use. Along with potential crop yields, farmer adaptation, agricultural policy, and world markets are important factors in the economic impact of climate change on the agricultural sector.

*Coastal Systems*: Coastal zones are ecologically and economically important. Settlement and economic activity have reduced the resilience and adaptability of coastal systems to climate variability and change, as well as to sea-level rise. Some coastal areas already are beneath mean sea level, and many others are vulnerable to storm surges. Areas most at risk include the Dutch, German, Ukrainian, and Russian coastlines; some Mediterranean deltas; and Baltic coastal zones. Storm surges, changes in precipitation, and changes in wind speed and direction add to the concern of coastal planners. In general, major economic and social impacts can be contained with relatively low investment. This is not true, however, for a number of low-lying urban areas vulnerable to storm surges, nor for ecosystems—particularly coastal wetlands—which may be even further damaged by protective measures.

*Human Settlements*: Supply and demand for cooling water will change. Energy demand may increase in summer (cooling) and decrease in winter (heating), and peak energy demand will shift. Infrastructure, buildings, and cities designed for cooler climates will have to be adjusted to warming, particularly heat waves, to maintain current functions. In areas where precipitation increases or intensifies, there are additional risks from landslides and river floods.

*Human Health*: Heat-related deaths would increase under global warming and may be exacerbated by worsening air quality in cities; there would be a reduction in cold-related deaths. Vector-borne diseases would expand. Health care measures could significantly reduce such impacts.

*Conclusions*: Even though capabilities for adaptation in managed systems in many places in Europe are relatively well established, significant impacts of climate change still should be anticipated. Major effects are likely to be felt through changes in the frequency of extreme events and precipitation, causing more droughts in some areas and more river floods elsewhere. Effects will be felt primarily in agriculture and other water-dependent activities. Boreal forest and permafrost areas are projected to undergo major change. Ecosystems are especially vulnerable due to the projected rate of climate change and because migration is hampered.

## 6.6. Latin America

Latin America includes all continental countries of the Americas from Mexico to Chile and Argentina, as well as adjacent seas. The region is highly heterogeneous in terms of climate, ecosystems, human population distribution, and cultural traditions. Several Latin American countries—especially those of the Central American isthmus, Ecuador, Brazil, Peru, Bolivia, Chile, and Argentina—are significantly affected with adverse socio-economic consequences by seasonal to interannual climate variability, particularly the ENSO phenomenon. Most production is based on the region's extensive natural ecosystems, and the impacts of current climate variability on natural resources suggest that the impacts of projected climate changes could be important enough to be taken into account in national and regional planning initiatives. Land use is a major force driving ecosystem change at present, interacting with climate in complex ways. This factor makes the task of identifying common patterns of vulnerability to climate change very difficult.

*Ecosystems*: Large forest and rangeland areas are expected to be affected as a result of projected changes in climate, with mountain ecosystems and transitional zones between vegetation types extremely vulnerable. Climate change could add an additional stress to the adverse effects of continued deforestation in the Amazon rainforest. This impact could lead to biodiversity losses, reduce rainfall and runoff within and beyond the Amazon basin (reduced precipitation recycling through evapotranspiration), and affect the global carbon cycle.

*Hydrology and Water Resources*: Climate change could significantly affect the hydrological cycle, altering the intensity and temporal and spatial distribution of precipitation, surface runoff, and groundwater recharge, with various impacts on different natural ecosystems and human activities. Arid and semi-arid areas are particularly vulnerable to changes in water availability. Hydropower generation and grain and livestock production are particularly vulnerable to changes in water supply, particularly in Costa Rica, Panama, and the Andes piedmont, as well as adjacent areas in Chile and western Argentina between 25°S and 37°S. The impacts on water resources could be sufficient to lead to conflicts among users, regions, and countries.

*Food and Fiber Production*: Decreases in agricultural production —even after allowing for the positive effects of elevated $CO_2$ on crop growth and moderate levels of adaptation at the farm level—are projected for several major crops in Mexico, countries of the Central American isthmus, Brazil, Chile, Argentina, and Uruguay. In addition, livestock production would decrease if temperate grasslands have to face substantial decreases in water availability. Extreme events (e.g., floods, droughts, frosts, storms) have the potential to adversely affect rangelands and agricultural production (e.g., banana crops in Central America). The livelihoods of traditional peoples, such as many Andean communities, would be threatened if the productivity or surface area of rangelands or traditional crops is reduced.

*Coastal Systems*: Losses of coastal land and biodiversity (including coral reefs, mangrove ecosystems, estuarine wetlands, and marine mammals and birds), damage to infrastructure, and saltwater intrusion resulting from sea-level rise could occur in low-lying coasts and estuaries in countries such as those of the Central American isthmus, Venezuela, Argentina, and Uruguay. Sea-level rise that blocks the runoff of flatland rivers into the ocean could increase the risks of floods in their basins (e.g., in the Argentine Pampas).

*Human Settlements*: Climate change would produce a number of direct and indirect effects on the welfare, health, and security of the inhabitants of Latin America. Direct impacts resulting from sea-level rise, adverse weather, and extreme climatic conditions (e.g., floods, flash floods, windstorms, landslides, and cold and heat outbreaks), as well as indirect effects through impacts on other sectors such as water and food supply, transportation, energy distribution, and sanitation services, could be exacerbated by projected climate change. Particularly vulnerable groups include those living in shanty towns in areas around large cities, especially where those settlements are established in flood-prone areas or on unstable hillsides.

*Human Health*: Projected changes in climate could increase the impacts of already serious chronic malnutrition and diseases for some Latin American populations. The geographical distributions of vector-borne diseases (e.g., malaria, dengue, Chagas') and infectious diseases (e.g., cholera) would expand southward and to higher elevations if temperature and precipitation increase. Pollution and high concentrations of ground-level ozone, exacerbated by increasing surface temperature, would have the potential to negatively affect human health and welfare, especially in urban areas.

*Conclusions*: Increasing environmental deterioration (e.g., changes in water availability, losses of agricultural lands, and flooding of coastal, riverine, and flatland areas) arising from climate variability, climate change, and land-use practices would aggravate socioeconomic and health problems, encourage migration of rural and coastal populations, and deepen national and international conflicts.

## 6.7. North America

This region consists of Canada and the United States south of the Arctic Circle. Within the region, vulnerability to and the impacts of climate change vary significantly from sector to sector and from subregion to subregion. This "texture" is important in understanding the potential effects of climate change on North America, as well as in formulating and implementing viable response strategies.

*Ecosystems*: Most ecosystems are moderately to highly sensitive to changes in climate. Effects are likely to include both beneficial and harmful changes. Potential impacts include northward shifts of forest and other vegetation types, which would affect biodiversity by altering habitats and would reduce the market and non-market goods and services they provide; declines in forest density and forested area in some subregions, but gains in others; more frequent and larger forest fires; expansion of arid land species into the great basin region; drying of prairie pothole wetlands that currently support over 50% of all waterfowl in North America; and changes in distribution of habitat for cold-, cool-, and warm-water fish. The ability to apply management practices to limit potential damages is likely to be low for ecosystems that are not already intensively managed.

*Hydrology and Water Resources*: Water quantity and quality are particularly sensitive to climate change. Potential impacts include increased runoff in winter and spring and decreased soil moisture and runoff in summer. The Great Plains and prairie regions are particularly vulnerable. Projected increases in the frequency of heavy rainfall events and severe flooding also could be accompanied by an increase in the length of dry periods between rainfall events and in the frequency and/or severity of droughts in parts of North America. Water quality could suffer and would decline where minimum river flows decline. Opportunities to adapt are extensive, but their costs and possible obstacles may be limiting.

*Food and Fiber Production*: The productivity of food and fiber resources of North America is moderately to highly sensitive to climate change. Most studies, however, have not fully considered the effects of potential changes in climate variability; water availability; stresses from pests, diseases, and fire; or interactions with other, existing stresses. Warmer climate scenarios (4–5°C increases in North America) have yielded estimates of negative impacts in eastern, southeastern, and corn belt regions and positive effects in northern plains and western regions. More moderate warming produced estimates of predominately positive effects in some warm-season crops. Vulnerability of commercial forest production is uncertain, but is likely to be lower than less intensively managed systems due to changing technology and management options. The vulnerability of food and fiber production in North America is thought to be low at the continental scale, though subregional variation in losses or gains is likely. The ability to adapt may be limited by information gaps; institutional obstacles; high economic, social, and environmental costs; and the rate of climate change.

*Coastal Systems*: Sea level has been rising relative to the land along most of the coast of North America, and falling in a few areas, for thousands of years. During the next century, a 50-cm rise in sea level from climate change alone could inundate 8,500 to 19,000 km$^2$ of dry land, expand the 100-year floodplain by more than 23,000 km$^2$, and eliminate as much as 50% of North America's coastal wetlands. The projected changes in sea level due to climate change alone would underestimate the total change in sea level from all causes along the eastern seabord and Gulf coast of North America. In many areas, wetlands and estuarine beaches may be squeezed between advancing seas and dikes or seawalls built to protect human settlements. Several local governments are implementing land-use regulations to enable coastal ecosystems to migrate landward as sea level rises. Saltwater intrusion may threaten water supplies in several areas.

*Human Settlements*: Projected changes in climate could have positive and negative impacts on the operation and maintenance costs of North American land and water transportation. Such changes also could increase the risks to property and human health and life as a result of possible increased exposure to natural hazards (e.g., wildfires, landslides, and extreme weather events) and result in increased demand for cooling and decreased demand for heating energy—with the overall net effect varying across geographic regions.

*Human Health*: Climate can have wide-ranging and potentially adverse effects on human health via direct pathways (e.g., thermal stress and extreme weather/climate events) and indirect pathways (e.g., disease vectors and infectious agents, environmental and occupational exposures to toxic substances, food production). In high-latitude regions, some human health impacts are expected due to dietary changes resulting from shifts in migratory patterns and abundance of native food sources.

*Conclusions*: Taken individually, any one of the impacts of climate change may be within the response capabilities of a subregion or sector. The fact that they are projected to occur simultaneously and in concert with changes in population, technology, economics, and other environmental and social changes, however, adds to the complexity of the impact assessment and the choice of appropriate responses. The characteristics of subregions and sectors of North America suggest that neither the impacts of climate change nor the response options will be uniform.

Many systems of North America are moderately to highly sensitive to climate change, and the range of estimated effects often includes the potential for substantial damages. The technological capability to adapt management of systems to lessen or avoid damaging effects exists in many instances. The ability to adapt may be diminished, however, by the attendant costs, lack of private incentives to protect publicly owned natural systems, imperfect information regarding future changes in climate and the available options for adaptation, and institutional barriers. The most vulnerable sectors and regions include long-lived natural forest ecosystems in the east and interior west; water resources in the southern plains; agriculture in the southeast and southern plains; human health in areas currently experiencing diminished urban air quality; northern ecosystems and habitats; estuarine beaches in developed areas; and low-latitude cool- and cold-water fisheries. Other sectors and subregions may benefit from opportunities associated with warmer temperatures or, potentially, from $CO_2$ fertilization—including west coast coniferous forests; some western rangelands; reduced energy costs for heating in the northern latitudes; reduced salting and snow-clearance costs; longer open-water seasons in northern channels and ports; and agriculture in the northern latitudes, the interior west, and the west coast.

### 6.8. *Small Island States*

With the exception of Malta and Cyprus in the Mediterranean, all of the small island states considered here are located within the tropics. About one-third of the states comprise a single main island; the others are made up of several or many islands. Low-lying island states and atolls are especially vulnerable to climate change and associated sea-level rise because in many cases (e.g., the Bahamas, Kiribati, the Maldives, the Marshall Islands), much of the land area rarely exceeds 3–4 m above present mean sea level. Many islands at higher elevation also are vulnerable to climate change effects, particularly in their coastal zones, where the main settlements and vital economic infrastructure almost invariably are concentrated.

*Ecosystems*: Although projected temperature rise is not anticipated to have widespread adverse consequences, some critical ecosystems, such as coral reefs, are very sensitive to temperature changes. Although some reefs have the ability to keep pace with the projected rate of sea-level rise, in many parts of the tropics (e.g., the Caribbean Sea, the Pacific Ocean) some species of corals live near their limits of temperature tolerance. Elevated seawater temperatures (above seasonal maxima) can seriously damage corals by bleaching and also impair their reproductive functions, and lead to increased mortality. The adaptive capacity of mangroves to climate change is expected to vary by species, as well as according to local conditions (e.g., the presence or absence of sediment-rich, macrotidal environments, the availability of adequate fresh water to maintain the salinity balance). The natural capacity of mangroves to adapt and migrate landward also is expected to be reduced by coastal land loss and the presence of infrastructure in the coastal zone. On some islands, ecosystems already are being harmed by other anthropogenic stresses (e.g., pollution), which may pose as great a threat as climate change itself. Climate change would add to these stresses and further compromise the long-term viability of these tropical ecosystems.

*Hydrology and Water Resources*: Freshwater shortage is a serious problem in many small island states, and many such states depend heavily on rainwater as the source of water. Changes in the patterns of rainfall may cause serious problems to such nations.

*Coastal Systems:* Higher rates of erosion and coastal land loss are expected in many small islands as a consequence of the projected rise in sea level. In the case of Majuro atoll in the Marshall Islands and Kiribati, it is estimated that for a 1-m rise in sea level as much as 80% and 12.5% (respectively) of total land would be vulnerable. Generally, beach sediment budgets are expected to be adversely affected by reductions in sediment deposition. On high islands, however, increased sediment yield from streams will help to compensate for sand loss from reefs. Low-lying island states and atolls also are expected to experience increased sea flooding, inundation, and salinization (of soils and freshwater lenses) as a direct consequence of sea-level rise.

*Human Settlements and Infrastructure:* In a number of islands, vital infrastructure and major concentrations of settlements are likely to be at risk, given their location at or near present sea level and their proximity to the coast (often within 1–2 km; e.g., Kiribati, Tuvalu, the Maldives, the Bahamas). Moreover, vulnerability assessments also suggest that shore and infrastructure protection costs could be financially burdensome for some small island states.

*Human Health*: Climate change is projected to exacerbate health problems such as heat-related illness, cholera, dengue fever, and biotoxin poisoning, and would place additional stress on the already over-extended health systems of most small islands.

*Tourism:* Tourism is the dominant economic sector in a number of small island states in the Caribbean Sea and the Pacific and Indian Oceans. In 1995, tourism accounted for 69%, 53%, and 50% of gross national product (GNP) in Antigua, the Bahamas, and the Maldives, respectively. This sector also earns considerable foreign exchange for a number of small island states, many of which are heavily dependent on imported food, fuel, and a range of other vital goods and services. Foreign exchange earnings from tourism also provided more than 50% of total revenues for some countries in 1995. Climate change and sea-level rise would affect tourism directly and indirectly: Loss of beaches to erosion and inundation, salinization of freshwater aquifers, increasing stress on coastal ecosystems, damage to infrastructure from tropical and extra-tropical storms, and an overall loss of amenities would jeopardize the viability and threaten the long-term sustainability of this important industry in many small islands.

*Conclusions:* To evaluate the vulnerability of these island states to projected climate change, a fully integrated approach to vulnerability assessments is needed. The interaction of various biophysical attributes (e.g., size, elevation, relative isolation) with the islands' economic and sociocultural character ultimately determines the vulnerability of these islands. Moreover, some islands are prone to periodic nonclimate-related hazards (e.g., earthquakes, volcanic eruptions, tsunamis); the overall vulnerability of these islands cannot be accurately evaluated in isolation from such threats. Similarly, vulnerability assessments for these small island states should take into consideration the value of nonmarketed goods and services (e.g., subsistence assets, community structure, traditional skills and knowledge), which also may be at risk from climate change. In some island societies, these assets are just as important as marketed goods and services.

Uncertainties in climate change projections may discourage adaptation, especially because some options may be costly or require changes in societal norms and behavior. As a guiding principle, policies and development programs which seek to use resources in a sustainable manner, and which can respond effectively to changing conditions such as climate change, would be beneficial to the small island states, even if climate change did not occur.

The small island states are extremely vulnerable to global climate change and global sea-level rise. A range of adaptation strategies are theorectically possible. On some small low-lying island states and atolls, however, retreat away from the coasts is not an option. In some extreme cases, migration and resettlement outside of national boundaries might have to be considered.

## 6.9. Temperate Asia

Temperate Asia includes countries in Asia between 18°N and the Arctic Circle, including the Japanese islands, the Korean peninsula, Mongolia, most parts of China, and Russian Siberia. The east-west distance of the area is about 8,000 km, and its north-south extent is about 5,000 km. Distinct subregions include arid/semi-arid, monsoonal, and Siberian regions.

*Ecosystems*: Although the area of potential distribution of temperate forests in Temperate Asia is, to a large extent, cleared and used for intensive agriculture, global climate change can be considered sufficient to trigger structural changes in the remaining temperate forests. The nature and magnitude of these changes, however, depend on associated changes in water availability, as well as water-use efficiency. Shifts in temperature and precipitation in temperate rangelands may result in altered growing seasons and boundary shifts between grasslands, forests, and shrublands. Some model studies suggest that in a doubled $CO_2$ climate there would be a large reduction in the area (up to 50%) and productivity of boreal forests (primarily in the Russian Federation), accompanied by a significant expansion of grasslands and shrublands. There also would be a decrease in the area of the tundra zone of as much as 50%—accompanied by the release of methane from deep peat deposits—and an increase (less than 25%), in $CO_2$ emissions.

*Hydrology and Water Resources*: Overall, most $2xCO_2$ equilibrium scenario simulations show a decrease in water supply, except in a few river basins. Warmer winters may affect water balances because water demands are higher in spring and summer. Equilibrium climate conditions for doubled equivalent $CO_2$ concentrations indicate that a decrease of as much as 25% in mountain glacier mass is possible by 2050. Initially, runoff from glaciers in central Asia is projected to increase threefold by 2050, but by 2100 glacier runoff would taper to two-thirds of its present value. Model results suggest that runoff in the northern part of China is quite vulnerable to climate change, mainly as a consequence of changes in precipitation in spring, summer, and autumn, especially during the flood season. To balance water supply with water demand, increasingly efficient water management is likely the best approach for Japan. In other parts of Temperate Asia, water-resource development will remain important; the central adaptation issue is how the design of new water-resource infrastructure should be adjusted to account for uncertainties resulting from climate change. The most critical uncertainties are the lack of credible projections of the effects of global change on the Asian monsoon or the ENSO phenomenon, which have great influence on river runoffs. Multiple-stress impact studies on water resources in international river basins are needed in the future.

*Food and Fiber Production*: Projected changes in crop yields using climate projections from different GCMs vary widely. In China, for example, across different scenarios and different sites, the changes for several crop yields by 2050 are projected to be: rice, -78% to +15%; wheat, -21% to +55%; and maize, -19% to +5%. An increase in productivity may occur if the positive effects of $CO_2$ on crop growth are considered, but its magnitude remains uncertain. A northward shift of crop zones is expected to increase agricultural productivity in northern Siberia but to decrease (by about 25%) grain production in southwestern Siberia because of a more arid climate. Aquaculture is particularly important to Temperate Asia. Greater cultivation of warm-water species could develop. Warming will require greater attention to possible oxygen depletion, fish diseases, and introduction of unwanted species, as well as to potential negative factors such as changes in established reproductive patterns, migration routes, and ecosystem relationships.

*Coastal Systems*: An increase in sea level will exacerbate the current severe problems of tectonically and anthropogenically induced land subsidence in delta areas. Saltwater intrusion also would become more serious. A sea-level rise of 1 m would threaten certain coastal areas—for example, the Japanese coastal zone, on which 50% of Japan's industrial production is located (e.g., Tokyo, Osaka, and Nagoya); in addition, about 90% of the remaining sandy beaches in Japan would be in danger of disappearing.

*Human Health*: Heat-stress mortality and illness (predominantly cardiorespiratory) are projected to more than double by 2050 resulting from an increase in the frequency or severity of heat waves under climate-change conditions projected by a transient GCM (GFDL X2, UKMO X6). Net climate change-related increases in the geographic distribution (elevation and latitude) of the vector organisms of infectious diseases (e.g., malarial mosquitoes, schistosome-spreading snails) and changes in the life-cycle dynamics of vectors and infective parasites would, in aggregate, increase the potential transmission of many vector-borne diseases. Increases in nonvector-borne infectious diseases—such as cholera, salmonellosis, and other food- and water-related infections—also could occur because of climatic impacts on water distribution, temperature, and microorganism proliferation. Disease surveillance could be strengthened and integrated with other environmental monitoring to design early warning systems; develop early, environmentally sound public health interventions; and develop anticipatory societal policies to reduce the risk of outbreaks and subsequent spread of epidemics.

*Conclusions*: The major impacts in Temperate Asia under global climate change are projected to be large shifts of the boreal forests, the disappearance of significant portions of mountain glaciers, and water supply shortages. The most critical uncertainty in these estimates stems from the lack of credible projections of the hydrological cycle under global climate change scenarios. The effects of climate change on the Asian monsoon and the ENSO phenomenon are among the major uncertainties in the modeling of the hydrological cycle. Projections of agricultural crop yields are uncertain, not only because of the uncertainty in the hydrological cycle but also because of the potential positive effects of $CO_2$ and production practices. Sea-level rise endangers sandy beaches in the coastal zones, but remains an anthropogenically induced problem in delta areas. Integrated impact studies considering multi-stress factors are needed.

## *6.10. Tropical Asia*

Tropical Asia is physiographically diverse and ecologically rich in natural and crop-related biodiversity. The present total population of the region is about 1.6 billion, and it is projected to increase to 2.4 billion by 2025. The population is principally rural-based, although in 1995, the region included 6 of the 25 largest cities in the world. The climate in Tropical Asia is characterized by seasonal weather patterns associated with the two monsoons and the occurrence of tropical cyclones in the three core areas of cyclogenesis (the Bay of Bengal, north Pacific Ocean, and South China Sea). Climate change will add to other stresses such as rapid urbanization, industrialization, and economic development, which contribute to unsustainable exploitation of natural resources, increased pollution, land degradation, and other environmental problems.

*Ecosystems*: Substantial elevational shifts of ecosystems in the mountains and uplands of Tropical Asia are projected. At high elevation, weedy species can be expected to displace tree species—though the rates of vegetation change could be slow compared to the rate of climate change and constrained by increased erosion in the Greater Himalayas. Changes in the distribution and health of rainforest and drier monsoon forest will be complex. In Thailand, for instance, the area of tropical forest could increase from 45% to 80% of total forest cover, whereas in Sri Lanka, a significant increase in dry forest and a decrease in wet forest could occur. Projected increases in evapotranspiration and rainfall variability are likely to have a negative impact on the viability of freshwater wetlands, resulting in shrinkage and desiccation. Sea-level rise and increases in sea-surface temperature are the most probable major climate change-related stresses on coastal ecosystems. Coral reefs may be able to keep up with the rate of sea-level rise but suffer bleaching from higher temperatures. Landward migration of mangroves and tidal wetlands is expected to be constrained by human infrastructure and human activities.

*Hydrology and Water Resources*: The Himalayas have a critical role in the provision of water to continental monsoon Asia. Increased temperatures and increased seasonal variability in precipitation are expected to result in increased recession of glaciers and increasing danger from glacial lake outburst floods. A reduction in average flow of snow-fed rivers, coupled with an increase in peak flows and sediment yield, would have major impacts on hydropower generation, urban water supply, and agriculture. Availability of water from snow-fed rivers may increase in the short term but decrease in the long run. Runoff from rain-fed rivers may change in the future. A reduction in snowmelt water will put the dry-season flow of these rivers under more stress than is the case now. Increased population and increasing demand in the agricultural, industrial, and hydropower sectors will put additional stress on water resources. Pressure on the drier river basins and those subject to low seasonal flows will be most acute. Hydrological changes in island and coastal drainage basins are expected to be relatively small in comparison to those in continental Tropical Asia, apart from those associated with sea-level rise.

*Food and Fiber Production*: The sensitivity of major cereal and tree crops to changes in temperature, moisture, and $CO_2$ concentration of the magnitudes projected for the region has been demonstrated in many studies. For instance, impacts on rice yield, wheat yield, and sorghum yield suggest that any increase in production associated with $CO_2$ fertilization will be more than offset by reductions in yield from temperature or moisture changes. Although climate change impacts could result in significant changes in crop yields, production, storage, and distribution, the net effect of the changes regionwide is uncertain because of varietal differences; local differences in growing season, crop management, etc.; the lack of inclusion of possible diseases, pests, and microorganisms in crop model simulations; and the vulnerability of agricultural areas to episodic environmental hazards, including floods, droughts, and cyclones. Low-income rural populations that depend on traditional agricultural systems or on marginal lands are particularly vulnerable.

*Coastal Systems*: Coastal lands are particularly vulnerable; sea-level rise is the most obvious climate-related impact. Densely settled and intensively used low-lying coastal plains, islands, and deltas are especially vulnerable to coastal erosion and land loss, inundation and sea flooding, upstream movement of the saline/freshwater front, and seawater intrusion into freshwater lenses. Especially at risk are large delta regions of Bangladesh, Myanmar, Viet Nam, and Thailand, and the low-lying areas of Indonesia, the Philippines, and Malaysia. Socioeconomic impacts could be felt in major cities and ports, tourist resorts, artisinal and commercial fishing, coastal agriculture, and infrastructure development. International studies have projected the displacement of several millions of people from the region's coastal zone, assuming a 1-m rise in sea level. The costs of response measures to reduce the impact of sea-level rise in the region could be immense.

*Human Health*: The incidence and extent of some vector-borne diseases are expected to increase with global warming. Malaria, schistosomiasis, and dengue—which are significant causes of mortality and morbidity in Tropical Asia—are very sensitive to climate and are likely to spread into new regions on the margins of presently endemic areas as a consequence of climate change. Newly affected populations initially would experience higher fatality rates. According to one study that specifically focused on climate influences on infectious disease in presently vulnerable regions, an increase in epidemic potential of 12–27% for malaria and 31–47% for dengue and a decrease of schistosomiasis of 11–17% are anticipated under a range of GCM scenarios as a consequence of climate change. Waterborne and water-related infectious diseases, which already account for the majority of epidemic emergencies in the region, also are expected to increase when higher temperatures and higher humidity are superimposed on existing conditions and projected increases in population, urbanization, declining water quality, and other trends.

*Conclusions*: The potential direct effects of climate change assessed here, such as changes in water availability, crop yields, and inundation of coastal areas, all will have further indirect effects on food security and human health. The suitability of adaptation strategies to different climatic environments will vary across the diverse subregions and land uses of the region. Adaptive options include new temperature- and pest-resistant crop varieties; new technologies to reduce crop yield loss; improvements in irrigation efficiency; and integrated approaches to river basin and coastal zone management that take account of current and longer-term issues, including climate change.

## 7. Research Needs

The gaps and deficiencies revealed in this special report suggest some priority areas for further work to help policymakers in their difficult task. These needs include:

- Better baseline data, both climatic and socioeconomic
- Better scenarios, especially of precipitation, extreme events, sulfate aerosol effects, and regional-scale changes
- Better understanding of the ecological and physiological effects of increasing $CO_2$ concentrations, taking account of species competition and migrations, soil and nutrients, acclimation, and partitioning between crop yields, roots, stems, and leaves
- Dynamic models of climate, biospheric processes, and other socioeconomic factors to take account of the developing, time-varying nature of global change
- Impact assessments across a range of scenarios and assumptions to enable the assessment of risk, particularly in regions comprised primarily of developing countries and small island states, where resources for research and assessment have been inadequate to date
- Analysis of adaptation options, including the need for development of new technologies and opportunities for adapting existing technologies in new settings
- Integrated assessments across sectors, from climate change to economic or other costs, across countries and regions, including adaptations, and including other socio-economic changes.

## Authors/Contributors

Robert T. Watson (USA), Marufu C. Zinyowera (Zimbabwe), Richard H. Moss (USA), Reid E. Basher (New Zealand), Martin Beniston (Switzerland), Osvaldo F. Canziani (Argentina), Sandra M. Diaz (Argentina), David J. Dokken (USA), John T. Everett (USA), B. Blair Fitzharris (New Zealand), Habiba Gitay (Australia), Bubu P. Jallow (The Gambia), Murari Lal (India), R. Shakespeare Maya (Zimbabwe), Roger F. McLean (Australia), M.Q. Mirza (Bangladesh), Ron Neilson (USA), Ian R. Noble (Australia), Leonard A. Nurse (Barbados), H.W.O. Okoth-Ogendo (Kenya), A. Barrie Pittock (Australia), David S. Shriner (USA), S.K. Sinha (India), Roger B. Street (Canada), Su Jilan (China), Avelino G. Suarez (Cuba), Richard S.J. Tol (The Netherlands), Laura Van Wie McGrory (USA), Masatoshi Yoshino (Japan)

# 1

# Introduction

# CONTENTS

## 1.1. Scope of the Assessment

Worldwide concern about possible climate change and acceleration of sea-level rise resulting from increasing concentrations of greenhouse gases has led governments to consider international action to address the issue, particularly through the development of the United Nations Framework Convention on Climate Change (UNFCCC). Because the extent and urgency of action required to mitigate the source of the problem—namely the emission of greenhouse gases by human activities—depends on the level of vulnerability, a key question for the Conference of the Parties (COP) to the Convention, and for policymakers in general, is the degree to which human conditions and the natural environment are vulnerable to the potential effects of climate change. Impact assessments are needed to establish the costs and benefits of climatic change as a guide to what adaptation and mitigation measures might be justified. Without such assessments, we run the risk of making uninformed, unwise, and perhaps unnecessarily costly decisions.

The foundation for policy formulation for the climate change problem is scientific information on greenhouse gas emissions, the climate system and how it may change, and the likely impacts on human activities and the environment. To provide the best available base of scientific information for policymakers and public use, governments have requested that the Intergovernmental Panel on Climate Change (IPCC) periodically assess and summarize the current scientific literature related to climate change. The most recent assessment is the Second Assessment Report (SAR), a comprehensive three-volume report completed in 1995 and published in 1996. This assessment involved extensive inputs from thousands of scientists and was reviewed by governments and leading experts. The SAR takes a global view of the impacts of climate change, organizing chapters by ecosystem type or socioeconomic sector (e.g., forests, grasslands, agriculture, and industry).

In making use of the SAR, the UNFCCC negotiators found a need for more explicit information on how different regions of the world might be affected, to better assess their degrees of vulnerability. Accordingly, the Subsidiary Body for Scientific and Technological Advice (SBSTA) of the UNFCCC requested that the IPCC prepare a report that provided a geographically explicit view of the problem, particularly the vulnerabilities for each region. Initially, a Technical Paper was planned (which, under the IPCC rules of procedure, limited the authors to citing only material included in the SAR), but in September 1996 the IPCC XIIth Plenary at Mexico City decided that a Special Report should be produced. This decision was taken to allow the inclusion and proper review of new material postdating the SAR—especially new work emerging from several country studies programs, as well as regional studies which were not included in the SAR due to its global scope.

The present report is the result of this process. This report provides assessments of vulnerability of climate change for 10 regions of the globe: Africa, the Arctic and the Antarctic (Polar Regions), Australasia, Europe, Latin America, Middle East and Arid Asia (Arid Western Asia), North America, Small Island States, Temperate Asia, and Tropical Asia. It also includes several annexes that provide information about climate observations, climate projections, vegetation distribution projections, and socioeconomic baseline assumptions used in the report.

## 1.2. Approach of the Assessment

This report should be read as an assessment of the scientific and technical literature related to the sensitivity, adaptability, and vulnerability of ecosystems and social and economic sectors in the 10 regions—not as a quantitative integrated assessment of impacts. The approach used in preparing the assessment was agreed by the lead authors at a series of scoping meetings held in Washington, DC, in May and September 1996, which set the direction of the assessment when it was being prepared as a technical paper. These meetings were used to review materials from the sectoral assessments of the SAR and organize them into regional analyses, and to identify common issues across the regions and standardize approaches to addressing them. After the paper was reprogrammed as a special report, a series of chapter-specific regional consultations and meetings of lead authors and other experts was held to refine the scope of each regional assessment and to identify studies and methods to use in addition to those used in the SAR. These meetings were held in Toronto, Canada (13–15 January 1997); New Delhi, India (23–25 January 1997); Harare, Zimbabwe (27–29 January 1997); Tarawa, Kiribati (10–13 February 1997); Montevideo, Uruguay (11–13 February 1997); and Amsterdam, The Netherlands (19–21 March 1997).

On the basis of these meetings, the lead authors set about preparing each chapter to provide an assessment of the vulnerability of natural ecosystems, socioeconomic sectors, and human health in the region. The definition of vulnerability used in the SAR was adopted for use by the lead authors in this report: "Vulnerability" is the extent to which climate change may damage or harm a system; it is a function of both the "sensitivity" of a system or structure to climate and the opportunities for "adaptation"[2] to new conditions. Sensitivity is defined as the degree to which a system will respond to a change in climatic conditions (e.g., the extent of change in ecosystem composition, structure, and functioning, including primary productivity, resulting from a given change in temperature or precipitation). The responses may result in either beneficial or harmful effects. Adaptation is defined as adjustments in practices, processes, or structures in response to projected or actual changes in climate. Adjustments can be either spontaneous or planned, reactive or anticipatory. In some cases (e.g., for many ecosystems), options for planned or anticipatory adaptation may not exist. Adaptations can reduce negative impacts or take advantage of new opportunities presented by changing climate conditions. It is in part because of the uncertainties associated with regional projections of climate change (these uncertainties are summarized in Section 1.3.2. and described more fully in Annex B) that this report takes the approach of assessing vulnerabilities, rather than quantitatively assessing expected impacts of climate change.

This report is based upon evidence found in the published literature, which uses a diverse range of methods and models. This diversity reflects current uncertainties regarding the functioning of complex natural and social systems and how they respond to changes in climate. The assessment did not include the performance of new research or computer model simulations by the authors to estimate impacts under common scenarios of greenhouse gas emissions or climate change. Such work was beyond the scope of the report. Because the available studies have not employed a common set of climate scenarios, and because of uncertainties regarding the sensitivity and adaptability of natural and social systems, the assessment of regional vulnerabilities is necessarily qualitative. Often only very general conclusions can be supported by the currently available evidence. In a number of instances, quantitative estimates of impacts of climate change are cited in the report. Such estimates are strongly dependent upon the specific assumptions made and models used. These estimates should not be interpreted as predictions of the most likely impacts, but rather as illustrations of the potential character and magnitude of impacts that may result from specific scenarios of climate change.

Many impacts studies use model simulations for the equilibrium climatic response to a carbon dioxide ($CO_2$) doubling, rather than more recent model simulations of climate change resulting from gradually increasing $CO_2$ concentrations and changing concentrations of aerosols and stratospheric ozone. Thus the level of warming used in many of the impacts studies may not be reached until several decades after 2100, rather than by that date. However, this does not necessarily mean that all impacts will be slowed; for example, the transient simulations exhibit larger land-sea temperature change contrasts, and this would be expected to alter atmospheric circulation and weather patterns in ways not predicted in the equilibrium simulations. Historical observations of the impacts of weather patterns—including droughts, floods, storms, and other extreme weather events—suggest that changes in climate variability could have important impacts on natural and social systems.

Some readers of the special report will be interested only in a particular region, whereas others will be interested in comparing information from different regions. To facilitate such comparison, a common structure, or template, for each regional chapter was developed. The main elements of this chapter template follow:

Executive Summary

Regional Characteristics

- Biogeography (countries, ecosystems, socioeconomic activities covered)
- Trends (key socioeconomic and resource-use information based on data from existing international sources, compiled by the Technical Support Unit in cooperation with World Resources Institute)
- Major climatic zones
- Observed trends for temperature and precipitation (based on IPCC, 1996, WG I, Chapter 3, extended and updated to cover a broader number of contiguous regions)
- Summary of available information on projections of future climate (based on IPCC, 1996, WG I, Chapter 6) and including updated material specific to the region used in regional impact assessments

Sensitivity, Adaptability, and Vulnerability

- Coverage of topics in this section will vary by region, depending on the most important sectors for each region; however, chapters organize the information into the following categories:
  - Ecosystems (including biodiversity)
  - Hydrology/water supply
  - Food and fiber for human consumption (agriculture, forestry, and fisheries)
  - Coastal systems
  - Human settlements and urbanization
  - Human health
  - Other topics particularly relevant to each region (e.g., energy, transport, tourism)

Integrated Assessment of Potential Impacts

- Assessments of illustrative case examples related to ecosystems, water supply/basin management, and socioeconomic activities
- Integrated model results, if available
- Lessons from past fluctuations/variability

This approach is broadly consistent with the seven-step method outlined by the IPCC in its *Technical Guidelines for Assessing Climate Change Impacts and Adaptations* (IPCC, 1994b). These steps are: 1) defining the problem; 2) selecting the method; 3) testing the method/sensitivity; 4) selecting scenarios; 5) assessing biophysical/socioeconomic impacts; 6) assessing autonomous adjustments; and 7) evaluating adaptation strategies.

## 1.3. Baseline Data and Climate Scenarios

### *1.3.1. Climate Observations*

Current trends in regional variations of temperature and precipitation also are important parts of the baseline against which the potential effects of climate change should be assessed. IPCC (1996, WG I, Chapter 3) provided time series plots and global maps depicting trends for temperature and precipitation. This information was extended and updated by one of the lead authors of the WG I assessment (T. Karl, USA). The information was provided to the regional assessment lead authors and is contained in Annex A of this special report, which describes the data sets used for depicting these trends. Additional information based on regional analyses has been added to several of the regional chapters by the lead authors.

### 1.3.2. Climate Scenarios

GCM-based scenarios are the most credible and frequently used projections of climate change. Other types of climate projections include synthetic scenarios and analogue scenarios. These approaches and their limitations are described in IPCC (1994b).

In the IPCC's second assessment (1996, WG I, Chapter 6), seven regions were identified for regional analysis of climate simulations. That analysis was based on transient runs with atmosphere-ocean general circulation models (AOGCMs) suitable for construction of regional climate scenarios, using additional regionalization techniques to improve the simulation of regional climate change. The team of lead authors that conducted that analysis, led by F. Giorgi and G. Meehl, prepared information on the simulation of regional climate change with global coupled climate models and regional modeling techniques for use by the regional assessment teams. That information, which is presented in Annex B of this report, is based entirely on the information included in the WG I contribution to the SAR. No new information has been added to the previous analysis.

The wide range of changes in temperature and precipitation indicated at the time of doubled $CO_2$ concentrations for each region is illustrated in Figures B-1 and B-2, which show large model-to-model differences. Annex B provides the following conclusion regarding the confidence that can be placed in regional climate projections:

> "Analysis of surface air temperature and precipitation results from regional climate change experiments carried out with AOGCMs indicates that the biases in present-day simulations of regional climate change and the inter-model variability in the simulated regional changes are still too large to yield a high level of confidence in simulated change scenarios. The limited number of experiments available with statistical downscaling techniques and nested regional models has shown that complex topographical features, large lake systems, and narrow land masses not resolved at the resolution of current GCMs significantly affect the simulated regional and local change scenarios, both for precipitation and (to a lesser extent) temperature (IPCC, 1996). This adds a further degree of uncertainty in the use of GCM-produced scenarios for impact assessments. In addition, most climate change experiments have not accounted for human-induced landscape changes and only recently has the effect of aerosols been investigated. Both these factors can further affect projections of regional climate change."

The wide range of projected changes in temperature and precipitation suggest that caution is required in treating any impact assessments based on GCM results as firm predictions. This uncertainty is why the term "climate scenarios" has been adopted in most impact assessments. Such scenarios should be regarded as internally consistent patterns of plausible future climates, not as predictions. Decisionmakers need to be aware of the uncertainties associated with climate projections so that they can weigh them in formulation of strategies to cope with the risks of climate change.

The review chapters in this report summarize impact studies based on a range of climate scenarios where they were available. Most studies were based on the older, mixed-layer GCM climate scenarios; results from coupled transient models have only recently become available, and studies using these scenarios are only beginning to be conducted. The older GCM runs estimate stable equilibrium conditions for $1xCO_2$ and $2xCO_2$ climates and generally show more global mean warming than recent transient model runs (see Table 1-1 for a list of equilibrium scenarios used in studies assessed in this special report). In the transient model runs (see Table 1-2 for a listing of transient scenarios cited), in which trace gases increase slowly over a period of years, the full effects of changes in temperature and precipitation lag the effects of changes in atmospheric composition by a number of decades. Thus, in impact studies using transient scenarios (e.g., model studies of potential climate change impacts on vegetation distribution), the positive effects of $CO_2$ on plant growth and other variables dependent upon plant production precede the full effects of changes in climate.

This complication does not mean that impact assessments based on older equilibrium GCM projections are of no value. Rather, it suggests that their results should be carefully interpreted. Where possible, the actual projected changes in temperature, precipitation, and so forth have been stated in the text, and climate scenarios representing the range of potential changes in temperature and precipitation have been used for regions where a range of scenarios is available. Space limitations prevent the presentation of fine detail, but the original source papers and reports are listed. Unfortunately, even some of the original material does not give as much precise information as might be desired.

At the very least, impact assessments based on older climate scenarios can be used to estimate the sensitivity of the various sectors to climate change. New transient GCMs based on improved coupling to the oceans; better scenarios of greenhouse gas and sulfate aerosol emissions; and better representation of processes of cloud formation, water vapor transport, ice/snow formation, vegetation feedbacks, and ocean circulation will produce quantitatively different results.

### 1.3.3. Socioeconomic and Resource-Use Baseline Data

The vulnerability of ecosystems and socioeconomic sectors will be affected by their baseline or initial conditions and by the other stresses to which they may be subjected. For this reason, it is important to examine the vulnerability of these systems and sectors in the context of existing and projected developments. To provide a consistent set of socioeconomic and resource-use data, the Technical Support Unit collated data requested by the authors from, among other sources, *World Resources 1996–97* (WRI/UNEP/UNDP/World Bank, 1996) (see Annex D for a complete list of sources). These data include information on:

- Population and related indicators (1995 population, current population density, projected population density for 2025, and urban and coastal populations)
- Economic indicators [gross domestic product (GDP) per capita; annual growth rate for GDP; and percentage of GDP from agriculture, industry, and services]
- Land cover and use (total land area and amount of land in several categories, including permanent cropland, permanent pasture, forest and woodland, and other land)
- Agricultural activities (amount of irrigated land, size of agricultural labor force, and livestock holdings)
- Water use (water resources per capita and annual water withdrawals for domestic uses, industry, and agriculture)
- Energy use (total commercial energy consumption and consumption of energy sources that are sensitive to changes in climate, including traditional fuels and hydroelectric production)
- Biodiversity (known and endemic mammal, bird, and plant species).

It is important to note that these data are intended simply to provide a consistent set of *assumptions* on important social

***Table 1-1:** The global mixed-layer atmosphere-ocean general circulation models (equilibrium 2x$CO_2$ simulations) used for impact assessment studies in this report.*

| Group | Experiment Acronym | Horizontal Resolution (# of waves or lat x long) | Global Surface Air Temperature Change (°C) | Reference(s) |
|---|---|---|---|---|
| GFDL | A1 | R 15 | 3.2 | Wetherald and Manabe, 1988 |
| GFDL | A2 | R 15 | 4.0 | Manabe and Wetherald, 1987 |
| GFDL | A3 | R 30 | 4.0 | Wetherald and Manabe, 1989 |
| OSU | B1 | 4°x5° | 2.8 | Schlesinger and Zhao, 1989 |
| MRI | C1 | 4°x5° | ~4.3 | Noda and Tokioka, 1989 |
| NCAR | D1 | R 15 | 4.0 | Washington and Meehl, 1984; Meehl and Washington, 1990 |
| NCAR | D2 | R 15 | 4.6 | Washington and Meehl, 1993 |
| CSIRO4 | E1 | R 21 | 4.0 | Gordon *et al.*, 1992; Gordon and Hunt, 1994 |
| CSIRO9 | F1 | R 21 | 4.8 | Whetton *et al.*, 1993; Watterson *et al.*, 1995 |
| GISS | G1 | 8°x10° | 4.8 | Hansen *et al.*, 1984 |
| UKMO | H1 | 5°x7.5° | 5.2 | Wilson and Mitchell, 1987 |
| UKMO | H2 | 5°x7.5° | 3.2 | Mitchell and Warrilow, 1987 |
| UKMO | H3 | 2.5°x3.75° | 3.5 | Mitchell *et al.*, 1989 |
| CCC | J1 | T 32 | 3.5 | Boer *et al.*, 1992; McFarlane *et al.*, 1992; Boer, 1993 |
| MPI | K1 | T 106[a] | – | Bengtsson *et al.*, 1995, 1996 |

**Note**: In general, the findings on impact assessment contained in this report are based on climate change scenarios inferred from the model experiments listed above and cited in IPCC's First Assessment Report (1990) and its supplement (1992).
[a]Time-slice experiments with atmosphere-only ECHAM3 T 106 model.

and economic factors that will influence demands on environmental goods and services (and hence the stresses to which environmental systems may be subjected), as well as the human and financial capacity of societies to adapt to potential climate change. They are not intended to be a definitive source of data on social and economic trends in any particular country. Projections of socioeconomic conditions such as population, incomes, land uses, technological change, economic activity by sector, demands for water and other resources, and other variables are at least as uncertain as regional projections of climate change; as with regional climate information, they should be used as scenarios of future conditions, not treated as predictions.

### 1.3.4. *Development of Integrated Socioeconomic and Climate Change Scenarios*

It is important for policymakers to be able to put climate change impacts in the context of other social, economic, and technological conditions, such as:

- Demographic change
- Land-use change
- Land degradation
- Air and water pollution
- Economic and social change, "development" (including technological change), and poverty.

***Table 1-2:** A brief description of the global coupled atmosphere-ocean general circulation models (transient simulations) used for impact assessment studies in this report.*

| Group | Model Name[a] | Experiment Acronym[b] | Horizontal Resolution (# of waves or lat x long) | GHG Scenario[c] | Global Surface Air Temperature Change at $CO_2$ Doubling (°C) | Reference(s) |
|---|---|---|---|---|---|---|
| BMRC | – | X1 (a) | R 21 | 1%/yr | 1.35 | Colman *et al.*, 1995 |
| GFDL | – | X2 (g) | R 15 | 1%/yr | 2.2 | Manabe *et al.*, 1991, 1992 |
| MRI | – | X3 (p) | 4°x5° | 1%/yr | 1.6 | Tokioka *et al.*, 1995 |
| NCAR | 5° Ocean | X4 (q) | R 15 | 1%/yr | 2.3 | Meehl *et al.*, 1993 |
| NCAR | 1° Ocean | X5 (r) | R 15 | 1%/yr | 3.8 | Meehl, 1996<br>Washington and Meehl, 1996 |
| UKMO | UKTR1 | X6 (s) | 2.5°x3.75° | 1%/yr | 1.7 | Murphy, 1995; Murphy and Mitchell, 1995; Senior, 1995 |
| UKMO | HADCM2 | X7 (z) | 2.5°x3.75° | 1%/yr + aerosols | ~2.5 | Mitchell and Johns, 1997 |
| MPI | ECHAM1+LSG | X8 (m) | T 21 | 1.3%/yr | 1.3 | Cubasch *et al.*, 1992 |
| MPI | ECHAM3+LSG | X9 (y) | T 21 | 1.3%/yr + aerosols | na | Hasselmann *et al.*, 1995 |
| CSIRO | – | X10 (d) | R 21 | 1%/yr | 2.0 | Gordon and O'Farrell, 1997 |
| CCC | CGCM1 | X11 (b) | T 32 | 1%/yr | 2.6 | Boer *et al.*, 1997; Flato *et al.*, 1997 |
| GISS | – | X12 (k) | 4°x5° | 1%/yr | 1.4 | Russell *et al.*, 1995 |

Note: In general, the climate change scenarios described in this document are based on those inferred from the model experiments listed above and reported in the IPCC Second Assessment Report (1996). The future regional projections for combined greenhouse gases (equivalent $CO_2$) and aerosol forcings (based on experiments X7 and/or X9) also have been discussed for some regions.
na = not available
[a]If different from group name.
[b]Parenthetical refers to experiment listed in Table 6.3 of the SAR Working Group I volume (also see Table B-1 in Annex B).
[c]The GHG scenario refers to the rate of increase of $CO_2$ used in the model experiments; most experiments use 1%/yr, which gives a doubling of $CO_2$ after 70 years (IS92a gives a doubling of equivalent $CO_2$ after 95 years).

Thus, each chapter in this report has a section on "integrated assessment," which attempts to draw together the interactions among sectors, countries, and forces of change. Integrated assessment has been tackled at various levels:

- Integrating the chain of effects from changes in atmospheric composition and climate to changes in biophysical systems to socioeconomic consequences (the "vertical" dimension)
- Including the interactions among systems, sectors, and activities (the "horizontal" dimension)
- Considering climate change in the context of other trends and changes in society (the "time" or "global change" dimension).

Some case study examples have been highlighted in the following chapters, but integrated assessment is in its infancy, and the development of new integrated scenarios of socioeconomic changes, emissions of greenhouse gases, and potential changes in climate was not possible in the time available for preparation of this report. This type of analysis is a priority for the IPCC, however; it currently is the focus of two related activities: a special report on emissions scenarios and a task group on climate scenarios for impact analysis. We expect that the Third Assessment Report (TAR) will be based on such an integrated set of scenarios.

The gaps and deficiencies revealed in this special report suggest some priority areas for further work to help policymakers in their difficult task. These needs include:

- Better baseline data, climate and socioeconomic
- Better scenarios, especially of precipitation, extreme events, sulfate aerosol effects, and regional-scale changes
- Better understanding of the ecological and physiological effects of increasing $CO_2$ concentrations, taking account of species competition and migrations, soil and nutrients, acclimation, partitioning between crop yields, roots, stems, and leaves
- Dynamic models of climate, biospheric processes, and socioeconomic factors to take account of the developing, time-varying nature of global change
- Impact assessments across a range of scenarios and assumptions to enable the assessment of risk—particularly in regions composed primarily of developing countries, where resources for research and assessment have been inadequate to date
- More and better integrated assessments across sectors, from climate change to economic or other costs, across countries and regions, including adaptations and other socioeconomic changes.

Clearly, impact assessments have not been made across all potentially affected sectors and regions, so many potential costs and benefits remain to be examined and, where possible, quantified. Nevertheless, we believe the present report summarizes a substantial body of work that, if carefully interpreted, may provide useful guidance to policymakers.

## References


**Bengtsson**, L., M. Botzet, and M. Esch, 1995: Hurricane-type vortices in a general circulation model. *Tellus*, **47A**, 1751–1796.

**Bengtsson**, L., M. Botzet, and M. Esch, 1996: Will greenhouse gas-induced warming over the next 50 years lead to higher frequency and greater intensity of hurricanes? *Tellus*, **48A**, 57–73.

**Boer**, G.J., N.A. McFarlane, and M. Lazare, 1992: Greenhouse gas-induced climate change simulated with the CCC second-generation general circulation model. *J. Climate*, **5**, 1045–1077.

**Boer**, G.J., 1993: Climate change and the regulation of the surface moisture and energy budgets. *Clim. Dyn.*, **8**, 225–239.

**Boer**, G.J., G. Flato, C. Reader, and D. Ramsden, 1997: A transient climate change simulation with historical and projected greenhouse gas and aerosol forcing (to be submitted).

**Colman**, R.A., S.B. Power, B.J. McAvaney, and R.R. Dahni, 1995: A non-flux corrected transient $CO_2$ experiment using the BMRC coupled atmosphere-ocean GCM. *Geophys. Res. Lett.*, **22**, 3047–3050.

**Cubasch**, U., K. Hasselmann, H. Hock, E. Maier-Reimer, U. Mikolajewicz, B.D. Santer, and R. Sausen, 1992: Time dependent greenhouse warming computations with a coupled ocean atmosphere model. *Clim. Dyn.*, **8**, 55–69.

**Flato**, G., G.J. Boer, W.G. Lee, N.A. McFarlane, D. Ramsden, C. Reader, and A. Weaver, 1997: The CCCma Global Coupled Model and its control climate (to be submitted).

**Gordon**, H.B. and B. Hunt, 1994: Climate variability within an equilibrium greenhouse simulation. *Clim. Dyn.*, **9**, 195–212.

**Gordon**, H.B., P.H. Whetton, A.B. Pittock, A.M. Fowler, and M.R. Haylock, 1992: Simulated changes in daily rainfall intensity due to enhanced greenhouse effect: implications for extreme rainfall events. *Clim. Dyn.*, **8**, 83–102.

**Gordon**, H.B. and S.P. O'Farrell, 1997: Transient climate change in the CSIRO coupled model with dynamical sea ice. *Mon. Wea. Rev.*, **125**, 875–901.

**Hansen**, J., A. Lacis, D. Rind, G. Russell, P. Stone, I. Fung, R. Ruedy, and J. Lerner, 1984: Climate Sensitivity: Analysis of feedback mechanisms. In: *Climate Processes and Climate Sensitivity* [Hansen, J.E. and T. Takahashi (eds.)]. American Geophysical Union, Washington, DC, 130–163.

**Hasselmann**, K., L. Bengtsson, U. Cubasch, G.C. Hegerl, H. Rodhe, E. Roeckner, H. von Storch, R. Voss, and J. Waszkewitz, 1995: *Detection of anthropogenic climate change using a fingerprint method.* Proceedings of Modern Dynamic Meteorology Symposium in honour of Aksel Wiin-Nielsen [Ditlevsen, P. (ed.)]. ECMWF Press.

**IPCC**, 1994b: *IPCC Technical Guidelines for Assessing Climate Change Impacts and Adaptations.* Prepared by IPCC Working Group II [Carter, T.R., M.L. Parry, H. Harasawa, and S. Nishioka (eds.)] and WMO/UNEP. University College, London, United Kingdom, and Center for Global Environmental Research, Tsukuba, Japan, 59 pp.

**Manabe**, S. and R.T. Wetherald, 1987: Large scale changes of soil wetness induced by an increase in atmospheric carbon dioxide. *J. Atmos. Sci.*, **44**, 1211–1235.

**Manabe**, S., R.J. Stouffer. M.J. Spelman, and K. Bryan, 1991: Transient responses of a coupled ocean atmosphere model to gradual changes of atmospheric $CO_2$. Part I — Annual mean response. *J. Climate*, **4**, 785–818.

**Manabe**, S., M.J. Spelman, and R.J. Stouffer, 1992: Transient responses of a coupled ocean-atmosphere model to gradual changes in atmospheric $CO_2$. Part II — Seasonal response. *J. Climate*, **5**, 105–126.

**McFarlane**, N.A., G.J. Boer, J.P. Blanchet, and M. Lazare, 1992: The Canadian Climate Centre second general circulation model and its equilibrium climate. *J. Climate*, **5**, 1013–1044.

**Meehl**, G.A. and W.M. Washington, 1990: $CO_2$ climate sensitivity and snow — sea ice albedo parameterization in an atmospheric GCM coupled to a mixed layer ocean model. *Climatic Change*, **16**, 283–306.

**Meehl**, G.A., G.W. Branstator, and W.M. Washington, 1993: Tropical Pacific interannual variability and $CO_2$ climate change. *J. Climate*, **6**, 42–63.

**Meehl**, G.A., 1996: Vulnerability of fresh water resources to climate change in the tropical Pacific region. *J. Water, Air and Soil Pollution*, **92**, 203–213.

**Meehl**, G.A. and W.M. Washington, 1996: El Nino-like climate change in a model with increased $CO_2$ concentrations. *Nature*, **382**, 56–60.

**Mitchell**, J.F.B. and D.A. Warrilow, 1987: Summer dryness in northern mid-latitude due to increased $CO_2$. *Nature*, **330**, 238–240.

**Mitchell**, J.F.B., C.A. Senior, and W.J. Ingram, 1989: $CO_2$ and climate: A missing feedback? *Nature*, **341**, 132–134.

**Mitchell**, J.F.B. and T.J. Johns, 1996: Sensitivity of regional and seasonal climate to transient forcing by $CO_2$ and aerosols. *J. Climate*, **10**, 245–267.

**Murphy**, J.M., 1995: Transient response of the Hadley Centre coupled model to increasing carbon dioxide. Part II — Analysis of global mean response using simple models. *J. Climate*, **8**, 496–514.

**Murphy**, J. and J.F.B. Mitchell, 1995: Transient response of the Hadley Centre coupled model to increasing carbon dioxide. Part II — Temporal and spatial evolution of patterns. *J. Climate*, **8**, 57–80.

**Noda**, A. and T. Tokioka, 1989: The effect of doubling the $CO_2$ concentration on convective and non-convective precipitation in a general circulation model coupled with a simple mixed layer ocean. *J. Met. Soc. Japan*, **67**, 1055–1067.

**Schlesinger**, M.E. and Z.C. Zhao, 1989: Seasonal climate changes induced by doubled $CO_2$ as simulated in the OSU atmospheric GCM/mixed layer ocean model. *J. Climate*, **2**, 459–495.

**Senior**, C.A., 1995: The dependence of climate sensitivity on the horizontal resolution of a GCM. *J. Climate*, **8**, 2860–2880.

**Tokioka**, T., A. Noda, A. Kitoh, Y. Nikaidou, S. Nakagawa, T. Motoi, S. Yukimoto, and T. Takata, 1995: A transient $CO_2$ experiment with the MRI CGCM—quick report. *J. Met. Soc. Japan*, **74(4)**, 817–826.

**Washington**, W.M. and G.A. Meehl, 1984: Seasonal cycle experiment on the climate sensitivity due to a doubling of $CO_2$ with an atmospheric general circulation model coupled to a simple mixed layer ocean model. *J. Geophys. Res.*, **89**, 9475–9503.

**Washington**, W.M. and G.A. Meehl, 1993: Greenhouse sensitivity experiments with penetrative convection and tropical cirrus albedo effects. *Clim. Dyn.*, **8**, 211–223.

**Washington**, W.M. and G.A. Meehl, 1996: High latitude climate change in a global coupled ocean-atmosphere-sea ice model with increased atmospheric $CO_2$. *J. Geophys. Res.*, **101**, 12795–12801.

**Watterson**, I.G., M.R. Dix, H.B. Gordon, and J.L. McGregor, 1995: The CSIRO 9-level atmospheric general circulation model and its equilibrium present and doubled $CO_2$ climates. *Aust. Met. Mag.*, **44**, 111–125.

**Wetherald**, R.T. and S. Manabe, 1988: Cloud feedback processes in a general circulation model. *J. Atmos. Sci.*, **45**, 1397–1415.

**Whetton**, P.H., A.M. Fowler, M.R. Haylock, and A.B. Pittock, 1993: Implications of climate change due to the enhanced greenhouse effect on floods and droughts in Australia. *Climatic Change*, **25**, 289–317.

**Wilson**, C.A. and J.F.B. Mitchell, 1987: A doubled $CO_2$ climate sensitivity experiment with a global climate model including a simple ocean. *J. Geophys. Res.*, **92**, 13315–13343.

# 2

# Africa


MARUFU C. ZINYOWERA (ZIMBABWE), BUBU P. JALLOW (THE GAMBIA), R. SHAKESPEARE MAYA (ZIMBABWE), H.W.O OKOTH-OGENDO (KENYA)

Lead Authors:

*L.F. Awosika (Nigeria), E.S. Diop (Senegal), T.E. Downing (UK), M. El-Raey (Egypt), D. Le Sueur (South Africa), C.H.D. Magadza (Zimbabwe), S. Touré (Cote d'Ivoire), C. Vogel (South Africa)*

Contributors:

*E.L. Edroma (Uganda), A. Joubert (South Africa), W. Marume (Zimbabwe), S.L. Unganai (Zimbabwe), D. Yates (USA)*

# CONTENTS

# EXECUTIVE SUMMARY

Several climate regimes characterize the African continent; the wet tropical, dry tropical, and alternating wet and dry climates are the most common. Many countries on the continent are prone to recurrent droughts; some drought episodes, particularly in southeast Africa, are associated with El Niño-Southern Oscillation (ENSO) phenomena. Deterioration in terms of trade, inappropriate policies, high population growth rates, and lack of significant investment—coupled with a highly variable climate—have made it difficult for several countries to develop patterns of livelihood that would reduce pressure on the natural resource base. Under the assumption that access to adequate financing is not provided, Africa is the continent most vulnerable to the impacts of projected changes because widespread poverty limits adaptation capabilities.

*Ecosystems*: In Africa today, tropical forests and rangelands are under threat from population pressures and systems of land use. Generally apparent effects of these threats include loss of biodiversity, rapid deterioration in land cover, and depletion of water availability through destruction of catchments and aquifers. Changes in climate will interact with these underlying changes in the environment, adding further stresses to a deteriorating situation. A sustained increase in mean ambient temperatures beyond 1°C would cause significant changes in forest and rangeland cover; species distribution, composition, and migration patterns; and biome distribution. Many organisms in the deserts already are near their tolerance limits, and some may not be able to adapt further under hotter conditions. Arid to semi-arid subregions and the grassland areas of eastern and southern Africa, as well as areas currently under threat from land degradation and desertification, are particularly vulnerable. Were rainfall to increase as projected by some general circulation models (GCMs) in the highlands of east Africa and equatorial central Africa, marginal lands would become more productive than they are now. These effects are likely to be negated, however, by population pressure on marginal forests and rangelands. Adaptive options include control of deforestation, improved rangeland management, expansion of protected areas, and sustainable management of forests.

*Hydrology and Water Resources*: Of the 19 countries around the world currently classified as water-stressed, more are in Africa than in any other region—and this number is likely to increase, independent of climate change, as a result of increases in demand resulting from population growth, degradation of watersheds caused by land-use change, and siltation of river basins. A reduction in precipitation projected by some GCMs for the Sahel and southern Africa—if accompanied by high interannual variability—could be detrimental to the hydrological balance of the continent and disrupt various water-dependent socioeconomic activities. Variable climatic conditions may render the management of water resources more difficult both within and between countries. A drop in water level in dams and rivers could adversely affect the quality of water by increasing the concentrations of sewage waste and industrial effluents, thereby increasing the potential for the outbreak of diseases and reducing the quality and quantity of fresh water available for domestic use. Adaptation options include water harvesting, management of water outflow from dams, and more efficient water usage.

*Agriculture and Food Security*: Except in the oil-exporting countries, agriculture is the economic mainstay in most African countries, contributing 20–30% of gross domestic product (GDP) in sub-Saharan Africa and 55% of the total value of African exports. In most African countries, farming depends entirely on the quality of the rainy season—a situation that makes Africa particularly vulnerable to climate change. Increased droughts could seriously impact the availability of food, as in the Horn of Africa and southern Africa during the 1980s and 1990s. A rise in mean winter temperatures also would be detrimental to the production of winter wheat and fruits that need the winter chill. However, in subtropical Africa, warmer winters would reduce the incidence of damaging frosts, making it possible to grow horticultural produce susceptible to frosts at higher elevations than is possible at present. Productivity of freshwater fisheries may increase, although the mix of fish species could be altered. Changes in ocean dynamics could lead to changes in the migratory patterns of fish and possibly to reduced fish landings, especially in coastal artisinal fisheries.

*Coastal Systems*: Several African coastal zones—many of which already are under stress from population pressure and conflicting uses—would be adversely affected by sea-level rise associated with climate change. The coastal nations of west and central Africa (e.g., Senegal, The Gambia, Sierra Leone, Nigeria, Cameroon, Gabon, Angola) have low-lying lagoonal coasts that are susceptible to erosion and hence are threatened by sea-level rise, particularly because most of the countries in this area have major and rapidly expanding cities on the coast. The west coast often is buffeted by storm surges and currently is at risk from erosion, inundation, and extreme storm events. The coastal zone of east Africa also will be affected, although this area experiences calm conditions through much of the year. However, sea-level rise and climatic variation may reduce the buffer effect of coral and patch reefs along the east coast, increasing the potential for erosion.

A number of studies indicate that a sizable proportion of the northern part of the Nile delta will be lost through a combination of inundation and erosion, with consequent loss of agricultural land and urban areas. Adaptation measures in African coastal zones are available but would be very costly, as a percentage of GDP, for many countries. These measures could include erection of sea walls and relocation of vulnerable human settlements and other socioeconomic facilities.

*Human Settlement, Industry, and Transportation*: The main challenges likely to face African populations will emanate from extreme climate events such as floods (and resulting landslides in some areas), strong winds, droughts, and tidal waves. Individuals living in marginal areas may be forced to migrate to urban areas (where infrastructure already is approaching its limits as a result of population pressure) if the marginal lands become less productive under new climate conditions. Climate change could worsen current trends in depletion of biomass energy resources. Reduced stream flows would cause reductions in hydropower production, leading to negative effects on industrial productivity and costly relocation of some industrial plants. Management of pollution, sanitation, waste disposal, water supply, and public health, as well as provision of adequate infrastructure in urban areas, could become more difficult and costly under changed climate conditions.

*Human Health*: Africa is expected to be at risk primarily from increased incidences of vector-borne diseases and reduced nutritional status. A warmer environment could open up new areas for malaria; altered temperature and rainfall patterns also could increase the incidence of yellow fever, dengue fever, onchocerciasis, and trypanosomiasis. Increased morbidity and mortality in subregions where vector-borne diseases increase following climatic changes would have far-reaching economic consequences. In view of the poor economic status of most African nations, global efforts will be necessary to tackle the potential health effects.

*Tourism and Wildlife*: Tourism—one of Africa's fastest-growing industries—is based on wildlife, nature reserves, coastal resorts, and an abundant water supply for recreation. Projected droughts and/or reduction in precipitation in the Sahel and eastern and southern Africa would devastate wildlife and reduce the attractiveness of some nature reserves, thereby reducing income from current vast investments in tourism.

*Conclusions*: The African continent is particularly vulnerable to the impacts of climate change because of factors such as widespread poverty, recurrent droughts, inequitable land distribution, and overdependence on rain-fed agriculture. Although adaptation options, including traditional coping strategies, theoretically are available, in practice the human, infrastructural, and economic response capacity to effect timely response actions may well be beyond the economic means of some countries.

## 2.1. Introduction

This chapter focuses on the potential impacts of climate change on ecosystems, natural resources, and various socioeconomic sectors of mainland Africa. To the extent permitted by the literature, it describes the functions and current status of a number of key resource sectors and ecosystems; the ways in which these systems would respond to changes in climatic conditions; options for adaptation to projected changes in climate; and the vulnerability of each system or sector, taking into account adaptation options as well as impediments to their implementation. Downing (1992, 1996) suggests that vulnerability is an aggregate measure of human welfare that integrates environmental, social, economic, and political exposure to a range of potentially harmful perturbations or threats. Vulnerability varies spatially and temporally for different communities, although they may face the same risk (Eele, 1996). Feasible strategies for coping with future climate changes therefore must be rooted in a full understanding of the complex structure and causes of present-day social vulnerability, through an understanding of vulnerability to climatic variability on seasonal to interannual time scales.

Although Africa, of all the major world regions, has contributed the least to potential climate change because of its low per capita fossil energy use and hence low greenhouse gas emissions, it is the most vulnerable continent to climate change because widespread poverty limits capabilities to adapt. The ultimate socioeconomic impacts of climate change will depend on the relative resilience and adaptation abilities of different social groups. In general, the commercial sector and high-income households in communal areas are better equipped to adjust adequately and in a timely fashion. Much will depend on the coping abilities and mechanisms used by governments and households over the next 50 years or so. Such abilities are determined by political stewardships. If the region manages to achieve reasonable economic growth, the prospects for proper adjustments to climate change are much better than if economic stagnation prevails (Hulme, 1996b).

### *2.1.1. Physical Geography*

Africa has a total land area of 30,244,000 km$^2$. Countries considered in this chapter are listed in Box 2-1, and socioeconomic data are provided in Annex D.

Africa's physical features include a series of plateaus, higher in the east and gradually declining toward the west. The general elevation is relieved by great shallow basins and their river systems; by the deep incision of the 6,400-km Great Rift Valley; and by often-magnificent volcanoes, fault blocks, and inselbergs. Figure 2-1 shows capitals, other major cities, and elevations. The highest point is Mount Kilimanjaro (5,894 m); the lowest point is in the Qattara Depression, at 132 m below sea level. Africa's vast plateaus are broken only by a few rather low mountain ranges—of which the outstanding ones are the Atlas, Ahaggar, Cameroons, Tibetsi, and Ethiopian and east African highlands, as well as the Drakensberg Mountains. In east Africa are (in addition to Kilimanjaro) Mount Kenya (5,199 m), the Ruwenzoris (5,120 m), and Mount Elgon (4,321 m) (Pritchard, 1985).

The African continent encompasses a rich mosaic of ecological settings. Together these ecosystems harbor a wealth of economically and biologically important resources, from individual species to productive habitats (Huq *et al.*, 1996). One quarter of Africa is hyper-arid desert; one third is in the humid climate zone; and the remainder of the continent is dryland, consisting of arid, semi-arid, and dry subhumid areas (UNEP, 1992). These drylands are home to about 400 million people—two-thirds of the continent's total population. Recurrent droughts have long been a permanent feature of life throughout the drylands of Africa. Over the past 30 years or so, however, unusually severe and/or prolonged droughts in these drylands have seriously affected agriculture and wildlife and caused many deaths and severe malnutrition. In some areas, desertification has accompanied these droughts, although the processes leading to desertification are much more varied than climate alone. Currently, 36 countries in Africa are affected by recurrent drought and some degree of desertification (UNEP, 1992). The risk of drought is highest in the Sudano-Sahelian belt and in southern Africa (Nicholson *et al.*, 1988).

**Box 2-1. The Africa Region**

| | |
|---|---|
| Algeria | Libya |
| Angola | Madagascar |
| Benin | Malawi |
| Botswana | Mali |
| Burkina Faso | Mauritania |
| Burundi | Morocco |
| Cameroon | Mozambique |
| Central African Republic | Namibia |
| Chad | Niger |
| Congo | Nigeria |
| Cote d'Ivoire | Reunion |
| Democratic Republic of Congo (DRC) (formerly Zaire) | Rwanda |
| | Senegal |
| | Sierra Leone |
| Djibouti | Somalia |
| Egypt | South Africa |
| Equatorial Guinea | Sudan |
| Eritrea | Swaziland |
| Ethiopia | Tanzania |
| Gabon | The Gambia |
| Ghana | Togo |
| Guinea | Tunisia |
| Guinea-Bissau | Uganda |
| Kenya | Zambia |
| Lesotho | Zimbabwe |
| Liberia | |

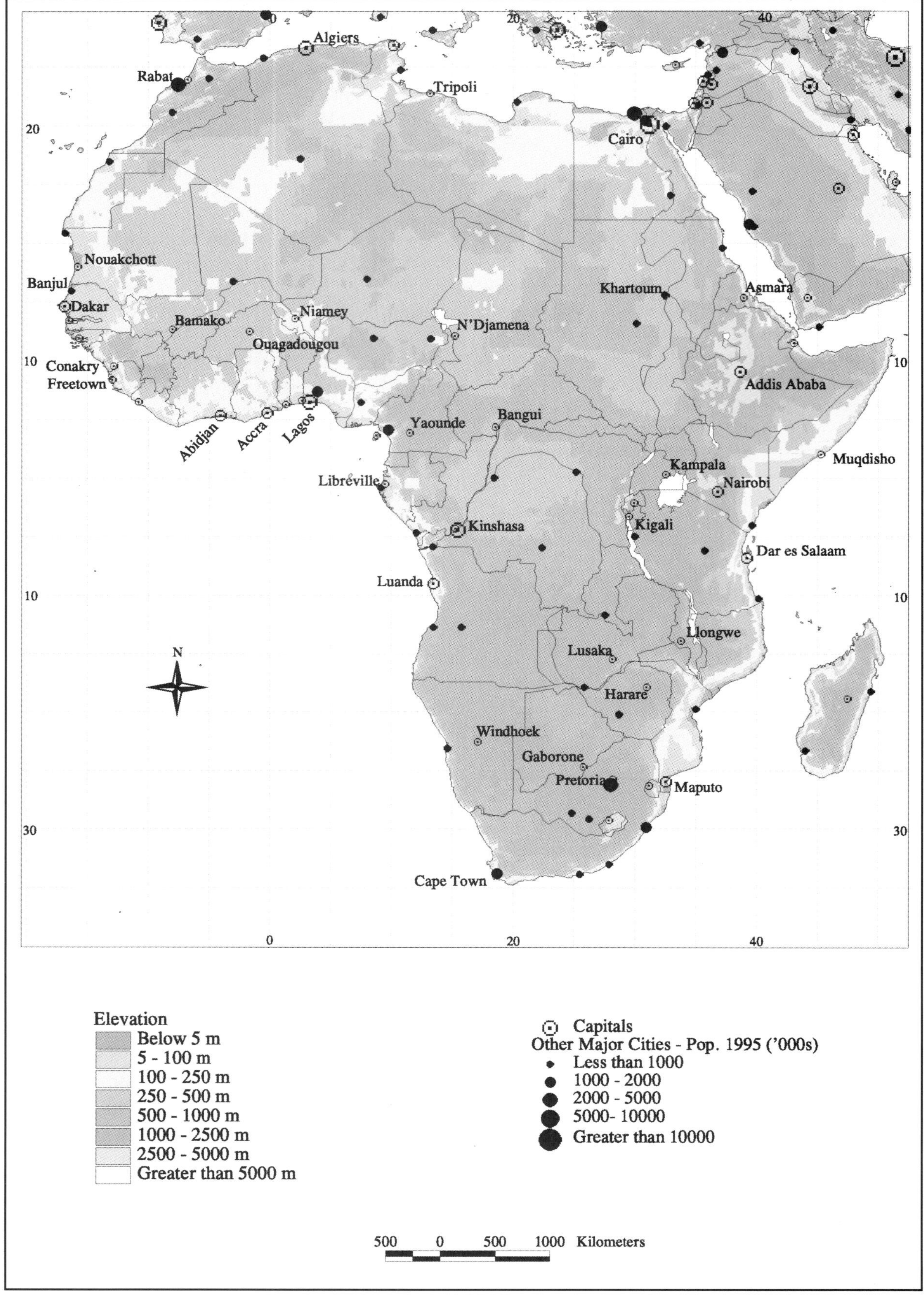


**Figure 2-1:** The Africa region [compiled by the World Bank Environment Department Geographic Information System (GIS) Unit; see Annex E for a color rendition].

### 2.1.2. *People and Natural Resources*

The population of Africa is approximately 650 million (World Bank, 1995b). About two-thirds of the population lives in rural areas and derives its main income from agriculture. In some countries—such as Burkina Faso, Ethiopia, Malawi, and Uganda—rural population makes up more than 80% of the total. Agriculture is the fundamental economic activity in these countries; although it accounts on average for only about 20–30% of GDP, agriculture contributes the largest share of total exports (Cleaver and Schreiber, 1994).

Cultivated land per capita varies considerably across the continent, reflecting the uneven distribution of population (for example, low density in the Congo basin but high in the east African Highlands) as well as low levels of technology and the unsuitability of wide areas for farming. Dependence on shifting cultivation makes the area needed for cropland high. Scarcity of good land, coupled with soil degradation and low levels of inputs and technology, results in increasing deficits in food production. The predominance of rain-fed subsistence agriculture and, across southern Africa, overdependence on (water-demanding) maize has helped ensure that food security for most of the continent is inextricably linked to the quality of each rainy season. In dryland regions, crop and livestock production are extremely susceptible to seasonal rainfall variability and, as a result, have shown considerable volatility in recent years. Drought shock, however, extends well beyond the confines of agriculture (crop) and livestock production, partly because of the important roles these two sectors play in the overall economies of many African states. Crop and livestock production are major employers and make significant contributions to GDP and export earnings. They also are major sources of raw materials for industries such as textiles, food processing, and fuel refining. On the other hand, they are major markets for other industries such as machinery, animal feeds, fuel, and fertilizer producers (Gibberd *et al.*, 1996).

Of the 30 poorest countries of the world, 22 are African (World Bank, 1996). The average income level in sub-Saharan Africa in 1993 was $520 per capita, and the average growth rate per capita was negative (-0.8% per year from 1980 to 1993). By the year 2000, the number of poor in Africa is projected to increase to an estimated 265 million—which would represent more than 40% of the continent's population (World Bank, 1990). In Africa, poverty is linked to the environment in complex ways, particularly in economies that are based on exploitation of natural resources. Degradation of these resources reduces the productivity of poor persons, who most rely on them, and makes poor communities even more susceptible to extreme events (whether meteorological, economic, or political in origin) (World Bank, 1996). Poverty is exacerbated by a demographic profile that continues to record an annual population growth rate of just under 3%, the highest in the world (WRI, 1994). As a result, the population in many African countries will continue to double every 20 to 30 years, well into the 21st century. However, the high incidence of incurable diseases—such as malaria, human immunodeficiency virus (HIV), and hepatitis B—may modify this estimate.

## 2.2. Regional Climate

The continent of Africa is characterized by several climatic regimes and ecological zones. All parts of the continent, except the Republic of South Africa, Lesotho, and the Mediterranean countries north of the Sahara, have tropical climates. These tropical climates may be divided into three distinct climatic zones: wet tropical climates, dry tropical climates, and alternating wet and dry climates (Huq *et al.*, 1996).

Several comprehensive descriptions of the climates of Africa exist, most notably those of Thomson (1965) and Griffiths (1972). Surveys of African rainfall have been carried out by Newell *et al.* (1972), Kraus (1977), Klaus (1978), Tyson (1986), and Nicholson (1994b). These researchers agree that summer rainfall maxima, which are dominant over most of Africa, are controlled primarily by the Inter-Tropical Convergence Zone (ITCZ). Over land, the ITCZ tends to follow the seasonal march

**Box 2-2. Africa's Natural Resources, Economy, and Political Environment**

African economies have made relatively low levels of investment in infrastructure and directly productive capital goods—and hence continue to rely heavily on natural capital (natural resources). This natural capital is at risk because poverty and high population growth often induce land degradation and deforestation—which in turn lead to growing food insecurity and loss of biodiversity. This pattern contributes to migration into rural areas that often are less suitable for agricultural expansion and to urban areas with inappropriate physical, social, and economic infrastructure. The process also contributes to population growth by creating an incentive for large families: Adding family labor becomes one way of coping with the increasing time costs of gathering fuel and water and clearing new land. The severity of this population-agriculture-environment nexus is compounded by low investment in human capital (human resource development), which often restricts individuals to continued reliance on unskilled labor and short-term exploitation of natural resources (the land) as the only feasible survival options.

Although Africa's youthful population represents future social capital, a great deal of investment in terms of education, training, and skills development will be required to ensure its full productivity in the next decade and beyond. Particular attention will need to be paid to capacity-building in information technology to better prepare African societies for efficient, sustainable management of the continent's fragile resources. In addition, the region's political structures, which determine decision making in resource allocation and consumption, will have to be stabilized.

of the sun and oscillates between the fringes of the Sahara in boreal summer and the northern Kalahari desert in the austral summer. The latitude zones of these arid and semi-arid deserts demarcate the tropics from the subtropics. Rainfall in the subtropics is modulated by mid-latitude storms, which may be displaced Equator-ward in winter. Further modification of these broad patterns is provided by natural features such as lakes and mountains, and by the influence of ocean currents. The poleward extremes of the continent have extratropical influences associated with mid-latitude synoptic disturbances, resulting in significant winter rainfall (Griffiths, 1972).

In general, surface air temperatures over most of Africa display a high degree of thermal uniformity, spatially and seasonally (Riehl, 1979). The extreme north and south of the continent, however, experience cold frontal systems that quasi-regularly introduce abrupt air mass changes. Temperatures there are more variable in response to a large annual cycle of insolation and the effects of seasonally varying air masses and winds. Mean temperature trends over the past 100 years, averaged over continental Africa, are shown in Figure 2-2. The highland areas of eastern and southern Africa are substantially cooler than lowland regions, and there is evidence that recent warming trends may have been exaggerated in these mountain areas (Hulme, 1996a).

Rainfall over Africa exhibits high spatial and temporal variability (see Figures 2-3 and 2-4). Mean annual rainfall ranges from as low as 10 mm in the innermost core of the Sahara to more than 2,000 mm in parts of the equatorial region and other parts of west Africa (Figure 2-4). The rainfall gradient is largest along the southern margins of the Sahara—the region known as the Sahel—where mean annual rainfall varies by more than 1,000 mm over about 750 km. This tight rainfall gradient means that relatively small changes in the position of the ITCZ can have large consequences for rainfall in the Sahel; thus, this region is a sensitive indicator of climate change in Africa. Coefficients of rainfall variability in Africa exceed 200% in the deserts; they are about 40% in most semi-arid regions, and between 5% and 20% in the wettest areas (Figure 2-3).

Most policymakers now recognize drought as a normal feature of Africa's climate and acknowledge that its recurrence is inevitable. Widespread occurrences of severe drought during the past three decades have repeatedly underscored the vulnerability of developed and developing societies to its ravages in Africa. Although the variability of African climate is inevitable, loss of human life and economic disruption associated with extreme climatic fluctuation can be lessened by advance warning. Recent scientific advances in understanding the climate system and breakthroughs in predictive capabilities on seasonal time scales provide an opportunity to reduce the vulnerability of human societies by planning for previously unexpected variations from mean climatic conditions. Current forecasting capabilities, although by no means perfect, provide a better indication of climatic conditions that are expected to prevail during the next season or two than simply assuming that rainfall and temperature will be normal (Bonkoungou, 1996).

### 2.2.1. *African Climate Trends and Projections*

Temperature and precipitation trends are reviewed in Annex A of this report. Rainfall trends—especially over the past 30 years or so—have had a very large bearing on socioeconomic development of the continent because most activities are agriculturally based (Serageldin, 1995).

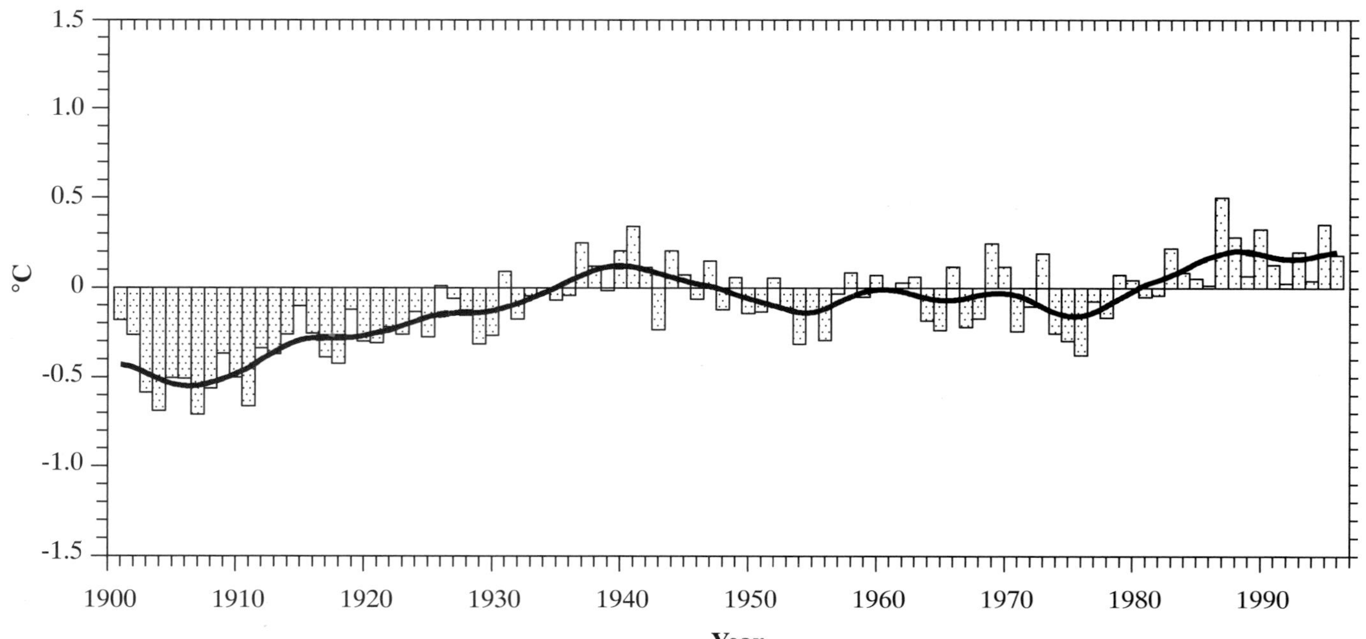


**Figure 2-2**: Observed annual temperature changes in Africa (see Annex A).

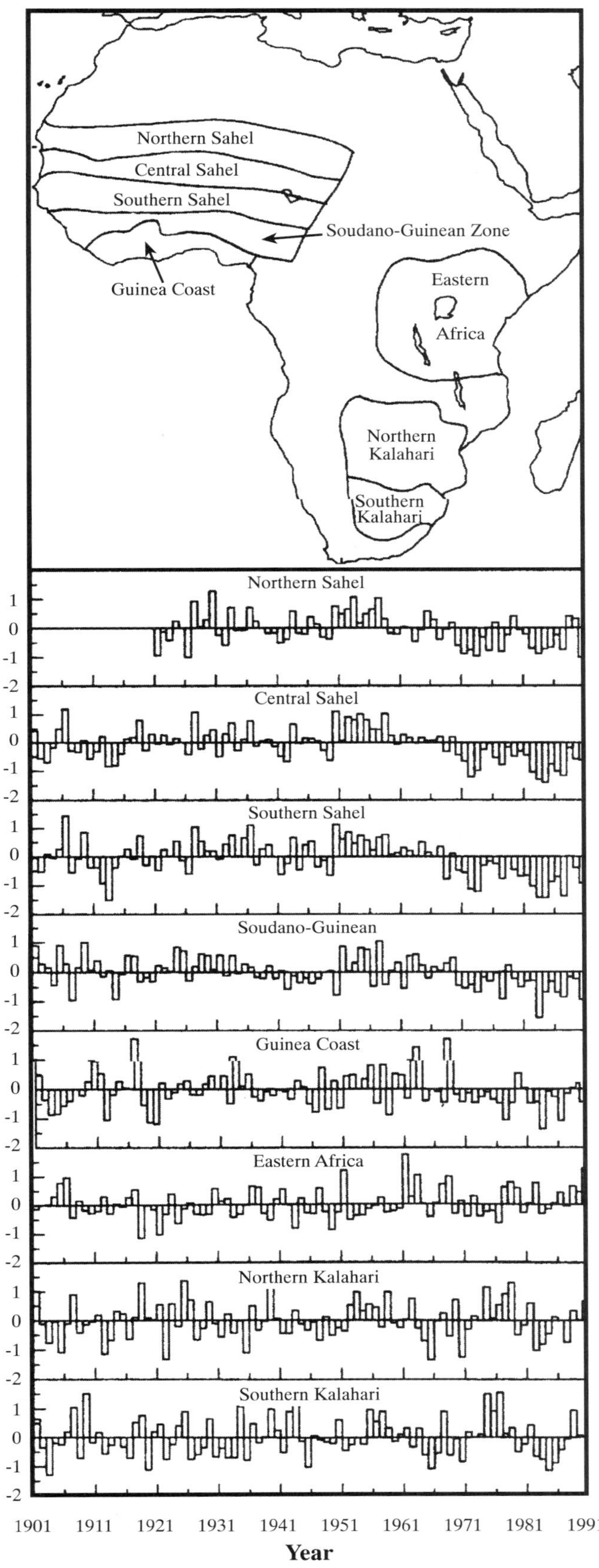


**Figure 2-3**: Rainfall fluctuations for select areas of Africa, expressed as a regionally averaged standard departure (Nicholson, 1993).

Uncertainties in GCMs make deriving regional climate change predictions impossible (see Annex B of this report for an assessment of regional projections of climate change). Therefore, it is important to interpret model outputs in the context of their uncertainties and to consider them as potential scenarios of change for use in sensitivity and vulnerability studies. In IPCC (1990), IPCC (1996), and Hernes *et al.* (1995), the Sahel (10–20°N, 20–40°E) was selected as a study region for purposes of inter-comparing GCM outputs. Since then, several approaches to subregional climate projection have been developed at national levels (e.g., Joubert, 1995; Ringius *et al.*, 1996); several others have been developed under the U.S. Country Studies Program (USCSP, 1996).

Although most initial climate change simulations used GCMs, an increasing number of climate-modeling centers have used regional climate models (RCMs). RCMs rely on similar physical representations of atmospheric processes as GCMs but operate at a much finer spatial resolution—typically 50 km—over limited domains. Little climate change work using RCMs nested within GCMs has been completed as yet for Africa (Ringius *et al.*, 1996), so it remains necessary to rely on extracting regional results for Africa from GCM climate change experiments. A selection of such results is summarized in Box 2-3.

## 2.3. Vulnerabilities and Potential Impacts for Key Sectors

### 2.3.1. *Terrestrial Ecosystems*

#### 2.3.1.1. *Introduction*

Numerous schemes have been used to describe Africa's vegetation and ecosystems. White's (1983) classification system is used here, aggregated (after Justice *et al.*, 1994) to show the rainforest in central Africa (unmarked) and two major categories of woodland savannas divided by moisture and nutrient level: broad-leafed, nutrient-poor, moist savannas; and fine-leafed, nutrient-rich, arid savannas (Figure 2-5). This aggregation summarizes current understanding of the role of soil nutrients and moisture on vegetation distribution in Africa, especially in savannas (Scholes and Walker, 1993). The broad-leafed savannas include the extensive miombo woodlands of central and southern Africa. Nutritionally poor soils support only low-quality grass for grazing, so the numbers of large herbivores is low in miombo and other broad-leafed savannas (Frost, 1996). The fine-leafed savannas include acacia-dominated thorn woodlands, which have higher-quality grass and so support large numbers of large herbivores; these areas constitute the main rangeland region.

Africa is composed essentially of woodlands and grasslands, with rainforests occupying only about 7% of the land area. Africa's rainforests represent slightly less than one-fifth of the total remaining rainforest in the world; Asia and Latin America contain the rest (Sayer *et al.*, 1992). Only about a third of Africa's historical forest extent remains, with west Africa's forests being lost faster than those of any other region. Annual deforestation rates average 0.7% per annum (FAO, 1997).

WRI (1996) indicates that only 8% (0.5 million km[2]) of Africa's regional forest remains as "frontier forest." (Frontier forest is essentially natural/primary forest of sufficient size to support ecologically viable populations of indigenous species.) More than 90% of west Africa's original forest has been lost, and only a small part of what remains qualifies as frontier forest. Of the remaining forest, 77% is under moderate or high threat from logging and commercial hunting to meet growing urban demand for bushmeat. Demands on forests also have escalated in some regions (e.g., as a result of civil unrest that has pushed hundreds of thousands of people into previously intact forest).

Many studies of African ecosystems emphasize particular vegetation types—savanna grasslands, miombo woodlands, mopane woodlands, rangelands, or rainforest—or particular regions: the Sahel, sub-Saharan Africa, or the Southern African Development Community (SADC). In general, the structure of Africa's vegetation is determined by climate at the large scale, then soil type (texture) and nutrient levels at the local scale (Scholes and Walker, 1993). Fire and herbivory are important disturbance factors. Increased moisture in drier areas will likely result in a complex set of feedbacks between nutrients, fire occurrence, decomposition, and competing vegetation.

Increased variability in rainfall and changes in temperature will likely disrupt key ecosystem processes such as phenology and will influence insect pests and diseases in mostly unknown ways. Direct effects on pests will involve disruption of insect life cycles or creation of more suitable conditions for new pests (or for old pests to expand their territory). Ticks, tsetse flies, and locusts are notable examples of serious insect pests in Africa. Although a lot of work has been done to study these insects, a lot remains to be done, especially in relation to how climate change may impact them.

### 2.3.1.2. *Climatic Driving Forces*

Of the many climatic factors that are important for plant growth, among the most significant in relation to climate change are temperature, water availability (determined by precipitation and soil characteristics, as well as other meteorological variables), and carbon dioxide ($CO_2$) concentrations. Consideration of the effects of climate change requires examination of the direct effects of changes in $CO_2$ concentrations and climate variables on the growth of plants, as well as the ways in which these direct effects are modified by soil feedbacks and biological interactions among different organisms.

The effects of temperature changes will vary in different subregions and ecosystems. An increase in temperatures will reduce the incidence of frost damage in areas where this damage occurs and widen the potential geographical range of species that are limited by minimum temperatures. The extent of effects of higher temperatures on African vegetation (e.g., effects on respiration rates, membrane damage) is largely uncertain. Temperature is known to interact with $CO_2$ concentration, so expected increases in respiration resulting from a temperature increase alone may be offset or even reduced by higher $CO_2$ concentrations (Wullschleger and Norby, 1992).

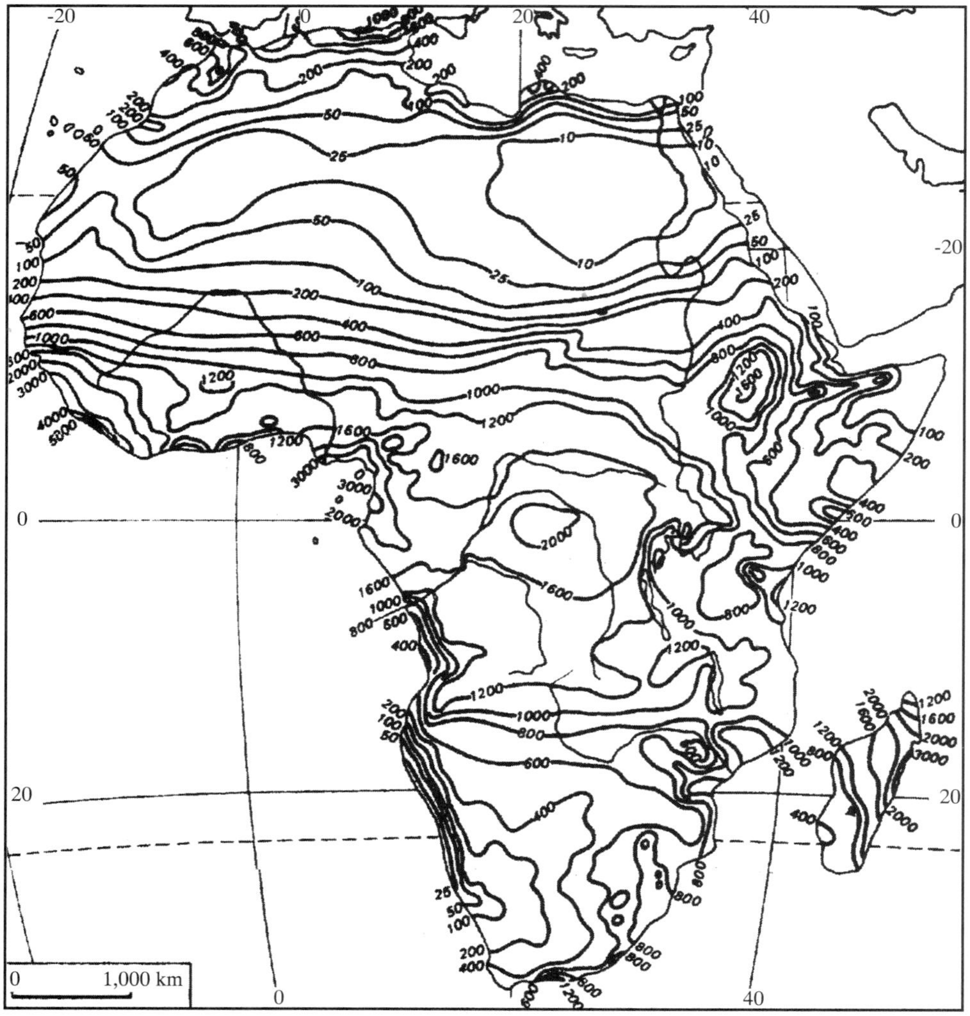

**Figure 2-4**: Mean annual precipitation (mm) in Africa (Martyn, 1992).

**Box 2-3. Climate Scenarios**

With respect to temperature, land areas may warm by 2050 by as much as 1.6°C over the Sahara and semi-arid parts of southern Africa (Hernes *et al.*, 1995; Ringius *et al.*, 1996). Equatorial countries (Cameroon, Uganda, and Kenya) might be about 1.4°C warmer. This projection represents a rate of warming to 2050 of about 0.2°C per decade. Sea-surface temperatures in the open tropical oceans surrounding Africa will rise by less than the global average (i.e., only about 0.6–0.8°C); the coastal regions of the continent therefore will warm more slowly than the continental interior.

Rainfall changes projected by most GCMs are relatively modest, at least in relation to present-day rainfall variability. In general, rainfall is projected to increase over the continent—the exceptions being southern Africa and parts of the Horn of Africa; here, rainfall is projected to decline by 2050 by about 10%. Seasonal changes in rainfall are not expected to be large; Joubert and Tyson (1996) found no evidence for a change in rainfall seasonality among a selection of mixed-layer and fully coupled GCMs. Hewitson and Crane (1996) found evidence for slightly extended later summer season rainfall over eastern South Africa (though nowhere else), based on a single mixed-layer model prediction. Great uncertainty exists, however, in relation to regional-scale rainfall changes simulated by GCMs (Joubert and Hewitson, 1997). Parts of the Sahel could experience rainfall increases of as much as 15% over the 1961–90 average. Equatorial Africa could experience a small (5%) increase in rainfall. These rainfall results are not consistent: Different climate models, or different simulations with the same model, yield different patterns. The problem involves determining the character of the climate change signal on African rainfall against a background of large natural variability compounded by the use of imperfect climate models.

Projected temperature increases are likely to lead to increased open water and soil/plant evaporation. Exactly how large this increased evaporative loss will be would depend on factors such as physiological changes in plant biology, atmospheric circulation, and land-use patterns. As a rough estimate, potential evapotranspiration over Africa is projected to increase by 5–10% by 2050. Little can be said yet about changes in climate variability or extreme events in Africa. Rainfall may well become more intense, but whether there will be more tropical cyclones or a changed frequency of El Niño events remains largely in the realm of speculation.

Changes in sea level and climate in Africa might be expected by the year 2050. Hernes *et al.* (1995) project a sea-level rise of about 25 cm. There will be subregional and local differences around the coast of Africa in this average sea-level rise—depending on ocean currents, atmospheric pressure, and natural land movements—but 25 cm by 2050 is a generally accepted figure (Joubert and Tyson, 1996). For Africa south of the Equator, simulated changes in mean sea-level pressure produced by mixed-layer and fully coupled GCMs are small (~1 hPa)—smaller than present-day simulation errors calculated for both types of models (Joubert and Tyson, 1996). Observed sea-level pressure anomalies of the same magnitude as simulated changes are known to accompany major large-scale circulation adjustments associated with extended wet and dry spells over the subcontinent.

In most of Africa, water availability is projected to have the greatest impact on plant processes. Individual species are adapted to particular water regimes and may perform poorly and possibly die out in conditions to which they are poorly adapted (e.g., Hinckley *et al.*, 1981). The effects of climate change will vary—depending, for example, on how particular plant types use water (water-use efficiency, WUE) or the amount of water available in the soil. Plants are grouped into $C_3$, $C_4$, and CAM plant types depending on how the process of photosynthesis takes place (see IPCC 1996, WG II, Section A.2.2). $C_3$ plants (which include most trees and crop species such as wheat, rice, barley, cassava, and potato) have relatively poor WUE, unlike $C_4$ plants (most of the tropical grasses and agricultural species such as maize, sugarcane, and sorghum). Higher $CO_2$ concentration will likely improve water-use efficiency and growth in $C_3$ plants in water-limited environments. $C_4$ and CAM plants (including desert plants such as cacti) are unlikely to be affected directly by changes in $CO_2$ concentration.

The amount of water available to plants over the course of the year affects plant growth and location across soil and climate types. Available soil water (in combination, of course, with other factors) generally controls the growth cycle (beginning and end) and other events such as when to leaf, shed leaves, set buds, and so forth. Water availability and temperature indices and parameters (maxima and minima, heat sums, cold sums) have been used to relate the distribution of vegetation formations to climatic factors (for more details on these plant biogeography models, see Section 2.3.1.4). However, large uncertainties in GCM precipitation projections constrain our ability to project ecosystem responses to changes in climate. Thus, improving climate modeling at the regional scale is a priority in most of Africa, where ecosystem processes are limited by moisture.

#### *2.3.1.3. Soils, Plant Growth, and Land Degradation*

Many African soils are agriculturally poor, because they are very old, badly leached, and often infertile. Laterites (the oldest soils) are agriculturally unproductive. Laterized red earths are younger and less leached and occur in regions of heavy rainfall, so they are quite agriculturally productive. Nonlaterized red earths, which are found in drier regions (e.g., savannas), are good agricultural soils. Upland red earths are an immature group that occasionally are intensively farmed. In regions of moderate rainfall, the most fertile soils are located in the high veld of southern Africa and parts of west Africa. The black soils—the vertisols—are very fertile but become adhesive during the rainy season and almost rock-like in the drought period. In arid regions, soil humus is very low; moisture often is drawn upward by capillarity and, on evaporation, deposits dissolved minerals in a crust at the surface. In Mediterranean regions, the summer drought, the absence of frost, and the small degree of chemical weathering has led to poorly formed soils.

The ability for soil to support particular natural or agricultural communities is fundamental in any future scenarios of ecosystem development. Soil development is slow and will likely lag climate and vegetation change. In the short term, changes in the soil-water regime and turnover of organic matter and related mineralization or immobilization of nitrogen and other nutrients will have the greatest effect on ecosystem functions. Among the factors that will affect these soil processes, fire and land use probably are the most important. Changes in fire regimes (e.g., frequency, intensity) will directly influence organic-matter processes in the soil and nutrient fluxes—and so will have a significant impact on how soils will function. Land use and land-use history on given sites influence nutrient dynamics and the potential for erosion damage.

Land degradation—defined as "a reduction in the capability of the land to support a particular use" (Blaikie and Brookfield, 1987)—is a major problem in Africa and the whole world. Support by African countries for the Convention on Desertification (United Nations, 1992)—which recognizes that 66% of the continent is desert or dryland, and 73% of the agricultural drylands already are degraded—clearly shows that most African governments are aware of this problem.

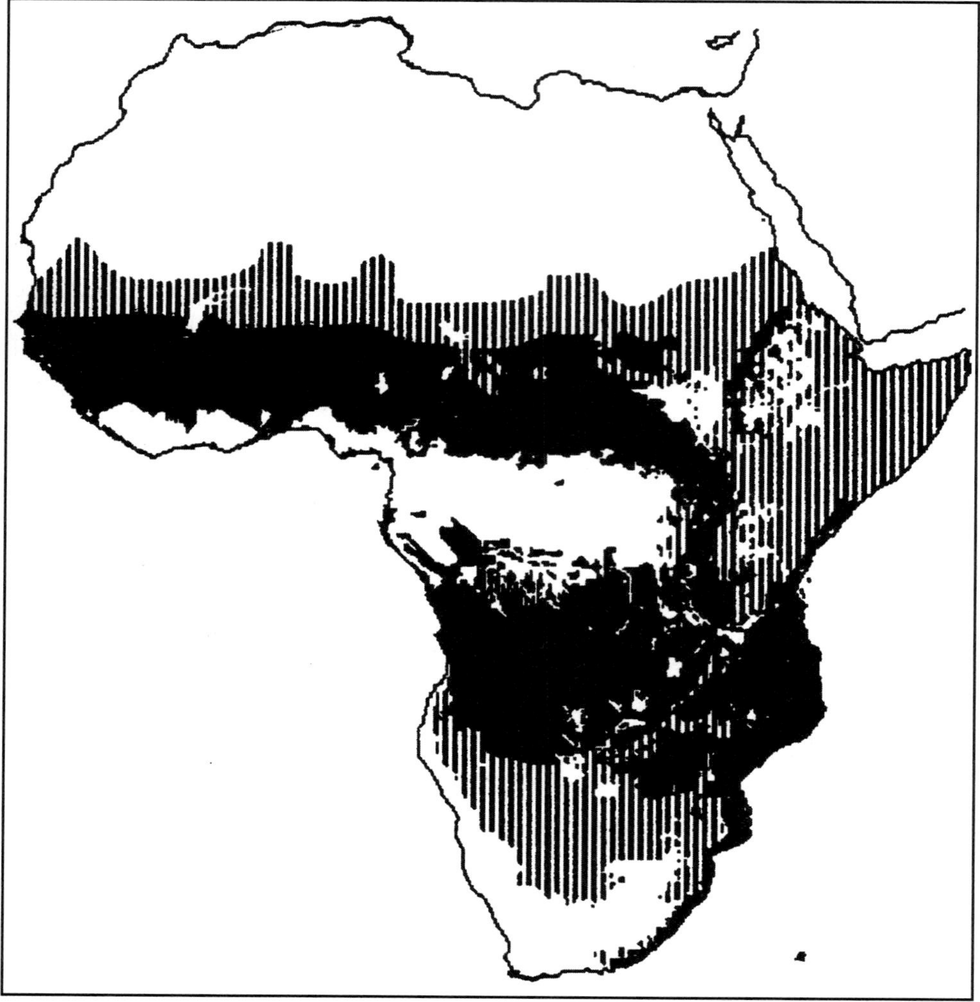

**Figure 2-5**: The distribution of broad-leafed and fine-leafed savannas in Africa. Dark-shaded areas are broad-leafed, nutrient-poor, moist savannas, and striped areas are fine-leafed, nutrient-rich, arid savannas. This map has been derived from White (1983) by reclassifying woodland and wooded grassland map units into one of the two savanna classes according to the dominant tree species.

Recognized forms of land degradation include soil erosion, salinization, soil contamination, loss of soil organic matter, decline in nutrient levels, acidification, and loss of soil structure. Low rainfall, long dry seasons, recurrent droughts, mobile surface deposits, skeletal soils, and sparse vegetation encourage desertification (Le Houerou, 1989; Dregne, 1983; Kassas, 1995). A combination of climatic variations and human land-management practices can lead to excessive land degradation, eventually leading to desertification. Thus, efforts to reduce vulnerability to climate change must take into account land management and the social and economic factors that drive people's use of the land.

Most studies of soil erosion have looked at soil loss from plot-based measurements and then extrapolated to estimate total soil loss per hectare. Although soil erosion clearly is a major problem in many parts of Africa, simple extrapolations from plots to whole countries and into the future can be misleading. Erosion is a major problem locally, and steps must be taken to combat soil erosion at the farm and catchment levels. Reij *et al.* (1996) argue for participatory approaches to soil and water conservation, rather than large-scale, top-down interventions that encompass technology alone. The social and economic context is critical for success.

### 2.3.1.4. Forest and Woodland Ecosystems

Forests in Africa are of great socioeconomic importance as sources of timber, fuel, and many nonwood products, as well as for the protection of water resources. Ecologically, they serve critical roles in water, carbon, and nutrient cycling. The impacts of climate change on forests at the continental scale will be assessed in very broad terms using biome distribution models.

In geological time scales, forest boundaries fluctuated a great deal during the Pleistocene epoch (Sayer *et al.,* 1992); the forests of Africa even now are considerably more extensive than they were during the most recent high-latitude glacial advance about 18,000 years ago. Environmental conditions in tropical Africa at about 18,000 before present (BP) are quite well known, thanks to a large number of pollen and plant microfossil studies (see Hamilton, 1988, for more details). During the severe arid period around 18,000 BP, core areas (centers of biotic diversity) were the main centers of forest survival. Two such principal core areas are located in Cameroon/Gabon and eastern Democratic Republic of Congo (DRC) (formerly Zaire); other, less-diverse core areas are in west Africa and near the east African coast (Sayer *et al.,* 1992). The core areas are not only rich in numbers of species and endemics but also are centers of distribution of disjunct species. Some of these species are unlikely to be able to disperse from core area to core area without continuous forest cover. For example, gorillas are disjunctly distributed across the Zaire basin (Figures 2-6 to 2-8). Although the forests between the two populations seem suitable for the species, some explanation is needed regarding how their ranges became fragmented. A likely explanation for the gorilla and

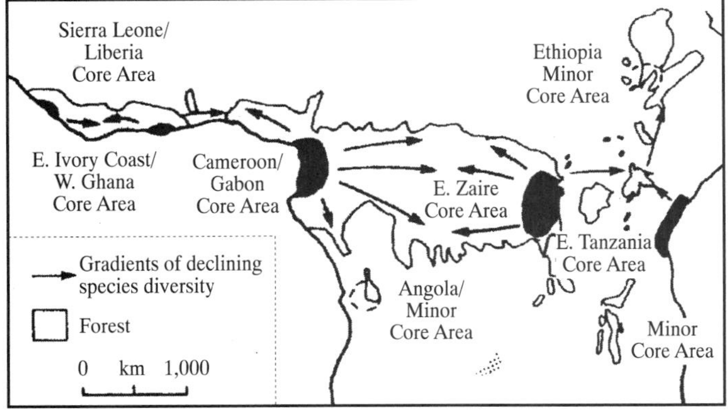


**Figure 2-6**: Distribution of forest, core areas, and gradients of decreasing biotic diversity in tropical Africa. Core areas are believed to approximate sites of forest refugia at the time of the last world glacial maximum (18,000 BP) (Hamilton, 1988).

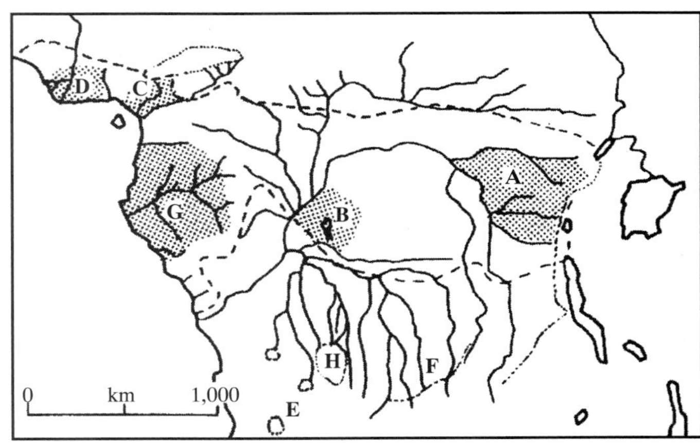


**Figure 2-7**: Forest refugia during arid periods in central Africa: A) Central refuge; D,C,G) Cameroon/Gabon refuge (D, Niger section; C, Cameroon section; G, Gabon or Ogoowe basin section); B) southern Zaire basin refuge; E) north Angola refuge; F) southern scarps of Zaire basin; and H) Lunda Plateau (Kingdon, 1980).

other obligate forest species is that their ranges became fragmented as a result of forest retraction at times of aridity and that these species subsequently have been slow to expand their range to include all potentially suitable habitat. Thus, animal species may not adapt fast enough to rapid changes in climate.

Current vegetation distribution can be studied using biome distribution models. Biome distribution models such as MAPSS (Neilson, 1995) and BIOME3 (Haxeltine and Prentice, 1996) simulate the distribution of potential global vegetation based on local vegetation and hydrological processes and the properties of plants. They simulate the mixture of life forms (such as trees, shrubs, and grasses) that can coexist at a site while in competition with each other for light and water. These models simulate the maximum carrying capacity, or vegetation density (in the form of leaf area), that can be supported at the site under the constraints of energy and water. A change in leaf area can be interpreted as a change in overall carrying capacity or standing crop of the site, regardless of whether it is potential natural vegetation or under cultivation. This carrying capacity potential is the basis for application of these models to all of Africa to indicate possible shifts in agricultural or vegetation potential.

MAPSS and BIOME3 outputs were generated (see Annex C) using several GCM scenarios, with and without $CO_2$ effects and aerosol emissions. Table 2-1 summarizes changes in areal coverage for four main biome types in Africa (the shrub/woodlands biome was combined with grasslands). The models indicate a net shift from more arid biomes (low leaf area index—LAI) to more mesic biomes (higher LAI) for the OSU, GFDL, and UKMO scenarios. Exact percentages of change and where this change would occur are highly uncertain because the models indicate potential, not actual, vegetation.

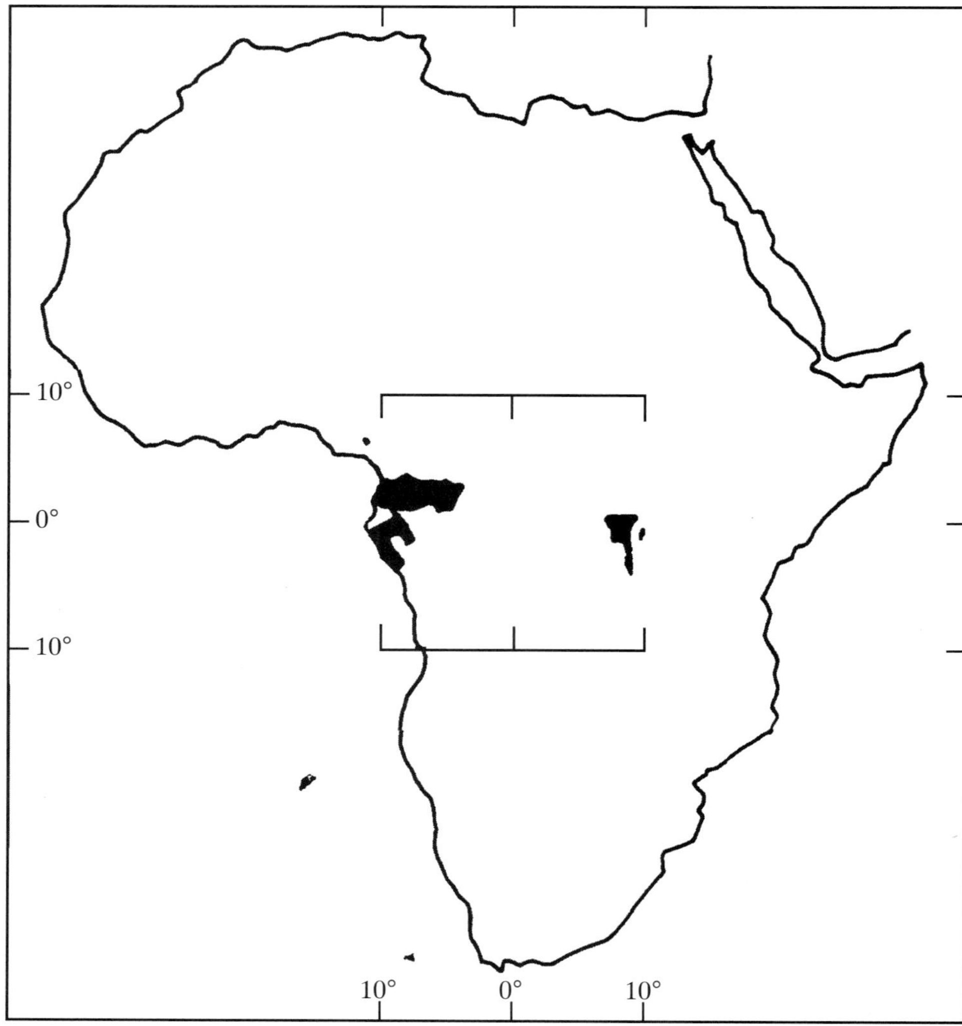


**Figure 2-8**: The distribution of gorillas in Africa (Harcourt *et al.*, 1989).

Biome distribution models simulate only static vegetation (i.e., equilibrium vegetation at a future date). Factors that affect vegetation dynamics, such as competition and disturbance (fire, herbivory), are not considered. These factors have not yet been incorporated in global vegetation models (but see Woodward and Steffen, 1996, for plans and developments in this area). An important factor to consider with regard to whether a vegetation type can respond as simulated is migration. Migrating species would require corridors of unmanaged land. The fragmented nature of remaining African vegetation (outside the rainforests) would make vegetation responses to climate change difficult. The destructive aspects of fires also would reduce migration. The dynamics of savannas and woodlands (such as miombo) are strongly linked to fires, so likely changes in fire intensity and frequency will have unknown consequences on vegetation.

The biome models are able to capture some overall divisions (e.g., rainforest versus woodland versus arid), but they do not yet capture fine-scale detail for Africa. This limitation is a result of the quality of the climate data used as inputs, coarse soils information, and the nature of the models—which were primarily designed to describe vegetation in temperate climates.

Some progress is being made in the development of models of vegetation dynamics that include the effects of fire and herbivory for African vegetation (e.g., Menaut *et al.,* 1991; Van Dalaan and Shugart, 1989; Desanker, 1996). However, these models have to be widely tested and validated before they can be used to evaluate impacts of climate change at the broad scale. Results of biogeochemical modeling in savannas using the CENTURY model of Parton *et al.* (1992) are discussed in Section 2.3.1.5.

***Table 2-1:*** *Changes in areal coverage (in 1,000 km²) of major biome types under current and GCM scenarios using MAPSS (with a direct $CO_2$ effect).*

| Biome Type[1] | Current[2] | OSU[3] | GFDL[3] | UKMO[3] |
|---|---|---|---|---|
| Tropical broad-leafed forest | 2,986 | 5,752 | 4,798 | 2,909 |
| Savanna/woodland | 8,845 | 8,662 | 9,462 | 11,449 |
| Lumped shrub and grass | 8,713 | 7,534 | 8,083 | 8,025 |
| Arid | 8,814 | 7,497 | 7,146 | 7,200 |

[1]Biome types are defined in the MAPSS model description in Annex C. Minor biome types have been ignored in this table; therefore, column totals will not be the same. Model experiments are $2xCO_2$ equilibrium scenarios and are described in Annex C.
[2]1961–90 average climate.
[3]All scenarios are $2xCO_2$ equilibrium scenarios.

### *2.3.1.5. Rangelands*

Rangelands in Africa (i.e., grasslands, savannas, and woodlands, which contain both grasses and woody plants) cover approximately $2.1 \times 10^9$ ha. Africa's livestock population of about 184 million cattle, 3.72 million small ruminants (sheep and goats), and 17 million camels extracts about 80% of its nutrition from these vast rangelands (IPCC 1996, WG II, Table 2-1). In addition, Africa's rangelands support a vibrant tourist industry that, in many countries, is the leading contributor to gross national product (GNP). Because Africa's population has been growing at about 3% annually, the rangelands recently have become an arena for intense human and animal conflict, leading to serious reduction in spatial distribution and diversity of species. This reduction is likely to be exacerbated by projected changes in climate.

From a land-use perspective, there are differences between west Africa and east Africa in the way rangelands are used. In arid and semi-arid areas of west Africa (rainfall 5–600 mm), millet (or another crop) is planted over a unimodal (one peak in rainfall per year) rainy season (3–4 months); then fields remain fallow over the 8- to 9-month dry season. Livestock eat crop residues. Land use is dominated by cultivation, with livestock playing a subsidiary role in the village economy (Ellis and Galvin, 1994). In east Africa, by contrast, areas with higher rainfall (up to 600 mm) are inhabited by pastoral people rather than farmers. The Ngisonyoka of northern Kenya, for example, are pastoral nomads, completely dependent on their livestock for food, livelihood, and life (Galvin, 1992). In dry parts of eastern Africa, cultivation is uncommon and occurs mainly where irrigation is possible or where water can otherwise be sequestered and stored for cropping (Ellis and Galvin, 1994). Rainfall is bimodal (two peaks in rainfall per year) in most east African rangelands, resulting in two plant-growing seasons. This pattern has important implications for natural vegetation and rain-fed agriculture (de Ridder *et al.,* 1982). According to Ellis *et al.* (1987), Turkana pastoralists in northern Kenya say that the best years for livestock production are not necessarily those with the greatest rainfall. Rather, years in which moderate rainfall extends over several months, resulting in a long period of foliage production and livestock milk production, are good years. Thus, the distribution and timing of rainfall will be at least as important as total annual amounts projected under climate change.

Rangelands are noted for high climatic variability and high frequency of drought events. They have a long history of human use. The combination of climatic variability and human land use make rangeland ecosystems more susceptible to rapid degeneration of ecosystem properties (Parton *et al.,* 1996). Because these systems develop under highly variable rainfall regimes, they are conducive to rapid changes in ecosystem structure given modifications in fire and grazing patterns (Archer *et al.,* 1994; Ojima *et al.,* 1994) and altered climate regimes (OIES, 1991).

At the very broad scale, simulations with the biome models (MAPSS and BIOME3) projected increases in the extent of rangelands, mainly as a result of a reduction in the area of arid and semi-arid desert resulting from the reduction of drought stress with projected higher rainfall. However, ecosystem process models are more appropriate in analysis of this specific ecosystem type. Ojima *et al.* (1996) used the CENTURY (Parton *et al.,* 1992) ecosystem process model of plant-soil interactions to analyze the impact of climate and atmospheric $CO_2$ changes on grasslands of the world, including 7 of 31 sites in Africa. Ojima *et al.* (1996) looked at the effects of increasing $CO_2$ and climate, using climate change scenarios based on the Canadian Climate Centre (CCC) and GFDL GCMs. They found that changes in total plant productivity were positively correlated to changes in precipitation and nitrogen mineralization (with nitrogen mineralization the most important factor). The response to nitrogen mineralization was consistent with the general observation that grasslands respond positively to addition of nitrogen fertilizer (Rains *et al.,* 1975; Lauenroth and Dodd, 1978). Plant responses to $CO_2$ were modified in complex ways by moisture and nutrient availabilities; their results generally indicated that $CO_2$ enrichment had a greater effect with higher moisture stress. However, nutrient limitations reduced $CO_2$ responses. Ojima *et al.* (1996) concluded that increased atmospheric $CO_2$ will offset the negative effects of periodic droughts, making grasslands more resilient to natural (and human-induced changes in) climate variability. The strength of this beneficial effect, however, is controlled by the availability of nitrogen and other nutrients, which tend to be limited in many African landscapes.

### *2.3.1.6. Deserts*

Deserts are an environmental extreme characterized by low rainfall that is highly variable intra-annually and interannually. Desert air is very dry; incoming solar and outgoing terrestrial radiation are intense, with large daily temperature fluctuations; and potential evaporation is high. Many organisms in the deserts already are near their tolerance limits (IPCC, 1996). The Sahara in north Africa and the Namib desert in southwest Africa are classified as the hottest deserts in the world—with average monthly temperatures above 30°C during the warmest months and extremes above 50°C. The diurnal temperature range often is large; winter nights in the Namib Desert sometimes are as cold as 10°C (IPCC 1996, WG II, Section 3.3.1) or lower. Extreme desert systems already experience wide fluctuations in rainfall and are adapted to coping with sequences of extreme conditions. Initial changes associated with climate change are less likely to create conditions significantly outside present ranges of tolerance; desert biota show very specialized adaptations to aridity and heat, such as obtaining their moisture from fog or dew (IPCC 1996, WG II, Section 3.4.2).

### *2.3.1.7. Mountain Regions*

Mountains usually are characterized by sensitive ecosystems and regions of conflicting interests between economic development and environmental conservation. In Africa, most mid-elevation ranges, plateaus, and high-mountain slopes are under considerable pressure from commercial and subsistence farming activities (Rogers, 1993). Mountain environments are potentially vulnerable to the impacts of global warming. This vulnerability has important ramifications for a wide variety of human uses—such as nature conservation, mountain streams, water management, agriculture, and tourism (IPCC 1996, WG II, Section 5.2).

There is a general picture of continuing ice retreat on the mountains. On Mount Kenya, the Lewis and Gregory glaciers have shown recession since the late 19th century (IPCC 1996, WG II, Box 5-3). Changes in climate (as projected in Greco *et al.*, 1994) could reduce the area and volume of seasonal snow, glacier, and periglacial belts—with a corresponding shift in landscape processes. The retreat of some glaciers on Kilimanjaro and Mt. Kenya would have significant impacts on downstream ecosystems, people, and their livelihoods because of moderation of the seasonal flow regimes of rivers upstream. Further reduction of snow cover and glaciers also could reduce the scenic appeal of African high mountain landscapes for tourists and thus have a negative impact on tourism.

Forest fires would increase in places where summers become warmer and drier. Prolonged periods of summer drought would transform areas already sensitive to fire into regions of sustained fire hazard. Mt. Kenya and mountains on the fringes of the Mediterranean Sea already subject to frequent fire episodes could be affected (IPCC 1996, WG II, Section 5.2.2.3).

#### 2.3.1.8. *Adaptation and Vulnerability*

There is potential for spontaneous and assisted adaptation in Africa. Many options will need to involve a combination of efforts to reduce land degradation and foster sustainable management of resources. This section highlights options for forestry and woodlands, rangelands, and wildlife.

A number of adaptive processes designed to prevent further deterioration of forest cover already are being implemented to some degree. Some of these measures involve natural responses when particular tree species develop the ability to make more efficient use of reduced water and nutrients under elevated $CO_2$ levels. Other adaptive measures involve human-assisted action programs (such as tree planting) designed to minimize undesirable impacts. These strategies will include careful monitoring and microassessment of discreet impacts of climate change on particular species. Low-latitude forest adaptation options, especially in west Africa, must include active vegetation and soil management. For example, Gilbert *et al.* (1995) have indicated that silvicultural practices, endangered species habitat management, watershed manipulation, and antidesertification techniques could be applied given current infrastructure in Cameroon and Ghana. These adaptive measures will help reduce climate change impacts on forest watersheds and semiarid woodlands. Smith and Lenhart (1996) have identified enhancement of forest seed banks as an adaptation policy option for maintaining access to a sufficient variety of seeds to allow the original genetic diversity of forests to be rebred. Genetic diversity also provides an assurance that benefits provided by forests are not lost forever (Smith and Lenhart, 1996) and is particularly relevant to the maintenance of the forests in the Sahel and other extremely sensitive regions of Africa where 20 years of recurrent drought have degraded the forests. Mwakifwamba (1997) asserts that adaptation strategies or measures in Tanzania should focus mainly on reducing high deforestation rates, protecting existing forests, and introducing new species or improving existing species.

For rangelands, Milton *et al.* (1994) present a conceptual model of arid rangeland degradation that suggests that degradation proceeds in steps—increasingly difficult and costly to reverse—and discusses adaptation options (see Box 2-4). Assisted management is a lot harder for wildlife in game reserves than for livestock. Monitoring is required to identify populations at risk (from deforestation), as well as reserved areas that are changing their vegetation types in response to climate, leaving some animals in habitat types that are not suitable. Massive fragmentation of previous forests and woodlands makes it difficult for wildlife to migrate along corridors to areas with more water and foliage. Close monitoring would identify groups of wildlife that are in danger, and steps can be taken to move them to suitable habitat.

At the institutional level, mechanisms need to be created (or improved upon) to facilitate the flow of scientific results into the decision-making and policy-making process. Joint planning of projects that would impact cross-boundary catchment areas will become increasingly important if the climate becomes more variable and water more scarce for many regions of Africa.

### 2.3.2. **Hydrology and Water Resources**

Water resources are inextricably linked with climate, so the prospect of global climate change has serious implications for water resources and regional development (Riebsame *et al.*, 1995). Efforts to provide adequate water resources for Africa will confront a number of challenges, including population pressure, problems associated with land use such as erosion/siltation, and possible ecological consequences of land-use change on the hydrological cycle. Climate change will make addressing these problems more complex.

#### 2.3.2.1. *Hydrological Systems*

Africa has several surface water bodies spread throughout the continent. Table 2-2 lists the 10 largest surface-water bodies in sub-Saharan Africa, along with basin countries and basin area (after Rangeley *et al.*, 1994). Other smaller water bodies exist within country boundaries. Africa has the greatest number of rivers and water bodies that cross or form international boundaries. The 10 river basins in Table 2-2 (including Lake Chad) have a total drainage area greater than 350,000 km$^2$, and they

**Box 2-4. A Conceptual Model of Arid Rangeland Degradation**

Overuse by a narrow suite of domesticated herbivores has led to progressive loss of secondary productivity and diversity in rangelands. Degraded rangelands may not return to their original state, even when they are rested for decades (Westoby *et al.,* 1989; O'Connor, 1991). Milton *et al.* (1994) develop the idea that the probability of reversing grazing-induced change may be inversely related to the amount of disturbance involved in the transition. They develop a step-wise model of rangeland degradation and show how the potential for recovery appears to be related to the function of the affected component. Their study stresses the need to recognize and treat degradation early because management inputs and costs increase for every step in the degradation process. Steps and management options are described below.

Similar models can be constructed for climate effects, to conceptualize potential impacts and points of intervention.

*Steps and management options for arid rangeland degradation.*

Stepwise degradation of arid or semi-arid rangelands. Symptoms describe the state of plant and animal assemblages; management options refer to actions that a manager could take to improve the condition of the range; and management level refers to the system (level of the food chain) on which management should be focused.

| | |
|---|---|
| **Step 0** | |
| Description: | Biomass and composition of vegetation varies with climatic cycles and stochastic events (e.g., droughts, diseases, hail, frost, fire) |
| Symptoms: | Perennial vegetation varies with weather |
| Management Option: | Adaptive management, involving timely manipulations of livestock densities |
| Management Level: | Secondary producers (i.e., grazers and herbivores) |
| **Step 1** | |
| Description: | Herbivory reduces reestablishment of palatable plants, allowing populations of unpalatable species to grow |
| Symptoms: | Demography of plant population changes (age-structural changes) |
| Management Option: | Strict grazing controls |
| Management Level: | Secondary producers |
| **Step 2** | |
| Description: | Plant species that fail to establish are lost, as are their specialized predators and symbionts |
| Symptoms: | Plant and animal losses, reduced capacity to support herbivores |
| Management Option: | Manage vegetation (e.g., add seed, remove plants) |
| Management Level: | Primary producers (i.e., vegetation) |
| **Step 3** | |
| Description: | Biomass and productivity of vegetation fluctuates as ephemerals and weed species benefit from loss of cover from perennial plants |
| Symptoms: | Perennial biomass reduced (short-lived plants and instability increase), resident birds decrease, nomadic bird species |
| Management Option: | Manage soil cover (e.g., mulching, erosion barriers, roughen soil surface) |
| Management Level: | Physical environment (soil) |
| **Step 4** | |
| Description: | Denudation and desertification involve changes in soil function and soil microbe activity |
| Symptoms: | Vegetation cover completely lost, erosion accelerated; soil salinization, aridification |
| Management Option: | Difficult to address; costs of restoration or rehabilitation too high; nonpastoral use of land only economic option |
| Management Level: | Difficult to identify |

*Table 2-2: The 10 largest surface-water bodies in sub-Saharan Africa (Rangeley et al., 1994).*

| Basin | No. of Countries | Basin Area (1,000 km²) | Basin Countries |
|---|---|---|---|
| Congo | 9 | 3,720 | Zaire, Central African Republic, Angola, Congo, Zambia, Tanzania, Cameroon, Burundi, Rwanda |
| Nile | 10 | 3,031 | Sudan, Ethiopia, Egypt, Uganda, Tanzania, Kenya, Zaire, Rwanda, Burundi |
| Niger | 9 | 2,200 | Mali, Nigeria, Niger, Guinea, Cameroon, Burkina Faso, Benin, Cote d'Ivoire, Chad |
| Lake Chad | 6 | 1,910 | Chad, Niger, Central African Republic, Nigeria, Sudan, Cameroon |
| Zambezi | 8 | 1,420 | Zambia, Angola, Zimbabwe, Mozambique, Malawi, Botswana, Tanzania, Namibia |
| Orange | 4 | 950 | South Africa, Namibia, Botswana, Lesotho |
| Okavango | 4 | 529 | Botswana, Angola, Namibia, Zimbabwe |
| Limpopo | 4 | 385 | South Africa, Botswana, Mozambique, Zimbabwe |
| Volta | 6 | 379 | Burkina Faso, Ghana, Togo, Cote d'Ivoire, Benin, Mali |
| Senegal | 4 | 353 | Mali, Mauritania, Senegal, Guinea |

combine to affect 33 sub-Saharan countries and Egypt. Sharma *et al.* (1996) assert that few of the transboundary river basins in the region are effectively jointly managed. Effective management would require treaties, political commitment, institutions, capacity, information, and finance. National interests often override regional objectives. The large number of countries belonging to multiple river and lake basins makes regional cooperation very difficult. Table 2-3 shows dependence on external surface water for selected countries. Coordinated action among African countries will determine whether countries in the region can effectively adapt to changes in the hydrology of African rivers and lakes.

The major effects of climate change on African water systems will be through changes in the hydrological cycle, the balance of temperature, and rainfall. A case study of the impacts of climate change on the Zambezi and Nile River basins follows, based on Riebsame *et al.* (1995). Additional literature on the Zambezi basin includes Calder *et al.* (1996), Pinay (1988), Balek (1977), Conway and Hulme (1993), Vorosmarty and Moore (1991), Vorosmarty *et al.* (1991), and du Toit (1983).

The Nile and Zambezi basins are the second and fourth largest river systems in Africa; key geographic characteristics are depicted in Figure 2-9 and key hydrological characteristics given in Table 2-4. Both the Nile and Zambezi have a low runoff efficiency and a high dryness index, indicating a high sensitivity to climate change. Analysis showed the Nile as very sensitive while the Zambezi was fairly sensitive. Although the severity of the impacts of climate change depended primarily on the magnitude of change, the different hydrological sensitivities of the basins are also important. The Nile and Zambezi are especially sensitive to climate warming: Runoff decreases

*Table 2-3: Dependence on external surface water—selected countries (after Gleick, 1993).*

| Country | % Total Flow Originating Outside Border | Ratio of External Water Supply to Internal Supply[1] |
|---|---|---|
| Egypt | 97 | 32.3 |
| Mauritania | 95 | 17.5 |
| Botswana | 94 | 16.9 |
| The Gambia | 86 | 6.4 |
| Congo | 77 | 3.4 |
| Sudan | 77 | 3.3 |
| Niger | 68 | 2.1 |
| Senegal | 34 | 0.5 |

[1]"External" represents river runoff originating outside national borders; "internal" includes average flows of rivers and aquifers from precipitation within the country.

in these basins even when precipitation increases, due to the large hydrological role played by evaporation.

There were striking responses in runoff for the Nile. Riebsame *et al.* (1995) conclude that despite potential adjustments, Nile flows throughout the basin are extremely sensitive to temperature and precipitation changes. GCM scenarios provide widely diverging pictures of possible future river flows, from a 30% increase to a 78% decrease. There are formal agreements between Egypt and Sudan on the allocation of flows from the Nile, now and under any future enhancements. However, any reductions over 20% would exceed the management capability of the agreements and

*Table 2-4: Hydrological characteristics for the Zambezi and Nile River basins (extracted from Riebsame et al., 1995).*

| Parameter | Zambezi | Nile | Blue Nile |
|---|---|---|---|
| Length (km) | 2,600 | 6,500 | 1,000 |
| Area (km$^2$ x 10$^3$) | 1,330 | 2,880 | 313 |
| Flow (m$^3$/sec) | 4,990 | 2,832 | 1,666 |
| Flow (10$^9$ m$^3$/yr) | 157 | 89 | 53 |
| Specific Discharge (l/sec-km$^2$) | 3.8 | 1.0 | 5.3 |
| Runoff (R) (mm) | 118 | 31 | 168 |
| Precipitation (P) (mm) | 990 | 730 | 784 |
| R/P | 0.12 | 0.04 | 0.21 |
| PET/P | 2.50 | 5.50 | 1.80 |

Note: PET = Potential evapotranspiration.

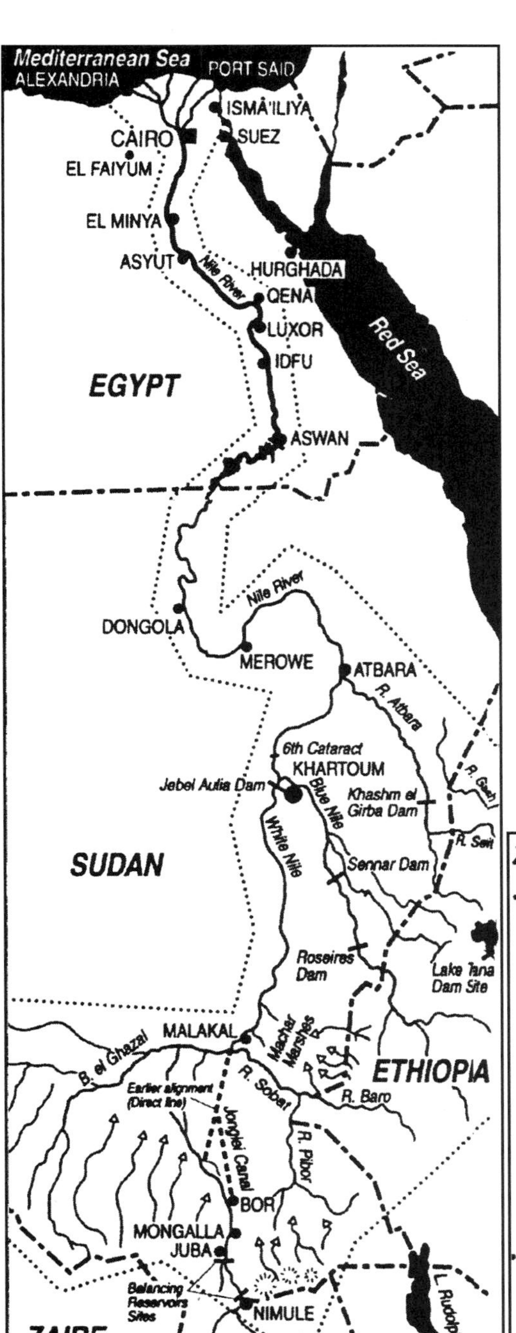


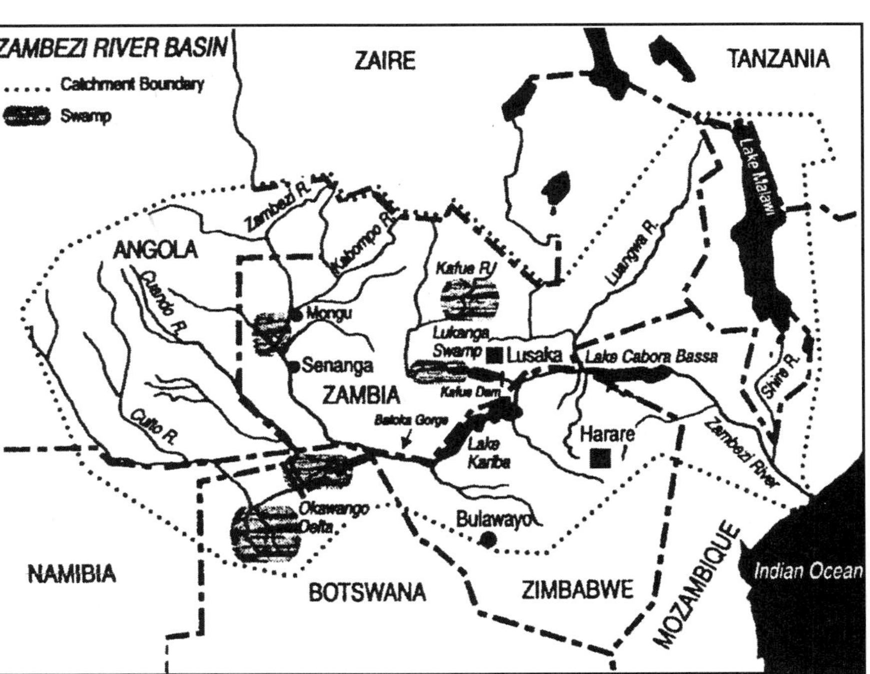


**Figure 2-9**: The Nile and Zambezi River basins. For the Zambesi River basin, climate change impacts were projected for the basin above Lake Kariba with the existing Kariba Dam and with a proposed new dam and reservoir at Batoka Gorge (Riebsame *et al.*, 1995).

would result in major social and economic impacts. Adjustments in response to climate change would either involve changes in water allocation or structural adjustments in the upper and lower basin. The large uncertainty in climate-change projections makes it very hard for basin managers to adopt any response policy. There is need for a regional climate modeling effort over the Nile to help reduce this uncertainty. It remains prudent to make capital investments in decreasing water demand via more efficient irrigation management as a very wise adaptation to climate change.

The seasonal runoff pattern for the Zambezi remained relatively unchanged; the river was sensitive, however, to temporal shifts of the rainy season. There was a net deficit in river flows due to higher surface temperatures, which increase the rate of evapotranspiration. Hydropower production at Kariba decreased slightly under the GISS and GFDL scenarios, while the cooler scenarios of UKMO and GISS led to small increases in power generation. Seasonality of flow had more marked effects on production, a function of storage capacity of the dams in relation to ability to store excess and regulate water flows. Under climate change, there would be less water entering Kariba, and this would likely lead to reduction in fish populations. Adaptation to climate change for the Zambezi basin was suggested to depend on better planning of water projects that consider hydrological inter-relationships of the basin as a whole, crossing many national boundaries. This requirement for countries to look beyond their own needs is a major factor in implementing adaptation options.

The Niger River runs over 4,000 km across west Africa, and its basin covers about one third of the subregion including Guinea, Cote d'Ivoire, Mali, Burkina Faso, Niger, Benin, Nigeria, Cameroon, and Chad. The pressure on this river basin is intense. For example, the Sahelian drought of the 1970s seriously affected hydropower generation from Nigeria's Kaiji Dam on the Niger River during the 1973-77 period. This caused a severe shortfall in power generation to consumers in Nigeria, Mali, Benin, and Chad.

There is some concern that the negative impacts of climate change on water supply could be larger (and the gains smaller) than those reported in current assessments. Many GCMs have not explicitly incorporated the influence of persistent drought in evaluating the impact of global warming. In particular, equilibrium models begin each year with no model memory of groundwater depletion in a preceding year. Yet the successive accumulation of back-to-back drought years often can have devastating effects on groundwater, runoff, reservoir storage, marginal agricultural activities, and water quality (Cline, 1992).

Despite relatively small climatic changes projected for the tropics, tropical lakes also may be quite sensitive to climate change (see Box 2-5). The level of Lake Victoria (in eastern Africa) rose rapidly in the early 1960s following only a few seasons with above-average rainfall and has remained high since (Sene and Pinston, 1994—as cited in IPCC 1996, WG II, Section 10.5.2).

**Box 2-5. African Lakes and Climate Change**

African Great Lakes are sensitive to climate variation on time scales of decades to millennia (Kendall, 1969; Livingstone, 1975; Haberyan and Hecky, 1987). Lake Victoria (the world's second-largest freshwater lake by area), Lake Tanganyika (the world's second-deepest lake), and Lake Malawi were closed basins for extended periods in the Pleistocene and Holocene epochs (Owen *et al.*, 1990). Lakes Malawi and Tanganyika were hundreds of meters below their current levels; Victoria dried out completely. Today these lakes are in delicate hydrological balance and are nearly closed. Only 6% of the water input to Tanganyika leaves at its riverine outflow (which was totally blocked when the lake was explored by Europeans) (Bootsma and Hecky, 1993).

Higher temperatures would increase evaporative losses, especially if rainfall also declined. Minor declines in mean annual rainfall (10–20%) for extended periods would lead to the closure of these basins even if temperatures were unchanged (Bootsma and Hecky, 1993). Tropical temperatures are increasing; temperatures in the 1980s were 0.5°C warmer than a century earlier and 0.3°C warmer than during the period 1951–1980. Concurrently, Lake Victoria's epilimnion was warmer by 0.5°C in the early 1990s than in the 1960s (Hecky *et al.*, 1994). Although current climate scenarios project only small increases in tropical temperatures, small changes in temperature and water balance can dramatically alter water levels, as well as mixing regimes and productivity.

Recent temperature and rainfall data and GCM simulations indicate increasing aridity in the tropics (Rind, 1995). Increases of 1–2°C in air temperatures could substantially increase the stability of stratification in permanently stratified Tanganyika and Malawi. Their deep waters are continuously warm, but the <1°C difference between surface and deep water in warm seasons maintains a density difference that prevents full circulation. Lake Tanganyika's deep water has been characterized as a "relict" hypolimnion that formed under a cooler climate within the past 1,000 years (Hecky *et al.*, 1994). Since then, warming has created a barrier to vertical circulation. Additional warming could strengthen this barrier and reduce the mixing of deep, nutrient-rich hypolimnetic water and nutrient-depleted surface layers; that mixing sustains one of the most productive freshwater fisheries in the world (Hecky *et al.*, 1981).

Source: IPCC 1996, WG II, Box 10-1.

### 2.3.2.2. Water Supply

Water supply undoubtedly is a most important resource for Africa's social, economic, and environmental well-being. Currently, about two-thirds of the rural population and one-quarter of the urban population are without safe drinking water, and even higher proportions lack proper sanitation. Climate change will likely make the situation more adverse. The greatest impact will continue to be felt by the poor, who have the most limited access to water resources. This section focuses mainly on sub-Saharan Africa (SSA).

Availability of water in SSA is highly variable. Only the humid tropical zones in central and west Africa have abundant water. Availability of water varies considerably within countries, too, influenced by physical characteristics and seasonal patterns of rainfall. According to Sharma *et al.* (1996), eight countries were suffering from water stress or scarcity in 1990; this situation is getting worse as a consequence of rapid population growth, expanding urbanization, and increased economic development. By 2000, about 300 million Africans risk living in a water-scarce environment. Moreover, by 2025, the number of countries experiencing water stress will rise to 18—affecting 600 million people (World Bank, 1995b). Figure 2-10 shows how countries will shift from water surplus to water scarcity as a result of population changes alone between 1990 and 2025, using a per capita water-scarcity limit of 1,000 $m^3$/yr. The scarcity statistics also can be associated with challenges to international water resources: Eight river basins already face water stress, and four face scarcity (Figure 2-11); Figure 2-12 illustrates water availability in the year 2025 (taking account of population increase alone) (Sharma *et al.*, 1996).

During the 1980s and 1990s, drought affected urban areas and industry very severely. Most water-dependent industries in southern Africa were forced to reduce their activities after water reservoirs fell to critical levels. Beverage companies, which use a lot of water to wash bottles, had to change to non-returnable aluminum cans (which require less water). Botswana's construction and textile industries had to retrench workers after operations were scaled down because of a severe shortage of water. Similar problems hit Bulawayo, the heart of Zimbabwe's industrial sector; companies were almost forced to pull out and relocate elsewhere because of a lack of water, and half of the small businesses crumbled. In South Africa, Swaziland, and Zimbabwe, sugarcane industries almost ground to a halt because there was no water for irrigation. Power rationing in Kenya in 1996–97 as a consequence of drought severely disrupted the country's manufacturing and engineering industries (UNEP, 1997).

Unfortunately, there are few assessments of how climate change or responses to it could affect local wetland biodiversity. The climate change scenarios of Greco *et al.* (1994) project that there could be less water in most of the large rivers in the Sahel over the next 30–60 years, with the possible exception of the major rivers flowing into Lake Chad. This shift would mean less available water in the large wetlands along these rivers, unless there are changes to the management of outflow from dams. Changes to the hydrology of smaller wetlands will depend not only on climate change but also on whether they are supplied with surface water or groundwater, as well as the extent of cropping in their catchment areas. The loss of small wetlands may lead to a significant risk of extinction for local populations of turtles and small birds (Gibbs, 1993), although

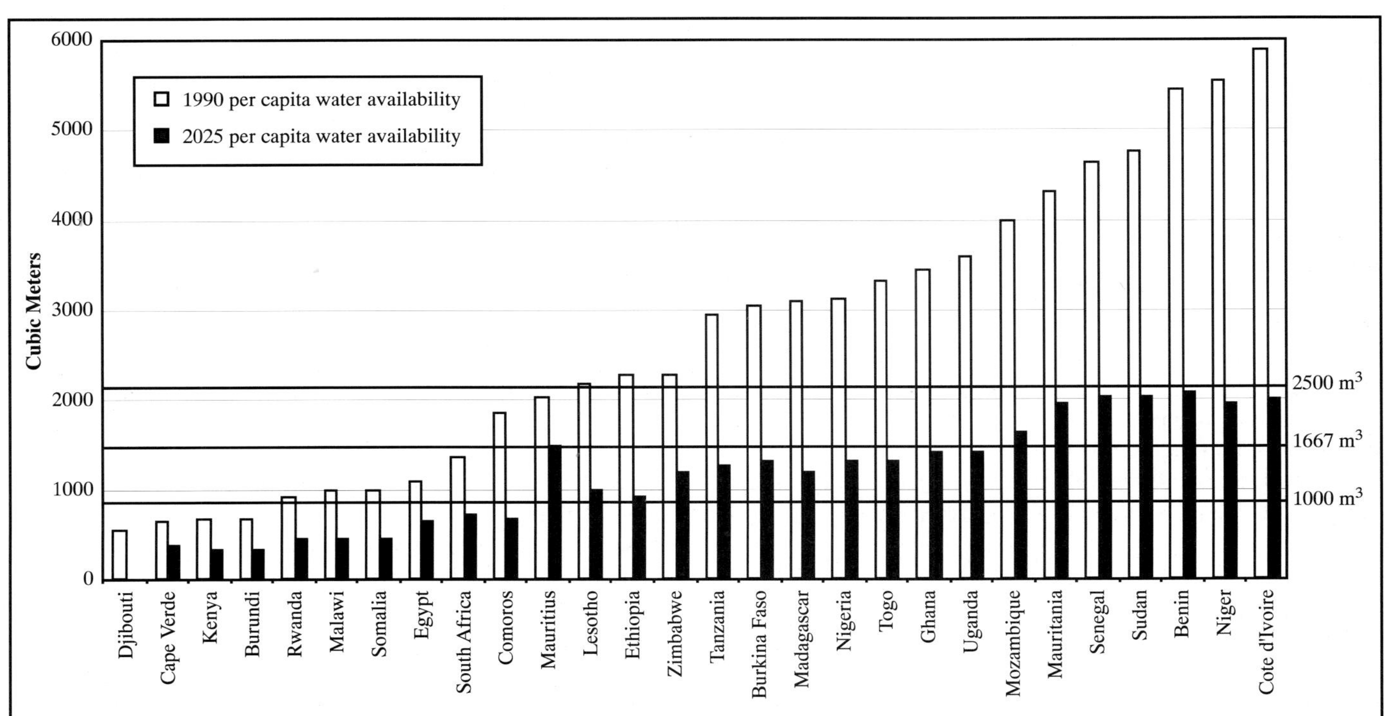


**Figure 2-10**: Water scarcity and people in Africa (Sharma *et al.*, 1996).

taxa that are easily transported as adults, eggs, cysts, larvae, and so forth would be subjected to less risk (Magadza, 1991). If wetlands in the eastern Sahel become drier, relatively mobile birds that are dependent on wetland habitat could move into wetlands further east (i.e., in Niger, northern Nigeria, northern Cameroon, Chad).

#### *2.3.2.3. Adaptation and Vulnerability*

Climate change will have various effects on water resources and water management in Africa. The large variability in projected climate scenarios over Africa's most vulnerable river basin systems (such as the Nile) makes any policy reformulation in anticipation of climate change difficult. However, improved efficiency in irrigation systems and water use are strongly recommended modes of action because they will benefit the region regardless of the degree and direction of climate change. Detailed studies of the river basins are essential to provide adequate information for planning and negotiation purposes in this area that will continue to generate tension across many borders.

Sharma *et al.* (1996) have evaluated the sub-Saharan African countries with respect to their degree of national commitment and planning to address water problems in general and have developed a list of country performance indicators (Table 2-5). Columns 4–7 describe the enabling environment; columns 1 and 2 are poverty indicators; and columns 1, 2, and 3 are risk indicators, where problems will call for either more water or more efficient management of existing stocks. The following points are critical:

- The extent of political stability, ownership of development efforts, and commitment to sustainable water resources management in each country
- The extent to which an enabling environment exists—consisting of transparent and accountable governance in the water sector, clear legislation and policy, strategies and investment programs, stakeholder participation, and the capacity for water resources management at all levels
- The extent to which information and knowledge exists to gauge water availability and quality, consumer demand, and sectoral needs (e.g., sanitation coverage, irrigation, hydropower).

Knowledge also is needed about multiple cross-sectoral linkages relating to a nation's water development (competing demands from agriculture, industry, and municipalities; reliance upon international waters). Depending on how a country fares with regard to the three critical points above, the types of efforts and interventions required by funding agencies and nations will vary. Countries that fare poorly in this analysis will be most vulnerable to climate changes because they will have less capacity to adapt.

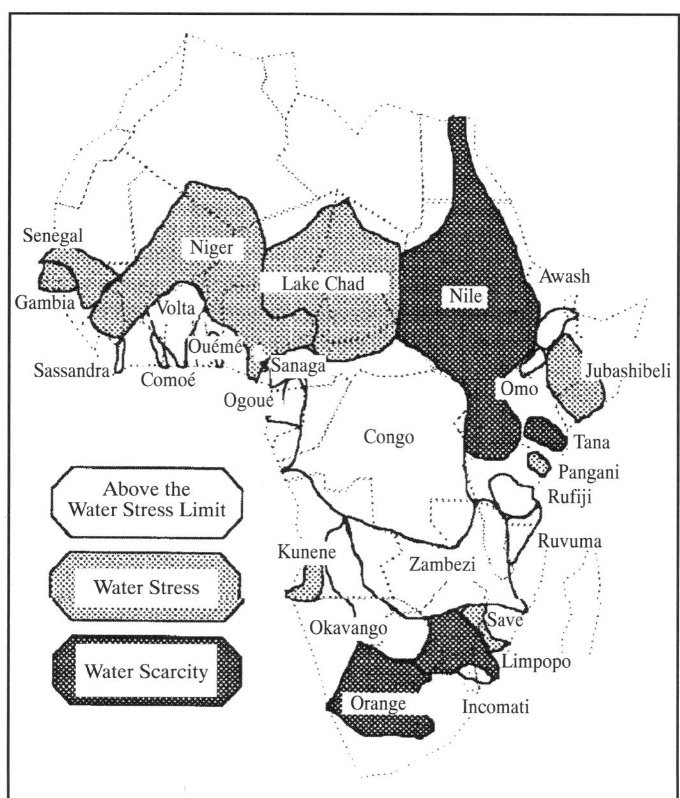


**Figure 2-11**: Water availability at river-basin level (1995) (Sharma *et al.*, 1996).

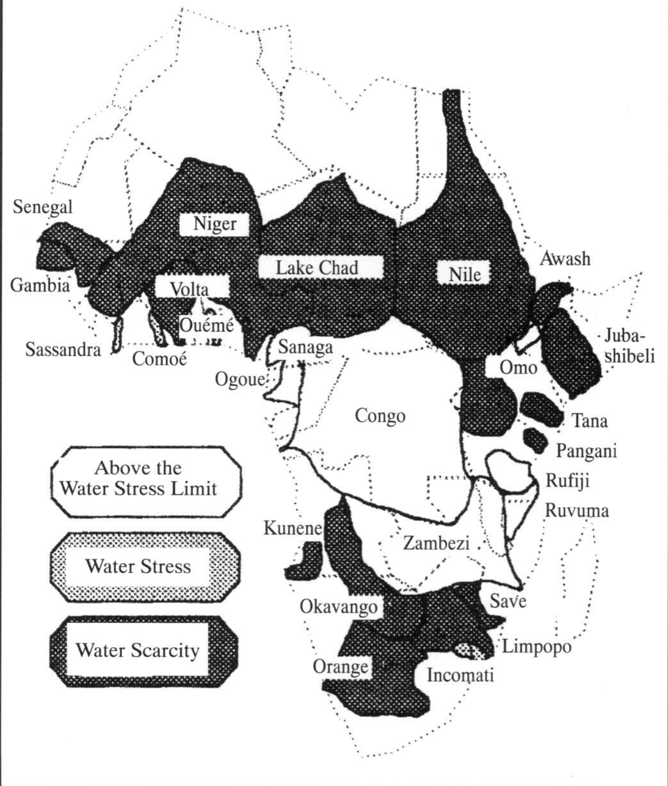


**Figure 2-12**: Water availability at river-basin level (2025) for projected population levels (Sharma *et al.*, 1996).

***Table 2-5:*** *Sub-Saharan Africa: Country indicators for water resources (after Sharma et al., 1996).*

| | (1) % Population with Safe Drinking Water | (2) % Population with Sanitation | (3) Irrigated Area as % of Potential | (4) Governance Environment | (5) Capacity for Resource Management | (6) Status of Water Legislation | (7) Status of Water Plant Policy |
|---|---|---|---|---|---|---|---|
| **Southern Africa** | | | | | | | |
| Angola | Low | Low | Low | Low | Low | Available | Nonavailable |
| Botswana | High | High | Low | High | High | Available | Available |
| Lesotho | Medium | Low | Low | Medium | High | Available | Available |
| Malawi | Medium | High | Low | High | Medium | Available | Available |
| Mozambique | Low | Low | Low | Low | | Available | Nonavailable |
| Namibia | | Low | | High | | Available | Available |
| South Africa | | | | High | Medium | Available | Available |
| Swaziland | Low | Medium | High | High | | Available | Nonavailable |
| Zambia | Medium | Medium | Low | Medium | | Available | Partial |
| Zimbabwe | Medium | Low | Medium | High | High | Available | Available |
| **Eastern Africa** | | | | | | | |
| Djibouti | Medium | Medium | | Low | Low | | |
| Eritrea | | | | Medium | | | |
| Ethiopia | Low | Low | Low | Medium | Medium | Available | Available |
| Kenya | Medium | Medium | Low | Low | Medium | Available | Available |
| Somalia | Low | Low | High | Low | Low | | |
| Sudan | Medium | Low | Medium | Low | Low | | |
| Tanzania | Medium | High | Low | Medium | Low | Available | Available |
| Uganda | Low | Low | Low | Medium | | Nonavailable | Available |
| **Central Africa** | | | | | | | |
| Burundi | Low | Medium | High | Low | Low | | |
| Cameroon | Medium | Medium | Low | Low | Low | Available | Nonavailable |
| Central African Republic | Low | Medium | Low | Low | Low | Available | Nonavailable |
| Chad | Low | | Low | Medium | Low | | |
| Comoros | High | | | Low | | | |
| Congo | Low | Low | Low | Medium | Low | | |
| Equitorial Guinea | Medium | Medium | | Low | Low | | |
| Gabon | High | | Low | Medium | Medium | Available | Nonavailable |
| Madagascar | Low | Low | High | Medium | Low | | |
| Mauritius | High | High | | High | Medium | | |
| Rwanda | Medium | Medium | Medium | Low | Low | | |
| Seychelles | High | | | High | | | |
| Zaire | Medium | Low | Low | Low | Low | | |
| **West-Central Africa** | | | | | | | |
| Benin | Medium | Low | Low | Medium | Medium | Available | Nonavailable |
| Burkina Faso | High | Low | Low | High | Medium | Available | Nonavailable |
| Cote d'Ivoire | High | High | Medium | Medium | Medium | Partial | Partial |
| Ghana | Medium | Low | Low | Medium | Medium | Available | Nonavailable |
| Niger | Medium | Low | Medium | Medium | | Nonavailable | Nonavailable |
| Nigeria | Medium | Low | Medium | Low | Low | Nonavailable | Nonavailable |
| Togo | Medium | Low | Low | Low | | | |
| **Western Africa** | | | | | | | |
| Cape Verde | Medium | Low | | Medium | | Available | Available |
| The Gambia | Medium | Medium | Medium | Medium | Low | Available | Available |
| Guinea | Low | Medium | Medium | Low | Low | Nonavailable | Nonavailable |
| Guinea-Bissau | Low | Low | | Low | | Available | Nonavailable |
| Liberia | Medium | Low | | Medium | Low | | |
| Mali | Low | Low | Medium | Low | | | |
| Mauritania | Medium | Low | Medium | Medium | | | |
| Sao Tome and Principe | Medium | Low | | Medium | Low | | |
| Senegal | Medium | Medium | Medium | Medium | | Partial | Partial |
| Sierra Leone | Low | Low | Medium | Low | Low | | |

Notes: Blank boxes indicate no data available. Columns 1 and 2: Low = 0–33%, Medium = 34–66%, High = 67–100%. Column 3: Low = 0–29%, Medium = 30–60%, High = 61–100%. Column 4 based on political and social stability. Column 5 based on efficiency of domestic resource mobilization and allocation. In Columns 6 and 7, "Partial" indicates draft bill/policy or obsolete laws.

### 2.3.3. *Agriculture and Food Security*

#### 2.3.3.1. *Socioeconomic Vulnerability*

Many indicators of human development highlight Africa's relative poverty and vulnerability (Table 2-6). With smaller holdings and little investment in agriculture, household production faces difficulties in meeting subsistence requirements or developing specialized export crops. Household expenditures on food are high—more than half of the annual budget, on average. Africa receives the largest amount of food aid of any continent. Low rates of female literacy and high rates of infant mortality are indicative of populations that have low status and inadequate infrastructure for education and health—two essential requirements for vigorous rural development. The high numbers of refugees highlight potential economic and political instability. Vulnerable populations include smallholder agriculturists with inadequate resources, pastoralists, rural landless laborers, and the urban poor. Rural populations are directly affected by climatic variations. Reduced food supplies and high prices immediately affect landless laborers who have little savings. The effect on agriculturists and pastoralists depends on how much surplus they produce and the relative terms of trade (e.g., between food and livestock). A dramatic increase in urban poverty has been noted in the past decade—one consequence of stagnant rural development and high population pressures. The urban poor are indirectly affected by climate change through changes in prices and regional investment.

#### 2.3.3.2. *Food and Fiber Production*

Agriculture constitutes a large share of African economies, with a mixture of subsistence and commercial production. Forestry is an important complement to agriculture in many rural areas, but managed forests are less significant. Fisheries are important in coastal areas and islands but a small component of the African economy as a whole. African agriculture is sensitive to present climatic variations. The effects of climate change are uncertain, but adverse impacts are likely in many regions. The future of African agriculture and food security depends on the outcome of climate change in Africa, indigenous responses to global change, development efforts in the next few decades, and global patterns of commodity production and demand (which also are affected by climate change and policy responses to global change).

##### 2.3.3.2.1. *Present agriculture*

African economies are highly dependent on agriculture: Arable land and permanent pasture occupy one-third of the land area of Africa. Agriculture constitutes approximately 30% of GDP (see Table 2-7 and Annex D). Almost three-fourths of the African population resides in rural areas, and almost all of the rural labor force is engaged in agriculture (including livestock, forestry, and fisheries). However, much of the land is of poor quality; less than 10% of Africa is actively cultivated. Annual

***Table 2-6:*** *Regional vulnerability to food crises in Africa.*

| | Expenditure on Food (% of consumption) | Food Aid (cereals) (kg per capita) | Refugees | Female Literacy (adult) (%) | Infant Mortality (per 1000) |
|---|---|---|---|---|---|
| **African Region**[1] | | | | | |
| Northern | 42 | 18 | 221,450 | 45 | 59 |
| Sudano-Sahelian | 42 | 13 | 974,800 | 17 | 119 |
| Gulf of Guinea | 39 | 6 | 819,750 | 28 | 109 |
| Central | 39 | 3 | 480,500 | 41 | 97 |
| Eastern | 37 | 4 | 1,408,150 | 43 | 102 |
| Indian Ocean | 57 | 12 | 0 | 73 | 66 |
| Southern | 57 | 15 | 1,793,800 | 53 | 85 |
| *Total* | *57* | *10* | *5,698,450* | *35* | *97* |
| **Comparison Country** | | | | | |
| Bangladesh | 59 | 12 | 245,300 | 22 | 108 |
| Thailand | 30 | 2 | 255,000 | 90 | 26 |
| Mexico | 35 | 3 | 47,300 | 85 | 35 |
| Greece | 30 | -1 | 1,900 | 89 | 8 |
| United Kingdom | 12 | -3 | 24,600 | X | 7 |

[1] **Northern**: Algeria, Egypt, Libya, Morocco, Tunisia; **Sudano-Sahelian**: Burkina Faso, Cape Verde, Chad, Djibouti, Eritrea, The Gambia, Mali, Mauritania, Niger, Senegal, Somalia, Sudan; **Gulf of Guinea**: Benin, Cote d'Ivoire, Ghana, Guinea, Guinea-Bissau, Liberia, Nigeria, Sierra Leone, Togo; **Central**: Angola, Cameroon, Central African Republic, Congo, Equatorial Guinea, Gabon, Sao Tome and Principe, Zaire; **Eastern**: Burundi, Ethiopia, Kenya, Rwanda, Tanzania, Uganda; **Indian Ocean**: Comoros, Madagascar, Mauritius, Seychelles; **Southern**: Botswana, Lesotho, Malawi, Mozambique, Namibia, South Africa, Swaziland, Zambia, Zimbabwe.
Source: WRI, 1994.

food production over the past few decades has grown by 2.8% per year for cereals, 2.9% for legumes, and 4.0% for roots and tubers, although the total cultivated area has grown by only 0.6%. Although population densities are relatively low compared with global averages, some rural areas are very densely populated, and population growth rates have yet to reach stability levels.

Throughout Africa, the staple crops are cereals—maize in particular. Millet and sorghum are widely grown as well; wheat and teff are common in some regions. Almost all agriculture is rain-fed, although irrigation is important in some regions. The absence of irrigation (less than 10% of the cultivated area is irrigated) increases the sensitivity of crop yields to climatic variations. Cash crops are important in every country but vary in their distribution and profitability. Coffee, tea, groundnuts, cocoa, tobacco, and palm oil are grown as cash crops. Other significant crops (at least in terms of household consumption) include cassava, yams, legumes, and horticultural crops. Agro-pastoralism and extensive nomadic pastoralism are common in semi-arid regions. Relying on grass and browse, pastoralism is particularly sensitive to long periods of drought when grazing resources are depleted by livestock and not renewed.

The regions of Africa have distinct characteristics. North Africa and the Indian Ocean islands rely on irrigated agriculture. In

***Table 2-7:*** *Regional agriculture in Africa.*

| Region[1] | Population Density (pop/km²) | Population Growth (%) | Crop Land (% of total) | Irrigated Land (% of total) | Average Yield of Cereals (kg/ha) | Fertilizer Use (kg/yr) | Food Prod. Index (1970=100) |
|---|---|---|---|---|---|---|---|
| ***Resources*** | | | | | | | |
| Northern | 226 | 2.25 | 5 | 27 | 1,973 | 94 | 115 |
| Sudano-Sahelian | 106 | 2.72 | 4 | 7 | 727 | 5 | 90 |
| Gulf of Guinea | 891 | 2.83 | 21 | 2 | 892 | 6 | 100 |
| Central | 145 | 2.70 | 4 | 1 | 923 | 2 | 87 |
| Eastern | 541 | 2.88 | 10 | 2 | 1,363 | 12 | 92 |
| Indian Ocean | 262 | 1.96 | 5 | 23 | 1,988 | 140 | 98 |
| Southern | 208 | 2.56 | 6 | 7 | 929 | 27 | 76 |
| *Total* | *253* | *2.65* | *6* | *8* | *1,098* | *25* | *92* |
| Bangladesh | 9,853 | 2.18 | 72 | 31 | 2,572 | 101 | 96 |
| Thailand | 1141 | 0.92 | 45 | 19 | 2,052 | 39 | 109 |
| Mexico | 491 | 1.55 | 13 | 21 | 2,430 | 69 | 100 |
| Greece | 795 | 0.07 | 30 | 31 | 3,700 | 172 | 101 |
| United Kingdom | 2,404 | 0.19 | 28 | 2 | 6,332 | 350 | 112 |

| | GNP per capita ($) | GDP in Agriculture (%) | GDP Growth Rate (%/yr) | Public Agricultural Investment ($) |
|---|---|---|---|---|
| ***Investment*** | | | | |
| Northern | 1,285 | 17 | 3.60 | 25 |
| Sudano-Sahelian | 860 | 34 | 2.36 | 7 |
| Gulf of Guinea | 760 | 39 | 1.87 | 15 |
| Central | 760 | 22 | 2.15 | 5 |
| Eastern | 593 | 47 | 3.05 | 13 |
| Indian Ocean | 280 | 22 | 3.85 | 6 |
| Southern | 333 | 21 | 3.38 | 7 |
| *Total* | *355* | *30* | *2.75* | *11* |
| Bangladesh | 205 | 37 | 4.20 | 68 |
| Thailand | 1,697 | 13 | 7.80 | 78 |
| Mexico | 2,971 | 8 | 1.50 | 129 |
| Greece | 6,530 | 17 | 1.60 | 25 |
| United Kingdom | 33,850 | 2 | 2.80 | 347 |

[1] Regions are as in Table 2-6.
Source: WRI, 1994.

west Africa, the gradient of climates from the Sahara to the humid coast determines the potential for agriculture. Subsistence agriculture and pastoralism dominate the Sudan and Sahelian regions; plantation agriculture is found along the Guinea coast. The highlands of east Africa are well known for productive agriculture that takes advantage of the two rainy seasons. The lowlands, however, are subject to erratic rainfall and poor soils. Coffee and tea are major cash crops in the highlands. The humid and sub-humid zones of central Africa, where drought is seldom a problem, are conducive to roots and tubers.

Most rural households engage in subsistence agriculture, although large commercial estates are found throughout Africa. Moreover, regionally diverse values, cultures, and practices in agriculture make for a truly unique region. In many African cultures, identity and the measure of personal wealth and value are determined by the amount of land one owns, the number of cattle in one's herd, or the amount of food produced for the community, rather than monetary wealth. These nuances make African agriculture a particularly important sector within the climate change debate.

Prolonged drought—lasting a season or longer over a widespread area—is the most serious climatic hazard affecting African agriculture, water supplies, and ecosystems. If droughts become more common, widespread, and persistent, many subhumid and semi-arid regions will have difficulty sustaining viable agricultural systems. Drought-prone environments already have been settled and land converted from extensive grazing and long-fallow cultivation to permanent cropping. Box 2-6 reviews the frequency of drought in Africa and its impacts.

### 2.3.3.2.2. *Impacts of climate change*

The effects of climatic variations on African agriculture have been well established through decades of field experiments, statistical analyses of observed yields, and monitoring of agricultural production. The most important climatic element is precipitation, particularly seasonal drought and the length of the growing season. The distribution of rainfall within the growing season also may affect yields. Local flooding and storms are minor problems. Low temperatures and radiation limit production in some high-elevation regions; frost is a hazard in South Africa. High temperatures can affect yields and yield quality in semi-arid and arid regions, although water is more important. Sea-level rise and coastal erosion will affect groundwater, irrigated agriculture, and low-lying coastal land in some areas.

The direct effects of $CO_2$ enrichment on plants tend to increase yields and reduce water use. Increased $CO_2$ concentrations increase the rate of photosynthesis and increase water-use efficiency (the efficiency with which plants use water to produce a unit of biomass or yield). The direct effects are strongest for plants with $C_3$ pathways, such as wheat, compared with $C_4$ plants like maize, sorghum, millet, and sugarcane—which are staples for much of sub-Saharan Africa. $CO_2$ enrichment also affects weeds, many of which are $C_4$ plants (Ringius *et al.*, 1996). According to the IPCC Second Assessment Report, the effect of a doubling in $CO_2$ concentrations (from the present) varies from a 10% increase to almost a 300% increase in biomass; WUE may increase by up to 50% (or more) (IPCC, 1996). Thus, the beneficial effects of increased concentrations of $CO_2$ are likely to offset some of the effects of decreased precipitation. However, the effect of $CO_2$ on crops in Africa—where nutrients often are a limiting factor and leaf temperatures are high—remains highly uncertain.

Unfortunately, regional projections of precipitation change diverge quite strongly in Africa. For example, scenarios of summer precipitation in the Sahel show a range of ±20% for nine atmosphere-ocean GCMs reported in Annex B. However, present trends in precipitation in Africa show a decrease in some regions. Recent transient scenarios report lower temperature changes, globally as well as for Africa. Thus, for agriculture, there is little confidence in present scenarios for precipitation—the most important aspect of climate change for African agriculture. However, a combination of the potential effects of increased $CO_2$ concentrations and lower temperatures (at time of doubling) using transient scenarios suggests that the impacts of these scenarios may be less severe than those of earlier equilibrium GCM model experiments. Nevertheless, even a small decrease in precipitation can be significant. Furthermore, few scenarios of drought risk and the distribution of rainfall within the growing season have been developed.

Although African agriculture clearly is sensitive to climatic variations, perhaps equally important is the gap between present and potential agricultural yields in Africa. For example, a serious impact of climate change might be a decrease of 20% in maize yields. Yet present yields among smallholders often are only half (or even one-tenth) of potential yields. The evaluation of potential impacts of climate change should not mask the enormous potential for more-productive agricultural systems in Africa (see Section 2.3.3.2.3).

At the national level, Figure 2-13 shows the variability in national maize yields for selected countries (Hulme, 1996a). The effects of the 1984–85 and 1991–92 droughts are clear (see Box 2-6). The coefficient of variation for annual maize yields varies from about 10% in central Africa to almost 50% in drier countries such as Botswana and Swaziland. A significant component of the variability is likely to be related to rainfall, although prices and market policies are influential. The role of precipitation in agricultural productivity was demonstrated dramatically in the Sahel and eastern and southern Africa during the drought period 1970–95 (Buckland, 1992). Water scarcity revealed widespread dependence on rain-fed agriculture and the lack of infrastructural development for supplemental irrigation and water resources. For example, Zimbabwe in 1991 and 1992 imported 800,000 tons of maize, 250,000 tons of wheat, and 200,000 tons of refined and semirefined sugar to make up the shortfall associated with reduced agricultural production as a result of rainfall shortages (Makarau,

1992). Studies of the role of climatic variability in African food security have a long tradition. At the local level, agroclimatic studies such as Akong'a *et al.* (1988), Downing *et al.* (1990), Mortimore (1989), and Sivakumar (1991, 1993) considered the effects of climatic variability on agriculture, with an emphasis on coping with drought. Back-to-back drought episodes in sub-humid and semi-arid zones led to the failure of crop production and dependency on other sources of income to buy food, or on famine relief.

National crop modeling studies that specifically address climate change now have been carried out for many countries for a variety of purposes (see Sivakumar, 1991, 1993; Eid, 1994; Muchena, 1994; Fischer and van Velthuizen, 1996; Makadho, 1996; Matarira *et al.*, 1996; Sivakumar *et al.*, 1996; USCSP, 1996). Recent studies sponsored by the United Nations Environment Programme (UNEP), the Global Environment Facility (GEF), the USCSP, and others are to be published soon. A few regional studies have been conducted (e.g., Hulme

**Box 2-6. African Drought: Episodes and Impacts**

Extensive droughts have afflicted Africa, with serious episodes since independence in 1965–1966, 1972–1974, 1981–1984, 1986–1987, 1991–1992, and 1994–1995 (WMO, 1995; Usher, 1997). The causes of African drought are numerous and vary among regions, seasons, and years. Local droughts occur every year; continental crises appear to occur once (or more recently twice) every decade. Major droughts tend to be connected to ENSO anomalies. It seems prudent to expect drought in Africa to continue to be a major climatic hazard.

The potential effect of climate change on drought in Africa is uncertain. At a local level, increased temperatures are likely to lead to increased moisture demand. The balance between rainfall and higher evapotranspiration implies more frequent water scarcity. However, a great deal depends on vegetation response to higher $CO_2$ concentrations and the timing of rainfall. The combination of higher evapotranspiration and even a small decrease in precipitation could lead to significantly greater drought risks. An increase in precipitation variability would compound temperature effects. For example, Hulme (1996b) reports that interannual variability increases on the order of 25% in much of southern Africa in the UKTR scenario for the 2050s. Within the region, however, some areas experience a similar decrease in variability. The temperature-precipitation-$CO_2$ forcing of seasonal drought probably is less significant than the prospect of large-scale circulation changes that drive continental droughts that occur over several years. A change in the frequency and duration of atmosphere-ocean anomalies, such as the ENSO phenomenon, could force such large-scale changes in Africa's rainfall climatology. However, such scenarios of climate change are not well developed at the global level, much less for Africa.

The effects of drought are cross-cutting, with severe direct impacts on agriculture, water resources, and natural vegetation and indirect effects on health, the economy, and institutions (see Benson and Clay, 1994, for an overview of drought impacts). The impacts of drought are confounded by environmental degradation, including soil erosion, water pollution, and deforestation. Intersectoral linkages, the diversity of the economy, the numbers of vulnerable people, the intensity of water use in the economy, the role of financial systems and public enterprises, and public revenue and expenditure affect the severity and distribution of drought impacts. Drought in the 1960s, 1970s, and 1980s triggered widespread starvation and loss of life, particularly in the Sahel and the Horn of Africa. Similar famine has been averted in the 1990s through more effective early warning systems and responses. The aggregate impact of drought on the economies of Africa can be large: 8–9% of GDP in Zimbabwe and Zambia in 1992, 4–6% of GDP in Nigeria and Niger in 1984 (Benson and Clay, 1994).

The 1991–1992 episode in southern Africa amply illustrates the impact of drought. In that episode, the SADC countries experienced the worst drought of the century: From central Zambia through central Malawi and Mozambique southward, there were seasonal deficits of as much as 80% of normal rainfall (Zinyowera and Unganai, 1993). Large sections of the SADC subregion received scanty rainfall—20–75% of normal—during the rainy season from October 1991 through April 1992. Abnormally high temperatures (47°C along the South Africa-Zimbabwe border) exacerbated the extreme dryness. Regional grain production fell 60% short of expected levels. Food stocks had been depleted, largely as a consequence of exports. Roughly five times more food had to be brought into southern Africa than had been delivered to the Horn of Africa during the famine of 1984–1985. Six different transport corridors were used to deliver food aid, and 11 countries assisted in trying to alleviate the crisis wrought by the drought. Even though 1992–1993 and 1993–1994 could be considered post-drought periods, recovery in the subregion was slow. Nutritional status was affected by crop failure, depending on alternative sources of income and drought responses. The number of food-insecure households among communal farmers in Zimbabwe more than doubled during the 1991–1992 drought, especially in semi-arid zones (Christensen and Stack, 1992). The level of the reservoir at Kariba Dam, which supplies power to Zambia and Zimbabwe, fell below the level required to generate hydroelectric power (see IUCN, 1994). Water shortages, electricity shortages and rationing, input supply difficulties, reduction in demand, and macroeconomic constraints led to a 9% reduction in manufacturing output in Zimbabwe, with a 6% loss in foreign currency receipts (Benson and Clay, 1994).

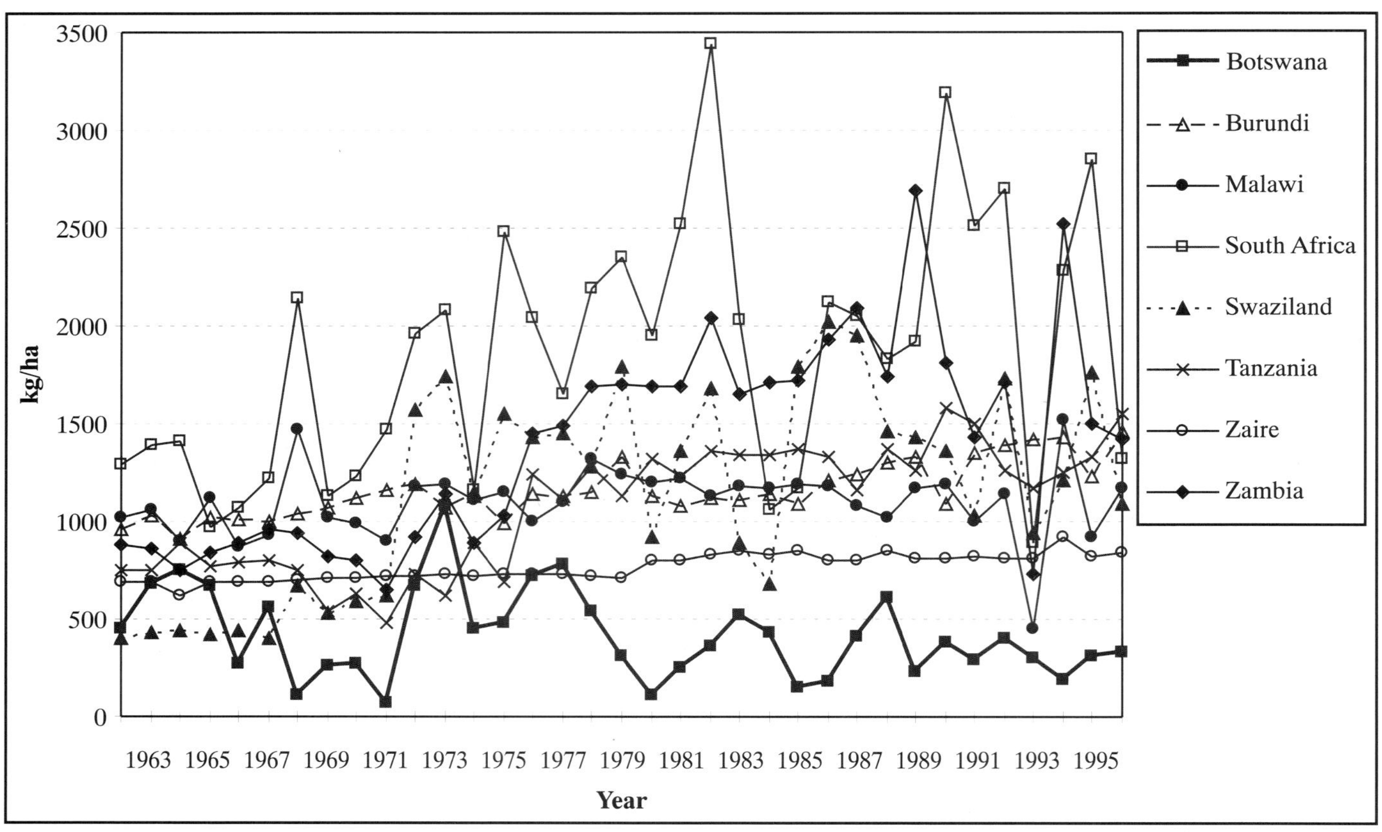


**Figure 2-13**: Maize yields in Africa (Hulme, 1996b), based on FAO data.

*et al.*, 1995; Hulme, 1996a; Ringius *et al.*, 1996), although an authoritative continental assessment has not been compiled. Global studies have included Africa—often, however, using poor data or inadequate spatial coverage and ignoring many critical issues of vulnerability and food security.

The general conclusion is that climate change will affect some parts of Africa negatively, although it will enhance prospects for crop production in other areas (see Downing, 1992, for case studies of agriculture in Kenya, Zimbabwe, and Senegal). Warmer climates will alter the distribution of agroecological zones. Highlands may become more suitable for annual cropping as a result of increased temperatures (and radiation) and reduced frost hazards. Although $C_3$ crops exhibit a positive response to increased $CO_2$ (as much as 30% with $2xCO_2$), the optimal productive temperature range is quite narrow. Some regions could experience temperature stress at certain growing periods—necessitating shifting of planting dates to minimize this risk. Expansion of agriculture is important in the east African highlands. For example, agroecological suitability in the highlands of Kenya would increase by perhaps 20% with warming of 2.5°C based on an index of potential food production (Downing, 1992). In contrast, semi-arid areas are likely to be worse off. In eastern Kenya, 2.5°C of warming results in a 20% decrease in calorie production. In some lowlands, high-temperature events may affect some crops. Growth is hindered by high temperatures, and plant metabolism for many cereal crops begins to break down above 40°C. Burke *et al.* (1988) found that many crops manage heat stress (with ample water supply) through increased transpiration to maintain foliage temperatures at their optimal range. Because a large portion of African agriculture is rain-fed, however, heat-related plant stress may reduce yields in several key crops—such as wheat, rice, maize, and potatoes. At the other extreme of the $C_3$ temperature spectrum, several crops (such as wheat and several fruit trees) require chilling periods (vernalization). Warmer night temperatures could impede vernalization in plants that require chilling, such as apples, peaches, and nectarines. Locations suitable for grapes and citrus fruit would shift to higher elevations. $C_4$ crops are more tolerant, in general, to climate variations involving temperature ranges between 25°C and 35°C. These crops most often are located in warmer, dryer climates; they are quite susceptible to water stress.

In Egypt, Eid (1994) evaluated wheat and maize sensitivity to warmer temperatures; Strzepek *et al.* (1995; see also Conway and Hulme, 1996) reported an integrated assessment of climate change impacts on coastal resources, agriculture, and water. The agricultural sector is highly sensitive to climate change, although different scenarios result in widely different impacts on irrigated agriculture.

In Kenya, a recent study by the International Institute for Applied Systems Analysis (IIASA) (Fischer and van Velthuizen, 1996) highlights the diverse effects of climate change. The Food and Agriculture Organization (FAO) Agroecological Zone model was used to delineate crop-growing regions and their suitability for a wide range of crops. Rising temperatures and increased plant water requirements would lead to dramatic reductions in agricultural production potential, especially in eastern and southern Kenya. In central and western Kenya, temperature increases would result in an extension of the land suitable for cultivation because some higher-elevation areas would become suitable for cropping. Along with higher cropping intensities in the highlands, this effect more than outweighs the effects of moisture stress in the lowlands. In humid regions (>270 days of growing periods), diminished wetness could reduce pest and disease constraints. The balance of increased evapotranspiration and precipitation in semi-arid regions determines the effect of climate change on agriculture and food security in the lowlands.

Using the ACRU/CERES hybrid model—one of the most sophisticated crop-climate models developed in Africa—Schulze *et al.* (1996) have evaluated the impact of climate change on maize in South Africa. The investigators divided the diverse geography of South Africa into 712 relatively homogeneous zones, each associated with vegetation, soil, and climate data. Daily values of temperature (minimum and maximum), rainfall, wind speed, and solar radiation are used in the crop evaluation, based on the CERES-Maize model. Recent scenario analysis of the model (see Hulme, 1996a) shows a wide range of potential maize yields in South Africa. For three scenarios of climate change (corresponding to the middle of the next century), yields decrease in the semi-arid west. For most of the country, however, potential yields would increase—generally by as much as 5 t/ha. The $CO_2$ enrichment effect counteracts the relatively modest changes in temperature and precipitation. In parts of the eastern highlands, particularly in Lesotho, dramatic increases in yields result from higher temperatures.

Hulme (1996b) presents an integrated view of climate impacts in southern Africa. Prospects for agriculture depend critically upon changes in precipitation. A "dry" scenario suggests less-suitable conditions in semi-arid and subhumid regions. With little decrease (or increases) in precipitation, agriculture should be able to cope with the average changes. However, shifts in drought risk need to be considered.

Ultimately, climate change is a global issue—even more so for traded commodities such as agricultural products. Some regions, for example, may be less competitive in national and global agricultural markets, with corresponding impacts on exports and imports. Africa, in particular, may be sensitive to changes in world prices and stocks because many countries rely on food imports. Several world-trade models have been tested with scenarios of climate change, with differing assumptions regarding economic growth, population growth, trade liberalization, and technological innovation (see Fischer *et al.*, 1994, 1996; Rosenzweig and Parry, 1994). Because they are global simulations, they can illustrate some of the dynamic adjustments in world prices and regional imports and exports that may result from climate change. However, Africa is not well represented in such assessments. Scenarios tend to be trend projections that discount the potential for dramatic improvements in agriculture or welfare. Moreover, the lack of uniform and accessible data on crop-climate sensitivity in Africa leads to large uncertainties in predicted impacts in Africa. A critical question is the extent to which climate change at the global level alters African exports (reflecting changes in comparative advantage) and food imports (reflecting the world price of cereals).

Most livestock in Africa are herded in nomadic areas, although significant numbers are kept in paddocks on farms. Domestic animals, especially cattle, also will be affected by climate change. In the cold highlands of Lesotho, for example, animals would benefit from warmer winters but could be negatively affected by a lowering of the already low nutritional quality of grazing. Heat stress also is a concern in warmer areas. The direct impacts of changes in the frequency, quantity, and intensity of precipitation and water availability on domestic animals are uncertain. However, increased droughts could seriously impact the availability of food and water—as in southern Africa during the droughts of the 1980s and 1990s (IPCC, 1996).

Agricultural pests, diseases, and weeds also will be affected by climate change. Little quantitative research on these topics has been undertaken in Africa, however. Perhaps the most significant shifts could occur in tsetse fly distributions and human disease vectors (such as mosquito-borne malaria). Tsetse fly infestation often limits where livestock can be kept or the expansion of extensive agriculture (Hulme, 1996a). Declining human health would affect labor productivity in agriculture.

African economies depend on natural resources, and the impact of changing natural resources affects several sectors. Perhaps more so than in many regions, the cross-sectoral impacts of climate change need to be understood. Agriculture depends on water resources, a healthy labor supply, and demand for its products. In turn, rural health, incomes, and development depend on viable agricultural economies. One example of the potential interactions is the role of drought. A small change in drought risk need not affect agriculture to a great extent, as long as food supplies and household income can be saved over several years. However, an increase in drought risk could affect regional water supplies, leading to rationing of water and energy and reduced irrigation. Increasing aridity and prolonged spells of severe drought could accelerate abandonment of the rural economy and migration to urban centers.

#### 2.3.3.2.3. *Adaptation and vulnerability*

Low investment in agriculture (see Table 2-7) means that African agriculture is extremely sensitive to climatic fluctuations. The current status of agriculture is not encouraging: It is characterized by stagnant yields, land degradation, and recurrent droughts (Box 2-7 suggests strategies to cope with current levels of climate variability and drought). Furthermore, political

### Box 2-7. Drought and Famine Responses

The loss of human life and economic disruption associated with extreme climatic fluctuation can be lessened by enhancing household coping strategies, preparedness, and advanced warning and timely responses. The initial aim in responding to drought is to maintain household livelihoods and mitigate the effects of rainfall shortages. Seasonal hunger, poverty, and household crises recur among vulnerable households throughout Africa. It is often difficult for governments, donors, and non-governmental organizations (NGOs) to distinguish between chronic conditions and an emerging crisis during a drought (or other shock). Recent scientific advances in understanding the climate system and predicting seasonal drought provide an opportunity to reduce the vulnerability of human societies (Rasmusson, 1987; Hastenrath, 1995; Bonkoungou, 1996). Predictions of seasonal rainfall already are operational in many regions of Africa, with lead times of up to several months. Longer-term forecasts also may be possible. However, drought predictions still need to be incorporated into African management of agriculture, water, energy, and national economies (see Gibberd *et al.*, 1996).

Increasing drought preparedness in Africa has resulted in a plethora of organizations that monitor climate, agriculture, vegetation, and resources; make available early-warning information; participate in multidisciplinary research; and promote preparedness plans. International early-warning systems such as the World Food Programme, the Food and Agriculture Organization, Save the Children Fund, U.S. Famine Early Warning System, and Agence Europèene pour le Développement et la Santé, among others, link local and national efforts with international responses. Following the devastating droughts of the 1970s in the Sahel, the AGRHYMET center was established in Niger. The SADC has a similar program on food security. Drought monitoring centers in Kenya and Zimbabwe and the African Centre for Meteorological Applications in Development were established in the 1990s. National departments, universities, and NGOs also have been active in promoting drought monitoring, mitigation, and preparedness. By monitoring key indicators of staple food production, national stocks, and import availability, a rough and ready national balance can be calculated using population estimates and per capita needs (SADC, 1996). Most famine early-warning systems (EWSs) predict the failure of food systems (Wilhite and Easterling, 1987). Increasingly, the focus of monitoring and intervention has been on food-insecure people who strive to subsist within food systems, rather than national food balances per se (Davies, 1996). Many EWSs now monitor upstream determinants of production (such as rainfall and soil moisture) as well as downstream outcome indicators (such as market prices for food and the nutritional status of those most vulnerable). Vulnerability assessment (and mapping) is one tool for integrating the climatic, production, and socioeconomic dimensions of drought in the context of sustainable development. Unfortunately, the focus often remains on food balances and food crises rather than solving chronic hunger and enhancing the livelihoods of the most vulnerable populations.

Early warnings do not guarantee effective responses. Early warnings by the drought monitoring centers and national meteorological services of a severe drought in 1991 were largely ignored by government and donor agencies (SADC, 1994). Economic structural adjustment programs failed to take rainfall variability into account in their design and emphasis on reform. Thus, numerous countries were poorly prepared, without a national disaster management program in place. There were conflicts over statistics used in emergency programs—where data from different sources were not reconciled—which led to delays in response. The response to the 1990s drought in Kenya was more positive; one month after the drought was apparent, large-scale food imports were made, and local food stocks were utilized (Longhurst, 1992; see Downing *et al.*, 1990, for a description of the similar success in 1984–1985). Preprogrammed drought responses also assisted in averting a major crisis: In northern Kenya, a decentralized early-warning system was well developed, and recommendations were translated rapidly into firm decisions to mitigate the impact of the drought.

Local coping strategies for drought and famine are well illustrated in the inner Niger delta of the Malian Sahel (Davies, 1996). Famine, although not entirely absent, is not a common occurrence in the Sahel. Periods of drought have been more frequent since the late 1960s, although there is little consensus regarding their origin or whether such drought is a permanent trend. Over the past 30 years or so, drought and famine have been part of a downward spiral of impoverishment, increasing vulnerability, destitution, and sometimes death. As a result of the Sahelian droughts of the early 1970s and the mid-1980s, fundamental changes are taking place in the livelihoods of people as they adapt to confront declining food security. Local livelihood systems have become less resilient and more sensitive (more vulnerable) to food stress with successive periods of drought. The nature and degree of vulnerability vary according to the livelihood system. Famines could be mitigated if policymakers can recognize and reinforce local coping strategies.

Lessons learned in responding to drought in Africa need to be widely implemented to reduce the threat of famine and enhance livelihood security. Promising developments in climate prediction and experience in early warning and responses lay the foundation for ending famine in Africa. Responses to drought also provide insight into effective adaptation to climate change. Reducing present vulnerability reduces the threat of catastrophic impacts. Institutional management of present climatic hazards should support adaptive learning to cope with future climate change.

conflicts have adversely affected food production, making the continent extremely vulnerable to climate change. Without a sound agricultural sector, Africa is unlikely to develop diversified economies that can cope with the impact of climate change. Thus, the impact of climate change on agriculture and food security in Africa over the next few decades will depend on progress in applied agricultural research and development.

There is a need for assessments of crops, agricultural systems, food economy, and food security at local, national, and regional levels, but data are not readily available (IPCC, 1996). In some African countries, such studies are underway.

Adaptation to climate change is not automatic or autonomous. The motivations, constraints, and domains of authority of decision-makers involved in shaping policy, implementing decisions, and coping with the consequences of changes in resources and hazards must be considered (see Grimble and Chan, 1995). The principal stakeholders in African agriculture range from vulnerable consumers and subsistence producers to national and international organizations charged with planning, research, and relief (see Gibberd *et al.*, 1996, for examples in the context of seasonal climate prediction). Consumers are the ultimate stakeholders in adapting to climate change. For particularly vulnerable groups (such as resource-poor farmers, landless laborers, the urban poor, the destitute and displaced, or refugee populations), the outcome of strategies to adapt to climate change and climatic hazards may alter their livelihoods. Failure to cope with adverse change could lead to significant deprivation, social disruption, and population displacement. Producers adapt to climatic variations every season but have varying interests in climate change. Subsistence farmers are unlikely to have the resources necessary to consider specific strategies to adapt to climate change. Commercial farmers are more likely to be linked to national markets and international agribusinesses and able to invest in agricultural technology. However, adopting specific strategies depends on cost effectiveness within the context of short-term enterprise goals. One of the key stakeholders in enacting forward-looking strategies is business—from local market traders to international commodity and research organizations. However, commodity traders are not likely to be affected directly by the consequences of climate change, as long as production is viable and trade required somewhere in the world. Incentives may be required to induce agribusiness to adopt a longer planning horizon and to develop and implement adaptive responses. At present, the bulk of responsibility for designing, evaluating, and implementing strategic responses falls upon national governments, national and international research centers, and aid organizations (particularly bilateral and multilateral groups, although some international and even local NGOs may take an interest in adaptation policies). These are the same actors charged with development; extending their purview to longer-term climate change should not be difficult.

**Box 2-8. Maize Adaptation to Climate Change in Zimbabwe**

Throughout the world there is great scope to adapt to the agricultural impacts of climate change. Simulations with the CERES-Maize crop model at four sites in Zimbabwe illustrate the impacts of climate change and the effect of changing planting dates (Makadho, 1996; see also Muchena, 1994; Hulme, 1996b). The CERES model simulates crop development and yield using specialized functions to calculate photosynthesis, phenological stage, evapotranspiration, and partitioning of biomass. Simulations were discrete, with the default initial soil water moisture set at the field capacity of the soils. Nitrogen stress and pests were not simulated. Average potential yields, as simulated with 40 years of climate data, range from more than 3.5 t/ha in Region II (Karoi) to barely 1 t/ha in Region V (Beit Bridge). There is a considerable gap between these potential yields and the yields realized by farmers in the region. In semi-arid zones (Region V), 500 kg/ha would be an above-average yield. Average commercial farm yields in the high veld are closer to the potential yield because of application of fertilizer. For the two equilibrium scenarios of climate change, yields decrease in the high veld (Karoi) and semi-arid region (Beit Bridge) but increase in the middle zones (Gweru and Masvingo). If the season start is adjusted, the yield decreases in the high veld, and semi-arid zones produce sizable increases. For example, planting on 15 October at Karoi results in 3.7 t/ha at present. With climate change and delaying planting to 1 November, yields would be 4.6 t/ha—a 25% increase. At Beit Bridge, the season would shift toward earlier planting to avoid high temperatures and water stress at the height of the summer.

*Grain yield (kg/ha) from CERES-Maize simulations for Zimbabwe.*[1]

| | P Date | Present | CCC | GFDL |
|---|---|---|---|---|
| Karoi | 15 Oct | 3,727 | 2,643 | 2,940 |
| | 1 Nov | 3,654 | 4,641 | 4,630 |
| Gweru | 15 Oct | 3,006 | 5,011 | 5,446 |
| Masvingo | 15 Oct | 3,006 | 3,493 | 3,097 |
| Beit Bridge | 15 Dec | 1,213 | 713 | 725 |
| | 1 Dec | 1,203 | 1,304 | 1,453 |
| | 1 Nov | 1,136 | 838 | 1,740 |

[1]P Date is simulated planting date; CCC is the Canadian Climate Centre equilibrium GCM experiment, and GFDL is the Geophysical Fluid Dynamics Laboratory equilibrium GCM experiment.
Source: Makadho, 1996; see also Matarira *et al.*, 1996.

What options are available to different stakeholders to begin preparing for climate change in Africa? The impacts of climate change could be serious, at least in some regions and for some stakeholders. Equally important, however, climate change presents new opportunities that may promote development. Table 2-8 lists generic agricultural adjustments that currently are practiced in Africa and may be appropriate to cope with climate change. Adjustments are grouped according to the level of the agricultural sector in which they might be applied: land management, crop variety and land use, crop husbandry, and

***Table 2-8:*** *Agricultural adjustments to climate change.*

| Strategy and Adjustment | Mechanisms | Costs | Implementation | Constraints and Issues |
|---|---|---|---|---|
| ***Land Management*** | | | | |
| Moisture management (conservation, irrigation, soil drainage, mulching, fallowing); soil management (mulching, tillage, crop rotation, land drainage) | Regulate soil water balance through incremental irrigation, drainage, control of evaporation, and runoff; enhance organic matter, use fertilizer, control soil erosion | Higher costs for additional irrigation works, water, operation, and maintenance; some additional labor and inputs | Gradual implementation with increased temperatures, often in response to drought | Water availability (surface and ground), water quality, terrain, alternative demands for water, investment capital and incentives |
| ***Crop Variety and Land Use*** | | | | |
| Cultivars; rotations; crop substitution; cropped area; crop location; conversion to/from crops or pasture; changes in specialization; livestock types and levels | Switch varieties, crops, or rotations (longer maturing varieties, heat- and drought-tolerant, requiring less vernalization); more flexible cropping system with seasonal forecasts, spread risk; switch location (regional or within farm) to new climates or soils; change specialization (e.g., arable/pasture production); change resource intensity (e.g., stocking rate) | Costs include development of cultivars, livestock breeding, and restructuring for different farming systems; marginal costs may be minimal if encompassed in normal agricultural investment | Costs are staged or incremental, but related to rate of climate change and possible effects of severe episodes; new cultivars require 10–15 years to develop | Lead time to develop new cultivars; soil suitability and terrain in conversion to agricultural uses; delayed response to new conditions; need for additional information and training; availability of genetic material |
| ***Crop Husbandry: Planting and Harvesting*** | | | | |
| Timing of planting and harvest; plant mixed varieties; planting depth; plant density | Earlier/later scheduling with changed growing season or to shift timing of heat stress; flexible cropping system; plant deeper in drier conditions; thin crop in dry years to lower plant density and reduce competition for moisture | Few additional costs; shifts in labor requirements during season | Gradual adjustment with little lead time; possibly greater flexibility in response to seasonal or monthly weather forecasts | Availability of cultivars, changes in winter season, frost, soils, field accessibility due to wet conditions may limit applicability |
| ***Crop Husbandry: Fertility and Pest Management*** | | | | |
| Herbicides; pesticides; fertilizer application; nitrogen-fixing crops | Control weeds to reduce competition for moisture, nutrients, and light; control pests and diseases that limit plant growth, yield, or yield quality; nature, quantity, and timing of fertilizer affect plant uptake | Input costs increase in general; considerable savings possible for some fertilizer/crop regimes, but increased costs in other regions | Gradual adjustment with short lead time and rapid responses, except for new crops and invading pests, diseases, and weeds | Toxic and ecological concerns with fertilizer leaching; information to respond to new pests, diseases, and weeds; new crops |

*Table 2-8 continued.*

| | Mechanisms | Costs | Implementation | Constraints and Issues |
|---|---|---|---|---|
| ***Economic Adjustments: Farm Level*** | | | | |
| Investment in agriculture (equipment and machinery); farm inputs; savings and storage; labor and employment; off-farm purchases; food consumption | Increased investment in agriculture to increase yields; increased food storage to reduce variability in supply; increased savings and purchases to supplement storage; off-farm employment to support increased investment and food purchases; altered food consumption to cope with seasonal shortages, shifts to new varieties, economic crises, labor demands | Infrastructure for storage and marketing; operation and maintenance costs for storage; opportunity cost of off-farm employment; costs of new technology; additional costs in dry years for purchases, replanting, etc. | Gradual but variable related to yields; storage facilities minor on-farm; gradual shifts in employment, but sudden with extreme episodes | Limited by finance, technology, type of agricultural production, surplus, access to regional and international economics and trade |

farm-level economic adjustments (macro-economic and sectoral planning also should be considered). Box 2-8 evaluates one such adjustment in Zimbabwe.

Beyond improving agriculture using available strategies, what more should be done to plan for climate change? Ringius *et al.* (1996) provided guidelines on adaptation for agriculture and water in Africa (see also Downing *et al.,* 1996). Anticipating climate change may be warranted for projects with long life spans (e.g., irrigation schemes); where the marginal cost of adaptation is small or brings benefits regardless of climate change; for protection against extreme events; and to prevent irreversible impacts (e.g., preservation of biodiversity). For example, the most certain aspect of climate change is increased $CO_2$ concentrations. Efforts to enhance positive $CO_2$ responses in new cultivars may be worth the investment in plant breeding and agricultural technology, irrespective of projected changes in moisture availability. Strategic food reserves to buffer potential increases in the variation of local and national production might be relatively inexpensive at the international level but not warranted for most countries or at local grain banks. Drought early warning and preparedness could build upon the considerable improvements that already are underway in many regions of Africa, including making better use of seasonal climate predictions.

Protection against irreversible impacts or losses of valued resources may be warranted in some situations. Thus, if coastal erosion and sea-level rise threaten valuable coastal resources, protection measures may be cost-effective. For example, groundwater pumping may be required to lower the water table if saline intrusion affects agriculture in low-lying areas. It is unlikely that such projects are a priority in the near future. However, protection measures should be considered in the design of new development.

Regulation of resource allocation and development is deficient in much of Africa. Institutional and regulatory reform may be warranted to preclude development in areas that become increasingly hazardous (such as coastal zones) and to protect vulnerable communities (e.g., economic restructuring). Establishing priorities for development based on future land capabilities may be premature for most regions, but flexibility in development priorities should be retained and new information taken into account. For example, the ability of community groups to manage rapid resource changes may warrant further support. Market structures often support crops with a high level of risk and fail to support markets for drought-tolerant crops. Regulations that constrain free trade may increase the volatility of local markets and food supplies in response to climatic variations.

If the present scope for adaptation is limited, investment in research and education are warranted to develop new solutions and stimulate behavioral changes to accommodate climate change. For example, development of cultivars that optimize responses to $CO_2$ enrichment and development and testing of new cultivars suitable for a range of likely climatic conditions require improved research. Education on environmental issues is warranted, although it is probably too soon to undertake specific campaigns designed to adapt to climate change. However, a broad capacity to address environmental issues and communicate understanding to stakeholders is urgently needed. This capacity is even more critical in linking greenhouse gas abatement with sustainable development issues.

Beyond specific adaptive actions, institutions may need to be strengthened—for example, to enhance the productivity of natural resources, to increase the capacity to respond to developmental pressures and resource crises, and to improve environmental quality. Concerning drought, institutions need to make better use of climate information to cope with climatic risks and reduce vulnerability. These are ongoing development objectives, but further assistance is warranted in light of the risk of changes in climatic hazards.

Ringius *et al.* (1996) evaluate four ensembles of adaptive strategies at different levels of planning:

1) Farm-level adaptive responses include substitution of agronomic practices (changes in sowing date, planting density, cultivars, etc.); altered inputs (e.g., fertilizer, pest and weed control, crop choice); and agricultural development (soil and water management that requires more substantial investment). The priority stakeholder is the smallholder farming sector. Commercial farms would be less likely to need assistance in these sorts of adaptive strategies. On the other hand, these strategies are less likely to be effective for agropastoralists and pastoralism in general. Such farm-level agronomic improvements can be effective, can be readily implemented, and have few substantial constraints to their adoption.
2) At the national level, three strategies commonly are proposed:
   - Maintaining strategic reserves allows the government (or marketing bodies) to dampen price fluctuations and release food in emerging crises. Large national reserves have been held in the past, but have been reduced under structural adjustment agreements because of their cost.
   - Markets and trading conditions could be adjusted to promote private-sector responses to climate change and climatic variability. This strategy might take the form of tax incentives for carry-over stocks or bonds to smooth income between adverse and good trading years. This approach would build upon present efforts to reduce trade barriers, with some specific adjustments to accommodate climate change.
   - Promoting agricultural development in general would close the gap between research and practice. The need to adapt to climate change could be used as one argument for fresh initiatives in promoting adaptive agricultural research and development in Africa.

   The primary beneficiaries of such national adaptive strategies are consumers and commercial producers who depend on markets for food consumption. Market adjustments may entail some trade-offs between consumers and producers or between relatively prosperous farmers and vulnerable smallholders who may not have access to inputs and markets. Yet the potential for multiple benefits is high (except for strategic reserves, which are a burden on the economy). As a group, these strategies would be reasonably effective in preparing for climate change.
3) At the global level, some policies to prepare for climate change may be justified:
   - Climate change may require additional trade to smooth out fluctuations in national production. Maintaining international prices within acceptable limits would benefit poorer countries that might not be able to afford large imports in times of scarcity. In the transition toward a new climate, such an international capacity to prevent food deficits from becoming survival emergencies would appeal to the humanitarian goals of ending famine and reducing hunger.
   - Encouraging free trade between and among countries should stimulate agricultural markets in regions with a comparative advantage. This trade may be a major benefit to some countries and a significant cost to others, as the impacts of climate change alter traditional markets. In principle, free trade allows national surpluses and deficits to be accommodated more efficiently. Thus, supply and price fluctuations are buffered at the global level, widening the potential pool of responses to climate change.
   - International mechanisms to promote agrotechnology transfers to developing countries might focus on basic foodstuffs: wheat, rice, and maize. International agencies might license new technologies developed by biotechnology firms for dissemination and use in developing countries.
4) Reducing drought risk and societal vulnerability would provide greater resilience in coping with climate change. Concerted action is required in three areas: mitigation to reduce vulnerability, monitoring drought and vulnerability, and preparedness to respond effectively to emerging crises. Considerable progress has been made in the past decade; a further decade of development might reap substantial rewards in efforts to eliminate widespread famine and enhance livelihood security, at least in times of drought (see Downing *et al.*, 1996; Gibberd *et al.*, 1996). The priority stakeholders should be the most vulnerable socioeconomic groups affected by drought crises—although many levels of local, national, and international actors are required to implement drought monitoring, mitigation, and emergency responses.

These four broad strategies would promote African resilience in the face of present climatic risks and enhance Africa's capacity to adapt to climate change. In the next few decades, Africa must manage the transition from vulnerable subsistence agriculture to resilient, adaptive systems. If real progress is achieved, most regions and populations could cope with climate change over the next few decades. Climate change on top of continued vulnerability and agricultural crises threatens regional security.

The fundamental requirement for adaptation in Africa is to promote sustainable agricultural development—closing the gap between experimental yields and farm yields, overcoming constraints in markets, and providing rural infrastructure (credit, inputs, transport, etc.), which will enhance the capacity of the agricultural sector to contribute to local and national economic development. The overall aim should be to make the best use of climate as a resource for agriculture by enhancing the capabilities of agriculturists, agribusiness, and organizations to respond to climate variations and climate change. Perhaps the clearest specific objective at present is to prepare for climatic hazards by reducing vulnerability by way of developing monitoring capabilities and by enhancing the responsiveness of the agricultural sector to forecasts of production variations and food crisis.

#### *2.3.3.3. Marine and Riverine Fisheries*

The African nations possess a variety of lacustrine, riverine, and marine habitats with more than 800 freshwater and marine species. Ten ichthyofauna regions, based largely on present-day drainage systems, have been delineated for Africa (Lowe-McConnell, 1987). These regions are dominated by the Niger, the Nile, the Congo, and the Zambezi River systems; they also include several inland drainage areas associated with lakes (Hlohowskyj *et al.*, 1996). Among the riverine systems, the Congo River (including its major tributaries) contains the most diverse fish fauna, with about 690 species (of which 80% are endemic) (Lowe-McConnell, 1987). Lacustrine systems in Africa (particularly the Rift Valley lakes) contain the most diverse and unique fish assemblages found anywhere in Africa, if not the world. For example, Lake Malawi has more than 240 identified fish species (of which more than 90% are endemic), and another 500 or more species awaiting taxonomic identification (Lewis *et al.*, 1986).

Fish make up a significant part of the food supply in Africa (Hersoug, 1995). FAO (1993) estimates the total fish harvest potential at around 10.5 million tons: 7.8 million in saltwater fisheries and 2.7 million in freshwater fisheries. In a densely populated country such as Nigeria, as much as one-third of the protein supply comes from fish (Hersoug, 1995). Consequently, any fluctuation in the fish stock will impact planning and management. A reduction in fish stocks will have the greatest effect on countries that are heavily dependent on fisheries and cannot diversify easily into other activities; Mauritania, Namibia, and Somalia are examples of African countries in this category (Clarke, 1993).

The productivity of freshwater and sea margins has become stressed mainly by economic activities rather than climate variability. For example, the artificial opening of the sand barrier at the mouth of the Cote d'Ivoire River to clear floating weeds has allowed seawater to enter the lower part of the river and has changed species dominance (IPCC 1996, WG II, Section 16.1.1). On the Nile, the Aswan Dam so thoroughly regulates flows that the delta has become degraded ecologically. Local sardine populations that once thrived and provided food for the region have collapsed with the decline in production that depended on the strong surges of floodwaters and their pulse nutrients. The Sahelian drought is causing increased salinity in the lower parts of Senegalese rivers, but a dam erected near the mouth of the Senegal River to stop the rising salinity and ease severe problems in local agriculture prevents fish migration (Binet *et al.*, 1995).

##### *2.3.3.3.1. Vulnerability of fisheries resources*

The vulnerability of fisheries to climate change depends on the nature of the climate change, the nature of the fishery, and its species and habitats. Changes in climatic conditions such as air temperature and precipitation affect fisheries by altering habitat availability or quality. Specifically, fisheries habitats may be affected by changes in water temperature; the timing and duration of extreme temperature conditions; the magnitude and pattern of annual stream flows; surface-water elevations; and the shorelines of lakes, reservoirs, and near-shore marine environments (Carpenter *et al.*, 1992).

Mean annual air temperature is the most important factor in predicting lake fish production across latitudes. Alterations in seasonal climate patterns should change the population distributions in larger lakes. Large-lake fish production could increase about 6% with a 1°C rise in average annual air temperature (Meisner *et al.*, 1987; IPCC 1996, WG II, Section 16.2.1). Warm-water lakes generally have higher productivity than cold-water lakes, and existing warm-water lakes will be in areas with the least change in temperature. It is reasonable to expect higher overall productivity from freshwater systems.

Although changing rainfall patterns and flood regimes may have profound effects on freshwater fish, marine fisheries are likely to be affected more by elevated temperatures (Hernes *et al.*, 1995). The impacts of elevated temperatures could include:

- A shift in centers of production and the composition of fish species as ecosystems move geographically and change internally. This is in contrast to freshwater fish species—particularly in small, shallow rivers and lakes—which will have limited possibilities to adapt to the changes through migration.
- Economic values can be expected to fall until long-term stability is reestablished. Rapid changes resulting from physical forcing favor smaller, low-priced, opportunistic species that discharge large numbers of eggs over long periods.
- Where ecosystems shift position, national fisheries will suffer if institutional mechanisms are not in place that enable fishermen to move within and across present exclusive economic zone boundaries. Subsistence and other small-scale fishermen (who dominate in Africa) probably will suffer disproportionally from such changes (Everett, 1994).

##### *2.3.3.3.2. Adaptation options for fisheries resources*

Adaptation to existing climate variability may demonstrate ways to deal with climate change. The following adaptation options are suggested for the fisheries industry (IPCC 1996, WG II, Section 16.3.1):

- Modify and strengthen fisheries management policies and institutions and associated fish population and catch-monitoring activities.
- Preserve and restore wetlands, estuaries, floodplains, and bottomlands—essential habitats for most fisheries.
- Cooperate more closely with forestry, water, and other resource managers because of the close interaction between land cover and maintenance of adequate fishery habitat. The adequacy of management practices in all sectors affecting fisheries (e.g., water resources, coastal management) needs to be examined to ensure that proper responses are made as climate changes.
- Promote fisheries conservation and environmental education among fishermen.
- In cases of species collapse and obvious ecosystem disequilibrium, restock with ecologically sound species and strains as habitat changes; great care is needed to avoid ecological damage.
- Develop aquaculture and tourism to make coastal communities better able to deal with uncertainties of climate change.

To reduce the possibility of fishery disruption, strict biological monitoring should be implemented, and properly enforced fishing controls must be instituted. These strategies would help keep stock-replacement levels stable in the face of physical stress caused by climate change and other environmental phenomena while meeting the growing demand for fish and fishery products by an ever-increasing population.

#### 2.3.3.4. Production Forestry

Plantations of exotic species are a major source of wood for timber and paper industries in Africa; indigenous species (hardwoods) are used mostly for fuelwood, for charcoal production, and in traditional construction. Some indigenous species have very specialized uses, including making furniture, musical instruments, railway sleepers, and carvings. In such cases, natural stands are harvested by removing large and viable trees, with residual stands left to recover. It is likely that natural stands would take 60 or more years before they produce trees of harvestable size. Exotic species, on the other hand, can take less than 10 years to about 30 years to produce poles and sawtimber-quality trees. The most common introduced species are pines (mostly *Pinus patula*), eucalypts, and cypress (*Cupressus spp.*).

In the 1980s, fast-growing exotic plantations were regarded as the solution to Africa's dwindling fuelwood resources. Many projects performed poorly, however, because of poor species choice, lack of species trials, limited site characterization, and unforeseen pests (such as goats and termites). Conversion of extensive areas of woodland, shrub, or savannas to forest plantations has had ecological consequences that have not been fully assessed for Africa. Elsewhere, extensive plantations lead to increased transpiration and reduced runoff, so catchment yields can be reduced. Although there are signs of this process in many parts of Africa, no data have been collected. Eucalypts, in particular, have been suspected of drying landscapes and lowering water tables (FAO, 1985).

In many countries in Africa, forest plantations are established in marginal areas (to avoid competing with agriculture), and often in complex terrain (mountainsides and plateaus). Soil erosion after timber harvesting leads to siltation problems in rivers and reservoirs and can cause serious problems in hydroelectric generation. Plantations traditionally have been managed to maximize financial returns. Erosion problems associated with clear-cutting and log extraction will need to be addressed in future silvicultural planning, especially where logging is being carried out in catchment areas that are important for drinking water or hydroelectric generation.

The widespread use of exotic conifer species has increased the incidence of exotic forest pests, such as aphid pests (Katerere, 1983). These include the Eurasian pine woolly aphid, *Pineus pini* (Macquart); the Holarctic pine needle aphid, *Eulachnus rileyi* (Williams); and the European cypress aphid, *Cinara cupressi* (Buckton). The cypress aphid was first reported in Africa in Malawi in 1986 (Chilima, 1991); by 1990, it had spread to several countries, including Kenya (Owuar, 1991). In Africa, the aphid attacks the exotic plantations of *Cupressus lusitanica*, as well as indigenous species such as *Juniperus procera* and *Widdringtonia nodiflora* (Mulanje cedar)—Malawi's national tree. In Kenya, about half of the approximately 150,000 ha of plantation forests is *Cupressus lusitanica*. The aphid thus poses a considerable threat to the timber industry (Orondo and Day, 1994). Data are being collected to assess the extent of damage and any correlations with climatic factors in eastern and southern Africa (Chilima, pers. comm.).

Plantations may be most vulnerable to climate change through increased stress resulting from drought, which makes conditions ideal for new or old pests and diseases. In a matter of a few years, an important species can be wiped out—before control measures are developed or new species found to replace it. No data exist on how plantation species in Africa might respond to increases in $CO_2$ concentrations. It would seem likely, though, that improved WUE associated with a $CO_2$ fertilization effect would boost productivity.

### 2.3.4. African Coastal Zones

The African coastal zone consists of a narrow, low-lying coastal belt. It also includes the continental shelf and coasts of 32 mainland countries. It is composed of a variety of ecosystems, including barrier/lagoons, deltas, mountains, wetlands, mangroves, coral reefs, and shelf zones. These ecosystems vary in width from a few hundred meters (in the Red Sea area) to more than 100 km, especially in the Niger and Nile deltas. In west Africa (Mauritania to Namibia), the coastal zone spans a broad range of habitats and biota and includes the pristine islands of Bijagos Archipelago; the offshore island nations of Cape Verde and Sao Tome and Principe; and the remote central Atlantic islands of San Helena and Ascension.

A large percentage of west Africa's urban population lives in coastal cities. In Nigeria, for example, about 20 million people (22.6% of the national population) live along the coastal zone; about 4.5 million Senegalese (66.6% of the national population) live in the Dakar coastal area. About 90% of the industries in Senegal are located within the Dakar coastal zone. In Ghana, Benin, Togo, Sierra Leone, and Nigeria, most of the economic activities that form the backbone of the national economies are located within the coastal zone. Coastal areas also form the food basket of the region. Offshore and inshore areas, as well as estuaries and lagoons, support artisinal and industrial fisheries accounting for more than 75% of fishery landings in the region.

The coastal zone of east Africa, including coastal wetlands, extends from Sudan to South Africa and includes the near-shore islands off the coast of Tanzania and Mozambique and the oceanic islands of Madagascar, the Seychelles, Comoros, Mauritius, and Reunion. The desert margins of the Red Sea feature some of the richest coral reefs in the world. Coral reefs further south, extending from Kenya to the Tropic of Capricorn, are well distributed around most of the oceanic islands. They buffer the coastline against the impact of wave breakers and the full force of storms and cyclones. Many principal east African cities are located inland. Despite their low densities, however, coastal cities like Dar es Salaam and Mombasa are experiencing annual population growth of 6.75% and 5%, respectively (World Bank, 1995a). Coastal tourism and fisheries represent large inputs into the GNP of east African states.

Coastal population pressures and increasing exploitation of coastal resources—utilizing conflicting exploitation methods—have led to coastal degradation. Coastal erosion, flooding, pollution (air, water, land), deforestation, saltwater intrusion, and subsidence are some of the environmental problems degrading large parts of the coastal area of Africa. Coastal erosion already has been reported to reach 23–30 m annually in some parts of coastal west Africa (Ibe and Quelennac, 1989). In Cote d'Ivoire, high erosion rates have been reported in areas off the Abidjan harbor. It also is estimated that about 40% of the mangroves in Nigeria had been lost by 1980 (WRI, 1990); about 60% of mangrove areas in Senegal also have been lost as a result of mangrove clearing, coastal erosion, and increases in the salinity of water and soil. Industrial pollution from oil spills and discharge of domestic untreated wastes is polluting large areas of the coast, including lagoons and near-shore areas. The Korle lagoon in Accra, Lagos lagoon, and Ebrie lagoon in Abidjan all have been polluted, resulting in loss of fisheries resources.

Under doubled $CO_2$, climate change is projected to adversely affect several physical, ecological/biological, and socioeconomic characteristics of the west African coastal zone and adjacent oceans that presently are under stress. At the same time, population pressures and conflicting policies of exploitation of coastal resources also have had adverse effects on coastal sustainability. Environmental problems degrading the coastal area are projected to increase as a result of either sea-level rise or an increase in extreme weather events (IPCC, 1996).

#### *2.3.4.1. Vulnerability and Impacts of Climate Change on Coastal Zones*

Climate change will exacerbate existing physical, ecological/biological, and socioeconomic stresses on the African coastal zone. Most existing studies focus on the extent to which rising sea level could inundate and erode low-lying areas or increase flooding caused by storm surges and intense rainstorms. The coastal nations of west and central Africa (e.g., Senegal, The Gambia, Sierra Leone, Nigeria, Cameroon, Gabon, Angola) have low-lying lagoonal coasts that are susceptible to erosion and hence are threatened by sea-level rise, particularly because most of the countries in this area have major and rapidly expanding cities on the coast (IPCC, 1996).

Africa's west coast often is buffeted by storm surges and currently is at risk from erosion, inundation, and extreme storm events. Inundation could be a significant concern (Awosika *et al.*, 1992; Dennis *et al.*, 1995; French *et al.*, 1995; ICST, 1996; Jallow *et al.*, 1996). Major cities such as Banjul (Jallow *et al.*, 1996), Abidjan, Tabaou, Grand Bassam, Sassandra, San Pedro (ICST, 1996), Lagos, and Port Harcourt (Awosika *et al.*, 1992)—all situated at sea level—would be very vulnerable. Finally, tidal waves, storm surges, and hazards also may increase and may modify littoral transport (Allersman and Tilsmans, 1993).

The coastal zone of east Africa also will be affected—although, unlike west Africa's Atlantic coast, this area experiences calm conditions through much of the year. Along the east coast of Africa, sea-level rise and climatic variation may decrease the attenuation of coral and patch reefs that have evolved along major sections of the continental shelf. The lessening of this buffer effect as a result of climate change would increase the potential for erosion of the east coast. Increases in population growth rates in the principal coastal cities of east Africa, combined with a likelihood of a 1-m sea-level rise, could create conditions for significant negative impacts on tourism-oriented economies, ecology, and natural habitats of this area.

Existing literature provides information about the implications of sea-level rise for Egypt, Nigeria, Senegal, Cote d'Ivoire, The Gambia, and Tanzania.

##### *2.3.4.1.1. Egypt*

Results from studies on various aspects of the impacts and possible responses to sea-level rise on the Egyptian coast (Broadus *et al.*, 1986; Milliman *et al.*, 1989; Sestini, 1989; Ante, 1990; El-Raey, 1990; El-Sayed, 1991; Khafagy *et al.*, 1992; Stanley and Warne, 1993) indicate that a sizable proportion of the northern part of the Nile delta will be lost to a combination of inundation and erosion, with consequent loss of agricultural land and urban areas. Furthermore, agricultural land losses will occur as a result of soil salinization (El-Raey *et al.*, 1995).

Khafagy *et al.* (1992) estimate that for a 1-m sea-level rise, about 2,000 km$^2$ of land in coastal areas of the lower Nile delta may be lost to inundation. Substantial erosion should be expected, possibly leading to land losses of as much as 100 km$^2$. A very rough estimate of the agricultural land area that might become unusable is 1,000 km$^2$ (100,000 ha). With an average land value of US$1.5/m$^2$, the value of land loss in the lower Nile delta as a result of flooding alone will be on the order of US$750 million (2,500 million Egyptian pounds) (Khafagy *et al.*, 1992). Outside the delta, erosion is expected to be quite limited. If average erosion were 20 m along 50% of the remaining coast (and assuming land values on the order of 5 Egyptian pounds per m$^2$), the total loss would be about US$60 million (200 million Egyptian pounds).

For the Governorate of Alexandria, two main economic areas appear most vulnerable: the Alexandria lowlands and the Alexandria beaches (El-Raey *et al.*, 1995). The Alexandria lowlands—on which the city of Alexandria originally developed—are vulnerable to inundation, waterlogging, increased flooding, and salinization under accelerated sea-level rise. The two surviving Alexandria beaches (Gleam and El Chatby) will be lost even with a 0.5-m rise in sea level. Based on the 0.5-m scenario, estimated losses of land, installations, and tourism will exceed US$32.5 billion. An average business loss is estimated at US$127 million/yr because most tourist facilities such as hotels, camps, and youth hotels are located within 200–300 m of the shoreline. It has been widely reported that 8 million people would be displaced in Egypt by a 1-m rise in sea level, assuming no protection and existing population levels (Broadus *et al.*, 1986; Milliman *et al.*, 1989). This estimate is based on the displacement of 4 million people in the Nile delta, as well as the entire population of Alexandria.

#### 2.3.4.1.2. *Nigeria*

If sea level rises, inundation could occur along more than 70% of the Nigerian coastline, placing land at risk many kilometers inland (Awosika *et al.*, 1992). In Nigeria, inundation is the primary threat for at least 96% of the land at risk (Awosika *et al.*, 1992; French *et al.*, 1995). With a 1-m rise in sea level, up to 600 km$^2$ of land would be at risk. This area includes parts of Lagos and other smaller towns along the coast. For the Mud Coast, a 1-m rise will place as much as 2,016 km$^2$ of land at risk. Even with no acceleration in sea-level rise, current rates of land loss through edge erosion alone could cause losses of as much as 250 km$^2$ by the year 2100. This land loss is equivalent to an average shoreline recession of 3 km. Erosion threatens a higher percentage of the land on the Strand Coast than in the delta (4.6–20.7% for the Strand Coast, versus 0.8–3.5% for the delta). Without consideration of oil wells in the Niger delta, the greatest value at risk is along the Barrier Coast—ranging from just over US$1.3 billion with a 0.2-m sea-level rise to almost US$14 billion with a 2-m rise. When the value of the oil fields is considered, the value at risk increases from US$81.4 million to almost US$2.2 billion with a 0.2-m rise and from US$6 billion to more than US$19 billion for a 2-m rise. On the Strand Coast, a 1-m rise will result in a value loss of more than US$18 billion; the largest contribution to this loss would come from the oil fields in the area.

In Nigeria, a potentially massive "environmental refugee" migration will occur. For a 1-m rise, more than 3 million people are at risk, based on the present population. The estimated number of people that would be displaced ranges from 740,000 for a 0.2-m rise to 3.7 million for a 1-m rise and 10 million for a 2-m rise (Awosika *et al.*, 1992).

#### 2.3.4.1.3. *Senegal*

In Senegal, a 1-m rise in sea level could inundate and erode more than 6,000 km$^2$ of land, most of which is wetlands (Dennis *et al.*, 1995). In general, inundation is responsible for more than 95% of the land loss, independent of the scenario considered. Dennis *et al.* (1995) have shown that, under a 1-m scenario, buildings with a total market value of at least US$499–707 million would be at risk. On a national basis, tourist facilities represent 20–30% of the total value at risk under the 1-m scenario. Under the 1-m scenario, it is estimated that at least 110,000–180,000 people—or 1.4–2.3% of the 1990 population of Senegal—are at risk (Dennis *et al.*, 1995). Nearly all of these people are located south of the Cape Verde peninsula; the bulk of the population at risk lives south of Rufisque.

#### 2.3.4.1.4. *Cote d'Ivoire*

In Cote d'Ivoire, a 1-m sea-level rise will lead to inundation of 1,800 km$^2$ of lowland. The rate of shoreline retreat as a result of erosion is estimated to vary from 4.5 m to 7.4 m per annum (ICST, 1996). The most threatened infrastructures on the coastal zone are the Autonomous Port of Abidjan (Port Autonome d'Abidjan, PAA) and the port of San Pedro.

#### 2.3.4.1.5. *The Gambia*

In The Gambia, inundation is projected to lead to a loss of 92 km$^2$ of land for a 1-m sea-level rise. Shoreline retreat is projected to vary between 6.8 m in cliff areas to about 880 m for flatter and sandier areas (Jallow *et al.*, 1996). If a 1-m sea-level rise were to occur as envisaged, without protective measures the whole capital city of Banjul would be lost in the next 50–60 years because a majority of the city is below 1 m. Preliminary analysis of data from the Gambian Department of Lands on the value of land and sample properties between Banjul and the Kololi Beach Hotel suggests that about 1,950 billion Dalasis (US$217 million) of property will be lost (Jallow *et al.*, 1996).

All of the structures located on land between Sarro and Banjul cemeteries and the whole of Banjul will be lost. (It was not possible to attach monetary value to this loss of public places.) According to Jallow *et al.* (1996), the entire population of Banjul (42,000 inhabitants) and people living in the eastern parts of Bakau and Cape St. Mary—as well as the swampy parts of Old Jeswang, Kanifing Industrial Estate, Eboe Town, Taldning Kunjang, Fagikunda, and Aruko—will be displaced.

###### *2.3.4.1.6. Tanzania*

Using sea-level rise scenarios of 0.5 and 1.0 m per century, Mwaipopo (1997) assessed the potential impacts of such a rise on the coastline of Tanzania. The results revealed that, with a 0.5-m and 1-m sea-level rise, about 2,090 km$^2$ and 2,117 km$^2$ of land would be inundated, respectively. With a 1-m sea-level rise, another 9 km$^2$ of land would be eroded. Projected damage is expected to be about Tsh. (Tanzanian shillings) 50 billion for a 0.5-m rise and Tsh. 86 billion for 1-m rise.

#### *2.3.4.2. Adaptation Strategies and Measures in the Coastal Zone*

Ibe (1990) has found that large-scale protective engineering measures are impractical in Africa because of the high costs to countries in the region. Instead, low-cost, low-technology, but effective measures—such as permeable, nonconcrete, floating breakwaters; artificial raising of beach elevations; and installation of rip-rap, timber groins, and so forth—are considered more sensible.

In Egypt, protection of beaches and associated infrastructure must depend upon continuous and periodic nourishment. The total cost of protection is estimated to be only US$21 million and $42 million for the 0.5-m and 1-m scenarios, respectively (El-Raey *et al.*, 1995). The city of Alexandria is built on three intermittent calcareous ridges, parts of which are leveled and thus constitute a potential pathway for rising water to reach the lowland south of the city. El-Raey *et al.* (1995) have suggested that maintaining the beaches in front of these pathways, using beach nourishment, will help to prevent surface water from reaching the lowland south of the city through these paths, particularly during storm surges. Construction of small dikes also may be necessary.

According to French *et al.* (1995), the potential cost of protection against a 1-m rise in sea level in Nigeria ranges from US$558–668 million if densely developed areas are protected (important areas protection) to US$1.4–1.8 billion if all shores are protected from erosion, inundation, and flooding (total protection). Under the "important areas protection" option, the 1-m sea-level rise scenario would require a total of 430 km of seawall and associated fill, costing nearly US$200 million. An additional 474 km of sheltered seawall and 180 km of open-coast seawall is required when total protection is considered. More than 600,000 villagers could be displaced by a 1-m rise in sea level, based on existing population (French *et al.*, 1995). It is impractical to protect these villages and populations at risk in Nigeria from a 1-m rise in sea level. Relocation probably is the most practical solution to the problem of inundation of the villages.

If the option considered in Senegal is "important areas protection," Dennis *et al.* (1995) suggest that about 70 km—or about 14.5%—of the open coastline would require protection. Total protection would require protection of 2,063 km, including 310 km of open-coast seawall and 1,680 km of sheltered-coast seawall. The cost of total protection is about 2.5–4 times higher than the costs of important areas protection.

With regard to The Gambia, Jallow *et al.* (1996) argue that if sea-level rise were to occur as envisaged, the most appropriate response (considering The Gambia's economic situation) would be to protect important areas. The projected cost includes US$3.1 million for construction of a 7-km low-cost seawall and US$3.9 million for construction of a revetment. About 16 km of dikes is required to protect villages bordering wetlands and swamplands from seasonal flooding. Four types of actions could fulfill required protection needs:

- Repair of groin systems
- Construction of breakwaters
- Construction of low-cost seawalls
- Construction of revetments.

In Cote d'Ivoire, the most essential response to a potential sea-level rise of 1 m in the next century would be to protect important areas—such as the two ports at Abidjan and San Pedro; the airport; tourism facilities; residential areas; and other areas of high economic importance, particularly in Grand Bassam, Abidjan, Grand Lahou, Sassandra, San Pedro, and Tabou (ICST, 1996). No cost estimates are attached to this protection.

Within an integrated approach, there is a great opportunity to anticipate problems associated with sea-level rise rather than simply react to change as, or after, it occurs (Nicholls and Leatherman, 1994). Furthermore, a well-planned response that seeks to anticipate the physical impacts of sea-level rise in a timely fashion will minimize unwise decisions and result in lower costs for reactive responses such as protection (Nicholls and Leatherman, 1995). Anticipatory responses include urban growth planning, building setbacks, wetland preservation and mitigation, public awareness, and integrated coastal zone management.

Policies and regulations concerning the use of the coastal zone for any form of human activity should include consideration of sea-level rise. Physical planning and building-control measures and regulations should be instituted and implemented. Allocation of land for any economically useful purpose in areas likely to be flooded or inundated should be avoided. The public should be informed of the risk of living in coastal and lowland areas that are threatened by sea-level rise. Timely public education about erosion, sea-level rise, and flooding risks could be a cost-effective means of reducing future expenditures. Where coastal infrastructures such as roads, fish land, and curing plants are approved and must be constructed, the authorities and owners of these infrastructures should make sure that marginal increases in the height of the structures are included to offset sea-level rise (Smith and Lenhart, 1996) and other related phenomena. People located in high-risk areas should be offered incentives to relocate out of these areas. Setbacks could be used as buffer zones to allow sea level to rise without threatening coastal development. French *et al.* (1995) recommended incorporating buffer zones between the shore and new coastal development in Nigeria.

Adaptation to climate change and concomitant adverse effects will involve an understanding of climate change parameters and dynamics, including monitoring and data analysis of climate change parameters. This strategy should lead to an African Climate Change Scenario (ACCS), upon which countries can base their adaptation options. Existing scenarios and adaptation measures for climate change and sea-level rise are built around Western experiences.

All of the adaptation strategies, options, and policies discussed above will reduce the risks from current climate variability and protect against potential climate change and sea-level rise. These measures should be combined in a coastal zone management plan (CZMP). The CZMP brings together all actors in the coastal area to address coastal-zone problems. The program should consist of a set of principles and plans to guide the use of coastal land for conservation, recreation, and development.

### 2.3.5. Human Settlements, Energy, Industry, and Transport

The pattern of distribution of human settlements often reflects the uneven nature of resource endowments and availability between regions and within individual communities. In Africa, as elsewhere, there are heavy concentrations of human settlements within 60 km of coastal zones, in areas of high economic potential, in river and lake basins, in close proximity to major transportation routes, and in places that enjoy hospitable climatic regimes. Changes in climate conditions would have severe impacts not only on the pattern of distribution of human settlements but also on the quality of life in particular areas. For example, wetter coasts or drier conditions in up-country areas could lead to spontaneous migrations as an adaptive option. Similarly, the pattern of energy use could change radically as a result of technological adaptations arising from climate change.

IPCC (1996) and UNFCCC (1992) acknowledge that developing countries' energy demands must increase to meet their needs for economic development. This increase must occur so these countries can respond to their development needs and to support the needs of growing populations. More of this economic development will be in industrial and transport sectors than in any other sector. It has been argued that the growth of the energy, industry, and transport sectors is needed as countries go through their economic transitions, which will decrease their vulnerability. Current high dependence on land-based production activities—such as agriculture and fisheries—only increases the vulnerability of African countries. The energy, industry, and transport sectors are thus important in discussing vulnerability and adaptation.

#### 2.3.5.1. *Human Settlements*

The main challenges likely to face African populations will emanate from the effects of extreme events such as tropical storms, floods, landslides, wind, cold waves, droughts, and abnormal sea-level rises that are expected as a result of climate change. These events are likely to exacerbate management problems relating to pollution, sanitation, waste disposal, water supply, public health, infrastructure, and technologies of production (IPCC, 1996).

Adaptation strategies lie mainly in relocating populations, efficient energy supply and use, introduction of adaptation technologies, and improved management systems. Because most of these strategies have high cost implications, existing economic constraints of African countries may present major obstacles. In addition, implementing some of these strategies may have aspects that go beyond costs; relocation of human settlements from low-lying coastal areas that are vulnerable to inundation, for example, is likely to create problems that go beyond cost implications and include changes in social structure—clear policies on land use, fortified by flexible land-tenure regimes, will be needed.

#### 2.3.5.2. *Energy*

The impacts of climate change on the energy sector will be felt primarily through losses or changes in hydropower potential for electricity generation and the effects of increased runoff (and consequent siltation) on hydrogeneration, as well as changes in the growth rates of trees used for fuelwood. The total primary energy use in 1990 in sub-Saharan Africa (including South Africa) was broken down in the following shares: biomass fuels (53%), petroleum (26%), coal (14%), large-scale hydro (3%), natural gas (2%), and other renewables (2%). The most vulnerable areas of the energy sector to climate change in Africa are the provision of energy services for rural areas and, to some extent, for urban low-income needs. Table 2-9 shows that millions of cubic meters of wood are harvested each year for energy purposes. The extent of biomass dependence for the African energy sector is high; this dependence is critical because the source of biomass is supported only by the natural regeneration of indigenous natural forests. In addition to the primary energy sources listed in Table 2-10, dependence on charcoal is high in east and southern Africa, in countries such as Zambia and Tanzania; in Zambia, where charcoal provides 80% of urban household energy needs, 3.5 million tons of charcoal are produced annually from indigenous forests.

***Table 2-9:*** *Relative extent of rural population in selected African countries and associated fuelwood production.*

| Subregion | Representative Country | Fuelwood Production ($10^3$ $m^3$) | Population ($10^3$) | Rural Population (%) |
|---|---|---|---|---|
| West Africa | Nigeria | 90,699 | 95,198 | 77 |
| | Ghana | 8,493 | 13,588 | 68 |
| East Africa | Kenya | 32,174 | 20,600 | 80 |
| | Ethiopia | 37,105 | 43,557 | 88 |
| Southern Africa | Zimbabwe | 5,988 | 8,777 | 75 |
| | Botswana | NA | 1,107 | 81 |
| North Africa | Sudan | 18,202 | 21,550 | 79 |
| | Egypt | 1,962 | 46,909 | 54 |
| Central Africa | Cameroon | 9,389 | 9,873 | 58 |
| | Chad | 3,137 | 5,018 | 73 |

Sources: Compiled from UNEP, 1990; ADB AEP, 1996.

In the household sector, fuelwood accounts for 97% of all energy consumed (Chidumayo, 1997). Although natural stocks of wood may be high, wood resources available to the majority of the rural population are very low in many areas. Brickmaking and tobacco and tea curing are major wood uses. In Zimbabwe, wood used for brickmaking is said to equal that used for cooking in rural areas (Bradley and Dewees, 1993); tobacco estates in Malawi account for 21% of total fuelwood consumption (Moyo *et al.*, 1993). In Botswana, the fencing of fields to keep out livestock consumes 1.5 times more wood than is used for cooking in farming households (Tietema *et al.*, 1991). Indigenous miombo and other woodlands in sub-Saharan Africa contribute significantly to the firewood harvested for consumption and conversion to charcoal. Stands are left to recover, with minimal active management. The ability of users to purchase alternative forms of energy (gas or electricity), as well as charcoal, depends on the economics of each family. Therefore, poor people (with limited buying power) are most vulnerable to reductions in fuelwood supply. Increasing populations also are contributing to depletion of resources. In relatively dense woodland areas, where population density is low, there usually is enough deadwood that can be collected and used for fuelwood. Increasing incidence of drought, however, leads to increased fire frequency—which, in turn, reduces deadwood material in woodlands. If current natural resource management systems are not changed, Africa could run the risk of depleting its forest resources used as biomass energy at a rate faster than the rate of population growth. The paucity of data on biomass depletion and regeneration rates makes meaningful assessment difficult and compounds the problems of possible reduced precipitation and subsequent lower regeneration rates by making it difficult to identify appropriate response options. There already are indications of a negative supply balance (e.g., extensive household utilization of agricultural and animal wastes for energy).

In 1992, Africa's electricity output was 312,000 GWh; thermal power provided 78% and hydroelectricity 19%, with a small amount (3%) from nuclear sources in South Africa (ADB AEP, 1996). Thermal power plants require huge volumes of water in their cooling systems; in a situation of reduced rainfall, loss of cooling-water resources will not only reduce generation capacity but also retard construction of new plants. It may be reasonably expected, therefore, that exploitation of the continent's massive coal reserves in areas with such resources would be inhibited by both the anti-coal lobby and shortages of cooling water. In the past (e.g., during the drought of 1991–92), declines in precipitation led to a significant loss of total hydropower energy, including losses of as much as 30% from the Kariba Dam (which supplies power for Zambia and Zimbabwe). It has been suggested that future hydropwer output could could be affected by climate change. Salewicz (1995) investigated the vulnerability of the Zambezi basin to climate change. He noted that 75% of the lower Zambezi waters flow into Kariba. Under climate change scenarios, this area is projected to experience increased rainfall and runoff into Lake Kariba. Although there may be shifts in the seasonal reliability of given discharges for the Lower Zambezi, it is possible that hydropower generation capacity would be adversely affected. Similar impacts could occur on the Congo, Nile, and Niger river hydropower systems, resulting in critical electricity supply shortfalls throughout the continent. In addition, the continent's massive hydroelectric potential of 150,000 GWh/yr would be significantly curtailed. Such a situation would lead to the introduction of major changes in fuel supply strategies in most countries. A case has been made for developing micro- and small-scale hydropower plants in Africa to overcome the cost of large-scale generation systems. This type of plant will require a defined minimum level of runoff. Reductions in precipitation could significantly reduce the number of viable sites for such micro-hydro installations.

***Table 2-10:*** *Estimates of primary energy supplies (%) in subregions' representative countries.*

| Subregion | Representative Country | Oil | Coal | Gas | Biomass | Electricity/Hydro |
|---|---|---|---|---|---|---|
| West Africa | Nigeria | 27 | 0.4 | 12.6 | 59 | 1 |
| | Ghana | 21 | – | – | 69 | 10 |
| East Africa | Kenya | 21 | 1 | – | 70 | 8 |
| | Ethiopia | 8 | – | – | 90 | 2 |
| Southern Africa | Zimbabwe | 10 | 50 | – | 25 | 15 |
| | Botswana | 17 | 6 | – | 73 | 4 |
| North Africa | Sudan | 19 | – | – | 80 | 1 |
| | Egypt | 54 | 2 | 21 | 15 | – |
| Central Africa | Cameroon | 19 | – | – | 67 | 14 |
| | Chad | 33 | – | – | 77 | – |
| **Average** | | | | | 62.5 | |

Sources: Compiled and computed from UNEP, 1990; ADB AEP, 1996.

### 2.3.5.3. *Industry*

Changes in future climate should be actively considered in developing a sustainable industrial development path for Africa. Vulnerability in African industry may relate more to the inhibiting effects of climate change on industrial expansion than to its effects on existing industrial installations and investments. The most serious impacts of climate change on this sector would be related to loss of competitiveness associated with increased costs of production resulting from changes or retrofitting of plants for cleaner production. Reduced surface-water supplies would lead to extended use of groundwater sources—which, in most cases, have to be treated on site to achieve desired water-quality standards for specific industrial applications. Other major effects will result from a lack of water for industrial processes and increased costs of cooling for temperature-controlled processes and storage; Africa's industry has a large number of agro-industrial operations that need large amounts of water.

Besides these direct effects, there will be indirect effects, such as rising water costs; in cases of severe and recurrent water shortages, this factor could lead to relocation of industrial plants. Electricity shortages, due to a drop in the water level which causes a decline in hydropower, also will affect industry—particularly the steel sector (including iron and steel), ferro-chrome production, cement production, textiles, and aluminum production. These industries are among some of the most advanced on the continent, but they are highly dependent on constant electricity supplies. Although there are no data to indicate the level of water shortages that may result from a decline in precipitation, it is obvious that water shortages that affect concentrated urban settlements also will have a debilitating effect on industrial production. Water demand in many states in southern and northern Africa already has exceeded or is expected to exceed water supply soon.

Although detailed assessments of the impact of sea-level rise on coastal industries have been made for Asia and other regions, little information is available for Africa. It can be assumed, however, that most of the impacts that would ensue in other coastal zones would apply equally in Africa. Most impacts would be related to relocation of industries. The extent of these impacts could not be assessed in any detail without a more complete assessment of coastal zone industrial locations.

### 2.3.5.4. *Transport*

The transport sector is based on long-term, immovable infrastructure such as roads, rails, and water. In most parts of the continent, road networks have tended to link industrial centers with major areas of agricultural activity; railways have been designed in the past primarily with a sea-route orientation to facilitate international shipments of primary products. Climate change may lead to industrial relocation, resulting either from sea-level rise in coastal-zone areas or from transitions in agro-ecological zones. This relocation would necessitate additional infrastructural investments. It also may render waterways dysfunctional, thereby necessitating additional road and rail investments to replace them. If sea-level rise occurs, the effect on the many harbors and ports around the continent will be quite devastating economically for many coastal-zone countries. Excessive precipitation, which may occur in some parts of Africa, is likely to have serious negative effects on road networks and air transport. On the other hand, if climate change leads to drier conditions, maintenance costs may be reduced. Typical road networks on the continent consist of gravel roads linking major urban centers, which have paved road systems. Swaziland has a reasonable percentage of paved roads (55% by 1990). The situation is much worse in other countries, as indicated in Table 2-11, which shows the quality of road networks for selected African countries.

Agricultural production activities involve extensive transportation on the farm, to ferry farm products and inputs and to move farm implements. These critical operations are easily disturbed by excessive rains. The effects of such disturbances are likely to be higher in the case of large-scale commercial farms, which are highly mechanized. During recent droughts on the continent, soils on roads became friable as a result of high temperatures and dry weather. Such roads were easily washed away when the rains finally came, resulting in serious soil erosion and high repair costs to farmers. For small-scale communal land farmers, this situation was represented by massive soil erosion along pathways on and off the farm. In many cases, these washed-out pathways became the source of gullies.

Impacts on air transport would be greatest for airports located in coastal areas or near other water bodies, such as rivers or lakes. Indirect effects also would be felt if increased precipitation resulted from climate change, as some models have suggested. In this case, existing air transport support systems—such as meteorological data bases—may be stretched to their limits because they were designed for certain prevailing climates in the continent. Bad weather, in general, would raise the potential for air traffic accidents.

#### 2.3.5.5. *Mitigation and Adaptation: Response Options*

The suggestion that a "solar revolution" (Kane, 1996) may replace the temporary fossil economy is rather simplistic. For African countries—which have yet to develop their infrastructure significantly and their basic industries—the need for centralized energy systems will continue for some time, although it may coexist with advances in solar installations. Besides this general transition, Africa's centralized energy systems—including hydroelectric, coal, and oil thermal systems—will need to benefit from cleaner, more efficient energy-conversion technologies. This transition will be based on autonomous efficiency improvements in the short to medium term, but Africa will not be in a position to drive trends toward such improvements, due to economic and technical trends.

Wind energy is a widely suggested option, based on average wind speeds of 5.8 m/s (Kane, 1996). Unless technologies are developed to generate electricity at lower speeds, the windmill option will remain limited in its application because wind speeds in the region generally are low, averaging 3–5 m/s.

New and renewable energy sources, however, offer other benefits in addition to being alternatives to large hydropower and thermal power generation systems. These alternative energy sources will generate more jobs overall in the economy. According to Kane (1996), these jobs will be high-quality jobs, mainly in systems design. For this reason, this benefit may remain limited in Africa because the continent lags other regions in the development of new technologies such as windmills, solar photovoltaics (PVs), and biogas digesters, which have been spearheaded in China and India. African countries should take immediate steps, with external support, to rectify this deficiency.

***Table 2-11:*** *Road networks and density for selected African countries.*

| Country | Paved Roads | % Paved | Total (km) | Density (km/1,000 km²) | Vehicles/km |
|---|---|---|---|---|---|
| Angola | 8,900 | 12.1 | 73,400 | 59 | 4 |
| Botswana | 3,740 | 20.4 | 18,330 | 32 | 4 |
| Ethiopia | 13,300 | 34.5 | 38,600 | 32 | 1 |
| Kenya | 9,800 | 14.5 | 67,800 | 117 | 5 |
| Lesotho | 540 | 11.2 | 4,840 | 161 | 5 |
| Malawi | 2,400 | 18.3 | 13,140 | 111 | 2 |
| Mauritius | 1,795 | 86 | 2,090 | 1,045 | 65 |
| Mozambique | 5,400 | 15.1 | 35,700 | 45 | 2 |
| Namibia | – | – | 6,500 | 8 | 21 |
| South Africa | – | 29.8 | – | – | – |
| Swaziland | 730 | 21.2 | 3,450 | 203 | 17 |
| Tanzania | 3,580 | 4 | 82,600 | 87 | 1 |
| Uganda (1993) | 8,342 | 21 | 40,057 | 166 | 2.4 |
| Zambia | 6,300 | 16 | 39,160 | 52 | 3 |
| Zimbabwe | 13,000 | 17 | 78,700 | 202 | 5 |
| **Average** | – | 23 | – | 166 | 10 |

Source: Zhou and Molcofi, 1997.

Energy efficiency improvements—which, in Africa, will have to be autonomous in most cases—offer a significant response option. Energy wastage in Africa is quite high; in some cases, savings of up to 40% can be achieved. In the electricity sector, total system losses sometimes exceed 30% in situations where universal standards are below 8%, including some systems in Africa (Davidson, 1992). Reducing these inefficiencies will provide a demand-side option for electricity supply. Demand-side management options, however, need strong support programs to overcome a number of implementation barriers.

With regard to industry, it has been suggested that "much of industry will have to mimic nature, reusing and recycling every chemical that it uses in cyclical processes." This approach implies "shifting to sustainable industries" (Kane, 1996). Before the current climate change debate, this strategy entailed using renewable materials for industrial raw materials—such as natural fibers as opposed to finite resources such as minerals. However, reduced productivity in agriculture—which is the current base for industrial raw materials in most African countries—limits the usefulness of this suggestion.

Small-scale mills have been suggested as ideal for developing countries because they could achieve recycling of resources and greater efficiencies (Kane, 1996). It would be more efficient, for example, to install small-scale mills in dispersed locations around the country to reduce the cost of moving scrap to large, centrally located mills. In Africa, this approach would be a relevant option to reduce the decline or depletion rate for natural ores and to disperse job opportunities around the country. Steel production is significant in Nigeria, Zimbabwe, and South Africa; plants in these countries, however, are traditional large-scale installations. The introduction of small-scale mills could enable the industry to spread to other locations across the continent.

### *2.3.6. Wildlife, Tourism, and Recreation*

Tourism—one of Africa's most promising and fastest-growing industries (about 15% per year)—is based on wildlife and water supply for recreation. Recurrent droughts in the past decade have depleted wildlife resources significantly. Permanent loss of such attractions would waste vast amounts of investment in tourism. The greatest impacts would occur in drought-prone areas of the Sahel, east Africa, and southern Africa.

High levels of floral and faunal species diversity exist in various reserved areas in relict, fragmented patches of natural vegetation. Most wildlife is in reserved areas surrounded by human land use (agriculture). This fragmentation and concentration of animals in specific areas make them highly vulnerable because vegetation (habitat) will not respond quickly enough to changed climate, and wildlife will be unable to migrate to more suitable climatic conditions because of limited corridors between wildlife reserves in different vegetation and climate types; moreover, wildlife would be slow in responding to a changing habitat boundary (see Section 2.3.1.4).

Climate change will impact the tourism industry indirectly through changes in water and vegetation, as well as through wider-scale socioeconomic changes—for example, fuel prices and patterns of demand for specific activities or destinations. Various indirect impacts also may derive from changes in landscape—the "capital" of tourism (Krippendorf, 1984)—which might lead potential tourists to perceive Africa as less attractive and consequently to seek new locations elsewhere. There also may be new competition from other tourist locations as climates change (particularly on seasonal time frames), especially in relation to northern vacation periods.

Tourist attractions such as Victoria Falls could become much less attractive as a result of reduced river discharge and alteration of the rainforest. Hydrology models for tropical savanna Africa suggest reduced runoff as a result of climate change (Hulme, 1996a). The tourist impacts of these changes will include alteration of characteristics of popular tourist destinations. In the 1992 drought period, Victoria Falls lost some of its attractiveness as a result of much reduced water discharge over the falls. Furthermore, the reduced flow resulted in reduction of the spray that maintains the rainforest that is part of Victoria Falls' aura—resulting in the death of flora around the falls.

In southern and eastern Africa, there are a variety of water-based tourist activities—such as sailing, skiing, angling, rafting, and so forth. Such activities could be affected by changes in river flows and the impacts of land use (especially agriculture) on water quality—such as eutrophication, which gives rise to objectionable blue-green algae and a proliferation of aquatic weeds (such as *Eichhornia* and *Salvinia*). These developments will compromise the aesthetic value of tourist destinations. Many reservoirs in Africa (such as Lake Victoria and the Nile Sudd) are under threat. In addition to their tourist value, these inland waters in Africa are a source of protein. Many water bodies in Africa show very high sensitivity to changes in runoff. Inland drainage systems—such as Lake Chad, the east African Rift Valley lakes (e.g., Lake Nakuru, Lake Naivasha), and other shallow water bodies such as Lake Chilwa and the Okovango delta—have a delicate hydrological balance. Complete drying of the lake recently occurred at Lake Chilwa in Malawi and Lake Nakuru in Kenya. Magadza (1984, 1996) has suggested that fish production in large reservoirs such as Lake Kariba can be significantly affected by decreased runoff because of reduced nutrient inflow.

The wetlands of Africa—such as the Okovango delta, the Kafue River floodplains, and Lake Bangweulu—have rich and varied wild fauna and are especially conspicuous for their avifauna. Magadza (1996) shows how the gradual drying of the Caprivi Strip wetlands resulted in a population reduction of the wetland vertebrates and the complete disappearance of some species from the area. Such aridification of wetlands has been followed by encroachment of human cultivation.

The destruction of coastal infrastructure, sandy beaches and barriers, and marine ecosystems would have negative impacts

**Box 2-9. Cape Cross: The Ecological Importance of Coastal Wetlands**

A number of African coastal areas, such as the Namibian coast, offer unique habitats for migratory bird species as well as a variety of local species (Williams, 1990). These wetlands are maintained by a complex hydrology of ephemeral rivers that flow onto the Namibian west coast. Changes in sea level would affect these western coastal wetlands through habitat loss.

Cape Cross, one of the earliest recorded (15th century) points of contact for European explorers in southern Africa, is best known for the large Cape Fur Seal (*Arctocephalus pusillus*) colony, numbering about 150,000 animals. To the south of the rocky promontory, a sand barrier has formed from sediments continually washed north from the Orange River. Behind this barrier is a series of lagoons, 5 $km^2$ in extent, comprising salt flats fed by seawater seepage and high-tide washovers (Williams, 1990). Wooden guano platforms were erected in these areas in the 1950s to allow cormorants to breed and roost (Rand, 1952). Guano is still harvested today from 30,000 pairs of Cape cormorants that nest there (Williams, 1990).

As many as 11,000 individuals of 28 species of coastal birds—including intra-African immigrants, palaearctic immigrants, and resident breeders—may be found on the lagoons (Williams, 1990). Among species listed as red data birds, more than 2,000 individuals of the southern African race of the black-necked grebe (*Podiceps nigricollis gurneyi)* may appear in these lagoons—about 16% of the estimated world population (Cooper and Hockey, 1981; Williams, 1990). The lagoons also regularly support between 1% and 3% of the southern African subcontinental population of greater flamingo (*Phoenicopterus ruber)* and lesser flamingo *(P. minor)* and up to 22% of the coastal population of Cape Teal (*Anus capensis)* (Williams, 1990). For palaearctic migrants, including the Curlew sandpiper (*Calidris ferruginea)*, the total number approaches 4,200 individuals; Cape Cross ranks about eighth in importance for these birds along the Namib coast (Williams, 1990). The Cape Cross seal colony falls within a national reserve and only limited disturbance occurs in the lagoons because of their inaccessibility. The area, therefore, suffers no current threat.

on tourism in these areas (Okoth-Ogendo and Ojwang, 1995). This effect could be exacerbated by disturbances in the pattern of human settlements in coastal zones and the general loss of environmental values.

Many tourist facilities (such as hotels) have been invested on inland lakeshores and reservoirs—such as Midmar Reservoir in South Africa, Lake Malawi, Lake Chad, Lake Victoria, and several other lakes in the Rift Valley in east Africa. In some

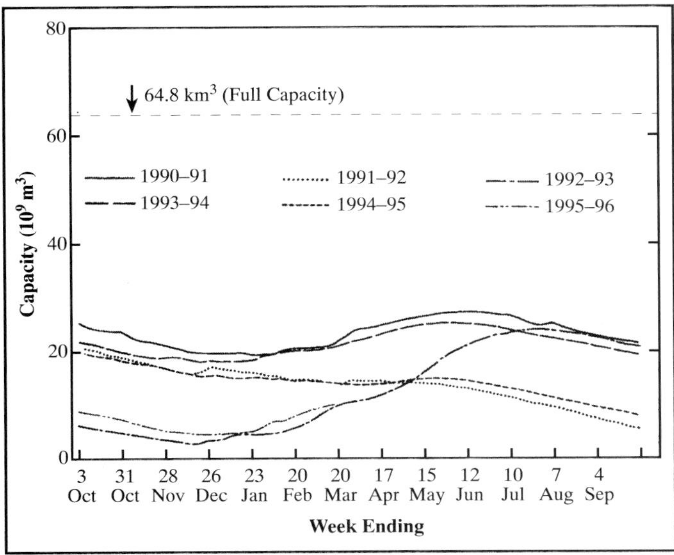


**Figure 2-14**: Lake Kariba water storage 1990–91 to 1995–96 (Hulme, 1996b).

cases, there also are downstream facilities, such as those on the Shire River and Liwonde. Past drought episodes have demonstrated that fluctuations in lake levels affect the quality of services that the lakeside resorts offer; the water level may recede a considerable distance from the facility. Lake Kariba currently is 13 m below its storage-capacity level; services such as dry-dock facilities have lain idle for several years. Figure 2-14 shows the changes in water level of Lake Kariba during the 1991–92 drought period. Where the lake is the primary source of an effluent river, river-based tourist facilities would be similarly prejudiced by lower lake levels. Such impacts would be more pronounced on reservoirs that combine other activities (e.g., irrigation or hydroelectric power generation) with tourism.

An assessment of how the whole natural environment and wildlife might be susceptible to climate change should be an integral part of environment planning (Mkanda, 1996). A number of sub-Saharan countries regard the conservation of biological diversity (wildlife) together with agriculture and tourism activities as an important win-win solution to the nature resource that also offers a major source of much-needed revenue (World Bank, 1996). In a report on wildlife conservancies in Zimbabwe, Price Waterhouse (1994) demonstrated that wildlife presented the best land-use options in semi-arid parts of southeast Zimbabwe. This application, in the Gonarezhou nature reserve, has been facilitated by an appropriate legal framework in Zimbabwe—based on the Parks and Wildlife Act, which essentially confers stewardship of wildlife on landowners.

### 2.3.7. *Human Health*

The links between climate and many environmental and vector-borne diseases (VBDs) are felt through the impacts of various climatic components (e.g., temperature, rainfall) on the physiology of pathogens and their vectors. Although there are crude atlases of disease distribution within Africa (Knoch and Schulze, 1956), accurate and verified models that translate these physiological climate-related processes into more detailed maps of disease distribution are scarce. Such maps and models are necessary to set the baseline of current levels and limits of transmission against which projected impacts of climate change can be measured. These changes may include shifts in the distribution of diseases into areas that previously were disease-free or a change in severity at a given location. Although such models now exist (Martens *et al.,* 1995a,b; le Sueur *et al.,* in preparation), most remain hypothetical and largely unverified. However, they may provide good starting points to illustrate the effects of projected climate change. In the case of malaria, a continental effort—Mapping Malaria Risk in Africa (MARA/ARMA)—is now underway; this effort will provide a data base of disease distribution and severity that can be used to verify climate-induced processes. No such parallel efforts, however, currently are underway for other diseases of the African continent that may be affected by climate change (e.g., arboviruses, trypansomiasis, and schistosomiasis).

In Africa, VBDs are major causes of illness and death. Table 2-12 provides global estimates of the number of people at risk from and the number of people who currently are infected by major VBDs. Currently, the distribution of most VBDs remains well within the climatic limits of their vectors. The extent to which disease transmission potential shifts geographically in response to shifts in vector distribution following climate change will depend partly on how human activities modify local ecosystems (McMichael *et al.*, 1996). Rodent-borne diseases that could be affected by climate change include plague and hantavirus pulmonary syndrome. In a warmer and more urbanized world, rodent populations—which act as pathogen reservoirs and as hosts for the relevant

***Table 2-12:*** *Major tropical vector-borne diseases and the likelihood of change in their distribution as a result of climate change.*

| Disease | Vector | Number at Risk (millions)[1] | Number Infected or New Cases/Year | Present Distribution | Likelihood of Altered Distribution with Climate Change |
|---|---|---|---|---|---|
| Malaria | Mosquito | 2,400 | 300–500 million | Tropics/subtropics | +++ |
| Schistosomiasis | Water snail | 600 | 200 million | Tropics/subtropics | ++ |
| Lymphatic filariasis | Mosquito | 1,094 | 117 million | Tropics/subtropics | + |
| African trypanosomiasis | Tsetse fly | 55 | 250,000–300,000 cases/yr | Tropical Africa | + |
| Dracunculiasis | Crustacean (copepod) | 100 | 100,000/yr | South Asia/ Middle East/ Central-West Africa | ? |
| Leishmaniasis | Phebotomine sand fly | 350 | 12 million infected, 500,000 new cases/yr[2] | Asia/South Europe/ Africa/Americas | + |
| Onchocerciasis | Blackfly | 123 | 17.5 million | Africa/Latin America | ++ |
| American trypanosomiasis | Triatomine bug | 100 | 18–20 million | Central-South America | + |
| Dengue | Mosquito | 2,500 | 50 million/yr | Tropics/subtropics | ++ |
| Yellow fever | Mosquito | 450 | <5,000 cases/yr | Tropical South America and Africa | ++ |

+ = likely; ++ = very likely; +++ = highly likely; ? = unknown.
[1]Top three entries are population prorated projections, based on 1989 estimates.
[2]Annual incidence of visceral leishmaniasis; annual incidence of cutaneous leishmaniasis is 1–1.5 million cases/yr.
Sources: PAHO, 1994; WHO, 1994, 1995; Michael and Bundy, 1996; WHO statistics.

arthropod vectors—will tend to increase. Thus, incidences of these diseases can be expected to rise (Shope, 1991).

Projected increases in the interannual variability of climate would have marked implications for the impact of seasonal epidemic diseases such as malaria. In general, control and mitigation activities for such diseases are planned around mean expected levels in any one year. Significant interannual variation impedes intervention and mitigation because of the impact on national budgets (which plan for mean circumstances) and lags that occur in relation to responses to climatically induced epidemic situations. In addition, such variation results in intermittent exposure of nonimmune populations—resulting in high levels of morbidity and mortality. The recent degree of variability is clearly illustrated in Table 2-13, which shows data for four southern African countries. This variability highlights the need for more climate-based forecasting systems capable of predicting such interannual variations with a lead time that allows health authorities to respond in a timely manner with preparatory/preventative measures (Jury, 1996; le Sueur and Sharp, 1996).

### 2.3.7.1. *Vulnerability and Adaptation*

It is projected that climate-related mortality will increase in the large cities of north Africa (IPCC, 1996), from direct effects as well as from indirect impacts of climate change. These impacts will include potential increases in the incidence of VBDs such as malaria, yellow fever, dengue fever, onchocerciasis, and trypanosomiasis arising from elevated temperatures and altered rainfall. High-elevation locations such as Nairobi or Harare may become vulnerable to malaria epidemics because the malaria parasite may be able to survive in the possibly warmer conditions (IPCC, 1996) at higher elevations than the current limits.

Minimum temperatures ($T_{min}$ or nighttime and winter temperatures) are crucial parameters for vector survival and affect the latitude and elevation of distribution, as well as the length of season permissive to transmission. $T_{min}$ is known to play an important role in limiting the distribution of malaria vector populations at a given locality where summer conditions are suitable for transmission (Craig and le Sueur, in preparation; Leeson, 1931). Thus, meteorological variables can create conditions conducive to disease spread or even to clusters of outbreaks (in the case of flooding or drought) (Epstein *et al.*, 1993). A drop in water level in dams and rivers also would affect the quality of household and industrial fresh water because reduced water volume would increase the concentration of sewage and other effluent in rivers—resulting in outbreaks of diseases such as diarrhea, dysentery, and cholera. During 1992 and 1993, cholera affected almost every country in the SADC region, claiming hundreds of lives. In many drought-affected areas in Zambia, Zimbabwe, Botswana, and South Africa, streams and rivers dried up. Villagers (mainly women) had to walk long distances—only to collect polluted water, which they shared with wild animals and livestock. SARDC (1994) notes that when a major cholera outbreak occurred in several countries in southern Africa in the mid-1980s, the region had just come out of another drought. Reduced water flow during these droughts reduced the capacity of rivers, streams, and swamps to dilute agrochemicals and process fertilizers in fields, adversely affecting soil ecosystems. These drought-related problems are likely to increase under projected climate change. Vulnerabilities and control measures will affect the impact, however.

The growing resistance of insects to insecticides and of microorganisms to antimicrobials, as well as the toxicity of pesticides to helpful (predator) insects and larger animals (including humans), will limit the effectiveness of these control measures. Thus, nurturing environmental conditions that reduce vulnerability to VBD spread (e.g., extensive land clearing and monocultures), and those that shore up generalized defenses (trees around plots and settlements to harbor birds that consume bugs) becomes increasingly important (Epstein, 1993, 1995).

In west Africa, the population of blackflies—which carry onchocerciases (river blindness)—may increase by as much as

***Table 2-13:*** *Interannual variability of malaria (number of cases) within the southern Africa region.*

| **Country** | **1992** | **1993** | **1994** | **1995** | **1996** |
|---|---|---|---|---|---|
| Botswana | | | | | |
| (confirmed) | 415 | 14,615 | 5,335 | 2,129 | 19,340 |
| (unconfirmed)[1] | 4,293 | 40,722 | 24,256 | 15,470 | 49,315 |
| Namibia[1] | 238,592 | 386,215 | 407,863 | 286,407 | 353,593[2] |
| South Africa | 2,886 | 13,330 | 10,298 | 9,287 | 29,206 |
| Zimbabwe[1] | 420,137 | 877,734 | 797,659 | 721,376 | 1,585,850 |

[1]Clinically diagnosed.
[2]Incomplete.

25%, according to precipitation projections from the GISS GCM. Schistosomiasis and malaria, both of which depend on water availability, also are likely to increase as a result of projected expansion of irrigation in hot climates. African trypanosomiasis (sleeping sickness), which is carried by the tsetse fly, also could proliferate because higher temperature would increase the range of the vector in areas prone to this infection. The distribution of tsetse flies may increase in east Africa (Rogers and Packer, 1993). Other minor health effects could include higher incidences of skin disorders or cancers, cataracts and similar forms of eye damage, and suppression of immune systems as a consequence of stratospheric ozone depletion—which leads to greater ultraviolet radiation (IPCC, 1996).

Nutritional status also is likely to be severely affected by droughts and associated crop failures, as in southern Africa during the droughts of the 1980s to early 1990s. This factor will further reduce the natural persistence of African communities and increase exposure to disease.

There is increasing evidence that many emergent and resurgent diseases may be related to ecosystem instability (Epstein, 1995). In many cases, this resurgence may be related not to climatic change but to other human-induced changes in the environment (e.g., lyme disease, dengue, hantavirus) (Levins *et al.,* 1994; Wenzel, 1994; Epstein, 1995). Other diseases with clear links to climate and climatic change include malaria (Bouma *et al.*, 1994; Loevinsohn, 1994) and cholera (Epstein, 1992, 1995; Epstein *et al.,* 1993). Nonclimatic environmental modifications often disrupt natural habitats through processes such as deforestation or afforestation, resulting in the provision of new habitats for vectors or pathogens. Climatically induced change, however, can include short-term impacts as a result of interannual variation or long-term change associated with factors such as global warming. The effects of such changes are likely to be most noticeable in areas that constitute the fringes of distribution of a particular disease. Changes in distribution often will cause susceptible populations (those with little or no immunity) to be exposed to diseases not previously encountered—and result in severe morbidity and mortality. In addition, areas of traditionally low risk may become more vulnerable. The implications of such changes, especially in populations that lack immunity, have severe consequences (individual, social, and economic) for exposed individuals and health authorities.

Rogers (1996) modeled the effect of projected future climate changes on the distribution of three important disease vectors—mosquitoes, tsetse flies, and ticks—in southern Africa. The human diseases relating to these vectors are malaria (mosquitoes) and human African trypanosomiasis (tsetse flies). Climatic change may alter not only the physiological constraints placed on the vector but also the ability of the parasite to survive within the vector (Molineaux, 1988).

Within Africa, little evidence exists of causal changes in disease transmission and climate. This lack of evidence does not mean that these changes do not exist; rather, it may reflect the lack of available epidemiological data as a result of poor or absent surveillance and health information systems. Within Africa, 71.3% of the burden of disease is attributed to infectious diseases; malaria is the single greatest contributor (10.8%). All other vector-borne, helminthic, and environmentally related diseases that are affected by climate contribute about 2% of the total burden of disease. With regard to environmentally related diseases in Africa, malaria contributes more than 80% of the cause of lost disability adjusted life years (World Bank, 1993; WHO, 1996b). These estimates exclude diarrhea but include cholera.

Malaria contributes the highest percentage (>80%) of the climate-related disease burden in Africa. The physiological relationships among climate, vectors, and pathogens are only partially understood. Malaria provides a good example of how potential climate change may affect environmental and vector-borne diseases. Surveillance systems and epidemiological data on malaria exist in some of the regions most susceptible to climate change, allowing future monitoring to move from speculative to causal relationships.

One of the deficient micronutrients in malnourished Africans is iron. Woman and children are disproportionately affected by anemia, and pregnant women are especially at risk. In addition, one of the major causes of death in acute and complicated malaria is anemia. Thus, the potential exists for exacerbated morbidity and mortality in regions in which climate change may decrease nutritional status and increase malaria transmission.

Climate-related impacts on health status are likely to increase most within Africa, largely as a result of the high burden of disease within the continent and its vulnerability in terms of mitigation and nutritional status. In contrast to other regions of the world, a significant proportion of the burden of disease in Africa is related to climatically affected infectious diseases. Typical of many of these diseases in Africa is the fact that their severity focuses within the tropics, and infection rates approach saturation. Thus, the marginal regions of distributions, where severity of transmission is restricted climatologically, are most likely to be affected by climatic change. It is important to distinguish between restrictions related to rainfall and those related to temperature. In cases where rainfall restricts distribution, increases in temperature may result in a decrease in disease incidence and distribution. Conversely, in areas where low temperature currently limits distribution, increased temperature may increase regional severity and cause distributional extension.

### 2.4. Integrated Assessment of Potential Impacts

Work on integrated assessment of climate change in the Africa region is in its infancy. Several different approaches are being applied in country studies and primary research. These approaches include assessments of illustrative case examples related to ecosystems, water supply/basin management, and socioeconomic activities such as agriculture. These case studies begin to integrate, for specific subregions (such as watersheds or agricultural production areas), the impacts of climate

**Box 2-10. The Potential Impact of Long-Term Climate Change on Vector-Borne Diseases: The Case of Malaria**

Malaria is a vector-borne environmental disease; as such, it is greatly influenced by macro- and microenvironmental changes. The impact of human-induced environmental change on malaria can be graded on a scale that commences at a global level and terminates at the level of an individual family homestead. In the investigation of environmentally induced changes in malaria transmission, it is impossible to restrict the study to localized environmental changes because they are nested within the changing macroenvironment. To understand the dynamics of these changes, as well as their interactions and their potential for prediction of occurrence, it is essential that we consider the role of macroenvironmental change (e.g., global warming) as well as microenvironmental change (e.g., deforestation or changes in land-use patterns, housing type, migration).

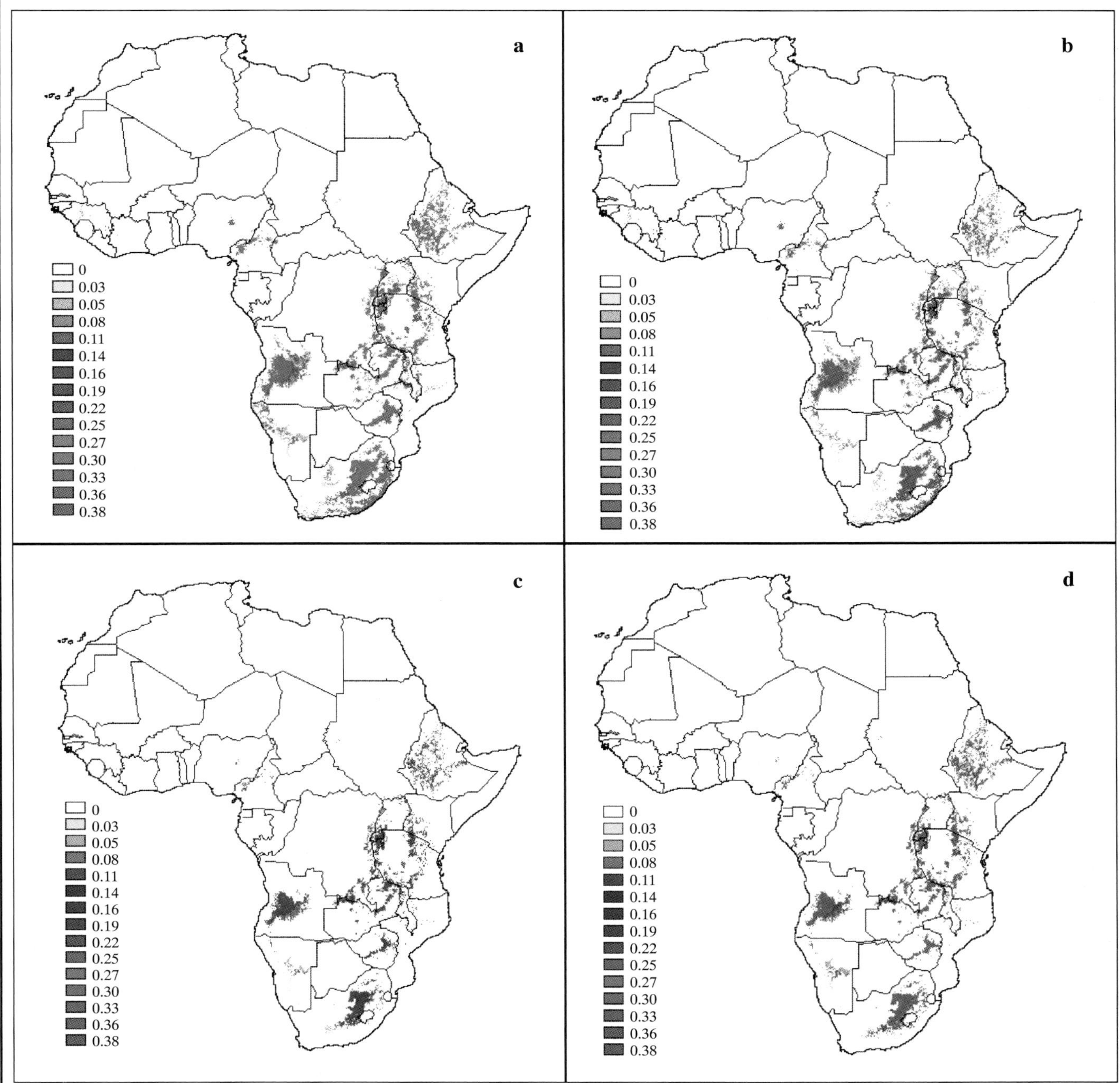


*Degree of Change: a) 5-month model with 1°C increase, b) 5-month model with 0.5°C increase, c) 5-month model with 0.5°C increase at elevation >1,300 m, and d) 5-month model with 1°C increase at elevation >1,300 m.*

**Box 2-10 continued**

There is little doubt that global warming is now a reality (IPCC 1990, Chapters 5, 7, 8). Over the past three decades there has been a marked increase in mean temperatures, especially at higher latitudes; in the tropics, the increase is estimated to be about 0.2°C (IPCC 1990, Chapter 5). Some attention has focused on global climate change and its implications for epidemic malaria transmission (Bouma *et al.,* 1994; Martens *et al.,* 1995a; Martin and Lefebvre, 1995). The impact of these climatic changes on malaria is likely to be greatest in regions where malaria transmission previously was limited physiologically by low temperatures, which limit the development of the vector and the parasite. In these areas, transmission is likely to be limited to specific seasons (when temperatures are favorable) or may not occur at all. Increased temperatures will lengthen the transmission season, resulting in a marked increase in incidence. Increased temperatures also can expand transmission into regions that previously defined the limits of malaria transmission because of the effects of latitude or elevation (or both).

Studies in Rwanda (Loevinsohn, 1994) and Kenya (Knight and Neville, 1991; Some, 1994) have investigated the extent to which small-scale land-use changes can be responsible for dramatic increases in highland malaria. The effects of short-term environmental changes, as well as changes in human behavioral patterns (e.g., migration, changes in control activities in adjacent regions), are superimposed on more general effects of macroenvironmental change and are likely to play an important role in the onset of epidemics. Thus, although climate change may mean certain regions become more susceptible to malaria transmission, the actual risk of epidemics may remain low because of the absence of local contributory risk factors.

The accompanying figures a, b, c, and d provide a delineation of regions that would be environmentally affected by 0.5°C and 1.0°C increases in temperature. Figures a and b limit the modeling process to the highland areas of Africa, whereas Figures c and d include regions in which malaria distribution previously was restricted by elevation and latitude. These "new" zones of malaria transmission would be intermediate between areas of known annual transmission and those where malaria has never occurred. With respect to the latter (the upper limit), the simulation errs on the side of caution: Malaria always is limited (cannot occur at present) and is not subject to interannual variation. It uses existing long-term, retrospective, climatological (temperature and rainfall) and topographic surfaces that already exist for Africa (Hutchinson *et al.,* 1995). Inherent in the prerequisite for such epidemic fringe malaria (elevation or latitude) is that in certain years a window of suitable environmental conditions will exist. This window consists of a number of consecutive months with suitable environmental conditions. The graphic provides a conservative estimate; it assumes a prerequisite of five consecutive months, whereas in some regions (such as the Sahel) this may be as short as three (Bagayako, pers. comm.). Despite the occurrence of such a window, transmission may not occur. Transmission then will be contingent on human activities that result in the introduction of the parasite into this suitable (short-term) climatological environment.

The premise that climate variables, on their own, may not always reliably predict epidemic risk has important implications for the forecasting of epidemics and changes in distribution. This model is not definitive but provides an illustration of how climate may impact upon disease distribution and how regions of impact may be delineated. Similarly, the change or shift in severity in existing malarial regions could be modeled. The figures illustrate that increased temperatures would significantly increase susceptible regions in areas of the Southern Hemisphere where distribution previously was constrained by latitude. In contrast, northern Africa is not affected because the distribution concurs with the Sahelian region, where the absence of rainfall and high temperatures limits the disease (similarly with Botswana). Conversely, in certain areas of southern Africa, especially South Africa, low temperatures previously were limiting—thus, a significant increase in distribution is likely.

change with the potential impacts of other factors such as land-use change, demographic change, land degradation, air and water pollution, and economic and social change (including factors such as changing resource demands resulting from economic development and technological change).

The use of integrated assessment models has not been widespread in the African context, although it is becoming an item on the agenda of several modeling groups and organizations. More widespread use of integrated assessment models is a priority for the IPCC in the Third Assessment Report, although it will require fundamental advances in the research literature. As a basis for integration, primary research on impacts on the potentially most vulnerable sectors and regions must be conducted so that interactions among the many potential costs and benefits can be assessed and, where possible, quantified.

## 2.5. Major Challenges Ahead for Africa

One of the first and most pressing scientific steps to address the shortage of data on African climate is the maintenance and (if possible) enhancement of the surface climate observing network. The network is important because, for the detection of long-term climate change, stable and continuous observing sites are necessary. The World Meteorological Organization (WMO) has a program to designate and maintain key sites as Reference Climate Stations. This program needs additional recognition and funding. The observing network also will help in the calibration of new satellite-based methods of observing the climate (WMO, 1992).

There is a continual need to evaluate the results of GCM experiments that simulate greenhouse gas- and aerosol-induced climate change to identify the likely subregional response within Africa to projected global-mean warming. This research should be seen as only one part of the much larger effort underway worldwide to narrow the uncertainties surrounding predictions of greenhouse gas-induced climate change. Determining whether recent desiccation in the Sahel is associated in some way with global air pollution also is of great importance (Hastenrath, 1995; Ringius *et al.*, 1996).

In addition to conducting more comprehensive assessments of the sensitivity and vulnerability of key resource sectors and systems, there is an urgent need to begin to apply existing and developing techniques for integration of potential climate change impacts on several dimensions, as suggested in the introduction to this report. These dimensions include:

- Integrating the chain of effects from changes in atmospheric composition and climate to changes in biophysical systems to the socioeconomic consequences (the "vertical" dimension)
- Including interactions among systems, sectors, and activities (the "horizontal" dimension)
- Considering climate change in the context of other trends and changes in society (the "time" or "global change" dimension)
- Considering the integrated impacts of current levels and episodes of climate variability.

In light of the large number and magnitude of socioeconomic and environmental changes projected for Africa, developing integrated assessments of climate change is an urgent priority for the region.

As this information on sectoral vulnerabilities and integrated assessment of potential climate change impacts is developed, it must penetrate more fully into national government organizations and international donor agencies. This penetration is necessary to ensure that what is known about past and present climate variability is properly taken into account in developing national economic and environmental plans (Sadowski *et al.*, 1996). Such sensitization of the policymaking process to climate variability also ensures that as knowledge about future climate change improves, it too can be sensibly used to guide drought/climate-related economic policy (OECD, 1996).

Within Africa, priority areas for environmental policy include securing sustainable water supply and quality; preventing and reversing desertification; combating coastal erosion and pollution; ensuring sustainable industrial development; making efficient use of energy resources; maintaining forests and wildlife resources; managing demographic change; and ensuring adequate food security. These priority areas highlight environmental and developmental concerns in the region that require immediate attention from the research and policy communities (Hulme, 1996a).

## References


**African Development Bank, African Development Fund/ Africa Energy Programme (ADB AEP)**, 1996: *Energy and Environmental Interactions.* Technical Paper No. EE1, Energy and Environment Technical Paper Series, Abidjan, Ivory Coast.

**Akong'a**, J., T.E. Downing, N.T. Konijn, D.N. Mungai, H.R. Muturi, and H.L. Potter, 1988: The effects of climatic variations on agriculture in central and eastern Kenya. In: *The Impact of Climatic Variations on Agriculture, Assessments in Semi-Arid Regions* [Parry, M.L., T.R. Carter, and N.T. Konijn (eds.).]. Kluwer Academic Press, Dordrecht, Netherlands, pp. 123–270.

**Allersman**, E. and W.K. Tilsmans, 1993: Coastal conditions in West Africa: A review. *Ocean and Coastal Management*, **19**, 199–240.

**Ante**, B., 1990: Impacts of climate change on the socioeconomic structure and activities in the Mediterranean region. In: *Changing Climate and the Coast,* volume 2 [Titus, J.G. (ed.)]. U.S. Environmental Protection Agency, Washington, DC, USA, pp. 127–138.

**Archer**, S., D.S. Schimel, and E.A. Holland, 1994: Mechanisms of shrubland expansion: Land use, climate or $CO_2$? *Climatic Change,* **29**, 91–99.

**Awosika**, L.F., G.T. French, R.T. Nicholls, and C.E. Ibe, 1992: The impacts of sea level rise on the coastline of Nigeria [O'Callahan, J. (ed.)]. In: *Global Climate Change and the Rising Challenge of the Sea.* Proceedings of the IPCC Workshop at Margarita Island, Venezuela, 9–13 March 1992. National Oceanic and Atmospheric Administration, Silver Spring, MD, USA, 690 pp.

**Bagayako**, M., 1996: Malaria Research and Training Centre, Bamako, Mali. (personal communication)

**Balek**, J., 1977: *Hydrology and Water Resources in Tropical Africa.* Elsevier Science Publishers, Amsterdam, Netherlands, 27 pp.

**Benson**, C. and E. Clay, 1994: The impact of drought on sub-Saharan African economies. *IDS Bulletin,* **25(4)**, 24–32.

**Binet**, D., L. LeReste, and P.S. Diouf, 1995. Influence des eaux de ruissellement et des rejets fluviaux sur les ecosystemes et les ressources vivantes des cotes d'Afrique occidentale. In: *FAO Fisheries Technical Paper, Effects of Riverine Inputs on Coastal Ecosystems.* Food and Agriculture Organization, Rome, Italy.

**Blaikie**, P.M. and H.C. Brookfield, 1987: *Land Degradation and Society.* Methuen, London, UK and New York, USA, 296 pp.

**Bonkoungou**, E.G., 1996: Drought, desertification and water management in sub-Saharan Africa. In: *Sustaining the Future: Economic, Social, and Environmental Change in Sub-Saharan Africa* [Benneh, G., W.B. Morgan, and J.I. Uitto (eds.)]. United Nations University Press, New York, NY, USA, pp. 165–180.

**Booth**, T. H., 1990: Mapping regions climatically suitable for particular tree species at the global scale. *Forest Ecology and Management*, **36**, 47–60.

**Bootsma**, H.A. and R.F. Hecky, 1993: Conservation of the African Great Lakes: a limnological perspective. *Conservation Biology*, **7**, 644–656.

**Bouma**, M.J., H.E. Sondorp, and H.J. van der Kaay, 1994: Climate change and periodic epidemic malaria. *Lancet,* **343**, 342.

**Bradley**, P. and P. Dewees, 1993: Indigenous woodlands, agricultural production and household economy in the communal areas. In: *Living with Trees: Policies for Forestry Management in Zimbabwe* [Bradley, P.N. and K. McNamara (eds.)]. The World Bank, Washington, DC, USA, pp. 63–137.

**Broadus**, J., S. Milliman, D. Edwards, D.G.Aubrey and F. Bable, 1986: Rising sea level and damming of rivers: possible effects in Egypt and Bangladesh. In: *Effects of Changes in Stratospheric Ozone and Global Climate,* volume 4 [Titus, J.G., (ed.)]. U.S. Environmental Protection Agency and UNEP, Washington, DC, USA, pp. 165–189.

**Buckland**, R.W., 1992: *Cereal Output in the SADCC Region During Dry Spells.* SADC Regional Workshop on Climate Change, Windhoek, Namibia.

**Burke**, J.J., J.R. Mahan, and J.L. Hatfield, 1988: Crop-specific thermal kinetic windows in relation to wheat and cotton biomass production. *Agronomy Journal,* **80**, 553–556.

**Calder**, I.R., R.L. Hall, H.G. Bastable, H.R. Gunston, O. Shela, A. Chirwa, and R. Kafundu, 1995: The impact of land use change on water resources in sub-Saharan Africa: a modelling study of Lake Malawi. *Journal of Hydrology,* **170**, 123–135.

**Carpenter**, S.R., S.G. Fisher, N.B. Grimm, and J.F. Kitchell, 1992: Global change and freshwater ecosystems. *Annual Review of Ecological Systems,* **23**, 119–139.

**Chidumayo**, E.N., 1997: *Miombo Ecology and Management.* IT Publications, London, United Kingdom (in press).

**Chilima**, C.Z., 1991: The status and development of conifer aphid damage in Malawi. In: *Exotic Aphid Pests of Conifers. A Crisis in African Forestry.* Proceedings of a Workshop at Maguga, Kenya, June 3-6, 1991. Food and Agriculture Organization, Rome, Italy, pp. 64–67.

**Chilima**, C.Z. 1997: Forest Research Institute of Malawi, Zomba, Malawi. (personal communication)

**Cleaver**, K.M. and G.A. Schreiber, 1994: *Reversing the Spiral: the Population, Agriculture and Environment Nexus in Sub-Saharan Africa: Directions on Development Series.* The World Bank, Washington, DC, USA, 293 pp.

**Cline**, W.R., 1992: *The Economics of Global Warming.* Institute for International Economics, Washington, DC, USA, 408 pp.

**Conway**, D. and M. Hulme, 1993: Recent fluctuations in precipitation and runoff over the Nile sub-basins and their impact on main Nile discharge. *Clim. Change,* **25**, 127–151.

**Conway**, D. and M. Hulme, 1996: The impacts of climate variability and future climate change in the Nile Basin on water resources in Egypt. *Water Resources Development,* **1(2/3)**, 277–296.

**Cooper**, J. and P.A.R. Hockey, 1981: *The Atlas and Site Register of South African Coastal Birds. Part 2, The Atlas.* Report to South African National Committee for Oceanographic Research, 324 pp.

**Craig**, M. and D. le Sueur (in preparation): *African Climatic Model of Malaria Transmission Based on Monthly Rainfall and Temperature.* 16 pp.

**Davidson**, O.R., 1992: Energy issues in sub-Saharan Africa: future directions. *Annual Review of Energy and the Environment.,* **17**, 359–404.

**Davies**, S., 1996: *Adaptable Livelihoods: Coping with Food Insecurity in the Malian Sahel.*

**Dennis**, K.C., I. Niang-Diop, and R.J. Nicholls, 1995: Sea level rise and Senegal: potential impacts and consequences. *J. Coastal Res.,* special issue **14**, 243–261.

**de Ridder**, N., L. Stroosnijder, A.M. Cisse, and H. van Keulen, 1982: PPS course book. Vol. I, Theory. *Productivity of Sahelian Rangelands: A Study of the Soils, the Vegetations and the Exploitation of That Natural Resource.* Department of Soil Science and Plant Nutrition, Wageningen Agricultural University, Wageningen, Netherlands, pp. 227–231.

**Desanker**, P.V., 1996: Development of a Miombo woodland dynamics model in Zambezian Africa using Malawi as a case study. *Climatic Change,* **36**, 279–288.

**Downing**, T.E., 1992: *Climate Change and Vulnerable Places: Global Food Security and Country Studies in Zimbabwe, Kenya, Senegal, and Chile.* Research Paper No. 1, Environmental Change Unit, University of Oxford, Oxford, United Kingdom, 54 pp.

**Downing**, T.E. (ed.), 1996: *Climate Change and World Food Security.* Springer-Verlag, Berlin, Germany, 662 pp.

**Downing**, T.E., S. Lezberg, C. Williams, and L. Berry, 1990: Population change and environment in central and eastern Kenya . *Environmental Conservation,* **17(2)**, 123–133.

**Downing**, T.E., M.J. Watts, and H.G. Bohle, 1996: Climate change and food insecurity: Toward a sociology and geography of vulnerability. In: *Climate Change and World Food Security* [Downing, T.E. (ed.)]. Springer-Verlag, Berlin, Germany, pp. 183–206.

**Dregne**, H.E., 1983: *Desertification of Arid Lands.* Harwood Academy, New York, NY, USA, 242 pp.

**du Toit**, R.F., 1983: Hydrological changes in the Middle-Zambezi system. *The Zambezi Science News,* **17(708)**, 121–126.

**Eele**, G., 1996: Policy lessons from communities under pressure. In: *Climate Change and World Food Security* [Downing, T.E. (ed.).]. Springer-Verlag, Heidelberg, Germany, pp. 184–206.

**Eid**, H.M., 1994: Impact of climate change on simulated wheat and maize yields in Egypt. In: *Implications of Climate Change for International Agriculture: Crop Modeling Study* [Rosenzweig, C. and A. Iglesias (eds.)]. U.S. Environmental Protection Agency, Washington, DC, USA, pp. 1–14.

**Ellis**, J.E. and K.A. Galvin, 1994: Climate patterns and land use practices in the dry zones of east and west Africa. *Bioscience,* **44(5)**, 340–349.

**Ellis**, J.E., K.A. Galvin, J.T. McCabe, and D.M. Swift, 1987: *Pastoralism and drought in Turkana District Kenya.* Report to the Norwegian Agency for International Development, Nairobi, Kenya.

**El-Raey**, M., 1990: Response to the impacts of greenhouse-induced sea-level rise on the northern coastal regions of Egypt. In: *Changing Climate and the Coast,* volume 2 [Titus, J.G. (ed.)]. U.S. Environmental Protection Agency, Washington, DC, USA, pp. 225–238.

**El-Raey**, M., S. Nasr, O. Frihy, S. Desouke, and K.H. Dewidar, 1995: Potential impacts of accelerated sea level rise on Alexandria Governate, Egypt. *J. Coastal Res.,* special issue **14**, 190–204.

**El-Sayed**, M.K., 1991: *Implications of Relative Sea Rise on Alexandria.* In: Proceedings of First International Meeting (Cities on Water), December, 1989, Venice, Italy, [Frassetto, R. (ed.)]. Marsilo Editori, pp. 183–189.

**Epstein**, P.R., 1992: Cholera and the environment. *Lancet,* **339**, 1167–1168.

**Epstein**, P.R., 1993: Algal blooms in the spread and persistence of cholera. *BioSystems,* **31(2-3)**, 209–221.

**Epstein**, P.R., 1995: Emerging diseases and ecosystem instability: new threats to public health. *American Journal of Public Health,* **85**, 168–172.

**Epstein**, P.R., T.E. Ford, and R.R. Collwell, 1993: Marine ecosystems. *Lancet,* **342**, 1216–1219.

**Everett**, G.V., 1994: Promoting industrial fisheries in West Africa, *FAO Fish. Circ.,* no. 857, Food and Agriculture Organization of the United Nations, Rome, Italy, 56 pp.

**FAO (Food and Agriculture Organization)**, 1985: The Ecological Effects of Eucalyptus. *FAO Forestry Paper 59,* Food and Agriculture Organization of the United Nations, Rome, Italy, 88 pp.

**FAO**, 1993: Yearbook Production: Volume 46, 1992. FAO Statistical Series No. 112, Food and Agriculture Organization of the United Nations, Rome, Italy.

**FAO**, 1994: *Review of Pollution in the African aquatic environment* [D. Calamari and H. Naeve (eds.)], CIFA Technical Paper 25, Food and Agriculture Organization of the United Nations, Rome, Italy.

**FAO**, 1997: *State of the World's Forests 1997,* Food and Agriculture Organization of the United Nations, Rome, Italy.

**Fischer**, G., K. Frohberg, M.L. Parry, and C. Rosenzweig, 1994: Climate change and world food supply, demand and trade. Who benefits, who loses? *Global Environmental Change,* **4(1)**, 7–23.

**Fischer**, G., K. Frohberg, M.L. Parry, and C. Rosenzweig, 1996: Impacts of potential climate change on global and regional food production and vulnerability. In: *Climate Change and World Food Security* [T.E. Downing (ed.)]. Springer, Heidelberg, Germany, pp. 115–160.

**Fischer**, G. and H.T. van Velthuizen, 1996: *Climate Change and Global Agricultural Potential Project: A Case Study of Kenya.* WP-96-071. International Institute for Applied Systems Analysis, Laxenburg, Austria, 50 pp.

**French**, G.T., L.F. Awosika, and C.E. Ibe, 1995: Sea level rise and Nigeria: potential impacts and consequences. *J. Coastal Res.,* special issue **14**, 224–242.

**Frost**, P.G.H., 1996: The ecology of miombo woodlands. In:*The Miombo in Transition: Woodlands and Welfare in Africa.* [Campbell, B. (ed.)]. CIFOR, Bogor, Indonesia, pp. 11–57.

**Galvin**, K.A., 1992: Nutritional ecology of pastoralists in dry tropical Africa. *American Journal of Human Biology,* **4(2)**, 209–221.

**Gibberd**, V., J. Rook, C.B. Sear, and J.B. Williams, 1996: *Drought Risk Management in Southern Africa: the Potential of Long Lead Climate Forecasts for Improved Drought Management.* Chatham Maritime, Natural Resources Institute, Chatham Maritime.

**Gibbs**, J.P., 1993: Importance of small wetlands for the persistence of local populations of wetland associated animals. *Wetlands,* **13**, 25–31.

**Gleick**, P.H., 1993: *Water in Crisis*. Oxford University Press, Oxford, UK, 473 pp.

**Greco**, S., R.H. Moss, D. Viner, and R. Jenne, 1994: *Climate Scenarios and Socio-Economic Projections for IPCC WGII Assessment.* Intergovernmental Panel on Climate Change, Washington, DC, USA, 67 pp.

**Griffiths**, J.F. (ed.), 1972: Climates of Africa. *World Survey of Climatology,* Volume 10. Elsevier, New York, NY, USA, 604 pp.

**Grimble**, R. and M.-K. Chan, 1995: Stake holder analysis for natural resource management in developing countries: Some practical guidelines for making management more participatory and effective. *Natural Resources Forum*, **19**, 113–124.

**Haberyan**, K.A. and R.E. Hecky, 1987: The Late Pleistocene and Holocene stratigraphy and paleolimnology of Lakes Kivu and Tanganyika. *Paleogeography, Paleoclimatology and Paleoecology*, **61**, 169–197.

**Hamilton**, A.C., 1988: Guenon evolution and forest history. In: *A Primate Radiation: Evolutionary Biology of the African Guenons* [Gautier-Hion, A., F. Bourliere, J.P. Gantier, and J. Kingdon (eds.)]. Cambridge University Press, Cambridge, United Kingdom, pp. 13–34.

**Harcourt**, A.H., K.J. Stewart, and J.M. Inahoro, 1989: Gorilla quest in Nigeria. *Oryx*, **23(1)**, 7–13.

**Hastenrath**, S., 1995: Recent advances in tropical climate prediction. *Bull. Amer. Meteor. Soc.,* **8**, 1519–1532.

**Haxeltine**, A. and I.C. Prentice, 1996: BIOME3: An equilibrium terrestrial biosphere model based on ecophysiological constraints, resource availability and competition among plant functional types. *Global Biogeochemical Cycles*, **10**, 693–710.

**Hecky**, R.E., F.W.B. Bugenyi, P. Ochumba, J.F. Talling, R. Mugidde, M. Gophen, and L. Kaufman, 1994: Deoxygenation of the deep water of Lake Victoria, East Africa. *Limnology and Oceanography*, **39**, 1476–1481.

**Hecky**, R.E., E.J. Fee, H.J. Kling, and J.M.W. Rudd, 1981: Relationship between primary productivity and fish production in Lake Tanganyika. *Transactions of the American Fisheries Society*, **110**, 336–345.

**Hernes**, H., A. Dalfelt, T. Berntsen, B. Holtsmark, L.O. Naess, R. Selrod, and A. Aaheim, 1995: *Climate Strategy for Africa.* Cicero Report No. 1995(3), Cicero, Oslo, Norway, 83 pp.

**Hinckley**, T.M., R.O. Teskey, F. Duhme, and H. Richter, 1981: Temperate hardwood forests. In: *Water deficits and plant growth. VI. Woody plant communities* [Kozlowski, T.T. (ed.)]. Academic Press, New York, NY, pp. 153–208.

**Hlohowskyj**, I., M.S. Brody, and R.T. Lackey, 1996: Methods for assessing the vulnerability of African fisheries resources to climate change. *Climate Research*, **6**, 97–106.

**Hulme**, M., 1996a: Climatic change within the period of meteorological records. In: *The Physical Geography of Africa* [Adams, W.M., A.S. Goudie, and A.R. Orme (eds.)]. Oxford University Press, Oxford, United Kingdom, 429 pp.

**Hulme**, M., 1996b: *Climate Change and Southern Africa: an Exploration of Some Potential Impacts and Implications in the SADC Region.* Report commissioned by WWF International and coordinated by the Climate Research Unit, UEA, Norwich, United Kingdom, 104 pp.

**Hulme**, M., D. Conway, P.M. Kelly, S. Subak, and T.E. Downing, 1995: *The Impacts of Climate Change on Africa.* Stockholm Environment Institute, Stockholm, Sweden, 46 pp.

**Huq**, S., L.J. Mata, I. Nemesova, and S. Toure, 1996: Chapter 4—Regional Summary. In: *Vulnerability and Adaptation to Climate Change: a Synthesis of Results from the U.S. Country Studies* [Lenhart, S., S. Huq, L.J. Mata, I. Nemesova, and S. Toure. (eds.).] U.S. Country Studies Program, Washington, DC, USA, pp. 69–108.

**Hutchinson**, M.F., H.A. Nix, J.P. McMahon, and K.D. Ord, 1995: Africa: a Topographic and Climatic Database. Version 1.0, October 1995. Centre for Resource and Environmental Studies, Australian National University, Canberra, Australia.

**Ibe**, A.C., 1990: Adjustments to the impact of sea level rise along the West and Central African coasts. In: *Changing Climate and the Coast, vol. 2* [Titus, J.G. (ed.)]. Proceedings of the first IPCC CZMS workshop, Miami, 27 November–1 December 1989, Environmental Protection Agency, Washington, DC, USA, pp. 3–12.

**Ibe**, A.C. and R.E. Quelennac, 1989: Methodology for assessment and control of coastal erosion in west Africa and central Africa. *UNEP Regional Sea Reports and Studies No. 107.* United Nations Environment Programme, New York, NY, USA.

**ICST (Ivorian Country Study Team)**, 1996: *Vulnerability of Coastal Zone of Côte d'Ivoire to Sea Level Rise and Adaptation Options.* Report on the Côte d'Ivoire/USA Collaborative Study on Climate Change in Côte d'Ivoire.

**IPCC**, 1990: *Climate Change: The IPCC Scientific Assessment.* [Houghton, J.T., G.J. Jenkins, and J.J. Ephraums (eds.)]. Cambridge University Press, New York, NY, USA, and Cambridge, United Kingdom, 365 pp.

- Mitchell, J.F., S. Manabe, T. Tokioka, and V. Meleshko, Chapter 5. *Equilibrium Climate Change —and its Implications for the Future,* pp. 131–172.
- Folland, C.K., T. Karl, and K.Y. Vinnikov, Chapter 7. *Observed Climatic Variations and Change,* pp. 195–238.
- Wigley, T.M.L. and T.P. Barnett, Chapter 8. *Detection of the Greenhouse Effect in the Observations,* pp. 239–256.

**IPCC**, 1996: *Climate Change 1995: The Science of Climate Change. Contribution of WGI to the Second Assessment Report of the Intergovernmental Panel on Climate Change* [Houghton, J.T., L.G. Meira Filho, B.A. Callander, N. Harris, A. Kattenberg, and K. Maskell (eds.)].Cambridge University Press, Cambridge, United Kingdom, 572 pp.

**IPCC**, 1996: *Climate Change 1995: Impacts, Adaptations and Mitigation of Climate Change: Scientific-Technical Analysis. Contribution of Working Group II to the Second Assessment Report of the Intergovernmental Panel on Climate Change* [Watson, R.T., M.C. Zinyowera, and R.H. Moss (eds.)]. Cambridge University Press, New York, NY, USA, and Cambridge, United Kingdom, 880 pp.

- Kirschbaum, M.U.F., Chapter A. *Ecophysiological, Ecological, and Soil Processes in Terrestrial Ecosystems: A Primer on General Concepts and Relationships*, p.57–74.
- Allen-Diaz, B., Chapter 2. *Rangelands in a Changing Climate: Impacts, Adaptations and Mitigation,* pp. 131–158.
- Noble, I.R. and H. Gitay, Chapter 3. *Deserts in a Changing Climate: Impacts,* pp. 159–189.
- Beniston, M. and D.G. Fox, Chapter 5. *Impacts of Climate Change on Mountain Regions,* pp. 191–213.
- Arnell, N., B. Bates, H. Lang, J.J. Magnuson, and P. Mulholland, Chapter 10. *Hydrology and Freshwater Ecology,* pp. 325–364.
- Everett, J., Chapter 16. *Fisheries,* pp. 511–537.

**Jallow**, B.P., M.K.A. Barrow, and S.P. Leatherman, 1996: Vulnerability of the coastal zone of The Gambia to sea level rise and development of response options. *Clim. Res.,* **6**, 165–177.

**Joubert**, A.M., 1995: Simulations of southern African climate by early-generation general circulation models. *South African J. Science,* **91**, 85–91.

**Jury**, M., 1996: *Malaria Forecasting Workshop: Report on Workshop on Reducing Climate Related Vulnerability in Southern Africa. Victoria Falls, Zimbabwe.* Office of Global Programs, National Oceanic and Atmospheric Administration, Washington, DC, USA, pp. 73–84.

**Justice**, C., B. Scholes, and P. Frost, 1994: *African Savannas and the global atmosphere: research agenda*:: IGBP Report 31. International Geosphere-Biosphere Programme, Stockholm, Sweden, 51 pp.

**Kane**, H., 1996: Shifting to Sustainable Industries. In: *State of the World 1996: a Worldwatch Institute report on progress toward a sustainable society.* W.W. Norton and Company, New York, USA and London, United Kingdom, pp. 152–167.

**Kassas**, M., 1995: Desertification: a general review. *Journal of Arid Environments*, **30(2)**, 115–128.

**Katerere**, Y., 1983: Insect pests of pine plantations in the eastern districts of Zimbabwe, 1. Preliminary list. *Zimbabwe Journal of Agricultural Research,* **21**, 101–105.

**Kendall**, R.L., 1969: An ecological history of the Lake Victoria basin. *Ecological Monographs*, **39**, 121–176.

**Khafagy**, A.A., C.H. Hulsbergen, and G. Baarse, 1992: Assessment of the vulnerability of Egypt to sea level rise. In: *Global Climate Change and the Rising Challenge of the Sea* [O'Callahan, J. (ed).]. Proceedings of the IPPCC Workshop, March 1992, Margarita Island, Venezuela. National Oceanic and Atmospheric Administration, Silver Spring, MD, USA.

**Kingdon**, J.S., 1980: The role of visual signals and face patterns in African forest monkeys (guenons) of the genus *Cercopithecus*. *Transactions of the Zoological Society, London*, **35**, 425–75.

**Klaus**, D., 1978: Spatial distribution and periodicity of mean annual precipitation south of the Sahara. *Arch. Meteor. Geophys. Bioklim. Ser. B.,* **26**, 17–27.

**Knight**, R., and C. Neville, 1991: *The Malaria Situation in Kenya, Seasonal Epidemic and Non-Endemic Ecozones.* Internal report to National Malaria Control Programme Subcommittee.

**Knoch**, K. and A. Schulze, 1956: Niederschlag. Temperatur und Schwule in Afrika. In: *World Atlas of Epidemic Diseases,* volume 2. Heidelberger and Akademie der Wissenschaften, Heidelberg, Germany.

**Kraus**, E.B., 1977: Subtropical droughts and cross-equatorial energy transports. *Mon. Wea. Rev.,* **105**, 1009–1018.

**Krippendorf**, J., 1984: The capital of tourism in danger. In: *The Transformation of Swiss Mountain Regions* [Brugger, E.A. et al. (eds.)]. Haupt Publishers, Bern, Switzerland, pp. 427–450.

**Lauenroth**, W.K. and J.L. Dodd, 1978: The effects of water- and nitrogen-induced stress on plant community structure in a semiarid grassland. *Oecologia*, **36**, 211–222.

**Leeson**, H.S., 1931: Anopheline mosquitoes in southern Rhodesia. *Mem. London Sch. Hyg. Trop. Med.,* **4**, 1.

**Le Houerou**, H.N., 1989: *The Grazing Land Ecosystems of the African Sahel.* Springer-Verlag, New York, NY, USA.

**le Sueur**, D., and B.L. Sharp, 1996: *Malaria Forecasting Project: Report on Workshop on Reducing Climate Related Vulnerability in Southern Africa. Victoria Falls, Zimbabwe.* Office of Global Programs, National Oceanic and Atmospheric Administration, Washington, DC, USA, pp. 65–73.

**Levins**, R., T. Awerbuch, U. Brinkman, I. Eckardt, P. Epstein, N. Makhoul, C. Albuququerque de Possas, C. Puccia, A. Spilman, and M.E. Wilson, 1994: The emergence of new diseases. *American Scientist,* **82**, 52–60.

**Lewis**, D., P. Reinthal, and J. Trendall, 1986: *A Guide to the Fishes of Lake Malawi National Park.* World Wildlife Fund, World Conservation Center, Gland, Switzerland.

**Livingstone**, D.A., 1975: Late quaternary climatic change in Africa. *Annual Review of Ecology and Systematics,* **6**, 249–280.

**Loevinsohn**, M.E., 1994: Climatic warming and increased malaria incidence in Rwanda. *Lancet,* **343**, 714–717.

**Lowe-McConnell**, R.H., 1987: *Ecological Studies in Tropical Fish Communities.* Cambridge University Press, Cambridge, United Kingdom.

**Magadza**, C.H.D., 1984: An analysis of plankton from Lake Bangweulu, Zambia. *Hydrobiologie du Bassin du Lake Bangweolo et du Luapula* [Symoens, J.J., (ed.)].

**Magadza**, C.H.D., 1991: Some possible impacts of climate change on African ecosystems. In: *Climate Change. Science, Impacts and Policy.* Proceedings of the Second World Climate Conference [Jager, J. and H.L. Ferguson (eds.)]. Cambridge University Press, Cambridge, United Kingdom, pp. 385–390.

**Magadza**, C.H.D., 1996: Climate change: some likely multiple impacts in southern Africa. In: *Climate Change and World Food Security* [Downing, T.E. (ed.)]. Springer-Verlag, Heidelberg, Germany, pp. 449–483.

**Makadho**, J.M., 1996: Potential effects of climate change on corn production in Zimbabwe. *Climate Research,* **6**, 147–151.

**Makarau**, A., 1992: *National Drought and Desertification Policies: the Zimbabwe Situation.* SADC Regional Workshop on Climate Change, 1992. Windhoek, Namibia.

**Martens**, W.J.M., T.H. Jetten, J. Rotmans, and L.W. Niessen, 1995a: Climate change and vector-borne diseases: a global modelling perspective. *Global Environmental Change,* **5**, 195–209.

**Martens**, W.J.M., L.W. Niessen, J. Rotmans, T.H. Jetten, and A.J. McMichael, 1995b: Potential impact of global climate change on malaria risk. *Environmental Health Perspectives,* **103**, 458–464.

**Martin**, P. and M. Lefebvre, 1995: Malaria and climate: sensitivity of malaria potential transmission to climate. *Ambio*, **24(4)**, 200–220.

**Martyn**, D., 1992: *Climates of the World, Developments in Atmospheric Science, 18.* Elsevier, New York, NY, USA, 435 pp.

**Matarira**, C.H., W. Kamukondiwa, F.C. Mwamuka, J.M. Makadho, and L.S. Unganai, 1996: Vulnerability and adaptation assessments in Zimbabwe. In: *Vulnerability and Adaptation to Climate Change: a Synthesis of Results from the U.S. Country Studies* [Lenhart, S., S. Huq, L.J. Mata, I. Nemesova, S. Toure, and J.B. Smith (eds.).] U.S. Country Studies Program, Washington, DC, USA, pp. 129–140.

**McMichael**, A. J., A. Haines, R. Sloof, and S. Kovats (eds.), 1996: *Climate Change and Human Health.* World Health Organisation, Geneva, 297 pp.

**Meisner**, J.D., J.L. Goodie, H.A. Regier, B.J. Shuter, and W.J. Christie, 1987: An assessment of the effect of climate warming on Great Lakes Basin fish. *J. Gr. L. Res.*, **13**, 340–352.

**Menaut**, J.C., J. Gignoux, C. Prado, and J. Clobert, 1991: The community dynamics in a humid savanna of the Côte d'Ivoire: modeling the effects of fire and competition with grass and neighbors. *J. Biogeogr.,* **17**, 471–481.

**Michael**, E. and D.A.P. Bundy, 1996: The global burden of lymphatic filariasis. In:*World Burden of Diseases* [Murray, C.J.L. and A.D. Lopez (eds.)]. World Health Organization, Geneva, Switzerland.

**Milliman**, J., J. Broadhaus, and F. Gable, 1989: Environmental and economic implications of rising sea level and subsiding deltas, the Nile and Bengal examples. *Ambio,* **18**, 340–345.

**Milton**, S.J., W. Richard, J. Dean, M.A. du Plessis, and W. Roy Siegfried, 1994: A conceptual model of arid rangeland degradation. *Bioscience*, **44**, 70–76.

**Mkanda**, F.X., 1996: Potential impacts of future climate change on nyala (*Tragelaphus angasi*) in Lengwe National Park, Malawi. *Clim. Res.,* **6**, 157–164.

**Molineaux**, L., 1988: The epidemiology of human malaria as an explanation of its distribution, including some implications for control. In: *Volume I, Malaria: Principles and Practises of Malariology.* Longman Group, United Kingdom, pp. 913–998.

**Mortimore**, M., 1989: *Adapting to Drought: Farmers, Famines and Desertification in West Africa.* Cambridge University Press, Cambridge, United Kingdom, 299 pp.

**Moyo**, S., P. O'Keefe, and M. Sill, 1993: *The Southern African Environment: Profiles of the SADC Countries.* ETC Foundation/Earthscan Publishers, Ltd., London, United Kingdom.

**Muchena**, P., 1994: Implications of climate change for maize yields in Zimbabwe. In: *Implications of Climate Change for International Agriculture: Crop Modeling Study* [Rosenzweig, C. and A. Iglesias (eds.)]. U.S. Environmental Protection Agency, Washington, DC, USA, pp. 1–9.

**Neilson**, R.P., 1995: A model for predicting continental scale vegetation distribution and water balance. *Ecol. Appl.*, **5**, 362–385.

**Newell**, R.E., J.W. Kidson, D.G. Vincent, and G.J. Boer, 1972: *The General Circulation of the Tropical Atmosphere and Interactions with Extra-Tropical Latitudes,* volume 1. MIT Press, Cambridge, MA, USA, 258 pp.

**Nicholls**, R.J. and S.P. Leatherman, 1994: Sea level rise and coastal management. In:*Geomorphology and Land Management in a Changing Environment* [McGregor, D. and D. Thompson (eds.)]. John Wiley and Sons, United Kingdom.

**Nicholls**, R.J. and S.P. Leatherman, 1995: The implications of accelerated sea-level rise for developing countries: a discussion. *J. Coastal Res.,* **14**, 303–323.

**Nicholson**, S.E., 1993: An overview of African rainfall fluctuations of the last decade. *Journal of Climate*, **6**, 1463–1466.

**Nicholson**, S.E., 1994a: A review of climate dynamics and climate variability in eastern Africa. In: *The Limnology, Climatology and Paleoclimatology of the East African Lakes* [Johnson, T.C. and E. Odada (eds.)]. Gordon and Breach, London, United Kingdom.

**Nicholson**, S.E., 1994b: Recent rainfall fluctuations in Africa and their relationship to past conditions. *Halocene*, **4**, 121–131.

**Nicholson**, S.E., J. Kim, and J. Hoopingarner, 1988: *Atlas of African Rainfall and its Interannual Variability.* Department of Meteorology, The Florida State University, Tallahassee, FL, USA, 237 pp.

**O'Connor**, T.G., 1991: Local extinction in perennial grasslands: a life-history approach. *Am. Nat.,* **137**, 735–773.

**OECD**, 1996: *Climate Change Policy Initiatives, the 1995/96 Update.* OECD Energy and the Environment Series, Paris, France, 185 pp.

**OIES (Office for Interdisciplinary Earth Studies)**, 1991: *Arid Ecosystems Interactions: Recommendations for Drylands Research in the Global Change Research Program.* OIES Report 6, Boulder, Colorado, USA.

**Ojima**, D.S., W.J. Parton, M.B. Coughenour, and J.M.O. Scurlock, 1996: Impact of climate and atmospheric carbon dioxide changes on grasslands of the world. In: *Global Change: Effects on Coniferous Forests and Grasslands* [Breymeyer, A.I., D.O. Hall, J.M. Mellilo, and G.I. Agren (eds.)]. Scientific Committee on Problems of the Environment (SCOPE), Chichester, NY, USA, Wiley, pp. 271–309.

**Ojima**, D.S., D.S. Schimel, W.J. Parton, and C.E. Owensby, 1994: Long- and short-term effects of fire on nitrogen cycling in tallgrass prairie. *Biogeochemistry*, **23**, 1–18.

**Okoth-Ogendo**, H.W.O. and J.B. Ojwang (eds.), 1995: *A Climate for Development: climate change policy options for Africa.* African Centre for Technological Studies, Nairobi, Kenya, 264 pp.

**Orondo**, S.B.O. and R.K. Day, 1994: Cypress aphid (*Cinara cupressi*) damage to a cypress (*Cupressus lusitanica*) stand in Kenya. *International Journal of Pest Management,* **40**, 141–144.

**Owen**, R.B., R. Crossley, T.C. Johnson, D. Tweddle, L. Kornfield, S. Davison, D.H. Eccles, and D.E. Engstrom, 1990: Major low levels of Lake Malawi and their implications for speciation rates in cichlid fishes. *Proceedings of the Royal Society*, London, UK, **240**, 519–553.

**Owuar**, A.L., 1991: Exotic conifer aphids in Kenya, their current status and options for management. In: *Exotic Aphid Pests of Conifers. A Crisis in African Forestry.* Proceedings of a Workshop at Maguga, Kenya, June 3-6, 1991. Food and Agriculture Organization, Rome, Italy, pp. 58–63.

**PAHO**, 1994: Leishmaniasis in the Americas. *Epidemiological Bulletin*, **15(3)**, 8–13.

**Parton**, W.J., M.B. Coughenour, J.M.O. Scurlock, and D.S. Ojima, 1996: Global grassland ecosystem modeling; development and test of ecosystem models for grassland ecosystems. In: *Global Change: Effects on Coniferous Forests and Grasslands* [Breymeyer, A.I., D.O. Hall, J.M. Mellilo, and G.I. Agren (eds.)]. Scientific Committee on Problems of the Environment (SCOPE), Chichester, NY, USA, Wiley, pp. 229–269.

**Parton**, W.J., B. McKeown, V. Kirchner, and D. Ojima, 1992: *CENTURY Users' Manual.* Natural Resource Ecology Laboratory, Colorado State University, Fort Collins, CO.

**Pinay**, G., 1988: *Hydrobiological Assessment of the Zambezi River System: a Review.* International Institute for Applied Systems Analysis (IIASA) Working Paper WP-88-089. IIASA, Laxenburg, Austria, 116 pp.

**Price Waterhouse**, 1994: *The Lowveld Conservancies: New Opportunities for Productive and Sustainable Land-Use.* Price Waterhouse Wildlife, Tourism, and Environmental Consulting, Save Valley, Bubiana and Chiredzi River Conservancies, Harare, Zimbabwe.

**Pritchard**, 1985: *Africa: a Study Geography for Advanced Students.* Longman, Zimbabwe.

**Rains**, J.R., C.E. Owensby, and K. Kemp, 1975: Effects of nitrogen fertilization, burning, and grazing on reserve constituents of big bluestem. *J. Range Manage.,* **28**, 358–362.

**Rangeley**, R., B.M. Thiam, R.A. Andersen, and C.A. Lyle, 1994: *International River Basin Organizations in Sub-Saharan Africa.* Technical Paper No. 250. The World Bank, Washington, DC, USA, 70 pp.

**Rasmusson**, E.M., 1987: Global climate change and vulnerability: effects on drought and desertification in Africa. In: *Drought and Hunger in Africa* [Glantz, M.H. (ed.)]. Cambridge University Press, Cambridge, United Kingdom, pp. 3–22.

**Reij**, C., I. Scoones, and C. Toulmin (eds), 1996: *Sustaining the Soil: Indigenous Soil and Water Conservation in Africa.* Earthscan Publications Ltd., London, United Kingdom, 260 pp.

**Riebsame**, W.E., K.M. Strzepek, J.L. Wescoat, Jr., R. Perrit, G.L. Graile, J. Jacobs, R. Leichenko, C. Magadza, H. Phien, B.J. Urbiztondo, P. Restrepo, W.R. Rose, M. Saleh, L.H. Ti, C. Tucci, and D. Yates, 1995: Complex river basins. In: *As Climate Changes, International Impacts and Implications* [Strzepek, K.M. and J.B. Smith (eds.)]. Cambridge University Press, Cambridge, United Kingdom, pp. 57–91.

**Rind**, D., 1995: Drying out the tropics. *New Scientist*, **146**, 36–40.

**Ringius**, L., T.E. Downing, M. Hulme, and R. Selrod, 1996: *Climate Change in Africa: Issues and Regional Strategy.* Cicero Report No. 1996(2), Cicero, Oslo, Norway, 154 pp.

**Rogers**, D., 1996: Changes in disease vectors. In: *Climate Change and Southern Africa: an Exploration of Some Potential Impacts and Implications in the SADC Region* [Hulme, M. (ed.)]. Report commissioned by World Wildlife Fund International and coordinated by the Climate Research Unit, UEA, Norwich, United Kingdom, pp. 49–55.

**Rogers**, D.J. and M.J. Packer, 1993: Vector-borne diseases, models and global change. *Lancet*, **342**, 1282–1284.

**Rogers**, P., 1993: What water managers and planners need to know about climate change and water resources management. In: *Proc. Conference on Climate Change and Water Resources Management* [Ballentine, T. and E. Stakhiv (eds.)]. U.S. Army Institute for Water Resources, Fort Belvoir, VA, pp. I/1–I/14.

**Rosenzweig**, C. and M.L. Parry, 1994: Potential impact of climate change on world food supply. *Nature*, 367(6450), 133–138.

**Sadowski**, M., S. Meyers, F. Mullins, J. Sathaye, and J. Wisnieski, 1996: Methods of assessing GHG mitigation for countries with economies in transition. *Environ. Management,* **20** (Supplement 1), S1–S118.

**Salewicz**, A., 1995: Impact of climate change on the operation of Lake Kariba hydropower scheme on the Zambezi river. In: *Water Resources Management in the Face of Climatic and Hydrologic Uncertainties* [Kaczmarek, Z. et al. (eds.)]. Kluwer Academic Publishers, Dordrecht, The Netherlands.

**SARDC**, 1994: *State of the Environment in Southern Africa.* A Report by the Southern African Research and Documentation Center (SARDC) in Collaboration with IUCN—The World Conservation Union—and the Southern African Development Community, Harare, Zimbabwe, 332 pp.

**Sayer**, J. A., C.S. Harcourt, and N.M. Collins, 1992: *The Conservation Atlas of Tropical Forests: Africa.* Macmillan, for International Union for Conservation of Nature and Natural Resources, 288 pp.

**Scholes**, R. J. and B.H. Walker, 1993: *An African Savanna. Synthesis of the Nylsvley Study.* Cambridge University Press, Cambridge, United Kingdom, 306 pp.

**Schulze**, R.E., G.A. Kiker, and R.P. Kunz, 1996: Global climate change and agricultural productivity in Southern Africa: thought for food and food for thought. In: *Climate Change and World Food Security* [Downing, T.E. (ed.)]. North Atlantic Treaty Organization ASI Series, Volume 137. Springer-Verlag, Berlin and Heidelberg, Germany, pp 421–447.

**Sene**, K.J. and D.T. Pinston, 1994: A review and update of the hydrology of Lake Victoria, East Africa. *Hydrological Sciences Journal*, **39**, 47–63.

**Serageldin**, I., 1995: *Nurturing Development (Aid and Cooperation in Today's Changing World).* The World Bank, Washington, DC, USA.

**Sestini**, G., 1989: *The Implication of Climatic Changes for the Nile Delta.* Report WG 2/14, UNEP/OCA, Nairobi, Kenya.

**Sharma**, N.P., T. Damhaug, E. Gilgan-Hunt, D. Grey, V. Okaru, and D. Rothberg, 1996*: African Water Resources: Challenges and Opportunities for Sustainable Development.* World Bank Technical Paper No. 331, The World Bank, Washington, DC, USA, 115 pp.

**Shope**, R., 1991: Global climate change and infectious diseases. *Environmental Health Perspectives*, **96**, 171–174.

**Sivakumar**, M.V.K., 1991: Climate change and implications for agriculture in Niger. *Climatic Change,* **20**, 297–312.

**Sivakumar**, M.V.K., 1993: Global climate change and crop production in the Sudano-Sahelian zone of west Africa. In: *International Crop Science,* volume I. Crop Science Society of America, Madison, WI, USA.

**Smith**, J.B. and S.S. Lenhart, 1996: Climate change adaptation policy options. *Clim. Res.,* **6**, 193–201.

**Some**, E.S., 1994: Effects and control of highland malaria epidemic in Uasin Gishu District, Kenya. *East African Medical Journal*, **71** **(1)**, 2–8.

**Stanley**, D.J. and A.G. Warne, 1993: Nile Delta: recent geological evaluation and human impact. *Science,* **260**, 628–634.

**Strzepek**, K.M., S.C. Onyeji, M. Saleh, and D.N. Yates, 1995: An assessment of integrated climate change impacts on Egypt. In: *As Climate Changes: International Impacts and Implications.* [Strzepek, K.M. and J.B. Smith (eds.)]. Cambridge University Press, Cambridge, United Kingdom, pp. 180–200.

**Tietema**, T., D.J. Tolsma, E.M. Veenendaal, and J. Schroten, 1991: Plant responses to human activities in the tropical savanna ecosystem of Botswana. *Vegetatio,* **115**, 157–167.

**Tyson**, P.R., 1986: *Climate Change and Variability in Southern Africa.* Oxford University Press, Oxford, United Kingdom, 220 pp.

**UNEP**, 1992: *World Atlas of Desertification* [Middleton, N.J. and D.S.G. Thomas (eds.)]. Edward Arnold Publishers, Sevenoaks, United Kingdom, 69 pp.

**UNFCCC**, 1992: *United Nations Framework Convention on Climate Change.* United Nations, New York, NY, USA, 29 pp.

**USCSP (U.S. Country Studies Program)**, 1996: *Vulnerability and Adaptation to Climate Change: a Synthesis of Results from the U.S. Country Studies* [Smith, J., S. Huq, S. Lenhart, L.J. Mata, I. Nemesova, and S. Toure (eds.).] U.S. Country Studies Program. Kluwer, Norwell, MA, USA, 366 pp.

**Usher**, P., 1997: *Comments on the 1996/97 Droughts in Kenya.* (unpublished).

**Van Dalaan**, J.C. and H.H. Shugart, 1989: OUTENIQUA — A computer model to simulate succession in the mixed evergreen forests of the Southern Cape, South Africa. *Landscape Ecology*, **24**, 255–267.

**Vorosmarty**, C.J. and B. Moore III, 1991: Modeling basin-scale hydrology in support of physical climate and global biogeochemical studies: an example using the Zambezi River. *Surveys in Geophysics,* **12**, 271–311.

**Vorosmarty**, C.J, B. Moore III, A. Grace, B.J. Peterson, E.B. Rastetter, and J. Mellilo, 1991: Distributed parameter models to analyze the impact of human disturbance of the surface hydrology of a large tropical drainage basin in Southern Africa. In: *Hydrology for the Water Management of Large River Basins.* Proceedings of the Vienna Symposium, August 1991. IAHS Publication No. 201 [van de Ven, F.H.M, D. Gutnecht, D.P. Loucks, and K.A. Salewicz (eds.)], pp. 233–244.

**Wenzel**, R.P., 1994: A new Hantavirus infection in North America. *New England Journal of Medicine,* **330**, 1004–1005.

**Westoby**, M., B. Walker, and I. Noy-Meir, 1989: Opportunistic management for rangelands not at equilibrium. *Journal of Range Management*, **42**, 26–274.

**White**, F., 1983: *The Vegetation of Africa*. UNESCO, Paris, France, 356 pp.

**WHO (World Health Organization)**, 1990: *Potential Health Effects of Climatic Change*: Report of a WHO Task Group. WHO/PEP/90.10, WHO, Geneva, Switzerland, 58 pp.

**WHO**, 1994: *Progress Report Control of Tropical Diseases.* CTD/MIP/94.4, unpublished document.

**WHO**, 1996a: *Climate Change and Human Health*. World Health Organization, Geneva, Switzerland, 297 pp.

**WHO**, 1996b: *Investing in Health Research and Development.* Report of the Ad Hoc Committee on Health Research Relating to Future Intervention Options. WHO, Geneva, Switzerland, 278 pp.

**Wilhite**, D.A. and W.E. Easterling, 1987: *Planning for Drought: Toward a Reduction of Societal Vulnerability.* Westview Press, Boulder, CO, USA and London, United Kingdom.

**Williams**, M., 1990: Understanding wetlands. In: *Wetlands: a threatened landscape* [Williams, M. (ed.)]. Basil Blackwell, Ltd., Oxford, UK, pp. 1–14.

**WMO (World Meteorological Organization)**, 1992: *The Third Long-Term Plan. Part II, Volume 1.* World Weather Watch Program. WMO, Geneva, Switzerland.

**WMO**, 1995: *Global Climate System Review: Climate System Monitoring.* WMO, Geneva, Switzerland.

**Woodward**, F.I. and W.L. Steffen, 1996: *Natural Disturbances and Human Land Use in Dynamic Global Vegetation Models.* International Geosphere-Biosphere Programme Report 38, Stockholm, Sweden.

**World Bank**, 1990: *World Development Report 1990: Poverty.* World Bank and Oxford University Press, New York, NY, USA, 272 pp.

**World Bank**, 1993: *World Development Report: Investing in Health.* Oxford University Press, New York, NY, USA, 329 pp.

**World Bank**, 1995a: *A Framework for Integrated Coastal Zone Management in Sub-Saharan Africa: Building Blocks for Environmentally Sustainable Development in Africa.* Paper No. 4, Africa. Technical Department. The World Bank, Washington, DC, USA.

**World Bank**, 1995b: *Toward Environmentally Sustainable Development in Sub-Saharan Africa.* The World Bank, Washington, DC, USA, 300 pp.

**World Bank**, 1996: *Toward Environmentally Sustainable Development in Sub-Saharan Africa: a World Bank Agenda.* The World Bank, Washington, DC, USA, 140 pp.

**WRI (World Resources Institute)**, 1990: *World Resources 1990–1991.* World Resources Institute/United Nations Environment Programme/United Nations Development Programme/The World Bank. Oxford University Press, New York, NY, USA.

**WRI**, 1994: *World Resources: A Guide to the Global Environment, 1994–1995.* World Resources Institute/United Nations Environment Programme/United Nations Development Programme/The World Bank. Oxford University Press, New York, NY, USA, 400 pp.

**WRI**, 1996: *World Resources: A Guide to the Global Environment, 1996–97.* World Resources Institute/United Nations Environment Programme/United Nations Development Programme/The World Bank. Oxford University Press, New York, NY, USA, 342 pp.

**Wullschleger**, S.D. and R.J. Norby, 1992: Respiratory cost of leaf growth and maintenance in white oak saplings exposed to atmospheric $CO_2$ enrichment. *Canadian Journal of Forest Research*, **22**, 1717–1721.

**Zinyowera**, M.C., and S.L. Unganai, 1993: Drought in southern Africa. An update on the 1991–92 drought. *Drought Network News Int.,* **4(3)**, 3–4.

# 3

# The Arctic and the Antarctic

JOHN T. EVERETT (USA) AND B. BLAIR FITZHARRIS (NEW ZEALAND)

Lead Author:
*B. Maxwell (Canada)*

Contributors:
*M. Beniston (Switzerland), P.A. Berkman (USA), H.S. Bolton (USA), J. Brown (USA), R. Brown (Canada), D. Demaster (USA), S. Diaz (Argentina), R. Dixon (USA), D. Drewry (UK), L. Dyke (Canada), K.A. Edwards (Kenya), A. Gunn (Canada), R. Hewitt (USA), R. Jefferies (Canada), S. Kim (Korea), P. Kuhry (Finland), J. Kuylenstierna (UN), W.J.M. Martens (Netherlands), M. McFarland (USA), A.D. McGuire (USA), A.J. McMichael (UK), U. Molau (Sweden), M. Oquist (Sweden), C.J. Peckham (USA), C.A. Rinaldi (Argentina), F. Roots (Canada), E. Sakshaug (Norway), M. Scott (USA), J.B. Smith (USA), R.S.J. Tol (The Netherlands), C. Tynan (USA), E.C. Weatherhead (USA), G. Weller (USA), E. Wolff (UK)*

Note: This draft is built mostly on the IPCC Second Assessment Report, with contributions from the peer-reviewed literature provided by those listed above. The location of the supporting text is referenced as appropriate.

## CONTENTS

# EXECUTIVE SUMMARY

## Main Points

- The Arctic is extremely vulnerable to projected climate change and its impacts. Over the period of this assessment, climate change is expected to contribute to major physical, ecological, sociological, and economic changes already begun in the Arctic. Because of a variety of positive feedback mechanisms, the Arctic is likely to respond to climate change rapidly and more severely than any other area on earth. The most direct and pronounced changes to the Arctic are likely to include changes in temperature and precipitation, with subsequent effects on sea ice and permafrost. Much smaller changes are likely for the Antarctic over the period of this assessment. The Antarctic is likely to respond relatively slowly to climate change.
- Substantial loss of sea ice is expected in the Arctic Ocean. If there is more open water, there will be impacts on the climate system of northern countries as temperatures moderate and precipitation increases. If warming occurs, there will be considerable thawing of permafrost—leading to changes in drainage, increased slumping, and altered landscapes over large areas.
- Polar warming probably should increase biological production but may lead to different species composition on land and in the sea. On land, there will be a tendency for poleward shifts in major biomes and associated animals. However, the Arctic Ocean geographically limits movement of the tundra, taiga, and boreal forest. In the sea, marine ecosystems will move poleward. Animals dependent on ice may be disadvantaged.
- Changes in the polar climate are likely to affect the rest of the world through increased sea levels from melting of the cryosphere, increased warming of lower latitudes from slowing of oceanic transport of heat, and increased greenhouse gas levels through carbon dioxide and methane emissions in the Arctic.
- Human communities in the Arctic will be affected by these physical and ecological changes. The effects will be particularly important for indigenous peoples leading traditional lifestyles.
- There will be economic benefits and costs. Benefits include new opportunities for shipping across the Arctic Ocean, lower operational costs for the oil and gas industry, lower heating costs, and easier access for tourism. Increased costs can be expected from several sources, including disruptions caused by thawing of permafrost and reduced transportation capabilities across frozen ground and water. Sea-ice changes in the Arctic have major strategic implications for trade and defense.

## Document Summary

- **Regional Climate.** Cold is the overwhelming characteristic of polar regions, with nine months of snow, ice, cold, and relative darkness. In the Arctic, there are only a few weeks of thaw, when much of the ground is boggy and awash with water. There also are episodes of extreme cold and storms. In Antarctica, temperatures mostly remain below freezing.
- **Climate Trends.** Parts of the Arctic and Antarctic have warmed over the past half-century, while some parts appear to have cooled. Precipitation seems higher. Trends in the overall ice balance of the world's major ice sheets in Greenland and Antarctica are uncertain. There is conflicting information on changes in the thickness and extent of sea ice globally. Ice seems to be thinning in the Arctic but not in the Antarctic.
- **Vulnerability and Impacts.** Systems in the Arctic and Antarctic Peninsula are extremely sensitive to temperature. The primary impacts will be on the physical environment, biota and ecosystems, and human activities. The Arctic Ocean has strong sensitivity to temperature because its exposed areal extent grows and shrinks by as much as 50% within a single year. Interannual variability in sea ice similarly shows extreme sensitivity to temperature. The Arctic terrestrial system, primarily based on permafrost, also is very sensitive to temperature because much of the permafrost is currently discontinuous. Strong positive feedback mechanisms within the Arctic suggest that climate change is likely to be more severe in the Arctic than in the rest of the world.
- **Terrestrial Ecological Systems.** Climate change may occur at a rapid rate relative to forest migration ability. If there is warming, there will be a poleward migration of the northern treeline. However, geographic limitations may restrict the migration of the boreal forest. If warming occurs and the forest cover changes, there will be major impacts on biological resources (e.g., bears, caribou, small mammals, amphibians, and insects). The effects of climate change have not been deeply investigated in the Antarctic terrestrial ecosystem, but its distribution and specific composition could be altered by global warming.
- **Marine Ecological Systems.** In general, productivity should rise. Warming should increase growth and development rates of nonmammals; however, ultraviolet-B (UV-B) radiation is still increasing, which may constrain primary productivity as well as fish productivity. Additional risks include the loss of sea-ice cover, upon which several marine mammals depend for food and protection. Also,

Arctic shipping, oil exploration and transport, and economic development could bring risks to many species.

- **Additional Stressors.** In addition to climatically driven changes, such as changes in sea ice and temperature, the Arctic currently is stressed from a variety of sources, and some of these stresses are likely to increase in the next few decades. Additional stresses that may confound the effects of climate change include hydrocarbons, radionuclides, acidification, heavy metals, persistent organic pollutants, and UV radiation, as well as increased human development, traffic, and potential oil spills.
- **Ice.** If there is warming, the Arctic Ocean, as well as Arctic lakes and streams, could experience a thinner and reduced ice cover. In contrast, vast Antarctica is so cold that any warming within the IPCC scenarios should have little impact except in the Dry Valleys and on the Antarctic Peninsula. In fact, ice could accumulate through greater snowfall, slowing sea-level rise.
- **Permafrost.** Permafrost covers all of Antarctica and most of the Arctic. Much of the Arctic permafrost is close to thawing, making it an area extremely sensitive to even small changes in temperature. Effects of thawing permafrost include large-scale slumping, erosion, and sinking of areas. All of these effects will disrupt current vegetation, ecosystems, water balance, and human structures.
- **Fisheries.** Warming could lead to a rise in production unless changes in water properties would disrupt the spawning grounds of fish in high latitudes. There could be a substantial redistribution of important fish species. Fisheries on the margin of profitability could prosper or decline. Fishing seasons will lengthen, but most stocks already are fully exploited.
- **Navigation and Transport.** If sea-ice coverage is reduced, coastal and river navigation will increase. Opportunities for water transport, tourism, and trade will increase. The Arctic Ocean could become a major trade route. Seasonal transport across once-frozen land and rivers may become more difficult or costly. Offshore oil production should benefit from less ice.
- **Sub-Antarctic Islands.** These small, isolated, usually uninhabited islands in the Southern Ocean have specialized flora and important marine mammals and birds. Many seem to have warmed over the past 50 years, and this trend is expected to continue. Climate change impacts are unlikely to be important for most animal and bird species, but plant communities could change. Glaciers will probably shrink.
- **Arctic Settlements.** If the climate ameliorates, conditions will favor the northward spread of agriculture, forestry, and mining, with an expansion of population and settlements. More infrastructure—such as marine, road, rail, and air links—would be required. Changes in the distribution and abundance of sea and land animals will impact on traditional lifestyles of native communities.
- **Infrastructure.** If the permafrost thaws, much infrastructure—including pipelines, airstrips, water supply and sewage systems, building foundations, roads, rail lines, and mining systems and structures—will be damaged. Thawing could disrupt petroleum production in the tundra. Redesign and replacement of many of these systems will be needed.
- **Integrated Analysis.** Direct effects could include ecosystem shifts, sea- and river-ice loss, and permafrost thaw. Indirect effects could include positive feedback to the climate system. There will be new challenges and opportunities for shipping, the oil industry, fishing, mining, tourism, infrastructure, and the movement of populations, resulting in more interactions and changes in trade and strategic balance. There will be winners and losers. As examples, a reduced and thinning ice cover will disadvantage polar bears, while sea otters will have new habitats; communities on new shipping routes will grow, while those built on permafrost will have difficulties. Native communities will face profound changes impacting on traditional lifestyles.

## 3.1. Introduction

Polar environmental changes are expected to be greater than for many other places on Earth (IPCC 1996, WG II, Section 7.5). The Arctic and Antarctic contain about 20% of the world's land area. Athough similar in many ways, the two polar regions are different in that the Arctic is a frozen ocean surrounded by land, whereas the Antarctic is a frozen continent surrounded by ocean. The Antarctic is thermally isolated by the polar vortex, whereas the Arctic is influenced by seasonal atmospheric transport from the surrounding continents.

The Antarctic, for the purposes of this document, comprises the Antarctic continent, the surrounding Southern Ocean, south of 55°S, and the Sub-Antarctic islands (e.g., Campbell Island, Heard Island, South Georgia Islands). It is the driest, windiest, coldest, and cleanest continent and covers around 14 million $km^2$ (UNEP, 1997). It is devoid of trees. Its boundary is sometimes taken to coincide with the Antarctic Convergence, which roughly parallels the mean February air isotherm of 10°C and is the northern boundary of the Antarctic marine ecosystem. The area is managed by the Consultative Parties to the Antarctic Treaty to the dedication of science and peace (UNEP, 1997.) The Arctic, for the purposes of this document, is defined as the area within the Arctic Circle; it includes the boreal forests and discontinuous permafrost, although some authors prefer to use the area north of the natural tree line—which coincides approximately with the mean July air isotherm of 10°C (Sugden, 1982). The Arctic areas of North America, Asia, and Europe are included here, rather than in other regional chapters.

The polar regions are a zone marginal for the distribution of many species; however, native organisms thrive in terrestrial and marine ecosystems. Apart from research bases, the Antarctic is virtually uninhabited by humans. It is the only continent without indigenous peoples (about 4,000 persons are there for prolonged periods, engaged in scientific research). The Arctic, however, has been populated for thousands of years by a variety of indigenous peoples who have developed ways of life to adapt to the harsh and changing climate, but at very low densities compared to the rest of the world. A number of urban outposts have developed in recent times. There is a distinct contrast and sometimes conflict between intrusive modern society and indigenous culture.

There is little resource use in the Antarctic and surrounding Southern Ocean apart from fishing and tourism. These industries have been increasing in activity over recent years and have considerable potential for growth. Although tourists generally make visits of shorter duration, the number of tourists now is about double the number of scientists. Antarctic tourism is growing rapidly; the expected number of tourists might exceed 10,000 persons in the 1997–98 season (IAATO, 1997). Some local fish populations have been depleted, but the krill population could become a food source even though the maximum harvest has only been on the order of 500,000 tons. There is a multinational approach to natural resources and environmental management, with minerals exploration and exploitation banned by international agreement. By contrast, the Arctic lies within the political boundaries of some of the world's richest and most powerful nations. There is considerable economic activity based around fishing, farming and herding, oil extraction, mining, and shipping. All of these activities are climate sensitive. The Arctic has been a critical strategic area, and there still are considerable defense establishments in the region.

## 3.2. Regional Climate and Past Variability

The overwhelming characteristic of polar regions, in terms of both intensity and duration, is the cold. The long winter night ensures very low temperatures in winter. However, warm North Atlantic water flows into the Arctic Ocean at about 500 m water depth at a temperature of +2 to +4°C in the Fram Strait area and remains near 0°C even after reaching the other side of the Arctic. The polar regions show large seasonal variations in incoming solar radiation, from none during the winter to 24 hours of sunlight at mid-summer. The poles receive less solar radiation annually than the equator, but in their respective mid-summers the daily totals are greater than at other places on earth. The high albedo of polar regions, from the persistent snow and ice and the large loss of long-wave radiation due to the exceptional clarity and dryness of the atmosphere, is a key factor in the surface energy budget and ensures a net loss of radiation in all or most of the months of the year. The loss is compensated through transport of sensible and latent heat from lower latitudes, usually within cyclones, and by heat carried within ocean currents. Because of the lack of transport of warm current to the Southern Ocean and the pressure of strong westerlies, which blocks heat supply over the Antarctic, the Antarctic is colder than the Arctic. Summer temperatures in most of the Antarctic continent remain well below freezing. In the Arctic, however, rapid and strong snowmelt produces a large influx of fresh water to the rivers and Arctic Ocean in the spring and summer and supports a burst of life during a brief and intense summer. Important circulation systems of the world's oceans are driven by sinking cold water at the periphery of polar regions.

After nine months of snow, ice, cold, and relative darkness, there are a few brief weeks of thaw when much of the Arctic ground is awash with water and boggy. Overland travel is easy when surfaces are firmly frozen but becomes more difficult in summer when they are not. Surface transport over ice in seas, lakes, and rivers much of the year must give way to transport over water in summer. This marked seasonal contrast provides two dramatically different environments, which are a challenge and constraint to traditional and modern human systems (Sugden, 1982).

The freeze-thaw threshold of 0°C is crucial in polar regions. Large changes in physical, biological, and human systems occur when temperature crosses this threshold. Therefore, any climate change that shifts the freeze-thaw line, whether in space or time, will bring about important impacts.

Temporary incursions of cold air at lower latitudes have their source in polar regions. Antarctic storms sometimes strongly affect South America and southern New Zealand and exert some influence on the weather and climate of Australia and much of the Southern Hemisphere. In the Northern Hemisphere, the northern part of the subtropical zone and the southern part of the temperate zone in winter are the most vulnerable. Episodes of extreme cold and blizzards are major climate concerns for circumpolar countries like Russia and Canada.

Ice cores from the Arctic and Antarctic provide a particularly valuable archive of past climate and are direct evidence for the amount of increases in carbon dioxide ($CO_2$), methane ($CH_4$), and nitrous oxide ($N_2O$). There is evidence from these records of rapid warming ~11,500 years ago, at the end of the last glacial period. Recent results (Cuffey *et al.*, 1995; Johnsen *et al.*, 1995) suggest that the temperature changes were larger than previously thought; that the coldest parts of the last glacial period could have been as much as 21°C colder than the present temperature in central Greenland; and that temperatures increased by more than 10°C in a few decades. There is evidence of an even more rapid change in the precipitation pattern, rapid reorganizations in atmospheric circulation, and periods of rapid warming during the past 20,000 years. Surface water salinity and temperature have exhibited parallel changes that resulted in reduced oceanic convection in the North Atlantic and in reduced strength of the global conveyor belt ocean circulation (IPCC 1996, WG I, Section 3.6.4).

There also are indications of rapid warm-cold oscillations during the last glacial period in the central Greenland records. Rapid warmings of ~10°C in a few decades were followed by periods of slower cooling over a few centuries and then a generally rapid return to glacial conditions. About 20 such intervals, each lasting between 500 and 2,000 years, occurred during the last glacial period (IPCC 1996, WG I, Section 3.6.3). However, the great ice sheet of Greenland has changed little in extent during this century (IPCC 1996, WG II, Section 7.4).

The exact sequence of events leading to rapid climate events within the glacial period is not well understood, but it has generally been believed that the mechanisms are related to the existence of large Northern Hemisphere ice sheets. Although there is some indication that a cooling event may have occurred within the last interglacial period (Maslin and Tzedadakis, 1996), the evidence from paleorecords for rapid and catastrophic events in interglacial periods remains a topic in need of further study.

There has been a tendency toward warmer temperatures in parts of the Arctic and Antarctic over the past half-century (Raper *et al.*, 1984; Jones *et al.*, 1986; Chapman and Walsh, 1993). Figures 3-1 and 3-2 depict winter temperature anomalies from 1961–1990 means for the Arctic and the Antarctic Peninsula (60° and higher latitude), respectively, for this century. The former depicts evidence of long-term fluctuation in winter temperatures, with about a 2°C warming trend over nearly 100 years of recorded data. This trend is supported by the annual temperature record of Figure A-5 in Annex A. Figure 3-2 is less definitive in depicting a clear warming trend in the Antarctic, but it should be noted that the data base for this record is more limited, consisting mostly of data from near the Antarctic Peninsula. Borehole temperature measurements show that permafrost is warming in some areas, though not everywhere. Later freeze-up and earlier break-up dates for river and lake ice are observed in some tundra and boreal lands.

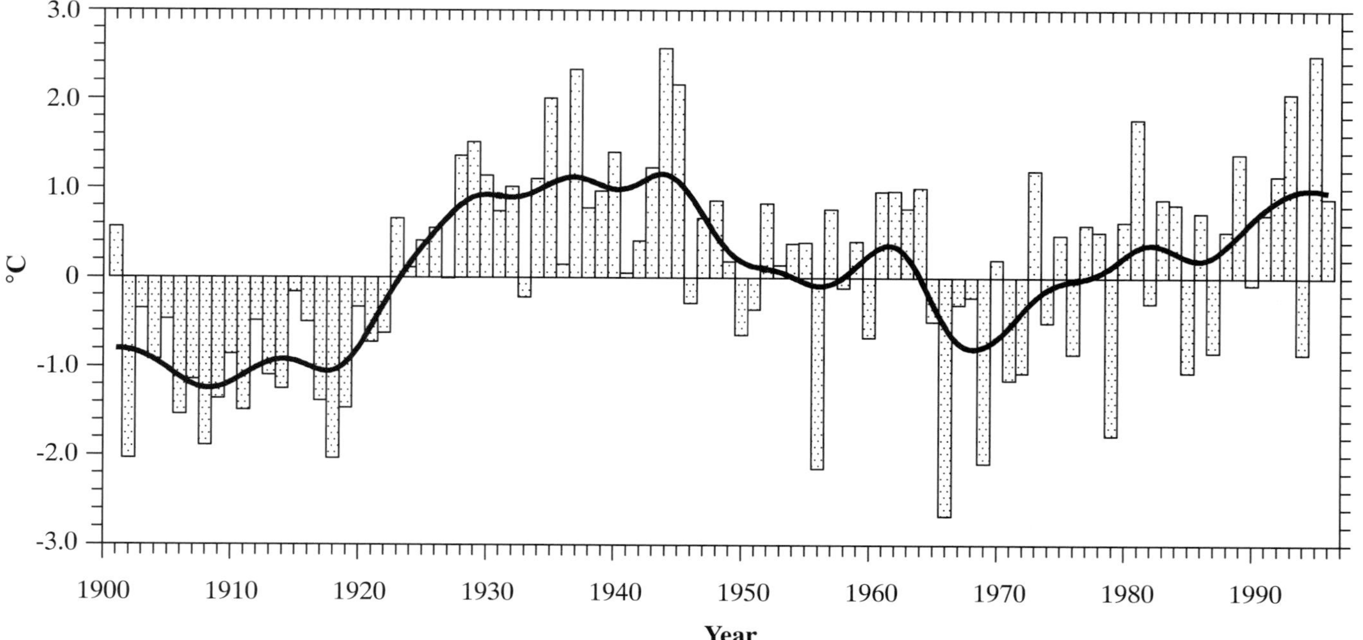


**Figure 3-1**: Observed annual winter (DJF) temperature anomaly over the Arctic during the period 1900–96.

These latter events are each at least a week different compared to the last century (IPCC 1996, WG II, Chapter 7 Executive Summary).

There is growing evidence of recent atmospheric warming on the Antarctic Peninsula. A warming trend of 0.056°C/yr—a total increase of ~2.5°C since 1945—has been recorded for Faraday Station on the west coast of the Antarctic Peninsula (King, 1994), and a 2.1°C increase in decadal average of the mean annual temperature between 1931–1940 and 1981–90 has been recorded from Orcadas Station on South Orkney Island (Hoffman *et al.*, 1997). The glacier retreat (Skvarca *et al.*, 1995) and the very recent collapses of the northernmost Larsen (Rott *et al.*, 1996) and Wordie Ice Shelves appear to be consistent with a warming trend in this region.

From the main part of the Antarctic continent, there have been discharges of enormous icebergs from the Filchner and Ross Ice Shelves (Rott *et al.*, 1996; Vaughan and Doake, 1996). However, the great ice sheets of Antarctica have changed little in extent during this century. The dynamic responses of the different ice sheets are influenced largely by whether they are marine (e.g., West Antarctic Ice Sheet) or land-based (e.g., East Antarctic Ice Sheet). Accumulation on the Antarctic continent has increased significantly (by as much as 5–10%) in the past few decades. The dynamic response times of land-based ice sheets (East Antarctica and Greenland) to climate change are on the order of thousands of years, so they are not necessarily in equilibrium with current climate. The response times of marine-based ice sheets (West Antarctica) probably are much shorter because they may be directly influenced by sea level and other environmental effects such as salinity, temperature, and currents. Observational evidence is insufficient to determine whether they are in balance or have decreased or increased in volume over the past 100 years (IPCC 1996, WG II, Section 7.4).

In the circum-Arctic, there has been a tendency for negative mass balances in ice caps and glaciers over the past 30 years or longer (IPCC 1996, WG II, Section 7.2.2). A recent Canadian study, however, measured no significant changes in either the mass balance of snow accumulation or ice melt over the past 32 years in the Canadian Arctic (Koerner and Lundgaard, 1995).

An increase has been found in the numbers of cyclones and anticyclones over the Arctic between 1952 and 1989 (IPCC 1996, WG I, Section 3.5.3.2). Station measurements indicate that annual snowfall has increased over the period 1950–90 by about 20% over northern Canada (north of 55°N) and by about 11% over Alaska. Total precipitation has increased in all of these regions (IPCC 1996, WG I, Section 3.3.2).

Projections of future polar climate face several difficulties. The reliability of the simulated climate change scenarios is not high, and there are considerable model-to-model differences. However, all or most general circulation models (GCMs) show the following features: greater warming over land than sea; reduced warming, or even cooling, in the high-latitude Southern Ocean and part of the northern North Atlantic Ocean; maximum warming in high northern latitudes in winter and little warming over the Arctic in summer; increased precipitation and soil moisture in high latitudes in winter; a reduction in the strength of the North Atlantic currents; and a widespread reduction in diurnal range of temperature (IPCC 1996, WG I, Section 6.2.5).

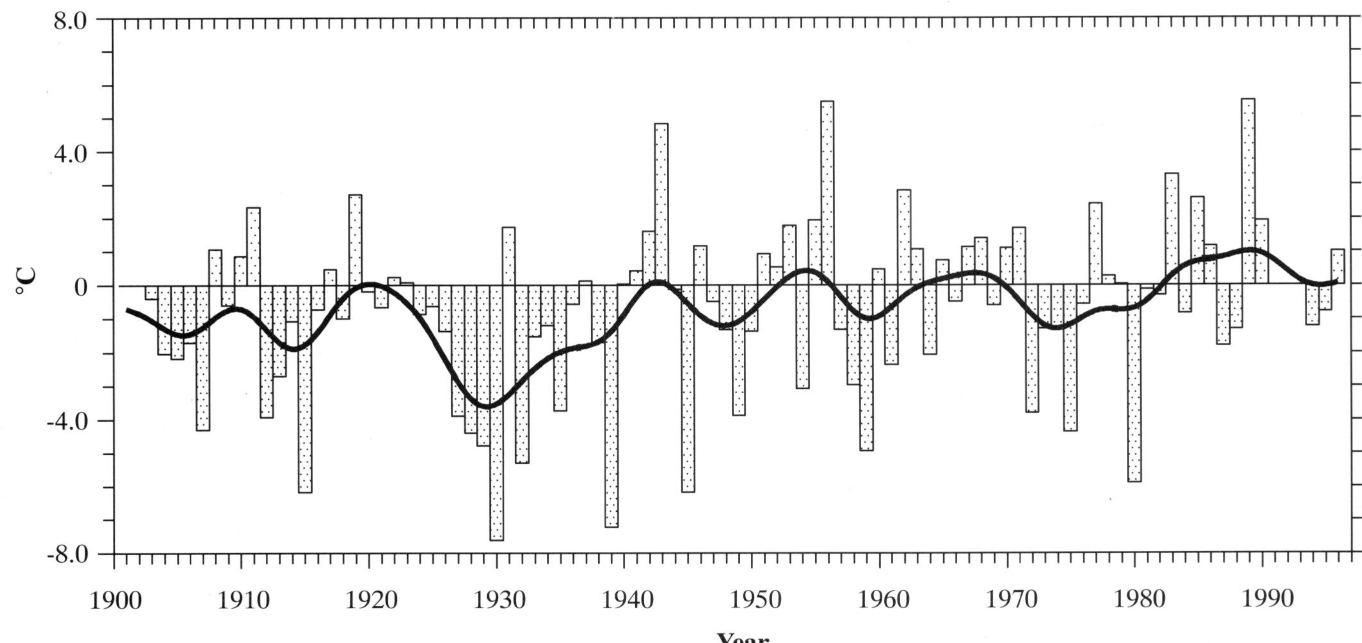


**Figure 3-2**: Observed annual winter (JJA) temperature anomaly over Antarctica during the period 1900–96.

## 3.3. Vulnerability and Potential Impacts

The polar systems are extremely sensitive to the variability of temperature, and several aspects of these systems will be affected by any further climate change. The primary impacts will be on the physical environment, including ice, permafrost, and hydrology; on biota and ecosystems, including fisheries and terrestrial systems; and on human activities, including social and economic impacts on settlements, on resource extraction and transportation, and on existing infrastructure.

### *3.3.1. Ice and Snow*

Sea ice covers about 11% of the world's ocean, depending on the season. It affects albedo, salinity, and ocean-atmosphere thermal exchange (IPCC 1996, WG II, Section 8.3.1.4). Studies on regional changes in the Arctic and Antarctic indicate trends of decadal length, often with plausible mechanisms proposed for periodicities of a decade or more (IPCC 1996, WG II, Section 8.3.1.4). At present, however, there is considerable year-to-year variation and no convincing evidence of long-term changes in the extent of global sea ice.

In Antarctica, temperatures are so low that comparatively little surface melting occurs on the continental ice sheet; ice loss is mainly by iceberg calving, the rates of which are determined by dynamic processes involving long response times (thousands of years). Even if Antarctica were to warm in the future, its mass balance is expected to become more positive: The rise in temperature would be insufficient to initiate melt but would increase snowfall (IPCC 1996, WG II, Section 7.4). Little change in Antarctic ice sheets is expected over the next 50 years, although longer-term behavior—including that of West Antarctic ice—remains uncertain, and some instability is possible. Some areas of Antarctica may show a pronounced change and dynamic response. The Antarctic Peninsula, for instance, receives 28% of the continent's snowfall and experiences warmer temperatures and summer melting at sea level. A rise in temperature would be expected to cause continued wasting of marginal ice shelves in the Antarctic Peninsula, but this melting has no direct effects on sea level, nor is it indicative of changes in the Antarctic ice sheets.

In the Antarctic, where the sea-ice cover is divergent and land boundaries are less important, it is more reasonable to suppose that the main effect of global warming will be a simple retreat of the ice edge southward. Even here, however, a complex set of feedback mechanisms comes into play when the air temperature changes. The balance of lead concentration, upper ocean structure, and pycnocline depth adjusts itself to minimize the impact of changes, tending to preserve an ice cover even though it may be thinner and more diffuse (IPCC 1996, WG II, Section 7.4.5). The modeling of Gordon and O'Farrell (1997) demonstrates an ultimate reduction in Antarctic sea ice of about 25% for a doubling of $CO_2$, with the ice retreating fairly evenly around the continent.

A recent study by Nicholls (1997) discounts the significance of air temperature-induced melting of the more massive ice shelves south of the Antarctic Peninsula in a climate-warming scenario. Studies of the seasonal water temperature changes of the sub-ice-shelf cavity of the Antarctic Filchner-Ronne Ice Shelf indicate that the flux of high-salinity shelf water (HSSW) is responsible for the net melting at the ice shelf's base. As rates of sea-ice formation decrease during warmer winters, the flux of HSSW beneath the ice shelf is reduced. Subjected to less warm water flux, the sub-ice cavity will cool. Nicholls concludes that a moderate warming of the climate could, in fact, lead to a basal thickening of the Filchner-Ronne Ice Shelf, perhaps increasing its longevity.

With climate warming, ice cover in lakes and rivers is expected to decrease, with large changes in lake water levels. In the largest Antarctic desert, the McMurdo Dry Valleys, closed-basin lakes fed by glacier meltwater streams rose as much as 10 m from 1970 to 1990, or almost 480 mm per year. In the Antarctic Dry Valley, ice cover has thinned for some permanently ice-covered lakes. Lake Hoare thinned by 20 cm/yr over a 10-year period beginning in 1977. Because light attenuation by the ice is a major limiting factor, these climate-related changes are expected to cause shifts in the biota of such lakes (IPCC 1996, WG II, Section 10.6.1.4).

The Greenland ice sheet, which has no floating ice shelves of consequence, is different from the Antarctic ice sheets. Ice loss from surface melting and runoff is of the same order of magnitude as loss from iceberg calving. Thus, climate change in Greenland could have immediate effects on the surface mass balance of the ice sheet through melting and runoff as well as through accumulation. If there is warming, both the melt rates at the margins and the accumulation rates in the interior should increase. The former rate is expected to dominate. Thus, the mass balance could become negative (IPCC 1996, WG II, Section 7.4). Nevertheless, changes in the general form of the total ice sheet over the next century are expected to be small.

Increased temperatures in the Arctic are likely to shorten the duration of ice cover on Arctic lakes. A longer open-water period, together with warmer summer conditions, will increase evaporative loss. Some patchy wetlands and shallow lakes owe their existence to a positive water balance and the presence of an impermeable permafrost substrate that inhibits deep percolation; enhanced evaporation and ground thaw will cause some to disappear (IPCC 1996, WG II, Section 7.4.4). If there is decreased precipitation and lower flood frequency, the Mackenzie Delta in the Canadian arctic could shrink in several decades (IPCC 1996, WG II, Section 10.5.2), although the pre-Pleistocene depositions of gravel in the delta are sufficiently large that a more likely scenario is a profound change in the delta's form and extent.

A major source of uncertainty about sea-level change relates to the future behavior of the polar ice sheets, which hold most of the nonoceanic water on Earth's surface. Most of their volumes lie on land above sea level. However, much of the West

Antarctic ice sheet is below sea level. The discovery of major recent changes in certain Antarctic ice streams has focused public attention on the possibility of "collapse" of this ice reservoir within the next century, with potential impacts on sea level (IPCC 1996, WG I, Section 7.3.3.1). The collapse of the West Antarctic ice sheet by the year 2100—with consequential major sea-level rise—is not impossible, but its likelihood is considered to be very low (IPCC 1996, WG I, Section 7.5.5).

Observed variations in sea-ice thickness are in accord with the predictions of numerical models that take account of ice dynamics and deformation as well as ice thermodynamics. The limited information available does not provide evidence of detectable change in the thickness of Antarctic sea ice (IPCC 1996, WG II, Section 7.2.5). Nevertheless, researchers have identified significant reductions in the summer sea-ice cover in the Bellingshausen and Admundsen Seas in the late 1980s and early 1990s that are consistent with a warming climate west of the Antarctic Peninsula (Jacobs and Comiso, 1993). Measurements from a series of submarine transects near the North Pole show large interannual variability in ice draft over the period 1979–1990. There is some evidence of a decline in mean thickness in the late 1980s relative to the late 1970s (IPCC 1996, WG II, Section 7.2.5). Between 1978 and 1994, the Arctic sea ice appears to have decreased by 5.5%.

Based on some global model scenarios for a doubling of $CO_2$, a large change in the extent and thickness of sea ice is possible, not only from warming but also from changes in circulation patterns of both atmosphere and oceans. There is likely to be substantially less sea ice in the Arctic Ocean (IPCC 1996, WG II, Chapter 7 Executive Summary). Major areas that are now ice-bound throughout the year are likely to have long periods during which waters are open and navigable. Some models even predict an ice-free Arctic, although most scenarios maintain significant summer ice centered on the North Pole. Melting of snow and glaciers will lead to increased freshwater influx, changing the chemistry and salinity of oceanic areas affected by the runoff.

However, the ability of existing GCMs to predict the extent of Arctic ice change is limited by the inadequacy of regional polar models to simulate the multiscale dynamics of sea ice. GCM experiments with simplified treatments of sea-ice processes produce widely varying results and do not portray the extent of and seasonal changes in sea ice for the current climate very well. Given these limitations in existing ice modeling, some authors estimate that with a doubling of greenhouse gases, sea ice would cover only about 50% of its present area; others project a sea-ice reduction of 43% for the Southern Hemisphere and 33% for the Northern Hemisphere. The global area of sea ice is projected to shrink by up to 17 x $10^6$ $km^2$ (IPCC 1996, WG II, Section 7.4.5).

Ramsden and Fleming (1995) suggest that Arctic sea ice would be little altered under doubled $CO_2$ conditions. More recent analyses by Gordon and O'Farrell (1997), who use a dynamic ice procedure within a transient-coupled-atmosphere-ocean model, predict a 60% loss in summer sea ice in the Northern Hemisphere by the time $CO_2$ has doubled. In their model, ice loss appeared earlier and proceeded more rapidly than it did in the Southern Hemisphere.

Although current polar models are subject to considerable uncertainty, they are the best available tools to explore possible future scenarios. A regional study of the Canadian sector of the Beaufort Sea proposed an increase from 60 days to 150 days in the open-water (ice-free) season (McGillivray *et al.*, 1993). There could also be reduced ice thickness. The maximum extent of open water in summer could increase from the current 150–200 km to 500–800 km offshore. At the same time, if longer fetches are available, wave heights would increase (e.g., the proportion of waves in excess of 6 m would rise from 16% to 39%) (IPCC 1996, WG II, Section 7.5.4). This increased open-water exposure of the ice-rich coastlines would most certainly result in increased erosion.

Using empirical ice growth-melt models, Wadhams (1990) predicts that in the Northwest Passage and Northern Sea Route, a century of warming would lead to a decline in winter fast-ice thickness from 1.8–2.5 m at present to 1.4–1.8 m and an increase in the ice-free season of 41–100 days. Other researchers, using another mix of models, find these ice-free day estimates a little high (Flato and Brown, 1996). This effect will be of great importance for the extension of the navigation season in the Russian Northern Sea Route and the Northwest Passage.

Predicting the future character of moving pack ice is a difficult problem because dynamics (ocean and wind currents), rather than thermodynamics (radiation and heat components), determine its average thickness. Wind stress acting on the ice surface causes the ice cover to open up to form leads (ice-free areas). Later, under convergent stress, refrozen leads and thinner ice elements are crushed to form pressure ridges. Exchanges of heat, salt, and momentum are all different from those that would occur in fast-ice cover. The area-averaged growth rate of ice is dominated (especially in autumn and early winter, when much lead and ridge creation take place) by the small fraction of the sea surface occupied by ice less than 1 m thick (IPCC 1996, WG II, Section 7.4.5).

In the coastal zones of the Arctic Ocean—such as off the Canadian Arctic Archipelago, where there is net convergence of currents—the mean ice thickness is very high (7 m or more) because of ridging. Here the mean thickness is determined by mechanical factors, largely the strength of the ice, and may not be as sensitive to global warming as in other regions (IPCC 1996, WG II, Section 7.4.5).

Feedback mechanisms in the Arctic often are strong and complex, and not all have been fully identified or quantified. Most are positive in nature. For example, as biomes migrate northward, taller plants will tend to lower albedo, especially where they protrude through the snow. This will lead to a further enhancement of warming. Similarly, decreases in the extent of snow and ice cover will lower albedo and act to warm the water and land. Increased precipitation, expected with many warming scenarios, could provide a possible negative feedback as some will contribute to increased snow thickness. As Arctic warming increases the open-water area, precipitation may further increase and cause thicker snow cover, including on sea ice. The growth rate of land-fast ice could be expected to decrease, as has been directly observed (Brown and Coté, 1992). However, if snow thickness is increased to the point where not all of it is melted in summer, more protection may be conferred on the ice surface—which could lead to an increase, rather than a decrease, in equilibrium ice thickness (IPCC 1996, WG II, Section 7.4.5). This scenario also would increase the stability of the upper mixed layer of the ocean, leading to more sea ice production.

### 3.3.2. *Permafrost, Hydrology, and Water Resources*

Permafrost—ground material that remains below freezing—underlies as much as 25% of the global land surface, including all of Antarctica and virtually all of the Arctic (IPCC 1996, WG II, Section 7.1). In continental areas in the tundra, as well as some boreal lands, it extends to considerable depths. It also is present under shallow polar seabeds, in ice-free areas in Antarctica, on some sub-Antarctic islands, and in many mountain ranges and high plateaus of the world (IPCC 1996, WG II, Section 7.2.3). Ice-rich permafrost may contain up to twice as much (frozen) water as the same soil in a thawed state (IPCC 1996, WG I, Section 7.3.4). It forms an impervious layer to deep infiltration of water, maintaining high water tables and poorly aerated soils.

By the year 2050, increases are expected in the thickness of the active layer of permafrost and in the loss of extensive areas of discontinuous permafrost in Arctic and sub-Arctic areas. Major changes in the volume and extent of deep, continuous permafrost are unlikely because it is very cold and reacts with longer time lags (IPCC 1996, WG II, Chapter 7 Executive Summary).

In areas of good drainage, significant increases in active layer depth or loss of permafrost are expected to cause drying of upper soil layers in most regions.

Widespread loss of permafrost will trigger erosion or subsidence of ice-rich landscapes (IPCC 1996, WG II, Chapter 7 Executive Summary). A critical factor influencing the response of tundra to warming is the presence of ground ice. Ground ice generally is concentrated in the upper 10 m of permafrost—the very layers that will thaw first as permafrost degrades. This loss is effectively irreversible because once the ground ice melts, it cannot be replaced for millennia, even if the climate were to cool subsequently. The response of the permafrost landscape to warming will be profound, but it also will vary greatly at the local scale, depending on ground-ice content. As substantial ice in permafrost is melted, there will be land subsidence. This process of thermokarst erosion in lowland areas will create many ponds and lakes and lead to coastal retreat and inland erosion (IPCC 1996, WG II, Section 7.5). These physical changes will result in major changes in ecosystem structure and landscape in the interior land masses of the sub-Arctic.

Peatlands will be extremely vulnerable to climate change if warmer temperatures lead to a thawing of the permafrost layer and affect their hydrology through changes in surface elevation, drainage, or flooding. These wetlands have a limited capacity to adapt to climate change because it is unlikely that new permafrost areas will form (IPCC 1996, WG II, Section 6.3.1). Permafrost is the key factor, generally, in maintaining high water tables in these peatland ecosystems (IPCC 1996, WG II, Chapter 7 Executive Summary).

A forerunner of future landscapes can be seen in areas of massive ground ice—such as in Russia, where past climatic warming has altered the landscape by producing extensive flat-bottomed valleys. Ponds within an area of thermokarst topography eventually grow into thaw lakes. These lakes continue to enlarge for decades to centuries because of wave action and continued thermal erosion of the banks. Liquefaction of the thawed layer will result in mudflows on slopes in terrain that is poorly drained or contains ice-rich permafrost. On steeper slopes there also will be landslides. Winter discharge of groundwater often leads to ice formation. This formation is expected to increase on hill slopes and in the stream channels of the tundra (IPCC 1996, WG II, Section 7.5).

Climatic warming would likely make notable changes in the hydrology of Arctic areas. The nival regime runoff patterns will weaken for many rivers in the permafrost region. The pluvial influence upon runoff will intensify for rivers along the southern margin of Arctic regions of Eurasia and North America. Should the climate continue to warm, the vegetation will likely be different from today. When lichens and mosses—which tend to be suppressers of evapotranspiration—are replaced by transpiring plants, evaporative losses will increase. Enhanced evaporation will lower the water table, which would be followed by changes in the peat characteristics as the extensive wetland surfaces become drier (IPCC 1996, WG II, Section 7.5.1).

Thawing of permafrost deepens the active layer, allowing greater infiltration and water storage, especially for rain that falls during the thawed period. Warming of the ground also will lead to the formation of unfrozen zones within the permafrost that provide porousness to enhance groundwater flow and increase chemical weathering and nutrient release. The chemical composition and amount of groundwater discharge may be changed as subpermafrost or intrapermafrost water is connected to the surface. In autumn and winter, more groundwater should be available to maintain baseflow, further extending the stream-flow season. With earlier snowmelt, the seasonality of river flows will be different (IPCC 1996, WG II, Section 7.5.1). Because many of the major river systems are north-flowing and cross several climate zones that may respond differently to climate warming, predictions are further complicated.

### 3.3.3. Terrestrial Ecological Systems

Climate change will affect terrestrial ecological systems through changes in permafrost as well as direct climatic changes, including changes in precipitation, snow cover, and temperature. Terrestrial ecoystems are likely to change from tundra to boreal forests, although vegetative changes are likely to lag climatic change. Vegetational ecosystem models suggest that the tundra will decrease by as much as one- to two-thirds of its present size (see Annex C). Climate change is expected to occur at a rapid rate relative to the speed at which forest species grow, reproduce, and reestablish themselves or to their ability to develop appropriate soils. The boreal forest covers approximately 17% of the world's land surface in a circumpolar complex of forested and partially forested ecosystems in northern Eurasia and North America (IPCC 1996, WG II, Section 1.6.1). If there is warming, the greatest forest changes are expected in high latitudes. Polar projections of this region's terrestrial ecological changes (Figure 3-3) derived from two different models' depictions of the future climate reinforce this conclusion, showing significant reduction in the tundra and taiga biomes and expansion of the boreal forest.

A poleward migration of the northern tree line would decrease winter albedo because the tree canopy has much lower albedo than exposed snow surfaces, affecting local climate through increased absorption of the sun's incoming energy (IPCC 1996, WG II, Chapter 6). Large losses are projected in the area of boreal forests despite their encroachment into current tundra. Shrinkage in total area because of the geographically limited poleward shift leads to a net loss of about 25% in boreal coverage (IPCC 1996, WG II, Section 1.3.4). On the other hand, vegetative ecosystem modeling suggests an expansion in the boreal forest ranging from 108% to 133% of the present size (see Annex C). The species composition of forests is likely to change; entire forest types may disappear, while new assemblages of species, and hence new ecosystems, may be established (IPCC 1996, WG II, Section 3.1). The rate of migration is critical, and it is likely that the loss from the southern margin will be greater than the gain in the northern margin. Projections for habitat changes are presented in Table 3-1.

By 1995, the Arctic contained 285 protected areas covering 2.1 million km$^2$ (UNEP, 1997). There will be considerable impacts of climate change on resource management in the tundra (IPCC 1996, WG II, Section 7.5). There is likely to be a change in the migration patterns of polar bears and numbers of caribou, along with other biological impacts (IPCC 1996, WG II, Section 7.5). High-arctic Peary caribou and musk-oxen may become extinct, and barren-ground caribou and musk-oxen elsewhere may be reduced (Gunn, 1995). Arctic island caribou migrate seasonally across the sea ice between many of the arctic islands in late spring and fall; changes in sea ice would disrupt those migrations, with unforeseen consequences to population survival and gene flow.

It is also anticipated that small mammals, from lemmings to aquatic furbearers (muskrats, beavers, mink), will face ecosystem alterations that will change their abundance and distribution. Amphibians and reptiles can be expected to increase their range, and the distribution and abundance of hundreds of arthropod species will alter. Changes in the timing and abundance of forage availability and parasite infestations may accumulate—driving populations into decline, with serious consequences for people still depending on them (Jefferies, 1992). Increased incidences of forest fires and insect outbreaks are likely, such as already observed in Alaska.

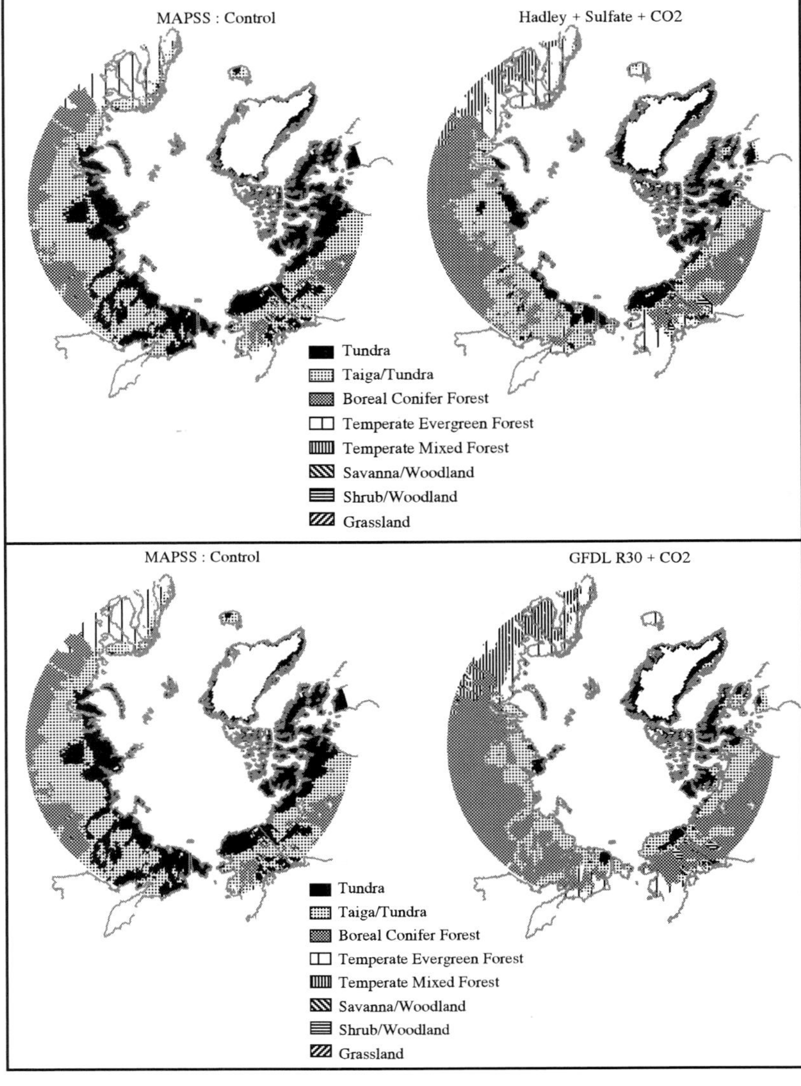


**Figure 3-3**: Ecological shifts derived from the MAPSS model in response to changes projected by two GCM simulations of future climate.

***Table 3-1:** Area of each biome (in 1,000 km²) north of the Arctic Circle, under current and future climate(s), as projected by three models discussed in Annex C.*

| | Present | MAPSS + $CO_2$ Effect OSU | GFDLR30 | UKMOS |
|---|---|---|---|---|
| Ice (perennial land-mass ice) | 1,745 | 1,437 | 1,328 | 1,532 |
| Tundra | 4,316 | 2,972 | 2,019 | 1,832 |
| Taiga/Tundra | 1,950 | 2,520 | 2,632 | 2,265 |
| Boreal Conifer Forest | 0 | 791 | 1,432 | 1,643 |
| Other | 178 | 468 | 778 | 918 |

Indirect temperature effects associated with changes in tundra thaw depth, nutrient availability, and vegetation will cause substantial changes in Arctic species composition, litter quality, and nutrient availability. Increased nutrient availability increases shrub abundance and decreases the abundance of mosses, an important soil insulator (IPCC 1996, WG II, Section 2.7.3).

Some alpine and tundra plants have no photoperiod requirements for initiating growth and therefore are affected by spring frosts. For example, in northern Sweden, increased length of the growing season and increased summer temperature will have a significant impact on reproduction and population dynamics in Arctic and alpine plants (IPCC 1996, WG II, Section 5.2.3.1). Experimental doubling of $CO_2$ concentration has relatively little multiyear effect on plant growth in Arctic tundra, presumably because of constraints of low nutrient supply (IPCC 1996, WG II, Section 2.2.2).

Major shifts in biomes will be associated with changes in microbiological (bacteria, algae, etc.) and insect communities. As a consequence, some may diminish while others prosper. All microbiologically facilitated processes are strongly affected by moisture and temperature. Annual soil respiration rates are likely to increase because of the lengthened season for breakdown of plant material and because increasing temperature strongly stimulates organic-matter decomposition, especially in Arctic regions subject to permafrost (IPCC 1996, WG II, Section A.3.1).

In the recent geologic past, the tundra was a carbon sink. Recent climatic warming in the Arctic, coupled with the concomitant drying of the active layer and the lowering of the water table, has shifted areas of the Arctic from sinks to sources of $CO_2$ (Oechel *et al.*, 1993). Extrapolating results from the Alaskan tundra to the circumpolar Arctic, regional warming could have caused a net flux from the land to the atmosphere of about 0.2 GtC/yr during the 1980s (IPCC 1996, WG I, Section 9.2.3.1). However, unpublished evidence from the Land-Atmosphere-Ice-Interactions component of ARCSS indicates that there may be substantial interannual variability in the source-sink relationship of Arctic tundra. This concern is reinforced by recent study by Myneni *et al.* (1997). Loss of a sizable portion of more than 50 Gt of carbon in Arctic soils and 450 Gt of carbon in soils of all tundra ecosystems could cause an appreciable positive feedback on the atmospheric rise of $CO_2$ (IPCC 1996, WG II, Section 7.4.3). It should be noted, however, that the preceding projections are based on an assessment for all high-latitude ecosystems, not just the tundra, and could be misleading as an indication of the potential for tundra ecosystems to act as a positive feedback. In addition, Schimel *et al.* (1994) and McGuire *et al.* (1995) have modeled the temperature sensitivity of soil carbon responses and find levels much less than 450 GtC.

Enhanced decomposition of soil organic matter leads to the release of trace gases and feedbacks on the global climate. A change in total $CH_4$ flux from Arctic wetlands can be expected if the areal extent of wetlands changes, the duration of the active period changes, or the per-unit-area production or oxidation of $CH_4$ changes (IPCC 1996, WG II, Section 6.3.3.2.2). The limited available data for the $CH_4$ content of permafrost has substantial variability. High $CH_4$ concentrations in ice-bonded sediments and gas releases suggest that pore-space hydrate may be found at depths as shallow as 119 m. These data raise the possibility that gas hydrates could occur at much shallower depths and may be more rapidly influenced by climate change than previously thought (IPCC 1996, WG II, Section 7.4.3).

In Antarctica, the terrestrial ecosystem is comparatively simple, constrained by an exposed land area that is very cold. Only 2% of the Antarctic surface is not covered by ice. This limited regime is mostly rocky areas where the temperature is below freezing except for periods of a few days or weeks during the Austral summer. These "Antarctic oases" provide a natural laboratory for assessing the vulnerability and response of this ecosystem to climate change. The mainland plays host to a number of microscopic plants that are found mainly in crevices and cavities of exposed rocks. Even the poorly developed soil of Antarctica harbors bacteria, algae, yeast and other fungi, lichens, and even moss spore (though usually in a dormant stage). The coastal region is particularly hospitable to the vegetation of lichens and mosses. Meltwater in the area helps to support herbaceous species such as grass. Some species of invertebrates survive in the harsh environment by super-cooling or anhydrobiosis mechanisms (Walton and Bonner, 1985). The Dry Valley's environmental conditions resemble those on Mars; this area is one of the world's most extreme desert regions. It was formed by the advances and retreats of glaciers through the coastal mountain ranges. The ecosystem there consists of microorganisms, microinvertebrates, mosses, and lichens. Climate change will impact on the physiology, distribution, and species composition of this terrestrial ecosystem.

### 3.3.4. Freshwater and Marine Ecological Systems

The ranges of many species in lakes and streams are likely to shift poleward by about 150 km for every 1°C increase in air temperature (IPCC 1996, WG II, Technical Summary, Section 3.3.1). This axiom is not useful for extrapolation in oceanic waters. If global warming effects continue, the sea surface temperature increases would be about 3°C in the North Pacific in 50 years. This change represents the same effect that would be seen in a present isotherm shift northward of about 500 km in mid-latitudes (Kim, 1995). Climate change or regime shifts might change distribution, species composition, and productivity in the North Pacific as well as adjacent subarctic areas.

Water temperatures have a direct impact on the productivity of fish species and the relative abundance of different fish species. Climatic changes are likely to affect not only water temperatures but also salinity and seasonal water cycles. Temperature influences biological production and can have a profound effect on growth and metabolic processes. Observations show that biological rates double or halve with a 10°C increase or decrease in temperature, respectively. Over the range of temperatures encountered in the ocean (-2 to +30°C), maximum growth rates of plankton vary by about a factor of 10. At high latitudes, where the temperature range is smaller (-2 to +3°C), temperature changes may have a relatively larger effect. Even small changes in temperature could have pronounced effects on biological rates of growth and development. Further, glacial meltwater is known to affect the nearshore zonation of Antarctic marine invertebrates (Berkman, 1994) and increased meltwater production may further impact species in the Antarctic coastal zone. Climate impacts on long-lived benthic invertebrate species such as sponges and some calcareous species, some of which have decadal and perhaps century-long lifespans, are unknown.

Greenhouse warming of the troposphere would be accompanied by cooling of the stratosphere. Changes in the stratospheric ozone layer have occurred over both the Arctic and the Antarctic. In the Antarctic, the occurrence of a persistent stratospheric ozone hole occurs during the spring and is due to ozone-depleting substances, including chlorofluorocarbons. The low human and species populations in the Antarctic have limited the biological damage, although there is evidence that significant changes in ecosystem function could occur. In the Arctic, where considerably more biological activity may be affected, ozone depletion and increased UV radiation have been observed over the past decade. Ozone depletion has occurred both as a steady decline and also with short, isolated areas of very low ozone (Weatherhead and Morseth, 1997). Climate change may further ozone depletion. The cooling of the stratosphere is likely to increase this depletion with current chlorine loading. However, chlorine loading can be expected to decline considerably in the future. Some of the episodes of low ozone observed in the Arctic are not associated with chemical depletion but are due to the influx of low-ozone air from lower latitudes (Taalas, 1993; Taalas *et al.*, 1995). Whether these episodes will increase or decrease will depend on stratospheric circulation patterns near the Arctic; thus, it also may be influenced by climate changes. The chemically induced and the dynamically induced episodes of low ozone occurring in the Arctic both appear to be increasing in frequency and severity (Taalas *et al.*, 1997; Weatherhead and Morseth, 1997). These depletion events are most prevalent in the spring, when biological activity is highly sensitive to UV-B radiation. Increased levels are likely to affect the human populations as well as the aquatic and terrestrial species and ecosystems (Taalas, 1993; SCOPE, 1996).

Sea ice is important in the development and sustenance of temperate-to-polar ecosystems. Ice conditions conducive to ice-edge primary production provide a primary food source in polar ecosystems. Ice-dependent activities of organisms ensure energy transfer from primary producers (algae and phytoplankton) to higher trophic levels (fish, marine birds, and mammals). As a consequence, the ice-dependent habitat maintains and supports abundant biological communities (IPCC 1996, WG II, Section 8.3.2).

One of the direct effects of global warming will be a change in the extent of sea ice—which will have the beneficial effect of increasing the extent of many aquatic species but may negatively impact one of the areas most important for primary production: the ice edge. For some species, this change would be beneficial because it would allow greater penetration of light and freer movement. The absence of ice over the continental shelf of the Arctic Ocean would produce a sharp rise in the productivity of this region, provided that a sufficient supply of nutrients is maintained. For other species, there appear to be significant disadvantages. For example, in southern latitudes, warming could affect pack-ice movement and timing and reduce its extent, in turn affecting the distribution and reducing the abundance and productivity of krill. Krill is the fundamental link in the food web of the Southern Ocean. Changes in wind strength, the effects of the Antarctic Convergence, and the west-east and east-west surface currents of the Southern Ocean also may affect productivity in the ice edge zone (IPCC 1996, WG II, Section 16.2.2.2).

In the polar regions, the seasonal development and regression of sea ice is of substantial biological importance (in addition to its profound role in global ocean-atmosphere exchange of heat, water, and gases). Blooms of phytoplankton at the ice edge, induced by upwelling of nutrient-rich water and/or the stabilizing effect of melting icewater, are significant at high latitudes in both hemispheres. Changes in the extent and thickness of ice affect the timing, magnitude, and duration of the seasonal pulse of primary production in polar regions. The composition of the marine ecosystem depends on the mode of sea-ice formation. These changes in turn affect the quantity and quality of food available to grazers and the timing and magnitude of vertical flux of carbon and other elements. Such series of events are particularly important in highly productive systems such as the Bering Sea, where a 5% reduction in sea ice has been observed over the past decade.

UV-B radiation may directly affect fish and their larvae, phytoplankton, and krill. The development and survival of some species are at apparent threshold levels at current UV-B levels. Significant changes in ecosystem function and functioning could occur as species sensitive to UV-B radiation are replaced by more resistant species (IPCC 1996, WG II, Section 16.2.7). Experimental results have shown damaging effects on aquatic organisms from polar regions, particularly on phytoplankton (photosystem damage, DNA damage, modifications in nutrient uptake, changes in species composition, etc.), but a series of internal and external protective mechanisms (synthesis of photoprotective pigments, DNA repair mechanisms, vertical mixing in the water column, etc.) may counteract these negative effects (Vincent and Roy, 1993). On the other hand, large-scale ice retreat due to global warming would increase the exposure to UV-B radiation in the water column during the maximum development of the ozone hole (September–October). However, the solar angle during the spring period is very low. This factor may minimize the penetration of UV-B radiation in the water column, thus reducing harmful effects on the biota (Arrigo, 1994; McMinn *et al.*, 1994). Thus, care should be taken in extrapolating experimental results to the real world. More *in situ* studies and modeling efforts are necessary to assess the effects of increased UV-B radiation during ozone hole episodes.

In the Arctic, mean levels of ozone have decreased. In addition, smaller, less well-understood ozone-depletion events, or mini-ozone holes, have occurred. Presently, Arctic ozone-depletion events are not sufficiently understood, and whether they will increase in severity also is not clear. However, there is some indication that ozone depletion in the Arctic is tied closely to global warming and the accompanying cooling of the Arctic stratosphere. Arctic ozone-depletion events are most prevalent in the spring when biological activity is highly sensitive to UV-B radiation. Increased levels are likely to affect human populations as well as aquatic and terrestrial species and ecosystems.

Warming, changes in upwellings, circulation variations, and wind changes would affect the distribution and characteristics of polynyas (ice-free areas) and ice edges that are vital to polar ecosystems. Predicted climate change also may have important impacts on marine mammals such as whales, dolphins, and seals and seabirds such as cormorants, penguins, storm petrels, and albatross. Geographic centers of food production will most certainly shift. Underlying primary productivity will change due to changes in upwelling, loss of ice-edge effects, and ocean temperatures. Changes in critical habitats such as sea ice (due to climate warming) and nesting and rearing beaches (due to sea-level rise) will occur. Diseases and production of oceanic biotoxins due to warming temperatures and shifts in coastal currents will increase (IPCC 1996, WG II, Section 8.3.2). However, it is presently impossible to predict the magnitude and significance of these impacts. Recent changes in the penguin population on the Antarctic Peninsula have been related to climate warming (Fraser *et al.*, 1992), but should also be interpreted in the context of penguin rookery variations that have occurred in the past (Baroni and Orombelli, 1994). Penguin are a good example for illustrating the vulnerability and complex responses of Antarctic ecosystems to climate, environmental, and ecosystem changes.

A number of marine mammals depend explicitly on ice cover. For example, the extent of the polar bear's habitat is determined by the maximum seasonal surface area of marine ice in a given year. The disappearance of or a major reduction in ice cover would threaten the very survival of the polar bear, as well as certain marine seals. Similarly, a reduction in ice cover would reduce food supplies for seals and walruses and increase their vulnerability to natural predators and human hunters and poachers. Other animals, such as the sea otter, could benefit by moving into new territories with reduced ice (IPCC 1996, WG II, Section 8.3.2).

Furthermore, many species of penguins and seals that are dependent upon krill production in the Southern Ocean use coastal areas as breeding sites. Year-class strength of krill is closely related to climate change or sea-ice distribution, and the growth, survival, and hatching rates of penguin chicks or seal pups at the rookery are directly influenced by krill abundance in the sea.

Animals that migrate great distances—as do most of the great whales and seabirds—are subject to possible disruptions in the distribution and timing of their food sources during migration. It remains unclear how the contraction of ice cover would affect the migration routes of animals that normally follow the ice front. The timing and order of whale migration, for instance, may be influenced by changes in sea-ice cover, with subsequent impacts on their energetics (e.g., feeding and reproductive biology). At least some migrating species may respond rapidly to new situations; for example, migrating ducks have altered their routes to take advantage of the recent exploding population of zebra mussels in the Great Lakes (IPCC 1996, WG II, Section 8.3.2).

Many pelagic marine mammals (e.g., the great whales) are able to locate and follow seasonal centers of food production that frequently change from year to year because of local oceanographic conditions. Similarly, their migrations may change to accommodate interannual differences in environmental conditions. However, some marine mammal stocks (e.g., seals and sea lions) have life histories that tie them to specific geographic features (e.g., pupping beaches or ice fields). Although they have some flexibility in their need for specific habitats, some marine mammals may be more severely affected than others by changes in the availability of necessary habitats and prey species that result from climate change.

With sea-level rise, marine mammal calving and pupping beaches may disappear from areas where there are no alternatives. Some species will be able to take advantage of increases in prey abundance and spatial/temporal shifts in prey distribution toward or within their primary habitats. Conversely, some populations of birds and seals will be adversely affected by climatic changes if food sources decline or are displaced away from regions suitable for breeding or rearing of young (IPCC 1996, WG II, Section 8.3.2).

Furthermore, recent evidence indicates that the extent of seasonal sea ice, at least in some areas of the Northern Hemisphere, is retreating. This information, coupled with predictions of warming from GCMs, suggests that current barriers to gene flow among marine mammal stocks in the Arctic may change dramatically in the next 50 years. Although this shift may not result in a reduction in abundance at the species level, it could very well change the underlying population structure of many species of Arctic whales and seals, which will greatly affect their management.

Although the impacts of these ecological changes are likely to be significant, they cannot be reliably forecast or evaluated. Climate change may have positive and negative impacts, even on the same species. Positive effects such as extended feeding areas and seasons in higher latitudes, more-productive high latitudes, and lower winter mortality may be offset by negative factors that alter established reproductive patterns, breeding habitats, disease vectors, migration routes, and ecosystem relationships (IPCC 1996, WG II, Section 8.3.2).

### *3.3.5. Fisheries*

The basic limiting factors for fish production in polar and subpolar regions are light and temperature. Warming in high latitudes should lead to longer growing periods, increased growth rates, and ultimately, perhaps, increases in the general productivity of these regions (IPCC 1996, WG II, Section 16.2.2.2). On the other hand, the probability of nutrient loss resulting from reduced deep-water exchange could result in reduced productivity in the long term. Again, this complexity highlights the importance of changes in temperature for patterns of circulation. Global warming could have especially strong impacts on the regions of oceanic subpolar fronts, where temperature increases in deep water could lead to a substantial redistribution of pelagic and benthic communities, including commercially important fish species (IPCC 1996, WG II, Section 8.3.2).

In polar regions, the number of dominant fish species is small; many species of low abundance are typical of tropical regions, with the exception of upwelling areas. Only 15–20 commercially important species in the Arctic or Antarctic Oceans are recorded, whereas the numbers increase to about 50 and 16–450 in the boreal and tropic areas, respectively (Laevastu, *et al.*, 1996). The poleward distribution of fish due to climate warming generally expands fishing areas. This expansion might produce better yields of fish production. In the higher latitudes, however, spawning grounds of cold-water species that are very sensitive to the temperature change might be destroyed by changes in water properties.

In some cases, fisheries on the margin of profitability could prosper or decline. For example, if there is a retreat of sea ice in Antarctica, the krill fishery—which is regulated by the current ice-free period—could become more attractive to nations not already involved (IPCC 1996, WG II, Section 16.2.2.3). Fishery statistics may be more valuable for the analyses of interannual and long-term fluctuations of marine populations than was previously thought. Time series of catch-per-unit-effort (CPUE) statistics from the commercial krill fishery operating around South Georgia during 1973–1993 have been used for considering the hypothesis that fluctuations in the abundance of krill in the Scotia Sea area are related to environmental changes. A consistent correlation has been found between the various CPUE indices and ice-edge positions: The further south the ice-edge occurred during the winter-spring season, the lower the CPUE values in the following fishing season. The most extreme expression of this relationship was the lack of a krill fishery in 1978 and 1984, when the ice did not extend far north during the previous winter. By contrast, in 1978 and 1984 the March ice-edge reached its northern limit at 50°S, preceding high CPUE values in 1979 and 1985. A consistent relationship also exists between CPUE and water temperature. Warm-water temperature in the South Georgia shelf area in January–February corresponded to lower CPUE values in the same year. There also is significant correlation between air transport in late spring and CPUE in the next year. For example, a prevalence of southerly meridional air transport precedes high CPUE values (Fedoulov *et al.,* 1996). It must be emphasized, however, that the physical regimes of the sub-Antarctic region in the vicinity of South Georgia are very complicated, and this model may not be applicable to the entire Antarctic.

Fedoulov *et al.* (1996) proposed the following mechanism as a hypothesis to explain how ice, ocean, and atmospheric components of the Southern Ocean affect krill distribution. Krill usually are more abundant in the southern Scotia Sea along the Weddell Scotia Confluence (WSC), so it is likely that the currents play a key role in krill transport to South Georgia from the Antarctic Peninsula. The WSC zone extends northward in the eastern Scotia Sea, and this colder water penetrates along the southeastern shelf of South Georgia. The position of the WSC is thought to be determined by the intensity of the Weddell gyre, which in turn is driven by the formation of dense and cold Weddell water. The main factor in the creation of the cold Weddell water is increased salinity resulting from ice formation. Hence, the dominance of a warm or cold year reflects the intensity of the Weddell gyre and consequently the general position of the WSC. It is reasonable to suppose that ice can start to influence krill distribution when it is close to or covers the area of the WSC. Ice cover modifies the mechanism of drift current formation and creates favorable (northern ice-edge position) or unfavorable (southern ice-edge position) conditions for krill transport to South Georgia.

In a recent study, Loeb *et al.* (1997) are documenting a more complex relationship between krill and salpa, a pelagic tunicate. In essence, extensive seasonal ice cover promotes early krill spawning, inhibits population blooms of pelagic salps, and favors the survival of krill larvae through their first winter. Salpa blooms affect adult krill reproduction and the survival of krill larvae. If a decrease in the frequency of winters with extensive sea-ice development accompanies the warming trend in the Antarctic Peninsula area, the frequency of krill recruitment failures would be expected to increase, and the krill population would decline. An increase in salpa blooms would further depress krill numbers. This codependency of competing species on changing climate variables has implications for the management of the krill fishery and for populations of vertebrate predators such as penguins, fish, and whales, which depend on krill.

### *3.3.6. Agriculture*

Agriculture in polar lands is severely limited by the harsh climate. Although agriculture is not practiced in Antarctica, some agriculture currently takes place in the Arctic. If temperature were to increase and result in earlier last-freeze dates and later first-frost dates, conditions for Arctic agriculture should become more favorable, although climate conditions will still make agriculture in the Arctic extremely difficult. There already are indications in Alaska, based on the last 70 years, for an increase in the length of the growing season (Sharratt, 1992). Although considerably more land will be available for farming if temperatures increase, moisture and nutrient problems will limit the productivity of these areas (Mills, 1994). The more immediate effects are likely to be on plants such as cotton grass that are important to caribou and reindeer populations (Kuropat, 1984).

### *3.3.7. Navigation*

Reduced sea-ice extent and thickness would increase the seasonal duration of polar navigation on rivers and in coastal areas that are presently affected by seasonal ice cover (IPCC 1996, WG II, Section 7.5). Improved opportunities for water transport, tourism, and trade at high latitudes are expected as a result. These activities will have important implications for the people, economies, and navies of nations along the Arctic rim (IPCC 1996, WG II, Chapter 7 Executive Summary). Reduced sea ice will provide safer approaches for tourist ships and new opportunities for sightseeing around Antarctica and the Arctic (IPCC 1996, WG II, Section 7.5.5). Increased calving of icebergs from the Antarctic Peninsula may, however, affect navigation and shipping lanes north of the Antarctic Convergence. Decreased sea-ice extent around Antarctica could make it easier for tourist vessels with less preparedness for sea-ice travel to visit the continent and surrounding islands. Some may be ill-prepared to navigate and respond to the extreme and highly variable environmental conditions in the Southern Ocean. There is no clear consensus, however, about whether the frequency of icebergs, and their danger to shipping, will change with global warming (IPCC 1996, WG II, Section 7.4). Increased precipitation may reduce the enthusiasm for tourism in some areas.

Projected reductions in the extent and thickness of the sea-ice cover in the Arctic Ocean and its peripheral seas could substantially benefit shipping, perhaps opening the Arctic Ocean as a major trade route (IPCC 1996, WG II, Technical Summary, Section 3.2.4). This projection would include the opening of both the Northwest Passage and the Russian Northern Sea Route for up to 100 days a year. One French experiment indicated that the use of the Northeast Passage in ice-free seasons shortened by about 3 weeks the shipping duration between Europe and Far East Asia compared with the present route (i.e., via the Suez Canal). Although a reduction of sea ice may be a boon to international shipping and consumers in East Asia, North America, and Western Europe, policies designed to limit the total burden of pollutants entering the Arctic environment from ports, ship operations, and accidents may have to be developed (IPCC 1996, WG II, Chapter 8 Executive Summary).

Less river ice and a shorter ice season in northward flowing rivers of Canada and Russia should enhance north-south river transport. Combined with less sea ice in the Arctic, this development would provide new opportunities for reorganization of transport networks and trade links. Ultimately, those changes could affect Northern Hemisphere trading patterns (IPCC 1996, WG II, Section 7.5.1).

A survey of the potential impacts on Canadian shipping suggested net benefits to Arctic and ocean shipping due to deeper drafts in ports and longer navigational seasons, with mixed results for lake and river shipping due to the opposing effects of a longer shipping season but lower drafts. Demands to maintain Arctic shipping may increase. In Siberia and Canada, many rivers are used as solid roads during winter. Warmer winters would require a shift to water transport or the construction of more all-weather roads. Other impacts on means of transport could arise from changes in snowfall or melting of the permafrost (IPCC 1996, WG II, Section 11.5.2.1).

Currently, ice-breaking efforts are an expensive aspect of navigation in the Arctic. Interannual variability prevents the elimination of these programs unless extreme changes in sea ice should occur. Some ice-breaking programs in some areas may be cut back with moderate warming of the Arctic. In other areas, costs may rise to keep newly available routes open longer. A disappearance of sea ice south of Labrador would eliminate Canadian Coast Guard ice-breaking requirements. This would mean an annual saving of CDN$15–20 million. Even larger savings also can be expected in the former Soviet Union if ice retreats from the shores of the Kara, Laptev, and Chukchi Seas. Similar savings would accrue along the Gulf of Bothnia with the absence of ice. The effect of annual warming on ice calving (simulated using a simple degree-day model) shows that for every 1°C of warming there would be a 1° latitude retreat of iceberg occurrence in the Atlantic Ocean. In the Southern Ocean, any effects of reduced sea ice will be economically less pronounced (IPCC 1996, WG II, Section 7.5.3).

Offshore oil and gas exploration and production conducted at high latitudes may be assisted by a longer ice-free season. A possible beneficial effect would be shorter winters to disrupt construction, exploration, and drilling programs. A decrease in thickness and extent of sea ice in the Arctic will extend the drilling seasons for floating vessels considerably. A reduction in sea ice and icebergs also will reduce "downtime" on offshore oil and gas drilling explorations. Currently such interruptions cost Canadian explorations more than CDN$40 million annually. The most critical factor could be ice movement during winter. If there were increased numbers and severity of storm surges and wave activity, design requirements for offshore structures and associated coastal facilities would increase, and oil spill clean-up could become more difficult (IPCC 1996, WG II, Section 7.5.4).

### 3.3.8. Sub-Antarctic Islands

The sub-Antarctic islands are small land areas surrounded by vast areas of the Southern Ocean. Some of these islands are sufficiently high to possess glaciers and small ice caps. Their climate is characterized by strong and persistent winds, little sunshine, many rain days, and cool temperatures. They are generally uninhabited, though they contain highly specialized flora and important marine mammal and bird populations.

Many of these islands have shown a tendency for warming over the last half of this century. Their future climate will be controlled by changes in the surface temperature of the Southern Ocean and the strength of Southern Hemisphere westerlies. The sub-Antarctic islands are expected to continue to warm. The impacts of climate change are unlikely to be important for most animal and bird species, but there could be changes in the species composition of plant communities. Glaciers will probably shrink.

### 3.3.9. Infrastructure, Settlements, and Health

The climate dictates many aspects of the way of life for all people living in the Arctic. Climate changes are likely to affect the current subsistence economies, habitability, and health of the Arctic people. If warming occurs, there will be striking changes in the landscapes at northern high latitudes. These changes may be exacerbated where they are accompanied by growing numbers of people and increased economic activity (IPCC 1996, WG II, Chapter 7 Executive Summary). There could be a northward spreading of agricultural, forestry, and mining activities—resulting in increased population and intensified settlement patterns—into Arctic areas. The rate of such spread is an issue of debate and will depend on social and economic factors, intervention strategies, and natural processes. Marine, road, rail, and air links would have to be expanded accordingly. Although this development would entail substantial extra capital and operating costs, it also would be an economic opportunity (IPCC 1996, WG II, Section 11.4.4).

Changes in landscape, sea-ice distribution, and river and lake ice could have a major impact on indigenous people who live in Arctic regions and depend upon traditional occupations, food gathering, and hunting. They depend directly on the living resources of the area and often travel on ice, so their livelihood may be widely affected. The resulting redistribution and abundance changes of terrestrial and marine animals that are vital to the subsistence lifestyles of Native communities may have major impacts. Ice roads and crossings commonly are used to link northern settlements. The greatest economic impact is likely to stem from decreases in ice thickness and bearing capacity, which could severely restrict the size and load limit of vehicular traffic (IPCC 1996, WG II, Section 7.5). Road maintenance related to permafrost thawing already is a major problem with high associated costs in many sub-Arctic areas. At the same time, further warming is likely to cause many small settlements, particularly along the coast and on small islands, to be abandoned because of permafrost loss and sea-level rise.

Anticipated hydrological changes and reductions in the areal extent and depth of permafrost could lead to large-scale damage to infrastructure. Some transportation systems, mining activities, and structures will be threatened by thawing (IPCC 1996, WG II, Section 7.5.1). For example, thawing could lead to disruption of existing petroleum production and distribution systems in the tundra unless mitigation techniques are adopted (IPCC 1996, WG II, Chapter 7 Executive Summary). This impact may require changes in the design of oil pipelines to avoid slumping, breaks, and leaks (IPCC 1996, WG II, Section 11.5.4). Fortunately, most of this technology already is in use south of the permafrost line.

Structures such as pipelines, airstrips, community water supply and sewage systems, and building foundations are susceptible to performance problems if existing frozen foundations or subgrades thaw, even minimally. Extensive measures would be needed to ensure the structural stability and durability of installations for tourism, mining, and telecommunication in permafrost areas affected by climate warming. In some cases, existing settlements would become uninhabitable because of permafrost changes. Transport links also could be affected, with serious disruption and increased maintenance costs from ground subsidence, side-slope slumpings, landslides, icings, and ice-mound growth.

On the other hand, many northern cities will spend less money on snow and ice clearance. More frequent periods of open water for rivers, lakes, and seas, however, will produce greater snowfall downwind. This factor will be important near Hudson Bay, the Barents Sea, and the Sea of Okhotsk (IPCC 1996, WG II, 7.4.1). Engineering design criteria will need to be modified to reflect changing snow and frost climates, deepening of the active layer over permafrost, and warming and ultimate disappearance of marginal or discontinuous permafrost. Present permafrost engineering commonly designs for the warmest year in the past 20 years of record; such criteria may need to be reviewed and revised (IPCC 1996, WG II, Section 7.5.1).

Sea-level rise will affect a few outlying Arctic communities, many of which are on the shoreline, as well as major coastal industrial facilities such as the Prudhoe Bay oilfields in Alaska. Sea-level rise also will allow ice-thrust events to be more damaging to the shoreline. Coastal permafrost will be thermally eroded, which will produce local slumping and coastal retreat.

Episodes of extreme cold and blizzards are major climate concerns for circumpolar countries like Russia and Canada (IPCC 1996, WG II, Section 12.4.2.5). However, the Polar regions will remain cold, so the direct effects of global warming are likely to have little effect on human health. Potential indirect effects, such as changes in infectious diseases and vector organisms, are largely unknown. UV-B radiation is increasing, which can damage the genetic (DNA) material of living cells (in an inverse relationship to organism complexity) and induce skin cancers in experimental animals. It also may affect human health: UV-B radiation is implicated in the causation of human skin cancer and lesions of the conjunctiva, cornea, and lens; it also may impair the body's immune system (IPCC 1996, WG II, Section 18.4).

## 3.4. Integrated Analysis of Potential Vulnerabilities and Impacts and Future Research Needs

Individual sectors and resources will not be affected in isolation but in interaction with one another. The direct effects of climate change include shifts in geographic distribution of ecosystems, changes in sea and river ice, thaw of permafrost, and changes in coastal areas. Indirect effects include feedbacks to the climate system through less ice and snow reflectivity, changes in sea ice and ocean circulation, and ecological sources and sinks of $CO_2$ and $CH_4$. Improvements are needed in modeling of future sea-ice distribution and thickness because of their importance as drivers for other physical and ecological systems, economics, and transportation. There is uncertainty about whether changes in tundra will cause it to act as a source or a sink for $CO_2$. This issue needs to be resolved.

There is a variety of positive and negative shifts in opportunities for shipping, the oil industry, fishing, mining, and tourism, as well as coastal infrastructure and the movement of populations. These impacts will lead to further interactions and potential changes in trade and strategic balance. The health consequences of these shifts will depend on human capacity to adapt.

The Arctic is more vulnerable than the Antarctic because of its sensitive and fragile ecosystems and the impacts on traditional lifestyles of indigenous peoples and because climate changes are expected to be greater. An integrated impact assessment using multiple stress models is required for the Arctic. When associated with baseline monitoring, it should be possible to distinguish natural variability from human impacts.

Further analysis should focus on particularly sensitive zones and activities, vulnerable species, marginal communities, and estimates of economic impacts. To improve biological components of regional and global models, it is important to understand how, when, and where productivity in the Southern Ocean will change with global warming (IPCC 1996, WG II, Section 8.3.2). More certainty is required regarding the future behavior of the marine-based West Antarctic Ice Sheet and ice balance of the continent. Changes have the potential to substantially alter sea level and southern hemispheric climate, but the time frame needs to be defined.

Most human and natural systems in the Arctic and the Antarctic Peninsula are extremely sensitive to temperature. Future global warming is expected to be greatest in these areas. Large reductions in sea ice, permafrost, and tundra will disrupt many natural systems and change species composition over land and in the polar oceans. These areas appear to be vulnerable to climate change, although the number of people directly affected would be relatively small. However, most are indigenous people, who lead traditional lifestyles and have few adaptive strategies that can be implemented. Antarctica is less vulnerable because the temperature changes envisioned over the next century are likely to have little impact, and few people are involved.

## References


**Arrigo**, K.R., 1994: Impact of ozone depletion on phytoplankton growth in the Southern Ocean: large-scale spatial and temporal variability. *Mar. Ecol. Prog. Ser.*, **114**, 1–12.

**Baroni**, C. and G. Orombelli, 1994: Abandoned penguin rookeries as Holocene paleoclimatic indicators in Antarctica. *Geology*, **22**, 23–26.

**Berkman**, P.A., 1994: Epizoic zonation on growing scallop shells in McMurdo Sound, Antarctica. *J. Exper. Marine Biology and Ecology*, **179**, 49–67.

**Brown**, R.D. and P. Coté, 1992: Interannual variability of landfast ice thickness in the Canadian High Arctic, 1950–1989. *Arctic*, **45**, 273–284.

**Chapman**, W.L. and J.E. Walsh, 1993: Recent variations of sea ice and air temperature in high latitudes. *Bull. Am. Met. Soc.*, **74**, 33–47.

**Cuffey**, K.M., G.D. Clow, R.B. Alley, M. Stuiver, E.D. Waddington, and R.W. Saltus, 1995: Large Arctic temperature changes in the Wisconsin-Holocene glacial transition. *Science*, **270**, 455–458.

**Fedoulov**, P.P., E. Murphy, and K.E. Shulgovsky, 1996: Environment-krill relations in the South Georgia marine ecosystem. *CCAMLR Science*, **3**, 13–30.

**Flato**, G.F. and R.D. Brown, 1996: Variability and climate sensitivity of landfast sea ice. *J. Geophysical Research*, **101(C10)25**, 767–25, 777.

**Fraser**, W.R., W.Z. Trivelpiece, D.G. Ainley, and S.G. Trivelpiece, 1992: Increases in Antarctic penguin populations: reduced competition with whales or a loss of sea ice due to environmental warming. *Polar Biology*, **11**, 525–531.

**Gordon**, H.B. and S.P. O'Farrell, 1997: Transient climate change in the CSIRO coupled model with dynamic sea ice. *Monthly Weather Review*, **25(5)**, 875–907.

**Gunn**, A., 1995: Responses of arctic ungulates to climate change In: *Human ecology and climate change*. [Peterson, D.L. and D. R. Johnson (eds.)]. Taylor and Francis, Washington, DC, USA, pp. 90–104.

**Hoffman**, J.A.J., S.E. Nuez, and W.M. Vargas, 1997: Temperature, humidity, and precipitation variations in Argentina and the adjacent sub-Antarctic region during the present century. *Meteorol. Zeitschrift*, **N.F. 6**, 3–11.

**IAATO**, 1997: *Overview of Antarctic tourism activities: A summary of 1996–1998 and five year projection 1997– 2002*. XXI ATCM/IP75, May 1997, Christchurch, New Zealand.

**IPCC**, 1996. *Climate Change 1995: The Science of Climate Change. Contribution of Working Group I to the Second Assessment Report for the Intergovernmental Panel on Climate Change* [Houghton, J.J., L.G. Meiro

Filho, B.A. Callander, N. Harris, A. Kattenberg, and K. Maskell (eds.)]. Cambridge University Press, Cambridge, United Kingdom and New York, NY, USA, 572 pp.

- Nicholls, N., G.V. Gruza, J. Jouzel, T.R. Karl, L.A. Ogallo, and D.E. Parker, Chapter 3. *Observed Climate Variability and Change,* pp. 133–192.
- Kattenberg, A., F. Giorgi, H. Grassl, G.A. Meehl, J.F.B. Mitchell, R.J. Stouffer, T. Tokioka, A.J. Weaver, and T.M.L. Wigley, Chapter 6. *Climate Models—Projections of Future Climate,* pp. 289–357.
- Warrick, R.A., C. Le Provost, M.F. Meier, J. Oerlemans, and P.L. Woodworth, Chapter 7. *Changes in Sea Level,* pp. 259–406.
- Melillo, J.M., I.C. Prentice, G.D. Farquhar, E.-D. Schulze, and O.E. Sala, Chapter 9. *Terrestrial Biotic Responses to Environmental Change and Feedbacks to Climate,* pp. 445–482.

**IPCC**, 1996. *Climate Change 1995: Impacts, Adaptations, and Mitigation of Climate Change: Scientific-Technical Analyses. Contribution of Working Group II to the Second Assessment Report for the Intergovernmental Panel on Climate Change* [Watson, R.T., M.C. Zinyowera, and R.H. Moss (eds.)]. Cambridge University Press, Cambridge, United Kingdom and New York, NY, USA, 880 pp.

- Kirschbaum, M.U.F. and A. Fischlin, Chapter 1. *Climate Change Impacts on Forests,* pp. 95–130.
- Allen-Diaz, B., Chapter 2. *Rangelands in a Changing Climate: Impacts, Adaptations, and Mitigation,* pp. 131–158.
- Noble, I.R. and H. Gitay. Chapter 3. *Deserts in a Changing Climate: Impacts,* pp. 159–170.
- Beniston, M. and D.G. Fox, Chapter 5. *Impacts of Climate Change on Mountain Regions,* pp. 191–214.
- Öquist, M.G. and B.H. Svensson, Chapter 6. *Non-Tidal Wetlands,* pp. 215–240.
- Fitzharris, B.B., Chapter 7. *The Cryosphere: Changes and Their Impacts,* pp. 241–266.
- Ittekkot, V., Chapter 8. *Oceans*, pp. 267–288.
- Arnell, N., B. Bates, H. Lang, J.J. Magnuson, and P. Mulholland, Chapter 10. *Hydrology and Freshwater Ecology,* pp. 325–364.
- Moreno, R.A. and J. Skea, Chapter 11. *Industry, Energy, and Transportation: Impacts and Adaptation,* pp. 365–398.
- Scott, M. J., Chapter 12. *Human Settlements in a Changing Climate: Impacts and Adaptation,* pp. 399–426.
- Everett, J.T., Chapter 16. *Fisheries,* pp. 511–538.

**Jacobs**, S.S. and J.C. Comiso, 1993: A recent sea-ice retreat west of the Antarctic Peninsula. *Geophysical Research Letters*, **20(12)**, 1171–74.

**Jefferies**, R.L., 1992: Tundra grazing systems and climatic change. In: *Arctic ecosystems in a changing climate an ecophysiological perspective* [Chapin, F.S., R.L. Jefferies, J.F. Reynolds, G.R. Shaver, J. Svoboda, and E.W. Chu (eds.)]. Academic Press Inc., New York, NY, USA, pp. 391–412.

**Johnsen**, S.J., D. Dahl-Jensen, W. Dansgaard, and N. Gundestrup, 1995: Greenland palaeotemperatures derived from GRIP bore hole temperature and ice core isotope profiles. *Tellus*, **47B**, 624–629.

**Jones**, P.D., S.C.B. Raper, and T.M.L. Wigley, 1986: Southern Hemisphere surface air temperature variations: 1851–1984. *Journal of Climate and Applied Meteorology*, **25**, 1213–1230.

**Kim**, S., 1995: Climate change and fluctuation of fishery resources in the North Pacific (written in Korean with English abstract). *Ocean Policy Res.*, **10**, 107–142.

**King**, J.C., 1994: Recent climate variability in the vicinity of the Antarctic Peninsula. *Int. J. Climatol.*, **14**, 357–369.

**Koerner**, F. and L. Lundgaard, 1995: Glaciers and global warming. *Geog. Phys. Quat.*, **49(3)**, 429–434.

**Kuropat**, P., 1984: Foraging behavior of caribou on the calving ground in northwestern Alaska. M.S. Thesis, University of Alaska, Fairbanks, 95 pp.

**Laevastu**, T., D.L. Alverson, and R.J. Marasco, 1996: *Exploitable marine ecosystems: their behavior & management*. Fishing News Books, 321 pp.

**Loeb**, V. V. Siegel, O. Holm-Hansen, R. Hewitt, W. Fraser, W. Trivelpiece, and S. Trivelpiece, 1997: Effects of sea-ice extent and krill or salp dominance on the Antarctic food web. *Nature*, **387**, 897–900.

**Maslin**, M. and C. Tzedadakis, 1996: Sultry last interglacial gets sudden chill. *EOS Transactions*, **77**, 353–354.

**McGillivray**, D.G., T.A. Agnew, G.A. McKay, G.R. Pilkington, and M.C. Hill, 1993: Impacts of climatic change on the Beaufort sea ice regime: implications for the Arctic petroleum industry. *Climate Change Digest*, **93-01**, pp. 1–17.

**McGuire**, A.D., J.M. Melillo, D.W. Kicklighter, and L.A. Joyce, 1995: Equilibrium responses of soil carbon to climate change - empirical and process-based estimates. *Journal of Biogeography*, **22**, 785–796.

**McMinn**, A., H. Heijnis, and D. Hogson, 1994: Minimal effects of UVB radiation in Antarctic diatoms over the past 20 years. *Nature*, **370**, 547–549.

**Mills**, P.F., 1994: The agricultural potential of Northwestern Canada and Alaska and the impact of climatic change. *Arctic*, **47(2)**, 115–123.

**Myneni**, R.B., C.D. Keeling, C.J. Tucker, G. Asrar, and R.R. Nemani, 1997: Increased plant growth in the northern high latitudes from 1981 to 1991. *Nature*, **386**, 698–702.

**Nicholls**, K.W., 1997: Predicted reduction in basal melt rates of an Antarctic ice shelf in a warmer climate. *Nature*, **388**, 460–461.

**Oechel**, W.C., S.J. Hastings, G. Vourlitis, M. Jenkins, G. Riechers, and N. Grulke, 1993: Recent change of Arctic tundra ecosystems from a net carbon dioxide sink to a source. *Nature*, **361(6412)**, 520–523.

**Ramsden**, D. and G. Fleming, 1995: Use of a coupled ice-ocean model to investigate the sensitivity of the Arctic ice cover to doubling atmospheric $CO_2$. *J. Geophys. Res.*, **100(C4)**, 6817–6828.

**Raper**, S.C.B., T.M.L. Wigley, P.R. Mayes, P.D. Jones, and M.J. Salinger, 1984: Variations in Surface Air Temperature Part 3: The Antarctic, 1957–82. *Monthly Weather Review*, **112**, 1342–1353.

**Rott**, H., P. Skvarca, and T. Nagler, 1996: Rapid collapse of northern Larsen Ice Shelf, Antarctica. *Science*, **271**, 788–792.

**Schimel**, D.S., B.H. Braswell, E.A. Holland, R. McKeown, D.S. Ojima, T.H. Painter, W.J. Parton, and A.R. Townsend, 1994: Climatic, edaphic, and biotic controls over storage and turnover of carbon in soils. *Global Biogeochemical Cycle*, **8**, 785–796.

**SCOPE**, 1996: *Effects of Increased Ultraviolet Radiation in the Arctic.* IASC report #2. Oslo, Norway, 58 pp.

**Sharratt**, B.S., 1992: Growing season trends in the Alaskan Climate Record. *Arctic*, **45(2)**, 124–127.

**Skvarca**, P., H. Rott, and T. Nagler, 1995: Satellite imagery, a baseline for glacier variation study on James Ross Island, Antarctica. *Ann. Glaciol.*, **21**, 291–296.

**Sugden**, D., 1982: *Arctic and Antarctic: a modern geographical synthesis*. Basil Blackwell, Oxford, United Kingdom and New York, NY, USA, 472 pp.

**Taalas**, P., 1993: Factors affecting the behavior of tropospheric and atmospheric ozone in the European Arctic and in Antarctica. *Report No. 10. Finnish Meteorological Institute Contributions*, Finnish Meteorological Institute, Helsinki, Finland, 138 pp.

**Taalas**, P., J. Damski, A. Korpela, T. Koskela, E. Kyro, and G. Braathen, 1995: Connections between atmospheric ozone, the climate system and ultraviolet-B radiation in the Arctic. In: *Ozone as a climate gas,* vol. 32 [Wang, W.C. and I.S.A. Isaksen (eds.)]. NATO ASI Series. Springer-Verlag, Heidelberg, Germany, pp. 411–426.

**Taalas**, P., J. Damski, E. Kyro, M. Ginzburg, and G. Talamoni, 1997: The effect of stratospheric ozone variations on UV radiation and on tropospheric ozone at high latitudes. *J. Geophys. Res.* (in press).

**UNEP**, 1997: *Global Environment Outlook: An Overview.* ISBN 92-807-1638-7, UNEP in collaboration with the Stockholm Environment Institute. Oxford University Press, New York and Oxford, 264 pp.

**Vaughan**, D.G. and C.S.M. Doake, 1996: Recent atmospheric warming and retreat of ice shelves on the Antarctic Peninsula. *Nature*, **379**, 328–331.

**Vincent**, W.F., and S. Roy, 1993: Solar ultraviolet-B radiation and aquatic primary productivity: damage protection and recovery. *Environ. Rev.*, **1**, 1–12.

**Wadhams**, P., 1990: Sea ice and economic development in the Arctic Ocean: a glaciologist's experience. In: *Arctic Technology and Economy. Present Situation and Problems, Future Issues*. Bureau Veritas, Paris, France, pp. 1–23.

**Walton**, D.W.H. and W.N. Bonner, 1985: Terrestrial habitats — Introduction. In: *Antarctica — Key Environments* [Bonner, W.N. and D.W.H. Walton (eds.)]. Pergamon Press, Oxford and New York, 381 pp.

**Weatherhead**, E.C. and C.M. Morseth, 1997: Climate change, ozone and UV radiation. In: *Arctic Monitoring and Assessment Program's Assessment Report*. Oslo, Norway, (in press).

# 4

# Australasia

REID E. BASHER (NEW ZEALAND) AND A. BARRIE PITTOCK (AUSTRALIA)

Lead Authors:

*B. Bates (Australia), T. Done (Australia), R.M. Gifford (Australia), S.M. Howden (Australia), R. Sutherst (Australia), R. Warrick (New Zealand), P. Whetton (Australia), D. Whitehead (New Zealand), J.E. Williams (Australia), A. Woodward (New Zealand)*

# CONTENTS

# EXECUTIVE SUMMARY

This region is defined here as Australia, New Zealand, and their outlying islands. Australia is a large, flat continent spanning the tropics to mid-latitudes, with relatively nutrient-poor soils and a very arid interior, whereas New Zealand is much smaller, mountainous, and well watered. Both have "Gondwanan" ecosystems and unique flora and fauna. They have been subject to significant human influences—particularly from fire, agriculture, deforestation, and introduced exotic plants and animals. The total land area is 8 million square kilometers, and the population is 22 million. In contrast to other Organisation for Economic Cooperation and Development (OECD) countries, commodity-based industries of agriculture and mining dominate the economies and exports. Tourism is a major and rapidly growing industry.

*Climate:* Australasia's climate is strongly influenced by the surrounding oceans. Key climatic features include tropical cyclones and monsoons in northern Australia; migratory mid-latitude storm systems in the south, including New Zealand; and the El Niño-Southern Oscillation (ENSO) phenomenon, which causes floods and prolonged droughts, especially in eastern Australia. Unfortunately, climate models at present cannot provide reliable predictions for these features under climate change.

*Climate trends:* The region's climatic trends are consistent with those of other parts of the world. Mean temperatures have risen by up to 0.1°C per decade over the past century; nighttime temperatures have risen faster than daytime temperatures; and the past decade has seen the highest mean annual temperatures ever recorded. Increases in the frequency of heavy rainfalls and average rainfall have been reported for large areas of Australia. Sea levels have risen on average by about 20 mm per decade over the past 50 years.

*Climate scenarios:* Climate modeling and climate change scenarios for the region are relatively well developed. For Australia, the recently revised (1996) Commonwealth Scientific and Industrial Research Organization (CSIRO) scenarios for 2030 indicate temperature increases of 0.3–1.4°C and rainfall changes of up to 10% in magnitude (decreases in winter, increases or decreases in summer, and overall a tendency for decreases). The projected changes for 2070 are about twice those of the 2030 changes. Increases in the intensity of heavy rainfall events are indicated. For New Zealand, the temperature increases are expected to be similar to those for Australia, but the recent revision indicates the possibility of an increase in westerly winds (unlike the decreases of previous scenarios)—and hence precipitation increases in the west and precipitation decreases in the east. The changes in scenarios serve to caution against overinterpretation of impact studies based on any single scenario.

*Ecosystems:* In responding to climate change, Australasia's biota may face a greater rate of long-term change than ever before. They also must respond in a highly altered landscape fragmented by urban and agricultural development. There is ample evidence for significant potential impacts. Alterations in soil characteristics, water and nutrient cycling, plant productivity, species interactions (competition, predation, parasitism, etc.), and composition and function of ecosystems are highly likely responses to increases in atmospheric carbon dioxide ($CO_2$) concentration and temperature and to shifts in rainfall regimes. These changes would be exacerbated by any increases in fire occurrence and insect outbreaks.

Aquatic systems will be affected by the disproportionately large responses in runoff, riverflow and associated nutrients, wastes, and sediments that are likely from changes in rainfall and rainfall intensity and by sea-level rise in estuaries, mangroves, and other low-lying coastal areas. Australia's Great Barrier Reef and other coral reefs are vulnerable to temperature-induced bleaching and death of corals, in addition to sea-level rise and weather changes. However, there is evidence that the growth of coral reef biota may be sufficient to adapt to sea-level rise. Our knowledge of climate change impacts on aquatic and marine ecosystems is relatively limited.

Prediction of climate change effects is very difficult because of the complexity of ecosystem dynamics. Although Australasia's biota and ecosystems are adapted to the region's high climate variability (exemplified in arid and ENSO-affected areas), it is unclear whether this will provide any natural adaptation advantage. Many species will be able to adapt through altered ecosystem relationships or migration, but such possibilities may not exist in some cases, and reduction of species diversity is highly likely. Climate change will add to existing problems such as land degradation, weed infestations, and pest animals and generally will increase the difficulties and uncertainty involved in managing these problems.

The primary human adaptation option is land-use management—for example, by modification of animal stocking rates in rangelands, control of pests and weeds, changed forestry practices, and plantings along waterways. Research, monitoring, and prediction, both climatic and ecological, will be necessary foundations to human adaptive responses. Active manipulation of species generally will not be feasible in the region's extensive natural or lightly managed ecosystems, except for rare and endangered species or commercially valuable species. In summary, it must be concluded that some of the region's ecosystems are very vulnerable to climate change.

*Hydrology and water resources:* The four hydrological situations of most concern to the region are drought-prone areas, flood-prone urban areas, low-lying islands, and alpine snowfields. Model simulations suggest changes of as much as ±20% in soil moisture and runoff in Australia by 2030, with considerable variation from place to place and season to season and with the possibility of an overall reduction in average runoff. Water shortages would sharpen competition among various uses of water, especially where large diversions are made for economic purposes. One study shows Australia's major Murray-Darling River system facing constraints on existing irrigation uses and/or harm to the riverine environment. More frequent high-rainfall events would enhance groundwater recharge and dam-filling events but also would increase the impacts of flooding, landslides, and erosion. A preliminary study for an urban area near Sydney showed a tenfold increase in the potential damage of the "100-year" flood under a doubled $CO_2$ scenario. Water supplies on atolls and low-lying islands are vulnerable to saltwater intrusion of groundwater from rising sea levels and to possible rainfall reductions. Reduced snow amounts and a shorter snow season appear likely and would decrease the amenity value of the mountains and the viability of the ski industry. The glaciers of New Zealand's Southern Alps are likely to shrink further.

Adaptation options include integrated catchment management, changes to water pricing systems, water efficiency initiatives, building or rebuilding engineering structures, relocation of buildings, urban planning and management, and improved water supply measures in remote areas and low-lying islands. The financial exposure, and cost involved in potential adaptations, indicate a high vulnerability with respect to hydrology.

*Agriculture:* Significant impacts on agriculture are likely, including crop and pasture performance increases from $CO_2$ rises; mixed effects of temperature rises; changes in soil fertility; changes in quality of grain and pasture nutrition; shifts in the suitability of districts for particular crops, such as kiwifruit; and possibly increased problems with weeds, pests, and diseases.

Impacts will vary widely from district to district, crop to crop, and decade to decade. Grain crops may gain in the first few decades because of the immediate beneficial effect of higher $CO_2$, but that advantage may be eroded as the delayed rise in temperature becomes greater and reduces the grain-filling period. In Australia's warm low latitudes, some crops are near their maximum temperature tolerance and are likely to suffer increasingly as the temperature increases. The possibility of overall decreased rainfall in Australia would negatively affect rangeland pastoralism and irrigated agriculture (which is a major source of production). Any changes in global production, and hence international food commodity prices, would have major economic impacts.

Farming in the region is well adapted to dealing with variability and change through a variety of natural and market factors—for example, by means of plant breeding, diversification, seasonal climate prediction, and so forth. Such techniques are likely to be sufficient to adapt to climate change over the next few decades; in some cases, these techniques may facilitate expanded production. As the time horizon extends, however, the changing climate is likely to become less favorable to agricultural production in Australia, leading to a long-term increase in vulnerability.

*Forestry:* Production forestry will be affected by changes in tree productivity; forest operational conditions; and weed, disease, and wildfire incidence. The net impact is not clear. The longer time to reach maturity results in a relatively large exposure to financial loss from extreme events, fire, or any rapid change in climate conditions.

*Fisheries:* A number of climate sensitivities have been identified, but the impacts cannot be predicted with any confidence. It appears that freshwater and near-shore fisheries will be more affected than open-ocean fisheries. The principal adaptation option is scientifically based integrated fisheries management and coastal zone management.

*Coastal zones:* The impacts on coastal zones of sea-level rise and climate change include inundation, riverine flooding, saline intrusion, erosion, and wave damage. The impacts of changes in weather conditions (winds, waves, storms, and storm surge) may be comparable to those of sea-level rise alone. Beaches, estuaries and coastal wetlands, and reefs—including the Great Barrier Reef—have adapted naturally to past changes in climate and sea level. However, in the future they are likely to face faster rates of change, and in many cases landward migration will be blocked by human land uses and infrastructure. Impacts will be complex, both physically and socioeconomically, and will vary greatly from place to place. There is potential for considerable damage to the region's low-lying coastal settlements and infrastructure where populations, tourism, and capital investment are large and growing. New Zealand also is exposed to impacts on its Pacific island territories, including the eventual possibility of environmental refugee fluxes. Some parts of Australasia's coastline, including a number of coastal cities and indigenous communities in low-lying coastal settlements and islands, are highly vulnerable.

Adaptation options include integrated coastal zone management; redesign, rebuilding, or relocation of capital assets; protection of beaches and dunes; development zone control; and retreat plans.

*Human settlements*: In addition to possible changes in the frequency and magnitude of "natural disaster" type events such as urban fringe fires, floods, and extreme sea-level events, urban areas and industry will experience a variety of direct climatic impacts on water and air quality, water supply and drainage, waste disposal, energy production and distribution, minerals production, transport operations, insurance, and tourism. The effects of the direct impacts are likely to be small relative to other economic influences, but the sectors are very large, and the impacts and necessary adaptations may represent major losses and costs. Thus, a moderate degree of vulnerability is present. Adaptations will include improved planning, zoning, and engineering standards that take climate change and sea-level rise into account.

*Human health*: Increases are expected in heat stress mortality (particularly in Australia), tropical vector-borne diseases such as dengue, and urban pollution-related respiratory problems. These impacts are small compared to the total burden of ill-health, but they have the potential to cause significant community impact and cost. Indigenous communities and economically disadvantaged persons appear to be more at risk. Adaptation responses include strengthening existing public health infrastructure and meeting the needs of vulnerable groups. A moderate degree of vulnerability is apparent with health.

*Integration:* Some preliminary studies of impacts for limited areas or issues have been attempted for the region, but comprehensive integrated analyses are not available. A cross-sector economic costing approach indicates substantial impacts on Australasia's gross domestic product (GDP) (several percent per annum for a doubling of $CO_2$), but the method involved is subject to considerable criticism. The New Zealand CLIMPACTS and Australian OzClim software packages have been developed as national-scale, integrated assessment methodologies to enable the progressive exploration of sensitivities, impacts, adaptation options, and so forth as knowledge develops.

*Research needs:* Research is needed on better climate scenarios; the dynamic responses of ecological systems; impacts on marine, aquatic, and coastal ecosystems; possible changes in the magnitude and frequency of "natural disasters"; impacts on settlements, industries, and health; cross-sector and multinational interactions; and potential adaptations and their ameliorating effects.

*Conclusion:* Australia's relatively low latitude makes it particularly vulnerable through impacts on its scarce water resources and on crops presently growing near or above their optimum temperatures, whereas New Zealand—a cooler, wetter, mid-latitude country—may gain some benefit from the ready availability of suitable crops and the likely increase in agricultural production. In both countries, however, there is a wide range of situations in which vulnerability is thought to be moderate to high—particularly in ecosystems, hydrology, coastal zones, settlements and industry, and health.

---

## 4.1. Background Characterization of Region

*Summary: This region is defined here as Australia, New Zealand, and their outlying islands. Australia is a large, flat continent spanning the tropics to mid-latitudes, with relatively nutrient-poor soils and a very arid interior, whereas New Zealand is much smaller, mountainous, and well watered. Both have "Gondwanan" ecosystems and unique flora and fauna. They have been subject to significant human influences—particularly from fire, agriculture, deforestation, and introduced exotic plants and animals. The total land area is 8 million square kilometers, and the population is 22 million. In contrast to other OECD countries, commodity-based industries of agriculture and mining dominate the economies and exports. Tourism is a major and rapidly growing industry.*

Australia is a large, ancient, eroded, and relatively flat continent, similar in size to Europe or the continental United States, with generally nutrient-poor soils. Situated between the Indian and Pacific Oceans, it is bordered to the north at 11°S by tropical waters and to the south at 44°S by the cold and windy Southern Ocean (Figure 4-1). It is the only major OECD country that lies largely in tropical and subtropical latitudes. In contrast, New Zealand is much smaller; it comprises two narrow, geologically young, and mountainous main islands located wholly in the mid-latitudes, from 35°S to 48°S. The total land area of the two countries is 8 million km$^2$. Small offshore islands greatly extend both countries' jurisdictions. Key geographical features of the region are its relative isolation in an oceanic hemisphere, Gondwanan ecosystems and unique flora and fauna, low human population densities, and significant climatic features such as the ENSO phenomenon, the very arid interior of Australia, and tropical cyclones.

The ecosystems of the region are extremely varied, owing to the range of climates arising from the large latitudinal range, the continentality of Australia, and the mountainous, maritime nature of New Zealand. Australia has mainly rainfall-limited ecosystems (both natural and agricultural), although high temperatures also are limiting to crop production in some more northern areas. New Zealand has generally ample rainfall and mainly low-temperature limited ecosystems. More important biomes include the interior Australian deserts; semi-arid shrublands and savannas; tropical and temperate grasslands; tropical and temperate rainforests; sclerophyll forests and woodlands; alpine zones in southeast Australia and New Zealand; freshwater and coastal wetlands; tropical and subantarctic islands; coral reefs; and deep ocean systems. Land-use is reported as being 6% cropland (with a large fraction irrigated), 55% permanent pasture (much of it semi-arid rangelands), 19% forest and woodland, and 20% other (see WRI, 1996, Table 9.1; see also Annex D of this report).

Many parts of the region have been subject to significant human influences, particularly from the use of fire, widespread agriculture, vegetation clearance, deforestation, and other land-use

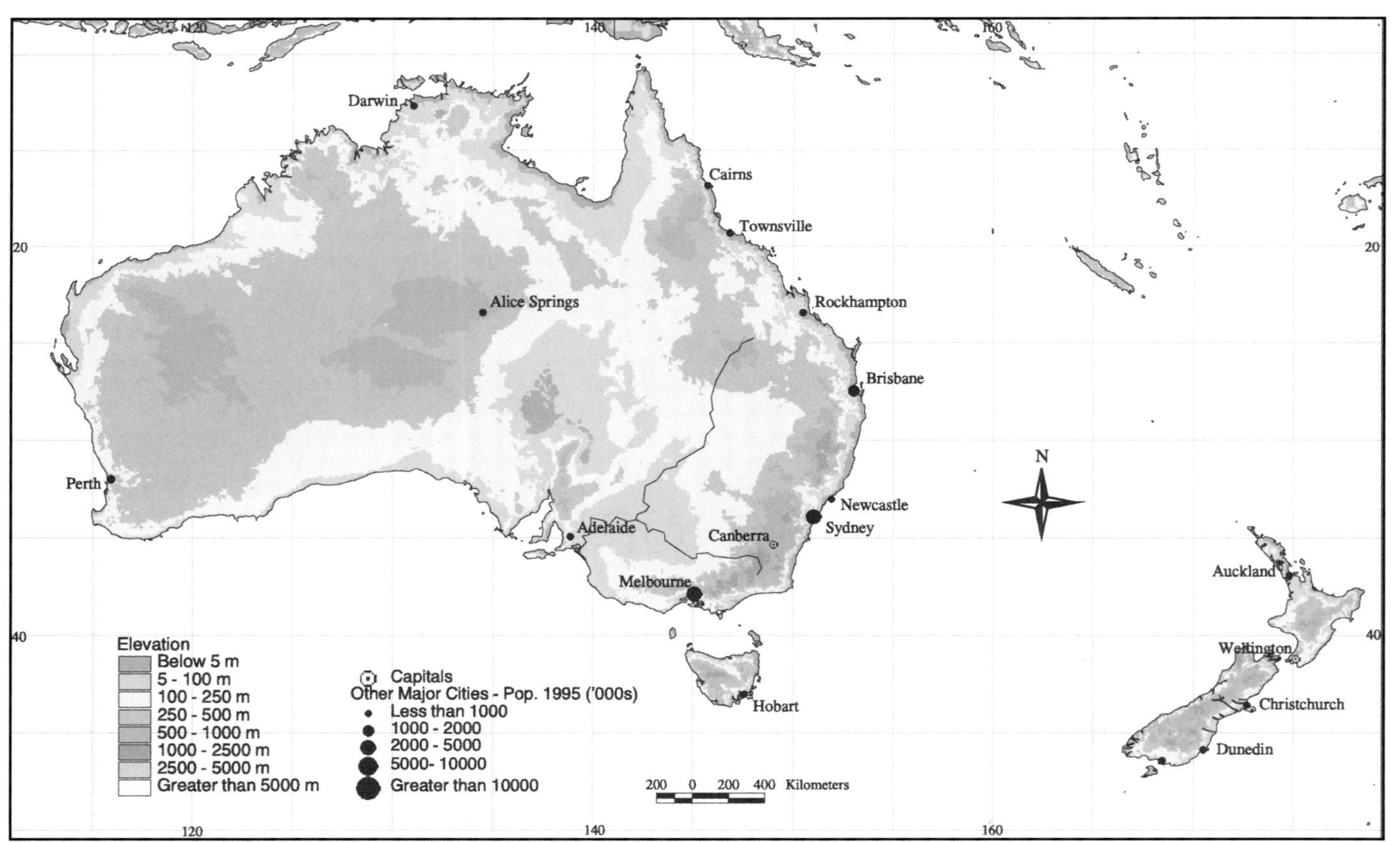


**Figure 4-1:** The Australasia region [compiled by the World Bank Environment Department Geographic Information System (GIS) Unit; see Annex E for a color rendition].

change and from the introduction of exotic plants and animals—particularly rabbits and foxes in Australia and rabbits, deer, and Australian possums in New Zealand. Owing to millions of years of isolation, the region has a very high proportion of endemic species (plants and animals found only in this region), and its ecosystems are extremely vulnerable to introduced pests, diseases, and weeds. Environmental pollution is relatively low, in line with the low human population. Local air and water pollution from urban industries, land transport, and intensive farming and related processes are of concern in some areas, and soil erosion, rising water tables, and dryland salinization (as a result of land clearing and irrigation) are a concern in Australia (SOEC, 1996).

The region's present population is about 18.5 million (Australia) and 3.5 million (New Zealand). Population growth of about 1.6% per annum is higher than the average for OECD countries, largely due to immigration. About 85% of the population live in urban and suburban areas, mostly near the coast. Half of Australia's population lives in just four coastal cities. Average population density is low; even when arid areas are excluded, it is only about 5 persons/km$^2$ for Australia (see Annex D). Both countries have strong relationships with Pacific island countries, in some cases extending to specific constitutional responsibilities, and they are home to sizable populations of Pacific Islanders. Large populations outside the region are dependent on the region's agricultural exports.

The two countries have open market economies, are members of the OECD, and in general have good access to the capital, technological, and human resources needed to adapt to the impacts of climate change. Per capita GDP is currently about US$19,000 in Australia and US$14,000 in New Zealand (the lower values in Annex D are calculated on a different, comparative purchasing power basis). Real annual growth rates have been about 3%, and inflation is low. Unemployment is about 8–10%.

Primary industries are a cornerstone of the region's economy. These industries include Australia's low-input pastoralism (grain, meat, and wool) and irrigated agriculture and horticulture and New Zealand's meat, wool, dairy, and horticulture industries, together with the associated processing industries in both countries. Forestry, fishing, and tropical crops also are important. Australia is a major exporter of coal, as well as energy-intensive iron, steel, and aluminium. Coal reserves are about 100 billion tonnes, with annual production of about 200 million tonnes, mostly for export. About two-thirds of New Zealand's electricity is hydro-generated. The bulk of the region's exports is raw or processed agricultural and mining products; in this respect the regional economies are anomalous among OECD countries, being more like the less developed and emerging economies (Crocombe *et al.*, 1991). Thus, regional exports are heavily exposed to fluctuations in world commodity prices and trading conditions.

Tourism in the region—which is largely dependent on landscape, biodiversity, and climate—is growing faster than the global average and now contributes about 13% of all export earnings, exceeding traditional export earners such as wool and wheat.

Socioeconomic trends in the region are similar to those in other developed countries. The population is aging, and there is a continuing shift of employment and population from lower value, commodity-related activities to higher value manufacturing and service activities. There is more intensive use of the less arid and more nutrient-rich land, especially for horticulture; in New Zealand especially there is rapid development of plantation forestry.

## 4.2. Regional Climate Information

### *4.2.1. Current Climate*

*Summary: Australasia's climate is strongly influenced by the surrounding oceans. Key climatic features include tropical cyclones and monsoons in northern Australia; migratory mid-latitude storm systems in the south, including New Zealand; and the ENSO phenomenon, which causes floods and prolonged droughts, especially in eastern Australia. Unfortunately, climate models at present cannot provide reliable predictions for these features under climate change.*

The region's climate is strongly influenced by its oceanic context. Northern Australia, lying just south of the Western Pacific oceanic "warm pool," experiences tropical conditions, with a summer monsoon (Sturman and Tapper, 1996). Tropical cyclones (averaging six per year) are a major concern for northern coastal regions. Cyclones can track southward as far as New Zealand, bringing very high rainfalls (Sinclair, 1993a, b). Western and central Australia experience generally clear, dry conditions owing to large-scale subsidence. In the winter half-year, eastward-moving anticyclones cross the continent, and northern areas are influenced by mild, dry, southeast tradewinds. In the summer half-year, the anticyclones cross at higher latitudes—bringing fine weather to southern Australia and to New Zealand.

South of about 35°S there are increasingly strong westerly winds, within which large-scale atmospheric waves develop. The resulting migratory cyclonic and anticyclonic systems are key features of the weather of New Zealand, Tasmania, and the southern coasts of Australia, especially during winter (Maunder, 1971; Sturman and Tapper, 1996). The eastward progression of these weather systems brings cycles of warm northerly conditions followed by depressions, fronts, and sometimes severe outbreaks of cold air from the Southern Ocean.

Australia's generally low relief (apart from Tasmania and the Great Dividing Range parallel to the east coast) does not significantly affect the weather regimes. Because of the size of the continent, however, the rain-bearing weather systems progressively dry out as they penetrate inland, resulting in a very arid central desert region (Sturman and Tapper, 1996). Australia is the driest populated continent; two-thirds of the land is classified as arid or semi-arid (<400 mm annual rainfall). About 87% of all rainfall is lost to evaporation from the land surface and vegetation, compared to 60% for the United States and Europe. Less than 5% of Australia has an average annual runoff greater than 250 mm; 26% of the land surface contributes more than 88% of total runoff, and half the land surface has no direct discharge to the ocean.

An important feature of the region's climate is the ENSO phenomenon, which causes high year-to-year variability, especially of rainfall in northern and eastern Australia. The ENSO phenomenon involves the interaction of the tropical Pacific Ocean with the global atmosphere and shifts in the major climatic patterns of the globe every 2–5 years or so (McBride and Nicholls, 1983; Rasmusson and Wallace, 1983; Bureau of Meteorology, 1989; McCreary and Anderson, 1991; Trenberth, 1991). During "El Niño" years—which are marked by unusually warm sea surface temperatures in the tropical eastern Pacific Ocean—Australia experiences reduced rainfall and extensive and often prolonged droughts, with severe impacts on rural communities and agricultural production. Major wildfires also can occur. At the opposite phase of the oscillation, during "La Niña" years, flooding is common. Australia has been described as *a land of drought and flooding rains*.

Owing to its generally low rainfall and low runoff and the large seasonal and ENSO-related variations, Australia's hydrological features are significantly different from almost all other regions and continents (Finlayson and McMahon, 1988; McMahon and Finlayson, 1991; Fleming, 1995; Thomas and Bates, 1997). The variability of Australian rainfall and runoff is among the greatest in the world: two to four times those of northwestern Europe and North America for the same climatic zones. Compared with other nations, the frequency and duration of drought is extreme, though these are interspersed with sequences of above-average rainfall. About 25% of the variance in annual rainfall is attributable to the ENSO phenomenon. Most of Australia's major rivers have ceased to flow at least once during the past 100 to 200 years, and groundwater levels show remarkable variations through time.

In contrast, New Zealand has a maritime climate; few places are more than 100 km from the coast. Rainfall is relatively well-distributed by region and season, and the influence of the ENSO phenomenon on year-to-year variations is less than in Australia. Mountains rising to 2000–3000 m transverse to the prevailing westerly winds result in very wet conditions in western areas and drier "rainshadow" conditions in eastern areas. In the west of the South Island, annual rainfall can be as much as 12,000 mm, with high runoff, steep torrents, extensive permanent snow fields at high altitudes, and fast-moving glaciers. Barely a few tens of kilometers to the east, annual rainfall may be as low as 500 mm. The eastern plains experience summer drought, which sometimes extends to spring and autumn.

### 4.2.2. *Trends in Climate, Sea Level, and $CO_2$*

*Summary: The region's climatic trends are consistent with those of other parts of the world. Mean temperatures have risen by up to 0.1°C per decade over the past century; nighttime temperatures have risen faster than daytime temperatures; and the past decade has seen the highest mean annual temperatures ever recorded. Increases in the frequency of heavy rainfalls and average rainfall have been reported for large areas of Australia. Sea levels have risen on average by about 20 mm per decade over the past 50 years.*

Climate exhibits a variety of natural variations—from those occurring over geological time (McGlone *et al.*, 1996) to those measured on decadal, year to year, and seasonal time scales. Long-term climate fluctuations in the Australasia region, assessed from historical data measured over the past century, are summarized in regional reviews by Hobbs (1988) and Salinger *et al.* (1996) and the regional-mean series in Figure A-6 of Annex A. The region's trends are broadly consistent with global trends reported in the recent Intergovernmental Panel on Climate Change (IPCC) second assessment (IPCC 1996, WG I, Chapter 3). The impacts of such long-term variations on land use in parts of southern Australia are documented by Heathcote (1994, 1996).

Air temperatures in the region have risen by 0.5–0.9°C since the beginning of the century (Salinger *et al.*, 1996), and New Zealand temperatures show a trend of +0.11°C per decade (Zheng *et al.*, 1997). These figures are somewhat higher than the global mean trend (IPCC 1996, WG I, Section 3.2.1). There are large variations in decadal averages (Figure A-6), presumably of natural origin. The largest increases have occurred since about 1950. Regional trends in intraseasonal and interannual temperature variability were mixed and generally not statistically significant (Plummer, 1996). Nineteenth-century Australian temperature data may be systematically too high, owing to instrumentation error, which may partially mask the real temperature rise for Australia derived from these early data (IPCC 1996, WG I, Section 3.2.2.1). Ocean temperatures in the region also are generally rising (Folland and Salinger, 1995; Salinger *et al.*, 1996; Holbrook and Bindoff, 1997; Zheng *et al.*, 1997).

Analyses have shown that, along with other parts of the world, the region's minimum daily temperatures have tended to rise faster than the maximum daily temperatures, which means that the day-night difference (the diurnal temperature range) has decreased noticeably in most places, by up to 1°C over the past 40 years (Karl *et al.*, 1992; Salinger, 1995; Torok and Nicholls, 1996; Zheng *et al.*, 1997). The trend in diurnal temperature range has been strongest since the 1950s and appears to be a result of increasing cloudiness—the source of which is unclear at present (IPCC 1996, WG I, Section 8.5.3). In Australia, there has been a decrease in the number of clear days since at least the 1950s, with the largest declines occurring in spring (Plummer *et al.*, 1997), and cloud cover has increased by 5% overall since 1910 (Jones, 1991).

As expected, the trend toward rising temperatures has resulted in an increase in the frequency of very warm days and nights and a decrease in the frequency of very cool days and nights. Plummer *et al.*'s (1997) study of Australian data for 1961 to 1995 showed that the decrease in the frequency of cool nights exhibited the strongest trend, whereas the decrease in the frequency of cool days was a relatively weak trend.

Rainfall is intrinsically more variable than temperature, both from place to place and time to time; thus, the detection of trends in rainfall is more difficult and uncertain. The picture for the region is not straightforward. There are seasonal and subregional differences, the trends are affected by the ENSO phenomenon, and the results depend on the period chosen for analysis. Some of the observed trends, notably those in recent decades in the southwest of Western Australia (Allan and Haylock, 1993), have been related to changes in regional atmospheric circulation. The time series of regional average annual rainfall is shown in Figure A-6. There are marked interdecadal variations, which are dominated by ENSO-induced variations in summer half-year rainfall over northern and eastern Australia.

Recent studies (Nicholls and Lavery, 1992; Suppiah and Hennessy, 1996, 1997; Lavery *et al.*, 1997) demonstrate an increase in heavy rainfall and average rainfall over large areas of Australia from 1910 to 1990. The largest increases have occurred along the east coast, particularly in New South Wales, but decreases are evident in southwest Western Australia and inland Queensland. In the summer half-year, the all-Australia average rainfall (based on areal weighting of station data) increased by 14%, heavy rainfall increased by 10–20%, and the number of dry days decreased by 4%. In the winter half-year, the changes were about half these figures. The trends in heavy rainfall are partially but not totally explained by ENSO fluctuations over recent decades.

The evidence on global trends in tropical cyclones, ENSO behavior, mid-latitude storms, and other atmospheric circulation is inconclusive (IPCC 1996, WG I, Sections 3.5.2.3, 3.5.3.1, 3.5.3.2). However, there is evidence of more frequent depressions along the east coast of Australia, which may be due to local circulation changes (Hopkins and Holland, 1997; Leighton *et al.*, 1997). Increased numbers of tropical cyclones over the past few decades have been reported in the southwest Pacific region (north of New Zealand), with the greatest increases for the stronger cyclones, but there are doubts about the homogeneity of the available database, partly owing to improvements in observation capabilities in recent decades (Thompson *et al.*, 1992; Radford *et al.*, 1996). In the Australian region (105–160°E), the total number of cyclones from the 1969–70 to the 1995–96 seasons has decreased, but the number of strong cyclones has increased slightly, reflecting an ENSO influence, as has the total duration of tropical cyclones (Nicholls *et al.*, 1997).

Although sea-level measurements are strongly influenced by local vertical land movement and other factors (and must be carefully interpreted), the evidence points to an average rise in sea level in the Australasia region over the past 50 years of about 2 mm per year (Rintoul *et al.*, 1996; Salinger *et al.*, 1996). This figure is within the range of the current estimate of global sea-level rise (IPCC 1996, WG I, Section 7.2.1).

Higher $CO_2$ concentration has direct biological effects on plants through the increased $CO_2$ "fertilization" effect and increased plant water-use efficiency (IPCC 1996, WG II, Chapter A), in addition to its effect on the climate system. For this reason, we note that present-day atmospheric $CO_2$ concentrations are about 30% higher than in pre-industrial times (IPCC 1996, WG I, Section 2.1.1; Manning *et al.*, 1996) and are currently increasing at about 0.4% per year.

### *4.2.3. Climate Scenarios*

*Summary: Climate modeling and climate change scenarios for the region are relatively well developed. For Australia, the recently revised (1996) CSIRO scenarios for 2030 indicate temperature increases of 0.3–1.4°C and rainfall changes of up to 10% in magnitude (decreases in winter, increases or decreases in summer, and overall a tendency for decreases). The projected changes for 2070 are about twice those of the 2030 changes. Increases in the intensity of heavy rainfall events are indicated. For New Zealand, the temperature increases are expected to be similar to those for Australia, but the recent revision indicates the possibility of an increase in westerly winds (unlike the decreases of previous scenarios)—and hence precipitation increases in the west and precipitation decreases in the east. The changes in scenarios serve to caution against overinterpretation of impact studies based on any single scenario.*

The IPCC has concluded that climate models at present provide useful predictions at the global and continental scale, but as yet allow little confidence at subcontinental scales (IPCC 1996, WG I, Section 6.6.3; Annex B of this report). A considerable amount of global climate model development, testing, and intercomparison has been done in the Australasia region (Garratt *et al.*, 1996; Gordon *et al.*, 1996; Whetton *et al.*, 1996b; Whetton *et al.*, 1996d). This research currently provides the best source of climate change scenarios for the region. Two factors that are important in the region's temperature changes are the moderating role of the extensive Southern Hemisphere oceans and the relative absence of aerosols that partially offset greenhouse gas warming in the northern hemisphere (Ayers and Boers, 1996; IPCC 1996, WG I, Section 6.2.2.2; Rintoul *et al.*, 1996).

To facilitate impact and policy studies, a series of scenarios for the region have been progressively developed in line with the development of climate modeling research. Scenarios for Australia were released in 1987 (Pittock, 1988), in 1990 (Whetton and Pittock, 1990), in 1991 (CSIRO, 1991; Whetton *et al.*, 1992), in 1992 (CSIRO, 1992; Mitchell *et al.*, 1994; Whetton *et al.*, 1996a), and in 1996 (CSIRO, 1996a). Similar scenarios have been released for New Zealand (Salinger *et al.*, 1987; Salinger and Hicks, 1990a, 1990b; Salinger and Pittock, 1991; Mullan, 1994—also described in Whetton *et al.*, 1996a). The integrated impact assessment models being developed for New Zealand ('CLIMPACTS,' Kenny *et al.*, 1995) and Australia ('OzClim,' CSIRO, 1996b) both include a scenario generator to enable the potential impacts of multiple scenarios to be investigated.

The majority of recently published regional impact studies are based on the scenarios of CSIRO (1992) for Australia and Mullan (1994) for New Zealand. These scenarios used spatial patterns of regional climate change derived from five general circulation model (GCM) experiments—namely, J1, F1, A3, H3 of Table 1-1, and the Bureau of Meteorology Research Center (BMRC) experiment of Coleman *et al.* (1994). For 2030, these gave warmings in Australia of 0.5–2.5°C for inland areas, 0–1.5°C for northern coastal areas, and 0.5–2.0°C for southern coastal areas, and warmings in New Zealand of 0.5–2.0°C in inland Canterbury and Otago and 0.5–1.5°C elsewhere. The Australian summer precipitation projections ranged from little change to as much as a 20% increase, and winter precipitation projections showed as much as 10% change—but the direction of change depended on the district. In New Zealand, regardless of season, precipitation increases from zero to about 20% were projected—except around Wellington, the east coast of the North Island, South Canterbury, and Otago. The projected changes for 2070 were about twice the magnitude of the 2030 changes.

The CSIRO (1992) and Mullan (1994) scenarios took into account the IPCC (1990) range of plausible greenhouse-gas emission scenarios and sensitivity of the global climate to increased greenhouse forcing, but the GCMs used were of a "slab ocean" type—that is, they did not include an interactive deep ocean. Very recently, the application of coupled ocean-atmosphere GCMs and the use of updated IPCC (1996a) global warming scenarios has led to significant shifts in the projected changes of climate (CSIRO, 1996a). The CSIRO (1992) and Mullan (1994) scenarios used the same five slab-ocean GCMs (except for an improved version of F1), and five coupled-ocean GCMs (X10, X2, X8, and X6 of Table 1-1, and another MPI experiment as in Lunkeit *et al.*, 1994). The marked differences in scenarios from the coupled models in the Australasian region highlight the need for caution in the use of climate scenario information for impact assessment.

In the new CSIRO (1996a) scenarios, the projections of warming for Australia have been reduced, mainly due to the downward revision of global warming estimates. The new scenarios for 2030 are 0.4–1.4°C for inland areas, 0.3–1.0°C for northern coastal areas, and 0.3–1.3°C for southern coastal areas. The use of coupled ocean-atmosphere models in addition to the older non-coupled "slab ocean" models has led to major changes in summer rainfall scenarios, which for 2030 now vary from –10% to +10%, unlike the increases projected by the slab models alone. This difference appears to be due to the coupled models' much slower rate of warming in the higher latitudes of the Southern Hemisphere (Whetton *et al.*, 1996b). Winter rainfall change scenarios for 2030 are similar in coupled and slab models and range from decreases of up to about 10% over inland areas to increases of a similar magnitude in the far south (mainly Tasmania). Related scenarios based on the results of a regional model (DARLAM) nested in the CSIRO slab-ocean GCM (experiment F1, Table 1-1) also have been developed (Hennessy *et al.*, 1997; Whetton *et al.*, 1997).

It should be noted that the CSIRO (1996a) scenarios continue to use results from the older slab-ocean climate models as well as from the newer coupled models. This is because there are still some doubts about the reliability of the scenarios from the coupled models (Whetton *et al.*, 1996b). These doubts partly relate to the observed latitudinal gradient of warming in the Southern Hemisphere (IPCC 1996, WG I, Section 3.2.2.3), which is more like that in the slab-ocean GCMs, and to the simulated rainfall decreases in summer over Australia—which are contrary to the observed small increases this century. This disagreement could be because the observed trends are not predominantly greenhouse-related and may be influenced by natural multidecadal fluctuations in the coupled ocean-atmosphere system.

Alternative scenarios for New Zealand based on the newer coupled model results are in preparation. Significant differences in the projected patterns of rainfall change are expected because the coupled models simulate an increase in the strength of the mid-latitude westerly winds, which is the reverse of the previous slab-model based scenarios. Stronger westerlies are likely to show a greater tendency for rainfall increases on the wet western side of the Southern Alps and decreases on the dry eastern plains.

Any overall increase in temperature will tend to cause an increase in the frequency of extremely high temperatures and a decrease in the frequency of extremely low temperatures (e.g., frost) (IPCC 1996, WG I, Section 6.5.7). A similar effect will apply to areas of overall rainfall increase, and *vice versa* for areas of overall rainfall decrease. In addition, global climate models tend to predict that in a warmer climate there will be more high-intensity rainfall events, with more severe droughts and floods in some places but less severe droughts and floods in other places (Fowler and Hennessy, 1995; IPCC 1996, WG I, Section 6.5.7). Where interannual variability in rainfall is high, there may be more sensitivity to changes in the frequency of extreme events than to small changes in the mean climate (IPCC 1996, WG II, Section 2.7.1).

Soil moisture will be directly affected by rainfall changes, including the replenishment of groundwater by more frequent high-rainfall events. Moreover, higher temperatures will increase the potential for evapotranspiration, causing faster drying of the soil.

Of particular concern for the Australasian region are possible changes to the region's major weather systems—especially the timing, intensities, and locations of the tropical monsoonal and cyclone systems; the locations and intensities of mid-latitude weather systems and the subtropical anticyclone belt; and the frequency, intensity, and bias of the El Niño-Southern Oscillation. Unfortunately, climate models do not at this stage provide firm projections for any of these key features (e.g., Pittock *et al.*, 1995; IPCC 1996, WG I, Sections 6.4.4, 6.5.4; Katzfey and McInnes, 1996; Pittock *et al.*, 1996; Suppiah *et al.*, 1996), although there are some tentative indications that tropical cyclones could increase their maximum potential intensities by as much as 10–20% under doubled $CO_2$ conditions (Henderson-Sellers and Zhang, 1997; Holland, 1997). The use of nested regional models may provide scenarios of tropical cyclone changes in the future (Suppiah *et al.*, 1996; Walsh and Watterson, 1997).

There are large differences between the results of the different global models, especially at the subcontinental scale (IPCC 1996, WG I, Section 6.6; Whetton *et al.*, 1996d), and any regional circulation changes could impose large changes in temperature and rainfall on top of the continental-scale changes indicated above.

## 4.3. Vulnerabilities and Potential Impacts for Key Sectors

### *4.3.1. Ecosystems*

*Summary: In responding to climate change, Australasia's biota may face a greater rate of long-term change than ever before. They also must respond in a highly altered landscape fragmented by urban and agricultural development. There is ample evidence for significant potential impacts. Alterations in soil characteristics, water and nutrient cycling, plant productivity, species interactions (competition, predation, parasitism, etc.), and composition and function of ecosystems are highly likely responses to increases in atmospheric $CO_2$ concentration and temperature and to shifts in rainfall regimes. These changes would be exacerbated by any increases in fire occurrence and insect outbreaks.*

*Aquatic systems will be affected by the disproportionately large responses in runoff, riverflow and associated nutrients, wastes and sediments that are likely from changes in rainfall and rainfall intensity and by sea-level rise in estuaries, mangroves, and other low-lying coastal areas. Australia's Great Barrier Reef and other coral reefs are vulnerable to temperature-induced bleaching and death of corals, in addition to sea-level rise and weather changes. However, there is evidence that the growth of coral reef biota may be sufficient to adapt to sea-level rise. Our knowledge of climate change impacts on aquatic and marine ecosystems is relatively limited.*

*Prediction of climate change effects is very difficult because of the complexity of ecosystem dynamics. Although Australasia's biota and ecosystems are adapted to the region's high climate variability (exemplified in arid and ENSO-affected areas), it is unclear whether this will provide any natural adaptation advantage. Many species will be able to adapt through altered ecosystem relationships or migration, but such possibilities may not exist in some cases, and reduction of species diversity is highly likely. Climate change will add to existing problems such as land degradation, weed infestations, and pest animals and generally will increase the difficulties and uncertainty involved in managing these problems.*

*The primary human adaptation option is land-use management—for example, by modification of animal stocking rates in rangelands, control of pests and weeds, changed forestry practices, and plantings along waterways. Research, monitoring, and prediction, both climatic and ecological, will be necessary foundations to human adaptive responses. Active manipulation of species generally will not be feasible in the region's extensive natural or lightly managed ecosystems, except for rare and endangered species or commercially valuable species. In summary, it must be concluded that some of the region's ecosystems are very vulnerable to climate change.*

#### *4.3.1.1. General*

Climate is a primary influence not only on the individual plant, animal, and soil components of an ecosystem but also on water and nutrient availability and cycling within the ecosystem, on fire and other disturbances, and on the dynamics of species interactions. Changes in climate therefore affect ecosystems both by directly altering an area's suitability to the physiological requirements of individual species and by altering the nature of ecosystem dynamics and species interactions (Peters and Darling, 1985). In addition, biota face an environment in which the rising atmospheric $CO_2$ concentration also will directly affect plants and soils.

The rate of climatic change may exceed any that the biota have previously experienced (IPCC 1996, WG II, Chapter A and Section 4.3.3). This rate of change poses a potentially major threat to ecosystem structure and function and possibly to the ability of evolutionary processes, such as natural selection, to keep pace (Peters and Darling, 1985). Although many of the biota and ecosystems in the region have adapted to high climate variability (exemplified in the region's arid and ENSO-affected areas), it is unclear whether this will provide any advantage in adapting to the projected changes in climate.

Furthermore, in contrast to the case of climate change over geological time scales, today the region's biota must respond in a landscape that has been highly modified by agricultural and urban development and introduced species (Peters, 1992). Considerable fragmentation of habitat has occurred in Australasia's forests, temperate woodlands, and rangelands. In the short term, land-use changes such as vegetation clearance are likely to have a much greater bearing on the maintenance of conservation values than the direct effects of climate change on biodiversity (Saunders and Hobbs, 1992). In the longer term, however, climate change impacts are likely to become increasingly evident, especially where other processes have increased ecosystem vulnerability (Williams *et al.*, 1994).

Australasia's isolated evolutionary history has led to a very high level of endemism (plants and animals found only in the region). For example, 77% of mammals, 41% of birds, and 93% of plant species are endemic (see Annex D). As one of the 12 recognized "mega-diversity" countries (and the only one that is an OECD member), Australia has a particular stewardship responsibility toward an unusually large fraction of the world's biodiversity. Many of New Zealand's endemic bird species are endangered. Species confined to limited areas or habitat, such as Australia's endangered Mountain Pygmy Possum (*Burramys parvus*)—which is only found in the alpine and subalpine regions of southeast Australia (Dexter *et al.*, 1995)—may be especially vulnerable to climate change.

Certain ecosystems have particular importance to the region's indigenous people, both for use as traditional sources of food and materials and for their cultural and spiritual significance. Selected climate change impacts on Australian Aborigines and New Zealand Maori are considered in Sections 4.3.3.1 and 4.3.4.2 respectively.

### *4.3.1.2. Soil Properties and Plant Growth*

The impact of climate change on soil is difficult to predict (Jenkinson *et al.*, 1991; Gifford *et al.*, 1996a; IPCC 1996, WG II, Chapters A and 4; McMurtrie and Comins, 1996; Tate *et al.*, 1996; Thornley and Cannell, 1996). Soil nutrient availability is particularly important in Australia, where a high proportion of the continent has soils of low nutrient content. Increased $CO_2$ concentration and temperature will change the carbon-to-nitrogen ratios of biomass and hence of decomposing organic material. Higher temperatures are likely to increase the rate of decomposition of organic material; this loss may be partially offset, however, by the small observed increases in primary production that can occur with increased atmospheric carbon dioxide levels. Increases in soil carbon as a result of improved stock and pasture management are likely to be significant in Australian rangelands (Ash *et al.*, 1996).

Soil water availability is likely to be the most important limiting factor for productivity in much of the region, especially in inland Australia and eastern districts of New Zealand. Soil water content will be affected by changes in rainfall (increases or decreases) and by increased evapotranspiration occurring as a result of increased temperature. Tropical forests that experience a long dry season—such as those in northern Australia—may be more sensitive to changes in rainfall than to changes in temperature.

Increased salinization and alkalization would occur in areas where evaporation increased or rainfall decreased (Varallyay, 1994); this development could have significant impacts on large areas of Australia's semi-arid zones. In areas where salinity is a result of recharge processes, salinization would increase if the upstream recharging rainfall increased (Peck and Allison, 1988). Increasing atmospheric $CO_2$ concentration can reduce the impact of salinity on plant growth (Nicolas *et al.*, 1993). An increased frequency of higher rainfall events would increase soil erosion (IPCC 1996, WG II, Section 4.2.1), which would be a concern in much of New Zealand's deforested hill country.

Rising levels of atmospheric $CO_2$ will have a considerable impact on the growth and morphology of plants, with likely increases in potential productivity through increases in carbon assimilation, water-use efficiency, and possibly nutrient-use efficiency. Increased water-use efficiency under higher $CO_2$ conditions will lead to higher productivity, especially in water-limited systems; but the magnitude of the response will depend on other limiting factors such as soil nitrogen (Eamus, 1991; Gifford, 1992). Some experiments have shown an "acclimatization" or "acclimation" effect, in which the growth response to higher $CO_2$ in the longer term is less than in short-term experiments (Gunderson and Wullschleger, 1994); whether this effect applies at the ecosystem level over many years remains untested, however.

In temperate zones, increased temperatures generally enhance the rate of plant and soil biochemical processes and lead to greater plant productivity. Thus, higher air temperatures are likely to increase plant growth in the mid-latitudes of New Zealand and southern Australia, where productivity is currently limited by lower temperatures. In Australia's tropical areas, however, higher temperature stresses (above 35–40°C over extended periods) may result in more frequent damage to the vegetation from desiccation and sunscald (IPCC 1996, WG II, Section 1.4.3.2). In systems where $C_3$ and $C_4$ species co-exist, their relative proportions may or may not change much, depending on the balance between increased photosynthesis in $C_4$ species at higher temperatures and increased photosynthesis in $C_3$ species arising from elevated $CO_2$ concentration (Campbell and Hay, 1993).

Cloudiness has a major influence on the amount of solar radiation reaching the Earth's surface. Any changes in cloudiness associated with changing weather patterns would directly affect the characteristics of photosynthetically active radiation and the amounts of solar UVB radiation, which is often detrimental to biota and to the productivity of crop plants and trees (Hunt *et al.*, 1996). Many parts of Australasia experience relatively high levels of solar radiation and solar UVB radiation. At present, climate models cannot provide reliable predictions of how cloudiness might change in the future.

### *4.3.1.3. Terrestrial Ecosystems*

Predictions based on scenarios of climate change suggest increased variability and unpredictability in productivity and community composition. Any changes that might occur in the location and seasonal patterns of rainfall and in ENSO characteristics would result in significant impacts and vulnerability (IPCC 1996, WG II, Section 2.7.2 and Boxes 2.2 and 2.8). Although migration to more suitable climatic regimes is an option for some biota (Mitchell and Williams, 1996), ecosystems are not expected to shift *en masse*. Differential rates of migration and survival of different species will inevitably change the abundance and distribution of species, community structure, and possibly ecosystem function at any given location. Altered species interactions—for example, predation, parasitism, and competition—also are likely and may eliminate formerly successful species even if the climate remains within their physiological tolerances (Peters, 1992). Ecosystems whose "keystone" species are particularly sensitive to climate will be more at risk. (A keystone species is one that has a central servicing role affecting many other organisms and whose demise is likely to result in the loss of a number of species and lead to major changes in ecosystem function.) Some species will be slow to migrate because of factors such as slow reproduction rate or limited seed dispersal mechanisms (IPCC 1996, WG II, Section 1.3.5); species with better capacity for dispersal and establishment, including weeds, will have an advantage. Disruption in forest composition is most likely to occur where fragmentation of the forest reduces the potential for dispersal of new, more suitable species (Whitehead *et al.*, 1992). Although many species will be able to adapt, climate change is expected to reduce biodiversity in individual ecosystems overall (IPCC 1996, WG II, Section 1.3.6).

Rising temperatures can be accommodated by moves toward higher elevations (if the terrain allows) because air temperature decreases by about 1°C for every 100–200 m increase in elevation (IPCC 1996, WG II, Section 5.2.3.3), or toward higher latitudes because temperatures generally decline as one moves poleward, especially in mid-latitudes. However, elevational migration is not an option for most of Australia's predominantly flat expanses, and latitudinal migration is a limited option in parts of tropical Australia where the latitudinal variation of temperature is small and poleward migration is limited by desert or land-use change; here the biota will increasingly experience temperatures never previously encountered. In contrast to North America and Eurasia, the oceanic boundary of southern Australia will restrict the poleward migration of terrestrial biota. Small offshore islands and mountain tops provide limited migration options, and escarpments, deserts, agricultural land, and urban areas present physical barriers to migration. Conversely, changes in climate thresholds, such as frost, may remove an existing limitation to species survival and performance. The survival of vulnerable species in refugia, particularly during drought, and then expansion into adjacent areas (e.g., Morton *et al.*, 1995) suggest that migration will be even more limited if there are increases in the frequency or intensity of climatic extremes such as drought.

More than half of the region comprises the arid and semi-arid ecosystems of grasslands, shrublands, and savanna. These areas exhibit a high degree of spatial variation, from small-scale variation in water and soils to larger-scale variation in climate patterns. In such rangeland ecosystems, the amount and timing of rainfall and nutrient limitations are the major determinants of plant community composition, distribution, and productivity (Mott *et al.*, 1985; IPCC 1996, WG II, Section 2.1). Possible increases in rainfall intensity with climate change would increase the proportion of rain that runs off particular landscape elements and runs onto others, resulting in changes in the temporal and spatial functioning of rangelands (Stafford Smith *et al.*, 1994). In general, rangelands are not in equilibrium but fluctuate between states according to rainfall, fire, grazing, and other factors (Westoby *et al.*, 1989; IPCC 1996, WG II, Section 2.1). They have adapted to the naturally high variation of rainfall associated with their low mean rainfall and ENSO-related interannual variability. Should ecosystem productivity and composition become more variable and unpredictable—as has been suggested by some climate change studies—then land management issues, such as the type of grazing activity, the use of fire for woody weed control, and pest animal management, will become more important (Stafford Smith *et al.*, 1994).

The responses of animals to climate change will be partly determined by the response of co-occurring plants and habitat. For example, migratory species may be affected by reduction of suitable habitats along their migration routes. Thermal stress, which is particularly evident in animals in a variety of Australian ecosystems (Nix, 1982), may affect the geographic range and the reproductive biology of species. Indeed, contractions in core climatic habitat were shown for more than 80% of threatened vertebrate species in Australia under three climate change scenarios based on CSIRO (1992) (Dexter *et al.*, 1995). Although some animal species may be physically capable of moving great distances, behavioral traits may restrict dispersal options (Peters, 1992). As with plants, animals that are less adaptable, less mobile, or physically restricted (such as fish in lakes) may decline or become extinct.

Forest and woodland are reported to cover 19% of Australia and 28% of New Zealand (WRI, 1996). Indigenous forests occupy nearly a quarter of New Zealand, mainly under conservation control and often in mountain lands. The slowly maturing trees of indigenous forests will be more vulnerable than other plants to long-term external change.

Alpine ecosystems in Australia, which occupy only a small area, have been identified as being particularly susceptible to climate change (Busby, 1988; Nias, 1992; Brereton *et al.*, 1995). Upslope areas required to cope with the predicted rise in temperatures may not be available for many of these ecosystems. In small alpine streams, any increase in water temperatures and reduction in water flow could be stressful for alpine stream animals. Model studies for Australian mountain vegetation show that there is potential for the expansion of woody vegetation and shrub communities, as well as rises in treelines (IPCC 1996, WG II, Section 5.2.3.4).

Pests, weeds, and diseases play significant roles in Australian and New Zealand ecosystems (and agriculture). The geographical distribution and severity of their impacts on host plants and ecosystems could be dramatically changed by the combination of changes in climate, atmospheric composition, and habitat (Sutherst *et al.*, 1996). For example, any changes in the timing and intensity of plant moisture stress will bring changes in the relative advantage of different types of herbivorous insects.

Fire plays an important role in the composition, function, and dynamics of many Australian ecosystems (IPCC 1996, WG II, Section 1.5.1). For example, the importance of the length of inter-fire interval has been repeatedly demonstrated (Christensen and Kimber, 1975; Burgman and Lamont, 1992; Gill and Bradstock, 1995; Morrison *et al.*, 1995; Keith, 1996). It has been strongly suggested that the risk of fire may increase as the climate changes (IPCC 1996, WG II, Section 1.3.1) because of factors such as higher temperatures and water stress—though there remains much speculation about future fire regimes owing to the uncertainty associated with the many climatic and ecosystem variables involved (Williams and Gill, 1995). If fires become more frequent under climate change (Beer and Williams, 1995; Pittock *et al.*, 1997), species composition and structure could be altered, at least in southern Australia. Changes in other components of the fire regime, such as intensity (which could be affected by changes in fuel loading) and season, likewise could affect species diversity. Any increased fire occurrence also could present increased risks to people and infrastructure in forested urban fringe areas of southeastern Australia.

#### 4.3.1.4. Aquatic and Marine Ecosystems

Aquatic ecosystems are exposed to the primary effects of local changes in temperature, sunshine, wind, and so forth and to a wide range of secondary effects, particularly from changes in hydrology and waterborne materials. Increased water temperatures, increased evaporation, and changes in inflows and flooding would change the thermal and chemical structures of rivers and lakes. These changes would directly affect the nutrient status of aquatic ecosystems; the survival, reproduction, and growth of organisms; the distribution and diversity of species; and overall ecosystem productivity (IPCC 1996, WG II, Sections 10.5, 10.6). Changes in rainfall produce disproportionate changes in runoff and in mean flows and levels of rivers and some lakes; these effects might be exacerbated if the climate were to become more variable, with more frequent flood and drought events. Sea-level rise is likely to have significant effects on lowland aquatic ecosystems near the coasts. (Coral reefs are discussed in Section 4.3.1.5.)

An increased frequency of more intense rainfalls would increase the intensity of runoff events and thus might contribute greater inputs of nutrients, organic material, agricultural waste, and sediment (IPCC 1996, WG II, Section 10.5). This could lead to reduced water quality, stress on some species, and potentially increased biological productivity forming nuisance growths. Waterways in the many intensively farmed areas of the region may be particularly vulnerable to increased pollution by fertilizers, animal waste, and agrochemicals, with greater potential for algal blooms and aquatic plant proliferation. Algal blooms and eutrophication already are a major problem in many of Australasia's inland waters (SOEC, 1996). Excessive plant growth in lowland streams can reduce drainage capacity and increase risks of flooding.

Sediment transport following heavy rainfall can smother extensive areas of estuarine habitat, resulting in loss of breeding habitat essential to many coastal fish species and affecting food supply for seabirds. Such an extreme event occurred in New Zealand's Whangapoua estuary in 1995. Any increase in extreme rainfall events and sedimentation would be likely to have major impacts on river, lake, estuarine, and coastal waters—particularly the Great Barrier Reef lagoon (Larcombe *et al.*, 1996)—and lead to reduced aesthetic values and reduced recreational and tourist use.

In the lowland coastal rivers and floodplains of northern Australia, the possibly lower rainfall projected by the revised climate change scenarios (CSIRO, 1996a) would most likely lead to greatly decreased biodiversity because it has been found that poor wet seasons reduce the extent and duration of inundation—which in turn has dramatic impoverishing effects on the abundance and diversity of the biota. Drier conditions would likely lead to a decrease in the problem of insect-borne diseases, however. These floodplains may be particularly prone to the impacts of sea-level rise (Steering Committee of the Climate Change Study, 1995; Waterman, 1995), whereby storm surges, cyclonic floods, and seawater intrusions could devastate the freshwater biota. The possible effect of climate change on the present invasion of floodplains by aquatic and semi-aquatic weeds (Lonsdale, 1994; Miller and Wilson, 1995), or on future invasions, is unclear.

The fauna of small lowland tropical streams in northeastern Australia appear to be susceptible to depletion by floods and to have a relatively low rate of recovery (Rosser and Pearson, 1995), which suggests that an increase in the frequency and magnitude of extreme events may lower the diversity of lowland rainforest streams. In contrast, upland rainforest streams have a high diversity (Pearson *et al.*, 1986) and appear to be relatively resilient (Benson and Pearson, 1987; Rosser and Pearson, 1995). A study in southeast Australia found that droughts deplete introduced trout but not the native galaxiids, resulting in an expansion of galaxiid populations downstream with the death of trout (Closs and Lake, 1996); thus, with increased drought the range of native fish in such small streams might increase.

The many ephemeral river and lake ecosystems in inland Australia (temporary rivers and lakes that only flow and fill occasionally) (Boulton and Lake, 1988; Lake, 1995) are attuned to high climate variability, but their resilience to long-term change in the frequency and intensity of events is less certain. Information on their biota is scanty. A survey of the small and intermittently flowing streams of the George Gill Range in central Australia found an unexpected diversity in species (Davis *et al.*, 1993). Although increased temperature and droughts may threaten the viability of fish populations in this region, such changes may not greatly alter the invertebrate biota—which appear to be well adapted to variability in water availability. Breeding cycles of water birds may be affected (Hassall and Associates, 1997; see also Box 4-1).

The natural effects of drought may be exacerbated by climate change in catchments like the Murray-Darling basin in southeastern Australia, with its present high human demands on water. Billabongs (cutoff meanders), which are water bodies of high diversity (Boon *et al.*, 1990; Hillman and Shiel, 1991), and off-channel wetlands may dry up for extensive periods and become depleted in terms of diversity. There is evidence that where species are eliminated by drought, recolonization may be slow (e.g., Boulton and Lake, 1992). The already increasing salinity of the Murray-Darling system (Macumber, 1990) could alter under a changed climate. Although much of the biota of its floodplain sections appears to have considerable salinity tolerance (Walker, 1992), above a salinity of 1 g/l many species of freshwater biota, especially invertebrates, are adversely affected (Hart *et al.*, 1991). In lowland streams of northern Victoria, saline pools that develop under normal low-flow conditions (McGuckin *et al.*, 1991) are increasing in number and are marked by a very low faunal diversity (Metzeling *et al.*, 1995).

In relation to its size, southwestern Australia has highly endemic fish and invertebrate fauna but low species richness (Bunn and Davies, 1990). The low diversity may be due to the isolation of the area, its history of aridity, and a very low level of primary productivity. If changes in climate were to eliminate species in this area, it is unlikely that they would be replaced by species colonizing from elsewhere.

Although a great deal is known about New Zealand's aquatic ecosystems, little research has been done on the potential impacts of climate change. A number of issues and possible impacts—such as the effects of rising sea level on estuaries and temperature effects on spawning—have been identified and discussed in preliminary reviews by Burns and Roper (1990), Glova (1990), Paul (1990), and McDowell (1992). McDowell concluded that scientific knowledge of the controls on New Zealand freshwater communities' species composition and interactions and of ecosystem function is very limited, and is inadequate to make specific predictions even if very detailed climate scenarios were available.

It is known, however, that New Zealand's lake ecosystems are very similar in composition and function despite the latitudinal range of 5°C in mean temperature, which suggests an insensitivity of these ecosystems to projected temperature rises. Furthermore, New Zealand's river ecosystems contain relatively simple generalist species with little diversity—presumably a response to the high degree of short-term hydrological variability of the country's short, steep rivers (IPCC 1996, WG II, Section 10.6.3.4). Thus, these river ecosystems have an existing resilience to hydrological variability, which may offer some degree of insensitivity to predicted increases in rainfall variability.

Specific effects on aquatic ecosystems cannot be predicted with confidence. A general rule appears to be that the drier the climate, the greater the sensitivity to climate change (IPCC 1996, WG II, Chapter 10)—which implies that sensitivities and impacts are likely to be more significant in Australia than in New Zealand. Any increase in climatic variability, especially in extremes of flooding and drought, would have greater ecological effects than a change in mean conditions (IPCC 1996, WG II, Sections 10.5, 10.6).

The region's marine environment extends from tropical to subantarctic waters and includes extensive areas of open ocean. Any changes in the structure and strength of the region's wind fields and ocean currents would alter nutrient upwelling, food networks, reproductive patterns, and species distribution (IPCC 1996, WG II, Section 16.2.2). In addition, the Australasian region can expect to be affected by major oceanic changes occurring in other regions because the waters of the globe's oceans circulate throughout all the major ocean basins. Changes in wind speed and direction over upwind land surfaces would alter wind-borne terrestrial nutrient input, including micronutrients such as iron. Paleo-oceanographic research on sediments east of New Zealand has shown a close relationship between ocean phytoplankton productivity—as represented by diatom content—and regional wind strength, as represented by terrestrial quartz content (Fenner *et al.*, 1992).

As with freshwater systems, our knowledge of the role of climate in marine systems and our ability to predict climate change outcomes for the marine environment is very limited, though some of the responses of species and ecosystems to climate change described above for terrestrial systems are expected to apply to marine systems as well. Migration is an option for fish populations where coasts have a generally north-south alignment and suitable habitats exist. However, along the northern and southern extremities of Australia and New Zealand, or where a population is in equilibrium with a specialized habitat such as an estuary, rocky reef, island, or submarine seamount, there may be more random losses and gains of faunal components. (Impacts on coastal systems are discussed further in Section 4.3.4.)

#### *4.3.1.5. Coral Reefs*

Australia's coral reefs, including the Great Barrier Reef, are among the region's most sensitive environments to sea-level rise and climate change—through potential inundation, flooding, erosion, saline intrusion, bleaching and death of corals, and possible changes in tropical cyclone occurrence (some of these issues are covered in Section 4.3.4). Climate change impacts will be compounded by the rapid growth in environmental stresses arising from existing population growth and increasing tourism (IPCC 1996, WG II, Sections 9.4.3, 9.4.5). Coral bleaching and algal invasions are sometimes observed on the Great Barrier Reef, but the extent to which they are natural or human-induced is unknown.

Coral bleaching is associated with several factors (including extreme temperatures and solar irradiance, subaerial exposure, sedimentation, freshwater dilution, contaminants, and diseases) acting singly or in combination (Glynn, 1996). During bleaching, corals expel the symbiotic single-celled algae that live within their tissues, and then sometimes die (Brown, 1997). This occurs when sea temperatures rise more than 2°C above normal, which is often associated with ENSO episodes. The frequency of high-temperature episodes will increase as mean temperatures gradually rise, which would result in more frequent and widespread damage to corals—especially those remote from a reliable supply of larvae of reef species or stressed through exposure to local climatic and/or human impacts such as riverine runoff (Larcombe *et al.*, 1996), high solar irradiance (especially ultraviolet wavelengths) (Glynn, 1996), and pollution (Dubinsky and Stambler, 1996). The effects of temperature rise also will depend on latitude, coral height in relation to sea level, and the direction and size of the prevailing wave climate.

There is evidence, however, that the temperature changes by themselves may be slow enough for the coral reef biota to adapt, through changes in their symbiotic partnerships (Brown, 1997) or genetically over a number of generations, as better-adapted genotypes settle and survive. At the scale of whole coral reefs and their communities, migration to higher latitudes may be a possibility. However, there is a host of historical, hydrodynamic, and ecological factors that determine current distributions of coral reef biota (Veron, 1995), and it could be centuries before substantial changes to species composition and increases in diversity are detected at Australia's southernmost coral reefs.

It has been predicted that, by itself, a rise in sea level might benefit many reefs because corals on reef tops would have a renewed and/or extended opportunity for vertical growth—unlike the last several thousand years, when they have been limited by sea level to exclusively horizontal growth (Wilkinson and Buddemeier, 1994; Wilkinson, 1996). Thus, the potential effect of sea-level rise over the next 50 years in tidally limited reef-top coral habitats is an increase in living coral cover and, in many cases, an increase in topographic relief. This scenario is quite likely to be played out in sheltered parts of many reefs, assuming that the majority of these reef-top corals are not killed by increased temperature or other causes (such as burial by sediments if high-intensity rainfall events were to increase). Although the vertical growth of most shallow corals will be able to keep pace with a rate of sea-level rise of up to 5 cm per decade, the mechanisms for reef island growth are not expected to keep pace with sea-level rise, and therefore many low islands may eventually become uninhabitable (Wilkinson, 1996; see also Section 4.3.4).

If reef tops do not keep pace with sea-level rise, there could be dramatic changes in the reef-top zonation of corals, other biota, and abiotic reef substrata because these are strongly controlled by the energy of the waves that pass over the reef (Done, 1983). The tendency would be for the plunging point for waves to gradually move toward the back of the reef, causing changes in the benthic zonation at the new location of the breaker zone and for several meters to tens of meters either side. Moreover, any significant change in the frequency or magnitude of tropical storms (Suppiah *et al.*, 1996; Henderson-Sellers and Zhang, 1997; Holland, 1997) would affect both the structure and growth of coral reefs (Lough, 1994; Wilkinson, 1996), since individual storms can cause considerable damage to reefs, from which it may take a decade or more to recover.

#### *4.3.1.6. Adaptation and Vulnerability*

Adaptations to climate change may be categorized as either "autonomous" adaptations—where biota and ecosystems respond and change of their own accord—or "planned" or "conscious" adaptations, undertaken by humans. The primary option for conscious adaptation measures in respect to near-natural ecosystems is land-use management. This includes modification of land clearing, forestry practices, fire management, rangeland animal stocking rates, pest animal management, control of herbaceous and woody weeds, development of corridors for species migration, resting and rehabilitating degraded areas, and the consideration of climate change in management plans for urban, coastal, catchment, and other zones. Implementation of riparian (riverside) management practices and selective planting of upslope source areas would mitigate many of the adverse impacts of land development and hydrological changes on water quality and habitat values of streams, rivers, and ultimately lake and estuarine habitats. Active manipulation of species generally will not be feasible in the region's extensive, lightly managed, natural ecosystems, though in some circumstances—such as for endangered or commercially important species—relocation to new and more suitable habitats may be desirable and feasible.

Continued research on natural ecosystems also is a necessary, integral part of adaptation responses. In many instances, our knowledge is insufficient to provide the guidance sought by managers and policymakers. Routine climatic, environmental, and ecological monitoring and prediction also are necessary.

In summary, it must be concluded that some of the region's species and ecosystems are rather vulnerable to climate change because of the likely continued increase of greenhouse gases; the likely rate of climate change; the fundamental nature of likely changes to biota and ecosystems; the large proportion of Australasia affected; the ecological isolation and fragmentation of ecosystems of the region; existing environmental stresses; and the limited available options for conscious adaptation.

### *4.3.2. Hydrology and Water Resources*

*Summary: The four hydrological situations of most concern to the region are drought-prone areas, flood-prone urban areas, low-lying islands, and alpine snowfields. Model simulations suggest changes of as much as ±20% in soil moisture and runoff in Australia by 2030, with considerable variation from place to place and season to season and with the possibility of*

**Box 4-1. Macquarie River Basin Study: An Integrated Assessment of Impacts**

A collaborative study by a consulting company, two New South Wales (NSW) government agencies, and the CSIRO has made a preliminary integrated assessment of the impacts of climate change on the management of the scarce water resources of the Macquarie River basin in northern NSW (Hassall and Associates, 1997). The catchment contains dry-land agriculture (mainly wheat) and pastoralism, irrigated agriculture (mainly cotton), several small towns, and an episodically flooded wetlands area known as the "Macquarie Marshes," which is a major breeding area for birds. Over the past decade, agricultural and pasture production of sheep, beef, wool, wheat, and cotton contributed 92% to the regional economy.

The study considered the impacts of "low change" and "high change" climate scenarios in the region by 2030, based on estimates from the CSIRO regional climate model nested in the CSIRO slab-ocean GCM (experiment F1, Table 1-1), and IPCC (1996a) ranges of uncertainty in global warming. Spatially and seasonally varying projected rainfall and temperature changes in the catchment ranged from about 0 to -15%, and 0.4 to 1.2°C, respectively. These were used in a catchment model to quantify possible changes in moisture, runoff, and water supplies. Output was then used in the IQQM river management model developed by the NSW Department of Land and Water Conservation. Consequences for the pastoral, agricultural, and wider economy of the region were then considered, using simple models of yield and income on climate. Consideration of the need for "environmental flows" in the river and to ensure wetland breeding habitat led to limitations on water diversions for irrigation. Results of this study showed that mean annual runoff to the Burrendong Dam (the main water-storage facility) was reduced by 11% in the low-change scenario and by 30% in the high-change scenario, with correspondingly less water available for irrigation under present river management rules. Using optimistic assumptions regarding the beneficial effects of increased $CO_2$ concentrations on crops and pastures, the study found aggregate losses to the agricultural economy in 2030 of 6% in the low-change case and 23% in the high case. Beneficial effects of increased $CO_2$ on cotton approximately balanced the effects of climate change, but this was less true for wheat. By far the biggest losses were for sheep, beef, and wool, which constitute 63% of total agricultural production in the region.

The NSW National Parks and Wildlife Service forecast that, if the rules were not changed, the loss of water supply would lead to reduced filling of the Macquarie Marshes and thus reduced or less frequent breeding of some bird species, with possible local or regional extinctions, depending on what changes occur in other breeding areas.

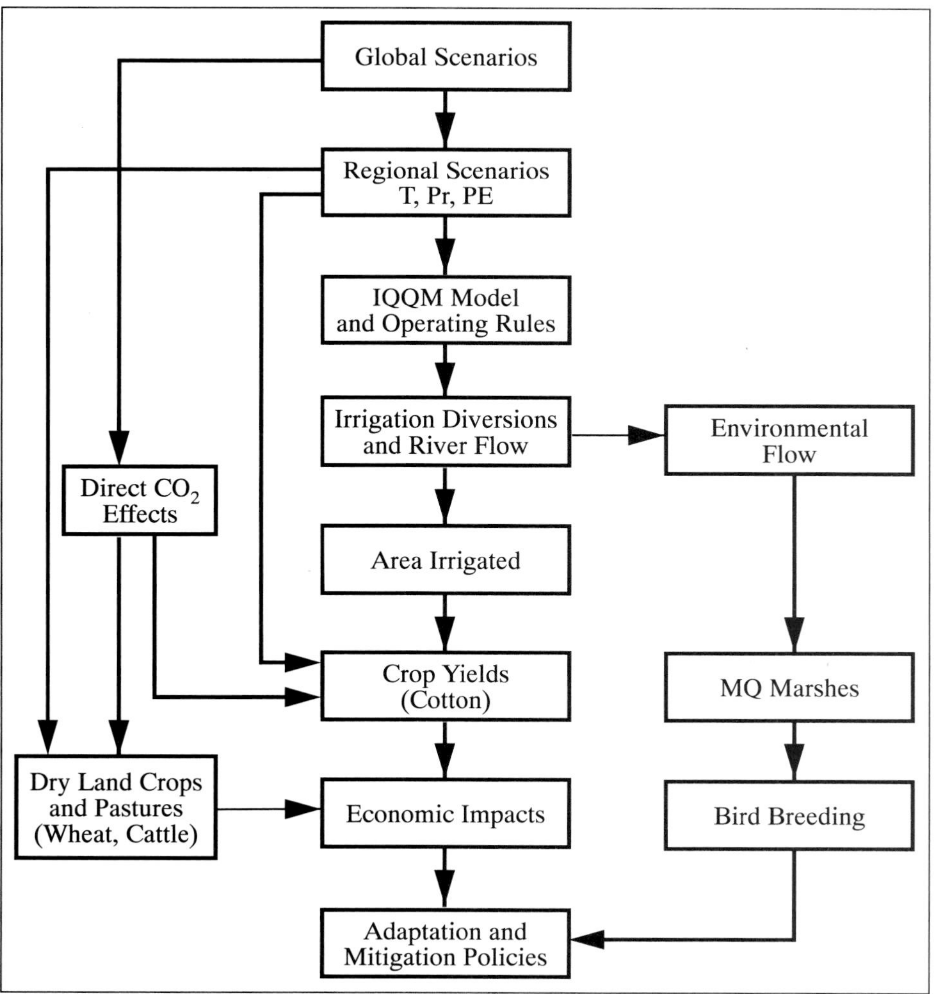


*Schematic of the integrated climate change impact assessment approach used in the Macquarie River Basin Study. T is temperature, Pr is precipitation, PE is potential evaporation, and IQQM is the "Integrated Quantity and Quality Model" used for modeling regulated river systems in New South Wales.*

*an overall reduction in average runoff. Water shortages would sharpen competition among various uses of water, especially where large diversions are made for economic purposes. One study shows Australia's major Murray-Darling River system facing constraints on existing irrigation uses and/or harm to the riverine environment. More frequent high-rainfall events would enhance groundwater recharge and dam-filling events but also would increase the impacts of flooding, landslides, and erosion. A preliminary study for an urban area near Sydney showed a tenfold increase in the potential damage of the "100-year" flood under a doubled $CO_2$ scenario. Water supplies on atolls and low-lying islands are vulnerable to salt-water intrusion of groundwater from rising sea levels and to possible rainfall reductions. Reduced snow amounts and a shorter snow season appear likely and would decrease the amenity value of the mountains and the viability of the ski industry. The glaciers of New Zealand's Southern Alps are likely to shrink further.*

*Adaptation options include integrated catchment management, changes to water pricing systems, water efficiency initiatives, building or rebuilding engineering structures, relocation of buildings, urban planning and management, and improved water supply measures in remote areas and low-lying islands. The financial exposure, and cost involved in potential adaptations, indicate a high vulnerability with respect to hydrology.*

#### *4.3.2.1. Hydrological Systems*

Hydrological systems are potentially very sensitive to changes in climate. The three key variables are soil moisture, which is a primary control on vegetation and ecosystems; groundwater recharge, which feeds groundwater reserves; and runoff, which feeds rivers and causes floods. Increased temperatures are expected to cause increased potential evaporation and less snow, and possible changes in mean rainfall, rainfall intensity, and rainfall seasonality would affect soil moisture, streamflow, and groundwater recharge and the occurrence of floods and droughts. More frequent high-intensity rainfall would tend to increase the occurrence of flooding. Specific effects will depend on the pattern of change, in rainfall particularly, and the characteristics of catchments and cannot be predicted with confidence. In general, the drier the climate, the greater the sensitivity to climate change (IPCC 1996, WG II, Chapter 10). Although the effects of rainfall changes and sea-level rise on groundwater resources are not adequately understood at present, they cannot be ignored (IPCC 1996, WG II, Section 10.3.6) and may be significant for inland and coastal aquifers in Australia (Ghassemi *et al.*, 1991).

Water resources in the region are strongly affected by the heavy rainfall of major weather events, such as tropical cyclones in northern Australia, and by the ENSO phenomenon, which is the main source of year-to-year variation and contributes both widespread heavy rainfall and widespread drought, depending on its phase. However, climate models are unable to represent these well as yet and therefore cannot represent the resulting major sources of surface runoff and groundwater recharge events. This situation presents a basic difficulty in assessing climate change impacts on hydrological systems and water resources.

A number of preliminary assessments and research studies of hydrological response in the region are available (e.g., Griffiths, 1990; Mosley, 1990; Bates *et al.*, 1994; Chiew *et al.*, 1995; Bates *et al.*, 1996; Fitzharris and Garr, 1996; Minnery and Smith, 1996; Schreider *et al.*, 1996). Most of the research work has been based on available regional scenarios of temperature and rainfall changes (see Section 4.2.3) and has focused on changes in water yield from unregulated rural catchments. Recent studies have begun to consider impacts on groundwater recharge (Green *et al.*, 1997) and on water resources systems as a whole (Hassall and Associates, 1997; see Box 4-1).

A wide range of significant changes in water yield and soil moisture was found by Chiew *et al.* (1995), who considered the potential impacts of CSIRO (1992) climate scenarios on 28 catchments that represent the large range of climatic, physical, and hydrological regimes experienced in Australia. By 2030, increases in annual runoff of up to 25% and 10% occurred for catchments in the wet tropics of northeastern Australia and in Tasmania, respectively. Decreases of up to 35% occurred for South Australia, and changes of ±20% and ± 50% occurred for southeastern Australia and the west coast, respectively. Changes in annual soil moisture levels ranged from -25% to +15%. Although the specific figures have high uncertainty (arising from the large scenario uncertainty), their magnitudes provide an indication of the size of changes that may conceivably occur.

Sizable changes in median monthly runoff during the wettest parts of the year and increases in annual maximum monthly runoff were found by Bates *et al.* (1996), who used the results of a single climate model (CSIRO9, experiment F1, Table 1-1) and a stochastic daily weather generator to represent climate and hydrological variability under current and doubled $CO_2$ conditions. The increases were due to the general increase in rainfall intensity and the increased frequency of heavy rainfall events indicated by the CSIRO9 model.

A similar shift toward greater variability was found by Schreider *et al.* (1996), who applied the "most wet" and "most dry" climate change scenarios for 2030 and 2070 (adapted from CSIRO, 1992) to historical daily rainfall and temperature series for 14 rivers in the Ovens and Goulburn Basins in southeastern Australia. Figure 4-2 shows that in scenarios in which rainfalls were projected to decrease ("most dry" scenario), the frequency of high-flow events substantially decreased, and the drought frequency (as indicated by a soil wetness index) increased. However, in scenarios in which rainfalls were projected to increase, there was little increase in average annual runoff, but the frequency of high flow events increased.

Changes in catchment vegetation—either from climate change directly or from adaptation responses (such as afforestation)—

would alter catchment hydrological characteristics, including evaporation, runoff, and extreme events (IPCC 1996, WG II, Sections 14.2, 14.4). The effects of changes in rainfall amount and timing on groundwater recharge can be amplified by the dynamic response of vegetation, according to a study by Green *et al.* (1997) that used a daily soil-vegetation-atmosphere model to determine changes in groundwater recharge beneath North Stradbroke Island in northeastern Australia. It was found that the net recharge increased consistently by amounts greater than the change in rainfall and that the recharge was more affected by vegetation type than by soil type.

The responses to climate change of the large, arid, ephemeral lake systems of interior Australia are difficult to predict. Significant water-level changes may occur for nonephemeral lakes in dry evaporative drainages or small basins where present evaporation is comparable with rainfall inputs (IPCC 1996, WG II, Section 10.3).

Rising sea levels will cause the tidal saltwater wedge to intrude further upstream in estuaries and rivers, with resulting changes in salinity affecting estuarine aquatic ecosystems (Chappell *et al.*, 1996; Waterman, 1996). Similarly, sea-level rise also may lead to greater saltwater intrusion into coastal groundwater, aquifers, and surface waters in some coastal systems (Ghassemi *et al.*, 1991).

Hydrological systems in the future also will be affected by other changes such as deforestation and urbanization, both of which tend to increase runoff amount and runoff speed—increasing the risks of flash flooding, high sediment loadings, and pollution. Changes to water pricing and allocation

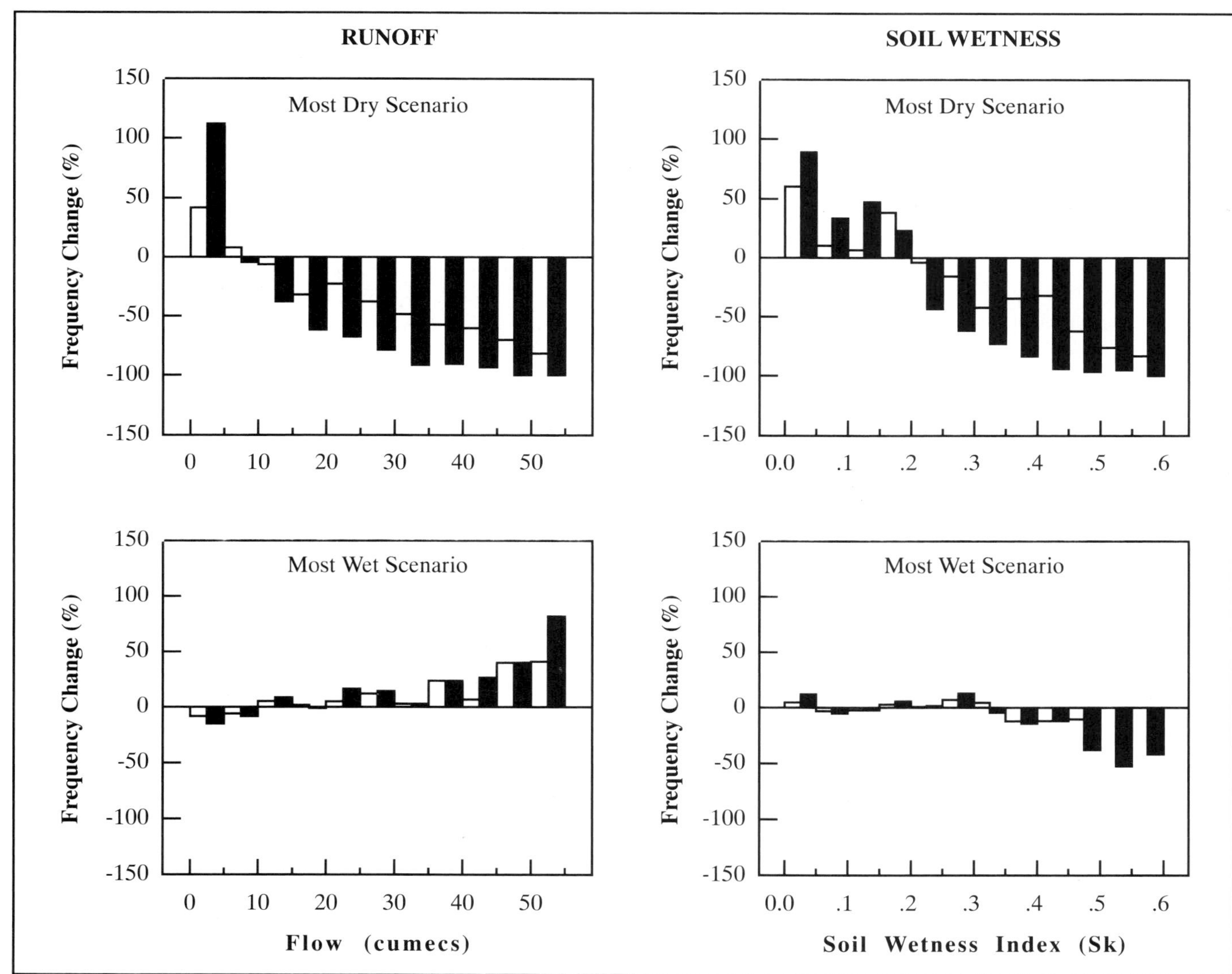


**Figure 4-2:** Change in frequencies of runoff and soil wetness for different climate scenarios. White bars are for 2030, and black bars are for 2070. Under the "most dry" scenario (top row), where rainfall decreases and warmings are relatively large, the frequencies of high runoff and high soil wetness substantially decrease. However, under the "most wet" scenario (bottom row), where rainfall increases and warmings are less, there is some increase in the highest runoffs but also some decreases in the highest soil wetness frequencies. For further details, see Schreider *et al.* (1996).

mechanisms also will affect patterns of water use and demand (Fenwick, 1995; McClintock, 1997), and indeed can be used as adaptation measures.

#### *4.3.2.2. Water Hazards and Water Supply*

If rainfall becomes more extreme, flooding, landslides, and erosion would become greater and/or more frequent (Fowler and Hennessy, 1995; IPCC 1996, WG II, Sections 4.2, 12.4). Locations most at risk include steep or unstable hill country, river valleys, and floodplains. Most of the region's settlements are adjacent to rivers and harbors. Furthermore, in low-lying coastal areas, sea-level rise will compound the problem by reducing river flood discharge rates and thus exacerbate flood heights and durations.

Where urban areas are exposed to flood risk, either directly or through failure of upstream river management works, the potential human and infrastructural vulnerability is extreme and has significant financial implications. Major costs to dam owners and public agencies would arise from any upward revision of Probable Maximum Flood estimates. The costs to private property and public buildings and infrastructure of floods under the existing climate are already high; for example, the damage caused by the 1974 floods in Brisbane was reported to be A$660 million (Minnery and Smith, 1996). A study by these authors of the Hawkesbury-Nepean corridor, an urban area near Sydney, estimated that the potential damage from the "100-year" flood rose more than tenfold for a doubled $CO_2$ scenario in which the flood return periods decreased by a factor of four (Table 4-1).

Climate change is expected to affect water demand, water supply, and water quality (IPCC 1996, WG II, Sections 14.2, 14.3). For many parts of the region, rainfall and water supply are generally adequate. However, the drier inland areas of Australia are vulnerable to potential water shortages during the seasonal minimum and during droughts; any additional shortages in these regions arising from climate change would sharpen competition among various economic, social, and environmental uses and hence increase the effective cost of water. Considerable demand arises from urbanization and the diversion of large amounts of water for economic purposes such as mining and irrigated agriculture. This competition may be exacerbated by trends toward population growth, higher valuation of natural waters, and possibly shifts to more intensive farming systems. If, on the other hand, there were an increase in water availability due to climate change, it might well encourage demand for more irrigation, with obvious short-term benefits—although in the longer-term this could lead to increased salinization in semi-arid regions.

**Table 4-1:** *Potential direct damage (in millions of Australian dollars) of a 1-in-100-year flood in the Hawkesbury-Nepean corridor for current and doubled $CO_2$ scenarios (Minnery and Smith, 1996).*

| | Current | Doubled $CO_2$ |
|---|---|---|
| Residential | 19 | 190 |
| Commercial | 8 | 140 |
| **Total** | 27 | 330 |

In central Australia, low rainfall and high evaporation forces the few towns—such as Alice Springs and Yulara—and other tourist centers, cattle stations (ranches), and Aboriginal and mining settlements to rely on fresh or desalinated brackish groundwater (Knott and McDonald, 1983; Jacobson *et al.*, 1989; Jacobson, 1996). Economic growth and population growth will put added pressure on these supplies, which are recharged in part by occasional heavy rainfall events. Rainfall changes would have complex impacts on this groundwater supply because any reduced recharge would lower water tables and water supplies, and *vice versa*. The effects on water quality are not clear.

The water resources of atolls and low-lying islands are usually restricted to rainwater or limited sub-surface supplies. These are sensitive to climatic variations and environmental impacts, and in many cases are already stressed by unsustainable demand and pollution. With climate change, the freshwater lenses beneath coral atolls are exposed to saltwater intrusion from rising sea levels and possibly increased storm events, as well as to possible rainfall reductions where these occur. The indigenous populations of the Torres Strait, the Cocos (Keeling) Islands, and Pacific islands associated with New Zealand are likely to be affected.

The major Murray-Darling River system, in southeastern Australia, is heavily regulated by dams and weirs and supplies 10 billion $m^3$ of water annually for human use, mainly for irrigation. This amounts to about 40% of the mean annual flow. Application of various scenarios (e.g., CSIRO, 1992) has suggested a possible combination of decreased mean rainfall, higher temperatures and evaporation, and a higher frequency of extreme events (i.e., more floods and droughts) (e.g., Schreider *et al.*, 1996; see Figure 4-2); if so, the inflows to the system may be reduced, and if irrigation demands remain fixed or increase, the remaining river flows would substantially decrease (see, e.g., Hassall and Associates, 1997, and Box 4-1).

The Asian-Pacific Integrated Model (AIM) team has simulated changes in drought and flood intensities for a wide region, including Australia and New Zealand; AIM (1997) shows results using the Geophysical Fluid Dynamics Laboratory (GFDL) Q-flux climate model (experiment A2 or A3). Based on the estimated magnitude of 10-year return period monthly discharge events, they show small and spatially-varying changes in flood discharge over Australia but apparently significant reductions in minimum flows in the Murray-Darling Basin. However, the rainfall simulation of experiment A2 over Australia is questionable (Whetton *et al.*, 1994).

Because of the higher variability of Australian rainfall, the storage capacities of Australian water supply dams are typically about six times larger than those of European dams for the same percentage of mean annual streamflow and probability of water shortfall. In sharp contrast, New Zealand's higher and more reliable rainfalls enable its hydro-electricity system to operate on a total reservoir capacity equal to only about six weeks of national demand—albeit with the attendant risk of a rapid decline in supply in times of drought, as was experienced during 1992 (Fitzharris, 1992).

#### *4.3.2.3. Snowfields and Hydroelectricity*

Increased temperatures will lead to a reduced fraction of precipitation falling as snow, higher snowlines, earlier spring snowmelt, and a shorter snow season in Australia (Whetton *et al.*, 1996c) and New Zealand (Garr and Fitzharris, 1994; Fitzharris and Garr, 1996). Figure 4-3 shows the results of a simulation of the progressive shortening of the snow season at Mt. Bogong—one of Australia's higher-altitude snow areas—over the first half of the next century, based on the CSIRO (1992) scenarios. Greater ablation of New Zealand's Southern Alps' glaciers can be expected, and the volumes of glaciers and total snowpack may decrease, depending on precipitation changes (note that the new scenarios for New Zealand referred to in Section 4.2.3 may result in increased precipitation in the Southern Alps). However, it has been estimated that even with a 10% increase in precipitation, a 2°C temperature rise would cause a 20% reduction in the Southern Alps' snow amount (IPCC 1996, WG II, Section 7.4.1). Reductions in relative snow amounts would change the seasonality of runoff by increasing winter runoff and decreasing spring runoff (IPCC 1996, WG II, Section 10.3.5). This would reduce spring flood risks and provide more seasonally smooth hydroelectricity generation (IPCC 1996, WG II, Section 7.5.2). Any changes in atmospheric circulation patterns also would cause changed seasonality of rainfall and river flows, but there is less certainty about the nature of these changes.

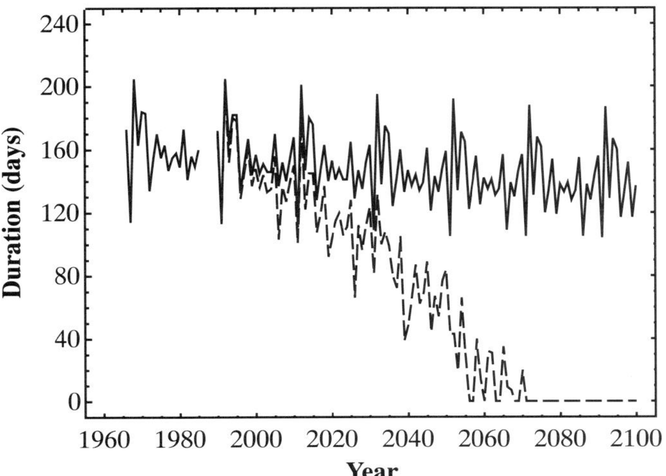


**Figure 4-3:** Simulation of the future decrease in the length of the snow season at Mt. Bogong in southeastern Australia, for best-case (solid) and worst-case (dashed) climate change scenarios. An indication of year-to-year variability is provided by superimposing the year-to-year climate anomalies for 1966–85. Further details can be found in Whetton *et al.* (1996c).

Under the scenario of more frequent extreme rainfalls, hydroelectricity systems would need to be managed more conservatively, to avoid the risk of overtopping of dams in floods and running out of water during droughts. Increased sediment transport would accelerate the reduction of the storage capacities of hydro lakes. Increases in the temperatures of rivers would reduce the rivers' capacity to cool thermal generating plants, with increased difficulty in meeting regulatory constraints on downstream river temperature (IPCC 1996, WG II, Section 14.3.3).

Decreased snow amounts would most likely reduce the amenity value of mountain landscapes for the local population and tourists and reduce the viability of the region's ski industry, whose options for relocation are very limited—by low altitude in Australia and by rugged terrain and conservation estate regulations in New Zealand (IPCC 1996, WG II, Section 7.5.5; Whetton *et al.*, 1996c).

#### *4.3.2.4. Adaptation and Vulnerability*

The hydrological situations expected to be of most concern in the region are the water-limited drought-prone areas, the exposure of the built environment to possible increases in flooding, the supply of potable water to remote indigenous populations in inland Australia and on low-lying islands, and the progressive loss of snowfields.

Integrated catchment management provides an adaptation framework for the long-term management of catchment water and surface properties and the short-term tradeoffs between competing demands for water. Already, major systems such as Australia's Murray-Darling River system are subject to intensive management, but generally this does not include consideration of possible decadal-scale changes or even of the predictable El Niño-related seasonal-to-annual variations. Water pricing and water efficiency initiatives may be used as an effective adaptation strategy (Fenwick, 1995; McClintock, 1997). The risk of landslides and soil erosion can be reduced through informed land management, particularly by avoiding vegetation clearance in vulnerable areas and by rehabilitating exposed areas that have been cleared.

Urban planning and management will be needed to deal with increased flooding risk—for example through provision of retention basins and zoning. Any slow change in mean rainfall or rainfall intensity, if well-enough known in advance, could be accommodated within the lead time of about 10 years needed for planning and constructing major facilities. Dams and flood-protection works can be redesigned and rebuilt and vulnerable buildings relocated, albeit at considerable cost. However, any slow increase in interannual rainfall variability may result in early stresses for supply systems designed to cope with existing rainfall variability; at such times, the adaptation options would be very limited, as was shown during the public water supply crisis in Auckland in 1994. Measures needed to improve the water supply situation of indigenous peoples in inland Australia and on Australia's low-lying coastal islands are discussed in Moss (1994).

Adaptations to change in snowfields and glaciers are very limited. Artificial snowmaking is a potential strategy for skifield operators, but only to a limited extent because of the strong impact of temperature increases on snow amount, the environmental impacts of the large water storage dams required, and the costs involved.

Although many management and engineering adaptation options exist, the magnitude of the financial exposure and the considerable time and expense involved in adaptations indicate a high vulnerability with respect to hydrology.

### 4.3.3. Food and Fiber Production

*Summary: Significant impacts on agriculture are likely, including crop and pasture performance increases from $CO_2$ rises; mixed effects of temperature rises; changes in soil fertility; changes in quality of grain and pasture nutrition; shifts in the suitability of districts for particular crops, such as kiwifruit; and possibly increased problems with weeds, pests, and diseases.*

*Impacts will vary widely from district to district, crop to crop, and decade to decade. Grain crops may gain in the first few decades because of the immediate beneficial effect of higher $CO_2$, but that advantage may be eroded as the delayed rise in temperature becomes greater and reduces the grain-filling period. In Australia's warm low latitudes, some crops are near their maximum temperature tolerance and are likely to suffer increasingly as the temperature increases. The possibility of overall decreased rainfall in Australia would negatively affect rangeland pastoralism and irrigated agriculture (which is a major source of production). Any changes in global production, and hence international food commodity prices, would have major economic impacts.*

*Farming in the region is well adapted to dealing with variability and change through a variety of natural and market factors—for example, by means of plant breeding, diversification, seasonal climate prediction, and so forth. Such techniques are likely to be sufficient to adapt to climate change over the next few decades; in some cases these techniques may facilitate expanded production. As the time horizon extends, however, the changing climate is likely to become less favorable to agricultural production in Australia, leading to a long-term increase in vulnerability.*

*Production forestry will be affected by changes in tree productivity; forest operational conditions; and pests, weeds, disease, and wildfire incidence. The net impact is not clear. The longer time to reach maturity results in a relatively large exposure to financial loss from extreme events, fire, or any rapid change in climate conditions.*

*A number of climate sensitivities have been identified for fisheries, but the impacts cannot be predicted with any confidence. It appears that freshwater and near-shore fisheries will be more affected than open-ocean fisheries. The principal adaptation option is scientifically based integrated fisheries management and coastal zone management.*

#### 4.3.3.1. *Agriculture and Horticulture*

The possible interactions among atmospheric $CO_2$, climate, soil, plants, animals, and humans in the agricultural system are very complex, which makes it difficult to draw firm conclusions about agricultural responses to projected climate changes (Warrick *et al.*, 1996). However, research has indicated that increased $CO_2$ fertilization, higher temperatures, and changed rainfall characteristics are likely to lead to changes in productivity and crop yield; changes in grain quality and animal pasture feed quality; changes in the incidence of diseases, pests, and weeds; and changes in the suitability of districts for particular crop types and farming systems (IPCC 1996, WG II, Chapter 13). Furthermore, it can be expected that climate change impacts on other countries' agriculture will lead to changes in global food markets and that these external changes may have as great an effect on the region's agricultural industries as internal changes.

The yield responses reported for studies of Australasian crops and pasture (IPCC 1996, WG II, Section 13.6.4 and Table 13-7) cover a wide range of values, depending on the situation studied and the research assumptions. Considering atmospheric $CO_2$ concentration and temperature rises (but not necessarily rainfall changes) and ignoring adaptive measures, it appears that pasture productivity may generally increase (e.g., Greer *et al.*, 1995), but annual crop yield may increase only slightly or remain about constant (e.g., Wang *et al.*, 1992; Rawson, 1995; Wang and Connor, 1996), or even reduce slightly (Muchow and Sinclair, 1991). As long as slower-maturing grain cultivars are available, crop yields may benefit, with the beneficial effects of higher $CO_2$ concentrations outweighing the reductions in yield expected from shorter grain-filling periods associated with warmings (Wang *et al.*, 1992; Wang and Connor, 1996). However, if sufficiently longer-maturing cultivars cease to become available due to greater warmings, yields may decrease. This would happen first at low latitudes, and the effect would increase as further warming occurs later in the next century (Pittock, 1995, 1996).

An empirically based study by Nicholls (1997) has attributed 30–50% of the observed increase in Australian wheat yield since 1952 (which was about 45%) to observed climatic trends, with increases in minimum temperature the dominant influence. The method excludes the direct physiological effects of increasing $CO_2$ concentration on yield but could be confounded if farmers used more fertilizer in climatically favorable years.

The interaction of climate change and increasing $CO_2$ concentrations on pasture quality and subsequent animal productivity remains uncertain (Owensby, 1993). Pasture plants grown under elevated $CO_2$ concentration always seem to have lower protein (Arp and Berendse, 1993) and higher nonstructural carbohydrate contents (Lutze, 1996), which implies an impact on the nutritional quality of the forage. In high-rainfall temperate regions, particularly early in the growing season, most forage is eaten as green sward (or products preserved from green material), in which case the increased soluble carbohydrate content would increase the digestibility and nutritional value of the forage (Smith *et al.*, 1997). Reductions in feed quality may occur through reductions in digestibility and increase in lignin content of tropical grasses grown at higher temperatures (Wilson, 1982) and increased stem-to-leaf ratios of plants grown under high $CO_2$ levels (e.g., Morison and Gifford, 1984). Thus, rangeland forage quality is likely to decrease at some times of year—suggesting an increased need for feed supplements or legume pastures. However, higher plant water-use efficiency under elevated $CO_2$ levels may result in more "green days" each year, partly offsetting the above effects (McCown, 1980).

In Australia's rangelands, scenarios of climate change suggest alterations in the temporal and spatial components that are likely to increase variability and unpredictability in plant productivity and community composition. Increased competition from woody weed species in mixed pasture may have a significant negative effect on grass growth (Scanlan, 1992) and hence livestock production (Stafford Smith *et al.*, 1994). It is anticipated that in temperate areas of Australia and New Zealand, encroachment of subtropical grasses into pastures may be exacerbated not only by the warming conditions but also by the strong response of $C_4$ growth during recurring periods of water stress (Campbell *et al.*, 1996). This would affect the types of possible grazing activity and would require greater consideration of land management issues, such as the use of fire for woody weed control and pest animal management (Stafford Smith *et al.*, 1994).

Aborigines have a large and increasing role in land management in the pastoral zone (Young *et al.*, 1991; Ross *et al.*, 1994). Aboriginal land supports a wide range of uses, including pastoralism, hunting and gathering, horticulture, conservation, cultural tourism, and mining. Climate change may affect native bush food supplies and sacred biota, as well as water resources for remote Aboriginal communities.

Temperature increases will very likely lead to an increase in pasture production in mid-latitudes, with corresponding increases in livestock production. However, if most of the increase in pasture growth comes in summer, winter feed may remain the limiting factor unless there is increased feed conservation or changes in livestock production systems. Increased pasture production may not occur if there is an increase in the frequency of extreme events, such as summer droughts or late-winter snowstorms in lowland sheep areas of New Zealand. Any increases in production also would depend on farmers' attitudes toward risk and on the signals reflected back to producers from the effect of economic and climate changes on customers and competitors (Martin *et al.*, 1991).

Livestock—which in this region are largely unhoused—are expected to benefit from warmer winters, possibly extended growing seasons, and possible minor improvements in feed quality in temperate high-rainfall zones. In some circumstances, however, they may be negatively affected by lower nutritional quality of feed (Owensby, 1993) and greater summer heat stress (Russell, 1991; McKeon *et al.*, 1993). There may be increases in diseases of tropical origin but decreases in cold-weather diseases (Sutherst *et al.*, 1996). In the New Zealand high country, it is not uncommon for huge losses of livestock to result from cold, wet weather and heavy snowfalls during winter and early spring. Warmer winters may lead to reductions in the numbers of animals lost.

Grain quality may decline as a result of increases in the frequency of high-temperature extremes during grainfill (Blumenthal *et al.*, 1991a,b), and increased atmospheric $CO_2$ may reduce grain protein contents under some circumstances (e.g., Gifford *et al.*, 1996). Any reduced protein content of grains would be detrimental for some grain products (like bread wheat), reducing their value, but advantageous for others (like brewing barley).

There is also the possibility of an increase in the incidence of pests and diseases and in some areas an increase in competition from weed growth (Farrow *et al.*, 1993; Sutherst *et al.*, 1996). However, Sutherst *et al.* (1996) noted the general lack of quantitative research on the effects of climate on weed species dynamics and the current inability to predict such interactions.

Temperature rises are expected to result in a general poleward shift in preferred crops and production systems, though there will be many subregional variations depending on the total climatic and soil requirements of a crop.

Higher temperatures will reduce the viability of existing temperate horticultural areas, especially with respect to winter chilling requirements for apples, stone fruit (Hennessy and Clayton-Greene, 1995), kiwifruit, and some vegetables and cereals such as winter wheat and barley. There are likely to be positive effects for tropical fruits and warm-climate crops such as maize and sweet corn. For these crops, the longer growing seasons, increased availability of "growing degree days," and reduced risk of early- or late-season frost damage will allow the use of later maturing cultivars with higher yield potential. Warmer conditions will accelerate crop development, allowing crops to avoid early- and late-season climatic risks and perhaps providing better timing of harvesting to meet market needs (e.g., Gifford *et al.*, 1996b). However, unless cultivars are changed, the shorter growth duration will allow less opportunity for yield accumulation and hence lower yield.

Recent work using a simple model developed by Salinger and Kenny (1995) shows that a critical change in kiwifruit production in New Zealand may occur around the year 2050 (Warrick *et al.*, 1996), although there are large uncertainties and caveats. Increased winter temperatures would reduce winter chilling, requiring the greater use of chemicals to manage the breaking of dormancy, but overseas market resistance may require reduction in the use of such chemicals in the future. Ongoing work using the CLIMPACTS model (see Section 4.4) also indicates that warmer conditions will lead to later maturity of kiwifruit. Changes in maturity dates, especially for temperate, perennial fruit crops, may make it more difficult to match the timing of harvesting to market needs.

Biological risks are a major concern for the region's agricultural industries. As long-isolated island countries, Australia and New Zealand are extremely vulnerable to invasion by foreign weeds, insects, and pathogens for which there are no natural local biological controls (Sutherst *et al.*, 1996). These invaders can directly affect the productivity and quality of pastoral lands, horticultural crops, and forest products and cause the closure of foreign markets to exports. Vulnerability to biological risks may alter under a changed climate, both from the direct effects on biota and ecosystems and indirectly from any adaptations that involve changes to crop types and farming systems.

Any changes in rainfall or rainfall reliability—for example, from changes in ENSO characteristics or other regional synoptic circulations—causing increased drought in drought-prone areas of either country could seriously affect pastoral and dairy production (e.g., McKeon *et al.*, 1993). Irrigated agriculture produces a significant part of Australia's agricultural income, and rainfall decreases would have a major impact on the demand for and the relative availability of irrigation water (Pittock, 1995; Schreider *et al.*, 1996; Hassall and Associates, 1997). Conversely, wetter conditions may result in the expansion of cropping and pastoralism (McKeon *et al.*, 1988). Some crops that require dry conditions at certain stages, like grapes, could be harmed by any increased frequency of unseasonal rainfall (Boag *et al.*, 1988).

There may be increased risk of soil erosion in agricultural areas from the expected increase in the frequency of intense rainfalls (Yu and Neil, 1995). This will require improved soil surface management to reduce runoff rates (McKeon *et al.*, 1988).

#### 4.3.3.2. *Production Forestry*

The production of wood and paper products is a major industry in the region. In New Zealand, the industry is primarily based on plantations of fast-growing exotic species, whereas in Australia there is still a considerable forest industry based on native eucalypt forests in addition to expanding plantation forests. A rising world population, combined with increased living standards, may (at current prices) demand more wood than can be supplied from global resources by the year 2050. Temperate forest products are expected to play an important role in meeting demand (IPCC 1996, WG II, Sections 15.2.3, 15.4.3). There also is an increasing trend to see plantation forests as a means of sequestering carbon in the region (Maclaren, 1996).

$CO_2$ fertilization is likely to increase growth, especially when trees are water limited. However, growth enhancements due to increased temperature may occur only in trees not subject to water limitations (Landsberg, 1996). As already noted, however, the $CO_2$ impact on tree growth decreases with time for many species (Gunderson and Wullschleger, 1994). Any overall reductions in rainfall or changes in seasonality that result in water limitations or prolongation of droughts would negatively affect production and plantation seedling establishment. On the positive side, in New Zealand forestry many of the worst exotic weeds are $C_4$ plants, so the competitive effects may be reduced if elevated $CO_2$ concentration favors growth of $C_3$ over $C_4$ plants in temperate zones. Again, however, this gain might be negated by the relative advantage of $C_4$ plants at higher temperatures. Changes in tree and forest water use would alter the catchment hydrological characteristics, particularly runoff and extreme events (IPCC 1996, WG II, Sections 14.2, 14.4).

The exposure of forests and forest operations to fire risk may increase, particularly in Australia, and there is potential for changes in the frequency and intensity of damaging events from wind and storm, particularly in New Zealand. More intense rainfall events would exacerbate soil erosion and pollution of streams during forestry operations and make these operations more difficult to carry out. Warmer and wetter conditions could provide the opportunity for increased incidence of arthropod pests and pine needle blight (*Dothistroma pini*). *Pinus radiata* constitutes 91% of the exotic plantation forests in New Zealand and 68% in Australia, so this blight is a major potential risk to production forestry.

An important distinguishing feature of forestry is the long time scale of the tree lifecycle and the very large investment in the standing crop relative to the annual yield. Whereas a wheat farmer stands to lose one year's production in a climatic disaster, a forester may have at risk a full 30 years' growth. Thus, more so than with agricultural crops, global change has the potential to adversely affect the substantial accumulated capital of a standing forest. Furthermore, the long time scale means there are less frequent opportunities to apply adaptation options to any particular forest. The slower growing indigenous forests will be even more affected than the fast-rotation exotic forest plantations.

#### 4.3.3.3. *Fisheries*

Relatively little is known about how climate affects marine fishes—particularly in the Southern Hemisphere, where data series tend to be short. Thus, it is extremely difficult to predict how climate change may affect Australasia's fish stocks and fishing industry, particularly in the context of the present stresses on fish stocks. There is some evidence, however, that climate impacts can be quite profound. The IPCC Second Assessment Report (SAR) concluded that although global marine fisheries production may remain about the same—despite possible changes in dominant species—there are likely to be collapses and expansions of specific regional fisheries (IPCC 1996, WG II, Section 16.2.2). These conclusions may likewise apply to large oceanic regions like Australasia, though current knowledge is not adequate to predict the impacts—positive or negative—on total productivity for the region.

In these circumstances, mobile high-seas fishing fleets are less likely to be affected, provided that access regulation is not tied to geographical areas. However, among the more localized small-scale fishers, who are dependent on specific in-shore fisheries, there may be large gains and large losses if fish populations shift their areas of abundance (IPCC 1996, WG II, Chapter 16). In addition, in-shore fisheries and marine stocks that need to reproduce in freshwater, estuaries, or mangroves may be negatively affected by changes in terrestrial and coastal processes, such as increased pollution and sediment discharge or loss of habitat. The economic impacts are unclear but could be significant for some parts of the industry.

Climate conditions are a factor in the outbreaks of bloom-forming algae and shellfish diseases that periodically occur in Australasian waters. These organisms include naturally occurring species and exotic species introduced by discharge of ships' ballast water. In the region's nontropical waters, some of the organisms are likely to be at the margins of suitable temperature conditions, and warming may give them increased opportunities to survive, spread, and form problem populations (IPCC 1996, WG II, Section 16.2.4).

Aquaculture and freshwater fisheries at mid-latitudes may benefit from longer growing seasons, lower natural winter mortality, and faster growth rates. Studies for New Zealand suggest that rising sea level and temperatures may increase oyster farm areas and productivity—but that sea temperatures may become excessive for salmon farming. Any increases in rainfall intensity over land are likely to increase coliform bacteria counts in runoff and result in more frequent closures of shellfish beds (IPCC 1996, WG II, Section 16.2.3).

#### *4.3.3.4. Adaptation and Vulnerability*

Agriculture is an intrinsically adaptive activity. Farming in the region is highly decentralized, technologically well-supported, and market-responsive and routinely deals with variability on a variety of time scales, arising from climatic, biological, and market factors. Operational decisions on the variety of crop and animal options are made annually or more frequently; structural investment decisions on the farm or orchard, and at processing plants, have a currency of a decade or so. Although adjusting production systems to changing climate will not be without cost and will require systematic awareness-raising and information dissemination (Stafford Smith *et al.*, 1994), it is very likely to be a smaller and slower influence than changes arising from markets, prices, and technology. Furthermore, gains in production in some areas may offset or even exceed costs.

Adaptation options include plant breeding and cultivar choice, adjustment of planting times to realign thermal and vernalization requirements, changes in crop sequences, improved soil management, diversification of crops, adoption of sustainable farming methods, monitoring and prediction of seasonal climate and associated crop and pasture/livestock modeling, and regional monitoring and management of drought (Gifford *et al.*, 1996b). Adaptation strategies will need to go hand in hand with mitigation strategies to reduce farm system emissions of greenhouse gases where there may be significant opportunities to act as a carbon sink (Ash *et al.*, 1996) or reduce methane emissions (Howden *et al.*, 1994).

In rangeland agriculture, more flexible management responses to enable adjustment to fluctuating forage supplies would be required if there were changes in the frequency or intensity of extreme events (McKeon *et al.*, 1993). The resulting land-use change may have considerable regional socioeconomic impacts—as well as raising issues relating to sustainable use of land and water resources. Approaches to soil erosion control, pest animal management, weed control, and tree-clearing may have to be adjusted as climate changes. Seasonal to interannual climate prediction, particularly of rainfall in the ENSO-affected agricultural areas, is an adaptation option that offers increasing potential to manage climate variability.

Existing trends of diversification and specialization also provide a basis for the development of adaptation strategies (Stafford Smith *et al.*, 1994). Many potential adaptation options already will exist in particular farming systems or particular localities, but their widespread use may require further research and coordination at different scales, from land managers to governments (McKeon *et al.*, 1993). Adaptations involving changes to crop types, farming systems, and adjacent ecosystems may change vulnerability to biological risks; such risks also may be minimized by the introduction of less vulnerable species or increased diversification in farming systems.

The existing diversity of uses of Aboriginal land will provide resilience in coping with change, though some traditional management approaches such as the use of fire and harvesting of native foods may need to be modified. There will be a need to develop awareness of climate change among Aboriginal managers, and to learn from the traditional Aboriginal management, which has survived past climatic changes (Stafford Smith *et al.*, 1994).

Recent policy changes by the region's governments have shifted a greater part of the responsibility for agricultural risk management to farmers and the private sector. This shift was effected partly to cut the cost and overheads of government payouts to farming communities during droughts and other weather-related disasters and partly to promote more economically and environmentally rational decisions in land use and farm investment; the goal was to encourage decisions that properly account for the long-run risks involved. Although this approach is sound in principle and should improve outcomes over the medium term, there probably will be many individual farmers who will be unable to appraise, cost, or insure against a widespread very extreme event, an increased frequency of extreme events, or the generally uncertain effects of climate change. These risks inevitably become shared by the whole community and so remain a responsibility for community cooperation and government leadership. Few farmers in the region make use of commercial crop insurance.

The capacity of the region's agricultural industry to adapt to climate change will depend on the magnitude of change and hence the time frame. Over the next few decades—when the warming will be relatively small and the rainfall perhaps little changed—adaptation techniques are likely to be sufficient to cope without great consequences, and vulnerability will be small. However, as the time horizon extends and the climate changes become larger, there is likely to be a trend toward reduced production and increasing aridity in many areas (mainly in Australia). The greatest vulnerability appears to lie in the context of this long-term outlook, in the uncertainty of possible changes in rainfall associated with synoptic weather patterns and the ENSO phenomenon, and in possible external market responses and biological risks.

The forestry industry has some degree of vulnerability to climate change but may also gain from productivity increases. Greater attention to forest management—particularly with respect to seedling establishment, soils, fire risk, and disease—may be a required adaptation. In exotic production forestry, the multidecadal crop cycle is still sufficiently short to allow some adaptation through choice of species and of areas for planting.

The principal adaptation option for all categories of fisheries is integrated management through international and national mechanisms as appropriate and including consideration of habitat and all life-cycle stages. The greatest vulnerability is expected for freshwater fisheries generally—owing to direct temperature and hydrological effects and limited adaptation possibilities in the confines of rivers and lakes—and for fisheries dependent on estuaries and mangroves that may become subject to sea-level rise, flooding, and pollution by organisms, chemicals, and sediment from runoff (IPCC 1996, WG II, Chapter 16).

### 4.3.4. Coastal Systems

*Summary: The impacts on coastal zones of sea-level rise and climate change include inundation, riverine flooding, saline intrusion, erosion, and wave damage. The impacts of changes in weather conditions (winds, waves, storms, and storm surge) may be comparable to those of sea-level rise alone. Beaches, estuaries and coastal wetlands, and reefs—including the Great Barrier Reef—have adapted naturally to past changes in climate and sea level. However, in the future they are likely to face faster rates of change, and in many cases landward migration will be blocked by human land uses and infrastructure. Impacts will be complex, both physically and socioeconomically, and will vary greatly from place to place. There is potential for considerable damage to the region's low-lying coastal settlements and infrastructure where populations, tourism, and capital investment are large and growing. New Zealand also is exposed to impacts on its Pacific island territories, including the eventual possibility of environmental refugee fluxes. Some parts of Australasia's coastline, including a number of coastal cities, and indigenous communities in low-lying coastal settlements and islands are highly vulnerable.*

*Adaptation options include integrated coastal zone management; redesign, rebuilding, or relocation of capital assets; protection of beaches and dunes; development zone control; and retreat plans.*

#### 4.3.4.1. Coastal Zones and Responses

The Australasian region contains diverse coastal zones spanning tropical to subantarctic latitudes and including continental coasts and small islands. The habitats of coastal zones and their adjacent small islands support a great range of ecosystems and complex food networks. A variety of economic activities are located in the coastal zone, which also provides important landscape and recreation values. Direct modification of the coastline through structural works is confined to limited areas of industrial, urban, and resort development. However, changes in sediment processes as a result of upstream land-use change and modification of areas adjacent to the shoreline and of coastal dune systems have resulted in some significant morphological changes in coastal areas. In addition, discharge of nutrients, sediments, and pollutants from agricultural and urban runoff and domestic and industrial wastes affects the composition and quality of some coastal and estuarine waters (Zann, 1995; Larcombe *et al.*, 1996).

The Australian coastline is about 70,000 km in length (at the 0.1 km scale); it includes 12,000 islands and 783 major estuaries, 415 of which are in the tropics. About 24% of the Australian coast consists of dunes and beaches, and there are large areas of intertidal and supratidal mud, alluvium, mangroves, and seagrass, as well as thousands of kilometers of coral reefs (Zann, 1995). The New Zealand coastline is approximately 15,000 km—of which 25% is eroding, 19% accreting, and 56% (mostly hard-rock cliffs) static (Gibb, 1980); it is more tectonically active than the Australian coast.

The potential impacts of climate change on the coastal zone include the direct effects of rising sea level; rising temperatures; and possible changes in weather, storminess, and wave characteristics, together with the indirect impacts of changes on adjacent lands and rivers—resulting in altered inputs of freshwater, floodwaters, sediments, nutrients, and biota to the coastal seas. Global sea level is projected to rise 15–110 cm above current levels by the year 2100, with a central estimate of about 50 cm (IPCC 1996, WG I, Section 7.5.2). The estimated rate of rise is about two to five times faster than that experienced over the past 100 years.

Actual rates of rise at specific coastal locations can vary substantially due to local rise or fall of the land, different rates of oceanic warming and expansion between regions, and changes in oceanic and atmospheric circulation under climate change (Pittock *et al.*, 1995; IPCC 1996, WG I, Section 7.2; IPCC 1996, WG II, Section 9.3.1). It is likely that storminess will increase in some regions and decrease in others, with parallel increases and decreases in impacts.

The coasts of Australia and New Zealand are not simple, passive systems but respond dynamically to sea level and climate in a range of ways depending on local circumstances. Weather elements and related processes such as wind speed and direction, rainfall intensities, ocean wave energies, and storm surge have major roles in coastal geomorphology with respect to the mean state and extreme events. Even small changes in weather patterns may cause impacts commensurate with those of sea-level rise alone (Cowell *et al.*, 1996; IPCC 1996, WG II, Box 9.4; McInnes and Hubbert, 1996). Recent reviews of climate change impacts and adaptations for the region may be found in Chappell *et al.* (1996) and Kay *et al.* (1996).

Global research indicates that estuaries and coastal wetlands have coped with historical sea-level rise—for example, by migration landward. Salt marshes and mangroves have survived where the rate of sedimentation approximates the rate of local sea-level rise; beaches have grown or decayed according to changes in prevailing winds and seas; and coral reefs have demonstrated the capacity to grow vertically in response to past sea-level rise (IPCC 1996, WG II, Section 9.4). However, these past rates of natural adaptation may be insufficient for higher rates of future sea-level rise; in many cases, landward migration will be blocked by human infrastructure such as causeways, flood protection levees, and urban development, leading to a reduction in the area of the wetland or mangrove.

Sea-level rise will increase the penetration of the saltwater wedge in the many tidal estuaries, leading to salinity changes and consequent effects on estuarine ecosystems (see Section 4.3.2.1). Some coastal aquifers also may be affected by saline intrusion (Ghassemi *et al.*, 1991).

Natural responses may be complex. For example, in macrotidal estuaries in northern Australia, channel widening from sea-level rise may contribute sediment to adjacent estuarine plains, enhancing vertical accretion. Backwater swamps and freshwater ecosystems on the estuarine plains may be endangered (IPCC 1996, WG II, Section 9.4.2). In these sediment-rich areas, where strong tidal currents redistribute sediments, mangrove communities have a better chance of survival (IPCC 1996, WG II, Section 9.4.4; see also Section 4.3.4.3 below). However, losses of coastal wetlands for Australia are expected to be greater than the global average (IPCC 1996, WG II, Section 9.5.2.3).

The potential impacts of climate change on Australia's coral reefs, including the Great Barrier Reef, have been discussed in Section 4.3.1.5. Coral reefs face the types of coastal impacts described above, questions about whether reef growth can keep pace with sea-level rise, and the problems of coral bleaching and algal invasions. For small coral islands, the higher sea level will erode inhabited coral cays and contaminate water tables to an extent that is likely to be prohibitively expensive to combat or rectify (Wilkinson and Buddemeier, 1994). On many of Australia's coral reef islands, the impacts are likely to be highly visible and would have significant consequences for tourism activities.

Climate change impacts on the coasts will be compounded by rapid growth in environmental stresses arising from existing population growth and increasing tourism.

#### *4.3.4.2. Coastal Infrastructure and Human Impacts*

The vast majority of the population and infrastructure of both Australia and New Zealand lies in the coastal zone (Resources Assessment Commission, 1993; Zann, 1995). Thus, potential damage to coastal settlements and infrastructure from rising sea level, coupled with possibly increased storminess and more intense flooding events, is a major concern in view of the magnitude of capital investment (IPCC 1996, WG II, Section 12.3.4) and the nature and potential cost of adaptation options. The exposure of such areas is increasing as coastal populations continue to grow. Ports, harbors, and low-lying tourism and recreational developments such as beach resorts and marinas are most at risk.

Shipping and offshore oil and gas facilities in Australia's Bass Strait and Northwest Shelf and off New Zealand's Taranaki coast are exposed to any increased storminess. Populated areas already are vulnerable to pollution from human activity of terrestrial and marine origin, including sewage wastes, industrial wastes, agricultural runoff, oil spills, and ballast water discharge (Zann, 1995). Changes in weather patterns, especially if accompanied by increased intensity of rainfall or increased frequency of damaging storms, and sea-level rise could increase the risk and magnitude of pollution events.

Vulnerable inhabited Australian islands include the Cocos (Keeling) Islands (Jacobson, 1976); Millingimby in the Northern Territory; and Boigu, Saibai, Masig, Poruma, and Waraber Islands in the Torres Strait (Mulrennan, 1992). Boigu and Saibai are low mud islands, the second of which suffered flooding during exceptional high tides in the 1940s; Masig, Poruma, and Waraber are low coral cays subject to active erosion and dune development.

In the tropical Pacific, New Zealand has responsibility for Tokelau, a populated atoll group, and constitutional ties with Niue and the Cook Islands. Potential impacts of climate change for such small islands are elaborated in Chapter 9 of this report; impacts for Tokelau are considered in Hooper (1990). Tropical cyclones or other severe weather events, even if unchanged, superimposed on rising sea levels may progressively stress the habitability of the Pacific islands and lead to impacts on New Zealand with respect to emergency disaster responses and migration.

Coastal areas and estuaries have particular importance to Maori (indigenous people of New Zealand) (Fraser, 1990), and much Maori land and economic enterprise is on or near low-lying areas. Coastal seafood is of immense cultural value and is important as an economic resource and for recreation. Land is the principal element of iwi (tribal/family) identity; together with coasts, rivers, and forests, it is central to the cultural, spiritual, and economic life of Maori. Detailed impact studies are not available, but any loss or damage to traditional land, sacred ancestral or archaeological sites, or food-gathering areas and other natural resources over which Maori have traditional domain would have large ramifications for the Maori (Fraser, 1990).

#### *4.3.4.3. Vulnerability Assessments and Case Studies*

In Australia, vulnerability assessments were carried out for Cocos (Keeling) Island (an Australian Territory), Geographe Bay in Western Australia, and Kiribati in response to the first version of the IPCC Common Methodology (IPCC, 1991). This led to recommendations for changes to the Common Methodology (McLean and Mimura, 1993).

Meanwhile, a major vulnerability study was carried out by the Port of Melbourne Authority (1992) on the impact of projected greenhouse changes on the 3,000 km of the Victorian coast. Quantitative assessment was carried out for selected beaches for 0.3- and 0.5-m rises in mean sea level, without considering other changes. The largest recession rates were indicated for beaches west of Cape Otway, due to the relatively fine sand and more severe wave conditions. Shoreline recession rates also were significant in parts of Port Phillip Bay and the Gippsland Lakes, due to generally narrower beaches. Detailed assessments identified beaches, barriers, estuaries, mudflats, swamps, and sedimentary cliffs that are particularly vulnerable.

A series of nine case studies initiated by the Australian Department of the Environment, Sport and Territories (DEST) on coastal vulnerability to climate change and sea-level rise was undertaken during 1994 and 1995 (Waterman, 1996). Impacts considered included sea-level rise, temperature and rainfall variation, changes in storm intensity and frequency, and saltwater intrusion into surface and groundwater supplies. Case studies were undertaken for Mackay, Queensland; Batemans Bay, New South Wales; Gippsland Lakes and Port Phillip Bay, Victoria; South Arm, Hobart, Tasmania; Northern Spencer Gulf, South Australia; Perth, Western Australia; and the Darwin and Alligator Rivers regions in the Northern Territory. Emphasis in the studies was placed on the wider socioeconomic, environmental, aesthetic, and cultural aspects, as well as strategies for integration of vulnerability assessment into coastal management and planning processes. The main aim in most of the studies, however, was to test the methodology rather than to produce definitive assessments of the vulnerability in the regions.

The studies found that vulnerability to climate change is difficult to determine in the context of the inherent variability of natural systems and the impacts of development on the coastal zone. In addition, the effects of change are poorly understood, and the implications for coastal management are not well appreciated by the wider community. The researchers concluded that community consultation, along with serious attempts to address the complex jurisdictional issues concerning coastal management, was essential.

Several of the studies, such as those for Port Phillip Bay and Hobart, noted the high degree of variability in potential impacts, even over distances of a few kilometers. The basis for identifying vulnerable areas was a geomorphological study that identified the main coastal landforms. At the broadest scale of generalization, the whole coast of Australia has been categorized by Chappell *et al.* (1996) according to several overlapping geomorphological and greenhouse hazard criteria, partly based on the work of Bird (1988, 1993) and Thom (1984). High wave energy coasts in the southern parts of the west and east of the continent are vulnerable to shifts in wave direction and energy, especially on beach coasts. Across most of northern Australia, rising sea level and changes in tropical cyclone activity are the greatest threats, whereas the Great Barrier Reef may benefit in some respects from moderate sea-level rises.

Severe limitations applied in quantitatively assessing likely impacts in the DEST case studies because of the wide range of uncertainty in the scenarios for sea-level rise, temperature and rainfall changes, and changes in storm intensity and frequency. Moreover, even in well-studied areas—such as the Alligator Rivers region in the Northern Territory, where rapid changes (such as saltwater intrusion into the Kakadu freshwater flood plains) (Knighton *et al.*, 1991; Woodroffe and Mulrennan, 1993) have occurred in historical times—processes of natural change are not well understood. Nevertheless, it is clear that Kakadu National Park, which is a World Heritage Area and major tourist attraction as well as Aboriginal hunting and fishing grounds, would be almost totally transformed into mangrove forest by a sustained 1-m rise in mean sea level.

In several of the areas studied, towns such as Port Augusta, Port Pirie, Lakes Entrance, and Mackay have experienced partial flooding from the sea due to extreme events in the recent past, so they are very vulnerable to effects of sea-level rise. This was highlighted for Mackay by Smith and Greenaway (1994), who note a high sensitivity of flood damage to any change in the climatology of storm surges.

Potential vulnerability to health effects, particularly due to increases in standing water (see Section 4.3.6.1), also were noted. The study of the Perth metropolitan coastline noted the natural "breakwater effect" of the offshore reef system and concluded that "climate change, natural or greenhouse-induced, appears to be more important to future changes ... than projected rises in mean-sea level." Saltwater penetration into the coastal aquifer also is an issue in the Perth region, as it is in some other coastal aquifers (Ghassemi *et al.*, 1991).

Small, low-lying coastal and island communities are particularly vulnerable to sea-level rise and changes in climatic forcing of extreme events. These communities include several of the Torres Strait islands (Mulrennan, 1992), the Australian territory of Cocos (Keeling) Islands, and several Aboriginal coastal communities across northern Australia. Unlike cities such as Darwin, these small communities usually do not abide by strict building codes or have evacuation procedures. Development of such codes and procedures therefore is necessary for effective adaptation.

The DEST project (Waterman, 1996) identified a number of issues concerning how policy and management decisions are made about the effects of climate change and sea-level rise by authorities and groups in coastal areas. Common methodological elements that should form part of a region-specific approach to vulnerability assessment were identified. Specific recommendations included capacity-building at governmental and community levels, including awareness raising; the development of a better appreciation of insurance, legal, and land-use planning implications, including professional and public liability; the development of a guide to best practice in integrated coastal zone management; and the development of benchmarks, monitoring programs, and databases for coastal zones.

In New Zealand, there has been no comprehensive national assessment of coastal vulnerability to climate and sea-level changes. Nonetheless, under the Resource Management Act (RMA) of 1991, local authorities are obligated to ensure the "avoidance or mitigation of natural hazards," including coastal flood and erosion hazards. Subsequent legal rulings have confirmed that future sea-level rise should be taken into account in coastal hazard analyses in connection with resource consent (permit) procedures as specified under the RMA. Furthermore, the New Zealand Coastal Policy Statement (Department of Conservation, 1994), which contains policies aimed at achieving the single purpose of the RMA (the "sustainable management of natural and physical resources"), includes specific reference to the possibility of a "rise in sea level" and its impacts on the subdivision, use, and development of the coast.

As a consequence of this policy framework, numerous coastal hazard studies have been commissioned by local authorities and developers in New Zealand for planning, policy development, and environmental impact assessment purposes. On the whole, these analyses—mostly in the form of unpublished consultants' reports—suggest that a considerable portion of the existing built environment of the New Zealand coast would be endangered by a future rise in sea level of the order of 0.5–1 m, especially along sandy beach and dune coasts. Where development pressures are high, such as in the Bay of Plenty, the potential increase in vulnerability is large unless adaptive measures are taken.

#### 4.3.4.4. *Adaptation and Vulnerability*

The specific options for adaptation are wide-ranging. Often the existing capital assets could be used to the end of their design life and rebuilt to suit higher sea levels and increased risks of coastal erosion and flooding. In some cases, however, there may be a need for costly raising or relocation of railways, ports, bridges, dams, highways, electricity lines, and so forth; in other cases, the preferred option may be to build protective dikes, seawalls, and pumping systems. "Softer" options include setbacks of development zones, beach nourishment, dune protection, and phased retreat plans—options that increasingly are perceived as viable alternatives, especially in coastal zones facing high development pressures. These options are set out in guidelines issued by the Australian Institution of Engineers (1991).

The time frame for response to sea-level rise is sufficient, in principle, to allow for the development of suitable coastal policies and management practices to adjust. At present, however, in Australia coastal hazards are generally not well accommodated within current coastal management planning frameworks. This was highlighted for Australia in the Report of the Coastal Zone Enquiry (Resource Assessment Commission, 1993), which recommended a National Coastal Action Plan. The approaches to coastal hazard management of the six Australian states are summarized in Kay *et al.* (1996). In most cases, there is some way to go to factor in the relatively long-term issue of climate change.

In New Zealand, the overarching national policies attempt to guide and encourage adaptation at the local level. The RMA requires that alternative options be systematically evaluated during local resource consent procedures (s.32). The New Zealand Coastal Policy Statement directs that a "precautionary approach" should be adopted and states that reliance on coastal protection works should be avoided; an emphasis should be placed instead on the enhancement of natural protective features of coastal systems and on the location and design of new developments to avoid hazardous situations. This policy framework in New Zealand is a potentially powerful mechanism for promoting adaptation to reduce the impacts of climate change in the coastal zone. However, it is too soon to assess its effectiveness because very few regional and district policies and plans have completed the lengthy public submission and appeal processes required.

For the future, a key strategy for the Australasia region will be to consider adaptation options within the context of integrated coastal zone management (IPCC 1996, WG II, Section 9.6.4). Within this increasingly accepted approach, the necessary detailed hazard mapping, monitoring, and policy development can occur, and a range of sociocultural, structural, legal, financial, economic, and institutional measures can be developed to deal with any additional climate-related risks. Such advance planning may substantially reduce the later costs of adaptation.

### 4.3.5. Human Settlements

*Summary: In addition to possible changes in the frequency and magnitude of "natural disaster" type events such as urban fringe fires, floods, and extreme sea-level events, urban areas and industry will experience a variety of other direct climatic impacts on water and air quality, water supply and drainage, waste disposal, energy production and distribution, minerals production, transport operations, insurance, and tourism. The effects of the direct impacts are likely to be small relative to other economic influences, but the sectors are very large, and the impacts and necessary adaptations may represent major losses and costs. Thus, a moderate degree of vulnerability is present. Adaptations will include improved planning, zoning, and engineering standards that take climate change and sea-level rise into account.*

#### 4.3.5.1. Energy

Rising temperatures will lead to reduced heating demand in winter and increased air conditioning demand in summer. Because of the temperate marine climates of coastal Australia and New Zealand, there are relatively smaller demands for heating and air conditioning than in continental climates, and the net impact on the economy is likely to be relatively small. In New Zealand, the expected smaller seasonal variation in demand for electricity would provide an operational gain for the electricity industry. In Australia, the summer seasonal peak from air conditioning would increase, requiring additional peak plant capacity. The efficiency of thermal generation plants decreases with increasing temperature, as does the maximum carrying capacity of transmission lines. Any increase in weather extremes would decrease system reliability (IPCC 1996, WG II, Section 11.5.3). Any changes in rainfall characteristics in hydroelectricity catchments would have impacts on electricity generation operations (see Section 4.3.2.3).

#### 4.3.5.2. Air and Water Pollution

Air pollution from vehicle emissions and industrial processes is primarily a problem of the region's urban areas. The climate affects the photochemical production of secondary pollutants—through sunlight, humidity, and temperature—and the dispersal of source pollutants and photochemical products through winds, rainfall, and atmospheric stability. The severity of photochemical smog episodes is increased by higher temperatures, more sunshine, and lighter winds. Reviews of the impacts of climate change on air pollution for New Zealand (Wratt, 1990) and Australia (SOEC, 1996) note a variety of possible effects, the most certain being the role of increased temperature.

Any increase (decrease) in high rainfall events and runoff would increase (decrease) the risk of breach in sewerage systems (which in some cities, such as Sydney, already are subject to overflow), as well as other waste collection facilities and chemical storage facilities. Conversely, any reduced water flows may result in increased concentrations of chemical and biological toxins in waterways, and vice versa. Nonurban water pollution has been discussed in previous sections; urban-rural interactions may be important with respect to water sources for cities and pollution of natural environments downstream of cities.

As with most other environmental issues, there are many factors involved in air and water pollution, many of which are either more significant or more rapidly changing than climate, and there is great uncertainty about likely climate impacts. In general, the technical means to appraise, monitor, and manage the potential impacts of climate change on air and water quality already are available, if not in place.

#### 4.3.5.3. Minerals

The minerals industry (predominantly in Australia) will experience some impacts—depending on the extent of climate change, the environmental conditions at the site, and the nature of the operations. Temperature has a role in process design and operation; rainfall and evaporation affect materials handling, the design and operation of civil structures (pits, drainage systems, tailings and mineral waste disposal, roads, and railways), supplies of surface and groundwater, the quality and disposal of wastewater, management of dust and air quality, and site rehabilitation and revegetation. Storm events (in coastal subtropical regions) and sea level have impacts on ports and other materials transport facilities. Positive and negative impacts may be expected. These impacts also will depend on minerals demand; government policy, especially in respect to fuels and climate change; and knowledge of the impacts and possible adaptation options. Because of the size of the industry, any impacts would have significant consequential effects on the Australian economy.

#### 4.3.5.4. Insurance

Because of the region's relatively high frequency of climatic and other natural hazards (Pittock *et al.*, 1997), insurance against extreme events is especially important; the cost of insured catastrophes in Australia is at least 12% of the costs of non-life insurance premiums, whereas globally the corresponding figure is only 2.5% (IPCC 1996, WG II, Section 17.5). However, in Australia, private insurance has never been available for flood damage to domestic dwellings and small businesses, presumably because the risks are too high—though assistance may be made available under National Disaster Relief arrangements. The cost of insurance in the region has risen already because of increasingly large losses in the international reinsurance market from recent, mainly weather-related, disasters. The common perception that there is a global trend toward increased frequency of severe climatic events is not well substantiated by meteorological evidence (IPCC 1996, WG II, Section 17.4.1), apart from the evidence of increases in the frequency of heavy rainfall (Karl *et al.*, 1995; Suppiah and Hennessy, 1996, 1997). However, there is some evidence for possible future increases in flood and fire risks (Pittock *et al.*, 1997; see also Sections 4.3.1.3 and 4.3.2.2).

It is likely that climate change will adversely affect property insurance, where such insurance is available, but current knowledge is insufficient to allow the industry to quantify its changed exposure (IPCC 1996, WG II, Section 17.4.2). Even if the mean long-run losses were to remain unchanged, the greater uncertainty in quantifying the risk would demand a higher premium. Adaptation options include primary risk management through hazard assessment (such as Australia's 52 zones for cyclone, thunderstorm, wildfire, and earthquake) (IPCC 1996, WG II, Section 17.3.2), land-use regulation, building design and construction permits, retrofitting in high-risk areas, and public education (IPCC 1996, WG II, Section 17.6).

An important feature of the region is the very high level of individual home ownership. Personal assets often are closely tied to the market value of a person's home and the land it is built on, and equity usually is highly leveraged with mortgages. Typically, the real long-term risks of urban flooding and coastal inundation in a locality are not transmitted into market values or into insurance premiums, and an adverse event can cause large personal and insurance losses, withdrawal of insurance coverage, and dramatic declines in property values and personal equity. By way of example, residential losses in Florida constituted 65% of the insurable losses from Hurricane Andrew (IPCC 1996, WG II, Section 12.3.4).

#### 4.3.5.5. *Transportation*

The transportation industry has long time frames for investment, is very sensitive to weather, and may be affected by climate change in various ways (IPCC 1996, WG II, Section 11.4.4). Any changes in production systems, such as in agriculture and mining, would alter the patterns of demand for transport services. Higher temperatures would reduce maximum takeoff payloads for aircraft and increase problems with buckling of railway lines and melting or softening of road tar surfaces—but also would reduce the risks of winter snow and ice on roads. Any changes in winds at flight levels would affect aircraft operations and economics. More intense rainfalls would cause more frequent landslides and blockages to rail and road routes and might increase flooding impacts on airports, roads, rail tracks, and bridges. Any increase in storm frequency or intensity would negatively affect all transport operations, especially shipping and airways.

Australia and New Zealand are particularly dependent on efficient, reliable transport within their relatively sparse domestic markets and for international trade. Impacts on transport services therefore could affect tourism and exports, in addition to the direct effects on transport safety and costs.

Because virtually all transport services are fossil fuel-driven, $CO_2$ emission management will result in further impacts on the industry. A detailed consideration of Australia's options with respect to transport and climate change mitigation (BTCE, 1996) noted several "no regrets" strategies that would have benefits for air quality and urban congestion. Adaptation to the potential direct impacts of climate change could be dealt with similarly through normal engineering design and operations responses, but some residual vulnerability may remain.

#### 4.3.5.6. *Tourism*

Tourism is the fastest-growing sector of world trade, and Australasia's share of this rising market has been increasing. It represents the most important economic use of the region's landscape, biodiversity, and climate—ahead of agriculture (it was worth about US$40 billion in 1995–96). The industry is vulnerable to climate change in several ways. Coastal resorts, particularly buildings and infrastructure, may be vulnerable to sea-level rise and changes in storm surge magnitude and frequency, as well as consequential changes in beach extent. Ski resorts could be adversely affected (see Section 4.3.2.3) by shorter snow seasons and an increased need for artificial snow, and resorts on the Queensland coast may be affected by any damaging impacts on the Great Barrier Reef from coral bleaching or any increased damage from tropical cyclones or riverine runoff (see Section 4.3.1.5). On the other hand, moderate rates of sea-level rise could invigorate coral reef growth, increasing the reefs' attractiveness.

Changes also could occur in the prevalence of insect pests and vector-borne diseases, affecting tourist health and comfort and resulting in stricter quarantine precautions. Human comfort also may change adversely, particularly by warmer and possibly more humid conditions, including the number of days of rain, and the length of the monsoon and tropical cyclone seasons in northern Australia. Any increase in heat and humidity discomfort would increase the demands for air conditioning at resorts and could lead to a shift of demand away from tropical resorts to cooler locations such as Tasmania or the South Island of New Zealand.

#### 4.3.5.7. *Adaptation and Vulnerability*

A wide range of potential "direct" climatic impacts on settlements and industry, largely located in urban areas, has been identified in this section. Several other vulnerabilities have been identified elsewhere in this chapter—notably, possible increased exposure to fires in the urban fringes (Section 4.3.1.3), increased riverine flooding (Section 4.3.2.2), and extreme sea-level events (4.3.4.3). The potential additional cost of these "natural disaster" events attributable to climate change is very difficult to quantify at present, but it may be very large, and the possibility heightens the need for effective current planning, zoning, and engineering standards and for improved disaster management.

Most of the direct impacts identified here are relatively small compared to other economic influences, and their likely slow development would allow adaptation through normal processes of planning, management, and engineering design. However, the sectors involved (e.g., energy, mining, transport, tourism, and insurance) are very large; therefore, small fractional impacts and small adaptation responses could represent substantial total losses and costs. On this basis, a moderate degree of vulnerability is present.

### 4.3.6. Human Health

*Summary: Increases are likely in heat stress mortality (particularly in Australia), tropical vector-borne diseases such as dengue, and urban pollution-related respiratory problems. These impacts are small compared to the total burden of ill-health, but they have the potential to cause significant community impact and cost. Indigenous communities and economically disadvantaged persons appear to be more at risk. Adaptation responses include strengthening existing public health infrastructure and meeting the needs of vulnerable groups. A moderate degree of vulnerability is apparent with health.*

#### 4.3.6.1. *Health Impacts*

Average life expectancy in Australia and New Zealand is relatively high, and access to medical care is relatively good by international standards. There are considerable inequalities in health status and access to services, however. Disadvantaged groups such as indigenous peoples and the poor are likely to be most at risk from the effects of climate change.

In parts of urban Australia, the frequency of very hot days (over 40°C) is expected to increase by 50–100% for a 2°C increase in mean temperature (Hennessy and Pittock, 1995). This is likely to lead to an increase in deaths, especially in cities such as Melbourne that are subject to wide variability in temperature, but insufficient information is available at present to quantify this impact. There may be fewer deaths in winter with warmer temperatures. For example, the winter excess in coronary heart disease mortality in Australia and New Zealand appears to be related in part to ambient temperature (Enquselassie *et al.*, 1993). However, studies in other parts of the world suggest that reduced mortality from cold extremes would probably only partly offset the heat stress effect (IPCC 1996, WG II, Section 18.2.1).

In both Australia and New Zealand, there are significant food- and waterborne diseases that occur more commonly in warmer conditions (IPCC 1996, WG II, Section 18.3.5). For example, in parts of Australia, an amoeba that causes meningoencephalitis proliferates in water pipes that are heated in summer as the water travels overland (IPCC 1996, WG II, Section 18.3.2). Hotter weather also will exacerbate urban air pollution due to photochemical oxidants (Woodward *et al.*, 1995), possibly leading in the major cities (Sydney, Melbourne, Auckland) to increased frequency of respiratory problems and deaths (IPCC 1996, WG II, Section 18.3.5). Australia and New Zealand have rates of asthma and other allergic conditions that are higher than elsewhere in the Pacific or in many parts of Europe, and these diseases may be exacerbated by warmer and more humid climates (Pearce *et al.*, 1993).

Any increase in extreme events and flooding, including those associated with sea-level rise, would increase deaths, injury, infectious diseases, and psychological disorders and may increase road accidents. Other waterborne illnesses—such as viral, bacterial, and protozoal diarrhoea and cyanobacterial poisoning—would be affected by any changes in water availability (due to flooding, droughts, and public water shortages). Some disease vectors also are influenced by changes in water availability. New Zealand currently is free of human arbovirus infections (i.e., viruses borne by arthropods—ticks, mites, etc.), but if temperature and rainfall alter it may become susceptible to infections that currently are common in Australia (Weinstein *et al.*, 1995). Vectors for dengue and malaria already exist in Australia, and southward expansion of the vector populations with temperature increases would increase the potential for insect-borne diseases to become more widely established. These diseases are very sensitive to climate variability—for example, outbreaks of dengue are much more common in the western Pacific islands during relatively warm and wet La Niña years (Hales *et al.*, 1996).

Conditions forecast under climate change could alter the distribution and proliferation of arthropod vectors and/or natural vertebrate hosts. Warmer and wetter conditions would lead to increased incidences of insect-borne infections such as Japanese Encephalitis virus, Murray Valley Encephalitis virus, Ross River virus, and dengue. In southeastern Australia, epidemics of Murray Valley Encephalitis and Ross River virus infection follow heavy rain in the Murray-Darling basin (Nicholls, 1986; IPCC 1996, WG II, Section 18.3.1.7), so additional cases might be expected if the frequency or duration of heavy rainfall events increased. However, the relationship between climate, vector distribution, and disease are complex. For example, although the incidence of infection with Ross River virus may increase with increased rainfall, the incidence of symptomatic disease may actually fall because individuals who are infected as children remain asymptomatic and are immune for life.

The indigenous peoples of Australia and New Zealand—Australian Aborigines, Torres Strait Islanders, and Maori—may be more vulnerable to climate change because their health status in general is worse than for other Australians (Australian Bureau of Statistics, 1997) and other New Zealanders (Pomare *et al.*, 1995). These health status differences result largely from social and economic disadvantage, which itself is a cause for susceptibility. For example, many Australian Aborigines live in remote areas, in poor housing, and are particularly susceptible to infections related to deficiencies in water supply and sewage disposal (Moss, 1994). Under some scenarios, these conditions could worsen due to higher temperatures, changes in rainfall, higher water tables, and rising sea levels. Reduced freshwater supply, increased standing water, or both could increase exposure to microbial and viral infections and vector-borne diseases (Kolsky, 1993; Hales *et al.*, 1995; Henderson *et al.*, 1995; Jacobson and Graham, 1996).

The health impacts of climate change will include economic and social costs. These costs are difficult to quantify, but the evidence suggests that the health impacts for the region are likely to be relatively small compared to impacts from other diseases and sources of mortality. At the same time, it should be cautioned that there is considerable uncertainty in our current knowledge of not only the likely direct impacts but also the complexities of social responses to change and interaction with other sectors. For example, there could be changes in the incidence of zoonoses (vertebrate animal-mediated diseases) such as leptospirosis, a concern in New Zealand, as farming systems and zoonoses transmission alter in response to climate change. Heat stress impacts will be affected by housing choices and changing attitudes to air conditioning. More difficult to estimate precisely—but possibly of greater importance to public health in the long term—are the indirect effects of climate change, such as any adverse social and economic effects, especially on vulnerable groups, arising from impacts on other sectors and any large population shifts of "environmental refugees" (for example, from Pacific atolls) (Moore and Smith, 1996).

#### 4.3.6.2. *Adaptation and Vulnerability*

Major adaptation responses include strengthening the existing public health infrastructure (such as disease monitoring, vector management, and primary health care); improving protective systems where deficiencies are apparent already (for example, comprehensive infectious disease surveillance and border quarantine controls to reduce introduction of exotic pathogens); and meeting the needs of the most vulnerable groups in the population who will be at greatest risk from new climate-related threats to health (Woodward, 1996), including through improvements in water supply and sewerage systems (Moss, 1994). Adaptations to deal with the indirect socioeconomic effects arising in other sectors may require significant policy responses at the national and international levels.

### 4.4. Integrated Analysis of Potential Impacts

*Summary: Some preliminary studies of impacts for limited areas or issues have been attempted for the region, but comprehensive integrated analyses are not available. A cross-sector economic costing approach indicates substantial impacts on Australasia's GDP (several percent per annum for a doubling of $CO_2$), but the method involved is subject to considerable criticism. The New Zealand CLIMPACTS and Australian OzClim software packages have been developed as national-scale, integrated assessment methodologies to enable the progressive exploration of sensitivities, impacts, adaptation options, and so forth as knowledge develops.*

*Research is needed on better climate scenarios; the dynamic responses of ecological systems; impacts on marine, aquatic, and coastal ecosystems; possible changes in the magnitude and frequency of "natural disasters"; impacts on settlements, industries, and health; cross-sector and multinational interactions; and potential adaptations and their ameliorating effects.*

*In summary, Australia's relatively low latitude makes it particularly vulnerable through impacts on its scarce water resources and on crops presently growing near or above their optimum temperatures, whereas New Zealand—a cooler, wetter, mid-latitude country—may gain some benefit from the ready availability of suitable crops and the likely increased agricultural production. In both countries, however, there is a wide range of situations in which vulnerability is thought to be moderate to high—particularly in ecosystems, hydrology, coastal zones, settlements and industry, and health.*

A comprehensive integrated analysis of impacts for the region is not possible at present, mainly because of the large uncertainties in each of the three stages of analysis: first, in the magnitude and direction of change in the climate elements; second, in the estimations of potential impacts for individual components of systems; and third, concerning possible interactions between components. Instead, we discuss preliminary attempts to develop integrated assessment methodologies and present results from two case studies.

Integrated assessments of climate change are attempts to integrate three components or dimensions of the problem (Weyant *et al.*, 1996; IPCC 1996, WG III, Chapter 10):

- The chain of effects, from changes in atmospheric composition and climate to changes in biophysical systems to socioeconomic consequences (the "vertical" dimension)
- The interactions between systems, sectors, and activities (the "horizontal" dimension)
- The context of other trends and changes in society, such as population, technology, land-use, and economics (the "time" dimension).

New Zealand was one of the first countries in the world to carry out such an assessment of the effects of climate change at the national scale, based on the work of expert panels and integrated through a national committee (Mosley, 1990a). In Australia, the idea of integrated climate impact assessments has been developed, to some extent, in the context of climatic variability rather than climate change, to better adapt Australian agriculture to the large year-to-year climatic variability associated with

**Box 4-2. Climate Change and Cattle Ticks in Australia: An Integrated Assessment**

Sutherst *et al.* (1996) carried out an integrated assessment of the national socioeconomic impact of climate change on the cattle tick (*Boophilus microplus*), which is an important parasite of cattle in northern Australia. There is concern that global warming will result in a spread of ticks and an increase in their numbers, as indicated by an earlier study for both New Zealand and Australia (Sutherst *et al.*, 1996). The cattle tick causes losses in productivity through feeding in large numbers and by transmitting a blood pathogen that causes high rates of mortality in nonimmune cattle. The southern limit of the ticks currently is maintained by a quarantine and eradication program at the border between NSW and Queensland, at a cost of several million dollars per year.

A population model of the cattle tick was used first to estimate current national losses of productivity of beef cattle due to the ticks in each part of the country. These impacts were costed using an economic model of the cattle industry, after which the exercise was repeated using the changed climate scenarios derived from the CSIRO9 global climate model (experiment F1, Table 1-1). Potential costs were estimated on the basis that the quarantine line would not be economically sustainable because of the recurrent development of strains of ticks that are resistant to the chemicals used to control them and the increasing difficulty of eradication under more favorable climatic conditions.

The maps in the accompanying graphic illustrate the estimated potential costs under current and 2070 climate scenarios. The number of ticks was estimated to increase in the southern part of the existing range—where they are limited by the shortness of the warm season, which allows reproduction, and by severe winter mortality. Both of these constraints were reduced by global warming; as a result, most of the estimated increases in costs would be sustained within the current tick-infested area in Queensland. There was a parallel increase in the potential area affected by ticks in NSW. The final estimates of potential national costs, in terms of net present value, ranged from A$18 million to A$192 million, but these estimates will vary with discount rates.

The analysis was unique in Australia in that it was one of the first to use a fully integrated hierarchical approach to assessing impacts of climate change on pests, diseases, and weeds, including climate change scenarios generated by CSIRO; simulation of the target species (ticks) with a population model; and calculation of economic impacts on the national beef industry, estimated using a beef industry model developed by the Australian Bureau of Agricultural and Resource Economics. To achieve these objectives, an interdisciplinary team of climate modellers, ecologists, and economists was needed. The results are available at http://www.modeling.ctpm.uq.edu.au/dest_public/ on the World Wide Web.

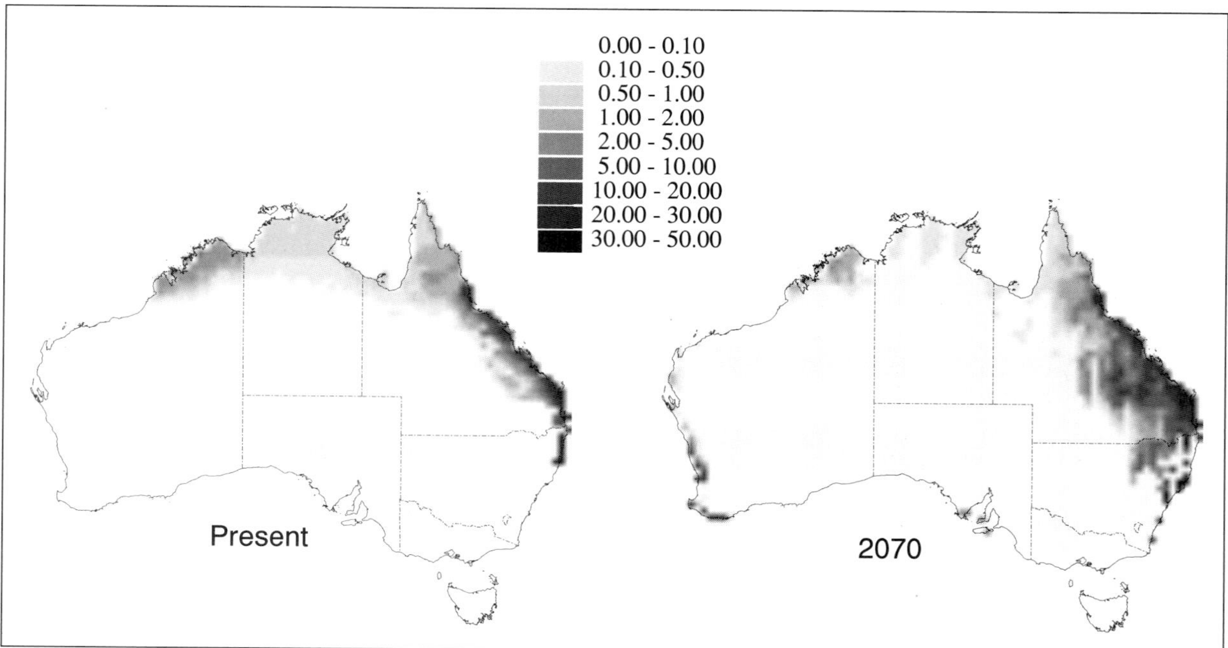


*Estimated costs of cattle tick, in Australian dollars per head of cattle, for current (left panel) and 2070 (right panel) climate, using a three-level integrated approach composed of climate scenario, tick/cattle model, and economic model of the national beef industry.*

ENSO fluctuations. Collaborative projects involving climatologists, agricultural scientists, rural economists, and stakeholders have led to the development of crop-, farm-, and industry-level models that facilitate the application of ENSO-derived seasonal climate forecasting to improve farm and industry management (Stone *et al.*, 1993; BRS, 1994; Stafford Smith *et al.*, 1994). The ability to better predict and manage this short-term climate variability is itself an important adaptation tool.

Two initial efforts at integrated assessment in Australia are described in Box 4-1, on the impacts of climate change on the management of the scarce water resources of the Macquarie River basin in northern NSW, and in Box 4-2, on the impacts of climate change on the cattle tick.

These integrated assessments in Australia and New Zealand are subject, however, to large uncertainties arising from continually evolving GCM-based scenarios of regional climate change. Indeed, the remarkable contrast between the rainfall change scenarios for Australia (and also by implication for New Zealand—see Section 4.2.3) emerging from simulations using slab-ocean GCMs and from the more comprehensive coupled-ocean models (Whetton *et al.*, 1996a) has increased the uncertainty regarding the "correct" scenario to use in impact analyses (CSIRO, 1996a) and has thrown considerable doubt on the predictions of many early impact studies (though they remain valuable as sensitivity studies). As a consequence, there is a growing recognition that integrated assessments for Australia and New Zealand need to be carried out within a versatile framework that allows the range of uncertainties to be explored and the effects of changing scenarios (due to scientific, economic, or policy changes) to be readily assessed in an iterative fashion. For this approach, integrated models are required.

Methods for integrated assessment have been explored at several recent workshops involving Australian and New Zealand participants (Pittock and Mitchell, 1994; Braaf *et al.*, 1995; Hennessy and Pittock, 1996a, b). In New Zealand, a major research program known as CLIMPACTS has been under way since 1993, involving more than seven different laboratories (Kenny *et al.*, 1995; Warrick *et al.*, 1996). This project has coupled the MAGICC model (Wigley, 1994)—which generates a global average warming curve for given input assumptions—to a regional pattern of climate change per 1°C global warming, which is based on a range of GCM simulations and local statistical interpolation techniques. A number of impacts models are now being coupled to this framework, in order to examine the sensitivity of the New Zealand environment to climate variability and change. First-order impacts and sensitivities have been assessed for grain maize, pasture, wheat, and kiwifruit.

In Australia, CSIRO has begun developing an Australian version of CLIMPACTS called OzClim (CSIRO, 1996b). This is part of a broad-based climate impacts "tool kit" that has been adopted as a goal of the CSIRO Climate and Atmosphere Sector. A comprehensive model for assessing the effect of climate change on pests and diseases also has been applied in Australia (Sutherst, 1995; Sutherst *et al.*, 1996) and could be part of an impacts tool kit.

These integrated models provide flexible, evolving tools for comprehensive assessments of vulnerability and adaptation options. They are especially useful in performing "diagnostic" analyses of environmental and socioeconomic sensitivities to climate change and variability under the shadow of the large scientific uncertainties in GCM-based predictions of regional climate change.

Another way to estimate integrated effects is to calculate a comprehensive monetary estimate, where all known impacts are expressed in a dollar value and then added (IPCC 1996, WG III, Chapter 6). There is considerable room for criticism of this method's assumptions, data, and results. Shortcomings aside, however, the estimates available for the Australasian region using this approach indicate very significant impacts on total GDP (See Box 4-3).

The Australasian region is subject to high climate variability, especially associated with the ENSO phenomenon. Therefore, a better understanding of how the region's unique ecosystems are affected by climate variability, extremes, and chaotic sequences of climate conditions—as well as small, slow changes in mean conditions—is needed to enable us to understand how ecosystems might respond to future changes in this variability. The issues of climate variability; time characteristics of physical and biological components of ecosystems, including hydrology, soils, and weeds; and the resulting dynamic, nonlinear behavior of ecosystems are key issues, both generally and for predicting the impacts of climate change.

Research also is needed on vulnerability to present-day climate fluctuations in the region and on how people and systems respond to and cope with these fluctuations. The emerging ability to predict interannual climate variations, especially those arising from the ENSO phenomenon, and to use those predictions to optimize management strategies (McKeon and Howden, 1993; Stone *et al.*, 1993) is an example of an approach that not only serves immediate needs but also can become a tool for adaptation to long-term change.

Australia and New Zealand also are extremely vulnerable to international influences arising from climate change: through their high levels of international trade in food commodities whose supply, demand, and prices will alter in a warmer world; from climate-mediated risks of imported biological pests and diseases that have the potential to decimate food production systems; from the consequences of international greenhouse gas mitigation policies on the region's coal industry, energy-intensive industries (transport fuels, smelting), and possibly agricultural industry; and through their special responsibilities to assist South Pacific island communities in the event of climatic disaster and loss of habitability.

To improve the prediction of climate change impacts for Australasia, additional research and information is needed on

**Box 4-3. Estimation of Integrated Economic Impacts of Climate Change**

In principle, the total potential economic impact of climate change on the world (or a region or country) can be found by deriving a comprehensive monetary estimate—where all of the known impacts, positive and negative, are expressed in a dollar value and combined (IPCC 1996, WG III, Section 6.1.2.). For marketed goods and services (for example, increased agricultural production, land loss due to sea-level rise, or energy savings in winter), this calculation is more or less straightforward because the prices are known. In the case of nonmarketed goods and services (for example, wetland loss or mortality changes), monetary estimates of impacts can be obtained through examining market transactions where such goods or services are implicitly traded (e.g., the higher price of a house with a good view) or through interviewing people for their preferences. That is, human preferences are expressed by people's willingness to pay to secure a benefit or their willingness to accept compensation in exchange for a loss. In OECD countries, such valuation techniques are well established and have been widely applied.

When applied to agriculture, energy, water supply, sea-level rise, mortality, and species losses, the method shows that with an equilibrium doubling of $CO_2$, the impacts amount to an annual reduction of several percent of GDP in many regions of the world (IPCC 1996, WG III, Section 6.2.15). The calculations are carried out under current economic conditions and include some account of international trade effects in agriculture but not of interactions between the different impacts or adaptation responses. It should be noted that these results have been strongly criticized—first on the basis of inadequate representation of the full range of impacts and second because many researchers hold the view that the importance of nonmarket goods, especially lives and quality of life, cannot be expressed in monetary terms.

Updated values of the SAR estimates of the annual impact for Australia and New Zealand combined fall in the range of -1.2% to -3.8% of GDP, depending on economic model assumptions (Fankhauser and Tol, 1997). The real uncertainty in these estimates is much greater than the range given. Most of the impact arises from health and environmental effects, which do not appear in national economic accounts. At face value, these figures indicate substantial negative effects on Australasia's economy. Furthermore, they may seriously underestimate the true total impact because they do not include a number of impacts that are potentially important to Australasia—such as changes in weeds, pests, and diseases; storm surges; and urban flooding. In addition, damages may increase nonlinearly with increased global warming, so the impacts and costs could rise faster than the rate of warming. On the other hand, adaptations to climate change, which will act to reduce total costs, also are not included in the calculations.

more accurate and region-specific climate scenarios; the dynamics and internal interactions of ecological systems, as noted above; impacts on marine, aquatic, and coastal ecosystems; possible changes to the magnitude and frequency of "natural disasters" such as droughts, floods, tropical cyclones, fire, and extreme sea-level events (Pittock *et al.*, 1997); impacts and costs for human activities (settlements, industries, and health); cross-sector and multinational interactions; and potential adaptation options (Henderson-Sellers, 1996) and their impact-ameliorating effects.

In broad summary, climate change impacts in Australia are likely to be dominated by the following factors. Some of these factors are more similar to those operating in low-latitude developing countries than to those in the more developed countries of the OECD:

- The effects of climate change on water resources, due to the arid and semi-arid nature of much of the country
- Potential changes in rainfall variability associated with the ENSO phenomenon, tropical cyclones, and the increased water-holding capacity of warmer air
- The fact that much of Australia lies in the tropics and subtropics, where many crops grow near or above their optimum temperatures
- The important role of fire in Australian ecosystems
- The lack of substantial cooler alpine or higher-latitude zones to which heat-intolerant Australian species, crops, or activities could retreat
- The fragmented nature of many of the remaining Australian native ecosystems
- The large role of commodity exports in the Australian economy, which makes Australia extremely vulnerable to changes in world trade and commodity prices—which in turn may be affected by climate change.

On the other hand, New Zealand is a mid-latitude country with relatively large alpine areas and greater water resources. Despite New Zealand's large dependence on export commodities—which may be affected by world commodity prices—general warming would allow adaptation through the introduction of more heat-tolerant crops or the migration of species and activities to higher elevations or latitudes. Increased agricultural production appears likely. One of the most obvious impacts of warming in New Zealand would be the retreat of snowfields and glaciers, which may have impacts on tourism, water resources, and hydroelectric power generation.

For both countries, it is clear that studies of potential impacts are still very preliminary, with little work to date based on the latest scenarios from the coupled-ocean GCMs. Further, there has been relatively little integration across sectors, few attempts to take account of other components of global change, and few systematic studies of potential adaptations.

Despite these serious shortcomings, preliminary indications are that in both countries many activities and ecosystems are quite sensitive to climate change, with positive and negative effects. Australia's relatively low latitude makes it particularly vulnerable through impacts on its scarce water resources and on crops presently growing near or above their optimum temperatures. The effects may be quite nonlinear, with small warmings having benign effects but larger warmings producing more negative impacts.

High priority must be given in the future to studies that take a more integrated approach and facilitate assessment of the impacts of a wide range of climatic, socioeconomic, and other scenarios, as well as potential adaptations. Until then, any quantitative estimates of the potential cost of climate change must be regarded as highly uncertain.

## References


**AIM**, 1997: *Asian-Pacific Integrated Model.* AIM Project Team, National Institute for Environmental Studies, Tsukuba, Japan, 83 pp.

**Allan**, R.J. and M.R. Haylock, 1993: Circulation features associated with the winter rainfall decrease in southwestern Australia. *J. Climate*, **6**, 1356–1367.

**Allden**, W.G., 1981: Energy and protein supplements for grazing livestock. In: *Grazing Animals* [Morley, F.H.W. (ed.)]. Elsevier, Amsterdam, Netherlands, pp. 89–307.

**Arp**, W. and F. Berendse, 1993: Plant growth and nutrient cycling in nutrient poor ecosystems. In: *Climate Change: Crops and Terrestrial Ecosystems* [van de Geijn, S.C., J. Goudriaan, and F. Berendse (eds.)]. Agrobiologische Themas **9**, CABO-DLO, Wageningen, The Netherlands, pp. 109–124.

**Ash**, A.J., S.M. Howden, and J.G. McIvor, 1996: Improved rangeland management and its implications for carbon sequestration. In: *Rangelands in a Sustainable Biosphere.* Vol. 1, Proceedings of the Fifth International Rangelands Congress [West, N.E. (ed.)], 23-28 July 1995, Salt Lake City, Utah, USA, pp. 19–20.

**Australian Bureau of Statistics**, 1997: *The Health and Welfare of Australia's Aboriginal and Torres Strait Islander Peoples.* Australian Bureau of Statistics Catalogue No. 4704.0, Canberra, 9 pp.

**Australian Institution of Engineers**, 1991: *Guidelines for Responding to the Effects of Climatic Change in Coastal Engineering Design.* National Committee on Coastal and Ocean Engineering, Barton, ACT, 446 pp.

**Ayers**, G.P. and R. Boers, 1996: Climate, Clouds and the sulfur cycle. In: *Greenhouse; Coping with Climate Change*, [Bouma, W.J., G.I. Pearman, and M.R. Manning (eds.)]. CSIRO Publishing, Collingwood, Victoria, Australia, pp. 27–41.

**Bates**, B.C., A.J. Jakeman, S.P. Charles, N.R. Summer, and P.M. Fleming, 1996. Impact of climate change on Australia's surface water resources. In: *Greenhouse: Coping with Climate Change* [Bouma, W.J., G.I. Pearman, and M.R. Manning (eds.)]. CSIRO Publishing, Collingwood, Victoria, Australia, pp. 248–262.

**BCTE**, 1996: *Transport and Greenhouse: Costs and Options for Reducing Emissions*, Bureau of Transport and Communications Economics, Australian Govt. Pub. Service, Canberra, 541 pp.

**Beer**, T. and A. Williams, 1995: Estimating Australian forest fire danger under conditions of doubled carbon dioxide concentrations. *Climatic Change*, **29**, 169–188.

**Benson**, L.J. and R.G. Pearson, 1987: The role of drift and the effect of season on macroinvertebrate colonization of implanted substrata in a tropical Australian stream. *Freshwater Biology*, **18**, 109–116.

**Bird**, E.C.F., 1988: Physiographic indications of a sea-level rise. In: *Greenhouse: Planning for Climate Change* [Pearman, G.I. (ed.)]. CSIRO, Melbourne, Australia, pp. 60–73.

**Bird**, E.C.F., 1993: *Submerging Coasts: The Effects of Rising Sea-Level on Coastal Environments.* John Wiley and Sons, Chichester, UK, 184 pp.

**Blumenthal**, C.S., I.L. Batey, F. Bekes, C.W. Wrigley, and E.W.R. Barlow, 1991a: Seasonal changes in wheat-grain quality associated with high temperatures during grain filling. *Australian Journal of Agricultural Research*, **42**, 21–30.

**Blumenthal**, C.S., F. Bekes, I.L. Batey, C.W. Wrigley, H.J. Moss, D.J. Mares, and E.W.R. Barlow, 1991b: Interpretation of grain quality results from wheat variety trials with reference to high temperature stress. *Australian Journal of Agricultural Research*, **43**, 325–334.

**Boag**, T.S., L. Tassie, and K. Hubick, 1988: The greenhouse effect: implications for the Australian grape and wine industry. *Australian and New Zealand Wine Industry Journal*, **3**, 30–36.

**Boon**, P., J. Frankenberg, T. Hillman, R. Oliver, and R. Shiel, 1990: Billabongs. In: *The Murray* [Mackay, N. and D. Eastburn (eds.)]. Murray Darling Basin Commission, Canberra, Australia, pp. 182–198.

**Boulton**, A.J. and P.S. Lake, 1988: Australian temporary streams — some ecological characteristics. *Verh. Internat. Verein. Limnol.* **23**, 1380–1383.

**Boulton**, A.J. and P.S. Lake, 1992: The ecology of two intermittent streams in Victoria, Australia. III. Temporal changes in faunal composition. *Freshwater Biology*, **27**, 123–138.

**Braaf**, R., A. Henderson-Sellers, G. Holland, and W. Howe (eds.), 1995: *Climate Change Prediction, Impacts Evaluation and Integrated Assessment.* Proceedings of Users of Climate Change Predictions Experts' Workshop, 31 May – 1 June 1995. Climatic Impacts Centre, Macquarie University, Sydney, Australia, 56 pp.

**Brereton**, R., S. Bennett, and I. Mansergh, 1995: Enhanced Greenhouse climate change and its potential effect on selected fauna of south-eastern Australia: a trend analysis. *Biological Conservation*, **72**, 339–354.

**Brown**, B.E., 1997: Adaptations of reef corals to physical environmental stresses. *Advances in Marine Biology*, **31**, 221–299.

**Brown**, B.E., 1997: Coral bleaching: causes and consequences. *Coral Reefs* (in press).

**BRS**, 1994: Climate and Risk. Bureau of Resource Sciences, special issue of *Agricultural Systems and Information Technology*, **6(2)**, 1–75.

**Bunn**, S.E. and P.M. Davies, 1990: Why is the stream fauna of south-western Australia so impoverished? *Hydrobiologia*, **194**, 169–176.

**Bureau of Meteorology**, 1989: *The Climate of Australia.* Australian Government Publishing Service, Canberra, 49 pp.

**Burgman**, M.A. and B.B. Lamont, 1992: A stochastic model for the viability of *Banksia cuneata* populations: environmental, demographic and genetic effects. *Journal of Applied Ecology*, **29**, 719–727.

**Burns**, N.M. and D.S. Roper, 1990: Estuaries. In: *Climatic Change: Impacts on New Zealand* [Mosley, M.P. (ed.)]. Ministry for the Environment, Wellington, pp. 81–84.

**Busby**, J.R., 1988. Potential impacts of climate change on Australia's flora and fauna. In: *Greenhouse: Planning for Climate Change* [Pearman, G.I. (ed.)]. CSIRO, Melbourne, pp. 387–398.

**Campbell**, B.D. and R.J.M. Hay, 1993: *Will subtropical grasses continue to spread throughout New Zealand*? Proceedings of the XVII International Grassland Congress, 13 February 1993. Palmerston North, New Zealand, pp. 1126–1128.

**Campbell**, B.D., G.M. McKeon, R.M. Gifford, H. Clark, D.M. Stafford-Smith, P. Newton, and J.L. Lutze, 1996: Impacts of atmospheric composition and climate change on temperate and tropical pastoral agriculture. In: *Greenhouse: Coping with Climate Change* [Bouma, W.J., G.I. Pearman, and M.R. Manning (eds.)]. CSIRO Publishing, Collingwood, Victoria, Australia, pp. 171–189.

**Ceulemans**, R. and M. Mousseau, 1994: Effects of elevated atmospheric $CO_2$ on woody plants. *New Phytologist*, **127**, 425–446.

**Chappell**, J., P.J. Cowell, C.D. Woodroffe, and I.G. Eliot, 1996: Coastal impacts of enhanced greenhouse climate change in Australia: implications for coal use. In: *Greenhouse: Coping with Climate Change* [Bouma, W.J., G.I. Pearman, and M.R. Manning (eds.)]. CSIRO Publishing, Collingwood, Victoria, Australia, pp. 220–234.

**Chiew**, F.H.S., P.H. Whetton, T.A. McMahon, and A.B. Pittock, 1995: Simulation of the impacts of climate change on runoff and soil moisture in Australian catchments. *J. Hydrol.,* **167**, 121–147.

**Christensen**, P.E. and P.C. Kimber, 1975: Effect of prescribed burning on the flora and fauna of south-west Australian forests. *Proceedings of the Ecological Society of Australia*, **9**, 85–106.

**Closs**, G.P. and P.S. Lake, 1996: Drought, differential mortality and the coexistence of a native and an introduced fish species in a south east Australian intermittent stream. *Env. Biol. Fish.*, **47**, 17–26.

**Coleman**, R.A., B.J. McAvaney, J.R. Fraser, and S.B. Power, 1994: Annual mean meridional energy transport modelled by a general circulation model for present and 2 x $CO_2$ equilibrium climates. *Climate Dynamics*, **10**, 221–229.

**Cowell**, P.J., P.S. Roy, T.Q. Zeng, and B.G. Thom, 1996: *Practical relationships for predicting coastal geomorphic impacts of climate change.* Proceedings of Ocean and Atmosphere Pacific [Aung, T.H. (ed.)]. National Tidal Facility, The Flinders University of South Australia, Adelaide, South Australia, pp. 16–21.

**Crocombe**, G.T., N.J. Enright, and M.E. Porter, 1991: *Upgrading New Zealand's Competitive Advantage*, Oxford University Press, Oxford, 235 pp.

**CSIRO**, 1991: *Current Climate Change Scenario for the Australian Region — 2030.* Climate Impact Group, CSIRO Division of Atmospheric Research, Melbourne, Australia, 4 pp.

**CSIRO**, 1992: *Climate Change Scenarios for the Australian Region.* Climate Impact Group, CSIRO Division of Atmospheric Research, Melbourne, Australia, 6 pp.

**CSIRO**, 1996a: *Climate Change Scenarios for the Australian Region.* Climate Impact Group, CSIRO Division of Atmospheric Research, Melbourne, Australia, 8 pp.

**CSIRO**, 1996b: *OzClim: A Climate Scenario Generator and Impacts Package for Australia.* Issued November 1996, 4 pp. Also available on the Internet at http://www.dar.csiro.au/pub/programs/climod/impacts/ozclim.htm

**Davis**, J.A., S.A. Harrington, and J.A. Friend, 1993. Invertebrate communities of relict streams in the arid zone: the George Gill Range, central Australia. *Aust. J. Mar. Freshwater Res.*, **44**, 483–505.

**Department of Conservation**, 1994: *New Zealand Coastal Policy Statement.* Department of Conservation, Wellington, New Zealand, 26 pp.

**Dexter**, E.M., A.D. Chapman, and J.R. Busby, 1995: *The Impact of Global Warming on the Distribution of Threatened Vertebrates.* Available on the Internet at http://www.erin.gov.au/life/end_vuln/animals/climate/climate_change/cc_exe.html

**Done**, T.J., 1983: Coral zonation: its nature and significance. In: *Perspectives on Coral Reefs. Australian Institute of Marine Science* [Barnes, D.J. (ed.)]. Brian Cluston Publisher, Manuka, ACT, Australia, pp. 107–147.

**Dubinsky**, Z. and N. Stambler, 1996: Marine pollution and coral reefs. *Global Change Biology,* **2**, 511–526.

**Eamus**, D., 1991: The interaction of rising $CO_2$ and temperature with water use efficiency. *Plant, Cell, and Environment*, **14**, 843–852.

**Enquselassie**, F., A.J. Dobson, H.M. Alexander, and P.L. Steele, 1993: Seasons, temperature and coronary disease. *International Journal of Epidemiology,* **137**, 331–341.

**Fankhauser**, S. and R.S.J. Tol, 1997: The social costs of climate change: the IPCC Second Assessment Report and beyond. *Mitigation and Adaptation Strategies for Global Change*, **1**, 385–403.

**Farrow**, R.A., G. McDonald, and P.D. Stahle, 1993: Potential impact of rapid climate change through the greenhouse effect on the pests of pastures in southeast Australia. In: *Pests of Pastures* [Delfosse, E.S. (ed)]. CSIRO, Melbourne, pp. 142–151.

**Fenner**, J., L. Carter, and R. Stewart, 1992: Late Quaternary paleoclimatic and paleooceanographic change over northern Chatham Rise, New Zealand. *Marine Geology*, **108**, 383–404.

**Fenwick**, T., 1995: Water management — a new age. In: *Jack Beale Water Resources Lecture Series 1990–1994.* Water Research Foundation of Australia, c/o Australian National University, Canberra, pp. 35–47.

**Finlayson**, B.L. and T.A. McMahon, 1988: Australia vs the world: A comparative analysis of streamflow characteristics. In: *Fluvial Geomorphology of Australia* [Warner, R.J. (ed.)]. Academic Press, Sydney, pp. 17–40.

**Fitzharris**, B.B., 1992: The 1992 electricity crisis and the role of climate and hydrology. *New Zealand Geographer*, **48(2)**, 79–83.

**Fitzharris**, B. and C. Garr, 1996: Climate, water resources and electricity. In: *Greenhouse: Coping with Climate Change* [Bouma, W.J., G.I. Pearman, and M.R. Manning (eds.)]. CSIRO Publishing, Collingwood, Victoria, Australia, pp. 263–280.

**Fleming**, P.M., 1995: Australian water resources are different. *Australasian Science*, **16(2)**, 8–10.

**Folland**, C.K. and M.J. Salinger, 1995: Surface temperature trends and variations in New Zealand and the surrounding ocean, 1871–1993. *International Journal of Climatology,* **15**, 1195–1218.

**Fowler**, A.M. and K.J. Hennessy, 1995: Potential impacts of global warming on the frequency and magnitude of heavy precipitation. *Natural Hazards*, **11**, 283–303.

**Fraser**, T., 1990: Maori and Their Resources. In: *Climate Change: Impacts on New Zealand*, N.Z. Ministry for the Environment, Wellington, pp. 215–218.

**Garr**, C.E. and B.B. Fitzharris, 1994: Sensitivity of mountain runoff and hydroelectricity to changing climate. In: *Mountain Environments in Changing Climates* [Beniston, M. (ed.)]. Routledge, London, United Kingdom, pp. 366–381.

**Garratt**, J.R., M.R. Raupach, and K.G. McNaughton, 1996: Climate and the terrestrial biosphere. In: *Greenhouse: Coping with Climate Change* [Bouma, W.J., G.I. Pearman, and M.R. Manning (eds.)]. CSIRO Publishing, Collingwood, Victoria, Australia, pp. 42–55.

**Ghassemi**, F., G. Jacobson, and A.J. Jakeman, 1991: Major Australian aquifers: potential climatic change impacts. *Water International*, **16**, 38–44.

**Gibb**, J.G., 1980: *Late Quaternary Shoreline Movements in New Zealand.* Ph. D. Thesis, Victoria University of Wellington, Wellington, 217+ pp.

**Gifford**, R.M., 1992: Interaction of carbon dioxide and growth-limiting environmental factors in vegetation productivity: Implications for the global carbon cycle. *Advances in Bioclimatology,* **1**, 24–58.

**Gifford**, R.M., D.J. Barrett, J.L. Lutze, and A.B. Samarakoon, 1996a: Agriculture and global change: Scaling direct carbon dioxide impacts and feedbacks through time. In: *Global Change and Terrestrial Ecosystems* [Walker, B., and W. Steffen (eds.)]. Cambridge University Press, Cambridge, UK, pp. 229–259.

**Gifford**, R.M., B.D. Campbell, and S.M. Howden, 1996b: Options for adapting agriculture to climate change: Australian and New Zealand examples. In: *Greenhouse: Coping with Climate Change* [Bouma, W.J., G.I. Pearman, and M.R. Manning (eds.)]. CSIRO Publishing, Collingwood, Victoria, Australia, pp. 399–416.

**Gill**, A.M. and R.A. Bradstock, 1995: Extinction of biota by fires. In *Conserving Biodiversity: Threats and Solutions*, [Bradstock, R.A., T.D. Auld, D.A. Keith, R.T. Kingsford, D. Lunney, and D.P. Sivertsen (eds.)] Surrey Beatty and Sons, Chipping Norton, New South Wales, Australia, pp. 309–322.

**Glova**, G.J., 1990: Freshwater fisheries. In: *Climatic Change: Impacts on New Zealand* [Mosley, M.P. (ed.)]. Ministry for the Environment, Wellington, pp. 78–80.

**Glynn**, P.W., 1996: Coral reef bleaching: facts, hypotheses and implications. *Global Change Biology,* **2**, 495–509.

**Gordon**, H.B., B.J. McAvaney, and J.L. McGregor, 1996: Perspectives on modelling global climate change. In: *Greenhouse: Coping with Climate Change,* [Bouma, W.J., G.I. Pearman, and M.R. Manning (eds.)]. CSIRO Publishing, Collingwood, Australia, pp. 56–80.

**Green**, T.R., B.C. Bates, P.M. Fleming, and S.P. Charles, 1997: Simulated impacts of climate change on groundwater recharge in the subtropics of Queensland, Australia. In: *Subsurface Hydrological Responses to Land Cover and Land Use Changes*, [Taniguchi, M. (ed.)]. Kluwer Academic, Boston, Massachusetts, pp. 187–204.

**Greer**, D.H., W.A. Laing, and B.D. Campbell, 1995: Photosynthetic responses of thirteen pasture species to elevated $CO_2$ and temperature. *Aust. Journal of Plant Physiology*, **22**, 713–722.

**Griffiths**, G.A., 1990: Water resources. In: *Climatic Change: Impacts on New Zealand* [Mosley, M.P. (ed.)]. Ministry for the Environment, Wellington, pp. 38–43.

**Gunderson**, C.A. and S.D. Wullschleger, 1994: Photosynthetic acclimation in trees to rising atmospheric $CO_2$: a broader perspective. *Photosynthetic Research*, **39**, 369–388.

**Hales**, S., P. Weinstein, and A. Woodward, 1996: Dengue fever epidemics in the South Pacific: driven by El Niño Southern Oscillation? *Lancet*, **348**, 1664–1665.

**Hales**, S., A. Woodward, and C. Guest, 1995: Climate change in the South Pacific: priorities for public health research. *Aust. Journal of Public Health*, **19**, 543–545.

**Hart**, B.T., P. Bailey, R. Edwards, K. Hortle, K. James, A. McMahon, C. Meredith, and K. Swadling, 1991: A review of the salt sensitivity of the Australian freshwater biota. *Hydrobiologia,* **210**, 105–144.

**Hassall and Associates**, 1997: *Climate Change Scenarios and Managing the Scarce Water Resources of the Macquarie River.* Report for the Department of the Environment, Sport and Territories, under the Climate Impacts and Adaptation Grants Program, Canberra (in press).

**Heathcote**, R.L., 1994: Australia. In: *Drought Follows the Plough* [Glanz, M.H. (ed.)]. Cambridge University Press, Cambridge, United Kingdom, pp. 91–102.

**Heathcote**, R.L., 1996: Settlement advance and retreat: a century of experience on the Eyre Peninsula of South Australia. In: *Climate Variability and Vulnerability: Causality and Response* [J.C. Ribot, A.R. Magalhães, and S.S. Panagides (eds.)]. Cambridge University Press, Cambridge, United Kingdom, pp. 109–122.

**Henderson**, G, P. McKenna, A. Kingsley, R. Little, A. Sorby, G. Jacobson, and G. Fisher, 1995: Hepatitus A and water supply in the Torres Strait Area. *Aboriginal and Torres Strait Islander Health Information Bulletin*, **21**, 48–58.

**Henderson-Sellers**, A., 1996: Adaptation to climatic change: its future role in oceania. In: *Greenhouse: Coping with Climate Change* [Bouma, W.J., G.I. Pearman, and M.R. Manning (eds.)]. CSIRO Publishing, Collingwood, Victoria, Australia, pp. 349–376.

**Henderson-Sellers**, A. and H. Zhang, 1997: *Tropical Cyclones and Global Climate Change.* Report from the WMO/CAS/TMRP Committee on Climate Change Assessment (Project TC-2), World Meteorological Organization, Geneva, Switzerland, 47 pp. (also accessible via Internet at http://www.bom.gov.au/bmrc/).

**Hennessy**, K.J. and K. Clayton-Greene, 1995: Greenhouse warming and vernalisation of high-chill fruit in southern Australia. *Climatic Change*, **30**, 327–348.

**Hennessy**, K.J. and A.B. Pittock, 1995: Greenhouse warming and threshold temperature events in Victoria, Australia. *Internat. Journal of Climatology*, **15**, 591–612.

**Hennessy**, K.J. and A.B. Pittock, 1996a: *Climate Impacts Assessment Workshop Report: Development and Application of Climate Change Scenarios.*, CSIRO Division of Atmospheric Research, Melbourne, Australia, 4–5 Dec. 1995, 47 pp.

**Hennessy**, K.J. and A.B. Pittock, 1996b: *Climate Impacts Assessment Workshop Abstracts: Development and Application of Climate Change Scenarios.* CSIRO Division of Atmospheric Research, Melbourne, Australia, 4–5 Dec. 1995, 117 pp.

**Hennessy**, K.J., P.H. Whetton, X. Wu, J.L. McGregor, J.J. Katzfey, and K. Nguyen, 1997: *Fine Resolution Climate Change Scenarios for New South Wales*, N.S.W. Environment Protection Authority, Sydney, Australia, 42 pp.

**Hillman**, T.J. and R.J. Shiel, 1991: Macro- and micro-invertebrates in Australian billabongs. *Verh. Internat. Verein. Limnol.*, **24**, 1581–1587.

**Hobbs**, J.E., 1988: Recent climate change in Australia. In: *Recent Climate Change* [Gregory, S. (ed.)]. Belhaven Press, London, United Kingdom, pp. 285–297.

**Holbrook**, N.J. and N.L. Bindoff, 1997: Interannual and decadal temperature variability in the southwest Pacific Ocean between 1955 and 1988. *J. Climate*, **10**, 1035 –1049.

**Holland**, G.J., 1997: The maximum potential intensity of tropical cyclones. *J. Atmospheric Science*, **54**, 2519–2541.

**Hooper**, A., 1990: Tokelau. In: *Climate change: Impacts on New Zealand*, N.Z. Ministry for the Environment, Wellington, pp. 238–242.

**Hopkins**, L.C. and G.J. Holland, 1997: Australian heavy-rain days and associated east coast cyclones: 1958–1992. *J. Climate*, **10**, 621–635.

**Howden**, M., D.H. White, G.M. McKeon, J.C. Scanlan, and J.O. Carter, 1994: Methods for exploring management options to reduce greenhouse gas emissions from tropical grazing systems. *Climatic Change*, **27**, 49–70.

**Hunt**, J.E., F.M. Kelliher, and D.L. McNeil, 1996. Response in chlorophyll *a* fluorescence of six New Zealand tree species to a step-wise increase in ultraviolet-B irradiance. *New Zealand Journal of Botany,* **34**, 401–410.

**IPCC**, 1990: *Climate Change: The IPCC Scientific Assessment* [Houghton, H.T., G.J. Jenkins, and J.J. Ephraums (eds.)]. Cambridge University Press, Cambridge, United Kingdom, 365 pp.

**IPCC**, 1991: *Common Methodology for Assessing Vulnerability to Sea-level Rise.* Intergovernmental Panel on Climate Change, Coastal Zone Management Group, Ministry of Transport and Public Works, The Hague, Netherlands.

**IPCC**, 1996a: Climate Change 1995: The Science of Climate Change. Contribution of Working Group I to the Second Assessment Report of the Intergovernmental Panel on Climate Change [Houghton, J.T., L.G. Meira Filho, B.A. Callander, N. Harris, A. Kattenberg, and K. Maskell (eds.)]. Cambridge University Press, Cambridge and New York, 572 pp.

- Schimel, D., D. Alves, I. Enting, M. Heimann, F. Joos, D. Raynaud, T.M.L. Wigley, E. Sanhueza, X. Zhou, P. Jonas, R. Charlson, H. Rodhe, S. Sadasivan, K.P. Shine, Y. Fouquart, V. Ramaswamy, S. Solomon, J. Srinivasan, D. Albritton, R. Derwent, Y. Isaken, M. Lal, and D. Wuebbles, Chapter 2. *Radiation Forcing of Climate Change,* pp. 65–131.
- Nicholls, N., G.V. Gruza, J. Jouzel, T.R. Karl, L.A. Ogallo, and D.E. Parker, Chapter 3. *Observed Climate Variability and Change,* pp. 133–192.
- Kattenberg, A., F. Giorgi, H. Grassl, G.A. Meehl, J.F.B. Mitchell, R.J. Stouffer, T. Tokioka, A.J. Weaver, and T.M.L. Wigley, Chapter 6. *Climate Models —Projections of Future Climate,* pp. 289–357.
- Warrick, R.A., C. Le Provost, M.F. Meier, J. Oerlemans, and P.L. Woodworth, Chapter 7. *Changes in Sea Level,* pp. 359–405.
- Santer, B.D., T.M.L. Wigley, T.P. Barnett, and E. Anyamba, Chapter 8. *Detection of Climate Change and Attribution of Causes*, pp. 407–443.

**IPCC**, 1996b: Climate Change 1995: Impacts, Adaptations, and Mitigation of Climate Change: Scientific-Technical Analyses. Contribution of Working Group II to the Second Assessment Report of the Intergovernmental Panel on Climate Change [Watson, R.T., M.C. Zinyowera, and R.C. Moss (eds.)]. Cambridge University Press, Cambridge and New York, 878 pp.

- Kirschbaum, M.U.F., Chapter A. *Ecophysiological, Ecological, and Soil Processes in Terrestrial Ecosystems: A Primer on General Concepts and Relationships*, p.57–74.
- Kirschbaum, M.U.F. and A. Fischlin, Chapter 1. *Climate Change Impacts on Forests,* pp. 95–130.
- Allen-Diaz, B., Chapter 2. *Rangelands in a Changing Climate: Impacts, Adaptations and Mitigation,* pp. 131–158.
- Bullock, P. and H. Le Houérou, Chapter 4. *Land Degradation and Desertification,* pp. 170–189.
- Beniston, M. and D.G. Fox, Chapter 5. *Impacts of Climate Change on Mountain Regions,* pp. 191–213.
- Fitzharris, B.B., Chapter 7. *The Cryosphere: Changes and their Impacts,* pp. 240–265.
- Bijlsma, L, Chapter 9. *Coastal Zones and Small Islands*, pp. 289–324.
- Arnell, N., B. Bates, H. Lang, J.J. Magnuson, and P. Mulholland, Chapter 10. *Hydrology and Freshwater Ecology,* pp. 325–364.
- Acosta-Moreno, R. and J. Skea, Chapter 11. *Industry, Energy, and Transportation: Impacts and Adaptation,* pp. 365–398.
- Scott, M.J., Chapter 12. *Human Settlements in a Changing Climate: Impacts and Adaptation,* pp. 399–426.
- Reilly, J., Chapter 13. *Agriculture in a Changing Climate: Impacts and Adaptation*, pp. 427–467.
- Kaczmarek, Z., Chapter 14. *Water Resources Management*, pp. 469–486.
- Solomon, A.M., Chapter 15. *Wood Production under Changing Climate and Land Use,* pp. 487–510.
- Everett, J., Chapter 16. *Fisheries,* pp. 511–537.
- Dlugolecki, A.F., Chapter 17. Financial Services, pp.539–560.
- McMichael, A., Chapter 18. *Human Population Health,* pp. 561–584.

**IPCC**, 1996c: Climate Change 1995: Economic and Social Dimensions of Climate Change. Contribution of Working Group III to the Second Assessment Report of the Intergovernmental Panel on Climate Change [Bruce, J.P., H. Lee, and E.F. Haites (eds.)]. Cambridge University Press, Cambridge and New York, 448 pp.

- Pearce, D.W., W.R. Cline, A.N. Achanta, S. Fankhauser, R.K. Pachauri, R.S.J. Tol, and P. Vellinga, Chapter 6. *The Social Costs of Climate Change: Greenhouse Damage and Benefits of Control*, pp. 179–224.
- Weyant, J., Chapter 10. *Integrated Assessment of Climate Change: An Overview and Comparison of Approaches and Results*, pp.367–396.

**Jacobson**, G., 1976: The freshwater lens on Home Island in the Cocos (Keeling) Islands. *BMR Journal of Australian Geology and Geophysics*, **1**, 335–343.

**Jacobson**, G., 1996: The interrelationship of hydrology and landform in central Australia. In: *Exploring Central Australia: Society, the Environment and the 1894 Horn Expedition* [Morton, S.R. and D.J. Mulvaney (eds.)]. Surrey Beatty and Sons, Chipping Norton, New South Wales, pp. 249–266.

**Jacobson**, G., G.E. Calf, J. Jankowski, and P.S. McDonald, 1989: Groundwater chemistry and palaeorecharge in the Amadeus Basin, Central Australia. *Journal of Hydrology*, **109**, 237–266.

**Jacobson**, G. and T. Graham, 1996: Groundwater discharge to coastal swamps in the Torres Strait Islands: high water table and mosquito habitat. In: *Groundwater Discharge in the Coastal Zone* [Buddemeier, R.W. (ed.)]. *LOICZ Reports and Studies*, **8**, pp. 55–61.

**Jenkinson**, D.S, D.E. Adams, and A. Wild, 1991: Model estimates of $CO_2$ emissions from soil response to global warming. *Nature*, **351**, 304–306.

**Jones**, P.A., 1991: Historical records of cloud cover and climate for Australia. *Australian Meteorological Magazine,* **39**, 181–189.

**Karl**, T.R., P.D. Jones, R.W. Knight, G. Kukla, N. Plummer, V. Razuvayev, K.P. Gallo, J. Lindseay, R.J. Charlson, and T.C. Peterson, 1993: Asymmetric trends of daily maximum and minimum temperature. *Bulletin American Meteorological Society*, **74**, 1007–1023.

**Karl**, T.R., R.W. Knight, and N. Plummer, 1995: Trends in high-frequency climate variability in the twentieth century. *Nature*, **377**, 217–220.

**Katzfey**, J.J. and K.L. McInnes, 1996: GCM simulations of eastern Australian cutoff lows. *Journal of Climate*, **9**, 2337–2355.

**Kay**, R., A. Kirkland, I. Stewart, J. Bailey, C. Berry, B. Caton, J. Dahm, I.E. Eliot, E.J.A. Kleverlaan, G. Morvell, D. Slaven, and P. Waterman, 1996: Planning for future climate change and sea-level rise induced coastal change in Australia and New Zealand. In: *Greenhouse: Coping with Climate Change* [Bouma, W.J., G.I. Pearman, and M.R. Manning (eds.)]. CSIRO Publishing, Collingwood, Victoria, Australia, pp. 377–398.

**Keith**, D., 1996: Fire-driven extinction of plant populations: a synthesis of theory and review of evidence from Australian vegetation. *Proceedings of the Linnean Society of New South Wales*, **116**, 37–78.

**Kenny**, G. J., R.A. Warrick, N.D. Mitchell, A.B. Mullan, and M.J. Salinger, (1995): CLIMPACTS: An Integrated Model for Assessment of the Effects of Climate Change on the New Zealand Environment. *Journal of Biogeography*, **22**, 883–895.

**Knighton**, A.D., K. Mills, and C.D. Woodroffe, 1991: Tidal creek extension and salt water intrusion in northern Australia. *Geology*, **19**, 831–834.

**Knott**, G.G. and P.S. McDonald, 1983: Groundwater for Central Australian Aboriginal communities. *Australian Water Resources Council, Conference Series*, **3**, 141–150.

**Kolsky**, P.J., 1993: Diarrhoeal disease: current concepts and future challenges. Water sanitation and diarrhoea; the limits of understanding. *Trans. Royal Society of Tropical Medicine and Hygiene*, **87**, 43–46.

**Lake**, P.S. 1995: Of floods and droughts: river and stream ecosystems of Australia. In: *Ecosystems of the World 22. River and Stream Ecosystems* [Cushing, C.E., K.W. Cummins, and G.W. Minshall (eds.)]. Elsevier, Amsterdam, Netherlands, pp. 659–694.

**Landsberg**, J.J., 1996: Impact of climate change and atmospheric carbon dioxide concentration on the growth of planted forests. In: *Greenhouse: Coping with Climate Change* [Bouma, W.J., G.I. Pearman, and M.R. Manning (eds.)]. CSIRO Publishing, Collingwood, Victoria, Australia, pp. 205–219.

**Landsberg**, J. and M. Stafford Smith, 1992: A functional scheme for predicting the outbreak potential of herbivorous insects under global atmospheric change. *Australian Journal of Botany*, **40**, 565–577.

**Larcombe**, P., K. Woolfe, and R. Purdon (eds.), 1996: *Great Barrier Reef: Terrigenous Sediment Flux and Human Impacts*, CRC Reef Research Centre, Current Research, Townsville, Queensland, Australia, 2nd ed., 174 pp.

**Lavery**, B.M., G. Joung, and N. Nicholls, 1997: An extended high-quality rainfall data set for Australia. *Austr. Met. Mag.*, **46**, 27–38.

**Leighton**, R.M., K. Keay, and I. Simmonds, 1997: Variation in annual cyclonicity across the Australian region for the 29-year period 1965–1993 and the effect on annual all-Australia rainfall. In: *Proceedings, Workshop on Climate Prediction for Agriculture and Resource Management*. Bureau of Resource Sciences, Canberra.

**Lonsdale**, W.M. 1994: Inviting trouble: introduced pasture species in northern Australia. *Aust. Journal of Ecolology,* **19**, 345–354.

**Lough**, J.M., 1994: Climate variation and El Niño — Southern Oscillation events on the Great Barrier Reef: 1958–1987. *Coral Reefs*, **13**, 181–195.

**Lunkeit**, F., R. Sausen, and J.M. Oberhuber, 1994: Climate simulations with the global coupled atmosphere-ocean model ECHAM/OPYC, part I: present-day climate and ENSO events. *MPI Report No. 132*, Max-Planck-Institut, Hamburg, Germany, 47 pp.

**Lutze**, J.L., 1996: *Carbon and nitrogen relationships in swards of Danthonia richardsonii in response to carbon dioxide enrichment and nitrogen supply.* Ph.D. Thesis, Australian National University, Canberra, Australia.

**Maclaren**, J.P., 1996: Plantation forestry: its role as a carbon sink. In: *Greenhouse: Planning for Climate Change* [Pearman, G.I. (ed.)]. E.J. Brill Publishers, Leiden, Netherlands and CSIRO Publishing, East Melbourne, Victoria, Australia, pp. 417–436.

**Macumber**, P.G., 1990: The salinity problem. In: *The Murray* [Mackay, N. and D. Eastburn (eds.)]. Murray Darling Basin Commission, Canberra, Australia, pp. 110–125.

**Manning**, M.R., G.I. Pearman, D.M. Etheridge, P.J. Fraser, D.C. Lowe, and L.P. Steele, 1996: The changing composition of the atmosphere. In: *Greenhouse; Coping with Climate Change* [Bouma, W.J., G.I. Pearman, and M.R. Manning (eds.)]. CSIRO Publishing, Collingwood, Victoria, Australia, pp. 3–26.

**Martin**, R.J., C.J. Korte, D.G. McCall, D.B. Baird, P.C.D. Newton, and N.D. Barlow, 1991: Impacts of potential change in climate and atmospheric concentration of carbon dioxide on pasture and animal production in New Zealand. *Proceedings of the New Zealand Society of Animal Production*, **51**, 25–33.

**Maunder**, W.J., 1971: The climate of New Zealand — physical and dynamic features. In: *World Survey of Climatology* [Gentilli, J. (ed.)]. Elsevier, Amsterdam, Netherlands, **13**, 213–227.

**McBride**, J. and N. Nicholls, 1983: Seasonal relationships between Australian rainfall and the Southern Oscillation. *Monthly Weather Review*, **111**, 1998–2004.

**McClintock**, 1997: Irrigation in the Murray-Darling Basin. *ABARE Update*, April 1997. Australian Bureau of Agricultural and Resource Economics, Canberra, pp. 14–15.

**McCown**, R.L., P. Gillard, L. Winks, and W.T. Williams, 1981: The climatic potential for beef cattle production in tropical Australia: Part II — Liveweight change in relation to agro-climatic variables. *Agricultural Systems,* **7**, 1–10.

**McCreary**, J.P. and D.L.T. Anderson, 1991: An overview of coupled ocean-atmosphere models of El Niño and the Southern Oscillation, *J. Geophys. Res.*, **96**, supplement, 3125–3150.

**McDowell**, R.M., 1992: Global climate change and fish and fisheries: what might happen in a temperate oceanic archipeligo like New Zealand. *GeoJournal*, **28**, 29–37.

**McGlone**, M., G. Hope, J. Chappell, and P. Barrett, 1996: Past climatic change in oceania and Antarctica. In: *Greenhouse: Coping with Climate Change* [Bouma, W.J., G.I. Pearman, and M.R. Manning (eds.)]. CSIRO Publishing, Collingwood, Victoria, Australia, pp. 81–99.

**McGuckin**, J., J.A. Anderson, and R.J. Gasior, 1991: *Salt-Affected Rivers in Victoria.* Arthur Rylah Institute for Environmental Research, Technical Report Series No. 118, Melbourne, Victoria, Australia.

**McInnes**, K.L. and G.D. Hubbert, 1996: *Extreme events and the impact of climate change in Victoria's coastline.* Environment Protection Authority and Melbourne Water, Melbourne, Australia, 69 pp.

**McKeon**, G.M. and S.M. Howden, 1993: Adapting the management of Queensland's grazing systems to climate change. In: *Climate Change: Implications for Natural Resource Conservation* [Burgin, S. (ed.)]. University of Western Sydney, Occasional Paper No. 1, pp. 123–140.

**McKeon**, G.M., S.M. Howden, N.O.J. Abel, and J.M. King, 1993: *Climate change: adapting tropical and subtropical grasslands.* Proceedings of the XVII International Grassland Congress, 13-16 February 1993, Palmerston North, New Zealand, pp. 1181–1190.

**McKeon**, G.M., S.M. Howden, D.M. Silburn, J.O. Carter, J.F. Clewett, G.L. Hammer, P.W. Johnston, P.L. Lloyd, J.J. Mott, B. Walker, E.J. Weston, and J.R. Willcocks, 1988: The effect of climate change on crop and pastoral production in Queensland. In: *Greenhouse: Planning for Climate Change* [Pearman, G.I. (ed.)]. E.J. Brill Publishers, Leiden, Netherlands and CSIRO Publishing, East Melbourne, Victoria, Australia, pp. 546–563.

**McLean**, R. and N. Mimura (eds.), 1993: *Vulnerability Assessment to Sea-Level Rise and Coastal Zone Management.* Proceedings IPCC Eastern Hemisphere Workshop, Tsukuba, Japan, 3–6 August 1993, 429 pp.

**McMahon**, T.A. and B.L. Finlayson, 1991: Australian surface and groundwater hydrology — regional characteristics and implications. In: *Water Allocation and the Environment* [Pigram, J.J. and B.P. Hooper (eds.)]. Centre for Water Policy Research, University of New England, Armidale, New South Wales, pp. 21–40.

**McMurtrie**, R.E. and H.N. Comins, 1996: The temporal response of forest ecosystems to doubled atmospheric $CO_2$ concentration. *Global Change Biology,* **2**, 49–57.

**Metzeling**, L., T. Doeg, and W. O'Connor, 1995: The impact of salinization and sedimentation on aquatic biota. In: *Conserving Biodiversity: Threats and Solutions* [Bradstock, R.A., T.D. Auld, D.A. Keith, R.T. Kingsford, D. Lunney, and D.P. Sivertsen (eds.)]. Surrey Beatty and Sons, Chipping Norton, New South Wales, Australia, pp. 126–136.

**Miller**, I.L. and C.G. Wilson, 1995: Weed threats to Northern Territory wetlands. In: *Wetland Research in the Wet-dry Tropics of Australia* [Finlayson, C.M. (ed.)]. Supervising Scientist for the Alligator Rivers Region, Canberra, pp. 190–195.

**Minnery**, J.R., and D.I. Smith, 1996: Climate change, flooding and urban infrastructure. In: *Greenhouse: Coping with Climate Change* [Bouma, W.J., G.I. Pearman, and M.R. Manning (eds.)]. CSIRO Publishing, Collingwood, Victoria, Australia, pp. 235–247.

**Mitchell**, C.D., K.J. Hennessy, and A.B. Pittock, 1994: *Regional Impact of the Enhanced Greenhouse Effect on the Northern Territory, Final Report.* Northern Territory Government and CSIRO Division of Atmospheric Research, Aspendale, Victoria, Australia, 85 pp.

**Mitchell**, N.D. and J.E. Williams, 1996: The consequences for native biota of anthropogenic-induced climate change. In: *Greenhouse; Coping with Climate Change* [Bouma, W.J., G.I. Pearman, and M.R. Manning (eds.)]. CSIRO Publishing, Collingwood, Victoria, Australia, pp. 308–324.

**Moore**, E.J. and J.W. Smith, 1996: Migration response to climate change. In: *Greenhouse: Coping with Climate Change* [Bouma, W.J., G.I. Pearman, and M.R. Manning (eds.)]. CSIRO Publishing, Collingwood, Victoria, Australia, pp. 325–348.

**Morison**, J.I.L. and R.M. Gifford, 1984: Plant growth and water use with limited water supply in high $CO_2$ concentrations. II Plant dry weight, partitioning and water use efficiency. *Aust. J. Plant Physiology*, **11**, 375–384.

**Morrison**, D.A., G. Cary, S.M. Pengelly, D.G. Ross, B.J. Mullins, C.R. Thomas, and T.S. Anderson, 1995: Effects of fire frequency on plant species composition of sandstone communities in the Sydney region: inter-fire interval and time-since-fire. *Australian Journal of Ecology,* **20**, 239–247.

**Morton**, S.R., D.M. Stafford Smith, M.H. Friedel, G.F. Griffin, and G. Pickup, 1995: The stewardship of arid Australia: ecology and landscape management. *Journal of Environmental Management*, **43**, 195–217.

**Mosley**, M.P. (ed.), 1990a: *Climatic Change: Impacts on New Zealand*, Ministry for the Environment, Wellington, 244 pp.

**Mosley**, M.P., 1990b: Water industry. In: *Climatic Change: Impacts on New Zealand* [Mosley, M.P. (ed.)]. Ministry for the Environment, Wellington, pp. 133–139.

**Moss**, I., 1994: *Water: A report on the provision of water and sanitation in remote Aboriginal and Torres Strait Islander communities.* Federal Race Discrimination Commissioner, Australian Government Publishing Service, Canberra, 487 pp.

**Mott**, J.J., J. Williams, M.H. Andrew, and A.N. Gillison, 1985: Australian savanna ecosystems. In: *Ecology and Management of the World's Savannas* [Tothill, J.C. and Mott, J.J. (eds.)]. Australian Academy of Science, Canberra, pp. 56–82.

**Muchow**, R.C. and T.R. Sinclair, 1991: Water deficit effects on maize yields modeled under current and "Greenhouse" climates. *Agronomy Journal*, **83**, 1052–1059.

**Mullan**, A.B., 1994: *Climate Change Scenarios for New Zealand: Statement for Greenhouse 94*, NIWA, Wellington, New Zealand, 9 pp.

**Mulrennan**, M.E., 1992: *Coastal management: Challenges and Changes in the Torres Strait Islands.* Discussion Paper No.5, North Australian Research Unit, Australian National University, Canberra, 40 pp.

**Nias**, R., 1992: *Global Climate Change and Biological Diversity.* World Wide Fund for Nature, Sydney, Australia, pp. 23.

**Nicholls**, N., 1986: A method for predicting Murray Valley Encephalitis in southeast Australia using the Southern Oscillation. *Australian Journal of Experimental and Biological Medical Science,* **64**, 587–594.

**Nicholls**, N., 1997: Increased Australian wheat yield due to recent climate trends. *Nature*, **387**, 484–485.

**Nicholls**, N. and B. Lavery, 1992: Australian rainfall trends during the twentieth century. *Int. J. Climate*, **12**, 153–163.

**Nicholls**, N., C. Landsea, and J. Gill, 1997: Recent trends in Australian region tropical cyclone activity. *Meteorology and Atmospheric Physics*, Special Edition on Tropical Cyclones (in press).

**Nicolas**, M.E., R. Munns, A.B. Samarakoon, and R.M. Gifford, 1993: Elevated $CO_2$ improves the growth of wheat under salinity. *Aust. Journal of Plant Physiology*, **20**, 349–360.

**Nix**, H.A., 1982: Environmental determinants of biogeography and evolution in Terra Australis. In: *Evolution of the Flora and Fauna of Arid Australia* [Barker, W.R. and P.J.M. Greenslande (eds.)]. Peacock Press, Adelaide, pp. 47–66.

**Owensby**, C.E., 1993: *Climate change and grasslands: ecosystem-level responses to elevated carbon dioxide.* Proceedings of the XVII International Grassland Congress, 13-16 February 1993, Palmerston North, New Zealand, pp. 1119–1124.

**Paul**, L.J., 1990: Marine fish, fisheries, and aquaculture. In: *Climatic Change: Impacts on New Zealand* [Mosley, M.P. (ed.)]. Ministry for the Environment, Wellington, pp. 85–94.

**Pearce**, N., S. Weiland, U. Keil, P. Langridge, H.R. Anderson, D. Strachan, et al., 1993: Self-reported prevalence of asthma symptoms in children in Australia, England, Germany and New Zealand: an international comparison using the ISAAC protocol. *Europ. Respir. Journal*, **6**, 1455–1461.

**Pearson**, R.G., L.J. Benson, and R.E.W. Smith., 1986: Diversity and abundance of the fauna of Yuccabine Creek, a tropical rainforest stream. In: *Limnology in Australia* [De Deckker, P. and W.D. Williams (eds.)]. CSIRO, Melbourne, Victoria, and W. Junk, Dordrecht, Netherlands, pp. 329–342.

**Peck**, A.J. and G.B. Allison, 1988: Groundwater and salinity response to climate change. In: *Greenhouse: Planning for Climate Change* [Pearman, G.I. (ed.)]. CSIRO Publishing, East Melbourne, Victoria, Australia, and E.J. Brill Publishers, Leiden, Netherlands, pp. 238–251.

**Peters**, R.L., 1992: Conservation of biological diversity in the face of climate change. In: *Global Warming and Biological Diversity* [Peters, R.L. and T.E. Lovejoy (eds.)]. Yale University Press, New Haven, CT, USA and London, United Kingdom, pp. 15–26.

**Peters**, R.L. and Darling, J.D.S., 1985: The greenhouse effect and nature reserves. *Bioscience*, **35**, 707–717.

**Pittock**, A.B., 1988: Actual and anticipated changes in Australia's climate, In: *Greenhouse: Planning for Climatic Change* [Pearman, G.I. (ed.)]. CSIRO Publishing, East Melbourne, Victoria, Australia, and E.J. Brill Publishers, Leiden, Netherlands, pp. 35–51.

**Pittock**, A.B., 1995: Report on reports "Climate Change" and "World Food Supply" and Special Issues of "Global Environment Change" and "Food Policy," *Environment*, **37(9)**, 25–30.

**Pittock**, A.B., 1996: *Adapting agriculture to climate change: a challenge for the 21st century*. Conference of Proceedings, Second Australian Conference on Agricultural Meteorology, University of Queensland, Brisbane, 1–4 October 1996, pp. 28–34.

**Pittock**, A.B., R.J. Allan, K.J. Hennessy, K.L. McInnes, R. Suppiah, K.J. Walsh, P.H. Whetton, H. McMaster, and R. Taplin, 1997: Climate change, climatic hazards and policy responses in Australia. In: *Climate, Change and Risk* [Downing, T.E., R.J.S. Tol, and A.A. Olsthoorn (eds.)]. Routledge, London, United Kingdom (in press).

**Pittock**, A.B., M.R. Dix, K.J. Hennessy, J.J. Katzfey, K.L. McInnes, S.P. O'Farrell, I.N. Smith, R. Suppiah, K.J. Walsh, P.H. Whetton, and S.G. Wilson, 1995: Progress towards climate change scenarios for the southwest Pacific. *Weather and Climate*, **15**, 21–46.

**Pittock**, A.B. and C.D. Mitchell, 1994: *Towards an Integrated Approach to Climate Change Impact Assessment*, Report of Workshop 26 April 1994, CSIRO Division of Atmospheric Research, Aspendale, Victoria, Australia, 72 pp.

**Pittock**, A.B., K.J. Walsh, and K.L. McInnes, 1996: Tropical cyclones and coastal inundation under enhanced greenhouse conditions. *Water, Air and Soil Pollution*, **92**, 159–169.

**Plummer**, N., 1996: Temperature variability and extremes over Australia: part 1 — recent observed changes. *Australian Meteorological Magazine*, **45**, 233–250.

**Plummer**, N., N. Nicholls, B.M. Lavery, R.M. Leighton, and B. Trewin, 1997: Twentieth century trends in Australian climate extremes indices. In: *Proceedings of CLIVAR/GCOS/WMO Workshop on Indices and Indicators for Climate Extremes*. 3-6 June 1997, Asheville, North Carolina, USA.

**Pomare**, E., V. Keefe-Ormsby, C. Ormsby, et al. (1995). *Hauora. Maori Standards of Health III. A Study of the Years 1970–1991*. Wellington. Te Ropu Rangahau a Eru Pomare. Wellington School of Medicine. 200 pp.

**Port of Melbourne Authority**, 1992: *Victorian Coastal Vulnerability Study*. Coastal Investigations Unit, Port of Melbourne Authority, Port Melbourne, Victoria, Australia, 170 pp.

**Radford**, D., R. Blong, A.M. d'Aubert, I. Kuhnel, and P. Nunn, 1996: *Occurence of Tropical Cyclones in the Southwest Pacific Region 1920–1994*. Greenpeace International, Amsterdam, Netherlands, 35 pp.

**Rasmusson**, E.M. and J.M. Wallace, 1983: Meteorological aspects of the El Niño/Southern Oscillation, *Science*, **222**, 1195–1202.

**Rawson**, H.M., 1995: Yield responses of two wheat genotypes to carbon dioxide and temperature in field studies using temperature gradient tunnels. *Australian Journal of Plant Physiology*, **22**, 23–32.

**Resource Assessment Commission**, 1993: *Coastal Zone Enquiry: Final Report*. Commonwealth of Australia, Australian Government Publishing Service, Canberra, 664 pp.

**Resource Management Act**, 1991, No. 69, Wellington, New Zealand Government. 382 pp.

**Rintoul**, S., G. Meyers, J. Church, S. Godfrey, M. Moore, and B. Stanton, 1996: Ocean processes, climate and sea level. In: *Greenhouse: Coping with Climate Change* [Bouma, W.J., G.I. Pearman, and M.R. Manning (eds.)]. CSIRO Publishing, Collingwood, Victoria, Australia, pp. 127–144.

**Ross**, H., E. Young, and L. Liddle, 1994: Mabo: an inspiration for Australian land management. *Australian Journal of Environmental Management*, **1**, 24–41.

**Rosser**, Z.C. and R.G. Pearson, 1995: Responses of rock fauna to physical disturbance in two Australian tropical rainforest streams. *Journal North American Benthol. Society*, **14**, 183–196.

**Russell**, J.S., 1991: Likely climate changes and their impact on the northern pastoral industry. *Tropical Grasslands*, **25**, 211–218.

**Salinger**, M.J., 1995: Southwest Pacific temperatures: trends in maximum and minimum temperatures. *Atmospheric Research*, **37**, 87–100.

**Salinger**, M.J., R. Allan, N. Bindoff, J. Hannah, N. Plummer, and S. Torok, 1996: Observed variability and change in climate and sea level in Australia, New Zealand and the South Pacific. In: *Greenhouse: Coping with Climate Change* [Bouma, W.J., G.I. Pearman, and M.R. Manning (eds.)]. CSIRO Publishing, Collingwood, Victoria, Australia, pp. 100–127.

**Salinger**, M.J., R.E. Basher, B.B. Fitzharris, J.E. Hay, P.D. Jones, J.-P. MacVeigh, and I. Schmidely-Leleu, 1995: Climate trends in the Southwest Pacific, *International Journal of Climatology*, **15**, 285–302.

**Salinger**, M.J. and D.M. Hicks, 1990a: The scenarios. In: *Climatic Change: Impacts on New Zealand*. N.Z. Ministry for the Environment, Wellington, pp. 12–18.

**Salinger**, M.J. and D.M. Hicks, 1990b: Appendix 1: Regional climate change scenarios. In: *Climate Change: Impacts on New Zealand*, N.Z. Ministry for the Environment, Wellington, pp. 238–242.

**Salinger**, M.J. and G.J. Kenny, 1995: Climate and kiwifruit cv. "Hayward," 2. Regions in New Zealand suited for production. *New Zealand Journal of Crop and Horticultural Science*, **23**, 173–184.

**Salinger**, M.J., A.B. Mullan, and J.G. Gibb, 1987: *Climate Change Scenario*. A working scenario prepared for the seminar Climate Change: The New Zealand Response, N.Z. Ministry for the Environment, Wellington, December 1987, 4 pp.

**Salinger**, M.J. and A.B. Pittock, 1991: Climate scenarios for 2010 and 2050 AD Australia and New Zealand. *Climatic Change*, **18**, 259–269.

**Saunders**, D.A. and R.J. Hobbs, 1992: Impact of biodiversity of changes in land-use and climate. In: *Biodiversity of Mediterranean Ecosystems in Australia* [Hobbs, R.J. (ed.)]. Surrey Beatty and Sons, Chipping Norton, New South Wales, pp. 61–75.

**Scanlan**, J.C., 1992: A model of woody-herbaceous biomass relationships in eucalypt and mesquite communities. *J. Range Management*, **45**, 5–80.

**Schreider**, S.Y., A.J. Jakeman, A.B. Pittock, and P.H. Whetton, 1996: Estimation of possible climate change impacts on water availability, extreme flow events and soil moisture in the Goulburn and Ovens basins, Victoria. *Climatic Change*, **34**, 513–546.

**Sinclair**, M.R., 1993a: Synoptic-scale diagnosis of the extratropical transition of a Southwest Pacific tropical cyclone. *Monthly Weather Review*, **121**, 941–960.

**Sinclair**, M.R., 1993b: A diagnostic study of the extratropical precipitation resulting from tropical cyclone Bola. *Monthly Weather Review*, **121**, 2690–2707.

**Smith**, D.I. and M.A. Greenaway, 1994: *Tropical Storm Surge, Damage Assessment and Emergency Planning: A Pilot Study for Mackay, Queensland*. Resource and Environmental Studies No. 8, Centre for Resource and Environmental Studies, Australian National University, Canberra, 59 pp.

**Smith**, K.P.H., F.M. Reed, and J.Z. Foot, 1997: An assessment of the relative importance of specific traits for the genetic improvement of nutritive value in dairy pasture. *Grass and Forage Science*, **52**, 167–175.

**SOEC**, 1996: *Australia, State of the Environment 1996*: An Independent Report Presented to the Commonwealth Minister for the Environment by the State of the Environment Advisory Council, CSIRO Publishing, Collingwood, Victoria, Australia, 539 pp.

**Stafford Smith**, M., 1994: *Sustainable production systems and natural resource management in the rangelands*. Proceedings of the ABARE Outlook Conference, Canberra, Australia, February 1994, pp. 148–159.

**Stafford Smith**, M., B. Campbell, W. Steffen, and S. Archer (eds.), 1994: State-of-the-science assessment of the likely impacts of global change on the Australian rangelands. *GCTE Working Document No. 14*, GCTE Core Project Office, Canberra, Australia, 72 pp.

**Steering Committee of the Climate Change Study**, 1995: *Climate Change Science: Current Understanding and Uncertainties*. Australian Academy of Technological Sciences and Engineering, Canberra, 100 pp.

**Stone**, R.C., G.L. Hammer, and D. Woodruff, 1993: *Assessment of risk associated with climate prediction in management of wheat in north-eastern Australia*. Proceedings of the 7th Australian Agronomy Conference, Adelaide, 19-24 September 1993. Australian Society of Agronomy, Parkville, Victoria, Australia, pp. 174–177.

**Sturman**, A.P. and N.J. Tapper, 1996: *The Weather and Climate of Australia and New Zealand*. Oxford University Press, Melbourne, 476 pp.

**Suppiah**, R., R. Allan, K. Hennessy, R. Jones, B. Pittock, K. Walsh, I. Smith, and P. Whetton, 1996: *Climate Change under Enhanced Greenhouse Conditions in Northern Australia. Second Annual Report, 1995–1996*. CSIRO Division of Atmospheric Research, Aspendale, Victoria, 63 pp.

**Suppiah**, R. and K.J. Hennessy, 1996: Trends in the intensity and frequency of heavy rainfall in tropical Australia and links with the Southern Oscillation. *Australian Meteorological Magazine*, **45**, 1–17.

**Suppiah**, R. and K.J. Hennessy, 1997: Trends in total rainfall, heavy rain events and number of dry days in Australia, 1910–1990. *International Journal of Climatology* (in press).

**Sutherst**, R.W., 1995: The potential advance of pests in natural ecosystems under climate change: implications for planning and management. In: *Impacts of Climate Change on Ecosystems and Species: Terrestrial Ecosystems* [Pernetta, J.C., R. Leemans, D. Elder, and S. Humphrey (eds.)]. IUCN, Gland, Switzerland, pp. 83–98.

**Sutherst**, R.W., T. Yonow, S. Chakraborty, C. O'Donnell, and N. White, 1996: A generic approach to defining impacts of climate change on pests, weeds and diseases in Australasia. In: *Greenhouse: Coping with Climate Change* [Bouma, W.J., G.I. Pearman, and M.R. Manning (eds.)]. CSIRO Publishing, Collingwood, Victoria, Australia, pp. 281–307.

**Tate**, K.R., D.J. Giltrap, A. Parshotam, A.E. Hewitt, D.J. Ross, G.J. Kenny, and R.A. Warrick, 1996: Impacts of climate change on soils and land system in New Zealand. In: *Greenhouse: Coping with Climate Change* [Bouma, W.J., G.I. Pearman, and M.R. Manning (eds.)]. CSIRO Publishing, Collingwood, Victoria, Australia, pp. 190–204.

**Tate**, K.R. and D.J. Ross, 1997: Elevated $CO_2$ and moisture effects on soil carbon storage and cycling in temperate grasslands. *Global Change Biology*, **3**, 225–235.

**Thom**, B.G., 1984: *Coastal Geomorphology in Australia*. Academic Press, Sydney, 349 pp.

**Thomas**, J.F. and B.C. Bates, 1997: *Responses to the variability and increasing uncertainty of climate in Australia.* Proceedings of the George Kovacs Colloquium, 19–21 September 1996, Paris, France. UNESCO/IHP, Cambridge University Press, Cambridge, UK (in press).

**Thompson**, C., S. Ready, and X. Zheng, 1992: *Tropical cyclones in the Southwest Pacific: November 1979 to May 1989*. New Zealand Meteorological Service, Wellington, New Zealand, 35 pp.

**Thornley**, J.H.M. and M.G.R. Cannell, 1996: Temperate forest responses to carbon dioxide, temperature and nitrogen: a model analysis. *Plant Cell and Environment,* **19**, 1331–1348.

**Torok**, S. J. and N. Nicholls, 1996: A historical annual temperature dataset for Australia. *Australian Meteorological Magazine*, **45**, 251–260.

**Trenberth**, K.E., 1991: General characteristics of El Niño — Southern Oscillation. In: *Teleconnections Linking Worldwide Climate Anomalies* [Glantz, M.H., R.W. Katz, and N. Nicholls (eds.)]. Cambridge University Press, Cambridge, United Kingdom, pp. 13–42.

**Varallyay**, G., 1994: Climate change, soil salinity and alkalinity. In: *Soil Responses to Climate Change* [Rounsevell, M.D.A. and P.J. Loveland (eds.)]. NATO ASI Series I, Global Environmental Change, Vol. 23. Springer-Verlag, Heidelberg, Germany, pp. 39–54.

**Veron**, J.E.N., 1995: *Corals in Space and Time. The Biogeography and Evolution of the Scleractinia.* University of New South Wales Press, Sydney, Australia, 321 pp.

**Walker**, K.F., 1992: The River Murray, Australia: A semiarid lowland river. In: *The Rivers Handbook. Volume 1* [Calow, P. and G.E. Petts (eds.)]. Blackwell Scientific Publications, Oxford, United Kingdom, pp. 472–492.

**Walsh**, K. and I.G. Watterson, 1997: Tropical cyclone-like vortices in a limited area model: comparison with observed climatology. *Journal of Climate*, **10**, 2240–2259.

**Wang**, Y.P and D.J. Connor, 1996: Optimal development for spring wheat at two locations in southern Australia under present and changed climate conditions. *Agricultural and Forest Meteorology*, **79**, 9–28.

**Wang**, Y.P., J.R. Handoko, and G.M. Rimmington, 1992: Growth sensitivity to air temperature, rainfall and ambient carbon dioxide concentration of a wheat crop in Victoria, Australia — a simulation study. *Climate Research*, **2**, 131–149.

**Warrick**, R.A., R.M. Gifford, and M.L. Parry, 1986: $CO_2$, climate change and agriculture. In: *The Greenhouse Effect, Climatic Change and Ecosystems* [Bolin, B., B.R. Doos, J. Jaeger, and R.A. Warrick (eds.)]. John Wiley and Sons (SCOPE 29), Chichester, UK, pp. 393–473.

**Warrick**, R.A., G.J. Kenny, G.C. Sims, N.J. Ericksen, Q.K. Ahmad, and M.Q. Mizra, 1996: Integrated model systems for national assessments of the effects of climate change: Applications in New Zealand and Bangladesh. *Journal of Water, Air and Soil Pollution,* **19**, 215–227.

**Waterman**, P., 1995: Assessing the vulnerability of the coastlines of the wet-dry tropics to natural and human-induced changes. In: *Wetland Research in the Wet-dry Tropics of Australia* [Finlayson, C.M. (ed.)]. Supervising Scientist for the Alligator Rivers Region, Canberra, pp. 218–226.

**Waterman**, P., 1996: *Australian Coastal Vulnerability Assessment Project Report.* 2 volumes and CD-ROM. Department of the Environment, Sport and Territories, Australian Government, Canberra.

**Weinstein** P., M. Laird, and L. Calder, 1995: Australian arboviruses: at what risk New Zealand? *Australia New Zealand Journal of Medicine*, **25**, 666–669.

**Westoby**, M., B. Walker, and I. Noy-Meir, 1989: Opportunistic management for rangelands not at equilibrium. *J. Range Management*, **42**, 266–274.

**Whetton**, P.H., M.H. England, S.P. O'Farrell, I.G. Watterson, and A.B. Pittock, 1996b: Global comparison of the regional rainfall results of enhanced greenhouse coupled and mixed layer ocean experiments: implications for climate change scenario development. *Climatic Change*, **33**, 497–519.

**Whetton**, P.H., A.M. Fowler, C.D. Mitchell, and A.B. Pittock, 1992: *Regional impact of the enhanced greenhouse effect on Victoria: annual research report 1990–91*, CSIRO Division of Atmospheric Research and EPA Victoria, Victorian Government, Melbourne, Australia, 68 pp.

**Whetton**, P.H., M.R. Haylock, and R. Galloway, 1996c: Climate change and snow-cover duration in the Australian Alps. *Climatic Change,* **32**, 447–479.

**Whetton**, P.H., K.J. Hennessy, X. Wu, J.L. McGregor, J.J. Katzfey, and K. Nguyen, 1997: *Fine Resolution Assessment of Enhanced Greenhouse Climate Change in Victoria*: *Part 2*, Department of Natural Resources and Environment, Victorian Government, Melbourne, Australia (in press).

**Whetton**, P., A.B. Mullan, and A.B. Pittock, 1996a: Climate change scenarios for Australia and New Zealand. In: *Greenhouse: Coping with Climate Change* [Bouma, W.J., G.I. Pearman, and M.R. Manning (eds.)]. CSIRO Publishing, Collingwood, Victoria, Australia, pp. 145–170.

**Whetton**, P.H. and A.B. Pittock, 1990: *Regional Impact of the Enhanced Greenhouse Effect on Victoria: Annual Research Report 1990*, CSIRO Division of Atmospheric Research and EPA Victoria, Victorian Government, Melbourne, 70 pp.

**Whetton**, P., A.B. Pittock, J.C. Labraga, A.B. Mullan, and A. Joubert, 1996d: Southern hemisphere climate: comparing models with reality. In: *Climate Change: Developing Southern Hemisphere Perspectives* [Giambelluca, T.W. and A. Henderson-Sellers, (eds.)]. John Wiley and Sons, Chichester, UK, pp. 89–129.

**Whetton**, P.H., P.J. Rayner, A.B. Pittock, and M.R. Haylock, 1994: An assessment of possible climate change in the Australian region based on an intercomparison of general circulation modeling results. *Journal of Climate*, **7**, 441–463.

**Whitehead**, D., J.R. Leathwick, and J.F.F. Hobbs, 1992: How will New Zealand's forests respond to climate change? Potential changes in response to increasing temperature. *New Zealand Journal of Forestry Science,* **22**, 39–53.

**Wigley**, T.M.L., 1994: *MAGICC: User's Guide and Scientific Reference Manual.* National Center for Atmospheric Research, Boulder, CO, USA. 23 pp.

**Wilkinson**, C.R., 1996: Global change and coral reefs: impacts on reefs, economies and human cultures. *Global Change Biology,* **2**, 547–558.

**Wilkinson**, C.R., R.W. Buddemeier, 1994: *Global Climate Change and Coral Reefs. Implications for People and Reefs.* Report of the UNE-IOC-ASPIE-IUCN Global Task Team on Coral Reefs. IUCN, Gland, Switzerland, 124 pp.

**Williams**, J.E. and A.M. Gill, 1995: *The Impact of Fire Regimes on Native Forests in Eastern New South Wales*. Forest Issues 1. Environmental Heritage Monograph Series No 2. New South Wales National Parks and Wildlife Service. 68 pp.

**Williams**, J.E., T.W. Norton, and H.A. Nix, 1994: *Climate Change and the Maintenance of Conservation Values in Terrestrial Ecosystems.* Report to the Climate Change and Marine Branch, Department of the Environment, Sport and Territories, Canberra, Australia (also available on the Internet at http://www.environment.gov.au/air/climate/clim_change/eco).

**Wilson**, J.R., 1982: Environmental and nutritional factors affecting herbage quality. In: *Nutritional Limits to Animal Production from Pastures* [Hacker, J.B. (ed.)]. CAB International, Farnham Royal, UK, pp. 111–131.

**Woodroffe**, C.D. and M.E. Mulrennan, 1993: *Geomorphology of the Lower Mary River plains, Northern Territory*. Australian National University, North Australian Research Unit and Conservation Commission of the Northern Territory, Darwin, 152 pp.

**Woodward**, A., 1996: What makes populations vulnerable to ill health? *New Zealand Medical Journal*, **109**, 265–267.

**Woodward**, A., C. Guest, K. Steer, A. Harman, R. Scicchitano, D. Pisaniello, I. Calder, and A. McMichael, 1995: Tropospheric ozone: respiratory effects and Australian air quality goals. *Journal of Epidemiology and Community Health*, **49**, 401–407.

**Wratt**, D.S., 1990: Atmosphere. In: *Climatic Change: Impacts on New Zealand* [Mosely, M.P. (ed.)]. Ministry of Environment, Wellington, pp. 29–33.

**WRI**, 1996: *World Resources: A Guide to the Global Environment, 1996–97*. World Resources Institute/United Nations Environment Programme/United Nations Development Programme/The World Bank. Oxford University Press, New York, 342 pp.

**Young**, E., H. Ross, J. Johnson, and J. Kesteven, 1991: *Caring for Country: Aborigines and Land Management*. Australian National Parks and Wildlife Service, Canberra, 215+ pp.

**Yu**, B. and D. Neil, 1995: Implications of a decreases in rainfall erosivity since the 1920s in the wet tropics of Queensland. *Australian Journal of Soil and Water Conservation*, **8(3)**, 38–43.

**Zann**, L.P., 1995: *Our Sea, Our Future: Major Findings of the State of the Marine Environment Report for Australia*. Department of Environment, Sport, and Territories, Canberra, ACT, Australia, 112 pp.

**Zheng**, X., R.E. Basher, and C.S. Thompson, 1997: Trend detection in regional-mean temperature: maximum, minimum, mean temperature, diurnal range and SST. *Journal of Climate*, **10**, 317–326.

# 5

# Europe

MARTIN BENISTON (SWITZERLAND)
AND RICHARD S.J. TOL (THE NETHERLANDS)

Lead Authors:
*R. Delécolle (France), G. Hoermann (Germany), A. Iglesias (Spain), J. Innes (Switzerland), A.J. McMichael (UK), W.J.M. Martens (The Netherlands), I. Nemesova (Czech Republic), R. Nicholls (UK), F.L. Toth (Germany)*

Contributors:
*S. Kovats (UK), R. Leemans (The Netherlands), Z. Stojic (Slovenia)*

# CONTENTS

# EXECUTIVE SUMMARY

## General Considerations

- *Climate Change.* Climate model projections suggest a general increase in temperature, greatest in northerly latitudes. Precipitation changes are considerably more uncertain, but one could expect generally wetter conditions in the north, drier conditions in the south, and increasingly drier conditions from west to east. Winter precipitation may be greater than today, while summer precipitation is likely to decrease.
- *Sensitive Regions.* As water is one of the main integrating factors for many environmental and economic systems in Europe, currently sensitive areas in terms of their hydrology include the Mediterranean region, the Alps, northern Scandinavia, certain coastal zones, and central and eastern Europe. A changing climate is likely to enhance water-related stresses in these already sensitive regions.

## Vulnerability and Potential Impacts

### *Hydrology, Snow and Ice, Water Supply and Demand*

- Evapotranspiration will increase in a warmer climate, with potential reductions in water availability; however, the response of hydrological systems depends on the distribution of precipitation (highly variable, as suggested above) and storage capacity.
- Many regions in the southern and interior parts of Europe could experience a general decrease in runoff, though the change in runoff may range between -5% and +12%.
- More droughts could be expected in southern Europe, and the potential for winter and springtime flooding could be greater in northern and northwestern Europe. However, this pattern is not the same for all general circulation models (GCMs).
- Intrusion of saline waters into coastal aquifers and the expected reduction in precipitation could aggravate the problem of freshwater supply in some areas.
- Snow and ice are likely to decrease in many places, with consequences for the timing and amount of runoff in river basins, as well as winter tourism.
- Demand for water could increase in summer. Supply could decrease, though there may be regional differences in which storage capacity plays an important role.
- Pollution is a major stress factor for many European rivers, and a decrease in discharge would increase pollutant concentrations, leading to reductions in water quality.
- Current national and international policies and practices for water resources management will be put under stress by climate change.

### *Ecosystems*

- With the exception of parts of Scandinavia and the Russian Federation, Europe has few genuine natural ecosystems. Natural ecosystems generally are confined to poor soils and are fragmented and disturbed; consequently, they tend to be more sensitive to climate change than agriculture, which occupies the most fertile soils.
- The reaction of European ecosystems to global change is difficult to predict because there are a number of interactions and feedback loops between increasing temperatures, decreasing availability of soil water, and increasing carbon dioxide ($CO_2$) concentrations.
- Increasing $CO_2$ concentration increases the productivity of plants with $C_3$ metabolism under laboratory conditions (for most agricultural plants, except maize and millet). However, many other factors come into play under field conditions, such as water and nutrient stress, increased respiration losses, and interactions between species. Therefore, the overall change in productivity can only be predicted if these interacting environmental conditions are taken into account. Many studies indicate that $CO_2$ increases alone may have relatively little impact under field conditions.
- The forests in many parts of Europe are affected by high deposition rates of nitrogen. Their productivity is not only a function of climatic factors but of the change in nitrogen deposition, which can both act as a fertilizer and cause disturbances to many processes within the ecosystem.

### *Agriculture*

- Crop mixes and production zones will be redistributed, and the use of water, fertilizers, herbicides, and pesticides will shift with them.
- Conflicting demands for water—for instance, between irrigation and domestic supply in southern Europe—will need to be taken into account.
- Changes in potential production translate in a complex way to farmer incomes and food prices, depending on technology, farmer adaptation, world markets, and agricultural policies.

### *Coastal Zones*

- Sea-level rise will place additional stress on coastal zones already stressed by other factors (urbanization, coastal developments, pollution, etc.).
- The level of impact will depend on the adaptation capacity (e.g., the ability of systems to move inland) and policies of individual countries (e.g., trade-offs between lands that are not considered important and those that need to be protected).
- Sensitive zones include areas already close to or below mean sea level (such as the Dutch and German North Sea coastlines, the Po River delta, and the Ukrainian Black Sea coast), areas with low intertidal variation (such as the coastal zones of the Baltic Sea and the Mediterranean), and coastal wetlands.
- Changes in the nature and frequency of storm surges, particularly in the North Sea, are likely to be of considerable importance for low-lying coastal areas.

### *Other Infrastructure, Activities, Settlements*

- *Energy.* Changing hydrology will impact those energy and industrial production sectors that depend on water for cooling. There is a potential for increased energy demand related to cooling in summer, and decreased energy demand related to heating in winter. Such changes would lead to shifts in peak energy demand.
- *Urbanization.* Infrastructure, buildings, and cities designed for cooler climates will have to be adjusted to warmer conditions, particularly heat waves, to maintain current functions.

### *Health*

- While there are fewer heat-related deaths in Europe than in some other parts of the world, the risk of heat-related deaths would probably increase with summer warming. The risk of cold-related deaths would probably decline with winter warming. It is not clear what the net change in risk would be for Europe.
- Warmer temperatures will exacerbate summer air pollution episodes and their health impacts in many cities.
- Some vector-borne infectious diseases will have the potential to extend their range; the adaptation capacity of individual countries will depend on their level of environmental management, public health surveillance, and health care.

## 5.1. Background Characterization/Current Baseline Conditions

### *5.1.1. Geography*

Although referred to as a continent, Europe constitues only the western fifth of the Eurasian landmass, which is made up primarily of Asia (see Figure 5-1 and Box 5-1). The Ural Mountains, the Ural River, and part of the Caspian Sea generally are recognized as forming the main boundary between Europe and Asia. The second smallest of the seven continents, Europe has an area of 10,525,000 km$^2$, but it has the second largest population of all the continents (about 685 million). The European mainland stretches from the North Cape in Norway (71°N) to Gibraltar (36°N). The western and eastern extremes are defined by the west coast of Iceland (24°W) and the Ural River in Russia (50°E).

Europe is a highly fragmented landmass consisting of a number of large peninsulas (e.g., Fennoscandia, Iberia, the Balkans, and Italy), as well as smaller ones such as Jutland (Denmark) and Brittany (France). It also includes a large number of islands—notably Iceland, Great Britain, and Ireland. Europe has coastlines on the Arctic Ocean and the North and Baltic Seas in the north; on the Caspian Sea to the southeast; on the Black Sea and the Mediterranean Sea in the south; and on the Atlantic Ocean to the west. A number of mountainous regions are located in Europe. The Alpine arc stretches from the Mediterranean coast of France to central Europe; the Alps include the highest summits of western Europe, with about 90 peaks exceeding an altitude of 4,000 m. The Pyrenees form the border of Spain and France; the Apennines are in Italy. Other mountain ranges include the Scandinavian range in Norway and Sweden, the Tatras in Slovakia, the coastal ranges of Slovenia and Croatia, and the Carpathians in Romania. Further east, the Caucasus range stretches northward from the Black Sea and forms a natural boundary between Europe and Asia. The highest point on the continent is located in this mountain range (Mt. Elbrus, 5,642 m).

One major geological region of Europe consists of a belt of sedimentary materials that sweeps in an arc from southeast England and southwest France into Belgium and the Netherlands and on into Germany, Poland, and western Russia. These sedimentary rocks, covered in places by a layer of till, form the Great European Plain. Some of the best soils of Europe are found here—particularly along the southern margin, which is rich in loess.

South of the Great European Plain are the central European Uplands—including the Jura, the Vosges and the Black Forest mountains, the Massif Central, and the Meseta; here, mountains alternate with plateaus and valleys.

Most European streams flow outward from the core of the continent. The Alps are the location of the headwaters of major

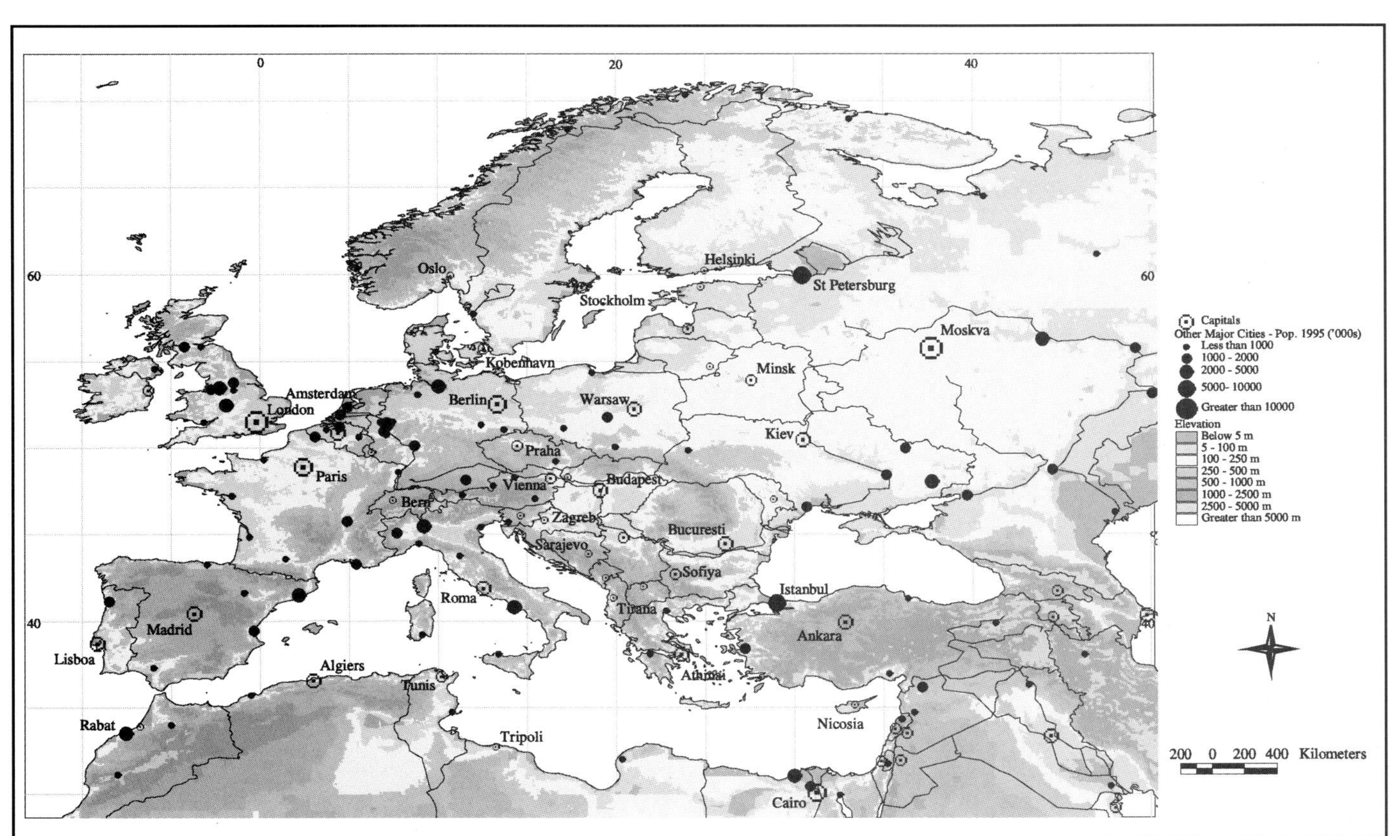


**Figure 5-1:** The Europe region [compiled by the World Bank Environment Department Geographic Information System (GIS) Unit; see Annex E for a color rendition].

**Box 5-1. The Europe Region**

| | |
|---|---|
| Albania | Liechtenstein |
| Andorra | Lithuania |
| Armenia | Luxembourg |
| Austria | Macedonia, Former Yugoslav Republic of |
| Azerbaijan | Moldova, Republic of |
| Belarus | Monaco |
| Belgium | Norway |
| Bosnia and Herzegovina | Poland |
| Bulgaria | Portugal |
| Croatia | Romania |
| Czech Republic | Russian Federation |
| Denmark | San Marino |
| Estonia | Slovak Republic |
| Finland | Slovenia |
| France | Spain |
| Georgia | Sweden |
| Germany | Switzerland |
| Greece | The Netherlands |
| Hungary | Ukraine |
| Iceland | United Kingdom |
| Ireland | Yugoslavia, Federal Republic of |
| Italy | |
| Latvia | |

rivers such as the Rhone, the Rhine, the Danube, and the Po. The longest river in Europe, the Volga, flows southward into the Caspian Sea, whereas the Danube (Europe's second longest) flows eastward to the Black Sea. The radial drainage pattern lends itself to the interconnection of rivers by canals. Lakes occur in mountainous areas—such as in Switzerland (the Lakes of Geneva and Constance are western Europe's largest bodies of fresh water), Italy, and Austria—and in lowland regions such as in Finland, Poland, and Hungary (Lake Balaton being central Europe's largest lake). Europe's largest freshwater lake is Lake Ladoga in northwest Russia.

### 5.1.2. *Ecology*

Although much of Europe—particularly the west—originally was covered by forest, natural vegetation patterns have been transformed by direct human interference through the clearing of land for agriculture and urbanization. Only in the most northerly mountains and in parts of northern and central European Russia has the forest cover been relatively unaffected by human activity. A considerable amount of the continent, however, is covered by woodland that has been planted or has reoccupied cleared lands.

The largest vegetation zone in Europe—cutting across the middle portion of the continent from the Atlantic to the Urals—is a belt of mixed deciduous and coniferous trees (oak, maple, and elm interspersed with pine, fir, and spruce). The Arctic coastal regions of northern Europe and the upper slopes of its highest mountains are characterized by tundra vegetation, which consists mostly of lichens, mosses, herbs, and shrubs. The milder but still cool temperatures of the inland parts of northern Europe create an environment favorable to a continuous cover of coniferous trees—especially spruce and pine, although birch and aspen also occur (Larsen, 1980). Much of the Great European Plain is covered with prairies—areas of relatively tall grasses. Further to the east, Ukraine is characterized by steppe—a flat and comparatively dry region with short grasses. The Mediterranean regions are covered by vegetation that has adapted to generally dry and warm conditions; natural vegetation tends to be more sparse in the southern and eastern reaches of the Mediterranean basin, reflecting regional differences in precipitation and temperature regimes.

At one time Europe was home to a large variety of wild mammals, including deer, elk, bison, boar, wolf, and bear. Many species of animals have become extinct or have been greatly reduced in number. Today, deer, elk, wolf, and bear occur in the wild in significant numbers only in northern Fennoscandia and Russia, as well as the Balkan Peninsula. Elsewhere, they exist mainly in protected preserves. Indigenous mountain animals have survived human encroachment on their habitats (to some extent); chamois and ibex are found in the higher elevations of the Pyrenees and Alps. Europe still has many smaller mammals and contains many indigenous bird species.

### 5.1.3. *Population and Demographic Trends*

Europe has the highest overall population density of all the continents. The most heavily populated area includes a belt originating in England and continuing eastward through Belgium and the Netherlands, Germany, the Czech Republic, Slovakia, Poland, and into the European part of Russia. Northern Italy also has a high population density. The average annual growth rate for the European population during the 1980s was only about 0.3%; by comparison, in the same period, the population of Asia grew by about 1.8% per year and that of North America by about 0.9% annually. At the same time, wide variations in growth rate occurred from country to country in Europe. For instance, during the late 1980s, Albania had a yearly growth rate close to 1.9% and Spain about 0.5%, while the population of Great Britain did not change appreciably, and that of Germany declined slightly (particularly in the former East Germany). The overall slow rate of population increase in the latter half of the 20th century has been the result primarily of a low birth rate.

Europeans generally enjoy some of the longest average life expectancies at birth: 75 years in most countries, compared with less than 60 years in India and most countries of Africa.

### 5.1.4. *Economics*

Europe does not have a homogeneous pattern of wealth and economic development. Western and northern Europe have some of the highest standards of living in the world in terms of gross domestic product (GDP) per capita, life expectancy, literacy rate, level of health care, or other common criteria. The standard of living in southern Europe today is close to that in most western European countries. The new democracies of eastern Europe face major economic difficulties related to the rapid transition from the planned economies that prevailed into the early 1990s to the free-market economies of today; the Czech Republic, Poland, and Hungary have been most successful in the transition to a free-market economy. Certain republics of the former Yugoslavia and Albania remain in political turmoil, which has severely damaged the national economies of these countries. The emerging republics of the former Soviet Union—such as the Baltic states, Ukraine, Belarus, and the Russian Federation—are experiencing economic difficulties following the collapse of industrial production and guaranteed markets in the former Eastern Bloc. GDP per capita ranges from close to US$40,000 per annum in Switzerland and Luxemburg to less than US$1,000 per annum in Albania and Macedonia. The countries of western Europe have a high level of supranational organization—with institutions such as the Organisation for Economic Cooperation and Development (OECD), the European Free Trade Association, and the European Union (EU). With the recent fall of communism, eastern European institutions have ceased to exist or have become ineffective, and most countries in this region now seek to join western institutions.

#### *5.1.4.1. Economic Sectors: Agriculture, Forestry, and Fishing*

Mediterranean agriculture is dominated by the production of wheat, olives, grapes, and citrus fruit. In most of these countries, farming plays a more important role in the national economy than in the northern countries. Throughout much of western Europe, dairy and meat production are major agricultural activities. To the east, crops become more important. In the nations of the Balkan peninsula, crops account for 60% of agricultural production; in Ukraine, wheat production overshadows all other agriculture. Europe as a whole is particularly noted for its great output of wheat, barley, oats, rye, corn, potatoes, beans, peas, and sugar beets. Besides dairy and beef cattle, large numbers of pigs, sheep, goats, and poultry are raised. Most of Europe is self-sufficient in basic farm products.

The boreal forests, which extend from Norway through northern European Russia, are the main sources of forest products in Europe. Sweden, Norway, Finland, and Russia all have relatively large forestry industries that produce pulpwood, wood for construction, and other products. In southern Europe, Spain and Portugal produce a variety of cork products from the cork oak.

All of the coastal European countries engage in some commercial fishing, but the industry is especially important in the northern countries—particularly Norway, Iceland, and Denmark. Spain, Russia, Great Britain, and Poland also are major fishing nations. The industry currently faces major problems, however, and is in a state of decline in many countries.

#### *5.1.4.2. Economic Sectors: Mining, Manufacturing, Energy Patterns, and Transportation*

International trade is important in Europe. Much of the trade is intracontinental, especially among members of the European Union, but Europeans also engage in large-scale trade with nations of other continents. Germany, France, Great Britain, Italy, and The Netherlands are among the world's greatest trading nations. A large portion of European intercontinental trade involves the export of manufactured goods and the import of raw materials.

Coal mining in areas such as the British Midlands, the Ruhr district of Germany, Ukraine, and the Silesian fields of Poland established industrial patterns that continue to exist today. Although coal mining is declining in Europe, it remains important in some countries. Iron ore is produced in large quantities in northern Sweden, eastern France, and Ukraine. A wide range of other minerals—such as bauxite, copper, manganese, nickel, and potash—are mined in substantial amounts. Oil and natural gas are mined in the North Sea and its bordering areas, as well as in the southern part of European Russia (notably the Volga River basin) and Romania.

Manufacturing of a wide variety of goods, ranging from bulk chemicals to high-tech equipment, is concentrated in England, eastern and southern France, northern Italy, Switzerland, Belgium, The Netherlands, Germany, Poland, the Czech Republic, Slovakia, southern Norway, Sweden, European Russia, and Ukraine.

Europe consumes great quantities of energy, though per capita levels of energy consumption are lower than in North America. The leading energy sources are coal, lignite, petroleum, natural gas, nuclear power, and hydropower. Norway, Sweden, France, Switzerland, Austria, Italy, and Spain all have major hydroelectric installations, which contribute much of the annual output of electricity. Nuclear power is important in France, Great Britain, Germany, the former Soviet republics, Belgium, Sweden, Switzerland, Finland, and Bulgaria.

Europe has highly developed transportation systems, which are densest in the central part of the continent; Fennoscandia, the former Soviet Union, and southern Europe have fewer transport facilities in relation to their land area. Europeans own large numbers of private cars, and much freight is transported by road. Rail networks are well maintained in most European countries and are important carriers of passengers as well as freight. High-speed train networks in France, Italy, and Germany make use of the most advanced technology. Water transport also plays a major role in the European economy. Several countries—such as Greece, Great Britain, Italy, France, Norway, and Russia—maintain large fleets of merchant ships. Rotterdam, in The Netherlands, is one of the world's busiest seaports. Much freight is carried on inland waterways; European rivers with substantial traffic include the Rhine, Elbe, Danube, Volga, and Dniepr. In addition, Europe has a number of important canals. Almost all European countries maintain national airlines; several are major worldwide carriers. Most transportation systems in European countries are government-controlled, although a recent tendency toward privatization and deregulation has come into effect in many sectors, including civil aviation.

## 5.2. Regional Climate Characteristics

### 5.2.1. *Current Climate*

Europe's particular distribution of land and sea—which includes several major inland seas such as the Mediterranean, the Baltic, and the Black Sea—and its long coastline facing the eastern North Atlantic ocean are shaping factors of the numerous regional climates of the continent. The presence of numerous regions of high mountains, which act as a physical barrier to atmospheric flows, is responsible for substantial regional differences in precipitation patterns.

Although much of Europe lies in the northern latitudes, the relatively warm seas that border the continent give most of central and western Europe a temperate climate, with mild winters and summers. European climate is determined essentially by the interactions among three pressure centers: the Icelandic Low, the Azores High, and continental highs (which predominate in winter) and lows (which generally are confined to summer months). Indeed, in recent years, there has been increasing interest in the relative strengths of these systems, as well as their persistence. This is exemplified particularly by the North Atlantic Oscillation index (e.g., Hurrell and van Loon, 1997), which is a measure of the strength of atmospheric flows over the North Atlantic and their links to temperature and precipitation patterns over Europe. The prevailing westerly winds, warmed in part by their passage over the North Atlantic ocean currents (the Gulf Stream), bring precipitation throughout most of the year. The strength of these winds varies, partly in response to the North Atlantic Oscillation. In the Mediterranean area (i.e., Spain, southern France, Italy, southern Croatia, Montenegro, Macedonia, Albania, and Greece), the summer months usually are hot and dry; almost all rainfall in this area occurs in the winter. From central Poland eastward, the moderating effects of the seas are reduced; consequently, drier conditions prevail, accompanied by a greater amplitude of annual variation of temperatures (i.e., hot summers and cold winters). Northwestern Europe is characterized by relatively mild winters, with abundant precipitation along the Scottish and Norwegian coasts and mountains, and much colder winters and generally drier conditions in Sweden and Finland. In mountain regions such as the Alps, winters generally are cool, and snow remains on the ground for several months of the year; summers typically are cool and moist. The mountains intercept precipitation driven by frontal systems and can trigger convective rainfall (summer thunderstorms) in the absence of major synoptic disturbances.

Table 5-1 provides some insight into the variety of European climates—which are determined not only by latitude or altitude

***Table 5-1:*** *Climatological statistics for selected European stations.*

| Station | Latitude | Longitude | Altitude (m) | January Temp (°C) | July Temp (°C) | Annual Range (°C) | Annual Precip (mm) |
|---|---|---|---|---|---|---|---|
| Sodankyla (Finland) | 67.3 N | 26.6 E | 178 | -13.5 | 14.7 | 28.2 | 508 |
| Bergen (Norway) | 60.4 N | 5.3 E | 43 | 1.5 | 15.0 | 13.5 | 1,958 |
| Göteborg (Sweden) | 57.7 N | 12.0 E | 40 | -1.1 | 17.0 | 18.1 | 670 |
| Moscow (Russia) | 55.7 N | 37.5 E | 156 | -9.9 | 19.0 | 28.9 | 575 |
| Copenhagen (Denmark) | 55.6 N | 12.7 E | 5 | 0.1 | 17.8 | 17.7 | 602 |
| Berlin (Germany) | 52.5 N | 13.5 E | 50 | -0.5 | 19.4 | 19.9 | 556 |
| Kiev (Ukraine) | 50.4 N | 30.5 E | 179 | -6.1 | 20.4 | 26.7 | 615 |
| Plymouth (UK) | 50.3 N | 4.1 W | 27 | 6.2 | 16.2 | 10.0 | 990 |
| Prague (Czech Rep.) | 50.1 N | 14.3 E | 380 | -2.6 | 17.9 | 20.5 | 508 |
| Paris (France) | 49 N | 2.5 E | 53 | 3.1 | 19.0 | 15.9 | 585 |
| Vienna (Austria) | 48.2 N | 16.3 E | 203 | -1.4 | 19.9 | 21.3 | 660 |
| Budapest (Hungary) | 47.5 N | 19.0 E | 118 | -1.1 | 22.2 | 23.3 | 630 |
| Zurich (Switzerland) | 47.3 N | 8.5 E | 569 | -1.1 | 17.6 | 18.7 | 1,137 |
| Säntis (Switzerland) | 47.3 N | 9.3 E | 2,496 | -9.0 | 5.6 | 14.6 | 2,488 |
| Zagreb (Croatia) | 45.8 N | 16.0 E | 156 | 0.2 | 22.0 | 21.8 | 864 |
| Marseille (France) | 43.5 N | 5.2 E | 20 | 5.5 | 23.3 | 17.6 | 546 |
| Barcelona (Spain) | 41.3 N | 2.1 E | 93 | 7.3 | 25.0 | 17.7 | 1,189 |
| Athens (Greece) | 38.0 N | 23.7 E | 107 | 9.3 | 27.6 | 18.3 | 402 |
| Almeria (Spain) | 36.8 N | 2.3 W | 21 | 11.4 | 25.3 | 13.9 | 226 |

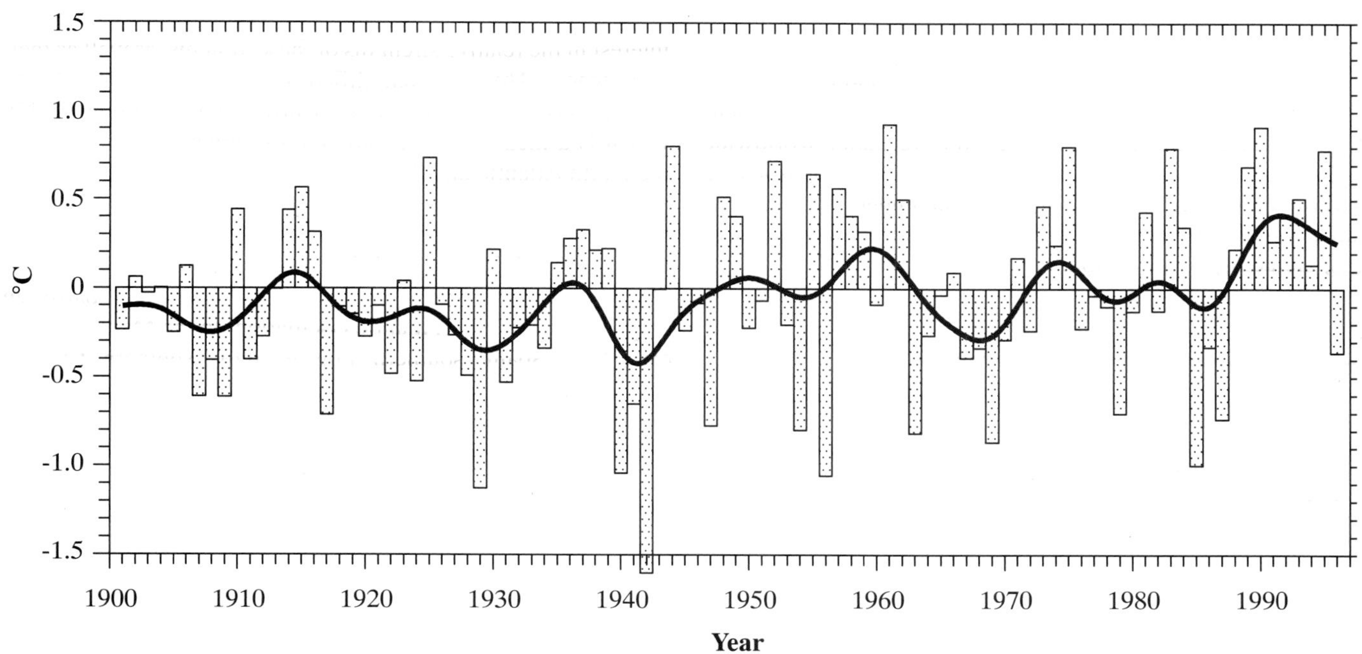


**Figure 5-2**: Annual winter (DJF) temperature anomaly over Europe during the period 1900–96.

but by proximity to the ocean or to one of the numerous inland seas. Annual temperature ranges vary from some 10°C in coastal regions of the United Kingdom and Ireland to about 30°C in Finland and Russia. Annual precipitation totals range from as low as 200 mm in southern Spain and Greece to more than 2,000 mm in coastal regions of Scotland and Norway and at some locations in the Alps.

### 5.2.2. *Climatic Trends in This Century*

Although regional differences are relatively high, most of Europe has experienced increases in temperature of about 0.8°C on average in this century (Schoenwiese *et al.*, 1993; Brazdil *et al.*, 1996; IPCC 1996, WG I, Chapter 3; Onate and Pou, 1996; Schuurmans, 1996). The increase has not been continuous throughout the century; at most stations, an increase to about 1940 was observed, followed by a leveling off or even a decrease until about 1970, and then a renewed warming to the present period. These features are most pronounced in middle to high latitudes. Some locations in southern Europe exhibit different trends—such as in Greece and parts of eastern Europe, where some stations show a cooling trend over much of the century (see Figure A-2 in Annex A). During the most recent decade (1981–1990), warming over most of Europe has been exceptionally great, with increases in yearly means of 0.25–0.5°C with respect to the long-term average. The warming is most apparent in a belt extending from Spain through central Europe into Russia. At some high-elevation sites in the Alps, temperature increases have been even more marked, exceeding 1°C in the 1980s (Auer and Boehm, 1994; Beniston and Rebetez, 1995). The 1980s have exhibited annual temperature anomalies that are systematically in excess of the long-term mean (see Figure A-7). Temperature rise has been most marked during the winter period (see Figure 5-2); much evidence suggests that minimum temperature increases have been far larger than changes in maximum temperatures (e.g., Beniston *et al.*, 1994; Brazdil *et al.*, 1996). In other words, the diurnal temperature range is decreasing, which is consistent with evidence from other regions of the world (e.g., Karl *et al.*, 1993). The geographical distribution of temperature trends emphasizes greater warming (2°C per century) in the southwestern part of Europe (Iberian Peninsula, south and central France) than in the British Isles or along the Baltic coastline (1°C per century). The northern and central parts of European Russia also have experienced greater warming than the European average—in some places exceeding 3°C per century (see Figure A-2).

Annual precipitation trends in this century are characterized essentially by enhanced precipitation in the nothern half of Europe (i.e., north of the Alps to northern Fennoscandia), with increases ranging from 10% to close to 50%. The region stretching from the Mediterranean through central Europe into European Russia and Ukraine, by contrast, has experienced decreases in precipitation by as much as 20% in some areas (see Figure A-1). In time-series analyses of precipitation averaged over the European region (see Figure A-7), it is difficult to determine a meaningful trend in precipitation, especially since the 1950s. The interannual variability seems to have decreased in the latter part of the record: The amplitude of departures in precipitation from long-term averages is far less than in the first half of the century. This pattern does not necessarily mean that the amplitude of interannual variability has decreased at the regional scale or at specific sites.

### 5.2.3. *Projections for the Future*

GCM projections, though undergoing continuous improvements, remain too uncertain over a region such as Europe to draw more than very tentative conclusions. Nested regional climate model (RCM) simulations also are too tenuous to provide firm conclusions about climatic change in Europe. Temperature changes in large parts of Europe will be a function not only of positive radiative forcing resulting from increases in the atmospheric concentrations of a number of greenhouse gases but also of the countereffects of aerosols. Although the magnitude of the aerosol effects is still quite uncertain, the regional negative forcing of sulfate aerosols in central Europe could offset almost half the positive radiative forcing of $CO_2$ (Mitchell *et al.,* 1995). The summary of results that follows is based on GCM simulations without aerosol forcing that were published in the Working Group I volume of the Second Assessment Report (SAR) (IPCC 1996, WG I, Chapter 6).

Most GCM-based projections for the European region indicate that there may be an overall increase in winter annual temperatures (IPCC 1996, WG I, Figure 6.32) and that this increase could be larger in boreal latitudes than in mid-latitude Europe. The diurnal temperature range, according to the model used, is 2.5–4.5°C for northerly latitudes, compared with 1.5–4.5°C for southern Europe. The range of summer temperatures forecast by the different models is larger than for the winter period, but the upper limit of the range is about 4.5°C increase in a $2xCO_2$ climate for southern and northern Europe.

Projected precipitation patterns are more uncertain. Most models show an increase in precipitation for Europe as a whole as a consequence of a higher content of water vapor in the atmosphere. Winter precipitation in high latitudes of Europe may increase by as much as 20% (IPCC 1996, WG I, Figure 6.32), according to most models. Rainfall during the summer months may remain unchanged in many parts of Europe. Some models show decreases in the Mediterranean region and in central and eastern Europe, though others show increases; in northern Europe, most models suggest an increase in summer precipitation. There is much uncertainty associated with future precipitation trends, however; for instance, GCM simulations incorporating the aerosol effect provide conflicting evidence for future precipitation trends in parts of Europe, compared with greenhouse-gas-only simulations.

## 5.3. Key Impacts

### 5.3.1. *Ecological Systems*

Europe has few genuine natural ecosystems. Natural ecosystems generally are confined to small areas; agriculture and forestry occupy most soils. Many sites with intermediate soils are occupied by seminatural vegetation types.

Vegetation responds to climate change directly and indirectly. Direct effects include responses to temperature; indirect effects occur primarily as soil-mediated phenomena, such as the influence of precipitation on soil moisture regimes. Indirect effects also may occur as a result of the responses of herbivores and pests to climate changes, changes in soil fauna, and changes in the frequency and severity of disturbances such as fire. In addition, vegetation responds directly and indirectly to atmospheric $CO_2$ concentrations. The responses are species-dependent; no two taxa respond to climatic change in exactly the same way (e.g., Stirling *et al.*, 1997). The main response of individual taxa to climatic change consists of changes in distribution; adaptive evolutionary changes are very rare (Huntley, 1991). At the ecosystem level, the impact of $CO_2$ on ecosystem processes remains very uncertain. Zaller and Arnone (1997) have documented an increase in surface-casting earthworm activity in Swiss grasslands exposed to 610 $\mu l$ $CO_2/l$, but such studies are extremely rare.

In a warmer climate, the pattern of species response will be extremely complex within Europe (Grime, 1996) because a variety of temperature and moisture gradients exist. Although there is a general temperature gradient from south to north and from low to high altitudes, east-west gradients in temperature and precipitation also exist; the latter are associated with increasing continentality toward central Europe. Further complications arise in the prediction of species responses to temperature changes because of the importance of the nature of the change. For example, a rise in late-summer temperatures will have different impacts than a rise in early-spring temperatures or an average rise in temperatures spread evenly throughout the year (Fitter *et al.*, 1995). Similarly, the occurrence of late frost may play an important role in restricting the responsiveness of small-genome species to mean temperature changes (MacGillivray and Grime, 1995). This factor makes the prediction of species responses difficult; a general northward shift in species distributions is now recognized as too simplistic a hypothesis. Successful migration depends on a number of factors—in particular, the range of tolerance of a given plant or tree to heat and moisture stress, environmental conditions at the new location, the rate of migration, the presence of competing species, and natural and human barriers to migration (Thompson, 1994; Malanson and Cairns, 1997). Anthropogenic barriers are especially important for large portions of western and central Europe, where land use is dominated by direct human intervention.

Short-term experiments suggest that many types of plants will respond positively to increases in $CO_2$ concentrations in the atmosphere (the so-called $CO_2$ fertilization effect), whereby their photosynthetic rates increase if other factors remain constant; this is particularly the case for $C_3$ plant types and all important European crops except maize (Semenov *et al.*, 1996; Wolf *et al.*, 1996). Most experiments have been undertaken using isolated plants with optimum nutrient supply. Such experimental conditions are relevant to horticultural and agro-industrial situations but are inapplicable to natural and semi-natural plant communities (Körner, 1996). A wide range of vegetation types may show little or no response to increasing $CO_2$ concentrations under field conditions (Körner, 1996). The responses, however, will be species-specific, especially when other factors such as enhanced nitrogen deposition are taken into account (Hättenschwiler and Körner, 1996a). The net primary productivity of other plants may increase (provided that they are not water limited), but there are many uncertainties regarding the long-term responses of plants to increased $CO_2$; studies around natural sources of $CO_2$ have not revealed any gradients in the growth rates or biomass of Mediterranean grassland species (Körner and Miglietta, 1994). Studies of trees growing around natural sources of $CO_2$ revealed no changes in stomatal density, but the guard cells were reduced in size (Miglietta and Raschi, 1993). Downy oak (*Quercus pubescens* Willd.) growing close to a $CO_2$ source had lower stomatal conductance than those further away, but Holm oak (*Quercus ilex* L.) showed no such trend (Tognetti *et al.*, 1996). In both species, the osmotic potential and apoplasmic fraction of water was elevated close to the $CO_2$, indicating that these trees were more tolerant of drought conditions (Chaves *et al.*, 1995; Tognetti *et al.*, 1996).

There is increasing evidence that traits other than photosynthetic metabolism are more important in determining the response to elevated $CO_2$ of different species under field conditions (e.g., Körner, 1993; Körner *et al.*, 1995; Diaz, 1995, 1996). For example, increased levels of $CO_2$ are likely to result in increased water-use efficiency in many species. Increased water-use efficiency may help many plants and trees resist the extremes of heat and drought that may occur more frequently in southern Europe and the Mediterranean region.

Studies of model ecosystems exposed to enhanced $CO_2$ and nitrogen deposition suggest the presence of nonlinear system-level adjustments (Hättenschwiler and Körner, 1996b). These adjustments include physiological down-regulation of photosynthesis at the leaf level, reduced leaf area index, and increasing strength of below-ground carbon sinks. At the same time, no aboveground growth stimulation was observed. Changes were observed between $CO_2$ concentrations of 280 and 420 $\mu$l/l; major changes in coniferous forest ecosystems may be underway already in response to increasing $CO_2$ concentrations.

The impact of climate change on biodiversity and the composition of ecosystems in Europe is extremely difficult to predict. A great deal depends on the impacts on ecosystem processes, such as the rates and magnitudes of disturbance. The resilience of many ecosystems to change also is very uncertain; many Norway spruce forests, for example, are likely to persist for several hundred years in the absence of any major disturbance (Sykes and Prentice, 1996). This inertia, along with the possibility of species acclimation to changed environmental conditions (Kellomäki and Wang, 1996), may delay the onset of many changes in natural ecosystems (Woodward, 1993). Models that currently are available for predicting such changes generally are restricted to the most important components of vegetation (e.g., trees in forests); little research has been done on possible future interactions between these and other ecosystem components under changed climatic and $CO_2$ conditions. The models suggest that in some cases, species arrivals may compensate for species losses, whereas in others, human-induced changes in land use may delay the arrival of new species, causing a reduction in diversity. Consequently, there is a need to combine models that simulate changes in species distributions with models developed to look at species turnover at specific sites (compare van der Maarel and Sykes, 1993; Økland, 1995a,b; Fröborg and Eriksson, 1997).

#### 5.3.1.1. *Forests*

The species composition of forests in Europe is determined more by past management activities than by natural factors (Ellenberg, 1986). Because there currently is a tendency toward a more ecological approach to forest management (e.g., Innes, 1993; Lämås and Fries, 1995), it will be difficult to separate the influence of changing climates from the influence of changing management practices. The trend toward tree species compositions in forests that more closely mimic those occurring naturally has been regarded as making the forests more adaptable to climate change. The evidence to support this hypothesis, however, is very limited (see, e.g., Fries *et al.*, 1997). Current species combinations are based on today's conditions; future communities are likely to be composed of species assemblages that do not necessarily occur today (Lindner *et al.*, 1997). The successful establishment today of species mixes that will be appropriate for future climatic conditions represents a major challenge for modern silviculture.

The range of European forests is limited primarily by climate, either through moisture availability or through temperature (both absolute amounts and seasonal distributions) (Berninger, 1997). A change in temperature or precipitation will affect the current distributions and productivity of forests (Bugmann, 1996; Lindner *et al.*, 1996; Kellomaki *et al.*, 1997). The response can be predicted with greater ease at the northern boundaries of forests in Fennoscandia and northern Russia, where an expansion of Norway spruce and Scotch pine into tundra regions is likely to occur under warmer conditions (Sykes and Prentice, 1996). These changes would be accompanied by a northward movement of the southern limit of these two species in Fennoscandia. Similar conclusions were reached by Kräuchi (1995), whose model predictions indicated the replacement of Norway spruce by beech and other broad-leaved species at a site in northern Germany (Solling), associated with an increase in mean annual temperature of 0.3°C/decade until the end of the 21st century.

The rate of northward extension of the forest limit and individual species is highly uncertain because it depends not only on the rate of climate change but also on associated rates of dispersal (Malanson and Cairns, 1997), soil development, species composition (e.g., spruce/pine/beech) (Sykes and Prentice, 1996), and age of the trees. Such changes in the distribution of species will be highly individualistic (Huntley and Webb, 1989; Huntley, 1991)—with some species expanding their ranges, some showing little or no change, and others contracting their ranges (Sykes and Prentice, 1995, 1996), possibly to the point of extinction. Climate projections suggest a displacement of climatic zones suitable for forests by 150–550 km over the next century (IPCC 1996, WG II, Chapter 1). This shift is faster than the estimated potential of many species to migrate (20–200 km/century) (Davis, 1981; Birks, 1989) or the capability of many soils to develop a new structure.

In mountain regions, certain species and communities could disappear altogether because the upward displacement of species living close to the upper reaches of mountains will be constrained by the lack of any place in which they can become established (Kienast, 1991; Kienast *et al.*, 1996). Modeling studies in Switzerland (Kräuchi and Kienast, 1993) suggest that a temperature increase of 3°C would result in deciduous broad-leaved trees invading the subalpine belt and coniferous trees invading the alpine zone. Within the deciduous zone, colline *Carpinion* forests would expand at the expense of submontane and low-montane beech forests (Brzeziecki *et al.*, 1995). The species composition of future forests would depend on changes in the continentality of each site. Such results have not yet been confirmed by empirical studies. Hättenschwiler and Körner (1995) found no indication of an upward movement in Scotch pine in the Swiss central Alps in response to summer temperatures during the period 1982–1991—when temperatures were, on average, 0.8°C warmer than those of the period 30 years before. They argued that the primary control on the altitudinal distribution of Scotch pine and Arolla pine (*Pinus cembra* L.) was more likely to be interspecific competition than temperature. However, some studies (e.g., Hofer, 1992; Grabherr *et al.*, 1994) have noted an upward shift in the distribution of some alpine species.

Changes in the forests of southern Europe are most likely to be driven primarily by changes in water availability, although changes in temperature (particularly reductions in frost occurrence) also may play a role in the expansion of some forest types (e.g., *Quercus pyrenaica* and *Quercus rotundifolia*). Changes in precipitation will determine the relative importance of sclerophyllous and deciduous species; water availability in the period April–June would be particularly important (Gavilán and Fernández-González, 1997).

Many changes in forests may occur as a consequence of subtle alterations in the competitive balance between species. For example, rising temperatures are likely to change the time interval between budburst and leaf fall, but the effects will differ among species. Kramer (1995) argues that the duration of the growing season is likely to decrease with increasing temperature for European larch (*Larix decidua* Miller) and pedunculate oak (*Quercus robur* L.), whereas it will increase for beech (*Fagus sylvatica* L.) and small-leaved lime (*Tilia cordata* Miller). There already is some evidence that a substantial (25–30%) proportion of the forests in Switzerland is poorly adapted to the expected natural species distribution under current climatic conditions—and that this proportion will increase with an increase in temperature (Kienast *et al.*, 1996). Similar patterns are likely elsewhere in Europe.

Increased climatic stress is likely to result in increased amounts of dead plant material within forests. At the same time, potential increased carbon-to-nitrogen ratios as a result of increased $CO_2$ may reduce decomposition rates; this effect remains uncertain, however: Some studies suggest no change (e.g., Norby *et al.*, 1995), whereas increased temperatures are likely to increase decomposition rates in areas such as northern Scandinavia (Johansson, 1994; Johansson *et al.*, 1995). The net effect of these different trends is largely unknown. Increased volumes of dead organic matter, combined with drier climatic conditions, would increase the fire hazard in many forests. The Mediterranean zone is likely to be particularly affected, but other areas also are likely to see an increase in the fire hazard. Boreal forests, which currently are mainly temperature-limited, could become soil moisture-limited if warmer temperatures in spring and summer lead to earlier melting of snowpack and drying of soil in summer. This shift would lead to increased risk of biomass loss through drought-driven mortality and fire. Reduced winter snowpacks also may increase the risk of winter damage to trees during periods when air temperatures are above zero and the ground remains frozen. Cold winters in boreal zones and some temperate regions have protected forests from many of the insects and fungi that are common further south; a warmer climate is likely to change this situation (Cammell and Knight, 1992; Straw, 1995).

Many forested regions of Europe have been cleared through human land-use practices—particularly in western, central, and southern Europe. In the Alps, for example, the timberline currently is as much as 200 m below its natural potential limit because of overgrazing by game and livestock and the removal of trees over several centuries to provide summer grazing areas (Holtmeier, 1994). If warming is accompanied by drier conditions in the Alps and Pyrenees, a northward and upward progression of Mediterranean ecotypes is likely to occur.

There is a large degree of uncertainty regarding the sensitivity of soil respiration to changing heat and moisture conditions, as well as to nitrogen deposition rates. As a result, nutrient availability and hence overall ecosystem productivity could either increase or decline (see, e.g., Heath and Kerstiens, 1997). A warmer climate, particularly in the form of warmer winters, may favor certain pests or pathogens—such as bark beetles, aphids (Straw, 1995), or fungi (Brasier, 1996). Because many of these species live in complex relationships with their hosts, as well as with predators and other animals, it is difficult to predict the broad-scale net effect of climate change on them.

Many European forests are subject to a variety of anthropogenic influences—including direct management, nitrogen deposition and availability (Kinney and Lindroth, 1997), sulfur emission (Posch *et al.*, 1996; Slovik, 1996), ozone injury, and changes caused by unbalanced game populations. These factors are likely to interact with climatic change to bring about a complex series of responses that will differ from place to place (Thornley and Cannel, 1996; Bugman, 1997; Young, 1997). Many studies have failed to take such interactions into account; this limitation should be considered when interpreting the results (Innes, 1994). There is some evidence, however, that the biomass of northern forests was growing during the last decades (Houghton, 1996; Lakida *et al.*, 1997; Lelyakin *et al.*, 1997; Myneni, 1997).

The interactions between a general increase in winter temperature and the occurrence of frost damage have been studied in detail. The timing of budburst is triggered by temperature; it is a function of the chilling requirement and the critical temperature sum (although precipitation, photoperiod, $CO_2$ concentration, and the nutritional status of the tree also may play roles). There are differences in the timing of budburst between species, between genotypes, and between phenotypes. Consequently, some models have predicted an increase in frost injury, such as in central and northern Finland (Bach, 1987; Kellomäki *et al.*, 1995), whereas others—such as that developed by Kramer (1994) for The Netherlands and Germany—suggest less injury. In Britain, Murray *et al.* (1989) have suggested that there may be no change in frost injury at lowland sites and an increase at upland sites for some species. There is likely to be considerable flexibility in budburst within a given population, providing some capacity for adaptation to climate change (Beuker, 1994). Because increased $CO_2$ concentrations may delay budburst, there appears to be an antagonistic effect between rising $CO_2$ concentrations and temperatures that may reduce the increase in frost risk for some species in some areas (Murray *et al.*, 1994).

The climatic changes that have been modeled are likely to have considerable direct and indirect impacts on plantation forests in Europe. Proe *et al.* (1996) have suggested, on the basis of an ecophysiological process model, that the general yield class (GYC) of Sitka spruce (*Picea sitchensis* (Bong) Carrière) in Scotland would increase by 2.8 $m^3$/ha/yr for each 1°C rise in temperature. This estimate was based on a uniform increase in temperature throughout the year. If there were greater warming during winter months, the GYC increase was 2.4 $m^3$/ha/yr. The reduction is attributable to the negative correlation between winter temperature and growth rate in this species, which is believed to be caused by the increased activities of pests in warmer winters and the winter chilling requirements of the species. Similar negative relationships between GYC and winter temperature have been identified in Scotland for Douglas fir (*Pseudotsuga menziesii* F. Mirb.) and Scotch pine (*Pinus sylvestris* L.), but not for Japanese larch (*Larix kaempferi* (L.) Carrière) (Tyler *et al.*, 1996). The winter chilling requirement seems to be a major factor in determining the distribution of many tree species (Sykes *et al.*, 1996).

The land area capable of supporting Sitka spruce with a GYC of more than 18 $m^3$/ha/yr in Scotland is expected to double within 20 years and to show a similar increase during the subsequent 40 years (Proe *et al.*, 1996). The model did not include potential effects of increased $CO_2$ because these effects are considered less important for growth responses than changes in temperature (Prentice *et al.*, 1991; Luxmoore *et al.*, 1993).

Similar changes in yield have been predicted in Finnish forests (Kellomäki and Kolström, 1994), although the nature of the changes varies with latitude. Yield increases for Scotch pine, silver birch (*Betula pendula* Roth) and downy birch (*Betula pubescens* Ehrh.) were predicted with increasing temperatures, with the increases being greater in northern Finland than in the southern part of the country. The productivity of Norway spruce was predicted to decrease in the south but to increase substantially in the north. These changes did not take into account any changes in precipitation, the nature of which remains uncertain. However, other modeling work (Talkkari and Hypén, 1996) has suggested that any effects of precipitation changes are likely to be much less important than the effects of temperature changes, mainly because of the morainic soils (which tend to retain moisture) that predominate over much of northern Scandinavia. On sandy soils in southern Finland, soil moisture deficits may become a problem under future climatic scenarios (Kellomäki, 1995; Kellomäki and Väisänen, 1996).

There is considerable evidence that predicted changes in yield already are occurring in European forests (Spiecker *et al.*, 1996). No systematic studies have been undertaken, but growth increases have been recorded in forests in Austria, Denmark, France, Germany, Slovenia, Sweden, and Switzerland, with possible increases also occurring in Portugal and Spain. The causes of these growth increases were not evaluated, although climatic change, increasing $CO_2$ concentrations, increased nitrogen deposition, and changes in management practices were cited as possible causes.

#### 5.3.1.2. *Grasslands and Rangelands*

Because of the likely shifts of phytoclimatic boundaries in a warmer world (based on GCM estimates for Europe of 3±1.5°C in the next century), there will be changes in the distributions of Mediterranean and boreal grassland species. Changes in individual species will depend on the nature of the climate change in a given area. Because the ranges of many species are determined primarily by soil moisture patterns, changes in the absolute amounts of precipitation and the seasonal distributions of precipitation will be important in determining vegetation responses. Some grassland species also may respond to increasing $CO_2$ concentrations—with the greatest responses among species characteristic of productive, undisturbed habitats and the lowest responses among species adapted to high environmental stress or rates of physical disturbance (Hunt *et al.,* 1991, 1993). Experiments with alpine grassland species suggest, however, that any effects will be apparent only if $CO_2$ increases are accompanied by climatic warming or enhanced nitrogen supply; increases in $CO_2$ concentrations alone have little effect (Schäppi and Körner, 1996).

In Mediterranean countries, displacement of grass and dwarf shrub steppes will occur at the expense of existing sclerophyllous shrubland. Extension of shrubland as a result of agricultural release is expected—with a parallel rise in wildfire episodes, loss of water through enhanced evapotranspiration, and a decrease in livestock grazing (with concomitant increases in game and other wildlife). Furthermore, shifts in carbon storage from soil to biomass are likely to occur.

In northern Europe, there is likely to be a reduction in the number and extent of mires and tundra and permafrost areas as forests expand into the tundra zone. In addition, changes in the concentration of atmospheric gases may alter competitive relationships in the plant community. A series of complex responses can be expected because of the interaction of different environmental factors. For example, an increase in temperature is likely to result in an increase in nutrient availability because of the greater mineralization of soil organic matter by soil microorganisms (Anderson, 1991; Bonan and Van Cleve, 1992). As with forests, the responses will be species-specific (Baxter *et al.*, 1994; Parsons *et al.*, 1994, 1995).

#### 5.3.1.3. Noncoastal Wetlands

Most of the major noncoastal wetland areas in Europe are confined to northern Scotland, Fennoscandia, and northern Russia (Hartig *et al.*, 1997). A changing climate is likely to have a significant impact on peat formation and ecological function in such regions. An increase in temperature by 1–2°C accompanied by decreases in soil moisture would lead to an estimated 25% reduction in peat formation.

Tundra peatlands will be extremely vulnerable because higher temperatures will result in thawing of the permafrost layer in areas with discontinuous permafrost, as well as an enlargement of the active layer in continuous permafrost. This shift will have considerable implications because permafrost is a key factor in maintaining high water levels in these systems. Further, it is unlikely that new permafrost areas will develop. As a result, hydrology and landscape patterns will be affected, leading to lowered water tables in some areas and the overflow of flooded thaw lakes in others. Such melting may shift bogs on permafrost back to fens, from which they originated after the warm mid-Holocene period (5000–6000 years BP).

A rapid rate of climatic change—implied by climate model simulations for northern latitudes—may cause degradation of the southern boundaries of wetlands and peatlands much faster than the northward expansion potential of their northern boundaries. Such an imbalance between biomass loss on the one hand and biomass increase on the other is likely to have implications for the carbon cycle, whereby the wetlands could undergo a reversal from sinks to sources of carbon. There also would be consequences for methane fluxes in these regions. A change in the total methane flux from northern wetlands can be expected if the areal extent of wetlands changes, the duration of the biologically active period is modified, or the production or oxidation of methane per unit area changes. Because of uncertainties in the changes in water regimes as projected by climate models, however, it is difficult to obtain reliable quantitative estimates for shifts in methane fluxes to the atmosphere.

Many peatlands may be subject to increasing pressure from afforestation operations as a result of changes in land capability classifications (Proe *et al.*, 1996). At the same time, degradation of peatlands is likely to increase the conservation value of the remaining intact areas, thereby creating the potential for more intense land-use conflicts.

Water draining from peatlands also is likely to be sensitive to climatic change, particularly in the form of summer droughts. Such droughts could increase autotrophy in the streams, leading to increases in the biofilm biomasses present in the water. Changes in nutrient content also may occur, with increases in inorganic nutrients and decreases in organic nutrients (Freeman *et al.*, 1994).

Studies indicate that wetlands in semi-arid regions of southern Europe can be very sensitive to climate warming; such warming has severe effects on their hydrological and ecological functions. They also are likely to be adversely affected by increased water demands. Biological reserves such as the Cota Donana in Spain are likely to come under increasing pressure.

#### 5.3.1.4. Freshwater Ecosystems

Many of the ecological impacts on freshwater systems depend on hydrological responses to climatic change, which are relatively uncertain given the current status of GCMs. The problem is one of scale: Many of the most significant impacts on freshwater ecosytems will result from hydrological changes at the scales of small catchments and drainage basins (which are unresolved by GCMs) and changes in hydrological variability rather than mean conditions, which is poorly taken into account by climate models.

Although the biodiversity of freshwater species may increase in a warmer climate—particularly in middle and high latitudes—there may be an initial reduction in species diversity in cool temperate and boreal regions (essentially Fennoscandia and northern Russia) if the northward migration of warm-water species cannot keep pace with the rate of loss of cool-water species because of physiological limitations, lack of north-south migration corridors, and limitations to genetic adaptation.

It is likely that the impacts of climate change will be more pronounced in the littoral zones of lakes than in the pelagic zones. Because aquatic macrophytes have access to the nutrients contained in sediments, they will be able to exploit higher temperatures; emergent vegetation is likely to benefit from higher $CO_2$ concentrations (Kankaala *et al.*, 1996).

Water levels in lakes and reservoirs are highly sensitive to weather conditions; in some regions of Europe, small lakes and reservoirs may fluctuate rapidly in response to changes in precipitation and evapotranspiration. Where water levels are likely to decline, inshore areas will change significantly. In shallow lakes and reservoirs in particular, inshore aquatic vegetation and surrounding wetlands would decrease in area. This decrease may result in changed habitats for aquatic biota, reduction of productivity, and even extinction of fish and invertebrate species that are dependent on these types of biomes. Where lakes have extensive bordering wetlands, declining water levels would reduce productivity and impact negatively on populations of fish and invertebrates that are dependent on these types of wetlands for their survival. Even if lake levels remain constant, decreasing throughflow may change the water quality of lakes.

Decreasing lake volumes and areas also would result in increased loading of nutrients (nitrogen, phosphorus, and others) from catchments per unit lake area or volume. Correspondingly, increased eutrophication can be expected—with high production of aquatic biomass, decreasing species diversity, deterioration of oxygen conditions, and adverse effects on water quality. In regions where most drinking water comes from surface sources, decreasing lake volumes and lower water quality may cause serious problems for human use.

In areas where increased snowfall and rainfall is expected, loading by acidifying pollutants from the atmosphere may increase. This effect would be critical for lakes located on bedrock with low neutralizing capacity (as in parts of the Alps, the Pyrenees, the Carpathians, and Fennoscandia). With further acidification of lakes, reductions in species and trophic-level diversity might occur. In the past decade, however, emission reduction measures in Europe have resulted in a decrease in atmospheric sulfur deposition of up to 50%; nitrogen deposition levels have remained unchanged. As a result, sulfate concentrations in freshwaters are decreasing at most European sites, and alkalinity values are increasing (particularly in the 1990s). Nitrate concentrations increased in European freshwaters during the 1980s but have remained stable or decreased in the 1990s. It has been postulated that the changes in the 1990s may reflect a climatic influence rather than changes in nitrogen deposition rates (UNECE, 1997).

In many parts of Europe, hydrological systems are heavily impacted by human activities (e.g., water use, sewage effluents, channel modification, removal of vegetation that acts as refugia for numerous plant and animal species, drainage, land-use changes, fertilization in catchment areas, soil erosion following deforestation). Such direct human interference, along with other environmental stress factors such as air pollution, is likely to exacerbate the impacts of climate change, with a consequent reduction of biodiversity in most parts of the continent.

#### 5.3.1.5. Littoral and Pelagic Zones

Littoral and pelagic zones often are forgotten in studies of climate change impacts. However, important changes have already been observed (e.g., a significant increase in average temperatures of the waters in the western Mediterranean basin over the past 20–30 years) (Francour *et al.*, 1994). Temperature changes are reflected in substantial changes in the relative abundance of thermophilic species; there have been increased catches of thermophilic fish species such as *Diplodus cervinus*, *Epinephus marginatus*, *Pomadasys incisus*, *Sphyraena sphyraena*, *Balistes carolinensis*, *Sardinella aurita*, and *Pomatomus saltatrix*.

#### 5.3.1.6. Adaptation Options

Natural ecosystems may adapt to climatic change in one way or another—but not necessarily in ways preferred by humans. Therefore, some kind of human intervention, in the form of management, may be necessary. Adaptation options for forests and plants could include the creation of refugia; migration corridors and/or assisted migration; and improvements in integrated fire-, pest-, and disease-management techniques. This approach poses problems in many parts of Europe, where ecosystems have been so fragmented and the population density is so high that some of these options may be impossible to implement. Reforestation would be a viable adaptation option in some cases, as would afforestation of abandoned agricultural land—increasing habitats and establishing corridors between fragmented nature reserves. For wetlands and peatlands, reducing the impacts of climate change could be achieved through wetland restoration or creation techniques; in boreal regions where permafrost is a major feature of wetland ecosystems, however, such techniques would not be of much use. Freshwater biological systems can be assisted in a number of ways that could help mitigate the impacts of climate change, particularly through the increase and protection of riparian vegetation and the restoration of river and stream channels to their natural, adaptive morphologies. Such restoration processes may take several decades.

For all ecological systems, the reduction of pollution and land-use stresses in more heavily populated regions of Europe would contribute to removing major stress factors. In some circumstances, this strategy might allow plants to adjust more easily to the negative effects of climatic change.

An increasingly determined effort toward protection and revitalization of freshwater ecosystems in developed countries already has shown positive achievements, particularly through improvements in the water quality of watersheds. The protection of threatened aquatic habitats such as wetlands has facilitated the reintroduction of several endangered vertebrate species. This has been the case particularly in central Europe as a consequence of increased awareness of environmental problems.

### 5.3.2. Agriculture and Fisheries

The impacts of climate change on agriculture can be defined at different scales—including crop yield, farm or sector profitability, regional economic activity, or hunger vulnerability. Impacts depend on biophysical and socioeconomic responses.

In the past decade, substantial amounts of research have focused on regional and national assessments of the potential effects of climate change on agriculture. Relatively little work, however, has included systematic assessments that identify vulnerable socioeconomic groups, integrate effects across sectors, describe impacts at different spatial and temporal scales, or address the efficacy of the range of practical responses (see Parry *et al.*, 1992). For the most part, studies have treated each region or nation in isolation (see IPCC, 1996, for a complete reference list of studies to date; see also Sirotenko *et al.*, 1997), without regard to changes in production in other places, and have not addressed in an integrated way interactions with other related systems (e.g., water resources, socioeconomy, policy). Global assessments of climate change impacts in agriculture and agricultural markets have been few to date (Parry *et al.*, 1988; Smit, 1989; Martin *et al.*, 1990; Kane *et al.*, 1992; Rosenberg and Crosson, 1991; UK Department of the Environment, 1991; Rosenzweig and Parry, 1994; Darwin *et al.*, 1995). In addition, a major shortcoming of most climate impact assessments has been their lack of in-depth treatment of adaptation, in part because of its complexity and in part because of the lack of a suitable methodological framework.

### 5.3.2.1. *Agricultural Production*

Although agriculture on a global basis could remain unchanged relative to base production in a changed climate—as a result of technological improvements and more efficient agricultural practices (IPCC 1996, WG II, Chapter 13)—there will be large regional discrepancies; these differences may be less acute in Europe, where agriculture is highly advanced.

Agricultural production can be described in terms of amount and quality. The reactions of an individual crop to global change will depend on the balance of shorter cycles resulting from increased air temperatures, shorter periods to accumulate yield products (at least in the case of determinate crops), higher potential yields resulting from increased assimilation of $CO_2$, and increased water-use efficiency resulting from enhanced regulation of transpiration with elevated $CO_2$. This equilibrium could be quite subtle in the highly mechanized context of European agriculture.

Projections of future European production derive from controlled-conditions experiments and from feeding crop simulation models, with climatic observations adjusted according to scenarios of future climate. Quantitative results of simulations therefore are highly dependent on the type of climate scenario used (especially in terms of available precipitation), though qualitative indications (trends) generally are constant among published results.

Most simulation and experimental studies so far have used expected fluctuations of mean values for climate variables, but increasing emphasis is being put on possible consequences of a more variable interannual and intra-annual climate (i.e., within and between years).

#### 5.3.2.1.1. *Annual crops*

In the following paragraphs, crops are classified according to their main growing period: winter (i.e., late autumn, winter, spring, and sometimes early summer) or summer (i.e., spring, summer, and early autumn). For Europe, winter crops include winter cereals, peas, and rapeseed; summer crops include maize (forage and grain), sunflowers, potatoes, sugar beet, and spring cereals. The results below need to be tempered by the physiological characteristics of crops—for instance, the shortening of the crop season resulting from increasing temperature is far less marked on nondeterminate crops (peas, potatoes) than on determinate crops (wheat).

*Suitable areas*

As a result of increasing air temperatures in winter, the risks associated with damaging frosts will be reduced as a whole. This factor will allow expansion of winter cereals and probably other winter crops (CLAIRE, 1996) in large areas such as southern Fennoscandia, western Russia, and the Alpine regions—topography, physiography, soils, and socioeconomic circumstances permitting. Increasing spring temperatures will accelerate soil temperature increases and extend suitable zones for most summer crops, allowing a reasonable length for their growth season. In the case of sunflowers (and probably for grain maize), the area of suitability would extend eastward to Belarus and northward to northern Germany and southern Fennoscandia for a specified climate change. As an illustrative example, the northern limit of reliable spring cereal cultivation in Finland has been estimated to shift northward by 100–150 km for each 1°C of warming (Carter *et al.*, 1996). For a representative range of temperature scenarios, the rate of northward shift is about 10–80 km per decade (Saarikko and Carter, 1996). Some northward shift also is simulated for vegetable crops like onion (CLAIRE, 1996). As far as temperature is concerned, there is no region that would become completely unsuitable for agriculture—even in southern Europe under high-temperature climate change scenarios. Studies currently are being carried out to determine what damage can occur in crops at extremely high temperatures.

*Crop duration*

Expected increases in temperature will cause faster rates of development as a whole (see "Adaptive Responses" below) and shorten the length of growing periods for determinate crops, consequently shortening the length of the grain-filling period. The total growing season for these crops may be reduced by 15–30 days, depending on the climate scenarios used; in this respect, crop duration could be more reduced in central and eastern Europe (4 weeks) than in western Europe (3 weeks). Cereal harvest dates therefore would occur sooner. Nevertheless, some indication is given by Miglietta *et al.* (1995) that a lack of cold days could reduce vernalization effects and consequently lengthen the first part of the growing season for winter cereals. Because of faster rates of development, nondeterminate crops would develop more potential harvestable organs (theoretically during a longer period) because no development event would stop the process of production;

increasing temperatures, however, probably would favor an increased senescence rate of the crop and tend to mitigate the beneficial effect of increases in potential organs by reducing the length of crop photosynthetic life.

Temperature increases in spring and summer will accelerate the course of crop development more crucially on short-cycle crops that are sown in spring than on winter crops. This general rule must be adapted, however, to local conditions (see, e.g., Saarikko and Carter, 1996)—who show by simulation that the sowing-to-leading phase in spring cereals in Finland declines by about 1 day per 1°C warming, and the heading-to-yellow ripening shortens by 2–4 days per 1°C warming). A simulation exercise on sunflowers shows a reduction of the crop cycle by 10–50 days with UKTR3140 scenario and 10–70 days with UKTR6675 (see Harrison and Butterfield, 1996 for extensive results on sunflower development under different GCM-based scenarios). This quite important change is likely to affect most of Europe, with a gradient from the southwest (low reduction in Spain and Italy, where cycles are already short) to the northeast (Poland and Russia). This pattern can be extrapolated to all determinate summer crops. Nondeterminate crops would experience a faster rate of development as well, which would induce earlier senescence.

*Crop yields*

Accounting for the enhancement of growth resulting from increasing $CO_2$ concentrations, the potential yield of winter crops (assuming that neither precipitation nor irrigation is limiting) would increase almost everywhere (with central or southern Europe experiencing the highest winter wheat yield boosts, depending on the climatic scenario). If water limitations are considered, crop response apparently would depend on the scenario chosen for the time evolution of $CO_2$ concentrations. In the case of winter wheat, there is some indication that the rate of increase in yields across Europe could be 0.2–0.36 T/ha/decade under the IS92a emission scenario and 0.13 T/ha/decade with the IS92d emission scenario, under both the UKTR3140 and the UKTR6675 climate scenarios. The largest increases would occur in central and eastern Europe (regardless of changes in management practices that may occur in some countries as a result of changes in economic structure) and in southern Europe (see Table 5-2). All winter crops probably would follow the pattern of winter wheat yield changes. The largest increases per country might occur in northern Europe because of increased possibilities for taking winter cereals into cultivation.

For summer crops, determinate crop yields would be affected by the shortened crop cycle and reduced time to assimilate supply and grain-filling periods. On the other hand, improvements in the rate of dry-matter production can be expected from enhanced $CO_2$ concentrations. This effect has been illustrated for sunflowers (CLAIRE, 1996), where simulation models suggest a compensation between the negative effects of temperature and the positive effects of carbon fertilization. Under the IS92a emission scenario, yield benefits occur in central and eastern Europe and losses occur in western Europe—but all of them are small. With IS92d emissions, the gains are even smaller, but the losses are amplified (see Table 5-3). The same pattern can be expected for all determinate summer crops (even if the fertilizing effect of atmospheric $CO_2$ is less beneficial to $C_4$ crops such as maize). Similar results were obtained with onion, whose simulated yield increased 4–8% with UKTR3140 and GFDL2534 (IS92a emissions) and 15% (resp. 7%) with UKTR6675 and GFDL5564, IS92a (resp. IS92c)—$CO_2$ effects again counteracting a negative effect of reduced growth period on yield (see CLAIRE, 1996). For nondeterminate crops, Peiris *et al.* (1996) expect potato yields to increase by as much as 35% under northern conditions as a result of the lengthening of the growing season, regardless of $CO_2$ fertilization effects.

*Climatic risks*

Climatic risks will be highly dependent on the expected pattern of time variability for weather variables. Increasing temperatures will promote the development rate of all winter crops, which therefore will face extreme events (cold spells) at a later stage (i.e., when they are more sensitive). Consequences depend as a whole on the probabilities of such extreme events and a higher intra-annual variability of minimum temperatures—yielding a

***Table 5-2:*** *Mean wheat yields (and standard deviation) simulated with EURO Wheat model for four predefined regions of Europe for baseline climate (1961–90) and climate change scenarios (in T/ha), reproduced from CLAIRE (1996).*

| **Emission Scenario ($CO_2$ ppmv)** | **GCM Scenario** | **Region[1]** Europe | E.U. | Northern E.U. | Southern E.U. |
|---|---|---|---|---|---|
| Base (353) | Base | 8.07 (2.34) | 7.77 (2.86) | 9.25 (1.22) | 5.94 (3.23) |
| 2 x $CO_2$ (560) | UKHI | 8.43 (2.55) | 8.50 (2.55) | 9.70 (1.37) | 6.79 (3.49) |
| IS92a (454) | UKTR3140 | 8.74 (2.50) | 8.61 (2.69) | 9.86 (1.33) | 6.91 (3.09) |
| IS92a (617) | UKTR6675 | 10.21 (2.22) | 10.17 (2.62) | 11.13 (1.37) | 8.83 (3.28) |
| IS92d (545) | UKTR6675 | 9.28 (2.30) | 9.21 (2.69) | 10.33 (1.32) | 7.64 (3.27) |

[1]Regions are defined as follows: Europe is the large region from Scandinavia to North Africa and from Ireland to the Black Sea; E.U. is the 15 countries of the European Union; Northern E.U. is all E.U. regions north of 45°N; and Southern E.U. is all E.U. regions south of 45°N.

***Table 5-3:*** *Mean sunflower yields (and standard deviation) simulated with EURO Sunflower model for four predefined regions of Europe for baseline climate (1961–90) and climate change scenarios (in T/ha), reproduced from CLAIRE (1996).*

| **Emission Scenario ($CO_2$ ppmv)** | **GCM Scenario** | **Region[1]** Europe | E.U. | Northern E.U. | Southern E.U. |
|---|---|---|---|---|---|
| Base (353) | Base | 1.53 (1.27) | 1.36 (1.22) | 2.41 (1.09) | 0.78 (0.84) |
| 2 x $CO_2$ (560) | UKHI | 0.93 (0.77) | 0.98 (0.83) | 1.46 (0.77) | 0.76 (0.74) |
| IS92a (454) | UKTR3140 | 1.37 (1.11) | 1.24 (1.05) | 1.94 (1.11) | 0.86 (0.78) |
| IS92a (617) | UKTR6675 | 1.59 (1.22) | 1.47 (1.16) | 2.15 (1.11) | 1.10 (1.01) |

[1]Regions are defined as follows: Europe is the large region from Scandinavia to North Africa and from Ireland to the Black Sea; E.U. is the 15 countries of the European Union; Northern E.U. is all E.U. regions north of 45°N; and Southern E.U. is all E.U. regions south of 45°N.

higher probability of crop failure from frost damage. The same problem arises with winter cereals, which face extreme temperature maxima in early summer during the grain-filling period. Recent investigations show that the probabilities of extreme temperatures increase under all climate scenarios (CLAIRE, 1996) and that thermal shocks on poorly adapted genotypes lead to losses in grain yield and quality.

Higher temperatures in summer should not be a real challenge to summer crops (except spring cereals, if subjected to elevated temperatures during the grain-filling period) because they are more resilient than winter crops. Drought could be a major concern in the future, however, particularly in the Mediterranean zone and in central Europe. This is a genuinely complex problem; GCM-based scenarios do not agree on the magnitude of changes in space of at least one component of the water budget (precipitation), and changes in another component—potential evapotranspiration (PET)—are extremely dependent on calculation methods. Le Houérou (1995) states that a 1°C increase in air temperature will induce 37 mm more PET south of 40°N (ECRASE, 1996)—an enormous 60% increase in PET in southern European countries. If simulated changes of -10% to -20% in summer precipitation for western, southern, and central Europe are reliable, fully irrigated crops may become even larger competitors to domestic and industrial users for water resources stored in aquifers and rivers (a 20% reduction represents a significant loss in currently moist areas).

*Adaptive responses*

Adaptive responses could be facilitated by increased knowledge of weather patterns and climate-related variability through the use of climate forecast information.

To abate the shortening effect of temperatures on crop cycles, changed sowing dates and later-maturing genotypes could be used whatever the type of crop. This approach, however, may cause a problem with winter cereals whose cycle length often is linked with cold temperature requirements (vernalization) that may be not completely fulfilled during warmer winters. Later-maturing crops also should face climatic risks in the early summer. On the other hand, the sowing dates of winter crops cannot be postponed to early fall because of the much higher probability of experiencing low temperatures at a sensitive stage and because of the cost of fungal disease control during periods in the early fall.

For summer crops, using earlier sowing dates or longer-maturing varieties would counteract the detrimental effects of climate change in all cases—as was demonstrated for sunflowers throughout Europe in CLAIRE (1996), for spring wheat in Finland in Saarikko and Carter (1996), and for maize in Spain in CLAIRE (1996). Choosing adequate sowing dates also could help to synchronize full canopy development and maximum radiation availability on maize-type crops in northern European regions (Delécolle *et al.*, 1996), which would enhance final production. Similarly, earlier sowing dates would allow the crop to develop during periods of lower PET demand, implying an improvement in global water-use efficiency and a reduction in irrigation demand—as simulated in Spain by CLAIRE (1996).

#### 5.3.2.1.2. *Perennial crops*

Perennial crops mostly include fruit trees, grapevines, and grasslands.

*Cycle length*

Assuming that climate variability does not increase, higher temperatures in late winter and early spring will hasten development stages. Concerning fruit trees, no experimental or simulation results are known, but it can be expected that budbreak will be promoted, as will later stages. On the other hand, higher temperatures could be detrimental to flowering quality for cold-requiring species and therefore could reduce fruit production.

With grapevines, simulation studies have shown that fruit biomass would decrease consistently in northern Italy under weather conditions foreseen by equilibrium climate scenarios (but would increase with some transient scenario outputs), even if the fertilizing effect of $CO_2$ is accounted for. The promoting effect of $CO_2$ has been experimentally demonstrated under current conditions by Bindi *et al.* (1996), who show that berry weight and total acidity increase, whereas total sugars decrease.

Higher temperatures probably will be beneficial to grasslands (see, e.g., Tchamitchian, 1994 for perennial ryegrass), at least early in the season, through increased early biomass production (earlier senescence also is expected). Higher temperatures during the summer may decrease the growth capabilities of grass. Higher $CO_2$ tends to abate the negative effects of temperature, as demonstrated experimentally by Soussana *et al.* (1994). The interaction between nitrogen level and carbon distribution in crops nevertheless is a key factor in forecasting the responses of grasses (Casella *et al.*, 1996). Some recent years in Europe have illustrated the immediate impact of dry periods on grassland production because of low rooting depth, shallow soils, etc., leading to a reduction in yield. Similar events may become more common in the future. Through simulation experiments, Rounsevell *et al.* (1996) show that grassland production in England and Wales is resilient to small increases in temperature and precipitation—and even stimulated at higher elevations. Experiments in northern and western Europe have shown positive responses in total biomass production across a range of grass species to increased temperature (+3°C) and elevated $CO_2$ concentrations (Jones *et al.*, 1996; Jones *et al.*, 1997). The interactive response may be less than additive, however, indicating a decline in response to elevated $CO_2$ as temperature increases. A significant part of the temperature response is manifest through an extended growing period. The response of below-ground growth to elevated $CO_2$ appears to be greater than that of above-ground growth, and there are differential responses between species—suggesting that sward conditions may change significantly in the future (Jones *et al.*, 1996; Jones *et al.*, 1997).

For woody species, the influence of changing climates will really depend primarily on how long increasing $CO_2$ concentrations enhance growth and production (i.e., whether there is any acclimation, whether farmers are capable of adapting their strategy to such situations, etc.) and secondarily on the influence of $CO_2$, temperature, and water shortage on fruit quality—which is the major issue in fruit production. No real information is available on irrigation needs for fruit trees under changing climate conditions. Although grasslands show very different responses according to species, they apparently display no acclimation to increasing $CO_2$.

### *Weeds, pests, and diseases*

Weeds are expected to benefit from higher $CO_2$ concentrations. The expected result of the crop-weed competition will depend on their respective reactions to climate and atmospheric fertilization. $C_3$ plant growth probably would be proportionally more enhanced than $C_4$ growth as a result of increasing $CO_2$ concentrations. This difference can lead to various consequences during the early crop vegetation period (fall or spring), depending on which species is weed and which is crop. Chemical control of weeds obviously remains possible, but this approach must be considered in the context of increasing incentives for environment-friendly agriculture in Europe.

Increasing precipitation and temperature in the northern half of Europe probably will be linked to increasing air humidity and possibly leaf wetness duration. All factors are favorable to early (fall, spring) disease outbursts for annual and perennial crops; the same holds for early pest attacks (Harrington *et al.*, 1994). Sophisticated agricultural systems probably can cope with increasing weeds, pests, and diseases in a sustainable manner (better-targeted and more efficient chemicals, risk forecast models, adapted machinery, etc.). The global result will depend on how widespread such agriculture will be in Europe around 2050. Whether controls will be possible in autumn and spring (i.e., whether working days will be available) also is questionable and must be studied in more detail. (Some model results are available for Finland.) The risk of crop damage from pests and diseases increases in all regions under a warming of climate. Northward shifts in the distribution of certain pests could be of similar magnitudes and rates as those estimated for cereal crops. Additional generations of multivariate species also can be expected. The damage potential of diseases such as potato late blight could increase at a similar rate as the potential increase in yields of the crop host. This analysis implies an increased requirement for pest and disease control, with associated consequences for the environment (Carter *et al.*, 1996).

None of the models that have been used to predict future agricultural yields have taken into account the possibility of future reductions in productivity associated with increased concentrations of ozone. Depending on the supply of precursors, ozone concentrations are likely to increase in warmer temperatures. For example, Legge *et al.* (1996) have shown that current yield losses of spring wheat in Hessen, Germany, may be as high as 15% in some years. Greater losses can be expected in parts of Europe where ozone concentrations are higher.

### *Soil erosion*

Increasing precipitation probably will induce greater risks of soil erosion, depending on the intensity of rain episodes (such information is not currently available from available climate scenarios). This possibility needs to be examined, as does the expected evolution of soil organic matter. If soil organic matter content decreases with increasing temperature as a result of a higher mineralization rate, soils will be more susceptible to slaking; consequently, early development of crops will be hampered. Some indication exists that the negative effect of temperature will be more or less compensated under European conditions by a positive impact of carbon fertilization (Balesdent *et al.*, 1994) and that the temperature-dependent mineralization rate is larger for clay and sand soils than for loam (Houot *et al.*, 1995). Increased rainfall amounts also could increase fertilizer leaching in already wet areas; for example, Peiris *et al.* (1996) have simulated decreasing wheat yields under Scottish conditions that are explained by such a lack of fertilizer. Changes in vegetation cover probably will play a role as well.

Concerning working days, Rounsevell and Brignall (1994) have suggested that opportunities for autumn soil tillage in Great Britain will be improved by global warming (as a result of increasing water demand), as long as the future precipitation increase is no more than 15%. This analysis must be verified for spring conditions and other areas.

#### 5.3.2.1.3. *General trends in agriculture*

The general trend of simulation results for coming decades is an increase in crop yields, at least in the northern half of Europe. As an example, a raw application of yield projections to presently suitable areas and yield levels of 1990 for winter wheat in the EU suggests an average increase of 9–11 million tons per year for the years 2013–2036 (IS92 a and d emission scenarios, UKTR3140 and GFDL2534 climate scenarios, equivalent years based on climate sensitivity) and 24–26 million tons per year for the years 2042–2100 (IS92a and d emission scenarios, UKTR6675 and GFDL5564 climate scenarios) (CLAIRE, 1996).

Increasing stress may be associated with a more variable climate (both intra-annual and interannual), with an increase in the probability of rare events. This effect is not obvious at all stages, however (e.g., winter temperature in high latitudes may become less variable where the meridional temperature gradient is reduced). Because European agriculture is highly adapted—that is, prepared to provide large outputs as a response to a restricted range of inputs—it could be severely impacted by more frequent sequences of such extreme events. As Semenov and Porter (1995) state, "Future climatic scenarios derived from GCMs have described changes in mean weather, but nonlinear crop models require explicit incorporation of changes in climatic variability to assess the risks of agricultural production from climate change." Breeding of new cultivars and planting of cultivars in previously less-suitable areas may be successful adaptation options, provided that farmers can distinguish between anomalous weather and climatic trends.

Increasing yields are related chiefly to increasing production of dry matter. Simulations show that only carbon weight is promoted in this dynamic, and the carbon-to-nitrogen ratio increases in dry matter. This effect often is linked with a loss in product quality (for bakeries, breweries, animal feeding, etc.).

A northward change in temperature patterns may not necessarily correspond to a simple shift in latitude of suitable areas for usual crops because many plants are sensitive to photoperiod and adapted to a combination of temperature and photoperiod ranges. New genotypes therefore might be necessary to meet this new agricultural frontier, provided that the available soils are suitable for the crop.

### 5.3.2.2. *Fisheries*

Any effects of climate change on fisheries will occur in a sector that already is characterized on a global scale by full utilization, massive overcapacities of usage, and sharp conflicts between fleets and among competing uses of aquatic ecosystems. Climate change impacts are likely to exacerbate existing stresses on fish stocks—notably overfishing, diminishing wetlands and nursery areas (perhaps aggravated by sea-level rise), pollution, and UV-B radiation.

The fishing industry is a significant income earner in Europe, but in recent decades the industry has experienced numerous problems related to fishing rights and quotas, dwindling stocks due to overfishing, and disputes as part of the European fishing fleet encroaches upon areas traditionally exploited by other nations (e.g., eastern Canada, the western African coast). Evidence in support of the view that environmental changes drive many changes in fish stocks has been accumulating in recent years (Mann and Lazier, 1991; Mann, 1992, 1993; Mann and Drinkwater, 1994; Polovina *et al.*, 1995). The question of whether overfishing, environmental change, or a combination of the two is responsible for major declines in fish stocks is still a matter for debate and is situation-specific.

About 70% of global fish resources depend on near-shore or estuarine habitats at some point in their life cycle (IPCC, 1990; Chambers, 1991). The growing rate of human occupation (with associated pollution) and the high property values of littoral areas, especially in western countries, will severely constrain the inland displacement of wetlands and other habitats as sea level rises. Fish production will suffer when wetlands and other habitats that serve as nurseries are lost (Costa *et al.*, 1994). Fish habitats are downstream of many impacts, and fish integrate the effects. Fish are symbolic of the health of ecosystems and our ability to manage our resources.

Growing demands for fish, water, and space; encroachment by large-scale fishing and aquaculture operations; population concentrations; urban expansion; pollution; and tourism already have harmed small-scale fishing communities in shallow marine waters, lakes, and rivers. Persons engaged in this economic sector have limited occupational or geographic mobility. With climate change, global and regional problems of disparity between catching power and the abundance of fish stocks will worsen—particularly with regard to the interaction between large mobile fleets and localized fishing communities. Aquaculture will develop in new areas—sometimes assisting, sometimes disrupting existing traditional fishing techniques.

Many studies have related historical changes in the abundance and distribution of aquatic organisms to climate changes. These studies reveal that relatively small changes in climate often produce dramatic changes in the abundance of species—sometimes of many orders of magnitude—because of impacts on water masses and hydrodynamics (e.g., Sharp and Csirke, 1983; Beukema *et al.*, 1990; Kawasaki *et al.*, 1991). The impacts of warming can be inferred from past displacements of transition zones—for example, the Russell cycle in the western English Channel (Southward, 1980) and simultaneous discontinuities in the local occurrence of pilchard and herring (Cushing, 1957, 1982). Opposing fluctuations in these fisheries in the Channel and the Bay of Biscay have followed long-term changes in climate for three centuries (Binet, 1988b).

A general poleward extension of habitats and range of species is likely, but an extension toward the Equator may occur in eastern boundary currents. For example, the range of *Sardina pilchardus* prior to World War II was from south Brittany to Morocco. In warm years following the war, sardines were fished up to the North Sea. During the 1970s, new fisheries developed off the Sahara and Mauritania, with small amounts landed as far south as Senegal—perhaps as a consequence of upwelling and ecological processes related to tradewind acceleration (Binet, 1988a). Changes in the circulation pattern are likely to induce changes in the larval advection/retention rates in and out of favorable areas and may explain changes in abundance and distribution (Binet, 1988a; Binet and Marchal, 1992). Areal overlaps in closely related species may change in unpredictable directions (Ntiba and Harding, 1993).

A poleward movement of species in response to climate warming is predictable on intuitive grounds (Shuter and Post, 1990). Habitat, food supply, predators, pathogens, and competitors, however, also constrain the distributions of species. Furthermore, there must be a suitable dispersion route, not blocked by land or some property of the water such as temperature, salinity, structure, currents, or oxygen availability. Movement of animals without a natural dispersal path may require human intervention; in the absence of intervention, such movement may take hundreds or thousands of years (Kennedy, 1990).

As stresses intensify, impacts that for a long time were limited to freshwaters and littoral areas now are observed in closed and semi-enclosed seas (FAO, 1989; Caddy, 1993). Some semi-enclosed seas—such as the Mediterranean, Black, Aegean, and northern Adriatic—already are eutrophic. The diversity of uses in these areas also has introduced some of the most pervasive and damaging anthropogenic impacts on the world's ecosystems in the form of the spread of nonindigenous species (Mills *et al.*, 1994).

#### 5.3.2.3. Adaptation Options

Most studies show that if climate variability remains the same as at present, adaptive strategies (change in sowing dates, genotypes, crop rotation) will continue to offset expected production losses, as long as such strategies are not counteracted by government or other policies in place. The outcome could be different if climate conditions become more variable; averaging across years then would be less appropriate.

In economic terms, most model studies suggest that on average Europe would benefit from climate change because of an overall increase in crop yields, potentially leading to lower consumer prices in a free market. Such studies have considered adaptation to climate change, in which adjustments at the farm level—including shifts in planting dates, changes in water use in irrigated areas, and changes in crop cultivars—are considered. Scenarios without such adaptations also have been undertaken; the results of the model investigations clearly point to the beneficial effects of early adaptation to climatic change.

Other adaptation options include technological advances and socioeconomic options, such as land-use planning, watershed management, improved distribution infrastructure, information dissemination (including improved climate and weather forecasts), adequate trade policy, and national agricultural programs.

The ultimate impacts of climate on the agricultural sector may be determined by nonclimate factors that control the system. In Europe, a major concern—and the main driving force behind agricultural policy—are laws derived from the Common Agricultural Policy (CAP) for the countries in the EU. These laws also affect other countries in Europe. A common aim in all countries is to reduce land degradation and environmental problems. In countries with water limitations, the growing competition for water, which leads to water-saving programs, is an additional concern. In Spain, for example, one of the main objectives of the future law on water use (Ley del Plan Hidrologico Nacional, MOPT, 1993) is to increase water-use efficiency to increase water availability; the specific water savings are reflected mainly in improvement of the irrigation systems and an efficient adjustment between crop water demand and supply.

Little concern has been devoted to the large percentage of the population working in agriculture in southern Europe. The consequences of climatic variations in agronomic systems that are highly regulated, such as the systems in the countries of the EU, are difficult to predict because crops are highly subsidized and therefore crop prices are artificially high. For example, although agricultural production in Spain decreased in 1994 (-4.3%) as a result of the drought conditions, the Spanish agrarian rent increased (+14%, the largest increase in the EU) (MAPA, 1995) because of the increase in Spanish commodity prices and direct subsidies to Spanish farmers. From the Spanish perspective, it is clear that agricultural regulations and the devaluation of the national currency (peseta) had an initial positive effect on agricultural prices and on farmers' subsidies (Matea, 1995).

Fisheries adaptation options could provide large benefits irrespective of climate change:

- Design and implement national and international fishery management institutions that recognize shifting species ranges, accessibility, and abundances and balance species conservation with local needs for economic efficiency and stability.
- Support innovation through research on management systems and aquatic ecosystems.

- Expand aquaculture to increase and stabilize seafood supplies, help stabilize employment, and conserve wild stocks.
- In coastal areas, integrate the management of fisheries with other uses of coastal zones.
- Monitor health problems (e.g., red tides, ciguatera, cholera) that could increase under climate change and harm fish stocks and consumers.

Genetic engineering has the potential to increase the production and efficiency of fish farming (Fischetti, 1991). However, resource managers are very concerned about accidental or intentional release of altered and introduced species that might harm natural stocks and gene pools. Around Fennoscandia, escapees and nonindigenous reproduction may have reached or exceeded the recruitment of salmon wild stocks (Ackefors *et al.*, 1991). Introduction of pathogenic organisms and antibiotic-resistant pathogens also is of concern; this problem needs to be addressed in any long-term planning.

### 5.3.3. *Cryosphere, Hydrology, Water Resources, and Water Management*

#### 5.3.3.1. *Snow and Ice*

As contributors to hydrological systems, snow and ice and their potential changes in a warmer global climate will have profound impacts on European streams and rivers. Mountains—in particular the Alps—are the source of most of Europe's major rivers; the timing and amount of flow in rivers such as the Rhine, the Rhone, and the Danube (via the Inn River) are strongly dependent on the seasonal accumulation and melting of snow and, during the summer and fall, on meltwater from mountain glaciers. Changes in the mountain cryosphere in the Alps and the Fennoscandian mountains would have significant consequences for the flow regimes of rivers originating in these mountains. Changes in hydrological regimes would particularly affect populations living downstream of the mountains that depend on the water the mountains provide for freshwater supply, industrial and energy usage, irrigation, and, in some cases, transportation—as on the Rhine and the Danube, both of which are particularly sensitive to flow changes. As the extensive flooding in Poland, Germany, and the Czech Republic in 1997 has demonstrated, many flood defense systems have a limited capacity, and any changes in hydrological regimes could have major impacts in floodplain areas.

Higher temperatures will push the snowline upwards by about 150 m for every 1°C rise; the seasonal patterns of snowfall are likely to change, with the snow season beginning later and ending earlier. The timing and amount of seasonal flow patterns currently experienced by European rivers also would change as a result of snowpack conditions in the mountains; peak flow would occur earlier in the season, and there could be shortfalls as a result of drier summer conditions over much of Europe and reduced river flow in the summer. The fact that winter runoff is likely to increase and spring runoff probably will decrease could benefit the hydropower industry, however, because it would be in a more favorable position to generate electricity at the peak demand period of the year.

Mountain glaciers generally have been shrinking and are projected to lose about 25% of their mass worldwide by the middle of the 21st century. In the European Alps, about half of the original ice volume has been lost since 1850; as much as 95% of the existing glacier mass could disappear over the next 100 years with anticipated warming, and many of the small glaciers could disappear altogether within decades (Haeberli and Hoezle, 1995). Large reductions in glacier mass would impact most hydrological basins, with significant reductions in flow regimes during the summer and early fall. Accelerated glacier shrinkage and permafrost degradation also is likely to cause increasing slope stability problems in areas above the present timberline (Haeberli *et al.*, 1997).

In terms of river and lake ice—which are particularly relevant to northern Sweden, Finland, and Russia, as well as high-altitude Alpine lakes—a warmer climate will lead to delayed freezing at the beginning of the winter season, while break-up will begin earlier. In Russia and Fennoscandia, the river-ice season could be shortened by up to one month. Many rivers in the temperate regions of central Europe will become ice-free or develop only intermittent or partial ice coverage.

#### 5.3.3.2. *Hydrology and Water Management*

The main components of the hydrological cycle are precipitation, evaporation, discharge, storage in reservoirs, groundwater, and soil. Because of many complex interactions between these factors in time and in space, changes in the hydrological cycle are more difficult to model and to analyze than temperature and precipitation data. A detailed discussion is beyond the scope of this paper (see IPCC 1996, WG I, Chapter 3; IPCC 1996, WG II, Chapter 10).

Potential changes in the hydrological cycle are easier to understand if some basic mechanisms are kept in mind. Generally, higher temperatures lead to higher potential evaporation and decreased discharge (which also is a function of precipitation, storage, and topography). The storage of water in the soil serves as a buffer; in winter and spring, increasing precipitation normally generates higher discharge because the buffer is full and evaporation is low. During the summer, storage is reduced by evapotranspiration and must be refilled before discharge begins. Changes in the hydrological cycle therefore are more variable than changes in other climatic factors. Seasonal-to-interannual variability in precipitation and temperature also accounts for some of the variability in hydrological characteristics in European river basins.

The hydrological cycle in Europe is strongly influenced by anthropogenic factors. During the past century, land-use patterns have changed, the drainage conditions of rivers and land have been altered, the proportion of impermeable (urban) areas has increased, and additional reservoirs have been built. Changes or trends in the discharge of rivers therefore are more difficult to analyze than temperature and precipitation trends, especially in densely populated areas (Forch *et al.*, 1996; Holt and Jones, 1996; Giakoumakis and Baloutsos, 1997).

GCM-based analyses for the European continent (IPCC 1996, WG II, Table 10-1) give a range of possible responses of river runoff in a warmer global climate—from decreases in some regions (e.g., Hungary, Greece) to increases in other regions (United Kingdom, Finland, Ukraine); these estimates are a function of precipitation, evapotranspiration, and soil moisture projections in the different GCMs. The uncertainties of climate model results, however, remain very large in terms of hydrological forecasting, particularly at the regional scale. This limitation is particularly critical for water management practices in the future because water resource impacts occur at the local scale, not at regional or larger scales.

The river basin is the natural spatial unit for water resource management; because neighboring river basins can have quite different climate change impacts, making quantitative statements about regional-scale climate impacts on water resources is difficult given that many regions include a wide range of hydroclimatic regions. The results of catchment-scale simulations with conventional hydrological models driven by GCM data therefore are highly variable. Arnell and Reynard (1996), for example, simulated changes of ±20% in annual runoff for 21 catchments in Great Britain—with a tendency toward lower amounts of discharge, especially in sensitive areas and during the summer months.

Another consequence of increasing temperatures is a change in the distribution of water in the landscape. Especially in flat regions (e.g., lowlands in The Netherlands and northern Germany), catchment areas depend on groundwater recharge. Changes in percolation therefore can change the size of catchments (Hoermann *et al.*, 1995). On a local catchment scale, distribution of water in the landscape can change even if annual discharge remains unchanged: Whereas hilltops are severely stressed by drought, areas with high groundwater levels may remain largely unaffected.

The effects on different regions of Europe can be classified according to the climatic gradients from north to south and from west to east (the maritime-to-continental gradient). According to current and projected distributions of rainfall, there may be an increase in summer drought. Observed precipitation shows a decrease in summer and an increase in winter and spring. Such a change is likely to affect mainly areas that already are sensitive to drought: the southern and continental parts of Europe, especially the Mediterranean region. Additional consumption of water for irrigation may create additional depletion of water reservoirs and groundwater.

Although the debate about changes in the frequency of floods is still open, an increase in rainfall during periods when soils are saturated (i.e., winter and spring), along with earlier snowmelt, could increase the frequency and severity in floods. An increase in large-scale precipitation might lead to increased flood risks on large river basins in western Europe in winter. The increased temperatures expected in summer could lead to higher local precipitation extremes and associated flood risks in small catchment areas.

Another unknown factor is the effect of increased plant water-use efficiency (WUE) resulting from increased $CO_2$ in the atmosphere. Under laboratory and greenhouse conditions, plants seem to use less water for constant yield or exhibit higher productivity. In the field, however, predictions of WUE at the catchment or even ecosystem level are not yet feasible. In water-limited systems (e.g., in the Mediterranean region), a higher WUE can increase or maintain productivity—whereas in the humid areas of northern Europe, it could theoretically decrease evaporation and therefore increase discharge.

Warmer average and extreme temperatures will enhance the demand for freshwater, particularly for agriculture and direct human consumption, although the water-quality requirements are different for these two types of consumption. Changes in precipitation patterns, particularly over regions that already are sensitive (e.g., the Mediterranean basin), may lead to increased demand for water for irrigation purposes, especially for soils with low water-storage capacities. If precipitation were to undergo a decline in the Mediterranean basin (which is not always the case suggested by GCM simulations), countries such as Spain, Italy, and Greece would face substantially increased risks of summer water shortages. In such a situation, increases in storage capacity would be needed to maintain existing water and energy supplies. The Netherlands could face the desiccation of most of its wetland areas or be forced increasingly to rely on the Rhine to maintain present water levels. Groundwater aquifers could be affected by increased saltwater intrusion as sea level rises. Given a possible 4°C temperature increase and a rise in the Alpine snow line of 500–700 m in the summer, the flow of the Rhine could decline by 10% during this season. A temperature increase of only 0.5°C and a daily rainfall decrease of a few percent would decrease runoff in Hungary by 25–30%. Despite projected wetter winters, drier summers and increased evaporation in southeastern England may reduce yields in impounding reservoirs by 8–15%. Aquifer yields, which are very important to the London area, may fall by 8%. Supplies of water during warmer, possibly drier, summers would need to be maintained through larger storage or transfers from wetter regions. Water quality may deteriorate because there would be less river flow to dilute contaminants.

In terms of water management, system yields of river basins are likely to decrease as a result of changes in flow patterns; these river basins certainly will be sensitive to expected increases in floods and droughts. There may be a potential for conflict over environmental and economic uses of freshwater, especially for river basins that are shared by several countries. The high population density in western Europe does not allow for much flexibility for major changes in flood-protection practices.

Floodplains in most countries already are overpopulated; thus, it will be extremely difficult to find the necessary areas for an effective flood-protection system.

#### 5.3.3.3. Adaptation Options

Of all the systems that are sensitive to climatic change, the cryosphere may be the most vulnerable; no adaptation measure can counter the disappearance of snow and ice in an environment that is likely to experience more frequent episodes of above-freezing temperatures than at present. In very restricted situations, the use of snowmaking equipment may help to extend the skiing season in particular areas, but this solution is expensive, and its environmental impact has not been fully evaluated.

In terms of water resource management, technological measures—including land-use criteria and erosion control, reservoirs and pipelines to increase the availability of freshwater supply, and improvements in the efficiency of water use—can be envisaged. Socioeconomic options should include direct measures to control water use and land use, as well as indirect measures such as incentives or taxes; institutional changes for improved resource management also may be needed. Specific examples of possible options include supplementing rain-fed agriculture with irrigation; water-conserving irrigation practices; enhanced coordination of surface and groundwater management; changes in cropping patterns; watershed management; structural and nonstructural flood-control management; and reallocation of water resources among water-use sectors and among nations.

### 5.3.4. Coastal Zones

Coastal zones are characterized by highly diverse ecosystems that are important as sources of food and habitats for many species. In many areas in Europe, population, economic activity, and arable land are concentrated in coastal zones, which has led to a decrease in their resilience and adaptability to variability and change. Some coastal areas—such as much of The Netherlands, the fens in eastern England, and the Po River plain (Italy)—already are beneath mean sea level; many more areas are vulnerable to flooding from storm surges. Fixed, rigid flood defenses and sea-level rise already are causing "coastal squeeze" (i.e., a decline in intertidal coastal habitats).

Sea-level rise and possible changes in the frequency and/or intensity of extreme events—such as temperature and precipitation extremes, cyclones, and storm surges—represent consequences of climate change that are of most concern to coastal zones. Except for sea-level rise itself, there currently is little understanding of the possible interaction of different aspects of climate change in the coastal zone. Other possible changes in climate could be costly. In The Netherlands, the costs of protection against an adverse 10% change in the direction and intensity of storms may be worse than the costs of a 60-cm rise in sea level (Peerbolte *et al.*, 1991; IPCC 1996, WG II, Chapter 9).

Under the IS92a scenario, global sea level is projected to rise by about 5 mm/yr (with an uncertainty range of 2–9 mm/yr). This increase is two to five times the rate experienced over the past century. Regional and local sea-level rise will not necessarily be the same as the global average because of vertical land movements (glacial isostatic rebound, tectonic activity, subsidence) and possible changes in ocean water characteristics (oceanic circulation, wind and pressure patterns, ocean water density). Other climatic change in Europe is uncertain. An increase in precipitation intensity seems likely, increasing the flood risk in low-lying coastal areas.

Without adaptation, a rise in sea level would inundate and displace wetlands and lowlands, erode shorelines, exacerbate coastal storm flooding, increase the salinity of estuaries, threaten freshwater aquifers, and otherwise impact water quality. The impacts would vary from place to place and would depend on coastal type and relative topography. Areas most at risk would be tidal deltas, low-lying coastal plains, beaches, islands (including barrier islands), coastal wetlands, and estuaries. Tidal range also is a key factor: In general, the smaller the tidal range, the greater the response to a given rise in sea level. This pattern suggests that the Mediterranean and Baltic coasts, with their low tidal range, may be more vulnerable to sea-level rise than the open ocean coasts.

Examples of susceptible coasts include the Rhone, Po, and Ebro deltas (Jimenez and Sanchez-Arcilla, 1997; Sanchez-Arcilla and Jimenez, 1997). These areas already are subsiding because of natural and sometimes human factors, and they are sediment-starved as a result of changes in catchment management. For example, the Ebro delta has lost 97% of its sand supply since the 1950s. Reduction or loss of these areas would impact important agricultural and natural values. Many of Europe's largest cities—such as London, Hamburg, St. Petersburg, Thessaloniki, and Venice—are built on estuaries and lagoons (Frasetto, 1991). Such locations are exposed to storm surges, and climatic change is an important factor to consider for long-term development. In Venice, a 30-cm rise in relative sea level this century has greatly exacerbated flooding and damage to this unique medieval city; permanent solutions to this problem are still being investigated. Beaches tend to erode given sea-level rise, which destroys a valuable resource and exposes human activities landward of the beach to increased wave and flood action. Intense recreational use of beaches in many coastal areas, particularly around the Mediterranean, makes this erosion a particular problem; some response to such changes often is essential. In higher latitudes, gravel beaches are more common than sand beaches (Carter and Orford, 1993). Gravel beaches often provide an important coastal protection function, which sea-level rise may disrupt.

The Global Vulnerability Analysis (GVA) was a first-order analysis that combined available global data sets with a number of assumptions to provide regional and global perspectives on vulnerability to accelerated sea-level rise for a limited number of parameters (Hoozemans *et al.*, 1993). GVA results are consistent with a number of national studies (Nicholls, 1995); the following results for Europe can be derived. In 1990, more than 30 million people were estimated to live below the 1,000-year storm surge level. These people generally are protected from flooding by structural measures and the low incidence of flooding. A 1-m rise in sea level relative to the 1990 conditions would increase the population living below the 1,000-year storm surge level by about 30%, to 40 million people. The incidence of flooding will depend on human response: Proactive measures that anticipate the rise in sea level will maintain a low incidence of flooding.

Europe is estimated to have at least 2,860 km$^2$ of salt marshes and 6,690 km$^2$ of other unvegetated intertidal habitat (these estimates exclude the former Soviet Union). Based on coastal morphological type, the GVA estimates that coastal wetlands will decline given a 1-m rise in sea level, with 45% of salt marshes and 35% of other intertidal areas lost. Considering coastal squeeze increases the losses to 62% and 41%, respectively. The characteristics of the surviving salt marsh and intertidal areas may be greatly altered compared with current conditions. The northern Mediterranean is most vulnerable to such losses; the Baltic Sea is most vulnerable to coastal squeeze. Such losses could have serious consequences for biodiversity in Europe, particularly for bird populations.

Table 5-4 presents what is known to date from national studies in the Intergovernmental Panel on Climate Change (IPCC) (IPCC 1996, WG II, Chapter 9) and U.S. Country Studies Program (Lenhart *et al.*, 1996) on European countries, supplemented by a national study of Germany (Sterr and Simmmering, 1996; Ebenhoe *et al.*, 1997). A regional study of East Anglia in the United Kingdom also is available (Turner *et al.*, 1995). Although the results vary from country to country, they all emphasize the large human and ecological values that could be impacted by sea-level rise. Other values that may be impacted include cultural and archeological resources at the coast (Fulford *et al.*, 1997). Historic cities, such as Venice, might be included in this category.

#### 5.3.4.1. *Adaptation*

People tend to respond to the threat of sea-level rise. Three broad response strategies can be distinguished: planned retreat, accommodation, and protection. In the latter strategy, land is maintained as it is now, albeit with a decline in natural functions and values. In the first strategy, coastal processes are allowed to occur naturally, but valuable developments are prevented or withdrawn in time. The strategy of accommodation is intermediate; the response in this case is to control the impacts of sea-level rise by changing land use. The strategy chosen depends on national circumstances, including the economic and ecological importance of the coastline, technical and financial capabilities, and the legislative and political structure of the countries concerned.

Because most low-lying coastal areas in Europe already are protected from flooding, a rise in sea level by itself will not cause serious inundation of human-occupied areas. Assuming other climatic factors remain constant, the protection offered by the flood defenses will decline as sea level rises, but its effectiveness will not disappear immediately (see, e.g., Kelly, 1991, for a discussion of the Thames Barrier, which protects London). Based on the GVA, the reduction in flood protection will be largest around the Mediterranean (Nicholls and

***Table 5-4:*** *Social and economic impacts of sea-level rise in The Netherlands, Germany, Poland, Estonia, and Ukraine.*

| Country | Sea-Level Rise (m) | People Affected # People (000s) | People Affected % Total | Capital Value at Loss Million US$[1] | Capital Value at Loss % GNP | Land at Loss km$^2$ | Land at Loss % Total | Wetland at Loss km$^2$ | Adaptation/ Protection Million US$[1] | Adaptation/ Protection % GNP |
|---|---|---|---|---|---|---|---|---|---|---|
| Netherlands[2] | 1.0 | 10,000 | 67 | 186,000 | 69 | 2,165 | 5.9 | 642 | 12,300 | 0.05 |
| Germany[3] | 1.0 | 3,200 | 4 | 7,500 | 0.05 | 13,900 | 3.9 | 2,000 | 23,500 | 0.2 |
| Poland[2] | 1.0 | 240 | 1 | 22,000 | 24 | 1,700 | 0.5 | 36 | 1,400 | 0.02 |
| Poland[4] | 0.1 | 40 | 0.2 | 10,000 | 11 | 845 | 0.2 | n.a. | 2,300+10 | 0.03 |
| Poland[4] | 2.5 | 235 | 1 | 75,000 | 82 | 2,203 | 0.6 | n.a. | 9,800+500 | 0.15 |
| Estonia[5] | 1.0 | n.a. | n.a. | n.a. | n.a. | 60 | n.a. | n.a. | n.a. | n.a. |
| Ukraine[6] | 0.5 | n.a. | n.a. | n.a. | n.a. | n.a. | n.a. | 32 | n.a. | n.a. |
| Ukraine[6] | 2.0 | n.a. | n.a. | n.a. | n.a. | n.a. | n.a. | 370 | n.a. | n.a. |

[1]All costs have been adjusted to 1990 US$.
[2]Source: IPCC 1996, WG II, Chapter 9. Results are for existing development. People affected, capital loss, land loss, and wetland loss assume no human response, whereas adaptation assumes protection except in areas with low population density.
[3]Sources: Sterr and Simmering, 1996; Ebenhoch *et al.*, 1997.
[4]Source: U.S. Country Studies Program. Adaptation costs are full protection plus yearly maintenance. See also Zeidler, 1997.
[5]Source: Kont *et al.*, 1997; estimates for two sites only.
[6]Estimates for wetland loss along Black Sea coastline only.

Hoozemans, 1996). Anticipating accelerated sea-level rise in the design of new flood defenses will counter this effect (U.K. Department of the Environment, 1996).

Analysts often employ simplifying assumptions regarding response strategies. The costs in Table 5-4, for example, assume that all densely populated areas will be protected. An alternative assumption is that the costs of protection (including the monetary values of nature) will be balanced against the costs of land loss. Fankhauser (1994) analyzed this assumption for countries in the OECD. He concludes that it is optimal to protect all harbors and cities. Open coasts and beaches need to be maintained and protected in densely populated areas (such as The Netherlands) and in areas where beach tourism is important (such as Greece, Italy, and Spain). On the other hand, areas with low population density (such as Norway and Iceland) and poorer areas (such as Turkey) have a lower optimal level of protection. Turner *et al.* (1995) analyzed protection in East Anglia, England. At the regional scale, protection can be justified for the entire coast. When the 113 individual flood compartments were evaluated, however, the study indicated that it may not be economically justifiable to continue to protect 20% of compartments, even with modest rates of sea-level rise. This analysis assumes that there is no interaction between flood compartments—which may not always be the case. However, it suggests that a range of responses, rather than a single response, may be appropriate at regional and national scales.

Responses may be hindered by resource constraints. In Cyprus, the coast is almost totally sediment-starved as a result of catchment regulation and management (Nicholls and Hoozemans, 1996), and there are no ready sources of sand available for beach nourishment. Yet maintaining the beach is critical to the tourist industry. Therefore, external (and hence costly) sources of sand may be required for beach nourishment. Many other Mediterranean islands appear to have similar problems.

Some adaptation that anticipates climate change already is being implemented, despite the costs (U.K. Department of the Environment, 1996). In The Netherlands, national policies that outlaw erosion and mandate maintenance to the present shoreline position already are in place (Koster and Hillen, 1995). In the Ebro, Rhone, and Po deltas, conceptual and quantitative models of deltaic responses to sea-level rise are being developed (Jimenez and Sanchez-Arcilla, 1997; Sanchez-Arcilla and Jimenez, 1997). The ultimate goal of this work is to harness natural processes within deltas to counter global and local (due to subsidence) sea-level rise. However, other coastal settings often are not being managed with climate change in mind (Nicholls and Hoozemans, 1996). This approach often ignores numerous opportunities to use existing changes, such as the redevelopment cycle of coastal urban areas, to adapt to climate change (IPCC 1996, WG II, Chapter 9). Planning for and responding to sea-level rise raise important questions of equity, such as who pays for eroded land (if retreat is a response). These issues remain to be addressed systematically.

A more strategic and integrated perspective regarding coastal management and shoreline protection has been triggered by climate change, although resulting efforts have to consider more than just climate change. In Britain, guidelines for estuary management (English Nature, 1993) and shoreline management (U.K. Ministry of Agriculture, Fisheries and Food *et al.*, 1995) have been published, and a number of shoreline management plans and estuary management plans have been completed or are being formulated. Recognition of sediment cells and the benefits of erosion and beach nourishment in terms of sediment supply to other parts of the cell are being more fully realized (Bray *et al.*, 1996). The flood-prevention benefits of coastal wetlands also are being assessed. In eastern Britain, some managed setback schemes have been carried out on a pilot basis. In these areas—which had previously been reclaimed for agriculture—the sea defense has been realigned and the seawall breached to recreate areas of salt marsh and intertidal habitat. This strategy is expected to assist in the restoration of natural balance in estuaries and to provide flood alleviation benefits. It is anticipated that larger-scale application of such defense realignment, in appropriate locations, may be required to counter expected losses of intertidal habitat and relieve pressure on artificial defenses from sea-level rise. Such a policy may find more widespread application in Europe and would help minimize wetland losses resulting from sea-level rise.

### 5.3.5. *Other Infrastructure/Activities/Settlement*

#### *5.3.5.1. Energy Demand*

Energy demand will be affected by warming, but the direction and strength of the impact are unclear. Air conditioning (cooling) is a relatively new but growing source of energy demand in northern Europe, and warmer summers would increase this demand. Such a tendency would enhance the urban heat island effect and thereby heat stress. Increased demand for irrigation water also would augment the demand for energy. On the other hand, warmer winters would reduce the demand for heating energy.

In the United Kingdom, peak demand for heating fuels might decline less than total annual demand (as a result of a shortened heating season), leading to a reduced demand load factor. On aggregate, UK demand for fossil fuels may decline by 5–10% and electricity demand by 1–3% for a 2.2°C temperature rise by 2050; Finnish electricity demand would fall by 7–23% for a temperature increase of 1.2–4.6°C; Russian fossil fuel demand would fall by 5% and electricity demand by 1% for a 2.0°C temperature increase (IPCC 1996, WG II, Chapter 11).

#### *5.3.5.2. Water Supply*

As discussed in Section 5.3.3, GCM-based analyses project a range of possible runoff trends—from slightly drier to slightly moister conditions in Europe in the future. Many European rivers are likely to have less flow in dry periods of summer, aggravating problems of water supply to major cities. Numerous public supplies depend on groundwater (e.g., 94% of Portuguese supplies); any decrease in winter recharge could have serious implications. Reduced runoff also would negatively affect cooling of electric-power and industrial plants, particularly if there are environmental constraints on waste-heat production. For example, several French nuclear power stations were forced to close down or operate well below design capacity during the 1991 drought (IPCC 1996, WG II, Chapter 14).

Anticipated climatic changes in Greece are likely to dramatically increase the risk of summer water shortages. Significant increases in storage capacity would be needed to maintain existing water and energy supplies. The Netherlands could face the desiccation of most of its wetland areas or be forced increasingly to rely on the Rhine to maintain present water levels. Groundwater aquifers could be affected by increased saltwater intrusion as sea level rises. Given a possible 4°C temperature increase and a rise in the Alpine snow line of 700 m in summer, the summer flow of the Rhine could decline by 10%. A temperature increase of only 0.5°C and a daily rainfall decrease of 0.08 mm would decrease runoff in Hungary by 25–30%. Despite projected wetter winters, drier summers and increased evaporation in southeastern England may reduce yields in impounding reservoirs by 8–15%. Aquifer yields (which are very important to the London area) may fall by 8%. Supplies of water during warmer, possibly drier, summers would need to be maintained through larger storages or transfers from wetter regions. Water quality may deteriorate because there would be less river flow to dilute contaminants (IPCC 1996, WG II, Chapter 10).

#### 5.3.5.3. *Water Demand*

Increasing temperatures intensify water demand—particularly for agriculture, human consumption (e.g., watering of gardens and lawns), cooling water for electric-power and industrial plants, and natural ecosystems (IPCC 1996, WG II, Chapter 14). Changing precipitation also modifies demands for irrigation, particularly in regions with soils of low water-storage capacity (e.g., northern Germany, Denmark, Poland).

#### 5.3.5.4. *Air Pollution*

More stable anticyclonic conditions in the summer would provide increased opportunities for ground-level ozone to build up and lead to deteriorating urban air quality. Existing problems in Mediterranean cities such as Athens will increase if conditions become drier. Summertime smog (largely ground-level ozone) has become a significant problem in many parts of Europe (IPCC 1996, WG II, Chapter 12). The extent of these problems in the future will very much depend on future transport policies in affected areas.

#### 5.3.5.5. *Construction and Infrastructure*

In general, risks to construction and infrastructure are caused by extreme events (e.g., droughts, high winds) rather than average conditions. In northern Europe, the greatest negative impacts on construction are likely to arise from river floods. Studies by Kwadijk and Middelkoop (1994) and Penning-Rowsell *et al.* (1996) show that small increases in winter precipitation could lead to drastic increases in flood depths. Unless adaptive measures are installed in time, flood damage also would increase dramatically. Increased summer drought would aggravate land subsidence induced by clay-shrinkage, affecting dwellings particularly in England and Poland (Brignall *et al.*, 1996). Increased summer drought also would heighten fire danger, especially in the Mediterranean area. In temperate and northern climates, higher temperatures will result in lower road maintenance costs, particularly if snowfall and the number of freeze-thaw cycles decrease (IPCC 1996, WG II, Chapter 11).

#### 5.3.5.6. *Insurance*

Insurance companies are vulnerable to climate change through changes in windstorms, droughts, and floods that could affect covered property (IPCC 1996, WG II, Chapter 17). Flood insurance can be bought on the free market only in the United Kingdom; other European countries have government-backed insurance or none at all (Albala-Bertrand, 1993). Thus, possible increases in flood magnitudes and frequencies are unlikely to seriously affect the insurance sector. Droughts affect insurance only through the impact of land subsidence on buildings. Damage to buildings from windstorms is widely covered by insurance in Europe. Dorland *et al.* (1996) show that an increase of a few percent in storm intensity (not inconsistent with GCM results) could double or triple storm damages, most of which are insured. This effect occasionally could lead to troubles in the insurance sector—witness the total insured losses of $10 billion (compared with $15 million total losses) during the first months of 1990. Storm damage insurance, however, accounts for only a tiny fraction of total property insurance (Tol, 1996).

Europe is home to the largest reinsurance companies in the world. Reinsurers insure other insurance companies from all over the world for large losses—for instance, losses resulting from natural disasters. Thus, increases in floods or tropical cyclones in highly developed areas may affect Europe's financial sector. Experience in the early 1990s shows that the reinsurance market is capable of rapid reform under stress—which indicates that climate change could well lead to incidental turmoil but is unlikely to cause structural problems (Tol, 1996).

### 5.3.6. **Human Health**

Global climate change over the coming decades could have various effects on the health of human populations within the European region. Because of the nature and scale of the exposures involved, such effects generally would apply to entire populations or communities rather than to small groups or individuals. Climate change could affect human health through six major pathways.

1) An increased frequency or severity of heat waves would cause an increase in heat-related mortality and illness. In contrast, less-severe cold weather would reduce the documented seasonal excess of deaths in winter. Many studies have shown that instances of heat hyperpyrexia and overall death rates rise during heat waves, particularly when the temperature rises above the local population's physiological threshold and the temperature increases are accompanied by high humidity. Several U.S. studies indicate that, on average, approximately one-third of these deaths would occur 2–3 weeks after the heat wave in susceptible persons, whereas other deaths apparently are unexpected. By applying these documented heat wave-associated risks to transient climate change scenarios associated with $CO_2$-doubling, U.S. researchers have estimated that the number of heat-related deaths may increase several-fold in very large U.S. cities by 2050 (Kalkstein, 1993). In such cities, this would represent up to several thousand additional deaths per year.

No European-equivalent estimates of additional heat-related deaths attributable to climate change, based on local empirical studies, are yet available. A clear rise in daily mortality was associated with the 1995 extreme heat wave in the United Kingdom; the increase was approximately twice as great within the urban London population (Rooney *et al.*, 1997). By simple extrapolation from a U.S. study, Fankhauser (1995) concluded that about 9,600 additional heat-related deaths per year may occur in the European Union for a 2.5°C temperature rise, and about 8,400 additional deaths may occur in the former Soviet Union. In both cases, full physiological acclimatization is assumed; population size, air conditioning, and medical care are assumed to be unchanged from present conditions.

This heat-related increase in deaths would be partially offset by reductions in cold-related deaths (predominantly cardiovascular). A recent study (Martens, 1997) estimated that in areas with temperate and cold climates—which includes large parts of the European continent—a globally averaged temperature increase of approximately 1°C could result in a reduction in winter cardiovascular mortality, especially in older people. Another British study (Langford and Bentham, 1995) has forecast that approximately 9,000 fewer winter-related deaths would occur annually by the year 2050 in England and Wales under a 2–2.5°C increase in average winter temperature. On the other hand, Fankhauser (1995) arrives (again by extrapolation from the United States) at about 800 fewer winter deaths for the entire EU. The EUROWINTER group (1997) assessed increases in mortality per 1°C decrease in temperature in various European regions. This study shows that mortality increased to a greater extent with a given decline in temperature in regions with warm winters, in populations with cooler homes, and among people who wore fewer clothers and were less active outdoors. Apparently there still is insufficient European information to quantify this trade-off between heat-associated losses and gains associated with milder winters. Furthermore, the balance will vary by location and adaptive response.

2) Changes in seasonal and daily temperatures and humidity are likely to affect the concentration of airborne materials that impinge on respiratory health. The production of photochemical smog proceeds more rapidly at higher temperatures. The concentrations and onset and duration of season of allergenic pollens and spores are related to cumulative temperatures and rainfall, though in a complex manner (Emberlin, 1994; Spieksma *et al.*, 1995). Alterations in the concentration of aeroallergens may affect the seasonality of certain allergic respiratory disorders. Relatively little research has been done on these processes and relationships within Europe or elsewhere, however. Interactive adverse effects on mortality have been reported from Athens in response to simultaneous high temperatures and high levels of air pollution (Katsouyanni *et al.*, 1993).

3) Extreme weather events (floods, storms, fogs, etc.) cause deaths, injury, certain infectious diseases, and mental health disorders. The number of victims of such events is relatively low in Europe (Alexander, 1993; IDNDR, 1994); for instance, the extremely stormy first quarter of 1990 caused "only" 225 casualties throughout Europe (Munich Re, 1993). Indications exist that flood incidence in the northern parts of Europe and storminess in the western and central parts of Europe (and hence health risks) may increase (Downing *et al.*, 1996); to date, no research has been reported that quantifies the impact on mortality and morbidity risks. Disease (e.g., hepatitis) may break out in areas affected by severe flooding, particularly if drinking water becomes contaminated by sewage.

4) Organisms and biological systems that determine the spread of infectious diseases typically are sensitive to climatic variables. Net climate change-related increases in the geographic distribution of vector organisms (e.g., ticks, mosquitoes, sand flies) of various infectious diseases, along with changes in the life-cycle dynamics of vectors and infectious parasites, would, in aggregate, increase the potential for transmission of many vector-borne diseases in Europe.

Summer conditions in Europe are warm enough in some countries for native mosquito species to transmit malaria. Indeed, until the third quarter of this century, the disease was present in southern Europe. Although malaria was successfully eradicated from most of Europe during the 1950s and 1960s, there have been recurrent outbreaks in Turkey and the former Soviet Union (Bruce-Chwatt and de Zulueta, 1980), and it is present in parts of North Africa. Different types of malaria are present; the severity of symptoms is related to the plasmodia involved. The most dangerous form is *Plasmodium falciparum* (other species are *P. vivax*, *P. ovalum,* and *P. malariae*). Climate change would increase the risk of malaria transmission (which currently is low). However, existing public-health resources—surface-water management, disease surveillance, and medical treatment—would make re-emergent malaria unlikely. Localized outbreaks of "airport malaria" may occur (because of increases in tourism, increased malaria in countries that Europeans visit, and warmer local conditions in Europe), so it is important to strengthen current policies. For example, about 2,000 cases of malaria were diagnosed in England in 1990, and there are 10–12 fatalities each year. A troublesome trend is the increasing resistance of the plasmodia to prophylactics, which increases the chances of disease transmission.

Several existing tick-borne infectious diseases are significant in Europe (e.g., Lyme disease, which is caused by the bacterium *Borrelia burgdorferi*, and tick-borne viral encephalitis). The tick's life cycle is sensitive to temperature and humidity, and tick populations depend on access to intermediate host animal species (especially rodents and deer). Unpublished data indicate that an earlier onset of spring and a later arrival of the winter season are related to increases in the incidence of tick-borne encephalitis in Sweden (Lindgren and Gustafson, submitted).

Leishmaniasis is endemic at a low level in all countries bordering the Mediterranean, as well as Portugal (Gradoni *et al.*, 1995). The visceral form of the disease, known as kala-azar (which can be fatal), is limited by the distribution of its sand fly vectors. Climate change is likely to expand the distribution of the sand fly vectors. Higher temperatures also would accelerate the maturation of the protozoal parasite, thereby increasing the risk of infection (Rioux *et al.,* 1985).

The mosquito *Ae. albopictus* is considered second only to *Ae. aegypti* in its importance as a disease vector of dengue and dengue hemorrhagic fever. The first sighting of the vector species in Europe (in 1979) came from Albania. However, it was only when *Ae. albopictus* was introduced into Italy in 1990—through the importation of used tires—and subsequently spread that the species was considered a threat to public health. By the end of 1995, *Ae. albopictus* infestations had been reported in 10 Italian regions and 19 provinces. Other countries where climatic conditions meet such criteria and that may be vulnerable to a potential introduction of *Ae. albopictus* include Spain, Portugal, Greece, Turkey, France, Albania, and the former Republic of Yugoslavia (Knudsen *et al.*, 1996).

5) Increases in non-vector-borne infectious diseases also would occur because of the effect of higher temperatures on microorganism proliferation. Many food-related infections are affected by ambient temperatures and have their annual peaks during the summer months (e.g., food poisoning from salmonellosis) (Bentham and Langford, 1995). Declining water availability and quality also would pose risks to health. Flooding and regional droughts would alter the mix of drinking water sources in some locations. Outbreaks of waterborne infectious diseases such as cryptosporidiosis—which have occurred in both the United Kingdom and the United States in recent years—may become more likely.

6) Climate change also could affect human health indirectly through other impacts. For example, a potentially important health impact would result from deterioration in social and economic circumstances that might arise from effects of climate change on patterns of employment, wealth distribution, and population mobility and settlement.

The actual impact of climate change on human health is modulated by the susceptibility of the exposed population and the human ability to adapt. In general, the most vulnerable populations, communities, or subgroups would be people living in poverty or with a high prevalence of malnutrition, chronic exposure to infectious disease agents, and inadequate access to social and physical infrastructures. Within Europe, in general, the vulnerability to many health impacts is limited by existing material resources, public health measures, and systems of health care. Heat waves, other forms of extreme weather, airborne respiratory hazards, and vector-borne infectious diseases may have an impact, however.

## 5.4. Integrated Assessment of Potential Vulnerabilities and Impacts

In the foregoing sections, impact categories have been treated separately. Obviously, however, the impacts of climate change interact with one another—through the use of resources such as water and land and through economy, society, and politics. An integrated assessment tries to take these interactions into account. In addition, an integrated assessment tries to compare the diversity of impacts with respect to their seriousness (from a human viewpoint) and evaluate overall vulnerability to climate change.

Three approaches to integrated assessment have been applied to Europe: monetization, integrated modeling, and coupled models and expert panels. Each method has its advantages and disadvantages, which are discussed at length in Weyant *et al.* (IPCC 1996, WG III, Chapter 10). The focus here is on the results. Note, however, that integrated assessment is a young field, without a great number of well-established methodologies, let alone results.

### *5.4.1. Integrated Assessment Using Monetization*

One way of integrating the range of potential impacts of climate change is to derive a comprehensive monetary estimate, which adds all impacts expressed in their dollar value. This approach allows for comparison of the seriousness of climate change with other problems, comparison of vulnerabilities to climate change among regions and sectors, and comparison of the impact of climate change with the impact of greenhouse gas emission reduction. Expressing effects on marketed goods and services (e.g., land loss resulting from sea-level rise, energy savings in winter) in monetary terms is relatively straightforward because the price is known. Expressing damage to nonmarketed goods and services (e.g., wetland loss, mortality changes) in monetary terms can be accomplished by examining market transactions where such

**Box 5-2. Integrated Regional Climate Impact Assessment in Brandenburg**

The Brandenburg project (Stock and Toth, 1996) was an integrated regional climate impact study. A variety of statistically derived climate change scenarios were used as inputs to various types of ecosystem and forest growth models, plant development simulations, and hydrological analyses. Socioeconomic impacts were limited to human health and energy use. Integration across sectors and climate impacts was limited, however.

The observed climate of 1937–1992 served as a reference scenario. Six additional scenarios were developed to capture possible changes in climate under various assumptions about the frequency, length, and type of anomalous weather conditions. A synthetic year was composed from a warm and wet spring, a hot and dry summer, a warm and wet fall, and a warm and wet winter. The frequency of these extreme years was increased by 20% in Scenario 2; Scenario 3 contained the same extreme year only. Scenarios 4 and 6 were created by modifying the frequency of extreme conditions by 20% in monthly patterns and summer weather conditions, respectively. Scenarios 5 and 7 contained only synthetic years derived from modified monthly and summer patterns, respectively.

Brandenburg is the driest region in Germany. This condition is aggravated by the predominance of sandy soils with low water-holding capacities and by the lack of substantial amount of water inflows to the region. Two regional water models were used to analyze the hydrological implications of climate change. Under all scenarios considered, critical low-flow values fall significantly. This result would create problems for water availability, water quality, and aquatic ecosystems.

Three different modeling approaches have been used to study possible impacts on the composition, productivity, and stability of forests and natural ecosystems. A global vegetation potential model projects a transition toward steppe as the dominant vegetation form as a consequence of increasing dryness; it is uncertain, however, that dryness will increase. On the other hand, two local forest succession models show that the simulated potential natural vegetation of beech will be replaced almost everywhere by mixed stands of oak, lime, and fir. Currently, forestry in Brandenburg relies mainly on fir stands; simulation results show that there is little chance for improved conditions at current sites even under optimistic projections of climates. Reduced water availability would result in substantially reduced yields. Unfavorable secondary impacts—such as more frequent forest fires and increased pest and insect damages—together with higher management costs and timber loss are likely to generate significant economic losses.

Additional components of the study looked at possible impacts on yields in agriculture, human health, and energy production and use. In addition to preliminary results, the major conclusion in all these areas was that additional and more detailed studies are required to improve our understanding of possible threats, as well as the benefits and disadvantages associated with possible adaptation strategies.

goods or services are implicitly traded (e.g., landscape beauty) or by interviewing people about their preferences. That is, human preferences are expressed by people's willingness to pay to secure a benefit or their willingness to accept compensation for a loss. In western Europe, valuation techniques are well established and widely applied. Numerous theoretical and empirical problems remain, however. Based on the SAR (IPCC 1996, WG II, Chapter 6), Fankhauser and Tol (1997) report best estimates for the annual impact resulting from a doubling of atmospheric concentrations of carbon dioxide of about -1.6% to -1.4% of the GDP in western Europe, using a mix of earlier GCM scenarios standardized to a 2.5°C increase in the global mean temperature. Best estimates for eastern Europe and the former Soviet Union vary between -0.4% and +0.4% (a benefit) of GDP. These figures compare with an estimated world impact of -1.8% to -1.2% of GDP. National differences may be hidden by the regional average, but only one country-specific study has been carried out to date. Using similar methods and scenarios as Fankhauser and Tol (1997), Kuoppomaeki (1996a, b) concludes that Finland may gain about 1% of GDP. Many assumptions underlie these best guesses; the uncertainties are large, yet unknown.

### 5.4.2. *Integrated Assessment Using Integrated Modeling*

An integrated assessment model combines climate change scenarios with models of various impacts of climate change in a single computer code (i.e., hard-linking of models). One advantage of this approach is that consistency is ensured (e.g., land or water used for agriculture is not used in the domestic sector). A disadvantage is that modeling and computational requirements entail simplifications of the state-of-the-art. Only two integrated assessment models are even somewhat useful for climate change impact assessment in Europe; the other twenty-odd models reviewed in IPCC (1996, WG III, Chapter 10) have other aims or are based on monetization. The first integrated assessment model is the ESCAPE model (Rotmans *et al.*, 1994), which was developed specifically for the EU. It includes a variety of impacts on natural and human systems; since its completion in 1991, however, the model has not been updated. The second

**Box 5-3. Integrated Study: Effects of Increasing Temperatures on the Ecosystems of the Bornhöved Lake District**

The Bornhöved lake district is located in the north of Germany, near the Baltic Sea; it includes agroecosystems, beech forest, alder bog, grassland, and a lake. The surface of the catchment is about 5 km$^2$, including 1 km$^2$ of lake surface. The dominating soil types are dystric arenosols on the hills.

All results are based on a climate data set of 60 years with a maximum increase of 2.7ºC; rainfall was assumed to be unchanged. Potential evapotranspiration (pET) would increase by about 100 mm, but actual evapotranspiration (aET) would differ only slightly from observed values because soil water storage would become a limiting factor. During summer, there may be considerably more drought stress. Soil moisture repletion in autumn and winter would take considerably longer: The profile would not be filled until January. As a result, farmers would have to change their land use from spring cereals to winter crops or install irrigation systems on the light soils.

The extent of the relatively flat catchment area depends on the amount and spatial distribution of groundwater recharge. Because of higher evapotranspiration, groundwater recharge would be reduced by 25% (100 mm/year), and the catchment area of the lake would decrease. Overall, the inflow into Lake Belau could decline by about 50%.

The discharge of the rivers could drop by about 30% compared with present conditions. The frequency of low water levels could increase, causing additional stress for flora and fauna in the rivers. Water quality in the eutrophic Lake Belau is determined by the load of the river that flows through the lake. As a result of decreasing discharge, there would be higher nutrient concentrations in the groundwater and the river water; the overall nutrient load will increase substantially.

Under unchanged management, nitrogen concentrations in the soil solution would rise, but the overall loss would decrease slightly. However, a change in the crop rotation and a reduction of fertilization from the present 340 kg N/ha to 163 kg N/ha would affect the nitrogen leaching much more.

Vegetation studies were conducted for primary production of alder, vegetation structure in grassland, soil respiration, changes in the length of the growth period in beech forest, and production of beech and *Carex acutiformis*. Zoological case studies were carried out for molluscs, birds, mice, several invertebrates, and the food chain of the robin. Particularly interesting is the decrease in the number of long-distance (trans-Sahara) migratory birds—a phenomenon that also has been observed in other regions of Germany. One possible explanation could be that with increasing temperatures, the nests and territories of these birds already have been occupied by nonmigratory birds and birds with short passage that can respond more quickly to the changing conditions of rising temperatures.

Changes in agricultural yield were simulated for wheat, corn, rape, grassland, and beans. Because of the light soils in the research area, corn yield will decrease dramatically under scenario conditions, and the risk of bad harvests will increase. An earlier seed time and the use of other varieties of corn has almost no effect on yields. Because the agriculture of the region is dominated by dairy production, food for cattle has the highest priority. To minimize the risk of bad harvests, farmers should irrigate or replace corn with grassland, which uses winter rainfall more efficiently.

Under scenario conditions, rivers and lakes will be more severely affected than terrestrial ecosystems. The structure of the landscape will be more pronounced between dry uphill patches and downhill regions with access to groundwater. The soil water storage capacity will become much more limiting and will increase the local risk of bad harvests; adaptation of land-use practices would be necessary. Despite numerous efforts to predict species composition in terrestrial and aquatic ecosystems, predictions on a local scale are not possible. Changes should be evaluated by a combination of models and monitoring programs (predictive monitoring).

Source: Hörmann *et al.*, 1995.

such model is the IMAGE2 model (Alcamo, 1994). This model is up-to-date. It models the whole world, with a focus on land use. The resolution is 0.5°x0.5°, with east and west Europe grouped into two economic regions. The socioeconomic modeling of agriculture and human infrastructure—the dominant land-use tupes in Europe—requires further work.

### *5.4.3. Integrated Assessment Using Coupled Models and Expert Panels*

A third form of integrated impact assessment is the use of coupled models and expert panels. The knowledge of a range of experts on various impacts—supported, where possible, by a suite of models—is combined; consistency is attempted through common scenarios, soft-linking of models (i.e., using output from one model as input to another), and discussion. This approach better reflects the full richness of the literature and the complexity of the issues (including nonmodeled parts), but it lacks the solid outcome of a model. This approach has been used for the United Kingdom (CCIRG, 1991, 1996), Finland (SILMU, 1996), the state of Brandenburg in Germany (see Box 5-2), the Bornhöved lake district in Germany (see Box 5-3), and land use in England and Wales (Parry *et al.*, 1996). Ensuring consistency only in scenarios, the UK and Finland studies are least ambitious in integration yet most comprehensive in areas and topics covered; relevant findings are discussed elsewhere in this chapter. The Brandenburg and Bornhöved studies consider only small areas, focusing on the impact of climate change on ecosystems. The study on England and Wales uses world food prices (from Rosenzweig and Parry, 1994) and climate change scenarios to investigate land-use changes; a major result is that, although the aggregate agricultural output of this region is unlikely to change drastically, the spatial distribution of activities may well change substantially.

## References


**Ackefors**, H., N. Johansson, and B. Walhlberg, 1991: The Swedish compensatory programme for salmon in the Baltic: an action plan with biological and economic considerations. In: *Ecology and Management Aspects of Extensive Mariculture: A Symposium Held in Nantes* [Lockwood, S. (ed.)]. ICES Mar. Sci. Symp., 248 pp.

**Albala-Bertrand**, J.M., 1993, *Political Economy of Large Natural Disasters*. Clarendon Press, Oxford, United Kingdom, 259 pp.

**Alcamo**, J., 1994: *IMAGE 2.0 — Integrated Modeling of Global Climate Change*. Kluwer Academic Publishers, Dordrecht, Netherlands,321 pp.

**Alexander**, D., 1993: *Natural Disasters*. University College, London Press, London, United Kingdom, 632 pp.

**Anderson**, J.M., 1991: The effects of climate change on decomposition processes in grassland and coniferous forests. *Ecological Applications*, **1**, 326–347.

**Arnell**, N.W. and N.S. Reynard, 1996: The effects of climate change due to global warming on river flows in Great Britain. *J. Hydrology*, **183(3–4)**, 397–424.

**Auer**, I. and R. Böhm, 1994: Combined temperature-precipitation variations in Austria during the instrumental period. *Theoretical and Applied Climatology*, **49**, 161–174.

**Bach**, W. 1987. Development of climate change scenarios: A. From general circulation models. In: *The impact of climatic variations on agriculture, vol. 1: Assessment in cool temperature and cold regions* [Parry, M.L., T.R. Carter, and N.T. Konijn (eds.)]. Kluwer Academic Publishers, Dordrecht, The Netherlands, pp. 125–157.

**Balesdent**, J., S. Houot, and S. Recous, 1994: Réponse des matières organiques des sols aux changements atmosphériques globaux. IV. Simulation des effets d'un réchauffement et d'une production carbonée accrue sur les stocks de carbone des sols cultivés. In: *Ecosystèmes et Changements globaux* [Perrier, A. and B. Saugier (eds.)]. INRA, Paris, France, **8**, 91–95.

**Baxter**, R., T.W Ashenden, T.H Sparks, and J.F. Farrar, 1994: Effects of elevated carbon dioxide on three montane grass species. I. Growth and dry matter partitioning. *Journal of Experimental Botany*, **45**, 305–315.

**Beniston**, M. and M. Rebetez, 1995: Regional behavior of minimum temperatures in Switzerland for the period 1979–1993. *Theor. and Appl. Clim.*, **53**, 231–243.

**Beniston**, M., M. Rebetez, F. Giorgi, and M.R. Marinucci, 1994: An analysis of regional climate change in Switzerland. *Theor. and Appl. Clim.*, **49**, 135–159.

**Bentham**, G. and I.H. Langford, 1995: Climate change and the incidence of food poisoning in England and Wales. *International Journal of Biometeorology*, **39**, 81–86.

**Berninger**, F., 1997: Effects of drought and phenology on GPP in Pinus sylvestris: A simulation study along a geographical gradient. *Functional Ecology*, **11(1)**, 33–42.

**Beukema**, J.L., W.J. Wolff, and J.J.W.M. Brouns (eds.), 1990: Expected effects of climatic change on marine coastal ecosystems. In: *Developments in Hydrobiology 57*. Kluwer Academic Publishers, Dordrecht, The Netherlands, 221 pp.

**Beuker**, E., 1994: Adaptation to climatic changes of the timing of bud burst in populations of *Pinus sylvestris* (L.) and *Picea abies* (L.) Karst. *Tree Physiology*, **14**, 961–970.

**Bindi**, M., L. Fibbi, B. Gozzini, S. Orlandini, and F. Miglietta, 1996: Modeling the impact of future climate scenarios on yield and yield variability of grapevine. *Climate Research*, **7**, 213–224.

**Binet**, D., 1988a: Rôle possible d'une intensification des alizés sur le changement de répartition des sardines et sardinelles le long de la côte ouest africaine. *Aquat. Living Resour.*, **1**, 115–132.

**Binet**, D., 1988b: French sardine and herring fisheries: a tentative description of their fluctuations since the XVIIIth century. In: *Long Term Changes in Marine Fish Populations* [Wyatt, T. and M.G. Larraneta (eds.)]. Symposium held 18–20 November 1986, Bayona Imprenta REAL, Vigo, Spain, pp. 253–272.

**Binet**, D. and E. Marchal, 1992: Le développement d'une nouvelle population de sardinelles devant la Côte d'Ivoire a-t-il été induit par un changement de circulation? *Ann. Inst. océanogr.*, **68(1–2)**, 179–192.

**Birks**, H.J.B., 1989: Holocene isochrone maps and patterns of tree-spreading in the British Isles. *Journal of Biogeography*, **16**, 503–540.

**Bonan**, G.B. and K. Van Cleve, 1992: Soil-temperature, nitrogen mineralization, and carbon source sink relationships in boreal forests. *Canadian Journal of Forest Research*, **22**, 629–639.

**Brasier**, C.M., 1996: Phytophthora cinnamomi and oak decline in southern Europe. Environmental constraints including climate change. *Annales Des Sciences Forestières*, **53(2–3)**, 347–358.

**Bray**, M.J., D.J. Carter, and J.M. Hooke, 1996: Littoral Cell Definition and Budgets for Central Southern England, *Journal of Coastal Research*, **11**, 295–570.

**Brazdil**, R., M. Budikova, I. Auer, R. Bohm, T. Cegnar, P. Fasko, M. Lapin, M. Gajiccapka, K. Zaninovic, E. Koleva, T. Niedzwiedz, Z. Ustrnul, S. Szalai, and R.O. Weber, 1996: Trends of maximum and minimum daily temperatures in central and southeastern Europe. *International Journal of Climatology*, **16(7)**, 765–782.

**Brignall**, A.P., M.J. Gawith, J. Orr, and P.A. Harrison, 1996: Towards an Index for Assessing the Potential Effects of Climate Change on Clay Shrinkage Induced Land Subsidence. In: *Climate Change and Extreme Events — Altered Risk, Socio-economic Impacts and Policy Responses* [Downing, T.E., A.A. Olsthoorn, and R.S.J. Tol, (eds.)]. Institute for Environmental Studies, Vrije Universiteit and Environmental Change Unit, University of Oxford, Amsterdam, Netherlands and Oxford, United Kingdom, pp. 35–50.

**Bruce-Chwatt**, L.J. and J. de Zulueta, 1980: *The rise and fall of malaria in Europe*. Butler and Tanner Ltd., London, United Kingdom, 240 pp.

**Brzeziecki**, B., F. Kienast, and O. Wildi, 1995: Modelling potential impacts of climate change on the spatial distribution of zonal forest communities in Switzerland. *Journal of Vegetation Science*, **6**, 257–268.

**Bugmann**, H., 1996: A simplified forest model to study species composition along climate gradients. *Ecology*, **77(7)**, 2055–2074.

**Bugmann**, H., 1997: Sensitivity of forests in the European Alps to future climatic change. *Climate Research*, **8(1)**, 35–44.

**Caddy**, J.F., 1993: Toward a comparative evaluation of human impacts on fishery ecosystems of enclosed and semi-enclosed seas. *Reviews and Fisheries Science*, **1(1)**, 57–95.M.25.

**Cammell**, M.E. and J.D. Knight, 1992: Effects of climate change on the population dynamics of crop pests. *Advances in Ecological Research*, **22**, 117–162.

**Carter**, T.R., R.A. Saarikko, and K.J. Niemi, 1996: Assessing the Risks and Uncertainties of Regional Crop Potential under a Changing Climate in Finland. *Agriculture and Food Science in Finland,* **5**, 329–320.

**Casella**, E., J.F. Soussana, and P. Loiseau, 1996: Long-term effects of $CO_2$ enrichment and temperature increase on a temperate grass sward. 1. Productivity and water use. *Plant and Soil,* **182(1)**, 83–99.

**CCIRG (Climate Change Impacts Review Group)**, 1991: *The Potential Effects of Climate Change in the United Kingdom.* HMSO, London.

**CCIRG**, 1996: *The Potential Effects of Climate Change in the United Kingdom.* HMSO, London.

**Chambers**, J.R., 1991: Coastal degradation and fish population losses. In: *Stemming the Tides of Coastal Fish Habitat Loss: Proceedings of the Marine Recreational Fisheries Symposium* [Stroud, R.H. (ed.)]. National Coalition for Marine Conservation, Savannah, GA, USA, pp. 45–51.

**Chaves**, M.M., J.S. Pereira, S. Cerasoli, J. Clifton-Brown, F. Miglietta, and A. Raschi, 1995: Leaf metabolism during summer drought in Quercus ilex trees with lifetime exposure to elevated $CO_2$. *Journal of Biogeography,* **22**, 255–259.

**CLAIRE**, 1996: *Climate Change and Agriculture in Europe: Assessment of Impacts and Applications* [Harrison, P., R. Butterfield, and T. Downing (eds.)]. Research Report No. 9, Environmental Change Unit, University of Oxford, 411 pp.

**Costa**, M.J., J.L. Costa, P.R. Almeida, and C.A. Assis, 1994: Do eel grass beds and salt marsh borders act as preferential nurseries and spawning grounds for fish? An example of the Mira estuary in Portugal. *Ecological Engineering,* **3**, 187–195.

**Cushing**, D.H., 1957: The number of pilchards in the Channel. *Fish. Invest. Ser. II,* **21(5)**, 1–27.

**Cushing**, D.H., 1982: *Climate and Fisheries.* Academic Press, London, United Kingdom, 373 pp.

**Darwin**, R., M. Tsigas, J. Lewandrowski, and A. Raneses, 1995: *World Agriculture and Climate Change: Economic Adaptations.* Economic Research Service, USDA, AER-703, Washington, DC, pp. 1–86.

**Davis**, M.B., 1981: Quaternary history and the stability of forest communities. In: *Forest succession* [West, D.C., H.H. Shugart, and D.B. Botkin (eds.)]. Springer-Verlag, New York, NY, USA, pp. 132–153.

**Delécolle**, R., F. Ruget, D. Ripoche, and G. Gosse, 1996: Possible effects of climate change on wheat and maize crops in France. Climate Change and Agriculture: Analysis of Potential International Impacts. *ASA Special Publications,* **59**, 241–257.

**Diaz**, S. 1995. Elevated-$CO_2$ responsiveness, interactions at the community level, and plant functional types. *Journal of Biogeography,* **22**, 289–295.

**Diaz**, S. 1996. The effects of elevated $CO_2$ on root symbionts mediated by plants. *Plant and Soil,* **187**, 309–320.

**Dorland**, C., R.S.J. Tol, A.A. Olsthoorn, and J.P. Palutikof, 1996: An Analysis of Storm Impacts in The Netherlands. In: *Climate Change and Extreme Weather —Altered Risk, Socio-economic Impacts and Policy Responses* [Downing, T.E., A.A. Olsthoorn, and R.S.J. Tol (eds.)]. Institute for Environmental Studies, Vrije Universiteit and Environmental Change Unit, University of Oxford, Amsterdam, The Netherlands and Oxford, United Kingdom, pp. 157–184.

**Downing**, T.E., A.A. Olsthoorn, and R.S.J. Tol, 1996: *Climate Change and Extreme Events — Altered Risk, Socio-economic Impacts and Policy Responses.* Institute for Environmental Studies, Vrije Universiteit and Environmental Change Unit, University of Oxford, Amsterdam, The Netherlands and Oxford, United Kingdom, 309 pp.

**Ebenhoe**, W., H. Sterr, and F. Simmering, 1997: *Potentielle Gefaehrdung und Vulnerabilitaet der deutschen Nord- und Ostseekueste bei fortschreitenden Klimawandel*, unpublished report, 138 pp.

**Ellenberg**, H., 1986: *Vegetation Mitteleuropas mit den Alpen.* Ulmer, Stuttgart, Germany, 4th ed., 989 pp.

**Emberlin**, J., 1994: The effects of patterns in climate and pollen abundance on allergy. *Allergy,* **94**, 15–20.

**English Nature**, 1993: *Strategy for the Sustainable Use of England's Estuaries.* English Nature, Peterborough, United Kingdom, 43 pp.

**European Climate Support Network (ECSN)**, 1995: *Climate of Europe — recent variation, present state and future prospects.* National Meteorological Services 1995. KNMI, De Bilt, The Netherlands, 73 pp.

**EUROWINTER Group**, 1997: Cold exposure and winter mortality from ischaemic heart diseases, cerebrovascular disease, respiratory disease, and all causes in warm and cold regions of Europe. *Lancet,* **349**, 1341–1346.

**Fankhauser**, S., 1994: Protection versus retreat: estimating the costs of sea-level rise. *Environment and Planning A.,* **27**, 299–319.

**Fankhauser**, S., 1995: *Valuing Climate Change — The Economics of the Greenhouse.* EarthScan Publications, London, United Kingdom, 180 pp.

**Fankhauser**, S. and R.S.J. Tol, 1997: The social costs of climate change: the IPCC Second Assessment Report and beyond. *Mitigation and Adaptation Strategies for Global Change,* **1**, 385–403.

**FAO**, 1989: Recent Trends in Mediterranean Fisheries, XIXth Session of the GFCM, Livorno (Italy), 27 February–3 March 1989 [Caddy, J.F., (ed.)]. GFCM/RM/7/89/3, FAO, Rome, Italy, 71 pp.

**Fischetti**, M., 1991: A feast of gene-splicing down on the fish farm. *Science,* **253**, 512–513.

**Fitter**, A.H., R.S.R. Fitter, I.T.B. Harris, and M.H. Williamson, 1995: Relationships between first flowering date and temperature in the flora of a locality in central England. *Functional Ecology,* **9**, 55–60.

**Forch**, G., F. Garbe, and J. Jensen, 1996: Climatic change and design criteria in water resources management — A regional case study. *Atmospheric Research,* **42(1–4)**, 33–51.

**Francour**, P., C.F. Boudouresque, J.G. Harmelin, M.L. Harmelin-Vivien, and J.P. Quignard, 1994: Are the Mediterranean waters becoming warmer? Information from biological indicators. *Marine Pollution Bulletin,* **28**, 523–526.

**Freeman**, C., R. Gresswell, H. Guasch, J. Hudson, M.A. Lock, B. Reynolds, F. Sabater, and S. Sabater, 1994: The role of drought in the impact of climatic change on the microbiota of peatland streams. *Freshwater Biology,* **32**, 223–230.

**Fries**, C., O. Johansson, P. Pettersson, and B. Simonsson, 1997: Silvicultural models to maintain and restore natural stand structures in Swedish boreal forests. *Forest Ecology and Management,* **94(1–3)**, 89–103.

**Fröborg**, H. and O. Eriksson, 1997: Local colonization and extinction of field layer plants in a deciduous forest and their dependence upon life history features. *Journal of Vegetation Science,* **8**, 395–400.

**Fulford**, M., T. Champion, and A. Long (eds.), 1997: *England's Coastal Heritage.* Royal Commission on the Historical Monuments of England and English Nature, Archaeological Report 15, London, United Kingdom, 268 pp.

**Gavilán**, R. and F. Fernández-González, 1997: Climatic discrimination of Mediterranean broad-leaved sclerophyllous and deciduous forests in central Spain. *Journal of Vegetation Science,* **8**, 377–386.

**Giakoumakis**, S.G. and G. Baloutsos, 1997: Investigation of trend in hydrological time series of the Evinos river basin. Hydrological Sciences Journal — *Journal Des Sciences Hydrologiques,* **42(1)**, 81–88.

**Grabherr**, G., M. Gottfried, and H. Pauli, 1994: Climate effects on mountain plants. *Nature,* **369**, 448.

**Gradoni**, L., A. Bryceson, and P. Desjeux, 1995: Treatment of Mediterranean visceral leishmaniasis. *Bulletin of the World Health* Organization. **73(2)**, 191–197.

**Grime**, J.P., 1996: The changing vegetation of Europe: What is the role of elevated carbon dioxide? Carbon Dioxide, Populations, and Communities. In: *Physiological Ecology — A Series of Monographs, Texts, and Treatises (1996),* pp. 85–92.

**Haeberli**, W. and M. Hoezle, 1995: Application of inventory data for estimating characteristics of regional climate change effects on mountain glaciers: a pilot study with the European Alps. *Annals of Glaciology,* **21**, 206–212.

**Haeberli**, W., M. Wegmann, and D. Vonder Muhll, 1997: Slope stability problems related to glacier shrinkage and permafrost degradation in the Alps. *Ecologae Geologicae Helvetiae* (in press).

**Harrington**, R., M. Tatchell, and J. Bale, 1994: Aphid problems in a changing climate. In: *Insects in a changing climate* [Harrington, R. and N.E. Stork (eds.)]. Royal Entomological Society, London, United Kingdom.

**Harrison**, P.A. and R.E. Butterfield, 1996: Effects of climate change on Europe-wide winter wheat and sunflower productivity. *Climate Research,* **7**, 225–241.

**Hartig**, E.K., O. Grozev, and C. Rosenzweig, 1997: Climate change, agriculture and wetlands in Eastern Europe: Vulnerability, adaptation and policy. *Climatic Change*, **36(1–2)**, 107–121.

**Hättenschwiler**, S. and C. Körner, 1995: Responses to recent climate warming of Pinus sylvestris and Pinus cembra within their montane transition zone in the Swiss Alps. *Journal of Vegetation Science*, **6**, 357–368.

**Hättenschwiler**, S. and C. Körner, 1996a: Effects of elevated $CO_2$ and increased nitrogen deposition on photosynthesis and growth of understory plants in spruce model ecosystems. *Oecologia*, **106**, 172–180.

**Hättenschwiler**, S. and C. Körner, 1996b: System-level adjustments to elevated $CO_2$ in model spruce ecosystems. *Global Change Biology*, **2**, 377–387.

**Hofer**, H.R., 1992: Veränderungen in der Vegetation von 14 Gipfeln des Berninagebietes zwischen 1905 und 1985. *Ber. Geobot. Inst. Eidg. Tech. Hochsch. Stift. Rübel Zür*, **58**, 39–54.

**Holt**, C.P. and Jones, J.A.A., 1996: Equilibrium and transient global, warming scenario implications for water resources in Wales. *Water Resources Bulletin*, **32(4)**, 711–721.

**Holtmeier**, F.-K., 1994: Ecological aspects of climatically-caused timberline fluctuations. Review and outlook. *Mountain environments and changing climates* [Beniston, M. (ed.)]. Routledge, London, United Kingdom, pp. 220–233.

**Hoozemans**, F.M.J., M. Marchand, and H.A. Pennekamp, 1993: *A Global Vulnerability Analysis – Vulnerability Assessment for Population, Coastal Wetlands and Rice Production on a Global Scale.* Delft Hydraulics and Rijkswaterstaat, Delft and The Hague, The Netherlands, 2nd ed., 184 pp.

**Hörmann**, G., C. Ebrecht, M. Herbst, K. Geffers, W. Kluge, E-W. Reiche, and P. Widmoser, 1995: Auswirkungen einer Temperaturerhoehung auf den Wasserhaushalt der Bornhoeveder Seenkette. *EcoSys*, **5**, 27–49.

**Houghton**, R.A., 1996: Terrestrial sources and sinks of carbon inferred from terrestrial data. *Tellus Series B — Chemical and Physical Meteorology*, **48(4)**, 420–432.

**Houot**, S., V. Bergheaud, J.N. Rampon, and J. Balesdent, 1995: Réponse des matières organiques des sols aux changements atmosphériques globaux. III. Thermodépendance de la minéralisation des fractions de matière organique de biodégradabilité différente. In: *Ecosystèmes et Changements globaux* [Perrier, A. and B. Saugier, (eds.)]. INRA, Paris, France, **8**, 87–90.

**Hunt**, R., D.W. Hand, M.A. Hannah, and A.M. Neal, 1991: Response to $CO_2$ enrichment in 27 herbaceous species. *Functional Ecology*, **5**, 410–421.

**Hunt**, R., D.W. Hand, M.A. Hannah, and A.M. Neal, 1993: Further responses to $CO_2$ enrichment in British herbaceous species. *Functional Ecology*, **7**, 661–668.

**Huntley**, B. 1991. How plants respond to climate change: migration rates, individualism and the consequences for plant communities. *Annals of Botany*, **67 (Supplement 1)**, 15–22.

**Huntley**, B. and T. Webb III, 1989: Migration: species responses to climatic variations caused by changes in the earth's orbit. *Journal of Biogeography*, **16**, 5–19.

**Hurrell**, J.W. and H. Van Loon, 1997: Decadal variations in climate associated with the North Atlantic oscillation. *Climatic Change*, **36**, 301–326.

**IDNDR**, 1994: *Natural Disasters in the World —Statistical Trends on Natural Disasters*, National Land Agency, Tokyo, Japan.

**Innes**, J.L., 1993: New perspectives in forestry: a basis for a future forest management policy in Great Britain? *Forestry*, **66**, 395–421.

**Innes**, J.L., 1994: Climatic sensitivity of temperate forests. *Environmental Pollution*, **83**, 237–243.

**IPCC**, 1990: *Climate Change, The IPCC Impacts Assessment.* Australian Government Publishing Service, Canberra, Australia, 268 pp.

**IPCC**, 1996. *Climate Change 1995: The Science of Climate Change. Contribution of Working Group I to the Second Assessment Report of the Intergovernmental Panel on Climate Change* [Houghton, J.J., L.G. Meiro Filho, B.A. Callander, N. Harris, A. Kattenberg, and K. Maskell (eds.)]. Cambridge University Press, Cambridge, United Kingdom and New York, NY, USA, 572 pp.

- Nicholls, N., G.V. Gruza, J. Jouzel, T.R. Karl, L.A. Ogallo, and D.E. Parker, Chapter 3. *Observed Climate Variability and Change*, pp. 133–192.
- Kattenberg, A., F. Giorgi, H. Grassl, G.A. Meehl, J.F.B. Mitchell, R.J. Stouffer, T. Tokioka, A.J. Weaver, and T.M.L. Wigley, Chapter 6. *Climate Models —Projections of Future Climate*, pp. 289–357.

**IPCC**, 1996. *Climate Change 1995: Impacts, Adaptations, and Mitigation of Climate Change: Scientific-Technical Analyses. Contribution of Working Group II to the Second Assessment Report of the Intergovernmental Panel on Climate Change* [Watson, R.T., M.C. Zinyowera, and R.H. Moss (eds.)]. Cambridge University Press, Cambridge, United Kingdom and New York, NY, USA, 880 pp.

- Kirschbaum, M.U.F. and A. Fischlin, Chapter 1. *Climate Change Impacts on Forests*, pp. 95–130.
- Bijlsma, L, Chapter 9. *Coastal Zones and Small Islands*, pp. 289–324.
- Arnell, N., B. Bates, H. Lang, J.J. Magnuson, and P. Mulholland, Chapter 10. *Hydrology and Freshwater Ecology*, pp. 325–364.
- Moreno, R.A. and J. Skea, Chapter 11. *Industry, Energy, and Transportation: Impacts and Adaptation*, pp. 365–398.
- Scott, M.J., Chapter 12. *Human Settlements in a Changing Climate: Impacts and Adaptation*, pp. 399–426.
- Reilly, J., Chapter 13. *Agriculture in a Changing Climate: Impacts and Adaptation*, pp. 427–467.
- Kaczmarek, Z., Chapter 14. *Water Resources Management*, pp. 469–486.
- Dlugolecki, A.F., Chapter 17. *Financial Services*, pp. 539–560.

**IPCC**, 1996. *Climate Change 1995: Economic and Social Dimensions of Climate Change. Contribution of Working Group III to the Second Assessment Report of the Intergovernmental Panel on Climate Change* [Bruce, J.P., H. Lee, and E.F. Haites (eds.)]. Cambridge University Press, Cambridge, United Kingdom and New York, NY, USA, 448 pp.

- Pearce, D.W., W.R. Cline, A.N. Achanta, S. Fankhauser, R.K. Pachauri, R.S.J. Tol, and P. Vellinga, Chapter 6. *The Social Costs of Climate Change: Greenhouse Damage and Benefits of Control*, pp. 179–224.
- Weyant, J., Chapter 10. *Integrated Assessment of Climate Change: An Overview and Comparison of Approaches and Results*, pp. 367–396.

**Jimenez**, J.A. and A. Sanchez-Arcilla, 1997: Physical impacts of climatic change on deltaic coastal systems (II): driving terms. *Climatic Change*, **35**, 95–118.

**Johansson**, M.-B., 1994: Decomposition rates of Scots pine needle litter related to soil properties, litter quality, and climate. *Canadian Journal of Forest Research*, **24**, 1771–1781.

**Johansson**, M.-B., B. Berg, and V. Meentemeyer, 1995: Litter mass-loss rates in late stages of decomposition in a climatic transect of pine forests. Long-term decomposition in a Scots pine forest. IX. *Canadian Journal of Botany*, **73**, 1509–1521.

**Jones**, M.B., M. Jongen, and T. Doyle, 1996: Effects of elevated carbon dioxide concentrations on agricultural grassland production. *Agricultural and Forest Meteorology*, **79(4)**, 243–252.

**Jones**, R. G., J.M. Murphy, M. Noguer, and A.B. Keen, 1997: Simulation of climate change over Europe using a nested regional-climate model. 2. Comparison of driving and regional model responses to a doubling of carbon dioxide. *Quarterly Journal of the Royal Meteorological Society* (in press).

**Kalkstein**, L.S., 1993: Health and climate change: direct impacts in cities. *Lancet*, **342**, 1397–1399.

**Kane**, S., J. Reilly, and J. Tobey, 1992: An empirical study of the economic effects of climate change on world agriculture. *Climatic Change*, **21**, 17–35.

**Kankaala**, P., A. Ojala, T. Tulonen, J. Haapamäki, and L. Arvola, 1996: Impact of climate change on carbon cycle in freshwater ecosystems. In: *The Finnish Research Programme on Climate Change. Final Report.* [Roos, J. (ed.)]. Publications of the Academy of Finland, Helsinki, April 1996, pp. 196–201.

**Karl**, T.R., P.D. Jones, R.W. Knight, G. Kukla, N. Plummer, V. Razuvayev, K.P. Gallo, J. Lindseay, R.J. Charlson, and T.C. Peterson, 1993: Asymmetric trends of daily maximum and minimum temperature. *Bull. American Meteorol. Soc.*, **74**, 1007–1023.

**Katsouyanni**, K., A. Pantazopoulu, G. Touloumi, I. Tselepidaki, K. Moustris, D. Asimakopoulos, G. Poulopoulou, and D. Trichopolous, 1993: Evidence of interaction between air pollution and high temperatures in the causation of excess mortality. *Arch. Env. Health*, **48**, 235–242.

**Kawasaki**, T., S. Tanaka, Y. Toba, and A. Taniguchi (eds.), 1991: *Long-Term Variability of Pelagic Fish Populations and Their Environment.* Pergamon Press, Tokyo, Japan, 402 pp.

**Kellomäki**, S., H. Hänninen, and M. Kolström, 1995: Computations on frost damage to Scots pine under climatic warming in Boreal conditions. *Ecological Applications,* **5**, 42–52.

**Kellomäki**, S., T. Karjalainen, and H. Vaisanen, 1997: More timber from boreal forests under changing climate? *Forest Ecology and Management,* **94(1–3)**, 195–208.

**Kellomäki**, S. and M. Kolström, 1994: The influence of climate change on the productivity of Scots pine, Norway spruce, Pendula birch and Pubescent birch in southern and northern Finland. *Forest Ecology and Management,* **65**, 201–217.

**Kellomäki**, S. and H. Väisänen, 1996: Model computations on the effect of rising temperature on soil moisture and water availability in forest ecosystems dominated by Scots pine in the Boreal zone in Finland. *Climatic Change,* **32**, 423–445.

**Kellomäki**, S. and K.-Y. Wang, 1996: Photosynthetic responses to needle water potentials in Scots pine after a four-year exposure to elevated $CO_2$ and temperature. *Tree Physiology,* **16**, 765–772.

**Kelly**, M.P., 1991: Global Warming: Implications for the Thames Barrier and Associated Defences. In: *Impact of Sea Level Rise on Cities and Regions.* Proceedings of the First International Meeting "Cities on Water," Venice, December 1989 [Frasetio, R. (ed.)]. Marsilio Editori, Venice, Italy, pp. 93–98.

**Kennedy**, V.S., 1990: Anticipated effects of climate change on estuarine and coastal fisheries. *Fisheries,* **15(6)**, 16–24.

**Kienast**, F., 1991: Simulated effects of increasing atmospheric $CO_2$ and changing climate on the successional characteristics of Alpine forest ecosystems. *Landscape Ecology,* **5**, 225–238.

**Kienast**, F., B. Brzeziecki, and O. Wildi, 1996: Long-term adaptation potential of Central European mountain forests to climate change: a GIS-assisted sensitivity assessment. *Forest Ecology and Management,* **80**, 133–153.

**Kinney**, K.K. and R.L. Lindroth, 1997: Responses of three deciduous tree species to atmospheric $CO_2$ and soil $NO_3$- availability. *Revue Canadienne de Recherche Forestière,* **27(1)**, 1–10.

**Knudsen**, A.B., R. Romi, and G. Majori, 1996: Occurrence and spread in Italy of Aedes albopictus, with implications for its introduction into other parts of Europe. *J. Am. Mosq. Control Assoc.,* **June 12(2 Pt 1)**, 177–183.

**Körner**, C., 1993: $CO_2$ fertilisation: the great uncertainty in future vegetation development. In: *Vegetation dynamics and global change* [Solomon, A.M. and H.H. Shugart (eds.)]. Chapman and Hall, New York, NY, USA, pp. 53–70.

**Körner**, C., 1996: The response of complex multispecies systems to elevated $CO_2$. In: *Global change and terrestrial ecosystems* [Walker, B.H. and W.L. Steffen (eds.)]. Cambridge University Press, Cambridge, United Kingdom, pp. 20–42.

**Körner**, C. and F. Miglietta, 1994: Long term effects of naturally elevated $CO_2$ on mediterranean grassland and forest trees. *Oecologia*, **99**, 343–351.

**Körner**, C., S. Pelaez-Riedl, and A.J.E. Van Bel, 1995: $CO_2$ responsiveness of plants: a possible link to phloem loading. *Plant, Cell and Environment,* **18**, 595–600.

**Koster**, M.J. and R. Hillen, 1995: Combat Erosion by Law: Coastal Defence Policy for The Netherlands. *Journal of Coastal Research*, **11**, 1221–1228.

**Kramer**, K., 1994: A modelling analysis of the effects of climatic warming on the probability of spring frost damage to tree species in the Netherlands and Germany. *Plant, Cell and Environment,* **17**, 367–377.

**Kramer**, K., 1995: Phenotypic plasticity of the phenology of seven European tree species in relation to climatic warming. *Plant, Cell and Environment,* **18**, 93–104.

**Kräuchi**, N., 1995: Application of the model FORSUM to the Solling spruce site. *Ecological Modelling,* **83**, 219–228.

**Kräuchi**, N. and F. Kienast, 1993: Modelling subalpine forest dynamics as influenced by a changing environment. *Water, Air, and Soil Pollution,* **68**, 185–197.

**Kuoppomaeki**, P., 1996a: *Impacts of Climate Change from a Small Nordic Country Perspective.* ETLA — The Research Institute of the Finnish Economy, Series B119, Helsinki, Finland, 156 pp.

**Kuoppomaeki**, P., 1996b: The Impacts of Climate Change on the Finnish Economy. In: *The Finnish Research Programme on Climate Change — Final Report* [Roos, J. (ed.)]. Publications of the Academy of Finland, April 1996, pp. 460–465.

**Kwadijk**, J. and H. Middelkoop, 1994: Estimation of Impact of Climate Change on the Peak Discharge Probability of the River Rhine. *Climatic Change*, **27**, 199–224.

**Lakida**, P., S. Nilsson, and A. Shvidenko, 1997: Forest phytomass and carbon in European Russia. *Biomass & Bioenergy,* **12(2)**, 91–99.

**Lämås**, T. and C. Fries, 1995: Emergence of a biodiversity concept in Swedish forest policy. *Water, Air, and Soil Pollution,* **82**, 57–66.

**Langford**, I.H. and G. Bentham, 1995: The potential effects of climate change on winter mortality in England and Wales. *International Journal of Biometeorology,* **38**, 136–145.

**Larsen**, J.A., 1980: *The Boreal Ecosystem.* Academic Press, New York, NY, USA.

**Leemans**, R., 1996: Incorporating land-use change in Earth system models illustrated by IMAGE 2. In: *Global change and terrestrial ecosystems.* [Walker, B.H. and W.L. Steffen (eds.)]. Cambridge University Press, Cambridge, United Kingdom, pp. 484–510.

**Legge**, A.H., L. Grünhage, M. Nosal, H.-J. Jäger, and S.V. Krupa, 1996: Ambient ozone and adverse crop response: an evaluation of North American and European data as they relate to exposure indices and critical levels. In: *Exceedance of critical loads and levels. Spatial and temporal interpretation of elements in landscape sensitive to atmospheric pollutants* [Knoflacher, M., J. Schneider, and G. Soja (eds.)]. Umweltbundesamt, Vienna, Austria, pp. 18–46.

**LeHouérou**, H.N., 1995: Climate change, drought and desertification. *Journal of Arid Environments,* **34(2)**, 133–185.

**Lelyakin**, A.L., A.O. Kokorin, and I.M. Nazarov, 1997: Vulnerability of Russian forests to climate changes. Model estimation of $CO_2$ fluxes. *Climatic Change,* **36(1–2)**, 123–133.

**Lenhart**, S., S. Huq, L.J. Mata, I. Nemesova, S. Toure, and J.B. Smith (eds.), 1996: *Vulnerability and Adaptation to Climate Change—A Synthesis of Results from the U.S. Country Studies Program.* Interim Report. January 1996, U.S. Country Studies Program, Washington, DC, USA, 366 pp.

**Lindgren**, E. and R. Gustafson, submitted: Climate and tick-borne encephalites in Sweden.

**Lindner**, M., H. Bugmann, P. Lasch, M. Flechsig, and W. Cramer, 1997: Regional impacts of climatic change on forests in the state of Brandenburg, Germany. *Agricultural and Forest Meteorology,* **84(1–2)**, 123–135.

**Lindner**, M., P. Lasch, and W. Cramer, 1996: Application of a forest succession model to a continentality gradient through Central Europe. *Climatic Change,* **34(2)**, 191–199.

**Luxmoore**, R.J., S.D. Wullschleger, and P.J. Hanson, 1993: Forest responses to $CO_2$ enrichment and climate warming. *Water, Air, and Soil Pollution,* **70**, 309–323.

**MacGillivray**, C.W. and J.P. Grime, 1995: Genome size predicts frost resistance in British herbaceous plants: implications for rates of vegetation response to global warming. *Functional Ecology,* **9**, 320–325.

**Malanson**, G.P. and D.M. Cairns, 1997: Effects of dispersal, population delays, and forest fragmentation on tree migration rates. *Plant Ecology,* **131**, 67–79.

**Mann**, K.H., 1992: Physical influences on biological processes: how important are they? Benguela trophic functioning [Payne, A.L., K.H. Mann, and R. Kilborn (eds.)]. *S. Afr. J. Mar. Sci.,* **12**, 107–121.

**Mann**, K.H. and K.F. Drinkwater, 1994: Environmental influences on fish and shellfish production in the Northwest Atlantic. *Envir. Rev.,* **2**, 16–32.

**Mann**, K.H. and J.R.N. Lazier, 1991: *Dynamics of Marine Ecosystems Biological Physical Interactions in the Oceans.* Blackwell Scientific Publications, Boston, MA, USA, 466 pp.

**Martens**, W.J.M., 1997: *Health Impacts of Climate Change and Ozone Depletion: An Eco-Epidemiological Modelling Approach.* Maastricht University, The Netherlands, 158 pp.

**Matea**, M.L., 1995: La reforma de la politica agricola comun: el sistema agromonetario. *Boletin Economico*, Banco de Espana, Mayo 1995, pp. 33–43.

**Miglietta**, F. and A. Raschi, 1993: Studying the effect of elevated $CO_2$ in the open in a naturally enriched environment in central Italy. *Vegetatio,* **104/105**, 391–402.

**Miglietta**, F., M. Tanasescu, and A. Marica, 1995: The expected effects of climate change on wheat development. *Global Change Biology,* **1(6)**, 407–415.

**Mills**, E.L., J.H. Leach, J.T. Carlton, and C.L. Secor, 1994: Exotic species and the integrity of the Great Lakes. *Biosci.,* **44(1)**, 666–676.

**Mitchell**, J.F.B., R.A. Davis, W.J. Ingram, and C.A. Senior, 1995: On surface temperature, greenhouse gases and and aerosols: models and observations. *J. Climate,* **10**, 2364–2386.

**MOPT**, 1993: *Plan Hidrologico Nacional. Memoria.* Ministerio de Obras Publicas y Transportes. Secretaria de estado para las Politicas del Agua y el Medio Ambiente. Madrid. *Anteproyecto de Ley del Plan Hidrologico Nacional.* Ministerio de Obras Publicas y Transportes. Secretaria de estado para las Politicas del Agua y el Medio Ambiente. Madrid, Spain.

**Munich Re**, 1993: *Winter Storms in Europe An Analysis of 1990 Losses and Future Loss Potential,* Muenchener Rueckversicherungs-Gesellschaft, Munich, Germany, 55 pp.

**Murray**, M.B., M.G.R. Cannell, and R.T. Smith, 1989: Date of budburst of fifteen tree species in Britain following climatic warming. *Journal of Applied Ecology*, **26**, 693–700.

**Murray**, M.B., R.I. Smith, I.D. Leith, D. Fowler, H.S.J. Lee, A.D. Friend, and P.G. Jarvis, 1994: Effects of elevated $CO_2$, nutrition and climatic warming on bud phenology in Sitka spruce (*Picea sitchensis*) and their impact on the risk of frost damage. *Tree Physiology*, **14**, 691–706.

**Myneni**, R.B., C.D. Keeling, C.J. Tucker, G. Asrar, and R.R. Nemani, 1997: Increased plant growth in the northern high latitudes from 1981 to 1991. *Nature*, **386**, 698–702.

**Nicholls**, R.J., 1995: Synthesis of Vulnerability Analysis Studies. In: *Preparing to Meet the Coastal Challenges of the 21$^{st}$ Century.* Proceedings of the World Conference, Noordwijk, November 1993, Rijkswaterstaat, The Hague, The Netherlands, pp. 181–216.

**Nicholls**, R.J. and F.M.J. Hoozemans, 1996: The Mediterranean: Vulnerability to Coastal Implications of Climate Change. *Ocean and Coastal Management,* **31**, 105–132.

**Norby**, R.J., E.G. O'Neill, and S.D. Wullschleger 1995: Belowground responses to atmospheric carbon dioxide in forests. In: *Carbon forms and functions in forest soils* [McFee, W.W. and J.M. Kelly (eds.)]. Soil Science Society of America, Madison, WI, USA, pp. 397–418.

**Ntiba**, M.J. and D. Harding, 1993: The food and the feeding habits of the long rough dab Hippoglossoides platessoides (Fabricius, 1780), in the North Sea. *Neth. J. Sea Res.,* **31**, 189–199.

**Økland**, R.H., 1995a: Persistence of vascular plants in a Norwegian boreal coniferous forest. *Ecography,* **18**, 3–14.

**Økland**, R.H., 1995b: Changes in the occurrence and abundance of plant species in a Norwegian boreal coniferous forest, 1988–1993. *Nordic Journal of Botany,* **15**, 415–438.

**Oliver**, J.E. and R.W. Fairbridge, 1987: *The encyclopedia of climatology.* Van Nostrand Reinhold Company, New York, NY, USA, 986 pp.

**Onate**, J.J. and A. Pou, 1996: Temperature variations in Spain since 1901: A preliminary analysis. *Int. Journal of Climatology,* **16(7)**, 805–815.

**Parry**, M.L., T.R. Carter, and N.T. Konijin (eds.), 1988: *The Impact of Climatic Variations on Agriculture. Vol. 2, Assessments in Semi-Arid Regions*. Kluwer Academic Publishers, Dordrecht, The Netherlands, 764 pp.

**Parry**, M.L., T.R. Carter, J.R. Porter, G.J. Kenny, and P.A. Harrison, 1992: *Climate Change and Agricultural Suitability in Europe*. Environmental Change Unit, University of Oxford. Report No. 1. Oxford, United Kingdom.

**Parry**, M.L., J.E. Hossell, P.J. Jones, T. Rehman, R.B. Tranter, J.S. Marsh, C. Rosenzweig, G. Fischer, I.G. Carson, and R.G.H. Bunce, 1996: Integrating global and regional analyses of the effects of climate change: a case study of land use in England and Wales. *Climatic Change,* **32**, 185–198.

**Parsons**, A.N., M.C. Press, P.A. Wookey, J.M. Welker, C.H. Robinson, T.V. Callaghan, and J.A. Lee, 1995: Growth responses of Calamgrostis lapponica to simulated environmental change in the Sub-arctic. *Oikos,* **72**, 61–66.

**Parsons**, A.N., J.M. Welker, M.A. Wookey, M.C. Press, T.V. Callaghan, and J.A. Lee, 1994: Growth responses of four sub-Arctic dwarf shrubs to simulated environmental change. *Journal of Ecology,* **82**, 307–318.

**Peerbolte**, E.B., J.G. de Ronde, L.P.M. de Vrees, M. Mann, and G. Baarse, 1991: *Impact of Sea Level Rise on Society: A Case Study for the Netherlands.* Delft Hydraulics and Rijkswaterstaat, Delft and The Hague, The Netherlands, 404 pp.

**Peiris**, D.R., J.W. Crawford, C. Grashoff, R.A. Jefferies, J.R. Porter, and B. Marshall, 1996: A simulation study of crop growth and development under climate change. *Agricultural and Forest Meteorology.* **79(4)**, 271–287.

**Penning-Rowsell**, E., J.W. Handmer, and S. Tapsell, 1996: Extreme Events and Climate Change: Floods. In: *Climate Change and Extreme Events — Altered Risk, Socio-economic Impacts and Policy Responses* [Downing, T.E., A.A. Olsthoorn, and R.S.J. Tol (eds.)]. Institute for Environmental Studies, Vrije Universiteit and Environmental Change Unit, University of Oxford, Amsterdam, The Netherlands and Oxford, England, pp. 97–128.

**Polovina**, J.J., G.T. Mitchum, and G.T. Evans, 1995: Decadal and basin-scale variation in mixed layer depth and the impact on biological production in the Central and North Pacific, 1960–88. In: *Deep Sea Research* (in press).

**Posch**, M., J.P. Hettelingh, J. Alcamo, and M. Krol, 1996: Integrated scenarios of acidification and climate change in Asia and Europe. *Global Environmental Change — Human and Policy Dimensions,* **6(4)**, 375–394.

**Prentice**, I.C., M.T. Sykes, and W. Cramer, 1991: The possible dynamic response of northern forests to global warming. *Global Ecology and Biogeography Letters,* **1**, 129–135.

**Proe**, M.F., S.M. Allison, and K.B. Matthews, 1996: Assessment of the impact of climate change on the growth of Sitka spruce in Scotland. *Canadian Journal of Forest Research,* **26**, 1914–1921.

**Rioux**, J-A., J. Boulker, G. Lanotte, R. Killick-Hendrick, and A. Martini-Dumas, 1985: Ecologie des leishmanioses dans le sud de France. 21 — Influence de la température sur le développement de Leishmania infantum Nicolle, 1908 chez Phlebotomus ariasi Tonnoir, 1921. Étude experimentale. *Annales de Parasitologie,* **60(3)**, 221–229.

**Rooney**, C., A.J. McMichael, R.S. Kovats, and M. Coleman, 1997: Excess Mortality in England and Wales, and in Greater London, During the 1995 Heatwave. *Journal of Epidemiology and Community Health* (in press).

**Rosenberg**, N.J. and P.R. Crosson, 1991: *Processes for Identifying Regional Influences of and Responses to Increasing Atmospheric $CO_2$ and Climate Change: the MINK Project, An Overview*. Resources for the Future. Dept. of Energy DOE/RL/01830T-H5. Washington, DC, USA, 35 pp.

**Rosenzweig**, C. and M.L. Parry, 1994: Potential impact of climate change on world food supply. *Nature,* **367**, 133–138.

**Rotmans**, J., M. Hulme, and T.E. Downing, 1994: Climate Change Implications for Europe: An Application of the ESCAPE Model. *Global Environmental Change*, **4(2)**, 97–124.

**Rounsevell**, M.D.A. and A.P. Brignall, 1994: The potential effects of climate change on autumn soil tillage opportunities in England and Wales. *Soil and Tillage Research,* **32**, 275–289.

**Rounsevell**, M., P. Brignall, and P. A. Siddons, 1996: Potential climate changes effects on the distribution of agricultural grassland in England and Wales. *Soil Use and Management,* **12(1)**, 44-51.

**Saarikko**, R.A. and T.R. Carter, 1996: Estimating Regional Spring Wheat Development and Suitability in Finland under Climate Warming. *Climate Research,* **7**, 243–252.

**Sanchez-Arcilla**, A. and J.A. Jimenez, 1997: Physical impacts of climatic change on deltaic coastal systems (I): an approach. *Climatic Change*, **35**, 71–93.

**Schäppi**, B. and C. Körner, 1996: Growth responses of an alpine grassland to elevated $CO_2$. *Oecologia,* **105**, 43–52.

**Schäppi**, B. and C. Körner, 1997: In situ effects of elevated $CO_2$ on the carbon and nitrogen status of alpine plants. *Functional Ecology,* **11(3)**, 290–299.

**Schoenwiese**, C.-D., J. Rapp, T. Fuchs, and M. Denhard, 1993: *Klimatrend-Atlas Europa 1891-1990.* ZUF-Verlag, Frankfurt, Germany, 210 pp.

**Schuurmans**, C.J.E., 1996: Climate variability in Europe. In: *Climate Variability and climate change — vulnerability and adaptation.* Proceedings of the Reg. Clim. Workshop, September 1996. Prague, Czech Republic, pp. 25–33.

**Semenov**, M.A. and J.R. Porter, 1995: Climatic variability and the modelling of crop yields. *Agricultural and Forest Meteorology,* **73**, 265–283.

**Sharp**, G.D. and J. Csirke (eds.), 1983: *Proceedings of the Expert Consultation to Examine the Changes in Abundance and Species Composition of Neritic Fish Resources 18–29 April 1983, San Jose, Costa Rica.* FAO Fish. Rep. Ser., Rome, Italy, **291(2–3)**, 1294.

**Shuter**, B.J. and J.R. Post, 1990: Climate, population viability, and the zoogeography of temperate fishes. *Trans. Amer. Fish. Soc.*, **119**, 314–336.

**SILMU**, 1996: *Climate Change and Finland — Summary of the Finnish Research Programme on Climate Change.* Helsinki University Press, Helsinki.

**Sirotenko**, O.D., H.V. Abashina, and V.N. Pavlova, 1997: Sensitivity of the Russian agriculture to changes in climate, $CO_2$ and tropospheric ozone concentrations and soil fertility. *Climatic Change,* **36(1–2)**, 217–232.

**Slovik**, S., 1996: Early needle senescence and thinning of the crown structure of Picea abies as induced by chronic $SO_2$ pollution. 1. Model deduction and analysis — 2. Field data basis, model results and tolerance limits. *Global Change Biology,* **2(5)**, 459–477.

**Soussana**, J.F., E. Casella, and P. Loiseau, 1994: Long-term effects of $CO_2$ enrichment and temperature increase on a temperate grass sward 2. Plant nitrogen budgets and root fraction. *Plant and Soil*, **182(1)**, 101–114.

**Southward**, A.J., 1980: The western English Channel: an inconstant ecosystem? *Nature*, **285**, 361–366.

**Spiecker**, H., K. Mielikäinen, M. Köhl, and J. Skovsgaard (eds.), 1996: *Growth trends in European forests. Studies from 12 countries.* Springer-Verlag, Berlin, Germany, 372 pp.

**Spieksma**, F.T.M., J. Emberlin, M. Hjelmroos, S. Jager, and R.M. Leuschner, 1995: Atmospheric birch (Betula) pollen in Europe: trends and fluctuations in annual quantities and the starting dates of the seasons. *Grana,* **34**, 51–57.

**Sterr**, H. and F. Simmering, 1996: Die Kuestenregionen im 21. Jahrhundert. In: *Beitrage zur aktuellen Kuestenforschung* [Sterr, H. and C. Preu (eds.)]. Vechtaer Studien zür Angewandten Geographie und Regionalwissenschaft (VSAG), **18**, 181–188.

**Stirling**, C.M., P.A. Davey, T.G. Williams, and S.P. Long, 1997: Acclimation of photosynthesis to elevated $CO_2$ and temperature in five British native species of contrasting functional type. *Global Change Biology*, **3(3)**, 237–246.

**Stock**, M. and F. Toth (eds.), 1996: *Moegliche Auswirkungen von Klimaaenderungen auf das Land Brandenburg.* PIK, Potsdam, Germany, 166 pp.

**Straw**, N.A., 1995: Climate change and the impact of the green spruce aphid, Elatobium abietinum (Walker), in the UK. *Scottish Forestry,* **49**, 134–145.

**Sykes**, M.T. and I.C. Prentice, 1995: Boreal forest futures: modelling the controls on tree species range limits and transient responses to climate change. *Water, Air, and Soil Pollution,* **82**, 415–428.

**Sykes**, M.T. and I.C. Prentice, 1996: Climate change, tree species distributions and forest dynamics: a case study in the mixed conifer/northern hardwoods zone of Northern Europe. *Climatic Change,* **34**, 161–177.

**Talkkari**, A. and Hypén, H., 1996: Development and assessment of a gap-type model to predict the effects of climate change on forests based on spatial forest data. *Forest Ecology and Management,* **83**, 217–228.

**Tchamitchian**, M., H. Colas, and J. Roy, 1994: Modélisation des effets à long terme du $CO_2$, de la température et de la fertilisation sur la photosynthèse de la feruille de ray-grass anglais (Lolium perenne, L.). In: *Ecosystèmes et Changements globaux.* [Perrier, A. and B. Saugier (eds.)]. *INRA,* Paris, France, **8**, 99–106.

**Thompson**, K., 1994: Predicting the fate of temperate species in response to human disturbance and global change. In: *Biodiversity, temperate ecosystems, and global change* [Boyle, T.J.B. and C.E.B. Boyle (eds.)]. Springer-Verlag, Berlin, Germany, pp. 61–76.

**Thornley**, J.H.M. and M.G.R. Cannell, 1996: Temperate forest responses to carbon dioxide, temperature and nitrogen: A model analysis. *Plant Cell and Environment,* **19(12)**, 1331–1348.

**Tognetti**, R., A. Giovannelli, A. Longobucco, F. Miglietta, and A. Raschi, 1996: Water relations of oak species growing in the natural $CO_2$ spring of Rapolano (central Italy). *Annales Des Sciences Forestières,* **53**, 475–485.

**Tol**, R.S.J., 1996: The Weather Insurance Sector. In: *Climate Change and Extreme Events — Altered Risk, Socio-economic Impacts and Policy Responses* [Downing, T.E., A.A. Olsthoorn, and R.S.J. Tol (eds.)]. Institute for Environmental Studies, Vrije Universiteit and Environmental Change Unit, University of Oxford, Amsterdam, Netherlands and Oxford, United Kingdom, pp. 209–250.

**Turner**, R.K., P. Doktor, and W.N. Adger, 1995: Assessing the Costs of Sea Level Rise. *Environment and Planning,* **A27**, 1777–1796.

**Tyler**, A.L., D.C. Macmillan, and J. Dutch, 1996: Models to predict the General Yield Class of Douglas fir, Japanese larch and Scots pine on better quality land in Scotland. *Forestry,* **69**, 13–24.

**UK Department of the Environment**, 1991: *The Potential Effects of Climate Change in the United Kingdom.* United Kingdom Climate Change Impacts Review Group, HMSO, London, United Kingdom, 124 pp.

**UK Department of the Environment**, 1996: *Review of the Potential Effects of Climate Change in the United Kingdom.* HMSO, London, United Kingdom, 247 pp.

**UK Ministry of Agriculture**, **Fisheries and Food**, The Welsh Office, Association of District Councils, English Nature and the National Rivers Authority, 1995: *Shoreline Management Plans: A Guide for Coastal Defence Authorities*, Ministry of Agriculture, Fisheries and Food, London, United Kingdom, 24 pp.

**UNECE**, 1997: *International cooperative programme on assessment and monitoring of acidification of rivers and lakes. The nine year report: Acidification of surface water in Europe and North America — long-term developments (1980s and 1990s).* Programme Centre, Norwegian Institute for Water Research, Oslo, Norway, 168 pp.

**van der Maarel**, E. and M.T. Sykes, 1993: Small-scale plant species turnover in a limestone grassland: the carousel model and some comments on the niche concept. *Journal of Vegetation Science,* **4**, 179–188.

**Wolf**, J., L.G. Evans, M.A. Semenov, H. Eckersten, and A. Iglesias, 1996: Comparison of wheat simulation models under climate change: 1. Model calibration and sensitivity analyses. *Climate Research,* **7(3)**, 253–270.

**Woodward**, F.I., 1993: The lowland-to-upland transition — modelling plant responses to environmental change. *Ecological Applications,* **3**, 404–408.

**Young**, J.W.S., 1997: A framework for the ultimate environmental index putting atmospheric change into context with sustainability. *Environmental Monitoring and Assessment,* **46(1–2)**, 135–149.

**Zaller**, J.G. and J.A. Arnone, 1997: Activity of surface-casting earthworms in a calcareous grassland under elevated atmospheric $CO_2$. *Oecologia,* **111**, 249–254.

# 6

# Latin America


OSVALDO F. CANZIANI (ARGENTINA) AND SANDRA DIAZ (ARGENTINA)

Lead Authors:
*E. Calvo (Peru), M. Campos (Costa Rica), R. Carcavallo (Argentina), C.C. Cerri (Brazil), C. Gay-García (Mexico), L.J. Mata (Venezuela), A. Saizar (Uruguay)*

Contributors:
*P. Aceituno (Chile), R. Andressen (Venezuela), V. Barros (Argentina), M. Cabido (Argentina), H. Fuenzalida-Ponce (Chile), G. Funes (Argentina), C. Galvao (Brazil), A.R. Moreno (Mexico), W.M. Vargas (Argentina), E.F. Viglizzo (Argentina), M. de Zuviría (Bolivia)*

# CONTENTS

# EXECUTIVE SUMMARY

Latin America includes all continental countries of the Americas from Mexico to Chile and Argentina, as well as adjacent seas. The region is highly heterogeneous in terms of climate, ecosystems, human population distribution, and cultural traditions. Most Latin American production activities are based on the region's extensive natural ecosystems. Land use is a major force driving ecosystem change at present; it interacts with climate in complex ways. This complexity makes the task of identifying common patterns of vulnerability to climate change very difficult. Major sectors in which the impacts of climate change could be important are natural ecosystems (e.g., forests, rangelands, wetlands), water resources, coastal zones, agriculture, and human health. The relative importance attributed to each of these projected impacts varies among countries.

Changes in climate over the past century have included a rise in the mean surface temperature, particularly at middle and high latitudes, and changes in precipitation rates and intensities in various countries of the region (e.g., southern Brazil, Paraguay, Argentina). Climate change could modify present conditions, with beneficial or adverse impacts—as presently occurs as a result of the El Niño-Southern Oscillation (ENSO) phenomenon. Natural climate variability at the time scale of seasons to several years has produced significant effects on Latin American countries, suggesting that climate change projections become an important element for national and regional planning. However, climate change should be considered not in isolation but in close interaction with other important factors for development, such as land-use practices and land-use change, population growth, economic situations, and community behavior.

Latin America's geographical location and geomorphology contribute to its large variety of climates, ranging from hyper-arid desert climates to humid tropical forest climates. The regional climate distribution is defined by interactions among the predominant atmospheric circulation patterns and the region's topographical features, radiation budgets, and heat and water balances—which, in turn, depend on the vast range of soil/vegetation types of the region. The extensive central portion of Latin America is characterized largely by humid, tropical conditions; important areas (e.g., in Brazil) are subject to drought, floods, and freezes. Atmospheric circulation and ocean currents are causal factors of extensive deserts in northern Mexico, Peru, Bolivia, Chile, and Argentina.

The relationship between the ENSO phenomenon and changes in precipitation and temperature has been well documented for countries of the Central American isthmus and South America. ENSO events associated with massive fluctuations in the marine ecosystems off the coasts of Ecuador, Peru, and northern Chile (which are among the richest fisheries in the world) would have adverse socioeconomic consequences on fishing and fishmeal production. Experimental El Niño forecasts have been applied, with remarkable success, in Peru and Brazil to reduce economic disruptions in agriculture. Climate variability also determines important changes in the distribution and intensity of rainfall and snow. This variability represents an additional stress on already limited freshwater availability in Chile and western Argentina at latitudes between 25°S and 37°S.

The surface area of Latin America is occupied by natural ecosystems whose genetic resources are among the richest in the world. The Amazon rainforest contains the largest number of animal and plant species in Latin America. Temperate and arid zones in this region—which, until recently, have received less attention—also contain important genetic resources, in terms of wild and domesticated genotypes.

The Latin American contribution to global emissions of greenhouse gases is low at present (approximately 4%). However, potential future impacts of climate and land-use changes could be large and costly for this region. In addition, the release of carbon to the atmosphere as a consequence of massive and continued deforestation in Latin America would have the potential to alter the global carbon balance. On the other hand, some studies suggest that technologically simple adaptation options could improve the capacity for carbon sequestration, as well as economic productivity, in some ecosystems.

Latin American forests—which occupy approximately 22% of the region and represent about 27% of global forest coverage—have a strong influence on local and regional climate, play a significant role in the global carbon budget, contain an important share of all plant and animal species of the region, and are economically very important for national and international markets. Vulnerability studies indicate that forest ecosystems in many countries (e.g., Mexico, countries of the Central American isthmus, Venezuela, Brazil, Bolivia) could be affected by projected climatic changes. Deforestation in the Amazon rainforest is likely to have a negative impact on the recycling of precipitation through evapotranspiration. Rainfall would be markedly reduced, leading to important runoff losses in areas within and beyond this basin.

Rangelands cover about one-third of the land area of Latin America. Rangeland productivity and species composition are directly related to the highly variable amount and seasonal distribution of precipitation; they are only secondarily affected by other climate variables (with the exception of high-temperature persistence in wildfire-prone areas). Temperate grasslands are vulnerable to drought; therefore, livestock production would drop drastically if precipitation decreased substantially or if higher temperatures led to increased evapotranspiration rates. An increased frequency of extreme events is likely to have larger impacts than changes in mean temperature or precipitation. The preservation of large-scale range management units and protected areas may assist migration and recolonization by native species in response to changing environmental conditions.

Mountain ranges and plateaus play an important role in determining Latin America's climate, hydrological cycle, and biodiversity. They are source regions of massive rivers (e.g., the tributary rivers of the Amazonas and Orinoco basins) and represent important foci of biological diversification and endemism—and they are highly susceptible to extreme events. The cryosphere in Latin America is represented by glaciers in the high Andes and three major ice fields in southern South America. Warming in high mountain regions could lead to the disappearance of significant snow and ice surfaces. In addition, changes in atmospheric circulation resulting from the ENSO phenomenon and climate change could modify snowfall rates—with a direct effect on the seasonal renewal of water supply—and surface and underground runoff in piedmont areas. This could affect mountain sports and tourist activities, which represent an important source of income in some economies. Glaciers are melting at an accelerated rate in the Venezuelan and Peruvian Andes; however, the largest glaciers in the Patagonian Andes would continue to exist into the 22nd century.

Approximately 35% of the world's continental water (freshwater) is found in Latin America. However, its distribution within and among countries is highly variable. Freshwater systems (i.e., rivers, lakes, reservoirs, nontidal wetlands) and their ecosystems are potentially very sensitive to climate change and vulnerable to short-term fluctuations in climate, such as those associated with the ENSO phenomenon.

In the humid tropics, extreme precipitation events would increase the number of reservoirs silting up well before their design lives have been reached. Other areas affected by the impact of climate change on water resources could be those that rely on freshwater ecosystems (i.e., lakes and inland wetlands and their biota), including commercial and subsistence fisheries. Climate change will interact strongly with anthropogenic changes in land use, waste disposal, and water extraction; regional water resources will become increasingly stressed by higher demands to meet the needs of growing populations and economies, as well as by temperature increases. Conflicts may arise among users and regions—and even among Latin American countries that share common river basins. The effects of climate change on agricultural demands for water, particularly for irrigation, will depend significantly on changes in agricultural potential, prices of agricultural produce, and water costs.

The vulnerability of Latin American countries to climate change strongly depends on the impacts of climate change on water availability, as shown by studies performed under the auspices of the U.S. Country Studies Program (USCSP) and the Global Environment Facility (GEF). Changes in precipitation rates have consequences for hydropower production in Costa Rica, Panama, and western Argentina. ENSO events during recent decades have led to a significant reduction in runoff and consequently to higher dependence on thermal energy production, especially in areas with few energy alternatives. Alterations in hydrological balance resulting from climate change could reduce already impaired water supply and distribution systems in major Latin American cities and rural areas. According to vulnerability studies carried out in Mexico and Peru, the combined impacts of global warming and population growth could result in major reductions in water availability in both countries in coming decades. Monitoring of the hydrological cycle, sensitivity studies, development planning, and improved water management practices are key elements in preparing for projected water shortages.

Studies of vulnerability to sea-level rise have suggested that countries of the Central American isthmus as well as Venezuela and Uruguay could suffer adverse impacts leading to losses of coastal land and biodiversity, saltwater intrusion, and infrastructure damage. Impacts likely would be multiple and complex, with major economic implications.

Agricultural lands (excluding pastures) represent approximately 19% of the land area of Latin America. Over the past 40 years, the contribution of agriculture to the gross domestic product (GDP) of Latin American countries has been on the order of 10%. Agriculture remains a key sector in the regional economy, however, because it occupies an important segment (30–40%) of the economically active population. It also is very important for the food security of the poorest sectors of the population. Studies in Brazil, Chile, Argentina, and Uruguay based on GCMs and crop models project decreased yields of a number of crops (e.g., barley, grapes, maize, potatoes, soybeans, wheat)—even when the direct effects of carbon dioxide ($CO_2$) fertilization and the implementation of moderate adaptation measures at the farm level are considered. Global warming also could enhance the negative impacts of animal and plant diseases and pests, with further negative effects on production.

Large alterations in Latin American ecosystems resulting from climate change impacts would have the potential to endanger the livelihoods of subsistence farmers and pastoral peoples, who make up a large portion of the rural populations of the Andean plateaus and tropical and subtropical forest areas.

Projected changes in climate could increase the impacts of already serious chronic malnutrition and diseases affecting a large sector of the Latin American population. The geographical distribution of vector-borne diseases (e.g., malaria, dengue, Chagas') and infectious diseases (e.g., cholera) would expand southward and to higher elevations if temperature and precipitation increase. Increasing pollution and high concentrations of ground-level ozone, aggravated by increasing surface temperatures and higher solar radiation rates, could negatively affect human health and welfare, especially in urban areas.

Although climate change may bring benefits for certain regions in Latin America, increasing environmental deterioration resulting from the misuse of land might be aggravated by the impacts of climate change on water availability and agricultural lands, as a result of coastal inundation stemming from sea-level rise and riverine and flatland flooding. Socioeconomic and health problems could be exacerbated in critical areas, fostering massive migrations of rural and coastal populations and deepening national and international conflicts.

Additional efforts in monitoring, research, and sensitivity analyses are needed to allow decision makers and other stakeholders to understand the potential consequences of climate change and variability, take advantage of potential benefits, minimize negative impacts, and seek adequate adaptation options. Specific priorities include organization and/or updating of biogeophysical observation networks and monitoring systems, development of regional circulation models, performance of sensitivity analyses of key sectors—from ecosystems to infrastructure—that appear to be most vulnerable in each country or subregion, and development of suitable adaptation options. These efforts include the creation of new technologies and the adaptation of existing technologies generated within or outside the region. Promising options include horizontal cooperation initiatives among countries of this region and assistance from countries of other regions of the world—particularly Canada and the United States—as well as countries of the temperate Southern Hemisphere that already are integrated into the Valdivia Group.

## 6.1. Regional Characteristics

The Latin American region spans a vast geographic and ecological range, from the subtropics of the Northern Hemisphere to the subpolar tip of the South American subcontinent; the region's largest portion lies in the tropical zone. The region consists of 20 independent states and the territory of French Guiana (an overseas department of France) (see Box 6-1 and Figure 6-1).

**Box 6-1. The Latin America Region**

| | |
|---|---|
| Argentina | Guyana |
| Belize | Honduras |
| Bolivia | Mexico |
| Brazil | Nicaragua |
| Chile | Panama |
| Colombia | Paraguay |
| Costa Rica | Peru |
| Ecuador | Suriname |
| El Salvador | Uruguay |
| French Guiana | Venezuela |
| Guatemala | |

The northern part of the region includes Mexico and the countries of the Central American isthmus, which are characterized by a broken relief of mountain ranges, tablelands or plateaus, shallow depressions, and numerous valleys—including desert (Baja California and Sonora) and semi-arid areas (Mexican highlands and coastal plains). South America is vertebrated by the Andean cordillera, a continuous mountain chain about 9,000 km long. The massive Andes host important glaciers and volcanoes, as well as a number of high plateaus. These high plateaus were the cradle of ancient civilizations and today host the region's largest rural population. This remarkable orographic barrier and the large oceans surrounding the subcontinent greatly influence the region's climate and land-use patterns. The lower eastern slopes of the Andes, together with the Guianas Highlands and the Brazilian Plateau, make up the habitats of the Colombia-Venezuela Llanos and the Amazon rainforest—the most important humid forest of the world.

Major biogeographical areas to the south of the Amazon forest include the woody Cerrado/Cerradinho and Chaco ecosystems and, further south, the Pampean region of Argentina. Extending southward from a latitude of 40°S, the Patagonian tableland—a region of vast steppe-like plains—rises westward from about 100 m on the coastline to about 1,000 m at the base of the Andes, with a surface area of about 670,000 km$^2$. Other important ecosystems in the region are the Yunga valleys in Bolivia and tropical/subtropical forests in Paraguay, Brazil, and Argentina. In recent decades, these forests have been subject to strong anthropogenic pressures to increase agricultural land area. For instance, only 4% of the tropical forest originally covering eastern Paraguay remained in the mid-1950s.

*Table 6-1: Amazon watershed area.*

| Country | Watershed (km$^2$) | % of National Territory | % of Watershed |
|---|---|---|---|
| Bolivia | 824,000 | 75.0 | 11.2 |
| Brazil | 4,982,000 | 58.5 | 67.8 |
| Colombia | 406,000 | 36.0 | 5.5 |
| Ecuador | 123,000 | 45.0 | 1.7 |
| Guyana | 5,870 | 2.7 | 0.08 |
| Peru | 936,751 | 74.4 | 13.0 |
| Venezuela | 53,000 | 5.8 | 0.7 |
| **Total** | 7,350,621 | | 100.0 |

Source: Commission on Development and Environment for Amazonia (CDEA), 1982.

South America also has important coastal and inland wetlands with very high biodiversity; the combined biodiversity of these ecosystems and of Latin America's tropical, subtropical, and temperate ecosystems represents the world's largest genetic pool. Annex D of this report provides the number of known and endemic mammal, bird, and flowering plant species in each of the Latin American countries. The most important freshwater wetlands are those of El Pantanal (Brazil) and Iberá (Argentina). These wetlands are associated with the large international Rio de la Plata basin (embracing about 5.1 million km$^2$), whose main component rivers—Paraguay, Paraná, and Uruguay—have a discharge of 79,400 m$^3$/s. Integration with the Orinoco (70,000 m$^3$/s) and Amazonas (180,000 m$^3$/s) basins makes this area the largest running surface-water system in the world (329,400 m$^3$/s), accounting for approximately 35% of global runoff and covering an area of about 12 million km$^2$. These very important river systems could be adversely affected by climate change and mismanagement of associated ecosystems, particularly further deforestation and, inter alia, deterioration of the buffer capacity of inland wetlands. The importance of the Amazon basin is depicted in Table 6-1. Table 6-2 provides the estimated deforestation rate of each country within this basin.

The total surface of the Latin American region is approximately 19.93 million km$^2$; the region's unevenly distributed population of approximately 446.2 million people has varied growth rates (see Annex D). Although the percentage of urban population is the largest in the developing world (73.6%), there are still countries—for example, Bolivia, Ecuador, Paraguay, and some in Central America—that have large rural populations. Countries with high percentages of population living in cities and megalopolises are Mexico (75.2%), Venezuela (85.8%), Chile (84.4%), Argentina (88.3%), and Uruguay (94.6%). Metropolitan Mexico City, for example, has about 24 million people, and greater Sao Paulo's population is expected to reach 23 million people by the end of this century (Book of the Year, 1995; Encyclopaedia Britannica, 1996). Socioeconomic and educational conditions vary widely among Latin American

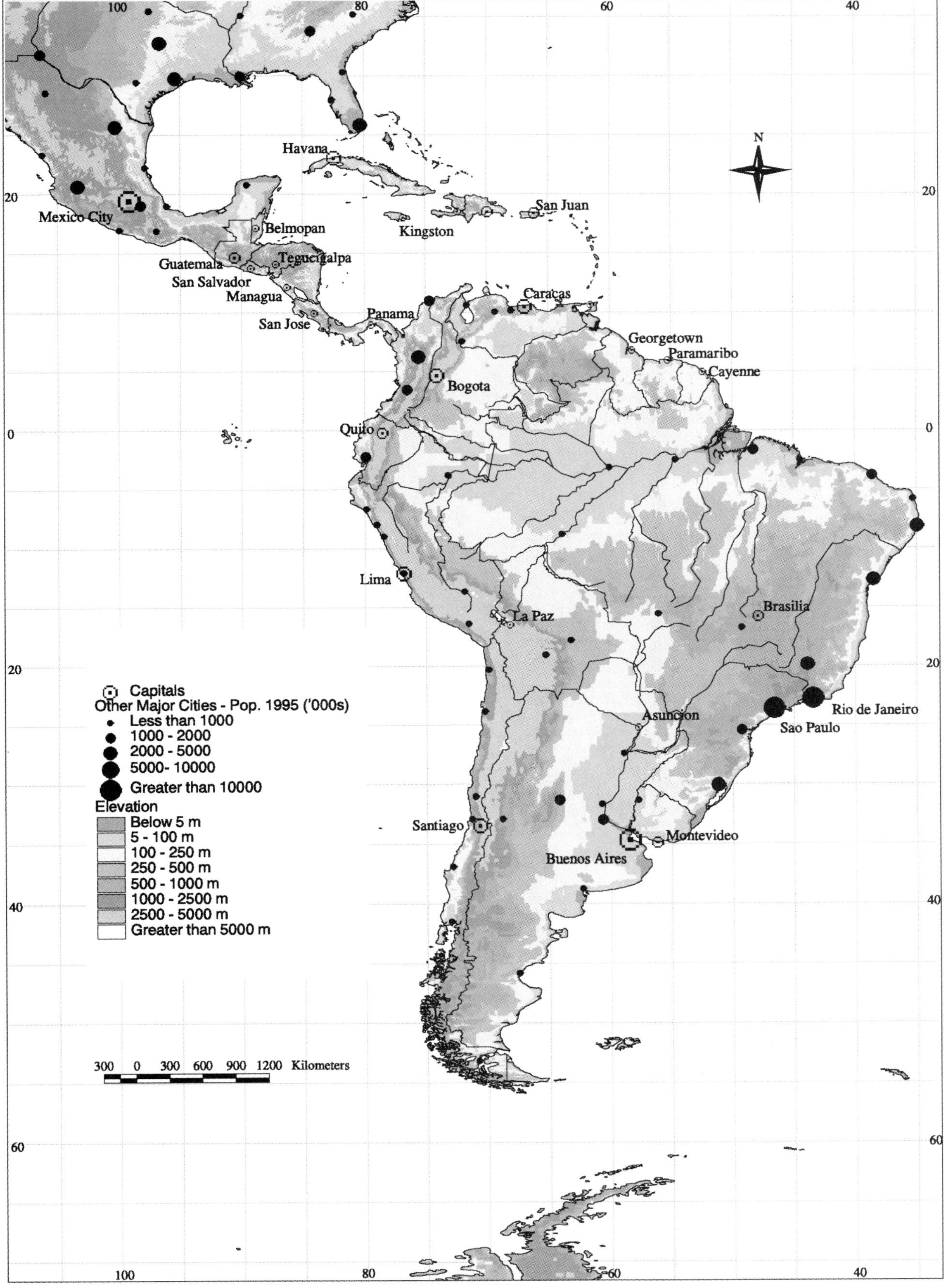


**Figure 6-1:** The Latin America region [compiled by the World Bank Environment Department Geographic Information System (GIS) Unit; see Annex E for a color rendition].

*Table 6-2: Estimated deforestation rates in the tropical rainforest of Amazonian countries, 1981–90.*

| Country | Remaining Rainforest Area 1990 (Mha) | Rate of Deforestation (Mha/yr) | Total Deforestation 1981–90 (Mha) | Total Deforestation 1981–90 as % of Remaining Forest |
|---|---|---|---|---|
| Bolivia[1] | 49,137 | 625 | 6,250 | 12.7 |
| Brazil | 291,597 | 1,012 | 10,120 | 3.5 |
| Colombia | 47,455 | 223 | 2,230 | 4.7 |
| Ecuador | 7,150 | 142 | 1,420 | 19.9 |
| Guyana | 13,337 | 0.0 | 0.0 | 0.0 |
| Peru | 40,358 | 114 | 1,140 | 2.8 |
| Venezuela | 19,602 | 147 | 1,470 | 7.5 |
| **Total** | 468,816 | 2,263 | 22,630 | 4.8 |

[1]Bolivia's total forest area and total deforestation rate are given, because FAO (1993) reports that Bolivia has no tropical rainforest zone.
Source: FAO, 1993.

countries. Poverty is widespread and may be aggravated by global warming if the impacts contribute to declines in the sustainability of Latin America's ecosystems, particularly in areas with marginal environmental and socioeconomic conditions (see Annex D).

Latin America's contribution to global greenhouse gas (GHG) emissions is relatively low (see Box 6-2). In fact, Mexico is the only Latin American country on the list of the world's 15 largest emitter countries; mainly as a result of thermal energy production, it accounts for approximately 2% of the carbon emitted by those countries (WRI, 1992). A large portion of Latin American emissions is caused by deforestation, mainly related to the expansion of agricultural lands (Gay-García *et al.*, 1995; Lebre-La Rovère, 1995; Massera *et al.*, 1996; Perdomo, 1996). On a country-by-country basis, however, deforestation is not always the main source of GHG emissions. For example, carbon emissions resulting from deforestation of Mexican tropical and temperate forests has been estimated at 52,000–100,000 Gg/yr (Massera *et al.*, 1996)—which represents less than 20% of the country's total emissions as estimated by Perdomo (1996) (see Box 6-2). The National Greenhouse Gas Inventories—recently undertaken by the majority of Latin American countries in response to UN Framework Convention on Climate Change (UNFCCC) requirements—will soon enable a better estimate of the real situation in this region, where data availability still is inadequate.

## 6.2. Regional Climate Information

### 6.2.1. *Current Climate*

Because Latin America spans a vast range of latitudes and contains important high-elevation mountain ranges, it is hardly surprising that it has a wide variety of climates. Although it is the only southern continent to reach high latitudes, its broadest extent is in the equatorial zone; thus, tropical conditions prevail over the larger portion of the region. The large-scale climatic features of Central and South America are defined by the

**Box 6-2. Latin America's Contribution to Global Greenhouse Gas Emissions**

Estimates of Latin America's contribution to global GHG emissions in 1990 are on the order of 902,000 Gg $CO_2$ equivalent (about 4.28%). Within the region, the largest emitter is Mexico, followed by Brazil, Argentina, Venezuela, and, to a lesser degree, Colombia and Chile (Perdomo 1996, on the basis of Oak Ridge National Laboratory information).

| Country | Total GHG Emission (Gg of $CO_2$-equiv) | % Distribution $CO_2$ | $CH_4$ | $N_2O$ |
|---|---|---|---|---|
| Bolivia | 69,987 | 78.6 | 21.0 | 0.4 |
| Costa Rica | 8,037 | 47.8 | 49.4 | 2.8 |
| Mexico | 531,906 | 81.5 | 17.9 | 0.5 |
| Peru | 138,573 | 74.3 | 24.3 | 1.4 |
| Venezuela | 269,951 | 70.6 | 28.8 | 0.55 |

Analysis of estimated emissions of $CO_2$, methane ($CH_4$), and nitrous oxide ($N_2O$) from the five countries for which detailed information is available revealed that most emissions are associated with the energy sector in Mexico, Venezuela, and Costa Rica and with land-use changes in Peru and Bolivia. Methane emissions from agriculture, especially from enteric fermentation and animal manure, occupy a second place in most cases (Perdomo, 1996).

region's predominant atmospheric circulation patterns and geomorphological features. The main circulation features are relatively low pressure at the equatorial belt (10°N–10°S), quasi-permanent high-pressure cells over the north and south Atlantic and southeast Pacific Oceans, and a belt of low pressure defining the westerly flow on the southern portion of the South American subcontinent. Mexico and Central America are affected by the penetration of cold fronts and tropical cyclones over the Atlantic and Pacific Oceans, whereas the Atlantic coast of South America is mostly free of high-intensity tropical storms.

The development of a thermal low located between 20°S and 30°S over the high, dry lands to the east of the Andes introduces a monsoonal circulation pattern, bringing seasonal precipitation on the high Andean Plateau and influencing the positioning of the Inter-Tropical Convergence Zone (ITCZ) in the region.

Latin America is characterized largely by humid, tropical conditions. However, important areas (e.g., northeastern Brazil) are subject to droughts and floods, and others are affected by freezes—all of which have negative impacts on agricultural production (Magalhães and Glantz, 1992).

Atmospheric circulation and cold ocean currents have remarkable influence on the weather and climate in the southern part of the region—giving origin to the Peruvian, Atacama, and Patagonian deserts, which receive less than 100 mm of mean annual precipitation. The cold Humboldt ocean current—which flows northward along the west coast of South America—brings to the coasts of Ecuador, Peru, and Chile large masses of phytoplankton that originate in the Antarctic Ocean, supporting one of the world's richest fisheries. This process is interrupted by the western displacement of the Humbolt current and the irruption (advection) of warmer waters caused by the weakening of the easterly surface winds (El Niño phenomenon), resulting in adverse conditions for fishing.

### 6.2.2. *Trends in Climate*

Climate trends over the past century have been investigated in the majority of countries that have the required amount of information (i.e., measured data and reliable proxy data). Studies are available on average and extreme values of temperature, humidity, and precipitation for different regions of Latin America and different periods of time.

Some of these studies reveal significant warming in southern Patagonia east of the Andes, with increases in maximum, minimum, and daily mean temperatures of more than 1°C. According to some researchers, no warming has been observed north of about 42°S. These observations are consistent with changes in vapor pressure and precipitation—which have increased (north of 40°S) since 1940 (Hoffman *et al.*, 1996).

Similar studies peformed in Chile indicate that mean surface temperatures showed no increasing trend before 1900—but that, during the period 1900–90, the temperature in the Southern Hemisphere increased by a total of 0.4°C, at a fairly constant rate (Rosenblüth *et al.*, 1997). These authors have reported a significant cooling in the southern half of Chile in 1991 and 1992, coinciding with the eruptions of the Pinatubo and Hudson's volcanoes (June and August 1991, respectively), as a result of the effect of the sulfate aerosols emitted into the stratosphere.

A trend analysis of 81 series of precipitation data in Central America (Brenes-Vargas and Trejos, 1994) shows that changes in general circulation influence precipitation rates and consequently the availabilty of water on the isthmus, with related problems in hydroenergy production (see Section 6.3.2).

Further, an analysis of precipitation trends in the southern portion of South America, east of the Andes cordillera, indicates that the mean annual precipitation in the humid Pampa has increased by about 35% in the past half-century (Forte-Lay, 1987; Castañeda and Barros, 1996).

Trends in annual precipitation (in percentages), mean annual temperatures (in degrees centigrade per hundred years), and variations in annual precipitation and temperatures are provided in Annex A.

### 6.2.3. *Climate Variability*

Climate variability means the alternation between the "normal climate" and a different, but recurrent, set of climatic conditions over a given region of the world. In the Latin American region, climatic variability is related, inter alia, to the Southern Oscillation (SO) and the El Niño phenomenon (EN). Studies on the SO effects, using an SO Index (SOI), have shown its connection with pressure, temperature, and rainfall, as well as with hydrometeorological anomalies (e.g., record river discharges and lake levels) (Aceituno, 1987).

Prominent among SO-related anomalies in South America is the well-documented tendency for anomalously-wet conditions along the otherwise arid coast of northern Peru and southern Ecuador during El Niño episodes (Rasmuson and Carpenter, 1982). The relationship between the SO and rainfall anomalies in northeastern Brazil has long been recognized (Walker, 1928; Doberitz, 1969; Caviedes, 1973; Hastenrath, 1976; Kousky *et al.*, 1984). Rainfall anomalies related to the SO in extratropical South America have been documented primarily for central Chile (Rubin, 1955). A tendency for precipitation in subtropical Chile to be exceptionally abundant during El Niño years has been noted. This relationship is consistent with the significant negative correlation between pressure differences (Tahiti minus Darwin) and annual precipitation in central Chile (Pittock, 1980) and snow accumulation in the southern Andes (Cerveny *et al.*, 1987). Rainfall in the Central America-Caribbean domain also is related to the SO, as revealed by the tendency for drought conditions to occur during warm episodes off the Peruvian coast (Hastenrath, 1976).

The strong El Niño episode in 1982–83, which coincided with a marked negative SO phase, was associated with extreme climatic conditions in various parts of South America. Examples include the convection regime associated with flooding in northern Peru (Horel and Cornejo-Garrido, 1986), droughts in northeastern Brazil (Rao *et al.*, 1986), and the remarkable precipitation and circulation anomalies over South America (Nobre and De Oliveira, 1986; Minetti and Vargas, 1995).

During recent decades, the influence of the ENSO phenomenon on the interannual variability of weather and climate in South America has been the research subject of other authors (Berlage, 1966; Burgos *et al.*, 1991; Santibañez and Uribe, 1994; Vargas and Bishoff, 1995; IPCC 1996, WG I, Chapter 4). During ENSO years, precipitation in some areas of northern South America is lower (Aceituno, 1987), increasing the likelihood of drought. ENSO events also can lead to higher precipitation and air temperature, as has occurred in the coastal deserts of Peru and Chile (Caviedes, 1973). In addition, these events are related to massive fluctuations in marine ecosystems of the southern Pacific and Atlantic Oceans, with adverse socioeconomic consequences for commercial fishing and fishmeal production in Chile, Peru, Brazil, and Argentina (Pauly and Tsukayama, 1987; Pauly *et al.*, 1987; Pauley *et al.*, 1989; Yañez, 1991; Bakun, 1993; Sharp and McLain, 1993).

Records of precipitation in countries of the Central American isthmus show an important reduction in precipitation during the ENSO period, particularly along the Pacific watershed. This reduction has considerable effects on the most important economic activities and sectors in these countries (Campos *et al.*, 1996).

It is unclear whether the ENSO phenomenon would change with long-term global warming—and what the consequences would be of overlapping sources of climatic variability on the Earth's systems. However, ENSO effects have been used by regional scientists to help define real scenarios that could be useful as analogs for climate change and for studies of potential responses of the countries and sectors affected (Campos *et al.*, 1996).

### 6.2.4. *Climate Change*

Annex B of this report provides information on the progress made with transient runs of atmosphere-ocean general circulation models (AOGCMs), which allow—within certain limits and with reliable complementary data—climate projections at a regional scale. Annex B also provides information on different regionalization techniques, which recently have been developed and tested to improve the coarse resolution in regional climate change simulation. However, the regions identified for this type of simulation do not include Latin America.

Whetton *et al.* (1996) analyzed the performance of models in South America to assess the ability of GCMs to reproduce key climate features of the Southern Hemisphere, as well as the major gaps in their simulation capability. The authors considered several "mixed-layer ocean" experiments and several "coupled" or "full dynamic ocean" experiments. The correlation between modeled and observed annual mean precipitation over the region ranges from 0.35 to 0.70 in mixed-layer GCM runs; slightly higher correlations are found in coupled GCM climate experiments.

In greenhouse climate experiments performed over South America and adjacent oceans (Labraga, 1997; Labraga and Lopez, 1997), mean temperature increases simulated in mixed-layer equilibrium experiments ranged between 1.5°C and 4.0°C at the time of $CO_2$ doubling; transient experiments with coupled GCMs simulated mean temperature increases of 1.2–1.7°C. According to the results of five coupled GCMs, temperatures in the semi-arid zone of central Chile and central-western Argentina during the Southern Hemisphere summer are simulated to increase by 1–3°C at the time of $CO_2$ doubling. During the same season, precipitation in the area was projected to decrease by 10–15% per degree of global warming, according to the same set of experiments. There is consensus among models that the semi-arid subtropical zone will experience intensified and extended dry conditions. In addition, the coupled GCMs all project increases in rainfall in the ITCZ in the eastern equatorial Pacific and the northwestern part of the continent, the South Atlantic Convergence Zone (SACZ) in the eastern part of Brazil and the adjacent Atlantic Ocean, and the southern tip of the continent.

Figures 6-2 and 6-3 reproduce maps showing expected seasonal changes in surface temperatures and precipitation, as projected by transient coupled AOGCM experiments performed at the Max Planck Institute (MPI, Germany) and the Bureau of Meteorology Research Centre (BMRC, Australia) (IPCC 1996, WG I, Chapter 6).

Because of the great climatic diversity of this region and the present limitations of GCMs in simulating Latin America's regional climate, many studies that have estimated vulnerability have used the GCMs as tools for obtaining reference scenarios, in order to perform sensitivity analyses (e.g., for different crops at specific sites and water resources for specific basins). These analyses have included detailed knowledge of the local climate founded on sufficient and reliable real data and correctly evaluated proxy data. A good example of this approach appears in a study on "Global Warming and Climate Change in Mexico" (Liverman and O'Brien, 1991), in which the authors note the importance of acknowledging differences among climate model projections—as well as between model simulations and observed climates—because these differences underscore the uncertainties involved in assessing the regional impacts of global warming. A range of possible outcomes is captured by using the results of different GCMs and undertaking a sensitivity analysis of the results. Some of the conclusions reached appear in the relevant sections of this chapter, which also provide information on the vulnerabilities of and potential impacts on natural and managed systems and human

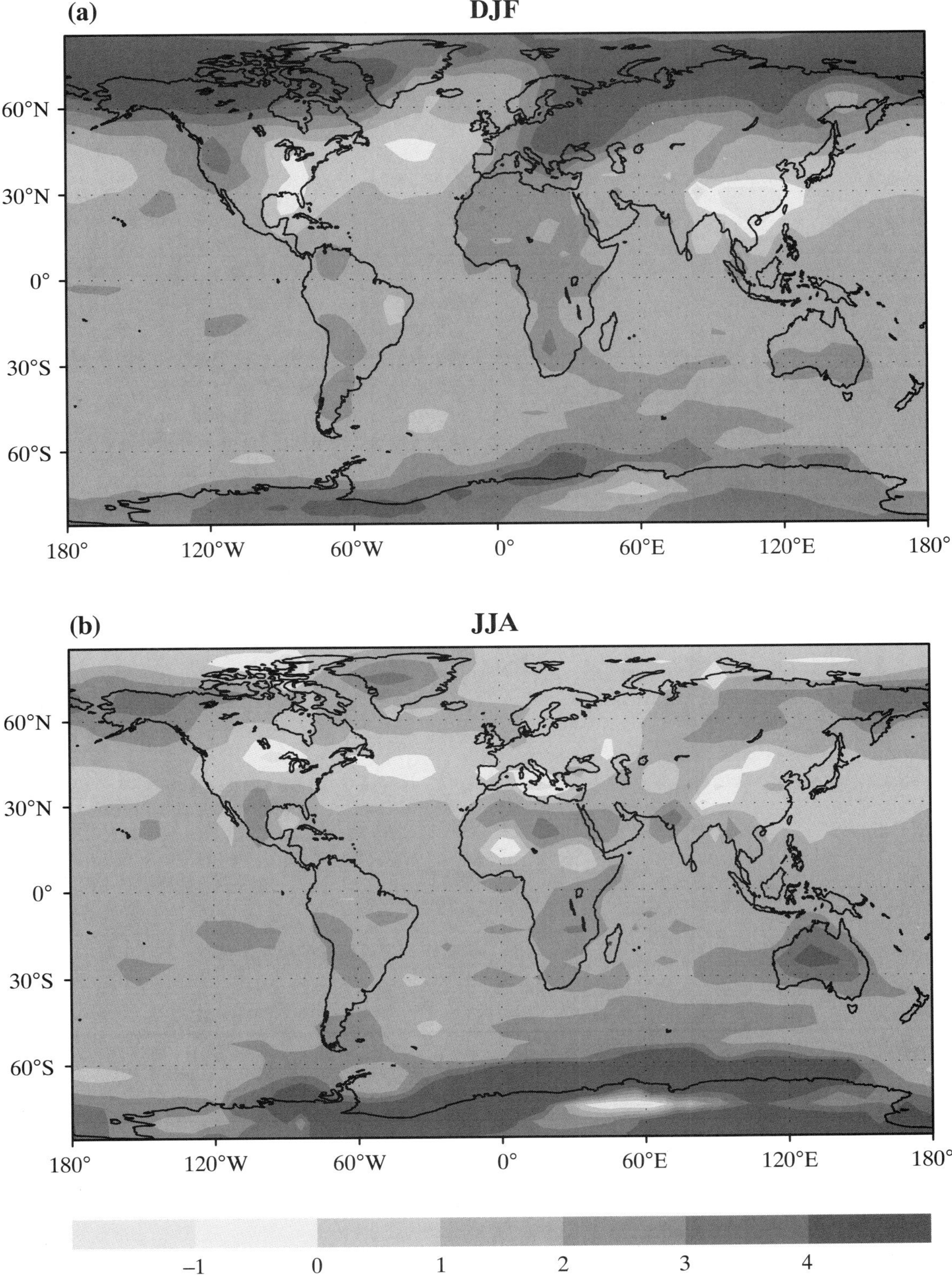


**Figure 6-2:** Seasonal change in surface temperature from 1880–1889 to 2040–2049 in simulations with aerosol effects included. Contours are at every 1°C (IPCC 1996, WG I, Figure 6.10) (see Annex E for a color rendition).

health in Latin America as a result of global warming. In this context, it is very important to realize that climate change is only one among many causes leading to the changes that have been observed in these systems during recent decades. In fact, many factors—such as limited access to technology, poor research capacity, economic crises, social inequality, and population growth—could give rise to larger effects than those stemming from atmospheric warming.

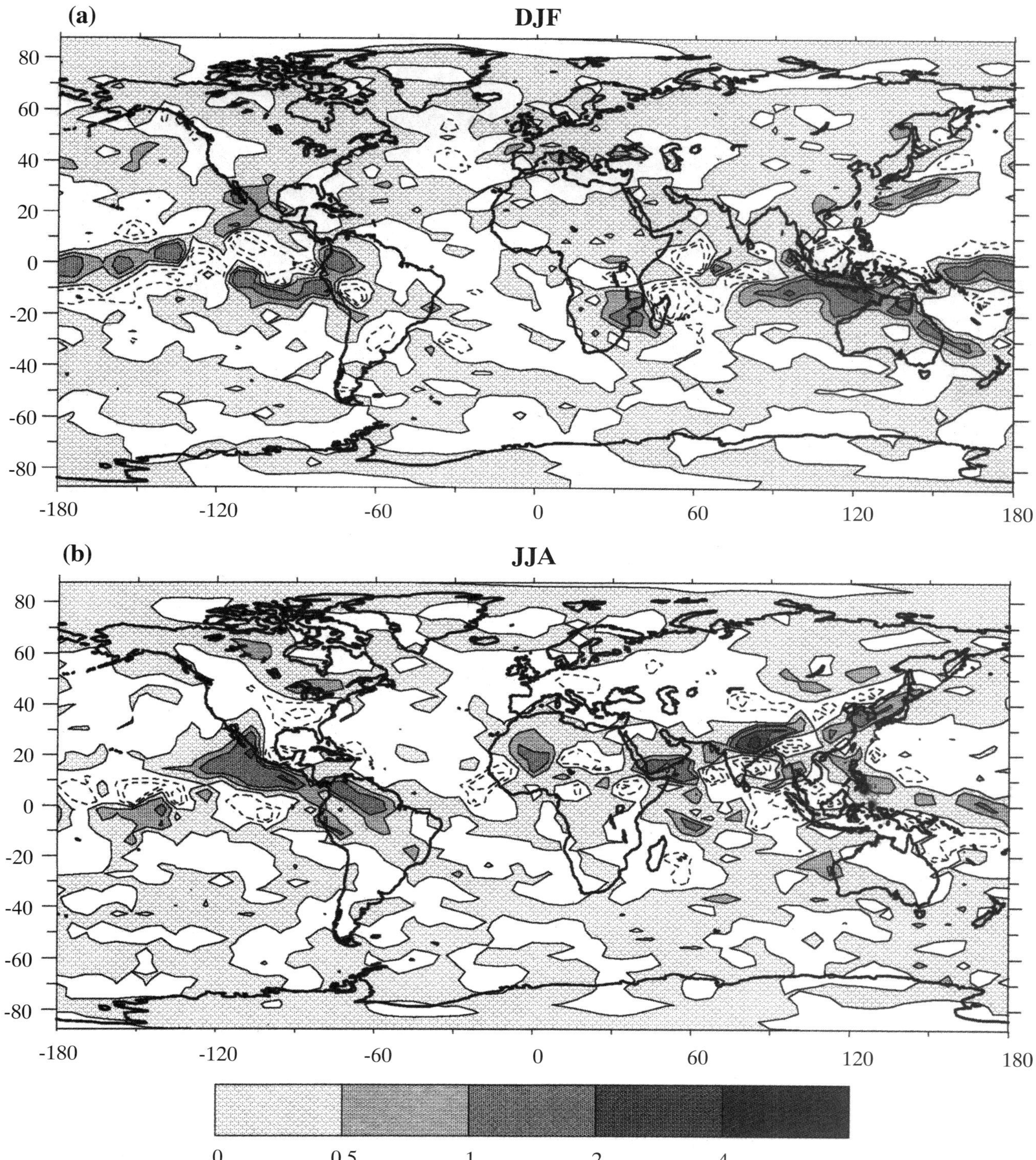


**Figure 6-3:** Seasonal changes in precipitation at the time of doubling $CO_2$ following a 1%/yr increase. Contours are at ±0.5, 1, 2, and 4 mm/day; negative contours are dashed and areas of increase areas stippled (IPCC 1996, WG I, Figure 6.11).

## 6.3. Sensitivity, Adaptability, and Vulnerability

As noted above, climate change is only one of the causes leading to global change. As happens when there are multiple stresses, different stresses may be prevalent in different situations. In this regard, the vulnerability of certain systems and activities—such as mountain regions with snow/ice cover, low coastal areas, agriculture, water resources management and hydropower generation, and human health—are recognized as vulnerable under projected climate change scenarios. Furthermore, climate change may worsen some existing problems—such as desertification and freshwater shortages (in certain regions)—and give rise to new problems, such as the expansion of the geographical and altitudinal range of some human diseases. Warming of the atmosphere also may result in some benefits, such as the enhancement of high-latitude agriculture and reductions in the effects of wintertime diseases. Politicians and decisionmakers need to be aware of the vulnerability and adaptability of ecosystems and activities in their countries and their region, as well as elsewhere around the world, to design and implement sustainable development initiatives that take advantage of beneficial climate changes and to build strategies for international commercial exchanges in light of differential climate change impacts around the world.

Regarding the assessment of climate change impacts on different economic sectors, climatic projections at scales relevant to the production and management fields are still inadequate (as indicated in Annex B). The resolution of current GCMs is too coarse to allow reliable projections or cost/benefit analyses of possible adaptation options for the individual countries of Latin America. In some cases, however, appropriate downscaling and other techniques are available, making possible the use of GCM outputs, and regional models suitable for impact and vulnerability analysis are now being developed for this region.

The largest impacts of climate change are likely to affect natural ecosystems and sectors related to primary production, such as agriculture, livestock raising, and fisheries. Water resources are at risk in many areas. Human health and human settlements, especially in coastal lowlands and environmentally and socioeconomically marginal areas, also are vulnerable. Impacts are expected to be less severe for industry, transportation, and infrastructure outside of flood-prone areas, although increased frequency or severity of extreme events may affect these sectors.

### 6.3.1. *Terrestrial Ecosystems: Vulnerability and Impacts*

Virtually all of the world's major types of ecosystems are present in Latin America (Figure 6-4), which is characterized by high species and ecosystem diversity. Some Latin American countries—such as Mexico, Venezuela, Colombia, Ecuador, Bolivia, Brazil, and Peru—are among the world's richest in terms of terrestrial plant and animal species (Annex D; see also WRI, 1990–91; LAC CDE, 1992). Forest biomes in this region include tropical rainforests, such as the Amazon; other tropical forests in eastern Mexico, Central America, and northern South America; and the endangered Mata Atlantica in Brazil. There also are important sectors of tropical deciduous forests in the Yucatán; in the Pacific watershed of Central America, Venezuela, and Ecuador; and on the Brazilian coast from about 7°S to the Tropic of Capricorn. Mid-latitude deciduous or temperate forests are established on low-elevation coastal mountains in southern Brazil and southern Chile and to a lesser extent in southern Argentina (in the piedmont area of the Patagonian Andes). Austral forests are located on the southernmost tip of the subcontinent and on Tierra del Fuego Island. Grasslands, shrublands, and deserts—the most extensive ecosystems in the region—are found on the Mexican Pacific and Venezuelan Caribbean coasts, in northeastern Brazil, and in inland areas between Brazil and Bolivia. The Gran Chaco ecosystem is located across parts of Bolivia, Paraguay, and Argentina. The chaparral biome is on the central Chilean coast. Mid-latitude grasslands occupy extensive areas in southern Brazil, Uruguay, and central and eastern Argentina; tropical grasslands and savannas are present in Central America, the Guyanas, Venezuela, Colombia, Brazil, Paraguay, and Argentina. Arid shrublands occupy the west of Argentina and Patagonia, and hyper-arid areas exist along the west coast of Peru and northern Chile, as well as in southern Bolivia and northwestern Argentina. The Latin American region includes almost 23% of the world's potentially arable land, 12% of the current cropland, and 17% of all pastures (Gómez and Gallopin, 1991). Table 6-3 lists the surface area of different biomes in South America, and Table 6-4 lists land-cover types by country; Figure 6-4 provides information on the actual vegetation cover in South America, based on satellite imagery (Stone *et al.*, 1994).

Biomass accounts for about 15% of the world's enegy use and 38% of the energy use in developing countries. In Latin America, however, most biomass is used inefficiently—mainly for cooking and heating—and often in much the same way it has been used for millenia (Velez *et al.*, 1990). Biomass can be converted into modern energy carriers, such as gaseous and liquid fuels and electricity, that can be widely used in more affluent societies (see Section 6.3.7), and it has a number of other benefits for developing countries: Its large-scale use for energy can provide a basis for rural development and employment in developing countries, and sites used for plantations in these countries are deforested areas or otherwise degraded lands. Revenues from the sale of plantation biomass crops grown on degraded lands can help finance the restoration of these lands.

The expected vulnerability of terrestrial ecosystems to climate change is high. Changes in extreme events, which have been increasing in frequency since the end of the 19th century (Karl *et al.*, 1995), are likely to be more disruptive than changes in mean values. Biogeographical models—based on local vegetation and ecophysiological and hydrological processes, as well as projections of future scenarios—have facilitated the simulation of changes in vegetation distribution in a $CO_2$-rich world. Annex C describes the MAPSS (Neilson, 1995) and BIOME 3 (Haxeltine and Prentice, 1996) biogeographical models, which have been run for the Latin American region under different scenarios. Under most scenarios, leaf area index (LAI)—an indicator of how much vegetation can live on the ground surface—decreases or remains constant over southeast and northwest Brazil, northeast Argentina, and Uruguay and increases slightly over central Brazil. Runoff decreases or remains constant over the southern portion of South America. It should be stressed that these models are based on potential vegetation only, and their projections do not incorporate land-use changes. At present, some land-use practices are a major threat, and

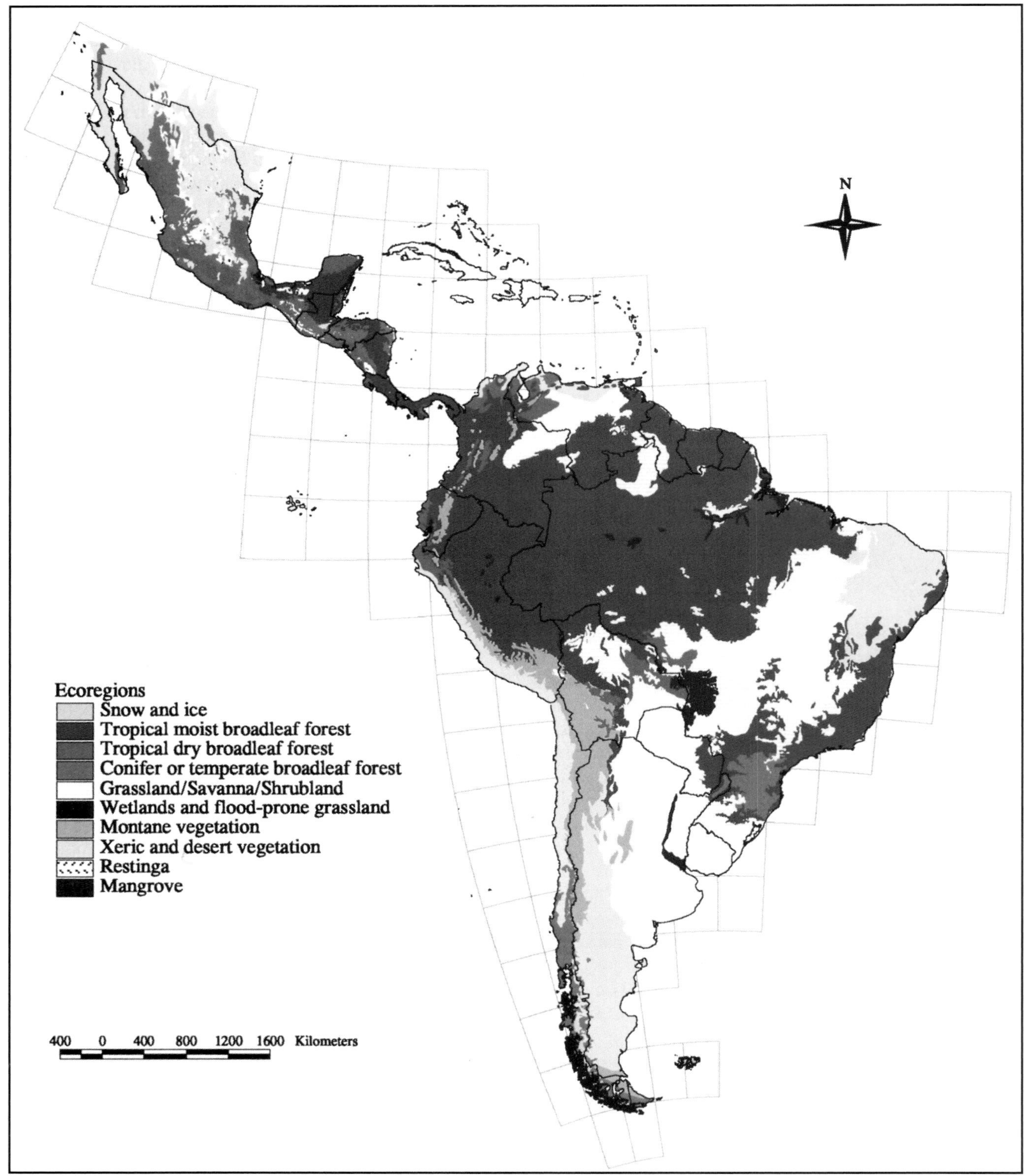


**Figure 6-4:** Major biomes in Latin America [compiled by the World Bank Environment Department Geographic Information System (GIS) Unit; see Annex E for a color rendition].

their effects are liable to be more severe than those of climate change (Brinkman and Sombroek, 1996). Studies developed in Costa Rica and Nicaragua (Halpin *et al.*, 1995a,b) observed that shifts may occur in climatic zones that are associated with particular vegetation types in these countries. The combined impacts of long-term climate change, land-use practices, and an increasingly variable climate on an interannual scale would pose greater stresses to Latin America's ecosystems. Therefore, designing and implementing adaptation measures is of primary importance for facing these combined impacts.

***Table 6-3**: Surface area of different biomes in South America.*

| **Biome** | **Area (km²)** | **Percent** |
|---|---|---|
| Tropical moist forest and semideciduous tropical moist forest | 5,858,100 | 33.13 |
| Seasonally deciduous woodlands (e.g., chaco in Argentina and cerrado in Brazil) | 2,300,100 | 13.01 |
| Savanna/grasslands and pasture | 2,296,900 | 12.99 |
| Secondary seasonal forest with agricultural activity | 979,000 | 5.54 |
| Cool deciduous scrublands (especially Argentina) | 905,000 | 5.40 |
| Xerophytic woodlands (thornforest or caatinga in Brazil) | 437,200 | 2.47 |
| Tropical seasonal or deciduous forest | 366,500 | 2.08 |
| Agriculture | 353,000 | 2.00 |
| Recently cleared tropical moist forest | 342,700 | 1.94 |
| Desert | 278,900 | 1.58 |
| Unclassified | 275,800 | 1.56 |
| Montane degraded grasslands (especially Bolivia and Peru) | 271,509 | 1.54 |
| Degraded seasonally deciduous woodlands | 266,700 | 1.51 |
| Montane grassland, tundra, or polar grasslands | 263,200 | 1.49 |
| Degraded xerophytic woodlands (thornforest) | 233,000 | 1.32 |
| Secondary forest in the tropical moist forest region | 220,800 | 1.25 |
| Wet vegetation (generally mixed water and vegetation) | 212,900 | 1.20 |
| Mixed pine with secondary forest and agriculture (southern Brazil) | 190,300 | 1.08 |
| Degraded grasslands or grasslands with agricultural activity | 184,100 | 1.04 |
| Cool deciduous woodlands | 173,800 | 0.98 |
| Montane degraded woodlands | 166,500 | 0.94 |
| Water (open) | 163,200 | 0.92 |
| Deciduous temperate forest | 121,600 | 0.69 |
| Montane woodlands | 120,800 | 0.68 |
| Xerophytic scrublands | 115,400 | 0.65 |
| Seasonally flooded grasslands (Pantanal) | 81,800 | 0.46 |
| Tropical open forest mixed | 77,000 | 0.44 |
| Cool deciduous forest | 66,600 | 0.38 |
| Montane forest | 64,100 | 0.36 |
| Inland salt marsh community | 52,700 | 0.30 |
| Snow and rock | 45,400 | 0.26 |
| Degraded tropical seasonal forest | 39,200 | 0.22 |
| Tropical gallery forests | 38,300 | 0.22 |
| Degraded temperate deciduous forest | 29,200 | 0.17 |
| Tropical moist forest with bamboo (in Acre, Brazil, and Bolivia) | 13,600 | 0.08 |
| Mangroves | 4,300 | 0.02 |
| Bare soil and rock | 3,500 | 0.02 |
| Urban regions | 2,200 | 0.01 |
| Xerophytic littoral vegetation (Venezuelan coast) | 700 | 0.00 |
| Montane degraded forest | 0 | 0.00 |
| **Total** | 17,680,200 | 100.00 |

Source: Stone *et al.*, 1994.

#### *6.3.1.1. Forests*

Tropical forests represent about 40% of the world's forested area and contain about 60% of global forest biomass. Latin American tropical forests—which represent about 22% of the global forest coverage—have a strong influence on local and regional climate (Salati and Vose, 1984; IPCC 1996, WG II, Section 1.4), play a significant role in the global carbon budget (Dixon *et al.*, 1994; IPCC 1996, WG II, Chapter 24), and contain a remarkably large share of all plant and animal species of the world (Annex D; see also Bierregaard *et al.*, 1992; Mabberley, 1992; Myers, 1992; Riede, 1993). Moreover, Latin American tropical and subtropical forests are economically very important, providing commercial products for national and international markets. Large numbers of people base their activities in these trades; many others, particularly indigenous people, subsist mainly on the forests' non-market products. Temperate forests in South America are

***Table 6-4:*** *Generalized land-cover types for South America, by country.*

| Country | Closed Trop. Moist Forest | Recently Degraded TMF | Closed Forest | Degraded Closed Forest | Wood-lands | Degraded Wood-lands | Savanna Grass-lands | Scrub-lands | Desert, Bare Soil | Water | Snow, Rock, Ice | Other |
|---|---|---|---|---|---|---|---|---|---|---|---|---|
| Argentina | 1.2 | 0.0 | 96.8 | 0.0 | 645.4 | 15.2 | 755.4 | 1,126.8 | 37.9 | 34.0 | 31.4 | 35.7 |
| Bolivia | 323.5 | 12.7 | 409.2 | 24.6 | 345.1 | 102.2 | 87.7 | 91.0 | 16.5 | 11.0 | 1.4 | |
| Brazil | 3,522.3 | 519.7 | 3,686.0 | 1,692.2 | 1,555.9 | 330.0 | 740.0 | 179.4 | 0.0 | 80.9 | 0.0 | 124.0 |
| Chile | 0.0 | 0.0 | 134.1 | 29.1 | 75.2 | 29.6 | 101.1 | 100.9 | 186.8 | 7.0 | 16.6 | 3.8 |
| Colombia | 581.6 | 5.4 | 622.5 | 11.4 | 116.3 | 14.5 | 255.5 | 64.0 | 0.0 | 3.1 | 0.0 | 22.8 |
| Ecuador | 115.5 | 1.7 | 121.0 | 1.7 | 33.7 | 4.3 | 41.9 | 16.5 | 2.5 | 0.6 | 0.0 | 0.8 |
| French Guiana | 78.8 | 0.0 | 79.8 | 2.4 | 0.6 | 0.0 | 0.2 | 0.0 | 0.0 | 0.1 | 0.0 | 1.0 |
| Guyana | 150.4 | 2.0 | 171.6 | 2.4 | 5.4 | 0.3 | 18.4 | 1.5 | 0.0 | 1.2 | 0.0 | 3.7 |
| Paraguay | 0.3 | 0.0 | 8.0 | 0.2 | 209.1 | 50.7 | 104.0 | 26.5 | 0.0 | 0.6 | 0.0 | 1.1 |
| Peru | 620.8 | 19.1 | 654.7 | 10.1 | 88.0 | 78.8 | 139.0 | 161.7 | 88.0 | 8.3 | 0.7 | 5.6 |
| Suriname | 136.0 | 2.5 | 128.5 | 10.0 | 0.5 | 0.3 | 1.2 | 0.4 | 0.0 | 1.1 | 0.0 | 3.3 |
| Uruguay | 0.0 | 0.0 | 2.1 | 0.0 | 0.9 | 0.0 | 154.1 | 11.0 | 0.0 | 3.0 | 0.0 | 5.9 |
| Venezuela | 370.1 | 0.2 | 415.5 | 0.0 | 33.9 | 40.2 | 243.3 | 109.2 | 0.0 | 11.4 | 0.0 | 8.4 |
| **Total** | 5,908.5 | 563.4 | 6,530.7 | 1,803.7 | 3,109.8 | 666.9 | 2,642.0 | 1,889.0 | 331.7 | 163.2 | 48.9 | 217.2 |

Notes: All values are given in 1,000s km$^2$; there are 314,200 km$^2$ unclassified; IMF includes tropical moist, semi-deciduous, and gallery forests; grasslands includes those seasonally flooded; closed forest includes TMF, montane forests, cool and temperate deciduous forest, and tropical seasonal forest; scrublands also include degraded savanna grasslands and agriculture; desert, bare soil includes salt marsh communities; and other includes wet vegetation and mangroves.
Source: Stone *et al.*, 1994.

important, to a much smaller extent, for the export economies of Chile and, to an even lesser extent, Argentina (IPCC 1996, WG II, Section 15.4.3).

Forest cover in Latin America declined from 992 million ha in 1980 to 918 million ha in 1990, with an annual deforestation rate of 0.8% over this period. Average annual deforestation rose from 5.4 million ha in 1970 to 7.4 million ha in 1990 (FAO, 1993). Between 1980 and 1990, deforestation reduced tropical forest cover from 826 million ha to 753 million ha—a decrease of 0.9% (UNEP, 1992). The tropical forests of the Pacific coast of Central America once covered 55 million ha; less than 2% of this forest now remains, although countries like Costa Rica have preserved and protected some of their forests under national park or reserve status. Similarly, only 4% of the original 100 million ha of the Atlantic forest of Brazil (also marginally present in Paraguay and northeastern Argentina) remains as relatively pristine forest. In Argentina, forests covered 106 million ha in 1914, but less than a third of that surface (32.3–35.5 million ha) remained by the 1980s (Di Pace and Mazzuchelli, 1993). High deforestation rates also have been observed in the subtropical Paranaense and Great Chaco forests, as well as in the Andean-Patagonian and Austral forests.

Latin American forests would face an additional threat from climate change, as indicated in vulnerability studies by USCSP and GEF projects. Unless appropriate action is taken, mismanagement of these ecosystems will, in turn, make climate change impacts more severe. Wood harvesting is expected to increase, especially in tropical and subtropical countries; local communities will face serious shortages in forest products required for subsistence and traditional trading (IPCC 1996, WG II, Section 15.2.3). Forest clearing also is expected to increase in response to an increasing need for agricultural land (DENR-ADB, 1990; Starke, 1994; Zuidema *et al.*, 1994).

Large-scale conversion of tropical forest into pasture will likely lead to changes in local climate through increased surface and soil temperatures, diurnal temperature fluctuations, and reduced evapotranspiration (Salati and Nobre, 1991; Cerri *et al.*, 1995). A considerable proportion of the precipitation over the Amazon basin originates from evapotranspiration (Molion, 1975; Salati and Vose, 1984), which could be reduced by continued and large-scale deforestation. Such large-scale forest clearing could reduce the enormous runoff of the Amazon river system and result in other far-reaching, undesirable impacts beyond the cleared areas (Gash and Shuttleworth, 1991). According to projections by Shukla *et al.* (1990), if tropical forests were replaced by degraded pastures, there would be significant increases in surface temperature and decreases in evapotranspiration and precipitation in the Amazon basin. Furthermore, increases in the length of the dry season would make reestablishment of forests difficult.

The global carbon cycle also could be altered. The potential of Latin American tropical forests to act as carbon sinks has been considered high (da Rocha, 1996; Massera *et al.*, 1996; Molion, 1996). According to Batjes and Sombroek (1997), however, the effect of climate change on carbon sequestration in tropical soils can be very complex; it depends on air temperature, $CO_2$ concentration, seasonal rainfall distribution, nitrogen deposition, and fires. Forest-to-pasture conversion during a 35-year sequence in Amazonia resulted in the creation of a net source of methane (Steudler *et al.*, 1996) and a net sink of $CO_2$; 9–18% more carbon was stored in the soil under pasture than under the original forest. However, the conversion process involves a large net organic carbon mineralization (Cerri *et al.*, 1995; Cerri *et al.*, 1996).

Tropical forests are likely to be more affected by changes in soil water availability (e.g., from seasonal droughts or soil erosion and nutrient leaching resulting from heavy rainfall events) and possibly by $CO_2$ fertilization than by changes in temperature (IPCC 1996, WG II, Section 1.4). Nutrient leaching, erosion, and timber harvesting also are likely to result in decreased biomass and biodiversity (Whitmore, 1984; Jordan, 1985; Vitousek and Sanford, 1986).

Global vegetation models do not agree on whether climatic change (in the absence of land-use change) will increase or decrease the total area of tropical forests in Latin America. These forests are projected to expand their geographical range in the MAPSS and BIOME3 models, although the results vary depending on the scenario (Annex C). When the IMAGE 2.0 model (Alcamo, 1994) has been used to simulate the joint effects of climatic and land-use changes (Leemans, 1992; IPCC 1996, WG II, Section 1.4), the projected net loss of tropical forest is not very large. This result may reflect the fact that, in that model, land-use changes are driven by the dynamics of human population growth and economic development. In reality, however, these variables are not necessarily directly related to natural resource exploitation in many Latin American countries, where harvests often are determined by needs outside the region (IPCC 1996, WG II, Chapter 15). Increased occurrence of droughts (as in the projections compiled by Greco *et al.*, 1994) would affect natural forests and plantations in low-precipitation areas.

According to climate change projections, approximately 70% of the current temperate forest in Mexico could be affected by climate change (Villers, 1995). Other vulnerability studies (Gay-García and Ruiz Suarez, 1996), carried out on the basis of CCC-J1 GCMs (Boer *et al.*, 1992; Boer, 1993; McFarlane *et al.*, 1992; GDFL-A3, Wetherald and Manabe, 1988), suggest that 10% of all vegetation types in northern Mexico's ecosystems—including forests and shrublands of southern Chihuahua, eastern Coahuila, northern Zacatecas, and San Luis Potosí—would be affected by drier and warmer conditions, resulting in the expansion of dry and very dry tropical forests and xerophytic shrublands. Changes in precipitation intensity and seasonality are likely to have larger impacts on the future composition and distribution of Costa Rican and Nicaraguan forests than annual temperature changes alone (Halpin *et al.*, 1995a,b). About 35 million ha in Venezuela have been projected to change from subtropical forest to tropical forest, and 40–50 million ha could shift from moist to dry or very dry forest by the time atmospheric $CO_2$ concentration doubles, according to studies performed on the basis of scenarios derived from a number of models: GFDL (Manabe and Wetherald, 1987), GISS (Hansen *et al.*, 1984), OSU (Schlesinger and Zhao, 1989), and UKMO-H2 (Mata, 1996; Mitchell and Warrilow, 1987).

In Bolivia, forests cover about 53.4 million ha (48% of the national territory). The deforestation rate over a period of 18 years has been estimated to be on the order of 168,000 ha per year, as a result of the expansion of agricultural frontiers and the selective extraction of species with high commercial value. Changes in Bolivian forest characteristics resulting from climate change have been assessed on the bases of Holdridge's model, and on the GISS-G1 (Hansen *et al.*, 1984) and UKMO-H3 (Mitchell *et al.*, 1989) GCMs, assuming a doubling of $CO_2$ concentrations. The resulting projections indicate reductions in humid subtropical forests (86.6% under GISS-G1, 95% under UKMO-H3), and expansions of humid tropical forests (61.8% under GISS-G1, 8.9% under UKMO-H3) (Tejada Miranda, 1996).

Middle- to high-latitude forests would be more strongly affected by temperature than tropical and subtropical forests; they also are believed to be more sensitive to the speed of climate change than to its magnitude (IPCC 1996, WG II, Section 1.5). In Latin America, therefore, climate warming per se should have its greatest impact on the Austral forests of southern Chile and Argentina, especially at their northern boundaries with other ecosystem types.

The foregoing information has very important implications for governments, local communities, and other stakeholders in Latin American countries: It is apparent that the consideration of feedbacks between changes in climate and land-use practices is essential in the design of sustainable development policies. The study of linkages between the present climate and projected climate, in the context of seasonal-to-interannual variability, is the first step in this design process.

#### *6.3.1.2. Rangelands*

Rangelands—grasslands, shrublands, savannahs, and hot and cold deserts, excluding hyper-arid deserts—cover 33% of the area of Latin America (IPCC 1996, WG II, Section 2.1). Rangeland productivity and species composition are directly related to the highly variable amounts and seasonal distribution of precipitation and only secondarily controlled by other climate variables. Because rangeland systems are driven by extremes, increases in the frequency and magnitude of extreme events (Easterling, 1990) may have a disproportionate effect on them (Westoby *et al.*, 1989). Because of their low productivity, rangeland management units usually are very large, often incorporating considerable spatial heterogeneity (Stafford Smith and Pickup, 1993).

Latin American rangelands sustain pastoralist activities, subsistence farming, and commercial ranching and are a key factor in the economy of many countries (e.g., Mexico, countries of the Central American isthmus, Brazil, Argentina, Uruguay). There are approximately 570 million animal units on the subcontinent, and over 80% of them feed from rangelands (Annex D).

Human activities may bring about more change in rangeland ecosystems than any other forces of global change and may interact strongly with climate change impacts, particularly in tropical and subtropical areas (IPCC 1996, WG II, Section 2.3.3). In Argentine Patagonia, for example, the introduction of unsustainable numbers of sheep along with inappropriate management policies has resulted in major changes in pastureland composition and even desertification (Soriano and Movia, 1986). This process is causing the loss of approximately 1,000 $km^2$ per year (LAC CDE, 1992); overall, 35% of the area's pastureland has been transformed into desert (Winograd, 1995). As a result, the number of sheep decreased by 30% between 1960 and 1988—representing a loss of about US$260 million (Paruelo and Sala, 1992).

Boundaries between rangelands and other biomes are likely to be affected by climate changes both directly (e.g., changes in species composition) and indirectly, through changes in fire regimes, opportunistic cultivation, or agricultural release of less-arid margins of the rangeland territory.

Concerning possible climate impacts of tropical deforestation, Salati and Nobre (1991) pointed out that the large-scale conversion of tropical forests into pastures will likely lead to changes in local climates of the Amazon region. This kind of land-use change increases surface and soil temperatures, the diurnal fluctuation of temperature, and the specific humidity deficit and reduces evapotranspiration—there is less available radiative energy at the canopy level because grass has a higher albedo than forests (Cerri *et al.*, 1995).

Climate characteristics affecting soil moisture conditions, relative humidity, or drought stress—in conjunction with changes in fire and grazing regimes—will have the greatest influence on the boundaries of grassland and woody species. In Mexico, approximately 70% of xerophytic shrublands have been projected to shift their geographical distribution as a result of climate change (Villers, 1995). This projection is roughly in agreement with outputs from MAPSS and BIOME 3 (Neilson, 1995; Haxeltine and Prentice, 1996; see Annex C), although these models suggest less dramatic suface reduction. The overall carrying capacity for herbivores may increase or decrease, depending on the balance between eventual increases in productivity and decreases in plant nutritional value. In some regions, warmer temperatures and increased summer rainfall, with fewer frost days, may facilitate the replacement of temperate grasses by tropical grasses (IPCC 1996, WG II, Section 2.6). Although some laboratory experiments on plants grown individually have shown that $C_3$ (temperate) plants tend to respond more positively than $C_4$ (tropical) plants to elevated $CO_2$, differential effects can be offset or even reversed in the field because traits other than photosynthetic pathways—such as architecture phenology, water, and nutrient-use efficiency—tend to play more decisive roles in the field (Bazzaz and McConnaughay, 1992).

The net effect of elevated $CO_2$ on forage quality also is uncertain. Elevated $CO_2$ usually results in higher C:N ratios in plants. This effect is reflected in increased herbivore consumption and decreased herbivore fitness in laboratory experiments involving insect populations and individual plants (Lindroth, 1996). It is uncertain whether similar patterns would occur in systems involving rangelands and large vertebrate herbivores (Díaz, 1995). Any alteration of the standing capacity of grasslands will be economically important, given the scale of livestock production in Latin American tropical and temperate grasslands.

Fire has been a factor in the evolution of grasslands and many other types of rangelands (Medina and Silva, 1990; Eskuche, 1992). Projected substantial increases in the frequency and severity of wildfires (Ottichilo *et al.*, 1991; Torn and Fried, 1992) could lead to vegetation and soil alterations (Ojima *et al.*, 1990).

Temperate grasslands and the animal production depending on them are vulnerable to drought. Therefore, livestock production could be negatively affected by higher temperatures or increased evapotranspiration rates. However, experience has shown that extreme events, such as large-scale floods or drought-erosion cycles, may pose the highest risks (Burgos *et al.*, 1991; Suriano *et al.*, 1992).

Because of their large extent and important capacity to sequester carbon, temperate grasslands play an important role in the maintenance of the composition of the atmosphere. Sala and Paruelo (1997) have estimated the value of maintaining native grasslands to be US$200/ha; they stress that, once grasslands have been transformed into croplands, the reverse process of abandonment of croplands and their slow transformation into native grasslands sequesters only a modest amount of carbon over relatively long periods of time. Fisher *et al.* (1994) have proposed increasing the carbon sequestration capacity of grasslands through reduced burning frequency, nutritional supplementation of soils, and introduction of deep-rooted grasses and legumes, in combination with controlled stocking rates.

#### *6.3.1.3. Deserts*

Extremely arid (<100 mm annual precipitation) deserts in Latin America (i.e., the Chihuahuan, Sonoran, Peruvian, Atacama, Monte, Patagonian deserts) occupy a large proportion of the region (see Figure 6-4); they have significant species richness and a high degree of endemism (IPCC 1996, WG II, Section 4.4.2). On the other hand, their contribution to the region's primary productivity and C and N pools is extremely low. Because they are driven by discrete events (e.g., rainfall) that occur at irregular intervals, deserts are described as pulse-driven ecosystems (IPCC 1996, WG II, Section 3.3.2); soil instability, dune systems, and shifting boundaries are some of their typical features. In some of the driest coastal deserts (e.g., in northern Chile and Peru), fog banks (camanchaca) are particularly frequent and provide the largest, if not the entire, moisture supply in most years (IPCC 1996, WG II, Figure 3-1 and Section 3.3.4). Harvesting of camanchaca—using frames with vertical nylon threads, either static or rotating at very low speeds—is a common activity among communities along the coast; such harvesting provides enough water to maintain some coastal rangelands and woodlands, which support seasonal grazing by sheep and goats.

Specific information on the vulnerability of Latin America's extreme deserts is rather poor, and further research is required. As with all extreme deserts, however, these systems already experience wide fluctuations in rainfall and are adapted to coping with the consequences of extreme conditions. Initial changes associated with climate change are unlikely to create conditions significantly outside the present range of variation (IPCC 1996, WG II, Section 3.2).

Changes in precipitation projected under a set of climate scenarios assembled for sensitivity studies in the Second Assessment Report (SAR) (Greco *et al.*, 1994) (e.g., a 25% increase in the present mean of 2 mm/yr rainfall) are unlikely to produce major ecosystem changes. Regarding changes in temperature, increases projected in the Greco *et al.* (1994) scenarios typically are in the range of 0.5–2.0ºC, with greater increases in the summer. A rise of 2ºC without an increase in precipitation would increase potential evapotranspiration by 0.2–2 mm per day (IPCC 1996, WG II, Sections 3.3.2, 4.2.1).

Human-induced desertification at the boundary between arid and extremely arid areas has the potential to counteract any ameliorating effects of climate change on most deserts, unless appropriate management actions are taken. Even if wetter conditions were to prevail, their effects may be overridden by pressure from resource exploitation (IPCC 1996, WG II, Section 4.4.2). Other arid and semi-arid areas in Latin America also suffer from desertification; in Mexico, for example, water erosion alone affects 85% of the territory (Gligo, 1995). Vulnerability studies undertaken in Mexico under the USCSP (Gay-García and Ruiz Suarez, 1996) show that drought in central Mexico could be severe, affecting various states in the area (such as Michoacan, where more than 50% of the land surface is highly vulnerable to desertification). Similar trends are expected in Jalisco, Colima, Nayarit, Queretaro, Hidalgo, and Guanajuato.

#### 6.3.1.4. *Mountain Ecosystems and Cryosphere*

Latin America's mountain chains strongly influence its climate, hydrological cycles, and biodiversity. They are source regions for massive rivers (e.g., the tributaries of the Amazonas and Orinoco basins) and represent major foci of biological diversification and endemism (IPCC 1996, WG II, Section 5.1.4). Mountain areas are exposed to extreme weather and climate events, such as unusually high or low temperatures or precipitation. The Agroclimatological Study of the Andean Zone (Frére *et al.*, 1978) provides detailed information on the particular limitations that climate imposes on development in the high South American plateaus and associated mountain ranges in the central Andes (Ecuador, Peru, and Bolivia).

Although the present importance of mountain ecosystems in the national market economies varies from country to country, Andean and extra-Andean mountain ranges have sustained traditional subsistence agriculture for centuries to millennia. Athough the human population density is very low in the northern and southern Andes, the central Andes support the largest percentage rural population in the region.

The cryosphere in the Latin American region is represented by glaciers in the Andes; Patagonia's ice fields (between 47°S and 52°S); and the Darwin ice field in Tierra del Fuego, at about 54ºS. Seasonal snowfall on the high Andes is critical for the subsistence of communities in central Chile and large piedmont communities in Argentina, where water supply depends almost entirely on snowmelt.

At present, GCMs do not provide sufficiently accurate regional projections; therefore, scenarios of climate change for Latin American mountain areas are highly uncertain. An added drawback is the scarcity of continuous and reliable meteorological and hydrological records in most areas. If climate changes as projected in Intergovernmental Panel on Climate Change (IPCC) scenarios (Greco *et al.*, 1994), the length of time that snowpacks remain will be reduced, altering the timing and amplitude of runoff from snow/ice and increasing evaporation—hence altering ecosystems at lower elevations and affecting human communities that depend on this runoff. If extreme events increase in frequency or intensity, landslides, flash flooding, and fires would become more likely, and soil instability would increase. Even in subhumid mountain environments, fires in the dry season followed by heavy rainfall events—before substantial vegetation recovery has taken place—usually lead to almost irreversible landscape modification (IPCC 1996, WG II, Section 5.1.3).

Warming in high-mountain regions would lead to the reduction or disappearance of significant snow and ice surfaces (IPCC 1996, WG II, Sections 7.4.1 and 7.4.2). Furthermore, changes in atmospheric circulation stemming from global warming may modify snowfall rates, with a direct effect on the seasonal renewal of water supply and the variability of runoff and underground water supplies in piedmont areas (Del Carril *et al.*, 1996). Glaciers are melting at accelerated rates in the Venezuelan and Peruvian Andes (Schubert, 1992; Hastenrath and Ames, 1995), as well as at the southern extreme of the subcontinent (Aniya *et al.*, 1992; Kadota *et al.*, 1992; Malagnino and Strelin, 1992). Nevertheless, larger glaciers, such as those in the Patagonian Andes, should continue to exist into the 22nd century. More water would be released from regions with extensive glaciers; as a result, some arid regions in their vicinity would benefit from additional runoff. Based on seasonal patterns and extreme events in these regions, however, this runoff could trigger major erosion, flooding, and sedimentation problems (IPCC 1996, WG II, Section 7.2.2). Melting of Andean glaciers and ice fields in southern Patagonia and Tierra del Fuego would increase runoff, as well as the water levels of neighboring rivers and lakes on both sides of the Andes.

Expected impacts of climate change on the biological diversity of mountain ecosystems would include loss of the coolest climatic zones toward the peaks of the mountains and shifting of remaining vegetation belts upslope, resulting in a net decrease in biodiversity. Mountaintops may become more vulnerable to genetic and environmental pressure (Bortenschlager, 1993; IPCC 1996, WG II, Section 5.2.2).

Viable areas for crop production in mountainous regions would likely change as a result of climate change. Elevational shifts in vegetation and altered hydrological patterns may have major implications for the use and conservation of multiple vegetation belts by traditional Andean peoples. These shifts may lead to competition between alternative land uses (e.g., conservation of endangered species versus expansion of subsistence agriculture) toward the mountaintops. Given the wide range of microclimates in Latin America's mountain areas that have been exploited through the cultivation of diverse crops, direct negative effects of climate change on crop yields may not be extremely serious. However, continued adaptation to varied climates may be possible only if the remarkable diversity of local genotypes is conserved. Development planners and decision makers should be fully aware that the protection of the wide variety of wild and domesticated genotypes of major crops in the Andes also will be crucial for the production of new crop varieties in the face of changing climatic conditions in other areas of the world. In 1960, for example, the discovery of two varieties of tomato in Peru provided the industry with economic benefits estimated at US$5 million per year (LAC CDE, 1992). Decision makers also should be aware that, unless appropriate adaptation measures are taken, climate change could completely disrupt lifestyles in mountain villages by altering already marginal food production and the availability of water resources.

Shifts in crop and rangeland elevation belts may be dramatic in rural, densely populated areas of the central Andes; deep socioeconomic changes have been brought about by past climatic pulses (Cardich, 1974; Frére *et al.*, 1978). Studies of the effects of climate change in Ecuador's central Sierra (Parry, 1978; Bravo *et al.*, 1988) have shown that crop growth and yields are controlled by complex interactions among different climatic factors and that specific methods of cultivation may permit crop survival in sites where microclimates otherwise would be unsuitable. Such specific details cannot be included in GCM-based impact assessments—which have suggested positive impacts, such as decreasing frost risks in the Mexican highlands (Liverman and O'Brien, 1991), and negative impacts, such as decreases in the productivity of upland agriculture (Parry *et al.*, 1990).

In many mountain regions, tourist resorts and large urban areas close to the mountains have spread into high-risk areas; these areas will be increasingly endangered by slope instability and flood risk, particularly as a consequence of extreme events aggravated by climate change. Such situations already have been observed in Andean cities (Bogota, La Paz, Mendoza, Quito); cities on the Serra do Mar mountain range (Rio de Janeiro, Sao Paulo, Santos) and their outskirts; and Guatemala City, in Central America.

Runoff changes resulting from snow/ice melting and from changes in winter snowfalls on high-altitude mountains would affect important sectors and activities (freshwater supply, agriculture, industry, and power generation) in the Andes piedmont areas (e.g., Cuyo region in Argentina, Elqui Valley in Chile). In view of the economic importance of these activities, adaptation measures are likely to be necessary (Fuenzalida *et al.*, 1993; Del Carril *et al.*, 1996; IPCC 1996, WG II, Section 5.2.4).

### 6.3.2. Hydrology, Water Resources, and Freshwater Fisheries

Latin America (particularly its tropical region) is rich in freshwater systems. Large river basins, lakes, freshwater wetlands, and reservoirs distributed throughout the region have facilitated the development of thousands of human settlements—with associated agricultural and industrial activities—and the harnessing of rivers for energy production. In addition, important segments of national and international rivers permit fluvial transportation of people and merchandise. The opening of the international Paraguay-Paraná Rivers Hydroway would expedite the increasing commercial traffic arising from the developing regional market, Mercosur. Freshwater systems also are sources of income from fisheries, aquaculture, and tourism. Latin America's extensive river basins, large numbers of lakes, and important freshwater wetlands (i.e., El Pantanal and Iberá) host a great number of fish species, amphibians, semiaquatic rodents, and reptiles, some of which—like the capybara (*Hydrochoerus hydrochoeris*) and yacaré (a type of *Caiman latirostris*), which are particular to this region—are important sources of income for local communities.

Tropical rivers in Latin America host about 1,500 species; in addition to their commercial value, these species are the basis for the yearly international Alto Parana river "dorado" fishing contest, an important tourist event. The wetlands also are stopover places for a number of bird species—especially waterfowl, which fly back and forth between the United States and Canada and Latin America. The longest waterfowl migration probably is that of the blue-winged teal (*Anas discors*), which nests as far north as 60°N in North America and winters south of 30°S, traveling a distance of more than 9,600 km.

Freshwater systems are potentially very sensitive to climate change (IPCC 1996, WG II, Section 10.3) and vulnerable to interannual fluctuations in climate, such as those associated with the ENSO phenomenon. Increased climatic variability is expected to have greater ecological effects than changes in average conditions. Associated flash floods and droughts would reduce the biological diversity and productivity of stream ecosystems (IPCC 1996, WG II, Section 10.6) and affect living conditions and welfare in flood-prone areas as well as in arid and semi-arid regions. Thus, monitoring of associated environmental variables is an urgent need.

Most Latin American countries, however, have inadequate hydrological and meteorological observation networks. Large, uninhabited areas have no surface observation systems, and the densities and operational practices of existing networks generally do not meet the recommended international and regional standards. As a result, records of variables that are important for the study of freshwater systems are scarce and suffer from incomplete areal and temporal coverage, particularly because of frequent closings of observation posts in recent decades. Under these conditions, water resources monitoring vis-à-vis climate change is very indequate. In fact, the 1992 Supplementary Report to the IPCC First Assessment Report (FAR) concluded that, although significant progress has been made in hydrological sensitivity analyses in developed countries, large gaps exist for less-developed countries.

About 35% of the world's continental waters (freshwater) are found in Latin America, but the distribution within and among countries is highly variable (see Annex D for data on water resources per capita and annual domestic and industrial water withdrawals). Many areas (e.g., northern Mexico, northeastern Brazil, coastal Peru, northern Chile) have great difficulty meeting their water needs. About two-thirds of Latin America is arid or semi-arid, including large portions of Argentina, Bolivia, Chile, Peru, northeastern Brazil, Ecuador, Colombia, and central and northern Mexico.

Climate change impacts on hydrological cycles are expected to be greatest in semi-arid and arid areas; on lakes and streams in high evaporative drainages; in basins with small catchments and relatively short retention times; in shallow lakes, streams, and rivers in which appropriate thermal refugia are not available; and on irrigation systems that rely on isolated reservoirs or snowmelt from seasonal snowfall (IPCC 1996, WG II, Chapters 7 and 10). The implications of climate warming on the hydrological cycle and its consequences for precipitation distribution, intensity, and timing; surface runoff; and underground water resources will be aggravated in certain areas by population growth and unsustainable development of water-consuming activities. Country studies undertaken to satisfy commitments under the UNFCCC indicate the vulnerability of Latin American communities that are dependent on water resource availability. In this respect, reduced precipitation has had serious consequences for hydropower production in Central American countries; for example, ENSO events during the past decades in Costa Rica led to significant reductions in runoff and consequently to a higher demand for thermal energy production (Campos *et al.*, 1996) (see Figure 6-5). On the other hand, increased precipitation (up to 35% in the province of La Pampa) has been observed in the Pampas—the most important area for grain and livestock production in Argentina (Canziani *et al.*, 1987; Forte-Lay, 1987; Vargas, 1987). The benefits from such an increase in precipitation have been offset by more frequent extreme precipitation events and higher risks of flooding.

Extreme precipitation events could increase the number of reservoirs in the humid tropics that silt up well before their design lives have been reached. Poor management of a basin (e.g., severe deforestation in drainage areas, especially in the case of rivers flowing along deep valleys) would increase erosion and interact with hydrological changes, causing siltation and decreasing the potential for hydroelectricity generation (Bruijnzeel and Critchley, 1994; Campos *et al.*, 1996).

Lakes have individualistic and often rapid responses to climate changes. Lake Titicaca, for example, experienced a 6.3-m rise in water level from 1943 to 1986 (IPCC 1996, WG II, Section 10.5.2). This increase exceeds by a factor of 40 the change in mean sea level estimated from global warming. Fluctuations in lakes with large temporal changes in water level are expected to be aggravated by climate change. Changes in mixing regimes also are projected as a result of climate change in temperate zones because of increases in winter air temperatures—with potentially large effects on biota.

Wetlands are distributed throughout the region but are more extensive in the tropics and subtropics. The effects of climate change on wetlands remain very uncertain (Gorham, 1991; IPCC 1996, WG II, Figure 6-1). Human intervention may be more critical than climate change in affecting these ecosystems.

Freshwater fisheries and aquaculture generally could benefit from climate change, though there could be some significant negative effects, depending on the species and on climate changes at the local level. Positive factors associated with warming and increased precipitation at higher latitudes include faster growth and maturation rates, lower winter mortality rates from cold or anoxia, and expanded habitats as a result of ice retreat. Offsetting negative factors include increased summer anoxia, increased demand for food to support higher metabolisms, possible negative changes in lake thermal structures, and reduced thermal habitat for cold-water species. Individual effects are difficult to integrate. However, warm-water lakes generally have higher productivity levels than cold-water lakes; because of the latitudinal distribution of projected warming, warm-water lakes will be in areas with the least change in temperature. It is reasonable to expect higher overall productivity from freshwater systems. Finally, fishery managers heavily manipulate freshwater fisheries, and aquaculture is expanding in many Latin American countries. If species mixes continue to be changed to support angler and market preferences and changing habitats, climate-change damages may diminish while benefits increase.

Impacts on biota—including commercial and subsistence fisheries—are expected to be most pronounced in isolated systems, such as high-elevation Andean lakes or lakes in the far south of

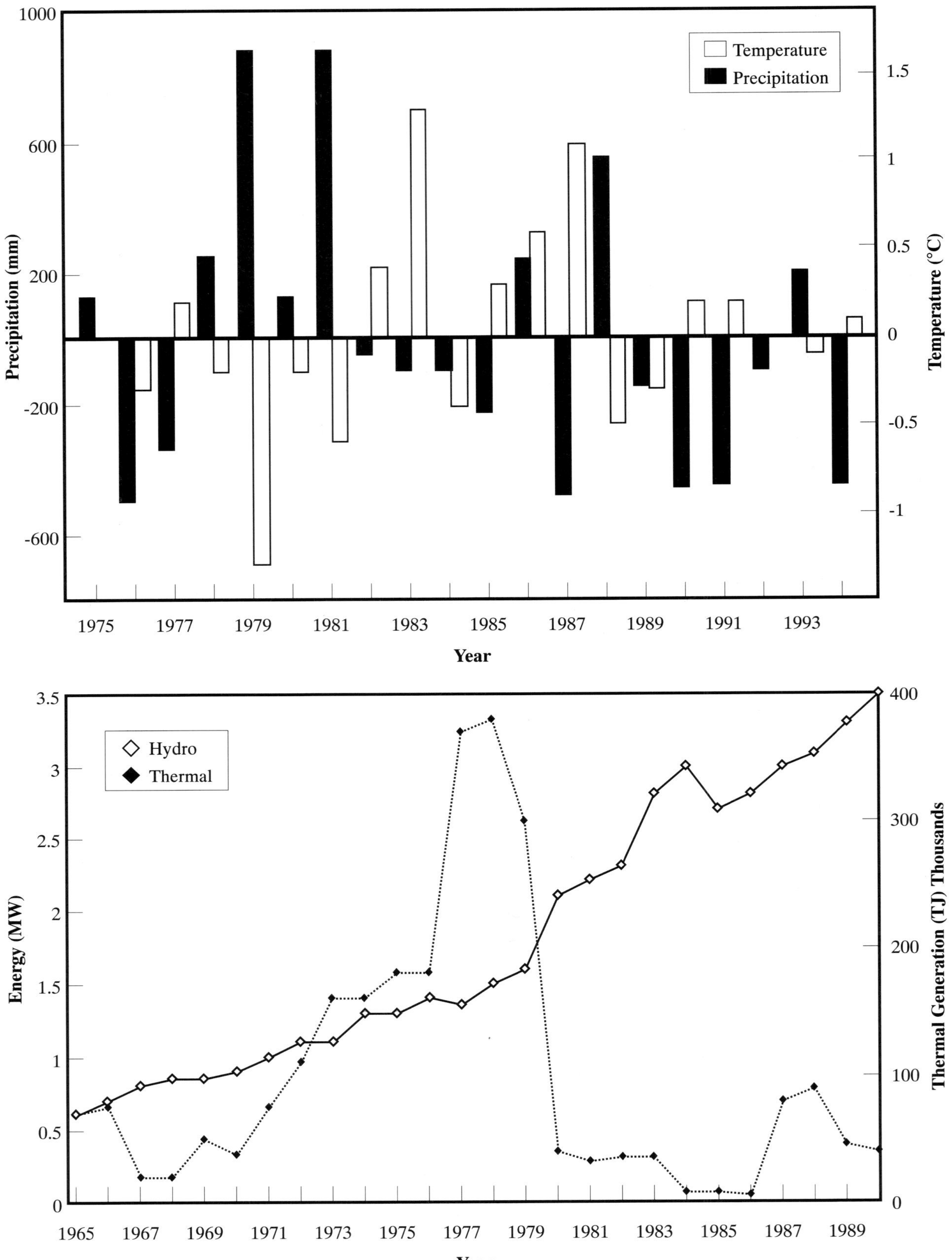


**Figure 6-5:** Precipitation and maximum temperature anomalies for the Pacific coast of Costa Rica resulting from ENSO phenomena and changes in electricity generation types.

Argentina and Chile, as well as in areas in which species are close to their geographical distribution limits. Within large drainage systems, flexibility for migratory shifts to compatible temperatures will be greater in north-south flowing systems (Meisner, 1990; IPCC 1996, WG II, Sections 10.6.1.2, 16.2.1).

The effects of climate change on freshwater ecology will interact strongly with anthropogenic changes in land use, waste disposal, and water extraction. Regional water resources will become increasingly stressed because of higher demands to meet the needs of growing populations and economies, as well as temperature increases. These anthropogenic effects are expected to increase in Latin America, and there is a risk that these impacts will extend to tropical streams and wetlands that, owing to their remoteness, have so far escaped the impacts of human activities (Armentano, 1990; IPCC 1996, WG II, Sections 6.3 and 10.2.2).

Conflicts may arise among users and regions and among Latin American countries that share common river basins. The effects of climate change on agricultural demands for water, particularly for irrigation, will depend significantly on changes in agricultural potential, prices of agricultural produce, and water costs (IPCC 1996, WG II, Chapter 14), as well as on other consumptive uses. Water availability in two countries of the region (Mexico and Peru; see Box 6-3) is expected to decline significantly with respect to the present climate. Furthermore, as has happened in the past, reductions in or a lack of freshwater availability, or cases of excess leading to flood conditions, may lead to interstate or international conflicts or invasion of neighboring territories by "campesinos" in distress.

### *6.3.3. Oceans, Saltwater Fisheries, and Coastal Zones*

#### *6.3.3.1. Oceans and Saltwater Fisheries*

The oceans function as regulators of the Earth's climate and sustain planetary biogeochemical cycles. Natural fluxes of $CO_2$ among the atmosphere, oceans, and land biota are much larger than anthropogenic perturbations (IPCC 1996, WG I, Section 2.1). The oceans are estimated to have absorbed about 30% of $CO_2$ emissions arising from fossil fuel use and tropical deforestation between 1980 and 1989 (Siegenthaler and Sarmiento, 1993), thus slowing the rate of greenhouse warming. The oceans also play a major role in the global hydrological cycle. Cycling of the oceanic freshwater fraction through advection, evaporation, precipitation and—in higher latitudes and elevations—the solid-ice phase is affected by changes in wind systems and oceanic current systems, as is the case with the Latin American climate (see Section 6.2). A changing pattern of rainfall over the oceans would cause changes in the rainfall pattern over land—which would, in turn, have a considerable effect on the salinity of marginal seas.

Living marine resources include fish, crustaceans and shellfish, marine mammals, and seaweed. In 1990, the world's fish catch was 97 million tons (14 Mt from inland sources and 83 Mt from marine sources) (FAO, 1992a). Latin America, which is located between two large oceans, has a recognized wealth of marine biotic resources, sustaining large schools of fish and enabling the development of some of the most important saltwater fisheries in the world. The average annual catch by Latin American countries during the period 1985–87 was about 13 Mt, or about 17% of the world's catch (WRI, 1992). Although these figures do not include catches made by other nations plying the region's seas, they provide a clear idea of the economic importance of the oceans in Latin America. Furthermore, krill—a pelagic member of the crustacean suborder *Euphausiacea*—is of great importance in the trophic chain of the oceans (as food for various fish, birds, and whales, particularly blue whales and finback whales). From January to April, swarms of krill (*Euphausiacea superba*) appear in the Antarctic Ocean and, because of the predominant sea currents, move northward into the Atlantic and Pacific Oceans surrounding South America. They have a large impact on fisheries along the eastern and western coasts of the subcontinent. It should be stressed that, because of their vast numbers and nutritive qualities, krill have been regarded by ecologists as a potential food source for humans. During the Southern Hemisphere summer, the open waters of the Antarctic Ocean may contain as much as 20 kg/m$^3$ of these animals.

**Box 6-3. Water Availability (m$^3$/yr) in Mexico and Peru under Present Climatic Conditions in 1990 and 2050 and for Three Scenarios for 2050**

The figures in column 2 show only the effect of population growth (no climate change is assumed). Column 3 summarizes the combined impact of population growth and climate change on water availability. Calculations are based on IPCC (1992) socioeconomic scenarios and the outputs of three transient GCM runs: GDFL-X2, UKMO-H3, and MPI-K1 (see Tables 1-1 and 1-2 in the Introduction).

| Country | Present Climate 1990 | Present Climate 2050 | Scenario Range 2050 |
|---|---|---|---|
| Mexico | 4,720 | 2,100 | 1,740–2,010 |
| Peru | 1,860 | 880 | 690–1,020 |

Currently, the annual rate of growth in the world marine catch is declining (FAO, 1992b) as a consequence of uncontrolled growth in fishing activities and overfishing of important stocks in the Atlantic Ocean, including the waters off the South American coasts. A combination of human activities (e.g., overfishing; pollution of estuaries and coastal oceans; and the destruction of habitat, especially wetlands and seagrasses) currently exerts a far more powerful effect on marine fisheries than declines expected from climate change (IPCC 1996, WG II, Section 8.2.2).

The oceans also are climate regulators: The combined effect of atmospheric and oceanic circulation defines particular weather and climate conditions throughout the region. The effect of sea currents is more evident during ENSO events, when resulting climatic variabilities have strong socioeconomic implications (see Section 6.2.3). The catastrophic effects of storms and storm surges are well-known, particularly for their role in exacerbating flooding in coastal areas and in erosion and restructuring of coastal formations (IPCC 1996, WG II, Section 8.4).

It is very probable that climate change would have significant negative impacts on human uses of the oceans—the most important resulting from impacts on biotic resources. It also is very likely that increased precipitation, river runoff, and atmospheric deposition from land-based activities would lead to increased loading of pollutants in coastal waters, with an adverse impact on fisheries, coral production, and tourism (IPCC 1996, WG II, Section 8.4).

Global warming may lead to poleward shifts in species ranges and migration patterns within the seas along the South American coast; this shift could lead to increased survival of economically valuable species and increased yields in marine fisheries. Such cases have been observed as a result of the large and intense 1983 El Niño event (Wooster and Fluharty, 1985), which may be considered a test for future climate scenarios. However, the impacts of El Niño are not the same for different regions of the Pacific Ocean. In fact, in areas where productivity depends on upwelling (e.g., the Pacific coast of Peru and Chile), the El Niño phenomenon leads to a decrease in ocean productivity (Jordan Sotelo, 1986; Budyko and Izrael, 1987; Lapenis *et al.*, 1990; IPCC 1996, WG II, Box 16.6).

Under IPCC scenarios, saltwater fisheries production should remain about the same—or undergo a significant increase, if management deficiencies are corrected. These conclusions depend on the assumption that interannual and decadal natural climate variability and the structure and strength of winds and ocean currents will remain about the same. Changes in any of these variables are expected to have significant effects on the distribution of major fish stocks, although not on overall production. Even without major changes in atmospheric and oceanic circulation, local shifts in centers of production and mixes of species in marine and fresh waters are expected to occur as ecosystems are displaced geographically and change internally.

Ecosystems and the services that they provide are sensitive to the rate and extent of changes in climate. The distribution of the world's major biomes, for example, is correlated with mean annual temperature and mean annual precipitation. The composition, dominances, and geographic distribution of many ecosystems will likely shift as individual species respond to environmental change driven by changes in climate: There will likely be changes in biological diversity and in the goods and services that ecosystems provide. Some ecological systems may not reach a new equilibrium for several centuries after the climate achieves a new balance.

Changes in the extent and duration of sea ice, combined with changes in the characteristics of sea currents, may affect the distribution, abundance, and harvesting of krill—an important link in the ocean fauna in the southern oceans. If there is a rapid retreat of sea ice in Antarctica or if the sea ice is reduced in extent, the krill fishery could become more attractive to nations involved in krill fishing (IPCC 1996, WG II, Section 16.2.2.2). Such a possibility could critically harm fishing activities in South America.

In tropical latitudes, most migratory organisms are expected to be able to tolerate climate warming, but the fate of sedentary species will be highly dependent on local climate changes. For example, corals are sensitive to changes in temperature; coral bleaching occurred during El Niño events in 1983 and 1987 (Glynn, 1989; Brown and Odgen, 1993), when the ocean temperatures were higher than normal. Therefore, under warmer environmental conditions resulting from climate change, the expectation is that corals and other sedentary species would be affected. Coral mortality is positively correlated with the intensity and length of warming episodes (Glynn, 1989; Glynn and Cruz, 1990); recent paleoclimatic investigations of ENSO phenomena show that coral records can be read as a proxy for increased sea-surface temperarures (SSTs), potentially over several thousand years (Cole *et al.*, 1992; Shen, 1993). Researchers recognize, however, that other environmental stresses—such as pollution, sedimentation, or nutrient influx—also would affect corals and other sedentary species (Maul, 1993; Milliman, 1993; IPCC 1996, WG II, Section 9.4).

Regarding the impacts of climate change on navigation in the southern oceans, there is no clear consensus—despite recent high rates of iceberg calving (Doake and Vaughan, 1991; Skvarca, 1993, 1994)—on whether the abundance of icebergs and their danger to shipping would change with global warming (IPCC 1996, WG II, Section 7.5.3).

#### 6.3.3.2. *Sea-Level Rise*

Projections of sea-level rise reported by IPCC Working Group I (scenario IS92a) indicate that sea level could rise, on average, about 5 mm/yr, within a range of uncertainty of 2–9 mm/yr. An important point to bear in mind is that the current best estimates represent a rate of sea-level rise that is about two to five times the rate experienced over the past 100 years (1.0–2.5 mm/yr). Changes in sea level at regional and local levels will not necessarily be the same as the global average change because vertical land movements affect sea level and there are dynamic effects resulting from oceanic circulation, wind and pressure patterns, and ocean-water density that cause variations in the level of the sea surface with respect to the geoid (IPCC 1996, WG II, Section 9.4).

Biogeophysical effects of sea-level rise will vary greatly in different coastal zones around the world because coastal landforms and ecosystems are dynamic; they respond to and modify the variety of external and internal processes that affect them. For instance, flooding conditions in the Pampas, in the province of Buenos Aires, would be exacerbated by any level of sea-level rise because the effectiveness of the Salado river as the only drainage system for this flatland would be reduced by sea-level rise. Some coastal sectors in Central America and on the Atlantic coast of South America would be subject to inundation risk. Flat areas—such as the Amazon, Orinoco, and Paraná river deltas—and the mouths of other rivers, such as the Magdalena in Colombia, would be affected by sea-level rise (IPCC, 1990). Estuaries like the Rio de la Plata also would suffer increasingly from saltwater intrusion, creating problems in freshwater supply. These effects will depend on the amount of sea-level rise and the characteristics of atmospheric and oceanic circulation.

Actual land loss resulting from sea-level rise may represent a small fraction of national territories, but it may have major impacts in those areas where large human settlements, tourist resorts, and other activities and infrastructure are located. Synthesized results of country case studies are presented in Table 6-5, including the estimated impacts of a 1.0-m sea-level rise. Consideration of particular cases, such as that of Uruguay, shows that although the amount of land lost on the Uruguayan coast would be rather small, the capital risk is very important. The national tourist industry, which creates more than US$200 million per year in revenues and attracts more than one million people each summer, might be seriously affected. Protected coastal areas, such as Laguna de Tacarigua in Venezuela, could be altered dramatically (Perdomo *et al.*, 1996). The only areas of Argentina that are vulnerable to sea-level rise are the coasts of the province of Buenos Aires and the Rio de la Plata estuary (Perillo and Picccolo, 1992). The coastal aquifers also would be affected by sea-level rise and would suffer from seawater intrusion, affecting the supply of freshwater on both margins of the Rio de la Plata, including the densely populated Buenos Aires metropolitan area.

Most global fish resources depend on near-shore or estuarine habitats at some point in their life cycles (IPCC, 1990; Chambers, 1991). Fish production would suffer if coastal wetlands and other habitats that serve as nurseries were lost as a consequence of sea-level rise (Costa *et al.*, 1994). It also is apparent that mangrove communities starve in microtidal, sediment-poor environments—such as around the Caribbean, where the lack of strong tidal currents does not permit sediment distribution (Parkinson *et al.*, 1994).

Coastal oceans already are under stress from a combination of factors, such as increased population pressure, habitat destruction, increased land-based pollution, and increased river inputs of nutrients and other pollutants (IPCC 1996, WG II, Section 9.2.2). Therefore, the effects of global climate change could represent a mixed, and probably synergistic, series of impacts on an already overstressed context. There also may be synergistic effects between climate change and overfishing. These combined stresses would reduce fish quality and stocks in Latin American seas, increasing the vulnerability of the fishing industry—particularly on regional seas plied by countries outside the region. In addition to the aforementioned coastal zone vulnerability, sea-level rise and increasing storm activity

***Table 6-5:*** *Synthesized results of country studies. Results are for existing development and a 1-m sea-level rise. People affected, capital value at loss, and wetland at loss assume no adaptive measures (i.e., no human response), whereas adaptation assumes protection except in areas with low population density. All costs have been adjusted to 1990 US$ (adapted from Nicholls, 1995).*

| Country | People Affected | | Capital Value at Loss | | Land at Loss | | Wetland at Loss | Adaptation/ Protection | |
|---|---|---|---|---|---|---|---|---|---|
| | # People (000s) | % Total | Million US$ | % GNP | km$^2$ | % Total | km$^2$ | Million US$ | % GNP |
| Argentina | — | — | >5,000[1] | >5 | 3,400 | 0.1 | 1,100 | >1,800 | >0.02 |
| Belize | 70 | 35 | — | — | 1,900 | 8.4 | — | — | — |
| Guyana | 600 | 80 | 4,000 | 1,115 | 2,400 | 1.1 | 500 | 200 | 0.26 |
| Uruguay | 13[2] | <1 | 1,700[1] | 26 | 96 | 0.1 | 23 | >1,000 | >0.12 |
| Venezuela | 56[2] | <1 | 330 | 1 | 5,700 | 0.6 | 5,600 | >1,600 | >0.03 |
| Venezuela | — | — | 153[3] | 0.46 | 124 | 0.22 | — | — | — |

[1]Minimum estimates—capital value at loss does not include ports.
[2]Minimum estimates—number reflects estimated people displaced.
[3]Including land and buildings.
Note: The two Venezuelan cases are based on different methodologies and refer to different places.
Source: IPCC 1996, WG II, Table 9-3.

would have adverse consequences for offshore extraction activities, marine transportation, human health, recreation, and tourism. An additional impact of sea-level rise on Latin American countries, particularly those with low population densities, would stem from political and cultural stresses arising from the relocation of immigrants from other regions (e.g., as a result of the inundation of low coastal zones and small islands) (Canziani, 1993).

In Central America, impacts associated with sea-level rise would have their greatest effects on infrastructure, agriculture, and natural resources along the coastline, with immediate effects on socioeconomic conditions in the isthmus countries. Sea-level rise would exacerbate the processes of coastal erosion and salinization of aquifers and increase flooding risks and the impacts of severe storms along the coastline (Campos *et al.*, 1997).

### *6.3.4. Agriculture*

Latin America has about 23% of the world's potential arable land, although—in contrast with other regions—it maintains a high percentage of natural ecosystems (UNEP, 1992). The total cropland (about 134 million ha), excluding pastures, represents less than 10% of the total area of the Latin American region (see Annex D). About 306 million ha (72.7%) of the agricultural drylands in South America (i.e., irrigated lands, rain-fed croplands, and rangelands) suffer from moderate to extreme deterioration (UNEP, 1992), and about 47% of the soils in grazing lands have lost their fertility (LAC CDE, 1992). These land degradation figures include erosion and soil degradation on hillsides and in mountain areas, as well as desertification by overgrazing, salinization, and alkalinization of irrigated soils in tropical pasture lands. The area of irrigated land has increased substantially: Data for the periods 1966–68 and 1986–88 show that irrigated land has increased from 8,674,000 ha to 14,040,000 ha (UNEP, 1992). A preliminary calculation by the Global Assessment of Soil Degradation (GLASOD, 1990) suggests that approximately 14% of the land area in South America is affected by some form of human-induced soil degradation. The percentage distribution of this land area, by type and degree of soil degradation, is summarized in Table 6-6.

***Table 6-6:*** *Type and degree of human-induced soil degradation in South America.*

| | Degree (%) | | | |
|---|---|---|---|---|
| **Type** | Slight | Moderate | Severe | Total |
| Water erosion | 14 | 29 | 4 | 47 |
| Nutrient decline | 12 | 17 | 5 | 34 |
| Wind erosion | 9 | 6 | | 15 |
| Waterlogging | 3 | | | 3 |
| Salinization | 1 | | | |
| **Total** | 39 | 52 | 9 | 100 |

Source: GLASOD, 1990; UNEP, 1992.

Historical shifts in land-use patterns are important factors to consider in the design of adaptive strategies aimed at maintaining or increasing production in the face of climate change while preventing soil erosion and other undesirable environmental consequences (Viglizzo *et al.*, 1995).

Studies of the dynamics of climate and land use during the past century have been performed by Viglizzo *et al.* (1997), with the aim of quantifying the impact of climate on land-use change and discussing adaptive land-use strategies for the Argentine Pampas. These studies show a high and positive degree of association between rainfall and crop growth in humid zones. This relationship declines, however, in the transition from humid to semi-arid zones. In these marginal environments, land-use change appears to be highly correlated to the crop yields (which are a combined function of climate and technology) and less correlated to the rainfall pattern (Viglizzo *et al.*, 1995; Viglizzo *et al.*, 1997).

The contribution of agriculture (10–12%) to the GDP of Latin American countries in the past 40 years has been secondary to other human activities (Baethgen, 1994). Agriculture is still a key sector in the region's economy, however, because it occupies an important proportion (30–40%) of the economically active population. Moreover, agriculture has generated the largest export income in countries having neither oil nor mineral production. Finally, in the smaller and poorer countries of the region, agriculture is the basis of subsistence lifestyles, the largest manpower user, and the main producing and exporting sector (Baethgen, 1994; IPCC 1996, WG II, Section 13.7).

Because agricultural production in lower-latitude and lower-income countries is more likely to be negatively affected by climate change (IPCC 1996, WG II , Section 13.8), large Latin American communities would be vulnerable to global warming. Climatic variations that result in shorter rainy seasons and/or increased frequency of rainless years would have extremely negative consequences for the region. However, yield impacts also depend on other factors, such as cultivars and specific environmental conditions.

In Mexico, any shift toward warmer, drier conditions could bring nutritional and economic disaster because agriculture already is stressed by low and variable rainfall. More than one-third of the Mexican population works in agriculture—a sector whose prosperity is critical to the nation's debt-burdened economy. Although only one-fifth of Mexico's cropland is irrigated, this area accounts for half of the value of the country's agricultural production, including many export crops (Liverman and O'Brien, 1991). The principal conclusion of Liverman and O'Brien's work is that Mexico is likely to be warmer and drier. According to each of the models used, potential evaporation is likely to increase; in most cases, moisture availability will decrease, even where the models project an increase in precipitation. Sensitivity analyses of evaporation and moisture-deficit calculations indicate that water availability would be expected to increase only if the model results produced much higher rainfall and relative humidities or significantly less solar

**Box 6-4. Assessment of the Vulnerability of Agricultural Activities in Belize to Climate Change**

Under the coordination of the Proyecto Centroamericano de Cambios Climáticos (PCCC) and with the support of the USCSP, vulnerability studies were undertaken on the impact of climate change on the agricultural sector in Central America. The methodology considered the acquisition and analysis of data; the use of agroclimatological models; and the performance of biophysical studies, involving simulations under base and modified climates and the evaluation of results (t/ha, crop duration, evapotranspiration). Scenarios selected included a range of temperature changes between +1°C and +2°C and precipitation changes ranging from +20 mm to -20 mm, at 10-mm intervals.

Impacts of Climate Change on Maize, Red Kidney Beans, and Rice Production in Belize

The field experiment took place in two places: one rain-fed area, for maize and beans; and one dryland area, for rice. The basic data included detailed soil information, maximum and minimum temperatures, daily sunshine hours, and precipitation; the model computed other required variables. The model also included management information (date, amount, type, fertilizer applications, depth and spacing of the plant, and planting date).

The conclusions of this assessment may be synthesized as follows: On the whole, simulated changes in crop yields are driven by two factors—changes in climate (temperature and precipitation) and $CO_2$ enrichment. The interactions of these factors on baseline crop growth are often complex. However, yield decreases are caused primarily by the increase in temperature, which shortens the duration of crop growth stages. New and fluctuating weather patterns could have a strong negative impact on economic activities in agriculture. The majority of Belizean people who are highly dependent on farming might well see their livelihoods destroyed by reduced rainfall and increased temperatures as a consequence of climate change. Agriculture continues to be a major part of the economy of Belize. Therefore, planning and evaluating strategies for adapting to climate change are important.

radiation and wind. Such changes are possible, of course, as the modeling of clouds and synoptic conditions improves; at present, however, it seems that a moisture decrease is more likely.

A series of studies in all of the countries of the Central American isthmus under the Central America Project on Climate Change—with the cooperation of the USCSP—estimated the vulnerability of agricultural resources. These studies were based on scenarios generated through a set of GCMs, including CCC-J1, UKMO-H3, GISS-G1, GFDL-A1, and GFDL-K2. The studies focused on specific crops (e.g., maize, rice, sorghum, beans)—surprisingly enough, however, not on export crops, such as bananas and coffee. A full summary of the results of these studies would be tedious, and they are not highly reliable (as indicated in the Executive Summary of the report, the aforementioned GCMs, in general, are not quite fitted for this type of specific study). Therefore, Box 6-4 presents a case study to provide information about the nature of the results achieved in this first exercise to fulfill the UNFCCC's national reporting requirement.

Regarding the relationship between agriculture and water resources, the studies mentioned above are limited to the known impacts of water deficits on agricultural activities. However, information about some cash crops in countries of the Central American isthmus indicates that, under current climate conditions, the productivity of banana crops is historically affected, particularly in areas already subject to flooding, by environmental conditions associated with tropical storms. This preexisting condition would indicate that, along the Central American-Caribbean watershed, these crops could be additionally stressed if climate change leads to increasing frequency of storms and heavy precipitation (Campos, 1996).

In climatic studies involving projections of GCMs and crop models of wheat, maize, barley, soybeans, potatoes, and grapes in Latin America (Table 6-7), crop yields decreased in the face of climate change—even when direct effects of elevated $CO_2$ were taken into account—in 9 of 12 studies. Experience available with regard to the development and spread of pests and diseases permits the inference that climate change would trigger many of them and extend their geographical ranges (Austin-Bourke, 1955; Omar, 1980; Pedgley, 1980).

One area that is highly vulnerable to climate change is the Brazilian northeast, which is strongly influenced by the ENSO phenomenon. Years with no rain are frequent; these periods are characterized by the occurrence of famine and large-scale migrations to metropolitan areas (Magalhães and Glantz, 1992; IPCC 1996, WG II, Section 13.7; IPCC 1996, WG III, Section 6.5.9). In the global agricultural model of Rosenzweig *et al.* (1993), yield impacts of climate change in Brazil are among the most severe for all regions. Under $2xCO_2$ scenarios, yields are projected to fall by 17–53%, depending on whether direct effects of $CO_2$ are considered. Similar reductions also are projected for Uruguay and Mexico (Conde *et al.*, 1996; IPCC 1996, WG II, Section 13.6.6).

Agroindustries that depend on primary production will be vulnerable to climate changes (see Section 6.3.7). Capital-intensive

***Table 6-7:*** *Agricultural yields estimated from different GCMs under current conditions of technology and management (see IPCC 1996, WG II, Chapter 13 for complete reference information).*

| Source | Scenario | Geographic Scope | Crop(s) | Yield Impact (%) |
|---|---|---|---|---|
| Baethgen (1992, 1994) | GISS<br>GFDL<br>UKMO[1] | Uruguay | Barley<br>Wheat | -40 to -30<br>-30 |
| Baethgen and Magrin (1994) | UKMO[1] | Argentina<br>Uruguay | Wheat | -10 to -5 |
| Siqueira *et al.* (1994)<br>Siqueira (1992) | GISS<br>GFDL<br>UKMO[1] | Brazil | Wheat<br>Maize<br>Soybean | -50 to -15<br>-25 to -2<br>-10 to +40 |
| Liverman *et al.* (1991, 1994) | GISS<br>GFDL<br>UKMO[1] | Mexico | Maize | -61 to -6 |
| Downing (1992) | +3°C<br>-25% precip. | Norte Chico, Chile | Wheat<br>Maize<br>Potatoes<br>Grapes | decrease<br>increase<br>increase<br>decrease |
| Sala and Paruelo (1992, 1994) | GISS<br>GFDL<br>UKMO[1] | Argentina | Maize | -36 to -17 |

[1]These studies also considered yield sensitivity to +2°C and +4°C, and -20% and +20% changes in precipitation.

livestock operations, which depend on grassland production, also are likely to be negatively affected (Parry *et al.*, 1988; Baker *et al.*, 1993; Klinedinst *et al.*, 1993). Impacts may be minor, however, for relatively intense livestock production systems (e.g., confined beef, dairy, poultry, swine) (IPCC 1996, WG II, Section 13.5).

Climate change also will affect the distribution and degree of infestation of insects indirectly through climatic effects on hosts, predators, competitors, and insect pathogens. There is some evidence that the risk of crop loss will increase as a result of poleward expansion of insect distribution ranges. Insect species characterized by high reproduction rates generally are favored (Porter *et al.*, 1991). Human alteration of conditions that affect host plant survival—irrigation, for example—also affects phytophagous (leaf-eating) insect populations.

The occurrence of plant fungal and bacterial pests depends on temperature, rainfall, humidity, radiation, and dew. Climatic conditions affect the survival, growth, and spread of pathogens, as well as the resistance of hosts. Friederich (1994) summarizes the observed relationship between climatic conditions and important plant diseases. In Latin America, warm, humid conditions lead to earlier and stronger outbreaks of late potato blight (*Phytophthora infestans*), as in Chile in the early 1950s (Austin-Bourke, 1955; Löpmeier, 1990; Parry *et al.*, 1990). Warmer temperatures would likely shift the occurrence of these diseases into presently cooler regions (Treharne, 1989).

As a result of these trends, farmers with limited financial resources and farming systems with few adaptive technological opportunities to limit or reverse the impacts of climate change may suffer significant disruption and financial loss from relatively small changes in crop yield and productivity (Parry *et al.*, 1988; Downing, 1992). Conflict is likely to arise between alternative uses of land areas under changing climate conditions—for example, competition for the same land may arise between expanding agriculture and other land uses (e.g., conservation, afforestation, population relocation).

Disparities in agricultural impact between developed and developing countries can be affected by international markets—which can moderate or reinforce local and national exchanges (Reilly *et al.*, 1994; Rosenzweig and Parry, 1994). Countries whose economies rely strongly on agricultural production would face major imbalances between production costs and international prices. According to Rosenzweig and Parry (1994), modeled yield changes in low-latitude countries are primarily negative, even though direct effects of $CO_2$ on plants, moderate levels of adaptation at the farm level, and production and price responses of the world food system were considered. Economic limitations, social conflicts (e.g., farmers' reluctance to abandon traditional practices), and environmental problems (e.g., salinization resulting from increased irrigation, which is not considered in the models) are likely to severely limit the capacity for adaptation and hinder the expansion of agricultural frontiers. Estimated net economic impacts of climate change on crops are negative for several Latin American countries analyzed by Reilly *et al.* (1994), even when modest levels of adaptation are considered. The only exception would be Argentina because, as a major exporter of grain, it should benefit from high world prices even if yields fall.

### 6.3.5. Human Health

Health has been defined by the World Health Organization (WHO) as "a state of complete physical, mental, and social well-being and not merely the absence of disease or infirmity." Different aspects of this well-being are related to weather and climate; primarily, however, it depends fully on the community's welfare. Because Latin America has a large tropical and subtropical environment, its inhabitants already are exposed to a number of infectious diseases and pests typical of these zones. The most vulnerable communities are those living in poverty, those with a high prevalence of undernutrition, and those with chronic exposure to infectious disease agents (IPCC 1996, WG II, Section 18.1.3). As a result, an increasing number of people who are living under these critical conditions in Latin America would be affected if, as expected, global warming aggravates disease and pest-transmission processes. Table 6-8 provides data on the estimated number of undernourished people in Latin America.

The major potential health impacts have been classified as "direct" and "indirect" impacts, according to whether they occur predominantly via the direct effect of exacerbated values of one or more climate variables (e.g., temperature, precipitation, solar radiation) on the human organism or are mediated by climate-induced changes in complex biogeochemical processes or climatic influences on other environmental health hazards.

#### 6.3.5.1. Direct Impacts of Climate Change

The direct impacts of climate change depend mainly on exposure to heat or cold waves or extreme weather events. The former involves an alteration of heat- and cold-related illnesses and deaths. Although Latin America was not included in the five regions identified by IPCC (1990) for analysis of regional climate change simulation, Kattemberg *et al.* (IPCC 1996, WGI, Chapter 6) made generalized tentative assessments concerning extreme events. Studies in temperate and subtropical countries have shown increases in daily death rates associated with extreme outdoor temperatures. Mortality increases much more steeply with rising temperatures than with falling temperatures (Kalkstein, 1993). The lowest mortality occurs within a range of intermediate comfortable temperatures and humidities (between 21°C and 26°C and below 60% relative humidity, in these countries).

***Table 6-8:*** *Estimated number of undernourished people in Latin America for 1969–71, 1979–81, and 1983–85.*

| Period | Population (millions) | Share of Population (%) |
|---|---|---|
| 1969–71 | 51 | 18 |
| 1979–81 | 52 | 15 |
| 1983–85 | 55 | 14 |

Source: UN World Food Council (WFC), Thirteenth Ministerial Session, Beijing, China, 1987.

No references to studies and research on direct health impacts from projected warming in Latin America are included in the SAR; however, extrapolation of investigations performed in cities in the United States, China, The Netherlands, and the Middle East indicates that morbidity and mortality also could increase in this region as a result of the expected increase in the number of days with high daily temperatures (i.e., the persistence of days with higher-than-normal maximum and minimum temperatures) (Haines *et al.*, 1993; Kalkstein, 1993; IPCC 1996, WG II, Section 18.2.1). The impacts would be exacerbated by high humidity rates, intense solar radiation, and weak winds. All of these factors affect the physiological mechanisms of human adaptation.

High temperatures and air pollutants, especially particulates, act synergistically to influence human mortality. This effect is occurring in large cities, such as Mexico City and Santiago, Chile, where such conditions enhance the formation of secondary pollutants (e.g., ozone) (Escudero, 1990; Katsouyanni *et al.*, 1993; Canziani, 1994).

Global warming could increase the number and severity of extreme weather events such as storms, floods, and droughts, and related landslides and wildfires. Such events tend to increase death and pathology rates—directly through injuries or indirectly through infectious diseases, as well as through social problems that stem from the dislocation of people, adverse psychological effects, and other stresses (IPCC 1996, WG II, Section 18.2.2). A number of slums and shanty towns located on hills, as well as human settlements located in flood-prone areas, are subject to periodic natural disasters that adversely affect human health (Section 6.3.6). These overcrowded and poorly-serviced peri-urban settlements also provide a potential breeding ground for disease hosts (e.g., rats, mice, cockroaches, flies) and disease organisms, increasing the population's vulnerability. Communities surrounded by these poverty belts also become more vulnerable to periodic disease outbreaks (WHO Commission on Human Health and Environment, 1992).

Climate variability also may aggravate diseases resulting from water contamination. Increases in *Salmonella* infections following a flood in Bolivia resulted from the El Niño event of 1983 (Telleria, 1986).

#### 6.3.5.2. *Indirect Impacts of Climate Change*

Infectious and parasitic diseases are important causes of morbidity and mortality in Latin America, and the main cause of death in children (PAHO, 1994). Some infectious diseases are more common in tropical and subtropical areas than in temperate or cold areas. Therefore, global warming would tend to extend their area of influence or increase the importance of outbreaks. Some of these diseases are food- or water-related infections; after they are introduced into a region, they show a tendency to spread over the whole region. Viral, bacterial, and protozoan agents of diarrhea can survive in water—especially in warmer waters—for long periods of time and thus spread at increased rates in rainfall periods, enhancing their transmissibility among people. An example is cholera, which was introduced into Peru in 1993. It produced an outbreak that spread to most of the South American subcontinent, including places as far as Buenos Aires (PAHO, 1994). Cholera's relation to the ENSO phenomenon was proposed by Colwell (1996). This disease and other diarrheas and dysenteries are associated with the distribution and quality of surface water, as well as with flooding and water shortages. These conditions alter the population dynamics of organisms, impede personal hygiene, and impair local sewage systems. Increases in coastal algal blooming also may amplify the proliferation and transmission of *Vibrio cholerae*.

Algal blooming also may be associated with biotoxin contamination of fish and shellfish (Epstein *et al.*, 1993). With ocean warming, temperature-sensitive toxins produced by phytoplankton could cause contamination of seafood more often, resulting in an increased frequency of poisoning. Thus, climate-induced changes in the production of aquatic pathogens and biotoxins may jeopardize seafood safety.

Other infectious diseases not included in the SAR have now achieved importance and could reach critical levels in South America. Some viruses have had unexpected outbreaks—such as arenaviruses in Argentina and Bolivia (PAHO, 1996) and hantaviruses in the south of Argentina; their relationship to climate change is not yet well understood. Fungi such as *Paracoccidiodes brasiliensis*—which require high humidity and generally are associated with rainfall regimes of 500–2,000 mm/yr and average temperatures of 14–30°C—are found in some areas of South America (e.g., Brazil, Venezuela, and northern Argentina) (Restrepo *et al.*, 1972), where this mycosis is becoming endemic. It may spread if climate change provides adequate conditions to start the epidemiological chain. In this connection, increasing surface traffic that has resulted from commercial activity stemming from the new regional common market—Mercosur—may call for the development of appropriate sanitary barriers (e.g., disinfection of vehicles and their contents to block transport of harmful fungi) at borders.

A special category of infectious diseases—a group known as vector-borne diseases (VBDs)—already affect a large number of people in Latin America. These diseases could expand their geographic and elevational ranges because conditions would be more favorable for viruses and other living agents, reservoirs, and vectors as a result of global warming. The most important VBDs in Latin America are listed below, with indications of some of their main vectors:

- Malaria: Vectors are several species of the mosquito genus *Anopheles.* Malaria's incidence is affected by temperature, surface water, and humidity (Carcavallo *et al.*, 1995).
- Dengue: Vectors are *Aedes* and other mosquito species. Dengue is expanding in Latin America (Koopman *et al.*, 1991; Herrera-Basto *et al.*, 1992). High temperatures, particularly in winter (Halstead, 1990), promote the spread of this disease.
- Yellow fever: Several species are vectors. Yellow fever has an urban epidemiology similar to dengue, but it also has cycles that develop in the wild (Martinez *et al.*, 1967).
- Chagas' disease or American trypanosomiasis: Vector is *Triatominae* bug of the order Hemiptera. About 100 million people in Latin America are at risk, and 18 million people are infected (Hack, 1955; Carcavallo and Martinez, 1972; WHO, 1995).
- Schistosomiasis: The vector is the water snail (Grosse, 1993).
- Onchocersiasis or river blindness: Vectors are several species of *Simulildae* or "blackflies" (WHO, 1985).
- Leishmaniasis: Vectors are several species of *Phlebotominae* or "sandfly" (Bradley, 1993).
- *Limphatic filariasis*: Vectors are several mosquito species (PAHO, 1994).

Other viruses also affect human beings in this region. One produces Venezuelan equine encephalitis; it is transmitted by several mosquito species, and cases have been reported in Colombia and Venezuela (WHO, 1996). Its relation to climate change has not yet been demonstrated; however, warming could affect the geographical distribution and dispersion, as well as some behaviors and patterns, of vertebrate reservoirs (mammals and birds) and vectors.

Infective agents and vectors are sensitive to environmental changes, especially those conditioned by temperatures and humidity. Vectors also are sensitive to wind, soil moisture, surface water, and changes in vegetation and forest distribution (Bradley, 1993). Temperatures and humidity influence the geographical and elevational dispersion of vectors (Burgos *et al.*, 1994; Curto de Casas *et al.*, 1994), as well as their population dynamics and behavior. Precipitation is an important factor for vectors with aquatic stages, such as mosquitoes and blackflies, because breeding places are increased and maintained by rainfall. Winds may contribute to the dispersion of some flying insects, such as mosquitoes, blackflies, and sandflies (Ando *et al.*, 1996).

Several years ago, VBDs were very critical in many areas of the region; during recent years, however, some of them have been almost entirely controlled or reduced in their endemicity. Malaria is prevalent in tropical and subtropical areas of the American continent, from south of Mexico to northern Argentina. Several species of mosquitoes of genus *Anopheles* are vectors of *Plasmodium vivax* and *Plasmodium falciparum,* which are causal agents of the disease that cannot survive at temperatures below 14–18°C, depending on the *Plasmodium* species (see Figure 6-6). The normal type of epidemiological curve of malaria in northeast Argentina has shown a peak in new cases every 1–4 years, particularly during the months of March to May or June, when *Anopheles darlingi* has been present. During the period 1991–93, however, there were cases every month of the year, and *Anopheles darlingi* were captured or reported sighted during the whole period. Thus, more studies are needed to assess the influence of new dams on the behavior of vector species and the epidemiology of the disease, as well as the influence of increases in minimum temperatures (Carcavallo *et al.*, 1995). Martens *et al.* (1995) have predicted that temperature increases of several degrees Celsius at higher elevations—which could occur in Andean ranges under projected climate change—may produce seasonal epidemic transmission in areas currently free of paludism.

In the recent past, irrigation schemes and agricultural changes in Central America, which have been associated with an increased resistance of vectors to insecticides, have brought an increase in new cases of malaria. The most severe outbreaks, however, are registered in Brazil (particularly in the Amazon

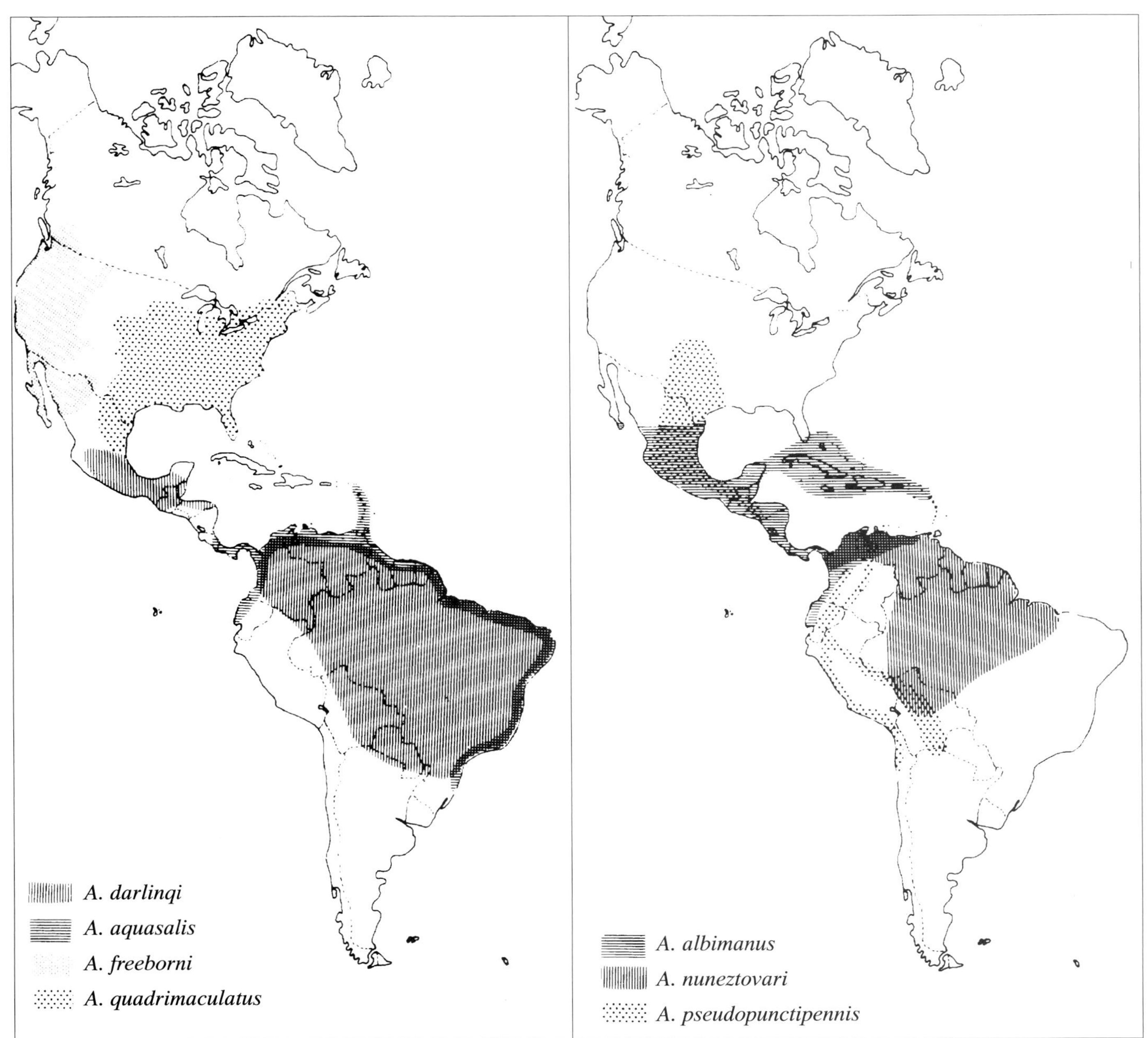


**Figure 6-6:** Distribution of principal malaria vectors in Latin America (WHO, 1996).

basin)—where more than 50% of malaria cases in the Americas are reported, including 6,000–10,000 deaths per year, as a consequence of new settlements and mining activities developed in unhealthy locations of the Amazon forest (WHO, 1993).

Dengue and yellow fever involve similar viruses and epidemiology; they are transmitted by the same urban vector, the *Aedes aegypti* mosquito. Although this species is recognized as the main vector, another (*Aedes albopictus*) has invaded the American continent, colonizing the United States and Brazil. In Latin America, dengue currently is more active than yellow fever. Dengue is a severe, influenza-like virus; it sometimes takes the form of hemorragic fever, and it is lethal about 15% of the time. Studies performed in Mexico have shown that an increase of 3–4°C in mean temperatures may double the transmission rate (Koopman *et al.*, 1991), and transmission is common in areas with temperatures below 20°C (Halstead, 1990). Cases of dengue at 1,700 m above sea level—an elevation never before attained—were found in the state of Guerrero (Herrera-Basto *et al.*, 1992) and have since been found at even higher elevations in other regions. Apparently, the thermal limit of *Aedes aegypti* could be a midwinter temperature of 10°C (Shope, 1991). Junin *et al.* (1995) reported re-infestation of the city of Buenos Aires by this species—after 30 years of formally declared erradication. Temperature is considered to be a predictor of dengue outbreaks. Martens *et al.* (1997) have identified epidemiological areas too cold for the transmission of the virus, including those at the current temperature of Mexico City.

Chagas' disease, or American trypanosomiasis, is a parasitic disease produced by the flagellate *Trypanosoma cruzi.* This infection is one of the most important public health scourges in Latin America. It affects the human digestive and nervous systems and especially the heart. Its vectors span latitudes from 42°N to 46°S, with an elevational dispersion ranging from sea level to about 4,100 m above sea level (Bejarano, 1967). However, the disease's endemic area is much smaller; it is confined mainly to tropical and subtropical regions, especially in poverty-stricken areas. *Triatoma infestans*—the most important vector of *Tripanosoma cruzi*—is found in Argentina, Bolivia, Brazil, Chile, Paraguay, Peru, and Uruguay. Most of the population of this vector lives within or around human habitats, particularly in poor rural dwellings that have mud walls and roofs made of tree branches, straw, or palm leaves. Colonies also exist outdoors in places with cold winters and warm summers (Carcavallo and Martinez, 1972). Climate change may influence the population dynamics of vectors (Hack, 1955; Burgos *et al.*, 1994), as well as the number of blood-feedings and, therefore, the possibilities of infective indirect contacts (Catala, 1991; Catala *et al.*, 1992).

The region also is affected by some rickettsial diseases, which are transmitted by ticks and have wild mammals as reservoirs. They could become increasingly important under climate change because they facilitate the coexistence of every link on the epidemiological chain.

Another indirect effect of climate change on human health would be the threat to food production (Rosenzweig *et al.*, 1993; Reilly *et al.*, 1994)—particularly among populations already living in marginal conditions, which have minimal capacity for adaptive changes (Leemans, 1992). For example, climate variability and change could increase the adverse effects of plant and animal pests and diseases on the quality and quantity of food production. The ENSO phenomenon has had similar effects on anchovy and sardine fishing and the related fishmeal industry, both of which are important for human and livestock diets.

### 6.3.6. *Human Settlements*

In Latin America, the region with the largest urbanization rate, a large spectrum of urban settlements has been established at different elevations (from sea level to mountain ranges exceeding 3,000 m). These locations have a wide variety of geographic and topographic characteristics (see Figure 6-1). Because weather and climate events already are affecting environmental conditions, climate change may be expected to have a host of direct and indirect effects on human settlements, even where urban migration is not a factor. However, such migration is significant in some countries of the region. In fact, large groups of people migrate internally and regionally from rural, drought-prone areas (Ezcurra, 1990) or poverty stricken communities (Canziani, 1996) to modern, well-developed cities. Unemployment rates indicate that this tendency will continue, particularly because cities and urban/industrial/commercial nuclei in Latin America offer better opportunities for employment than do those in other regions. Such cities may become the kernels of Latin American megalopolises, attracting people from less-developed neighboring countries as well as from within. Population displacements are likely to have serious socioeconomic and cultural as well as health implications (also see Section 6.3.5). More than one poverty belt, such as those in the metropolitan areas of some Latin American cities, may arise around a single city.

Latin American cities already suffer from the impacts of sea-level rise, adverse weather, and extreme climate conditions (e.g., floods, flash floods, windstorms, landslides, and cold and heat outbreaks). They also suffer from indirect effects through impacts on other sectors, such as water supply, energy distribution, transportation, agriculture, and sanitation services. Thresholds beyond which impacts escalate quickly are unique to local situations and tend to depend on the degree of preparedness for adaptive response (e.g., warning and alert systems and procedures; traffic rerouting procedures; flexible welfare systems) (IPCC 1996, WG II, Section 12.2).

Global warming can be expected to further affect the availability of water resources and biomass (for charcoal and fuel production), both of which are important energy sources in many Latin American countries. Water and biomass resources already are under stress in many of these areas as a result of rising demand, which will increase because of the urban migration trend. The increasing urban population in the region also has generated difficulties in providing water of adequate quality to urban residents. In most cities, piped drinking water and sewage services are not available to everyone. In the Buenos Aires metropolitan area, about 55% of the population obtains its drinking water from underground sources—some of which have serious levels of contamination from industrial waste dumps, agrochemicals, and, particularly, precarious sewage systems consisting of *pozos negros* (excreta and fecal holes) (González, 1990; De Filippi *et al.*, 1994).

In mountain regions, hundreds of thousands of people live in precarious settlements on potentially unstable hillsides that are especially vulnerable to climatic impacts. In recent decades, hundreds of people have been killed or seriously injured and thousands left homeless by landslides in Rio de Janeiro, Guatemala City, Medellín, Mendoza, Mexico City, Santos, and Sao Paulo (WHO Commission on Human Health and Environment, 1992; Aguilar and Sanchez, 1993), as well as Caracas (Hardoy *et al.*, 1992). Shantytowns (*barriadas*, *favelas*, or *villas miseria*) surrounding large cities in the region sometimes are established in the drainage valleys of rivers and streams—where flooding frequency already is increasing as a result of climatic variability (Canziani, 1996) and might be exacerbated as a result of global warming. Flooding and landslides have adverse effects on the welfare and health conditions of poorer communities.

Nonclimatic effects may be more important than climate change. Local environmental and socioeconomic situations are changing rapidly; the living standards of millions of people in Latin America are lower now than they were in the early 1970s. Poverty has become an increasingly urban phenomenon in recent decades. Migrants tend to live in informal, peri-urban settlements with serious infrastructure problems—ranging from unhealthy environments and water supplies and lack of sewage systems to difficult access to energy, transportation, communications, and even decent shelter. Some informal settlements around large Latin American cities, especially around the capital cities, are homes for many hundreds of thousands of people, leading to impacts on human health (see Section 6.3.5).

### 6.3.7. Industry, Energy, and Transportation

Weather and climate play decisive roles in many artificial systems; climate variability and climate change would affect these activities, with beneficial and adverse impacts. These effects include direct impacts on the planning, implementation, and operation of industry; energy production; and transportation, as well as on the economics of corresponding systems. Climate affects markets for goods and services and the natural resources on which such economic activity depends (IPCC 1996, WG II, Section 11.5). Activities directly sensitive to climate include construction, fossil fuel production (including offshore drilling), water-dependent manufacturing, hydropower generation, space heating and air conditioning, tourism, and recreation. Industry located in coastal zones and flood-prone areas also would likely be affected by climatic variations. Other activities that depend on climate-sensitive resources include agro-industries (food/drink, forestry-related activities, and textiles), production of energy from biomass fuels (see Section 6.3.4), and activities associated with the production of other renewable energy.

Agro-industry is of relatively greater significance in Latin American countries where activities associated with agriculture constitute the bulk of the economy, than elsewhere. Agro-industries that depend on the production and transportation of products such as grain, sugar, and rubber are vulnerable to changes in precipitation patterns and the frequency and intensity of extreme weather events. Globalization undoubtedly will better define the future activities of some Latin American countries. The strategy of "dumping" industrialized products—which has been carried out by more developed countries and some of the Far East's emerging economies—suggests that Latin American countries might well reorient their development toward agro-industry to take advantage of their natural resources and secure a place in international markets. The inception of regional markets would facilitate a combination of agricultural and agro-industrial production capacities within the region, assuming that available natural resources are evaluated, through monitoring and research, in the light of projected climate change. This route could lead to the region's sustainable development and the improvement of its productive capacity to satisfy the demands of an increasing global population.

An important domain within these agro-industries is the development of energy sources through biomass exploitation (see Section 6.3.4). In Brazil, the Fuel-Alcohol Program—one of the largest commercial biomass-to-energy programs—has facilitated the substitution of ethanol for gasoline in passenger cars and light vehicles (Goldenberg *et al.*, 1993). Technological advances, including more efficient production and processing of sugarcane, are responsible for the availability and low price of ethanol. The transition to ethanol fuel has reduced Brazil's dependence on foreign oil (thus lowering its import-export ratio), created significant employment opportunities, and greatly improved urban air quality. In addition, because sugarcane-derived ethanol is a renewable resource (cane is replanted at the same time that it is harvested), the combustion of ethanol adds virtually no net $CO_2$ to the atmosphere—and thus helps reduce the threat of global warming. Hydropower production would be influenced by changes in precipitation; whether these changes would be beneficial depends heavily on the relationship between seasonal patterns of precipitation and electricity demand. In Costa Rica and the Cuyo region in Argentina, for example, energy production already is affected by climate variability. Whereas hydroelectricity contributed 14.5% of the world's generated electricity in 1986, hydroelectric sources satisfied 53% of Latin America's electrical energy demands in that year (of the rest, 34% was supplied through thermal generation and the remaining 13% was shared between nuclear and geothermal generation). Of the world's technically exploitable hydroelectric potential, Latin America (excluding Mexico) accounts for 3,500 TWh per year. The region has an installed capacity of 86.69 GW, which in 1988 generated 367 TWh (Moreira and Poole, 1992); thus, about 20% of the potentially exploitable hydroenergy was used. Under these circumstances, the region would be in a position to increase its hydropower production to satisfy increasing demands during the coming century, even if climate change results in some reductions in power generation. (This hypothesis assumes the operation of new dams and stations, starting in the early 1990s—particularly in South America—and the installation of interconnected transmission networks among countries in the region.)

As noted in Section 6.5, sensitivity studies and monitoring of water resources are needed to prevent serious adverse climate change impacts. However, other factors affecting hydropower generation may have more weight than those resulting from climate change—such as social and environmental impacts, which may be the principal constraints and sources of uncertainty affecting hydropower development. Large storage reservoirs exacerbate many problems, particularly in tropical regions. The impacts of hydropower development can be categorized as direct—such as those that arise while constructing the dam, filling the reservoir, and changing the river flow—or indirect, such as effects on the health and well-being of the community and local ecosystems. However, many of these problems can be reduced by improved planning, in coordination with national or regional communities and their participatory social movements [e.g., non-governmental organizations (NGOs)].

Although the energy, industry, and transportation sectors are of great economic importance and clearly are sensitive to climate change, the capacity for autonomous adaptation is relatively high, as long as climate change takes place gradually. The lifetimes of most assets are relatively short compared with projected time scales for climate change (Campos *et al.*, 1996).

## 6.4. Integrated Analysis of Potential Impacts and Adaptation Options

To fulfill the commitments arising from the UNFCCC, Conference of the Parties (COP) members in Latin America have initiated the preparation of national GHG inventories, with the assistance of the USCSP and the GEF. These inventories generally are associated with the performance of vulnerability studies aimed at defining the potential impacts of climate change on natural and human systems on which the economies and the well-being of the respective national communities rely. The resulting information is designed to assist Latin American governments in better understanding the urgency of developing more and better information on impacts and adaptation strategies; as a tool for minimizing negative impacts and taking the best possible advantage of new opportunities under changing climatic conditions; and for achieving sustainable development practices. Vulnerability and impact assessments and adaptation options should consider interactions and feedbacks among different sectors. Some of the outputs of country studies are reports of vulnerability to climate change in selected areas in different countries. Examples are the case study for Belize (reported in Box 6-4); other case studies cited in the bibliography, such as country studies in Venezuela (Perdomo *et al.*, 1996) and Peru (Teves *et al.*, 1996); and information included in the preliminary report of the National GHG Inventory in Mexico (Gay-García *et al.*, 1996). Another interesting approach, which may be considered as an integrated analysis study with implications for some adaptation options, is summarized in Box 6-5.

**Box 6-5. Integrated Analysis and Possible Adaptation Options: Land Use and Climate Change**

Considering that, under changing climate conditions, land use is a key factor for agricultural production as well as for environmental preservation, effective land-use regulations would be necessary to reduce the vulnerability of this type of human system to climate change. In this connection, an analysis of trade-offs among productivity, stability, and sustainability in low-output agricultural systems was undertaken by Viglizzo and Roberto (1997). It showed that the productivity of herbage and grain (primary products) clearly exceeds that of milk and beef (secondary products). However, the secondary activities involve processes that are less affected by environmental disturbances, whereas primary product processes are more extractive in terms of soil nutrients and, consequently, less sustainable in the long term. This analysis suggests that different combinations of cropping and cattle activities on agricultural lands may need to be regulated to prevent undesirable effects in regions with different vulnerability to climate change, like Argentina's Pampas.

Climate change under $2xCO_2$ scenarios is projected to produce economic losses of 0.9–3.1% of Latin America's GDP. These large losses—compared with global (1.1–1.8%) and Organisation for Economic Cooperation and Development (OECD) (1.3–1.9%) loss projections—are explained by the fact that primary production accounts for a higher share of GDP in Latin America (Fankhauser and Tol, 1997); therefore, a large output is directly exposed to climatic influences (IPCC 1996, WG III, Section 6.5.9). It should be noted, however, that many assumptions underlie these best guesses—and that large uncertainties remain.

Many of the impacts of climate change may not be directly revealed in the marketplace. Nonmarket impacts on biodiversity, subsistence agriculture, traditional land-use patterns, and so forth may be no less important than market impacts (IPCC 1996, WG III, Table 6-1 and Section 6.2.1). In addition, the impact of climate change on Latin American ecosystems needs to be considered in parallel with impacts caused by unsustainable land-management practices and the effects of increasing population. In most cases, it is impossible to separate the effects of these impacts, and land-use impacts and population growth are expected to result in more severe changes than are changes in climate. Desertification, for example, is a widespread problem in Latin America arising from human abuse of the land and adverse climate conditions. Even if wetter conditions were to prevail, human-induced desertification could increase the vulnerability of land to desertification and escalate the desertification process (see Section 6.3.1.3).

In most managed ecosystems and human systems, adaptation to climate change in Latin America is not as much an issue of scientific knowledge or technical feasibility as of socioeconomic and cultural factors and political decisions. It is extremely difficult at this stage to analyze the viability of different options because their cost/benefit ratio depends on ecological and social conditions and the integration of local and external markets. In virtually all cases, political measures (e.g., credits, subsidies); educational programs; and interactive communication among experts, policymakers, and stakeholders are needed—especially within local communities—to implement viable adaptation options. Table 6-9 presents a very preliminary summary of possible adaptation options for different sectors within Latin America.

## 6.5. Monitoring and Research Needs

Latin America has the expertise needed for the study of climate, climate variability, and climate change—as is apparent from the relatively ample bibliography of research by local authors. This expertise is important because recent observations have shown that the Southern Hemisphere, where the largest portion of the region lies, is warming more than the Northern Hemisphere. The very heavily-populated land masses in the north produce more atmospheric pollution (in particular, sulfate aerosols and particulates), which dampens the heating capacity of solar radiation on the Earth. This situation is quite different from that in the oceanic Southern Hemisphere, where the distribution and density of air pollutants are remarkably lower. As a result, the absorption and reflection of incoming solar radiation do not counteract the background release of GHGs to the extent that they do in the north.

Other characteristics further differentiate the two hemispheres. The geographical features of the Southern Hemisphere (and Latin America in particular) consist mainly of a solid continent on the South Pole (the Antarctic); a very large oceanic mass with a well-known thermal moderating capacity; a subcontinent (South America) extending to the southernmost latitudes; and a vertebral mountain chain (the Cordillerra de los Andes) running from north to south over the entire subcontinent—all sandwiched between the two largest oceans of the world. Another important difference affecting the middle atmosphere over this region is the seasonal and regional formation of the Antarctic ozone hole. This remarkable stratospheric ozone depletion, combined with dynamic changes in middle atmospheric circulation, has the double impact of creating a window for infrared radiation—which globally compensates about 25% of GHG warming effects—and increasing the amount of ultraviolet-B (UV-B) radiation that reaches the ground, with direct effects on the region's terrestrial and marine ecosystems.

Thus, there is room not only for improving mathematical models to encompass hemispheric peculiarities (as is being done in Australasia) but also for developing appropriate regional climate models that should have the capacity, inter alia, to provide appropriate climate scenarios for running hydrological models for the region's river basins.

This analysis once again underscores the fact that the region suffers from insufficiently dense and reliable observation networks and that other basic information—biological, economic, and social—necessary to build up complete and coherent regional scenarios is missing. Last but not least, appropriate coordination with the Australasian countries, as well as with the United States and Canada, is becoming more and more critical because of common factors affecting climate variability (e.g., the ENSO phenomenon) and climate change over these regions.

In summary, the situation in Latin America calls for specific research initiatives; comprehensive ecological and socioeconomic databases; appropriate data validation techniques; proxy data capture, particularly regarding the neighboring Antarctic continent (which began keeping regional records only after the 1958 geophysical year); and appropriate training of personnel, particularly regarding integrated assessments and development of appropriate adaptation option methodologies. More fundamental research on ecosystem functioning, hydrographic systems, and interactions between land use and technological approaches by different communities is indispensable for the region to thoroughly undertake integrated assessments of potential impacts of climate change and develop appropriate adaptation and mitigation alternatives. Therefore, combined actions with countries in neighboring regions, such as those already implemented through joint activities with the USCSP, and the organization of activities within the Valdivia Group—which associates Latin American countries in the temperate Southern Hemisphere with activities of common interest in Australia, New Zealand, and South Africa—are in order. The participation of grassroots organizations and NGOs also is essential for achieving sustainable development in the region vis-à-vis climate change.

## Acknowledgments

We wish to acknowledge the valuable comments and kind cooperation of S. Basconcelo (Argentina), J.L. Buizer (USA), R. Basher (New Zealand), Rodolfo Dirzo (México), V. Falczuk (Argentina), D. Gray (World Bank), R. Neilson (USA), N. Pérez Harguindeguy (Argentina), A.B. Pittock (Australia), R. Street (Canada), R. Tol (The Netherlands), F.S. Vendramini (Argentina), and the entire staff of the IPCC WG II Technical Support Unit.

***Table 6-9:*** *Summary of adaptation options in response to climate change in Latin America, based primarily on the IPCC Technical Paper on Technologies, Policies, and Measures for Mitigating Climate Change (IPCC, 1996b).*

| Sector | Adaptation Option | Other Benefits | Difficulties to be Considered |
|---|---|---|---|
| *Forests* | Reduction of social pressure driving land conversion | Soil and biodiversity conservation, watershed benefits | National and international socioeconomic conflict |
| | Large tree plantations on highly degraded areas; short-term rotation plantations for local fuel needs | Proper site and species selection for soil conservation and watershed benefits | Costs vary between countries (US$4–31/tC) |
| | Assisted migration | Biodiversity conservation | High costs, uncertain success |
| | Low-impact harvest practices | Soil and biodiversity conservation, watershed benefits | |
| *Rangelands* | Preservation of an extensive spatial scale in management units | Preservation of traditional organization patterns of rural communities, soil and biodiversity conservation | Possible socioeconomic conflicts |
| | Active selection of plant species and control of animal stocking rates | Increased productivity, biodiversity, soil conservation | Land tenure and market problems, cultural difficulties |
| | Increase of the area devoted to capital-intensive improved pastures | Alleviation of pressure on larger areas of rangeland | High costs |
| | Agroforestry, particularly involving legume tree species | Increased productivity, biodiversity, and soil conservation | |
| *Mountains* | Conservation of traditional cultivation practices and genotypes | Conservation of local biodiversity and world's genetic resources; promotion of indigenous knowledge | Market problems |
| | Adjustment of infrastructure (dams, pipelines, erosion protection, etc.) | | High costs |
| *Agriculture* | Expansion of agricultural land area | | Competition with other uses, high environmental impacts in forest areas, threatening for subsistence lifestyles |
| | Changes in agricultural practices (sowing dates, tillage, irrigation, fertilization, crop varieties, species) | Reduced soil erosion, increased yields in some cases | Market problems, including marketing difficulties in adopting new practices, and environmental impacts in case of irrigation and fertilization |
| *Freshwater Systems* | Assisted dispersal of ecologically and/or economically important species to isolated locations | | High costs |
| | Restoration of rivers and stream channels to more natural morphologies; large-scale hydrological engineering in floodplains | | High costs, conflicting interests among stakeholders, ecological and cultural impacts |

*Table 6-9 continued*

| Sector | Adaptation Option | Other Benefits | Difficulties to be Considered |
|---|---|---|---|
| *Freshwater Systems (cont.)* | Augmentation of riparian vegetation to reduce negative effects of warming; decreased loading of nutrients to reduce eutrophication processes (which are believed to be exacerbated by increasing water temperature) | | |
| *Hydro-power* | Construction of new hydropower plants | Level supply and demand curve | Very high costs, increased need for international loans, potentially high ecological and sociological impacts |
| | Reduced consumption; improved use efficiency; improved electricity transmission | Lower energy costs | |
| | Raise reservoir capacity | Use of secondary water for other purposes (leisure activities) | |
| *Coastal Zones and Saltwater Fisheries*[1] | Structural protection measures (dikes, seawalls, breakwaters, beach groins) in heavily populated areas | | Very high costs |
| | Retreat | | Subject to land availability inland, socio-cultural conflict and high environmental impact likely |
| | Design and implementation of national and international fishery-management policies that recognize shifting species ranges, accessibility, and abundances and balance species conservation with local needs | | Possible internal and international interests |
| | Expansion of aquaculture to increase and stabilize seafood supplies, help stabilize employment, and carefully augment wild stocks | | |
| *Human Population* | Introduction of protective technologies (e.g., insulated buildings, air-conditioning, strengthened sea defenses, disaster warning systems); education efforts aimed at high-risk groups | | High costs |
| | Environmental management of ecosystems (e.g., freshwater resources, wetlands, and agricultural areas sensitive to invasion by vectors) | | High costs, poorly understood consequences for other components of ecosystems |
| | Improved primary health care for vulnerable populations, and public health surveillance and control programs (especially for infectious diseases) | | |

*Table 6-9 continued*

| Sector | Adaptation Option | Other Benefits | Difficulties to be Considered |
|---|---|---|---|
| *Human Settlements* | Decentralization of basic infrastructure to mitigate immigration into cities | | High costs, conflict of interests, cultural problems |
| | Better designed urban infrastructure (buildings, recreation areas, water delivery systems, etc.) | | High costs |
| | Improved treatment of sewage and industrial residues; fines to heavy polluters | | |
| *Industry* | Diversification of agroindustrial production | | Socioeconomic, marketing, and cultural problems |
| | Alternative sources of energy | Reduction of air pollution in some cases | Cost-benefit ratios have to be evaluated case by case |

[1]Adaptation to the impact of climate change on open oceans is limited by the nature of these changes, and their scale.

## References


**Aceituno**, P., 1987: *Aspectos tri-dimensionales del funcionamiento de la Oscilación del Sur en el sector sudamericano.* Anales del Segundo Congreso Interamericano de Meteorología, November 30–December 4, 1987, Centro Argentino de Meteorólogos, Buenos Aires, Argentina, pp. 5.1.1–5.1.6.

**Aguilar**, A.G. and M.L. Sanchez, 1993: Vulnerabilidad y riesgo en la ciudad de Mexico. *Ciudades,* **17**, 31–39.

**Alcamo**, J. (ed.), 1994: *IMAGE 2: Integrated Model of Global Change.* Kluwer Academic Publishers, Dordrecht, The Netherlands, 318 pp.

**Ando**, M., R.U. Carcavallo, P.R. Epstein, 1996: Climate Change and Human Health [McMichael, A.M., A. Haines, R. Sloof, and S. Kovats (eds.)]. WHO, Geneva, Switzerland, 297 pp.

**Aniya**, M., R. Naruse, M. Shizukuishi, P. Skvarca, and G. Casassa, 1992: Monitoring recent glacier variations in the southern Patagonia Icefield, utilizing remote sensing data. *International Archives of Photogrammetry and Remote Sensing,* **29(B7)**, 87–94.

**Armentano**, T.V., 1990: Soils and ecology: tropical wetlands. In: *Wetlands: a Threatened Landscape* [Williams, M. (ed.)]. The Alden Press, Ltd., Oxford, United Kingdom, pp. 115–144.

**Austin-Bourke**, P.M., 1955: *The Forecasting from Weather Data of Potato Blight and Other Plant Diseases.* WMO TN 42, Technical Report, WMO, Geneva, Switzerland.

**Baethgen**, W.E., 1994: Impact of climate change on barley in Uruguay: yield changes and analysis of Nitrogen management systms. In: *Implications of Climate Change for International Agriculture*, [Rosenzweig, C and A. Iglesias (eds.)], U.S. Environmental Protection Agency, Washington, DC, pp. 1-15.

**Baker**, B.B., J.D. Hanson, R.M. Bourdon and J.B. Eckert, 1993: The potential effects of climate change on ecosystems processes and cattle production on U.S. rangelands. *Climate Change*, **23**, pp. 27-47.

**Bakun**, A. 1993: The California Current, Benguela Current, and Southwestern Atlantic Shelf ecosystems: a comparative approach to identifying factors regulating biomass yields. In: *Large Marine Ecosystems: Stress. Mitigation and Sustainability* [Sherman, K., L.M. Alexander, and B.D. Gold (eds.)]. AAAS Press, Washington, DC, pp. 199–221.

**Batjes**, N.H. and W.G. Sombroek, 1997: Possibilities for carbon sequestration in tropical and subtropical soils. *Global Change Biology*, **3**, pp. 161-173.

**Bazzaz**, F.A., and K.D.M. McConnaughay, 1992: Plant interactions in elevated $CO_2$ environments. *Australian Journal of Botany*, **40**, pp. 547-563.

**Bejarano**, J.F.R., 1967: Lucha contra las enfermedades transmisibles por artropodos. Servicios especificos y servicios generales de salud. *Segundas Jorn. Entomoep. Argentinas,* **3**, 471–507.

**Bierregaard**, J.R., T.E. Lovejoy, V. Kapos, A.A. dos Santos, and R.W. Hutchings, 1992: The biological dynamics of tropical rainforest fragments. *Bioscience,* **42**, 859–865.

**Boer**, G.J., 1993: Climate change and the regulation of the surface moisture and energy budgets. *Clim. Dyn.*, **8**, 225–239.

**Boer**, G.J., N.A. McFarlane, and M. Lazare, 1992: Greenhouse gas-induced climate change simulated with the CCC second-generation general circulation model. *J. Climate*, **5**, 1045–1077.

**Bortenschlager**, S., 1993: Das höchst gelegene Moor der Ostalpen "Moor am Rofenberg" 2760 m Festschrift Zoller, Diss. Bot. **196**, pp. 329-334.

**Bradley**, D.J., 1993: Human tropical diseases in a changing environment. In: *Environment Change and Human Health. CIBA Foundation Symposium.* CIBA Foundation, London, United Kingdom, pp. 146–162.

**Bravo, A.**, *et al.*, 1988:The effects of climatic variations on agriculture in the Central Sierra of Ecuador. In: *The Impact of Climate Variations on Agriculture Impact, Vol 2. Assessment in Semi-arid Regions* [Parry, M.L, T.R. Carter, and N.T. Konijn (eds.)]. Kluwer Academic Publishers, Dordrecht, The Netherlands, pp. 381–493.

**Brenes-Vargas**, A., and V.F. Savorido Trejos, 1994: Changes in General Circulation and their influence on precipitation trends in Central America. *Costa Rica Ambio,* 87–90.

**Brinkman**, R. and W.G. Sombroek, 1996*: The Effects of Global Change Conditions in Relations to Plant Growth and Food Production.* In: F. Bazzaz and W.G. Sombroek (eds), Global Climatic Change and Agricultural Production: Direct and Indirect Effects of Changing Hydrological, Soil, and Plant Physiological Processes. John Wiley, Chichester, UK.

**Brown**, B.E. and J.C. Odgen, 1993: Coral bleaching. *Scientific American,* **268**, 44–50.

**Bruijnzeel**, L.A. and W.R.S. Critchley, 1994: *Environmental Impacts of Logging Moist Tropical Forests.* IHP Humid Tropics Programme, Series 7, UNESCO.

**Budyko**, M.I. and Y.U.A. Izrael (eds.), 1987: Anthropogenic climate change. *Leningrad Hydromteoizdat,* 378 pp. (in Russian)

**Burgos**, J.J., S.I. Curto de Casas, R.U. Carcavallo, and I. Galindez Giron, 1994: Global climate change influence in the distribution of some pathogenic complexes (malaria and Chagas' disease). *Entomol. Vect.* **1**, 69–78.

**Burgos**, J.J., J. Fuenzalida, and C.B. Molion, 1991: Climate change predictions for South America. *Climate Change,* **18**, 223–229.

**Campos**, M., A. Sanchez, and D. Espinoza, 1996: *Adaptation of Hydropower Generation in Costa Rica and Panama to Climate Change.* Central American Project on Climate Change, Springer-Verlag.

**Campos**, M., C. Hermiosilla, J. Luna, M. Marin, J. Medrano, G. Medina, M. Vives, J. Diaz, A. Gutierrez, and M. Dieguez, 1997: *Global Warming and the Impacts of Sea Level Rise for Central America: an Estimation of Vulnerability.* Workshop organized by U.S. EPA, Chinese Taipei, USCSP, Government of The Netherlands, and NOAA, February 24–28, 1997,Taipei, Taiwan.

**Campos**, M., 1996: *Estimación de la Vulnerabllidad de los Recursos Hidricos, Marinos-Costeros y Agricolas en Centroamerrica, ante un Potencial Cambio Climático*, USCSP/Proyecto Centroamericano sobre Cambio Climático (in press).

**Canziani**, O.F., 1993: Cambio globales: posibles efectos socioeconómicos en la Argentina. In: *Elementos de Política Ambiental.* Honorable Cámara de Diputados de la Provincia de Buenos Aires, pp. 779–789.

**Canziani**, O.F., 1994: *La Problemática Ambiental Urbana. Seminbario sobre Gestión Municipal de Residuos Urbanos.* UNDP/IEIMA, La Plata, Argentina, pp. 11–49.

**Canziani**, O.F., 1996: *Urbanization and Environmental Problem (with Emphasis in Mercosur Countries).* Seminar on European Community, Mercosur and Environment, February 1996, Brussels, Belgium, Environmental European Bureau, Brussels, Belgium.

**Canziani**, O.F., R.M. Quintela, J.A. Forte-Lay, and A. Troha, 1987: *Estudio de Grandes Tormentas en la Pampa Deprimida, en la Provincia de Buenos Aires y su Incidencia en el Balance Hidrológico.* Proceedings of the Symposium on Hydrology of Large Plains, 1983, Olavarria, Argentina, UNESCO, Buenos Aires, Argentina.

**Carcavallo**, R.U., S.I. Curto de Casas, and J.J. Burgos, 1995: Blood-feeding Diptera: epidemiological significance and relation to climate change. I: General aspects, genera *Anopheles* and *Aedes*. Experimental and field data. *Entomol. Vect.* **2**, 35–60.

**Carcavallo**, R.U. and A. Martinez, 1972: Life cycles of some species of *Triatoma* (Hemipt. Reduviidae). *Canadian Entomology,* **104**, 699–704.

**Cardich**, A., 1974: Los yacimientos de la etapa agrìcola de Lauricocha, Perú, y los límites superiores del cultivo. *Relaciones de la Sociedad Argentina de Antropología,* **8**, 27–48.

**Castañeda**, M.E. and V. Barros, 1996: *Sobre las causeas de las tendencias de precipitación en el Cono Sur de America del Sur al este de los Andes.* Report 26-1996, Center for Ocean-Atmosphere Studies.

**Catala**, S.S., 1991: The biting rate of *Triatoma infestans* in Argentina. *Med. Vet. Entomol.,* **5**, 325–333.

**Catala**, S.S., D.E. Gorla, and M.A. Basombrio, 1992: Vectorial transmission of *Trypanosoma cruzi* an experimental field study with susceptible and inmunized hosts. *American J. Trop. Med. Hyg.,* **47**, 20–26.

**Caviedes**, C.N., 1973: Secas and El Niño, two simultaneous climatic hazards in South America. *Proc. Assoc. Amer. Geogr.,* **5**, 44–49.

**Cerri**, C.C., M. Bernoux, and B.J. Feigl, 1995: *Deforestation and Use of Soil as Pasture: Climatic Impacts.* Interdisciplinary Research on the Conservation and Sustainable Use of the Amazonian Rainforest and its Information Requirements, Brasilia, Brazil, November 20–22, 1995, [Lieberei, R., C. Reisdorff, and A.D. Machado, (eds.)] MCT-CNPQ, pp. 177-186.

**Cerri**, C.C., M. Bernoux, B. Volkoff, and J.L Moraes, 1996: Dinamica do carbono nos solos da Amazonia. In: *O solo nos grandes dominios morfoclimáticos do Brasil e desenvolvimento sustentado* [Alvarez, V.H., L.E. Fontes, and M.P.F. Fontes (eds.)]. Visosa, pp. 61-69.

**Cerveny**, R.S., B.R. Skeeter, and K.F. Dewey, 1987: A preliminary investigation of a relationship between South American snow cover and the Southern Oscillation. *Mon. Wea. Rev.,* **115**, 620–623.

**Chambers**, J.R., 1991: Coastal degradation and fish population losses. In: *Stemming the Tides of Coastal Fish Habitat Loss* [Stroud, R.H. (ed.)]. Proceedings of the Marine Recreational Fisheries Symposium National Coalition for Marine Conservation, Savannah, GA, USA, pp. 45–51.

**Cole**, J.E, G.T. Shen, and M. Moore, 1992: Coral monitors of El Niño-Southern Oscillation dynamics aspects across equatorial Pacific. In: *El Niño: Historical and Paleoclimatic Aspects of the Southern Oscillation.* Cambridge University Press, New York, NY, USA, and London, United Kingdom, pp. 349–375.

**Colwell**, R.R., 1996: Global climate and infectious disease: the cholera paradigm. *Science*, **274**, 2025–2031.

**Conde**, X., D. Liverman, M. Flores, and T.C. Ferrere, 1996: *Vulnerabilidad del cultivo de maiz de temporal en México ante el Cambio Climatico.* Taller de Vulnerabilidad, 1996, Montevideo, Uruguay.

**Costa**, M.J., J.L. Costa, P.R. Almeida and C.A. Assis, 1994: Do eel grass beds and salt marsh borders act as preferential nurseries and spawning grounds for fish? An example of the Mira estuary in Portugal. *Ecological Engineering*, **3**, pp. 187-195.

**Curto de Casas**, S.I., R.U. Carcavallo, C.A. Mena Segura, and I. Galindez Giron, 1994: Bioclimatic factors of *Triatominae* distribution. Useful techniques for studies on climate change. *Entomol. Vect.* **1**, 51–68.

**da Rocha**, H.R., 1996: $CO_2$ flux over the Brazilian tropical rain forest and Cerrado vegetation. A review of recent measurements, and modeling data. In: *Greenhouse Gas Emissions Under Developing Countries Point of View* [Pinguelli, R.L. and M.A. dos Santos (eds.)]. COPPE, Federal University of Rio de Janeiro, Brazil, pp. 68–75.

**De Filippi** *et al.*, 1994: Los residuos sólidos urbanos en ciudades pequefias y medianas. Seminario de Gestión Municipal de los Residuos Urbanos. La Plata, Provincia de Buenos Aires, Instituto de Estudios e Investigaciones sobre el Medio Ambiente, IEIMA, Buenos Aires, pp. 51-62.

**Del Carril**, A., M. Doyle, V. Barros, and M.N. Nuñez, 1996: *Sobre el Mínimo en la Descarga de los Rios Cuyanos, a Principios de la Década del 70.* USCSP Workshop on Vulnerability Studies in Latin America, February 1996, Montevideo, Uruguay.

**DENR-ADB**, 1990: *Master Plan for Forestry Development.* Department of Environment and Natural Resources and the Asian Development Bank, Manila, Philippines, 523 pp.

**Díaz**, S., 1995: Elevated $CO_2$ responsiveness, interactions at the community level and plant functional types. *J. Biogeogr.* **22**, 289–295.

**Di Pace**, M.J. and S.A. Mazzuchelli, 1993: Desarrollo sustentable en la Argentina. Implicaciones regionales. In: *Elementos de Política Ambiental* [Goin, F. and R. Goñi (eds.)]. Honorable Cámara de Diputados, Buenos Aires, pp. 869–890.

**Dixon**, R.K., S. Brown, R.A. Houghton, A.M. Solomon, M.C. Trexler, and J. Wisneiwski, 1994: Carbon pools and flux of global forest ecosystems. *Science,* **263**, 185–190.

**Doake**, C.S.M. and D.G. Vaughan, 1991: Rapid disintegration of the Wordie ice shelf in response to atmospheric warming, *Nature*, **35(6316)**, pp. 328-330.

**Doberitz**, R., 1969: Cross spectrum and filter analysis of monthly rainfall and wind data in the Tropical Atlantic region. *Bonner Meterol. Abhandl.,* **11**, 1–43.

**Downing**, T.E., 1992: *Climate Change and Vulnerasble places: Global Food Security and country studies in Zimbawe, Kenya, Senegal and Chile.* Research Report 1, Environmental Change Unit, University of Oxford, United Kingdom, 54 pp.

**Easterling**, W.E., 1990: Climate trends and prospects. In: *Natural Resources for the 21st Century* [Sampson, R.N. and D. Hair (eds.)]. Island Press, Washington, DC, USA, pp. 32–55.

**Epstein**, P.R., T.E. Ford, and R.R. Colwell, 1993: Marine ecosystems. *Lancet,* **342**, 1216–1219.

**Escudero**, J., 1990: *Control Ambiental en Grandes Ciudades: Caso de Santiago de Chile.* Seminario Latinoamericano sobre Medio Ambiente y Desarrollo, Instituto de Estudios e Investigaciones sobre el Medio Ambiente, October 1990, Bariloche, Argentina, IEMA, Buenos Aires, Argentina, pp. 229–236.

**Eskuche**, U., 1992: Sinopsis cenosistematica preliminar de los pajonales mesofilos seminaturales del nordeste de la Argentina, incluyendo pajonales pampeanos y puntanos. *Phytoccenologia,* **21**, 287–312.

**Ezcurra**, E., 1990: The basin of Mexico. In: *The Earth as Transformed by Human Action: Global and Regional Changes in the Biosphere over the Past 300 Years* [Turner, B.L, *et al.* (eds.)]. Cambridge University Press, New York, NY, USA, pp. 577–588.

**Fankhauser**, S. and R.S. J. Tol, 1997: The Social Costs of Climate Change: The IPCC Second Assessment Report and Beyond. *Mitigation and Adaptation Strategies for Global Change,* **1**, pp. 385-403.

**FAO**, 1982: *Conservation and Development of Tropical Forest Resources.* FAO Forest Paper 37, Food and Agriculture Organization of the United Nations, Rome, Italy, p. 122.

**FAO**, 1992a, *Yearbook of Fishery Statistics. Catches and Landings 1990*, **70**, Department of Fisheries, UN, Rome, Italy.

**FAO**, 1992b, *Review of the Status of World Fish Stocks*, Department of Fisheries, UN, Rome, Italy.

**FAO**, 1993: *Forest Resource Assessment 1990: Tropical Countries.* FAO Forest Paper 112, Food and Agriculture Organization at the United Nations, Rome, Italy, pp. 59.

**Fisher**, M.J., I.M. Rao, M.A. Ayerza, C.E. Lascano, J.I. Sanz, R.J. Thomas, and R.R. Vera, 1994: Carbon storage by introduced deep-rooted grasses in the South American savannas. *Nature,* **266**, pp. 236-248.

**Forte-Lay**, J.A, 1987: *Evolución de las Características Hidrometeorológicas de la Llanura Pampeana Argentina.* Seminario Internacional Hidrología de Grandes Llanuras. UNESCO Programa Hidrológico Internacional, Argentina. UNESCO, Buenos Aires, Argentina.

**Frére**, M., J.O. Rijks and J. Rea, 1978: *Agroclimatological Study of the Andean Zone.* Inter-institutional Project FAO/UNESCO and WMO, WMO TN 161, Geneva, Switzerland, 298 pp.

**Friederich**, S., 1994: Wirkung veräderter Klimatischer factorem auf pflanzenschaedlinge. In: *Klimaveraenderungen und Landwirtschaft, Part II Landbauforsc,* [Brunnert, H. and U. Dämmgen (eds.)] Vlkenrode, **148**, pp. 17–26.

**Fuenzalida**, P.H., B. Rosenblüth, and R. Sanguineti, 1993: *Temperature Variations in Chile and Austral South America During the Present Century and its Relation With Rainfall.* The Quaternary of Chile International Workshop, November 1–9, 1993, University of Chile.

**Gallopin**, G., M. Winograd, and Y. Gomez, 1991: *Ambiente y Desarrollo en América Latina y el Caribe: Problemas, Oportunidades y Prioridades.* GASE, Bariloche, Argentina.

**Gash**, J.H.C. and W.J. Shuttleworth, 1991: Tropical deforestation: albedo and the surface-energy balance. *Climatic Change,* **19**, 123–133.

**Gay-García**, C., L.G. Ruiz Suarez, 1996: UNEP Preliminary *Inventory of GHG Emissions: Mexico.* (In press)

**GLASOD**, 1990, *Global Assessment of Soil Degradation*, Project UNEP/ISRIC/Intnl Reference and Information Centre, Part 3: Natural Resources, UNEP 1992 Environmental Data Report, Nairobi, Kenya.

**Gligo**, N., 1995: Situación y perspectivas ambientales en América Latina y el Caribe. *Revista de la CEPAL,* **55**, 107–122.

**Glynn**, P.W. 1989: Coral mortality and disturbances to coral reefs in the tropical eastern Pacific. In: *Global Ecological Consequences of 1982–1983 ENSO* [Glynn, P.W., (ed.)]. Elsevier Science Publishers, New York, NY, USA, pp. 55–126.

**Glynn**, P.W. and L.D. Cruz, 1990: Experimental evidence of high temperature stress causes of El Niño-coincident coral mortality. *Coral Reefs*, **8**, 181–191.

**Goldenberg**, J., L.C. Monaco, and I.C. Macedo, 1993: *The Brazilian Fuel-Alcohol Program. Renewable Energy, Sources of Fuels and Electricity, UNCED* [De Johansson, T.B., H. Kelly, A.K.N. Reddy, and R.H. Williams (eds.)]. Island Press, Washington, DC, pp. 841–863.

**Gómez**, I.A. and G.C. Gallopin, 1991: Estimacion de la productividad primaria neta de ecosistemas terrestres del mundo en relacion a factores ambientales. *Ecologia Austral,* **1**, 24–40.

**González**, N., 1990: *La Contaminación del Agua. Política Ambietal y Gestión Municipal.* Instituto de Estudios e Investigaciones sobre el Medio Ambiente, Buenos Aires, Argentina.

**Gorham**, E., 1991: Northern peatlands: role in the carbon cycle and probable responses to climate warming. *Ecol. Applic.,* **1(2)**, 182–195.

**Greco**, S., R.H. Moss, D. Viner, and R. Jenne (eds.), 1994*: Climate Scenarios and Socioeconomic Projections for IPCC WG II Assessment.* Intergovernmental Panel on Climate Change Working Group II, IPCC-WMO and UNEP, Washington, DC, 67 pp.

**Grosse**, S., 1993: *Schistosomiasis and Water Resources Development: A Reevaluation of an Important Environment-Health Linkage.* Working paper of the Environment and Natural Resources Policy and Training Project, EPAT/MUCIA, Technical Series No. 2., May 1993, University of Michigan Press, Ann Arbor, MI, USA, 32 pp.

**Hack**, W.H., 1955: Estudios sobre biologia del *Triatoma infestans* (Klug, 1834) (Hemiptera, Reduviidae). *An. Inst. Med. Reg.,* **4**, 125–134.

**Haines**, A., P.R. Epstein, P.R., and A. McMichael, 1993: Global health watch: monitoring impacts of environment change. *Lancet,* **342**, 1464–1469.

**Halpin**, P.N., P.M. Kelly, C.M. Secrett, and T.M. Smith, 1995a: *Climate Change and Central America Forest System.* Background paper of the Nicaragua Pilot Project.

**Halpin**, P.N., P.M. Kelly, C.M. Secrett, and T.M. Smith, 1995b: *Climate Change and Central America Forest System.* Background paper of the Costa Rica Pilot Project.

**Halstead**, S.B., 1990: Dengue. In: *Tropical and Geographical Medicine* [Warren, K. and A.A.F. Mahmoud (eds.)]. McGraw-Hill, New York, NY, USA, 2nd ed., pp. 675–685.

**Hansen**, J., A. Lacis, D. Rind, G. Russell, P. Stone, I. Fung, R. Ruedy, and J. Lerner, 1984: Climate Sensitivity: Analysis of feedback mechanisms. In: *Climate Processes and Climate Sensitivity* [Hansen, J.E. and T. Takahashi (eds.)]. American Geophysical Union, Washington, DC, 130–163.

**Hardoy**, J.E., D. Mitlin, and D. Satterthwaite, 1992: *Environmental Problems in Third World Cities.* Earthscan, London, United Kingdom, 302 pp.

**Hastenrath**, S., 1976: Variations in the low altitude circulation and extreme climatic events in tropical America. *J. Atm. Sc.,* **33**, 202–215.

**Hastenrath**, S. and A. Ames, 1995: Recession of Yanamarey Glacier in Cordillera Blanca, Peru, during the 20th century. *Journal of Glaciology,* **41(137)**, 191–196.

**Haxeltine**, A. and C.I. Prentice, 1996: BIOME3: An equilibrium terrestrial biosphere model based on ecophysiological constraints, resource availability and competition among plant functional types. *Global Biogeochemical Cycles* **10**, pp. 693-710.

**Herrera-Basto**, E., D.R. Prevots, M.L. Zarate, J.L. Silva, and J.S. Amor, 1992: First reported outbreak of classical dengue fever at 1,700 meters above sea level in Guerrero State, Mexico, June 1988. *American Journal of Tropical Medicine,* **46(6)**, 649–653.

**Hoffman**, J.J.A., W.M. Vargas, and S. Nunez, 1996: Temperature, humidity and precipitation variations in Argentina and adjacent sub-Antarctic region during the present century. *Meteorologisches Zeitschrift,* (in press)

**Horel**, J.D. and A.G. Cornejo-Garrido, 1986: Convection along the Coast of Northern Perú, during 1983: Spatial and Temporal variations of clouds and rainfall. *Mon. Wea. Rev.,* **114**, 2091–2105.

**IPCC**, 1990: *Climate Change: The IPCC Scientific Assessment* [Houghton, H.T., G.J. Jenkins, and J.J. Ephraums (eds.)]. Cambridge University Press, Cambridge, United Kingdom, 365 pp.

**IPCC**, 1992: *Climate Change 1992: The Supplementary Report to The IPCC Scientific Assessment.* Prepared by IPCC Working Group I [Houghton, J.T., B.A. Callander, and S.K. Varney (eds.)] and WMO/UNEP. Cambridge University Press, Cambridge, United Kingdom, 200 pp.

**IPCC**, 1996: *Climate Change 1995: The Science of Climate Change. Contribution of Working Group I to the Second Assessment Report of the Intergovernmental Panel on Climate Change* [Houghton, J.J., L.G. Meiro Filho, B.A. Callander, N. Harris, A. Kattenberg, and K. Maskell (eds.)]. Cambridge University Press, Cambridge, United Kingdom and New York, NY, USA, 572 pp.

- Schimel, D., D. Alves, I. Enting, M. Heimann, F. Joos, D. Raynaud, T.M.L. Wigley, E. Sanhueza, X. Zhou, P. Jonas, R. Charlson, H. Rodhe, S. Sadasivan, K.P. Shine, Y. Fouquart, V. Ramaswamy, S. Solomon, J. Srinivasan, D. Albritton, R. Derwent, Y. Isaken, M. Lal, and D. Wuebbles, Chapter 2. *Radiation Forcing of Climate Change,* pp. 65–131.
- Dickinson, R.E., V. Meleshko, D. Randall, E. Sarachik, P. Silva-Dias, and A. Slingo, Chapter 4. *Climate Processes,* pp. 193–227.
- Kattenberg, A., F. Giorgi, H. Grassl, G.A. Meehl, J.F.B. Mitchell, R.J. Stouffer, T. Tokioka, A.J. Weaver, and T.M.L. Wigley, Chapter 6. *Climate Models —Projections of Future Climate,* pp. 289–357.

**IPCC**, 1996: *Climate Change 1995: Impacts, Adaptations, and Mitigation of Climate Change: Scientific-Technical Analyses. Contribution of Working Group II to the Second Assessment Report of the Intergovernmental Panel on Climate Change* [Watson, R.T., M.C. Zinyowera, and R.H. Moss (eds.)]. Cambridge University Press, Cambridge, United Kingdom and New York, NY, USA, 880 pp.

- Kirschbaum, M.U.F. and A. Fischlin, Chapter 1. *Climate Change Impacts on Forests,* pp. 95–130.
- Allen-Diaz, B., Chapter 2. *Rangelands in a Changing Climate: Impacts, Adaptations and Mitigation,* pp. 131–158.

- Noble, I.R. and H. Gitay, Chapter 3. *Deserts in a Changing Climate: Impacts,* pp. 159–189.
- Bullock, P. and H. Le Houérou, Chapter 4. *Land Degradation and Desertification,* pp. 170–189.
- Beniston, M. and D.G. Fox, Chapter 5. *Impacts of Climate Change on Mountain Regions,* pp. 191–213.
- Öquist, M.G. and B.H. Svensson, Chapter 6. *Non-Tidal Wetlands,* pp. 215–239.
- Fitzharris, B.B., Chapter 7. *The Cryosphere: Changes and their Impacts,* pp. 240–265.
- Ittekkot, V., Chapter 8. *Oceans,* pp. 266–288.
- Bijlsma, L, Chapter 9. *Coastal Zones and Small Islands,* pp. 289–324.
- Arnell, N., B. Bates, H. Lang, J.J. Magnuson, and P. Mulholland, Chapter 10. *Hydrology and Freshwater Ecology,* pp. 325–364.
- Acosta-Moreno, R. and J. Skea, Chapter 11. *Industry, Energy, and Transportation: Impacts and Adaptation,* pp. 365–398.
- Scott, M.J., Chapter 12. *Human Settlements in a Changing Climate: Impacts and Adaptation,* pp. 399–426.
- Reilly, J., Chapter 13. *Agriculture in a Changing Climate: Impacts and Adaptation*, pp. 427–467.
- Kaczmarek, Z., Chapter 14. *Water Resources Management*, pp. 469–486.
- Solomon, A.M., Chapter 15. *Wood Production under Changing Climate and Land Use,* pp. 487–510.
- Everett, J., Chapter 16. *Fisheries,* pp. 511–537.
- McMichael, A., Chapter 18. *Human Population Health,* pp. 561–584.
- Brown, S., Chapter 24. *Management of Forests for Mitigation of Greenhouse Gas Emissions*, pp. 773–797.

**IPCC**, 1996: *Climate Change 1995: Economic and Social Dimensions of Climate Change. Contribution of Working Group III to the Second Assessment Report of the Intergovernmental Panel on Climate Change* [Bruce, J.P., H. Lee, and E.F. Haites (eds.)]. Cambridge University Press, Cambridge, United Kingdom and New York, NY, USA, 448 pp.

- Pearce, D.W., W.R. Cline, A.N. Achanta, S. Fankhauser, R.K. Pachauri, R.S.J. Tol, and P. Vellinga, Chapter 6. *The Social Costs of Climate Change: Greenhouse Damage and Benefits of Control*, pp. 179–224.

**IPCC**, 1996b: *Technologies, Policies, and Measures for Mitigating Climate Change: IPCC Technical Paper 1.* Intergovernmental Panel on Climate Change, Working Group II [Watson, R.T., M.C. Zinyowera, and R.H. Moss (eds.)]. World Meterological Organization, Geneva, Switzerland, 84 pp.

**Junin**, B., H. Grandinetti, J.M. Marconi, and R.U. Carcavallo, 1995: Vigilania del *Aedes aegypti* en la ciudad de Buenos Aires (Argentina). *Entomol. Vect.,* **2**, 71–75.

**Kadota**, T., R. Naruse, P. Skvarca, and M. Aniya, 1992: Ice flow and surface lowering of Tyndall Glacier, southern Patagonia. *Bulletin of Glacier Research,* **10**, 63–68.

**Kalkstein**, A., 1993: Health and climate change: direct impacts in cities. *Lancet,* **342**, 1397–1399.

**Karl**, T.R., R.W. Knight, and N. Plummer, 1995: Trends in high frequency climate variability in the XX Century. *Nature,* **377**, 217–220.

**Katsouyanni**, K., A. Pantazopoulu, and G. Touloumi, 1993: Evidence for interaction between air pollution and high temperature in the causation of excess mortality. *Arch. Environm. Health,* **48**, 235–242.

**Klinedinst**, P.L., D.A. Wilhite, G.L. Hahn and K.G. Hubbard, 1993: The potential effects of climate change on summer season dairy cattle milk production and reproduction. *Climate Change*, **23(1)**, pp. 21–36.

**Koopman**, J.S., D.R. Prevots, M.A.V. Marin, H.G. Dantes, M.L.Z. Aqino, I.M. Longini, and J.S. Amor, 1991: Determinants and predictors of dengue infection in Mexico. *American Journal of Epidemiology,* **133**, 1168–1178.

**Kousky**, V.E., M.T. Kagano, and I.A. Cavalcanti, 1984, A review of the Southern Oscillation: oceani-atmospheric circulation change and related rainfall anomalies. *Tellus,* **36A**, 490–504.

**Labraga**, J.C., 1997: The climate change in South America due to a doubling in the $CO_2$ concentration: intercomparison of general circulation model equilibrium experiments. *International Journal of Climatology,* **17**, 377–398.

**Labraga**, J.C. and M. Lopez, 1997: A comparison of the climate response to increased carbon dioxide simulated by general circulation models with mixed-layer and dynamic ocean representation in the region of South America. *International Journal of Climatology* (in press).

**LAC CDE**, 1992: *Our Own Agenda. Latin America and Caribbean Commission on Development and Environment.* UNDP and IDB, in collaboration with ECLAC and UNEP, Santiago, Chile.

**Lapenis**, A.G., N.S. Oskina, M.S. Barash, N.S. Blyum, and Y.V. Vasileva, 1990: The late quaternary changes in ocean biota productivity. *Okeanologiya/Oceanology,* **30**, 93–101. (in Russian)

**Lebre-La Rovère**, 1995: Mitigation measures for reduction of GHG emissions in Brazil. *Revista de la Facultad de Ingeniería de la Universidad Central de Venezuela,* **10(1–2)**.

**Leemans**, R., 1992: Modelling ecological and agricultural impacts of global change on a global scale. *Journal of Scientific and Industrial Research,* **51**, 709–724.

**Lindroth**, R.L., 1996: $CO_2$-mediated changes in tree chemistry and tree-lepidopteran interaction. In: *Carbon Dioxide and Terrestrial Ecosystems* [Koch, G.W. and H.A. Mooney (eds.)]. Academic Press, San Diego, CA, USA, pp. 105–120.

**Liverman**, D.M. and K.L. O'Brien, 1991: Global warming and climate change in Mexico. *Global Environmental Change,* **1(4)**, 351–363.

**Löpmeier**, F.J., 1990: Klimaimpaktforschung aus agrarmeteorologischer sicht. *Bayer. Landw. Jarhb.,* **67(1)**, 185–190.

**Mabberley**, D.J., 1992: *Tropical Rainforest Ecology.* Blackie and Son Ltd., Glasgow and London, United Kingdom, 2nd ed., 300 pp.

**Magalhães**, A.R. and M.H. Glantz (eds.), 1992: *Socioeconomic Impacts of Climate Variations and Policy Responses in Brazil.* UNEP, SEPLAN, Esquel Brasil Foundation, Brasilia, Brazil.

**Malagnino**, E. and J. Strelin, 1992: Variations of Upsala glacier in Southern Patagonia, since the late Holocen to the present. In: *Glaciological Research in Patagonia (1996)* [Naruse, R. and M. Aniya (eds.)] pp. 61–85.

**Manabe**, S. and R.T. Wetherald, 1987: Large scale changes of soil wetness induced by an increase in atmospheric carbon dioxide. *J. Atmos. Sci.*, **44**, 1211–1235.

**Martens**, W.J.M., T.H. Jetten, and D.A. Focks, 1997: Sensitivity of malaria, aschistosomiasis and dengue to global warming. *Climate Change,* **35**, 145–156.

**Martens**, W.J.M., L.W. Niessen, J. Rotmans, T.H. Jetten, and A.J. McMichaels, 1995: Potential impact of global climate chance on malaria risk. *Environm. Health Perspect.,* **103**, 458–464.

**Martinez**, A., R.U. Carcavallo, and A.F. Prosen, 1967: Los mosquitos posibles transmisores amarilicos en Argentina. *Segundas Jorn. Entomoep. Argentinas,* **1**, 7–26.

**Massera**, O.R., M.J. Ordoñez, and R. Dirzo, 1996: Carbon emissions from Mexican forests: current situation long-term scenarios. *Climate Change* (in press).

**Maul**, G.A. (ed.), 1993: *Climatic Change in the Intra-Americas Sea.* UNEP/IOC, E. Arnold, London, United Kingdom, 384 pp.

**McFarlane**, N.A., G.J. Boer, J.P. Blanchet, and M. Lazare, 1992: The Canadian Climate Centre second general circulation model and its equilibrium climate. *J. Climate*, **5**, 1013–1044.

**Medina**, E. and J. Silva, 1990: Savannas of the northern South America: a steady state regulated by water-fire interactions on a background of low nutrient availability. *Journal of Biogeography,* **17**, 403–413.

**Meisner**, J.D., 1990: Effect of climatic warming on the southern margins of the native range of brook trout, Salvelinus fontinalis. *Canadian Journal of Fisheries and Aquatic Sciences,* **47**, 1067–1070.

**Milliman**, J.D., 1993: Coral reefs and their response to global climate change. In: *Climatic Change in the Intra-Americas Sea* [Maul, G.A. (ed.)]. UNEP/IOC, E. Arnold, London, United Kingdom, pp. 306–322.

**Minetti**, J.L. and W.M. Vargas, 1995: *Trends and Jumps in South American Annual Precipitation South of Parallel 15º S,* (in press).

**Mitchell**, J.F.B., C.A. Senior, and W.J. Ingram, 1989: $CO_2$ and climate: A missing feedback? *Nature*, **341**, 132–134.

**Mitchell**, J.F.B. and D.A. Warrilow, 1987: Summer dryness in northern midlatitude due to increased $CO_2$. *Nature*, **330**, 238–240.

**Molion**, L.C.B., 1975: *A Climatonomy Study of the Energy and Moisture Fluxes of the Amazon Basin with Consideration of Deforestation Effects.* Diss., University of Wisconsin.

**Molion**, L.C.B., 1996: Global climate impacts of Amazonia deforestation. In: *Greenhouse Gas Emissions Under Developing Countries Point of View* [Pinguelli, R.L. and M.A. dos Santos (eds.)]. COPPE, Federal University of Rio de Janeiro, Brazil, pp. 78–89.

**Moreira**, J.R. and A.D. Poole, 1992*: Hydropower and its constraints. Renewable Energy. Sources of Fuels and Electricity, UNCED* [De Johansson, T.B., H. Kelly, A.K.N. Reddy, and R.H. Williams (eds.)]. Island Press, Washington, DC, USA, pp. 73–120.

**Myers**, N., 1992: Synergisms: joint effects of climate change and other forms of habitat destruction. In: *Global Warming and Biological Diversity* [Peters, R.L. and T.E. Lovejoy (eds.)]. Yale University Press, New Haven, CT, USA and London, United Kingdom, pp. 344–354.

**Neilson**, R.P., 1995: A model for predicting continental-scale vegetation distribution and water balance. *Ecological Applications,* **5**, pp. 362–385.

**Ojima**, D.S., W.J. Parton, D.S. Schimel, and C.E. Owensby, 1990: Simulated impacts of annual burning on prairie ecosystems. In: *Fire in North American Tallgrass Prairies* [Collins, S.L. and L.L. Wallace (eds.)]. University of Oklahoma Press, Norman, OK, USA, pp. 99–118.

**Omar**, M.H., 1980: *Meteorological Factors Affecting the Epidemiology of the Cotton Leaf Worm and the Pink Bollworm.* WMO TN 532, WMO, Geneva, Switzerland.

**Ottichilo**, W.K., J.H. Kinuthia, P.O. Ratego, and G. Nasubo, 1991: *Weathering the Storm: Climate Change and Investment in Kenya.* ACTS Press, Nairobi, Kenya.

**PAHO**, 1976: *Report of the Director.* Washington, DC.

**PAHO**, 1994: *Health Conditions in the Americas,* Vols 1 and 2. Washington, DC.

**PAHO**, 1996: *Report of the Director.* Washington, DC.

**Parkinson**, R.W., R.D. De Laum, and J.R.White, 1994: Holocene sea level rise and the fate of mangrove forests within the Wider Caribbean region. *Journal of Coastal Research,* **10**, 1077–1086.

**Parry**, M.L., 1992: *The Potential Socio-Economic Effects of Climate Change in South-East Asia.* United Nations Environment Programme, Nairobi, Kenya.

**Parry**, M., T.R. Carter, and N.T. Konijn (eds.), 1988*: The Impact of Climatic Variations on Agriculture. Vol. 2, Assessment in Semi-Arid Regions.* Kluwer Academic Publishers, Dordrecht, The Netherlands, 764 pp.

**Parry**, M.L., J.H. Porter and T.R. Carter, 1990: Agriculture: climate change and its implications. *Trends in Ecology and Evolution*, **5**, pp. 318-322.

**Paruelo**, J.M. and O.E. Sala, 1992: El impacto de la desertificación sobre la capacidad de carga de las estepas patagonicas: sus consecuencias económicas. II Congreso Latinoamericano de Ecologia, Caxambú, Mina Gerais, Brazil.

**Pauly**, D., D.P. Much, J. Mendo, and J. Tsukayama, 1989: *The Peruvian upwelling ecosystem: dynamics and interactions. ICLARM Conference Proceedings 18.* GTZ Gmbh. Eschbom, FRG, and International Center for Living Aquatic Resources Management, Manilla, Phillipines, 438 pp.

**Pauly**, D., M.L. Palomares, and F.C. Gayanilo, 1987: VPA estimates of the monthly population length composition, recruitment, mortality, biomass and related statistics of Peruvian anchoveta, 1953 to 1981. In: *ICLARM Studies and Reviews 15,* Instituto del Mar del Peru (IMARPE), Callao, Peru; Deutsche Gesellschaft für Technische Zusammenarbeit (GTZ), GmbH, Eschborn, Germany; and International Center for Living Aquatic Resources Management (ICLARM), Manila, Philippines.

**Pauly**, D. and I. Tsukayama, 1987: The Peruvian anchoveta and its upwelling ecosystem: three decades of change. In: *ICLARM Studies and Reviews 15,* Instituto del Mar del Peru (IMARPE), Callao, Peru; Deutsche Gesellschaft für Technische Zusammenarbeit (GTZ), GmbH, Eschborn, Germany; and International Center for Living Aquatic Resources Management (ICLARM), Manila, Philippines, 351 pp.

**Pedgley**, D.E., 1980: *Weather and Airborne Organisms.* WMO TN 562, Geneva, Switzerland.

**Perdomo**, M., 1996: Regional synthesis chapter for Latin America. In: *Greenhouse Gas Emission Inventories — Interim Results from the U.S. Country Studies Program* [Braatz, B.V., S. Jallow, S. Molnar, D. Murdiyarso, M. Perdomo, and J.F. Fitzgerald (eds.)].

**Perdomo**, M., M.L. Olivo, Y. Bonduki, and L.J. Mata, 1996: Vulnerability and adaptation assessments for Venezuela. In: *Vulnerability and Adaptation to Climate Change* [Smith, J.B., S. Huq, L.J. Lenhart, L.J. Mata, I. Nemesova, and S. Toure (eds.)]. Kluwer Academic Publishers, Dordrecht, Netherlands, pp. 347–366.

**Perillo**, G.M. and M.C. Piccolo, 1992: *Impact of SLR on the Argentinan Coastline.* Proceedings of the International Workshop on Coastal Zones, March 9–13, 1992, Isla Margarita, Venezuela.

**Pittock**, A.B., 1980: Patterns of climatic variation in Argentina and Chile. Precipitation, 1931–1960. *Mon. Wea. Rev.,* **108**, 1347–1361.

**Porter**, J.H., M.L. Parry and T.R. Carter, 1991: The potential effects of climate change on agricultural insect pests. *Agricultural/Forest Meteorology*, **57**, pp. 221-240.

**Rao**, V.B., P. Satyamurty, and J.I.B. Brito, 1986: On the 1983 drought in Northeast Brazil. *J. Clim.,* **6**, 43–51.

**Rasmuson**, E.M. and T.H. Carpenter, 1982: Variations in tropical sea surface temperature and surface windfields associated with the Southern Oscillation/El Niño. *Mon. Wea. Rev.,* **109**, 1163–1168.

**Reilly**, J., N. Hohmann, and S. Kane, 1994: Climate change and agricultural trade: who benefits, who loses? *Global Environmental Change,* **4(1)**, 24–36.

**Restrepo**, M., A., Greer, D.F., and Moncada, L.H., 1972: *Relationship between the environment and the paracoccidioimycosis.* Proc. Symp. Paracoccidioimycosis, PAHO Publication 254, 84 pp.

**Riede**, K., 1993: Monitoring biodiversity: analysis of Amazonian rainforest sounds. *Ambio,* **22**, 546–548.

**Rosenblüth**, B.H., H. Fuenzalida-Ponce, and P. Aceituno, 1997: Recent temperature variations in southern South America. *Intl. J. of Climatology,* **17**, 1–19.

**Rosenzweig**, C., M.L. Parry, G. Fischer, and K. Frohberg, 1993: *Climate Change and World Food Supply.* Research Report No. 3, Environmental Change Unit, Oxford University, Oxford, United Kingdom, 28 pp.

**Rosenzweig**, C. and M.L. Parry, 1994: Potential impact of climate change on world food supply. *Nature,* **367**, 133–138.

**Rubin**, M.J., 1955: An analysis of pressure anomalies in the Southern Hemisphere. *Notos,* **4**, 11–16.

**Sala**, O.E. and J.M. Paruelo, 1997: Ecosystem services in grasslands. In: G.C. Daily (de.), *Nature's Services: Societal Dependence on Natural Ecosystems*, pp. 237-252. Island Press, Washington, DC.

**Salati**, E. and C.A. Nobre, 1991: Possible climatic impacts of tropical deforestation. *Climate Change,* **19**, 177–196.

**Salati**, E. and P.B. Vose, 1984: Amazon Basin: a system in equilibrium. *Science,* **225**, 129–138.

**Santibañez**, F. and J. Uribe, 1994: *El Clima y la Desertificacion en Chile.* En Taller en Nacional del Plan Nacional de Accion para Combatir le Desertificacion, Universidad de Chile, Santiago, pp. 17–24.

**Schlesinger**, M.E. and Z.C. Zhao, 1989: Seasonal climate changes induced by doubled $CO_2$ as simulated in the OSU atmospheric GCM/mixed layer ocean model. *J. Climate*, **2**, 459–495.

**Schubert**, C., 1992: The glaciers of the Sierra Nevada de Merida (Venezuela): a photographic comparison of recent deglaciation. *Erdkunde,* **46**, 59–64.

**Sharp**, G.D. and D.R. McLain, 1993: Fisheries, El Niño-Southern Oscillation and upper ocean temperature records: an eastern Pacific example. *Oceanog.,* **5(3)**, 163–168.

**Shen**, G.T., 1993: Reconstruction of El Niño history from reef coral. *Bull. of Intl. France études andean,* Vol 112, pp. 125–158.

**Shope**, R.E., 1991: Global climate change and infectious diseases. *Environm. Health Perspect.,* **96**, 171–174.

**Shukla**, J., C. Nobre, and P. Sellers, 1990: Amazon deforestation and climate change. *Science* **247**, pp. 1322–1325.

**Siegenthaler**, U. and J.L. Sarmiento, 1993: Atmospheric carbon dioxide and the ocean. *Nature,* **365**, 119–125.

**Skvarca**, P., 1993: Fast recession of the northern Larsen Ice Shelf monitored by space images. *Annals of Glaciology,* **17**, 317–321.

**Skvarca**, P., 1994: Changes and surface features of the Larsen Ice Shelf, Antarctica, derived from the Landsat and Kosmos mosaics. *Annals of Glaciology,* **20**, 6–12.

**Soriano**, A. and C. Movia, 1986: Erosón y desertización en la Patagonia. *Interciencia* **11**, pp. 77–83.

**Stafford Smith**, D.M. and G. Pickup, 1993: Out of Africa, looking in: understanding vegetation change. In: *Range Ecology at Disequilibrium* [Behnke, R., I. Scoones, and C. Keerven (eds.)]. Overseas Development Institute, London, United Kingdom, pp. 196–244.

**Starke**, L. (ed.), 1994: *State of the World 1994A Worldwatch Institute Report on Progress Toward a Sustainable Society.* W.W. Norton and Company, New York, NY, USA, and London, United Kingdom, 265 pp.

**Stone**, T.A., P. Schlesinger, R.A. Houghton, and G.M. Woodwell, 1994: *A Map of Vegetation of South America Based on Satellite Imagery.* American Society for Photogrammetric Remote Sensing, Woods Hole Research Center, Woods Hole, MA, USA, pp. 541–551.

**Suriano**, J.M., L.H. Perpozzi, and D.E. Martinez, 1992: El cambio global: tendencias climticas en la Argentina y el mundo. *Ciencia Hoy,* **3**, 32–39.

**Tejada Miranda**, F., 1996: *Evaluación de la vulnerabilidad de los bosques al cambio climático en Bolivia.* Climate Change National Programme, Ministry of Sustainable Development and Environment, La Paz, Bolivia.

**Telleria**, A.V., 1986: Health consequences of floods in Bolivia in 1982. *Disasters,* **10**, 88–106.

**Teves**, N., G. Laos, C. San Roman, S. Carrasco, and L. Clemente, 1996: Vulnerability and adaptation assessments for Venezuela. In: *Vulnerability and Adaptation to Climate Change* [Smith, J.B., S. Huq, S. Lenhart, L.J. Mata, I. Nemesova, and S. Toure (eds.)]. Kluwer Academic Publishers, Dordrecht, Netherlands, pp. 347–366.

**Torn**, M.S. and J.S. Fried, 1992: Predicting the impacts of global warming on wildland fire. *Climatic Change,* **21**, 257–274.

**Treharne**, K., 1989: The implications of the "greenhouse effect" for fertilizer and agrochemicals. In: *The Greenhouse Effect and UK Agriculture 19* [de Bennet, R.M. (ed.)]. Centre for Agricultural Strategy, University of Reading, United Kingdom, pp. 67–78.

**UNEP**, 1992: *Environmental Data Report 1991–1992,* prepared by GEMS Monitoring and Assessment Research Centre, London, UK, in co-operation with World Resources Institute, Washington, DC and UK Department of the Environment, London. B. Blackwell Ltd, Oxford, UK, 408 pp.

**Vargas**, W.M., 1987: El Clima y sus Impactos, Implicancias en las inudaciones del Noroeste en Buenos Aires. *Boletin Informative Techint,* **250**, Buenos Aires, 12 pp.

**Vargas**, W.M. and S. Bishoff, 1995: Statistical study of climatic jumps in the regional zonal circulation over South America. *Journal of the Meteorological Society of Japan,* **73(5)**, 849–856.

**Vargas**, W.M. and J.L. Minetti, 1996: Trends and jumps in South America annual precipitation south of the 15°S parallel. *Climate Change* (submitted).

**Velez**, S.A., N. Nassif, J. Togo, and S. Wottips, 1990: *Energetico socio-alimentario.* Seminario sobre Latino-America, Medio Ambiente y Desarrollo, Instituto de Estudios e Investigaciones sobre el Medio Ambiente, October 1990, Bariloche, Argentina, IEIMA, Buenos Aires, Argentina.

**Viglizzo**, E.F. and Z.E. Roberto, 1997: On trade-offs in low input agroecosystems. *Agricultural Systems* (in press).

**Viglizzo**, E.F., Z.E. Roberto, M.C. Filippin, A.J. Pordomingo, 1995: Climate variability and agroecological change in the Central Pampas of Argentina. *Agriculture, Ecosystems and Environment,* **55**, pp. 7-16.

**Viglizzo**, E.F., Z.E. Roberto, F. Lértora, Gay E. López, J. Bernardos, 1997: Climate and land-use change in field-crop ecosystems of Argentina. *Agriculture, Ecosystems and Environment.* (In press)

**Villers**, L., 1995: *Vulnerabilidad de los Ecosistemas Forestales.* Country Study Mexico Report 6.

**Vitousek**, P.M. and R.L. Sanford, Jr., 1986: Nutrient cycling in moist tropical forest. *Annual Review of Ecology and Systematics,* **17**, 137–167.

**Walker**, G.T., 1928: Cerá (Brazil) famines and the general air movement. *Beitr. Phys. d. Frein. Atmosph.,* **14**, 88–93.

**Westoby**, M., B. Walker, and I. Noy-Meir, 1989: Opportunistic management for rangelands not at equilibrium. *Journal of Range Management,* **42**, 26–274.

**Wetherald**, R.T. and S. Manabe, 1988: Cloud feedback processes in a general circulation model. *J. Atmos. Sci.,* **45**, 1397–1415.

**WFC**, 1987: *Proceedings of the Thirteenth Ministerial Session.* Beijing, China.

**Whetton**, P.H., A.B. Pittock, J.C. Labraga, A.B. Mullan, and A. Joubert, 1996: Southern hemisphere climate comparing models with reality. In: *Climate Change, Developing Southern Hemisphere Perspectives* [Henderson-Sellers, A. and T. Giambelluca, (eds.)]. Wiley and Sons, Chichester, United Kingdom, pp. 89–130.

**Whitmore**, 1984: *Tropical Rain Forests in the Far East.* Clarendon Press, Oxford, United Kingdom, 2nd ed., 352 pp.

**WHO**, 1985: *Ten Years of Onchocerciasis Control in West Africa: Review of Work of the OCP in the Volta River Basin Area from 1974 to 1984.* OCP/GVA/85.1B, World Health Organization, Geneva, Switzerland, 113 pp.

**WHO Commission on Human Health and Environment**, 1992: *Report of the Panel on Urbanization.* World Health Organization, Geneva, Switzerland, pp. 127.

**WHO**, 1993: *A Global Strategy for Malaria Control*, Geneva, Switzerland.

**WHO**, 1995: *The World Health Report 1995: Bridging the Gaps.* World Health Organization, Geneva, Switzerland, 118 pp.

**WHO**, 1996: *The World Health Report 1996: Fighting Disease Fostering Development.* World Health Organization, Geneva, Switzerland.

**Winograd**, M., 1995: *Environmental Indicators for Latin America and the Caribbean: Toward Land Use Sustainability.* GASE, in collaboration with OAS, IICA/GTZ, and WRI.

**Wooster**, W.S. and D.L. Fluharty (eds.), 1985: *El Niño North: Niño Effects in the Eastern Subarctic Pacific Ocean.* Washington Sea Grant Program, Seattle, WA, USA.

**WRI**, 1990–1991: *Special Focus on Latin America*, Oxford University Press, New York, NY, pp. 33–64.

**WRI**, 1992: *World Resources 1992–93: A Guide to the Global Environment.* Oxford University Press, New York, NY, USA, 385 pp.

**WRI**, 1996: *World Resources 1996–97: A Guide to the Global Environment.* World Resources Institute/United Nations Environment Programme/ United Nations Development Programme/The World Bank. Oxford University Press, New York, 342 pp.

**Yañez**, E., 1991: Relationships between environmental changes and fluctuating major pelagic resources exploited in Chile (1950–1988). In: *Long-Term Variability of Pelagic Fish Populations and Their Environment* [Kawasaki, T., S. Tanaka, Y. Toba, and A. Taniguchi (eds.)]. Pergamon Press, Tokyo, Japan, pp. 301–309.

**Zuidema**, G., G.J. van den Born, J. Alcamo, and G.J.J. Kreileman, 1994: Determining the potential distribution of vegetation, crops and agricultural productivity. *Water, Air, and Soil Pollution,* **76(1/2)**, 163–198.

## For Further Reading

**Aceituno**, P., 1988: On the functioning of the Southern Oscillation in the South America sector. Part I: Surface climate. *Mon. Wea. Rev.,* **116**, 505–524.

**Anderson**, J.M., 1992: Responses of soils to climate change. *Advances in Ecological Research,* **22**, 163–210.

**Biggs**, G.R., 1993: Comparison of coastal wind and pressure trends over the tropical Atlantic: 1946–1988. *Intl. J. Climatol.,* **13**, 411–421.

**Binet**, D., 1988: Rôle possible d'une intensification des alizés sur le changement de repartition des sardines et sardinelles le long de la côte ouest africaine. *Aquat. Living Resources,* **1**, 115–132.

**Brinkman**, R. and W.G. Sombroek, 1993: *The Effects of Global Change on Soil Conditions in Relation to Plant Growth and Food Production.* Expert Consultation Paper on Global Climate Change and Agricultural Production: Direct and Indirect Effects of Changing Hydrological, Soil and Plant Physiological Processes. Food and Agriculture Organization, Rome, Italy, December 1993, 12 pp.

**Busby**, J.R., 1988: Potential implications of climate change on Australia's flora and fauna. In: *Greenhouse Planning for Climate Change* [Pearman, G.I. (ed.)]. CSIRO, Melbourne, Australia, pp. 387–398.

**Campbell**, B.D. and R.M. Hay, 1993: Will subtropical grasses continue to spread through New Zealand? In: *Proceedings of the XVII International Grassland Congress,* pp. 1126–1128.

**Caviedes**, C.N., 1984: El Niño 1982–1983. *Geographical Review,* **74**, 268–290.

**Cole**, J.E., R.G. Fairbanks, and G.T. Shea, 1993: Recent variability in the Southern Oscillation. Isotopic results from Tarawa Atoll coral. *Science,* **260**, 1790–1793.

**D'Oliveira**, A.S. and C.A. Nobre, 1986: *Interactions Between Frontal Systems in South America and Tropical Convection over the Amazon.* Preprints of the Second International Conference on Southern Hemisphere Meteorology, 1986, Wellington, New Zealand, AMS.

**Eakin**, C.M., 1995: *Post-El Niño Panamanian reefs: Less Accretion, More Erosion and Damselfish Protection.* Proceedings of the Seventh Coral Reef Symposium, vol. 1, June 22–27, 1992, Guam, University of Guam Press, Manqilao, Guam, USA, pp. 387–396.

**Ehleringer**, J.R. and C.B. Field (eds.), 1993: *Scaling Physiological Processes: Leaf to Globe.* Academic Press, New York, NY, USA, 388 pp.

**Epstein**, P.R., 1992: Cholera and the environment. *Lancet,* **339**, 1167–1168.

**Flenley**, J.R., 1979: The late quaternary vegetational history of the equatorial mountains. *Progress in Physical Geography,* **3**, 488–509.

**Frasier**, G.W., J.R. Cox, and D.A. Woolhiser, 1987: Wet-dry cycle effects on warm-season grass seedling establishment. *Journal of Range Management,* **40**, 2–6.

**Garms**, R., J.F. Walsh, and J.B. Davis, 1979: Studies on the reinvasion of the Onchocerciasis Control Programme in the Volta River Basin by *Sumulium damnosum* s.l. with emphasis on the Southwestern areas. *Tropical Medicine and Parasitology,* **30**, 345–362.

**Gilles**, H.M., 1993: Epidemiology of malaria. In *Bruce-Chwatt's Essential Malariology* [Giles, H.M. and D.A. Warrell (eds).]. Edward Arnold, London, United Kingdom, pp. 124-163.

**Henderson-Sellers**, A. and A. Hansen, 1995: *Atlas of Results from Greenhouse Model Simulations.* Model Evaluation Consortium for Climate Assessment, Climate Impacts Centre, Macquarie University, Kluwer Academic Publishers, Dordrecht, Netherlands.

**Henderson-Sellers**, A. and K. McGuffie, 1994: Land surface characterisation in greenhouse climate simulations. *International Journal of Climatology,* **14**, 1065-1094.

**Horel**, J.D. and J.M. Wallace, 1981: Planetary scal eatmospheric phenomenas associated with the Southern Oscillation. *Mon. Wea. Rev.,* **109**, 813–823.

**Mar del Peru (IMARPE)**, Callao, Peru; Deutsche Gesellschaft für Technische Zusammenarbeit (GTZ), GmbH, Eschborn, Federal Republic of Germany; and International Center for Living Aquatic Resources Management (ICLARM), Manila, Philippines, pp. 142–178.

**Martinelli**, L.A., A.V. Krusche, and R.L. Victoria, 1996: Changes in carbon stock associated to land-use/cover changes and emissions of $CO_2$. In: *Greenhouse Gas Emissions Under Developing Countries Point of View* [Pinguelli, R.L. and M.A. dos Santos (eds.).]. COPPE, Federal University of Rio de Janeiro, Brazil, pp. 90–101.

**Masera**, O.R., 1995: Mitigation options in forest sector with reference to Mexico. Latin America GHG emissions and mitigation options. *Revista de la Facultad de Ingenieria de la Universidad Central de Venezuela,* **10(1–2)**.

**Mata**, L.J., 1996: A study of climate change impacts on the forests of Venezuela. In: *Adapting to Climate Change: Assessment and issues* [Smith, J., N. Bhatti, G. Menzhulin, R. Benioff, M. Budyko, M. Campos, B. Jallow, and F. Rijsberman (eds.)]. Springer-Verlag.

**Medina**, E., 1982: Physiological ecology of neotropical savanna plants. In: *Ecology of Tropical Savannas* [Hyntley, B.J. and B.H. Walker (eds.)]. Springer-Verlag, Berlin, Germany, pp. 308–335.

**Mills**, D.M., 1995: A climatic water budget approach to blackfly population dynamics. *Publications in Climatology,* **48**, 1–84.

**Myers**, N., 1993: Questions of mass extinction. *Biodiversity and Conservation,* **2**, 2–17.

**Parker**, C. and J.D. Fryer, 1975: Weed control problems causing major reductions in world food supplies. *FAO Plant Protection Bulletin,* **23**, 83–95.

**Peters**, R.L., 1992: Conservation of biological diversity in the face of climate change. In: *Global Warming and Biological Diversity* [Peters, R.L. and T.E. Lovejoy (eds.)]. Yale University Press, New Haven, CT, USA, and London, United Kingdom, pp. 59–71.

**Rauh**, W., 1985: The Peruvian-Chilean deserts. In: *Ecosystems of the World: Hot Deserts and Arid Shrublands*, vol. 12A [Evenari, M., I. Noy Meir, and D. Goodall (eds.)]. Elsevier Science Publishers, Amsterdam, Netherlands, pp. 239–267.

**Reibsame**, W.E., 1990: Anthropogenic climate change and a new paradigm of natural resource planning. *Professional Geographer,* **42(1)**, 1–12.

**Reilly**, J. and N. Hohmann, 1993: Climate change and agriculture: the role of international trade. *American Economic Association Papers and Proceedings,* **83**, 306–312.

**Reynolds**, R.E. 1987: Breeding duck population, production and habitat surveys, 1979–1985. *Trans. N. Am. Wildl. Nat. Resour. Conf.,* **52**, 186–205.

**Reynolds**, J.F. and P.W. Leadley, 1992: Modelling the response of arctic plants to changing climate. In: *Arctic Ecosystems in a Changing Climate, an Ecophysiological Perspective* [Chapin III, F.S., R.L. Jefferies, J.F. Reynolds, G.R. Shaver, and J. Svoboda (eds.)]. Academic Press, San Diego, CA, USA, pp. 413–438.

**Riebsame**, W.E., 1989: *Assessing the Social Implications of Climate Fluctuations.* United Nations Environment Programme, Nairobi, Kenya.

**Riebsame**, W.E., 1990: The United States Great Plains. In: *The Earth as Transformed by Human Action: Global and Regional Changes in the Biosphere over the Past 300 Years* [Turner, B.L.

**Riebsame**, W.E. *et al.*, 1995: Complex river basins. In: *As Climate Changes: International Impacts and Implications* [Strzepek, K.M. and J.B. Smith (eds.)]. Cambridge University Press, Cambridge, United Kingdom, pp. 57–91.

**Riebsame**, W.E., W.B. Meyer, and B.L. Turner II, 1994: Modeling land use and cover as part of global environmental change. *Climatic Change,* **28**, 45–64.

**Sharkey**, T.D., 1985: Photosynthesis in intact leaves of C3 plants: physics, physiology and rate limitations. *Botanical Review,* **51**, 53–105.

**Smayda**, T.J., 1990: Novel and nuisance phytoplankton blooms in the sea: evidence for a global epidemic. In: *Toxic Marine Phytoplankton* [Graneli, E. *et al.* (eds.)]. Elsevier Science Publishers, New York, NY, USA, pp. 29–40.

**Smith**, T.M. and H.H. Shugart, 1993: The transient response of terrestrial carbon storage to a perturbed climate. *Nature,* **361**, 523–526.

**Solley**, W.B., C.F. Merk, and R.P. Pierce, 1988: *Estimated Use of Water in the United States in 1985.* Circular 1004, U.S. Geological Survey, Washington, DC.

**Solomon**, A.M., 1992a: A global biome model based on plant physiology and dominance, soil properties and climate. *Journal of Biogeography,* **19**, 117–134.

**Solomon**, A.M., 1992b: The nature and distribution of past, present and future boreal forests. In: *A Systems Analysis of the Global Boreal Forest* [Shugart, H.H., G.B. Bonan, and R. Leemans (eds.)]. Cambridge University Press, Cambridge, United Kingdom, pp. 291–307.

**Tol**, R.S.J., 1994: The damage costs of climate change: A note on tangibles and intangibles applied to DICE. *Energy Policy,* **22**, 436–438.

**Troadec**, J.-P., 1989: Elements pour une autre strategie. In: *L'homme et les ressources halieutiques, essai sur l'usage d'une ressource commune renouvelable* [Troadec, J.-P. (sous la dir.)]. IFREMER, Paris, France, pp. 747–795.

**UNEP**, 1997: *Global Environmental Outlook.* United Nations Environmental Programme, WRI, USA, Distributed by Oxford University Press and UNEP, 265 pp.

**UNESCO**, 1978: *Tropical Forest Ecosystems: A State-of-Knowledge Report.* UNESCO, UNEP, and FAO, Paris, France, 683 pp.

**Van der Hammen**, T., 1974: The Pleistocene changes of vegetation and climate in tropical South America. *Journal of Biogeography,* **1**, 3–26.

**Walsh**, J.F., J.B. Davis, and R. Garms, 1981: Further studies on the reinvasion of the Onchocerciasis Control Programme by *Simulium damnosum* s.l. *Tropical Medicine and Parasitology,* **32(4)**, 269–273.

**Warren**, C.R., 1994: Against the grain; a report on the Pio XI glacier Patagonia. *Geographical Magazine,* **66(9)**, 28–30.

**WHO**, 1992: *Our Planet, Our Health.* World Health Organization, Geneva, Switzerland.

**WHO**, 1996: *Climate Change and Human Health.* Fide,[ McMichael, A.J., A. Haines, R. Siloff and S. Kovats (eds.)] WHO/EHG/96.7. World Health Organization, Geneva, Switzerland, 297 pp.

# 7

# Middle East and Arid Asia

HABIBA GITAY (AUSTRALIA) AND IAN R. NOBLE (AUSTRALIA)

Lead Author:
*O. Pilifosova (Kazakstan)*

Contributors:
*B. Alijani (Iran) and U.N. Safriel (Israel)*

# CONTENTS

# EXECUTIVE SUMMARY

The Middle East and Arid Asia includes 21 countries of the predominantly arid and semi-arid region of the Middle East and central Asia. The region extends from Turkey in the west (26°10'E) to Kazakstan in the east (86°30'E) and from Yemen in the south (12°40'N) to Kazakstan in the north (50°30'N). Although the relief is mostly low, there are several peaks of 7,500 m or more in the Hindu Kush and Tien Shan mountain ranges; the lowest point is the Dead Sea in Israel (-395 m). Many of the region's countries are landlocked. The total population is 433 million, with half living in urban centers; six of the 21 countries have urban populations of over 80%. Countries of the region vary greatly in gross national product (GNP)—they include relatively resource rich, oil exporting countries and several poorer countries.

*Climate*: Average monthly mean daily temperatures in the region range from -10°C to 25°C in January to 20°C to >35°C in July. The rainfall in the region is low but highly variable. Average monthly rainfall ranges from 0 mm to 200 mm in January and from 0 mm to 500 mm in July.

*Climate trends:* Records of annual temperatures during the period 1900–96 show almost no change for most of the Middle East region, but a 1–2°C/century increase for central Asia. There was a 0.7°C increase from 1900 to 1996 for the region as a whole. There was no discernible trend in annual precipitation during 1900–95 for the region as a whole, nor in most parts of the region—except in the southwestern part of the Arabian peninsula, where there was a 200% increase. This increase, however, is in relation to a very low base rainfall (<200 mm/yr).

*Climate scenarios*: Climate models project that temperatures in the region will increase by 1–2°C by 2030–2050, with the greatest increases in winter in the northeast and in summer in part of the southwest. Precipitation is projected to increase slightly in the winter throughout the region and in the summer to remain the same in the northeast and increase in the southwest (i.e., the southern part of the Arabian peninsula). These precipitation projections vary from model to model and are unlikely to be significant. Because of projected increases in temperatures, higher evaporation is expected. Soil moisture is projected to decrease in most parts of the region, which may lead to increased areas of soil degradation.

Because of the arid nature of the region, some sectors will be particularly affected by climate change. The impacts on these sectors are summarized below.

*Ecological systems*: The region is mostly arid and semi-arid and is dominated by grasslands, rangelands, deserts, and some woodlands. Vegetation models project little change in most arid (or desert) vegetation types under climate change projections. The impacts may be greater in the semi-arid lands of the region than in the arid lands, especially in composition and distribution of vegetation types. The projected small increase in precipitation is unlikely to improve land conditions in the next century, partly because soil conditions take a long time to improve and partly because human pressure on these systems may contribute to land degradation. Improved water-use efficiency by some plants under elevated carbon dioxide ($CO_2$) may lead to some improvement in plant productivity and changes in ecosystem composition. Grasslands, livestock, and water resources are likely to be most vulnerable to climate change in this region because they are located mostly in marginal areas. Management options, such as better stock management and more integrated agroecosystems, could improve land conditions and counteract pressures arising from climate change.

Forests/woodlands are important resources, although they cover only a small area. They will have to be safeguarded, given the heavy use of wood for fuel in some countries.

The region has a large area dominated by mountains, such as the Hindu Kush, Karakoram, and Tien Shan mountains in the eastern part of the region. Mountain areas are under pressure from human use, which is leading to land degradation in some areas. Some of the mountains have permanent glaciers, which will be affected by climate change. Glacial melt is projected to increase under climate change, leading to increased flows in some river systems for a few decades; this will be followed, however, by a reduction in flow as the glaciers disappear—creating larger areas of arid, interior deserts in low- and mid-lying parts of central Asia.

A tenth of the world's known species of higher plants and animals occur in this region. Many countries are centers of origin for crop and fruit tree species of critical importance to world food production; thus, they are important sources of genes of wild relatives. Few studies have assessed the impact of climate change on biodiversity in the region. Current human activity, however, is causing a loss of biodiversity.

*Water resources*: Water shortages already are a problem in many countries of this predominantly arid region, and are unlikely to be reduced and may be exacerbated by climate change. Projected precipitation increases are small, and temperatures and evaporation are projected to rise. Rapid development is threatening some water supplies through salinization and pollution, and expanding populations are increasing the demand for water. Adaptation strategies might include more efficient organization of water supply, treatment, and delivery systems for urban areas and, in arid Asia, increased use of groundwater. Measures to conserve or reuse water already have been implemented in some countries; such strategies may overcome some shortages, especially if they are adopted widely throughout the region. Changes in cropping practices and improved irrigation practices could reduce water use significantly in some countries, especially those of the former Soviet Union (FSU).

*Food and Fiber*: Land degradation problems and limited water supplies constrain present agricultural productivity and threaten the food security of some countries. Though there are few projections of the impacts of climate change on food and fiber production for the region, studies in Kazakstan and Pakistan have suggested some negative impacts on wheat yields. There are also projected increases (e.g., winter wheat in Kazakstan). Many of the options available for combating existing problems will contribute to reducing the anticipated impacts of climate change. Food and fiber production concentrated on more intensively managed lands could lead to greater reliability in food production and reduce the detrimental impacts of extreme climatic events, such as drought, on rangeland systems. Implementation of more flexible risk-management strategies (e.g., long-term and appropriate stocking rates, responding to variations in precipitation by changing animal numbers annually)—along with the use of a wider variety of domestic animals, game ranching, and multiple production systems—would provide greater food security to the region.

*Health:* Human health in the region is variable, reflecting the economies of the countries. Some countries, where poverty is high, have high infant mortality rates and low life expectancies. The impacts of climate change are likely to be detrimental to the health of the population, mainly through heat stress and possible increases in vector-borne (e.g., dengue fever and malaria) and waterborne diseases. Decreases in water availability and food production (especially if there is a shortage of water for irrigation) would lead to indirect impacts on human health associated with nutritional and hygiene issues.

*Integration:* Countries of the FSU are undergoing major economic changes, resulting in changes in agricultural systems and management. This transition is likely to provide significant "win-win" opportunities for the conservation of resources, to offset the impacts of climate change. Opportunities to change crop types and introduce more efficient irrigation are among the most promising. Human activity can exacerbate the effect of climate change in this arid/semi-arid region, leading to long-term detrimental effects on ecosystems and human health. The Aral Sea is an illustrative example of the multiplicative effects of resource overuse, which can lead to local environmental and even climate change. Extensive redirection of water from feeder rivers to irrigated agriculture since 1960 has led to a reduction in the surface area of the lake, as well as damage to the surrounding wetlands and the species that depend on them. Air temperature in the vicinity of the lake has increased. Saline and polluted dust from the exposed lake bed has been implicated in significant health problems and increases in infant mortality.

There are some obvious research needs. Clearly, many basic physiological and ecological studies of the effects of changes in atmospheric and climatic conditions are necessary. The most pressing need over much of the region is for sound assessment and monitoring programs to establish current baselines and identify rates of change.

## 7.1. Regional Characterization and Baseline Conditions

This chapter addresses 21 countries of the predominantly arid and semi-arid region of the Middle East and central Asia (see Figure 7-1 and Box 7-1). The region extends from Turkey in the west (26°10'E) to Kazakstan in the east (86°30'E), and from Yemen in the south (12°42'N) to Kazakstan in the north (50°30'N). The relief is mostly low. The highest point is Communism Peak in Tajikistan (7,495 m); the lowest point is the Dead Sea in Israel (-395 m). Many countries in the region are landlocked.

**Box 7-1. Middle East and Arid Asia**

| | |
|---|---|
| Afghanistan | Pakistan |
| Bahrain | Qatar |
| Iran, Islamic Republic of | Saudi Arabia |
| Iraq | Syria |
| Israel | Tajikistan |
| Jordan | Turkey |
| Kazakstan | Turkmenistan |
| Kuwait | United Arab Emirates |
| Kyrgyz Republic | Uzbekistan |
| Lebanon | Yemen |
| Oman | |

Although it is common to separate semi-arid, arid, and extreme arid parts of a country/region, they are amalgamated for most purposes in this chapter and (for conciseness) referred to as "arid" unless a specific distinction is needed. In the Second Assessment Report (SAR), much of this region was discussed in the chapters on deserts (see IPCC 1996, WG II, Chapter 3) and rangelands (see IPCC 1996, WG II, Chapter 2).

Because there are few statistics specifically for arid parts of the region, statistics for whole countries have been used. Data collection is sparse for many countries in the region, for a range of social and physical reasons. Economic and social statistics often are unavailable for countries that were part of the FSU. In addition, with the exception of Kazakstan, few countries have studied the impacts of climate change in this region. The brevity of this chapter reflects the lack of available published literature. This lack of data would have to be addressed in the near future, especially for some sectors (e.g., fisheries).

The region is vulnerable to climate change because it is dry and water availability is thus limited. In some countries, the ability to adapt to the impacts of climate change will be reduced by a lack of infrastructure.

The region covers approximately 9% of the world's land area. It is dominated by arid (50%) and semi-arid (11%) lands. The total population is 433 million, or 8% of the global population.

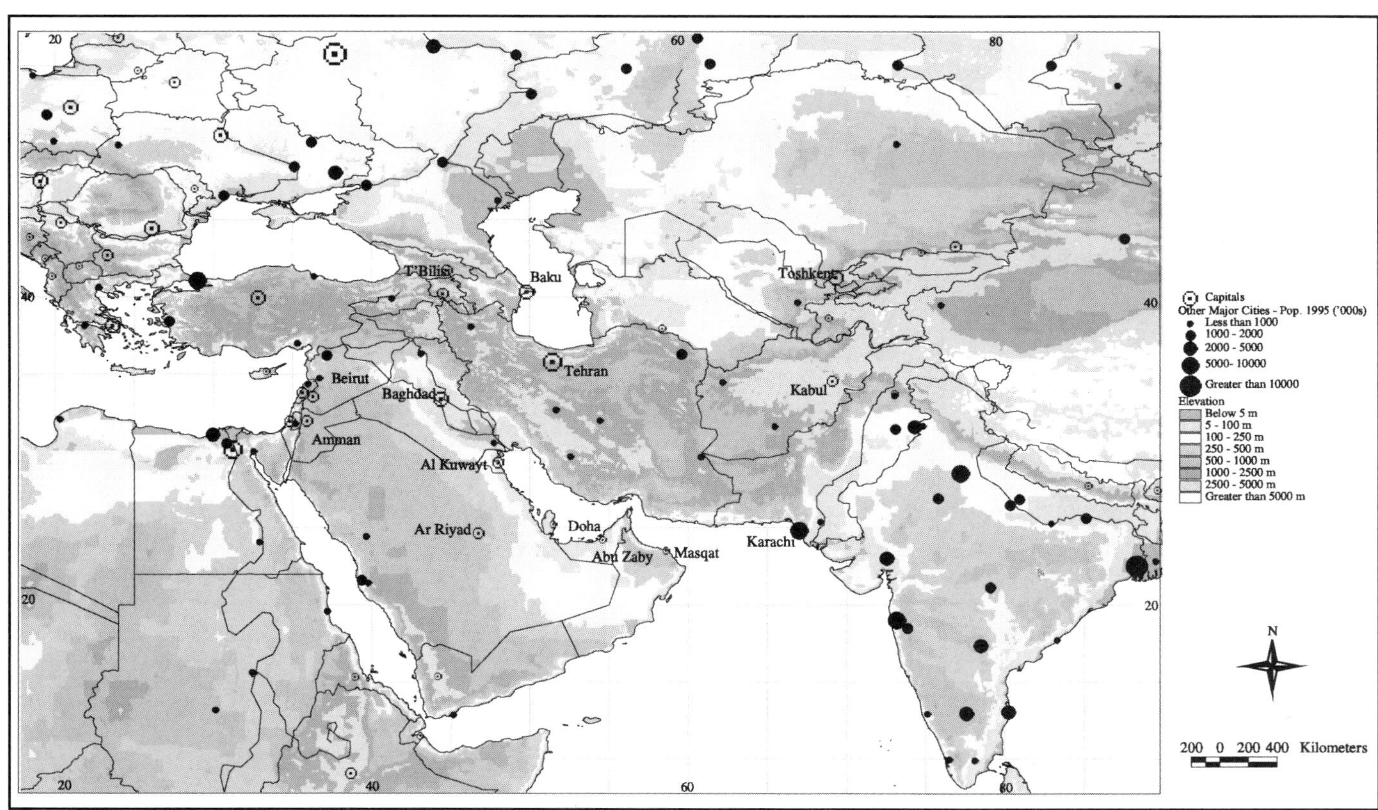


**Figure 7-1:** The Middle East and Arid Asia region [compiled by the World Bank Environment Department Geographic Information System (GIS) Unit; see Annex E for a color rendition].

Annual population growth rates range from -6.5% (Kuwait) to +5.8% (Afghanistan), compared with a world average of 1.6% for 1990–95. Half of the population lives in urban centers (compared with a world average of 45%), and 6 of the 21 countries are more than 80% urbanized. The population density varies between 6 persons/km$^2$ (Kazakstan) and 831 persons/km$^2$ (Bahrain); the regional average of 38 persons/km$^2$ is lower than the world's mean density (44 persons/km$^2$).

A regional summary of various economic, social, and environmental statistics from the World Resources Institute (WRI, 1994, 1996) is presented in Table 7-1 (see Annex D-5 for country-based data on most of the variables listed).

Annual water consumption in the region is 1,710 m$^3$ per capita, compared with the world average of 645 m$^3$ per capita. In many countries, the dominant water use is for irrigation to support small—but economically important—permanent pasturelands and croplands. The fraction of available water that is withdrawn annually for consumption varies from as little as 10% in Syria to, effectively, more than 100% in countries heavily reliant on desalinization.

The region is heterogeneous in terms of the countries' economies. Because it includes some of the richest and some of the poorest countries in the world, regional average economic performance statistics are misleading (IPCC 1996, WG II, Section 13.7). Industry and services contribute 82% of the gross domestic product (GDP), and agriculture accounts for the remainder. Per capita GNP has fallen substantially over the past decade, in part as a result of declining oil prices and political disruptions (IPCC 1996, WG II, Section 13.7).

## 7.2. Regional Climate

Two-thirds of the region can be classified as hot or cold desert. In the northern part of the region, a steppe climate prevails, with cold winters and hot summers. A narrow zone contiguous to the Mediterranean Sea is classified as a Mediterranean zone, with wet and moderately warm winters and dry summers. Permafrost zones exist in high mountain areas in the southeast part of the region.

### 7.2.1. *Observed Temperature and Future Projections*

Temperatures in the region range from -10°C to 25°C (January) to 20°C to >35°C (July) (Oxford World Atlas, 1994). The observed change in annual temperature in the region from 1955–74 to 1975–94 was 0.5°C (IPCC 1996, WG I, Figure 3.4). Temperature changes were smallest in December–February (0 to -0.25°C) and largest in September–November (~1°C). Annual temperatures in most of the Middle East region showed almost no change during the period 1901–96, but a 1–2°C/century increase was discernible in central Asia (based on the 5°x5° grid; see Annex A, Figure A-2). There was a 0.7°C increase during 1901–96 in the region as a whole (see Annex A, Figure A-9).

Climate models that include the effects of sulfate aerosols (GFDL and CCC) (IPCC 1996, WG I, Figure 6.7) project that the temperature in the region will increase 1–2°C by 2030–2050. The greatest increases are projected for winter in the northeast and for summer in part of the region's southwest (IPCC 1996, WG I, Figure 6.10).

### 7.2.2. *Observed Precipitation and Future Projections*

Rainfall is low in most of the region, but it is highly variable seasonally and interannually. Average monthly rainfall ranges from 0 mm to 200 mm in January and from 0 mm to 500 mm in July (Oxford World Atlas, 1994). There was no discernible trend in annual precipitation during 1901–95 for the region as a whole (see Annex A, Figure A-9), nor in most parts of the region—except in the southwestern part of the Arabian peninsula, where there was a 200% increase (see Annex A, Figure A-1). This increase, however, is in relation to a very low base rainfall (<200 mm/yr). Precipitation tends to be very seasonal; in the Middle East countries, for example, precipitation occurs during winter, and the summer dry period lasts for 5–9 months (UNEP, 1997).

Winter precipitation is projected to increase slightly (<0.5 mm/day) throughout the region; summer precipitation is projected to remain the same in the northeastern part of the region and increase (0.5–1 mm/day) in the southwest (i.e., the southern part of the Arabian peninsula). These projected changes vary, however, from model to model and are unlikely to be significant.

### 7.2.3. *Observed Moisture Availability and Future Projections*

Soil moisture is projected to decrease in most parts of the region because projected precipitation increases are small and evaporation will increase with rising temperatures (IPCC 1996, WG I, Figure 6.12).

The ratio of precipitation to potential evapotranspiration (P:PET) is a measure of moisture availability and an indicator of the type of vegetation that an area can support. The Middle East and Arid Asian region has a P:PET ratio in the range of 0.01–1.6; most of the area has a value of <0.45—typical of semi-arid and arid climates that support grasslands, shrublands, and some woodlands. Projected changes in P:PET for different model runs show no consistent trend (IPCC 1996, WG II, Chapter 14): Some models project a decrease (greater aridity) and some an increase. The lack of data specifically from the arid and semi-arid regions may contribute to this uncertainty.

## 7.3. Sensitivities, Adaptations, and Vulnerabilities of Key Sectors

This section discusses the region's major vulnerabilities—particularly those that apply to large parts of the region; many impacts are common to the region as a whole. Later in the

***Table 7-1:*** *Summary of socioeconomic, land, and biological data for the Middle East and Arid Asia region.*

| | | Middle East and Arid Asia Region | | | | |
|---|---|---|---|---|---|---|
| **Variable** | **World** | as % of the World | Average | 25th Percentile | 75th Percentile | Number of Countries Included |
| GNP total million US$ | | | 48,940 | 8,584 | 57,767 | 16 |
| GDP PPP % growth 1983–93 | | | 3.1 | 1.5 | 4.4 | 12 |
| GDP per capita PPP 1992 INT$ | | | 5,729 | 3,347 | 8,561 | 17 |
| Distribution of GDP 1993, industry (%) | | | 37 | 28 | 46 | 15 |
| Distribution of GDP 1993, services (%) | | | 45 | 41 | 53 | 15 |
| Distribution of GDP 1993, agriculture (%) | | | 18 | 5 | 28 | 15 |
| Population 1995 (thousands) | 5,716,426 | 7.6 | 20,622 | 3,009 | 20,141 | 21 |
| Annual average population change 1990–95 (%) | 1.6 | | 2.7 | 2.3 | 3.7 | 18 |
| Population density (#/km$^2$) (mean-weighted) | 44 | | 38 | 24 | 79 | 21 |
| Life expectancy 1990–95 | 64.7 | | 66.4 | 66.8 | 69.9 | 20 |
| Infant mortality (per 1,000 live births 1990–95) | 64.0 | | 49.4 | 30.0 | 52.5 | 20 |
| % urban population | 45 | | 50 | 39 | 85 | 21 |
| % in absolute poverty in rural areas | | | 28 | 15 | 30 | 9 |
| % urban population with access to safe water | | | 91 | 92 | 100 | 18 |
| % rural population with access to safe water | | | 77 | 70 | 100 | 17 |
| Total land area (Mha) | 13,098,404 | 8.8 | 54,704 | 8,360 | 65,209 | 20 |
| % arid land (mean-weighted) | | 50 | 77 | 59 | 95 | 15 |
| % arid land with soil constraints (mean-weighted) | | 90 | | | | 13 |
| % semi-arid land (mean-weighted) | | 9 | | | | 15 |
| % semi-arid with soil constraints (mean-weighted) | | 99.8 | | | | 12 |
| % forest/woodland | 31.8 | 1.4 | 4.7 | 0.8 | 4.9 | 16 |
| % pastureland 1991–93 (mean-weighted) | 25.7 | 42.8 | | 6.8 | 44.7 | 20 |
| % cropland (mean-weighted) | 11.1 | 11.8 | | 2.3 | 17.1 | 20 |
| % domesticated land (% of land area) | 38.0 | | 40.3 | 23.5 | 57.0 | 19 |
| % irrigated land (% of cropland) | 17.0 | | 41.1 | 15.0 | 60.0 | 18 |
| Coastline (km) | | | | 6.5 | 1677 | 19 |
| % grain-fed livestock | | | 36.1 | 24.5 | 46.0 | 16 |
| Annual average livestock numbers (000) | 3,268,787 | 407,450 | 20,369 | 1,334 | 16,330 | 20 |
| % of land area conserved under IUCN category I-V | 7.1 | | 2.5 | 0.3 | 3.8 | 19 |
| Total # of known species of vertebrates and higher plants | 295,299 | 10.3 | 1,697 | 157 | 2,126 | 18 |
| % of these species classed as threatened | | | 10.8 | 1.0 | 17.8 | 18 |
| % of these species classed as endemic | | | 5.3 | 0.0 | 5.0 | 18 |
| % water use for agriculture (1987) | 69.0 | | 78.7 | 79.0 | 91.5 | 19 |
| % water use by industry (1987) | 23.0 | | 8.6 | 4.5 | 9.5 | 19 |
| % domestic water use (1987) | 8.0 | | 1,271 | 3.3 | 13.5 | 19 |
| % annual withdrawal of water (1975–89) | 8.0 | | 63.0 | 24.0 | 70.0 | 18 |
| Annual renewable water resource per capita 1995 (m$^3$) | 7,176 | | 4,613 | 637 | 5,507 | 19 |
| Total internal renewable water resource 1995 (km$^3$) | 41,022 | 3.5 | 76.3 | 3.4 | 1,05.3 | 19 |
| Annual water withdrawal per capita (m$^3$) | 645 | | 1,710 | 466 | 2,374 | 19 |
| Total energy production 1993 (PJ) | 337,518 | 13.6 | 2,859 | 68 | 2,904 | 16 |
| Traditional fuel consumption as % of total energy consumption | 6 | | 5.9 | 0.0 | 1.0 | 17 |

**Notes**:

1) Data from World Resources 1996–97 (WRI, 1996).
2) Data for the following countries (maximum 21), when available, were included in the above table: Afghanistan, Bahrain, Iran, Iraq, Israel, Jordan, Kazakstan, Kuwait, Kyrgyz Republic, Lebanon, Oman, Pakistan, Qatar, Saudi Arabia, Syria, Tajikistan, Turkey, Turkmenistan, United Arab Emirates, Uzbekistan, and Yemen.
3) Country-by-country data are presented in Annex D.

section, some sector-specific impacts (e.g., those affecting agriculture or uplands), adaptations, and vulnerabilities for smaller parts of the region are highlighted.

### 7.3.1. *General Impacts*

The region is mostly semi-arid and arid, with significant areas of extreme aridity (i.e., deserts, as defined in IPCC 1996, WG II, Chapter 3). All areas experience wide fluctuations in rainfall, and their native plants and animals are adapted to coping with sequences of extreme climatic conditions. In many of these ecological systems, the initial climatic changes are unlikely to create conditions significantly outside the present range of variation (IPCC 1996, WG II, Chapter 3); thus, impacts from climate change may not be apparent for several decades.

Models that project changes in global vegetation in response to a doubled-$CO_2$-driven climate suggest little change in desert (arid) communities. One study, which used four general circulation model (GCM) climate projections, concluded that deserts were the most stable of the 16 vegetation types considered (IPCC 1996, WG II, Chapter 3); it estimated that only 59–66% of all combined vegetation types for the world would remain in the same category, whereas 82–92% of existing deserts would retain this classification. The impact of climate change may be greater in the semi-arid areas of the region than in the arid areas. Agriculture, natural grasslands, livestock, and water resources in marginal areas are most likely to be vulnerable to climate change (Smith *et al.*, 1996).

Although precipitation is projected by some models to increase slightly, this increase will have little impact because most of the region will remain arid. In rangelands or semi-arid lands, increased precipitation eventually may lead to improved soil conditions, including larger accumulations of organic matter, improved soil field capacity, enhanced moisture availability, and changed runoff patterns (IPCC 1996, WG II, Chapter 3). This process is slow (taking 400 years or more), however, so little advantage will accrue to these ecosystems during the next century. A more important process may be the demonstrated increase in the water-use efficiency of plants when high levels of $CO_2$ are present in the atmosphere (Bazzaz *et al.*, 1996). In atmospheres rich in $CO_2$, plants can take up the $CO_2$ necessary for photosynthesis with less water loss from their stomatal pores. This efficiency gain has been demonstrated to be most advantageous in plants growing in water-limited circumstances—and thus may lead to higher productivity throughout much of the region. The response of individual species and communities varies considerably, however. In high-$CO_2$ conditions, some species conserve water by using less water to maintain previous rates of productivity (i.e., photosynthesis); other species increase their productivity and maintain relatively high rates of water use. There is little understanding about how whole communities of plants respond, and, unfortunately, few studies have been done within the region. Nevertheless, the net result of increased water-use efficiency should be increased plant productivity in arid and semi-arid ecosystems; this effect probably will outweigh any direct effects stemming from increased precipitation resulting from climate change.

The processes associated with changed climatic and atmospheric conditions are expected to lead to changes in the composition of many plant communities—such as the mix between grasses and shrubs or weeds and non-weeds (IPCC 1996, WG II, Chapters 2 and 3). There also is evidence that the forage quality (e.g., the ratio of nitrogen to carbon) and the protein content of some cereals may decline. Although it has not been possible to test the effects of high $CO_2$ atmospheres on extensive areas of crops and pastures with large grazing animals (Diaz, 1995), it is possible that they may affect farm management systems significantly.

Even where wetter conditions prevail, however, any "greening" of the deserts and rangelands often might be counteracted by pressure from the expanding human population and associated desertification problems. Climatic conditions and social pressures vary greatly in a region that ranges from Mediterranean climate to extreme deserts, with economies as varied as those of the oil producing states, Israel, and the countries of the FSU. In the latter group of countries, the impacts of climate change are likely to be relatively small compared with the impacts of economic transitions on agricultural systems and the environment (see Section 7.3.4.1).

### 7.3.2. *Ecosystems*

- *The region is characterized by arid and semi-arid lands, on which it depends for much of its economic activity (e.g., agriculture and livestock grazing in rangelands). These ecosystems are under extreme pressure from human activities and current climatic conditions (e.g., high temperatures and prolonged droughts). Because of the current marginality of soil-water and nutrient reserves, some ecosystems in semi-arid regions may be among the first to show the effects of climate change. However, a number of management options (e.g., better stock management, more-integrated agroecosystems) could improve land conditions.*
- *The relatively hospitable mountain regions are under particular pressure from human settlements and commercial cultivation, which have led to land degradation and adverse effects on water supply. Glacial melt is forecast to increase under climate change, which would lead to increased summer flows in some river systems for a few decades, followed by a reduction in flow as the glaciers disappear.*
- *A tenth of the world's known species of higher plants and animals occurs in this region. Some countries (particularly Turkey and Tajikistan) are centers of origin for many crop and fruit-tree species; as such, they are important sources of genes for the wild relatives. Few studies are available to assess the impact of climate change on the biodiversity of the region; given its importance, however, there is a need for such studies. Biodiversity is being lost in the region because of human activities, especially land degradation and the overuse of resources.*

#### 7.3.2.1. Grasslands, Rangelands, Woodlands, and Deserts

Half of the region (see Table 7-1) is classified as arid land, much of which is true desert (i.e., extremely arid, as defined in IPCC 1996, WG II, Chapter 3). Another tenth of the region is classified as semi-arid rangeland, dominated by grasslands or shrublands. About 10% of the arid and semi-arid land is classified as having some soil constraints, indicating either that it shows significant soil degradation or that it is desertified (see Table 7-1). The region's major deserts and some of their characteristics are listed in Table 7-2, along with their classification (i.e., semi-arid, arid, or extremely arid, as defined in IPCC 1996, WG II, Chapter 3). For comparative purposes, deserts in adjoining regions (e.g., the Gobi desert from the Temperate Asia region) are included. Deserts are classified as cold deserts (e.g., upland deserts of Middle Asia that are 1,200 m above sea level) or hot deserts (e.g., those in the Middle East). The seasonality of the deserts' rainfall varies; some receive predominantly winter rain, whereas others receive mostly summer rain. Deserts with very sporadic rain events are classified as aseasonal.

Most semi-arid lands in the region are classified as rangelands, with a cover of grassland or shrubland. Moister areas are covered mostly by woodlands; the few forested areas account for only 5% of the land area in the region. Together, these areas represent 1% of the world's woodlands/forests (see Table 7-1). The region's forests and woodlands nevertheless are important resources that will have to be safeguarded, given the heavy use of wood as fuel in some countries.

Many desert organisms already are near their limits of temperature tolerance, and some may not be able to persist under hotter conditions. Because of the current marginality of soil-water and nutrient reserves, some ecosystems in semi-arid regions may be among the first to show the effects of climate change (IPCC 1996, WG II, Chapter 4). Higher temperatures in arid regions that experience cold winters are likely to allow spring growth to begin earlier. In some cases, this effect may result in earlier depletion of water reserves accumulated over the cooler winter period—leading to an even longer period of potential drought (IPCC 1996, WG II, Chapter 3) and a potential reduction in animal productivity, including wool production.

The influence of climate change on large-scale events, such as wildfires, can be inferred only indirectly. However, some authors have suggested that an increase in climate variability may increase the frequency and severity of wildfires in grasslands and rangelands (IPCC 1996, WG II, Chapter 2). Even without an increase in climate variability, fire frequencies may increase in some areas as a result of warmer weather. Fires may further reduce local precipitation because fire-emitted aerosols increase the number of cloud condensation nuclei, producing smaller cloud droplets that are less likely to fall as rain (IPCC 1996, WG II, Section 1.4.2).

In arid and semi-arid rangelands, annual net primary productivity is strongly influenced by fire and climate. Fires alter the structure of vegetation and affect nutrient cycling. By making nitrogen and other nutrients more available to soil microorganisms, fire may result in enhanced emissions of nitrous oxide ($N_2O$) from soil. Most of these emissions are short-term effects that occur immediately following the fire (IPCC 1996, WG II, Chapter 2). Parts of the region that become subject to more frequent fires also could suffer long-term effects, such as decreases in soil organic matter, changes in species composition, and long-term losses of nitrogen through volatilization (i.e., loss to the atmosphere during the fire) and immobilization (as very stable compounds in the soil). These effects would exacerbate any nitrogen limitation, which often is a factor in arid and semi-arid areas (IPCC 1996, WG II, Chapter 2).

The MAPSS and BIOME3 models of vegetation cover classify the region as arid lands dominated by shrub/woodland/grasslands; these models project little change in cover under the climate scenarios used (see Figure C-1), although they project a 1–10% decrease in leaf area index (see Figure C-4). However, these models—which are based on a simple matching of vegetation to climate—are unable to take into account many of the more complex interactions described above.

A vulnerability assessment of the productivity of semi-desert rangelands in the Aral Sea region shows that this region is very sensitive to changes in precipitation. Under a temperature increase of 2°C and a precipitation increase of 0.2–0.3 mm/day, grassland productivity is projected to increase by 20–55%. Under a temperature increase of 0.5°C and reduced precipitation, however, productivity could be reduced by 6–32%; temperature increases of 2–3°C, combined with reduced precipitation, could reduce grassland productivity by 40–90% (Smith *et al.*, 1996).

#### 7.3.2.2. Mountain Regions

Although most of the region is relatively flat, mountains (e.g., Hindu Kush, Karakoram, and Tien Shan) exist in the eastern part. Afghanistan, Kazakstan, Pakistan, Tajikistan, and Uzbekistan all have mountains of up to 5,000 m; the highest point in the region is Communism Peak in Tajikistan. The Kazak uplands in Kazakstan reach 1,560 m. Much of the Arabian peninsula has upland areas of 1,000–2,000 m, and some areas in Yemen and Oman reach 3,700 m and 3,000 m, respectively. Areas in western Pakistan are above 2,000 m, and the uplands in Iran and Turkey reach 2,000 m and 3,000 m, respectively. Countries with very few or no uplands (i.e., areas above 500 m) include Iraq, Kuwait, and the United Arab Emirates.

In this arid region, the mountainous areas are wet, cool, and hospitable for human dwelling and commercial cultivation. Human encroachment in mountain regions has reduced vegetation cover, which has increased soil-moisture evaporation, erosion, and siltation—with adverse effects on water quality and other resources (IPCC 1996, WG II, Section 5.1.5).

Some mountains in the region have permanent glaciers that will be affected by climate change. There will be pronounced

***Table 7-2:*** *Geophysical characteristics of deserts in the Middle East and Arid Asia region.*

| Desert (Country) | Aridity | Main Rainy Season | Temperature of Coldest Month/ Absolute Min (°C) | Temperature of Warmest Month/ Absolute Max (°C) | Latitude | Elevation (m) | Precipitation (mm/yr) | Area (000s km²) |
|---|---|---|---|---|---|---|---|---|
| **Arabian Desert** (includes Rub al Khali and Nafud) | Extreme Arid/ Arid/Semi-Arid | Aseasonal/ Winter | 0–10/10–20/ 20–30 | 20-0/>30 | 15–31°N | 0–1,200 | 25–150 | 800 |
| **Iranian Desert** (includes Dasht-i-Margo, Dasht-e-Naomid, Dasht-e-Kavir, Dasht-e-Lut) | Extreme Arid Semi-Arid/Arid | Winter | 0–10/–10 | 20-30/+45 | 27–36°N | 200–800 | 50–100 | 135 |
| **Negev** (Israel) | Arid | Winter | 10–20 | 20-30/>30 | | | | |
| **Syria, Iraq, N. Arabian, Jordanian Deserts** | Arid | Winter | 0–10/–11 | 20-0/+470 | 31–37°N | 200–800 | 100–150 | |
| **Registan** (Afghanistan) | Extreme Arid | Winter | /-19 | /+42 | 29–32°N | 500–1,500 | 50–100 | 40 |
| **Middle Asia[1] and Kazakstan** | | | | | | | | |
| Garagum | Extreme Arid/Arid | Biseasonal/Spring | -5–10/-35 | +27 +30/+ 50 | 37–42°N | 100–500 | 70–100 | 350 |
| Ustjurt and Mangyshlak | Arid/Semi-Arid | Biseasonal/Spring | -5–10/-40 | +26+28/+42 | 42–45°N | 200–300 | 80–150 | 200 |
| Qizilkum | Extreme Arid/ Arid/Semi-Arid | Biseasonal/ Spring | -4–8/-32 | +28+30/+45 | 42–44°N | 50–300 | 70–180 | 300 |
| Priaralski Garagum | Arid | Biseasonal/Spring | -11–14/-42 | +26+28/+42 | 46–48°N | 400 | 130–200 | 35 |
| Betpaqdala | Arid | Biseasonal/Spring | -12–13/-38 | +26+28/+43 | 44–46°N | 300–350 | 100–150 | 75 |
| Mujunkum | Arid/Semi-Arid | Biseasonal | -2–3/-45 | +24+28/+40 | 43–44°N | 100–660 | 170–300 | 40 |
| Moinkum | Semi-Arid | Biseasonal/Spring | -7–11 | +25+27 | 43–45°N | 300–700 | 250–300 | 80 |
| Saryesik-Atyrau | Arid | Biseasonal | -14–15 | +23+25 | 45–46°N | 300–500 | 150–200 | |
| Bolshie and Malye Barsuki | Arid | Biseasonal/Spring | -12–15 | +25+27 | 46–48°N | 100–200 | 150–200 | 40 |
| Naryn-Peski | Semi-Arid | Biseasonal | -9–12 | +23+25 | 46–50°N | 0–50 | 250 | 200 |
| **Others Adjacent to the Region** | | | | | | | | |
| Gobi (China, Mongolia)[2] | Semi-Arid/Extreme Arid | Biseasonal | /-40 | /+45 | 42–47°N | 900–1,200 | 50–200 | 1050 |
| Ordos (China, Mongolia) | Arid/Semi-Arid | Summer | /-21 | /+42 | 38–40°N | 1,100–1,500 | 150–300 | 95 |
| Sinai (Egypt) | Extreme Arid/Arid | Aseasonal | 10–20 | 20-30 | | | | |
| Takla Makan (China)[2] | Semi-Arid/Extreme Arid | Biseasonal | -10–20 /-27 | /+37 | 36–43°N | 800–1,500 | 50–75 | 271 |
| Thar (India, Pakistan)[2] | Arid/Semi-Arid | Summer/Aseasonal | 10–20/-1 | >30/+48 | 24–31°N | 0–800 | 150–500 | 300 |
| Thal (Pakistan) | Arid | Summer | /-2 | /+49 | 30–32°N | 100–200 | 50–200 | 26 |

Sources: McGinnies *et al.*, 1968; Wilson, 1976; Walter *et al.*, 1983; Evenari *et al.*, 1985; Babaev *et al.*, 1986.

[1]Includes FSU.

[2]Not in this region.

alterations in glacier-melt runoff as climate warms in the region. Glaciers will provide extra runoff as ice melts. In most mountain regions, this increase in runoff will last for a few decades, then decrease as the glaciers disappear. For countries with very large glaciers, the extra runoff may persist for a century or more and substantially increase regional water resources. Tentative estimates have been made for central Asia, based on mass balances from a small number of Tien Shan glaciers for the period 1959–1992 (IPCC 1996, WG II, Section 7.4.2). Extrapolation to the whole of central Asia suggests that glacier mass has decreased by 804 $km^3$ over that time, representing a 15% increase in glacial runoff.

A decrease in snowfall and glacier ice would influence the seasonality of river flow by reducing the production of meltwater in the warm season. The expected smoothing of the annual runoff amplitude could be beneficial (e.g., energy production in winter, reduction of summer flood peaks) and detrimental (e.g., reduced water supply for summer irrigation in dry areas). As mountain glaciers begin to disappear, the volume of summer runoff eventually will be reduced as a result of loss of ice resources. Consequences for downstream agriculture, which relies on this water for irrigation, will be unfavorable in some places. For example, low- and mid-lying parts of central Asia are likely to change gradually into more-arid, interior deserts (IPCC 1996, WG II, Section 7.5.1).

In parts of central Asia, regional increases in temperature will lead to an increased probability of events such as mudflows and avalanches that could affect human settlements (Iafiazova, 1997).

#### 7.3.2.3. *Biodiversity*

About 10% of the world's known species of higher plants and animals occur in this region; half of these species are found only within the region. About 10% of the region's known species are listed as threatened. In 1995, approximately 2.5% of the land in the region was protected under some conservation status (IUCN categories I-V; Table 7-1).

Some of the countries in the region (particularly Turkey and Tajikistan) are centers of origin for many crop and fruit-tree species; as such, they are important sources of genes for the wild relatives of these species, and even for new varieties that may be resistant to drought or disease. These species also provide a living laboratory for looking at possible new varieties, which may arise in response to climate change (World Bank, 1995b).

There is a general lack of detailed and coordinated information on biodiversity and environment in the region, resulting in temporal and spatial gaps and a lack of quality control in the available information (UNEP, 1997). Most of the available information is on vertebrates, with emphasis on birds and mammals. Some countries of the Middle East are rectifying information gaps by setting up database networks. In addition, the Middle Eastern countries have set up joint programs to protect biodiversity across political boundaries and to carry out biodiversity inventories.

In the Middle East, biodiversity is being lost as a result of development activities; land degradation (especially overgrazing and deforestation, leading to loss of plant cover); marine pollution; overfishing; hunting; and the overuse of freshwater, which affects the plants and animals of oases and wetlands (UNEP, 1997). Factors that are threatening biodiversity in central arid Asia include rapid changes in land use, extensive but poorly managed irrigation, more-intensive use of rangelands, medicinal and food-plant collection, dam building, and fuelwood collection (Bie and Imevbore, 1995; Kharin, 1995; World Bank, 1995b).

Wetlands, especially ephemeral wetlands, are an important part of this region culturally and economically, as well as in terms of its biodiversity (IPCC 1996, WG II, Chapter 3). The importance of particular wetlands to birds in semi-arid areas varies greatly from year to year, depending on local and regional conditions (IPCC 1996, WG II, Section 6.5.2), especially because the source of the water in some wetlands is thousands of kilometers away (e.g., the Tigris and Euphrates). In recent years, irrigation and artificial storage of water has created new habitats for water birds in central Asia, leading to an increase in their population (Bie and Imevbore, 1995). Few data are available to assess the impact of climate change on these systems; because the region has such limited rainfall, however, the wetlands may be among the most sensitive ecosystems to changes in the amount and seasonality of rainfall and evaporation. Such changes may lead to local extinction of some populations (IPCC 1996, WG II, Section 6.5.2).

Many countries (especially in the Middle East) are aware of these pressures and are implementing programs to conserve their biodiversity (UNEP, 1997).

### 7.3.3. *Water Resources*

- *In an area dominated by arid and semi-arid lands, water is a very limited resource. Droughts, desertification, and water shortages are permanent features of life in many countries in the region. Rapid development is threatening some water supplies through salinization and pollution, and increasing standards of living and expanding populations are increasing demand. Water is a scarce resource—and will continue to be so in the future. Projections of changes in runoff and water supply under climate change scenarios vary. Some countries are developing programs to conserve and reuse water or to achieve more efficient irrigation. Some countries are vulnerable to reductions in runoff; because they rely heavily on hydropower production, their energy supply is likely to be affected.*

In semi-arid and arid environments, rainfall is short-lived and often very intense. Because soils tend to be thin, much of the rainfall runs directly off of the surface, only to infiltrate deeper soils downslope or along river beds (IPCC 1996, WG II, Section 10.2.1). Thus, water availability is a major concern in most countries of the region (Middle East Water Commission, 1995; UNEP, 1997). Some countries (e.g., Syria, Iraq, Jordan, Lebanon) have reliable sources of surface water; the majority, however, depend either on groundwater or on desalinization for their water supply—both of which enable them to use water in amounts far exceeding the estimated renewable fresh water in the country (World Bank, 1995c; IPCC 1996, WG II, Chapter 4; GEO, 1997). Wetlands can retain water, especially during dry periods, and thus can enhance recharge to major aquifers (IPCC 1996, WG II, Section 6.3.3.2.3). However, in semi-arid and arid areas, where groundwater recharge occurs after flood events, changes in the frequency and magnitude of rainfall events will alter the number of recharge events (IPCC 1996, WG II, Section 10.3.6). Water from glacial melt is an important contribution to the flow of some river systems, and changes in seasonality and amount from this source are likely to occur as a result of climate change (see Section 7.3.2.2). Fossil aquifers are important water sources in many deserts; they will not be affected on a time scale relevant to humans (IPCC 1996, WG II, Chapter 3).

As a result of rapid increases in the income of some countries—with resulting increases in living standards—the demand for water has increased, such that many countries in the Middle East will experience chronic water shortages (IPCC 1996, WG II, Section 14.2.2; UNEP 1997). The list of nations with water supply problems is likely to expand as a consequence of the accelerated pace of urbanization. In some countries (e.g., Kuwait), monitoring programs for water quality have been established in an attempt to maintain water quality in accordance with World Health Organization (WHO) standards (UNEP, 1997). It is projected that the population without safe drinking water will almost double by 2030, assuming a "business-as-usual" scenario (IPCC 1996, WG II, Section 14.2.2).

Many countries in the region have highly urbanized populations, for whom it is easier and more efficient to organize water supply, treatment, and delivery systems than it is for rural populations. Moreover, demand can be more easily managed to promote water-use efficiency in urban areas. In most countries in the region, a large proportion of the population has running water in their homes. Although municipal and industrial water use will grow, per capita domestic use is likely to decrease and the quality of drinking water to increase with centralized treatment. However, future urban water demands are likely to compete with the irrigated agricultural sector (IPCC 1996, WG II, Section 14.2.2); 15 of the countries in the region use more than 75% of their water for irrigation (see Table 7-1).

There are other problems associated with some water bodies in the region, especially in areas of high human density. In these areas, habitat degradation often is important, causing many semi-enclosed water bodies to become eutrophic (Kharin, 1995; IPCC 1996, WG II, Section 16.1.1). Dryland salinization also is having an impact on water quality in some countries where groundwater is contaminated with salt; in other countries (e.g., Oman and United Arab Emirates), seawater has intruded into freshwater aquifers (UNEP, 1997).

Efforts are being made in some countries to use wastewater (Sarikaya and Eroglu, 1993; Abdelrahman and Alajmi, 1994; Shelef *et al.*, 1994; Shelef and Azov, 1996)—and to use water more efficiently, especially for agriculture. The Middle East Water Commission (1995) suggests that, for some countries, a reduction of 30% in water use for agriculture would alleviate some water crises. Increased recycling of water is being explored in areas where there are shortages of water, either perennially (e.g., Israel and Oman) or during some seasons (e.g., Turkey in the summer, when shortages are exacerbated by tourist demand) (Middle East Water Commission, 1995; IPCC 1996, WG II, Section 14.2.2). In many of the countries of the FSU, plans are underway to increase the efficiency of irrigation practices, with savings of over 20% expected in some regions (see Section 7.4.1).

Populations in many countries of the region are vulnerable because they depend on water supplies from outside their political boundaries. Formal agreements are in place, or are being developed, to share the water resources of rivers flowing across political boundaries (e.g., Euphrates, Tigris, Jordan, and Indus); such agreements would reduce the potential for future conflicts over water (Blaikie *et al.,* 1994; Middle East Water Commission, 1995).

River-basin runoff is very sensitive to small variations in climatic conditions and in the vegetation cover of its catchment. Riebsame *et al.* (1995) used two GCMs (GISS and GFDL) to assess the effect of climate change on various rivers throughout the world, including the Indus. The GCM climate scenarios projected an increase of 11–16% in annual runoff for the Indus. The authors also assessed the effect of a 2°C temperature increase with +20%, 0%, and -20% changes in precipitation and found that runoff is more sensitive to changes in precipitation than to changes in temperature.

The effect of climate change on runoff has been studied for some river basins in the region. Based on two GCM runs (CCC-J1 and GFDL-A3) with 2x$CO_2$ equilibrium climate scenarios, several mountain and plains watersheds experienced 30–35% reductions in runoff in years of water shortages and 20–25% reductions in years of excess water. A projection based on a third GCM run (GFDL-X2g transient model) confirms the tendency for a reduction in water resources during years of plentiful water supply for mountainous watersheds (Pilifosova *et al.,* 1996). In the short term, however, some studies suggest increases in water resources. For example, the above simulation projects increases of up to 12% in the water resources of some watersheds in 2000 and 2030 (Golubtsov *et al.,* 1996).

Some countries in the region produce large amounts of hydropower; Tajikistan, for example, is the third-highest producer in the world (World Bank, 1995b). Changes in runoff to the system could have a significant effect on the power output of these countries. One of the bigger river systems in the region, the Euphrates and Tigris, has a number of dams that are used for irrigation and water supply as well as for hydropower. To date, no studies have assessed the effect of climate change on these systems. However, if there is a reduction in total runoff as a result of climate change, the increased demand for agricultural and hydropower activities could place more pressure on water resources.

### 7.3.4. Food and Fiber for Human Consumption

- *Many countries in the region are highly dependent on agriculture. Land degradation problems and limited water supplies restrict present agricultural productivity and threaten the food security of some countries. There are few projections of the impacts of climate change on food and fiber production for the region. However, many of the options available for combating existing problems will contribute to reducing the anticipated impacts of climate change. A shift in reliance toward more suitable and more intensively managed land areas for food and fiber production—along with selected plant and animal introductions—could be beneficial, allowing greater reliability in food production; such approaches also would reduce the detrimental impacts of extreme climatic events, such as drought, on rangeland systems. Implementation of more flexible risk-management strategies (e.g., long-term and appropriate stocking rates, responding to variations in precipitation by changing animal numbers annually)—along with the use of various kinds of domestic animals, game ranching, multiple production systems, commercial hunting, and tourism—would facilitate socioeconomic activity and provide some food security to the region. These strategies can be viewed as "no regrets" options that might be adopted by many countries in the region because they would be beneficial even without any climate change.*

#### 7.3.4.1. Food and Fiber Production Systems

The climate of the region greatly limits the portion of land presently suitable for livestock production and crop agriculture. Two-thirds of the domestic livestock are supported on rangelands (IPCC 1996, WG II, Table 2-1), although in some countries, large percentages of animal fodder come from crop residues (e.g., 70% in Pakistan) (Pakistan Country Report, 1994). In many countries, crop agriculture is highly dependent on irrigation because rainfall is low and highly variable (Khan, 1985; UNEP, 1997). Only 8 of the 21 countries of the region have more than the world average area of arable land per person (0.26 ha/person); some countries have effectively no arable land. (Kazakstan is an exception, with more than 2 ha of arable land per person.) Nevertheless, agriculture is an important sector in the region as a whole, contributing a high proportion of the GDP in about half of the countries (>10% net added value in agriculture as a percent of GDP, compared with the world average of about 5%; World Bank, 1997). About a third of the GDP of the Kyrgyz Republic, Syria, Tajikistan, and Turkmenistan can be attributed to agriculture.

Many countries in the region (e.g., the countries of the FSU, Syria, Israel), are highly dependent on local agriculture for food; although some (e.g., Turkey and Kazakstan) are major food exporters, most countries are net importers of food (World Bank, 1997). Some countries (e.g., Tajikistan) rely heavily on imports of grain, both for human consumption and for livestock production (World Bank, 1995b). Land degradation and the consequential need to move to even more marginal lands threaten food security and the economies of countries that are highly dependent on agriculture.

In some countries, up to 89% of the croplands are irrigated (41% for the region as a whole; see Table 7-1). Improved irrigation practices (e.g., wider use of drip and underground irrigation) could save up to 50% of the water used in conventional irrigation systems (IPCC 1996, WG II, Chapter 4). Adoption of such conservation techniques could become increasingly important if climate change leads to reduced water availability. The expansion of winter-growing crops that demand much less water (and would use more of the projected precipitation in the region) may be another option (IPCC 1996, WG II, Chapter 4). Some of the FSU countries are making major changes in agricultural practices and would be able to incorporate new varieties that might respond better to low-water conditions (e.g., cotton in Tajikistan) (World Bank, 1995b).

Increasing populations in areas with limited arable land have led some countries (e.g., Jordan, Iraq, Syria, and Yemen) to use irrigated farming throughout the year, as well as to increase their use of fertilizers and pesticides (UNEP, 1997); these trends may add to the land degradation problem.

Little quantitative work has been done on the impacts of climate change on this sector. National and local assessments providing a detailed understanding of crop-specific responses and regional impacts for this region are still lacking (IPCC 1996, WG II, Section 13.6.1).

General assessments have suggested a range of positive and negative impacts of climate change on agriculture. Positive examples include decreasing frost risks and more productive upland agriculture, provided water availability does not decline (or irrigation is available) and appropriate cultivars are used. Increasing populations of pests and disease-causing organisms—many of which have distributions that are climatically controlled—may have a negative impact (IPCC 1996, WG II, Section 5.2.4.1). A number of factors affect the vulnerability of agricultural systems:

- Agro-industry, biomass production, and renewable energy sources depend heavily on climate-sensitive resources and therefore are potentially vulnerable to climate change.

- Climate change could have a severe impact on countries that are dependent on single crops. Diversification of economic activity could be an important precautionary response.
- The burden of climate change may affect lower-income households disproportionately.

Studies for sub-Saharan Africa suggest that there are critical thresholds related to precipitation and the length of the growing season that may affect agricultural practices. The yield of cereals in the Mediterranean climatic region is projected to decline as a consequence of increased drought resulting from the combination of increased temperature and decreased precipitation (or precipitation increases that are insufficient to counter higher evapotranspiration) (IPCC 1996, WG II, Section 13.6.7). In Kazakstan, where wheat is a major crop, spring wheat yields are projected to decrease by almost 60% under CCC-J1 and GFDL-A3 2x$CO_2$ equilibrium climate scenarios; winter wheat yields are projected to increase by about 20%. Because only 1% of the total cropland is suitable for winter wheat, this increase is not likely to significantly affect overall wheat yields, and the overall effect would be a decrease in wheat yield (Pilifosova *et al.,* 1996).

Studies in Pakistan project a decline in wheat yields under GISS and UKMO scenarios with 2x$CO_2$ equivalent, while under a GFDL scenario with 2x$CO_2$ equivalent an increase is projected (Qureshi and Iglesias, 1994). However, the results are highly dependent on assumptions made about the responsiveness of crops to high $CO_2$ levels.

If there is a decrease in rangeland productivity in the region as a result of temperature increases, the overall contribution of the agricultural industry to national economies would decline. Such a decrease also would have serious implications for the food policies of many countries—and for the lives of thousands of pastoral people (IPCC 1996, WG II, Chapter 2).

Some options to minimize the adverse effects of climate change are available. For example, rangelands in Kazakstan support sheep and goats, which are important for wool production. According to the CCC-J1 and GFDL-A3 2x$CO_2$ equilibrium climate scenarios, wool production in this country is projected to decrease by 10–25% (Pilifosova *et al.,* 1996). Possible adaptations identified from modeling studies include changing the current lambing and sheep shearing dates; however, this approach would be feasible only after considerable reeducation efforts were undertaken in the region.

In the rangelands of the Middle East and Arid Asia region, active selection of forage plants and greater control of animal stocking rates are the most promising management options for decreasing the negative impacts of projected climate changes (IPCC 1996, WG II, Chapter 2). Introducing or selecting some legume species in grassland systems—which would reduce reliance on fertilizer inputs and improve the nutritive value of forage—could make these systems more sustainable and resilient under climate change (IPCC 1996, WG II, Chapter 2). Use of various kinds of domestic animals, game ranching, multiple production systems, and commercial hunting and tourism (IPCC 1996, WG II, Chapter 4) are management options that also may facilitate socioeconomic activity in the region. These strategies could be viewed as "no regrets" options by many countries in the region because they would be beneficial even without climate change.

A shift in reliance toward more suitable and more intensively managed land areas for food and fiber production could have the dual benefits of allowing greater reliability in food production and reducing the detrimental impacts of extreme climatic events, such as drought, on rangeland systems (IPCC 1996, WG II, Chapter 2). Runoff farming—a form of water management that originated about 3,000 years ago in parts of the region—has been largely abandoned in some countries. It may provide opportunities, however, for sustainable cropping and horticulture under changed climates (IPCC 1996, WG II, Chapter 4).

In some countries, marine and/or freshwater fish are important sources of protein. The demand for fish is increasing along with increasing populations; this demand is expected to at least double by 2020, compared with 1995 levels. The increase in demand is expected to be met through inland aquaculture in earthen ponds, through mariculture in the Mediterranean (Mires, 1996), and through imports (e.g., Israel already imports 60% of its fish protein).

Many freshwater wetlands of great importance to agriculture and wildlife occur in the floodplains of lakes and rivers (IPCC 1996, WG II, Section 6.5.2). Water levels that fluctuate as a result of either human activity or climate-related changes may cause lakes to become separated from their bordering wetlands more frequently. Because a number of lake fish use these wetlands for spawning and nursery areas (IPCC 1996, WG II, Section 10.6.2.2), fish productivity may be affected—thus affecting the whole food chain.

#### *7.3.4.2. Land Degradation and Other Environmental Problems Affecting Food and Fiber Production*

Land degradation and desertification are major problems that are caused by natural factors (e.g., prolonged droughts) and by human activities—particularly overgrazing, uncontrolled cultivation, fuelwood gathering, inappropriate use of irrigation, uncontrolled urbanization, and tourism development (Kharin, 1995; IPCC 1996, WG II, Chapter 2; Schreiber and Shermuchamedov, 1996; UNEP, 1997). Urbanization and related activities (e.g., road construction) have resulted in losses of permanent pasture and increases in the agricultural use of marginal lands, leading to further degradation (UNEP, 1997). Dryland salinity and waterlogging, especially in low-lying countries in the Middle East and parts of central Asia (e.g., around the Aral Sea), also are contributing to land degradation (Kharin, 1997; UNEP, 1997).

The full extent of land degradation in the region is not known, although an estimated 10% of the arid and semi-arid land is classified as having some soil constraints (see Table 7-1). Future erosion risk is more likely to be influenced by increases in population density, intensive cultivation of marginal lands, and the use of resource-based and subsistence farming techniques than by changes in climate (IPCC 1996, WG II, Chapter 4).

Various options are available for reducing soil degradation by improving the carbon and water storage capacity of the soil (see below). Unfortunately, because of low surface-water availability, some of these practices might not be helpful for all of the countries in the region. However, other soil conservation and soil protection measures can be implemented in these areas.

*Agro-ecosystems.* Water availability is the main determinant of productivity; the use of reduced tillage and mulching to increase available water and reduce surface erosion can promote increased soil carbon. In areas where cropping and livestock are closely integrated, more-efficient use of manure and commercial fertilizer also can increase productivity and soil carbon. In temperate parts of the region, eliminating or reducing summer fallow through better water management in non-irrigated areas could significantly increase carbon and decrease soil erosion in semi-arid croplands (IPCC 1996, WG II, Section 23.2.2).

*Reduction in animal density.* In much of the semi-arid area, the predominant land use is pastoral, and grazing control is an important option for the maintenance of soil carbon. Reduction in animal numbers can increase carbon storage by enhancing plant cover. This practice can have a positive effect on the ecosystem if there is sufficient rainfall, but reductions in animal numbers on rangelands may require alternative sources of food for humans—and thus changes in national or regional food production policies.

*Changing animal distribution.* Changing animal distribution through salt placement, development of water sources, or fencing can lead to the more even use of pastures and consequently to increases in overall plant cover, improved status of the root system (as a consequence of less-intense grazing), and increased carbon sequestration. Appropriate animal-management practices will be specific to local and regional production systems; for example, fencing or salt placement may not be useful in herding systems and may interfere with wildlife migration.

*Watershed-scale projects.* Practices involving the development of dams with large-scale water-storage capacity may increase long-term carbon storage by improving animal management and food production systems. Such projects are expensive, however, and they can result in social and cultural dislocation, local extinctions of wildlife, and increases in human and animal population density. Because 6–89% of the region's cropland area already is irrigated (see Table D-5), watershed management with additions of dams may not be a suitable response.

#### *7.3.4.3. Climate Change and Desertification*

Desertification may occur in any area where the potential evapotranspiration is greater than 70% of the total precipitation—as is the case for parts of this region. It is feasible that desertification itself can exacerbate climate change. Changes in land-use practices can affect surface temperatures through boundary changes in vegetation cover, which lead to differences in albedo and thus to temperature differences (IPCC 1996, WG II, Chapters 3 and 4). It has been suggested that where these changes occur over large areas, they could affect global mean temperatures. The effect is mediated in the short term by surface soil moisture and in the longer term by changes in soil conditions and in carbon sequestration, which lead to changed emissions of $N_2O$ and methane. There are no specific data from this region to support or refute this possibility.

There also is a great deal of variation in assessments of the nature and severity of the desertification problem, mainly because of the lack of adequate data. Many large-scale studies have suggested that the annual rate of expansion of desertified lands in central Asia ranges from 0.5% to 0.7% of the arid zone. Approximately 570 million ha in the region are classified as arid land. Assuming a conservative rate of expansion of 0.5% per annum, 3 million ha of land are becoming desertified every year. Arid-land desertification could occur faster if human and livestock populations continue to increase and particularly if the resilience of arid regions is negatively affected by climate change. Some recent research, however, has not supported the expanding-desert hypothesis. During favorable or "wet" years, there is little evidence to suggest that the desertification front has expanded. The different views regarding the importance of human abuse versus adverse climatic conditions in causing desertification and the lack of agreement on the scale of the problem point to the need for a better understanding of the problem and, in particular, better and more comprehensive data on its nature and extent. There is little doubt that desertification is an important environmental problem that needs to be addressed urgently—and that climate change will have an impact on it (IPCC 1996, WG II, Chapter 4), with environmental and economic repercussions.

Some of the Middle Eastern countries recently have launched action plans to reduce the effects of desertification (UNEP, 1997), though the results of these actions remain to be seen. There also are many technical programs in the Middle East to assist in desertification problems (UNEP, 1997).

### *7.3.5.* ***Coastal Systems***

- *Coastal systems in the region are under threat from pollution and development, resulting in the deterioration of fish populations in some countries. No scenarios suggest that projected changes in sea level will have significant effects on the region as a whole. Certain coastlines of the region (e.g., around the Caspian Sea and Karachi) are predicted to be threatened by development.*

Seven of the 21 countries in the region are landlocked; the others have a total of 21,000 km of coastline, mainly on the Mediterranean Sea and the Gulf of Oman. Coastal zones in this region are not identified as significantly affected by sea-level changes (IPCC 1996, WG II, Chapter 12). The major pressures on them will be related to development rather than the direct result of climate change.

Part of the Mediterranean Sea, the northwestern part of the Gulf of Oman, the southern part of the Caspian Sea, and the coast of Karachi are classified as coastlines under high potential threat from development (WRI, 1996; McMichael *et al.*, 1996). This development includes large cities (>100,000 people), major ports, roads, and pipelines—all of which contribute to pollution in coastal areas.

Many countries in the region are major oil producers. Oil production is mostly land-based and is not as vulnerable to the impacts of climate change as offshore oil production in other regions. Transportation of oil also is less likely to be affected than some higher-latitude areas, which are projected to experience an increase in storm activity and icebergs (IPCC 1996, WG II, Section 8.2). However, coastal oil production areas are likely to be affected by storm surges (IPCC 1996, WG II, Sections 8.3.1 and 9.3.1).

Coastal pollution is a major problem in some countries of the Middle East. Some countries (e.g., Syria, Lebanon) have integrated coastal management programs to overcome some of these problems (UNEP, 1997). Despite such programs, the Red Sea and the Kuwait/Oman areas sustain more oil pollution than anywhere else in the world. In addition, coastal zones of the Middle Eastern countries experience oil spills from ships and pipelines, as well as land-based pollution discharges (UNEP, 1997)—all of which lead to the deterioration of coastal areas.

Fishery industries are important in some countries (e.g., the Persian Gulf countries in the Middle East). Overfishing and marine pollution have led to a decrease in fish catches. The loss of mangroves and intertidal areas (e.g., from dredging/infilling) probably has added to the problem by affecting breeding grounds (UNEP, 1997).

Preliminary studies and estimates show that if climate change were to cause an increase in flow from the Volga river, the resulting rise in the level of the Caspian Sea would significantly increase the vulnerability of its coastal zone, requiring some adaptation measures (Golubtsov and Lee, 1995).

### 7.3.6. Human Settlements and Urbanization

- *The region is highly urbanized, and rapid development of the urban areas has led to increased air and water pollution. In many countries, urbanization and development have been driven by the oil industry, although there also has been migration from rural to urban areas as a result of land degradation. Although no specific studies of the effects of climate change on urbanization have been done for this region, it is likely that increased land degradation and intensification of agricultural systems will increase population movement to urban centers.*

A high proportion (a mean of 50%; see Table 7-1) of the region's population lives in urban areas; the region contains 2 of the world's 25 megacities (Karachi, with a population of 9.9 million, and Istanbul, with a population of 7.5 million) (WRI, 1996). In many countries in the Middle East, rapid urbanization has increased the demand for resources, along with waste generation and management problems (UNEP, 1997). With increasing urbanization, problems of urban poverty, access to clean water and sanitation, food security, and air pollution—and thus general health—are issues that will have to be considered. In Middle Eastern countries, most urban houses have running water (approximately 90% of the population surveyed), as well as some sort of sewer and septic system for sewage disposal (WRI, 1996); rural areas, however, are not so well served.

Industrial development has resulted in major pollution problems in some countries, especially in the Middle East. Water and air pollution have affected the health of the population. Pesticide and herbicide use in agriculture also have increased, leading to contamination of food and water.

Political unrest has caused large populations to migrate to marginal lands, leading to further land degradation and water resource problems. In addition, the expansion of urban areas has encroached on some of the most productive lands, especially in countries in the Middle East—thus increasing the agricultural use of marginal lands. The introduction of modern production techniques, combined with industrialization, has had a negative impact on the lifestyles of nomadic populations in parts of the region. As a result of land degradation, there has been an increase in migration to urban areas, which has been detrimental to rural and urban areas (UNEP, 1997).

Although no specific studies of the effects of climate change on urbanization have been done for this region, it is likely that increased land degradation and intensification of agricultural systems will continue to increase population movement to urban centers.

#### 7.3.6.1. Domestic Energy Consumption

In general, decreasing rainfall will lead to a decrease in the production of biomass, which remains a major component of fuel use in some countries of the region. In some countries, population growth and land degradation may make adaptation necessary even before climate change becomes perceptible. Adaptation could take the form of either switching to new fuels or more efficient production and conversion of biomass (IPCC 1996, WG II, Section 16.2.3). It has been suggested that saline lands in coastal zones and arid regions could produce biomass using halophyte species. Although halophytes could assist in slowing degradation or rehabilitating degraded arid lands, their productivity is too low for them to be a significant source of biomass (IPCC 1996, WG II, Section 25.3.2).

#### 7.3.6.2. *Energy, Industry, and Transportation*

The region contains major fossil fuel reserves; it produced 14% of the world's energy in 1993 (Table 7-1). The energy sector in the region encompasses a range of activities, including coal, oil, and natural gas production; coke manufacture; production of refined petroleum products; and production of biomass fuels and renewable energy. Traditional fuels (e.g., fuelwood, animal dung, and crop residues) account for only 6% of the energy used in the region as a whole, compared with the world average of 12–15%. Nevertheless, up to 70% of the domestic energy consumption in some countries of the region (e.g., Afghanistan) is derived from traditional fuels.

In terms of the future economy of the region, oil exports from the Middle East are projected to decline absolutely but to grow as a percentage of global oil consumption—from about 20% in 1990 to more than 25% in 2025 and 33% in 2100. Total energy exports from the Middle East will double between 1990 and 2050, before declining to their 1990 level by the year 2100 (IPCC 1996, WG II, Figure 19-14). The Middle East is expected to increase its exports of natural gas and hydrogen derived from natural gas and solar electricity via electrolysis (IPCC 1996, WG II, Section 19.2.5).

The sensitivity of industry to climate change is widely believed to be low in relation to that of natural ecosystems and agriculture, and its adaptability is high.

### 7.3.7. **Human Health**

- *Human health in the region is variable, reflecting the economies of the different countries. In countries where poverty is prevalent, infant mortality rates are high and life expectancies are low. The population of the region is increasing by 2.7% per year; this trend may affect human health (e.g., through the spread of waterborne diseases associated with high population densities). The impacts of climate change are likely to be detrimental to the health of the population, mainly through heat stress and possible increases in vector-borne (e.g., dengue fever and malaria) and waterborne diseases. Decreases in water availability and food production (especially if there is a shortage of water for irrigation) would indirectly affect the health of the population.*

Human health in the region is variable, reflecting the economies of the different countries. The proportion of a population living in absolute poverty is considered to be an indication of human health. Nearly half of the developing world's poor—and nearly half of those in extreme poverty—live in south Asia; the next-largest numbers are (in order) in sub-Saharan Africa, the Middle East, and north Africa (IPCC 1996, WG II, Section 12.2.1.3). Estimates of rural poverty are available for nine countries of this region; they range from 6% (Oman) to 60% (Afghanistan). Access to safe drinking water in urban areas varies from 39% in Afghanistan to almost 100% in 13 countries in the region; in rural areas, the equivalent figures are 5% in Afghanistan to almost 100% in several countries (WRI, 1996). Because a high percentage of the population lives in urban areas, a high proportion of the population has access to safe drinking water, which diminishes the risk of waterborne infectious diseases.

Infant mortality rates and the prevalence of infectious diseases also are considered to be indicators of human health (WRI, 1996). Figures for infant mortality for 1970–75 varied between 23 deaths per 1,000 live births (Israel) to 194 deaths per 1,000 live births (Afghanistan), with a mean of 94 (the world average was 93). The equivalent figures for 1990–1995 varied from 9 to 163 (for the same countries), with a mean of 50 (the world average is 64). These statistics suggest that the health of the population has improved since the 1970s, in part as a result of better primary health care (Bener *et al.,* 1993). During 1990–95, life expectancy varied between 43.5 years (Afghanistan) and 76.5 years (Israel), with a mean of 66.6 (compared with the world average of 52.8).

Generally, the health of the population in the region has been improving faster than the world average—with a few exceptions related to civil unrest (e.g., Afghanistan). Given the natural wealth, in terms of petroleum deposits, of many countries in the region, this trend should continue in the near future. However, political tensions remain high in parts of the region, and much of the region is subject to major earthquakes; both of these factors could cause major setbacks in public health.

The population of the region is increasing by an average of 2.7% annually; as the population increases, air pollution, waste management, and sanitation will become important issues. Given global trends, there may be an increase in motor vehicle ownership, which is likely to add to the problem of air pollution. Photochemical smog produced by the reaction of sunlight with ozone and photochemical oxidants—such as peroxyacetal nitrate (PAN) from nitrogen oxides ($NO_x$) and hydrocarbon emissions—is particularly prevalent in cities in semi-arid regions with a high density of industrial pollution-producing industries (IPCC 1996, WG II, Section 12.3.3). This smog may lead to respiratory problems in the urban populations.

Climate change will have direct impacts (e.g., through heat stress) as well as indirect effects (e.g., through reductions in food, leading to poor nutrition and increased susceptibility to diseases) on human health. The overall impact of climate change on human societies will vary, depending on many factors—such as the amount of low-lying or arid land they occupy and their degree of dependence on agriculture or aquatic resources (IPCC 1996, WG III, Section 2.2.3).

Climate change is projected to increase the frequency of very hot days. Extensive research has shown that heat waves cause excess deaths (Larsen, 1990; McMichael *et al.,* 1996). Recent analyses of concurrent meteorological and mortality data in cities in the Middle East provide evidence that overall death rates rise during heat waves, particularly when the temperature rises above the local population's threshold value. Therefore, it can be predicted that climate change would cause additional heat-related deaths and illnesses in the region via increased exposure to heat waves (IPCC 1996, WG II, Section 18.2.1).

In some areas—especially where access to safe drinking water is poor—waterborne gastrointestinal diseases related to fecal contamination (e.g., *Giardiasis*, diarrhea) are a problem. Such diseases often lead to high infant mortality in Saudi Arabia (Jarallah *et al.*, 1993; Altukhi *et al.,* 1996); Afghanistan, Jordan, and Pakistan (Azim and Rahaman, 1993; Nazer *et al.,* 1993; Chavasse *et al.,* 1996); and Turkey. Waterborne diseases also occur in tourist resorts (Kocasoy, 1995) in Israel (Lowenthal, 1993), Tajikistan (World Bank, 1995b), and Turkmenistan (World Bank, 1995c). Some countries of the region have made efforts to control these diseases through various programs—some implemented in the early 1970s. If flooding occurs as a result of more intense rainfall events, waterborne diseases may become frequent, mostly because of the overloading of sewage systems (McMichael, 1997).

Other waterborne diseases not related to fecal contamination (e.g., *Schistosomiasis*) are linked to dams and irrigation projects (Hairston, 1973; Grosse, 1993; Hunter *et al.,* 1993). These types of projects may increase, depending on water management responses in the region.

Under current climate change projections, vector-borne diseases—particularly dengue fever—are expected to increase globally. The *Aedes* mosquito, which transmits dengue (McMichael *et al.,* 1996), is sensitive to temperature and rainfall; the incidence of the disease has been increasing globally in recent years, probably in response to changes in climate (Hales *et al.*, 1996). Dengue fever was reported on the eastern edge of the region in 1995; its spread will depend on whether climate changes produce moister and warmer conditions that favor the vector—even if only seasonally. Dengue can survive where the winter isotherm is above 10°C, and it can become epidemic when temperatures are above 20°C (Shope, 1991).

Malaria is endemic in certain parts of the region (e.g., in humid areas), even in arid countries such as Saudi Arabia (Annobil *et al.,* 1994). In parts of the arid/semi-arid areas, malaria epidemics are associated with high rainfall and floods (Akhtar and McMichael, 1996); in Pakistan, malaria epidemics are associated with El Niño events (McMichael *et al.,* 1996). Based on a climate scenario using the ECHAM1-A GCM, with a global mean temperature rise of 1.16°C, Martens *et al.* (1995) did not find indications of immediate increases in malaria cases in the region—but did project an expansion of malaria into adjoining areas by 2100. In addition, the present nonmalarial, higher-elevation areas in the region may experience seasonal epidemics.

The sand fly is a vector for sand fly fever and *Leishmaniasis*. Sand flies occur in parts of the region (e.g., the northwestern part of the Arabian peninsula and southwest Asia). With a temperature increase of 1°C, Cross and Hyams (1996) predict that there will be a seasonal and spatial (e.g., in uplands) expansion of sand flies—and thus potential increases in the diseases that they transmit. Monitoring the vector could form a basis for detecting the impact of climate change on human health.

There also are populations in the region that rely on nonirrigated (rain-fed) agriculture, which is likely to be vulnerable to climate change. Changes in nutrition may make these populations more vulnerable to disease, with an indirect effect on their health. No data are available to estimate the size of this vulnerable population.

### *7.3.8. Socioeconomic Impact of Climate Change*

- *Very little socioeconomic information about the impacts of climate change is available for the region. The best available estimates indicate that impacts on GDP may be higher than the world average.*

One way of expressing the potential impacts of climate change is to derive a comprehensive monetary estimate, which sums all known impacts and is expressed as a monetary value. Estimating effects on marketed goods and services (e.g., commercial or residential land loss resulting from sea-level rise, energy savings in winter) in monetary terms is relatively straightforward because the prices are known. Estimating damage to nonmarketed goods and services (e.g., wetland loss, mortality changes) in monetary terms is possible either through examining market transactions where such goods or services are implicitly traded (e.g., sites of landscape beauty) or through interviewing people to determine their preferences (i.e., their willingness to pay to secure a benefit or willingness to accept compensation for a loss). In Organisation for Economic Cooperation and Development (OECD) countries, such valuation techniques are relatively well established and have been widely applied, although the results often are contentious. Estimates for non-OECD countries are based on extrapolation from studies in the OECD. The best estimate of the annual impact of a doubling of the $CO_2$ concentration in the atmosphere is about 0.9–5.5% of GDP for the Middle East, compared with 1.4–1.9% of GDP for the world. Many assumptions underlie these best guesses, however, and large uncertainties remain. Nevertheless, the impact on this region appears to be greater than that on the world as a whole (Fankhauser and Tol, 1977).

## 7.4. Possible Approaches to Integration

### *7.4.1. Economies in Transition*

- *Countries of the FSU are undergoing major economic changes and consequential changes in agricultural systems and management. This transition is likely to provide "win-win" opportunities to offset the impacts of climate change, particularly in the areas of agriculture and resource management.*

The countries of the FSU are undergoing radical economic transitions, with great impacts on many sectors likely to be affected by climate change. Much of the agriculture of the Middle East and Arid Asia region developed in response to the agricultural plans of the FSU. Irrigated cotton production, for example, expanded from the 1960s through the 1980s, with significant impacts on water resources, the environment, and human health. Water resources were overcommitted to agriculture and were transferred unsustainably from several river systems. The drying of the Aral Sea (see Section 7.4.2) was one consequence; the salinization of many cropping areas was another. The heavy use of fertilizers and overuse of pesticides has left a legacy of polluted soils and water supplies. Recent assessments (World Bank, 1995a-d) conclude that agricultural practices are likely to change because agricultural products now are being traded on world markets. Changes include more targeted use of fertilizers, better use of integrated pest management, more efficient irrigation systems in existing systems, and changes in crop species and varieties. These changes will dominate policy development in this region, but they also offer opportunities to adapt to climate change by reducing water use and adopting new crop varieties.

In particular, targets for improved efficiency in irrigation could lead to savings in water use, which would be more important than any changes likely to result from climate change over the next few decades. Throughout the states of the FSU, options to reduce water use include lining more irrigation canals to reduce seepage losses (up to 40% of diverted water is lost in arterial channels) and reducing the area of crop and pasture irrigated by inefficient flooding methods while increasing the area of more-valuable fruit and vegetable crops irrigated by efficient drip and below-ground irrigation systems. In Turkmenistan, for example, cotton irrigation currently requires 12,000 $m^3$ water/ha; more modern techniques would require only 7,000 $m^3$/ha (World Bank, 1995a-d). More than one-third of all water used in Turkmenistan is applied to irrigated cotton; thus, modernizing techniques could save 20% of the country's current water use. Uzbekistan plans to reduce water consumption in agriculture by 20% during 1990–2005, and similar opportunities for such large savings probably apply in Tajikistan.

In many countries of the FSU, there also has been a decrease in pressure on the environment. These effects are apparent in the case of water resources, as well as in other sectors; for example, the total livestock population in Kazakstan (many existing in rangelands) decreased by 50% between 1990 and 1996 (Mizina *et al.*, 1997).

### 7.4.2. Impact of Human Activity and Climate Change on Ecosystems

- *The Aral Sea is a case study of the multiplicative effects of resource overuse, which can lead to local environmental and even climate change. Extensive redirection of water from feeder rivers to irrigated agriculture since 1960 has led to a reduction in the surface area of the lake and damage to the surrounding wetlands and the species that depend on them. Air temperature in the vicinity of the lake has increased. Saline and polluted dust from the exposed lake bed has been implicated in significant health problems and increases in infant mortality.*

No integrated assessments have been carried out on the impacts of human activity and climate change on the natural, economic, or social systems of the region. However, the recent history of the Aral Sea may serve as an illustrative case study of some of the multiple factors involved and the multiple effects that are likely to result.

The Aral Sea is a dramatic example of how inappropriate human activities—exacerbated by adverse impacts of climate change—can affect natural ecosystems and coastal systems, water supply, food production, human settlements, and human health (Popov and Rice, 1997). Since 1960, extensive irrigation development projects and intensive use of agricultural chemicals have resulted in regional environmental deterioration. The water level has dropped 16 m since 1960, and the open-water area has decreased by almost 50%. In the past 36 years, the Aral coastline has retreated by 50–100 km (Schreiber and Shermuchamedov, 1996). Schreiber and Shermuchamedov have detailed the resulting degradation in the region and suggested some measures to minimize the effects.

Surface water resources of the Aral basin originate from several large river basins; the most important are the Amudaria and Syrdaria basins. A period of low precipitation, which has reduced the flow of river water, has caused a 26% decrease in the water level of the Aral Sea. Runoff water regulations and intensive use of irrigation water for cotton and rice, along with the inflow of polluted water (i.e., sewage, industrial effluent, and mineralized pollutants) into the rivers, has changed the historic function of the Aral Sea, which used to be the main water and salt accumulation basin for Middle Asia. Currently, the waters of the Amudaria and Syrdaria Rivers are used primarily for irrigation; runoff into the Aral Sea has practically halted. Regulation of river runoff and its extensive use for irrigation also have caused dramatic changes in the rivers' hydrological regimes.

General degradation of floodplain and delta soils has taken place in the region since river flow regulations were implemented. The drying off process and subsequent desertification and salinization of soils have accelerated, resulting in an observed temperature increase of 1.5°C within 100–150 km of the edge of the sea. The process of desertification is accompanied by significant losses of soil organic matter (through wind erosion), particularly in marsh soils. The water in the main rivers of the region is contaminated by salts; other pollutants, such as nitrates, pesticides, organic and oil products; and increased bacterial contaminants and is not presently acceptable for drinking or irrigation use. The social, economic, and ecological consequences of the Aral Sea catastrophe are extensive: The infant mortality rate in the area is 46 deaths per 1,000 live births, and 80% of all human diseases in the region near the Aral Sea have been linked directly with drinking polluted water. The changes also are leading to a major loss of biodiversity of the region (Bie and Imevbore, 1995).

Because water is such a valuable resource in the area and affects all ecosystems—as well as human food and fiber production and health—it will have to be managed carefully to reduce the catastrophic and possibly long-term negative impacts in situations such as those illustrated by the Aral Sea. Several protective measures have been implemented in the area, especially to reduce wind erosion (which has decreased in recent years). Water quality monitoring, pollution control, and water protection zones have been set up in an attempt to minimize some of these effects.

Nissenbaum (1994) provides a similar reconstruction of rapid environmental degradation and impacts on the surrounding population—in an account of the collapse of the cities of the southern basin of the Dead Sea 4,000 years ago.

## 7.5. Research Needs

There are some obvious research needs. Clearly, many basic physiological and ecological studies of the effects of changes in atmospheric and climatic conditions are necessary. Plant and animal species show a great deal of variation in their responses to these changes, and local research at this level is always necessary. However, the most pressing need over much of the region is for sound assessment and monitoring programs to establish current baselines and identify rates of change.

## Acknowledgments

Thanks to Tony McMichael, Sandra Diaz, M. Mirza, and the government review teams for their constructive comments and to Alison Saunders and Margo Davies for help with the literature.

## References

**Abdelrahman**, H.A. and H. Alajmi, 1994: Heavy metals in some water-irrigated and wastewater-irrigated soils of Oman. *Communications in Soil Science and Plant Analysis,* **25**, 605–613.

**Akhtar**, R. and A. J. McMichael, 1996: Rainfall and malaria outbreaks in western Rajasthan. *Lancet,* **48**, 1457–1458.

**Altukhi**, M.H., M.N. Alahdal, J.P. Ackers, and W. Peters, 1996: Prevalence of *Giardia lamblia* infection in the city of Riyadh, Saudi Arabia. *Saudi Medical Journal,* **17**, 482–486.

**Annobil**, S.H., T.C. Okeahialam, G.A. Jamjoom, and W.A. Bassuni, 1994: Malaria in children — experience from Asir region, Saudi-Arabia. *Annals of Saudi Medicine,* **14**, 467–470.

**Azim**, S.M.T. and M.M. Rahaman, 1993: Home management of childhood diarrhoea in rural Afghanistan — a study in Urgun, Paktika province. *Journal of Diarrhoeal Diseases Research,* **11**, 161–164.

**Babaev**, A.G., N.N. Drozdov, I.S. Zonn, and Z.G. Friekin, 1986: Deserts. In: *Word Nature.* Mysl, Moscow, Russia, 318 pp.

**Bazzaz**, F.A., S.L. Bassow, G.M. Berntson, and S.C. Thomas, 1996: Elevated $CO_2$ and terrestrial vegetation: implications for and beyond the global carbon budget. In: *Global Change and Terrestrial Ecosystems* [Walker, B.H. and W.L. Steffen (eds.)]. Cambridge University Press, Cambridge, United Kingdom, pp. 43–76.

**Bener**, A., S. Abdullah, and J.C. Murdoch, 1993: Primary health care in the United-Arab-Emirates. *Family Practice,* **10**, 444–448.

**Bie**, S.W. and A.M.A. Imevbore, 1995: Executive summary. In: *Biological Diversity in the Drylands of the World* [Bie, S.W. and A.M.A. Imevbore (eds.)]. United Nations, Washington, DC, USA, pp. 5–21.

**Blaikie**, P., T. Cannon, I. Davis, and B. Wisner, 1994: *At Risk, Natural Hazards, People's Vulnerability and Disasters.* Routledge Publishers, London, United Kingdom, pp.57–79.

**Chavasse**, D., N. Ahmad, and T. Akhtar, 1996: Scope for fly control as a diarrhoea intervention in Pakistan — a community perspective. *Social Science and Medicine,* **43**, 1289–1294.

**Cross**, E.R. and K.C. Hyams, 1996: The potential effect of global warming on the geographic and seasonal distribution of *Phlebotomus papatasi* in southwest Asia. *Environmental Health Perspectives,* **104**, 724–727.

**Diaz**, S., 1995: Elevated $CO_2$ responsiveness, interactions at the community level and plant functional types. *Journal of Biogeography,* **22**, 289–295.

**Evanari**, M., I. Noy Meir, and D. Goodall (eds.), 1985: Hot deserts and arid shrublands. In: *Ecosystems of the World, Volume 12A.* Elsevier Press, Amsterdam, The Netherlands.

**Fankhauser**, S. and R.S.J. Tol, 1997: The social costs of climate change: the IPCC Second Assessment Report and beyond. *Mitigation and Adaptation Strategies for Global Change,* **1**, 385–403.

**Golubtsov**, V.V. and V.I. Lee, 1995: On calculation of the Caspian Sea level with regard to potential climate change. *Hydrometeorology and Ecology,* **1**, 28–38 (in Russian).

**Golubtsov**, V.V., V.I. Lee, and I.I. Scotselyas, 1996: Vulnerability assessment of the water resources of Kazakstan to anthropogenic climate change and the structure of adaptation measures. *Water Resources Development,* **12**, 193–208.

**Grosse**, S., 1993: *Schistosomiasis and Water Resources Development: a Re-evaluation of an Important Environment-Health Linkage.* The Environment and Natural Resources Policy and Training Project, Working Paper, EPAT/MUCIA, Technical Series No. 2, Washington, DC, USA.

**Hairston**, N.G., 1973: The dynamics of transmission. In: *Epidemiology and Control of Schistosomiasis* [Ansari, N. (ed.)] Karger, Basel, Switzerland, pp. 250–336.

**Hales**, S., P. Weinstein, and A. Woodward, 1996: Dengue fever epidemics in the South Pacific: driven by El Niño Southern Oscillation? *Lancet,* **348**, 1664–1665.

**Hunter**, J.M.L., L. Rey, K.Y. Chu, E.O. Adekolu-John, and K.E. Mott, 1993: *Parasitic Diseases in Water Resources Development: the Need for Intersectoral Negotiation.* World Health Organization, Geneva, Switzerland.

**Iafiazova**, R.K., 1997: Climate change impact on mud flow formation in Trans-Ili Alatay mountains. *Hydrometeorology and Ecology,* **3**, 12–23. (in Russian)

**IPCC**, 1996: *Climate Change 1995: The Science of Climate Change. Contribution of Working Group I to the Second Assessment Report of the Intergovernmental Panel on Climate Change* [Houghton, J.J., L.G. Meiro Filho, B.A. Callander, N. Harris, A. Kattenberg, and K. Maskell (eds.)]. Cambridge University Press, Cambridge, United Kingdom and New York, NY, USA, 572 pp.

- Nicholls, N., G.V. Gruza, J. Jouzel, T.R. Karl, L.A. Ogallo, and D.E. Parker, Chapter 3. *Observed Climate Variability and Change,* pp. 133–192.
- Kattenberg, A., F. Giorgi, H. Grassl, G.A. Meehl, J.F.B. Mitchell, R.J. Stouffer, T. Tokioka, A.J. Weaver, and T.M.L. Wigley, Chapter 6. *Climate Models — Projections of Future Climate,* pp. 289–357.

**IPCC**, 1996: *Climate Change 1995: Impacts, Adaptations, and Mitigation of Climate Change: Scientific-Technical Analyses. Contribution of Working Group II to the Second Assessment Report of the Intergovernmental Panel on Climate Change* [Watson, R.T., M.C. Zinyowera, and R.H. Moss (eds.)]. Cambridge University Press, Cambridge, United Kingdom and New York, NY, USA, 880 pp.

- Kirschbaum, M.U.F. and A. Fischlin, Chapter 1. *Climate Change Impacts on Forests,* pp. 95–130.
- Allen-Diaz, B., Chapter 2. *Rangelands in a Changing Climate: Impacts, Adaptations and Mitigation,* pp. 131–158.
- Noble, I.R. and H. Gitay, Chapter 3. *Deserts in a Changing Climate: Impacts,* pp. 159–189.
- Bullock, P. and H. Le Houérou, Chapter 4. *Land Degradation and Desertification,* pp. 170–189.
- Beniston, M. and D.G. Fox, Chapter 5. *Impacts of Climate Change on Mountain Regions,* pp. 191–213.
- Öquist, M.G. and B.H. Svensson, Chapter 6. *Non-Tidal Wetlands,* pp. 215–239.
- Fitzharris, B.B., Chapter 7. *The Cryosphere: Changes and their Impacts,* pp. 240–265.
- Ittekkot, V., Chapter 8. *Oceans,* pp. 266–288.
- Bijlsma, L, Chapter 9. *Coastal Zones and Small Islands*, pp. 289–324.
- Arnell, N., B. Bates, H. Lang, J.J. Magnuson, and P. Mulholland, Chapter 10. *Hydrology and Freshwater Ecology,* pp. 325–364.
- Scott, M.J., Chapter 12. *Human Settlements in a Changing Climate: Impacts and Adaptation,* pp. 399–426.
- Reilly, J., Chapter 13. *Agriculture in a Changing Climate: Impacts and Adaptation*, pp. 427–467.
- Kaczmarek, Z., Chapter 14. *Water Resources Management*, pp. 469–486.
- Everett, J., Chapter 16. *Fisheries,* pp. 511–537.
- McMichael, A., Chapter 18. *Human Population Health,* pp. 561–584.
- Ishitani, H. and T.B. Johansson, Chapter 19. *Energy Supply Mitigation Options*, pp. 587–648.
- Cole, V., Chapter 23. *Agricultural Options for Mitigation of Greenhouse Gas Emissions*, pp. 745–771.
- Leemans, R., Chapter 25. *Mitigation: Cross-Sectoral and Other Issues*, pp. 799–819.

**IPCC**, 1996: *Climate Change 1995: Economic and Social Dimensions of Climate Change. Contribution of Working Group III to the Second Assessment Report of the Intergovernmental Panel on Climate Change* [Bruce, J.P., H. Lee, and E.F. Haites (eds.)]. Cambridge University Press, Cambridge, United Kingdom and New York, NY, USA, 448 pp.

- Arrow, K.J., J. Parikh, and G. Pillet, Chapter 2. *Decision-Making Frameworks for Addressing Climate Change*, pp. 53–124.

**Jarallah**, J.S., S.A. Alshammari, T.A. Khoja, and M. Alsheikh, 1993: Role of primary health care in the control of *Schistosomiasis* — the experience in Riyadh, Saudi Arabia. *Tropical and Geographical Medicine,* **45**, 297–300.

**Khan**, S.M., 1985: Management of river and reservoir sedimentation in Pakistan. *Water International,* **10**, 18–21.

**Kharin**, N.G., 1995: Change of biodiversity in ecosystems of central Asia under the impacts of desertification. In: *Biological Diversity in the Drylands of the World* [Bie, S.W. and A.M.A. Invebore (eds.)]. United Nations, Washington, DC, USA, pp. 23–31.

**Kocasoy**, G., 1995: Waterborne disease incidences in the Mediterranean region as a function of microbial pollution and t-90. *Water Science and Technology,* **32**, 257–266.

**Larsen**, U., 1990: The effects of monthly temperature fluctuations on mortality in the United States from 1921 to 1985. *International Journal of Biometeorology,* **38**, 141–147.

**Lowenthal**, M. N., 1993: Water quality, waterborne disease and enteric disease in Israel, 1976–92. *Israel Journal of Medical Sciences,* **29**, 783–790.

**Martens**, W.J.M., L.W. Niessen, J. Rotmans, T.H. Jetten, and A.J. McMichael, 1995: Potential impact of global change on malarial risk. *Environmental Health Perspectives,* **103**, 458–464.

**McGinnies**, W.G., B.J. Goldman, and P. Paylore (eds.), 1968: Introduction. In: *Deserts of the World: an appraisal of Research into their Physical and Biological Environments.* University of Arizona Press, Tucson, AZ, USA, pp. 3–17.

**McMichael**, A.J., 1997: *Changes in Incidence of Vector-Borne Diseases due to Climate Change.* Paper presented at Asia Pacific Network Workshop on Human Dimension, 20–23 January 1997, Delhi, India.

**McMichael**, A.J., A. Haines, R. Slooff, and S. Kovaks, 1996: *Climate Change and Human Health.* World Health Organization, Geneva, Switzerland, 297 pp.

**Middle East Water Commission**, 1995: Observations regarding water sharing and management: an intensive analysis of the Jordan River Basin with references to long distance transfers. *International Journal of Water Resources Development,* **11**, 351–376.

**Mires**, D., 1996: Expected trends in fish consumption in Israel and their impact on local production. *Israeli Journal of Aquaculture (Bamidgeh),* **48**, 186–191.

**Mizina**, S.V., O.V. Pilifosova, and E.F. Gossen, 1997: GHG mitigation potential in non-energy sector of Kazakstan. *Hydrometeorology and Ecology,* **3**, 30–41. (in Russian)

**Nazer**, H., I. Aljobeh, H. Qubain, and D.A. Latif, 1993: Diarrhoea — a continuing health problem in Jordanian infants. *Journal of Tropical Pediatrics,* **39**, 195–196.

**Nissenbaum**, A., 1994: Sodom, Gomorrah and the other lost cities of the plain — a climatic perspective. *Climatic Change,* **26**, 435–446.

***The Oxford World Atlas***, 1994: Oxford University Press. Oxford, United Kingdom.

**Pakistan Country Report**, 1994: *Regional Study on Global Environmental Issues*. Asian Development Bank, Manila, Philippines.

**Pilifosova**, O., *et al.,* 1996: Vulnerability and adaptation assessment for Kazakstan. In: *Vulnerability and Adaptation to Climate Change: Interim Results from the U.S. Country Studies Program* [Smith, J.B., S. Huq, S. Lenhart, L.J. Mata, I. Nemesova, and S. Toure (eds.)]. Kluwer Academic Publishers, Boston, MA, USA, pp. 161–181.

**Popov**, Y.M. and T.J. Rice, 1997: Aral Sea. 1. Ecological and economic problems of the Aral Sea Region during transition to a market economy. *Ecology and Hydrometeorology,* **2**, 24–36.

**Qureshi**, A. and A. Iglesias, 1994: Implications of global climate change for agriculture in Pakistan: impacts on simulated wheat production. In: *Implications of Climate Change for International Agriculture: Crop Modeling Study*. U.S. Environmental Protection Agency, EPA 230-B-94-003, Washington, DC, USA.

**Riebsame**, W.E. *et al.,* 1995: Complex river basins. In: *As Climate Changes: International Impacts and Implications* [Strzepek, K.M. and J.B. Smith (eds.)]. Cambridge University Press, Cambridge, United Kingdom, pp. 57–78.

**Sarikaya**, H.Z. and V. Eroglu, 1993. Wastewater reuse potential in Turkey — legal and technical aspects. *Water Science and Technology,* **27**, 131–137.

**Schreiber**, W. and P. Shermuchamedov, 1996: *Turkestan —Our Common Home. Ecological and Environmental Problems in the Region of Central Asia*. Konrad Adenauer Foundation and the International Commitee for Ecology and Environmental Conservation of the Central Asia Region.

**Shelef**, G., Y. Azov, A. Kanarek, G. Zac, and A. Shaw, 1994: The Dan region sewerage wastewater treatment and reclamation scheme. *Water Science and Technology,* **30**, 229–238.

**Shelef**, G. and Y. Azov, 1996: The coming era of intensive wastewater reuse in the Mediterranean region. *Water Science and Technology,* **33**, 115–125.

**Shope**, R.E., 1991: Global climate change and infectious diseases. *Environmental Health Perspectives,* **96**, 171–174.

**Smith**, J.B., S. Huq, S. Lenghart, L.J. Mata, I. Nemesova, and S. Toure, 1996: Water resources. 1. In: *Vulnerability and Adaptation to Climate Change: a Synthesis of Results from the U.S. Country Studies Program.* U.S. Country Studies Program, Washington, DC, USA, pp. 161–181.

**UNEP**, 1997: *Global Environmental Outlook.* United Nations Environmental Programme and Oxford University Press, Oxford, UK, pp. 106–115 and 201–210.

**Walter**, H., E.O. Box, and W. Hilbig (eds.), 1983: The deserts of central Asia. In: *Ecosystems of the World, Volume 5, Temperate Deserts and Semi-Deserts.* Elsevier Press, Amsterdam, The Netherlands, pp. 193–236.

**Wilson**, A.W., 1976: *The Sonoran Desert: the Impact of Urbanization.* Canberra, ACT, Australia.

**World Bank**, 1995a: *World Bank Country Study—Kazakhstan: the Transition to a Market Economy*. World Bank, Washington, DC, USA, 233 pp.

**World Bank**, 1995b: *World Bank Country Study—Tajikistan*. World Bank, Washington, DC, USA, 240 pp.

**World Bank**, 1995c: *World Bank Country Study—Turkeministan*. World Bank, Washington, DC, USA, 244 pp.

**World Bank**, 1995d: *World Bank Country Study—Uzbekistan: an Agenda for Economic Reform.* World Bank, Washington, DC, USA, 318 pp.

**World Bank**, 1997: *World Development Indicators*. World Bank, Washington, DC, USA (CD-ROM).

**WRI**, 1994: *World Resources*. Basic Books, New York, USA.

**WRI**, 1996: *World Resources: A Guide to the Global Environment, 1996–97*. World Resources Institute/United Nations Environment Programme/ United Nations Development Programme/The World Bank. Oxford University Press, New York, 342 pp.

# 8

# North America

DAVID S. SHRINER (USA) AND ROGER B. STREET (CANADA)

Lead Authors:
*R. Ball (USA), D. D'Amours (Canada), K. Duncan (Canada), D. Kaiser (USA), A. Maarouf (Canada), L. Mortsch (Canada), P. Mulholland (USA), R. Neilson (USA), J.A. Patz (USA), J.D. Scheraga (USA), J.G. Titus (USA), H. Vaughan (Canada), M. Weltz (USA)*

Contributors:
*R. Adams (USA), R. Alig (USA), J. Andrey (Canada), M. Apps (Canada), M. Brklacich (Canada), D. Brooks (USA), A.W. Diamond (Canada), A. Grambsch (USA), D. Goodrich (USA), L. Joyce (USA), M.R. Kidwell (Canada), G. Koshida (Canada), J. Legg (Canada), J. Malcolm (Canada), D.L. Martell (Canada), R.J. Norby (USA), H.W. Polley (USA), W.M. Post (USA), M.J. Sale (USA), M. Scott (USA), B. Sohngen (USA), B. Stocks (Canada), W. Van Winkle (USA), S. Wullschleger (USA)*

# CONTENTS

# EXECUTIVE SUMMARY

Within the North American region (defined for the purposes of this report as the portion of continental North America south of the Arctic Circle and north of the U.S.-Mexico border), vulnerability to climate change varies significantly from sector to sector and from subregion to subregion. Recognition of this variability or subregional "texture" is important in understanding the potential effects of climate change on North America and in formulating viable response strategies.

The characteristics of the subregions and sectors of North America suggest that neither the impacts of climate change nor the response options will be uniform. This assessment suggests that there will be differences in the impacts of climate change across the region and within particular sectors. In fact, simply considering the relative climate sensitivity of different sectors or systems within a particular subregion (i.e., climate-sensitive, climate-insensitive, or climate-limited) would suggest differentiated impacts. This diversity also is reflected in the available response options. Sectors and subregions will need to adopt response options to alleviate negative impacts or take advantage of opportunities that not only address the impacts but are tailored to the needs and characteristics of that subregion.

Comprising most of Canada and the contiguous United States, this large area is diverse in terms of its geological, ecological, climatic, and socioeconomic structures. Temperature extremes range from well below -40°C in northern latitudes during the winter months to greater than +40°C in southern latitudes during the summer. The regional atmospheric circulation is governed mainly by upper-level westerly winds and subtropical weather systems, with tropical storms occasionally impacting on the Gulf of Mexico and Atlantic coasts during summer and autumn. The Great Plains (including the Canadian Prairies) and southeastern U.S. experience more severe weather—in the form of thunderstorms, tornadoes, and hail—than any other region of the world.

Our current understanding of the potential impacts of climate change is limited by critical uncertainties. One important uncertainty relates to the inadequacy of regional-scale climate projections relative to the spatial scales of variability in North American natural and human systems. This uncertainty is compounded further by the uncertainties inherent in ecological, economic, and social models—which thereby further limit our ability to identify the full extent of impacts or prescriptive adaptation measures. Given these uncertainties, particularly the inability to forecast futures, conclusions about regional impacts are not yet reliable and are limited to the sensitivity and vulnerability of physical, biological, and socioeconomic systems to climate change and climate variability.

Within most natural and human systems in North America, current climate—including its variability—frequently is a limiting factor. Climate, however, is only one of many factors that determine the overall condition of these systems. For example, projected population changes in North America and associated changes in land use and air and water quality will continue to put pressure on natural ecosystems (e.g., rangelands, wetlands, and coastal ecosystems). Projected changes in climate should be seen as an additional factor that can influence the health and existence of these ecosystems. In some cases, changes in climate will provide adaptive opportunities or could alleviate the pressure of multiple stresses; in other cases, climate change could hasten or broaden negative impacts, leading to reduced function or elimination of ecosystems.

Virtually all sectors within North America are vulnerable to climate change to some degree in some subregions. Although many sectors and regions are sensitive to climate change, the technological capability to adapt to climate change is readily available, for the most part. If appropriate adaptation strategies are identified and implemented in a timely fashion, the overall vulnerability of the region may be reduced. However, uncertainties exist about the feasibility of implementation and efficacy of technological adaptation.

Even when current adaptive capability has been factored in, long-lived natural forest ecosystems in the east and interior west; water resources in the southern plains; agriculture in the southeast and southern plains; human health in areas currently experiencing diminished urban air quality; northern ecosystems and habitats; estuarine beaches in developed areas; and low-latitude cold-water fisheries will remain among the most vulnerable sectors and regions. West coast coniferous forests; some western rangelands; energy costs for heating in the northern latitudes; salting and snow clearance costs; open-water season in northern channels and ports; and agriculture in the northern latitudes, the interior west, and west coast may benefit from opportunities associated with warmer temperatures or potentially from carbon dioxide ($CO_2$) fertilization.

The availability of better information on the potential impacts of climate change and the interaction of these impacts with other important factors that influence the health and productivity of natural and human systems is critical to providing the lead time necessary to take full advantage of opportunities for minimizing or adapting to impacts, as well as for allowing adequate opportunity for the development of the necessary institutional and financial capacity to manage change.

## Key Impacts to Physical, Biological, and Socioeconomic Systems

**Ecosystems: Nonforest Terrestrial** (Section 8.3.1). The composition and geographic distribution of many ecosystems will shift as individual species respond to changes in climate. There will likely be reductions in biological diversity and in the goods and services that nonforest terrestrial ecosystems provide to society.

*Increased temperatures could reduce sub-arctic (i.e., tundra and taiga/tundra) ecosystems. Loss of migratory wildfowl and mammal breeding and forage habitats may occur within the taiga/tundra, which is projected to nearly disappear from mainland areas.* This ecozone currently is the home of the majority of the Inuit population. It also provides the major breeding and nesting grounds for a variety of migratory birds and the major summer range and calving grounds for Canada's largest caribou herd, as well as habitat for a number of ecologically significant plant and animal species critical to the subsistence lifestyles of the indigenous peoples. Current biogeographic model projections suggest that tundra and taiga/tundra ecosystems may be reduced by as much as two-thirds of their present size, reducing the regional storage of carbon in the higher latitudes of North America—which may shift the tundra region from a net sink to a net source of $CO_2$ for the tundra region.

*The relatively certain northward shift of the southern boundary of permafrost areas (projected to be about 500 km by the middle of the 21st century) will impact ecosystems, infrastructure, and wildlife in the altered areas through terrain slumping, increased sediment loadings to rivers and lakes, and dramatically altered hydrology; affected peatlands could become sources rather than sinks for atmospheric carbon.* Projections suggest that peatlands may disappear from south of 60°N in the Mackenzie Basin; patchy arctic wetlands currently supported by surface flow also may not persist.

*Elevated $CO_2$ concentrations may alter the nitrogen cycle, drought survival mechanisms (e.g., the rate of depletion of soil water by grasses), and fire frequency—potentially decreasing forage quality and impacting forage production on rangelands.* Increases in $CO_2$ and changes in regional climate could exacerbate the existing problem of loss of production on western rangelands related to woody and noxious plant invasions by accelerating the invasion of woody $C_3$ plants (many crop and tree species) into mostly $C_4$ (tropical grasses, many weed species ) grasslands. Mechanisms include changes in water-use efficiency (WUE), the nitrogen cycle (increase in carbon-to-nitrogen ratio and concentrations of unpalatable and toxic substances), drought survival mechanisms, and fire frequency. Growth and reproduction of individual animals could decrease as $CO_2$ concentrations rise, without dietary supplementation. However, the data are ambiguous, and production may increase in some grassland ecosystems. Uncertainty exists in our ability to predict ecosystem or individual species responses to elevated $CO_2$ and global warming at either the regional or global scale.

*Arid lands may increase.* Current biogeographical model simulations indicate up to a 200% increase in leaf area index in the desert southwest region of North America and a northern migration and expansion of arid-land species into the Great Basin region of North America. Although uncertainty exists in predictions of regional climate changes and simulations of ecosystem responses to elevated $CO_2$ and global warming, long-term change in ecosystem structure and function is suggested.

*Landslides and debris flows in unstable Rocky Mountain areas and possibly elsewhere could become more common* as winter wet precipitation increases, permafrost degrades, and/or glaciers retreat. Water quality would be affected by increased sediment loads. Fish and wildlife habitat, as well as roads and other artificial structures, could be at increased risk.

**Ecosystems: Forested** (Section 8.3.2). Changes are likely in the growth and regeneration capacity of forests in many subregions. In some cases, this process will alter the function and composition of forests significantly.

*Forests may die or decline in density in some regions because of drought, pest infestations, and fire; in other regions, forests may increase in both area and density.* Models suggest that total potential forest area could increase by as much as 25–44%. For some individual forest types, however, range expansions could be preceded by decline or dieback over 19–96% of their area while the climate and ecosystems are adjusting, but before an equilibrium is attained. Even though total forest area could increase, northward shifts in distribution could produce losses in forest area in the United States.

*Geographic ranges of forest ecosystems are expected to shift northward and upward in altitude, but forests cannot move across the land surface as rapidly as climate is projected to change.* The faster the rate of climate change, the greater the probability of ecosystem disruption and species extinction. Climate-induced dieback could begin within a few decades from the present and might be enhanced by increases in pest infestations and fire. Alternatively, forest growth might increase in the early stages of global warming, followed by drought-induced forest dieback after higher temperatures have significantly increased evaporative demand. Migration into colder areas may be limited by seed dispersal (e.g., barriers may exist because of urbanization and changing land-use patterns), seedling establishment, and poor soils. As forests expand or contract in response to climate change, they will likely either replace or be replaced by savannas, shrublands, or grasslands. Imbalances between rates of expansion and contraction could result in a large pulse of carbon to the atmosphere during the transition.

*Longer fire seasons and potentially more frequent and larger fires are likely.* Because of decades of fire suppression—resulting in higher forest densities and increased transpiration—forests in the continental interior are experiencing increased drought stress; pest infestations; and catastrophic, stand-replacing fires, potentially resulting in changes in species composition. Future climate could result in longer fire seasons and potentially more frequent and larger fires in all forest zones (even those that currently do not support much fire), due to more severe fire weather, changes in fire management practices, and possible forest decline or dieback.

**Hydrology and Water Resources** (Section 8.3.3). Water is a linchpin that integrates many subregions and sectors. Water quantity and quality will be directly affected by climate change. Available water supplies also will be affected by changes in demand from multiple sectors competing for water resources. Changes in the hydrological cycle will cause changes in ecosystems—which will, in turn, affect human health (e.g., by altering the geographic distribution of infectious diseases) and biological diversity.

*Increases or decreases in annual runoff could occur over much of the lower latitudes and in midcontinental regions of mid and high latitudes.* Increases in temperature lead to a rise in evapotranspiration—which, unless offset by large increases in precipitation or decreases in plant water use, results in declines in runoff, lake levels, and groundwater recharge and levels. The greatest impact of declines in supply will be in arid and semi-arid regions and in areas with a high ratio of use relative to available renewable supply, as well as in basins with multiple competing uses. Alternatively, regions that experience substantial increases in precipitation are likely to have substantial increases in runoff and river flows.

*Climate projections suggest increased runoff in winter and early spring but reduced flows during summer in regions in which hydrology is dominated by snowmelt. Glaciers are expected to retreat, and their contributions to summer flows will decline as peak flows shift to winter or early spring.* In mountainous regions, particularly at mid-elevations, warming leads to a long-term reduction in peak snow-water equivalent; the snowpack builds later and melts sooner. Snow- or glacier-fed river and reservoir systems that supply spring and summer flow during the critical periods of high agricultural and municipal demand and low precipitation may tend to release their water earlier in the year, which would reduce supplies during summer droughts. Water supplies and water quality, irrigation, hydroelectric generation, tourism, and fish habitat, as well as the viability of the livestock industry, may be negatively impacted. The Great Plains of the United States and prairie regions of Canada and California are particularly vulnerable.

*Altered precipitation and temperature regimes may cause lower lake levels, especially in midcontinental regions and, along with the seasonal pattern and variability of water levels of wetlands, thereby affect their functioning—including flood protection, water filtration, carbon storage, and waterfowl/wildlife habitat.* The response of an affected wetland varies; it might include migration along river edges or the slope of a receding lake and/or altered vegetation species composition. Long-term lake levels would decline to or below historic low levels in the Great Lakes under several climate change scenarios. Prairie pothole lakes and sloughs may dry out more frequently in the north-central regions of North America. These wetlands currently yield 50–75% of all waterfowl produced annually in North America. In the Mackenzie delta of arctic Canada, many lakes could disappear in several decades because of decreased flood frequency and less precipitation.

*Ice-jam patterns are likely to be altered.* In New England, the Atlantic provinces, the Great Lakes, and central Plains areas, as well as northern regions susceptible to spring flooding, changes in late winter-early spring precipitation patterns could result in diminished frequency of ice jams and flooding. Damages caused by these events currently are estimated to cost Canadians CAN$60 million and Americans US$100 million annually, though northern deltas and wetlands appear to depend on the resulting periodic recharge. Depending on the specific pattern of altering climate, mid-latitude areas where ice jams presently are uncommon—such as the prairies; central Ontario and Quebec; and parts of Maine, New Brunswick, Newfoundland, and Labrador—may suffer from an increase in frequency and/or severity of winter breakup and associated jamming.

*Increases in hydrological variability (larger floods and longer droughts) are likely to result in increased sediment loading and erosion, degraded shorelines, reductions in water quality, reduced water supply for dilution of point-source water pollutants and assimilation of waste heat loads, and reduced stability of aquatic ecosystems.* Projected changes in snowfall and snowmelt—as well as suggested increases in warm-period rainfall intensity—could shift the periodicity of the flood regime in North America, possibly stressing the adequacy of dams, culverts, levees, storm drains, and other flood prevention infrastructures. The impacts of flooding are likely to be largest in arid regions, where riparian vegetation is sparse; in agricultural areas during winter, when soils are more exposed; and in urban areas with more impervious surfaces. Increases in hydrological variability may reduce productivity and biodiversity in streams and rivers and have large impacts on water resources management in North America, with increased expenditures for flood management. Increases in water temperature and reduced flows in streams and rivers may result in lower dissolved oxygen concentrations, particularly in summer low-flow periods in low- and mid-latitude areas.

*Projected increases in human demand for water would exacerbate problems associated with the management of water supply and quality.* Managing increased water demands will be particularly problematic in regions experiencing increases in variability and declines in runoff. Improved management of water infrastructure, pricing policies, and demand-side management of supply have the potential to mitigate some of the impacts of increasing water demand.

**Food and Fiber: Agriculture** (Section 8.3.4). As the climate warms, crop patterns will shift northward. Most studies of these shifts have focused on changes in average climate and assume farmers effectively adapt. They have not fully accounted for changes in climate variability, water availability, and imperfect responses by farmers to changing climate. Future consideration of these factors could either increase or decrease the magnitude of changes projected by these earlier studies.

*Climate modifications that lead to changes in daily and interannual variability in temperatures and, in particular, precipitation will impact crop yields.* Although changes in average temperature and precipitation can be expected to impact agriculture, few studies have considered the effects of increased climate variability on crop and livestock production. Increased variability in daily and interannual temperature and precipitation are likely to be as important or more important than the effects of mean changes in climate. Droughts, floods, and increased risks of winter injury will contribute to a greater frequency and severity of crop failure. An increased reliance on precision farming has increased vulnerability to climate variability outside a narrow range of change. These impacts are projected to be both site- and crop-specific; reliable forecasts for such occurrences, however, are not yet regionally available.

*The direct effects of a doubling of $CO_2$ on crop yields are largely beneficial.* Food and fiber production for crops like cotton, soybean, and wheat are expected to increase an average of 30% (range -10% to +80%) in response to a doubling of $CO_2$ concentration. The magnitude of this response will be highly variable and will depend on the availability of plant nutrients, temperature, and precipitation.

*Crop losses due to weeds, insects, and diseases are likely to increase and may provide additional challenges for agricultural sector adaptation to climate change.* Less severe winters due to climate change may increase the range and severity of insect and disease infestations. Increasing pressure to reduce chemical inputs (i.e., pesticides) in agriculture will necessitate a greater emphasis on concepts of integrated pest management and targeted application of agricultural chemicals through precision agricultural technologies.

*Recent analyses of issues of long-run sustainability associated with agricultural adaptation to climate change from an arbitrary doubling of equivalent $CO_2$ concentrations have concluded that there is considerably more sectoral flexibility and adaptation potential than was found in earlier analyses.* Much of this reassessment arises from a realization that the costs and benefits of climate change cannot be adequately evaluated independently of behavioral, economic, and institutional adjustments required by changing climate. Although scientific controversy over the nature and rate of climate change remains, most existing scenarios suggest gradual changes in mean climate over decades—providing ample opportunities for adaptation measures to be implemented within vulnerable subregions of North America. However, uncertainties remain about the implications of changes in climate variability, as well as crop responses to increases beyond a doubling of equivalent atmospheric $CO_2$ concentrations.

*Existing studies that have looked at changes in mean temperature and precipitation suggest that climate change is not likely to harm agriculture enough to significantly affect the overall economy of North America.* The economic consequences of climate change to U.S. agriculture are expected to be both positive and negative, depending on the nature of temperature and precipitation changes that occur in specific subregions. Subregions of North America that are dependent on agriculture may be more vulnerable than areas offering economic diversity. The Great Plains area, for example, relies heavily on crop and livestock production and, as a result, is potentially vulnerable to climate change, with negative consequences projected for southern extremes and potential positive impacts in northern areas as temperatures rise. Warmer temperatures at northern latitudes may lessen the adverse effects of frost damage, but the risk of early- and late-season frost will remain a barrier to the introduction of new crops.

*Consumers and producers could gain or lose; the long-term stability of the forest-products market is uncertain.* Consumer prices could increase by 100–250% with severe forest dieback, producing losses of 4–20% of the net value of commercial forests. Alternatively, consumer prices could decrease with increased forest growth and harvest in Canada, and producers could sustain economic losses. With exports from Canada to the United States, however, the net changes (consumers plus producers) could be negative for Canadians and positive for the U.S. market.

**Food and Fiber: Production Forestry** (Section 8.3.5). *The most intensively managed industry and private forestlands may be least at risk of long-term decline from the impacts of climate change because the relatively high value of these resources is likely to encourage adaptive management strategies.* Private forest managers have the financial incentive and the flexibility to protect against extensive loss from climate-related impacts. They can use several available techniques: short rotations to reduce the length of time that a tree is influenced by unfavorable climate conditions; planting of improved varieties developed through selection, breeding, or genetic engineering to reduce vulnerability; and thinning, weeding, managing pests, irrigating, improving drainage, and fertilizing to improve general vigor. Such actions would reduce the probability of moisture stress and secondary risks from fire, insects, and disease. However, the more rapid the rate of climate change, the more it may strain the ability to create infrastructure for seeding or planting of trees, or to support the supply of timber if there is a large amount of salvage. A fast rate of warming also may limit species constrained by slow dispersal rates and/or habitat fragmentation, or those that are already stressed by other factors, such as pollution.

**Food and Fiber: Fisheries and Aquatic Systems** (Section 8.3.6). Aquatic ecosystem functions will be affected by climate change, although the effects are likely to vary in magnitude and direction depending on the region.

*Projected increases in water temperature, changes in freshwater flows and mixing regimes, and changes in water quality could result in changes in the survival, reproductive capacity, and growth of freshwater fish and salmonid and other anadromous species.* In larger, deeper lakes—including the Great Lakes and many high-latitude lakes—increases in water temperature may increase the survival and growth of most fish species. In smaller, mid-latitude lakes and streams, however, increased water temperatures may reduce available habitat for some cold-water and cool-water species. Increased production rates of food (e.g., plankton) with warmer water temperature (e.g., plankton production increases by a factor of 2–4 with each 10°C increase) also may increase fish productivity. However, shifts in species composition of prey with warming may prevent or reduce productivity gains if preferred prey species are eliminated or reduced. Warmer freshwater temperatures and changes in the pattern of flows in spawning streams/rivers could reduce the abundance of salmon, although individual size may increase from improved growth in the warmer water. Increases in temperature in freshwater rearing areas and increased winter flows may increase mortality for stocks in southern rivers on the west coast.

*Freshwater species distributions could shift northward, with widespread/subregional species extinction likely at the lower latitudes and expansion at the higher latitudes of species ranges.* For example, a 3.8°C increase in mean annual air temperature is projected to eliminate more than 50% of the habitat of brook trout in the southern Appalachian mountains, whereas a similar temperature increase could expand the ranges of smallmouth bass and yellow perch northward across Canada by about 500 km. Whether fish are able to move or will become extinct in response to changes in or loss of habitat will depend on the availability of migration routes.

*Recreational fishing is a highly valued activity that could incur losses in some regions resulting from climate-induced changes in fisheries.* The net economic effect of changes in recreational fishing opportunities is dependent on whether the gains in cool- and warm-water fish habitat offset the losses in cold-water fish habitat. The loss of fishing opportunities could be severe in some parts of the region, especially at the southern boundaries of fish species' habitat ranges. Although gains in cool- and warm-water fishing opportunities may offset losses in cold-water fishing opportunities, distributional effects will cause concern.

*There will likely be relatively small economic and food supply consequences at the regional/national level as a result of the impacts on marine fisheries; however, impacts are expected to be more pronounced at the subregional and community levels.* The adaptability of fisheries to current climate variability and the relatively short time horizons on capital replacement (ships and plants) will minimize the regional- and national-level impacts of projected climate change. At the subregional and community levels, however, positive and negative impacts can be significant as a result of suggested shifts in the centers of production and ensuing relocation of support structures, processors, and people.

*Projected changes in water temperatures, as well as salinity and currents, can affect the growth, survival, reproduction, and spatial distribution of marine fish species and the competitors and predators that influence the dynamics of these species.* Growth rates, ages of sexual maturity, and distributions of some marine fish species are sensitive to water temperatures (e.g., cold temperatures typically result in delayed spawning, whereas warm temperatures result in earlier spawning), and long-term temperature changes can lead to expansion or contraction of the distribution ranges of some species. These changes generally are most evident near the northern or southern species boundaries (i.e., warming resulting in a distributional shift northward, and cooling drawing species southward).

*The survival, health, migration, and distribution of many North American marine mammals and sea turtles are expected to be impacted by projected changes in the climate through impacts on their food supply, sea-ice extent, and breeding habitats.* Although some flexibility exists in their need for specific habitats, some marine mammals and sea turtles may be more severely affected than others by projected changes in the availability of necessary habitat, including pupping and nesting beaches; in food supplies; and in associated prey species. Concerns are the result primarily of projected changes in seasonal sea-ice extent and accelerated succession or loss of coastal ecosystems as a result of projected rises in sea level.

**Coastal Systems** (Section 8.3.7). The implications of rising sea level are well understood, in part because sea level has been rising relative to the land along most of the coast of North America for thousands of years. Some coastal areas in the region will experience greater increases in sea level than others. Adaptation to rising seas is possible, but it comes at ecological, economic, and social costs.

*In the next century, rising sea level could inundate approximately 50% of North American coastal wetlands and a significant portion of dry land areas that currently are less than 50 cm above sea level. In some areas, wetlands and estuarine beaches may be squeezed between advancing seas and engineering structures.* A 50-cm rise in sea level would cause a net loss of 17–43% of U.S. coastal wetlands, even if no additional bulkheads or dikes are erected to prevent new wetland creation as formerly dry lands are inundated. Furthermore, in the United States, 8,500–19,000 km$^2$ of dry land are within 50 cm of high tide, 5,700–15,800 km$^2$ of which currently are undeveloped. Several states in the United States have enacted regulations to adapt to climate change by prohibiting structures that block the landward migration of wetlands and beaches. The mid-Atlantic, south Atlantic, and Gulf coasts are likely to lose large areas of wetlands if sea-level rise accelerates.

*Coastal areas in the Arctic and extreme North Atlantic and Pacific are less vulnerable, except where sea ice and/or permafrost currently is present at the shoreline.* Recent modeling suggests that projected increases in ocean fetches as a result of decreases in the period and extent of sea-ice cover could increase wave heights by 16–40% and therefore increase coastal erosion during the open-water season. Maximum coastal erosion rates are expected to continue in those areas where permafrost contains considerable pore, wedge, or massive ice or where the permafrost shoreline is exposed to the sea.

*Rising sea level is likely to increase flooding of low-lying coastal areas and associated human settlements and infrastructure.* Higher sea levels would provide a higher base for storm surges; a 1-m rise would enable a 15-year storm to flood many areas that today are flooded only by a 100-year storm. Sea-level rises of 30 cm and 90 cm would increase the size of the 100-year floodplain in the United States from its 1990 estimate of 50,500 $km^2$ to 59,500 $km^2$ and 69,900 $km^2$, respectively. Assuming that current development trends continue, flood damages incurred by a representative property subject to sea-level rise are projected to increase by 36–58% for a 30-cm rise and by 102–200% for a 90-cm rise. In Canada, Charlottetown, Prince Edward Island appears to be especially vulnerable, with some of the highest-valued property in the downtown core and significant parts of the sewage systems at risk.

*Saltwater is likely to intrude further inland and upstream.* Higher sea level enables saltwater to penetrate farther upstream in rivers and estuaries. In low-lying areas such as river deltas, saltwater intrusion could contaminate drinking water and reduce the productivity of agricultural lands.

**Human Settlements and Industry** (Section 8.3.8). Climate change and resulting sea-level rise can have a number of direct effects on human settlements, as well as effects experienced indirectly through impacts on other sectors.

*Potential changes in climate could have positive and negative impacts on the operation and maintenance costs of North American land and water transportation systems.* Higher temperatures are expected to result in lower maintenance costs for northern transportation systems, especially with fewer freeze-thaw cycles and less snow. However, some increased pavement buckling is a possibility because of projected longer periods of intense heat. Problems associated with permafrost thawing in the Bering Sea region could be particularly severe and costly. River and lake transportation could be somewhat more difficult, with increases in periods of disruption as a result of projected decreases in water levels (e.g., the Mississippi River and the Great Lakes-St. Lawrence Seaway system). Increases in the length of the ice-free season could have positive impacts for commercial shipping on the inland waterways and in northern ports (e.g., Arctic Ocean ports).

*Projected changes in climate could increase risks to property and human health/life as a result of changes in exposure to natural hazards (e.g., wildfire, landslides, and extreme weather events).* A large and increasing number of people and their property in North America are vulnerable to natural hazards. Projected changes in wildfires and landslides could increase property losses and increase disruptions and damages to urban and industrial infrastructure (e.g., road and rail transportation and pipeline systems). Although some questions remain regarding the extent and regional reflections of changes in extreme weather events as a result of climate changes, projected changes in the frequency or intensity of these events are of concern because of the implications for social and economic costs in a number of sectors. For example, extreme weather events can cause direct physical harm to humans; disrupt health infrastructure, causing contamination of water systems and creating breeding sites for insects or favorable conditions for rodents that carry diseases; and affect construction costs, insurance fees and settlement costs, and offshore oil and gas exploration and extraction costs.

*Climate warming could result in increased demand for cooling energy and decreased demand for heating energy, with the overall net effect varying among geographic regions. Changes in energy demand for comfort, however, are expected to result in a net saving overall for North America.* Projected increases in temperature could reduce energy use associated with space heating [e.g., a 1°C increase in temperature could reduce U.S. space-heating energy use by 11% of demand, resulting in a cost saving of $5.5 billion (1991$US)]. It also has been projected that a 4°C warming could decrease site energy use for commercial-sector heating and cooling by 13–17% and associated primary energy by 2–7%, depending on the degree to which advanced building designs penetrate the market. If peak demand for electricity occurs in the winter, maximum demand is likely to fall as a result of projected temperature changes, whereas if there is a summer peak, maximum demand will rise.

*The technological capacity to adapt to climate change is likely to be readily available in North America, but its application will be realized only if the necessary information is available (sufficiently far in advance in relation to the planning horizons and lifetimes of investments) and the institutional and financial capacity to manage change exists.* Some adaptations can be made without explicit climate predictions through increasing the resilience of systems, such as greater flood control, larger water reservoirs, and so forth, but these approaches are not without social and economic costs. Rapid changes in climate and associated acceleration of sea-level rise would limit adaptation options, thereby putting considerable strain on social and economic systems and increasing the need for explicit adaptation strategies.

**Human Health** (Section 8.3.9). Climate can have wide-ranging and potentially adverse effects on human health through direct pathways (e.g., thermal stress and extreme weather/climate events) and indirect pathways (e.g., disease vectors and infectious agents, environmental and occupational exposures to toxic substances, and food production).

*Direct health effects include increased heat-related mortality and illness and the beneficial effects of milder winters on cold-related mortality.* Under a warmer North America, current models indicate that by the middle of the next century, many major cities could experience as many as several hundred to thousands of extra heat-related deaths annually. The elderly, persons with preexisting health conditions, and the very young (0–4 years) are most vulnerable to heat stress. Gradual acclimatization to increasing temperatures, the use of air conditioners, and an adequate warning system for heat waves may help reduce heat-related deaths. Conversely, it has been suggested that winter mortality rates may decrease in the future with warmer winter temperatures.

*Climate warming may exacerbate respiratory disorders associated with reduced air quality and affects the seasonality of certain allergic respiratory disorders.* Concurrent hot weather and exposure to air pollutants can have synergistic impacts on health. Recent studies show a positive correlation between ground-level ozone and respiratory-related hospital admissions in the United States and Canada. Increased temperatures under climate change could lead to a greater number of days on which ozone levels exceed air quality standards. Global warming also may alter the production of plant aero-allergens, intensifying the severity of seasonal allergies.

*Changing climate conditions may lead to the northward spread of vector-borne infectious diseases and potentially enhanced transmission dynamics due to warmer ambient temperatures.* Vector-borne infectious diseases (e.g., malaria, dengue fever, encephalitis) and waterborne diarrheal diseases currently cause a large proportion of global fatalities. Temperature increases under climate change are expected to enlarge the potential transmission zones of these vectors into temperate regions of North America. Some increases in waterborne diseases may occur due to changes in water distribution, temperature, and microorganism proliferation under climate change. However, the North American health infrastructure likely would prevent a large increase in the actual number of vector-borne and waterborne disease cases.

**Integrative Issues** (Section 8.4). Taken individually, responses to any one of the impacts discussed here may be within the capabilities of a subregion or sector. The fact that they are projected to occur simultaneously and in concert with changes in population, technology, and economics and other environmental and social changes, however, adds to the complexity of the impact assessment and the choice of appropriate responses.

This assessment highlights a number of the uncertainties that currently limit our capability to understand the vulnerability of subregions and sectors of North America and to develop and implement adaptive strategies to reduce that vulnerability. The following research and monitoring activities are considered key to reducing these uncertainties:

- Improve regional and subregional projections of climate change that consider the physiographic characteristics that play a significant role in the North American climate (e.g., the Great Lakes, the nature of the coasts, and mountain ranges), and incorporate biosphere-atmosphere feedbacks.
- Improve projections of changes in weather systems and variability, including extremes.
- Develop a better understanding of physiological and ecosystem processes, with particular emphasis on direct $CO_2$ effects and how the $CO_2$ effects might be enhanced or diminished by nitrogen-cycle dynamics.
- Identify sensitivities and relative vulnerabilities of natural and social systems, including the availability of the necessary physical, biological, chemical, and social data and information.
- Identify beneficial impacts or opportunities that may arise as a result of climate change.
- Develop integrated assessments of impacts.
- Define viable response options that recognize the differentiated and integrative nature of the impacts and response options and the specific needs of sectors and subregions of North America.

## 8.1. Regional Characterization

North America, for the purposes of this regional assessment, is defined as continental North America north of the border between the United States and Mexico and south of the Arctic Circle. Comprising most of Canada and the United States, this area totals approximately 19.42 million km$^2$, with a combined population of approximately 292.7 million in 1995 (Annex D, Table D-6). Canada is the second-largest country in the world but one of the most sparsely populated, with nearly 90% of the population located along the border with the United States (Figures 8-1 and 8-2). The United States is the world's fourth-largest country in both area and population. Approximately 75% of the North American population is urban. North America is geologically and ecologically diverse and spans a full spectrum of land cover types and physiography (Figure 8-3 and Annex C). About 12% of the land area of the North American region is cropland, and 32% is forest and woodlands; the rest is divided among rangelands and other lands, including mountains, desert, wetlands and lakes, and wilderness. As such, management of croplands, forests, and rangelands within North America is a key part of sustainable development.

Canada and the United States rank among the wealthiest countries in the world in terms of per capita income and natural resources. In fact, there is a strong link between the region's economic prosperity and well-being and that of its natural resources. For example, Canada, more than most industrialized nations, depends on the land for its economic well-being, with one in three workers employed directly or indirectly in agriculture, forestry, mining, energy generation, and other land-based activities (Government of Canada, 1996). In the United States, although dependence on primary production is lower, agricultural production and marketing account for 16% of employment, and almost half of the total land area (excluding Alaska) is dedicated to agriculture-related purposes (PCSD, 1996).

North America has abundant energy resources—including uranium, oil, natural gas, and coal—and leads the world in the production of hydropower. It also is the world's largest consumer of energy, but because of more rapid growth in other regions of the globe, emissions from North America have shrunk from 45.1% of the global total in 1950 to 24.3% in 1994. The region historically and currently leads the world in greenhouse gas emissions, contributing 1509 million metric tons of carbon in 1994 (Canada, 122 million metric tons; the United States, 1387 million metric tons) (Marland and Boden, 1997).

Water availability and quality are among the most common concerns expressed by North Americans (UNEP, 1997). Despite an overall abundance, water shortages occur periodically in some localities (e.g., arid sections of the western United States, the Canadian prairies, and some of the interior valleys of the Rocky Mountains). Compared with people in most other regions, North Americans are among the world's leaders in per capita water consumption, and the region enjoys relatively good water quality. Nevertheless, the availability of safe water remains a problem in a number of areas, particularly in rural and remote areas. Improper agricultural practices and by-product and waste disposal practices in some areas have contributed to impaired water quality of rivers, lakes, and estuaries. For example, large concentrations of industrial capacity and agricultural production (nearly 25%

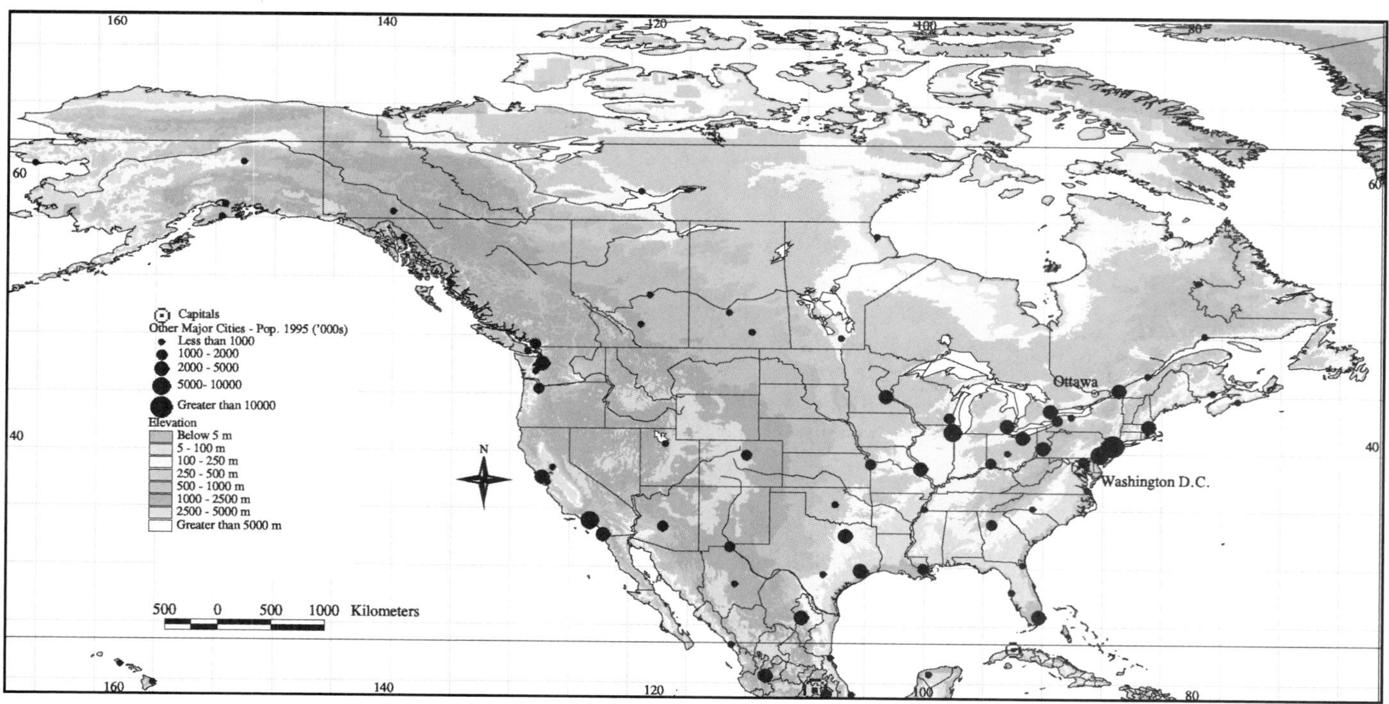


**Figure 8-1:** The North America region [compiled by the World Bank Environment Department Geographic Information System (GIS) Unit; see Annex E for a color rendition].

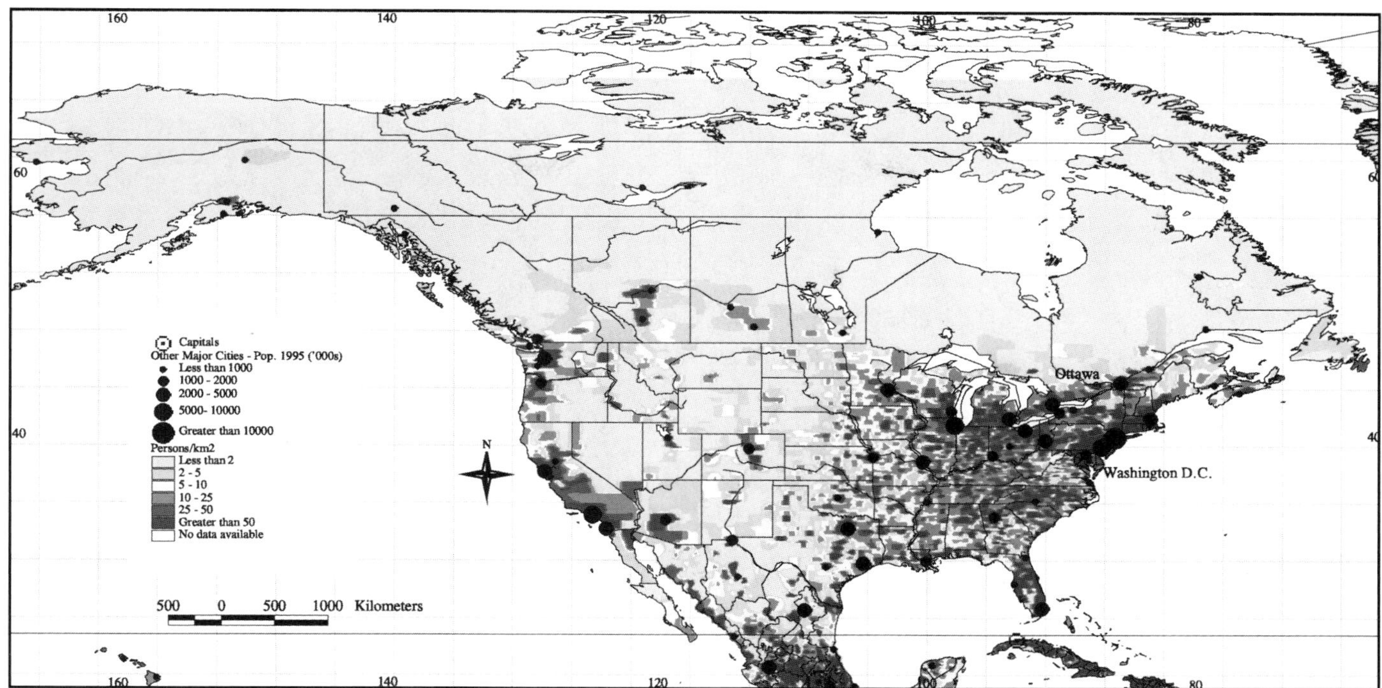


**Figure 8-2:** North American population density [compiled by the World Bank Environment Department Geographic Information System (GIS) Unit; see Annex E for a color rendition].

of the total Canadian agricultural production and 7% of the U.S. production) are located within the Great Lakes–St. Lawrence basin and have contributed to the reduced quality of the water contained within that basin (U.S. EPA and Environment Canada, 1995).

As in other regions of the globe, a number of pressures are being brought to bear on North America that are affecting the region's progress toward sustainability (e.g., population dynamics, land-use changes, changes in the global and regional economy, air and water pollution, consumption, and technological changes). Modern social and ecological systems, for the most part, have evolved and adapted to the prevailing local climate and its natural variability. Climate change, however, is an additional factor that will affect the evolution and adaptation (i.e., sustainability) of these systems. It acts in combination with these other pressures, resulting in either a negative impact or, in some cases, an opportunity that could benefit an area or sector. Climate change impacts and the ability of North America to adapt, therefore, must be assessed within the broader context of these other changes and development trends.

Another factor for consideration is the relative sensitivity of any activity, resource, or area of the region to climate and how these other changes and pressures will affect that sensitivity. Population growth, changing demographics, and the movement of a large proportion of that population to coastal communities are projected to increase the sensitivity of North Americans to climate change and variability. For example, by the year 2000, more than 75% of the U.S. population (PCSD, 1996) and approximately 25% of the Canadian population (Government of Canada, 1996) will reside in coastal communities. Projected increasing demands on water resources—in terms of absolute amounts and multiplicity of demands—are expected to increase the climate sensitivity of these resources. On the other hand, increased energy efficiency and related technological advances are projected to decrease climate sensitivity through their positive impacts on reducing energy demands (e.g., energy for lighting, heating, and cooling).

Some of the key issues that need to be considered when interpreting the results of this assessment or deciding how to respond are as follows.

*There are uncertainties associated with climate change and with the responses of natural and social systems to climate change, particularly at the regional and subregional scales (*IPCC 1996, WG I, Summary for Policymakers*).* Because of inherent uncertainties in our knowledge of the processes affected, our understanding of the magnitude of the responses is equally, or more, uncertain. These uncertainties are compounded further by uncertainties about how landowners and other decision makers will respond to associated risks. Assumptions about how people will respond to risks associated with climate change can significantly affect estimates of associated socioeconomic impacts.

*Differentiated impacts can occur across North America.* As this assessment notes, particular areas and sectors within North America are projected to experience negative impacts, whereas other areas and sectors could benefit from the projected changes. Similarly, because of differences in adaptive capacity within North America, different areas and sectors will be better able to respond to climate change.

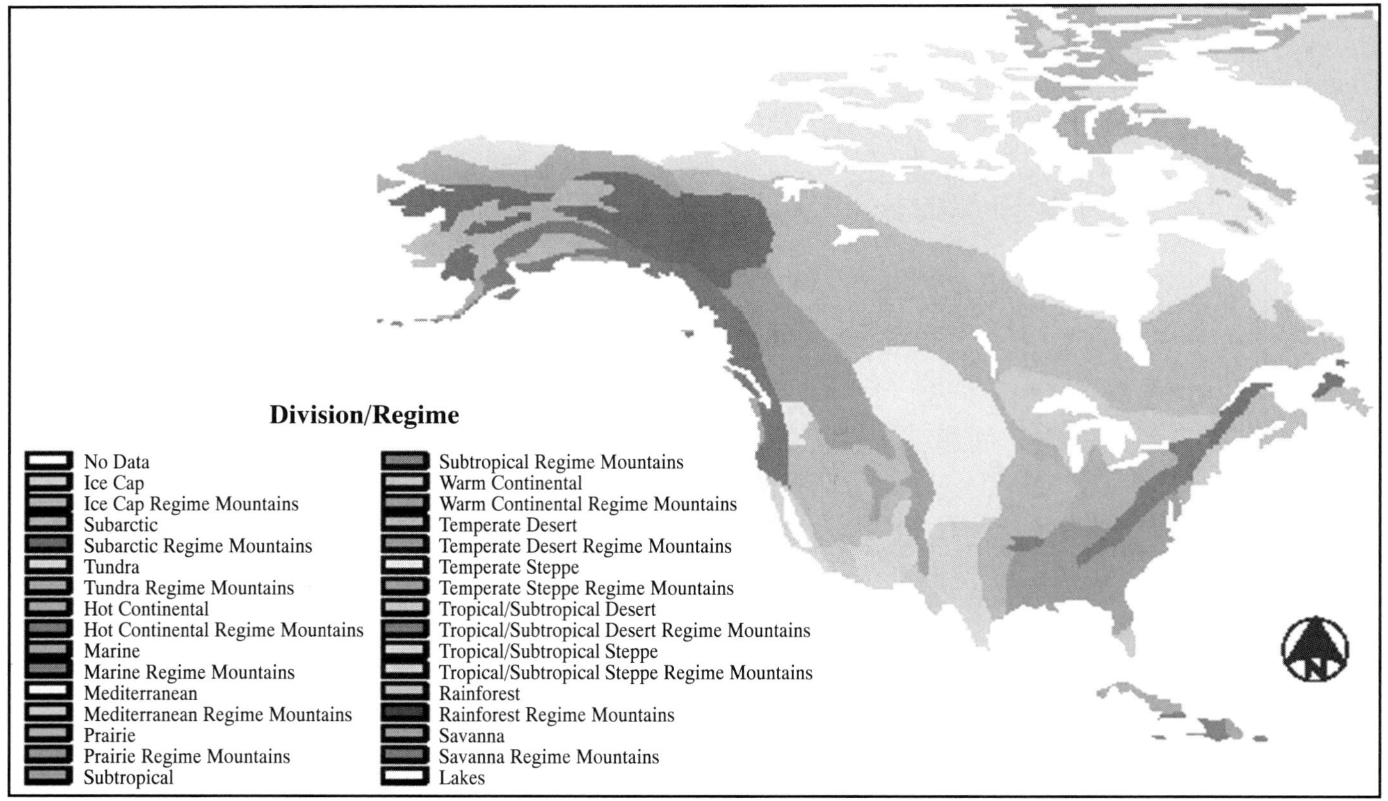


**Figure 8-3:** Ecoregions for North America (courtesy of R.A. Washington-Allen/ORNL), based on data provided by the National Geophysical Data Center (see Annex E for a color rendition).

*Climate change is a dynamic process that will occur over time.* Although most of the literature on impacts, including the analysis in this assessment, has focused on a $2xCO_2$ world, it is reasonable to expect that some of the kinds of impacts that have been discussed would begin to manifest themselves before atmospheric concentrations of $CO_2$ doubled. It also is important to recognize that $2xCO_2$ is not a magic concentration; it is likely that, unless action is taken, we will pass $2xCO_2$ on our way to even higher concentrations. It is not unreasonable to suggest that at these higher levels the negative and beneficial impacts (e.g., the benefits of $CO_2$ fertilization effects on vegetation) could be significantly different, even a nonfactor, in a $4xCO_2$ world.

*Many of the changes in climate that currently are being observed are consistent with the changes one would expect with greenhouse gas-induced climate change (*IPCC 1996, WG I, Summary for Policymakers*).*

*Many of the actions that might be taken today to address existing stresses might also help reduce vulnerability to potential climate change and variability.* Options include, but are not limited to, actions such as the following:

- Improve management of water infrastructure, pricing policies, and demand-side management of supply to support the competing needs of domestic water supply, agriculture, industrial cooling, hydropower, navigation, and fisheries and habitat.
- Introduce accommodating adaptation options in the transportation sector, as the industry copes with other significant changes, through new technologies, markets, and products.
- Introduce effective land-use regulations that also help reduce vulnerabilities and direct population shifts away from vulnerable locations such as floodplains, steep hillsides, and low-lying coastlines.

*Failure to consider climate change when making long-range decisions to manage stress response in any sector could increase the risk of taking actions that would prove ineffective or even counterproductive in the long run.* Examples where consideration of climate change could prove prudent are:

- Actions taken currently that would reduce weather-related deaths associated with heat waves in urban areas also could reduce vulnerability to potential climate change that might increase the frequency and intensity of heat waves (IPCC 1996, WG I, Summary for Policymakers).
- Introducing sustainable water-supply management concepts today, whether to deal with shortages or excesses of water, could reduce vulnerability to potential climate change.

*The concept of "effective" adaptation by any given sector assumes that those affected have the ability and the foresight to discern changing climate trends from short-term weather patterns and to make strategic anticipatory adaptations accordingly.* It is not clear that, if changes in climate and weather patterns or variables fall outside people's experience, they will be able to adapt effectively in the near term. If they are not able, there may be short-term transitional impacts on those individuals and decision makers (e.g., resource managers, farmers, fishermen, loggers, or ranchers).

## 8.2. Regional Climate Information

### *8.2.1. Current Climate*

North America possesses a multitude of diverse regional climates as a consequence of its vastness, its topography, and its being surrounded by oceans and seas with widely varying thermal characteristics. The North American region as analyzed in this report (see Figure A-3 in Annex A) extends latitudinally from approximately the Arctic Circle to the Tropic of Cancer and longitudinally from the Aleutian Islands in the west to the Canadian maritime provinces in the east. The regional atmospheric circulation is dominated by disturbances (waves) in the upper-level westerly winds. The development of these waves defines the position of the main upper-level jet stream over the continent and thus the position of the so-called Polar Front at the surface, which generally separates colder, drier air to the north from warmer, moister air to the south. In the colder half of the year, the position of the Polar Front can vary greatly, from southern Canada to the southern reaches of the United States. Such large shifts in the Polar Front are associated with long, high-amplitude waves that often cause one part of the continent to experience warm, moist, southerly airflow while another part experiences a blast of dry and cold Arctic air (meridional flow). These conditions may persist for lengthy periods because of the typically slow movement of these longer waves. At other times, however (mainly in the fall and spring), shorter, weaker waves move more quickly across the continent—producing highly variable weather with rapidly changing, but not extremely high or low, temperatures and short wet and dry periods. In the summer, the Polar Front retreats well into Canada, for the most part, and two oceanic semipermanent high-pressure systems tend to dominate the North American weather; as a result, there typically are fewer and weaker synoptic-scale disturbances in the westerlies. In summer and autumn, tropical storms of Atlantic, Caribbean, or Gulf of Mexico origin occasionally impact the Atlantic and Gulf coasts.

The temperature regime over North America varies greatly. Over all seasons, mean temperatures generally increase from the extreme north along the Arctic Ocean to the southern United States. Mean annual and wintertime temperatures along the west coast of the continent generally are higher than at equivalent latitudes inland or on the east coast because of the warming influence of Pacific air. During the winter in the far north, the long polar nights produce strong radiative cooling over the frozen Arctic Ocean and the typically snow-covered reaches of Alaska and Canada. This results in very cold surface temperatures and a temperature inversion that acts to inhibit cloud development, creating a positive feedback on the radiational cooling process. In this way, vast pools of cold, dense air (Arctic high-pressure systems) are formed and move over central and eastern North America; they sometimes move southward as far as the Gulf of Mexico. These extreme cold air outbreaks usually are confined to areas east of the Rocky Mountains; they often can produce temperatures below -40°C in the heart of the continent, with attendant sea-level pressure readings in excess of 1050 mb. To the west of the Rockies, warmer maritime airflow off the Pacific Ocean produces milder winters along the coast; the western cordillera effectively restricts this mild air from reaching and thus modifying temperatures in the interior. The eastern maritime regions of the continent enjoy much less warming influence from the Atlantic Ocean during these cold air outbreaks because the prevailing air flow is off the land (Schneider, 1996). Nevertheless, in winter the east and west coastal regions of Canada and the United States usually are warmer than inland regions, with the Pacific and Gulf coasts and Florida experiencing the shortest and mildest winters (Schneider, 1996).

In summertime, the large amount of insolation received over the very long days in the northern reaches of North America acts to raise temperatures there so that these areas are more in line with much of the rest of the continent, thus decreasing the north-south temperature gradient. The coldest areas are found in the western Canadian mountains and along the Labrador coast (Schneider, 1996). The highest continental temperatures are found in the U.S. desert southwest and southern plains states, where temperatures routinely exceed 38°C (~100°F). Occasionally, extreme summer heat waves spread over much of the central United States and parts of central and eastern Canada. These conditions can persist for days or weeks when occasional blocking high-pressure ridges form; these ridges may extend from the central United States to the western Atlantic. The hot air can be extremely humid because of low-level southerly airflow off warm Gulf of Mexico waters. The combination of heat and humidity produces dangerous health conditions that have resulted in significant numbers of fatalities (e.g., the July 1995 heat wave over the midwestern United States).

Annual precipitation amounts over North America show large spatial variations. The wettest regions lie along the Pacific coast, extending generally from Oregon to southern Alaska—with mean annual totals exceeding 300 cm at several Canadian locations (Environment Canada, 1995). The other main continental maximum in annual precipitation is located in the southeastern United States. It is centered mainly along the central Gulf coast states during winter, spring, and autumn and over Florida in the summer (Higgins *et al.*, 1997). Mean annual precipitation amounts along the central Gulf coast exceed 150 cm.

Another precipitation maximum typically is observed over the midwestern United States (centered roughly over Missouri and Iowa) (Higgins *et al.*, 1997) in the summer months, where mean rainfall (mainly convective in nature) typically exceeds 25 cm. This feature is associated with convection that often is fueled by a strong low-level southerly jet stream bringing abundant moisture from the Gulf of Mexico. The active convection often begins in the spring and continues through the summer, causing severe local- to regional-scale flooding. The convective activity observed over the Midwest, the Great Plains, and the southeastern United States also is responsible for the fact that this part of the United States experiences more severe weather (in the form of thunderstorms, tornadoes, and hail) than any other part of the world. Although North American annual precipitation is much more climatologically consistent than in many other parts of the world (e.g., northern Africa and eastern Australia), extremely damaging large-scale droughts and floods sometimes occur, often in association with blocking patterns in the large-scale circulation.

### 8.2.2. *Climate Trends*

A number of studies have examined long-term (century-scale) records of climate variables over the North American region. Most of this work has pertained to analyses of near-surface air temperature and precipitation. Gridded analyses of annual near-surface North American air temperatures for the period 1901–96 (see Figure A-2 in Annex A) show trends toward increasing temperatures over most of the continent. Temperature increases over land are greatest over an area extending from northwestern Canada, across the southern Canada/northern United States region, to southeastern Canada and the northeastern United States. These increases range mainly from 1–2°C/100 years. Decreases in annual temperature on the order of 1°C/100 years are observed along the Gulf coast and on the order of 0.5°C/100 years off the northeast coast of Canada (Environment Canada, 1995). Sea-surface temperatures appear to have warmed off both the west and east coasts of the continent, especially in the Gulf of Alaska.

The time series of anomalies in mean annual temperature for the entire North American region is depicted in Figure A-10 in Annex A. The record reveals temperatures increasing through the 1920s and 1930s, peaking around 1940, and then gradually decreasing through the early 1970s. From this point through the late 1980s, temperatures increased to levels similar to the 1940 era; they have remained mainly above normal since, with the exception of 1996. The more recent warmth has been accompanied by relatively high amounts of precipitation (see below), unlike the dry and warm 1930s. The value of the overall linear trend for 1901–96 is 0.57°C/100 years, a trend significant at better than the 99% confidence level.

The generally increasing temperatures of recent decades, both around the globe and across North America, have been found to result mainly from increases in daily minimum temperature ($T_{min}$); increases in daily maximum temperature ($T_{max}$) have less influence on the observed increase in the daily mean temperature (Karl *et al.*, 1993; Horton, 1995). This trend has caused the diurnal temperature range (DTR) to decrease in many areas. Over North America, Karl *et al.* (1993) found that $T_{min}$ increased greatly over the western half of the continent from 1951 to 1990—in many locales by as much as 2–3°C/100 years. Increases in $T_{max}$ were smaller, for the most part, with $T_{max}$ actually decreasing somewhat in the desert Southwest. The combined effect of these changes resulted in decreases in DTR of 1–3°C/100 years for much of western North America over the period 1951–90. [Trends are reported in Karl *et al.* (1993) as degrees per century to allow for direct comparison between regions with slightly different periods of record and should not be construed as representing actual trends over the past century.] Environment Canada (1995) also found that, over a longer period of record (1895–1991), maximum and minimum temperatures for Canada have been changing at different rates (Figure 8-4), with the minimum temperatures rising more than twice as much as maximum temperatures for the country as a whole.

Annual precipitation amounts from 1901 to 1995 over North America as a whole show evidence of a gradual increase since the 1920s, reaching their highest levels in the past few decades (see Figure A-10 in Annex A). Figure A-1 (Annex A) indicates that the regions experiencing the largest increases are portions of northwestern Canada (>20%), eastern Canada (>20%), and the Gulf coast of the United States (10–20%). The increases in eastern Canada shown in Figure A-1 are corroborated by Groisman and Easterling (1994) and Environment Canada (1995). The analysis of U.S. Historical Climatology Network data by Karl *et al.* (1996) for 1900–94 reveals the increases along the U.S. Gulf coast and also shows 10–20% increases over the central and northern Plains states, much of the Midwest and Northeast, and over the desert Southwest.

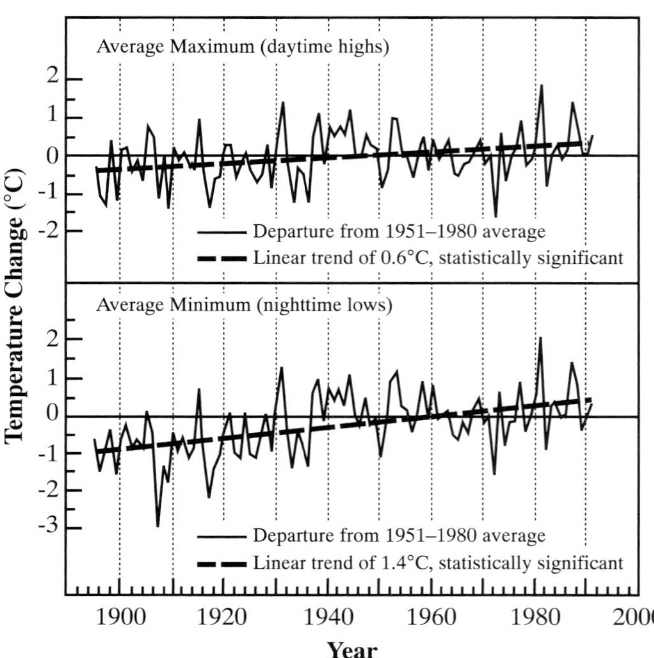


**Figure 8-4:** Canadian annual average maximum and minimum temperature trends for 1895–1991 (adapted from Environment Canada, 1995).

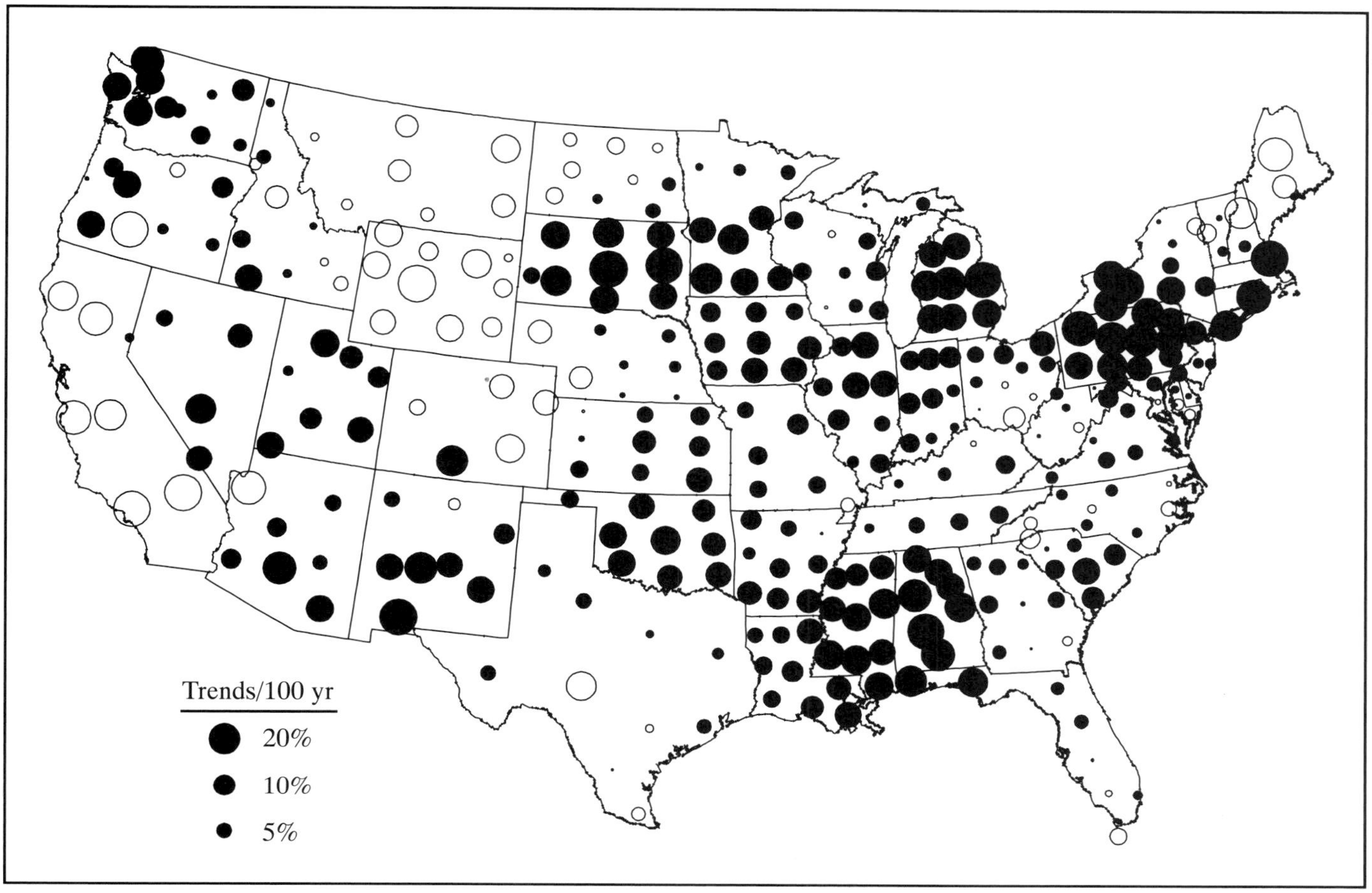

**Figure 8-5:** Conterminous U.S. precipitation trends for 1900–94 (converted to %/century), centered within state climatic divisions. The trend magnitude for each climatic division is reflected by the diameter of the circle. Solid circles represent increases, and open circles decreases.

Decreases of 10–20% are apparent over California and the northern Rocky Mountain states (Figure 8-5).

As part of an effort to monitor extreme weather events around the globe, some recent studies have examined the intensity of rainfall events. Karl *et al.* (1995) found a trend toward higher frequencies of extreme (>50.8 mm) 1-day rainfalls over the United States. The results pertained to the period 1911–92; the increasing frequency of such events was found to be a product mainly of heavier warm-season rainfall. Karl *et al.* (1996) also found a steady increase from 1910 to 1995 in the percentage area of the contiguous United States with a much above-normal (defined as the upper decile of all daily precipitation amounts) proportion of total annual precipitation coming from these extreme 1-day events (Figure 8-6). This area increased by 2–3%, and it was determined that there is less than 1 chance in 1000 that this change could occur in a quasi-stationary climate. To date, however, there is no similar evidence of an increase in the proportion of Canadian precipitation from extreme 1-day events (Hogg and Swail, 1997).

Although sea-level rise usually is not considered a climatic variable, it is arguably one of the most important potential impacts of global climate change in terms of environmental and social consequences (IPCC 1996, WG I, Section 7.1). Therefore, a brief summary of sea-level trends is appropriate. Global mean sea level is estimated to have risen 10–25 cm over the past 100 years (IPCC 1996, WG I, Section 7.2). These

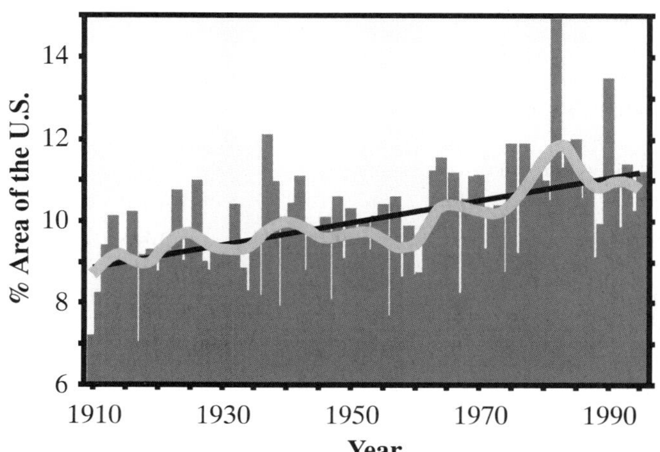


**Figure 8-6:** Percentage of the conterminous U.S. area with a much above normal proportion of total annual precipitation from 1-day extreme (more than 2 in. or 50.8 mm) events (Karl *et al.*, 1996).

estimates are based on tide gauge records; the increase is thought to result largely from the concurrent increase observed in global temperatures, which causes thermal expansion of the ocean and contributes to the melting of glaciers, ice caps, and ice sheets. In general, there is broad agreement that both thermal expansion and glaciers have contributed to the observed sea-level rise, but there are very large uncertainties regarding the role of the ice sheets and other hydrological factors (IPCC 1996, WG I, Section 7.4). There are differences in century-scale sea-level trends across regions of the globe because of vertical land movements such as "postglacial rebound." Figure 8-7 depicts sea-level trends for several North American sites. Sea level has risen 2.5–3.0 mm/year along parts of the U.S. Gulf coast and along the Atlantic coast south of Maine. Along the Texas-Louisiana coasts, sea level has been increasing about 10 mm/year as a result of rapid land subsidence in this region. Sea level is stable or dropping along much of the Canadian and Alaskan coasts because of postglacial rebound.

### 8.2.3. *Climate Scenarios*

As discussed in IPCC (1996, WG I, Section 6.6), output from transient runs of atmosphere-ocean general circulation models (hereafter referred to simply as GCMs) has become available that can be used as the basis for improved regional analysis of potential climate change. The main emphasis of current analyses is on the simulation of seasonally averaged surface air temperature and precipitation. Climate scenario information for North America is available from several GCMs. In IPCC (1990, WG I), one of the five regions identified for analysis of regional climate change simulation was central North America (35–50°N, 85–105°W). Output for this region from different coupled model runs with dynamic oceans was analyzed by Cubasch *et al.* (1994) and Kittel *et al.* (1998). Results for central North America, as well as the other identified regions, are depicted in Figure B-1 (Annex B), which shows differences between region-average values at the time of $CO_2$ doubling and the control run, as well as differences between control run averages and observations (hereafter referred to as bias) for winter and summer surface air temperature and precipitation. These model results reflect increasing $CO_2$ only and do not include the effects of sulfate aerosols. The biases in Figure B-1 (Annex B) are presented as a reference for interpretation of the scenarios because it can be generally expected that the better the match between control run and observed climate (i.e., the lower the biases), the higher the confidence in the simulated change scenarios. A summary of these transient model experiments is given in Table B-1 (Annex B). Most experiments use a rate of $CO_2$ increase of 1%/year, yielding a doubling of $CO_2$ after 70 years.

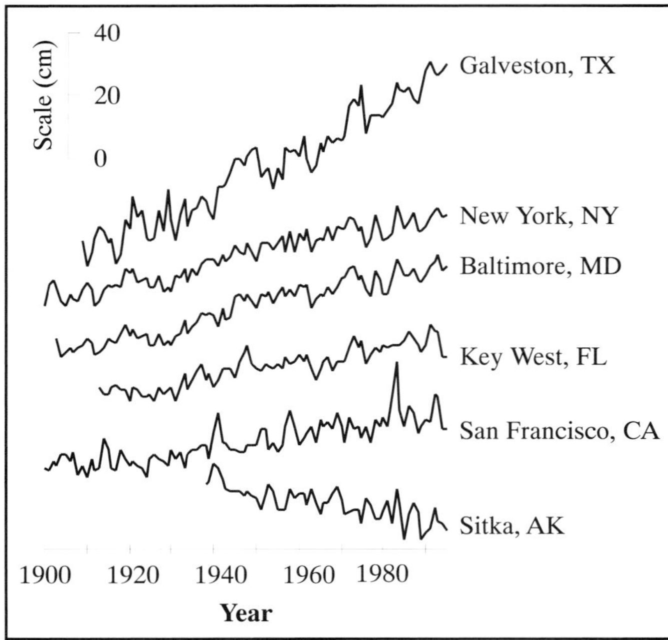


**Figure 8-7:** Relative sea-level trends for selected North American sites (adapted from Titus, 1997).

Scenarios produced for central North America by these transient experiments vary quite widely among models for temperature but less so for precipitation. GCM simulations also have been conducted that consider the effect of combined greenhouse gas- and direct sulfate aerosol-forcing on temperature, precipitation, and soil moisture (see Annex B). For central North America, the inclusion of sulfate aerosols results in a projected warming of 0–0.5°C in the summer and 1.4–3.4°C in the winter by the year 2100. In the case of precipitation, the inclusion of sulfate aerosol-forcing has little effect on the projections (see Annex B).

Using the Canadian Climate Centre (CCC) GCM (see Annex B), Lambert (1995) found a 4% decrease in cyclones in the Northern Hemisphere, though the frequency of intense cyclones increased. Lambert hypothesized that the latent heat effect is responsible for the greater number of intense storms. No change in storm tracks was evident. A few areas showed increased frequencies, such as off Cape Hatteras, over Hudson Bay, and west of Alaska. These results are similar to those of Rowntree (1993), who found a 40% increase in Atlantic gales, though fewer intense storms over eastern North America. Hall *et al.* (1994) and Carnell *et al.* (1996) found an intensification and northward shift of storm tracks.

Regarding sea-level rise scenarios, for IPCC Scenario IS92a, global mean sea level is projected to be about 50 cm higher by 2100 than today, with a range of uncertainty of 20–86 cm (IPCC 1996, WG I, Section 7.5). It is possible that for much of the North American coastline, future sea-level rise will be greater than the global average, given the higher historical rates of sea-level rise along the Gulf of Mexico and Atlantic coasts (see Section 8.2.2). By contrast, future sea-level rise along the Pacific coast may be less than the global average rise because of this region's generally lower historical rates. Even less sea-level rise might be expected in extreme northern North America, given the historical drop in sea levels at many locations (Titus and Narayanan, 1996).

## 8.3. Impacts and Adaptation

### 8.3.1. Ecosystems: Nonforest Terrestrial

#### 8.3.1.1. Distribution and Sensitivities

Nonforest terrestrial ecosystems are the single largest type of land surface cover (>51%) in North America. They are extremely diverse and include nontidal wetlands (bogs, fens, swamps, and marshes), ecosystems of the polar domain (tundra and taiga), traditional rangeland ecosystems (grasslands, deserts, and savannas), and improved pastures. These ecosystems are major components of every region of North America; they constitute about 80% of the land cover of western North America and nearly 100% of the land cover above the 75th parallel. They provide forage for 80 million cattle, sheep, and goats and 25 million deer, elk, antelope, caribou, and buffalo, as well as most of the breeding and feeding grounds for waterfowl in North America (Child and Frasier, 1992; WRI, 1996). Nonforest ecosystems are the source of most surface flow and aquifer recharge in the western Great Plains and the extreme northern regions of North America. Municipal, agricultural, and industrial sectors in these regions depend on nonforest ecosystems for the quantity and quality of water required for economic sustainability. The quality and quantity of water derived from nonforest ecosystems depend on the management these lands receive and the average annual and extreme climatic events they encounter.

Nontidal wetlands in North America include a variety of ecosystems such as bogs, fens, swamps, marshes, and floodplains. Classification systems are many and varied. These wetlands are distributed throughout North America, principally in a band extending from the New England states to Alaska. There are additional significant areas in the Mississippi Valley, the "Prairie Pothole" region, the many coastal wetlands (e.g., the Mississippi River delta, the Everglades, and the Okefenokee), the Atlantic coastal marshes and Fraser River estuary, the former Great Kankakee and Great Blackwater swamps, the Hudson Bay Lowlands, the Peace-Athabasca-Slave delta, the Mackenzie delta, and the Queen Maud Gulf on the Arctic Ocean (Mitsch and Gosselink, 1986; Ecological Stratification Working Group, 1995; IPCC 1996, WG II, Chapter 6). See Section 8.3.3 for a more detailed discussion of wetlands impacts.

Rangelands are characterized by native and introduced vegetation—predominantly grasses, grasslike plants, forbs, shrubs, and scattered trees. These lands are extremely varied: They include the tallgrass, mixed, and shortgrass prairie regions of central North America; tundra and taiga areas in the polar domain; annual grasslands of California; chaparral regions of Arizona and California; sagebrush shrub steppe and pinyon-juniper woodlands in the intermountain region of western North America; and the Chihuahuan, Sonoran, and Mojave deserts in the southwestern portion of North America. The associated ecosystems are complex and are affected by many interacting biotic and abiotic components, and their health depends on the interaction of climate, soils, species competition, fire, grazing, and management. These ecosystems provide a wide array of goods and services, including forage, water, and habitat for wildlife and domesticated livestock and open space for recreational activities, and they are the source of many of the raw materials needed to sustain our industrial society (i.e., pharmaceuticals, precious metals, minerals, construction materials, natural gas, oil, and coal) (Heady and Child, 1994).

Although some rangelands are fragile and easily disturbed by anthropogenic activity (Belnap, 1995), others are resistant to change. Semi-arid and arid ecosystems are considered among the most sensitive because these ecosystems often are water-limited and have marginal nutrient reserves (OIES, 1991; IPCC 1996, WG II, Chapter 2).

Current levels of uncertainties associated with the functioning and adaptive capacity of nonforest ecosystems under variable and changing climate and the possibility of critical thresholds limit our ability to identify the relative sensitivities of these ecosystems (and the potential impacts of changing climates). It is understood, however, that these ecosystems are sensitive to climate variability and that the impacts can vary depending on the resilience and resistance of the ecosystem to the stresses applied (e.g., changes in precipitation, $CO_2$, temperature, fire, land use, and land cover and management). Researchers also believe that the impacts of $CO_2$ enrichment and shifts in temperature and precipitation regimes are likely to be greatest when they are reinforced by other destabilizing forces. Lack of information about how these other factors interact with climate change also limits our understanding of ecosystem response. Also of concern are the relative sensitivities of species at the ecotones between vegetation types, such as between grasslands and woodlands and between woodlands and forests (Polley, 1997).

#### 8.3.1.2. Impacts, Vulnerabilities, and Adaptation

*The projected northward shift of the southern boundary of permafrost areas will alter ecosystem structure and functioning, with subsequent impacts on associated infrastructure and wildlife through terrain slumping, increased sediment loadings in rivers and lakes, and dramatically altered hydrology.*

Approximately half of the wetland areas of North America are located in Alaska, the Northwest Territories, and the Yukon (Table 8-1). Most of these wetlands rest on continuous or discontinuous permafrost, the distribution of which would be altered by climate warming.

The northward shift of the southern boundary of discontinuous and continuous permafrost areas is projected to be about 500 km by the middle of the 21st century (Anisimov and Nelson, 1996; IPCC 1996, WG II, Chapter 7; Prowse, 1997). This shift would have profound effects within the altered areas (as summarized by Prowse, 1997). The melting of widespread ground ice will result in downslope soil movement, bank failure, and massive terrain slumping, leading to increases in sediment loads to rivers and lakes. This process will in turn affect spawning areas, oxygen levels, and stream/wetland sediment budgets. A deeper active layer will reduce overland flow as infiltration and active layer storage capacity increase. Peatlands are projected to disappear from south of 60°N in the Mackenzie Basin (Cohen, 1997a); patchy arctic wetlands currently supported by surface flow would not persist. Lakes and ponds, which have permafrost hydrological divides, are more likely to drain laterally or to groundwater systems.

Landscape alteration on this scale has serious implications for hydrology, wildlife, cultural values, and lifestyles. The effects will likely extend to infrastructure and transportation—including the integrity of foundations (pipelines, bridges, and buildings), water-control structures, ice-roads, and tailings. Altered flooding patterns and sediment loadings will impact internationally significant wetland habitat such as the Peace-Athabasca-Slave delta, the Mackenzie delta, and habitats associated with Hudson Bay and Queen Maud Gulf lowlands.

*Many northern peatlands could become sources rather than sinks for atmospheric carbon.*

A primary impact of future climate change in nonforest terrestrial ecosystems is the projected reduction of subarctic (tundra/taiga) ecosystems (IPCC 1996, WG II, Chapter 2). Neilson *et al.* (see Annex C) estimate that tundra and taiga ecosystems may be reduced by as much as one- to two-thirds of their present size. This reduction will have an impact on regional storage of carbon in the higher latitudes of North America and may result in a shift from a net sink to a net source of $CO_2$ for the tundra region (Anderson, 1991; Oechel *et al.*, 1993). Climate warming also may cause reduction in total species biodiversity and total surface area covered by tundra vegetation, as well as decreased releases of methane from tundra plant communities as a result of alterations in the hydrological cycle, drier surface soils, and an increase in surface oxidation (IPCC 1996, WG II, Chapter 2).

*Loss of migratory wildfowl and mammal breeding and forage habitats will occur within the southern Arctic ecozone, which is projected to nearly disappear from mainland areas.*

The Queen Maud Gulf lowlands contain one of the largest sites (over 6 million ha) designated under the Ramsar Convention on Wetlands of International Importance. They are part of the southern Arctic ecozone, a transitional area between the taiga forest to the south and the treeless arctic tundra to the north. The ecozone includes the major summer range and calving grounds for Canada's largest caribou herds, as well as habitat for bear, wolf, moose, arctic ground squirrels, and lemmings. It is a major breeding and nesting ground for a variety of migratory birds, including yellow-billed, arctic, and red-throated loon; whistling swan; snow goose; oldsquaw; gyrfalcon; ptarmigan; and snowy owl (Ecological Stratification Working Group, 1995). It also is home to Canada's Inuit, whose subsistence lifestyle includes a diet dependent on this wildlife diversity. According to the U.S. Department of Agriculture (USDA) Forest Service MAPSS biome model using a variety of GCM simulations, this ecozone will nearly disappear from mainland North America under a climate brought about with $CO_2$ doubling (Neilson, 1993a,b,, 1995; Neilson and Marks, 1994; see also Annex C in this report).

***Table 8-1:** Estimated wetland area in North America.*

| | **Area (Mha)** | **Reference** |
|---|---|---|
| Alaska | 90.0 | OTA, 1984 |
| Yukon and NWT | 29.3 | Environment Canada, 1986 |
| *Subtotal* | 119.3 | |
| | | |
| U.S. Lower 48 | 40.1 | Frayer *et al.*, 1983 |
| Rest of Canada | 97.9 | Environment Canada, 1986 |
| *Subtotal* | 138.0 | |

*Elevated $CO_2$ concentrations may have a negative influence on forage quality and species diversity within North American rangeland ecosystems.*

Based on studies of plants grown in $CO_2$-enriched environments (Owensby *et al.*, 1993; IPCC 1996, WG II, Chapter 2), it has been suggested that forage quality in rangeland ecosystems may decrease with increasing $CO_2$ levels as a result of associated increases in carbon-to-nitrogen ratios and in concentrations of unpalatable and toxic substances, as well as decreases in mineral concentrations in the forage. There is evidence from field studies that low soil nitrogen—a common constraint on rangelands (McNaughton *et al.*, 1988; Seastedt *et al.*, 1991)—can limit plant growth responses to elevated $CO_2$ (Owensby *et al.*, 1994; Schäppi and Körner, 1996). Several studies have suggested that litter produced at increased levels of $CO_2$ will be nitrogen poor or that increased $CO_2$ concentrations slow nitrogen mineralization and reduce nitrogen availability to plants (Díaz *et al.*, 1993; Morgan *et al.*, 1994; Gorissen *et al.*, 1995). This decreased decomposition, mineralization, and uptake of nitrogen could initiate a negative feedback on nitrogen availability that reduces plant growth and forage production.

Without dietary supplementation, the growth and reproduction of individual animals could decrease as $CO_2$ concentrations rise (Owensby *et al.*, 1996). Rates of nitrogen input, litter quality, and frequency of events (like fire) that promote substantial nitrogen loss all mediate how quickly nitrogen accumulation and cycling approach the maximum at any given $CO_2$ concentration (Aber *et al.*, 1991). Potential production and forage quality on many rangelands therefore may be constrained by management practices that promote nitrogen loss or preclude nitrogen accumulation by limiting species change (Polley, 1997).

The existing data on the effect that rising $CO_2$ concentrations will have on the nitrogen cycle are ambiguous. Rising $CO_2$ may increase nitrogen input to rangelands directly or indirectly by promoting nitrogen fixation. Some of the most successful woody invaders in grasslands are legumes (e.g., species of the *Acaia* and *Prosopis* genera). Symbiotic fixation of nitrogen in these species can be significantly stimulated (+400%) by elevated $CO_2$ concentrations (Polley *et al.*, 1994; Polley *et al.*, 1997). By adding fixed nitrogen to rangelands in litterfall or root turnover, these woody invaders could potentially increase productivity if the nitrogen gains are not offset by losses or conversion of nitrogen to recalcitrant forms (Polley, 1997). Kemp *et al.* (1994) reported that the ambient $CO_2$ concentration at which plants were grown had little effect on the decomposition of standing dead material of three tallgrass prairie species. O'Neill and Norby (1996) concluded that decomposability of naturally abscissed leaf litter was not greatly affected by the $CO_2$ concentration at which the litter was produced.

On grasslands where the effects of increased WUE are not negated by an increase in leaf temperature and leaf area, rising $CO_2$ concentration should slow the depletion of soil water by grasses and potentially favor shrubs/woody plants that otherwise might succumb to water stress (Polley *et al.*, 1997). There is evidence that rising $CO_2$ also could contribute to species change by altering seedling survival rates during drought. Polley *et al.* (1996) found that more than twice as many honey mesquite tree seedlings survived soil-water depletion at elevated $CO_2$ levels than at ambient $CO_2$ levels. In the absence of fire or browsing, woody plants likewise would be expected to increase in size and abundance by exploiting the greater availability of soil water (Polley, 1997). Grass production and transpiration often decline following woody plant invasion (Vallentyne, 1971; Scifres, 1980; Knoop and Walker, 1985; Sala *et al.*, 1989).

By increasing the growth rates of woody seedlings or improving their ability to survive drought, rising $CO_2$ could act as a positive feedback to overgrazing in promoting woody plant invasion (Polley *et al.*, 1997). An increase in woody plants may result in decreased forage availability in mid-latitude grassland ecosystems and increased soil erosion, leading to a nearly irreversible loss of productive potential of the soil (Parton *et al.*, 1993).

Since the turn of the century, mesic and arid grasslands in North America have had increases in $C_3$ woody plants (Branson, 1985; IPCC 1996, WG II, Chapter 2). This change often has been accompanied by changes in runoff, accelerated soil erosion, and loss of the grazing resource (Rauzi and Fly, 1968; Spaeth *et al.*, 1996). Increases in $CO_2$ and changes in regional climate could exacerbate the existing problem of loss of production on western rangelands by accelerating the invasion of woody plants. Mayeux *et al.* (1991) discuss evidence that $C_4$ grasslands are being increasingly invaded by $C_3$ woody plants—a process that may have been abetted by the rise in $CO_2$ and changes in WUE over the past two centuries. However, conclusive evidence for this effect is not available (IPCC 1996, WG II, Chapter 2).

Consistent with trends from individual plants, many natural ecosystems (including grasslands) show little or no increase in standing crop or production at elevated $CO_2$ when temperatures are low or nutrients are limiting (Oechel *et al.*, 1993; Fredeen *et al.*, 1995; Schäppi and Körner, 1996). In the Arctic tundra there is little expected change in plant growth from increased $CO_2$, although there is an expected decrease in insect-pollinated forbs (IPCC 1996, WG II, Chapter 2). However, elevated $CO_2$ has been shown to increase aboveground net primary productivity in tallgrass prairies, shortgrass steppe, and coastal salt marshes (Curtis *et al.*, 1989; Owensby *et al.*, 1993; Hunt *et al.*, 1996) and root biomass in grasslands (Owensby *et al.*, 1993; Jongen *et al.*, 1995; Newton *et al.*, 1995) when essential elements like nitrogen are plentiful or water begins to limit growth and the positive effects of $CO_2$ on water relations are expressed.

Changes in species composition emerge as a major unknown with the potential to affect ecosystem processes (productivity, forage quality, and nitrogen cycling) in ways that are not evident from studies that consider the direct effects of elevated $CO_2$ concentration alone. Given that the geographic distribution of rangeland vegetation and aboveground net primary productivity are highly correlated with precipitation, temperature, nutrient status, and soil-water availability on rangelands (Sala *et al.*, 1988; Stephenson, 1990; Bailey, 1996; IPCC 1996, WG II, Chapter 2; Myneni *et al.*, 1997; Polley *et al.*, 1997), interactions among global warming, changes in precipitation, grazing, fire, rising $CO_2$ concentration, and species competition must be more clearly understood before we will be able to predict with confidence changes in forage quantity for North American rangelands.

On improved pastures, the alteration of species composition through reseeding with adapted grass species or the introduction of legumes to grass-dominated pastures is the most likely method to reduce the impacts of climate change. This approach would have the additional benefit of improving forage value for livestock while possibly reducing the average methane emission per head of livestock because of improved forage quality (IPCC 1996, WG II, Chapter 2). For native rangelands, active interventions to reduce impacts from increases in temperature or $CO_2$ or changes in precipitation frequency and amount are limited because of the large areal extent of rangelands and the low economic return per acre of land. Introduction of nonnative adapted species may be able to compensate for the loss of some forage production and watershed protection if native species decrease. However, the application and extent of this technology on federal lands may be limited by existing rules, regulation, and pending and future court cases.

*Climate-induced variability and extreme events will increase the complexity of managing rangelands.*

Rangeland vegetation is found where precipitation, temperature, and soil development provide suitable habitat for grasses, forbs, shrubs, and open stands of trees. Generally, these lands are characterized by extremes in temperature or in the timing, intensity, and amount of precipitation the site receives; these extremes drive rangeland ecosystems (Griffin and Friedel, 1985; Westoby *et al.*, 1989). Precipitation is the major determinant of the structure, function, and sustainability of natural ecosystems (Branson *et al.*, 1981). Current human activity on rangelands significantly alters plant species abundance and distribution and the hydrological cycle, accelerates erosion rates, and can overwhelm any change in regional or global climate (Thurow *et al.*, 1986; Weltz and Wood, 1986; IPCC 1996, WG II, Chapter 2; Williams and Balling, 1996). Small changes in the frequency or extent of extreme events may have a disproportionate effect on what management must cope with to sustain rangeland ecosystems (IPCC 1996, WG II, Chapter 2). Short-term variations in local or regional precipitation—upon which management planning often is based—are greater than the predicted change in the mean value of precipitation for North America (Shuttleworth, 1996). With the addition of climate change to existing stresses on rangelands, they may become more sensitive to extreme events such as drought, 100-year floods, and insect outbreaks that could reduce their long-term sustainability and escalate the desertification process in arid and semi-arid lands in North America (IPCC 1996, WG II, Chapters 2 and 4).

The most promising adaptive approach is to provide incentives to use management techniques that reduce the risk of these lands becoming degraded during extreme climatic events (i.e., droughts). The most cost-effective strategy is to improve lands already under stress and to strengthen their resistance to future extreme events. This approach could include changes in livestock type and number, changes in season of use, complete rest, or the development of additional infrastructure (new watering locations, fencing, etc.) to achieve proper stocking density.

A second promising avenue of adaptation is to provide arid and semi-arid land managers with more accurate predictions of regional precipitation on a seasonal to interannual basis. This information is particularly important prior to droughts or extremely wet years. Provision of these types of predictions is becoming more likely because recent research indicates that improvements in coupled ocean-atmosphere models make it possible to predict climatic conditions related to the El Niño-Southern Oscillation (ENSO) phenomenon more than a year before the event (Chen *et al.*, 1995). Predicted precipitation patterns can be used as inputs to ecological and hydrological models and thereby could provide the capability to assess the impacts of changing rainfall on flood frequency, surface hydrology, soil erosion, and forage and crop production and allow managers to develop mitigation plans to reduce degradation to rangeland ecosystems by altering grazing systems and purchasing supplemental feed before the onset of droughts.

*Arid lands may increase.*

Lane *et al.* (1994) reported that trend analyses for the period 1901–87 suggest that mean annual temperatures increased globally at the rate of 0.5°C per century, in the United States at 0.3°C per century, and in the southwestern desert region at about 1.2°C per century. Early climate change predictions suggested that a temperature increase of up to 17% in desert lands could occur in North America (Emanuel *et al.*, 1985). VEMAP Members (1995) considered climate change and doubled $CO_2$ from computer simulations of three different biogeographic models (see Annex C) and three different climate scenarios; a general result was that grasslands would contract and move eastward into the broadleaf forest and that shrublands would decrease within the United States. Potentially large increases (185%) in subtropical arid shrublands could occur in the southwest region of North America. Depending on the model and climate scenario, however, there could be a potential decrease (-56%) in subtropical arid shrublands (VEMAP Members, 1995). The most recent projections from the MAPSS model (Annex C) indicate up to a 200% increase in leaf area index in the desert southwest region of North America and a northern migration and expansion of arid-land species into the Great Basin region of North America. Various combinations of vegetation redistribution and altered biogeochemical cycles could result in novel plant communities and increases in arid regions.

Desertification is a function of human activities and adverse climate conditions. Recovery of desert soils from disturbance and desertification is a slow process. More than 50 years may be required to reestablish the nitrogen fixation capability of the soil, and 200 years may be required to completely reestablish the vegetation community in the arid southern region of North America (Belnap, 1995). During the recovery period, the site is at increased risk of wind and water erosion (Belnap, 1995; IPCC 1996, WG II, Chapter 4). Although long-term measures may need to be developed to cope with climate change, research that deals with annual and interannual fluctuations in precipitation must continue to receive attention because precipitation fluctuations directly affect North American strategic food and fiber supplies (grain production and forage for wildlife and livestock) (Oram, 1989).

**Box 8-1. Examples of Effects on Birds**

The distribution of birds in Canada shows strong correlation with habitat distribution, which in turn is influenced by climate as well as human land-use practices. Thus, changes in climate are expected to have significant effects on breeding and winter distributions. A combination of temperature and moisture considerations is the best predictor of the beginning of the nesting period, defining a "climate space" that can be extended to describe the limits of the breeding range (James and Shugart, 1974). In winter, the northern boundaries of many species coincide with January isotherms, reflecting daily energy requirements. For marine birds, shifts in the distribution of water masses of different temperature and salinity characteristics—supporting different species of prey—are expected to generate the most obvious responses to changing climates (Brown, 1991). For arctic nesting birds, such as geese and many shore birds, the timing of snowmelt is a critical variable that drives the success of nesting, as well as its timing (Boyd, 1987).

**Box 8-2. Examples of Effects on Wildlife**

Changes in the amount, type, and timing of winter precipitation will have considerable consequences for large ungulates (e.g., moose, caribou, elk, deer, and bison), as well as their most important predators (wolves and coyotes). Current distributions of ungulate communities correspond well with their adaptations to the type and depth of snow in the regions in which they occur. Shifts in winter climate could, based on past experience, lead to shifting suitability of ranges for these species. Predator-prey relations also could shift with changing distributions of snow types and amounts; wolves flounder in deep snow that caribou can cross, but moose are no better suited to travel through soft snow.

*Landslides in mountain areas could be more frequent.*

Catastrophic geomorphic processes in mountain terrain are heavily influenced by climate (e.g., precipitation). As a result, the occurrence of these processes—which include landslides and outburst floods—is sensitive to climate change. The frequency of debris flows and other landslide types can be expected to increase under conditions of increased precipitation, debuttressing of mountain slopes due to glacier ice losses, and the decay of mountain permafrost during recent and projected warming. As in the past, these events should be expected to impact on settlements, infrastructural elements, resources, and the environment, resulting in human and financial losses. Water quality would be affected by increased sediment loads. Fish and wildlife habitat, as well as roads and other structures, could be at increased risk. The nature of the landslide response is complicated by factors such as forest harvesting and other land-use changes (Eybergen and Imeson, 1989; IPCC 1996, WG II, Chapter 5; Evans and Clague, 1997).

### *8.3.2. Ecosystems: Forested*

The forests of North America north of Mexico occupy about 732 million hectares, representing about 17% of all global forested lands. Approximately 60% of the North American forests are Canadian (Brooks, 1993). The United States has about 13% of the world's temperate forests and almost half of the world's coastal temperate rainforest (Brooks, 1993). Nearly half of U.S. forests are privately owned, compared to only about 6% in Canada (Brooks, 1993).

Conifers constitute nearly 70% of the world's commercial timber harvest. In North America, conifer species dominate the boreal forests of Canada, Alaska, and the Pacific Northwest and share dominance with hardwoods in the southeastern and northeastern United States. Wood-based manufacturing accounts for about 2% (US$129 billion) of the U.S. gross domestic product (GDP) and about 3% (CAN$23 billion) of the Canadian GDP (Canadian Forest Service, 1996; International Monetary Fund, 1996; U.S. Department of Commerce, 1996).

Forests provide habitats for wildlife and fish, store and regulate freshwater supplies, are the repository of substantial plant and animal genetic resources, and satisfy aesthetic and spiritual values. Recreation activity associated with forests contributes to income and employment in every forested region of North America. Nontimber commodities gathered in forests are sources of income and recreation.

Forests hold about 62–78% of the world's terrestrial biospheric carbon (Perruchoud and Fischlin, 1995), about 14–17% of which is in the forests of North America; about 86% of that is in the boreal forest (Apps *et al.*, 1993; Heath *et al.*, 1993; Sampson *et al.*, 1993).

Forests play a large role in global water and energy feedbacks (Bonan *et al.*, 1995) and account for most of the world's terrestrial evapotranspiration, which is about 64% of the precipitation (Peixoto and Oort, 1992; Neilson and Marks, 1994). Most of the world's freshwater resources originate in forested regions, where water quality is directly related to forest health.

#### *8.3.2.1. Distribution and Sensitivities*

Three broad forest types are recognized in this assessment of North American forests: boreal, temperate evergreen, and temperate mixed forests. The boreal forest (Annex C, Figure C-1) is constrained by cold temperatures to the north that limit tree height and reproduction (Lenihan and Neilson, 1993; Starfield and Chapin, 1996). The southern limits of the boreal forest generally are defined by their juxtaposition with temperate forests or with interior savanna-woodlands and grasslands. Boreal tree species generally are not limited from growing further south. Rather, temperate hardwoods and conifers are limited by cold temperatures from spreading further north; where temperate species can flourish, they outcompete boreal species. Fire and herbivore browsing also are important constraints on forest distribution and species composition (Bergeron and Dansereau, 1993; Landhauser and Wein, 1993; Suffling, 1995; Starfield and Chapin, 1996). Wildfire and insect outbreaks limit forest productivity and can produce considerable mortality: Annual tree mortality losses from insect outbreaks in Canada are about 1.5 times the losses from wildfire and amount to about one-third of the annual harvest volume (Fleming and Volney, 1995). Annual losses from insects and fire in the United States also are about one-third of the annual harvest (Powell *et al.*, 1993). Warming-induced changes in the timing of spring frosts may be important in ending or prolonging outbreaks. Increased drought stress also may enhance insect outbreaks, and changes in climate could extend the ranges of some insects and diseases.

Temperate evergreen forests (Annex C), such as in the Pacific Northwest, tend to occur in areas that are warm enough for photosynthesis during the cool parts of the year but often are too cold for deciduous species to fix sufficient carbon during the frost-free season (Woodward, 1987). Areas with dry summers also tend to favor conifers or hardwoods with water-conserving leaves (Waring and Schlesinger, 1985; Neilson, 1995). Summer drought and winter chilling for frost hardiness are critical climate factors, rendering these forests sensitive to global warming (Franklin *et al.*, 1991). Northwest conifers are long lived and need only successfully reproduce once during their life span for population sustainability (Stage and Ferguson, 1982; Parker, 1986; Savage *et al.*, 1996). With global warming, however, establishment periods could become rare in some areas; after harvest, some forests may not be able to regenerate, even if mature trees could still survive the climate. Winter chilling may be required for adequate seed set or to confer frost-hardiness in some species (Burton and Cumming, 1995); because of the well-recognized spatial variation in the genetics of these species, however, such chilling requirements may not hold everywhere. Fire suppression in interior pine forests has left them in a sensitive condition with respect to drought, fire, and pests (Agee, 1990; Sedjo, 1991). Climate change could exacerbate all of these stressors (Williams and Liebhold, 1995). For example, increased drought stress could facilitate insect outbreaks; drought and infestation could lead to more fuel, increasing the risk of catastrophic fire.

Temperate mixed forests (mixed hardwood and conifer) are bound by cold temperatures to the north and subtropical dry regions to the south (the Caribbean coast in North America) and tend to occur in areas that are wet in both winter and summer. Temperate hardwood species also benefit from cold-hardening; with warmer winter conditions and less insulating snow cover, they can be sensitive to spring frost damage, which can kill roots and further sensitize the species to drought stress and widespread mortality (Auclair *et al.*, 1996; Kramer *et al.*, 1996). Southeastern U.S. pines within this type are among the most important commercial species on the continent. Natural southeastern pine stands historically relied on fire to maintain their composition (Komarek, 1974; Sedjo, 1991) but now are largely controlled by harvest. Compared to northwestern forests, southeastern conifers have a short rotation, which might confer more rapid adaptive capability through establishment of new genotypes.

Elevated $CO_2$ affects the physiology of trees, possibly increasing productivity, nitrogen-use efficiency and WUE (reduced transpiration per carbon fixed, conferring some drought resistance), and other responses (Bazzaz *et al.*, 1996; IPCC 1996, WG II, Section A.2.3). A review of 58 studies indicated an average 32% increase in plant dry mass under a doubling of $CO_2$ concentration (Wullschleger *et al.*, 1995). Norby (1996) documented an average 29% increase in annual growth per unit leaf area in seven broadleaf tree species under $2xCO_2$ scenarios over a wide range of conditions. WUE, examined in another review and indexed by reductions of leaf conductance to water vapor, increased about 30–40% (Eamus, 1991). If such responses were maintained in forests over many decades, they would imply a substantial potential for increased storage of atmospheric carbon, as well as conferring some increased tolerance to drought. However, some species or ecosystems exhibit acclimation to elevated $CO_2$ by downregulating photosynthesis (Bazzaz, 1990; Grulke *et al.*, 1990; Grulke *et al.*, 1993); others do not exhibit acclimation (Bazzaz, 1990; Teskey, 1997). Understanding the sources of large uncertainties in the linkages between forest physiology and site water balance is a research need; no two models simulate these complex processes in the same way.

Most of the early $CO_2$ research was done on juvenile trees in pots and growth chambers, which may limit the usefulness of some conclusions. New research is beginning to emerge that focuses on larger trees or intact forested ecosystems. Recent reviews of this newer literature (Curtis, 1996; Eamus, 1996a) indicate that acclimation may not be as prevalent when roots are unconstrained; that leaf conductance may not be reduced; and that both responses depend on the experimental conditions, the length of exposure, and the degree of nutrient or water stress. These results imply that forests could produce more leaf area under elevated $CO_2$ but may not gain a benefit from increased WUE. In fact, with increased leaf area, transpiration should increase on a per-tree basis, and the stand would use more water (Eamus, 1996a). Elevated temperatures would increase transpiration even further, perhaps drying the soil and inducing a drought effect on the ecosystem (Eamus, 1996a).

Nitrogen supply is prominent among the environmental influences that are thought to moderate long-term responses to elevated $CO_2$ (Kirschbaum *et al.*, 1994; McGuire *et al.*, 1995; Eamus, 1996b). Unless $CO_2$ stimulates an increase in nitrogen mineralization (Curtis *et al.*, 1995; VEMAP Members, 1995), productivity gains in high $CO_2$ are likely to be constrained by the system's nitrogen budget (Körner, 1995). Increased leaf area production is a common $CO_2$ response; however, nitrogen limitations may confine carbon gains to structural tissue rather than leaves (Curtis *et al.*, 1995). Thus, in areas receiving large amounts of nitrogen deposition, a direct $CO_2$ response could result in large increases in leaf area, increasing transpiration and possibly increasing sensitivity to drought via rapid soil-water depletion. Early growth increases may disappear as the system approaches carrying capacity as limited by water or nutrients (Körner, 1995). Shifts in species composition will likely result from different sensitivities to elevated $CO_2$ (Körner, 1995; Bazzaz *et al.*, 1996).

North American forests also are being subjected to numerous other stresses, including deposition of nitrogen and sulfur compounds and tropospheric ozone, primarily in eastern North America (Lovett, 1994). The interactions of these multiple stresses with elevated $CO_2$ and climate change and with large pest infestations (of, for example, the balsam wooly adelgid, gypsy moth, spruce budworm, and others) are very difficult to predict; however, many efforts are under way to address these questions (Mattson and Haack, 1987; Loehle, 1988; Fajer *et al.*, 1989; Taylor *et al.*, 1994; Winner, 1994; Williams and Liebhold, 1995). Anthropogenic nitrogen fixation, for example, now far exceeds natural nitrogen fixation (Vitousek, 1994). Atmospheric nitrogen deposition has likely caused considerable accumulation of carbon in the biosphere since the last century

(Vitousek, 1994; Townsend *et al.*, 1996). However, nitrogen saturation in soils also can be deleterious, possibly causing forest dieback in some systems (Foster *et al.*, 1997). Tropospheric ozone also can damage trees, causing improper stomatal function, root death, membrane leakage, and altered susceptibility to diseases (Manning and Tiedemann, 1995). Such ozone-induced changes can render trees more sensitive to warming-induced drought stress (McLaughlin and Downing, 1995). There are many other stress interactions, and researchers think that, in general, multiple stresses will act synergistically, accelerating change due to other stresses (Oppenheimer, 1989).

Assessments of possible consequences of climate change rely on linked atmospheric, ecological, and economic models. Significant uncertainties are associated with each model type, and these uncertainties may amplify as one moves down the line of linked models. The model capabilities of GCMs have improved significantly from the older (IPCC 1990, WG I, Chapter 3) to the newer (IPCC 1996, WG I, Chapter 6) scenarios, resulting in somewhat lower estimates of the potential $2xCO_2$ climate sensitivity and shifting much of the burden of uncertainty to the ecological and economic models. Ecological models still carry large uncertainties in the simulation of site water balance (among many other issues), particularly with respect to the role of elevated $CO_2$ on plant responses to water stress, competition, and nutrient limitations. Economic models carry uncertainties with respect to future management and technology changes, future per-capita income and available capital, GDP, international trade, and how to couple land-use management with ecological model output, among others. Ecological and economic models are rapidly being enhanced to narrow these uncertainties; improving the linkages between the many different model types necessary to permit fully time-dependent simulations for integrated regional assessments is an ongoing research need.

#### 8.3.2.2. Key Impacts on Forested Ecosystems of North America

*Forest gains as well as forest dieback and decline are projected, with regional differences in the expected response.*

Biogeography models—including a direct physiological $CO_2$ effect under three of the IPCC's First Assessment Report (FAR) $2xCO_2$ equilibrium GCM scenarios (Annex B)—simulate forest area gains of up to 20% over the conterminous United States under the cooler (least warming) or wetter scenarios and forest area losses of as much as 14% under the hottest scenario (VEMAP Members, 1995). The models produced similar forest redistribution patterns, including some conversion of northwest conifers to broadleaf deciduous under potential future equilibrium climates. The models have equal skill in simulating potential natural forest distribution under the present climate; although they diverge to some extent under future climates, they produce similar spatial responses and likely bound the range of forest responses to global warming. Extending these results with the FAR scenarios from the conterminous United States to all of North America using one of the biogeography models indicates that total forest area could increase as much as 32%—but that regions of forest decline or dieback (partial or total loss of trees) could range

***Table 8-2:*** *Potential future forest area (percentage of current) in North America simulated by the MAPSS and BIOME3 biogeography models under three older (IPCC 1990, WG I) equilibrium $2xCO_2$ GCM scenarios and under three newer (IPCC 1996, WG I) transient simulations from which $2xCO_2$ scenarios were extracted (Annex C). The reported ranges include both ecological models under several GCM scenarios. The baseline area estimates are outputs from each model. Because BIOME3 does not differentiate Taiga/Tundra from Boreal Forest, two different aggregations are presented. The Boreal Conifer and Total Forest summaries are MAPSS data only; the "Boreal + Taiga/Tundra" and "Total Forest + Taiga/Tundra" summaries are from both models. Numbers in parentheses are VEMAP results for the conterminous U.S. only, indicating some scenarios with losses in forest area over the U.S., and are based on MAPSS and BIOME2 output (VEMAP Members, 1995).*

| **Forest Type** | **Baseline Area (Mha)** MAPSS | BIOME3 | **With $CO_2$ Effect** FAR Scenarios | SAR Scenarios | **Without $CO_2$ Effect** SAR Scenarios |
|---|---|---|---|---|---|
| Boreal Forest + Taiga/Tundra | 594 | 620 | 65–105% | 64–87% | 28–86% |
| Boreal Conifer Forest | 295 | | 87–150% | 115–116% | 110–112% |
| Temperate Evergreen Forest | 127 | 110 | 130–180% | 78–182% | 82–129% |
| | (82) | (86) | (58–157%) | | |
| Temperate Mixed Forest | 297 | 383 | 107–141% | 146–198% | 129–159% |
| | (245) | (260) | (88–116%) | | |
| *Total Temperate Forest* | 424 | 493 | 114–153% | 137–171% | 121–142% |
| Total Forest + Taiga/Tundra | 1,019 | 1,113 | 102–116% | 107–118% | 99–105% |
| **Total Forest** | 719 | | 125–132% | 142–144% | 121–124% |
| | (327) | (346) | (86–123%) | | |

Note: FAR = First Assessment Report (IPCC 1990, WG I); SAR = Second Assessment Report (IPCC 1996, WG I).
Sources: Mitchell and Warrilow, 1987; Schlesinger *et al.*, 1989; IPCC, 1990; Bengtsson *et al.*, 1995; Mitchell *et al.*, 1995; Bengtsson *et al.*, 1996; IPCC, 1996, WG I, Chapters 5 and 6; Johns *et al.*, 1997.

***Table 8-3:*** *Percentage area of current forests that could undergo a loss of leaf area (i.e., biomass decrease) as a consequence of global warming under various older (IPCC 1990, WG I) and newer (IPCC 1996, WG II) GCM scenarios, with or without direct $CO_2$ effect (see Table 8-2 for details), as simulated by the MAPSS and BIOME3 biogeography models (ranges include both models). Losses in leaf area generally indicate a less favorable water balance (drought).*

| | **With $CO_2$ Effect** | | **Without $CO_2$ Effect** |
|---|---|---|---|
| **Forest Type** | FAR Scenarios | SAR Scenarios | SAR Scenarios |
| Boreal Forest + Taiga/Tundra | 19–40% | 0–9% | 4–45% |
| Boreal Conifer Forest | 37–80% | 14–19% | 79–89% |
| Temperate Evergreen Forest | 20–70% | 2–14% | 41–69% |
| Temperate Mixed Forest | 42–96% | 0–7% | 12–76% |

Sources: Mitchell and Warrilow, 1987; Schlesinger *et al.*, 1989; IPCC, 1990; Bengtsson *et al.*, 1995; Mitchell *et al.*, 1995; Bengtsson *et al.*, 1996; IPCC, 1996, WG I, Chapters 5 and 6; Johns *et al.*, 1997.

from 19–96% of the area of any individual forest type (Tables 8-2 and 8-3). Under hotter scenarios, forests die back from large areas of the conterminous United States (expressing declines in U.S. forest area) but expand into northern Canada and Alaska—so that total North American potential forest area actually increases. These simulated potential forest distributions do not include current or possible future land-use patterns, which will affect actual forest distribution. The models also do not simulate the differential rates of dieback and migration, which could produce near-term losses in total forest area in North America and a large pulse of carbon to the atmosphere.

Using more recent climate change scenarios (IPCC 1996, WG I, Chapter 6; Annex B and Annex C of this report), forest decline and dieback ranged only from 0–19% of the individual forest areas (Table 8-3) when a direct $CO_2$ effect was included. However, if the direct $CO_2$ effect is withheld under the newer scenarios, dieback could be quite extensive in all forest types—with a range of 12–89% of the forest area—with large range contractions in all major forest zones. Results without a $CO_2$ effect under the IPCC's Second Assessment Report (SAR) scenarios are similar in magnitude to results with a $CO_2$ effect under the FAR scenarios (Annex B). Because the full $2xCO_2$ effect may be less than fully realized, the potential forest response is bounded by simulating forests with and without a $CO_2$ effect. It is worth noting that by including sulfate aerosols in one SAR scenario, warming over the temperate mixed forest type in eastern North America is reduced, and there is less simulated forest dieback.

The SAR scenarios were produced from GCMs operated in a fully transient mode with gradual increases in greenhouse gases while coupled to a dynamic three-dimensional ocean. The biogeography models can only simulate equilibrium conditions, so average climate statistics were extracted from the simulations for a current-climate period and a period representing the time of $2xCO_2$ forcing. At the time of $2xCO_2$ forcing in the SAR scenarios, however, the GCMs had not equilibrated because of thermal inertia of the oceans and had only attained about 50–80% of their potential equilibrium temperature change. Thus, the SAR scenarios tend to be somewhat cooler than the FAR scenarios, which are run to equilibrium over a mixed-layer ocean. Nevertheless, when allowed to equilibrate to $2xCO_2$ forcing, the SAR scenarios exhibit a global temperature sensitivity that is similar to the FAR scenarios (Annex B). Thus, the less deleterious or even beneficial impacts simulated under the SAR scenarios may simply be precursors to the more severe impacts simulated under the equilibrium FAR scenarios.

Assessments using gap models under the equilibrium FAR GCM scenarios or sensitivity analyses have been completed for eastern North America (Solomon, 1986), the southeastern United States (Urban and Shugart, 1989), British Columbia (Cumming and Burton, 1996), central Canada (Price and Apps, 1996), and the Great Lakes region (Pastor and Post, 1988; Botkin *et al.*, 1989). Gap models differ from equilibrium biogeography models by simulating temporal dynamics of forests at a point; they therefore are able to simulate forest decline, dieback, or enhanced growth, as well as changes in species composition. The different models showed either increases or decreases in biomass depending on the method used for water balance calculations (Bugmann *et al.*, 1996). Slight variations in model structure, small differences in soil texture, or the method for implementing direct $CO_2$ effects can affect the magnitude and direction of change in simulated productivity and biomass storage (Martin, 1992; Post *et al.*, 1992).

Gap model results using the FAR GCM scenarios indicate the potential for significant forest dieback related to high temperatures, throughout the eastern United States—comparable to the more severe biogeography model simulations (Solomon, 1986; Pastor and Post, 1988; Botkin *et al.*, 1989; Urban and Shugart, 1989). Given cooler SAR climate scenarios and improved gap model technology, however, these older dieback results may be too severe (Fischlin *et al.*, 1995; Bugmann *et al.*, 1996; Loehle and LeBlanc, 1996; Martin, 1996; Oja and Arp, 1996; Pacala *et al.*, 1996; Post and Pastor, 1996; Shugart and Smith, 1996). Moreover, most of the gap model results do not include a direct $CO_2$ effect.

Studies using a regional forest-growth model suggest that forests in the northeastern United States might grow more (Aber *et al.*, 1995), while forests in the southeastern United

States could experience considerable forest dieback (McNulty *et al.*, 1996). These simulations incorporate a direct $CO_2$ effect but do not consider vegetation redistribution.

Gap model simulations in the boreal forest indicate increased biomass under cooler scenarios (less warming) but some dieback or shifts to drier types under the warmer scenarios (Price and Apps, 1996). Within the temperate evergreen zone of British Columbia (and throughout North America), gap models indicate that forests could shift upward in elevation and possibly disappear from some zones because of the lack of winter cooling for forest regeneration, increased sensitivity to spring frosts, and drought stress (Cumming and Burton, 1996). Where simulated dieback occurs in any of the forest zones, it tends to occur within a few decades of initial warming—with, for example, a 30% reduction in biomass within 50 years in the Douglas fir zone (Cumming and Burton, 1996).

Thus, two contrasting scenarios of the North American forest future must be considered: one with considerable forest dieback, another with much enhanced forest growth. These contrasting scenarios represent endpoints on a spectrum of possible responses. In general, however, the enhanced-growth scenarios occur under the least amount of simulated global warming, whereas the severe decline or dieback scenarios occur under the greatest projected global warming. With the incorporation of direct $CO_2$ effects, small temperature increases can produce increased growth, but larger temperature increases still produce declines. Without a direct $CO_2$ effect, forest decline simulations are far more widespread, even under the least warming scenarios. These results suggest the possibility that early forest responses to global warming could exhibit enhanced growth; later stages could produce widespread decline or dieback. Most combinations of scenarios and $CO_2$ effects produce intermediate scenarios, with a regional mosaic of forest dieback and enhanced forest growth. When coupled with economic models, these internally consistent but potentially opposite regional responses provide the basis for regional, national, and globally integrated assessments. Also, it is not clear that greenhouse gases will stabilize at the equivalent of $2xCO_2$ forcing; they could increase to $3xCO_2$ or $4xCO_2$ (IPCC 1996, WG I, Section 2.1.3).

*Forests cannot move across the land surface as rapidly as the climate can. The faster the rate of climate change, the greater the probability of ecosystem disruption and species extinction.*

Were temperature-induced drought dieback to occur, it likely would begin shortly after observable warming; if accompanied by short-term precipitation deficits, it could occur very rapidly (Solomon, 1986; King and Neilson, 1992; Martin, 1992; Smith and Shugart, 1993; Vose *et al.*, 1993; Elliot and Swank, 1994; Auclair *et al.*, 1996; Martin, 1996). That is, dieback could begin within a few decades from the present and might include potential increases in secondary impacts from pests and fire. Alternatively, forest growth might increase in the early stages of global warming, only to revert to widespread and rapid drought-induced forest dieback after higher temperatures have significantly increased evaporative demand. Vegetation change in areas of enhanced growth, especially previously unforested areas, would be more gradual (decades to hundreds of years), constrained by dispersal, establishment, and competition.

Under global warming the physical and biotic components of most animal habitats will likely change at different rates (Davis, 1986; Dobson *et al.*, 1989; Malcolm and Markham, 1996; Markham, 1996). The faster the rate of change, the greater the disequilibrium between physical and biotic habitat components and the higher the probability of substantial ecosystem disruption and species extinctions (Malcolm and Markham, 1996; Markham, 1996). However, species will respond differently than biomes (Neilson, 1993a,b; Lenihan and Neilson, 1995). The relative mixtures of species in forest communities will change—and under either forest expansion or contraction, some important species could be at risk.

*Forest ecosystems are expected to shift northward and upward in altitude, but expansion may be limited by dispersal and poor soils.*

All three major forest types within North America expand north and forested areas, with a few exceptions, increase under all scenarios with or without a direct $CO_2$ effect (biogeography models of potential natural forests under equilibrium conditions). Total forest area increases by as much as 25–32% as projected under the FAR $2xCO_2$ GCM scenarios (including a direct, physiological $CO_2$ effect)—much less than the 42–44% under the SAR scenarios with a direct physiological $CO_2$ effect. However, the projected forest-area increases under the SAR scenarios are reduced to 21–24% when a direct physiological $CO_2$ effect is not included. In the long term, more carbon would be sequestered by forests under these scenarios. Before equilibrium conditions are reached, however, the processes of forest redistribution could cause a temporary reduction in forest area and a carbon pulse to the atmosphere.

Boreal forests displace most of the taiga/tundra region and increase in area under the SAR scenarios but are projected to increase or decrease under the FAR scenarios (Table 8-2). It has been projected that the temperate evergreen forests shift northward into Canada and Alaska and expand under the climate projected by the FAR scenarios (Annex C, Figures C-2 to C-5; Table 8-2). Temperate evergreen forests may expand or contract in area—due in part to conversion from conifers to broadleaf deciduous forests and in part to severe forest dieback under some scenarios (Annex C, Figures C-2 to C-5; Table 8-2; VEMAP Members, 1995). The temperate mixed forest is projected to invade the boreal forest to the north and experience gains in area under all simulations. Smaller gains in area of both temperate forest types occur under more xeric scenarios; forest expansions to the north are balanced by forest dieback in the southern zones. Because dieback in the southern zones might occur more rapidly than northward advances, there could be a short-term reduction in the area of important temperate and boreal forests (King and Neilson, 1992; Smith and Shugart, 1993).

Changes in leaf area can be used to infer changes in forest biomass (Annex C, Figures C-6 to C-9; Tables 8-3 and 8-4). Because taiga/tundra is a low-density, ecotonal region, expansion of forests into that region always produces a dramatic increase in biomass. Under earlier FAR scenarios with a direct $CO_2$ effect, all three forest types have subregions (ranging from only 3% to slightly over half of the area) that undergo an increase in forest biomass (Table 8-4). Under newer SAR scenarios, however, the area of increased forest biomass is always well over 50%—ranging to as high as 97% in the temperate mixed forest (with a direct $CO_2$ effect). Without a direct $CO_2$ effect under the newer scenarios, areas of increased forest biomass range only from 2–49% (usually in the lower end of the range) and are similar to those projected under the earlier scenarios, which included a $CO_2$ effect.

All studies agree that where forests are limited by cold they will expand beyond current limits, especially to the far north. Whether forests will expand into the drier continental interior or contract away from it, however, depends on hydrological factors and remains uncertain. Vegetation distribution models that incorporate a direct physiological $CO_2$ effect indicate considerable expansion of all forest types into drier and colder areas and much enhanced growth over most areas—under the newer climate scenarios as well as some of the older studies. Under most of the FAR scenarios with a $CO_2$ effect and under the SAR scenarios without a $CO_2$ effect, however, forests would contract away from the continental interior because of increased drought stress.

*Longer fire seasons and potentially more frequent and larger fires are likely.*

Fire mediates rapid change and could increase in importance for vegetation change. Future climate scenarios could result in longer fire seasons and potentially more frequent and larger fires in all forest zones (even those that currently do not support much fire) because of more severe fire weather, changes in fire-management practices, and possible forest decline or dieback (Fosberg, 1990; Flannigan and Van Wagner, 1991; King and Neilson, 1992; Wotton and Flannigan, 1993; Price and Rind, 1994; Fosberg *et al.*, 1996).

Fire suppression during much of the 20th century has allowed biomass in many interior forests to increase by considerable amounts over historic levels (Agee, 1990). With increased biomass, forests transpire almost all available soil water; they become very sensitive to even small variations in drought stress and are very susceptible to catastrophic fire, even without global warming (Neilson *et al.*, 1992; Stocks, 1993; Stocks *et al.*, 1996). Forests in the interior of North America are experiencing increased frequencies of drought stress; pest infestations; and catastrophic, stand-replacing fires (Agee, 1990). This sequence of events is a reasonable analog for what could happen to forests over much larger areas in the zones projected by biogeography models to undergo a loss of biomass or leaf area as a consequence of temperature-induced transpiration increases and drought stress (Annex C, Figures C-6 to C-9; Table 8-3) (Overpeck *et al.*, 1990; King and Neilson, 1992).

*Enhanced fire and drought stress will facilitate changes in species composition and may increase atmospheric carbon contributions from forests.*

Given the ownership patterns and remote nature of much of the boreal forest lands, they are generally managed as natural systems. Even highly managed temperate forests are of such large extent that a rapid, large-scale management response would be logistically quite difficult and expensive.

On managed lands, harvesting of dead or dying trees, more rapid harvesting or thinning of drought-sensitive trees, and planting of new species could reduce or eliminate species loss or productivity declines. However, identification of which species to plant (and when) under a rapidly changing climate will be difficult management issues. The more rapid the rate of climate change, the more it may strain the ability to create infrastructure for seeding or planting of trees or support the supply of timber if there is a large amount of salvage. The fast rate of warming may limit some species that have slow dispersal rates or are constrained by human barriers, habitat fragmentation, or lack of suitable habitat—or already are stressed by pollution.

As fire-management agencies operate with increasingly constrained budgets, it is likely that any increases in fire frequency

***Table 8-4:*** *Percentage area of current forests that could undergo a gain of leaf area (i.e., biomass increase) as a consequence of global warming under various older (IPCC 1990, WG I) and newer (IPCC 1996, WG II) GCM scenarios, with or without direct $CO_2$ effect (see Table 8-2 for details), as simulated by the MAPSS and BIOME3 biogeography models (ranges include both models). Gains in leaf area generally indicate a more favorable water balance.*

| Forest Type | With $CO_2$ Effect FAR Scenarios | With $CO_2$ Effect SAR Scenarios | Without $CO_2$ Effect SAR Scenarios |
|---|---|---|---|
| Boreal Forest + Taiga/Tundra | 57–70% | 74–91% | 47–51% |
| Boreal Conifer Forest | 16–42% | 53–54% | 1–3% |
| Temperate Evergreen Forest | 11–49% | 52–79% | 7–16% |
| Temperate Mixed Forest | 3–53% | 92–97% | 2–49% |

Sources: Mitchell and Warrilow, 1987; Schlesinger *et al.*, 1989; IPCC, 1990; Bengtsson *et al.*, 1995; Mitchell *et al.*, 1995; Bengtsson *et al.*, 1996; IPCC, 1996, WG I, Chapters 5 and 6; Johns *et al.*, 1997.

or severity will result in a disproportionately large increase in area burned (Stocks, 1993). More and larger boreal fires will result in a reevaluation of protection priorities, with likely increased protection of smaller, high-value areas and reduced protection over large expanses. If forests die back from drought, infestations, or fire in extensive, remote regions, the impacts could include large-scale changes in nutrient cycling and carbon sequestration, as well as loss of value for future timber harvests or as habitat for wildlife and biodiversity. Some adaptive practices—such as harvesting dead or dying trees or thinning—could impact biodiversity, soil erosion, stream quality, and non-market forest products, producing potentially conflicting management options.

Markham and Malcolm (1996) have concluded that ecological resiliency can be increased by conserving biological diversity, reducing fragmentation and degradation of habitat, increasing functional connectivity among habitat fragments, and reducing anthropogenic environmental stresses. They also indicate that adaptation strategies should include redundancy of ecological reserves, reserves with much structural heterogeneity, and the flexibility to spatially relocate habitat protection depending on shifts in future climate (Peters and Darling, 1985; McNeely, 1990). Current habitat fragmentation patterns and human barriers may hinder species migration. Thus, management of the "seminatural matrix" may play an increasing role in fostering species redistribution (Peters and Darling, 1985; Bennett *et al.*, 1991; Franklin *et al.*, 1991; Parsons, 1991; Simberloff *et al.*, 1992).

### *8.3.3. Hydrology and Water Resources*

#### *8.3.3.1. Hydrological Trends and Variability*

Several reports of recent trends in precipitation and streamflow have shown generally increasing values throughout much of the United States; in Canada, total precipitation trends indicate an increase, but monthly streamflow analyses show varying seasonal changes. Lettenmaier *et al.* (1994) analyzed data over the period 1948–88 and found generally increasing trends in precipitation during the months of September to December and increasing trends in streamflow during the months of November to April, particularly in the central and north-central portions of the United States. Similarly, Lins and Michaels (1994) report that streamflow has increased throughout much of the conterminous United States since the early 1940s, with the increases occurring primarily in autumn and winter. In Louisiana, precipitation and simulated runoff (streamflow per unit drainage area) have increased significantly over the past 100 years (Keim *et al.*, 1995).

Mekis and Hogg (1997) analyzed annual and seasonal precipitation (total, rain and snow) trends for periods from 1948–96 to 1895–1996 for regions of Canada and noted significant increases in total annual precipitation and snow for most regions. In Ontario, 41 hydrometric stations with a minimum of 30 years of data ending in 1990 were analyzed by Ashfield *et al.* (1991). Mean monthly flows increased for the period September to January in more than 50% of the stations; approximately 25% of the stations show a downward trend in flow for the April to September period. Anderson *et al.* (1991) analyzed low-, average-, and maximum-flow time series for 27 stations (unregulated flow) across Canada; the data show a decrease in summer low flows and an increase in winter average and low flows but little trend in seasonal maximum flows. Burn (1994) analyzed the long-term record of 84 unregulated river basins from northwestern Ontario to Alberta for changes in the timing of peak spring runoff. In the sample, the more northerly rivers exhibited a trend to earlier spring snowmelt runoff; the observed impacts on timing were more prevalent in the recent portion of data. These trends generally are consistent with climate models that produce an enhanced hydrological cycle with increasing atmospheric $CO_2$ and warmer air temperatures, although some of the streamflow trends also may be the result of water-management or land-use changes that reduce surface infiltration and storage.

Recent investigations have shown how natural modes of variability at scales from seasons to years (e.g., ENSO, Pacific Decadal Oscillation) affect hydrological variability in different regions of North America and thereby have underlined the importance of increasing our understanding of the roles these features play in influencing hydrological characteristics. The ENSO phenomenon, a predictable climate signal, affects precipitation and streamflow in the northwestern, north-central, northeastern, and Gulf coast regions of the United States (Kahya and Dracup, 1993; Dracup and Kahya, 1994). For example, La Niña events (the cold phase of the ENSO phenomenon) produce higher than normal precipitation in winter in the northwestern United States, whereas El Niño events (the warm phase of the ENSO phenomenon) cause drier winters in the Northwest on roughly a bidecadal time scale. Precipitation over a large region of southern Canada extending from British Columbia through the prairies and into the Great Lakes shows a distinct pattern of negative precipitation anomalies during the first winter following the onset of El Niño events; positive anomalies occur in this region with La Niña events. On the other hand, the northern prairies and southeastern Northwest Territories show significant positive precipitation anomalies with El Niño events (Shabbar *et al.*, 1997). Variability in ENSO phenomena contributes natural variations in hydrology at decadal and longer time scales that are problematic for $CO_2$ climate change analysis. Changes in ENSO behavior related to increasing $CO_2$ are highly uncertain but could produce enhanced variability in precipitation and streamflow for the regions most sensitive to ENSO fluctuations (IPCC 1996, WG I, Section 6.4.4).

Wetlands in North America traditionally have been viewed as wasted land available for conversion to more productive use. This opinion has contributed to the loss of millions of wetland hectares that have been drained or filled for agriculture, highways, housing, and industry. In Canada, where wetlands occupy an estimated 14% of the landscape, 65–80% of Atlantic coastal marshes, southern Ontario wetlands, prairie potholes, and the Fraser River delta have been lost—largely to agriculture (Environment Canada, 1986, 1988). Figures for the United States indicate that approximately 53% of the original wetland area in the lower 48 states has been lost, mostly (87% of this figure) to agriculture (Maltby, 1986). These losses are accompanied by the loss of ecological, hydrological, and cultural functions wetlands provide, including water purification, groundwater recharge/discharge, stormwater storage/flood control, sediment and pollutant sequestering, carbon storage, cycling of sulfur, and wildlife habitat (Mitsch and Gosselink, 1986; IPCC 1996, WG II, Chapter 6).

Socioeconomically, wetlands provide direct benefits through the harvesting of timber, wild rice, cranberries, and horticultural peat—as well as through recreational activities such as hunting, fishing, and bird watching. The cultures and spiritual values of many First Nation peoples are linked to the health of wetlands.

#### 8.3.3.2. Impacts, Adaptations, and Vulnerabilities

Important vulnerabilities of water resources to potential climate change scenarios involve changes in runoff and streamflow regimes, reductions in water quality associated with changes in runoff, and human demands for water supplies.

*Seasonal and annual runoff may change over large regions as a result of changes in precipitation or evapotranspiration.*

Runoff is simply the area-normalized difference between precipitation and evapotranspiration; as such, it is a function of watershed characteristics, the physical structure of the watershed, vegetation, and climate. Although most climate change models show increases in precipitation over much of North America, rates of evaporation and perhaps transpiration also are likely to increase with increasing temperatures. Therefore, regions in which changes in precipitation do not offset increasing rates of evaporation and transpiration may experience declines in runoff and consequently declines in river flows, lake levels, and groundwater recharge and levels (Schindler, 1997). Alternatively, regions that experience substantial increases in precipitation are likely to have substantial increases in runoff and river flows.

Projected changes in annual discharge (summarized in Table 8-5) for some river basins in North America using various climate

***Table 8-5:*** *Summary of annual runoff impacts from climate change scenarios*

| Region/River Basin | Scenario Method | Hydrological Changes (annual) | Reference(s) |
|---|---|---|---|
| *East-Central Canada* | | | |
| St. Lawrence, Ontario and Quebec | GCM: CCC92 | -34% | Croley (1992) |
| Opinaca-Eastmain, Quebec | GCM: GISS84, GFDL80 | +20.2%, +6.7% | Singh (1987) |
| La Grande, Quebec | GCM: GISS84, GFDL80 | +15.6%, +16.5% | Singh (1987) |
| Caniapiscau, Quebec | GCM: GISS84, GFDL80 | +13.0%, +15.7% | Singh (1987) |
| Moise, Quebec | GCM: CCC92 | -5% | Morin and Slivitzky (1992) |
| Grand, Ontario | GCM: GISS87, GFDL87, CCC92 | -11%, -21%, -22% | Smith and McBean (1993) |
| *Canadian Prairie* | | | |
| Saskatchewan | GCM: GISS87[1] | +28%, +35% | Cohen *et al.* (1989); Cohen (1991) |
| | GCM: GFDL87[1] | -27%, -36% | |
| | GCM: OSU88[1] | +2%, -4% | |
| *Northwest Canada* | | | |
| Mackenzie | GCM: CCC92, GFDL-R30 | -3 to -7% | Soulis *et al.* (1994) |
| | analog | +7% | |
| *Mid-Atlantic USA* | | | |
| Delaware | GCM: GISS, GFDL, OSU | -5 to -38% (soil moisture index) | McCabe and Wolock (1992) |
| *Western USA* | | | |
| Upper Colorado | GCM: GISS, GFDL, UKMO | -33 to +12% | Nash and Gleick (1993) |

[1]Includes low and high irrigation.

Sources: CCC92 (Boer *et al.*, 1992; McFarlane *et al.*, 1992), OSU88 (Schlesinger and Zhao, 1988), GFDL87 (Manabe and Wetherald, 1987), GISS87 (Cohen, 1991), GISS84 (Cohen, 1991), GFDL80 (Cohen, 1991).

change scenarios indicate potential increases as well as declines. (Many of these hydrological impact assessments, however, were developed using older climate change scenarios of somewhat larger increases in global air temperature than the most recent scenarios that include regional aerosol-cooling effects.) Seasonal changes in runoff also could be substantial. Most climate change scenarios suggest increased winter precipitation over much of North America, which could result in increased runoff and river flows in winter and spring. Several climate change scenarios show declines in summer precipitation in some regions (e.g., the southeastern United States; IPCC 1996, WG I, Figure 6.11) or declines in summer soil-moisture levels (e.g., over much of North America; IPCC 1996, WG I, Figure 6.12), which could result in significant declines in summer and autumn runoff in these regions. However, climate change scenarios showing summer declines in precipitation or soil-moisture levels in these regions generally are produced from simulations with doubled $CO_2$ forcing alone; when aerosol forcing is included, summer precipitation and soil-moisture levels increase only slightly. This pattern highlights the large uncertainty in climate change projections of runoff.

Although large increases in annual runoff will affect flooding and flood management, large reductions may pose more serious threats to uses such as potable drinking water, irrigation, assimilation of wastes, recreation, habitat, and navigation. The greatest impact of declines in supply are projected to be in arid and semi-arid regions and areas with a high ratio of use relative to available renewable supply, as well as in basins with multiple competing uses. For example, reductions in outflow of 23–51% from Lake Ontario from assessments using four GCM scenarios suggest impacts on commercial navigation in the St. Lawrence River and the port of Montreal, as well as hydropower generation (Slivitzky, 1993). Lower flows also may affect the ecosystem of the river by allowing the saltwater wedge to intrude further upstream.

*Seasonal patterns in the hydrology of mid- and high-latitude regions could be altered substantially, with runoff and streamflows generally increasing in winter and declining in summer.*

Higher air temperatures could strongly influence the processes of evapotranspiration, precipitation as rain or snow, snow and ice accumulation, and melt—which, in turn, could affect soil moisture and groundwater conditions and the amount and timing of runoff in the mid- and high-latitude regions of North America. Higher winter temperatures in snow-covered regions of North America could shorten the duration of the snow-cover season. For example, one climate change scenario (CCC, Annex B) indicates up to a 40% decrease in the duration of snow cover in the Canadian prairies and a 70% decrease in the Great Plains (Boer *et al.*, 1992; Brown *et al.*, 1994). Warmer winters could lead to less winter precipitation as snowfall and more as rainfall, although increases in winter precipitation also could lead to greater snowfall and snow accumulation, particularly at the higher latitudes. Warmer winter and spring temperatures could lead to earlier and more rapid snowmelt and earlier ice break-up, as well as more rain-on-snow events that produce severe flooding, such as occurred in 1996–97 (Yarnal *et al.*, 1997).

Damages to structures, hydropower operations, and navigation and flooding caused by late-winter and spring ice-jam events are estimated to cost CAN$60 million annually in Canada and US$100 million in the United States. About 35% of flooding in Canada is caused by ice jams—principally in the Atlantic Provinces, around the Great Lakes, in British Columbia, and in northern regions (Beltaos, 1995). Northern deltas and wetlands, however, depend upon flooding for periodic recharge and ecological sustainability (Prowse, 1997). The $2xCO_2$ GCM simulations (IPCC 1996, WG I, Summary for Policymakers) suggest milder winters in higher latitudes and a general pattern of increased precipitation, with high regional variability. Where warmer winters result in reduced ice thickness, less severe breakups and reduced ice-jam flooding can be expected. However, major changes in precipitation patterns also are predicted. In some regions, there is an increased likelihood of winter or early spring rains. These climatic factors trigger sudden winter thaws and premature breakups that have the greatest potential for damage. Thus, although average conditions may be improved, the severity of extreme events in some regions appears likely to increase. In the more southern latitudes commonly affected by spring ice jams—such as the lower Great Lakes and central Great Plains areas, parts of New England, Nova Scotia, and British Columbia—there may be a reduction in the duration and thickness of the ice cover on rivers, as well as in the severity of ice jamming. In the north, similar effects are expected. In the intermediate latitudes—such as the prairies; much of Ontario and Quebec; and parts of Maine, New Brunswick, Newfoundland, and Labrador—spring jamming may become more common and/or severe. Such events are presently rare or completely unknown in some of these areas (Van Der Vinne *et al.*, 1991; Beltaos, 1995).

In mountainous regions, particularly at mid-elevations, warming could lead to a long-term reduction in peak snow-water equivalent, with the snowpack building later and melting sooner (Cooley, 1990). Glacial meltwater also is a significant source of water for streams and rivers in some mountainous regions, with the highest flows occurring in early or midsummer (depending on latitude). For example, glacial meltwater contributes an average of 85% of the August flow in the Mistaya River near Banff, Alberta (Prowse, 1997). Accelerated glacier melt caused by temperature increases means more runoff in the short term, but loss of glaciers could result in streams without significant summer flow in the future (IPCC 1996, WG II, Sections 7.4.2 and 10.3.7). Late-summer stream discharge could decrease suddenly within only a few years. A steady pattern of glacial retreat is apparent in the southern Rocky Mountains below central British Columbia and Alberta. Water supplies in small communities, irrigation, hydroelectric generation, tourism, and fish habitat could be negatively impacted (IPCC 1996, WG II, Chapter 7; Brugman *et al.*, 1997; Prowse, 1997).

In Arctic regions, permafrost maintains lakes and wetlands above an impermeable frost table and limits subsurface water storage. As described in Section 8.3.1, discontinuous and continuous permafrost boundaries are expected to move poleward as a result of projected changes in climate. Thawing of permafrost increases active-layer storage capacity and alters peatland hydrology. Although climatic warming could have a large effect on Arctic hydrology, the changes are highly uncertain at this time.

In general, increases in winter and early spring temperatures under a $2xCO_2$ climate could shift hydrological regimes toward greater flows in winter and early spring and lower flows in summer in the mid- and high-latitude regions of North America (Ng and Marsalek, 1992; Soulis *et al.*, 1994). River and reservoir systems that are fed by snowmelt or rely on glacier melt for spring and summer flow during critical periods of high agricultural and municipal demand and low precipitation may have critical supply-demand mismatches. California and the Great Plains and prairie regions of Canada and the United States could be particularly vulnerable (Cohen *et al.*, 1989; Gleick, 1993).

*Altered precipitation and temperature regimes will affect the seasonal pattern and variability of water levels of wetlands, thereby affecting their functioning—including flood protection, carbon storage, water cleansing, and waterfowl/wildlife habitat.*

It is difficult to generalize about the sensitivity of wetlands to climate change (IPCC 1996, WG II, Chapter 6). It appears, however, that climate change will have its greatest effect through alterations in hydrological regimes—in terms of the nature and variability of the hydroperiod (the seasonal pattern of water level) and the number and severity of extreme events (Gorham, 1991; Poiani and Johnson, 1993). However, other variables related to changing climate may drive a site-specific response. Such variables include increased temperature and altered evapotranspiration, altered amounts and patterns of suspended sediment loadings, fire, oxidation of organic sediments, and the physical effects of wave energy (Mitsch and Gosselink, 1986; IPCC 1996, WG II, Chapter 6).

There are many highly significant social and economic threats to wetlands, but there is insufficient information on the precise nature of anticipated local changes in climate. This difficulty prevents accurate assessment of risks or opportunities to adapt. The responses of affected wetlands are expected to vary; they might include migration of the wetland area along river edges or the slope of a receding lake and/or altered species composition. More serious effects would include altered physical characteristics; degradation to a simpler, less diverse form; or complete destruction. There also could be a loss of desired attributes, such as their ability to provide suitable habitat for particular species; their ability to act as a feeding or breeding area in support of an adjacent open-water commercial or recreational fishery; or their ability to buffer occasional flooding (Mitsch and Gosselink, 1986; IPCC 1996, WG II, Chapter 6). Altering climate and acid depositions can cause declining levels of dissolved organic carbon (DOC) in wetlands—thus increasing the water volumes, sediment areas, and associated organisms exposed to harmful ultraviolet-B (UV-B) irradiation. Potential effects include changes in aquatic communities and photoinhibition of phytoplankton (Schindler *et al.*, 1996; Yan *et al.*, 1996).

*Additional losses of prairie pothole wetlands could reduce migratory waterfowl and wildlife populations.*

Occupying depressions in the landscape in dry climates with small watershed areas, prairie pothole wetlands are highly susceptible to a lack of moisture occurring through the effects of decreased snowpack and associated spring recharge, droughts, and increased climatic variability. Already strained by losses of 71% in Canada (Environment Canada 1986, 1988) and 50–60% in the United States (Leitch, 1981), this area yields 50–75% of all the waterfowl produced in any year in North America (Leitch and Danielson, 1979; Weller, 1981). Trends in Canadian duck abundance already reflect the interactions between changing wetness regimes and landscape alterations (Bethke and Nudds, 1995). Any additional stress would be of great concern and could be accommodated only through active programs to protect, enhance, and increase wetland areas in this region.

*Increases in the frequency or magnitude of extreme hydrological events could result in water quality deterioration and water management problems.*

Hydrological variability (i.e., the frequency and magnitude of extreme events) is an extremely important issue for the management of water resources. Under a warmer climate, the hydrological cycle is projected to become more intense, leading to more heavy rainfall events (IPCC 1996, WG I, Section 6.5.6). Several $2xCO_2$ GCM simulations have indicated an increase in the magnitude of mean rainfall events, particularly for central and northwest North America, even with small changes in mean annual rainfall (Cubasch *et al.*, 1995; Gregory and Mitchell, 1995; Mearns *et al.*, 1995; IPCC 1996, WG I, Section 6.5.7). In addition, these simulations indicate increases in the length of dry spells (consecutive days without precipitation). However, few model simulation analyses have addressed the issue of variability in daily precipitation and increases in the frequency or severity of extreme hydrological events; as a result, issues surrounding variability and extreme hydrological events remain highly uncertain at this time.

In many regions, projected increases in hydrological variability would result in greater impacts on water resources than changes in mean hydrological conditions (IPCC 1996, WG II, Chapter 10 Executive Summary). Increases in the frequency or magnitude of extreme rainfall events would likely have their greatest impacts on water resources in the winter and spring, when the ground is frozen or soil moisture levels are high; severe flooding may be more likely. More severe or frequent floods could result in increased erosion of the land surface, as well as stream channels and banks; higher sediment loads and increased sedimentation of rivers and reservoirs; and increased loadings of nutrients and contaminants from agricultural and urban areas (IPCC 1996, WG II, Section 10.5.5). Longer dry spells would likely have their greatest impact in the summer, when streamflows generally are low. Increases in the severity of summer droughts could result in reduced water quality (e.g., lower dissolved oxygen concentrations, reduced dilution of effluents) and impaired biological habitat (e.g., drying of streams, expansion of zones with low dissolved oxygen concentrations, water temperatures exceeding thermal tolerances) (IPCC 1996, WG II, Sections 10.5.3 and 10.5.4).

Projected increases in hydrological variability (e.g., more frequent or larger floods) could lead to increased expenditures for flood management and disaster assistance (IPCC 1996, WG II, Section 14.4.3). Flood-control structures might require modifications to accommodate larger probable maximum-flow events. Alternatives to structural flood-control measures can be instituted to reduce risk at a lower cost to society, but these strategies require significant political will. Even with the high frequency of extreme events that have occurred recently (and their attendant costs), changes to less-costly and more effective nonstructural methods of risk reduction are slow in gaining acceptance. Navigation also might be impaired by changes in hydrological variability, as illustrated by severe restrictions on navigation in the Mississippi River during the drought of 1988 (IPCC 1996, WG II, Section 14.3.4). Hydropower generation could be severely restricted during droughts. It is estimated that hydropower production might decline 20% during the peak-load summer months in northern California as a result of more severe droughts (IPCC 1996, WG II, Section 14.3.3). More severe summer droughts also could increase agricultural irrigation demands (IPCC 1996, WG II, Section 14.3.1) and exacerbate current drinking water supply problems in some large urbanized areas, such as California and Houston, Texas (IPCC 1996, WG II, Section 12.3.5).

*Lake water levels could decline in regions such as the Arctic and the Great Lakes where precipitation increases do not compensate for warming-enhanced evaporation rate increases.*

Lake levels are sensitive to changes in precipitation and evaporation, which lead to changes in streamflow and groundwater flow. Changes in lake levels will depend on regional changes in temperature and precipitation (the latter of which is highly uncertain). In areas where climate change scenarios suggest that precipitation or soil moisture levels could decline, lake levels are likely to decline or to fluctuate more widely (IPCC 1996, WG II, Section 10.5.2). Decreases in precipitation and/or soil moisture are indicated for the southeastern United States and much of the midcontinental region of North America in several 2x$CO_2$ GCM simulations with only $CO_2$ forcing. Water-level declines would be most severe in lakes and streams in dry evaporative drainage basins and basins with small catchments. Semipermanent prairie sloughs are fed by groundwater in addition to precipitation and spring snowmelt. Severe droughts deplete groundwater storage and cause these sloughs to dry out—resulting in turn in a decline of bird habitats (Poiani and Johnson, 1991, 1993). In the north-central United States, some drainage lakes and seepage lakes are highly responsive to precipitation; lake levels declined substantially during the late-1980s drought (Eilers *et al.*, 1988).

High-latitude lakes also may be particularly vulnerable to changes in precipitation and temperature. For a 2x$CO_2$ climate change scenario with temperature increases of 3–5°C and precipitation increases of 10–15%, lake levels in the Mackenzie delta of arctic Canada fluctuate more widely. If precipitation were to decline by 10% with these temperature increases, however, many lakes could disappear within a decade as a consequence of decreased flood frequency (Marsh and Lesack, 1996).

The Great Lakes of North America are a critically important resource, and potential climate change effects are of great concern. Based on 2x$CO_2$ scenarios from several GCMs that indicated seasonal temperature increases of 2.6–9.1°C and seasonal precipitation changes of -30% to +40% (generally summer/autumn declines and winter increases), the following lake level declines could occur: Lake Superior -0.2 to -0.5 m, Lakes Michigan and Huron -1.0 to -2.5 m, and Lake Erie -0.9 to -1.9 m; the regulation plan for Lake Ontario cannot meet the minimum downstream flow requirements and maintain lake levels (Croley, 1990; Hartmann, 1990; Mortsch and Quinn, 1996). Using the Canadian Climate Centre (CCC) GCM II scenario (which generally has drier summer and autumn conditions than other GCMs for this region), the surface area of Lake St. Clair decreases by 15%; its volume is reduced by 37%; the water level declines 1.6 m; and the shoreline may be displaced 1–6 km lakeward, exposing lake bottom (Lee *et al.*, 1996). These Great Lakes water-level changes are based on climate change scenarios from models that produced global temperature increases that are at least twice as large and precipitation changes that generally are greater than the most recent climate change simulations with aerosols included. Nonetheless, although highly uncertain at this time, the potential declines in lake water levels shown in these analyses could have large effects on wetlands, fish spawning, recreational boating, commercial navigation, and municipal water supplies in the Great Lakes area. Also of concern is the exposure of toxic sediments and their remediation with declines in lake levels (Rhodes and Wiley, 1993).

Responses to adapt to these large changes in lake levels in developed areas could be costly. Changnon (1993) estimated the costs for dredging, changing slips and docks, relocating beach facilities, and extending and modifying water intake and sewage outfalls for a 110-km section of the Lake Michigan shoreline including Chicago to range from $298–401 million for a 1.3-m decline and $605–827 million (1988 dollars) for a 2.5-m decline.

*Water quality could deteriorate during summer low flows in regions experiencing reduced summer runoff.*

Changes in water quality as well as changes in hydrological regimes could occur as a result of climate warming. Increases in water temperature in streams and rivers reduce oxygen solubilities and increase biological respiration rates and thus may result in lower dissolved oxygen concentrations, particularly in summer low-flow periods in low- and mid-latitude areas (IPCC 1996, WG II, Section 10.5.4). Although temperature increases also may stimulate photosynthesis via increased nutrient cycling and thus prevent dissolved oxygen declines during the day, sharp nighttime declines could occur. Summer dissolved oxygen concentrations in the hypolimnion of lakes, particularly more eutrophic lakes, also may decline, and areas of anoxia may increase because of increased respiration rates in a warmer climate (IPCC 1996, WG II, Section 10.5.4). However, reduction in the length of winter ice cover may reduce the incidence of winter anoxia in more northerly lakes and rivers. Increases in water temperature also will impact industrial uses of water, primarily in the low and mid-latitudes, by reducing the efficiency of once-through cooling systems (IPCC 1996, WG II, Section 14.3.3). Increases in water temperature will have a positive impact on navigation in the mid- and high latitudes, especially in the Great Lakes, by increasing the length of the ice-free season (IPCC 1996, WG II, Section 14.3.4)—perhaps compensating for reduced cargo capacity due to low water levels.

Changes in the seasonality of runoff also may affect water quality. In the middle and high latitudes, the shift in the high-runoff period from late spring and summer to winter and early spring might reduce water quality in summer under low flows. Extended droughts in boreal regions have been shown to result in acidification of streams due to oxidation of organic sulfur pools in soils (Schindler, 1997). However, acidic episodes associated with spring snowmelt in streams and lakes in the northeastern United States and eastern Canada might be reduced under a warmer climate with lower snow accumulation and lower discharges during the spring melt (IPCC 1996, WG II, Section 10.5.3; Moore *et al.*, 1997). In general, water-quality problems (particularly low dissolved oxygen levels and high contaminant concentrations) associated with human impacts on water resources (e.g., wastewater effluents) will be exacerbated more by reductions in annual runoff than by other changes in hydrological regimes (IPCC 1996, WG II, Section 14.2.4).

*Increases in competition for limited water under a warmer climate could lead to supply shortfalls and water-quality problems, particularly in regions experiencing declines in runoff.*

Under a warmer climate, more intensive water resource management will be required because population growth, economic development, and altered precipitation patterns will lead to more intense competition for available supplies (IPCC 1996, WG II, Sections 12.3.5 and 14.4). Managing increased and diversified water demands will be particularly problematic in regions that currently have the lowest water availability (e.g., western-central North America) and those that will experience declines in runoff with climate change.

National water summaries by the U.S. Geological Survey provide comprehensive data on water availability and demand. Agriculture and steam electric generation account for approximately 75% of total water withdrawals in the United States; agricultural uses are most dominant west of the 100th meridian, where evaporation generally exceeds precipitation. When seasonal and interannual variability of regional climates are considered, the most inadequate water supplies within the United States (70% depletion of available supplies by off-stream uses) are in the southwest—including the lower Colorado River basin, the southern half of California's Central Valley, and the Great Plains river basins south of the Platte River.

A warmer climate will likely increase the demand for irrigation water by agriculture (IPCC 1996, WG II, Section 14.3.1) and for industrial cooling water at the same time that urban growth will be increasing the demand for municipal water supplies. In addition, higher water temperatures will reduce the efficiency of cooling systems (Dobrowolski *et al.*, 1995), and might make it increasingly difficult to meet regulatory constraints defining acceptable downstream water temperatures, particularly during extremely warm periods (IPCC 1996, WG II, Section 14.3.3). Furthermore, growing instream flow requirements to protect aquatic ecosystems also will reduce effective water supplies. However, improved management of water infrastructure, pricing policies, and demand-side management of supply have the potential to mitigate some of the impacts of increasing water demand (Frederick and Gleick, 1989; IPCC 1996, WG II, Section 12.5.5).

### 8.3.4. *Food and Fiber: Agriculture*

#### 8.3.4.1. *Description of the Resource*

Agricultural land represents about 12% of the land area of North America. Approximately 3% of the population and 1.7% of the annual growth in gross national product (GNP) are related to agriculture. Agricultural land use comprises a total of approximately 233 million ha. Irrigated farmland represents 21 million ha in the United States, with much of this along the Mississippi River, the central Great Plains, and the western states. North America is characterized by an abundance of fertile soils and a highly productive agricultural sector that leads the world in the production of small grains. Within the United States, there are 10 farm production regions, with 6 corresponding regions in Canada (Adams *et al.*, 1995b; Brklacich *et al.*, 1997a).

Agriculture in North America has a long history of sensitivity to climate variability (e.g., the timing and magnitude of droughts and floods, extremes in heat and cold) and is subject to a wide array of other factors that can limit potential productivity (e.g., tropospheric ozone, pests, diseases, and weeds). Agriculture has an equally long history of developing strategies to cope with the many factors capable of limiting production. Climate change is an additional factor that could enhance or reduce the sensitivity of the agricultural sector to these current stress factors. As world population grows, the demand for North American agricultural products is expected to increase, with possible increases in agricultural commodity prices (IPCC 1996, WG II, Section 13.6.8). Should increased demand lead to further intensification of agriculture in North America, increased emphasis on sustainable agriculture is likely (Matson *et al.*, 1997).

#### 8.3.4.2. Potential Impacts of Climate Change on Agriculture

Potential impacts of climate change on agriculture will be reflected most directly through the response of crops, livestock, soils, weeds, and insects and diseases to the elements of climate to which they are most sensitive. Soil moisture and temperature are the climate factors likely to be most sensitive to change across large agricultural areas of North America. The differential response of species to elevated $CO_2$ concentrations is expected to show a generally positive but variable increase in productivity and WUE for annual crops; limited evidence suggests less of a growth response for perennial crop species. Many weed species are expected to benefit from $CO_2$ "fertilization" and increased WUE, and increased temperatures may facilitate the expansion of warm-season weed species to more northerly latitudes (IPCC 1996, WG II, Section 13.2). Insect pests and fungal and bacterial pathogens of importance to agricultural production are sensitive to climate change through the direct effects of changes of temperature and moisture on the pest or pathogen, on host susceptibility, and on the host-parasite interrelation (IPCC 1996, WG II, Section 13.4). Livestock is sensitive to climate through impacts on feed and forage crops, through the direct effects of weather and extreme events on animal health, and through changes in livestock diseases (IPCC 1996, WG II, Section 13.5).

Long-term crop management strategies that increase soil organic matter will benefit agricultural lands by increasing soil nutrient status and water-holding capacity while increasing soil carbon storage (Matson *et al.*, 1997).

#### 8.3.4.3. Climate Variability and Extreme Events

*Changes in mean temperature and precipitation will likely affect agricultural crop and livestock production. Climate modifications that lead to changes in daily and interannual variability in temperatures and, in particular, precipitation also will impact crop yields.*

Mearns *et al.* (1996) used the Clouds and Earth's Radiant Energy System (CERES)-Wheat model to demonstrate the impact of daily temperature variability on simulated wheat yields at two sites in Kansas. A doubling of daily temperature variability contributed to increased crop failures and lower yields as a consequence of cold damage and winter kill. Simulated wheat yields also decreased as variability in precipitation increased, although absolute reductions in yield were dependent on soil type and associated moisture-holding capacity. Although these simulations illustrate the potential sensitivity of wheat production to increased variability in temperature and precipitation, they do not incorporate the beneficial role that elevated $CO_2$ may play in modifying these responses, nor are extreme events considered in these analyses. Extreme events like drought, flooding, hail, hurricanes, and tornadoes also will impact agriculture, but reliable forecasts of such occurrences are not yet regionally available.

#### 8.3.4.4. Direct and Indirect Effects

*The results of a large number of experiments designed to examine the effects of elevated $CO_2$ concentrations on crops have generally confirmed high confidence in a net beneficial effect of $CO_2$ fertilization, up to some level. Sustained plant response under field conditions to concentrations beyond $2xCO_2$ would likely be dependent on species as well as water and nutrient status and is highly uncertain.*

A mean value yield response of $C_3$ crops (most crops except maize, sugar cane, millet, and sorghum) to doubled $CO_2$ is reported to be approximately +30% (range -10% to +80%). There is reason to expect, however, that this value represents an upper estimate unlikely to be achieved under field conditions. Factors known to affect the magnitude of $CO_2$ response in crops include the availability of plant nutrients, the crop species, temperature, precipitation, and other environmental factors, such as air pollution, soil quality, weeds, insect pests, and diseases (IPCC 1996, WG II, Section 13.2.1). Increased WUE is a result of elevated $CO_2$ as well, though in many regions of North America, higher temperatures associated with elevated $CO_2$ can be expected to increase evaporative demand and transpiration, resulting in minimal benefit from the increase in WUE (Brklacich *et al.*, 1997b).

Changes in soils (e.g., loss of soil organic matter, leaching of soil nutrients, salinization, and erosion) are likely consequences of climate change for some soils in some climatic zones. Cropping practices such as crop rotation, conservation tillage, and improved nutrient management are technically effective in combating or reversing such deleterious effects (IPCC 1996, WG II, Section 13.3; Matson *et al.*, 1997).

Livestock production could be affected by changes in grain prices, changes in the prevalence and distribution of livestock pests, and changes in grazing and pasture productivity. Livestock are sensitive to stress from warmer, drier conditions, as well as reduced range forage quality and water availability. Warmer winter temperatures may enhance winter survival of range livestock. Taking action to improve forage quality or water supply could benefit livestock. Analyses indicate that intensively managed livestock systems such as those in North America have more potential for adaptation than crop systems because of their mobility in terms of access to food and water (IPCC 1996, WG II, Section 13.5).

*The risk of losses due to weeds, insects, and diseases is sensitive to temperature and moisture (including rainfall, humidity, and dew); the risk is likely to increase in subregions where these factors become more favorable for specific disease organisms but may decrease under drier conditions. Increased climate variability may provide additional challenges for pest-management adaptation to climate change.*

Elevated $CO_2$ levels may enhance the growth of $C_3$ weeds, based on the results of controlled exposure experiments. Evidence also exists, however, that other factors determining plant productivity may be more important in controlling plant response in the field (e.g., water- and nutrient-use efficiency) than the differential $CO_2$ response (Bazzaz and McConnaughay, 1992). There currently is little experimental evidence to directly evaluate the effects of elevated $CO_2$ on weed infestation, insect pests, or plant diseases under field conditions. Less severe winters may increase the range and severity of insect and disease infestations. Temperature and moisture are critical to the spread and development of many plant diseases (IPCC 1996, WG II, Section 13.4.3). Successful disease development requires convergence of a susceptible host, a virulent pathogen, and suitable environmental conditions. Increased variability of precipitation, for example, could affect the host-parasite interaction positively or negatively, leading to more or less disease development (Shriner, 1980). Increased climate variability also could render less effective disease-forecasting models currently used to manage some diseases and require increased reliance on pesticides. North American agriculture will need to address these concerns in the context of increasing pressure on agriculture to reduce chemical inputs.

### 8.3.4.5. *Yield and Production Changes by North American Subregion*

Previous studies that have simulated the impact of climate change on the North American agriculture sector have taken a variety of approaches. Tables 13-11 and 13-12 in the SAR (IPCC 1996, WG II) outline the range and variability of yield impacts that have been suggested across a number of studies looking at climate change impacts for the United States and Canada. For these studies (IPCC 1996, WG II, Section 13.6.8), when biophysical and economic impacts were combined, market adjustments were found to lessen the impacts of negative yield changes. More recent projections of increases in global mean surface temperatures in the future are lower than past estimates. These lower estimates are derived from new, transient GCM scenarios that take into account the interactions between the atmosphere and oceans and the cooling effects of sulfate and other aerosols in the troposphere (Darwin, 1997). Most of the impact studies currently available for review have not utilized the projections from these more recent climate model simulations as a background; as a result, they may overestimate the magnitude of expected temperature impacts.

The outcome of the net economic impact summarized in the SAR (IPCC 1996, WG II, Section 13.6.8) was sensitive to assumptions about population, income, trade barriers, and institutions and ranged from negative to positive. For Canada, the vulnerability of the agricultural sector derives from the importance of agriculture to subregional (e.g., the prairies) and rural economies, the location of agriculture in a marginal climate with regard to temperature and precipitation, and limitations to

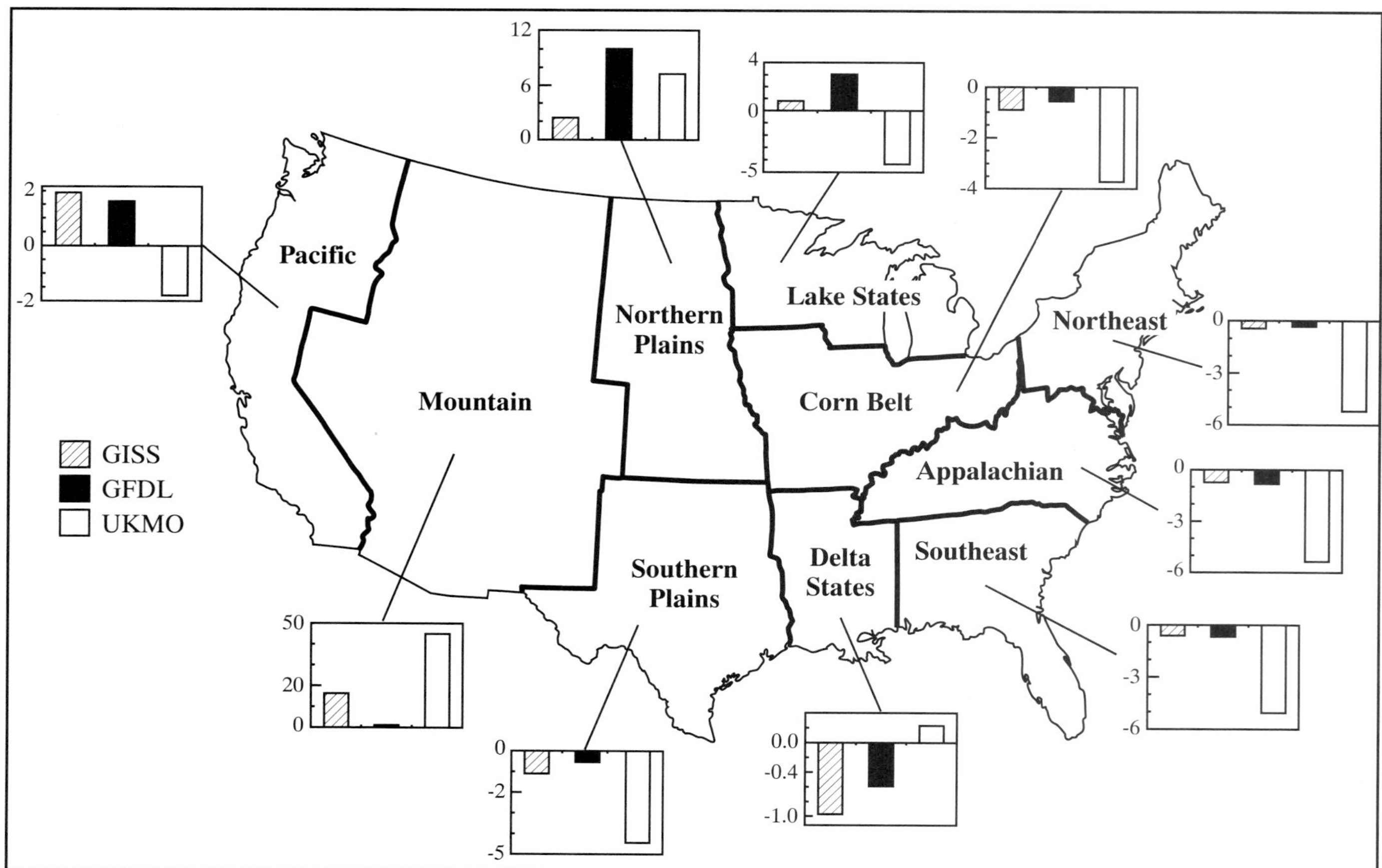


**Figure 8-8:** Effects of climate change on regional economic welfare (percentage change in total welfare from base, assuming 555 ppm $CO_2$ and changes in export demand) (adapted from Adams, 1995b). See page 287 for details of the climate scenarios used.

northward shifting of cropping by poor soil quality (Cohen *et al.*, 1992).

*Economic welfare may improve for more northerly farm production regions—with potential benefits indicated for the lake states, the northern Plains, the mountain region, and the Pacific region.*

Evaluation of the direct and indirect effects of climate on yield at the farm, regional, or higher level of aggregation requires integrated models that consider system interactions. Changes in crop production, crop water demand, and regional water resources will arise as a consequence of climate change, although the impacts of these changes on agriculture will be modified by trends in world food production and commodity exports. These interactions can be modeled to estimate the economic consequences of climate change for regions of the United States. Adams *et al.* (1995a,b) used three GCM scenarios to evaluate the economic consequences to crop production and regional welfare if climate changes of similar magnitude were to occur. Figure 8-8 illustrates the subregional variability in economic welfare suggested by climate change, including a $CO_2$ fertilization effect (and accounting for changes in export demand). Based on these analyses, eastern, southeastern, and corn belt regions yielded estimates of negative impacts, while positive effects were projected for northern plains and western regions.

*Subregional shifts in economic welfare with climate change will likely arise as a consequence of impacts on crops with specific temperature requirements for growth.*

Citrus production, for example, may shift slightly northward as temperatures rise in southern states. Yields are predicted to decline in southern Florida and Texas as a consequence of higher-than-favorable temperatures during the winter (Rosenzweig *et al.*, 1996). Warmer temperatures experienced at slightly more northern locations may lessen the chance of freeze damage, but the overall risk of early- and late-season frost will remain a major factor in crop loss. It is worth noting that as a woody perennial, citrus offers little possibility of short-term adaptation to climate change through management because the timing of phenological events in citrus is not under the control of the orchard manager. Other simulation studies show that corn and soybean yields may decrease across much of the U.S. corn belt with a 2°C rise in temperature (Phillips *et al.*, 1996) and that potato yields may decrease on average by 22% across sites from Maine to Washington as temperatures increase 1.5–5°C (Rosenzweig *et al.*, 1996).

Experimental results in combination with knowledge of physical and biological processes, along with modeling of these basic physiological processes, provide information on potential changes in yield, which can in turn provide input to an agricultural sector model capable of simulating sector-level impacts of alternative climate change scenarios (Adams *et al.*, 1995a,b).

Recent work has extended previous studies regarding the economic effects of climate change on agriculture to address some of the limitations in earlier studies. This new work (Adams *et al.*, 1995b) looks at U.S. agriculture sector impacts by incorporating other crops such as fruits and vegetables into the regional crop alternatives for the southeastern region and other southerly locations; considering the impacts of farmer adaptations to climate change; allowing for crop migration into regions where those crops are not currently being grown; incorporating changes in forage production and livestock performance; and assessing the potential for technological change, as manifested in present and future yields, to offset climate change.

The analyses by Adams *et al.* (1995b) used estimates from uniform incremental climate change scenarios and from two GCM-based analyses to assess a wide range of potential temperature and precipitation changes, as well as alternative levels for atmospheric $CO_2$ concentrations. The scenarios included 16 combinations of alternative temperature and precipitation changes (0°C, 1.5°C, 2.5°C, and 5°C for temperature and -10%, 0%, +7%, and +15% for precipitation). In addition, the analysis considered four alternative levels for atmospheric $CO_2$ concentrations (355, 440, 530, and 600 ppm). This scheme produced a total of 64 incremental scenarios, which were evaluated for 1990 and 2060 conditions. These scenarios represented the data set for estimation of a climate change response function for 1990 and 2060 conditions of technological development and agricultural demand. In addition to these uniform scenarios, the two GCM-based analyses used climate forecasts from the Goddard Institute for Space Studies (GISS) and the Geophysical Fluid Dynamics Laboratory (GFDL)-R30 GCMs (see Table 1-2). These analyses provided points of comparison with previous studies, as well as tests of the reasonableness of the economic effect response functions developed from the 64 uniform scenarios.

Estimation of the economic consequences of these scenarios required predictions of the impacts of climate change on yield levels for crops and forage, animal grazing requirements and performance, crop migration potentials, technology-based changes in yields, and changes in water resource availability. These predicted changes were then used in an economic model of U.S. agriculture. The economic model provided estimates of changes in social welfare, crop prices and quantities, resource use, and other measures of economic performance arising from the climate scenarios (Adams *et al.*, 1995b).

Under the majority of the climate change scenarios evaluated in the study, net welfare increased; only 34 of the 128 scenarios showed welfare losses. In general, increases in precipitation and $CO_2$ increase welfare. Slight-to-moderate increases in temperature also could increase welfare. The response function analysis also showed that increases in $CO_2$ and precipitation could offset the potentially negative effects of large temperature increases (Adams *et al.*, 1995b).

*The magnitude of welfare changes showed modest gains for an optimistic case and modest loss to modest gain from an adverse case for present and future technology and economic growth cases, respectively.*

The values represented very small percentage changes in total agriculture value (< 3% of the value of the base model solution). As a result, in the aggregate, Adams *et al.* (1995b) concluded that climate appears to be a relatively small stress to agriculture in the United States. Welfare losses from adverse climate change also tended to be smaller than previous estimates, whereas gains from favorable climate change tend to be larger because of the more comprehensive treatment of adjustment possibilities such as the inclusion of new crops and migration possibilities that were used in the analysis. This study concludes, as others have, that climate change of the type evaluated does not appear to be a food security issue for the United States (Adams *et al.*, 1995b). On the other hand, this study did not incorporate the costs of adaptations or the consequences of changes in subregional agricultural production for the economies of those subregions; it only addressed the consequences of change in mean climate, without increased climate variability; and it did not evaluate the possible consequences of potential secondary effects on pests, pathogens, or soils that might result from climate change. These shortcomings represent important research needs with regard to integrated modeling of climate change impacts for the agricultural sector.

Sensitivity analyses performed in conjunction with the Adams *et al.* (1995b) analyses indicated that potential farmer adaptations to climate change can play a major role in mitigating adverse effects of climate change, suggesting the importance of technology and related assumptions in future analyses. Sensitivity analyses of export assumptions reinforced the importance of world trade (exports) on the welfare of the U.S. agricultural sector. Global climate change is likely to increase the demand for U.S. commodities, with possible increases in welfare. Because of the importance of North American agriculture to world food production, trade issues are an important interregional consideration.

When the GCM-based analyses were considered, climate changes according to the GISS forecasts led to net welfare increases of $12 billion for the 1990 base—approximately 20% larger than previous analyses of GISS climate change using the Agriculture Sector Model (Adams *et al.*, 1995b). The increase was related primarily to changes in the model allowing other mitigation activities that permit the agricultural sector to exploit more fully new climate conditions. The GFDL-R30 analysis revealed losses of more than $14 billion (measured against the 1990 base). These losses arise from harsher climate conditions under the GCM. Results from the GCM cases did not compare well with estimates generated by response functions resulting from the uniform change scenarios, indicating the importance of regional differences in climate as determinants of estimates of national economic consequences.

*Several global climate change scenarios imply modest improvements in the agroclimatic potential for high-latitude agriculture in North America.*

Recent studies (Mills, 1994; Brklacich *et al.*, 1996, 1997a) have investigated the extent to which global climate change might shift the frontier for Canadian agriculture northward. About 1.2 million ha in the Peace River region of northern Alberta and British Columbia currently are devoted to spring-seeded cereals, forages, and pasture crops. Soils capable of supporting agricultural production north of the Peace River region (i.e., above 58°N latitude) are abundant, but these regions currently are too cool and too remote from markets.

Climate change scenarios derived from the GFDL and CCC $2xCO_2$ models were estimated to relax current constraints imposed by short, cool, frost-free seasons. However, it also was estimated that these benefits would be offset somewhat by declines in summer precipitation, with concomitant increases in crop moisture deficits (most notably under the CCC scenario).

The CERES-Wheat model was used to estimate the combined effects of increasing $CO_2$ levels to 555 ppm and the GFDL and CCC climatic change scenarios on spring wheat yields. Under the GFDL scenario, modest increases in spring-seeded wheat yields were estimated. Increases in crop moisture stress under the CCC scenario were estimated to offset the benefits of elevated $CO_2$ levels and longer, warmer, frost-free seasons; overall, modest declines in spring wheat yields were estimated.

Winter temperature increases were estimated at 3–4°C and 4°C under the GFDL and CCC scenarios, respectively. Assessments using CERES-Wheat indicated that these temperature increases would be sufficient to support winter wheat production at several locations within the Peace River region. It was also estimated that, for lands north of the Peace River region (i.e., north of the current climatic frontier for commercial agriculture), these temperature increases would not be sufficient to remove the risk of winter crop damage. In these northern prairie regions, soil capability to sustainably support agriculture may become more limiting than temperature.

Development of the capability to simulate the agricultural impacts of multiple transient climate scenarios is a research need that must be met to deal credibly with the cost of adaptation, about which there is significant uncertainty. Socioeconomic adjustment must be modeled to treat key dynamic processes—such as how the expectations of farmers change; whether farmers can easily detect climate change against a background of high natural variability; and how current investments in equipment, education, and training will affect the costs of adjustment.

#### *8.3.4.6. Adaptation*

Historically, farming systems have adapted to changing economic conditions, technology, and resource availability and have kept pace with a growing population (CAST, 1992; Rosenberg, 1992). Evidence exists that agricultural innovation responds to economic incentives such as factor prices and can relocate geographically (Hayami and Ruttan, 1985; CAST, 1992). A number of studies indicate that adaptation and adjustment at all levels—but especially at the farm level—will be important to limit losses or to take advantage of improving climatic conditions (IPCC 1996, WG II, Section 13.9). Examples of technological options for adaptation by agriculture include seasonal changes in sowing dates; different crop varieties or species; new crop varieties; water supply and irrigation systems; management adjustments with fertilizer, tillage, and so forth; and improved short-term climate prediction (IPCC 1996, WG II, Section 13.9.1; Darwin *et al.*, 1995). Socioeconomic options for adaptation include improved training and general education of populations dependent on agriculture; assessment of currently successful strategies for responding to climate variability; improved agricultural research to increase the robustness of new farming strategies; interactive communication to bring research results to farmers and farmers' problems to researchers; improved preservation and maintenance of genetic material critical to adaptation; and food programs to buffer against local supply changes. Transportation, distribution, and market integration provide additional flexibility for regions to respond to climate variability, and changes in policies could increase the adaptive capacity of agriculture (IPCC 1996, WG II, Section 13.9.2).

*Recent analyses of issues of long-run sustainability associated with agricultural adaptation to climate change from arbitrary doubling of equivalent $CO_2$ concentrations have concluded that there is considerably more sectoral flexibility and adaptation potential than was found in earlier analyses.*

Schimmelpfenning *et al.* (1996) concluded that the costs and benefits of climate change cannot be adequately evaluated independent of behavioral, economic, and institutional adjustments required by changing climate. Smit *et al.* (1996) and Brklacich *et al.* (1997b), in their research into agricultural adaptation to climatic variability and change in Ontario, reached a similar conclusion and urged that future research into agriculture and climatic change be reframed to explicitly consider agricultural decision making and adaptation processes. Although scientific controversy remains over the nature and rate of climate change and the importance of climate variability, most scenarios suggest gradual changes in mean temperature and precipitation over decades, providing opportunities for farms and other parts of the sector to adapt. In addition, the time scale of 80–100 years makes other profound social changes inevitable. Income and population growth and technological innovation will accelerate or decelerate, depending on global location, at the same time that adaptation to climate is taking place. Social and cultural factors may influence the rate at which adaptation measures are implemented within some subregions of North America. There may be time lags between decisions to follow an adaptive strategy and subsequent adjustments in the agricultural system. The costs and time required for such adjustments in infrastructure will need to be considered in planning adaptation options. Although none of these factors can be considered in isolation, recent research shows that the negative effects of climate change on agriculture probably are overestimated by studies that do not account for economic adjustments or consider the broader economic and environmental implications of such changes. However, uncertainties remain about the implications of changes in climate variability, as well as crop responses to increases beyond a doubling of equivalent atmospheric $CO_2$ concentrations.

#### *8.3.4.7. Vulnerabilities*

*Vulnerability to climate change-induced hunger or severe economic distress for the overall economy of North America as a result of climate change impacts on the agricultural sector is relatively low.*

The United States and Canada have high GNP per capita; the agricultural population is a small share of the total population; and agriculture is, in general, a small share of the economy. These areas are important for world food production. Midcontinental areas of the United States and Canada are prone to drought, which would be exacerbated if climate change reduced moisture availability or increased the demand for water (as occurs in several GCM scenarios). Economic dislocation is likely to be limited to the agricultural sector or to subregions highly dependent on agriculture (IPCC 1996, WG II, Section 13.7). Evidence suggests that yields of crops grown at the margin of their climatic range or in climates where temperature or precipitation could easily exceed threshold values during critical crop growth periods are more vulnerable (Matthews *et al.*, 1994a,b; IPCC 1996, WG II, Section 13.7). A regional economy that offers only limited employment alternatives for workers dislocated by the changing profitability of farming is relatively more vulnerable than those that are economically diverse. As an example, the Great Plains area of North America is most dependent on agriculture and thus might be the most economically vulnerable to climate change (Rosenberg, 1993).

### *8.3.5. Food and Fiber: Production Forestry*

Timber is one of the most valuable agricultural crops produced in North America. The forest products sector in the United States employs some 1.5 million people (1990) nationwide and adds approximately $80 billion to the GDP. Forest products are especially important to the economies of the Pacific Northwest and the southern United States. Two-thirds of U.S. forestlands (195 million ha) are considered productive enough to potentially support timber management and accessible to harvest. Less than one-third of U.S. timberlands are publically owned; approximately 15% is owned by the timber industry, and 57% is held by other private landowners. Timberland ownership differs greatly by region, with western forests largely on public lands and eastern forests largely in private ownership (OTA, 1993).

In Canada, direct employment in 1996 for the forest sector was 363,000; the sector contributed $16.177 billion to Canada's GDP. In terms of contribution to GDP, the sector is important to British Columbia, Quebec, and New Brunswick. Thirty-five percent of Canadian forestland is considered productive and accessible enough for harvest, though there may be some constraints on harvesting. Ownership is 71% provincial, 23% federal (including territorial land), and 6% private.

See section 8.3.2 for a detailed discussion of the impacts of climate change on forest ecosystems.

*Consumers and producers could gain or lose, and the long-term stability of the forest-products market could be jeopardized.*

Enhanced forest growth scenarios and extensive forest dieback scenarios have been analyzed with respect to market processes. Under the most severe ecological scenarios—where forest dieback occurs relatively early and there are long-term reductions in timber inventories and harvest—consumer prices would increase; producers would benefit from higher prices, but overall benefits to society would decrease. Under scenarios of more moderate dieback or forest expansion, consumers could benefit from lower prices; producers might benefit or lose, depending on local forest responses to dieback, local demand, and access to national and international markets. Alternative management practices must be carefully considered if benefits are to be gained under moderate forest dieback scenarios. However, the long-term sustainability of forests and any benefits could vary considerably under different scenarios.

Analyses of the economic consequences of climate change have been based on a number of quite different ecological assessments and a variety of economic models, but all are based on the equilibrium FAR scenarios or on sensitivity analyses. One study coupled the FASOM forest sector model—incorporating flexible pricing and forest management—to output from forest gap models over the conterminous United States (Callaway *et al.*, 1995; Adams *et al.*, 1996). The results, with severe forest dieback, indicated consumer price increases of 100–250%, with economic losses of 4–20% of the net value of commercial forests.

A study by Van Kooten and Arthur (1989) concluded that forest productivity could tend to increase in Canada but could increase or decrease in the United States. Consumer prices could decrease (largely as a consequence of increases in Canadian forest growth and harvest). Producers could sustain economic losses, but with exports from Canada to the United States, net changes (consumers plus producers) could be negative for Canadians and positive for the U.S. market.

Timber growth in the conterminous United States in one study generally increased over a 50-year projection period, pushing prices down by 6–35% (Joyce *et al.*, 1995). A study using a global trade model produced similar results, with general increases in global forest productivity (Perez-Garcia *et al.*, 1997).

A study using a different forest-sector model in the United States incorporated management strategies that optimized pre- and post-dieback forest management practices (Sohngen and Mendelsohn, 1997). The value of the market (consumer plus producer surpluses) could increase by 1–11% under scenarios of either enhanced forest growth or moderate forest dieback. The increased flow of green or salvage trees into the market depressed prices, but overall losses to producers were minimized by shifting biomes or altered yield functions. Economically optimal management strategies (such as thinning, salvage logging, and species transplanting), however, might be restricted by social and ecological constraints.

*The most intensively managed industry and private forestlands may be least at risk of long-term decline resulting from the impacts of climate change because the relatively high value of these resources is likely to encourage adaptive management strategies (OTA, 1993).*

Private forest managers have the financial incentive and the latitude to protect against extensive loss from climate-related impacts. They can use several available techniques: short rotations to reduce the length of time a tree is influenced by unfavorable climate conditions; planting of improved varieties developed through selection, breeding, or genetic engineering to reduce vulnerability; and thinning, weeding, managing pests, irrigating, improving drainage, and fertilizing to improve general vigor. Such actions would reduce the risk from moisture stress and secondary risks from fire, insects, and disease. Thinning, for example, reduces competition for moisture and can effectively increase tolerance to drought; it also may speed development of a climate-adapted forest by removing trees that are growing poorly (OTA, 1993). However, some adaptive measures such as thinning or harvesting dead or dying trees could impact biodiversity, soil erosion, stream quality, and nonmarket forest products, generating potentially conflicting management options.

Binkley and van Kooten (1994) found that overall impacts on the Canadian forest sector would not be significant. This finding was attributed to the ability of the production forest sector to adapt to whatever species prevail during and after climate change; to salvage-cut dying stands; to plant cut areas with species that are better adapted to the projected climate; and to move to locations where resources are more plentiful. Long-run sustainable yield levels, however, may be reduced as a result of increased losses to fire and insect outbreaks. For example, in the Mackenzie Basin Impact Study, a general decline in forest production has been suggested for the basin because of a combination of factors—including increased area burned, increased susceptibility to pests, and drought-related die offs (Cohen, 1997a).

*A healthy mixed-species, mixed-age forest probably is less susceptible to insect infestation than extensive areas of even-aged forest stands.*

Planting single-species forests might seem to pose increased threats of loss from insect pests or disease because of limited genetic diversity (Perry and Maghembe, 1989). However, commercial tree species show a great deal of genetic diversity among individuals—even among trees from the same parents (Kitzmiller, 1993). This inherent diversity could make trees less likely to succumb to a single pest or disease than most agricultural crops (Kellison and Weir, 1987). Forest managers should attempt to ensure diversity in the seedlings they use to establish their forest stands even if they are planting single-species forests (OTA, 1993).

*Once a decline in forest health begins, less intensively managed forests may face greater fire and pest damage. At particular risk will be forests already subject to moisture stress and fire hazard.*

Less-managed forests may not be at any greater inherent risk than actively managed forests, however. Once they are subjected to stress, wilderness forests and National Parks may be at elevated risk of substantial decline because of policy restrictions imposed on silvicultural and pest-management activities. Similarly, because management currently is limited on most National Forest lands and less-productive nonindustrial private lands, those forests could be at risk of unchecked loss. If the general health of these forests declines, their vulnerability to large-scale mortality could increase (OTA, 1993).

*Forests maintained for the production of wood products and fiber would benefit from any near-term or long-term increase in productivity. Reduced growth or increased mortality would have a damaging effect. Managers of industry forests and other private timberlands can be expected to respond with adaptive measures if and when they perceive changes in climate and market conditions.*

Although no timber company is altering forest practices today, some are actively preparing for the types of risks posed by climate change. Weyerhaeuser, for example, is conducting experimental silvicultural programs to examine the effects of thinning practices in ameliorating the effects of droughts (OTA, 1993). It also is sponsoring research on the genetics, physiology, and biotechnology of heat- and drought-tolerant seedlings. Such technological development should help protect the timber industry and future wood supplies (OTA, 1993).

*Despite the possibility of some adaptive management responses, climate change could be very costly to the timber industry.*

In the southern United States, declining timber volumes could lead to $300 million in lost annual revenues, whereas the increased management measures needed to compensate for poorer conditions could add $100 million to the annual costs of production (Regens *et al.*, 1989; Hodges *et al.*, 1992). A sea-level rise could force the movement of coastal pulp and paper mills, further increasing the costs of climate change. Some of these mills would cost as much as $1 billion to replace. For the Pacific Northwest, an expanded upslope range of Douglas fir forests might add 5% to the regional timber harvest (Hodges *et al.*, 1992). However, the increased costs of logging at higher elevations could offset much of this potential gain.

A report by the Office of Technology Assessment of the U.S. Congress (OTA, 1993) identified a number of adaptation strategies that should be considered to maintain and enhance commercial forest productivity under climate change. These strategies are largely measures that have value to the forest industry even without climate change but would serve to ameliorate the impacts of climate change as well:

- Establish an expanded forest seed-bank program to ensure maintenance of genetic diversity.
- Prepare to respond to major forest declines through improved forest health maintenance and measures to minimize risk of fire and pest and disease outbreaks.
- Develop new management strategies focused on adaptation to climate change.
- Improve incentives for maintaining and protecting private forestland.

### 8.3.6. *Food and Fiber: Fisheries and Aquatic Systems*

Although there is considerable uncertainty about the physical changes and response of the various freshwater and marine species, it is possible to suggest how certain species may respond to projected climate changes over the next 50–100 years. The uncertainties highlight the importance of research to separate the impacts of changing climate from natural population fluctuations and fishing effects. Many commercial finfish populations already are under pressure (e.g., overexploited), and global change may be of minor concern compared with the impacts of ongoing and future commercial fishing and human use or impacts on the coastal zone. Further, changes in the variability of climate may have more serious consequences on the abundance and distribution of fisheries than changes in mean conditions alone (Katz and Brown, 1992), and changes in future climate variability are poorly understood at this time.

Fish, including shellfish, respond directly to climate fluctuations, as well as to changes in their biological environment (predators, prey, species interactions, disease) and fishing pressures. Although this multiforcing sometimes makes it difficult to establish unequivocal linkages between changes in the physical environment and the responses of fish or shellfish stocks, some effects are clear (see reviews by Cushing and Dickson, 1976; Bakun *et al.*, 1982; Cushing, 1982; Sheppard *et al.*, 1984; Sissenwine, 1984; and Sharp, 1987). These effects include changes in the growth and reproduction of individual fish, as well as the distribution and abundance of fish populations. In terms of abundance, the influence occurs principally through effects on recruitment (how many young survive long enough to potentially enter the fishery) but in some cases may be related to direct mortality of adult fish.

Fish carrying capacity in aquatic ecosystems is a function of the biology of a particular species and its interrelationship with its environment and associated species. Specific factors that regulate the carrying capacity are poorly known for virtually all species, but some general statements can be made with some confidence. Fish are affected by their environment through four main processes (Sheppard *et al.*, 1984):

- *Direct physiological effects, including metabolic processes influenced by temperature, salinity, and oxygen levels*—Fish often seek optimal temperature or salinity regimes or avoid suboptimal conditions. Thus, ocean and freshwater changes as a result of projected climate changes can lead to distributional changes. In suboptimal conditions, performance is reduced, leading to starvation or increased predation.
- *Diseases*—Certain environmental conditions are more conducive to diseases than others (e.g., warm waters can trigger disease outbreaks; likewise, cold temperatures can limit them).
- *Food*—The environment affects feeding rates and competition, as well as abundance, quality, size, timing, spatial distribution, and concentration of food.
- *Predators*—The environment affects predation through influences on the abundance and distribution of predators.

Fish are influenced not only by temperature and salinity conditions but also by mixing and transport processes (e.g., mixing can affect primary production by promoting nutrient replenishment of the surface layers; it also can influence the encounter rate between larvae and prey organisms). Ichthyoplankton (fish eggs and larvae) can be dispersed by the currents, which may carry them into or away from areas of good food production, or into or out of optimal temperature or salinity conditions—and perhaps, ultimately determine whether they are lost to the original population.

Climate is only one of several factors that regulate fish abundance. Managers attempt to model abundance trends in relation to fishing effects in order to sustain fisheries. In theory, a successful model could account for global warming impacts along with other impacts without understanding them. For many species of fish, the natural mortality rate is an inverse function of age: Longer-lived fish will be affected by natural changes differently than shorter-lived fish. If the atmosphere-freshwater-ocean regime is stable for a particular time, it is possible to estimate the age-specific mortality rates for a species of interest. However, at least some parts of the atmosphere-freshwater-ocean system are prone to oscillations on a decadal scale, which may not be cyclical. These natural changes occur globally; thus, they will have impacts on the freshwater and marine ecosystems that support North American fish populations. Under natural conditions, it may be expected that the different life histories of these fish will result in different times of adjustment to a new set of environmental conditions.

Any effects of climate change on fisheries are expected to be most pronounced in sectors that already are characterized by full utilization, large overcapacities of harvesting and processing, and sharp conflicts among users and competing uses of aquatic ecosystems. Climate change impacts, including changes in natural climate variability on seasonal to interannual time scales, are likely to exacerbate existing stresses on fish stocks. The effectiveness of actions to reduce the decline of fisheries depends on our ability to distinguish among these stresses and other causes of change and on our ability to effectively deal with those over which we have control or for which we have adaptation options. This ability is insufficient at present; although the effects of environmental variability are increasingly recognized, the contribution of climate change to such variability is not yet clear.

*Recreational fishing is a highly valued activity that could incur losses in some regions as a result of climate-induced changes in fisheries.*

Recreational fishing is a highly valued activity within North America. In the United States, for example, 45 million anglers participate annually; they contribute to the economy through spending on fishing and related activities (US$24 billion in 1991). The net economic effect of changes in recreational fishing opportunities as a result of climate-induced changes in fisheries is dependent on whether projected gains in cool- and warm-water fisheries offset losses in cold-water fisheries. Work by Stefan *et al.* (1993) suggests mixed results for the United States, ranging from annual losses of US$85–320 million to benefits of about US$80 million under a number of GCM projections. A sensitivity analysis (U.S. EPA, 1995) was conducted to test the assumption of costless transitions across these fisheries. This analysis assumed that best-use cold-water fishery losses caused by thermal changes were effectively lost recreational services. Under this assumption, all scenarios resulted in damages, with losses of US$619–1,129 million annually.

#### 8.3.6.1. Freshwater Ecosystem Impacts, Adaptations, and Vulnerabilities

Commercial and recreational freshwater fisheries are important to the economy of many regions, as well as the well-being of native populations. In many aquatic ecosystems, freshwater fish also are important in maintaining a balance in other aquatic populations lower in the food web (via predatory and other effects). In broader terms, aquatic ecosystems are important as recreational areas, as sources of water for domestic and industrial use, and as habitat for a rich assemblage of species, including some that are threatened or endangered.

Several studies have indicated that projected climate change will have important impacts on North American freshwater fisheries and aquatic ecosystems. It must be noted, however, that most studies to date have used results from earlier climate model simulations that gave air temperature increases under a $2xCO_2$ climate that were as much as twice as large for the same time period as more recent estimates that include aerosol forcing—thus overestimating the effects of temperature increases, particularly in the summer.

*Changes in survival, reproductive capacity, and growth of freshwater fish and the organisms and habitats on which they depend result from changes in water temperature, mixing regimes, and water quality.*

In North America, freshwater fish have been grouped into three broad thermal groups (cold-water, cool-water, and warm-water guilds) based on differences in the temperature optima of physiological and behavioral processes. In simulations of deep, thermally stratified lakes in the mid- and high latitudes, including the Laurentian Great Lakes, winter survival, growth rates, and thermal habitat generally increase for fish in all three thermal guilds under the 2x$CO_2$ climate (DeStasio *et al.*, 1996; IPCC 1996, WG II, Sections 10.6.1.2 and 10.6.3.2; Magnuson and DeStasio, 1996). However, in smaller mid-latitude lakes, particularly those that do not stratify or are more eutrophic, warming may reduce habitat for many cool-water and cold-water fish because deep-water thermal refuges are not present or become unavailable as a consequence of declines in dissolved oxygen concentrations (IPCC 1996, WG II, Section 10.5.4). For example, Stefan *et al.* (1996) examined the effect of temperature and dissolved oxygen changes in lakes in Minnesota; they projected that under a 2x$CO_2$ climate (from a GISS GCM that projected a 3.8°C air temperature increase in northern Minnesota), cold-water fish species would be eliminated from lakes in southern Minnesota, and cold-water habitat would decline by 40% in lakes in northern Minnesota.

Changes in the productivity and species composition of food resources also may accompany climatic warming and, in turn, influence fish productivity. Production rates of plankton and benthic invertebrates increase logarithmically with temperature; rates increase generally by a factor of 2–4 with each 10°C increase in water temperature, up to 30°C or more for many organisms (Regier *et al.*, 1990; Benke, 1993; IPCC 1996, WG II, Section 10.6.1.1). Although this effect generally should increase fish productivity, shifts in species composition of fish prey with warming might prevent or reduce productivity gains. Biogeographic distributions of aquatic insects are centered around species thermal optima, and climate warming may alter species composition by shifting these thermal optima northward by about 160 km per 1°C increase in temperature (Sweeney *et al.*, 1992; IPCC 1996, WG II, Section 10.6.3.1). If species range shifts lag changes in thermal regimes because of poor dispersal abilities or a lack of north-south migration routes (e.g., rivers draining northward or southward) or if species adaptation is hindered by limited genetic variability, climatic warming might result initially in reductions in the preferred prey organisms of some fish (IPCC 1996, WG II, Section 10.6.3.3).

Climatic warming may result in substantial changes in the thermal regimes and mixing properties of many mid- and high-latitude lakes. In the mid-latitudes, some lakes that presently are dimictic (mixing in spring and autumn) may no longer develop winter ice cover and may become monomictic (mixing during fall, winter, and spring), with a longer summer stratification period. At high latitudes, some lakes that presently are monomictic and mix during summer may stratify in summer and mix twice a year, in autumn and spring (IPCC 1996, WG II, Section 10.5.4). Changes in lake mixing properties may have large effects on hypolimnetic dissolved oxygen concentrations (affecting available fish habitat) and on epilimnetic primary productivity, although these effects are likely to depend greatly on the morphometric characteristics of individual lakes and are difficult to predict (IPCC 1996, WG II, Section 10.5.4). For example, longer summer stratification and higher water temperature result in more severe hypolimnetic oxygen depletion in lakes in Minnesota under a 2x$CO_2$ climate simulation (Stefan *et al.*, 1993). In other lakes, reduction in the duration or lack of winter ice cover might reduce the likelihood of winter anoxia (IPCC 1996, WG II, Section 10.6.1.4). At high latitudes, development of summer stratification under a warmer climate might increase lake primary productivity by maintaining algae for longer periods within the euphotic zone. Climate changes that result in decline in runoff also may have substantial effects on the mixing properties of smaller lakes that are heavily influenced by fluxes of chemicals from their catchments. For example, the surface mixed layer of boreal lakes at the Experimental Lakes Area in northwest Ontario has deepened over the past 20 years as a result of a long-term drought that reduced inputs of DOC from the catchment and thus increased water clarity (IPCC 1996, WG II, Section 10.5.3 and Box 10-2; Schindler *et al.*, 1996).

Long-term research and monitoring of key physical, chemical, and ecological properties (particularly water temperature and mixing properties; concentrations of nutrients, carbon, and major ions; acid/base status; populations of key organisms; primary production; and organic-matter decomposition) remain key research needs to reduce uncertainties in projections of freshwater fisheries' responses to climate change.

*Climate warming may result in general shifts in freshwater species' distributions northward, with widespread/subregional species extinction at the lower latitudes and expansion at the higher latitudes of species' ranges.*

Climatic warming may produce a general shift in species distribution northward. Species extinction and extirpations are likely to occur at the lower latitude boundaries of species distributions, and range expansion likely will occur at the higher latitude boundaries of species distributions (IPCC 1996, WG II, Section 10.6.3). For example, a 3.8°C increase in mean annual air temperature is projected to eliminate more than 50% of the habitat for brook trout and result in severe fragmentation of its distribution in the southern Appalachian Mountains in the southeastern United States (Meisner, 1990). In contrast, a 4°C increase in mean air temperature is projected to expand the ranges of smallmouth bass and yellow perch northward across Canada by about 500 km (Shuter and Post, 1990).

In streams and rivers, particularly at low and mid-latitudes, the distributions of many fish species may contract because of limitations on availability of thermal refuges and migratory routes during periods of high temperatures and lower streamflow in the summer (IPCC 1996, WG II, Section 10.6.3.2). Eaton and

Scheller (1996) project that the suitable habitat for cold-, cool-, and even many warm-water fish species would be reduced by about 50% in streams of the lower 48 states in the United States by summer mean air temperature increases of 2–6°C (derived from a CCC GCM under a $2xCO_2$ climate). In the North Platte River drainage of Wyoming, Rahel *et al.* (1996) project losses from the present geographic range of cold-water fish species of from 7–16% for a 1°C increase to 64–79% for a 5°C increase in summer air temperatures, with considerable fragmentation of remaining populations. In a national study of climate-induced temperature effects on fish habitat, complete elimination of cold-water habitat is projected for sites modeled in 5 to 10 states and severe reductions in 11 to 15 states, depending on the GCM climate projection used (U.S. EPA, 1995). It must be noted, however, that most studies of warming effects on stream fish populations have used mean temperature increases that generally are greater than those produced by current models for the same time period; these later models include aerosol forcing and show minimal summer daytime changes and more cloud cover in summer for many regions.

Whole ecosystem experiments that alter the thermal, hydrological, or mixing regimes in small lakes and streams or in large-scale mesocosms (e.g., lake enclosures or artificial streams) are needed to determine the responses of organisms, processes, and habitats to global change. Additional work also is needed in the area of comparative studies of populations or ecological processes across latitudinal and hydrological gradients and ecosystem types to enable us to better understand climate-induced temperature effects in the context of natural seasonal and interannual variability.

In addition, qualitative projections of the consequences of climate change on the fish resources of North America will require good regional atmospheric and oceanic models of the response of the ocean to climate change; improved knowledge of the life histories of the most vulnerable species for which projections are required; and a further understanding of the roles that the environment, species interactions, and fishing play in determining the variability of growth, reproduction, and abundance of fish stocks.

*If climate changes result in lower water levels, reduced runoffs, and increased hydrological variability, the productivity of some freshwater species may decline.*

In areas of North America that experience significant reductions in runoff, lower water levels in some lakes may eliminate or reduce the productivity of fish species dependent on shallow near-shore zones or adjacent wetlands as spawning or nursery areas (IPCC 1996, WG II, Section 10.6.2.2). Some shallow lakes with relatively short water residence times may disappear entirely with reduced annual runoff. For example, Marsh and Lesack (1996) project with a hydrological model that under a $2xCO_2$ climate, many lakes in the Mackenzie delta in the Canadian arctic could disappear in several decades as a result of decreased precipitation and flooding frequency. In some lake districts, connectivity among lakes would be decreased by the cessation of flow in connecting streams—possibly eliminating species such as northern pike from some shallow-water lakes because they no longer have connections to deep-water lakes for winter habitat (IPCC 1996, WG II, Section 10.6.2.2).

In many areas, increases in flow variability are likely to produce larger effects on biota than changes in mean flows (IPCC 1996, WG II, Section 10.6.2.1), which could result in some changes in ecosystem productivity and organism abundance (including positive and negative effects). In arid-land streams, more intense storms and longer periods of drought may produce severe streambank erosion, lower biomass and productivity, and a decline in biological interactions (IPCC 1996, WG II, Box 10-3). In humid regions, more intense or clustered storms could reduce the abundance of many stream organisms via scouring of streambeds, although greater frequency of floodplain flooding also might increase the productivity of many river and stream organisms. Longer periods of drought in humid regions, particularly in summer, could increase the probability that streams will cease flowing and become dry; reductions in annual runoff also could increase the probability that streams will dry. In a regional analysis of U.S. streams, Poff (1992) projected that nearly one-half of perennial runoff streams in the eastern United States may become intermittent with only a 10% decline in annual runoff. Even if streams do not become intermittent, longer droughts and lower summer baseflows could result in more severe water quality deterioration (low dissolved oxygen concentrations, high concentrations of contaminants), which will reduce available habitat and eliminate intolerant species from streams (IPCC 1996, WG II, Section 10.5.4).

#### 8.3.6.2. Oceans

An early review (Wright *et al.*, 1986, summarized by Mann, 1993) projected that for the northern Atlantic, some of the consequences of global warming could include:

- A rise in the average sea surface temperature, causing an increase in evaporation and a more vigorous hydrological cycle of precipitation, runoff, and so forth
- The greatest increase in evaporation in mid-latitudes, leading to increased precipitation in northern regions, increased river runoff, increased stability of the water column, and increased strength of buoyancy-driven currents such as the Labrador current
- An increase in the north-south gradient in salinity
- A decrease in the thickness and extent of ice cover
- A reduction of the north-south temperature gradient and possibly a reduction in average wind stress over the whole of the north Atlantic, which could lead to a decrease in the strength of wind-driven currents such as the Gulf Stream.

In their summary of the Symposium on Climate Change and Northern Fish Populations, Sinclair and Frank (1995) described the variability of the northern Pacific in circulation and mixing and linked that variability in part to shifts in atmospheric circulation—specifically, the changes in the location and level of the Aleutian low-pressure system. Existing models have not been able to shed light on the most probable responses of the northern Pacific to a doubling of atmospheric $CO_2$.

Mann (1993) briefly considered various sources of data for the wind-driven coastal upwelling system off California. He suggested that available data could be used to support the hypothesis that coastal upwelling increases during global cooling but decreases during global warming.

### 8.3.6.3. Impacts, Adaptations, and Vulnerabilities of Ocean Fisheries Resources

*Overall, there likely will be relatively small economic and food supply consequences at the regional/national level; however, impacts are expected to be more pronounced at the subregional level.*

Natural climate variability—for example, changes in ocean temperatures and circulation patterns associated with the El Niño phenomenon and with the northern Pacific gyre—affects the distribution and composition of fisheries. Because interannual and decadal-scale natural variability is so great relative to global change and the time horizon on capital replacement (ships and plants) is so short, impacts on fisheries can be easily overstated; there likely will be relatively small economic and food supply consequences in the United States and Canada at the national level. At the state or regional level, impacts (positive and negative) will be more pronounced, particularly when a center of production shifts sufficiently to make one fishing port closer to a resource while a traditional port becomes more distant. Over time, fishing vessels and their support structure will relocate, followed by processors and eventually families as well. Community impacts can be significant.

Changes in primary production levels in the ocean as a result of climate change may affect fish stock productivity. As a first step in assessing the role of changes in primary production on fish productivity, global primary production in the ocean has been estimated by Longhurst *et al.* (1995) using satellite measurements of near-surface chlorophyll fields. Annual global primary production was estimated at 45–50 Gt carbon (C)/year. This annual global primary production is the sum of the annual primary production in 57 biogeochemical provinces covering the world ocean. More than 10 such provinces border North America. For example, the total primary production is estimated at 0.37 Gt C/year in the "California Upwelling Coastal" province and 1.08 Gt C/year in the "Northwest Atlantic Continental Shelf" province.

Exactly how climate-induced changes in primary production would affect the next trophic link, zooplankton, remains a matter of debate (e.g., Banse, 1995). However, changes in zooplankton biomass are known to affect fish stock productivity. Brodeur and Ware (1995) identified a twofold increase in salmonid biomass in the eastern subarctic Pacific since the 1950s, coincident with a large-scale doubling of the summer zooplankton biomass in the same region. Beamish and Bouillon (1995) examined trends in marine fish production off the Pacific coast of Canada and the United States. They concluded that the carrying capacity for fish in the northern North Pacific Ocean and the Bering Sea fluctuates in response to long-term trends in climate.

*Projected changes in water temperatures, salinity, and currents can affect the growth, survival, reproduction, and spatial distribution of marine fish species and of the prey, competitors, and predators that influence the dynamics of these species.*

Environmental conditions have a marked effect on the growth of many fish species. For example, mean bottom temperatures account for 90% of the observed (10-fold) difference in growth rates between different Atlantic cod (*Gadus morhua*) stocks in the north Atlantic (Brander, 1994, 1995). Warmer temperatures lead to faster growth rates. Regional studies have shown similar results (Fleming, 1960; Shackell *et al.*, 1995). In the northwest Atlantic, the largest cod typically are found on Georges Bank—where a 4-year-old fish, on average, is five times bigger than one off Labrador and Newfoundland. Temperature accounts not only for differences in growth rates between cod stocks but also year-to-year changes in growth within a stock.

In addition to growth, the environment affects the reproductive cycle of fish and shellfish. For example, the age of sexual maturity of certain fish species is determined by ambient temperature. Atlantic cod off Labrador and the northern Grand Banks mature at age 7 and in the northern Gulf of St. Lawrence and the eastern Scotian Shelf at age 6; in the warmer waters off southwest Nova Scotia and on Georges Bank, they mature at 3.5 years and 2 years, respectively (Myers *et al.*, 1996).

Spawning times also are influenced by temperature. Generally, cold temperatures result in delayed spawning (Hutchinson and Myers, 1994a), whereas warm temperatures result in earlier spawning. Marak and Livingstone (1970) found that a 1.5–2°C temperature change produced a difference in the spawning time of haddock on Georges Bank by a month, with earlier spawning and a longer duration in warmer years.

Temperature is one of the primary factors, along with food availability and suitable spawning grounds, that determine the large-scale distribution patterns of fish and shellfish. Because most fish species or stocks tend to prefer a specific temperature range (Coutant, 1977), long-term changes in temperature can lead to expansion or contraction of the distribution range of certain species. These shifts generally are most evident near their northern or southern boundaries; warming results in a distributional shift northward, and cooling draws species southward.

Changes in distribution also were observed during a warming trend in the 1940s in the Gulf of Maine—which produced a northward shift in the abundance and distribution of Atlantic mackerel, American lobster, yellowtail flounder, Atlantic menhaden, and whiting, as well as the range extension of more southern species such as the green crab (Taylor *et al.*, 1957).

Frank *et al.* (1990) projected a northward shift of the southern extensions of important fisheries such as Atlantic cod, Atlantic halibut, American plaice, and redfish out of the Gulf of Maine and into Canadian waters. Also projected is the northern extension of more southern fish species such as Atlantic menhaden, butterfish, and redhake further northward into the Gulf of Maine. Coutant (1990) suggests a northward shift of summer stocks of striped bass, with losses occurring for Virgina, Delaware, and New York and gains projected for Massachusetts, New Hampshire, Maine, and New Brunswick.

Understanding recruitment variability has been the number one issue in fisheries science in this century. Since the advent of intensive fishing, it has become increasingly difficult to sort out the relative importance of fishing versus environment as the cause of recruitment variability. Still, recruitment levels frequently have been associated with variations in temperature during the first years of life of the fish (Drinkwater and Myers, 1987). American lobster landings increased steadily during the 1980s and into the 1990s, to all-time historic highs. However, the temperature/landing relationships for lobster are not consistent with an expected positive linear relationship—suggesting that more than one variable can control the relationship, and a different variable may be the dominant one at any given time.

Climate also can affect the fishery through its influence on availability (fish available to be caught) and catchability (difficulty to catch), both of which depend not only on the abundance of fish but on when and how they are distributed. If cod traps are located in waters that are too cold, catches are low. Only when the temperature is warm enough do catches increase. In the case of lobster catchability, when temperatures are low, lobster are known to move slowly, reducing the potential for encountering lobster traps and hence reducing catchability (McLeese and Wilder, 1958).

Climate change can be expected to result in distributional shifts in species, with the most obvious changes occurring near the northern or southern boundaries of species' ranges. Migration patterns will shift, causing changes in arrival times along the migration route. Growth rates are expected to vary (with the amplitude and direction species-dependent). Recruitment success could be affected by changes in time of spawning, fecundity rates, survival rate of larvae, and food availability. Another possibility associated with climate change is a change in stratification (as a result of differences in heating, freshwater, and vertical mixing rates), which may lead to changes in the ratio of pelagic to groundfish abundance (Frank *et al.*, 1990). If stratification were to increase, more production would be expected to be recycled within the upper layers of the oceans, and less would reach the bottom.

Evidence of environmental control on the distribution of marine fish is abundant. For example, Welch *et al.* (1995) have identified critical temperatures defining the southern boundaries of salmonid species. The authors suggest that future temperature changes in the northern Pacific therefore could have a direct impact on the production dynamics of Pacific salmon. Impacts of global warming in the ocean, however, will be difficult to separate from natural shifts in ocean carrying capacity. A general warming of the ocean will have an impact on predators and prey distributions. In the Strait of Georgia, there was an abrupt decline in marine survival after the 1976–77 regime shift, but the mechanisms responsible remain unknown. On the west coast, warm periods after the 1989–90 climate change resulted in an influx of predators that caused large increases in juvenile mortalities. It is impossible to forecast the actual changes in the marine ecosystems; thus, the degree to which chinook marine survival may be affected is unknown. The abruptness of change in the Strait of Georgia and the west coast is of concern because it indicates that signals of change need to be detected quickly and managed effectively (Beamish *et al.*, 1997).

*Because salmonid species (and other anadromous species such as striped bass) rely on marine and freshwater aquatic systems at different points in their life cycles, projected changes in marine and freshwater water temperatures, ocean currents, and freshwater flows are more likely to impact growth, survival, reproduction, and spatial distribution of these species than of other fish species.*

Because of their anadromous life history, pink salmon are affected by changes in freshwater and changes in the ocean—and the impacts in each of these habitats are equally important. Recent research has shown that trends in pink salmon productivity shift in response to climate-driven changes in the ocean. Because the mortality of young pink salmon is so high (95–98%) shortly after they first enter the ocean, small changes in marine survival can result in large changes in adult returns.

Warmer freshwater and oceans and changes in the pattern of Fraser River flows probably will reduce the abundance of pink salmon, although individual size may increase because of improved growth in the warmer water. Warmer temperatures will reduce incubation time, and the longer period in fresh water will improve growth. In the smaller rivers, where flows are a function of winter precipitation, increased precipitation may increase water flows—resulting in higher egg and alevin mortality. Dracup *et al.* (1992) examined the effect of climate change in altering the timing of streamflow regimes; they suggest that these changes may increase mortality and reduce fish population in the Sacramento-San Juaquin chinook salmon fishery.

Marine effects obviously are relevant to hatchery-reared fish. Reduced coastal productivity resulting from reduced upwelling may reduce the total carrying capacity for pink salmon, and it may not be possible to build stocks to historic levels in a poor-productivity regime by producing more fry.

In recent years, research has shown that chum salmon productivity follows trends that shift in relation to climate-related changes in the ocean. Thus, changes in upwelling and the intensity of winds may reduce the carrying capacity for chum in the ocean to levels below what might occur during natural changes.

Increases in temperature in freshwater rearing areas and increased winter flows may increase freshwater chum mortalities for stocks in the Fraser River and other southern rivers. Chum are a very adaptable species, however, and spawning tends to be in the lower portion of rivers and streams; thus, the changes in saltwater may be more influential than changes in fresh water. It is possible that earlier and larger spring flows in rivers may improve survival in the ocean, if the initiation of the spring bloom occurs at a more favorable time. In recent years, relatively large numbers of age-0 ocean chum salmon have remained in the Strait of Georgia until late in the year, even though the surface temperatures have increased over the past 20 years. This pattern may indicate that the timing of plankton production is more favorable as a result of larger flows in April (Beamish *et al.*, 1997).

In the south, warmer river water and reduced flows in late summer may increase mortalities and reduce spawning success. Warmer waters in the winter will accelerate incubation and hatching and cause alevins to enter lakes earlier. Henderson *et al.* (1992) concluded that warming of sockeye rearing lakes would lower plankton production and reduce the size of smolts going to sea. These smaller smolts also may encounter reduced food when they enter the ocean, and the resulting slower growth may expose juveniles to predation longer and increase mortality in the early marine period. Welch *et al.* (1995) proposed that global warming would increase winter temperatures sufficiently that sockeye juveniles would migrate out of the northern Pacific into the Bering Sea, effectively reducing the winter feeding area. It is known that there are large interannual fluctuations in survival (Burgner, 1991) and large, natural, decadal shifts in marine survival (Hare and Francis, 1995; Adkison *et al.*, 1996; Beamish *et al.*, 1997). The mechanisms involved are not understood, but the shifts in abundance clearly show that changes in the ocean environment have profound impacts on the productivity of the stocks.

It is possible that changes affecting the northern stocks may not have a major impact on the stocks in the next 50 years. This speculation is based on the cumulative effects of freshwater and marine events in the early 1990s that have produced historic high returns to some of the northern sockeye stocks in Canada and Alaska.

Projected changes in climate can affect the timing of the return of anadromous species to fresh water to spawn in some of the smaller streams. Changes in the timing of spawning can change the behavior (e.g., select for later-spawning fish); however, it is not anticipated that large numbers of stocks would be adversely affected in the next 50 years. Warmer rivers will shorten the incubation time, which may result in a longer growing season in fresh water. Although fish may feed longer and grow to larger sizes, they also may enter the ocean earlier. This shift may change the percentage of life history types that survive more than overall survival because there already is an extended period of entry into saltwater for the various rearing types.

*Aquaculture potential will be affected by projected changes in climate and climate variability and could take advantage of extended favorable conditions in currently marginal areas.*

Most of the recent growth in total fisheries production is from aquaculture, which has grown rapidly during the past few decades and accounts for about 10% of total world fish production—mostly of higher-valued products. Aquaculture contributes to the resiliency of the fisheries industry, tending to stabilize supply and prices. Advancements are unevenly distributed across regions, farming systems, and communities. Growth in the United States is about 5% annually. The marine component is growing rapidly, but freshwater aquaculture is still dominant. Aquaculture will not rapidly solve the scarcity of natural fish, and current industry growth will fulfill the demand only for certain commodities, regions, and consumer groups.

Genetic engineering holds great promise to increase the production and efficiency of fish farming (Fischetti, 1991). However, fishers and resource managers are very concerned about accidental or intentional release of altered and introduced species that might harm natural stocks and gene pools. Around Scandinavia, escapees and nonindigenous reproduction may have reached or exceeded the recruitment of salmon wild stocks (Ackefors *et al.*, 1991). Other concerns associated with aquaculture are the discharge of excess nutrients into surrounding waters that can add to eutrophication; the heavy use of antibiotics and contamination with pesticides, potentially leading to disease outbreaks; and the introduction of pathogenic organisms and antibiotic-resistant pathogens.

Ranching (in which young fish are released to feed and mature at sea) and fish farming, like their equivalents on land, have self-generated and imposed impediments to success. The activities can compete for coastal space with other uses, and continued expansion can jeopardize the quality and quantity of fish habitat (e.g., through loss of mangroves and wetlands, competition for food with wild stocks, or other factors) (NCC, 1989).

Climate variability is important to aquaculture. Decreasing temperatures may cause low minimum temperatures through the year—possibly causing mass mortalities, especially along the east coast. Long-term temperature trends will affect what species of fish or shellfish are suitable, as well as the expansion or contraction of suitable aquaculture sites. General warming may allow aquaculture sites to expand into regions previously unavailable because water temperatures were too cold or there was a presence of sea ice. Growth rates of fish or shellfish and their food requirements are temperature dependent. Aquaculturists also are interested in projections of wind mixing, which contributes to flushing (i.e., the exchange of water between the aquaculture site and surrounding waters). Low flushing can lead to decreased oxygen; greater potential for the spread of diseases; and, in the case of filter feeders such as mussels, reduced food availability.

*The survival, health, migration, and distribution of many North American marine mammals and sea turtles are expected to be impacted by projected changes in the climate through impacts on their food supply, sea-ice extent, and breeding habitats.*

In North American waters, approximately 125 extant species of marine mammals (e.g., whales, dolphins, seals, sea lions, polar bears, and marine otters) are known to occur at least some time during the year. Although reliable abundance estimates for these mammals in North America are limited, there are endangered or threatened species among these mammals (e.g., 28 species are listed as either endangered or threatened under the U.S. Endangered Species Act or depleted under the U.S. Marine Mammal Protection Act); many are recovering from commercial harvesting and overexploitation.

Many marine mammals (e.g., the great whales) are able to locate and follow seasonal centers of food production, which frequently change from year to year depending on local oceanographic conditions. Similarly, their migrations may change to accommodate interannual differences in environmental conditions. However, some marine mammals (e.g., seals and sea lions) have life histories that tie them to specific geographic features (e.g., pupping beaches or icefields). Although there is some flexibility in their need for specific habitats, some marine mammals may be more severely affected than others by changes in the availability of necessary habitats and prey species that result from climate change.

Seasonal sea-ice extent, at least in some areas of the Northern Hemisphere, is retreating. This information, coupled with projections of warming, suggests that current barriers to gene flow among marine mammal stocks in the Arctic may change dramatically in the next 50 years. Although this shift may not result in a reduction in abundance at the species level, it could very well change the population structure of many species of Arctic whales and seals, which will greatly affect their management.

Coastal wetlands and beaches may be eliminated in some areas by rising sea level. As a result, marine mammal calving and pupping beaches may disappear from areas where there are no alternatives. Affected marine mammals could include, for example, all of the temperate and tropical seals and sea lions, coastal whales and dolphins, and manatees in estuarine habitats.

Six species of sea turtles (all of which are listed as endangered or threatened under the U.S. Endangered Species Act) regularly spend all or part of their lives off North American coasts and in U.S. territorial waters of the Caribbean Sea and Pacific Ocean. The loss of nesting beaches that would result from the combination of coastal development and projected sea-level rise is a threat to all marine turtle species.

### 8.3.7. Coastal Systems

Rising sea level is gradually inundating wetlands and lowlands; eroding beaches; exacerbating coastal flooding; threatening coastal structures; raising water tables; and increasing the salinity of rivers, bays, and aquifers (Barth and Titus, 1984). The areas most vulnerable to rising seas are found along the Gulf of Mexico and the Atlantic Ocean south of Cape Cod. Although there also are large low areas around San Francisco Bay and the Fraser delta (British Columbia), most of the Pacific coast is less vulnerable than the Atlantic and Gulf coasts. Because of a combination of rocky shores, lower rates of sea-level rise, higher elevations, and less shorefront development, most of the Canadian coast is much less vulnerable to the direct effects of rising sea level (Shaw *et al.*, 1994) than the low, sandy and muddy shores of the United States.

This section focuses primarily on the impacts of sea-level rise, which is the most thoroughly studied effect of global warming on coastal zones. Nevertheless, global climate change also is expected to alter coastal hydrology, the frequency and severity of severe storms, and sea-ice cover. Moreover, the impacts of regional climate change on inland areas also will affect coastal zones—particularly the estuaries into which most of the continent drains.

#### 8.3.7.1. Physical Effects and Their Implications

The implications of rising sea level are well understood, in part because sea level has been rising relative to the land along most of the coast of North America (and falling in a few areas) for thousands of years. For the most part, the relative rise and fall of sea level has been caused by adjustments of the earth's crust to the glacial mass that was removed from the land surfaces after the end of the last ice age (Grant, 1975). Change in the volume of water in oceans was also of importance. Water locked up in ice caps during ice ages lowered the volume of water in oceans, thus lowering sea level. The changes discussed here have occurred over geologic time (Holocene Epoch—last 10,000 years). The land is rising (i.e., relative sea level is falling) in the northern areas that had been covered by the ice sheet; land is subsiding in nearby areas that were not covered by the glaciers, such as the Canadian maritime provinces and U.S. middle Atlantic states.

*A 50-cm rise in sea level would inundate approximately 50% of North American coastal wetlands in the next century; many beaches would be squeezed between advancing seas and engineering structures, particularly along estuarine shores.*

Coastal marshes and swamps generally are found between the highest tide of the year and mean sea level. Coastal wetlands provide important habitat and nourishment for a large number of birds and fish found in coastal areas. Wetlands generally have been able to keep pace with the historic rate of sea-level rise (Kaye and Barghoorn, 1964). As a result, the area of dry land just above wetlands is less than the area of wetlands. If sea level rises more rapidly than wetlands can accrete, however, there will be a substantial net loss of wetlands (Titus, 1986; Park *et al.*, 1989). Because the current rate of sea-level rise is greater than the rate that prevailed over the past several thousand years

(IPCC 1996, WG I), some areas—such as Blackwater National Wildlife Refuge (NWR) along the Chesapeake Bay—are already experiencing large losses of coastal wetlands (Kearney and Stevenson, 1985). Blackwater NWR is also a victim of herbivory by an introduced rodent, which makes interpretation of the role of sea-level rise difficult.

Coastal development is likely to increase the vulnerability of wetlands to rising sea level. In many areas, development will prevent the wetland creation that otherwise would result from the gradual inundation of areas that are barely above today's high-water level (Titus, 1986, 1988). In Louisiana, flood control levees, navigation infrastructure, and other human activities have disabled the natural processes by which the Mississippi delta otherwise could keep pace with rising relative sea level; as a result, Louisiana currently is losing about 90 $km^2$ (35 $mi^2$) of wetlands per year (Gagliano *et al.*, 1981; Penland *et al.*, 1997).

Louisiana is expected to experience the greatest wetland loss from rising sea level, although most of these losses are predicted to occur even with the current rate of relative sea-level rise. The mid-Atlantic, south Atlantic, and Gulf coasts also are likely to lose large areas of wetlands if sea-level rise accelerates. A 50-cm rise in sea level would cause a net loss of 17–43% of the wetlands, even if no additional bulkheads or dikes are erected to prevent new wetland creation. Table 8-6 presents estimated losses in U.S. wetlands by region. Similar comprehensive assessments are unavailable for Canada. Nevertheless, regional studies suggest that the most vulnerable area is likely to be the salt marsh coast of the Bay of Fundy. Because 85% of these wetlands are enclosed by a system of dikes, the risk is not so much the direct submergence by higher water levels but rather the possibility that unless the dikes are fortified, an increased storm surge could overtop and breach the dikes. Many of the wetlands around San Francisco are similarly vulnerable.

In estuaries, sandy beaches may be even more vulnerable than vegetated wetlands to being squeezed between rising sea level and development. A 1-cm rise in sea level generally erodes beaches about 1 m (Bruun, 1962). Thus, because estuarine beaches usually are less than 5 m wide (Nordstrum, 1992), even a 5-cm rise in sea level can eliminate these systems in areas where adjacent land is protected with structures. Moreover, the environmental regulations that protect wetlands generally have not been applied to protect estuarine beaches

***Table 8-6:*** *Regional and national wetland losses in the U.S. for the trend and 1-m global sea-level rise scenarios (% loss of current area).*

| Region | Current Wetland Area ($mi^2$) | Trend | 1-m Shore Protection Policy Total[a] | Developed[a] | None[a] |
|---|---|---|---|---|---|
| Northeast | 600 | 7 | 16 | 10 | 1[d] |
| Mid-Atlantic | 746 | -5 | 70 | 46 | 38 |
| South Atlantic | 3,814 | -2 | 64 | 44 | 40 |
| South/Gulf Coast of Florida | 1,869 | -8 | 44 | 8[d] | 7[d] |
| Louisiana[b] | 4,835 | 52 | 85 | 85 | 85 |
| Florida panhandle, Alabama, Mississippi, and Texas | 1,218 | 22 | 85 | 77 | 75 |
| West[c] | 64 | -111 | 56 | -688 | -809 |
| United States | 13,145 | 17 | 66 | 49 | 50 |
| *Confidence Intervals* | | | | | |
| 95% Low | — | 9 | 50 | 29 | 26 |
| 95% High | — | 25 | 82 | 69 | 66 |

[a] The "total" protection scenario implies that all shorelines are protected with structures; hence, as existing wetlands are inundated, no new wetlands are formed. "Developed" implies that only areas that are currently developed will be protected; "no protection" assumes that no structures will be built to hold back the sea.
[b] Evaluation of management options currently contemplated for Louisiana (e.g., restoring natural deltaic processes) was outside the scope of this study.
[c] This anomalous result is from small sample size. The impact on nationwide results is small.
[d] Results are not statistically significant; sampling error exceeds estimate of wetlands lost.
Source: Titus *et al.*, 1991.

***Table 8-7:*** *Loss of dry land from sea-level rise (95% confidence interval, mi²).*

| | Baseline | Rise in Sea Level (cm) 50 | 100 | 200 |
|---|---|---|---|---|
| If no shores are protected | NR | 3,300–7,300 | 5,100–10,300 | 9,200–15,400 |
| If developed areas are protected | 1,500–4,700 | 2,200–6,100 | 4,100–9,200 | 6,400–13,500 |

NR = not reported.
Source: Titus *et al.*, 1991.

(Titus, 1997), which are important for recreation, navigation, and habitat for several endangered species (Nordstrum, 1992).

*A 50-cm rise in sea level could inundate 8,500–19,000 km² of dry land, even if currently developed areas are protected.*

The dry land within 1 m above high tide includes forests, farms, low parts of some port cities, communities that sank after they were built and that now are protected with levees, parts of deltas, and the bay sides of barrier islands. The low forests and farms generally are in the Mid-Atlantic and Southeast. Major port cities with low areas include Boston, New York, Charleston, Miami, and New Orleans. New Orleans' average elevation is about 2 m below sea level; parts of Texas City, San Jose, and Long Beach, California, are about 1 m below sea level. In the United States, 8,500–19,000 km² (3,300–7,300 mi²) of dry land are within 50 cm of high tide—5,700–16,000 km² (2,200–6,100 mi²) of which currently are undeveloped (Table 8-7) (Titus *et al.*, 1991). Approximately 100 km² of land in the Fraser delta (British Columbia) also is within 1 m of sea level.

*Many islands are at risk. The low bay sides of developed barrier islands could be inundated while their relatively high ocean sides erode. Undeveloped barrier islands will tend to migrate landward through the overwash process.*

The most economically important vulnerable areas are recreational resorts on the coastal barriers—generally long and narrow islands or spits (peninsulas) with the ocean on one side and a bay on the other—of the Atlantic and Gulf coasts. Typically, the oceanfront block is 2–5 m above high tide; the bay sides often are <0.5 m above high water.

Erosion threatens the high ocean sides of these densely developed islands; this oceanfront erosion generally is viewed as a more immediate problem than inundation of the islands' low bay sides. Shores currently are eroding at a rate of 0.25–0.5 m/yr in many areas. Studies using the "Bruun (1962) rule" have estimated that a 1-cm rise in sea level will cause beaches to erode 0.5–1 m from New England to Maryland, 2 m along the Carolinas, 1–10 m along the Florida coast, and 2–4 m along the California coast (Bruun, 1962; Kana *et al.*, 1984; Everts, 1985; Kyper and Sorensen, 1985; Wilcoxen, 1986). Because many U.S. recreational beaches are less than 30 m wide at high tide, even a 30-cm rise would threaten homes in these areas.

Canada's longest barrier coast is in New Brunswick along the Gulf of St. Lawrence; the narrow barrier islands and spits generally are undeveloped. Rising sea level tends to cause narrow islands to migrate landward through the overwash process (Leatherman, 1979). Although the barriers themselves are undeveloped, there are important recreational areas along the mainland coast behind the barriers, as well as environmentally sensitive freshwater bogs and woodlands.

Other types of islands also may be vulnerable to sea-level rise. In the Chesapeake Bay, several islands populated by a traditional subculture of fishermen are likely to be entirely submerged (Toll *et al.*, 1997). The coast of Prince Edward Island, except for some parts along the Northumberland Strait, is highly erodible because of its bedrock cliffs, sandy barriers, coastal dunes, salt marshes, and intertidal flats. The heart of the island's tourist industry, along the Gulf of St. Lawrence, is likely to experience increased beach erosion, which would threaten shorefront buildings.

*Rising sea level would increase flooding and storm damage. Regional climate change could offset or amplify these effects, depending on whether river flows and storm severity increase or decrease.*

Changing climate generally is increasing the vulnerability of coastal areas to flooding both because higher sea level raises the flood level from a storm of a given severity and because rainstorms are becoming more severe in many areas. It also is possible that hurricanes could become more intense, thus producing greater storm surges; IPCC (1996) concluded, however, that the science currently is inadequate to state whether or not this is likely. Existing assessments in coastal areas generally focus on the impact of rising sea level.

Because higher sea level provides a higher base for storm surges, a 1-m rise in sea level (for example) would enable a 15-year storm to flood many areas that today are flooded only by a 100-year storm (Kana *et al.*, 1984; Leatherman, 1984). Many coastal areas currently are protected with levees and seawalls. Because these structures have been designed for current sea level, however, higher storm surges might overtop seawalls, and erosion could undermine them from below (National Research Council, 1987). In areas that are drained artificially, such as New Orleans, the increased need for pumping could exceed current pumping capacity (Titus *et al.*, 1987).

The U.S. Federal Emergency Management Agency (FEMA, 1991) has examined the nationwide implications of rising sea level for the National Flood Insurance Program. The study estimated that rises in sea level of 30 cm and 90 cm would increase the size of the 100-year floodplain in the United States from 51,000 km$^2$ (19,500 mi$^2$) in 1990 to 60,000 km$^2$ and 70,000 km$^2$ (23,000 mi$^2$ and 27,000 mi$^2$), respectively. Assuming that current development trends continue, flood damages incurred by a representative property subject to sea-level rise are projected to increase by 36–58% for a 30-cm rise and 102–200% for a 90-cm rise.

Because of its higher elevations, the Canadian coastal zone is less vulnerable to flooding than the U.S. coast. Nevertheless, flooding appears to be a more serious risk to Canada than the loss of land from erosion or inundation. Some communities (e.g., Placentia, Newfoundland) already are vulnerable to flooding during high astronomical tides and storm surges, sometimes exacerbated by high runoff. In Charlottetown, Prince Edward Island, some of the highest-value property in the downtown core and significant parts of the sewage system would experience increased flooding with a 50- to 100-cm rise in sea level. According to Clague (1989), a rise of a few tens of cm would result in flooding of some waterfront homes and port facilities during severe storms in British Columbia, forcing additional expenditures on pumping.

Coastal flooding also is exacerbated by increasing rainfall intensity. Along tidal rivers and in extremely flat areas, floods can be caused by storm surges from the sea or by river surges. Washington, D.C., and nearby Alexandria, Virginia, were flooded twice by Hurricane Fran in 1996: first by a storm surge in the Chesapeake Bay and lower Potomac River, then three days later by the river surge associated with intense precipitation over the upper Potomac River's watershed. Higher sea level and more intense precipitation could combine synergistically to increase flood levels by more than the rise in sea level alone in much of coastal Louisiana and Florida, as well as in inland port cities along major rivers (such as Portland and Philadelphia). The direct effect of higher sea level also could be exacerbated throughout the coastal zone if hurricanes or northeasters become more severe—a possibility that has been suggested but not established (IPCC 1996, WG I).

*Rising sea level would increase salinities of estuaries and aquifers, which could impair water supplies, ecosystems, and coastal farmland. As with flooding, regional climate change could offset or amplify these effects, depending on whether river flows increase or decrease.*

Rising sea level also enables saltwater to penetrate farther inland and upstream in rivers, bays, wetlands, and aquifers; saltwater intrusion would harm some aquatic plants and animals and threaten human uses of water. Increased drought severity, where it occurs, would further elevate salinity. Increased salinity already has been cited as a factor contributing to reduced oyster harvests in Delaware Bay (Gunter, 1974) and the Chesapeake Bay and as a reason that cypress swamps in Louisiana are becoming open lakes (Louisiana Wetland Protection Panel, 1987).

Higher salinity can impair both surface and groundwater supplies. New York, Philadelphia, and much of California's Central Valley get their water from portions of rivers that are slightly upstream from the point at which the water is salty during droughts. If saltwater is able to reach farther upstream in the future, the existing intakes would draw salty water during droughts.

The aquifers that are most vulnerable to rising sea level are those that are recharged in areas that currently are fresh but could become salty in the future. Residents of Camden and farmers in central New Jersey rely on the Potomac-Raritan-Magothy aquifer, which is recharged by a portion of the Delaware River that is rarely salty even during severe droughts today but would be salty more frequently if sea level were to rise 50–100 cm or droughts were to become more severe (Hull and Titus, 1986). Miami's Biscayne aquifer is similarly vulnerable; the South Florida Water Management District already spends millions of dollars each year to prevent the aquifer from becoming salty (Miller *et al.*, 1992).

A second class of vulnerable aquifers consists of those in barrier islands and other low areas with water tables close to the surface, which could lose their freshwater lens entirely (see IPCC 1990, WG II, Figure 6.3; also Chapter 9 in this report).

Finally, rising sea level tends to make some agricultural lands too saline for cultivation. In areas where shorefront lands are cultivated, the seaward boundary for cultivation often is the point where saltwater from ground and surface waters penetrates inland far enough to prevent crops from growing. As sea level rises, this boundary penetrates inland—often rendering farmland too salty for cultivation long before inundation converts the land to coastal marsh (see, e.g., Toll, 1997).

*Coastal areas in the Arctic and extreme north Atlantic and Pacific are less vulnerable, except where sea ice and/or permafrost currently is present at the shoreline.*

Sea-level rise and storm surges along the tundra coastline of Alaska and Canada are likely to cause erosion, flooding, and inundation through mechanisms similar to those for other parts of the North American coast. Several additional factors, notably sea-ice effects and coastal permafrost degradation, also will come into play. Projected changes in sea ice include a 35% decrease in winter ice thickness, along with significant retreat of the southern limit of sea ice and complete absence of summer sea ice among the Arctic Islands (Maxwell and Barrie, 1989). These decreases in the period and extent of sea-ice cover will result in larger ocean fetches and greater wave attack on the coastal zone (Lewis, 1974), with attendant erosion. Subsequent modeling suggested that the wave energy during the open-water season may increase wave heights by 16–40% (McGillivray *et al.*, 1993).

Rates of erosion of permafrost also can be expected to increase. The Alaska and Yukon coasts already experience significant erosion during the annual thaw. According to Lewellen (1970), erosion rates in the mid-1960s and early 1970s ranged from a few decimeters to as much as 10 m per year. Maximum erosion occurred in areas where permafrost contained considerable pore, wedge, or massive ice (Lewis, 1974) or where the permafrost shoreline was exposed to the sea (Lewellen, 1970).

#### *8.3.7.2. Adapting to Sea-Level Rise*

*Adaptive responses focus on protection of shores or allowing them to retreat, with subsequent loss of existing shoreline systems and structures.*

Several U.S. government agencies have started to prepare for rising sea level. The U.S. Coastal Zone Management Act requires state coastal programs to address rising sea level, and a few states have modified coastal land-use policies to address rising sea level. The U.S. Army Corps of Engineers is required to consider alternative scenarios of future sea-level rise in its feasibility studies. These anticipatory measures have been implemented in part because assessments have identified measures whose costs are less than the benefits of preparation—even when future benefits are discounted by an economic rate of return.

##### *8.3.7.2.1. Erecting walls to hold back the sea*

Most assessments of North American response strategies to future sea-level rise have concluded that coastal cities will merit protection with bulkheads, dikes, and pumping systems (National Research Council, 1987; Titus *et al.*, 1991). Bulkheads, seawalls, and rock revetments already are being used to halt erosion to protect land that is well above sea level. Dikes and pumping systems are used to protect urban areas such as New Orleans that are below sea level, and other areas that are below flood levels.

Although structural measures can protect property from rising water levels, the resulting loss of natural shorelines may have adverse environmental, recreational, and aesthetic effects. Wetland and shallow-water habitats already are being lost because protective structures prevent those systems from migrating inland. In other areas, sandy and muddy beaches are being eliminated—impairing the ability of some amphibious species to move between the water and the land and directly removing the habitat of species that inhabit these beaches. The elimination of natural beaches may harm recreational and fishing navigation by removing locations from which small craft can be launched or beached in an emergency; the loss of beaches also impairs the ability of the public to move along the shore for fishing, recreation, and other uses. In the past 15 years, the state of Maryland alone has lost the use of 500 km (300 mi) of shorelines through the issuance of permits for bulkheads and revetments (Tidal Waters Division, 1978–93).

##### *8.3.7.2.2. Elevating land surfaces and beaches*

The effects of rising sea level can be offset by elevating beaches, land surfaces, and structures as sea level rises. A key benefit of this approach is that the character of the shore is not altered. Rapidly subsiding communities such as Galveston, Texas, have used fill to raise land elevations; some authors have suggested that it will be necessary to elevate Miami because the soils are too permeable for effective pumping (e.g., Walker *et al.*, 1989). Regulations along San Francisco Bay require projects along the shore or on newly reclaimed land to be either protected by a dike or elevated enough to accommodate accelerated sea-level rise.[1]

The practice of elevating land surfaces is most applicable to recreational barrier islands, where environmental and aesthetic factors (such as natural beaches and waterfront views) can be as important as property values and shore-protection costs (Gibbs, 1984; Howard *et al.*, 1985; Titus, 1990). Figure 8-9 illustrates possible responses to sea-level rise for barrier islands: building a dike, elevating the land surface, engineering a landward retreat, and no protection. A case study of Long Beach Island, New Jersey, concluded that any of the three protection options would be less costly than the current value of the threatened land (Titus, 1990). Although dikes have a lower direct cost than elevating land and structures, the latter approach is least disruptive to existing land uses and can be implemented gradually over time.

##### *8.3.7.2.3. Protecting natural shorelines by allowing shores to retreat*

Several planning measures have been proposed to enable some shorelines to remain in roughly their natural state as sea level rises, rather than be replaced with structures. For the most part, these measures apply to areas that are not yet developed. They broadly fall into two categories: setbacks, which are regulations that prevent development of areas likely to be inundated, and rolling easements—which allow development today, but only with the explicit condition that the property will not be protected from rising water levels (Titus, 1997).

Setbacks currently are used to ensure that homes are safe from current flood risks. Several U.S. states currently require an additional erosion-based setback, in which new houses are set back an additional 20 to 60 times the annual erosion rate (Klarin and Hershman, 1990; Marine Law Institute *et al.*, 1995). Eventually, however, the shore will erode to any setback

[1]San Francisco Bay Conservation and Development Commission. Resolution 88-15. Adoption of Bay Plan Amendment No. 3-88 Concerning Sea Level Rise Findings and Policies.

line. Moreover, it is economically inefficient, and sometimes unconstitutional, to prevent the use of property now solely to avoid an adverse impact in the future (Titus, 1991).

Many of these problems are avoided with rolling easements—a planning measure in which coastal development is allowed in return for the property owner agreeing not to build structures or otherwise artificially stop the natural inland migration of wetlands and beaches. This option requires neither a specific estimate of future sea-level rise nor large public land purchases, and it is economically efficient because it does not prevent owners from using their land unless or until the sea rises enough to inundate it. The ability of the government to prevent property owners from eliminating the shore is grounded in the "public trust doctrine," under which the public has always owned tidal waters and either owned or had an access right along all intertidal beaches (Slade, 1990). If this approach were implemented in the next decade, ensuring the continued survival of natural wetland and beach shores in U.S. areas that are still undeveloped would cost approximately $400–1,200 million (Titus, 1997).

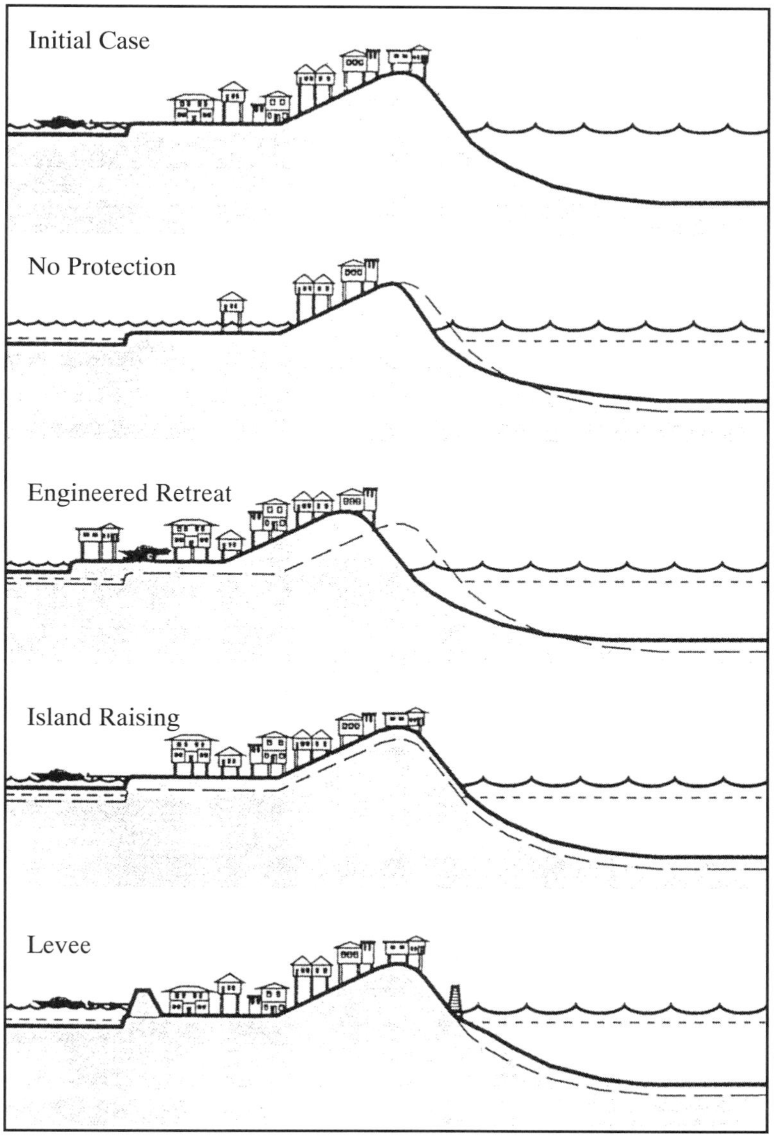

Initial Case

No Protection

Engineered Retreat

Island Raising

Levee

**Figure 8-9:** Responses to sea-level rise on developed barrier islands. Lightly developed islands may have no practical choice other than the "no protection" option, which would result in ocean-side erosion and in some cases bayside inundation. Under the "engineered retreat" option, a community might tolerate ocean-side erosion but move threatened structures to newly created bayside lands, imitating the natural overwash process that occurs with narrow undeveloped islands. A more common response is likely to be to raise entire islands as well as their beaches; although the sand costs are much higher than with an engineered retreat, existing land uses can be preserved. Finally, wide urbanized islands may choose to erect seawalls and levees (dikes); the loss of beach access and waterfront views, however, make this option less feasible for recreational barrier island resorts.

Texas common law recognizes rolling easements along its Gulf coast beaches. Maine and Rhode Island have issued regulations that prohibit structures that block the inland migration of wetlands. South Carolina's Beachfront Management Act, passed in response to the risks of a 1-ft rise in sea level, originally required setbacks along the coast, but in the aftermath of a trial court ruling that was eventually upheld by the U.S. Supreme Court (*Lucas* v. *South Carolina Coastal Council*), the statute was modified to require rolling easements in some locations[2] (South Carolina Beachfront Management Act, 1988). Because Canada inherited the same common law from England as the United States, all of these approaches could be applicable to Canada if its coastal zone becomes densely developed in the next century.

#### 8.3.7.2.4. *National assessments of adaptive responses*

Several nationwide assessments have been conducted in the United States, mostly focusing on the potential loss of wet and dry land and the cost of holding back the sea. These studies have recognized that the impact of sea-level rise ultimately depends on whether—and how—people hold back the sea; they generally estimate impacts assuming alternative policies for protecting coastal land. A rise of 50 cm would inundate 8,600–19,000 km$^2$ (3,300–7,300 mi$^2$) of dry land if no shores are protected and 5,700–16,000 km$^2$ (2,200–6,100 mi$^2$) if currently developed areas are protected (Table 8-7). The loss of coastal wetlands would be 17–43% if no shores are protected and 20–45% if currently developed areas are protected—but 38–61% if all shores were protected. These results

[2]The rolling easements are called "special permits." SC Code 48-39-290 (D)(1).

suggest that efforts to mitigate wetland loss from sea-level rise could exempt existing development and focus on areas that are still undeveloped (Titus *et al.*, 1991).

Studies generally estimate that the cumulative cost of a 50-cm rise in sea level through the year 2100 would be $20–200 billion; the cost of a 1-m rise would be approximately twice that amount. Titus *et al.* (1991) estimated that for a 50-cm rise, barrier islands could be protected by placing sand on eroding beaches and the low bay sides, at a cost of $15–81 billion; elevating houses and roads, at a cost of $29–36 billion; and protecting mainland areas with dikes and bulkheads, at a cost of $5–13 billion—for a total cost of $55–123 billion.

Yohe (1990) estimated that if shores were not protected, a 50-cm rise would inundate land and structures worth $78–188 billion; Yohe *et al.* (1996) estimated that the cost would only be $20 billion. Their lower estimate appears to have resulted from two differences in their study: First, rather than assuming that all developed areas would be protected, Yohe *et al.* assessed the value of land and structures and assumed that only areas that could be economically protected would be protected. Second, the two studies appear to make different assumptions regarding the area of developed barrier islands in the United States. The Yohe *et al.* (1996) analysis was based on a sample of the entire coast, which included five densely developed ocean beach resorts; Titus *et al.* (1991) based their estimates on an assessment by Leatherman (1989)—who examined every beach community between New Jersey and the Mexican border, as well as in California, along with one site in each of the other states.

### 8.3.8. Human Settlements and Industry

Housing, industry, commerce, and the major components of infrastructure that support settlements—energy, water supply, transportation, waste disposal, and so forth—have varying degrees of vulnerability to climate change. They can be affected directly through projected changes in climate (temperature, precipitation, etc.) and indirectly through projected impacts on the environment, natural resources, and agriculture. Indirect pathways to impacts include expected changes in the availability of natural resources, geographic shifts in climate-sensitive resource industries, effects on environmental quality and health from changes in ecosystems, and other effects resulting from changes in environmental service functions. Furthermore, these effects on human settlements in theory could lead to tertiary impacts—such as altering land use and redistributing population and activities to other regions—resulting in further changes in natural resources and other activities. Such effects, however, are largely speculative at the current state of knowledge.

Climate directly affects the quality of life; alters patterns of settlement and human activities; subjects humans to risks to their health, safety, and property (e.g., due to extreme events); and, therefore, has costs and benefits for individuals and for the private and public sectors. As such, changes in climate are expected to have positive and negative impacts.

Climate change will have direct impacts on economic activity in the industry, energy, and transportation sectors; impacts on markets for goods and services; and impacts on natural resources on which economic activity depends. Activities directly sensitive to climate include construction, transportation, offshore oil and gas production, manufacturing dependent on water, tourism and recreation, and industry that is located in coastal zones and permafrost regions. Activities with markets sensitive to climate include electricity and fossil fuel production for space heating and air conditioning, construction activity associated with coastal defenses, and transportation. Activities dependent on climate-sensitive resources include agro-industries (food/drink, forestry-related activity, and textiles), biomass production, and other renewable energy production.

Impacts occurring in the distant future are difficult to predict in detail because the context of human settlement patterns and technologies cannot be forecast accurately. Concomitantly, there are substantial opportunities for adaptation to changed climates in conjunction with the development of future housing and infrastructure facilities, depending in part on our capability to forecast climate changes. Many types of impacts on human facilities have the potential to be partially or completely reduced or eliminated through adaptation, though this usually will increase their costs.

#### 8.3.8.1. Impacts on Transportation

*Projected changes in climate will have both negative and positive impacts on the operation and maintenance costs of transportation systems.*

Studies in temperate and northern climates generally have indicated that higher temperatures will result in lower maintenance costs, especially with fewer freeze-thaw cycles and less snow (e.g., Walker *et al.*, 1989; Daniels *et al.*, 1992). Black (1990) points out, however, that increased pavement buckling caused by longer periods of intense heat is a possibility. Lewis (1988) and Hirsch (1988) cite such cases from the great North American summer heat wave of 1988.

In moderate climates, water transport would be affected by changes in river navigability. Reductions in rainfall, which are possible during the summer in mid-latitudes in North America, could adversely affect waterborne transportation. During the 1988 drought, industries that relied on bulk transportation of raw materials and finished products by barge on the Mississippi River found that low water kept more than 800 barges tied up for several months. In 1993, by contrast, floods in the upper Mississippi valley also disrupted the barge transportation system, and in 1997 increased siltation associated with floods prevented ships from reaching the port of New Orleans for several days. To the extent that industry is moving toward just-in-time production systems, it will become more vulnerable to interruptions for these and other reasons.

In colder regions, the most significant direct impact of warming is likely to be on inland and coastal water transportation. A longer season for Arctic shipping is likely for locations like Prudhoe Bay, Alaska, which depends on the short ice-free season to barge in modular loads too large to go by truck. Increased wave activity and increased frequency of extreme weather events might have a more significant effect on coastal transportation operations, but little research has been conducted on this topic. A survey of the potential impacts on Canadian shipping suggested net benefits to Arctic and ocean shipping as a consequence of deeper drafts in ports and longer navigational seasons (IBI Group, 1990).

Winter roads on ice constitute an important part of the transportation network in parts of Canada's north. For example, about 10–15% of the total annual flow of goods in the Mackenzie Valley moves over winter roads, some of which cross major rivers. As the name implies, winter roads (or ice roads) are functional in the winter only; they are made of snow, ice, or a mixture of soil and snow/ice and can be created on the frozen surface of lakes and rivers. Lonergan *et al.* (1993) found a substantial reduction in the length of the "ice road" season based on climate change projections.

Further south, there would be a greater number of ice-free days for inland waterways such as the Great Lakes and St. Lawrence Seaway (IBI Group, 1990). Inland waterways, however, may suffer loss of depth from greater periods of seasonal drought, reducing their usefulness for commercial shipping even if the ice-free season is lengthened (Black, 1990). A similar study showed that reduced ice cover compensated for lower water levels in two of three climate change scenarios but that dredging costs generally increased in the six Great Lakes ports examined (Keith *et al.*, 1989). Other climate impacts could arise from changes in snowfall or melting of the permafrost (IBI Group, 1990).

*Changes in the location and nature of agricultural activities, as well as other climate-dependent industries, could have a large impact on the freight transport system.*

Existing assessments of transportation impacts have recognized the potential significance of changes in geographical patterns of economic activity on the transportation network. Black (1990) notes that even gradual, long-term global warming could cause a major disruption of the movement of goods and people in North America. The IBI Group (1990) suggests that there probably would be a northward spreading of agricultural, forestry, and mining activities—resulting in increased population and intensified settlement patterns in Canada's mid-north and even in Arctic areas. Marine, road, rail, and air links would have to be expanded accordingly.

### *8.3.8.2. Recreation and Tourism*

Climate creates opportunities and limitations for outdoor recreation. It is a major influence on the economic viability of some recreation enterprises. Several studies have projected shorter North American skiing seasons as a result of climate change. In a study of the implications of an effective $CO_2$ doubling on tourism and recreation in Ontario, Canada, Wall (1988) projected that the downhill ski season in the South Georgian Bay region could be eliminated. This outcome assumed a temperature rise of 3.5–5.7°C and a 9% increase in annual precipitation levels. Some of these losses would be offset by an extended summer recreational season. Lamothe and Périard (1988) examined the implications of a 4–5°C temperature rise throughout the downhill skiing season in Quebec. They projected a 50–70% decrease in the number of ski days in southern Quebec; ski resorts equipped with snowmaking devices probably would experience a 40–50% reduction in the number of ski days. This change in winter recreational traffic would have direct implications for road traffic (down) and requirements for snow-removal and road repair (also probably down). On the other hand, Masterton *et al.* (1976) have noted that low temperatures are a limiting factor on recreation activity in the northern part of the prairie provinces. The summer recreation season in many areas may be extended (Masterton *et al.*, 1976; Staple and Wall, 1994). Warmer temperatures may offset some of the costs of sea-level rise for recreational barrier islands.

### *8.3.8.3. Extreme Weather Events*

*Human settlements and infrastructure are especially vulnerable to several types of extreme weather events, including droughts, intense precipitation, extreme temperature episodes, high winds, and severe storms. Hence, there could be impacts should the frequency or intensity of these extreme events increase or decrease with climate warming.*

Weather-related natural disasters (wildfires, hurricanes, severe storms, ice, snow, flooding, drought, tornadoes, and other extreme weather events) are estimated to have caused damages in the United States averaging about $39 billion per year during the years 1992–96 (FEMA, 1997). Those losses included damages to structures (buildings, bridges, roads, etc.) and losses of income, property, and other indirect consequences.

As indicated in Section 8.2.3 and IPCC (1996, WG I, Section 6.5), the ability to predict changes in the frequency or intensity of extreme weather events using global and regional models has been limited by their lack of small-scale spatial and temporal resolution and uncertainties about representation of some processes.[3] Historical changes in frequencies of extreme events

[3]Most hydrological studies of flooding and water resources now use scenarios based on GCM simulations, but there are important uncertainties in this use: "Weaknesses of models in coupling the land surface and atmospheric hydrologic cycles and in GCM simulations of regional climate and extremes, particularly with regard to precipitation. Weaknesses in using GCM simulations to define climate-change scenarios at the spatial and temporal resolution required by hydrological models. The spatial resolution of current GCMs is too coarse for their output to be fed directly into hydrological models." (IPCC 1996, WG II, Section 10.2.2).

also provide some insights on possible changes, but there is debate about which changes are significant and which are unambiguously attributable to climate warming. However, some indications of directions of change have been inferred from observations and model simulations for North America, particularly regarding increased variability of precipitation. Beyond those inferences, a number of vulnerabilities of resources to extreme events have been identified should such events increase in frequency or magnitude. Conversely, decreases in extreme events could reduce levels of damages currently experienced. Additional research is needed to better understand the sensitivity and vulnerabilities of North American human settlements and infrastructure to extreme events, including factors beyond climate that are changing those vulnerabilities.

*Flooding may be a very important impact because of the large amount of property and human life potentially at risk in North America, as is evident from historical disasters. There have been relatively few studies addressing the change in risk directly because of the lack of credible climate change scenarios at the level of detail necessary to predict flooding.*

The evidence for an increasing trend in warm-period rainfall intensities in the United States (discussed in Section 8.2.2) suggests the potential for a shift in the periodicity of the flood regime in North America. More frequent or more extreme flooding could cause considerable disruption of transportation and water supply systems.

Increases in heavy rainfall events (e.g., suggested changes in frequency of intense subtropical cyclones) (Lambert, 1995) and interactions with changes in snowmelt-generated runoff could increase the potential for flooding of human settlements in many water basins. Changes in snowmelt runoff may add to or subtract from rainfall events, depending on basin characteristics and climate changes for a basin. Extreme rainfall events can have widespread impacts on roads, railways, and other transportation links. As long as rainfall does not become more intense, impacts on urban roads and railways in temperate, tropical, and subtropical zones are likely to be modest.

Some areas in North America may experience changed risks of wildfire, land slippage, and severe weather events in a changed climate regime. Although this increase in risk is predicated on changes in the frequency or intensity of extreme weather events—about which there is controversy—considering these risks in the design of long-lived infrastructure may prove cost-effective in some circumstances. Human settlement infrastructure has increasingly concentrated in areas vulnerable to wildfire, such as the chaparral hillsides in California. Settlements in forested regions in many areas are vulnerable to seasonal wildfires. Areas of potentially increased fire danger include broad regions of Canada (Street, 1989; Forestry Canada, 1991) and seasonally dry Mediterranean climates like the state of California in the United States. It is possible that fuel buildup under drought conditions would decrease, decreasing fire intensities. Although generally less destructive of life than in many developing world locations, landslides triggered by periodic heavy rainfall events threaten property and infrastructure in steep lands of the western United States and Canada. Relict landslides occur in much of northern Europe and North America (Johnson and Vaughan, 1989). Although stable under present natural conditions, these landslides are reactivated by urban construction activities and are triggered by heavy rains (Caine, 1980). Lands denuded of vegetation by wildfire or urban development also are vulnerable.

Although there has been an apparent downward trend in Atlantic hurricanes in recent years (e.g., Landsea *et al.*, 1996), not all authors agree (Karl *et al.*, 1995b). What is certain is that the amount of property and the number of people in areas known to be vulnerable to hurricanes is large and increasing in low-lying coastal areas in much of the United States Atlantic and Gulf coasts. For example, although data on the amount or proportion of national physical assets exposed to climate hazards are not readily available, it is known that in the United States about $2 trillion in insured property value lies within 30 km of coasts exposed to Atlantic hurricanes (IRC, 1995).

Most authors have found increases in seasonal minimum temperatures in North America, but not in seasonal maximums (IPCC 1996, WG I, Chapter 3). These results would suggest reduced incidence of cold-related problems without a concomitant increase in heat-related problems. However, increases in regional cold outbreaks occurred from the late 1970s to the mid-1980s. There has been little evidence of an increase in danger from tornadoes in the region (Grazulis, 1993; Ostby, 1993).

Offshore oil and gas exploration and production would be influenced by change in extreme events. In the south, an increase in extreme storm events in the Gulf of Mexico may mean increasing fixed and floating platform engineering standards (i.e., more expensive platforms) and more frequent and longer storm interruptions. In terms of interruptions, weather-related production shutdowns result in losses to production companies in the range of $1 million dollars per day—$10,000–50,000 per facility where evacuation is necessary. The industry defers millions of dollars annually in royalties (approximately $7 million each day for offshore Gulf of Mexico facilities) paid for hydrocarbon produced from fields owned by the public.

#### *8.3.8.4. Energy Supply Systems*

The energy sector is diverse, but a few generalizations can be made. Many components of conventional energy supply systems that involve fossil and nuclear energy—including onshore extraction (with exceptions), land transportation of fuels, conversion, and end-use (except for space conditioning)—are largely independent of climate. However, exploration and well servicing offshore and in tundra and boreal regions, particularly in wet springs in boggy areas, are dependent on the climate regime; if climate conditions change (wetter or drier), the duration of the servicing/exploration season could change, with economic impacts in those sectors. Water transportation, activities in the arctic and mountainous areas, cooling systems for thermal power generation, and energy demand for space conditioning also may be affected to some degree by changes in climate, positively or negatively. Many renewable energy sources—such as hydropower, solar, wind, and biomass—are strongly affected by climate in positive or negative ways. Only large-scale hydropower and some biomass currently make a large contribution to the North American energy supply. However, this sector may become more dependent on renewable energy in the future, and hence more vulnerable to climate change, especially if greenhouse gas controls constrain the use of fossil fuels and current barriers to nuclear growth continue.

Thermal electric generating plants and nuclear energy plants are susceptible to hydrological and water resource constraints that affect their cooling water supply. Power plant output may be restricted because of reduced water availability or thermal pollution of rivers with a reduced flow of water. Events such as these have occurred during droughts in several parts of the world, including the United States (Energy Economist, 1988). Under more extreme temperature conditions, some nuclear plants might shut down to comply with safety regulations (Miller *et al.*, 1992). Future power plants are less likely to depend on once-through cooling and may be designed to deal with anticipated shortages of cooling water supplies.

Hydroelectricity, which provides 20% of the region's electricity (and is the primary energy source in some areas of North America, such as the Pacific Northwest and Quebec), depends on the quantity and seasonal distribution of precipitation. Greater annual precipitation overall in the North American region is projected, with the greatest increases in winter and spring. For the north, this likely will mean greater snowfall to be added to the spring runoff, which would put greater demand on reservoirs to even out electricity supply. For southern hydroelectric facilities, climate projections suggest greater seasonal variation—unfortunately not coincidental with anticipated increased demand for summer air conditioning. Some areas (particularly the southwest of the continent) may experience lower rainfalls in the summer and fall, which, along with increased demand for air-conditioning, would exacerbate peak power requirements. However, GCMs are less reliable in simulating regional precipitation than temperature, and these predictions currently are not sufficiently reliable as a basis for hydroelectric and water resource planning.

Local energy distribution will not be affected, but long-distance transmission lines and pipelines may be subject to land disturbances, particularly in the western mountains where increased precipitation may induce slope instability. In the north, the permafrost, which normally provides a solid base for construction and transportation, is expected to degrade or thaw faster in some areas, producing stress on structures that may have been designed for a permafrost regime. Projected changes include not only melting but also decreases in the strength properties of the permafrost and increases in frost heaving. The vulnerability of pipelines as a result of projected changes in underlying permafrost (Nixon *et al.*, 1990) are expected to be particularly acute in discontinuous permafrost areas and in the southern reaches of continuous permafrost. As a result, modifications or repairs to pipelines may be necessary, and some concerns have been raised regarding the potential of increased risk of environmental contamination.

Small-hydropower—usually located in nondammed streams—may provide more power in periods of peak runoff. Solar energy is highly dependent on cloud cover, which may increase with the expected intensification of the hydrological cycle; the exception might be the south-central area of North America, where increased insolation is expected (and where it would coincide with increased electricity demand for space cooling). The wind—not yet a significant contributor to North America's energy supply—is a highly variable source. Biofuels, currently primarily wood waste and grains, provide about 4% of the region's primary energy supply; changes in the availability of these fuels are possible as a result of projected changes to forest growth and productivity (see Section 8.3.2) and projected changes in the availability (absolutely and regionally) of grains, mainly corn for ethanol (see Section 8.3.4.1). However, future growth in biofuels is likely to involve dedicated energy farms utilizing short-rotation, highly managed crops.

#### 8.3.8.5. *Energy Demand*

*Climate warming would result in increased demand for cooling and decreased demand for heating energy, with the overall net effect varying with geographic region; however, changes in energy demand for comfort are expected to result in a net saving overall for North America.*

Space heating and cooling are the most climate-sensitive uses of energy; they account for about 14% of energy use in North America, based on U.S. estimates extended to include Canada (see Table 8-8). The demand for summer cooling is likely to increase with projected warming. On the other hand, winter heating demand will be reduced. Rosenthal *et al.* (1995) concluded that a 1.8°C global warming would reduce total U.S. energy use associated with space heating and air conditioning by 1 exajoule (EJ)—11% of demand—in the year 2010; costs would be reduced by $5.5 billion (1991 dollars). Belzer *et al.* (1996) found that a 4°C warming would decrease total site energy use for commercial sector heating and cooling by 0.5–0.8 EJ (13–17%) and associated primary energy by 0.1–0.4 EJ (2–7%), depending on the degree to which advanced building designs penetrate the market. (This analysis was based on projected buildings in the year 2030, though the assumed temperature increase is much greater than Intergovernmental Panel on Climate Change (IPCC) projections for that period.)

The seasonal occurrence of peak demand for electricity is an important factor. If peak demand occurs in winter, maximum demand is likely to fall, whereas if there is a summer peak,

*Table 8-8: Use of energy in buildings in U.S., 1989–90.*[1]

| Energy Source/Use | Residential | Commercial[2] | Total |
|---|---|---|---|
| Electricity | | | |
| – Air conditioning | 1.5 | 0.9 | |
| – Space heating | 0.9 | 0.3 | |
| – Ventilation | — | 0.8 | |
| Natural Gas | | | |
| – Space heating | 3.4 | 1.3 | |
| Fuel Oil | 1.0 | 0.4 | |
| LPG | 0.3 | — | |
| District Heat | — | 0.6 | |
| **Total** | 7.1 | 4.3 | 11.4 |

[1]All values based on *Buildings and Energy in the 1980s*, Energy Information Administration, DOE/EIA-0555(95)/1, June 1995.
[2]Commercial values are from 1989.
Note: Total energy resource consumption within the U.S. in 1990 was 80 quads, so space conditioning accounted for about 14% of total.

maximum demand will rise. The precise effects are strongly dependent on the climate zone (Linder and Inglis, 1989). Climate change may cause some areas to switch from a winter peaking regime to approach a summer peaking regime (Scott *et al.*, 1994). Fewer studies have estimated the possible impact of climate change on investment requirements in electricity supply. An exception is the "infrastructure" component (Linder and Inglis, 1989) of the U.S. national climate effects study (Smith and Tirpak, 1989), which estimated that with a 1.1°C warming, peak demand would increase by 29 gigawatts (GW), or 4% of the baseline level for that year. Decreases in investment for heating supply fuels have not been estimated.

Although usually smaller in total magnitude of energy demand, the use of electricity and fuels for irrigation pumping and the use of fuels for drying of agricultural crops also can be significant weather-sensitive demands in some regions (Darmstadter, 1993; Scott *et al.*, 1993). Pumping would tend to increase in regions suffering a decrease in natural soil moisture—for example, Goos (1989) found a 20% increase in energy demand for irrigation in the province of Alberta—whereas drying energy would decrease where humidity decreases. Automobile fuel efficiency may decrease slightly as a result of greater use of air-conditioning. With a 4°C warming, autos would consume an additional 47 liters of fuel per 10,000 km driven, for a total cost of $1–3 billion per year at current energy prices (Titus, 1992).

*The technological capacity to adapt to climate change is likely to be readily available in North America. However, its application will be realized only if the necessary information is available, the institutional and financial capacity to manage change exists, and the benefits of adaptive measures are considered to be worth their costs. Therefore, to increase the potential for adaptation and to reduce costs, it is essential that information about the nature of climate change is available sufficiently far in advance in relation to the planning horizons and lifetimes of investments.*

Some adaptation will occur if better information concerning the risks posed by climate change is available and the appropriate signals are available from the marketplace. Over periods of half a century or more, many sectors will change significantly, and new products, markets, and technologies will emerge. In the transportation industry, for example, nominal replacement cycles are 10–20 years for transit vehicles and 35–70 years for most infrastructure (National Council on Public Works Improvement, 1988). Turnover of capital stock provides the opportunity to adapt easily if information is available. Given uncertainties, however, autonomous adaptation signaled by the marketplace cannot be relied upon entirely for long-lived public transportation and other infrastructure. Governments may have to set a suitable policy framework, disseminate information about climate change, and act directly in relation to vulnerable infrastructures. For example, effective land-use regulation (zoning and building codes) can help reduce vulnerabilities by directing population shifts away from vulnerable locations such as floodplains, steep hillsides, and low-lying coastlines. Research is needed to better understand the factors that affect effective adaptive capacity and how those factors vary within North America.

Some forms of adaptation, such as those in the areas of fire prevention and water supply-and-demand balancing, will likely be advantageous regardless of climate regime. Examples would include major adaptive responses in fire-control systems in rural areas—such as controlled burns to limit fuel, fire breaks, aerial fire retardant delivery, and rural fire departments and "smoke jumpers"—and short-term activity controls during high fire-danger weather, such as prohibitions on open burning, commercial activities such as logging, and recreation activities such as hiking, hunting, or use of off-road vehicles. If the frequency or intensity of fire danger warrants, other potential adaptive responses involve improved spatial planning of communities and some longer-term land-use controls. These strategies would provide better isolation of fires and could limit damage to human settlements even under current conditions.

In the energy supply sector, anticipation of possible regional climate changes will be important in the design of and site selection for solar and wind energy systems, as well as energy transportation systems. For example, possible thawing of permafrost in Arctic regions may require changes in the design of oil pipelines to avoid slumping, breaks, and leaks (Brown, 1989; Anderson *et al.*, 1994). Systems with long lifetimes, such as large hydroelectric impoundment systems, will have difficulty adjusting in the absence of long-term predictions. Future biomass energy farms, however, are likely to be intensively managed and have short crop rotation; therefore, they would be better able to adapt to changing conditions through choices of crops and management techniques.

For energy demand, building design can assist adaptation. Increased building-shell efficiency and changes to building design that reduce air-conditioning load show promise (Scott *et al.*, 1994). Though effective, however, adaptive strategies are not implemented without costs (Loveland and Brown, 1990). Reducing the size of space-heating capacity in response to warmer climate would be a logical adaptive response in more temperate and polar countries and may free up investment funds for other purposes, even within the energy sector. Community design to reduce heat islands (through judicious use of vegetation and light-colored surfaces) (Akbari *et al.*, 1992), reducing motor transportation, and taking advantage of solar resources also may be viable and would have sustainability benefits.

### 8.3.9. Human Health

Climate change is likely to have wide-ranging and mostly adverse impacts on human health. These impacts would arise by direct pathways (e.g., exposure to thermal stress and extreme weather events) and indirect pathways (increases in some air pollutants, pollens, and mold spores; malnutrition; increases in the potential transmission of vector-borne and waterborne diseases; and general public health infrastructural damage) (IPCC 1996, WG II Sections 18.2 and 18.3, and Figure 18-1). Climate change also could jeopardize access to traditional foods garnered from land and water (such as game, wild birds, fish, and berries), leading to diet-related problems such as obesity, cardiovascular disorders, and diabetes among northern populations of indigenous peoples as they make new food choices (Government of Canada, 1996).

#### 8.3.9.1. Thermal Extremes

Temperate regions such as North America are expected to warm disproportionately more than tropical and subtropical zones (IPCC 1996, WG I). The frequency of very hot days in temperate climates is expected to approximately double for an increase of 2–3°C in the average summer temperature (CDC, 1989; Climate Change Impacts Review Group, 1991). Heat waves cause excess deaths (Kilbourne, 1992), many of which are caused by increased demand on the cardiovascular system required for physiological cooling. Heat also aggravates existing medical problems in vulnerable populations—particularly the elderly, the young, and the chronically ill (CDC, 1995; Canadian Global Change Program, 1995). For example, mortality during oppressively hot weather is associated predominately with preexisting cardiovascular, cerebrovascular, and respiratory disorders, as well as accidents (Haines, 1993; IPCC 1996, WG II, Section 18.2.1). In addition to mortality, morbidity such as heat exhaustion, heat cramps, heat syncope or fainting, and heat rash also result from heat waves. People living in hot regions, such as the southern United States, cope with excessive heat through adaptations in lifestyle, physiological acclimatization, and adoption of a particular mental approach (Ellis, 1972; Rotton, 1983). In temperate regions, however, periods of excessive heat occur less frequently, and populations accordingly are less prepared with responsive adaptive options (WHO, 1996).

Data in cities in the United States and Canada show that overall death rates increase during heat waves (Kalkstein and Smoyer, 1993), particularly when the temperature rises above the local population's temperature threshold. In addition to the 1980 heat wave that resulted in 1,700 heat-related deaths, heat waves in 1983 and 1988 in the United States killed 566 and 454 people, respectively (CDC, 1995). More recently, in July 1995, a heat wave caused as many as 765 heat-related deaths in the Chicago area alone (Phelps, 1996). Tavares (1996) examined the relationship between weather and heat-related morbidity for Toronto for the years 1979–89 and found that 14% of the variability for all morbidity in persons 0–65 years of age was related to weather conditions.

Death rates in temperate and subtropical zones appear to be higher in winter than in summer (Kilbourne, 1992). Comparative analyses of the causes of differences between summer versus winter weather-related mortalities are lacking, however. The United States averaged 367 deaths per year due to cold in the period 1979–94 (Parrish, 1997), whereas the annual average number of Canadians dying of excessive cold is 110 (Phillips, 1990). It has been suggested that winter mortality rates, which appear to be more related to infectious diseases than to extremely cold temperatures, will be little impacted by climate change. Any global warming-induced increases in heat-related mortality, therefore, are unlikely to be of similar magnitude to decreases in winter mortality (Kalkstein and Smoyer, 1993).

Mortality from extreme heat is increased by concomitant conditions of low wind, high humidity, and intense solar radiation (Kilbourne, 1992). In Ontario, the number of days annually with temperatures above 30°C could increase fivefold (from 10 to 50 days per year) under doubled $CO_2$ scenarios (Environment Canada *et al.*, 1995).

Several studies (e.g., WHO, 1996) have found that future heat-related mortality rates would significantly increase under climate change. Table 8-9 shows projected changes in heat-related deaths for selected cities in North America under two climate change scenarios. Acclimatization of populations, however, may reduce the predicted heat-related morbidity and mortality. Kalkstein *et al.* (1993) found that people in Montreal and Toronto might acclimatize somewhat to global warming conditions. People in Ottawa, on the other hand, showed no signs of potential acclimatization. It is important to note that acclimatization to increasing temperatures occurs gradually, particularly among the elderly, and may be slower than the rate of ambient temperature change.

***Table 8-9:*** *Total summer heat-related deaths in selected North American cities: current mortality and estimates of future mortality under two different climate change scenarios with global mean temperature increases of ~0.53°C or ~1.16°C.*[1]

| City | Present Mortality[2] | GFDL89 Climate Change Scenario ~0.53°C no acc | GFDL89 ~0.53°C acc | GFDL89 ~1.16°C no acc | GFDL89 ~1.16°C acc | UKTR Climate Change Scenario ~0.53°C no acc | UKTR ~0.53°C acc | UKTR ~1.16°C no acc | UKTR ~1.16°C acc |
|---|---|---|---|---|---|---|---|---|---|
| *United States* | | | | | | | | | |
| Atlanta | 78 | 191 | 96 | 293 | 147 | 247 | 124 | 436 | 218 |
| Dallas | 19 | 35 | 28 | 782 | 618 | 1364 | 1077 | 1360 | 1074 |
| Detroit | 118 | 264 | 131 | 419 | 209 | 923 | 460 | 1099 | 547 |
| Los Angeles | 84 | 205 | 102 | 350 | 174 | 571 | 284 | 728 | 363 |
| New York | 320 | 356 | 190 | 879 | 494 | 1098 | 683 | 1754 | 971 |
| Philadelphia | 145 | 190 | 142 | 474 | 354 | 586 | 437 | 884 | 659 |
| San Francisco | 27 | 19 | 40 | 104 | 85 | 57 | 47 | 76 | 62 |
| *Canada* | | | | | | | | | |
| Montreal | 69 | 121 | 61 | 245 | 124 | 460 | 233 | 725 | 368 |
| Toronto | 19 | 36 | 0 | 86 | 1 | 289 | 3 | 563 | 7 |

[1]Figures represent average summer-season heat-related deaths for each city under each climate change scenario. Figures assume no change in population size and age distribution in the future.
[2]Raw mortality data.
Source: Modified from Kalkstein *et al.* (1997) in WHO (1996), p. 57.

Air conditioning and adequate warning systems also may reduce heat-related morbidity and mortality in a warmer North America. It has been suggested that air conditioning could reduce heat-related deaths by 25% (Phelps, 1996). A warning system such as the Philadelphia Hot Weather-Health Watch/Warning System (PWWS) that alerts the public when oppressive air masses (e.g., extended periods of extreme high temperatures, high humidity, moderate to strong southwesterly winds, and high pressure) may occur might further reduce heat-related mortality (Kalkstein and Smoyer, 1993). The PWWS is a three-tiered system that produces a health watch, health alert, or health warning and then accordingly initiates a series of interventions, including media announcements, promotion of a "buddy system," home visits, nursing and personal care intervention, increased emergency medical service staffing, and provision of air-conditioned facilities (Kalkstein *et al.*, 1995).

#### 8.3.9.2. Air Quality and Ground-Level Ozone

*Projected climate changes could lead to exacerbation of respiratory disorders associated with reduced air quality in urban and rural areas and effects on the seasonality of certain allergic respiratory disorders.*

It is well established that exposure to single or combined air pollutants has serious public health consequences. For example, ozone at ground level has been identified as causing damage to lung tissue, particularly among the elderly and children—reducing pulmonary function and sensitizing airways to other irritants and allergens (Beckett, 1991; Schwartz, 1994; U.S. EPA, 1996). Ground-level ozone affects not only those with impaired respiratory function, such as persons with asthma and chronic obstructive lung disease, but also healthy individuals. Even at relatively low exposure levels, healthy individuals can experience chest pain, coughing, nausea, and pulmonary congestion as a result of exposure to ground-level ozone.

Researchers also recognize that concurrent hot weather and air pollution can have synergistic impacts on health (Katsouyanni *et al.*, 1993). For example, warmer temperatures can accelerate production and increase concentrations of photochemical oxidants in urban and rural areas and thus exacerbate respiratory disorders (Shumway *et al.*, 1988; Schwartz and Dockery, 1992; Dockery *et al.*, 1993; Katsouyanni *et al.*, 1993; Pope *et al.*, 1995; Phelps, 1996).

Few large-scale studies have been performed to assess the implications of climate change on air quality or population exposures to high concentrations of ground-level ozone. This limitation is related to difficulty in devising a defensible scenario of future climate change for a specific location, the previous focus on acute short-term effects rather than long-term effects, and the expense involved in modeling atmospheric chemistry. There is a limited number of studies, however, that shed some light on possible impacts of climate change on air quality and associated health implications.

Emberlin (1994), for example, has suggested that global warming may affect the seasonality of certain allergic respiratory disorders by altering the production of plant aero-allergens. Asthma and hay fever can be triggered by aero-allergens that cause high seasonal morbidity. The severity of allergies may be intensified by projected changes in heat and humidity, thereby contributing to breathing difficulties (Environment Canada *et al.*, 1995; Maarouf, 1995).

Ozone concentrations at ground level continue to be the most pervasive air pollution problem in North America. The U.S. population exposed to unhealthy levels of ozone has fluctuated over the past 10–20 years—reaching a peak in 1988, when 112 million people lived in areas with higher than acceptable concentrations. In addition, recent studies (U.S. EPA, 1996) provide evidence of a positive correlation between ground-level ozone and respiratory-related hospital admissions in several cities in the United States. Such hospital admissions in the province of Ontario strongly relate to ambient levels of sulfur dioxide and ozone and to temperature (Canadian Public Health Association, 1992).

Research has shown that ground-level ozone formation is affected by weather and climate. Many studies have focused on the relationship between temperature and ozone concentrations (Wolff and Lioy, 1978; Atwater, 1984; Kuntasal and Chang, 1987; Wackter and Bayly, 1988; Wakim, 1989). For example, the large increase in ozone concentrations at ground level in 1988 in the United States and in parts of southern Canada can be attributed, in part, to meteorological conditions; 1988 was the third-hottest summer in the past 100 years. In general, the aforementioned studies suggest a nonlinear relationship between temperature and ozone concentrations at ground level: Below temperatures of 22–26°C (70–80°F), there is no relationship between ozone concentrations and temperature; above 32°C (90°F), there is a strong positive relationship.

Regression analyses have revealed that high temperatures are a necessary condition for high ozone concentrations at ground level; other meteorological variables often need to be considered, however. Weather variables that have been included in regression equations include temperature, wind speed, relative humidity, and sky cover (Wakim, 1990; Korsog and Wolff, 1991); however, other variables that could be included are wind direction, dew-point temperature, sea-level pressure, and precipitation.

Studies of ground-level ozone concentrations in which emissions and other weather factors are held constant (Smith and Tirpak, 1989) suggest the following impacts on ground-level ozone as a result of a 4°C warming:

- In the San Francisco Bay area, maximum ozone concentration could increase by about 20% and could approximately double the area that would be out of compliance with the National Ambient Air Quality Standard (NAAQS).
- In New York, ground-level ozone concentrations could increase by 4%.
- In the Midwest and Southeast, changes in ground-level ozone levels could range from a decrease of 2.4% to an increase of 8%, and the area in exceedance of the ozone standard could exhibit nearly a threefold increase.

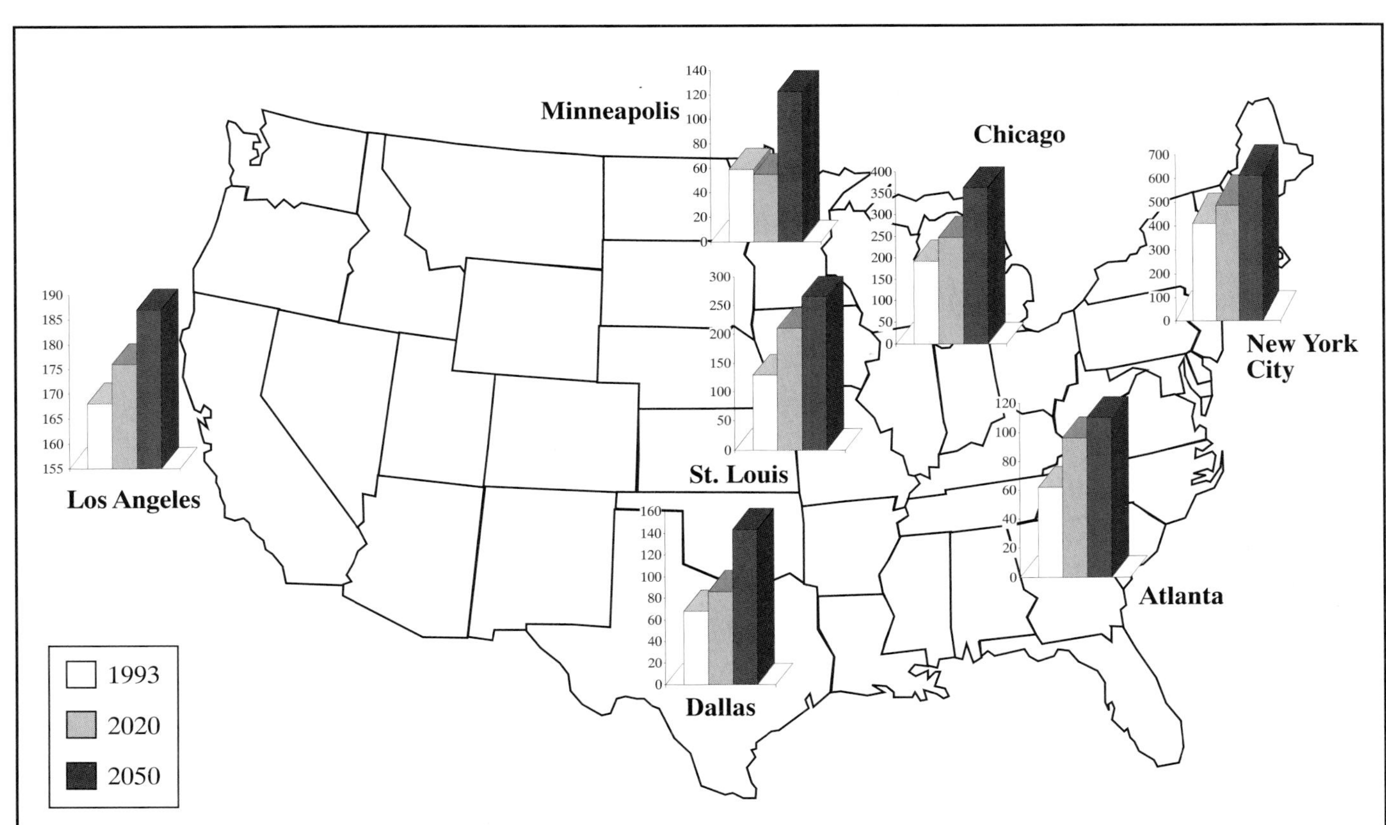


**Figure 8-10:** Average annual weather-related mortality for 1993, 2020, and 2050 climate (Kalkstein and Greene, 1997), based on 1980 population and the GFDL89 climate change scenario. Annual estimates were obtained by adding summer and winter mortality. The projections do not account for population growth, nor do they fully account for air-conditioning use; however, they do assume acclimation to changed climate.

In Canada, a projected fivefold rise in the frequency of hot days (i.e., those with temperatures >30°C) could lead to a greater number of days with levels of ground-level ozone considered to be a health risk for sensitive individuals in the population (Environment Canada *et al.*, 1995).

#### 8.3.9.3. *Extreme Weather Events*

In the United States, 145 natural disasters resulted in 14,536 deaths from 1945 to 1989. Of these events, 136 were weather disasters; these extreme weather events caused 95% of all disaster-related deaths. Floods are the most frequent type of disaster.

More frequent extreme weather events are predicted to accompany global warming (see Figure 8-10), in part as a consequence of projected increases in convective activity. More intense rainfall events accompanying global warming would be expected to increase the occurrence of floods, and warmer sea-surface temperatures could strengthen tropical cyclones (IPCC 1996, WG I).

Climate models are unable to predict extreme events because they lack spatial and temporal resolution. In addition, there is no clear evidence that sustained or worldwide changes in extreme events have occurred in the past few decades. Nonetheless, such events cause loss of life and endanger health by increasing injuries, infectious diseases, stress-related disorders, and adverse health effects associated with social and environmental disruptions and environmentally enforced migration. Because each extreme weather event is unique in scale and location, and population vulnerability varies considerably, it is not possible to quantify the health impacts that would be associated with potential changes in extreme weather events.

Recent floods in the United States (e.g., Mississippi River flooding in 1993) were caused primarily by unusually high precipitation combined with soil saturation from earlier precipitation (Kunkel *et al.*, 1994). In the United States, flash floods currently are the leading cause of weather-related mortality. In addition to causing deaths by drowning, flooding can lead to widespread destruction of food supplies and outbreaks of disease as a result of breakdowns in sanitation services. Flooding also may result in the release of dangerous chemicals from storage sites and waste disposal sites into floodwaters. Increased runoff from agricultural lands during periods of heavy precipitation also can threaten water supplies. The 1993 Mississippi River flooding, for example, caused wide dispersal of microorganisms and chemicals from agricultural lands and industrial sites (Changnon, 1996).

#### 8.3.9.4. *Biological Agents: Vector- and Waterborne Diseases*

##### 8.3.9.4.1. *Vector-borne diseases*

*Changing climate conditions may lead to the northward spread of vector-borne infectious diseases and potentially enhanced transmission dynamics as a result of warmer ambient temperatures.* Vector-borne diseases (primarily carried by arthropod or small mammal "vectors") and waterborne diarrheal diseases represent a large proportion of infectious diseases, which are the world's leading cause of fatalities. Projected changes in climate almost certainly would make conditions less suitable for the transmission of several vector-borne diseases (e.g., plague and some forms of encephalomyelitis) in much of their current North American range. Other diseases (e.g., Saint Louis encephalitis and western equine encephalomyelitis) might extend their range northward or exhibit more frequent outbreaks. The crucial factor is the availability of appropriate habitats for vectors and (in the case of zoonotic diseases) vertebrate "maintenance" hosts. Although projected changes in climate might provide opportunities for diseases to extend their range, the North American health infrastructure may prevent a large increase in disease cases; providing this protection, however, could increase the demands on and costs of the current public health system.

The transmission of many infectious diseases is affected by climatic factors. Infective agents and their vector organisms are sensitive to factors such as temperature, surface water, humidity, wind, soil moisture, and changes in forest distribution (IPCC 1996, WG II, Chapter 18).

Malaria: Climatic factors, which increase the inoculation rate of *Plasmodium* pathogens and the breeding activity of *Anopheles* mosquitoes, are considered the most important factors contributing to epidemic outbreaks of malaria in nonendemic areas. A temperature relationship for sporadic autochthonous malaria transmission in the temperate United States has been observed in New York and New Jersey during the 1990s (Layton *et al.*, 1995; Zucker, 1996). Common to these two outbreaks was exceptionally hot and humid weather, which reduced the development time of malaria sporozoites enough to render these northern anopheline mosquitoes infectious. Such temperature sensitivity of parasite development also has been observed in the laboratory (Noden *et al.*, 1995).

Martens *et al.* (1995) estimated that an increase in global mean temperature of several degrees by the year 2100 would increase the vectorial capacity of mosquito populations 100-fold in temperate countries. In these countries, however, continued and increased application of control measures—such as disease surveillance and prompt treatment of cases—probably would counteract any increase in vectorial capacity. Similarly, Duncan (1996) showed that projected increases in mean daily temperatures may allow for the development of malaria in Toronto. It was not suggested, however, that climate alone would permit the spread of malaria because many other factors must be considered.

Malaria once prevailed throughout the American colonies and southern Canada (Russell, 1968; Bruce-Chwatt, 1988). By the middle of the 19th century, malaria extended as far north as 50°N latitude. In Canada, malaria disappeared at the end of the 19th century (Bruce-Chwatt, 1988; Haworth, 1988). Serious malaria control measures were first undertaken in the southern United States in 1912 (Bruce-Chwatt, 1988). By 1930, malaria had disappeared from the northern and western United States and generally caused fewer than 25 deaths per 100,000 people in the South (Meade *et al.*, 1988). In 1970, the World Health Organization Expert Advisory Panel on Malaria recommended that the United States be included in the WHO official register of areas where malaria had been eradicated. The history of malaria in North America reinforces the suggestion that although increased temperatures may lead to conditions suitable for the reintroduction of malaria to North America, socioeconomic factors such as public health facilities will play a large role in determining the existence or extent of such infections.

Arboviruses: Dengue fever and dengue hemorrhagic fever (DHF) periodically have occurred in Texas, following outbreaks in Mexico, during the past two decades (Gubler and Trent, 1994; PAHO, 1994). Because of the sensitivity of dengue to climate, especially ambient temperature, it has been suggested that this disease may increase in the United States if a sustained warming trend occurs. However, due to high living standards, this disease is not likely to increase in incidence or geographic distribution in the United States, even if there is a sustained warming trend. Dengue viruses occur predominantly in the tropics, between 30°N and 20°S latitude (Trent *et al.*, 1983); freezing kills the eggs, larvae, and adults of *Aedis aegypte*, the most important vector (Chandler, 1945; Shope, 1991). It should be noted, however, that the eggs of *Ac. Alhopictus*, also a vector, are not killed by freezing.

Jetten and Focks (1997) analyzed the impact of a 2°C and a 4°C temperature rise on the epidemic potential for dengue, including the impacts for cities in temperate areas (Figure 8-11). Their analysis shows that areas adjacent in latitude or elevation to current endemic zones may become more receptive to viral introductions and enhanced transmission. Furthermore, their study shows that the proportion of the year when transmission can occur in North America could significantly increase under warming scenarios.

Encephalitides: Of reported encephalitis cases in North America, many are mosquito related, including Saint Louis encephalitis, which has occurred as far north as Windsor, Ontario (1975); LaCrosse encephalitis; and western, eastern, and Venezuelan equine encephalomyelitis (Shope, 1980). The elderly are at highest risk for Saint Louis encephalitis, and children under 16 years are at greatest risk of LaCrosse encephalitis.

Although mosquito longevity diminishes as temperatures rise, viral transmission rates (similar to dengue) rise sharply at higher temperatures (see Figure 8-11) (Hardy, 1988; Reisen *et al.*, 1993). From field studies in California (Reeves *et al.*, 1994), researchers have suggested that a 3–5°C temperature increase could cause a northern shift in western equine and Saint Louis encephalitis outbreaks, with the disappearance of western equine encephalitis in southern endemic regions. Also to be considered in these types of impact assessments is the impact of projected climate change on mosquito habitat (e.g., freshwater hardwood swamps for the eastern equine encephalomyelitis vector *Culiseta melanura*—which may well be eliminated from the southeast United States).

Outbreaks of Saint Louis encephalitis are correlated with periods of several consecutive days in which temperature exceeds 30ºC (Monath and Tsai, 1987). For example, the 1984 California epidemic followed a period of extremely high temperatures. In addition, eastern equine encephalitis has been associated with warm, wet summers along the east coast of the United States (Freier, 1993). Computer analysis of monthly climate data has demonstrated that excessive rainfall in January and February, combined with drought in July, most often precedes outbreaks of eastern equine encephalitis (Bowen and Francy, 1980). Such a pattern of warm, wet winters followed by hot, dry summers resembles many of the GCM projections for climate change over much of the United States.

Tickborne diseases: Ticks transmit Lyme disease—the most common vector-borne disease in the United States, with more than 10,000 cases reported in 1994—along with Rocky Mountain spotted fever (RMSF), and Ehrlichiosis. Involved tick and mammal host populations are influenced by land use and land cover, soil type, and elevation, as well as the timing, duration, and rate of change of temperature and moisture regimes (Mount *et al.*, 1993; Glass *et al.*, 1994). The relationships between vector life-stage parameters and climatic conditions have been verified experimentally in field and laboratory studies (Goddard, 1992; Mount *et al.*, 1993). *Ixodes scapularis*—an important hard-backed tick vector in North America—will not deposit eggs at temperatures below 8°C, and larvae will not emerge from eggs at temperatures below 12°C; the nymphal molt requires approximately 35 days at 25°C, and the adult molt requires 45 days at 25°C. Temperature also affects the activity of ticks; a minimum threshold for activity is 4°C. Ticks also are highly dependent on a humid environment. Climate change, therefore, could be expected to alter the distribution of these diseases in both the United States and Canada (Grant, 1991; Canadian Global Change Program, 1995; Environment Canada *et al.*, 1995; Hancock, 1997). For example, any tendency toward drying would suggest a reduction in the incidence of these diseases.

#### *8.3.9.4.2. Waterborne diseases*

Freshwater: Diarrheal diseases in North America can be caused by a large variety of bacteria (e.g., *salmonella*, *shigella,* and *campylobacter*), viruses (e.g., rotavirus), and protozoa (e.g., *giardia lamblia*, *toxoplasma*, and *cryptosporidium*). Climatic effects on the distribution and quality of surface water, including increases in flooding or water shortages, can impede personal hygiene and impair local sewage systems. For example, extreme precipitation contributed to an outbreak of toxoplasmosis in British Columbia in 1995 when excessive runoff contaminated a reservoir with oocysts from domestic and wild cats (British Columbia CDC, writ. comm. 1995).

Cryptosporidiosis, which causes severe diarrhea in children and can be fatal to immunocompromised individuals, is the most prevalent waterborne disease in the United States (Moore *et al.*, 1995). Natural events (e.g., floods, storms, heavy rainfall, and snowmelt) often can wash material of fecal origin, primarily from agricultural nonpoint sources, into potable water. The Milwaukee cryptosporidiosis outbreak in 1993 resulted in 403,000 reported cases; it coincided with unusually heavy spring rains and runoff from melting snow (MacKenzie *et al.*, 1994).

Factors enhancing waterborne cryptosporidiosis will depend on hydrological responses to climate change and the degree of flooding in water catchment areas. Flushing from heavy rains may be more important than actual flooding, especially for private wells influenced by surface water. Land-use patterns also determine contamination sources (e.g., agricultural activities) and therefore must be considered.

In addition, intensification of heavy rainfall events (as suggested by some scenarios) could lead to more rapid leaching from hazardous-waste landfills, as well as contamination from agricultural activities and septic tanks. This leaching or contamination represents a potential health hazard—particularly at times of extensive flooding, which can lead to toxic contamination of

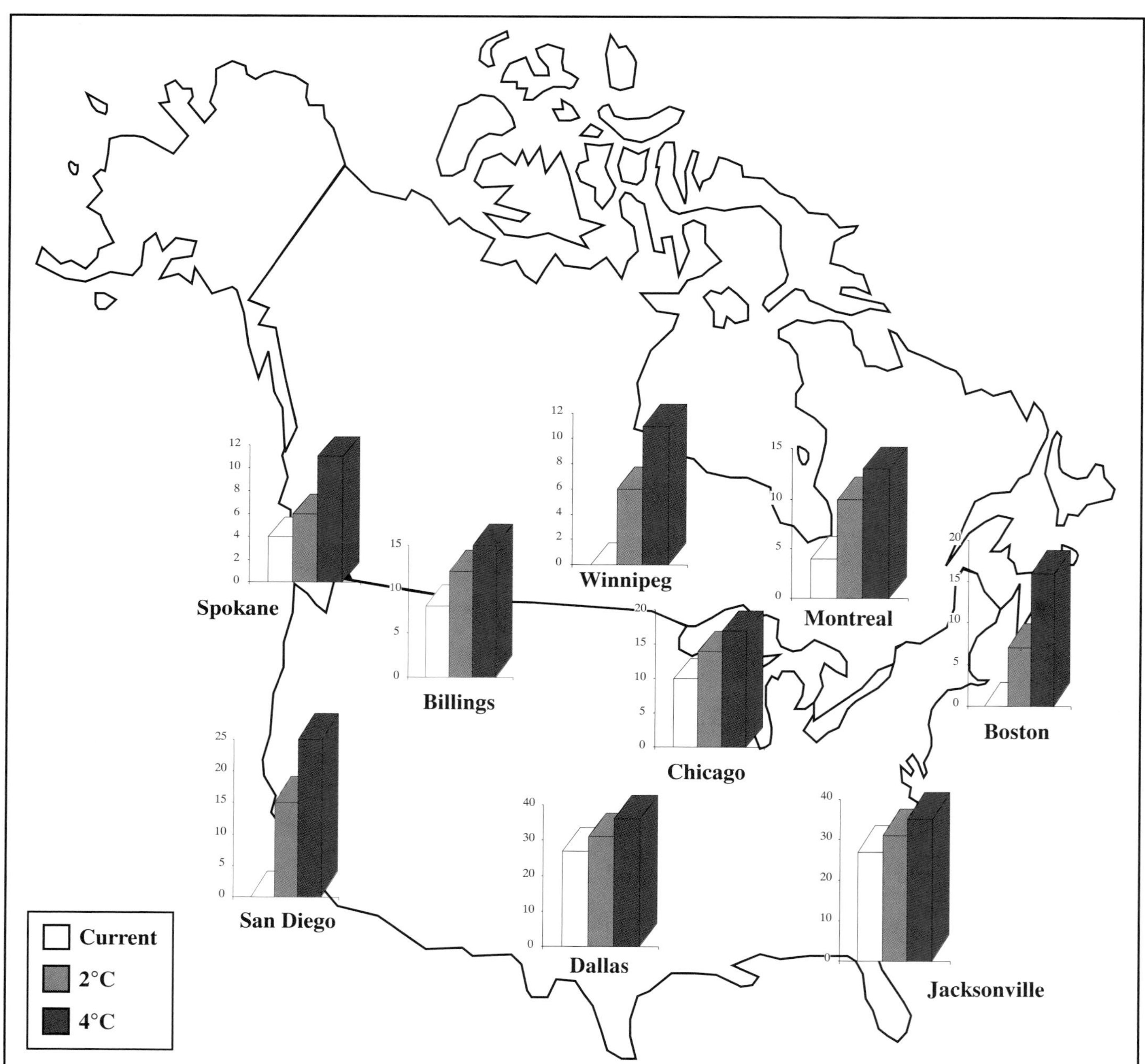


**Figure 8-11:** Weeks of potential dengue transmission under current temperature and 2°C and 4°C warming (adapted from Focks *et al.*, 1995; Jetten and Focks, 1997). Presence of dengue virus, mosquito vector, and exposed human populations are required for disease transmission.

groundwater or surface drinking water. Improvements in water treatment facilities and technologies could help ameliorate this situation.

Marine: Warm water favors the growth of toxic organisms such as red tides, which cause paralytic shellfish poisoning, diarrheic shellfish poisoning, and amnesic shellfish poisoning. For example, one species of toxic algae previously confined to the Gulf of Mexico (*Gymnodinium breve)* extended northward in 1987 after a "parcel of warm Gulf Stream water" reached far up the east coast, resulting in human neurologic shellfish poisonings and substantial fish kills (Tester, 1991). Domoic acid, a toxin produced by *Nitzchia pungens* diatom that causes amnesic shellfish poisioning, appeared on Prince Edward Island for the first time in 1987. The outbreak coincided with an El Niño year, when warm eddies of the Gulf Stream neared the shore and heavy rains increased nutrient-rich runoff (Hallegraeff, 1993).

Zooplankton, which feed on algae, can serve as reservoirs for *Vibrio cholerae* and other enteric pathogens, particularly gram-negative rods. Quiescent forms of *V. cholerae* have been found to persist within algae; these quiescent forms can revert to a culturable (and likely infectious) state when nutrients, pH, and temperature permit (Huq *et al*., 1990). *V. cholerae* occur in the Gulf of Mexico and along the east coast of North America. With warmer sea surface temperatures, coastal algal blooms therefore could facilitate cholera proliferation and transmission.

## 8.4. Integrative Issues

This chapter has discussed the impacts of climate change on the North American region largely in the context of sector-by-sector assessments of plausible impacts. Several common characteristics among sectors can be identified, however. Also, viewed collectively, interactions between sectors and subregions can be assessed, and insights about the integrated nature of the effects of climate change can emerge.

### *8.4.1. Limitations of Climate Scenarios for Regional Analyses*

Most impact studies have assessed how systems would respond to climate change resulting from an arbitrary doubling of equivalent $CO_2$ concentrations. These so-called $2xCO_2$ scenarios are limited for regional-scale analyses to the extent that they inadequately correspond to the spatial scales of variability in North American natural and human systems. They also do not permit an examination of the effects of climate variability on physical, biological, and socioeconomic systems. Very few studies have considered dynamic responses to steadily increasing concentrations of greenhouse gases. Consequently, important insights about the ability of systems to respond to changing climate over time are lost. This lack of information is of particular concern because the ability of natural ecological systems to migrate often may be much slower than the predicted rate of climate change. Even fewer studies have examined the consequences of increases beyond a doubling of equivalent atmospheric concentrations.

### *8.4.2. Regional Texture of Impacts*

All of the potential impacts of climate change exhibit a regional texture. Variations in the regional distribution of impacts need to be clearly articulated for policymakers. Failure to do so can lead to misleading impressions about the potential changes in social welfare as a result of climate change and alternative policy responses. A simple look at aggregate impacts on U.S. agriculture, for example, might suggest that climate change is not likely to harm agriculture enough to significantly affect the overall U.S. economy; policymakers might be left with the erroneous impression that no policy-relevant problems exist. Distributional differences emerge, however, upon examination of the regional texture of agricultural impacts.

Different adaptation strategies and options will be necessary to deal with these regional and sectoral differences. In areas where production significantly increases—such as the northern edge of agricultural production in North America—additional adaptation may be necessary in the development of infrastructure to support expanded population and transportation requirements associated with growth. The texture of the distribution of sectors and their biological, physical, and social components across the North American landscape cannot adequately be captured at a fine enough scale to be relevant to long-range planning at the present time; these are essential elements of future assessment needs.

It is also recognized—but poorly understood because of limited research—that climate change may have some benefits (e.g., it may reduce stress or provide opportunities) for certain areas or sectors within North America (e.g., expanded agriculture, reduced heating costs) or have a neutral effect on climate-insensitive sectors. If one examines any one particular climate impact, it is likely that there will be "winners" and "losers" either across subregions or within a subregion (e.g., across demographic groups). Nevertheless, the weight of evidence suggests that when all potential impacts are considered collectively, every subregion will incur some negative impacts of climate change.

### *8.4.3. The Role of Adaptation*

Some future climate change is inevitable. Strategies for technological and behavioral adaptation offer an opportunity to reduce the vulnerability of sensitive systems to the effects of climate change and variability. Some adaptive strategies can be undertaken in anticipation of future climate change; others are reactive and can be undertaken as the effects of climate change are realized.

Four points must be kept in mind when considering the extent to which adaptive strategies should be relied upon. First, adaptation is not without cost. Scarce natural and financial resources must be diverted away from other productive activities into adaptive practices. These costs must be carefully weighed when considering the tradeoffs among adapting to the change, reducing the cause of the change, and living with the residual impacts. Second, the economic and social costs of adaptation will increase the more rapidly climate change occurs. Third, although many opportunities exist for technological and behavioral adaptation, uncertainties exist about potential barriers and limitations to their implementation. Fourth, uncertainties exist about the efficacy and possible secondary effects of particular adaptive strategies.

### 8.4.4. *Water as a Common Resource Across Sectors and Subregions*

Water is a linchpin that integrates many subregions and sectors. Available water supplies will be directly affected by climate change, but they also are affected by changes in demand from the many sectors that rely upon the water. Water is a scarce resource used in the agriculture, forest, and energy sectors. It is used in urban areas and in recreational activities. It also is essential for the survival of wetlands, nonforest ecosystems, wildlife, and other ecological systems.

Assessments of the potential impacts of climate change and variability on any of these systems and sectors must account for the inherent competition for water supplies and the need for water of varying qualities in various activities. For example, in an assessment of the potential impacts of climate change on agriculture, an assumption that farmers will be able to adapt to changing climatic conditions through a reliance on irrigation is valid only to the extent that water is available under future climate scenarios. In many cases, the scarcity of available water supplies will increase because of the direct effects of climate change on water, as well as increased demands for available water supplies.

### 8.4.5. *Systemic Nature of the Problem*

In evaluating the implications of climate change impacts on North America, one must consider that although there are regional differences in response by sector and by subregion, the scale of anticipated changes is such that there may be adjustments taking place in every sector and subregion simultaneously. Any one of the impacts (whether beneficial or detrimental) that has been discussed for North America may appear well within the capability of existing structures and policies to adapt. However, the fact that they are occurring simultaneously may pose a significant challenge to resource managers and policymakers. The systemic nature of impacts and issues raises important questions about society's ability to manage the aggregate/cumulative risks posed by climate change.

This systemic problem also must be placed into the larger context of the multiple stressors that are and will be acting on North American resources. Many stressors (environmental, social, and economic) influence natural and human systems and pose significant challenges for decisionmakers and policymakers. The challenge of coping with the cumulative risks of climate change adds to the complexity. What must be kept in mind is that changing climate is not the only—nor necessarily the most important—factor that will influence these systems and that it cannot be isolated from the combination of other factors determining their future welfare.

### 8.4.6. *Integrated Nature of the Problem*

A complete assessment of the effects of climate change on North America must include a consideration of the potential interactions and feedbacks between sectors and subregions. Changes in the climate system can affect natural and human systems in a chain of consequences (see Figure 8-12). Some of these consequences are the results of direct effects of climate change and variability on physical, biological, and socioeconomic systems; some result from indirect links between climate-sensitive systems and related social and economic activities; some result from feedbacks between human activities that affect the climate system, which in turn can lead to further impacts (e.g., human activities affecting the climate system—which, in turn, can lead to further impacts on human health, the environment, and socioeconomic systems).

Most existing studies of potential impacts have focused on the more narrow direct pathway between climate change and climate-sensitive systems and sectors. These effects include direct climate impacts on human health (e.g., heat stress), environmental processes (e.g., impacts of runoff and streamflow on the hydrological cycle, coastal damages caused by sea-level rise, changes in biodiversity), market activities that are linked to the environment (e.g., agriculture, commercial timber, waterborne transport), and human behavior (e.g., changes in air conditioning use as a result of changes in the frequency of very hot days).

Fewer studies have captured the more indirect effects of climate change, which may take many different forms. Many of

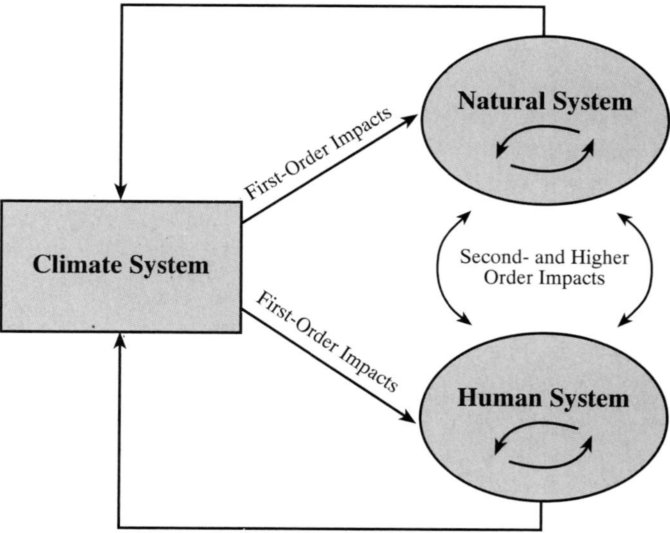


**Figure 8-12:** Chain of consequences.

the primary determinants of human health (adequate food, clean water, secure shelter) are related to outputs from sectors such as agriculture, water resources, and fisheries. The potential spread of infectious diseases is indirectly related to climate change through changes in ecosystems and the hydrological cycle. Therefore, it is important to integrate these relevant systems into a human health assessment.

Other indirect effects include secondary impacts on market activities that are dependent upon sectors directly affected by climate change. For example, climate change will directly affect crop yields and hence agricultural production and prices. These effects, in turn, will influence the prices of goods and services that use agricultural commodities in their production, which will feed back to the agricultural sector and agricultural prices. Shifts in agricultural production could have a large impact on freight transport patterns and may require adjustments in the transportation network—with marine, road, rail, and air links potentially needing expansion into areas not currently serviced. One study of the U.S. economy suggests that the direct effects of climate change on U.S. agriculture, energy use, and coastal protection activities could lead to price

**Box 8-3. Mackenzie Basin Impact Study**

The Mackenzie Basin Impact Study (MBIS) was a 6-year climate change impact assessment focusing on northwestern Canada (supported by Environment Canada and other sponsors) to assess the potential impacts of climate change scenarios on the Mackenzie Basin region, its lands, its waters, and the communities that depend on them (Cohen, 1997a). The MBIS was designed to be a scientist-stakeholder collaborative effort, with 30 research activities on various topics—ranging from permafrost and water levels to forest economics and community response to floods.

The MBIS integration framework included several integration modeling exercises—such as resource accounting, multiregional input-output modeling and community surveys of the nonwage economy of an aboriginal community, a multiobjective model focusing on scenarios of changing land utilization, and a land assessment framework (ILAF) with goal programming and an analytic hierarchy process. MBIS researchers identified six main policy issues related to climate change as another form of "vertical integration": interjurisdictional water management, sustainability of native lifestyles, economic development opportunities, buildings, transportation and infrastructure, and sustainability of ecosystems (Cohen, 1997a). Integration also was attempted through information exchange (scenarios and data) while study components were in progress and a series of workshops that provided opportunities for scientists and stakeholders to express their views on how climate change might affect the region and to react to research results (Cohen, 1997a,b).

The main result of the integrated assessment was that most participating stakeholders saw climate impacts scenarios as a new and different vision of the future for their region, and that adaptation measures alone might not be enough to protect the region from adverse impacts.

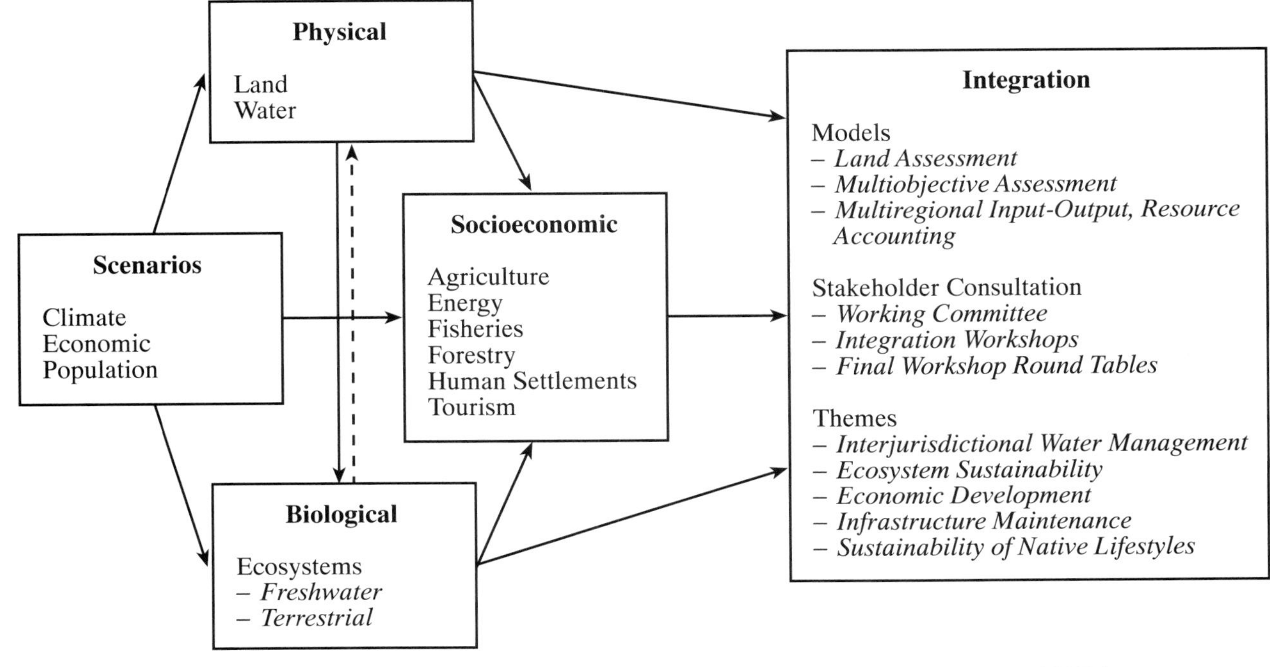


*Integration research framework for the Mackenzie Basin Impact Study (Cohen, 1997a).*

**Box 8-4. Great Lakes-St. Lawrence Basin Project**

The Great Lakes-St. Lawrence Basin (GLSLB) Project on "adapting to the impacts of climate change and variability" is a binational project initiated in 1992 by Environment Canada to improve our understanding of the complex interactions among climate, environment, and society so that regional adaptation strategies could be developed in response to potential climate change and variability (Mortsch and Mills, 1996). The GLSLB Project identified four climate-sensitive theme areas upon which to focus the research: water use and management, land use and management, ecosystem health, and human health. Another prime focus of the GLSLB Project is adaptation, which provides the conceptual framework for integrating research efforts. GLSLB Project studies contain common components that will facilitate the integration of research findings, including study objectives; policy issue foci; a multidisciplinary approach; scenarios tested; impact and adaptation assessment methods; the concept of adaptation; and scales of time, space, and human activity (Mortsch and Mills, 1996). A number of more pragmatic integration concepts also have been incorporated into the GLSLB Project, including the use of "end-to-end studies" assessing biophysical and socioeconomic impacts and developing and evaluating potential adaptation strategies, economic modeling (LINK model), the use of Geographic Information Systems (GIS), and common communication goals (Mortsch and Mills, 1996).

increases for all economic sectors, causing a reallocation of spending and the sectoral composition of output.

Other indirect effects include changes in nonmarket activities as a result of projected impacts of climate change on ecosystems (e.g., changes in recreational fishing as a result of projected impacts of climate change on aquatic ecosystems). For example, the loss of fishing opportunities could be severe in some parts of the region, especially at the southern boundaries of fish species' habitat regions. The loss of fishing opportunities may result in economic losses for the fishing industry. In turn, related industries such as the food, transportation, and lodging industries will be affected. All of these examples illustrate how each sector that is directly or indirectly affected by climate change can adversely affect others.

As this chain of consequences illustrates, the task of assessing various impacts and the feedbacks among them is enormously complex and requires a number of simplifying assumptions. Although there are complex macroeconomic models to assess the costs and consequences of various mitigation policies, the state of the art in impact work at present limits the insights that can be gained from this kind of "top-down" modeling. The dominant approach has been "bottom-up"—aggregating direct and indirect impacts into a single overall estimate, without much attention to feedbacks among various sectors. Nevertheless, the complex, integrated nature of the climate change problem suggests the need for integrated assessments that incorporate many aspects of the region. Sectoral assessments alone would not be sufficient.

Examples of broad, integrated approaches to climate impact assessment are two regional studies in North America: the Mackenzie Basin Impact Study and the Great Lakes-St. Lawrence Basin Project.

These two efforts have tried to account for some of the synergies and interactions among sectors that make each region unique. Each represents a learning experience that ultimately will lead to improvements in how regional assessments and integration are done.

## References


**Aber**, J.D., J.M. Melillo, K.J. Nadelhoffer, J. Pastor, and R.D. Boone, 1991: Factors controlling nitrogen cycling and nitrogen saturation in northern temperate forest ecosystems. *Ecological Applications,* **1**, 303–315.

**Aber**, J.D., S.V. Ollinger, C.A. Federer, P.B. Reich, M.L. Goulden, D.W. Kicklighter, J.M. Melillo, and R.G. Lathrop, 1995: Predicting the effects of climate change on water yield and forest production in the northeastern United States. *Climate Research*, **5(3)**, 207–222.

**Adams**, D., R. Alig, J.M. Callaway, and B.A. McCarl, and S.M. Winnet, 1996: *Forest and agricultural sector optimization model (FASOM): model structure and policy applications.* Research Paper PNW-RP-495. U.S. Department of Agriculture, Pacific Northwest Research Station, Portland, OR, USA, 60pp.

**Adams**, R.M., R.A. Fleming, C.-C. Chang, and B.A. McCarl, 1995a: A reassessment of the economic effects of global climate change on U.S. agriculture. *Climatic Change,* **30(2)**, 147–167.

**Adams**, R.M., B.A. McCarl, K. Segerson, C. Rosenzweig, K.J. Bryant, B.L. Dixon, R. Connor, R.E. Evenson, and D. Ojima, 1995b: *The economic effects of climate change on U.S. agriculture.* Final report. Electric Power Research Institute, Climate Change Impacts Program, Palo Alto, CA, 45 pp.

**Adkison**, M.D., R.M. Peterman, M.F. Lapointe, D.M. Gillis, and J. Korman, 1996: Alternative models of climatic effects on sockeye salmon (*Oncorhynchus nerka*), productivity in Bristol Bay, Alaska, and the Fraser River, British Columbia. *Fish. Oceanogr.* **5(3/4)**, 137–152.

**Agee**, J.K., 1990: The historical role of fire in Pacific Northwest forests. In: *Natural and prescribed fire in Pacific Northwest forests* [Walstad, J.D., S.R. Radosevich, and D.V. Sandberg (eds.)]. Oregon State University Press, Corvallis, OR, pp. 25–38.

**Akbari**, H., S. Davis, S. Dorsano, J. Huang, and S. Winnett, 1992: *Cooling our communities: a guidebook on tree planting and light-colored surfaces.* EPA 22P-2001, LBL-31587, U.S. Government Printing Office, Washington, DC, 217 pp.

**Anderson**, J.M., 1991: The effects of climate change on decomposition processes in grassland and coniferous forests. *Ecological Applications*, **1**, 226–247.

**Anderson**, J., S. Shiau, and D. Harvey, 1991: Preliminary investigation of trend/patterns in surface water characteristics and climate variations. In: *Using hydrometric data to detect and monitor climatic change* [Kite, G. and K. Harvey (eds.)]. Proceedings of NHRI Workshop, National Hydrology Research Institute, Saskatoon, Saskatchewan, pp. 189–201.

**Anderson**, W.P., R. DiFransesco, and M. Kilman, 1994: Potential impacts of climate change on petroleum production in the Northwest Territories. In: *Mackenzie Basin Impact Study Interim Report #2* [Cohen, S.J. (ed.)]. Environment Canada, Downsview, Ontario, pp. 433–441.

**Anisimov**, O.A. and F.E. Nelson, 1996: Permafrost distribution in the Northern Hemisphere under scenarios of climatic change. *Global and Planetary Change,* **14(1)**, 59–72.

**Apps**, M.J., W.A. Kurz, R.J. Luxmoore, L.O. Nilsson, R.A. Sedjo, R. Schmidt, L.G. Simpson, and T.S. Vinson, 1993: Boreal forests and tundra. *Water, Air, and Soil Pollution*, **70**, 39–53.

**Ashfield**, D., R. Phinney, H. Belore, and T. Goodison, 1991: Criteria for identifying streamflow stations applicable to climate change studies. In: *Using hydrometric data to detect and monitor climatic change* [Kite, G. and K. Harvey (eds.)]. Proceedings of NHRI Workshop, National Hydrology Research Institute, Saskatoon, Saskatchewan, pp. 165–180.

**Atwater**, M.A., 1984: Influence of meteorology on high ozone concentrations. In: *Air Pollution Control Association International Specialty Conference on Evaluation of the Scientific Basis for Ozone/Oxidants Standard.* Proceedings of a conference at Houston, TX, November 27–30, 1984. Air Pollution Control Association, Pittsburgh, PA, USA, 393 pp.

**Auclair**, A. N.D., J.T. Lill, and C. Revenga, 1996: The role of climate variability and global warming in the dieback of Northern Hardwoods. *Water, Air, and Soil Pollution*, **91**, 163–186.

**Bailey**, R.G., 1996: *Ecosystem Geography.* Springer-Verlag, New York, NY, 204 pp.

**Bakun**, A., J. Beyer, D. Pauly, J.G. Pope and G.D. Sharp, 1982: Ocean sciences in relation to living marine resources: a report. *Canadian Journal of Fisheries and Aquatic Sciences*, **39**, 1059–1070.

**Banse**, K., 1995: Zooplankton: pivotal role in the control of ocean production. *ICES J. Mar. Sci.*, **52**, 265–277.

**Barth**, M.C., and J.G. Titus (eds.), 1984: *Greenhouse effect and sea level rise: A challenge for this generation.* Van Nostrand Reinhold Co. Inc., New York, NY, USA, 325 pp.

**Bazzaz**, F.A., 1990: The response of natural ecosystems to the rising global $CO_2$ levels. *Annual Review of Ecology and Systematics*, **21**, 167–196.

**Bazzaz**, F.A., S.L. Bassow, G.M. Berntson, and S.C. Thomas, 1996: Elevated $CO_2$ and terrestrial vegetation: implications for and beyond the global carbon budget. In: *Global Change and Terrestrial Ecosystems* [Walker, B. and W. Steffen (eds.)]. Cambridge University Press, Cambridge, United Kingdom, pp. 43–76.

**Bazzaz**, F.A. and K.D.M. McConnaughay, 1992: Plant-plant interactions in elevated $CO_2$ environments. *Australian Journal of Botany*, **40**, 547–563.

**Beamish**, R.J. and D.R. Bouillon, 1995: Marine fish production trends off the Pacific coast of Canada and the United States. In: *Climate change and northern fish populations* [Beamish, R.J. (ed).]. *Can. Spec. Pub. Fish. Aquat. Sci.,* **121**, 585–591.

**Beamish**, R.J., M. Henderson, and H.A. Regier, 1997: Impacts of climate change on the fishes of British Columbia. In: *Responding to Global Climate Change in British Columbia and Yukon* [Taylor, E. and B. Taylor, (eds.)]. Volume I of the Canada Country Study: Climate Impacts and Adaptation, Environment Canada and B.C. Ministry of Environment, Lands and Parks, Vancouver, British Columbia, pp. 12-1 to 12-16.

**Beckett**, W.S., 1991: Ozone, air pollution and respiratory health, *Yale J Biol Med*, **64(2)**, 167–175.

**Belnap**, J., 1995: Surface disturbances: their role in accelerating desertification. *Environmental Monitoring and Assessment*, **37**, 39–57.

**Beltaos**, S., (ed.), 1995: *River ice jams.* Water Resource Publications, Highlands Ranch, CO, 372 pp.

**Belzer**, D.B., M.J. Scott, and R.D. Sands, 1996: Climate change impacts on U.S. commercial building energy consumption: An analysis using sample survey data. *Energy Sources,* **18(1)**, 177–201.

**Bengtsson**, L., M. Botzet, and M. Esch, 1995: Hurricane-type vortices in a general circulation model. *Tellus*, **47A**, 1751–1796.

**Bengtsson**, L., M. Botzet, and M. Esch, 1996: Will greenhouse gas-induced warming over the next 50 years lead to higher frequency and greater intensity of hurricanes? *Tellus*, **48A**, 57–73.

**Benke**, A.C., 1993: Concepts and patterns of invertebrate production in running waters. *Berh. Internat. Verein. Limnol.*, **25**, 15–38.

**Bennett**, S., R. Brereton, I. Mansergh, S. Berwick, K. Sandiford, and C. Wellington, 1991: *Enhanced greenhouse climate change and its potential effect on selected victorian fauna.* Arthur Rylah Institute Technical Report No. 132, Department of Conservation and Environment, Victoria, Australia.

**Bergeron**, Y. and P.R. Dansereau, 1993: Predicting the composition of Canadian southern boreal forest in different fire cycles. *Journal of Vegetation Science*, **4**, 827–832.

**Bethke**, R.W. and T.D. Nudds, 1995: Effects of climate change and land use on Duck abundance in Canadian prairie-parklands. *Ecological Applications*, **5(3)**, 588–600.

**Binkley**, C.S. and G.C. Van Kooten, 1994: Integrating climate change and forests: economic and ecologic assessments. *Climatic Change*, **281(1-2)**, 91–110.

**Black**, W.R., 1990: Global warming impacts on the transportation infrastructure. *TR News*, **150**, 2–34.

**Boer**, G.J., N. McFarlane, and M. Lazare, 1992: Greenhouse gas-induced climate change simulated with the CCC second-generation general circulation model. *Journal of Climatology,* **5**, 1045–1077.

**Bonan**, G.B., F.S. Chapin, and S.L. Thompson, 1995: Boreal forest and tundra ecosystems as components of the climate system. *Climatic Change*, **29(2)**, 145–168.

**Botkin**, D.A., R.A. Nisbet, and T.E. Reynales, 1989: Effects of climate change on forests of the Great Lake states. In: *The potential effects of global climate change on the United States: Appendix D — Forests* [Smith, J.B. and D.A. Tirpak (eds.)]. U.S. Environmental Protection Agency, Washington, DC, USA, pp. 2-1 to 2-31.

**Bowen**, S.G. and B.D. Francy, 1980: Surveillance. In: *St. Louis Encephalitis* [Monath, T.P. (ed.)]. American Public Health Association, Washington DC, 680 pp.

**Boyd**, H., 1987: Do June breeding temperatures affect the breeding success of Dark-bellied Brent Geese? *Bird Study*, **34**, 155–159.

**Brander**, K.M., 1994: Patterns of distribution, spawning, and growth in North Atlantic cod: the utility of inter-regional comparisons. *ICES marine science symposium*, **198**, 406–413.

**Brander**, K.M., 1995: The effects of temperature on growth of Atlantic cod (*Gadus morhua* L.). *ICES Journal of Marine Science*, **52**, 1–10.

**Branson**, F.A., 1985: *Vegetation changes on western rangelands.* Range Monograph No. 2. Society for Range Management, Denver, CO, USA, 76 pp.

**Branson**, F.A., G.F. Gifford, K.G. Renard, and R.F. Hadley, 1981: *Rangeland hydrology*. Society for Range Management, Denver, CO, USA, 340 pp.

**British Columbia Center for Disease Control**, 1995: written communication.

**Brklacich**, M., P. Curran, and D. Brunt, 1996: The application of agricultural land rating and crop models to $CO_2$ and climatic change issues in northern regions: the Mackenzie Basin case study. *Agricultural and Food Science in Finland*, **5**, 351–365.

**Brklacich**, M., P. Curran, D. Brunt, and D. McNabb, 1997a: $CO_2$-induced climatic change and agricultural potential in the Mackenzie Basin. In: *Mackenzie Basin Impact Study Final Report* [Cohen, S.J. (ed.)]. Atmospheric Environment Service, Environment Canada, Downsview, Ontario, Canada, pp. 242–246.

**Brklacich**, M., D. McNabb, C. Bryant, and J. Dumanski, 1997b: Adaptability of agricultural systems to global climatic change: a Renfrew county, Ontario, Canada pilot study. In: *Agricultural restructuring and sustainability: a geographic perspective* [Ilbery, B., T. Rickard, T., and Q. Chiotti (eds.)]. Commonwealth Agricultural Bureau International, Wallingford, UK, pp. 185–200.

**Brodeur**, R.D. and D. Ware, 1995: Interdecadal variability in distribution and catch rates of epipelagic nekton in the Northwest Pacific Ocean. In: *Climate change and northern fish populations* [Beamish, R.J. (ed.)]. *Can. Spec. Pub. Fish. Aquat. Sci.,* **121**, 329–356.

**Brooks**, D.J., 1993: *U.S. forests in a global context.* Gen. Tech. Rep. RM-228. USDA Forest Service, Rocky Mountain Research Station, Fort Collins, CO, USA, pp. 1–24.

**Brown**, H.M., 1989: *Planning for climate change in the Arctic—the impact on energy resource development.* Symposium on the Arctic and Global Change, Climate Institute, 25-27 October 1988, Ottawa, Ontario, Canada, pp. 96–97.

**Brown**, R.G.B., 1991: *Marine birds and climatic warming in the northwest Atlantic.* Canadian Wildlife Service Special Publication No. 68, pp. 49–54.

**Brown**, R., M. Huges, and D. Robinson, 1994: Characterizing the long term variability of snow cover extent over the interior of North America. *Annals of Glaciology,* **21**, 45–50.

**Bruce-Chwatt**, L., 1988: History of Malaria From Prehistory to Eradication. In: *Malaria: principles and practices of malariology* [Wernsdofer, W. (ed.)]. Churchill Livingstone, Edinburgh, United Kingdom, pp. 913–999.

**Brugman**, M.M., P. Raistrick, and A. Pietroniro, 1997: Glacier related impacts of doubling atmospheric carbon dioxide concentrations on British Columbia and Yukon. In: *Responding to global climate change in British Columbia and Yukon* [Taylor, E. and B. Taylor (eds.)]. Volume I, Canada Country Study: Climate Impacts and Adaptation. British Columbia Ministry of Environment, Lands and Parks and Environment Canada, Vancouver, British Columbia, pp. 6-1 to 6-9.

**Bruun**, P., 1962: Sea level rise as a cause of shore erosion. *J. of Waterways and Harbors Division (ASCE),* **88**, 116–130.

**Bugmann**, H.K.M., X. Yan, M.T. Sykes, P. Martin, M. Lindner, P.V. Desanker, and S.G. Cumming, 1996: A comparison of forest gap models: model structure and behaviour. *Climatic Change*, **34(2)**, 289–313.

**Burgner**, R.L., 1991: Life history of Sockeye salmon (*Oncorhynchus nerka*). In: *Pacific Salmon Life Histories* [Groot, C. and L. Margolis (eds.)]. University of British Columbia Press, Vancouver, BC, pp. 1–117.

**Burn**, D., 1994: Hydrologic effects of climate change in west-central Canada. *Journal of Hydrology,* **160**, 53–70.

**Burton**, P.J. and S.G. Cumming, 1995: Potential effects of climatic change on some western Canadian forests, based on pheological enhancements to a patch model of forest succession. *Water, Air, and Soil Pollution*, **82**, 401–414.

**Caine**, T.N., 1980: The rainfall intensity duration control of shallow landslides and debris flows. *Geografiska Annaler A*, **62A**, 23–35.

**Callaway**, M., J. Smith, and S. Keefe, 1995: *The economic effects of climate change for U.S. forests*. Final Report. U.S. Environmental Protection Agency, Adaptation Branch, Climate Change Division, Office of Policy, Planning, and Evaluation, Washington, DC, USA, 103 pp.

**Canadian Forest Service**, 1996: *Selected forestry statistics: Canada, 1995*. Inf. Rep. E-X-48, Natural Resources Canada, Ottawa, Ontario, Canada, 184 pp.

**Canadian Global Change Program**, 1995: *Implications of global change for human health: Final report of the health issues panel of the Canadian Global Change Program.* The Royal Society of Canada, Ottawa, Ontario, Canada, 30 pp.

**Canadian Public Health Association**, 1992: *Human and ecosystem health: Canadian perspectives, Canadian action*. Canadian Public Health Association, Ottawa, Ontario, 21 pp.

**Carnell**, R.E., C.A. Senior, and J.F.B. Mitchell, 1996: Simulated changes in Northern Hemisphere winter storminess due to increasing $CO_2$. *Climate Dynamics*, **12(7)**, 467–476.

**CAST**, 1992: *Preparing U.S. agriculture for global change*. Task Force Report No. 119, Council for Agricultural Science and Technology, Ames, IA, USA, 96 pp.

**CDC**, 1989: Heat-related deaths — Missouri, 1979–1989, *Morbidity and Mortality Weekly Report*, **38**, 437–439.

**CDC**, 1995: Heat-related illnesses and deaths — United States, 1994–1995, *Morbidity and Mortality Weekly Report,* **44(25)**, 465–468.

**Chandler**, A.C., 1945: Factors influencing the uneven distribution of *Aedes aegypti* in Texas cities. *American Journal of Tropical Medicine*, **25**, 145–149.

**Changnon**, S.A., 1993: Changes in climate and levels of Lake Michigan: shoreline impacts at Chicago. *Climatic Change,* **23(3)**, 213–230.

**Changnon**, S.A., 1996: Defining the flood: a chronology of key events. In: *The Great Flood of 1993: causes, impacts and responses* [Changnon, S.A. (ed.)]. Westview Press, Boulder, CO, USA, pp. 3–28.

**Chen**, D, S.E. Zebiak, A. Busalacchi, and M.A. Cane, 1995: An improved procedure for El Niño forecasting: implications for predictability. *Science*, **269**, 1699–1702.

**Child**, R.D. and G.W. Frasier, 1992: ARS range research. *Rangelands*, **14**, 17–32.

**Clague**, J.J., 1989: Sea levels on Canada's Pacific coast: past and future trends. *Episodes*, **12**, 29–33.

**Climate Change Impacts Review Group**, 1991: *The potential effects of climate change in the United Kingdom*. First Report, HMSO, London, UK, 124 pp.

**Cohen**, S.J., 1991: Possible impacts of climatic warming scenarios on water resources in the Saskatchewan River sub-basin, Canada. *Climatic Change*, **19(3)**, 291–317.

**Cohen**, S.J. (ed.), 1997a: *Mackenzie Basin Impact Study Final Report*. Environment Canada, Downsview, Ontario, Canada, 372 pp.

**Cohen**, S.J., 1997b: Scientist-Stakeholder collaboration in integrated assessment of climate change: lessons from a case study of northwest Canada. *Environmental Modelling and Assessment* (submitted).

**Cohen**, S.J., L.E. Welsh, and P.Y.T. Louie, 1989: *Possible impacts of climate warming scenarios on water resources in the Saskatchewan River subbasin.* CCC Report No. 89-9, Atmospheric Environment Service, National Hydrology Research Centre, Saskatoon, Saskatchewan, 87 pp.

**Cohen**, S.J., E.E. Wheaton, and J. Masterton, 1992: *Impacts of Climatic Change Scenarios in the Prairie Provinces: A Case Study from Canada*. SRC Publication No. E-2900-4-D-92, Saskatchewan Research Council, Saskatoon, Saskatchewan, 157 pp.

**Colwell**, R.R., 1996: Global Climate and Infectious Disease: The Cholera Paradigm. *Science,* **274**, 2025–2031.

**Cooley**, K.R., 1990: Effects of $CO_2$-induced climatic changes on snowpack and streamflow. *Hydrol. Sci.*, **35**, 511–522.

**Coutant**, C.C., 1977: Compilation of temperature preference data. *Journal of the Fisheries Research Board of Canada*, **34**, 739–745.

**Coutant**, C.C., 1990: Temperature-oxygen habitat for freshwater and coastal striped bass in a changing climate. *Transactions of the American Fisheries Society*, **119**, 240–253.

**Croley II**, T.E., 1990: Laurentian Great Lakes double-$CO_2$ climate change hydrological impacts. *Climatic Change*, **17(1)**, 27–47.

**Croley II**, T.E., 1992: CCC GCM 2x$CO_2$ hydrological impacts on the Great Lakes. In: *Climate, Climate Change, water level forecasting and frequency analysis, Supporting Documents, Volume 1. Water supply scenarios*. Task Group 2, Working Committee 3, International Joint Commission Levels Reference Study, Phase II, Washington DC, USA, 33 pp.

**Cubasch**, U., G. Meehl, and Z.C. Zhao, 1994: *IPCC WG I Initiative on Evaluation of Regional Climate Change*. Summary Report, 12 pp.

**Cubasch**, U., J. Waszkewitz, G. Hegerl, and J. Perlwitz, 1995: Regional climate changes as simulated in time-slice experiments. *Climatic Change,* **31(2–4)**, 273–304.

**Cumming**, S.G. and P.J. Burton, 1996: Phenology-mediated effects of climatic change on some simulated British Columbia forests. *Climatic Change*, **34(2)**, 213–222.

**Curtis**, P.S., 1996: A meta-analysis of leaf gas exchange and nitrogen in trees grown under elevated carbon dioxide. *Plant, Cell and Environment*, **19**, 127–137.

**Curtis**, P.S., B.G. Drake, P.W. Leadley, W.J. Arp, and D.F. Whigham, 1989: Growth and senescence in plant communities exposed to elevated $CO_2$ concentrations on a estuarine marsh. *Oecologia*, **78**, 20–26.

**Curtis**, P.S., C.S. Vogel, K.S. Pregitzer, D.R. Zak, and J.A. Teeri, 1995: Interacting effects of soil fertility and atmospheric $CO_2$ on leaf area growth and carbon gain physiology in *Populus* X *euramericana* (Dode) Guinier. *New Phytologist*, **129**, 253–263.

**Cushing**, D.H., 1982: *Climate and fisheries*. Academic Press, London, United Kingdom, 373 pp.

**Cushing**, D.H. and R.R. Dickson, 1976: The biological response in the sea to climatic changes. *Advances in Marine Biology*, **14**, 1–122.

**Daniels**, R.C., V.M. Gornitz, A.J. Mehta, S.C. Lee, and R.M. Cushman, 1992: *Adapting to sea-level rise in the U.S. southeast: the influence of built infrastructure and biophysical factors on the inundation of coastal areas*. ORNL/CDIAC-54, Oak Ridge National Laboratory, Oak Ridge, TN, USA, 268 pp.

**Darmstadter**, J., 1993: Climate change impacts on the energy sector and possible adjustments in the MINK region. *Climatic Change*, **24(1–2)**, 117–131.

**Darwin**, R., 1997. World agriculture and climate change: current questions. *World Resource Review*, **9**, 17–31.

**Darwin**, R., M. Tsigas, J. Lewandrowski, and A. Raneses, 1995: *World Agriculture and Climate Change: Economic Adaptations*. Economic Research Service, USDA, AER-703, Washington, DC.

**Davis**, M.B., 1986: Climatic instability, time lags, and community disequilibrium. In: *Community Ecology* [Diamond, J. and T.J. Case (eds.)]. Harper and Row, New York, NY, USA, pp. 269–284.

**DeStasio, Jr.**, B.T., D.K. Hill, J.M. Kleinhans, N.P. Nibbelink, and J.J. Magnuson, 1996: Potential effects of global climate change on small north-temperate lakes: physics, fish, and plankton. *Limnology and Oceanography*, **41**, 1136–1149.

**Díaz**, S., J.P. Grime, J. Harris, and E. McPherson, 1993: Evidence of a feedback mechanism limiting plant response to elevated carbon dioxide. *Nature*, **364**, 616–617.

**Dobson**, A., A. Jolly, and D. Rubenstein, 1989: The greenhouse effect and biological diversity. *Trends in Ecology and Evolution*, **4**, 64–68.

**Dockery**, D.W., C. Pope, X. Xu, J.D. Spengler, J.H. Ware, M.E. Fay, B.G. Ferris, and F.E. Speizer, 1993: An association between air pollution and mortality in six U.S. cities. *New England Journal of Medicine*, **329**, 1753–1759.

**DOE/IEA**, 1995: *Buildings and Energy in the 1980s*. Energy Information Administration, DOE/EIS-0555, 95(1), June.

**Dracup**, J.A., and E. Kahya, 1994: The relationships between U.S. streamflow and La Niña events. *Water Resources Research*, **30**, 2133–2141.

**Dracup**, J.A., S.D. Pelmulder, R. Howitt, J. Horner, W.M. Hanemann, C.F. Dumas., R. McCann, J. Loomis, and S. Ise, 1992: *The economic effects of drought impacts on California resources: implications for California in a warmer world.* National Institute for Global Climate Change, University of California, Davis, CA, USA, 112 pp.

**Drinkwater**, K.F. and R.A. Myers, 1987: Testing predictions of marine fish and shellfish landings from environmental variables. *Canadian Journal of Fisheries and Aquatic Sciences*, **44**, 1568–1573.

**Duncan**, K., 1996: Anthropogenic greenhouse gas-induced warming: suitability of temperatures for the development of *vivax* and *falciparum* malaria in the Toronto Region of Ontario. In: *Great Lakes-St. Lawrence Basin Project Progress Report #1: Adapting to the Impacts of Climate Change and Variability* [Mortsch, L.D. and B.N. Mills (eds.)]. Environment Canada, Burlington, Ontario, Canada, pp. 112–118.

**Eamus**, D., 1991: The interaction of rising $CO_2$ and temperatures with water use efficiency. *Plant, Cell and Environment*, **14**, 843–852.

**Eamus**, D., 1996a: Responses of field grown trees to $CO_2$ enrichment. *Commonwealth Forestry Review*, **75**, 39–47.

**Eamus**, D., 1996b: Tree responses to $CO_2$ enrichment: $CO_2$ and temperature interactions, biomass allocation and stand-scale modeling. *Tree Physiology*, **16**, 43–47.

**Eaton**, J.G., and R.M. Scheller, 1996: Effects of climate warming on fish thermal habitat in streams of the United States. *Limnology and Oceanography,* **41**, 1109–1115.

**Ecological Stratification Working Group**, 1995: *A National Ecological Framework for Canada*. Minister of Supply and Services Canada, Ottawa, Ontario, 125 pp.

**Eilers**, J.M., D.F. Brakke, and D.H. Landers, 1988: Chemical and physical characteristics of lakes in the Upper Midwest, United States. *Environmental Science and Technology*, **22**, 164–172.

**Elliott**, K.J. and W.T. Swank, 1994: Impacts of drought on tree mortality and growth in a mixed hardwood forest. *Journal of Vegetation Science*, **5**, 229–236.

**Ellis**, F.P., 1972: Mortality from heat illness and heat-aggravated illness in the United States. *Environmental Research*, **5**, 1–58.

**Emanuel**, W.R., H.H. Shugart, and M.P. Stevenson, 1985: Climate change and the broad-scale distribution of terrestrial ecosystem complexes. *Climatic Change*, **7(1)**, 29–41.

**Emberlin**, J., 1994: The effects of patterns in climate and pollen abundance on allergy. *Allergy*, **49**, 15–20.

**Energy Economist**, 1988: It is not just farmers who suffer from drought. *Financial Times Business Information*, July 4–7.

**Environment Canada**, 1986: *Wetlands in Canada: a valuable resource*. Lands Directorate Fact Sheet 86-4. Environment Canada, Ottawa, Ontario, Canada, 8 pp.

**Environment Canada**, 1988: *Wetlands of Canada*. Ecological Land Classification Series #24. Ottawa, Ontario, Canada, 452 pp.

**Environment Canada**, 1995: *The state of Canada's climate: monitoring variability and change.* State of Environment Rep. No. 95-1. Minister of Supply and Services, Ottawa, Ontario, Canada, 52 pp.

**Environment Canada**, Smith and Lavender Environmental Consultants, and Sustainable Futures, 1995: *Climate Change Impacts: an Ontario Perspective* [Mortsch, L. (ed.)]. Environment Canada, Burlington, Ontario, Canada, 76 pp.

**Evans**, S.G. and J.J. Clague, 1997: The impact of climate change on catastrophic geomorphic processes in the mountains of British Columbia, Yukon and Alberta. In: *Responding to Global Climate Change in British Columbia and Yukon* [Taylor, E. and B. Taylor (eds.)]. Volume I of the Canada Country Study: Climate Impacts and Adaptation, Environment Canada and B.C. Ministry of Environment, Lands and Parks, Vancouver, British Columbia, pp. 7-1 to 7-16.

**Everts**, C.H., 1985: Effect of sea level rise and net sand volume change on shoreline position at Ocean City, MD. In: *Potential impact of sea level rise on the beach at Ocean City, Maryland* [Titus, J.G. (ed.)]. U.S. Environmental Protection Agency, Washington, DC, USA, 176 pp.

**Eybergen**, J. and F. Imeson, 1989: Geomorphological processes and climate change. *Catena*, **16**, 307–319.

**Fajer**, E.D., M.D. Bowers, and F.A. Bazzaz, 1989: The effects of enriched carbon dioxide atmospheres on plant-insect herbivore interactions. *Science*, **243**, 1198–1200.

**FEMA**, 1991: *Projected impact of relative sea level rise on the National Flood Insurance Program*. Report to Congress. Federal Insurance Administration, Washington, DC, 172 pp.

**Fischlin**, A., H. Bugmann, and D. Gyalistras, 1995: Sensitivity of a forest ecosystem model to climate parametrization schemes. *Environmental Pollution*, **87**, 267–282.

**Flannigan**, M.D. and C.E. Van Wagner, 1991: Climate change and wildfire in Canada. *Canadian Journal of Forest Research*, **21**, 66–72.

**Fleming**, A.M., 1960: Age, growth and sexual maturity of cod (*Gadus morhua* L.) in the Newfoundland area, 1947-1950, *Journal of the Fisheries Research Board of Canada*, **17**, 775–809.

**Fleming**, R.A. and W.J.A. Volney, 1995: Effects of climate change on insect defoliator population processes in Canada's boreal forest: some plausible scenarios. *Water, Air, and Soil Pollution*, **82**, 445–454.

**Focks**, D.A., E. Daniels, D.G. Haile, and J.E. Keesling, 1995: A simulation model of the epidemiology of urban dengue fever: literature analysis, model development, preliminary validation, and samples of simulation results. *American Journal of Tropical Medicine and Hygiene,* **53**, 489–506.

**Forestry Canada**, 1991: *Selected Forestry Statistics, Canada, 1991*. Information Report E-X-46, Forestry Canada, Ottawa, Ontario, Canada.

**Fosberg**, M.A., 1990: Global change — a challenge to modeling. In: *Process modeling of forest growth responses to environmental stress* [Dixon, R.K., R.S. Meldahl, G.A. Ruark, and W.G. Warren (eds.)]. Timber Press, Inc., Portland, OR, USA, pp. 3–8.

**Fosberg**, M.A., B.J. Stocks, and T.J. Lynham, 1996: Risk analysis in strategic planning: fire and climate change in the boreal forest. In: *Fire in ecosystems of boreal Eurasia* [Goldammer, J.G. and V.V. Furyaev (eds.)]. Kluwer Academic Publishers N.V., Dordrecht, Netherlands, pp. 495–504.

**Foster**, D.R., J.D. Aber, J.M. Melillo, R.D. Bowden, and F.A. Bazzaz, 1997: Forest response to disturbance and anthropogenic stress. *BioScience*, **47**, 437–445.

**Frank**, K.T., R.I. Perry, and K.F. Drinkwater, 1990: Predicted response of Northwest Atlantic invertebrate and fish stocks to $CO_2$-induced climate change. *Transactions of the American Fisheries Society*, **119**, 353–365.

**Franklin**, J.F., F.J. Swanson, M.E. Harmon, D.A. Perry, T.A. Spies, V.H. Dale, A. McKee, W.K. Ferrell, J.E. Means, S.V. Gregory, J.D. Lattin, T.D. Schowalter, and D. Larsen, 1991: Effects of global climatic change on forests in northwestern North America. *The Northwest Environmental Journal*, **7**, 233–254.

**Frayer**, W.E., T.J. Monahan, D.C. Bowden, and F.A. Graybill, 1983: *Status and trends of wetlands and deepwater habitat in the coterminous United States, 1950s to 1970s*. Department of Forest and Wood Sciences, Colorado State University, Fort Collins, CO, USA, 32 pp.

**Fredeen**, A.L., G.W. Koch, and C.B. Field, 1995: Effects of atmospheric $CO_2$ enrichment on ecosystem $CO_2$ exchange in a nutrient and water limited grassland. *Journal of Biogeography*, **22**, 215–219.

**Frederick**, K.D., 1991: *Water resources. Report IV of Influences of and Responses to Increasing Atmospheric $CO_2$ and Climate Change: The MINK Project*. U.S. Department of Energy, Washington, DC, USA, 153 pp.

**Frederick**, K.D., and P.H. Gleick, 1989: Water resources and climate change. In: *Greenhouse Warming: Abatement and Adaptation* [Rosenberg, N.J., W.E. Easterling III, P.R. Crosson, and J. Darmstadter (eds.)]. Resources for the Future, Washington, DC, USA, pp. 133–143.

**Freier**, J., 1993: Eastern equine encephalomyelitis. *Lancet*, **342**, 1281–1282.

**Gagliano**, S.M., K.J. Meyer-Arendt, and K.M. Wicker, 1981. Land loss in the Mississippi River deltaic plain, *Transactions of the Gulf Coast Association of Geological Societies*, **31**, 295–300.

**Gibbs**, M., 1984: Economic analysis of sea level rise: Methods and results. In: *Greenhouse effect and sea level rise: a challenge for this generation* [Barth, M.C. and J.G. Titus (eds.)]. Van Nostrand Reinhold Co. Inc., New York, NY, USA, pp. 215–251.

**Glass**, G.E., F.P. Amerasinge, J.M. Morgan III, and T.W. Scott, 1994: Predicting *Ixodes scapularis* abundance on white-tailed deer using Geographic Information Systems. *American Journal of Tropical Medicine and Hygiene,* **51(5)**, 538–544.

**Gleick**, P.H., 1993: *Water in Crisis*. Oxford University Press, Oxford, United Kingdom, 473 pp.

**Goddard**, J., 1992: Ecological studies of adult *Ixodes scapularis* in Central Mississippi: questing activity in relation to time of year, vegetation type and meteorological conditions. *Journal of Medical Entomology*, **29(3)**, 501–506.

**Goos**, T.O., 1989: *The effects of climate and climate change on the economy of Alberta*. A summary of an Acres International Limited report: the environment/economy link in Alberta and the implications under climate change. Climate Change Digest CCD 89-05, Atmospheric Environment Service, Environment Canada, Downsview, Ontario, 9 pp.

**Gorham**, E., 1991: Northern peatlands: role in the carbon cycle and probable responses to climatic warming. *Ecological Applications*, **1(2)**, 182–195.

**Gorissen**, A., J.H. van Ginkel, and J.A. van Veen, 1995: Grass root decomposition is retarded when grass has been grown under elevated $CO_2$. *Soil Bio. Biochem*, **27**, 117–120.

**Government of Canada**, 1996: *The State of Canada's Environment Report*. Government of Canada, Ottawa. Available on the Internet at http://199.212.18.12/folio.pgi/soereng/query=*/doc

**Grant**, D.R., 1975: Recent coastal submergence of the Maritime Provinces. *Proceedings, Nova Scotian Institute of Science*, **27**, 83–102.

**Grant**, L., 1991: Human health effects of climate change and stratospheric ozone depletion. In: *Global climate change: health issues and priorities*. Health and Welfare Canada, Ottawa, Ontario, pp. 72–77.

**Grazulis**, T.P., 1993: A 110-Year Perspective of Significant Tornadoes. In: *The Tornado: its structure, dynamics, prediction, and hazards* [Church, C., D. Burgess, C. Doswell, and R. Davies-Jones (eds.)]. American Geophysical Union, Washington, DC, pp. 467–474.

**Gregory**, J.M., and J.F.B. Mitchell, 1995: Simulation of daily variability of surface temperature and precipitation over Europe in the current and $2xCO_2$ climates using the UKMO climate model. *Quart. J. R. Met. Soc.*, **121**, 1451–1476.

**Griffin**, G.F. and M.H. Friedel, 1985: Discontinuous change in central Australia: some implications of major ecological events for land management. *Journal of Arid Environments*, **9**, 63–80.

**Groisman**, P.Y. and D.R. Easterling, 1994: Century-scale series of annual precipitation over the contiguous United States and Southern Canada. In: *Trends '93: a compendium of data on global change* [Boden, T.A., D.P. Kaiser, R.J. Sepanski, and F.W. Stoss (eds.)]. ORNL/CDIAC-65, Carbon Dioxide Information Analysis Center, Oak Ridge National Laboratory, Oak Ridge, TN, USA, pp. 770–784.

**Grulke**, N.E., J.L. Hom, and S.W. Roberts, 1993: Physiological adjustment of two full-sib families of ponderosa pine to elevated $CO_2$. *Tree Physiology*, **12**, 391–401.

**Grulke**, N.E., G.H. Riechers, W.C. Oechel, U. Hjelm, and C. Jaeger, 1990: Carbon balance in tussock tundra under ambient and elevated atmospheric $CO_2$. *Oecologia*, **83**, 485–494.

**Gubler**, D.J. and D.W. Trent, 1994: Emergence of epidemic dengue/dengue hemorrhagic fever as a public health problem in the Americas, *Infectious Agents Diseases*, **2(6)**, 383–393.

**Gunter**, G., 1974: *An example of oyster production decline with a change in the salinity characteristics of an estuary—Delaware Bay, 1800–1973*. Proceedings of the National Shellfisheries Association, **65**, 3.

**Haines**, A., 1993: The possible effects of climate change on health. In: *Critical Condition: Human Health and the Environment* [Chivian, E., M. McCally, H. Hu, and A. Haines (eds.)]. MIT Press, Cambridge, MA, USA, pp. 151–170.

**Hall**, N.M.J., B.J. Hooskins, P.J. Valdes, and C.A. Senior, 1994: Storm tracks in a high resolution GCM with doubled $CO_2$. *Quarterly Journal of the Royal Met. Society*, **120**, 1209–1230.

**Hallegraeff**, G., 1993: A review of harmful algal blooms and their apparent increase. *Phycologica*, **32(2)**, 79–99.

**Hancock**, T., 1997: Catastrophic atmospheric events/Where science stands on climate variability, atmospheric change and human health. In: *Climate variability, atmospheric change and human health proceedings* [Holmes, R., K. Olgilvie and Q. Chiotti (eds.)] November 4–5, 1996, Downsview, Ontario. Pollution Probe Foundation, Toronto, Ontario, Canada, pp. 7–51.

**Hardy**, J.L., 1988: Susceptibility and resistance of vector mosquitoes. In: *The Arboviruses: Epidemiology and Ecology* [Monath, T.P. (ed.)]. CRC Press, Boca Raton, FL, USA.

**Hare**, S.R. and R.C. Francis, 1995: Climate change and salmon production in the northeast Pacific Ocean. In: *Climate Change and Northern Fish Populations* [R.J. Beamish (ed.)]. *Can. Spec. Publ. Fish. Aquat. Sci.*, **121**, pp. 357–372.

**Hartmann**, H.C., 1990: Climate change impacts on Laurentian Great Lakes levels. *Climatic Change*, **17(1)**, 49–68.

**Haworth**, J., 1988: The global distribution of malaria and the present control effort. In: *Malaria: Principles and Practices of Malariology* [Wernsdofer, W. (ed.)]. Churchill Livingstone, Edinburgh, United Kingdom, pp. 79–99.

**Hayami**, Y. and V.W. Ruttan, 1985: *Agricultural development: an international perspective*. The Johns Hopkins University Press, Baltimore, MD, USA, 506 pp.

**Heady**, H., and R.D. Child, 1994: *Rangeland ecology and management*. Westview Press, Boulder, CO, USA, 519 pp.

**Heath**, L.S., P.E. Kauppi, P. Burschel, H.D. Gregor, R. Guderian, G.H. Kohlmaier, S. Lorenz, D. Overdieck, F. Scholz, H. Thomasius, and M. Weber, 1993: Contribution of temperate forests to the world's carbon budget. *Water, Air, and Soil Pollution*, **70**, 55–69.

**Henderson**, M.A., D.A. Levy, and D.J. Stockner, 1992: Probable consequences of climate change on freshwater production of Adams River sockeye salmon (*Oncorhynchus nerka*). *GeoJournal*, **2**, 51–59.

**Higgins**, R.W., Y. Yao, E.S. Yarosh, J.E. Janowiak, and K.C. Mo, 1997: Influence of the Great Plains low-level jet on summertime precipitation and moisture transport over the Central United States. *Journal of Climate*, **10**, 481–507.

**Hirsch**, J., 1988: As streets melt, cars are flummoxed by hummocks. *The New York Times,* **137(45799)**, B1, B5.

**Hodges**, D.G., F.W. Cubbage and J.L. Regens, 1992: Regional forest migrations and potential economic effects, *Environmental Toxicology and Chemistry*, **11**, 1129–1136.

**Hogg**, W.D. and V.R. Swail, 1997: *Climate extremes indices in northern climates*. Proceedings of a workshop on indices and indicators for climate extremes, Asheville, NC, June 3–6, 1997, 14 pp.

**Horton**, E.B., 1995: Geographical distribution of changes in maximum and minimum temperatures. *Atmos. Res.*, **37**, 102–117.

**Howard**, J.D., O.H. Pilkey, and A. Kaufman, 1985: Strategy for beach preservation proposed. *Geotimes*, **30**, 15–19.

**Hull**, C.H.J. and J.G. Titus (eds.), 1986: *Greenhouse effect, sea level rise, and salinity in the Delaware Estuary*. U.S. Environmental Protection Agency and Delaware River Basin Commission, Washington, DC, USA, 88 pp.

**Hunt**, H.W., E.T. Elliot, J.K. Detling, J.A. Morgan, and D.-X. Chen, 1996: Responses of a $C_3$ and a $C_4$ perennial grass to elevated $CO_2$ and temperature under different water regimes. *Global Change Biology*, **2**, 35–47.

**Huq**, A., R.R. Colwell, R. Rahman, A. Ali, M. Chowdhury, S. Parveen, D. Sack, and E. Russek-Cohen, 1990: Detection of Vibrio cholerae 01 in the aquatic environment by fluorescent-monoclonal antibody and culture methods. *Applied and Environmental Microbiology*, **56**, 2370–2373.

**IBI Group**, 1990: *The implications of long-term climatic change on transportation in Canada*. CCD 90-02, Atmospheric Environment Service, Environment Canada, Downsview, Ontario, Canada, 8 pp.

**International Monetary Fund**, 1996: *International financial statistics yearbook*. International Monetary Fund, Washington, DC, 827 pp.

**IPCC**, 1990: *Climate Change: The IPCC Scientific Assessment* [Houghton, H.T., G.J. Jenkins, and J.J. Ephraums (eds.)]. Cambridge University Press, Cambridge, United Kingdom, 365 pp.

- Cubasch, U. and R.D. Cess, Chapter 3. *Processes and Modelling*, pp. 69–91.
- Mitchell, J.F.B., S. Manabe, V. Meleshko, and T. Tokioka, Chapter 5. *Equilibrium Climate Change — and its Implications for the Future*, pp. 131–172.

**IPCC**, 1996. *Climate Change 1995: The Science of Climate Change. Contribution of Working Group I to the Second Assessment Report of the Intergovernmental Panel on Climate Change* [Houghton, J.J., L.G. Meiro Filho, B.A. Callander, N. Harris, A. Kattenberg, and K. Maskell (eds.)]. Cambridge University Press, Cambridge, United Kingdom and New York, NY, USA, 572 pp.

- *Summary for Policymakers* and *Technical Summary*, pp. 1–49.
- Schimel, D., D. Alves, I. Enting, M. Heimann, F. Joos, D. Raynaud, T.M.L. Wigley, E. Sanhueza, X. Zhou, P. Jonas, R. Charlson, H. Rodhe, S. Sadasivan, K.P. Shine, Y. Fouquart, V. Ramaswamy, S. Solomon, J. Srinivasan, D. Albritton, R. Derwent, Y. Isaken, M. Lal, and D. Wuebbles, Chapter 2. *Radiation Forcing of Climate Change,* pp. 65–131.
- Nicholls, N., G.V. Gruza, J. Jouzel, T.R. Karl, L.A. Ogallo, and D.E. Parker, Chapter 3. *Observed Climate Variability and Change*, pp. 133–192.
- Gates, W.L., A. Henderson-Sellers, G.J. Boer, C.K. Folland, A.Kitoh, B.J. McAvaney, F. Semazzi, N. Smith, A.J. Weaver, and Q.-C. Zeng, Chapter 5. *Climate Models —Evaluation*, pp. 235–284.
- Kattenberg, A., F. Giorgi, H. Grassl, G.A. Meehl, J.F.B. Mitchell, R.J. Stouffer, T. Tokioka, A.J. Weaver, and T.M.L. Wigley, Chapter 6. *Climate Models —Projections of Future Climate,* pp. 289–357.
- Warrick, R.A., C. Le Provost, M.F. Meier, J. Oerlemans, and P.L. Woodworth, Chapter 7. *Changes in Sea Level,* pp. 359–405.
- Santer, B.D., T.M.L. Wigley, T.P. Barnett, and E. Anyamba, Chapter 8. *Detection of Climate Change and Attribution of Causes*, pp. 407–443.

**IPCC**, 1996. *Climate Change 1995: Impacts, Adaptations, and Mitigation of Climate Change: Scientific-Technical Analyses. Contribution of Working Group II to the Second Assessment Report of the Intergovernmental Panel on Climate Change* [Watson, R.T., M.C. Zinyowera, and R.H. Moss (eds.)]. Cambridge University Press, Cambridge, United Kingdom and New York, NY, USA, 880 pp.

- Kirschbaum, M.U.F., Chapter A. *Ecophysiological, Ecological, and Soil Processes in Terrestrial Ecosystems: A Primer on General Concepts and Relationships*, pp. 57–74.
- Allen-Diaz, B., Chapter 2. *Rangelands in a Changing Climate: Impacts, Adaptations and Mitigation,* pp. 131–158.
- Bullock, P. and H. Le Houérou, Chapter 4. *Land Degradation and Desertification,* pp. 170–189.
- Beniston, M. and D.G. Fox, Chapter 5. *Impacts of Climate Change on Mountain Regions,* pp. 191–213.
- Öquist, M.G. and B.H. Svensson, Chapter 6. *Non-Tidal Wetlands,* pp. 215–239.
- Fitzharris, B.B., Chapter 7. *The Cryosphere: Changes and their Impacts,* pp. 240–265.
- Arnell, N., B. Bates, H. Lang, J.J. Magnuson, and P. Mulholland, Chapter 10. *Hydrology and Freshwater Ecology,* pp. 325–364.
- Scott, M.J., Chapter 12. *Human Settlements in a Changing Climate: Impacts and Adaptation,* pp. 399–426.
- Reilly, J., Chapter 13. *Agriculture in a Changing Climate: Impacts and Adaptation*, pp. 427–467.
- Kaczmarek, Z., Chapter 14. *Water Resources Management*, pp. 469–486.
- McMichael, A., Chapter 18. *Human Population Health,* pp. 561–584.

**IRC**, 1995: *Coastal exposure and community protection—Hurricane Andrew's legacy*. Insurance Research Council, Wheaton, IL, USA, 48 pp.

**James**, F.C. and H.H. Shugart, 1974: The phenology of the nesting season of the American robin (*Turdus migratorius)* in the United States. *Condor*, **76**, 159–168.

**Jetten**, T.H. and D.A. Focks, 1997: Changes in the distribution of dengue transmission under climate warming scenarios, *Am J Trop Med Hyg* (in press).

**Johns**, T.C., R.E. Carnell, J.F. Crossley, J.M. Gregory, J.F.B. Mitchell, C.A. Senior, S.F.B. Tett, and R.A. Wood, 1997: The second Hadley Centre coupled ocean-atmosphere GCM: Model description, spinup and validation. *Climate Dynamics* (submitted).

**Johnson**, R.H. and R.B. Vaughan, 1989: The Cows Rocks landslide. *Geological Journal*, **24**, 354–370.

**Jongen**, M., M.B. Jones, T. Hebeisen, H. Blum, and G. Hendrey, 1995: The effects of elevated $CO_2$ concentrations on the root growth of *Lolium perenne* and *Trifolium repens* grown in a FACE system. *Global Change Biology,* **1**, 261–371.

**Joyce**, L.A., J.R. Mills, L.S. Heath, A.D. McGuire, R.W. Haynes, and R.A. Birdsey, 1995: Forest sector impacts from changes in forest productivity under climate change. *Journal of Biogeography*, **22(4–5)**, 703–713.

**Kahya**, E. and J.A. Dracup, 1993: U.S. streamflow patterns in relation to the El Niño/Southern Oscillation. *Water Resources Research*, **29**, 2491–2503.

**Kalkstein**, L.S. and J.S. Greene, 1997: An evaluation of climate/mortality relationships in large U.S. cities and the possible impacts of a climate change. *Environmental Health Perspectives*, **105(1)**, 84–93.

**Kalkstein**, L.S., J.S. Greene, M.C. Nichols and C.D. Barthel, 1993: A new spatial climatological procedure. In: *Proceedings of the eighth conference on applied climatology*, January 17–22, 1993, Anaheim, California, American Meteorological Society, Boston, MA, pp. 169–174.

**Kalkstein**, L.S., P.F. Jamason, J.S. Greene, J. Libby, and L. Robinson, 1995: The Philadelphia hot weather-health watch/warning system: development and application, summer 1995. *Bulletin of the American Meteorological Society*, **77(7)**, 1519–1528.

**Kalkstein**, L.S. and K.E. Smoyer, 1993: The impact of climate change on human health: some international implications. *Experiencia*, **49**, 469–479.

**Kalkstein**, L.S., et al., 1997: The impacts of weather and pollution on human mortality. EPA 230-94-019. US EPA Office of Policy, Planning and Evaluation, Washington, DC, USA (in press).

**Kana**, T.W., J. Michel, M.O. Hayes, and J.R. Jensen, 1984: The physical impact of sea level rise in the area of Charleston, South Carolina. In: *Greenhouse effect and sea level rise: A challenge for this generation* [Barth, M.C. and J.G. Titus (eds.)]. Van Nostrand Reinhold Co. Inc., New York, NY, USA, pp. 105–144.

**Karl**, T.R., P.D. Jones, R.W. Knight, G. Kukla, N. Plummer, V. Razuvayev, K.P. Gallo, J. Lindseay, R.J. Charlson, and T.C. Peterson, 1993: A new perspective on recent global warming: Asymmetric trends of daily maximum and minimum temperature. *Bulletin of the American Meteorological Society*, **74**, 1007–1023.

**Karl**, T.R., R.W. Knight, D.R. Easterling, and R.G. Quayle, 1995a: Trends in U.S. climate during the 20th century. *Consequences*, **1**, 3–12.

**Karl**, T.R., R.W. Knight, and N. Plummer, 1995b: Trends in high-frequency climate variability in the twentieth century. *Nature*, **377**, 217–220.

**Karl**, T.R., R.W. Knight, D.R. Easterling, and R.G. Quayle, 1996: Indices of climate change for the United States. *Bulletin of the American Meteorological Society*, **77**, 279–303.

**Katsouyanni**, K., A. Pantazopoulou, and G. Touloumi, 1993: Evidence for interaction between air pollution and high temperature in the causation of excess mortality. *Archives of Environmental Health*, **48**, 235–242.

**Katz**, R.W. and B.G. Brown, 1992: Extreme events in a changing climate: variability is more important than averages. *Climatic Change*, **21(3)**, 289–302.

**Kaye**, A., and E.S. Barghoorn, 1964: Lake quaternary sea level change and coastal rise at Boston, MA. *Geology Society of America Bulletin*, **75**, 63–80.

**Kearney**, M.S. and J.C. Stevenson, 1985: Sea level rise and marsh vertical accretion rates in Chesapeake Bay. In: *Coastal zone '85* [Magoon, O.T., H. Converse, D. Miner, D. Clark, and L.T. Tobin, (eds.)]. American Society of Chemical Engineers, New York, NY, pp. 1451–1461.

**Keim**, B.D., G.E. Faiers, R.A. Muller, J.M. Grymes III, and R.V. Rohli, 1995: Long-term trends of precipitation and runoff in Louisiana, USA. *International Journal of Climatology*, **15**, 531–541.

**Keith**, V.F., J.C. De Avila, and R.M. Willis, 1989: Effect of climate change on shipping within Lake Superior and Lake Erie. In: *Potential Effects of Global Climate Change on the United States: Appendix H — Infrastructure* [Smith, J.B. and D.A. Tirpak (eds.)]. U.S. Environmental Protection Agency, Washington, DC, USA.

**Kellison**, R.C. and R.J. Weir, 1987: Breeding strategies in forest tree populations to buffer against elevated atmospheric carbon dioxide levels. In: *The Greenhouse effect, climate change and U.S. forests* [Shands, W.E. and J.S. Hoffman (eds.)].

**Kemp**, P.R., D.G. Waldecker, C.E. Owensby, J.F. Reynolds, and R.A. Virginia, 1994: Effects of elevated $CO_2$ and nitrogen fertilization pretreatments on decomposition of tallgrass prairie leaf litter. *Plant Soil*, **165**, 115–127.

**Kilbourne**, E.M., 1992: Epidemiological Statistics. *WHO Reg Publ. Eur. Series*, **42**, 5–25.

**King**, G.A. and R.P. Neilson, 1992: The transient response of vegetation to climate change: a potential source of $CO_2$ to the atmosphere. *Water, Air and Soil Pollution*, **64**, 365–383.

**Kirschbaum**, M.U.F., D.A. King, H.N. Comins, R.E. McMurtrie, B.E. Medlyn, S. Pongracic, D. Murty, H. Keith, R.J. Raison, P.K. Khanna, and D.W. Sheriff, 1994: Modelling forest response to increasing $CO_2$ concentration under nutrient-limited conditions. *Plant, Cell and Environment*, **17**, 1081–1099.

**Kittel**, T.G.F., F. Giorgi, and G.A. Meehl, 1998: Intercomparison of regional biases and doubled $CO_2$ sensitivity of coupled atmosphere-ocean general circulation model experiments. *Climate Dynamics*. 26 manuscript pages.

**Kitzmiller**, J.H., 1993: Genetic considerations in propagating diverse tree species. In: *Proceedings, Western Forest Nursery Association, September 12-14, 1992, Fallen Leaf Lake, CA* [Landis, T.D. (ed.)]. General Technical Report RM-221, U.S. Department of Agriculture, Forest Service, Rocky Mountain Forest and Range Experiment Station, Fort Collins, CO, USA, 151 pp.

**Klarin**, P. and M. Hershman, 1990: Response of the coastal zone management programs to sea level rise in the United States, *Coastal Management*, **18(2)**, 143–166.

**Knoop**, W.T. and B.H. Walker, 1985: Interactions of woody and herbaceous vegetation in a southern Africa savanna. *Journal of Ecology*, **73**, 235–253.

**Komarek**, E.V., 1974: Effects of fire on temperate forests and related ecosystems: Southeastern United States. In: *Fire and Ecosystems* [Kozlowski, T.T. and C.E. Ahlgren (eds.)]. Academic Press, New York, NY, USA, pp. 251–277.

**Körner**, C., 1995: Towards a better experimental basis for upscaling plant responses to elevated $CO_2$ and climate warming. *Plant, Cell and Environment*, **18**, 1101–1110.

**Korsog**, P.E. and G.T. Wolff, 1991: An examination of tropospheric ozone trends in the northeastern U.S. (1973–1983) using a robust statistical method. *Atmospheric Environment*, **25B**, 47–51.

**Kramer**, K., A. Friend, and I. Leinonen, 1996: Modelling comparison to evaluate the importance of phenology and spring frost damage for the effects of climate change on growth of mixed temperate-zone deciduous forests. *Climate Research*, **7**, 31–41.

**Kunkel**, K.E., S.A. Changnon, and J.R. Angel, 1994: Climatic aspects of the 1993 Upper Mississippi River basin flood. *Bulletin of the American Meteorological Society*, **75(5)**, 811–822.

**Kuntasal**, G. and T.Y. Chang, 1987: Trends and relationship of $O_3$, $NO_x$, and HC in the south coast air basin of California. *Journal of the Air Pollution Control Association*, **37**, 1158–1163.

**Kyper**, T., and R. Sorensen, 1985: Potential impacts of selected sea level rise scenarios on the beach and coastal works at Sea Bright, New Jersey. In: *Coastal zone '85* [Magoon, O.T., H. Converse, D. Miner, D. Clark, and L.T. Tobin (eds.)]. American Society of Chemical Engineers, New York, NY, pp. 2645–2655.

**Lambert**, S.J., 1995: The effect of enhanced greenhouse warming on winter cyclone frequencies and strengths. *Journal of Climate*, **8**, 1447–1452.

**Lamothe and Périard**, 1988: *Implications of climate change for downhill skiing in Quebec*. CCD 88-03, Environment Canada, Downsview, Ontario, Canada, 12 pp.

**Landhauser**, S.M. and R.W. Wein, 1993: Postfire vegetation recovery and tree establishment at the Arctic treeline: climate-change-vegetation-response hypotheses. *Journal of Ecology*, **81**, 665–672.

**Landsea**, C.W., N. Nicholls, W.M. Gray, and L.A. Avila, 1996: Quiet early 1990s continues trend of fewer intense atlantic hurricanes. *Journal of Climate*, **5**, 435–453.

**Lane**, L.J., M.H. Nichols, and H.B. Osborn, 1994: Time series analysis of global change data. *Environmental Pollution*, **83**, 63–68.

**Layton**, M., M.E. Parise, C.C. Campbell, R. Advani, J.D. Sexton, E.M. Bosler and J.R. Zucker, 1995: Mosquito transmitted malaria in New York, 1993, *Lancet*, **346(8977)**, 729–731.

**Leatherman**, S.P., 1979: Migration of Assateague Island, Maryland, by inlet and overwash processes, Geology, **7**, 104–107.

**Leatherman**, S.P., 1984: Coastal geomorphic responses to sea level rise: Galveston Bay, Texas. In: *Greenhouse effect and sea level rise: a challenge for this generation* [Barth, M.C. and J.G. Titus, (eds.)]. Van Nostrand Reinhold Co. Inc., New York, NY, pp. 151–178.

**Leatherman**, S.P., 1989: National assessment of beach nourishment requirements associated with accelerated sea level rise. In: *The potential effects of global climate change on the United States: Appendix B — Sea level rise* [Smith, J.B. and D.A. Tirpak (eds.)]. U.S. Environmental Protection Agency, Washington, DC, USA.

**Lee**, D.H., R. Moulton, and B.A. Hibner, 1996: *Climate change impacts on western Lake Erie, Detroit River, and Lake St. Clair water levels*. Great Lakes Environmental Research Laboratory, Ann Arbor, MI and Environment Canada, Burlington, Ontario, Canada, 44 pp.

**Leitch**, J.A., 1981: The wetlands and drainage controversy–revisited. *Agricultural Economist 26*. University of Minnesota, St. Paul, MN, USA, 5 pp.

**Leitch**, J.A. and L.E. Danielson, 1979: *Social, economic and institutional incentives to drain or preserve prairie wetlands*. Department of Agriculture and Applied Economics, University of Minnesota, St. Paul, MN, USA, 78 pp.

**Lenihan**, J.M. and R.P. Neilson, 1993: A rule-based vegetation formation model for Canada. *Journal of Biogeography*, **20**, 615–628.

**Lenihan**, J.M. and R.P. Neilson, 1995: Canadian vegetation sensitivity to projected climatic change at three organizational levels. *Climatic Change*, **30(1)**, 27–56.

**Lettenmaier**, D.P., E.F. Wood, and J.R. Wallis, 1994: Hydro-climatological trends in the continental United States, 1948–1988. *Journal of Climate*, **7**, 586–607.

**Lewellen**, R., 1970: *Permafrost erosion along the Beaufort sea coast*. University of Denver, Geography and Geology Department, Denver CO, USA, 25 pp.

**Lewis**, C.P., 1974: *Sediments and sedimentary processes, Yukon-Beaufort sea coast*. Geological Survey of Canada Paper 75-1, Part B. Geological Survey of Canada, Ottawa, Ontario, Canada, pp. 165–170.

**Lewis**, N., 1988: Two more heat records fall as summer of 1988 boils on. *The Washington Post*, **111(257)**, A1, A10, A11, August 18.

**Linder**, K.P. and M.R. Inglis, 1989: The potential effects of climate change on regional and national demands for electricity. In: The potential impacts of global climate change on the United States. In: *Potential Effects of Global Climate Change on the United States: Appendix H — Infrastructure* [Smith, J.B. and D.A. Tirpak (eds.)]. U.S. Environmental Protection Agency, Washington, DC, USA.

**Lins**, H.F., and P.J. Michaels, 1994: Increasing U.S. streamflow linked to greenhouse forcing. *EOS Transactions*, **75**, 281–283.

**Loehle**, C., 1988: Forest decline: endogenous dynamics, tree defenses, and the elimination of spurious correlation. *Vegetatio*, **77**, 65–78.

**Loehle**, C. and D. LeBlanc, 1996: Model-based assessments of climate change effects on forests: a critical review. *Ecological Modelling*, **90**, 1–31.

**Lonergan**, S., R. DiFrancesco, and M. Woo, 1993: Climate change and transportation in Northern Canada: an integrated impact assessment. *Climatic Change*, **24(4)**, 331–351.

**Longhurst**, A., S. Sathyendranath, T. Platt, and C. Caverhill, 1995: An estimate of global primary production in the ocean from satellite radiometer data. *J. Plank. Res.* **17**, 1245–1271.

**Louisiana Wetland Protection Panel**, 1987: *Saving Louisiana's wetlands: The need for a long-term plan of action*. U.S. Environmental Protection Agency and Louisiana Geological Survey, Washington DC, 102 pp.

**Loveland**, J.E. and G.Z. Brown, 1990: *Impacts of climate change on the energy performance of buildings in the United States*. OTA/UW/UO, Contract J3-4825.0, Office of Technology Assessment, United States Congress, Washington, DC, USA, 58 pp.

**Lovett**, G.M., 1994: Atmospheric deposition of nutrients and pollutants in North America: an ecological perspective. *Ecological Applications*, **4**, 629–650.

**Lucas v. South Carolina Coastal Council**, ___ U.S. ___, 112 S.Ct. 2886 (1992).

**Maarouf**, A., 1995: Human Health. In: *The case for contributions to the Global Climate Observing System (GCOS)* [Canadian Climate Program Board (ed)]. Canadian Climate Program Board, Ottawa, Ontario pp. 105–108.

**MacKenzie**, W.R., N.J. Hoxie, M.E. Proctor, and M.S. Gradus, 1994: A massive outbreak in Milwaukee of cryptosporidum infection transmitted through public water supply. *New England Journal of Medicine,* **331(3)**, 161–167.

**Magnuson**, J.J., and B.T. DeStasio, 1996: Thermal niche of fishes and global warming. In: *Global warming — implications for freshwater and marine fish* [Wood, C.M. and D.G. McDonald (eds.)]. Society for Experimental Biology Seminar Series 61, Cambridge University Press, Cambridge, United Kingdom, pp. 377–408.

**Malcolm**, J.R. and A. Markham, 1996: Ecosystem resilience, biodiversity, and climate change: setting limits. *Parks*, **6**, 38–49.

**Maltby**, E., 1986: *Waterlogged wealth*. Earthscan Publications International Institute, Washington, DC, USA, 200 pp.

**Manabe**, S. and R.T. Wetherald, 1987: Large-scale changes in soil wetness induced by an increase in carbon dioxide. *J. Atmos. Sci.*, **44**, 1211–1235.

**Mann**, K.H., 1993: Physical oceanography, food chains and fish stocks, a review. *ICES J. Mar. Sci.*, **50**, 105–119.

**Manning**, W.J. and A.V. Tiedemann, 1995: Climate change: potential effects of increased atmospheric carbon dioxide ($CO_2$), ozone ($O_3$), and ultraviolet-B (UV-B) radiation on plant diseases. *Environmental Pollution*, **88**, 219–245.

**Marak**, R.R. and R. Livingstone Jr., 1970: Spawning date of Georges Bank haddock. *ICNAF Research Bulletin,* **7**, 56–58.

**Markham**, A., 1996: Potential impacts of climate change on ecosystems: a review of implications for policymakers and conservation biologists. *Climate Research*, **6**, 179–191.

**Markham**, A. and J. Malcolm, 1996: Biodiversity and wildlife conservation: adaptation to climate change. In: *Adaptation to climate change: assessment and issues* [Smith, J., N. Bhatti, G. Menzhulin, R. Benioff, M. Campos, B. Jallow, and F. Rijsberman (eds.)]. Springer-Verlag, New York, NY, USA, pp. 384–401.

**Marland**, G., and T.A. Boden, 1997: Global, Regional, and National $CO_2$ Emissions. In: *Trends: a compendium of data on global change*. Carbon Dioxide Information Analysis Center, Oak Ridge National Laboratory, Oak Ridge, TN, USA.

**Marsh**, P., and L.F.W. Lesack, 1996: The hydrologic regime of perched lakes in the Mackenzie Delta: potential responses to climate change. *Limnology and Oceanography*, **41**, 849–856.

**Martens**, W., T. Jetten, J. Rotmans, et al., 1995: Climate change and vector-borne diseases: a global modelling perspective, *Global Environmental Change,* **5(3)**, 195–209.

**Martin**, C., 1992: Climatic perturbation as a general mechanism of forest dieback. In: *Forest Decline Concepts* [Manion, P.D. and D. Lachance (eds.)]. APS Press, St. Paul, MN, USA, pp. 38–58.

**Martin**, P.H., 1996: Climate change, water stress, and fast forest response: a sensitivity study. *Climatic Change*, **34(2)**, 223–230.

**Masterton**, J.M., R.B. Crowe, and W.M. Baker, 1976: *The tourism and outdoor recreation climate of the Prairie Provinces*. REC-1-75, Meteorological Applications Branch, Environment Canada, Toronto, Ontario, 221 pp.

**Matson**, P.A., W.J. Parton, A.G. Power and M.J. Swift, 1997: Agricultural intensification and ecosystem properties, *Science*, **277**, 504–509.

**Matthews**, R.B., M.J. Kropff, and D. Bachelet, 1994a: Climate change and rice production in Asia. *Entwicklung und Ländlicherraum,* **1**, 16–19.

**Matthews**, R.B., M.J. Kropff, D. Bachelet, and H.H. van Laar, 1994b: *The impact of global climate change on rice production in Asia: a simulation study*. Report No. ERL-COR-821, U.S. Environmental Protection Agency, Environmental Research Laboratory, Corvallis, OR, USA, 289 pp.

**Mattson**, W.J. and R.A. Haack, 1987: The role of drought in outbreaks of plant-eating insects. *BioScience*, **37(2)**, 110–118.

**Maxwell**, J.B. and L.A. Barrie, 1989: Atmospheric and climatic change in the Arctic and Antarctic. *Ambio,* **18(1)**, 42–49.

**Mayeux**, H.S., H.B. Johnson, and H.W. Polley, 1991: Global change and vegetation dynamics. In: *Noxious Range Weeds* [James, L.F., J.O. Evans, M.H. Ralphs, and R.D. Childs (eds.)]. Westview Press, Boulder, CO, USA, pp. 62–74.

**McCabe, Jr.**, G.J. and D.M. Wolock, 1992: Effects of climatic change and climatic variability on the Thornthwaite moisture index in the Delaware River basin. *Climatic Change*, **20(2)**, 143–153.

**McFarlane**, N.A., G.J. Boer, J.-P. Blanchet, and M. Lazare, 1992: The Canadian Climate Centre second-generation general circulation model and its equilibrium climate. *J. Clim.*, **5**, 1013–1044.

**McGillivray**, D.G., T.A. Agnew, G.A. McKay, G.R. Pilkington, and M.C. Hill, 1993: *Impacts of climatic change on the Beaufort sea-ice regime: implications for the arctic petroleum industry*. CCD 93-01, Environment Canada, Downsview, Ontario, 17 pp.

**McGuire**, A.D., J.M. Melillo, and L.A. Joyce, 1995: The role of nitrogen in the response of forest net primary production to elevated atmospheric carbon dioxide. *Annual Review of Ecology and Systematics*, **26**, 473–503.

**McLaughlin**, S.B. and D.J. Downing, 1995: Interactive effects of ambient ozone and climate measured on growth of mature forest trees. *Nature*, **374**, 252–254.

**McLeese**, D.W. and D.G. Wilder, 1958. The activity and catchability of the lobster (*Homarus americanus*) in relation to temperature. *Journal of the Fisheries Research Board of Canada*, **15**, 1345–1354.

**McNaughton**, S.J., R.W. Ruess, and S.W. Seagle, 1988: Large mammals and process dynamics in African ecosystems. *BioSciences*, **38**, 794–800.

**McNeely**, J.A., 1990: Climate change and biological diversity: policy implications. In: *Landscape ecological impacts of climate change* [Boer, M.M. and R.S. de Groot (eds.)]. IOS Press, Amsterdam, Netherlands, pp. 406–428.

**McNulty**, S.G., J.M. Vose, and W. Swank, 1996: Potential climate change effects on loblolly pine forest productivity and drainage across the southern United States. *Ambio*, **25**, 449–453.

**Meade**, M., J. Florin, and W. Gesler, 1988: *Medical Geography*. The Guildford Press, New York, NY, 340 pp.

**Mearns**, L.O., F. Giorgi, L. McDaniel, and C. Shields, 1995: Analysis of daily variability of precipitation in a nested regional climate model: comparison with observations and doubled $CO_2$ results. *Global and Planetary Change*, **10**, 55–78.

**Mearns**, L.O., C. Rosenzweig, and R. Goldberg, 1996: The effects of changes in daily and interannual climatic variability on CERES-wheat: A sensitivity study. *Climatic Change*, **32(1)**, 257–292.

**Meisner**, J.D., 1990: Effect of climatic warming on the southern margins of the native range of brook trout, *Salvelinus fontinalis*. *Canadian Journal of Fisheries and Aquatic Sciences*, **47**, 1067–1070.

**Mekis**, E., and W. Hogg, 1997: Rehabilitation and analysis of Canadian daily precipitation time series. In: *10th conference on Applied Climatology, 20–24 October 1997*, Reno, Nevada, USA (in press).

**Miller**, B.A., et al., 1992: *Impact of incremental changes in meteorology on thermal compliance and power system operations.* Report WR28-1-680-109. Tennessee Valley Authority Engineering Laboratory, Norris, TN, USA, 32 pp.

**Mills**, P., 1994: The agricultural potential of northwestern Canada and Alaska and the impact of climatic change. *Arctic*, **47**, 115–213.

**Mitchell**, J.F.B., T.C. Johns, J.M. Gregory, and S. Tett, 1995: Climate response to increasing levels of greenhouse gases and sulphate aerosols. *Nature*, **376**, 501–504.

**Mitchell**, J.F.B. and D. A. Warrilow, 1987: Summer dryness in northern mid latitudes due to increased $CO_2$. *Nature*, **330**, 238–240.

**Mitsch**, W.J. and J.G. Gosselink, 1986: *Wetlands*. Van Nostrand Reinhold Co. Inc., New York, NY, USA, 539 pp.

**Monath**, T.P. and T.F. Tsai, 1987: St. Louis encephalitis: lessons from the last decade. *American Journal of Tropical Medicine and Hygiene,* **37**, 40S–59S.

**Moore**, M.V., M.L. Pace, J.R. Mather, P.S. Murdoch, R.W. Howarth, C.L. Folt, C.Y. Chen, H.F. Hemond, P.A. Flebbe, and C.T. Driscoll, 1997: Potential effects of climate change on freshwater ecosystems of the New England/Mid-Atlantic Region. *Hydrol. Processes*, (in press). Hydrologic Processes, **11**, 925–947.

**Moore**, R., S. Tzipori, J.K. Griffiths, and K. Johnson, 1995: Temporal changes in permeability and structure of piglet ileum after site-specific infection by cryptosporidium parvium. *Gastroenterology,* **108(4)**, 1030–1039.

**Morgan**, J.A., H.W. Hunt, C.A. Monz, and D.R. Lecain, 1994: Consequences of growth at two carbon dioxide concentrations and two temperatures for leaf gas exchange in Pascopyrum smithii ($C_3$) and Bouteloua gracilis ($C_4$). *Plant Cell Environment,* **17**, 1023–1033.

**Morin**, G. and M. Slivitzky, 1992: Impacts de changements climatiques sur le régime hydrologique: le cas de la rivière Moisie. *Revue des Sciences de l'Eau*, **5**, 179–195.

**Mortsch**, L. and F. Quinn, 1996: Climate change scenarios for Great Lakes Basin ecosystem studies. *Limnology and Oceanography*, **41(5)**, 903–911.

**Mortsch**, L.D. and B.N. Mills (eds.), 1996: Great Lakes-St. Lawrence Basin Project progress report #1: adapting to the impacts of climate change and variability. Environment Canada, Burlington, Ontario, 140 pp. + appendix. 160 pp.

**Mount**, G.A., D.G. Haile, D.R. Barnard, and E. Daniels, 1993: New Version of LSTSIM for computer simulation of *Amblyomma americanum* (Acari: Ixodes) population dynamics. *Journal of Medical Entomology,* **30(5)**, 843–857.

**Myers**, R.A., J.A. Hutchings and N.J. Barrowman, 1996: Hypotheses for the decline of cod in the North Atlantic. *Marine Ecology Progress Series*, **138**, 293–308.

**Myneni**, R.B., C.D. Keeling, C.J. Tucker, G. Asrar, and R.R. Nemani, 1997: Increased plant growth in the northern high latitudes from 1981 to 1991. *Nature*, **386**, 698–702.

**Nash**, L.L. and P.H. Gleick, 1993: *The Colorado River basin and climate change: the sensitivity of streamflow and water supply to variations in temperature and precipitation.* EPA-230-R-93-009, U.S. Environmental Protection Agency, Washington, DC, USA, 92 pp.

**National Council on Public Works Improvement**, 1988. *Fragile foundations — Final report to the President and Congress.* Washington, DC, USA, 226 pp.

**National Research Council**, 1987: *Responding to changes in sea level.* National Academy Press, Washington, DC, USA, 148 pp.

**Neilson**, R.P., 1993a: Vegetation redistribution: a possible biosphere source of $CO_2$ during climate change. *Water, Air and Soil Pollution*, **70**, 659–673.

**Neilson**, R.P., 1993b: Transient ecotone response to climatic change: some conceptual and modelling approaches. *Ecological Applications*, **3**, 385–395.

**Neilson**, R.P., 1995: A model for predicting continental-scale vegetation distribution and water balance. *Ecological Applications,* **5**, 362–385.

**Neilson**, R.P., G.A. King, and G. Koerper, 1992: Toward a rule-based biome model. *Landscape Ecology*, **7**, 27–43.

**Neilson**, R.P. and D. Marks, 1994: A global perspective of regional vegetation and hydrologic sensitivities from climatic change. *Journal of Vegetation Science*, **5**, 715–730.

**Newton**, P.C.D., H. Clark, C.C. Bell, E.M. Glasgow, K.R. Tate, D.J. Ross, G.W. Yeates, and S. Saggar, 1995: Plant growth and soil processes in temperate grassland communities at elevated $CO_2$. *J. Biogeography,* **22**, 235–240.

**Ng**, H., and J. Marsalek, 1992: Sensitivity of streamflow simulation to changes in climatic inputs. *Nordic Hydrology*, **23(4)**, 257–272.

**Nixon**, J.F., K.A. Sortland, and D.A. James, 1990: Geotechnical aspects of northern gas pipeline design. In: *5th Canadian Permafrost Conference.*, Collection Nordicana No. 54, Laval University, Montreal, Quebec, Canada, pp. 299–307.

**Noden**, B.H., M.D. Kent, and J.C. Beier, 1995: The impact of variations in temperature on early Plasmodium falciparum development in *Anopheles stephensi. Parasitology,* **111**, 539–545.

**Norby**, R.J., 1996: Forest canopy productivity index. *Nature*, **381**, 564.

**Nordstrum**, K.F., 1992. *Estuarine Beaches: an introduction to the physical and human factors affecting use and management of beaches in estuaries, lagoons, bays and fjords.* Elsevier Science Publishers, Essex, UK, 225 pp.

**NRC Committee on Rangeland Classification**, 1994: *Rangeland health: new methods to classify, inventory and monitor rangelands.* National Academy Press, Washington, DC, USA.

**Oechel**, W.C., S.J. Hastings, G. Vourlitis, and M. Jenkins, 1993: Recent change of arctic tundra ecosystems from net carbon dioxide sink to a source. *Nature*, **361**, 520–523.

**OIES**, 1991: *Arid Ecosystems Interactions.* Report OIES-6 UCAR, UCAR Office for Interdisciplinary Research, Boulder, CO, USA.

**Oja**, T. and P.A. Arp, 1996: Nutrient cycling and biomass growth at a North American hardwood site in relation to climate change: ForSVA assessments. *Climatic Change*, **34(2)**, 239–251.

**O'Neill**, E.G. and R.J. Norby, 1996: Litter quality and decomposition rates of foliar litter produced under $CO_2$ enrichment. In: *Carbon dioxide and terrestial ecosystems* [Koch, G.W. and H.A. Mooner (eds.)]. Academic Press, San Diego, CA, USA, pp. 87–103.

**Oppenheimer**, M., 1989: Climate change and environmental pollution: physical and biological interactions. *Climatic Change*, **15(1-2)**, 255–270.

**Oram**, P.A. 1989: *Climate and food security.* International Rice Research Institute, Manila, Philippines. 602 pp.

**Ostby**, F.P., 1993: The changing nature of tornado climatology. In: *Preprints, 17th Conference on Severe Local Storms, October 4–8, 1993, St. Louis, Missouri*, American Meteorological Society, Boston, MA, pp. 1–5.

**OTA**, 1984: *Wetlands: their use and regulation.* U.S. Congress OTA-O-206, Washington, DC, USA, 208 pp.

**OTA**, 1993: *Preparing for an uncertain climate - Volume I.* OTA-0-567. Government Printing Office, Washington, DC, USA, 372 pp.

**Overpeck**, J.T., D. Rind, and R. Goldberg, 1990: Climate-induced changes in forest disturbance and vegetation. *Nature*, **343**, 51–53.

**Owensby**, C.E., L.M. Auen, and P.I. Coyne, 1994: Biomass production in a nitrogen-fertilized, tallgrass prairie ecosystem exposed to ambient and elevated levels of $CO_2$. *Plant Soil,* **165**, 105–113.

**Owensby**, C.E., R.C. Cochran, and L.M. Auen, 1996: Effects of elevated carbon dioxide on forage quality for ruminants. In: *Carbon dioxide, populations, and communities* [Körner, C. and F.A. Bazzaz (eds.)]. Academic Press, San Diego, CA, pp. 363–371.

**Owensby**, C.E., P.I. Coyne, J.M. Ham, L.A. Auen, and A.K. Knapp, 1993: Biomass production in a tall grass prairie ecosystem exposed to elevated $CO_2$. *Ecological Applications,* **3**, 644–653.

**Pacala**, S.W., C.D. Canham, J. Saponara, J.A. Silander, R.K. Kobe, and E. Ribbens, 1996: Forest models defined by field measurements: estimation, error analysis and dynamics. *Ecological Monographs*, **66**, 1–43.

**PAHO**, 1994: *Dengue and dengue hemorrhagic fever in the Americas: guidelines for prevention and control.* Pan American Health Organization, Washington, DC, 98 pp.

**Park**, R.A., M.S. Treehan, P.W. Mausel, and R.C. Howe, 1989: The effects of sea level rise on U.S. coastal wetlands. In: *The potential effects of global climate change on the United States: Appendix B — Sea level rise* [Smith, J.B. and D.A. Tirpak (eds.)]. U.S. Environmental Protection Agency, Washington, DC, USA, 255 pp.

**Parker**, A.J., 1986: Environmental and historical factors affecting red and white fir regeneration in ecotonal forests. *Forest Science*, **32**, 339–347.

**Parrish**, G., 1997: Impact of Weather on Health. In: *Workshop on the Social and Economic Impacts of Weather* [Pielke, Jr., R.A. (ed.)]. April 2–4, 1997, Boulder, CO, USA, pp. 87–91.

**Parsons**, D.J., 1991: Planning for climate change in national parks and other natural areas. *The Northwest Environmental Journal*, **7**, 255–269.

**Parton**, W.J., J. M.O. Scurlock, D.S. Ojima, T.G. Gilmanov, R.J. Scholes, D.S. Schimel, T. Kirchner, J.C. Menaut, T. Seastedt, E. Gracia Moya, A. Kamnalrut, and J.L. Kinyamario, 1993: Observations and modeling of biomass and soils organic matter dynamics for grassland biome worldwide. *Global Biogeochemical Cycles*, **7**, 785–809.

**Pastor**, J. and W.M. Post, 1988: Response of northern forests to $CO_2$-induced climate change. *Nature*, **334**, 55–58.

**PCSD**, 1996: *Sustainable America: a new consensus for prosperity, opportunity, and a healthy environment for the future.* U.S. Government Printing Office, Washington, DC, USA, 186 pp.

**Peixoto**, J.P. and A.H. Oort, 1992: *Physics of climate.* American Institute of Physics, New York, NY, USA, 520 pp.

**Perez-Garcia**, J., L.A. Joyce, C.S. Binkley, and A.D. McGuire, 1997: Economic impacts of climatic change on the global forest sector: an integrated ecological/economic assessment. *Critical Reviews of Environmental Science and Technology* (in press).

**Perruchoud**, D.O. and A. Fischlin, 1995: The response of the carbon cycle in undisturbed forest ecosystems to climate change: a review of plant-soil models. *Journal of Biogeography*, **22(4–5)**, 759–774.

**Perry**, D.A. and J. Maghembe, 1989: Ecosystem concepts and current trends in forest management: time for reappraisal, *Forest Ecology and Management*, **26**, 123–140.

**Peters**, R.L. and J.D.S. Darling, 1985: The greenhouse effect and nature reserves. *BioScience*, **35**, 707–717.

**Phelps**, P., 1996: *Conference on human health and global climate change: summary of the proceedings.* September 11–12, 1995, Washington, DC, National Academy Press, Washington, DC, USA, 64 pp.

**Phillips**, D., 1990: *The Climates of Canada*. Environment Canada, Downsview, Ontario, Canada, 176 pp.

**Phillips**, D.L., J. Lee, and R.F. Dodson, 1996: Sensitivity of the U.S. corn belt to climate change and elevated $CO_2$: I. Corn and soybean yields. *Agricultural Systems*, **52**, 481–502.

**Poff**, J.L., 1992: Regional hydrologic response to climate change: an ecological perspective. In: *Global climate change and freshwater ecosystems* [Firth, P. and S.G. Fisher (eds.)]. Springer-Verlag, New York, NY, USA, pp. 88–115.

**Poiani**, K.A. and W.C. Johnson, 1991: Global warming and prairie wetlands: potential consequences for waterfowl habitat. *Bioscience*, **41**, 611–618.

**Poiani**, K.A. and W.C. Johnson, 1993: Potential effects of climate change on a semi-permanent prairie wetland. *Climatic Change*, **24 (37)**, 213–232.

**Polley**, H.W., 1997: Implications of rising atmospheric carbon dioxide concentrations for rangelands. *Journal of Range Management,* (in press).

**Polley**, H.W., H.B. Johnson, and H.S. Mayeux, 1994: Increasing $CO_2$: comparative responses of the $C_4$ grass *Schizachyrium* and grassland invader. *Prosopis*. *Ecology*, **75**, 976–988.

**Polley**, H.W., H.B. Johnson, H.S. Mayeux, C.R. Tischler, and D.A. Brown, 1996: Carobon dioxide enrichment improves growth, water relations and survival of droughted honey mesquite (*Prosopis glandulosa*) seedlings. *Tree Physiology,* **16**, 817–823.

**Polley**, H.W., H.S. Mayeux, H.B. Johnson, and C.R. Tischler, 1997: Viewpoint: atmospheric $CO_2$, soil water, and shrub/grass ratios on rangelands. *Journal of Range Management*, **50**, 278–284.

**Pope**, C., D. Bates, and M. Raizenne, 1995: Health effects of particulate air pollution: time for reassessment? *Environmental Health Perspectives*, **103**, 472–480.

**Post**, W.M. and J. Pastor, 1996: LINKAGES — An individual-based forest ecosystem model. *Climatic Change*, **34(2)**, 253–261.

**Post**, W.M., J. Pastor, A.W. King, and W.R. Emanuel, 1992: Aspects of the interaction between vegetation and soil under global change. *Water, Air, and Soil Pollution*, **64**, 345–363.

**Powell**, D.S., J.L. Faulkner, D.R. Darr, Z. Zhu, and D.W. MacCleery, 1993: *Forest resources of the United States, 1992*. USDA Forest Service, General Technical Report, RM-234, Fort Collins, CO, USA, 132 pp.

**Price**, C. and D. Rind, 1994: Possible implications of global climate change on global lightning distributions and frequencies. *Journal of Geophysical Research*, **99**, 10823.

**Price**, D.T. and M.J. Apps, 1996: Boreal forest responses to climate-change scenarios along an ecoclimatic transect in Central Canada. *Climatic Change*, **34(2)**, 179–190.

**Prowse**, T.D., 1997: Climate change effects on permafrost, freshwater ice and glaciers. In: *Ecosystem Effects of Atmospheric Change*. Proceedings of a colloquium, March 5–6, 1996, Pointe Claire, Quebec. National Water Research Institute, Environment Canada, Burlington, Ontario, pp. 105–136.

**Rahel**, F.J., C.J. Keleher, and J.L. Anderson, 1996: Potential habitat loss and population fragmentation for cold water fish in the North Platte River drainage of the Rocky Mountains: response to climatic warming. *Limnology and Oceanography*, **41**, 1116–1123.

**Rauzi**, F. and C.L. Fly, 1968: *Water Intake on Midcontinental Rangelands as Influenced by Soil and Plant Cover*. Technical Bulletin No. 1390. USDA-ARS, Washington, DC, USA, 58 pp.

**Reeves**, W.C., J.L. Hardy, W.K. Reisen, and M.M. Milby, 1994: The potential effect of global warming on mosquito-borne arboviruses. *Journal of Medical Entomology,* **31(3)**, 323–332.

**Regens**, J.L., F.W. Cubbage and D.G. Hodges, 1989: Greenhouse gases, climate change and U.S. forest markets, *Environment*, **31(4)**, 45–51.

**Regier**, H.A., J.A. Holmes, and D. Panly, 1990: Influence of temperature changes on aquatic ecosystems: an interpretation of empirical data. *Transactions of the American Fisheries Society,* **119**, 374–389.

**Reisen**, W.K., R.P. Meyer, S.B. Presser, and J.L. Hardy, 1993: Effects of temperature on the transmission of Western Equine encephalomyelitis and St. Louis encephalitis viruses by *Culex tarsalis* (Diptera: Culicidae). *Journal of Medical Entomology,* **30**, 151–160.

**Rhodes**, S.L. and K.B. Wiley, 1993: Great Lakes toxic sediments and climate change: implications for environmental remediation. *Global Environmental Change*, **3(3)**, 292–305.

**Rosenberg**, N.J., 1992: Adaptation of agriculture to climate change. *Climatic Change*, **21(4)**, 385–405.

**Rosenberg**, N.J., 1993: *Towards an integrated assessment of climate change: The MINK Study*. Kluwer Academic Publishers, Boston, MA, USA, 173 pp.

**Rosenthal**, D.H., H.K. Gruenspecht, and E. Moran, 1995: Effects of global warming on energy use for space heating and cooling in the United States. *Energy Journal*, **16(2)**, 77–96.

**Rosenzweig**, C., J. Phillips, R. Goldberg, J. Carroll, and T. Hodges, 1996: Potential impacts of climate change on citrus and potato production in the U.S. *Agricultural Systems*, **52**, 455–479.

**Rotton**, J., 1983: Angry, sad, happy? Blame the weather. *U.S. News and World Report*, **95**, 52–53.

**Rowntree**, P., 1993: *Workshop on socio-economic and policy aspects of change of incidence and intensity of extreme weather events*. Institute for Environmental Studies, W93/15, Free University, Amsterdam, The Netherlands, June 24–25, 1993.

**Russell**, P., 1968: The United States and malaria: debits and credits. *Bulletin of the New York Academy of Medicine,* **44**, 623–653.

**Sala**, O.E., R.A. Golluscio, W.K. Lauenroth, and A. Soriano, 1989: Resource partitioning between shrubs and grasses in the Patagonian steppe. *Oeceologia*, **81**, 501–505.

**Sala**, O.E., W.J. Parton, L.A. Joyce, and W.K. Lauenroth, 1988: Primary production of the central grasslands region of the United States. *Ecology*, **69**, 40–45.

**Sampson**, R.N., M. Apps, S. Brown, C.V. Cole, J. Downing, L.S. Heath, D.S. Ojima, T.M. Smith, A. M. Solomon, and J. Wisniewski, 1993: Workshop summary statement: terrestrial biospheric carbon fluxes — quantification of sinks and sources of $CO_2$. *Water, Air, and Soil Pollution*, **70**, 3–15.

**Savage**, M., P.M. Brown, and J. Feddema, 1996: The role of climate in a pine forest regeneration pulse in the southwestern United States. *Ecoscience*, **3**, 310–318.

**Schäppi**, B. and C. Körner, 1996: Growth responses of an alpine grassland to elevated $CO_2$. *Oecologia,* **105**, 43–52.

**Schimmelpfenning**, D.E., 1996: Uncertainty in economic models of climate-change impacts. *Climatic Change*, **33(2)**, 213–234.

**Schimmelpfenning**, D.E., J. Lewandrowski, J.M. Reilly, M. Tsigas, and I. Parry, 1996: *Agricultural Adaptation to Climate Change: Issues of Long Run Sustainability*. Natural Resource and Environment Division, Economic Research Service, USDA, AER-740, Washington, DC, 57 pp.

**Schindler**, D., 1997: Widespread effects of climatic warming on freshwater ecosystems in North America. *Hydrologic Processes*, **11**, 1043–1067.

**Schindler**, D.W., S.E. Bayley, B.R. Parker, K.G. Beaty, D.R. Cruikshank, E.J. Fee, E.U. Schindler, and M.P. Stainton, 1996: The effects of climatic warming on the properties of boreal lakes and streams at the Experimental Lakes Area, northwestern Ontario. *Limnology and Oceanography*, **41**, 1004–1017.

**Schindler**, D.W., P.J. Curtis, P. Parker, and M.P. Stainton, 1996: Consequences of climatic warming and lake acidification for UVb penetration in North American boreal lakes. *Nature*, **379**, 705–708.

**Schlesinger**, M.E. and Z.C. Zhao, 1988: *Seasonal Climate Changes Induced by Doubled $CO_2$ as Simulated by the OSU Atmospheric GCM/Mixed-Layer Ocean Model*. Oregon State University Climate Research Institute Report, Corvallis, OR, USA.

**Schlesinger**, M.E. and Z.C. Zhao, 1989: Seasonal climatic change introduced by double $CO_2$ as simulated by the OSU atmospheric GCM/mixed-layer ocean model. *Journal of Climate*, **2**, 429–495.

**Schneider**, S.H., 1996: *Encyclopedia of climate and weather*. Oxford University Press, New York, NY, USA, 929 pp.

**Schwartz**, J., 1994: Air pollution and daily mortality: a review and meta-analysis. *Environmental Research*, **64**, 36–52.

**Schwartz**, J. and D. Dockery, 1992: Increased mortality in Philadelphia associated with daily air pollution concentrations. *American Review of Respiratory Diseases*, **142**, pp. 600–604.

**Scifres**, C.J., 1980: *Brush Management*. Texas A&M University Press, College Station, TX, USA, 360 pp.

**Scott**, M.J., D.L. Hadley, and L.E. Wrench, 1994: Effects of climate change on commercial building energy demand. *Energy Sources*, **16(3)**, 339–354.

**Scott**, M.J., R.D. Sands, L.W. Vail, J.C. Chatters, D.A. Neitzel, and S.A. Shankle, 1993: *The effects of climate change on Pacific Northwest water-related resources: summary of preliminary findings*. PNL-8987, Pacific Northwest Laboratory, Richland, WA, USA, 46 pp.

**Seastedt**, T.R., J.M. Briggs, and D.J. Gibson, 1991: Controls of nitrogen limitation in tallgrass prairie. *Oecologia*, **87**, 72–79.

**Sedjo**, R.A., 1991: Climate, forests, and fire: a North American perspective. *Environment International*, **17**, 163–168.

**Shabbar**, A., B. Bonsal, and M. Khandekar, 1997: Canadian precipitation patterns associated with the Southern Oscillation. *Journal of Climate*, **10(11)**.

**Shackell**, N.L., K.T. Frank, W.T. Stobo and D. Brickman, 1995: *Cod (Gadus morhua) growth between 1956 and 1966 compared to growth between 1978 to 1985, on the Scotian Shelf and adjacent areas.* ICES Paper CM 1995/P:1, 18 pp.

**Sharp**, G.D., 1987: Climate and fisheries: cause and effect or managing the long and short of it all. *South African Journal of Marine Science*, **5**, 811–838.

**Shaw**, J., R.B. Taylor, D.L. Forbes, M.H. Rux and S. Solomon, 1994: *Sensitivity of the Canadian coast to sea-level rise*. Geological Survey of Canada Open File Report 2825. Natural Resources Canada, Ottawa. 114 pp.

**Sheppard**, J.G., J.G. Pope and R.D. Cousens, 1984: Variations in fish stocks and hypotheses concerning their links with climate. *Rapports et procès-verbaux des réunions, Conseil International pour l'Exploration de la Mer*, **185**, 255–267.

**Shope**, R.E., 1980: Arbovirus-related encephalitis. *Yale Journal of Biology and Medicine,* **53**, 93–99.

**Shope**, R.E., 1991: Global climate change and infectious diseases. *Environmental Health Perspectives,* **96**, 171–174.

**Shriner**, D.S., 1980: Vegetation surfaces: a platform for pollutant/parasite interactions. In: *Polluted Rain* [Toribara, T.Y., M.W. Miller, and P.E. Morrow (eds.)]. Plenum Publishers, New York, NY, pp. 259–272.

**Shugart**, H.H. and T.M. Smith, 1996: A review of forest patch models and their application to global change research. *Climatic Change*, **34(2)**, 131–153.

**Shumway**, R., A. Azari, and Y. Pawitan, 1988: Modelling mortality fluctuations in Los Angeles as functions of pollution and weather effects. *Environmental Research*, **45**, 224–241.

**Shuter**, B.J. and J.R. Post, 1990: Climate, population viability and the zoogeography of temperate fishes. *Transactions of the American Fisheries Society*, **119**, 316–336.

**Shuttleworth**, W.J., 1996: The challenges of developing a changing world. *EOS Transactions,* **77(36)**, 347.

**Simberloff**, D., J.A. Farr, J. Cox, and D.W. Mehlman, 1992: Movement corridors: conservation bargains or poor investments? *Conservation Biology*, **6**, 493–504.

**Sinclair**, M. and K. Frank, 1995: Symposium summary. In: *Climate change and northern fish populations* [Beamish, R.J. (ed.)]. *Can. Spec. Pub. Fish. Aquat. Sci.*, **121**, 735–739.

**Singh**, B., 1987: *The impacts of $CO_2$-induced climate change on hydro-electric generation potential in the James Bay territory of Quebec.* Proceedings of the influence of climate change and climatic variability on the hydrologic regime and water resources [Soloman, S.I., M. Beran, and W. Hogg (eds.)]. IAHS Publication no. 168, Vancouver, British Columbia, pp. 403–418.

**Sissenwine**, M.P., 1984: Why do fish populations vary? In: *Exploitation of Marine Communities*, [May, R.M. (ed.)]. Springer-Verlag, Berlin, Germany.

**Slade**, D.C. (ed.)., 1990: *Putting the public trust doctrine to work.* Coastal Sates Organization, Washington, DC, 361 pp.

**Slivitzky**, M., 1993: *Water management: water supply and demand the St. Lawrence River*. Proceedings of the Great Lakes-St. Lawrence Basin Project workshop on adapting to the impacts of climate change and variability [Mortsch, L., G. Koshida, and D. Tavares (eds.)]. Environment Canada, Downsview, Ontario, pp. 32–34.

**Smit**, B., D. McNabb, and J. Smithers, 1996: Agricultural adaptation to climatic variation. *Climatic Change*, **33(1)**, 7–29.

**Smith**, J., and E. McBean, 1993: The impact of climate change on surface water resources. In: *The impact of climate change on water in the Grand River Basin, Ontario* [Sanderson, M. (ed.)]. Department of Geography, University of Waterloo, Publications Series No. 40, Waterloo, Ontario, pp. 25–52.

**Smith**, J.B. and D.A. Tirpak (eds.), 1989: *The potential effects of global climate change on the United States*. EPA-230-05-89, Office of Policy, Planning and Evaluation, U.S. Environmental Protection Agency, Washington, DC, 411 pp. + app., 689 pp.

**Smith**, T.M. and H.H. Shugart, 1993: The transient response of terrestrial carbon storage to a perturbed climate. *Nature*, **361**, 523–526.

**Sohngen**, B. and R. Mendelsohn, 1997: A dynamic model of carbon storage in the United States during climate change. *Critical Reviews of Environmental Science and Technology* (in press).

**Solomon**, A.M., 1986 Transient response of forests to $CO_2$-induced climate change: simulation modeling experiments in eastern North America. *Oecologia*, **68**, 567–579.

**Soulis**, E.D., S.I. Solomon, M. Lee, and N. Kouwen, 1994: Changes to the distribution of monthly and annual runoff in the Mackenzie Basin using a modified square grid approach. In: *Mackenzie Basin Impact Study interim report #2* [Cohen, S.J. (ed.)]. Environment Canada, Downsview, Ontario, pp. 197–209.

**South Carolina**, **State of**, 1988: *South Carolina Beachfront Management Act*. SC Code §4839-250 et seq. (1988) and SC Code §4839-290(D)(1) (1990).

**Spaeth**, K.E., T.L. Thurow, W.H. Blackburn, and F.B. Pierson, 1996: Ecological dynamics and management effects on rangeland hydrologic processes. In: *Grazingland Hydrology Issues: Perspectives for the 21st Century*. Society for Range Management, Denver, CO, USA, pp. 25–51.

**Stage**, A.R. and D.E. Ferguson, 1982: Regeneration modeling as a component of forest succession simulation. In: *Forest succession and stand development research in the northwest* [Means, J.E. (ed.)]. Forest Research Laboratory, Corvallis, OR, USA, pp. 24–30.

**Staple**, T. and G. Wall, 1994: Implications of climate change for water-based recreation activities in Nahanni National Park Reserve. In: *Mackenzie Basin Impact Study Interim Report #2* [Cohen, S.J. (ed.)]. Environment Canada, Downsview, Ontario, pp. 453–455.

**Starfield**, A.M. and F.S. Chapin III, 1996: Model of transient changes in arctic and boreal vegetation in response to climate and land use change. *Ecological Applications*, **6**, 842–864.

**Stefan**, H.G., M. Hondzo, and X. Fang, 1993: Lake water quality modeling for projected future climate scenarios. *J. Environ. Quality*, **22**, 417–431.

**Stefan**, H.G., M. Hondzo, X. Fang, J.G. Eaton, and J. McCormick, 1996: Simulated long-term temperature and dissolved oxygen characteristics of lakes in the north-central United States and associated fish habitat limits. *Limnology and Oceanography*, **41(5)**, 1124–1135.

**Stephenson**, N.L., 1990: Climate control of vegetation distribution: the role of the water balance. *American Naturalist*, **135**, 649–670.

**Stocks**, B.J., 1993: Global warming and forest fires in Canada. *The Forestry Chronicle*, **69(3)**, 290–293.

**Stocks**, B., B.S. Lee, and D.L. Martell, 1996: Some potential carbon budget implications of fire management in the boreal forest. In: *Forest ecosystems, forest management and the global carbon cycle* [Apps, M.J. and D.T. Price (eds.)]. NATO ASI Series, Subseries 1, Global Environmental Change, Vol. 40, Springer-Verlag, Berlin, Germany, pp. 89–96.

**Street**, R.B., 1989: *Climate change and forest fires in Ontario.* Proceedings, 10th Conference on Fire and Forest Meteorology, Forestry Canada, Ottawa, Ontario, pp. 177–182.

**Suffling**, R., 1995: Can disturbance determine vegetation distribution during climate warming? A boreal test. *Journal of Biogeography*, **22**, 2387–2394.

**Sweeney**, B.W., J.K. Jackson, J.D. Newbold, and D.H. Funk, 1992: Climate change and the life histories and biogeography of aquatic insects in eastern North America. In: *Global Climate Change and Freshwater Ecosystems* [Firth, P.A. and S.G. Fisher (eds.)]. Springer-Verlag, New York, NY, USA, pp. 143–176.

**Tavares**, D., 1996: Weather and heat-related morbidity relationships in Toronto (1979–89). In: *Great Lakes-St. Lawrence Basin Project progress report #1: adapting to the impacts of climate change and variability* [Mortsch, L.D. and B.N. Mills (eds.)]. Environment Canada, Burlington, Ontario, pp. 110–112.

**Taylor**, C.C., H.B. Bigelow and H.W. Graham, 1957: Climate trends and the distribution of marine animals in New England. *Fisheries Bulletin*, **57**, 293–345.

**Taylor**, G.E., D.W. Johnson, and C.P. Andersen, 1994: Air pollution and forest ecosystems: a regional to global perspective. *Ecological Applications*, **4**, 662–689.

**Teskey**, R.O., 1997: Combined effects of elevated $CO_2$ and air temperature on carbon assimilation of *Pinus taeda* trees. *Plant, Cell and Environment*, **20**, 373–380.

**Tester**, P., 1991: Red Tide: effects on health and economics. *Health and Environ Digest*, **5**, 4–5.

**Thurow**, T.L., W.H. Blackburn, and C.A. Taylor, Jr., 1986: Hydrologic characteristics of vegetation types as affected by livestock grazing systems, Edwards Plateau, Texas. *Journal of Range Management*, **39**, 505–509.

**Tidal Waters Division**, Maryland Department of Natural Resources, 1978–93. Annual report. Department of Natural Resources, Annapolis, MD, USA.

**Titus**, J.G., 1986: Greenhouse effect, sea level rise, and coastal zone management. *Coastal Management*, **14(3)**, 147–171.

**Titus**, J.G., 1988: *Greenhouse effect, sea level rise, and coastal wetlands*. U.S. Environmental Protection Agency, Washington, DC, USA, 152 pp.

**Titus**, J.G., 1990: Greenhouse effect, sea level rise, and barrier islands. *Coastal Management*, **18(1)**, 65–90.

**Titus**, J.G., 1991: Greenhouse effect and coastal wetland policy: How Americans could abandon an area the size of Massachusetts at minimum cost. *Environmental Management*, **15**, 39–58.

**Titus**, J.G., 1992: The costs of climate change to the United States. In: *Global climate change: implications, challenges and mitigation measures* [Majumdar, S.K., L.S. Kalkstein, B. Yarnal, E.W. Miller, and L.M. Rosenfeld (eds.)]. Pennsylvania Academy of Science, Philadelphia, PA, USA, pp. 385–409.

**Titus**, J.G., 1997: Rising seas, coastal erosion, and the takings clause: how to save wetlands and beaches without hurting property owners. *Maryland Law Review* (in press).

**Titus**, J.G., C.Y. Kuo, M.J. Gibbs, T.B. LaRoche, M.K. Webb, and J.O. Waddell, 1987: Greenhouse effect, sea level rise, and coastal drainage systems. *J. of Water Resources Planning and Management*, **113**, 2.

**Titus**, J.G., R.A. Park, S. Leatherman, R. Weggel, M.S. Greene, M. Treehan, S. Brown, C. Gaunt, and G. Yohe, 1991: Greenhouse effect and sea level rise: The cost of holding back the sea. *Coastal Management*, **19**, 171–204.

**Titus**, J.G. and V. Narayanan, 1996: The risk of sea level rise: A delphic Monte Carlo analysis in which twenty researchers specify subjective probability distributions for model coefficients within their respective areas of expertise. *Climatic Change*, **33(2)**, 151–212.

**Toll**, J. et al., 1997: Chesapeake Bay at the crossroads: final report of a conference held at Chestertown, Maryland, October 18–19, 1996. Climate Institute, Washington, DC, USA, 15 pp.

**Townsend**, A.R., B.H. Braswell, E.A. Holland, and J.E. Penner, 1996: Spatial and temporal patterns in terrestrial carbon storage due to deposition of fossil fuel nitrogen. *Ecological Applications*, **6**, 806–814.

**Trent**, D.W., J.A. Grant, L. Rosen, and T.P. Monath, 1983: Genetic variations among dengue 2 viruses of different geographic origin. *Virology,* **128**, 271–284.

**UNEP**, 1997: *Global environment outlook*. Oxford University Press, New York, NY, USA, 264 pp.

**Urban**, D.L. and H.H. Shugart, 1989: Forest response to climatic change: a simulation study for southeastern forests. In: *The potential effects of global climate change on the United States: Appendix D — Forests* [Smith, J.B. and D.A. Tirpak (eds.)]. U.S. Environmental Protection Agency, Washington, DC, USA, 176 pp.

**U.S. Department of Commerce**, 1996: *Annual survey of manufacturers, 1994*. Economics and Statistics Administration, Bureau of the Census, Washington, DC, USA.

**U.S. EPA**, 1995: *Ecological impacts from climate change: an economic analysis of freshwater recreational fishing*. U.S. Environmental Protection Agency, Washington, DC, USA.

**U.S. EPA**, 1996: *National air quality and emissions trends report, 1995*. U.S. Environmental Protection Agency, Washington, DC, USA.

**U.S. EPA and Environment Canada**, 1995: *The Great Lakes: an environmental atlas and resource book.* Government of Canada, Toronto. U.S. Environmental Protection Agency, Great Lakes National Program Office, Chicago, IL, USA, 3rd ed., 46 pp.

**Vallentyne**, J.F., 1971: *Range development and improvements*. Brigham Young University Press, Provo, UT, USA, 516 pp.

**Van Der Vinne**, G., T.D. Prowse, and G. Andres, 1991: Economic aspects of river ice jams in Canada. In: *Northern hydrology: selected perspectives* [Prowse, T.D. and C.S.L. Ommanney (eds.)]. NHRI Symposium, National Hydrology Research Institute, Environment Canada, Saskatoon, Saskatchewan, pp. 333–352.

**Van Kooten**, G.C. and L.M. Arthur, 1989: Assessing economic benefits of climate change on Canada's boreal forest. *Canadian Journal of Forest Research*, **1**, 463–470.

**VEMAP Members**, 1995: Vegetation/ecosystem modeling and analysis project: comparing biogeography and biogeochemistry models in a continental-scale study of terrestrial ecosystem responses to climate change and $CO_2$ doubling. *Global Biogeochemical Cycles*, **9**, 407–437.

**Vitousek**, P.M., 1994: Beyond global warming: ecology and global change. *Ecology*, **75**, 1861–1876.

**Vose**, J., B.D. Clinton, and W.T. Swank, 1993: *Fire, drought, and forest management influences on pine/hardwood ecosystems in the southern Appalachians.* Proceedings of the 12th International Conference on Fire and Forest Meteorology, Society of American Foresters, 26–28 October, 1993, Jekyll Island, GA, USA, pp. 232–238.

**Wackter**, D.J. and P.V. Bayly, 1988: The effectiveness of emission controls on reducing ozone levels in Connecticut from 1976 through 1987. In: *The scientific and technical issues facing post-1987 ozone control strategies* [G.T. Wolff, J.L. Hanisch, and K. Schere (eds.)]. Air and Waste Management Association, Pittsburgh, PA, USA, pp. 398–415.

**Wakim**, P.G., 1989: Temperature-adjusted ozone trends for Houston, New York and Washington, 1981–1987. In: *Paper 89-35.1, Presented at the 82nd Annual Meeting and Exhibition of the Air and Waste Management Association*, Anaheim, CA, USA, June 25–30, 1989.

**Wakim**, P.G., 1990: 1981 to 1988 ozone trends adjusted to meteorological conditions for 13 metropolitan areas. In: *Paper 90-97.9, Presented at the 83rd Annual Meeting and Exhibition of the Air and Waste Management Association*, Pittsburgh, PA, USA, June 24–29, 1990.

**Walker**, J.C., T.R. Miller, G.T. Kingsley, and W.A. Hyman, 1989: Impact of global climate change on urban infrastructure. In: *The Potential Effects of Global Climate Change on the United States: Appendix H — Infrastructure* [Smith, J.B. and D.A. Tirpak (eds.)]. U.S. Environmental Protection Agency, Washington, DC, USA, 176 pp.

**Wall**, G., 1988: *Implications of climate change for tourism and recreation in Ontario*. CCD 88-05, Environment Canada, Downsview, Ontario, 16 pp.

**Waring**, R.H. and W.H. Schlesinger, 1985: *Forest ecosystems: concepts and management*. Academic Press, New York, NY, USA, 345 pp.

**Welch**, D.W., A.I. Chigirinsky, and Y. Ishida, 1995: Upper thermal limits on the oceanic distribution of Pacific salmon (*Oncorhynchus*) in the spring. *Can. J. Fish. Aquat. Sci.*, **52**, 489–503.

**Weller**, M.W., 1981: *Freshwater Marshes*. University of Minnesota Press, Minneapolis, MN, 146 pp.

**Weltz**, M.A., and M.K. Wood, 1986: Short duration grazing in Central New Mexico: effect on sediment production. *J. Soil and Water Conserv.*, **41**, 262–266.

**Westoby**, M., B. Walker, and I. Noy-Meir, 1989: Opportunistic management for rangelands not at equilbrium. *Journal of Range Management*, **42**, 266–274.

**WHO (World Health Organization)**, 1996: *Climate Change and Human Health* [McMichael, A.J., A. Haines, R. Slooff, and S. Kovats (eds.)]. World Health Organization, Geneva, Switzerland, 297 pp.

**Wilcoxen**, P.J., 1986: Coastal erosion and sea level rise: Implications for ocean beach and San Francisco's Westside Transport Project. *Coastal Zone Management Journal*, **14**, 3.

**Williams**, D.W. and A.M. Liebhold, 1995: Herbivorous insects and global change: potential changes in the spatial distribution of forest defoliator outbreaks. *Journal of Biogeography*, **22**, 665–671.

**Williams**, M.A.J., and R.C. Balling, Jr., 1996: *Interactions of desertification and climate*. World Meteorological Organization, United Nations Environment Programme, Wiley, New York, NY, USA, 270 pp.

**Winner**, W.E., 1994: Mechanistic analysis of plant responses to air pollution. *Ecological Applications*, **4**, 651–661.

**Wolff**, K. and P. T. Lioy, 1978: An empirical model for forecasting maximum daily ozone levels in the northeastern U.S. *Journal of the Air Pollution Control Association*. **28**, 1034–1038.

**Woodward**, F.I., 1987: *Climate and plant distribution*. Cambridge University Press, London, United Kingdom, 174 pp.

**Wotton**, B.M. and M.D. Flannigan, 1993: Length of fire season in a changing climate. *The Forestry Chronicle*, **69**, 187–192.

**WRI**, 1996: *World Resources 1996–1997*. Oxford University Press, New York, NY, USA, 365 pp.

**Wright**, D.G., R.M. Hendry, J.W. Loder, and F.W. Dobson, 1986: Oceanic changes associated with global increases in atmospheric carbon dioxide: a preliminary report for the Atlantic coast of Canada. *Can. Tech. Rep. Fish. Aquat. Sci.*, **1426**, vii + 78 pp.

**Wullschleger**, S.D., W.M. Post, and A.W. King, 1995: On the potential for a $CO_2$ fertilization effect in forest trees: an assessment of 58 controlled-exposure studies and estimates of the biotic growth factor. In: *Biotic feedbacks in the global climate system* [Woodwell, G.M. and F.T. Mackenzie (eds.)]. Oxford University Press, New York, NY, USA, pp. 85–107.

**Yan**, N.A., W. Keller, N.M. Scully, D.R.S. Lean and P.J. Dillon, 1996: Increased UV-B penetration in a lake owing to drought-induced acidification. *Nature*, **381**, 141–143.

**Yarnal**, B., D.L. Johnson, B.J. Frakes, G.I. Bowles, and P. Pascale, 1997: The Flood of '96 and its socioeconomic impacts in the Susquehanna River Basin. *Journal of the American Water Resource Association*, (in press).

**Yohe**, G., 1990: The cost of not holding back the sea. *Coastal Management*, **18**, 403–432.

**Yohe**, G., J. Neumann, P. Marshall, and H. Ameden, 1996: The economic cost of greenhouse-induced sea-level rise for developed property in the United States. *Climatic Change*, **32(4)**, 387–410.

**Zucker**, J.R., 1996: Changing patterns of autochtonous malaria transmission in the United States: a review of recent outbreaks, *Emerging Infectious Diseases*, **2(1)**, 37–43.

# 9

# Small Island States

LEONARD A. NURSE (BARBADOS), ROGER F. McLEAN (AUSTRALIA),
AND AVELINO G. SUAREZ (CUBA)

Contributors:
*M. Ali (Maldives), J. Hay (New Zealand), G. Maul (USA),
G. Sem (Papua New Guinea)*

# CONTENTS

# EXECUTIVE SUMMARY

## Introduction

All of the small island states considered in this report are located within the tropics, with the exception of Malta and Cyprus in the Mediterranean. About one-third of the states comprise a single main island; the others are made up of several or many islands. Although some states are experiencing relative declines in sea level, others—primarily low-lying island states and atolls—are especially vulnerable to climate change and associated sea-level rise because, in many cases (e.g., The Bahamas, Kiribati, Maldives, Marshall Islands), much of the land area is only 3–4 m above the present mean sea level. Islands at higher elevations also are vulnerable—particularly in coastal zones, where settlements, economic infrastructure, and vital services tend to be concentrated.

## Regional Characteristics

The ocean exerts a strong influence on small islands. Island climate is moderated by the maritime influence—which, given the islands' mainly tropical location, results in uniformly high temperatures (20°C and above) throughout the year. However, other climate variables often exhibit distinct seasonal patterns—particularly rainfall distribution, which results in wet and dry seasons. Some small island states are subject to tropical cyclones (i.e., hurricanes or typhoons); those that are outside the main storm tracks also are affected by high seas and swells associated with such events. In the Pacific, large interannual rainfall variations resulting from the El Niño-Southern Oscillation (ENSO) phenomenon are an important climate characteristic; ENSO effects also are felt on islands in the Caribbean Sea and Indian Ocean.

Economic activities in small island states frequently are dominated by agriculture (e.g., sugar and bananas for export; subsistence farming for local consumption) and by tourism, both of which are sensitive to external forces and are strongly influenced by climatic factors. Fisheries, although largely artisanal, also support an important economic activity. Although total population numbers are low, settlements commonly are concentrated in the capital city or on the capital island, where population densities often are very high. Human demands on coastal and marine resources are continuing to increase; the potential impacts of climate change, added to these pressures, almost certainly will result in the degradation or loss of some natural ecosystems that are important to the economies of small island states.

## Observed Climate Trends

Caribbean islands experienced an increase in mean annual temperature of more than 0.5°C during the period 1900–1995. During the same period, mean annual total rainfall decreased by about 250 mm, though throughout the rainfall record has been characterized by large variability. In the Pacific islands, the increase in average annual temperature has been less than 0.5°C since 1900. Rainfall records for the Pacific (1900–1995) reveal no clear trend; they show decadal fluctuations of ±200 mm for mean annual rainfall and ±50–100 mm for seasonal rainfall.

## Climate Model Projections

Because simulations using ocean-atmosphere general circulation models (GCMs) are not presently conducted at fine horizontal resolution, the ability to generate climate change scenarios for the small island states is somewhat limited. However, because of the strong influence of the surrounding oceans on the climate of these islands and because the oceans are projected to warm in the future [1–2°C for the Caribbean Sea and the Atlantic, Pacific, and Indian Oceans, with a doubling of carbon dioxide ($CO_2$)], small islands also are expected to experience moderate warming in the future. Mean rainfall intensity also is projected to increase by about 20–30% over the tropical oceans (the main locations of the small island states) with doubled $CO_2$. On the other hand, simulations conducted with combined greenhouse gas (GHG) and aerosol forcings project a decrease in mean summer precipitation over the Mediterranean Sea region (the location of Malta and Cyprus).

At this stage, there is much uncertainty in climate model projections with respect to possible changes in the distribution, frequency, and intensity of tropical cyclones and ENSO events. The most significant climate-related projection for small islands is sea-level rise. Current estimates of future global sea-level rise of 5 mm/yr (with a range of 2–9 mm/yr) represent a rate that is 2 to 4 times higher than what has been experienced globally over the past 100 years. Considerable local and regional variations in the rate, magnitude, and direction of sea-level change can be expected as a result of thermal expansion, tectonic movements, and changes in ocean circulation. However, although the level of vulnerability will vary from island to island, it is expected that practically all small island states will be adversely affected by sea-level rise.

## Sensitivity, Adaptation, and Vulnerability

### *Marine Ecosystems*

The impact of a climate change-related increase in air temperature on small island states has not been investigated in any detail. A rise in temperature (of the magnitude projected) is not anticipated to have widespread adverse consequences, though some critical ecosystems (e.g., coral reefs) are highly sensitive to temperature changes. Although reefs have the potential to keep up or catch up with the projected rate of sea-level rise, in many parts of the tropics (e.g., the Caribbean Sea, the Pacific Ocean) some species of corals live near their limits of tolerance to temperature (about 25–29°C). Thus, even relatively small projected increases in sea-surface temperature could have an adverse impact on the viability of some of these organisms. An increase in the incidence of bleaching—associated with the elevation of water temperatures above seasonal maxima—similarly would pose a threat to coral reef ecosystems.

The natural capacity of mangroves to adapt and migrate landward in response to projected sea-level rise is expected to be reduced by associated land loss, land-use practices, and the presence of infrastructure in the coastal zone. Survival of mangroves appears likely where the rate of sedimentation approximates the local rise in sea level, but landward migration commonly is inhibited by topography and infrastructure, which will constrict the mangrove belt.

Some ecosystems already are seriously affected by anthropogenic stresses—which, in some islands, may pose as great a threat as climate change itself. Where this situation exists, the natural capacity of ecosystems to adapt to the effects of climate change will be substantially reduced.

### *Coastal Systems*

Many islands are likely to experience increased coastal erosion and land loss as a consequence of sea-level rise. Beaches are expected to be affected by a reduced supply of sediment from adjacent reefs; on high islands, however, increased sediment yields from stream catchments may compensate for this effect, at least in the short term. In addition, increased sea flooding and inundation (as have been projected for the Marshall Islands and Kiribati) are expected in most low-lying islands and atolls.

### *Human Settlements and Infrastructure*

In many small island states, the largest settlements, much critical infrastructure, and major economic activities and services are located close to present sea level and therefore will be at risk from sea-level rise. Vulnerability assessment studies have shown that the costs of shoreline and other infrastructure protection could be burdensome for some islands.

### *Tourism*

Tourism is the dominant economic sector in many small island states in the Caribbean Sea and the Pacific and Indian Oceans. This sector is the single largest contributor to gross national product (GNP) in many countries. In 1995, for example, tourism accounted for 69%, 53%, and 50% of GNP in Antigua, the Bahamas, and the Maldives, respectively. The tourism sector also earns considerable foreign exchange for a number of small island states, many of which depend heavily on imported food, fuel, and a range of other goods and services. For example, in 1995, the tourist industry in the Maldives earned US$181 million in foreign exchange—amounting to more than 70% of the country's total foreign exchange earnings for that year. In the Bahamas, the industry earned US$1.3 billion in foreign exchange in 1995, which was equivalent to more than 50% of government revenues for the year. Climate change and sea-level rise would affect tourism directly and indirectly. The loss of beaches to erosion and inundation, increasing stress on coastal ecosystems, damage to infrastructure, and an overall loss of amenities would jeopardize the viability of the tourist industry in many small islands.

## Integration and Adaptation

Proper evaluation of the overall risk of island states to projected climate change effects requires adoption of a fully integrated approach to vulnerability assessments. The sensitivity of small islands to climate change cannot be attributed to any single factor (e.g., size or elevation); the cumulative and synergistic result of these and related biophysical attributes, combined with their economic and sociocultural character, ultimately determines the vulnerability of these islands. Moreover, some islands are prone to periodic non-climate-related hazards (e.g., earthquakes, volcanic eruptions, tsunamis); the degree of vulnerability of such islands to climate change therefore should not be evaluated in isolation from these threats. In addition, vulnerability assessments for small island states should take into consideration the value of nonmarketed goods and services (e.g., subsistence assets, community structure, traditional skills and knowledge), which also may be at risk. In some island societies, these assets are just as important as marketed goods and services.

Numerous adaptation measures for responding to climate change and sea-level rise theoretically are available to small island states; given their small size and limited human and financial resources, however, the costs of adaptation often are prohibitive. Shore protection, for instance, is very expensive—and in the past, the design of structures has not always been appropriate for the coastal environments of tropical small islands. The use of more flexible, easily replaceable, traditional shore protection measures could be explored, particularly in island states of the Pacific and Indian Oceans. Moreover, in planning for adaptation, efforts should be made to maximize the use of traditional knowledge and skills because island peoples have had to cope with a variety of environmental stresses in the past.

On low-lying islands and atolls where sea-level rise is projected to be a threat, retreat away from the coast rarely is an option, given the limited physical size of many islands. As a result, migration and resettlement outside national boundaries may have to be considered in extreme cases.

Uncertainties in climate change projections are likely to be a disincentive for implementing various adaptive measures, especially because some options may be costly or require changes in societal norms and behavior. For example, efforts to enhance the health and resilience of natural ecosystems—including coral reefs, mangroves, and beaches—by reducing activities that increase their vulnerability would be a vital adaptation strategy, but there would be monetary and social costs associated with their implementation. However, policies that seek to incorporate sound principles of integrated coastal management would be beneficial to the small island states in the long term—even in the absence of projected climate change and sea-level rise.

## 9.1. Introduction: General Characteristics of Small Islands

The majority of the world's small island states are concentrated in four tropical regions: the tropical Pacific Ocean, Indian Ocean, and Caribbean Sea and the Atlantic Ocean off the coast of west Africa. A few small islands are found outside these areas—for example, Malta and Cyprus, which are located in the Mediterranean Sea. This chapter principally addresses independent small island states; it does not focus on islands that are more appropriately considered to be part of another region (e.g., Singapore in Tropical Asia, Bahrain in the Middle East); the many islands that constitute part of a larger country (e.g., the thousands of small islands of Indonesia); or those that are administered by a metropolitan country, such as the United Kingdom, the United States, France, or The Netherlands. Nevertheless, many of the climate change projections and impacts detailed in this chapter apply equally to these other types of islands.

The small island states on which the discussion in this chapter focuses are depicted in Figure 9-1. Box 9-1 provides a listing of the islands for which socioeconomic data were compiled in Annex D.

Although these islands are by no means homogeneous politically, socially, or culturally—or in terms of physical size or stage of economic development—many tend to share a number of common characteristics. Island states generally have small land areas and high population densities, and large exclusive economic zones (EEZs) (for mid-Pacific states, these zones are 1,000 times larger than the land area); they are located predominantly in tropical and subtropical regions of the Caribbean Sea and the Indian and Pacific Oceans (Table 9-1). The climates of most islands generally are moderated by the maritime influence, with lower maximum and higher minimum temperatures than continental land masses at the same latitude. Most islands also have distinct seasonal patterns of rainfall and temperature. Although many islands are located outside the normal tropical storm tracks, some states in the Pacific and Indian Oceans and the Caribbean Sea periodically are subject to the devastating inflences of tropical cyclones (i.e., hurricanes and typhoons).

**Box 9-1. Small Island States**
**[United Nations Member States and Members of the Alliance of Small Island States (AOSIS)]**

| | |
|---|---|
| Antigua and Barbuda | Malta |
| The Bahamas | Marshall Islands |
| Barbados | Mauritius |
| Cape Verde | Nauru |
| Comoros | Palau |
| Cook Islands | Saint Kitts and Nevis |
| Cuba | Saint Lucia |
| Cyprus | Saint Vincent and the Grenadines |
| Dominica | Samoa |
| Dominican Republic | Sao Tome and Principe |
| Federated States of Micronesia | Seychelles |
| Fiji | Solomon Islands |
| Grenada | Tonga |
| Haiti | Trinidad and Tobago |
| Jamaica | Tuvalu |
| Kiribati | Vanuatu |
| Maldives | |

Small island states may comprise a single island (e.g., Barbados, Malta, and Nauru); a few islands, such as Tuvalu (9 islands), Vanuatu (12), and Cape Verde (15); or numerous islands—for instance, the Seychelles (115), Tonga (180), and the Maldives (1,200). Physiographically and geologically, island states may comprise low oceanic islands, including atolls and reef islands; high volcanic or limestone islands; or continental islands (Nunn, 1994). In the first group of islands, species diversity and endemism often are low. In the second group, species diversity may be high or low, but endemism usually is high. Continental islands—particularly those located on the broad shelves of neighboring land masses—tend to have high species diversity and low endemism, which are functions of their continental proximity.

The ecology of small islands generally is characterized by a limited range of terrestrial and coastal ecosystems, surrounded by a vast expanse of ocean. Vegetation usually consists of groups of easily dispersed species, which have a tendency to be restricted in their distribution. Forests (including strands and mangroves), coral reefs, and sea grass communities provide a range of food and other resources. Biodiversity is highly variable and depends on a combination of physical and other factors (e.g., location, area, geology). For example, because of the coralline nature of most low-lying atolls, soils usually are poorly developed, exhibit immature profiles, are highly alkaline and usually deficient in nutrients, and have low water-retention capacity.

To the extent that island species are restricted in range and distribution, they are potentially more endangered and prone to extinction than the flora and fauna of continental land masses. Communities of terrestrial fauna on small islands are poorly developed and include few indigenous land mammals. Avian fauna are dominated mainly by migratory sea birds, with a relatively small number of indigenous species (Table D-7), representing a limited number of families.

The economies of small island states are sensitive to external market forces over which they have little control. The economies generally are dominated by agriculture, fisheries, tourism, and international transport activities (air and sea). On some islands, remittances from expatriate nationals are an important component of family income. In the case of Tuvalu, remittances mainly from workers in the phosphate mines in Nauru, seamen, and contract workers in New Zealand exceed by far export earnings (SPREP, 1996). Primary production also is an important source of export earnings for many islands;

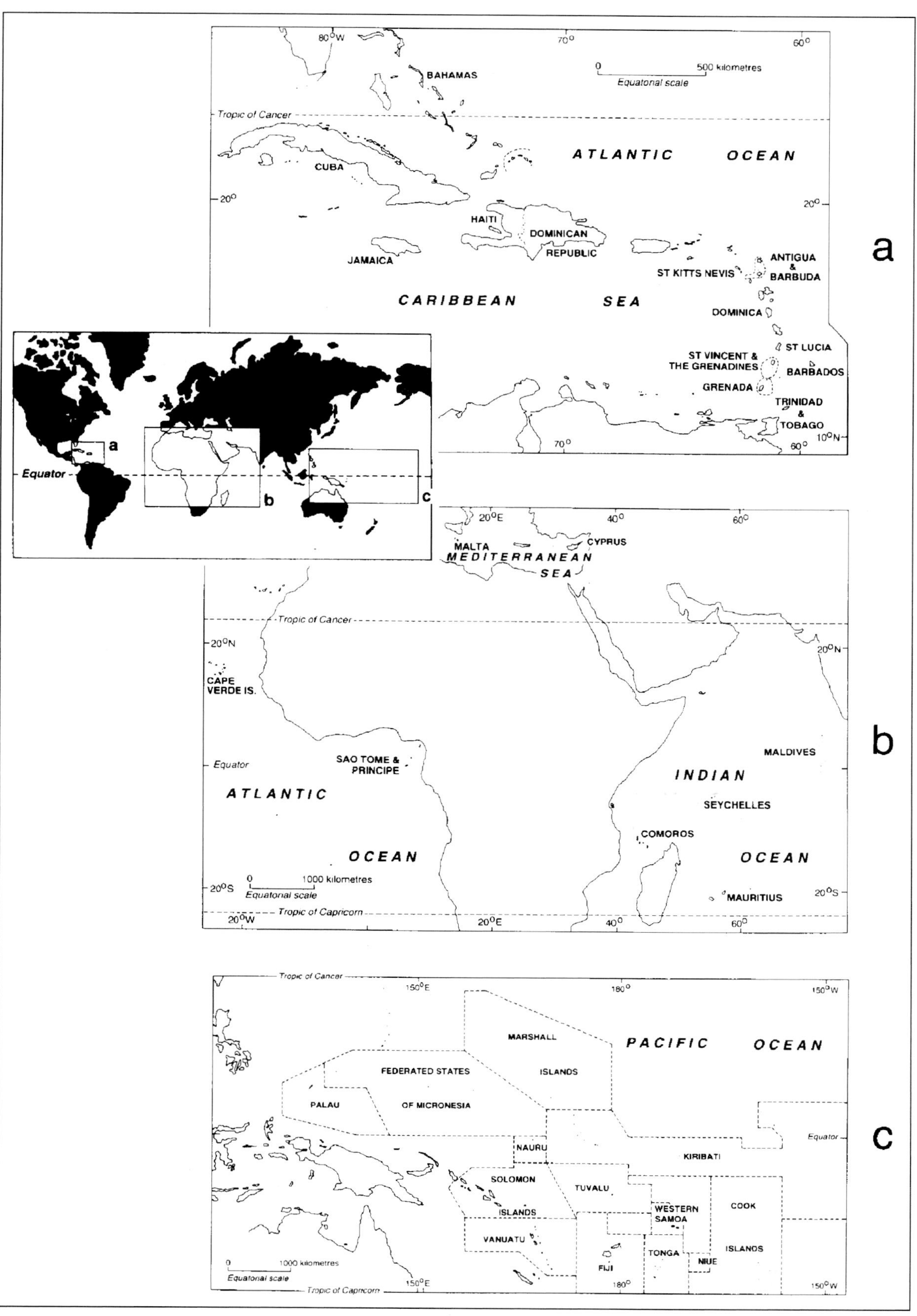


**Figure 9-1:** Main regions of the world in which small island states are located.

important products include sugar (Barbados, Cuba, Dominican Republic, Fiji, Mauritius), bananas (Dominica, St. Lucia, Tonga), copra (Cook Islands, Palau, Seychelles, Vanuatu), forestry (Solomon Islands), and mining (Cuba, Cyprus, Jamaica, Nauru). Subsistence agriculture also is an important economic activity, providing some level of food security and saving valuable foreign exchange that otherwise would be spent on food and other imports. Tourism also is an important foreign exchange earner in many small island states.

Infrastructure and services often are not well developed and frequently are concentrated in areas of high population density; on

***Table 9-1:*** *Land area and population data (1995) for selected small island states.*

| Country | Land Area ($km^2$) | Population (000s) | Pop. Density (persons/$km^2$) | Coastline Length (km) | EEZ (000s $km^2$) |
|---|---|---|---|---|---|
| *Atlantic Ocean* | | | | | |
| Cape Verde | 4,033 | 392 | 97 | 965 | 734 |
| Sao Tome and Principe | 960 | 133 | 139 | 209 | x |
| *Caribbean Sea* | | | | | |
| Antigua and Barbuda | 280 | 66 | 236 | 153 | 110 |
| Bahamas | 13,935 | 276 | 20 | 3,542 | 759 |
| Barbados | 431 | 262 | 607 | 97 | 167 |
| Cuba | 110,861 | 11,041 | 100 | 6,073 | x |
| Dominica | 750 | 71 | 95 | 148 | 15 |
| Dominican Republic | 48,442 | 7,823 | 161 | 940 | x |
| Grenada | 312 | 92 | 295 | 121 | 27 |
| Haiti | 27,750 | 7,180 | 259 | 370 | x |
| Jamaica | 10,991 | 2,447 | 223 | 1,022 | 297 |
| St. Kitts and Nevis | 269 | 41 | 152 | 135 | 11 |
| St. Lucia | 616 | 150 | 244 | 158 | 16 |
| St. Vincent and the Grenadines | 389 | 112 | 288 | 84 | 32 |
| Trinidad and Tobago | 5,128 | 1,306 | 255 | 3,760 | x |
| *Indian Ocean* | | | | | |
| Comoros | 2,171 | 653 | 292 | 340 | x |
| Maldives | 300 | 254 | 854 | 644 | x |
| Mauritius | 1,850 | 1,117 | 547 | 177 | 1,000 |
| Seychelles | 280 | 73 | 261 | 491 | 15,000 |
| *Mediterranean Sea* | | | | | |
| Cyprus | 9,251 | 742 | 80 | x | x |
| Malta | 316 | 366 | 1,159 | x | x |
| *Pacific Ocean* | | | | | |
| Cook Islands | 236 | x | x | 120 | x |
| Federated States of Micronesia | 720 | x | x | 6,112 | 2,978 |
| Fiji | 18,272 | 784 | 43 | 1,129 | 1,290 |
| Kiribati | 728 | 79 | 109 | 1,143 | 3,550 |
| Marshall Islands | 181 | x | x | 370 | 2,131 |
| Nauru | 21 | 11 | 523 | 30 | 320 |
| Palau | 497 | x | x | x | x |
| Samoa | 2,842 | 171 | 61 | 403 | 120 |
| Solomon Islands | 28,446 | 378 | 13 | 5,313 | 1,340 |
| Tonga | 697 | 98 | 141 | 419 | 700 |
| Tuvalu | 26 | 10 | 385 | 24 | 900 |
| Vanuatu | 14,763 | 169 | 14 | 2,528 | 680 |

x = no data available.
Sources: Wilkinson and Buddemeier, 1994; FAO, 1995; UN Population Division, 1995.

some islands, however, infrastructure for services and tourism is well developed. In most island states, the greatest concentration of people tends to occur in the capital city or on the capital island—where, in some cases, more than 50% of the country's total population lives. Nevertheless, in many cases (e.g., Pacific island communities), there are strong traditional and cultural ties to the home island, village, or rural area; there also may be strong religious links, which may have an important influence on perceptions of and attitudes toward climate change and sea-level rise (McLean and d'Aubert, 1993).

Some low-lying small island states—such as the atoll nations of the Pacific and Indian Oceans—are among the most vulnerable to climate change, seasonal-to-interannual climate variability, and sea-level rise. Much of their critical infrastructure and many socioeconomic activities tend to be located along the coastline, in many cases at or close to present sea level (Nurse, 1992; Pernetta, 1992; Hay and Kaluwin, 1993). Coastal erosion, saline intrusion, sea flooding, and land-based pollution already are serious problems that many of these countries face. Supplies of high-quality potable water often are inadequate, and water supply systems tend to lack adequate storage capacity. Moreover, valuable coastal ecological systems currently are stressed by human activities in the coastal zone and inland, reducing their capacity to adapt to the effects of climate change and sea-level rise. Pressures on these resources continue to increase in most small islands, partly as a consequence of high population growth rates, increasing urbanization, tourism development, and greater demand for material goods and services.

In addition, these island states are sensitive to changes in precipitation in terms of absolute amounts as well as spatial and temporal distribution. Similarly, island systems would be extremely vulnerable to any changes in the frequency or intensity of extreme events (e.g., droughts, floods, hurricanes, and storm surges). Indeed, vulnerability to these and other natural hazards—including some that may not be influenced by climate change (e.g., tsunamis, volcanoes)—contributes to the cumulative vulnerability of many small island states (Maul, 1996).

## 9.2. Regional Climate

### *9.2.1. Some Common Influences*

Although there is much climatic variation between localities, some factors and characteristics are common to most small islands—mainly as a result of their insular natures and tropical locations. For instance, it is generally true that:

- The ocean exerts a strong influence on the climate of islands.
- Temperatures usually are high, with mean annual values of 20°C and above.
- Diurnal and seasonal variations in temperature are low, with values around 5°C and below.
- Many small island states are influenced by tropical storms and cyclones (i.e., hurricanes and typhoons).
- In the tropical Pacific, most islands are strongly influenced by the ENSO phenomenon and associated high interannual variations in rainfall and sea level (Hay *et al.*, 1993a). The ENSO phenomenon also has an influence on the weather and climate of islands in the Caribbean Sea (Centella *et al.*, 1996), as well as in the Indian Ocean.

### *9.2.2. Observed Trends*

#### *9.2.2.1. Temperature and Precipitation*

Global and regional temperature and precipitation trends are presented in Annex A of this report. Figure 9-2a shows that, for the Caribbean islands, average annual temperatures have increased by more than 0.5°C over the period 1900–1995; the seasonal data are consistent with this overall trend. In the specific case of Cuba, for which a study of observed temperature trends has been undertaken, Centella *et al.* (1996) found that mean air temperature has risen by 0.6°C during the past 45 years. Rainfall data for the same period show much greater seasonal, interannual, and decadal-scale variability, although a declining trend in average annual rainfall—on the order of 250 mm—is evident (see Figure 9-2b). Average annual temperature also has increased since 1900 in the Pacific islands; the magnitude of the increase in this area, however, is less than 0.5°C, and seasonal trends are not coherent, nor do they track the annual average. Similarly, the rainfall data show considerable decadal-scale fluctuations—on the order of 200 mm for annual average rainfall and 50–100 mm for seasonal rainfall (Figure 9-2d,e).

Further details of climate trends for the southwest Pacific region have been reported by Hay *et al.* (1993a,b) and Salinger *et al.* (1995); several subregions with coherent trends and variability were identified. These studies indicate that the region as a whole has warmed at a rate of about 0.2°C per decade and that the 1981–1990 decade was the warmest on record. Within the region, there has been a steady increase in temperature to the south of the South Pacific Convergence Zone (SPCZ)—an area that includes Fiji and Tonga—since the 1880s. A rapid increase has occurred north of the SPCZ since the 1970s, after an earlier cooling trend from the 1940s; this region includes Tuvalu, Kiribati, Western Samoa, and the northern Cook Islands.

Rainfall in the southwest Pacific has been more variable, both temporally and spatially, over the past 100 years, and long-term trends are difficult to ascertain. In Kiribati, Tuvalu, and the northern Cook Islands, wetter-than-average conditions have prevailed since 1975; conditions have been drier in Fiji, Tonga, and Western Samoa. Rainfall patterns in the region clearly are associated with the ENSO phenomenon (Salinger *et al.*, 1995), although the precise nature of the relationship is not clearly understood.

#### *9.2.2.2. Tropical Cyclones*

The annual number of tropical cyclones for the greater Caribbean over the last 100 years and for the southwest Pacific

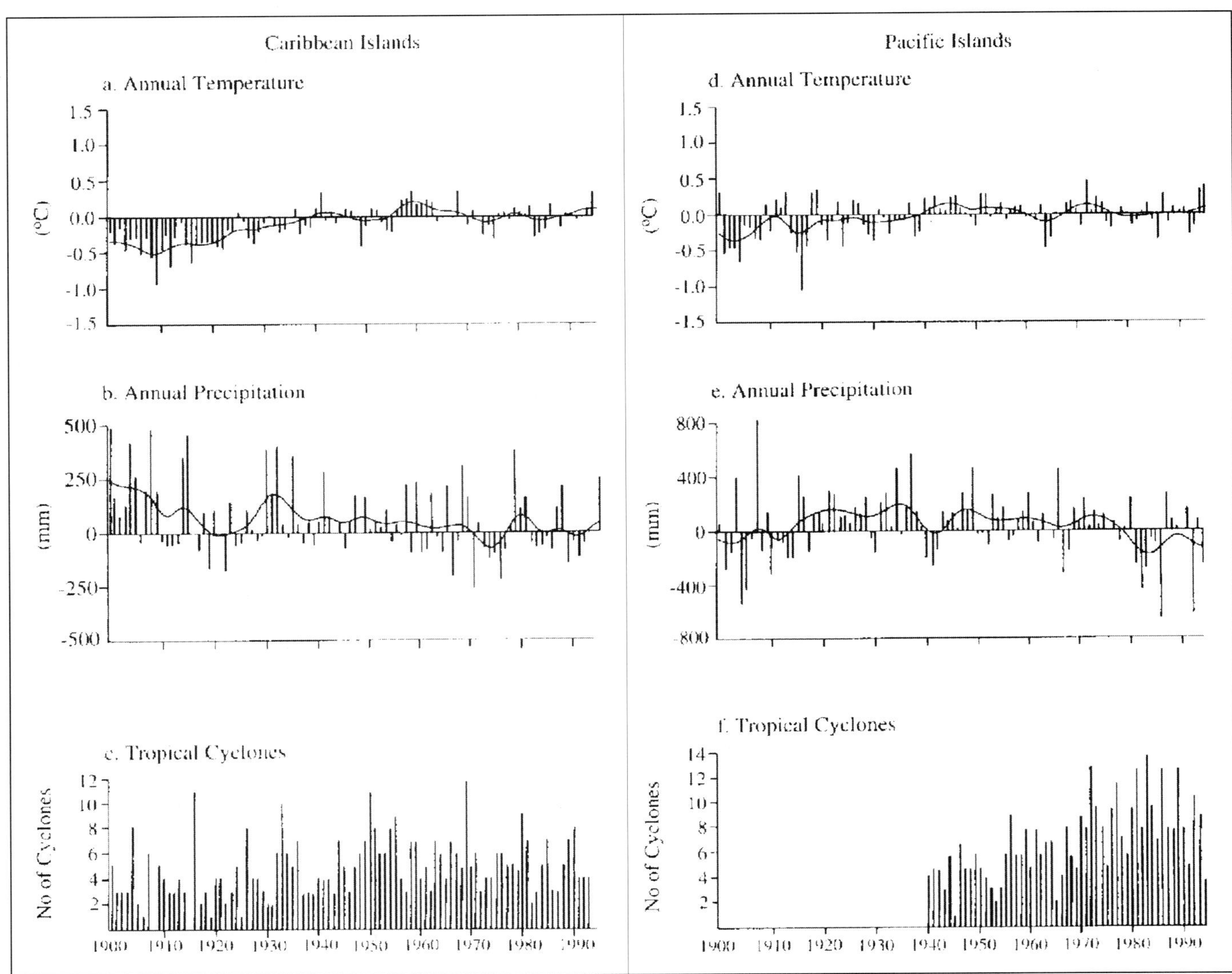


**Figure 9-2:** Time series of climate observations in the Caribbean islands (a–c) and Pacific islands (d–f). Observed annual temperature anomalies (a,d) and observed annual precipitation anomalies (b,e) both relative to 1961–90 means. Longer term variation of annual anomalies are emphasized by the smooth curve using a nine-point binomial filter. Number of Atlantic hurricanes including those in the Caribbean region (c), and annual numbers of tropical cyclones in the southwest Pacific region (f). Note that while there was an increase in the number of tropical cyclones detected since the advent of weather satellites in 1969, there is no clear evidence that tropical cyclones have been changing in number or intensity over the years.
Sources: Annex A (a,b,d,e); Environment Division, Ministry of Health and Environment, Barbados (c); Climate Research and Information Services, National Institute of Water and Atmospheric Research, Wellington, New Zealand (f).

cyclone belt for the past 50 years is given in Figure 9-2c,f. In both regions, high interannual and subdecadal variations are evident with no long-term overall trend.

### 9.2.3. *Model Projections*

#### 9.2.3.1. *Temperature, Precipitation, and Evaporation*

Analysis of climate model results (including results from a number of global coupled atmosphere-ocean models) suggests that global mean surface air temperatures can be expected to rise in the future. Unfortunately, current computing requirements for atmosphere-ocean GCMs (AOGCMs) virtually prohibit the modeling community from conducting simulations at fine horizontal resolution; because many of the small island states fall within a grid box of the models, they are not appropriately resolved. This constraint severely limits any ability to generate future projections of climate change for small islands. Nonetheless, because the climate of these islands is influenced by the surrounding oceans, and the oceans are expected to warm in the future (albeit at a slower rate than land masses), the small island states also are likely to experience moderate warming. Pittock *et al.* (1995) have made some progress toward developing climate change scenarios for the southwest Pacific.

An annual mean rise in tropical sea-surface temperature of about 1.0°C is projected by AOGCMs (with a doubling of $CO_2$), with no appreciable differences between the two seasons. This projection is in agreement with the analysis of Wigley and Santer (1993)—who were confident that, although the magnitude of the warming in the Caribbean is uncertain, the sign of the projection is correct. The models also suggest some differential mean warming of the eastern and western Pacific Ocean. Small reductions also are projected in mean diurnal temperature ranges.

One possible consequence of generally warmer temperatures would be an increase in evaporation in the tropics. As in other regions, however, the rates of increase in evaporation in the latitudes of small island states would not be uniform because evaporation is influenced by factors other than temperature (e.g., pressure). Thus, changes in evaporation rates would be expected to vary spatially and temporally, across and within regions.

Mean rainfall intensity is projected to increase by approximately 20–30% over the tropical oceans (the main locations of the small island states) at the time of doubling of $CO_2$. However, simulation experiments with combined GHG and aerosol forcings suggest that mean precipitation might decrease during summer over the Mediterranean Sea region, leading to a higher probability of dry soils in islands located in this region (i.e., Cyprus and Malta). Relatively greater increases in mean rainfall are projected for the central equatorial Pacific Ocean region (see IPCC 1996, WG I, Section 6.2).

Inter-model agreement (for equilibrium experiments, as well as for coupled model experiments) on rainfall change is variable. Agreement is poor for the western tropical Atlantic Ocean/Caribbean Sea and Mediterranean Sea regions; it is moderate for the Indian Ocean and fairly good over the Pacific Ocean region (see Maul, 1993; Whetton *et al.*, 1996).

### 9.2.3.2. *Extreme Events and Interannual Variability*

Various studies have attempted to assess possible climate change-related changes in extreme events, but few assessments have focused specifically on small island states. Several models, for instance, project an increase in precipitation intensity, runoff, and possible flooding for latitudes within which many of the islands are located. In a few cases, the possibility of change in the occurrence of droughts also has been projected. Given the constraints of scaling and resolution in the models with respect to small islands (see Section 9.2.3.1.) and the current status of inter-model agreement, however, such projections must be considered inconclusive at this stage.

A major ongoing concern for many small island states is whether global warming will lead to changes in the character and pattern of tropical cyclones (i.e., hurricanes and typhoons). Unfortunately, model projections suggest no clear trend, so it is not possible to state whether the frequency, intensity, or distribution of tropical storms and cyclones will change (IPCC 1996, WG I, Sections 3.5.2.3, 3.5.3.1, 3.5.3.2). Within the past few decades, an increase in the number of tropical cyclones also has been reported in the southwest Pacific, with the greatest increase in stronger cyclones (see Figure 9-2f). It should be noted, however, that some doubts exist about the homogeneity of the data base, due in part to recent improvements in observation capabilities (Thompson *et al.*, 1992; Radford *et al.*, 1996).

Gray (1993) reports that, based on historical data for the Caribbean Sea, increases in sea-surface temperature of 1.5°C have been associated with an increase in hurricane frequency. Although there is some uncertainty in the data base, projected increases of this magnitude represent approximately 40% more hurricane activity than normal. More recently, Holland (1997) projects that the intensity of tropical cyclones could increase by 10–20% under $2xCO_2$ conditions. In contrast, observed data suggest that, in the vicinity of longitudes 105–160°E (in the region of Australia), the total number of tropical cyclones between 1969–70 and 1995–96 has decreased, although there were slight increases in the number of strong systems and the duration of the cyclones (Nicholls *et al.*, 1997).

### 9.2.3.3. *Sea-Level Rise: Projections and Implications*

As a result of global warming, the penetration of heat into the ocean leads to the thermal expansion of the water; this effect, coupled with the melting of glaciers and ice sheets, results in a rise in sea level. Sea-level rise will not be uniform globally but will vary with factors such as currents, winds, and tides—as well as with different rates of warming, the efficiency of ocean circulation, and regional and local atmospheric (e.g., tectonic and pressure) effects.

For the IS92a emission scenario, it is estimated that sea level would rise, on average, about 5 mm/yr, within a range of uncertainty of 2–9 mm/yr (IPCC 1996, WG I, Section 7.5.2.4). On regional and local scales, sea level is expected to vary significantly from this global projection, as a result of factors such as vertical land movements (Aubrey and Emery, 1993; Hendry, 1993; Maul, 1996) and dynamic effects due to ocean circulation, wind and pressure patterns, and ocean-water density.

Although these projections are lower than the IPCC (1990) estimates, it should be emphasized that the current best estimates for sea-level rise represent a rate approximately two to four times higher than the rate experienced in the past 100 years (i.e., 1.0–2.5 mm/yr). Model runs also show that sea level would continue to rise beyond the year 2100 (because of lags in the climate response), even with assumed stabilization of global GHG emissions (IPCC 1996, WG II, Section 9.3.1.1). Recent estimates by Pittock *et al.* (1995) indicate a sea-level rise of 28–32 cm for the southwest Pacific region at the time of $CO_2$ doubling—the figure representing only the contribution from thermal expansion. These projections are of considerable concern to small islands (Pernetta, 1992; Hay and Kaluwin, 1993; Maul, 1993): Many islands and atolls in the Pacific and Indian Oceans rarely exceed 3–4 m above mean sea level in elevation and therefore could be vulnerable to changes of this magnitude. The extent of this threat, however, would depend on several factors, including the rate of sand production.

## 9.3. Sensitivity, Adaptation, and Vulnerability

The ecological systems of small islands—and the functions they perform—are sensitive to the rate and the magnitude of changes in climate. These systems provide food, medicine, and energy; process and store carbon and other nutrients; assimilate wastes; purify water and regulate runoff; and provide opportunities for recreation and tourism (see IPCC 1996, WG II, Section 9.2).

Although GCMs provide little insight at present into the extent of climate change impacts on island ecosystems, it is widely postulated that some effects could be quite dislocating for the inhabitants of many small islands (Pernetta, 1988; Roy and Connell, 1991; Holthus *et al.*, 1992; Maul, 1993, 1996; Nicholls, 1995). Various socioeconomic sectors, including tourism, infrastructure, agriculture, water resources, and human health—all of which are sensitive to fluctuations in rainfall, temperature, and sea level—also could be negatively affected.

### *9.3.1. Marine Ecosystems*

#### *9.3.1.1. Coral Reefs*

Coral reefs represent one of the most important resources of tropical islands. They perform valuable functions, including supplying sand to beaches and playing a critical role in the formation and maintenance of reef islands; they are habitats for a variety of marine communities; and they serve as spawning and nursery grounds for numerous species of reef fish. Reefs also function as protective barriers for beaches and coasts by reducing incident wave energy through the processes of wave reflection, dissipation, and shoaling; they also are significant contributors to the economic resource base of many small island states.

Given current projected rates of increase, sea-level rise per se is not expected to have widespread adverse effects on coral reefs. Indeed, some researchers argue that a rising sea level actually may be beneficial because the new conditions would be favorable for inducing vertical growth; in contrast, reef growth has been largely horizontal in the recent past, as a consequence of lower sea levels (Wilkinson and Buddemeier, 1994). Edwards (1995) further suggests that even slowly accreting reef flats should be able to cope with projected sea-level rise, in the absence of other negative forces (e.g., elevated seawater temperature and anthropogenic stresses).

The climate change effect of greatest potential significance to coral reefs is likely to be an increase in seawater temperature. Corals have narrow temperature tolerances (approximately 25–29°C) and salinity tolerances (about 32–36 ppm) (Bellairs Research Institute, 1990). In some islands, some species of corals currently live at or near their threshold of temperature tolerance (Goreau, 1992). Corals respond to the combined effects of irradiance and water temperature elevation by paling in color, or bleaching. However, corals generally do not bleach simply as a result of rapid fluctuation in water temperature but rather as a result of departures in temperature above their seasonal maximums. If the temperature elevation is substantial over an extended period (e.g., 3–4°C for >6 months), significant coral mortality is likely (Brown and Suharsono, 1990). On the other hand, if the temperature increase is relatively small (e.g., 1–2°C) for a short period, bleached corals may recover, though with reduced growth and impaired reproductive capabilities (Goreau, 1992; Brown and Ogden, 1993). A comprehensive review of the causes and consequences of coral bleaching has recently been completed (Brown, 1997).

Notwithstanding their adaptive capacity, corals—like other biotic communities—will continue to be subjected to increasing human stresses (e.g., nutrient loading and other types of chemical pollution, sedimentation from land-based activities, damage from anchoring of boats). These pressures inevitably will limit the innate capacities of these organisms to adapt to the effects of climate change.

ENSO events already have been associated with extensive coral bleaching in the Caribbean and the Pacific in the early 1990s. Although it is not yet clear how climate change will influence the incidence of ENSO events in the future, events that cause high temperature excursions will lead to coral bleaching and, possibly, to mortality (Goreau, 1992; Wilkinson and Buddemeier, 1994; see also Glynn, 1993).

#### *9.3.1.2. Mangrove Communities*

The capacity of mangrove forests to cope with sea-level rise is higher where the rate of sedimentation approximates or exceeds the rate of local sea-level rise. Indeed, Hendry and Digerfeldt (1989) have shown that mangrove communities in western Jamaica were able to keep pace with mid-Holocene sea-level rise (ca. 3.8 mm/yr). However, the adaptive capacity of mangroves and other coastal wetlands to sea-level rise (usually by landward migration) is now severely limited in many localities by increasing human activities. It has been suggested, for instance, that a 1-m rise in sea level in Cuba will drastically affect the viability of 333,000 ha of these wetland communities (approximately 93% of Cuba's mangroves) (Perez *et al.*, 1996). Additionally, adaptive capacity will vary among species; some species of mangroves appear to be more robust and resilient than others to the effects of climate change and sea-level rise (Ellison and Stoddart, 1991; Aksornkaoe and Paphavasit, 1993).

Some ecologists believe that mangrove communities are more likely to survive the effects of sea-level rise in macrotidal, sediment-rich environments—such as northern Australia, where strong tidal currents redistribute sediment (Semeniuk, 1994;

Woodroffe, 1995)—than in microtidal, sediment-starved environments like those in many small islands (e.g., in the Caribbean) (Parkinson *et al.*, 1994). Most small islands fall within the latter classification; therefore, they are expected to suffer reductions in the geographical distribution of mangroves. Furthermore, where the rate of shoreline recession increases, mangrove stands are expected to become compressed and suffer reductions in species diversity in the face of rising sea levels.

On the other hand, Snedaker (1993) argues that mangroves in the Caribbean are more likely to be affected by changes in precipitation than by higher temperatures and rising sea levels because they require large amounts of fresh water to reach full growth potential. He hypothesizes that a decrease in rainfall in the Caribbean would reduce mangroves' productive potential and increase their exposure to full-strength seawater. Thus, peat substrates would subside as a result of anaerobic decomposition by sulfate-reducing microorganisms, leading to the elimination of mangroves in affected areas (Snedaker, 1993).

#### *9.3.1.3. Seagrasses*

It has been postulated that seagrass meadows—which exist in shallow, intertidal coastal environments—are the ecosystems most likely to be negatively affected by climate change effects, particularly sustained elevation of sea-surface temperature or increases in freshwater runoff from land (Edwards, 1995). There is a growing consensus, however, that the main threats to seagrasses in the future are likely to come not from the effects of climate change but from anthropogenic disturbances—such as dredging, overfishing, water pollution, and land reclamation.

### *9.3.2. Terrestrial Ecosystems*

#### *9.3.2.1. Forests*

Although significant land clearance has been a feature of many small island states over decades of settlement, extensive areas of some islands still are covered by forests and other woodlands. For instance, forest and woodland cover in the Solomon Islands, Vanuatu, Dominica, and Fiji is more than 60% of the total land area (see Table 9-2 and Annex D, Table D-7).

Although the tropical forests on small islands are not as critical to the global carbon budget as the tropical rainforests of the South American and African continents, their influence on local and regional climates is no less important (IPCC 1996, WG II, Box 1-5). Forests also are of great socioeconomic importance as sources of timber, fuel, and many nonwood products. Furthermore, forests provide a broad range of other economic and social goods, even though their true value may be difficult to quantify. For example, they provide a basis for ecotourism, habitats for wildlife, and reservoirs for conservation of biological diversity; they also reduce soil erosion. Moreover, forests are of spiritual importance to many indigenous peoples (IPCC 1996, WG II, Section 1.1), many of whom inhabit small islands.

It is possible that tropical forests will be affected more by anthropogenic forces than by climate change per se, as long as deforestation continues at its current rate (IPCC 1996, WG II, Box 1-5). Tropical forests are likely to be affected more by changes in soil water availability (caused by the combined effects of changes in temperature and rainfall) than by changes in temperature alone. Forests are particularly vulnerable to extremes of water availability (drought or flooding) and will decline rapidly if conditions move toward one of these extremes. Increasing temperature and extreme events also may increase the incidence of pests and pathogens, as well as the frequency and intensity of fires.

On the other hand, increasing amounts of $CO_2$ may enable some forest species to use water and nutrients more efficiently (IPCC 1996, WG II, Section 1.3.7). $CO_2$ fertilization may have the greatest effect in the tropics, where it may lead to a gain in net carbon storage in undisturbed forests, especially in the absence of nutrient limitations (IPCC 1996, WG II, Section 1.4.3).

### *9.3.3. Biodiversity and Small Islands*

The biodiversity of islands also could be adversely affected by climate change. A wide range of changes might be expected, including alterations in population size, species distribution, species composition, and the geographical extent of habitats and ecosystems, as well as an increase in the rate of species extinction. One of every three known threatened plants are island endemics; among birds, approximately 23% of island species are threatened, compared with only 11% of the global bird population (McNeely *et al.*, 1993). Although this situation

***Table 9-2:*** *Forest and woodland cover for selected small island states, 1993.*

| Country | Forest/Woodland Cover (%) |
|---|---|
| Bahamas | 32 |
| Comoros | 18 |
| Cuba | 24 |
| Cyprus | 13 |
| Dominica | 67 |
| Dominican Republic | 12 |
| Fiji | 65 |
| Jamaica | 17 |
| Mauritius | 22 |
| St. Vincent and the Grenadines | 36 |
| Solomon Islands | 88 |
| Trinidad and Tobago | 46 |
| Vanuatu | 75 |

Source: UNFAO, 1995 (FAOSTAT-PC, FAO, Rome).

is believed to be linked closely to increasing population pressures and habitat alteration, additional stressors—such as projected climate change effects—could further adversely affect island biodiversity.

Small islands are variable in their marine, coastal, and terrestrial biodiversity. Some are very rich; for example, coral reefs have the highest biodiversity of any marine ecosystem, with some 91,000 described species of reef taxa. Table D-7 of Annex D shows that endemism among terrestrial flora is high in Fiji (58%), Mauritius (46%), Dominican Republic (36%), Haiti (35%), and Jamaica (34%). In the case of Cuba, 42.6% of the known flora and fauna is endemic (Vales *et al.*, 1996). In contrast, other island ecosystems such as low-reef islands tend to have both low biodiversity and low endemism.

The impacts of climate change, in association with human-induced stresses, probably would result in a loss of biodiversity. Bleaching of coral reefs as a result of changes in sea-surface temperature may deplete one of the world's most species-rich ecosystems (IPCC 1996, WG II, Section 9.4.5). In addition, the capacity of species and ecosystems, such as mangroves, to shift their ranges and locations in response to climate change will be hindered by land-use practices that have fragmented existing habitats. The establishment of nature reserves (terrestrial as well as marine) therefore is worth consideration as a viable option for arresting the decline in terrestrial, marine, and coastal biodiversity.

### *9.3.4. Hydrology and Water Resources*

It is highly probable that the effects of climate change will lead to adjustments in the global hydrological cycle, which could affect the distribution and availability of regional water resources. Climate variability on seasonal to interannual time scales can cause changes in precipitation—which can affect the magnitude, rate, and timing of runoff and the frequency and intensity of floods and droughts. Temperature variations can result in changes in evapotranspiration, soil moisture, and infiltration rates (IPCC 1996, WG II, Sections 10.1, 10.3, 10.4).

In many small island states (such as the islands of the eastern Caribbean), the annual rainfall regime often is characterized by pronounced wet and dry seasons. In some countries (e.g., Antigua and Barbuda, Barbados, Grenada), as much as 65% of the annual rainfall may occur during the wet season (June to December); this rainfall is associated largely with the northerly migration of the Inter-Tropical Convergence Zone (ITCZ) and the passage of major weather systems such as easterly waves, tropical depressions, storms, and hurricanes (Nurse, 1985; Gray, 1993). Therefore, to the extent that the availability of water resources in these islands is dependent on heavy rainfall events, changes in the occurrence of these phenomena inevitably will impact water supplies. In the south Pacific, the SPCZ plays a similar role (Salinger *et al.*, 1995).

There is growing evidence that hydrological variability might be associated with the occurrence of mega-scale climate anomalies, such as those associated with the ENSO phenomenon (IPCC 1996, WG II, Section 10.2.1). In the tropics and low-latitude regions of the Southern Hemisphere, the ENSO phenomenon is a major factor in year-to-year climate variability, with a marked effect on rainfall patterns (Pittock, 1984; Philander, 1990; Whetton *et al.*, 1996). Similarly, floods experienced in the Gulf states, Cuba, and other parts of the Americas in the early 1980s and hurricanes in Tahiti and Hawaii during the same period have been linked to a major ENSO event in 1982–83 (Shea, 1994). Possible climate changes associated with ENSO events could have serious consequences for water supplies and agriculture in many nations. As with sea-level rise, atoll communities dependent on rainfall for fresh water could be at risk from precipitation variability associated with anomalies such as the ENSO phenomenon (Meehl, 1994).

Coral islands and atolls are particularly sensitive to changes in groundwater recharge because almost all of their water supply comes from groundwater sources. In The Bahamas, for instance, freshwater lenses are the only exploitable groundwater resources; these lenses are affected periodically by salinity intrusions caused by overpumping and excess evapotranspiration (Cant, 1996). Sea-level rise may precipitate the intrusion of saltwater into the freshwater lens, reducing the quality and quantity of potable water, if the recharge rate or the width of the island is reduced. In contrast, if recharge and island-width remain constant or expand with rising sea level, the freshwater lens may increase in size (Buddemeier and Oberdorfer, 1990).

Various options have been suggested for minimizing the effects of climate change on water resources (IPCC 1996, WG II, Section 12.5.5), and many of these strategies are worthy of consideration by small island states. Options that these countries may wish to evaluate include the harvesting of rainwater, more efficient and extensive use of surface water, artificial recharge of aquifers with rainwater or treated wastewater, and more efficient management of existing supplies and associated infrastructure (e.g., use of various water-saving devices, reduction of leaks, replacement of worn pipes, and recycling).

### *9.3.5. Agriculture and Fisheries*

#### *9.3.5.1. Agriculture*

The climate-sensitive agricultural sector is important to the subsistence economies of many island states (SPREP, 1994a, 1994b, 1996). On many islands, agricultural production systems already are stressed as a consequnce of high population densities and growth rates. Few studies have been conducted specifically on the effects of climate change on agriculture in small islands (IPCC 1996, WG II, Section 13.6.4). However, those investigations that have been undertaken suggest that although $CO_2$ fertilization might have a beneficial effect—mainly on $C_3$ crops—the net effect of climate change is unlikely to be beneficial (Singh *et al.*, 1990; Singh, 1994; SPREP, 1996).

Examination of the agricultural impacts of climate variability associated with seasonal to interannual climate phenomena offers valuable insight into the potential effects of climate change on agriculture. Climate change and climate variability effects that alter the rainfall regime, increase evaporation, or reduce soil moisture would affect agricultural production—possibly with adverse consequences for food security and nutrition in many countries. Crop yields in small island states in the Pacific are projected to decrease as a consequence of reduced solar radiation (resulting from increased cloudiness), higher temperatures (causing shorter growth duration and increased sterility in some cultivars), and changes in water availability (because of changes in the frequency of droughts and floods, as well as changes in their spatial and temporal distribution) (Singh *et al.*, 1990). Seawater intrusion also would degrade the fertility of coastal soils and consequently contribute to a reduction in yields (IPCC 1996, WGII, Sections 13.6.4 and 13.7). Some of these adverse consequences of climate change may be partially offset, however, by the beneficial effects of increased atmospheric $CO_2$ (SPREP, 1996).

Singh (1994) projects that, for small islands in the Pacific, an extension of the dry season by 45 days would lead to a decrease in maize yields of 30–50%. Similarly, yields from sugarcane and taro would be reduced by 10–35% and 35–75%, respectively. On the other hand, Singh *et al.* (1990) found that significantly increased rainfall (>50%) during the wet season on the windward side of the larger islands would cause taro yields to increase by 5–15%, but would reduce rice yields by approximately 10–20% and maize yields by 30–100%.

In Tuvalu, contraction of the freshwater lens on the atolls (as a result of increasing water demand and sea-level rise) is expected to reduce crop yields severely. At locations such as Funafuti atoll, where there already is heavy demand on the water supply, the effects of sea-level rise will exacerbate water scarcity. The crop expected to be most seriously affected is pulaka (giant taro). Reduced yields from pulaka pits are "likely to have significant cultural ramifications, given the central role of this crop in Tuvaluan society" (SPREP, 1996, Section 6.4.1). The cultivation of taro, another important subsistence crop in the Pacific islands, would be similarly threatened (see Box 9-2).

Recent research has attempted to assess changes in world prices for selected crops, based on GCM projections. Reilly *et al.* (1994) estimate that, under the GFDL scenario, world prices for most crop commodities would fall relative to baseline prices; the exceptions are rice and sugar, which would rise by approximately 30% and 15%, respectively. Should such a projected increase in the price of world sugar occur, positive economic benefits (in the form of increased foreign exchange earnings) would accrue to the economies of sugar-exporting small island states. The estimated price increases are somewhat less for scenarios that include adaptation. In the GISS scenario, prices for soybeans, groundnuts, cotton, and tobacco fell by about 10–15% with adaptation. Prices for other crop commodities considered in the analysis rose by as much as 20% (IPCC 1996, WG II, Section 13.8.2).

**Box 9-2. Taro Cultivation and Sea-Level Rise**

The pit cultivation of taro is particularly susceptible to changes in freshwater quality. Taro is grown in depressions and pits that have been excavated down to the freshwater lens and partly filled with composting organic matter. Leaves of many species—including *Guettarda speciosa*, *Tournefortia argentea*, *Artocarpus altilis*, *Triumfetta procumbens*, and *Hibiscus tiliaceus*—are used to form the organic soil. In Kiribati, taro (*C. Chamissonis*) is planted in 20-m x 10-m pits, 2–3 m deep, with the taro corm placed in "organic baskets" of *Pandanus* and *Cocos nucifera* and anchored in holes 60 cm below the water level. In Puluwat Atoll, in contrast, the taro is planted in organic-matter bundles 0.5 m above the water level. Taro swamps also are sites of high evapotranspiration; increased loss of freshwater will increase the risk of saltwater intrusion.

A rise in sea level will have a serious impact on atoll agroforestry and the pit cultivation of taro. Erosional changes in the shoreline will disrupt populations, and the combined effects of freshwater lens loss and increased storm surges will stress freshwater plants and increase vulnerability to drought (Wilkinson and Buddemeier, 1994).

Although these scenarios do not address small island states specifically, some of the results would be relevant to any analysis of climate change impacts on agriculture in these nations. To the extent that island states depend heavily on food imports and export some agricultural products, such projected price changes would have a considerable impact on the ability of these countries to earn much-needed foreign exchange.

### 9.3.5.2. *Fisheries*

Fishing, although largely artisinal or small-scale commercial, is an important activity on most small islands (Blommestein *et al.*, 1996; Mahon, 1996). Although marine fisheries only account for approximately 1% of the global economy, many coastal and island states are far more dependent on the sector than this statistic would suggest (IPCC 1996, WG II, Section 16.1.1). For example, the annual yield of lobsters from the shelves and banks of the Caribbean islands (excluding the U.S. Virgin Islands) has a retail value in restaurants of approximately US$40 million (Vicente, 1996). Similarly, marine fish account for 16% of global animal protein consumption, but the contribution to protein intake is much greater in developing countries, where animal protein tends to be relatively expensive (Lauretti, 1992; Weber, 1993, 1994; IPCC 1996, WG II, Section 16.2.4).

The modest temperature increases projected for these regions are not anticipated to have a widespread adverse effect on small island fisheries. However, a temperature rise could have a negative effect on productivity in areas—such as shallow lagoons—where hypersalinity may occur, especially if juveniles are sensitive to salinity or temperature (Alm *et al.*, 1993). On the other hand, warming-induced changes in current patterns might increase upwellings at sea, bringing more nutrients to the surface and providing more food for the fish (Aparicio, 1993; Bakun, 1993; Chakalall, 1993; Maul, 1993).

There is some evidence that clam and sea-turtle fisheries could be sensitive to sea temperature changes. Clams, like corals, are known to suffer from bleaching with the expulsion of symbiotic algae, as a result of excesses in temperature or radiation (Gomez and Belda, 1988; Lucas *et al.*, 1988). Bleaching would cause "loss of productivity or devastation in stocks of growout clams on the reef-flats" (Wilkinson and Buddemeier, 1994). One possible option to minimize this possibility in locations where commercial clam mariculture is practiced (e.g., the Solomon Islands, Palau, Samoa) in deeper water, where light and temperature fluctuations would be reduced. This measure, however, will result in lower growth rates and reduced returns to the grower (Lucas *et al.*, 1988; Wilkinson and Buddemeier, 1994). In the case of sea turtles, most species nest during the summer, when temperatures are close to the organisms' upper thermal limit. It is uncertain whether turtles will be able to adapt to warmer temperatures by nesting in the cooler months (Limpus, 1993). If such adaptation is not possible, increased temperatures are likely to result in a higher ratio of females to males because the sex ratio of sea turtles is directly determined by temperature (Wilkinson and Buddemeier, 1994).

Generally, fisheries in the small island states are not expected to be adversely affected by sea-level rise per se. A higher sea level would be a critical factor for fisheries only if the rate of rise were far more rapid than current projections suggest. In such circumstances, the natural succession of coastal ecosystems on which some species depend (e.g., mangroves, seagrasses, corals) would be disrupted (IPCC 1996, WG II, Section 16.2.2.1). In tropical islands, these ecosystems function as nurseries and forage sites for a variety of commercially important species. Fish production obviously would suffer if these habitats were endangered or lost (Costa *et al.*, 1994).

### 9.3.6. Coastal Systems

#### 9.3.6.1. Sea-Level Rise and Coastal Changes

Major coastal impacts will result from accelerated sea-level rise; these effects will include coastal erosion, saline intrusion, and sea flooding, among other impacts. Impact studies have confirmed that low-lying deltaic and barrier coasts, low reef islands, and coral atolls are especially vulnerable to the potential impacts of sea-level rise (Maul, 1993). Some small islands could suffer land loss and experience increased beach erosion, inundation, and flooding from a sea-level rise of between 50 cm and 1 m. Leatherman (1994) suggests that sea-level rise could convert many islands in the Maldives to sandbars and significantly reduce available dry land on larger, more heavily populated islands. In Majuro Atoll (Marshall Islands), computation of land loss from a 1-m rise in sea level, based on the Bruun rule, suggests that approximately 60 ha of dry land (8.6% of the total land area) would be lost to erosion. It also is estimated that more than 115 ha of Majuro Atoll would be inundated if a 1-m sea-level rise were superimposed on present-day flooding from wave runup and overtopping (Holthus *et al.*, 1992). In the case of Kiribati, Woodroffe and McLean (1992) have suggested that 12.5% of the total land area would be vulnerable with a 1-m rise in sea level.

Notwithstanding these projections, the response of islands to the impacts of climate change will vary within regions and even within countries. Islands are not passive systems; they will respond dynamically in variable and complex ways to sea-level and climate changes (Aalbersberg and Hay, 1993; McLean and d'Aubert, 1993; McLean and Woodroffe, 1993). For example, the extent to which relative sea-level rise will affect coastal recession rates will depend on many factors, including (though not limited to) the rate of sediment supply relative to submergence; the width of existing fringing reefs; the rate of reef growth; whether islands are anchored to emergent rock platforms; whether islands are composed primarily of sand or coral rubble; the presence or absence of natural shore-protection structures, such as beachrock or conglomerate outcrops; the presence or absence of biotic protection, such as mangroves or other strand vegetation; the health of coral reefs; and, especially, the tectonic history of the island.

### 9.3.7. Human Settlement

#### 9.3.7.1. Infrastructure and Settlement

Generally, the largest concentrations of settlements on small islands occur no further than 1–2 km from the coast, and sometimes much less. In most of the eastern Caribbean states, for instance, more than 50% of the population resides within 2 km of the coast; the corresponding figure in Barbados is estimated to be in the region of 60% (Nurse, 1992). Similarly, large coastal populations are the norm in the Pacific and Indian Ocean islands—especially the atoll states, where settlement areas may even be sited on the beach itself or on the sand terrace (e.g., Tuvalu, Kiribati, Maldives). Clearly, such settlements are at risk from projected sea-level rise—which, in all likelihood, would be accompanied by inundation, increased flooding, coastal erosion, and consequently land loss.

On many small islands, critical infrastructure tends to be located in or near coastal areas; this infrastructure will be highly vulnerable to the effects of projected sea-level rise, especially during extreme events. Similarly, significant infrastructural damage could result from any increase in the frequency or intensity of extreme events such as floods, tropical storms, and storm surges (Pernetta, 1992; Alm *et al.*, 1993). Moreover, because island populations tend to congregate in the few urban centers where most of the infrastructure and services are located, damage to important infrastructure (e.g., coastal roads, bridges, seawalls) would be disruptive to several types of economic, social, and cultural activities. In Malta, for example, vital desalinization facilities on the coast would be at risk in such circumstances (Sestini, 1992). Social and economic dislocation would be especially severe among communities with high population densities—such as Eauripik, Federated States of Micronesia (950 persons/km$^2$); Majuro, Marshall Islands (2,188 persons/km$^2$); and Male, Republic of Maldives (5,000 persons/km$^2$).

The costs of protecting the shoreline and other infrastructure will vary, depending on the kind of protection needed, the length of area to be protected, design specifications to be adopted, and the availability of construction materials. There is concern, however, that the overall costs of infrastructure protection will be beyond the financial means of many island nations. Vulnerability studies conducted for selected small islands suggest that the costs of coastal protection ("hard" options) would be a significant proportion of GNP (see IPCC 1996, WG II, Table 9-3). In Malta and Cyprus, it is estimated that approximately US$550 million and US$190 million, respectively, would be needed to provide adequate shore protection works against a 20–30 cm rise in relative sea level (Sestini, 1992). Although these estimated costs might not be excessive in the context of large economies, they represent considerable financial resources that these small island states would have to reallocate.

The cost of insurance is another important factor that must be taken into consideration in any assessment of climate change impacts on infrastructure. Property insurance costs are extremely sensitive to the effects of catastrophic events such as hurricanes, floods, and earthquakes. High-risk locations therefore could face high insurance premiums—and even, in extreme cases, withdrawal of coverage (Box 9-3).

Like other countries, small islands might wish to consider implementing appropriate strategies to reduce the potential dislocation that may result from climate change-related infrastructural damage. As part of an overall response strategy, these countries may consider the gradual replacement of infrastructure in nonthreatened locations, where this proves feasible. Other options could include adoption of building codes and other design and construction standards, construction of mandatory building setbacks in coastal areas, and diversification of economic activities to the maximum extent possible. In the worst-case scenario, resettlement of entire communities also may have to be considered (see Box 9-4).

**Box 9-4. Sea-Level Rise and Settlement in the Caribbean**

The relocation of vulnerable coastal settlements is a difficult option for many Caribbean island states to consider. However, large populations and supporting infrastructure presently are located close to mean sea level; when the combined effects of increased flooding, salinization, coastal land loss, and infrastructural damage are contemplated, some resettlement of coastal populations seems inevitable. This resettlement would be socially dislocating for the people themselves and require an injection of funds that few countries in the region can easily afford.

The implications of resettlement will be even more serious for the smallest, low-lying states, for whom relocation within their own national boundaries may be physically impossible (see Nurse, 1992).

Successful implementation of these and other adaptation measures will depend largely on the extent to which island states are able to overcome associated constraints. Some possible constraining factors for many small islands include technology and human resources capability; financial limitations; cultural and social acceptability of adaptation measures; and the availability of adequate political, legal, and institutional support. On the positive side, many island societies possess a reserve of important traditional knowledge and skills—by virtue of having had to adapt, over generations, to a range of other stresses (Holthus, 1996; SPREP, 1996). This knowledge base is a vital resource that these countries may wish to exploit in designing and implementing appropriate adaptation strategies.

**Box 9-3. Hurricanes and Insurance in the Caribbean**

The Caribbean region suffered considerable damage from severe hurricanes (e.g., David, Hugo, Gilbert, Gabrielle, Luis, Marilyn) in the 1980s and 1990s. As a direct result, many insurance and reinsurance companies withdrew from the market. Those that remained imposed onerous conditions for coverage—including very high deductibles; separate, increased rates for windstorms; and insertion of an "average" clause to eliminate the possibility of underinsurance (see Murray, 1993; Saunders, 1993).

#### 9.3.7.2. *Tourism*

Tourism is one of the most important sectors of the economies of many small island states in the Caribbean Sea and the Pacific and Indian Oceans; the industry is equally important to the Mediterranean island of Malta. In some Caribbean countries, the annual number of long-staying visitors (i.e., excluding cruise ship arrivals) far exceeds the size of the resident population; in some cases, the ratio of tourists to residents is greater than 2:1 (e.g., approximately 3:1 in Antigua and almost 6:1 in the Bahamas; see Caribbean Tourism Organization, 1996). Tourism also is the largest single contributor to GNP in many islands. In 1995, for instance, this sector accounted for

69% and 53% of GNP in the Caribbean islands of Antigua and the Bahamas, respectively, and more than 10% in most other islands in the region. Tourism also earns valuable foreign exchange for these islands; the sector earned US$180 million in 1995 in the Maldives—more than 70% of the country's total foreign exchange earnings for that year. Moreover, tourism generates significant employment in many islands, such as the Bahamas—where three-quarters of the labor force is employed, directly or indirectly, in providing services to the industry (Box 9-5).

Tourism is so vital to many small island societies that when there is contraction in the industry (and, hence, reduced earnings), the rate of national economic growth often declines. In such circumstances, provision of many essential services would be jeopardized, and other vital sectors (e.g., health, education, and welfare)—whose budgetary allocations may be influenced by tourism earnings—also may be affected.

Climate affects tourism in many ways, directly and indirectly. Loss of beaches to erosion; inundation; degradation of ecosystems and related impacts (e.g., loss of coral reefs to bleaching, saline intrusion); and damage to critical infrastructure are only a few consequences that could undermine the tourism resource base of vulnerable small island states (Alm *et al.*, 1993). Although some of these impacts also can be triggered by non-climate-related factors, there is a growing consensus that climate change is likely to precipitate such changes, and that they would be disruptive (Holthus *et al.*, 1992; Pernetta, 1992; Sestini, 1992; IPCC, 1996, WG II, Box 9-3; SPREP, 1996). There is evidence that any such dislocation in the tourism sector would have severe repercussions for the economic, political, and sociocultural life of many small islands.

### 9.3.8. Human Health

Climate change-related disturbances of physical systems (e.g., weather patterns, sea level, water supplies) and ecosystems (e.g., agro-ecosystems, disease-vector habitats) could pose risks to the health of human populations. Some health impacts would occur via direct pathways (e.g., death from heat waves and other extreme weather events, such as hurricanes); others would occur via indirect pathways (e.g., changes in the geographical range of vector-borne diseases). Researchers generally agree that most of the impacts of climate change on human health are likely to be adverse. The vulnerability of populations to such threats will depend on many factors, including present health status, quality of available health care services and associated infrastructure, and availability of financial and technical resources (McMichael, 1993).

Should global warming increase the frequency and/or severity of extreme weather events such as droughts, floods, landslides, and tropical cyclones, it is likely that more deaths, injuries, infectious disease cases, and psychological disorders could result (McMichael and Martens, 1995; IPCC 1996, WG II, Section 18.2.2). Elevated global mean temperatures also could lead to a greater frequency of heat waves—and consequently a higher incidence of related illness (predominantly cardiorespiratory) and mortality. Inevitably, the most vulnerable countries will be those with reduced capacity to respond to or mitigate these threats. Because of their present social and economic circumstances, many small islands fall into this category.

The transmission of many infectious diseases is affected by climatic factors. Vector-borne diseases such as malaria, dengue, and yellow fever are sensitive to factors such as temperature, rainfall, and humidity. Climate change could increase the geographical range of disease agents; changes in the life-cycle dynamic of vectors and infective agents, in aggregate, could result in more efficient transmission of many vector-borne diseases (McMichael *et al.*, 1996). Mathematical models project that an additional 50–80 million cases of malaria would result from a mean temperature increase of 3.0ºC (Martens *et al.*, 1995). Although most of this projected increase is expected to be in temperate-zone countries, some increase in the tropics—the latitudinal zone in which the majority of small islands and other developing countries are located—cannot be ruled out (see Table 9-3). Moreover, even a marginal increase in the incidence of disease would place great stress on the public health systems of many small islands, where these facilities often are not well developed.

Results fom a recent study of dengue fever in island states of the South Pacific have shown a strong link between the incidence

**Box 9-5. Importance of Tourism in the Economies of the Bahamas and the Maldives**

**Bahamas**

- Accounts for 40% of gross domestic product (GDP)
- Earns approximately US$1.3 billion in foreign exchange
- Estimated to account for more than 50% of all government revenues
- Employs approximately 50% of labor force directly, another 25% in related services.

**Maldives**

- Accounts for 18% of GDP
- Earns approximately US$181 million in foreign exchange
- Accounts for 32.5% of all government revenues
- More than 25% of labor force employed directly in tourism.

Sources: Ministry of Tourism, Commonwealth of the Bahamas; Ministry of Tourism, Republic of the Maldives (1995 data).

***Table 9-3:** Estimated impacts of climate change on tropical vector-borne diseases.*

| Disease | Population at Risk (millions) | Prevalence or Incidence of Infection | Present Distribution | Possible Change of Distribution due to Climate Change[1] |
|---|---|---|---|---|
| Malaria | 2,400 | 300–500 million | Tropics/Subtropics | +++ |
| Schistosomiasis | 600 | 200 million | Tropics/Subtropics | ++ |
| Lymphatic filiariasis | 1,094 | 117 million | Tropics/Subtropics | + |
| African trypanosomiasis | 55 | 250,000–300,000 new cases/yr | Tropical Africa | + |
| Dracunculiasis | 100 | 100,000/yr | Tropics (Africa/Asia) | ? |
| Visceral leishmaniasis | 350 | 12 million (prevalence) + 500,000 new cases/yr | Asia/S. Europe/ Africa/South America | + |
| Onchocerciasis | 123 | 17.5 million | Africa/Latin America | ++ |
| American trypanosomiasis | 100 | 18 million | Central and South America | + |
| Dengue | 2,500 | 50 million/yr | Tropics/Subtropics | ++ |
| Yellow fever | 450 | <5,000 cases/yr | Africa/Latin America/ East and Southeast Asia | ++ |

[1] += likely; ++ = very likely; +++ = highly likely; ? = unknown.
Source: WHO, 1996.

of outbreaks of the disease and the ENSO phenomenon (Hales *et al.*, 1996). Although climate models do not, at present, give reliable projections of ENSO events with global warming, ENSO phenomena are known to be associated with interannual warming and rainfall anomalies (Hay *et al.*, 1993a; Salinger *et al.*, 1995; Centella *et al.*, 1996). It therefore appears highly probable that changes in ENSO patterns would affect the incidence of dengue fever.

An increase in the incidence of waterborne and food-borne infectious diseases, particularly in tropical regions, could result from a rise in temperature, microorganism proliferation, and a higher incidence of flooding and water shortages. Algal blooms—which occur in warm tropical waters and are associated with biotoxin contamination of fish and shellfish—also could become more frequent, resulting in the proliferation and increased transmission of cholera. Recent research has shown a clear relationship between increased incidence of cholera outbreaks and sea warming, owing to the ability of zooplankton to function as a reservoir for a dormant state of the pathogen (Colwell, 1996). In general, biotoxins such as ciguatera are known to be temperature sensitive; therefore, they could become more abundant with ocean warming, resulting in a higher incidence of seafood contamination and biotoxin poisoning. Microbiological contamination already is a problem in the coastal areas of many small islands, where sewage and solid-waste disposal systems currently are inadequate. These combined circumstances would place additional stress on already overextended health sectors of most small islands and may even threaten the nutritional status of some of these countries.

Among the strategies that could be adopted to ameliorate some of the negative effects of climate change on the health sector are the mounting of an effective public education program aimed at improving personal behavior and hygiene; upgrading, extension, and maintenance of existing health-care facilities and services; implementation and maintenance of appropriate, cost-efficient sewage and solid-waste management systems; adoption of appropriate disaster preparedness and management plans; and wider application of available protective technologies, such as water purification and sewage treatment (McMichael *et al.*, 1996). In addition, Health Early Warning Systems (HEWS)—which incorporate seasonal and interannual climate forecast information—could be used to detect ecological conditions conducive to outbreaks of climate-sensitive diseases such as cholera, malaria, and dengue. Lessening the vulnerability of small islands to such health threats through these types of improved planning measures and implementation of timely strategies would reduce human suffering and economic losses.

### 9.3.9. *Cultural Integrity*

In many small island states, a number of factors, including isolation and close traditional ties to the land and sea, have contributed to the development of a unique set of cultural traits on different islands or groups of islands. In Tuvalu, for instance—as in other small Pacific atoll states—attachment to land and sea is a critical component of local cosmology (SPREP, 1996). Any force that poses a threat to this attachment would be culturally and socially disruptive in these traditional societies.

Ethnic, linguistic, social, and religious differences among and between the peoples of Polynesia, Melanesia, and Micronesia in the South Pacific illustrate the cultural diversity of the island states. The unique cultures that have developed over millenia on the resource-rich and diverse high-volcanic and limestone islands in the region, such as Vanuatu, Fiji, and Samoa, are unlikely to be seriously threatened by climate change. On the other hand, resource-poor, low-reef islands and atolls, which have developed equally distinctive traditional identities over centuries—such as the Tuvaluan, Kiribati, Marshallese, and Maldivian cultures—are more at risk. The fragility of these low islands and their sensitivity to sea-level change and storms suggest that the future existence of such islands and their cultural diversity could be seriously threatened (Roy and Connell, 1991).

## 9.4. Integrated Analysis of Potential Vulnerabilities and Impacts

Because of their strong dependence on economic sectors that are highly sensitive to climate change effects (e.g., coastal tourism and agriculture), small island states clearly are a vulnerable group of countries (IPCC 1996, WG III, Sections 6.5.10, 6.5.11). Briguglio (1993) developed a mean "vulnerability index" for different categories of nations, based on three selected variables: export dependence, insularity and remoteness, and proneness to natural disasters. Although there may be some limitations associated with this index (e.g., the restricted criteria), it supports the widely held view that small island states will be more vulnerable than any other group of countries to projected climate change impacts. On a scale from 0 (lowest vulnerability) to 1 (highest vulnerability), a score of 0.590 (the highest index) was derived for the small island group. A lower index, 0.539, was calculated for other developing countries; the index for all developing countries as a group was 0.417. Indeed, based on Briguglio's index, 9 of the 10 most vulnerable countries are small islands. Thus, though the index has its limitations, it draws attention to the high vulnerability of small island states in relation to all other regional groups.

Pernetta (1988) ranks Pacific islands in terms of their vulnerability to sea-level rise, taking other factors—such as elevation—into consideration. Based on his classification, states such as the Marshall Islands, Tuvalu, and Kiribati would suffer "profound" impacts, including disappearance in the worst-case scenario; "severe impacts," resulting in major population displacement, would be experienced by the Federated States of Micronesia, Nauru, and Tonga; "moderate to severe impacts" would be felt by Fiji and the Solomon Islands; and "local severe to catastrophic" effects would be experienced by Vanuatu and Western Samoa.

It must be emphasized, however, that the sensitivity of small islands to the projected effects of climate change cannot be attributed to any single factor (e.g., size, elevation, remoteness, or any other) or to a select group of factors. Rather, the level of vulnerability of these islands is determined by the cumulative and synergistic result of these and related biophysical attributes (including the degree of natural adaptive capacity), combined with the islands' economic and sociocultural characters (including current and future levels of anthropogenic stress) (see, e.g., SPREP, 1996). Moreover, because many small islands already are prone to other hazards (e.g., tropical cyclones and storm surges) that invariably have adverse effects, climate change impacts on longer time scales could render these countries extremely vulnerable.

The Caribbean countries are a case in point. Some face an annual threat from hurricanes (cyclones); others, such as St. Vincent and Montserrat, are prone to disruptive volcanic activity; and still others are affected by periodic earthquakes and tsunamis (Maul, 1996). Most have extensive, vulnerable, low-lying coastal plains; some (e.g., Barbados, Antigua, St. Kitts, Bahamas) are heavily dependent on groundwater supplies; and for many, tourism is the most vital economic sector. A higher incidence of flooding and inundation, beach and coastal land loss, reef damage, salinization of the freshwater lens, and disruption of tourism and infrastructure would create economic and social crises in a number of these islands. Thus, many Caribbean countries must be classified as vulnerable to the effects of climate change and sea-level rise—not simply because of their size or elevation alone but because of strong linkages between these and other physical characteristics, natural resources, and socioeconomic structures.

Moreover, for small islands with limited resources, it is absolutely essential for integrated assessment models to include the value of nonmarketed goods and services that also will be at risk. Commodities such as cultural and subsistence assets (e.g., community structures), recreational values, traditional skills and knowledge, and natural values (e.g., the capacity of mangroves to filter nitrate and phosphate and thus reduce nutrient loading to the marine zone) are just as important to some small islands as marketed goods and services. Such nonmarketed goods and services often are not incorporated into integrated assessment models. From the perspective of small island states, the integration of these assets is an important and necessary challenge facing the modeling community. Some recent attempts to develop an appropriate methodology for vulnerability assessment, incorporating these factors, have been made by Yamada *et al.* (1995) for southwest Pacific islands.

**Box 9-6. A Small Island Response to Climate Change**

Climate change will impose diverse and significant impacts on small island states. Impediments to responses to climate change in small islands include:

- Lack of definitive projections of temperature, rainfall, and sea level
- Ambiguous statements on ENSO and tropical storms, hurricanes, and typhoons
- Many vulnerability assessments undertaken using methodologies "poorly harmonized with local conditions"
- Limited adaptive capacity of small island states.

Possible Regional Responses to *Facilitate* Adaptation to Climate Change for Small Island States

**Policy Responses**

- A policy of regional cooperation and coordination
- A policy of "owning" the issue of climate change and variability
- A policy of maximizing the benefits of climate change
- A policy to base plans and actions on factual understanding of climate change
- A policy of mainstreaming climate change responses in national planning through adaptation of sustainable management practices
- A policy of enhancing capacities to respond to the consequences of anticipated change in climate
- A policy of enhancing regional water, soil, habitat, and food security

**Priority Action Strategies**

- A strategy for capacity building, awareness-raising, and technical capacity enhancement
- A strategy for development and application of appropriate methodologies and information sources
- A strategy to identify, assess, and implement investment instruments relevant to adaptation
- A strategy to support optional management responses to climate change at the national level
- A strategy for regional support for integrated coastal zone management
- A strategy to undertake regionally relevant research on impacts and adaptations to climate change

Source: Modified from Hay, 1997.

Clearly, it would be inappropriate to assess the sensitivity of small islands to climate change impacts in isolation from other factors that contribute to their overall vulnerability. Constraining factors such as size, elevation, limited resources (natural, financial, and technological), proneness to natural hazards, dependence on external markets, and generally high population growth rates enhance the vulnerability of these island states (Alm *et al.*, 1993). Only when the effects of such factors are evaluated in combination with the threat of climate change impacts can a meaningful vulnerability index for small island states be developed and appropriate adaptation options pursued (see Box 9-6).

## References

**Aalbersberg**, B. and J. Hay, 1993: *Implications of Climate Change and Sea Level Rise for Tuvalu.* SPREP Reports and Studies Series No. 54, South Pacific Regional Environment Programme, Apia, Western Samoa, 80 pp.

**Aksornkaoe**, S. and N. Paphavasit, 1993: Effect of sea-level rise on the mangrove community in Thailand. *Malaysian Journal of Tropical Geography*, **24**, 29–34.

**Alm**, A., E. Blommestein, and J.M. Broadus, 1993: Climatic changes and socio-economic impacts. In: *Climatic Change in the Intra-Americas Sea* [Maul, G.A. (ed.)]. Edward Arnold, London, United Kingdom, pp. 333–349.

**Aparicio**, R., 1993: Meteorological and oceanographic conditions along the southern coastal boundary of the Caribbean Sea. In: *Climatic Change in the Intra-Americas Sea* [Maul, G.A. (ed.)]. Edward Arnold, London, United Kingdom, pp. 100–114.

**Aubrey**, D.G. and K.O. Emery, 1993: Recent global sea levels and land levels. In: *Climate and Sea Level Change: Observations, Projections and Implications* [Warrick, R.A., E.M. Barrow, and T.M.L. Wigley (eds.)]. Cambridge University Press, Cambridge, United Kingdom, pp. 45–56.

**Bakun**, A., 1993: Global greenhouse effects, multi-decadal wind trends, and potential impacts on coastal pelagic fish populations. *ICES Marine Science Symposium,* **195**, 316–325.

**Bellairs Research Institute**, 1990: *A Survey of Marine Habitats Around Anguilla, with Baseline Community Descriptors for Coral Reefs and Seagrass Beds.* Report prepared for the Department of Agriculture and Fisheries, Government of Anguilla. Funded by British Development Division in the Caribbean, Bridgetown, Barbados, 177 pp.

**Blommestein**, E., B. Boland, T. Harker, S. Lestrade, and J. Towle, 1996: Sustainable development and small island states of the Caribbean. In: *Small Islands: Marine Science and Sustainable Development* [Maul, G.A. (ed.)]. American Geophysical Union, Washington, DC, USA, pp. 385–419.

**Briguglio**, L. 1993: *The Economic Vulnerabilities of Small Island Developing States.* Report to U.N. Conference on Trade and Development (UNCTAD), Geneva, Switzerland,

**Brown**, B.E., 1997: Coral bleaching: Causes and Consequences. *Coral Reefs*, **16**, pp. 129-138.

**Brown**, B.E. and J.C. Ogden, 1993: Coral bleaching. *Scientific American,* **268**, 64–70.

**Brown**, B.E., and Suharsono, 1990: Damage and recovery of coral reefs affected by El Niño-related seawater warming in the Thousand Islands, Indonesia. *Coral Reefs,* **8**, 163–170.

**Buddemeier**, R.W. and J.A. Oberdorfer, 1990: Climate change and groundwater reserves. In: *Implications of Expected Climatic Changes in the South Pacfic Region: An Overview* [Pernetta, J.C. and P.J. Hughes (eds.)]. UNEP Regional Seas Reports and Studies No. 128, United Nations Environment Programme, Nairobi, Kenya, pp. 56–67.

**Cant**, R.V., 1996: Water supply and sewerage in a small island environment: the Bahamian experience. In: *Small Islands: Marine Science and Sustainable Development* [Maul, G.A. (ed.)]. American Geophysical Union, Washington, DC, USA, pp. 385–419.

**Caribbean Tourism Organization**, 1996: Caribbean Tourism Statistical Report, 1995, Bridgetown, Barbados, 249 pp.

**Centella**, A., L.R. Naranjo, and L.P. Paz, 1996: *Variations and Climate Changes on Cuba.* Technical Report, National Climate Center, Institute of Meteorology, Havana, Cuba, 58 pp. (in Spanish)

**Chakalall**, B., 1993: Policy issues for sustainable fisheries development in the Caribbean. In: *Report and Proceedings of the Meeting on Fish Exploitation Within the Exclusive Economic Zones of the English Speaking Caribbean Countries* [Chakalall, B. (ed.)]. FAO Fisheries Report No. 483, Food and Agriculture Organization, Rome, Italy, pp. 1–43.

**Colwell**, R.R., 1996: Global climate and infectious disease: the cholera paradigm. *Science*, **274**, 2025–2031.

**Costa**, M.J., J.L. Costa, P.R. Almeida, and C.A. Assis, 1994: Do eel grass beds and salt marsh borders act as preferential nurseries and spawning grounds for fish? An example of the Mira estuary in Portugal. *Ecological Engineering,* **3**, 187–195.

**Edwards**, A.J., 1995: Impact of climate change on coral reefs, mangroves and tropical seagrass ecosystems. In: *Climate Change: Impact on Coastal Habitation* [Eisma, D. (ed.)]. Lewis Publishers, Boca Raton, FL, USA, pp. 209–234.

**Ellison**, J.C. and D.R. Stoddart, 1991: Mangrove ecosystem collapse during predicted sea-level rise: Holocene analogues and implications. *Journal of Coastal Research,* **7**, 151–165.

**Glynn**, P.W., 1993: Coral reef bleaching: ecological perspectives. *Coral Reefs,* **12**, 1–17.

**Gomez**, E.D., and C.A. Belda, 1988: Growth of giant clams in Bolinao, Philippines. In: *Giant Clams in Asia and the Pacific* [Copland, J.W. and J.S. Lucas (eds.)]. Australian Centre for International Agricultural Research Monograph No. 9, Canberra, pp. 178–182.

**Goreau**, T.J., 1992: Bleaching and reef community change in Jamaica: 1951–1991. *American Zoology,* **32**, 683–695.

**Gray**, C.R., 1993: Regional meteorology and hurricanes. In: *Climatic Change in the Intra-Americas Sea* [Maul, G.A. (ed.)]. Edward Arnold, London, United Kingdom, pp. 87–99.

**Hales**, S., P. Weinstein, and A. Woodward, 1996: Dengue fever epidemics in the South Pacific: driven by El Niño Southern Oscillation. *Lancet,* **348**, 1664–1665.

**Hay**, J.E. 1997: A Pacific Response to Climate Change. *Tiempo*, **23**, pp. 1-10.

**Hay**, J.E. and C. Kaluwin (eds.), 1993: *Climate Change and Sea Level Rise in the South Pacific Region.* Proceedings of the Second SPREP meeting, Nomea, 6–10 April 1992, South Pacific Regional Environment Programme, Apia, Western Samoa, 238 pp.

**Hay**, J.E., C. Kaluwin, and N. Koop, 1993b: Implications of climate change and sea level rise for small island nations of the South Pacific: a regional synthesis. *Weather and Climate,* **16**, 1–20.

**Hay**, J., J. Salinger, B. Fitzharris, and R. Basher, 1993a: Climatological "seesaws" in the Southwest Pacific. *Weather and Climate,* **13**, 9–21.

**Hendry**, M.D., 1993: Sea-level movements and shoreline changes in the wider Caribbean region. In: *Climatic Change in the Intra-Americas Sea* [Maul, G.A. (ed.)]. Edward Arnold, London, United Kingdom, pp. 152-161.

**Hendry**, M.D. and G. Digerfeldt, 1989: Palaeogeography and palaeoenvironments of a tropical coastal wetland and adjacent shelf during Holocene submergence, Jamaica. *Palaeogeography, Palaeoclimatology. Palaeoecology,* **73**, 1–10.

**Holland**, G.J., 1997: The maximum potential instability of tropical cyclones. *Journal of Atmospheric Science* (in press).

**Holthus**, P.F., 1996: Coastal and marine environments of the Pacific Islands: ecosystem classification, ecological assessment and traditional knowledge for coastal management. In: *Small Islands: Marine Science and Sustainable Development* [Maul, G.A. (ed.)]. American Geophysical Union, Washington, DC, USA, pp. 341–365.

**Holthus**, P., M. Crawford, C. Makroro, and S. Sullivan, 1992: *Vulnerabilitry Assessment for Accelerated Sea Level Rise Case Study: Majuro Atoll, Republic of the Marshall Islands.* SPREP Reports and Studies Series No. 60, South Pacific Regional Environment Program, Apia, Western Samoa, 107 pp.

**IPCC**, 1996: *Climate Change 1995: The Science of Climate Change. Contribution of Working Group I to the Second Assessment Report of the Intergovernmental Panel on Climate Change* [Houghton, J.J., L.G. Meiro Filho, B.A. Callander, N. Harris, A. Kattenberg, and K. Maskell (eds.)]. Cambridge University Press, Cambridge, United Kingdom and New York, NY, USA, 572 pp.

- Nicholls, N., G.V. Gruza, J. Jouzel, T.R. Karl, L.A. Ogallo, and D.E. Parker, Chapter 3. *Observed Climate Variability and Change,* pp. 133–192.
- Kattenberg, N., F. Giorgi, H. Grassl, G.A. Meehl, J.F.B. Mitchell, R.J. Stouffer, T. Tokioka, A.J. Weaver, and T.M.L. Wigley, Chapter 6. *Climate Models: Projections of Future Climate,* pp. 285–357.
- Warrick, R.A., C. Le Provost, M.F. Meier, J. Oerlemans, and P.L. Woodworth, Chapter 7. *Changes in Sea Level,* pp. 359–405.

**IPCC**, 1996: *Climate Change 1995: Impacts, Adaptations and Mitigation of Climate Change: Scientific-Technical Analyses. Contribution of Working Group II to the Second Assessment Report of the Intergovernmental Panel on Climate Change* [Watson, R.T., M.C. Zinyowera, and R.H. Moss (eds.)]. Cambridge University Press, Cambridge, United Kingdom and New York, NY, USA, 880 pp.

- Kirschbaum, M.U.F. and A. Fischlin, Chapter 1. *Climate Change Impacts on Forests,* pp. 95–129.
- Bijlsma, L., Chapter 9. *Coastal Zones and Small Islands,* pp. 289–324.
- Arnell, N., B. Bates, H. Lang, J.J. Magnuson, and P. Mulholland, Chapter 10. *Hydrology and Freshwater Ecology,* pp. 325–363.
- Scott, M.J., Chapter 12. *Human Settlements in a Changing Climate: Impacts and Adaptation,* pp. 399–426.
- Reilly, J., Chapter 13. *Agriculture in a Changing Climate,* pp. 427–469.
- Everett, J.T., Chapter 16. *Fisheries,* pp. 511–537.
- McMichael, A.J., Chapter 18. *Human Population Health,* pp. 561–584.

**IPCC**, 1996. *Climate Change 1995: Economic and Social Dimensions of Climate Change. Contribution of Working Group III to the Second Assessment Report of the Intergovernmental Panel on Climate Change* [Bruce, J.P., H. Lee, and E.F. Haites (eds.)]. Cambridge University Press, Cambridge, United Kingdom and New York, NY, USA, 448 pp.

- Pearce, D.W., W.R. Cline, A.N. Achanta, S. Fankhauser, R.K. Pachauri, R.S.J. Tol, and P. Vellinga, Chapter 6. *The Social Costs of Climate Change: Greenhouse Damage and the Benefits of Control,* pp. 179–224.

**Lauretti**, E. (comp.), 1992: Fish and Fishery Products. World Apparent Consumption Statistics Based on Food Balance Sheets (1961–1990). FAO Fisheries Circular 821 (Rev. 2), Food and Agriculture Organization, Rome, Italy, 477 pp.

**Leatherman**, S.P., 1994: Rising sea levels and small island states. *EcoDecision,* **11**, 53–54.

**Limpus**, C., 1993: A marine resource case study: climate change and sea level rise probable impacts on marine turtles. In: *Climate Change and Sea Level Rise in the South Pacific Region* [Hay, J.E. and C. Kaluwin (eds.)]. Proceedings of the Second SPREP Meeting, Noumea, New Caledonia, 6-10 April 1992, South Pacific Regional Environment Programme, Apia, Western Samoa, p. 157.

**Lucas**, J.S., R.D. Braley, C.M. Crawford, and W.J. Nash, 1988: Selecting optimum conditions for ocean-nursery culture of Tridacna gigas. In: *Giant Clams in Asia and the Pacific* [Copland, J.W. and J.S. Lucas (eds.)]. Australian Centre for International Agricultural Research Monograph No. 9, Canberra, pp. 129–132.

**Mahon**, R., 1996: Fisheries of small island states and their oceanographic research and information needs. In: *Small Islands: Marine Science and Sustainable Development* [Maul, G.A. (ed.)]. American Geophysical Union, Washington, DC, USA, pp. 298–322.

**Martens**, W.J.M., T.H. Jetten, J. Rotmans, and L.W. Neissen, 1995: Climate change and vector-borne diseases: a global modelling perspective. *Global Environmental Change,* **5(3)**, 195–209.

**Maul**, G.A. (ed.), 1993: *Climatic Change in the Intra-Americas Sea*. Edward Arnold, London, United Kingdom, 389 pp.

**Maul**, G.A. (ed.), 1996: Small Islands: Marine Science and Sustainable Development. American Geophysical Union, Washington, DC, USA, 467 pp.

**McLean**, R. and A.M. d'Aubert, 1993: *Implications of Climate Change and Sea Level Rise for Tokelau.* SPREP Reports and Studies Series No. 61, South Pacific Regional Environment Program, Apia, Western Samoa, 53 pp.

**McLean**, R.F. and C.D. Woodroffe, 1993: Vulnerability assessment of coral atolls: the case of Australia's Cocos (Keeling) Islands. In: *Vulnerability Assessment to Sea Level Rise and Coastal Zone Management* [McLean, R. and N. Mimura (eds.)]. Proceedings of the IPCC /WCC '93 Eastern Hemisphere Workshop, 3–6 August 1993, Tsukuba, Japan, Department of Environment, Sport and Territories, Canberra, Australia, pp. 99–108.

**McMichael**, A.J., 1993: Global environmental change and human population health: a conceptual and scientific challenge for epidemiology. *International Journal of Epidemiology,* **22**, 1–8.

**McMichael**, A.J., A. Haines, R. Slooff, and S. Kovats (eds.), 1996: Climate Change and Human Health: An Assessment Prepared by a Task Group on Behalf of the World Health Organization, the World Meteorological Organization and the United Nations Environment Programme (WHO/EHG/96.7). WHO, Geneva, Switzerland, 297 pp.

**McMichael**, A.J. and W.J.M. Martens, 1995: The health impacts of global climate change: grapling with scenarios, predictive models, and multiple uncertainties. *Ecosystem Health,* **1(1)**, 23–33.

**McNeely**, J.A., M. Gadgil, C. Leveque, C. Padoch, and K. Redford, 1993: Human influences on biodiversity. In: *Global Biodiversity Assessment* [Heyword, V.H. and R.T. Watson (eds.)]. United Nations Environmental Programme, Cambridge University Press, Cambridge, United Kingdom, pp. 711–821.

**Meehl**, G.A., 1994: Possible changes of ENSO effects due to increased $CO_2$ in global coupled-ocean-atmosphere climate models. In: *Climate Change Implications and Adaptation Strategies for the Indo-Pacific Island Nations: Workshop Proceedings* [Rappa, P., A. Tomlinson, and S. Ziegler (eds.)]. Proceedings of the Workshop at Honolulu, Hawaii, September 26–30, 1994, U.S. Country Studies Program, p. 35.

**Murray**, C., 1993: Catastrophe reinsurance crisis in the Caribbean. In: *Catastrophe Reinsurance Newsletter,* 6, 1993. DYP Group Ltd., London, United Kingdom, pp. 14–18.

**Nicholls**, N., C. Landsea, and J. Gill, 1997: Recent trends in Australian region tropical cyclone activity. Submitted to *Meteorology and Atmospheric Physics*, special edition on tropical cyclones. (In press).

**Nicholls**, R.J., 1995: Synthesis of vulnerability analysis studies. In: *Preparing to Meet the Coastal Challenges of the 21st Century, vol. 1.* Proceedings of the World Coast Conference, November 1–5, 1993, Noordwijk, Netherlands, CZM-Centre Publication No. 4, Ministry of Transport, Public Works and Water Management, The Hague, Netherlands, pp. 181–216.

**Nunn**, P.D., 1994: *Oceanic Islands.* Blackwell Publishers, Oxford, United Kingdom, 413 pp.

**Nurse**, L.A., 1985: A review of selected aspects of the geography of the Caribbean and Their implications for environmental impact assessment. In: *Proceedings of the Caribbean Seminar on Environmental Impact Assessment* [Geoghegan, T. (ed.)]. May 27–June 7, 1985, Centre for Resource Management and Environmental Studies, University of the West Indies; Caribbean Conservation Association; Institute for Resource and Environmental Studies-Dalhousie University; and the Canadian International Development Agency, Bridgetown, Barbados, pp. 10–18.

**Nurse**, L.A., 1992: Predicted sea-level rise in the Wider Caribbean: likely consequences and response options. In: *Semi-Enclosed Seas* [Fabbri, P. and G. Fierro (eds.)]. Elsevier Applied Science, Essex, United Kingdom, pp. 52–78.

**Parkinson**, R.W., R.D. de Laune, and J.R. White, 1994: Holocene sea-level rise and the fate of mangrove forests within the Wider Caribbean region. *Journal of Coastal Research*, **10**, 1077–1086.

**Perez**, A.L., C. Rodriguez, and I. Salas, 1996: *Evaluation of Risks of Coastal Flooding, Cuba*. Institute of Physical Planning, Institute of Meteorology, Havana, Cuba, 20 pp. (in Spanish)

**Pernetta**, J.C., 1988: Projected climate change and sea-level rise: a relative impact rating for the countries of the Pacific basin. In: *Potential Impacts of Greenhouse Gas Generated Climate Change and Projected Sea Level Rise on Pacific Island States of the SPREP Region* [Pernetta, J.C. (ed.)]. ASPEI Task Team, Split, Yugoslavia, pp. 1–10.

**Pernetta**, J.C., 1992: Impacts of climate change and sea-level rise on small island states: national and international responses. *Global Environmental Change*, **2**, 19–31.

**Philander**, S.G., 1990: *El Niño, La Niña and the Southern Oscillation.* Academic Press, San Diego, CA, USA, 289 pp.

**Pittock**, A.B., 1984: On the reality, stability and usefulness of Southern Hemisphere teleconnections. *Australian Meteorological Magazine*, **32**, 75-82.

**Pittock**, A.B., M.R. Dix, K.J. Hennessy, J.J. Katzfey, K.L. McInnes, S.P. O'Farrel, I.N. Smith, R. Suppiah, K.J. Walsh, P.H. Whetton and S.G. Wilson, 1995: Progress towards climate change scenarios for the southwest Pacific. *Weather and Climate,* **15(2)**, 21–46.

**Radford**, D., R. Blong, A.M. d'Aubert, I. Kuhnel, and P. Nunn, 1996: *Occurrence of Tropical Cyclones in the southwest Pacific Region 1920–1994.* Greenpeace International, Amsterdam, Netherlands, 35 pp.

**Reilly**, J., N. Hohmann, and S. Kane, 1994: Climate change and agricultural trade: who benefits and who loses? *Global Environmental Change,* **4(1)**, 24–36.

**Roy**, P. and J. Connell, 1991: Climate change and the future of atoll states. *Journal of Coastal Research,* **7**, 1057–1075.

**Salinger**, M.J., R. Basher, B. Fitzharris, J. Hay, P.D. Jones, I.P. Macveigh, and I. Schmideley-Lelu, 1995: Climate trends in the south-west Pacific. *International Journal of Climatology,* **15**, 285–302.

**Saunders**, A., 1993: Underwriting Guidelines. Paper presented at the Thirteenth Caribbean Insurance Conference, Association of British Insurers, London, United Kingdom, 7 pp. (unpublished)

**Semeniuk**, V., 1994: Predicting the effect of sea-level rise on mangroves in Northwestern Australia. *Journal of Coastal Research,* **10**, 1050–1076.

**Sestini**, G., 1992: Sea-level rise in the Mediterranean region: likely consequences and response options. In: *Semi-Enclosed Seas* [Fabbri, P. and G. Fierro (eds.)]. Elsevier Applied Science, Essex, United Kingdom, pp. 79–109.

**Shea**, E., 1994: Climate change reality: current trends and implications. In: *Climate Change Implications and Adaptation Strategies for the Indo-Pacific Island Nations. Workshop Proceedings* [Rappa, P., A. Tomlinson, and S. Ziegler (eds.)]. Proceedings of the Workshop at Honlulu, Hawaii, September 26–30, 1994, U.S. Country Studies Program, pp. 26–28.

**Singh**, U., 1994: *Potential Climate Change Impacts on the Agricultural Systems of the Small Island Nations of the Pacific.* Draft paper, IFDC-IRRI, Los Banos, Philippines, 28 pp.

**Singh**, U., D.C. Godwin, and R.J. Morrison, 1990: Modelling the impact of climate change on agricultural production in the South Pacific. In: *Global Warming-Related Effects of Agriculture and Human Health and Comfort in the South Pacific* [Hughes, P.J. and J. McGregor (eds.)]. South Pacific Regional Environmental Program and United Nations Environment Programme, University of Papua New Guinea, Port Moresby, Papua New Guinea, pp. 24–40.

**Snedaker**, S.C., 1993: Impact on mangroves. In: *Climatic Change in the Intra-Americas Sea* [Maul, G.A. (ed.)]. Edward Arnold, London, United Kingdom, pp. 282–305.

**SPREP**, 1994a: *Integrated Coastal Zone Management Programme for Western Samoa and Fiji Islands —Assessment of Coastal Vulnerability and Resilience to Sea-Level Rise and Climate Change, Case Study: Savai'i Island, Western Samoa, Phase II: Development of Methodology.* South Pacific Regional Environment Programme; Environment Agency, Government of Japan; and the Overseas Environmental Cooperation Center, Japan, Apia, Western Samoa, 23 pp.

**SPREP**, 1994b: *Integrated Coastal Zone Management Programme for Western Samoa and Fiji Islands —Assessment of Coastal Vulnerability and Resilience to Sea-Level Rise and Climate Change, Case Study: Yasawa Islands, Fiji, Phase II: Development of Methodology.* South Pacific Regional Environment Programme; Environment Agency, Government of Japan; and the Overseas Environmental Cooperation Center, Japan, Apia, Western Samoa, 118 pp.

**SPREP**, 1996: *Integrated Coastal Zone Management Programme for Fiji and Tuvalu —Coastal vulnerability and resilience in Tuvalu —Assessment of Climate Change Impacts and Adaptation, Phase IV.* South Pacific Regional Environment Programme; Environment Agency, Government of Japan; and the Overseas Environmental Cooperation Center, Japan, Apia, Western Samoa, 130 pp.

**Thompson**, C., S. Ready, and X. Zheng, 1992: *Tropical Cyclones in the Southwest Pacific: November 1979 to May 1989.* New Zealand Meteorological Service, Wellington, New Zealand, 35 pp.

**Vales**, M.A., A. Alvarez, L. Montes, A. Avila, and H. Ferras (eds.), 1996: Country Study for the Biological Diversity of the Republic of Cuba (in Spanish). UNEP-CITMA, Havanna, Cuba, 433 pp.

**Vicente**, V.P., 1996: Littoral ecological stability and economic development in small island states. In: *Small Islands: Marine Science and Sustainable Development* [Maul, G.A. (ed.)]. American Geophysical Union, Washington, DC, USA, pp. 266–283.

**Weber**, A.P., 1993: *Abandoned Seas: Reversing the Decline of the Oceans.* World Watch Paper 116, World Watch Institute, Washington, DC, USA, 66 pp.

**Weber**, A.P., 1994: *Net Loss: Fish, Jobs and the Marine Environment.* World Watch Paper 120, World Watch Institute, Washington, DC, USA, 76 pp.

**Whetton**, P.H., M.H. England, S.P. O'Farrell, I.G. Walterson, and A.B. Pittock, 1996: Global Comparison of the regional rainfall results of enhanced greenhouse coupled and mixed layer ocean experiments: Implications for climate scenario development. *Climatic Change*, **33**, pp. 497-519.

**WHO**, 1996: *Climate Change and Human Health* [McMichael, A.J., A. Haines, R. Slooff, and S. Kovats (eds.)]. WHO, Geneva, Switzerland, 279 pp.

**Wigley**, T.M.L. and B.D. Santer, 1993: Future climate of the Gulf/Caribbean Basin from the global circulation models. In: *Climatic Change in the Intra-Americas Sea* [Maul, G.A. (ed.)]. Edward Arnold, London, United Kingdom, pp. 31–54.

**Wilkinson**, C.R. and R.W. Buddemeier, 1994: *Global Climate Change and Coral Reefs: Implications for People and Reefs.* Report of the UNEP-IOC-ASPEI-IUCN Global Task Team on the Implications of Climate Change on Coral Reefs. International Union for Conservation of Nature and Natural Resources (IUCN), Gland, Switzerland, 124 pp.

**Woodroffe**, C.D., 1995: Response of tide-dominated mangrove shorelines in Northern Australia to anticipated sea-level rise. *Earth Surface Processes and Landforms,* **20**, 65–86.

**Woodroffe**, C.D., and R.F. McLean, 1992: Kiribati Vulnerability to Accelerated Sea-Level Rise: A Preliminary Study. Department of the Arts, Sport, Environment and Territories, Canberra, Australia, 82 pp.

**Yamada**, K., P.D. Nunn, N. Mimura, S. Machida and M. Yamamoto, 1995: Methodology for the Assessment of Vulnerability of South Pacific Island Countries to Sea-Level Rise and Climate Change. *Journal of Global Environmental Engineering*, **1**, pp. 101-125.

# 10

# Temperate Asia

MASATOSHI YOSHINO (JAPAN) AND SU JILAN (CHINA)

Lead Authors:
*Byong-Lyol Lee (Korea), H. Harasawa (Japan), K.I. Kobak (Russia), Lin Erda (China), Liu Chunzen (China), N. Mimura (Japan)*

Contributors:
*Ding Yihui (China), M. Lal (India)*

# CONTENTS

# EXECUTIVE SUMMARY

## Introduction

Temperate Asia is composed of three regions: monsoon Asia, excluding its tropical subregion; the inner arid/semi-arid regions; and Siberia. The region includes countries in Asia between 18°N and the Arctic Circle, including the Japanese islands, the Korean peninsula, Mongolia, most parts of China, and Siberia in Russia. The east-west distance of the area is about 8,000 km, and its north-south extent is about 5,000 km. The largest plateau in the world—the Qing-Zang plateau (Tibetan plateau), with an average elevation of more than 4,000 m—is located in southwest China. Inner Siberia, with a mean monthly temperature in January of about -50°C, is the coldest part of the northern hemisphere in winter; this area is called the "cold pole." On the other hand, extremely dry, hot climate prevails in the Taklamakan Desert in China.

Human activities through the ages have brought profound changes to the landscape of this area. Except for boreal forests in Siberia, other natural forests in the region have long been destroyed. Broad plains have been cultivated and irrigated for thousands of years, and natural grasslands have been used for animal husbandry. In recent years, many countries in this region have shown marked economic development. Their gross national products (GNPs) and populations are increasing at an extremely rapid rate. The region's population is expected to grow from 1.42 billion in 1995 to 1.72 billion by 2025; the environment in this region already is under great stress. The impacts of expected climate change may exacerbate existing environmental problems.

## Climate Characteristics

Climate differs widely within Temperate Asia. It has a tropical monsoon climate in the far south; a humid, cool, temperate climate in the north; and a desert climate or steppe climate in the west and northwest. In the rest of the area—where most of the population of the region is concentrated—a humid, temperate climate prevails.

The east Asian monsoon has great influence on temporal and spatial variations in the hydrological cycle over parts of the region. For example, the summer monsoon accounts for 70% of the total annual runoff in China; for northern China, this precipitation often concentrates in a few storms during the flood season. At time scales longer than 100 years, the summer monsoons generally are stronger during (globally) warmer periods, leading to wetter conditions in northern China. On the other hand, drier conditions prevail over most of the monsoon-affected area during (globally) colder periods.

Tropical cyclones (typhoons) frequent the coastal regions. They are important not only because they cause disasters along the coasts but also because they are beneficial carriers of water resources to inland areas. The frequency, path, and intensity of typhoons vary greatly from year to year, with clear differences between El Niño-Southern Oscillation (ENSO) and non-ENSO years.

Over the past century, the average annual temperature in Temperate Asia has increased by more than 1°C. This increase has been most evident since the 1970s; seasonally, the warming is evident mainly in winter. Subregionally, over the past 100 years, there has been a 2–4°C temperature increase in eastern and northeastern Temperate Asia and a 1–2°C temperature decrease in some parts of southeasten China, except for the coastal area. These trends also are reflected in corresponding seasonal temperature distributions, except that summer temperatures in central Siberia actually are decreasing. In terms of precipitation, large decadal variability seems to have masked a smaller positive trend in annual precipitation. Over the past 100 years, there has been a 20–50% increase in east Siberia; a 10–20% increase across the Korean peninsula, northeast China, the Huaihe River Basin, and the Yellow River Basin; and a 10–20% decrease in Japan and the southern half of east China, including Taiwan. The increasing trend in annual precipitation in northeast China is evident mainly in spring and summer rainfall; in south China, annual winter precipitation shows a slight positive trend, in contrast to a negative trend in total annual precipitation.

## Future Climate Projection

Numerical experiments with coupled atmosphere-ocean general circulation models (AOGCMs) that include transient increases in greenhouse gases [i.e., increases in equivalent carbon dioxide ($CO_2$) at a rate of 1% per year] project a warming of between 2°C and 3°C over the annual mean of the region at the time of $CO_2$ doubling. Warming was more pronounced in the arid/semi-arid and Siberian regions than in the coastal monsoon region. Recent simulations, wherein GCMs also include the offsetting effects of sulfate aerosols, project a temperature rise of about 0.8°C over the eastern part of the region, about 1°C over most parts of eastern China, and close to 2°C in the Siberian region on an annual mean basis by the middle of the next century. It should be noted that the projections that account for sulfate aerosols also are highly uncertain.

In equilibrium and transient-response numerical experiments with GCMs, precipitation is projected to marginally increase (<0.5 mm/day) at the time of $CO_2$ doubling during the winter (DJF) throughout the region. In the summer (JJA), the spatial pattern of projected changes in precipitation is not uniform over the region. Model projections suggest that precipitation will increase slightly (0.5–1.0 mm/day) in the northern part of the region (Siberia), and by more than 1 mm/day over the Korean peninsula, the Japanese islands, and the southwestern part of China. In contrast, precipitation may decline in the northern, western, and southern parts of China. The projected decline in rainfall over most of China is substantial in numerical experiments that include the effects of sulfate aerosols.

### Summary Points

- For boreal forests, which are concentrated mainly in the Russian Federation, models suggest large shifts in distribution (e.g., area reductions of up to 50%) and productivity. In the boreal region, grasslands and shrublands may expand significantly, whereas the tundra zone may decrease by up to 50%, according to model predictions. Climatic warming would increase the release of methane from deep peat deposits—particularly from tundra soils, since they would become wetter. The release of $CO_2$ is expected to increase, though not by more than 25% of its present level.

- It is likely that by 2050, up to a quarter of the existing mountain glacier mass would disappear. For areas with very large glaciers, the extra runoff may persist for a century or more. By 2050, the volume of runoff from glaciers in central Asia is projected to increase threefold. Eventually, however, glacial runoff will taper off or even cease. Projected future glacier runoff is about 68 $km^3$/yr in 2100, compared with the present value of 98 $km^3$/yr. Permafrost in northeast China is expected to disappear if the temperature were to increase by 2°C or more. Over the Qing-Zang plateau, estimates of the impact of climate change on permafrost range from its complete disappearance as a result of a temperature increase of 2°C to a raising of its elevation limit to 4,600 m as a result of a warming of 3°C.

- Hydrologically, model results suggest that the areas most vulnerable to climate change would be in the northern part of China. Projected changes in runoff are due mainly to changes in precipitation in spring, summer, and autumn because of the strong influence of the monsoon climate. The most critical uncertainty is the lack of credible projections of the effect of climate change on either the Asian monsoon or the ENSO phenomenon, both of which strongly influence river runoff. In moderately and extremely dry years, the projected potential water deficiency caused by climate change—although less than that caused by population growth and economic development—may exacerbate seriously the existing water shortage.

- Different GCMs' estimated impacts on agricultural yields vary widely in range. Possible large negative impacts on rice production as a consequence of climate change would be of concern in the face of expected population increases. In China, across different scenarios and different sites, yield changes for several crops by 2050 are projected to be -78% to +15% for rice, -21% to +55% for wheat, and -19% to +5% for maize. An increase in productivity may occur if the positive effects of $CO_2$ on crop growth are considered, but the magnitude of the fertilization effect remains uncertain.

- Climate change is projected to have favorable impacts on agriculture in the northern areas of Siberia and to cause a general northward shift of crop zones. Grain production in southwestern Siberia is projected to fall by about 20% as a result of a more arid climate.

- Projections for 2010 indicate that the need for industrial roundwood could increase by 38% (southern Temperate Asia) to 96% (eastern Temperate Asia). These requirements may result in serious shortages of boreal industrial roundwood, placing further stress on boreal forests.

- Along most of the continental coast, relative sea level (i.e., sea level in relation to land) is an important factor for coastal environments. Deltaic coasts in China face severe problems from relative sea-level rise as a result of tectonically and anthropogenically induced land subsidence. In the next 50 years, the expected worldwide sea-level rise due to climate change will not be a major factor in relative sea-level rise for the Old Huanghe and Changjiang deltas in China, although it may be for the Zhujiang delta. Sea-level rise due to global warming will, however, exacerbate problems in all three deltas, along with saltwater intrusion problems in deltaic regions and coastal plains. China, however, has a long history in defenses against sea encroachment.

- Tokyo, Osaka, and Nagoya are all located in the coastal zone; together, they account for more than 50% of Japan's industrial production. In these metropolitan areas, already about 860 $km^2$ of coastal land—an area supporting 2 million people and with physical assets worth $450 billion—are below mean high-water level. With a 1-m rise in mean sea level, the area below mean high-water level would expand by a factor of 2.7, embracing 4.1 million people and assets worth more than $900 billion. The same sea-level rise would expand the flood-prone area from 6,270 $km^2$ to 8,900 $km^2$. The cost of adjusting existing protection measures has been estimated at about $80 billion.

- One of the potential threats that sea-level rise poses is exacerbated beach erosion. Sandy beaches occupy 20–25% of the total length of the Japanese coast. About 120 $km^2$ of these beaches have been eroded over the past 70 years. An additional 118 $km^2$ of beaches—57% of the remaining sandy beaches in Japan today—would disappear with a 30-cm sea-level rise. This percentage would increase to 82% and 90% if the sea level rose by 65 cm or 100 cm, respectively.

- An increase in the frequency or severity of heat waves would cause an increase in (predominantly cardiorespiratory) mortality and illness. Studies of urban populations in Temperate Asia indicate that the number of heat-related deaths would increase several-fold in response to modeled climate change scenarios for 2050.

- North China—including Beijing, Tianjin, the four provinces (Hebei, Henan, Shandong, Shanxi), part of Anhui province, and part of inner Mongolia—is an economic center of the country. It also is a topographic and climatological entity. Because this region already is at risk from normal climate variability, it also is likely to be quite vulnerable to long-term secular shifts.
  1) *Water resources:* Water resources in north China are vulnerable to climate change because of already low levels of available per capita water supplies, water projects that already are highly developed, large changes in river runoff related mainly to variability in flood season, and rapid economic development. Water resources also are sensitive to climate change because of the critical dependence of floods on the Asian monsoon and the ENSO phenomenon.
  2) *Agriculture:* This region appears to be especially sensitive to climate change because of potential increases in the soil moisture deficit. Warming and increased evapotranspiration, along with possible declines in precipitation, would make it difficult to maintain the current crop pattern in areas along the Great Wall and would limit the present practice of cultivating two crops in succession in the Huang-Hai Plains. Although climate warming may cause northward shifts of subtropical crop areas, frequent waterlogging in the south and spring droughts in the north would inhibit the growth of subtropical crops.
  3) *Forests:* Demands for agriculture as a result of population increases and changes in the characteristics of arable lands due to climate change will likely result in large reductions of forest area.
  4) *Coastal zones:* Climate change will exacerbate the already serious problem of relative sea-level rise because of tectonic subsidence and heavy groundwater withdrawal. Defense against sea encroachment would be the only viable response because of the high concentration of population and economic activities. Contamination of groundwater by seawater intrusion would further worsen the water resource shortage problem.

- *Research needs:* (i) Projection of regional climatic scenarios with high spatial and temporal resolution; (ii) improved hydrological models with appropriate land-surface parameters under nonstationary climatic conditions; (iii) multiple-stress impact studies on water resources in international river basins; (iv) implementation strategies for integrated coastal zone management, taking climate change into consideration; (v) studies of the interactive effects of stresses on human health from hot weather and high levels of air pollution; and (vi) integrated impact studies that consider different sectors and their response to adaptation strategies.

- *Conclusion:* The major impacts of global warming on Temperate Asia will likely take the form of large shifts of the boreal forests, the disappearance of significant portions of mountain glaciers, and shortages in the water supply. The most critical uncertainty in these estimates stems from the lack of credible projections of the hydrological cycle under global warming scenarios. The effects of climate change on the Asian monsoon and the ENSO phenomenon are among the major uncertainties in the modeling of the hydrological cycle. Projections of agricultural crop yields are uncertain not only because of the uncertainty in the hydrological cycle but also because of the potential positive effects of $CO_2$ and production practices. In the coastal zones, sea-level rise endangers sandy beaches but remains an anthropogenically induced problem in deltaic areas. Integrated impact studies that consider multiple stress factors are needed.

## 10.1. Regional Characteristics

### 10.1.1. Geography and Demography

This report covers countries in Temperate Asia between 18°N and the Arctic Circle, including the Japanese islands, the Korean peninsula, Mongolia, most parts of China, and Siberia in Russia (see Figure 10-1). Geographically, the region is located on the northeastern part of the Eurasian continent—the world's largest continent—and borders the Pacific, the world's largest ocean. The east-west distance of the area is about 8,000 km, and its north-south extent is about 5,000 km. The world's largest plateau—the Qing-Zang plateau (Tibetan plateau), with an average elevation of more than 4,000 m—is located in southwest China. Mt. Qomolangma (formerly Mt. Everest), the highest peak in the world (8,848 m), lies near the southern border of the plateau. The lowest point of the region (-154 m) is found in the Turfen Depression, northeast of the Taklamakan Desert.

Temperate Asia is composed of three regions: so-called monsoon Asia, excluding its tropical subregion; the inner arid/semi-arid regions; and Siberia (West Siberia, East Siberia, and Far East), which is covered largely by boreal forests and tundra. Tropical cyclones (typhoons) frequent the coastal regions. Inner Siberia, with a mean monthly temperature in January of about -50°C, is the coldest part of the northern hemisphere in winter; this area is called the "cold pole." On the other hand, extremely dry, hot climate prevails in the Taklamakan Desert.

Human activities through the ages have brought profound changes to the landscape of the region: Except for the boreal forests in Siberia, for example, natural forests in the region have long been destroyed. Only in mountain areas do secondary forests remain (Numata, 1974). Broad plains have been cultivated and irrigated for thousands of years, and natural grasslands have been used for animal husbandry.

The present population and projected population in 2025 for each country or region in Temperate Asia, except Siberia and Taiwan, are listed in Table 10-1, based on statistics from United Nations Population Division (1993) and FAO (1995). The region's population is expected to grow from 1.42 billion in 1995 to 1.72 billion by 2025. Table 10-1 also shows the current land use in each country or region.

In recent years, many countries in this region have shown marked economic development. Their GNPs and populations are increasing at an extremely fast rate. Thus, the environment

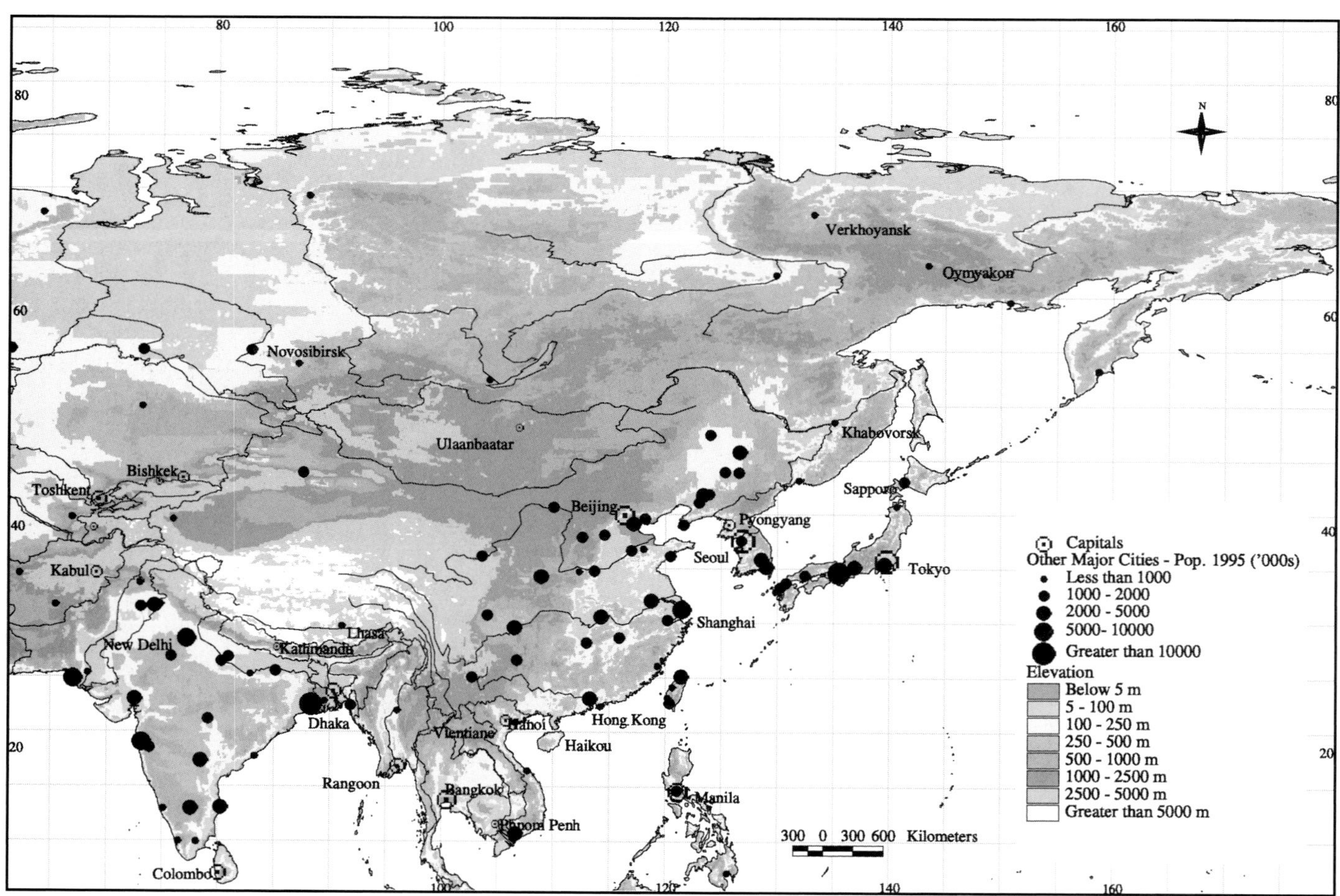


**Figure 10-1:** The Temperate Asia region [compiled by the World Bank Environment Department Geographic Information System (GIS) Unit; see Annex E for a color rendition].

**Box 10-1. Temperate Asia (Country/Region)**

China
Democratic People's Republic of Korea
Japan
Mongolia
Republic of Korea
Siberia (West Siberia, East Siberia, and Far East)
Taiwan, China

in this region already is under great stress. The impacts of expected climate change may exacerbate existing environmental problems.

### *10.1.2. Regional Climate Information*

The Tibetan plateau, rising to the 600-hPa level of the troposphere, strongly influences the atmospheric general circulation over the region, both thermally and dynamically. Development of polar frontal zones and cyclones in Temperate Asia is closely connected to activities associated with the westerly jet stream and the east Asian monsoon, both of which are significantly affected by the plateau (Kim *et al.*, 1973; Fukui, 1977; Yeh and Gao, 1979).

Climate differs widely within Temperate Asia. The region has a tropical monsoon climate in the far south, a humid, cool, temperate climate in the north, and a desert climate or steppe climate in the west and the northwest. In the rest of the area—where most of the population of the region is concentrated—a humid, temperate climate prevails. Mean monthly and annual temperatures and mean annual precipitation at selected sites are shown in Table 10-2.

China has kept proxy data, old documents, diaries, and other data throughout its long history. Likewise, Korea and Japan have kept relatively homogeneous proxy data for several hundred years. Such data are useful for analyzing climatic changes and their impacts on human activities.

#### *10.1.2.1. Past and Present Climate Characteristics*

Permafrost distribution in this region has changed substantially during warming periods of the Quaternary Period. The southern limit of lowland permafrost moved northward at a rate of 60 km/°C. In the alpine areas of Tibet, the boundary of lower-elevation permafrost changed by 160 m/°C; on the northern slope of the Himalayas, this sensitivity is about 80 m/°C (IPCC 1996, WG II, Section 7.2.3).

The length of many mountain glaciers in Pamirs, Tian Shan, and Altay has decreased by up to 4 km during the past two centuries. Fluctuations of 224 glaciers in central Asia from the 1950s to the 1980s can be summarized as retreating (73%), advancing (15%), and stable (12%). The mean equilibrium-line altitude—at which snow accumulation is equal to snow ablation for a glacier—is estimated as being 50–80 m lower today than during the Little Ice Age in the 18th century and the first half of the 19th century (IPCC 1996, WG II, Box 5-1 and Section 7.2.2).

Over the past century, the average annual temperature in Temperate Asia has increased by more than 1°C (Figure 10-2). This increase has been most evident since the 1970s. It is a reflection primarily of the warming of the winter season since that time, although temperatures in all the other seasons also show a slight increase (see Annex A). Decadal time scale variability is evident from long-term variations of annual and seasonal temperatures. In terms of precipitation, large decadal variability seems to have masked a smaller positive trend (see Figure 10-2 and Annex A).

Subregionally, over the past 100 years, there has been a 2–4°C temperature increase in eastern and northeastern Temperate Asia and a 1–2°C temperature decrease in the eastern half of south China, except for the coastal area. These trends also are reflected in corresponding seasonal temperature distributions, except that the summer temperature in central Siberia exhibits a negative trend. Over the same period, there has been a 20–50% increase in annual precipitation in east Siberia; a

***Table 10-1:*** *Population (1995 and 2025) and land use (1993) in Temperate Asia.*

| Country or Region[1] | Land Area (Mha) 1993 | Population (000s) 1995 | Population (000s) 2025 | Urban Population (000s) 1995 | Arable and Cropland (Mha) 1993 | Pastureland (Mha) 1993 | Forest and Woodland (Mha) 1993 |
|---|---|---|---|---|---|---|---|
| China | 932,641 | 1,221,462 | 14,96,722 | 369,736 | 95,975 | 400,000 | 130,496 |
| Hong Kong | 99 | 5,865 | 6,306 | 5,574 | 7 | 1 | 22 |
| Japan | 37,652 | 125,095 | 124,249 | 97,120 | 4,463 | 661 | 25,100 |
| Korea, Dem People's Rep | 12,041 | 23,917 | 33,522 | 14,650 | 2,000 | 50 | 7,370 |
| Korea, Rep | 9,873 | 44,995 | 52,946 | 36,572 | 2,055 | 90 | 6,460 |
| Mongolia | 156,650 | 2,410 | 4,421 | 1,468 | 1,401 | 125,000 | 13,750 |

Source: UN Population Division, FAO FAOSTAT.
[1]Data not available for Siberia and Taiwan.

***Table 10-2:*** *Temperature and precipitation at selected sites in Temperate Asia.*

| | Latitude (N) | Longitude (E) | Height (m) | Temperature (°C) January | Temperature (°C) July | Temperature (°C) Annual | Annual Precipitation (mm) |
|---|---|---|---|---|---|---|---|
| Verkhoyansk | 67°33' | 133°23' | 137 | -48.9 | 15.3 | -15.6 | 155 |
| Oymyakon | 63°16' | 143°09' | 740 | -50.1 | 14.5 | -16.5 | 193 |
| Novosibirsk | 44°12' | 37°48' | 37 | 2.5 | 23.6 | 12.7 | 638 |
| Khabarovsk | 48°31' | 135°07' | 86 | -22.7 | 21.0 | 1.2 | 569 |
| Hotan | 37°08' | 79°56' | 1,375 | -5.7 | 25.5 | 12.1 | 35 |
| Beijing | 39°48' | 116°28' | 51 | -4.7 | 26.0 | 11.6 | 683 |
| Shanghai | 31°10' | 121°26' | 5 | 4.2 | 27.8 | 16.1 | 1,135 |
| Haikou | 20°02' | 110°21' | 14 | 17.1 | 28.4 | 23.8 | 1,690 |
| Seoul | 37°03' | 126°58' | 86 | -4.9 | 24.5 | 11.2 | 1,259 |
| Sapporo | 43°03' | 141°20' | 17 | -4.9 | 20.2 | 8.0 | 1,158 |
| Tokyo | 35°41' | 139°46' | 5 | 4.7 | 25.2 | 15.3 | 1,460 |
| Lhasa | 29°42' | 91°08' | 3,658 | -2.3 | 14.9 | 7.5 | 454 |

10–20% increase across the Korean peninsula, northeast China, the Huaihe River Basin, and the Yellow River Basin; and a 10–20% decrease in Japan and the southern half of east China, including Taiwan. The increasing trend of annual precipitation in northeast China is manifested mainly in spring and summer rainfalls, whereas in south China the winter precipitation shows a certain degree of positive trend, in contrast to the trend in annual precipitation (see Annex A).

The temperature trend in east China (east of 105°E) differs from the global warming tendency. The peak warming period occurred in the 1940s. Since then, there has been a general cooling trend in this area (particularly in the southwestern area)—in contrast to the linear warming trend in Mongolia and the northern part of China. Similarly, the mean temperature in south China has decreased by about 0.8°C from the 1950s to the 1980s (Ding and Dai, 1994; Yatagai and Yasunari, 1994).

Rainfall over Mongolia concentrates mainly in the summer. Annual precipitation in this country is 100–400 mm over the steppe and less than 150 mm in the southern Gobi. In the Gobi area, summer rainfall decreased over the period 1970–1990; in particular, the number of days with relatively heavy rainfall (greater than 3 mm/day) dropped significantly (Mijiddorj *et al.*, 1992).

The east Asian monsoon greatly influences temporal and spatial variations in the hydrological cycle over many parts of the region. For example, the summer monsoon accounts for 70% of the total annual runoff in China; for northern China, it often concentrates in a few storms during the flood season. The characteristics of the hydrological cycle are significantly different north and south of the Huaihe River. South of the river, the ratio of evaporation to precipitation is 0.51, whereas to the north it is 0.75. The evaporation-to-precipitation ratio is higher than 0.8 for the Yellow River and the Hailuan River (DH-MWR, 1987).

At a time scale longer than 100 years, summer monsoons generally are stronger during (globally) warmer periods, leading to wetter conditions in northern China. On the other hand, drier conditions prevail over most of the monsoon-affected area during (globally) colder periods. At the 10–100 year time scale, however, such a global-regional relationship is not obvious (Yan and Marienicole, 1995).

Tropical cyclones (typhoons) are important not only because they cause disasters along the coasts south of 40°N but also because they are beneficial carriers of water resources to inland areas. The frequency, path, and intensity of typhoons vary greatly from year to year, with clear differences between ENSO and non-ENSO years (Nishimori and Yoshino, 1990). It is difficult to extrapolate impacts on ENSO behavior under global warming, however.

Because of the rising population, rapid industrialization, and increasing level of air pollution, heat-island effects are clearly evident in recent urban climate records of this region. The difference between average city temperatures during two periods—1961–90 and 1931–60—correlates closely with the urban population. For cities in Japan with a population of 6–7 million, the temperature difference is about 0.8°C; for Beijing, it is only about 0.4°C (Japan Meteorological Agency, 1996). Cities located in the northern part of South Korea show significantly higher warming trends than those in the southern part, especially in the anomalies of monthly mean minimum temperature. Thus, even if the global warming rate is lowered in the future, warming trends of the urban climate due to heat-island effects should be considered.

#### *10.1.2.2. Future Climate Projections*

Global climate models project that the mean annual surface temperature will rise by about 1.0–3.5°C by the year 2100. On regional scales, confidence in future climate projections remains low. There is more confidence in temperature projections than in precipitation changes. The degree to which regional climate variability will change also remains uncertain. Projections of future changes in temperature and rainfall for the Temperate Asia region, based on available scenarios from various GCM experiments, are described below.

*10.1.2.2.1. Temperature*

GCM equilibrium response experiments suggested an annual mean warming of 2–5°C over the region as a result of a doubling of $CO_2$ (experiment acronyms A3, F1, G1, and H2 in Table 1-1 of the Introduction). Warming was projected to be more pronounced during winter than in summer. Subsequent numerical experiments with coupled AOGCMs, which included transient increases in greenhouse gases (i.e., increases in equivalent $CO_2$ at a rate of 1% per year), projected a slightly lower degree of warming over the region of between 2°C and 3°C at the time of $CO_2$ doubling on an annual mean basis (experiment acronyms X2, X5, and X7 in Table 1-2 of the Introduction). Warming was more pronounced in arid/semi-arid and Siberian regions than in the coastal monsoon region. Recent simulation experiments (experiment acronyms X6 and X8 in Table 1-2 of the

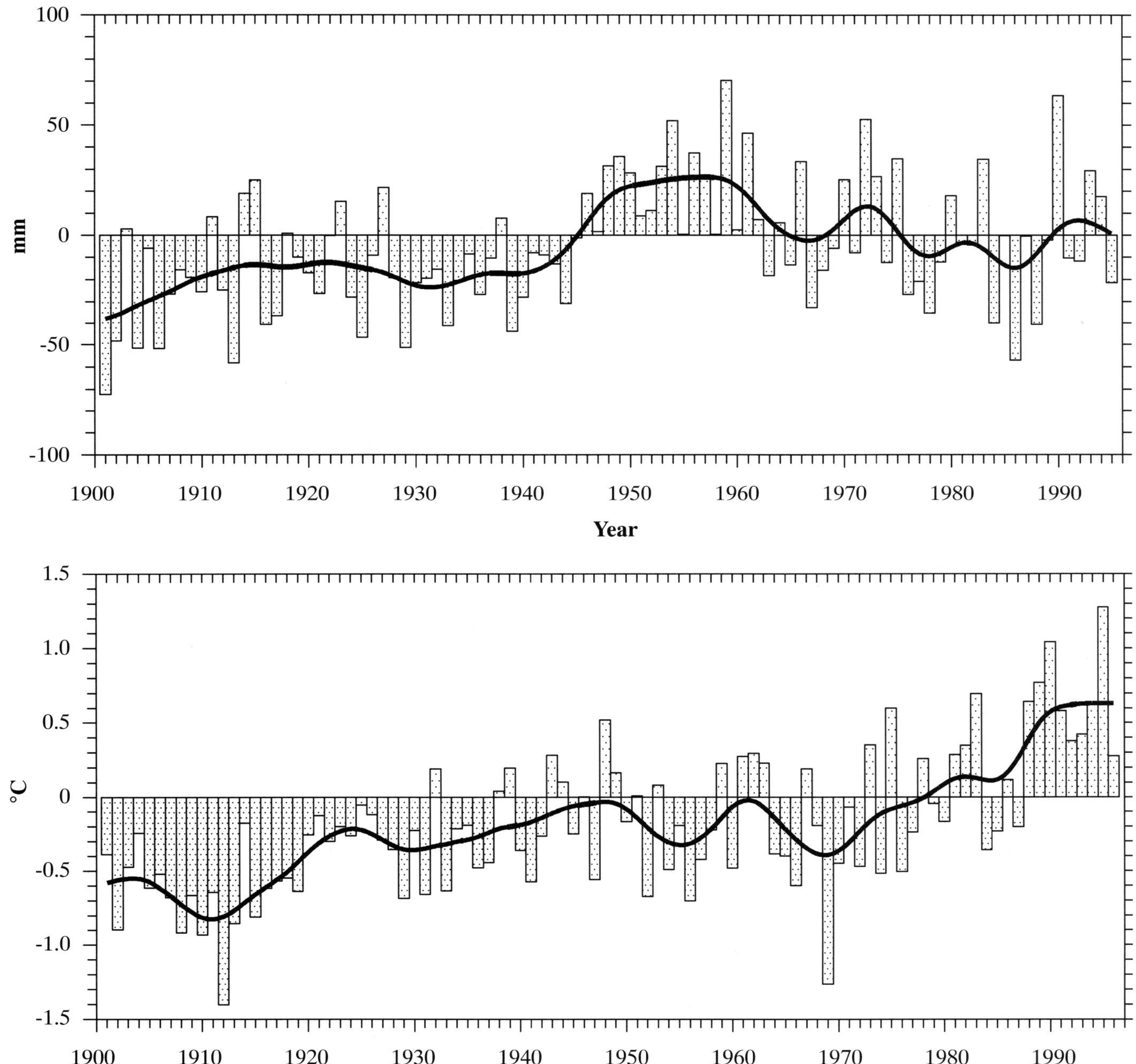


**Figure 10-2**: Trends in average precipitation (top) and temperature (bottom) in the Temperate Asia region, 1901–96 (see Annex A).

Introduction), wherein GCMs also include the offsetting effects of sulfate aerosols, project a temperature rise of about 0.8°C over the eastern part of the region, about 1°C over most parts of eastern China, and close to 2°C in the Siberian region on an annual mean basis by the middle of the next century. During winter, the projected changes are marginally higher for all of the regions; during summer, however, the projected warming is considerably less, and most parts of China show a cooling. It should be noted that the projections that account for sulfate aerosols also are highly uncertain because our current understanding of the indirect effects of sulfate aerosols is poor (IPCC 1996, WG I, Section 6.2.2).

*10.1.2.2.2. Precipitation*

In equilibrium and transient-response numerical experiments with GCMs, precipitation is projected to increase marginally (<0.5 mm/day) at the time of doubling of $CO_2$ during the winter (DJF) throughout the region. In summer (JJA), the spatial pattern of projected changes in precipitation is not uniform over the region. Model projections suggest that precipitation will increase slightly (0.5–1.0 mm/day) in the northern part of the region (Siberia) and by more than 1 mm/day over the Korean peninsula, the Japanese islands, and the southwestern part of China. In contrast, precipitation changes show a decline in the northern, western, and southern parts of China (IPCC WG I, Section 6.2.2). The projected decline in rainfall over most of China is substantial in numerical experiments that include the effects of sulfate aerosols.

*10.1.2.2.3. Possible changes in sea ice over the Sea of Okhotsk*

Tokioka et al. (1995) studied possible changes in sea ice over the Sea of Okhotsk based on future climate projections using a coupled GCM (MRI-X3). Sea ice in the Sea of Okhtosk is most vulnerable to climate, and it influences climate, ecosystems, fisheries, and transportation around the area. Because this area is the southernmost ocean in the Northern Hemisphere where sea ice forms, a sea ice-albedo feedback may occur early and effectively. These results suggest an early disappearance of sea ice and thus a large sea-surface temperature (SST) rise over the Sea of Okhotsk. Therefore, it is plausible that this area will suffer large impacts of climate change in coming decades.

## 10.2. Sensitivity, Adaptability, and Vulnerability

### *10.2.1. Ecological Systems (including Cryosphere)*

*10.2.1.1. Forests, Grasslands, and Tundra*

Although the area of potential distribution of temperate forests in Temperate Asia is, to a large extent, cleared and used for intensive agriculture, global warming can be considered sufficient to trigger structural changes in remaining temperate forests. The nature and magnitude of these changes depend on associated changes in water availability, as well as in water-use efficiency. Shifts in temperature and precipitation in temperate rangelands may result in altered growing seasons and boundary shifts among grasslands, forests, and shrublands (IPCC 1996, WGII, Sections 1.5 and 2.6).

According to simulation results of potential vegetation change by Neilson (see Annex C), temperate mixed forest areas, temperate evergreen forest, and grassland areas would expand in Temperate Asia under $2xCO_2$ equilibrium scenario conditions. However, there would be practically no change in the leaf area index (LAI) in almost all parts of the region (see Figures C-2 and C-4 of Annex C). Uncertainty remains about the time lag between climate change and migration of forest species at high latitudes.

For boreal forests, which are mainly concentrated in the Russian Federation, climate models (UKMO-H3, GFDL-A2) suggest large shifts in distribution (area reductions of up to 50%) and productivity (Dixon *et al.*, 1996). All components of boreal forest ecosystems would be affected, including water resources, soil systems, and wildlife, and the combined effect could be even stronger as a result of interacting factors.

Grasslands and shrublands in boreal regions may expand significantly, whereas the tundra zone may decrease by up to 50%, according to model projections. Climatic warming also would increase the release of methane from deep peat deposits, particularly from tundra soils, because they would become wetter. It is expected that the release of $CO_2$ would increase, though not by more than 25% of its present level (IPCC 1996, WG II, Section 7.4.2).

The most widely distributed coniferous forests in Siberia are the larch forests: West of the Yenissei River, *Larix russica* predominates; to the east, *Larix gmelini* prevails. The latter grows in the north of eastern Siberia, where the annual temperature range reaches about 100°C (-64°C to +38°C), as shown by mean long-term meteorological data from 1937 to the present from Yakutsk weather station. *Larix gmelini* has a specialized root system: Its apex central root dies off at the permafrost border, and a root system develops in the upper soil layers. The larch is vulnerable to damage by fires and insects, which may occur more frequently under global warming. Increased steppe area also may be expected in the southern part of eastern Siberia (Kobak *et al.*, 1996).

The biomass densities of larch (*Larix sibirica*), scotch pine (*Pinus silvestris*), Siberian pine (*Pinus sibirica*), and birch (*Betula platyphylla*) are projected to decrease by 27.7, 4.3, 28.5, and 2.6 t/ha, respectively, under a $2xCO_2$ equilibrium scenario (Ulziisaikhan, 1996). Such decreases seem to be caused by warming air temperature and reduced rainfall during the summer season. Gobi and steppe areas in Mongolia would therefore expand.

The numbers of all known and endemic species of mammals, birds, and flowering plants in Temperate Asia are shown in

Table 10-3. Existing factors causing the loss of biodiversity include natural factors, such as storms and floods; population pressure; soil erosion and desertification; deforestation and timber cutting; and overcollection of wild plant resources (Davis *et al.*, 1995; Heywood, 1995). Although global warming will play an important role in future changes in biodiversity, human influences also may have the potential to affect the climate system (Mooney et al., 1995). Such influences have taken the form of rapid, large, and frequent changes in land and resource use; increased frequency of biotic invasions; reductions in species numbers; creation of novel stresses; and the potential for change in the climate system.

Climate change may affect the biodiversity in boreal forests of Temperate Asia through a myriad of processes and effects: local mortality of boreal species and replacement by northern hardwoods or prairies, depending on locale and soil type; migration of boreal species northward and coastward, also depending on locale and soil type; increased probability of fire; increased or decreased soil nutrient availability, depending on permafrost, soil water-holding capability, and locale; increased emissions of greenhouse gases—particularly methane—from wetlands; and increased probability of outbreaks of pests, particularly insects, to drought-stressed trees (Pastor *et al.*, 1996).

Changes in local vegetation zones as a result of global warming could be important in marginal areas of the region. For example, in the Kamtchatka Peninsula, five ecological zones may be classified (Kojima, 1992): the *Larix Kamtchatka* zone, the *Picea ajanensis* zone, the *Betula ermanii* zone, the *Pinus pumila* zone, and the alpine tundra zone. These zones coincide with the division by continentality (or oceanity) defined by types of seasonal change, annual range, and other characteristics of air temperature. The first three zones are boreal forest, under three different climate types: subhumid continental, humid continental, and superhumid maritime, respectively. The fourth zone is subalpine, corresponding to a humid maritime climate zone. Under global warming, the most vulnerable region would be the *Larix Kamtchatka* zone because of the increasing aridity of soils or increasing frequency of forest fires. Major alterations in vegetation could be expected, especially in the mountains of the northern boreal subzone and the subarctic forest-tundra ecotone in northeast Siberia. In the middle and southern boreal subzones, vegetation changes may be more limited because of more-resistant species interaction in the forest communities of the continental area and the isolated islands of the Kuril, Shantar, and Kommander groups (Grishin, 1995).

Plant phenology is an important indicator of vegetation change. It also is quite useful and necessary for farmers of intensively cultivated small-hold rice fields in this region because their cultivation calendar depends on the year-to-year change of seasons, which is reflected in local plant phenology. For example, the flowering of cherry blossoms in spring begins about 3–4 days earlier per 1°C increase in the mean temperature in March (Yoshino and Ono, 1996). Leaf-color-change dates of *Ginko biloba* and *Acer palmatum* in autumn are delayed for 2–7 days per 1°C increase in monthly mean temperature (Kai *et al.*, 1996).

### *10.2.1.2. Deserts and Desertification*

Cold deserts and semi-deserts are widespread in west-central Temperate Asia. With projected shifts to warmer and drier conditions in Mongolia, the Gobi would change to warm temperate desert scrub, and the area of cool temperate desert scrub would expand to the Khangai mountains, supplanting forest areas (Ulziisaikhan, 1996). At the same time, low- and midland parts of central Asia are likely to change gradually into a more arid interior desert as the disappearance of mountain glaciers reduces the volume of summer runoff. Projection of potential changes in the total extent of cool semi-deserts varies among models: Whereas BIOME 1.1 suggests an almost 50% reduction in its extent, IMAGE 2.0 projects no change. However, desertification arises from adverse climatic conditions and from human abuse of the land; the latter can accelerate the desertification process (IPCC 1996, WG II, Section 2.6).

Evapotranspiration rates over the grasslands in inner Mongolia of China are two to three times higher in midsummer than in spring or autumn, and the heat budget of the grassland under such a desertification process is close to that of a sand dune (Harazono *et al.*, 1993). In northwestern China, the number of days each year when windstorms (greater than 17.2 m/sec)

***Table 10-3:*** *Number of species (all known and endemic) of mammals, birds, and plants in Temperate Asia, 1990s.*

| | **Mammals** | | **Birds** | | **Flowering Plants** | |
|---|---|---|---|---|---|---|
| **Country or Region**[1] | Known (All) | Endemic | Known (All) | Endemic | Known (All) | Endemic |
| China | 394 | 77 | 1,244 | 67 | 30,000 | 18,000 |
| Japan | 132 | 38 | 583 | 21 | 4,700 | 2,000 |
| Korea, Dem People's Rep | — | 0 | 390 | 0 | 2,898 | 107 |
| Kore, Rep | 49 | 0 | 372 | 0 | 2,898 | 224 |
| Mongolia | 134 | 0 | 290 | 0 | 2,272 | 229 |

Source: World Conservation Monitoring Center.
[1]Data not available for Siberia and Taiwan.

occur averages 20–40 days in arid regions and 10–25 days in semi-arid regions—but reaches more than 100 days in areas most affected (Xia, 1993; Zhao, 1996). These numbers are expected to increase under global warming because dryer and stronger-wind conditions are expected, resulting in accelerated soil degradation.

### *10.2.1.3. Lakes, Streams, and Wetlands*

Significant shortening of the duration of ice cover in lakes is expected. Enhanced evaporation, as well as ground thaw, would cause some arctic lakes to disappear. Lakes in northern latitudes may change from a vertically homogeneous state to a stratified state (vice versa for those in southerly areas). These changes in mixing regimes are likely to have very important effects on biota (IPCC 1996, WG II, Section 7.4.4).

Lake water levels are very sensitive to climate. In many regions, lake levels are likely to decline or fluctuate more widely because of changes in precipitation or increases in evapotranspiration. In lakes experiencing declines or rapid changes in water level, vegetation and habitat characteristics of highly productive inshore littoral areas may change significantly—with profound but highly lake-specific effects on food webs and productivity. Water-level declines in lakes with extensive bordering wetlands are likely to reduce lake productivity and populations of fish and invertebrates dependent on these wetlands for spawning and nursery grounds (IPCC 1996, WG II, Sections 10.5.2 and 10.6.2).

Climate change would result in a wide range of changes in the hydrological regimes of streams, with associated impacts on their ecosystems. For snowmelt streams, for example, a shift toward more rainfall and less snowfall in winter would result in a shift of the seasonality of stream discharge regimes toward higher winter flows and lower spring and summer flows. These changes may alter the timing of biogeochemical fluxes that could produce very important, but highly uncertain, effects on the biota in streams, lakes, and estuaries (IPCC 1996, WG II, Section 10.3.5).

Model results based on six different climate change scenarios (precipitation increasing or decreasing by 10% and temperature increasing 1°C, 2°C, and 3°C) suggest that the areal extent of herbaceous wetlands in eastern China would decline as a result of reductions in total rainfall and increases in evapotranspiration (IPCC 1996, WG II, Section 6.3.3).

Local extinctions and extirpations of cold-water and cool-water species of fish and invertebrates may be expected as a consequence of warming in the temperate zone. This effect will be most pronounced in shallow lakes, streams, and rivers in which appropriate thermal refugia (well-oxygenated deep waters, groundwater vents, access to higher elevations) are not available. Although biodiversity theoretically should increase with warming—particularly at mid- and high latitudes—initially there may be a loss in total diversity in cool temperate and boreal regions, if northward migration of warm-water species cannot keep pace with the rate of loss of cool-water and cold-water species due to limitations on dispersal (physiological limitations, lack of north-south corridors) and genetic adaptation (IPCC 1996, WG II, Sections 10.6.1 and 10.6.3).

Productivity and biodiversity in streams and rivers will likely be reduced in humid parts of Temperate Asia, which are expected to experience reductions in total rainfall and/or increases in evapotranspiration that produce longer, more severe droughts during the warm season. Declines in biodiversity would result from severe water quality deterioration (e.g., low dissolved oxygen levels, high concentrations of toxic substances, high temperatures) during extended low-flow periods in the summer, as well as from drying of previously perennial streams. These effects will be exacerbated in systems that are strongly affected by human activities (IPCC 1996, WG II, Sections 10.3.7, 10.4.2, 10.6.2, and 10.6.4).

### *10.2.1.4. Mountain Regions*

Subalpine conifers may expand into alpine regions under the influence of global warming—though this expansion would not disturb the upper vegetation zone, which already is occupied by well-developed creeping pine (*Pinus pumila*) scrub (Omasa *et al.*, 1996). Similarly, the distribution area of beech (*Fagus crenata*) on the Sea of Japan (East Sea) side of Honshu could shift upwards on the mountain slopes, but its extent will be affected by snow-cover status, which provides different vegetation dynamics.

The impacts of global warming on mountain regions in Temperate Asia are summarized in Table 10-4.

### *10.2.1.5. Cryosphere*

If projections of climate change for the year 2050 (UKMO-X5) are realized, a number of impacts on the cryosphere are likely. Projected warming of the climate would reduce the area and volume of the cryosphere. This reduction would have significant impacts on related ecosystems, people, and their livelihoods. There could be pronounced reductions in seasonal snow, permafrost, glaciers, and periglacial belts, with a corresponding shift in landscape processes. Pronounced alterations in glacier-melt runoff also are likely as the climate changes (IPCC 1996, WG II, Section 7.4).

Snow cover in the temperate regions ranges from a few meters to a few centimeters, but the snow often is close to its melting point; consequently, continental and alpine snow covers are very sensitive to climate. As warming occurs for many Asian mountains, there is a tendency for rainfall to occur at the expense of snowfall, although the extent of this shift depends on location. Less snow will accumulate at low elevations, although there may be more snow above the freezing level as a result of increased precipitation (IPCC 1996, WG II, Sections 7.3.1 and 7.4.1).

***Table 10-4:*** *Possible impacts of global warming on mountain regions in Temperate Asia.*

| | Field | Impacts | References |
|---|---|---|---|
| *Physical System* | Hydrology | a) Increase in air temperature and decrease in precipitation in summer<br>b) Increase in rate of rainfall/snow and decrease in depth and duration of snow accumulation<br>c) Upward shift of maximum precipitation zone on mountain slopes | Nakatsugawa, 1995; IPCC 1996, WG II, Chapter 10 |
| | Cryosphere | a) Direct effect of elevated temperature<br>b) Increasing density of snow accumulation and longer snow-free duration | IPCC 1996, WG II, Chapter 7 |
| | Abnormal Weather | a) Significant increase in frequency of strong winds and torrential rains<br>b) Warm and dry spring and summer, causing increased fire risk | Nakashizuka and Iida, 1995; IPCC 1996, WG II, Chapters 1 and 12 |
| | Topographical Process | a) Increase in precipitation maximum over mountain slopes, causing increased landslide, mud flow, soil erosion, groundwater pressure, etc.<br>b) Thawing of permafrost, causing increased grade and frequency of landslide | IPCC 1996, WG II, Chapters 5 and 7 |
| | Biological World/ Vegetation | a) Change of main species of mountain plants and animals, causing increased stress on mountain ecosystems<br>b) Easier upward shift of main species of mountain plants, due to short shift distance and less stress of adaptation to light condition within shift elevation<br>c) Effects of changing snow accumulation on plants and animals | Mooney *et al.*, 1995; IPCC 1996, WG II, Chapter 5 |
| *Socio-Economic System* | Mountain Agriculture | a) Decrease in existing crops<br>b) Change of cultivation calendar along mountain slopes | IPCC 1996, WG II, Chapters 5 and 13 |
| | Hydrological Power | a) Change of available water power due to seasonal change of hydrology<br>b) Increased demand for electric power in summer and decreased demand in winter | IPCC 1996, WG II, Chapter 10 |
| | Forestry | a) Change of economically predominant species<br>b) Increased damage by wildfire, pests, virus, disease, etc. | Kojima, 1992; Grishin, 1995; Mooney *et al.*, 1995; Nakashizuka and Iida, 1995; IPCC 1996, WG II, Chapters 1 and 12 |
| | Tourism | a) Change of elements dominating mountain landscape<br>b) Decreased length of skiing season | IPCC 1996, WG II, Chapter 11 |
| | Transportation | a) Increased accessibility due to amount and period of snow accumulation in winter<br>b) Possibility of accidents and damages by freezing for traffic and transmission systems | IPCC 1996, WG II, Chapter 11 |

Glaciers are more sensitive to changes in temperature than to any other climatic element. For Asian glaciers, where precipitation occurs mainly during the summer monsoon season, temperature rise has a double effect. The first effect is an increase in the absorption of solar radiation caused by a lowering of the surface albedo, as snowfall is converted to rainfall. The second effect is an increase in energy exchange between the atmosphere and the glacier surface (IPCC 1996, WG II, Section 7.3.1). Both effects lead to an increase in glacial melting.

Empirical and energy-balance models indicate that a large fraction (about one-third to one-half) of the world's existing mountain glacier mass could disappear with anticipated warming over the next 100 years. By 2050, up to a quarter of mountain glacier mass may have melted. However, the largest alpine glaciers—such as those found in the Pamirs, the Tien Shan, and the Himalayas—are expected to continue to exist into the 22nd century. Other studies are based on calculations of ablation and equilibrium-line altitude, according to given climatic scenarios. Ablation would intensify in central Asia with climate warming because conditions there could become even more continental (IPCC 1996, WG II, Section 7.4.2).

The broad sensitivity of permafrost to past climate change is well documented in Siberia and China. There would be a poleward shift of permafrost zones, although deep-seated, ice-rich permafrost will be resistant to changes. Permafrost in northeast China is expected to disappear if temperature increases by 2°C or more. Over the Qing-Zang plateau, estimates of climate change impacts on permafrost range from complete disappearance for a temperature increase of 2°C to the raising of its elevation limit to 4,600 m for a warming of 3°C (IPCC 1996, WG II, Sections 7.3.3 and 7.4.3).

### 10.2.2. Hydrology and Water Resources

#### 10.2.2.1. Hydrological Systems

Global warming affects stream hydrology in many ways. In mid- and high-latitude regions—where warmer temperatures and a shift toward more rainfall and less snowfall in winter are projected—the seasonality of discharge regimes of snowmelt streams would shift toward higher winter flows and lower spring and summer flows. In semi-arid regions, streams and rivers may experience large increases in hydrological variability, with more frequent and larger floods with high sediment loads and longer droughts. In humid regions, streams and rivers are likely to experience reductions in total rainfall and/or increases in evapotranspiration that produce longer, more severe droughts during the warm season (IPCC 1996, WG II, Sections 10.3.5, 10.3.7, and 10.4.2).

In mountain regions, glaciers will provide extra runoff as the ice disappears. In general, the extra runoff would persist for a few decades; in areas with very large glaciers, it may last for a century or more. By 2050, the volume of runoff from glaciers in central Asia is projected to increase threefold. Tentative estimates have been made for central Asia based on mass balances from a small number of Tien Shan glaciers for the period 1959–1992. Extrapolation to the whole area of central Asia suggests that its glacier mass may have decreased by 804 $km^3$ over that time, representing a 15% increase in glacial runoff. Eventually, however, glacial runoff will taper off or even cease. Projected glacier runoff in 2100 is about 68 $km^3$/yr, compared with the present value of 98 $km^3$/yr (IPCC 1996, WG II, Section 7.4.2).

Many rivers within temperate regions would tend to become ice-free or develop only intermittent or partial ice coverage. Ice growth and thickness would be reduced. In colder regions, the present ice season could be shortened by up to a month by 2050. Warmer winters would cause more midwinter breakups as rapid snowmelt, initiated particularly by rain-on-snow events, became more common. Warmer spring air temperatures may affect breakup severity, but the results would be highly site-specific because breakup is the result of a complex balance between downstream resistance (ice strength and thickness) and upstream forces (flood wave). Although thinner ice produced by a warmer winter would tend to promote thermal breakup, this effect might be counteracted to some degree by the earlier timing of the event, reducing breakup severity (IPCC 1996, WG II, Section 7.4.4).

Widespread loss of permafrost over extensive continental and mountain areas would trigger erosion or subsidence of ice-rich landscapes and change hydrological processes. Revegetation of terrain following deglaciation is slow in high-mountain areas. This lag leaves morainic deposits unprotected against erosion for extended periods (decades to centuries), which can result in increased sediment loads in alpine rivers and accelerated sedimentation in lakes and artificial reservoirs at high altitudes. On slopes steeper than about 25–30 degrees, stability problems such as debris flows will develop in freshly exposed or thawing nonconsolidated sediments (IPCC 1996, WG II, Section 7.4.3 and Box 7-2).

#### 10.2.2.2. Water Supply

Overall, most 2x$CO_2$ equilibrium scenario simulations show a decrease in water supplies throughout Temperate Asia, except in a few river basins. Warmer winters may affect water balances because water demands are higher in spring and summer. Glacial melt may lead to nonsustainable supplies of surface water (IPCC 1996, WG II, Chapter 14).

To balance water supply with water demand, more efficient water management is likely to be the approach taken by Japan. In other parts of Temperate Asia, water resource development will remain important. There, the central adaptation issue is how climate change might affect the design of new water resource infrastructure.

Because many river basins are international, climate change would exacerbate current international conflicts over water use and potentially cause new conflicts (IPCC 1996, WG II, Section 14.2.3). Clearly, multiple-stress impact studies on water resources in international river basins are needed.

***Table 10-5:*** *Impact of climate change on annual runoff of seven river basins in China, in % change of annual runoff (Liu, 1997).*

| | General Circulation Model | | | |
|---|---|---|---|---|
| | LLNL[1] | UKMO-H3 | OSU-B1 | GISS-G1 |
| Dongjiang | 0.4 | -3.1 | 8.1 | -4.9 |
| Hangjiang | -7.7 | -2.6 | 4.4 | -0.7 |
| Huaihe | -14.7 | -6.3 | 7.8 | -3.9 |
| Huanghe | -7.2 | -4.6 | -2.6 | 4.0 |
| Haihe | -16.0 | 7.2 | 1.0 | -7.3 |
| Liaohe | -14.0 | 17.4 | -2.3 | 5.9 |
| Songhuajiang | 3.4 | 12.1 | -7.5 | 3.8 |

[1]L. Gates, pers. comm.

### 10.2.2.3. Water Resources in China

Table 10-5 shows the impacts of climate change on the annual runoff of seven river basins in China, located in different climate zones from south to north. For all four GCMs, projected changes in runoff are the result mainly of changes in precipitation in spring, summer, and autumn because of the strong influence of the monsoon climate (Liu, 1997).

The model results from the four GCMs suggest that the most sensitive and vulnerable areas to climate change would be in the northern part of China:

- Within climatic regions, runoff is likely to be more sensitive to climate change in the plains than in mountain areas.
- In the Huaihe River basin and the area to its north, reduced runoff would occur mainly in the spring, summer, and autumn (most significantly in the summer as a result of high temperatures and less rainfall), which may unfavorably affect reservoir storage.
- For the Songhuajiang River, Liaohe River, Haihe River, upper reach of the Yellow River, Huaihe River, and Dongjiang River, increases in runoff would occur mainly in the spring, summer, and autumn, necessitating greater flood control.
- In moderately and extremely dry years, the projected potential water deficiency caused by climate change in the Huaihe River basin and the area to its north—although less than that caused by population growth and economic development—may seriously exacerbate existing water shortages and thus badly affect socioeconomic development in these areas.
- Single isolated reservoir systems are less adaptable than integrated multiple reservoir systems to reduced water resources.

Adaptation strategies for water-resource management in China could include:

- Strengthening water-resource management infrastructure
- Increasing investment in water-saving technology and enhancing public consciousness of water saving
- Strengthening the management of existing water projects
- Strengthening protection of water sources and developing new water sources
- Implementing step-by-step interbasin water-transfer projects
- Promoting water conservation and establishing water statutes
- Formulating contingent response measures for extraordinary conditions.

The most critical area of uncertainty is the lack of credible projections of the regional climate in Temperate Asia. In particular, the effect of climate change on the Asian monsoon or the ENSO phenomenon is unknown. Presently, there are two diametrically opposite projections: one showing a strengthening and the other a weakening of the Asian monsoon, which would result in completely opposite impacts on hydrology and water resources. Other uncertainties are introduced in the downscaling of precipitation from the GCM grid scale to the scale of hydrological models for river basins, through the stochastic or interpolation method. There also is considerable uncertainty in the translation of climate change into hydrological effects through hydrological models: The parameters in hydrological models, which are calibrated by historical data, are not transferable either geographically or in terms of a changing environment. In addition, the impact of climate change on evapotranspiration and groundwater—both of which are important for long-term projections of water resources—so far cannot be estimated appropriately.

Research and monitoring needs include projection of regional climate scenarios with high spatial and temporal resolution, improved hydrological models with appropriate land surface parameters under nonstationary climatic conditions, integrated impact studies that consider different sectors and their responses to adaptation strategies, and enhanced hydrological monitoring networks.

## 10.2.3. Food and Fiber Production

### 10.2.3.1. Agriculture

Table 10-6 shows major rice, wheat, maize, and soybean production in the countries of Temperate Asia (except Siberia) for 1990, 1993, and 1996. The increasing production of wheat, maize, and soybeans in China reflects better management practices, particularly in the northeast provinces. Rice production in the whole region seems to have leveled off. The drop in rice production in 1993 reflects the anomalously cool and dry summer of that year (Yoshino, 1997; see also Annex A).

Agricultural yields estimated using different GCMs are shown in Table 10-7; the impacts clearly vary widely. Nevertheless, possible large negative impacts on rice production as a result of climate change would be of concern in the face of expected population increases. In China, across different scenarios and different sites, yield changes for several crops by 2050 are projected to be -78% to +15% for rice, -21% to +55% for wheat, and -19% to +5% for maize. An increase in productivity may occur if the positive effects of $CO_2$ on crop growth are considered, but the magnitude of the fertilization effect remains uncertain.

Six areas are most likely to be negatively affected by climate change: the area around the Great Wall lying southeast of the transition belt between crop agriculture and animal husbandry; the Huang-Hai Plains, where dryland crops like wheat, cotton, corn, and fruit trees are grown; the area north of Huaihe River—including east Shandong—that lies along the south edge of the temperate crop zone; the central and southern areas of Yunnan plateau; the middle and lower reaches of the Yangtze River; and the Loess plateau. Except for the Yunnan plateau, these areas

***Table 10-6:*** *Major production of rice, wheat, and maize (million tons) in Temperate Asia for 1990, 1993, and 1996.*

| | China | Japan | Korea, Dem People's Rep | Korea, Rep | Mongolia |
|---|---|---|---|---|---|
| **Rice** | | | | | |
| 1990 | 191.59 | 13.12 | 3.08 | 7.72 | — |
| 1993 | 179.79 | 9.79 | 2.30 | 6.51 | — |
| 1996 | 190.10 | 13.00 | 2.80 | 6.28 | — |
| **Wheat** | | | | | |
| 1990 | 98.23 | 0.95 | 0.12 | — | 0.60 |
| 1993 | 106.39 | 0.64 | 0.12 | — | 0.45 |
| 1996 | 109.01 | 0.55 | 0.10 | — | 0.26 |
| **Maize** | | | | | |
| 1990 | 97.21 | — | 2.38 | 0.12 | — |
| 1993 | 103.11 | — | 1.96 | 0.08 | — |
| 1996 | 117.35 | — | 2.00 | 0.07 | — |
| **Soybean** | | | | | |
| 1990 | 11.08 | 0.22 | 0.46 | 0.23 | — |
| 1993 | 15.32 | 0.10 | 0.38 | 0.17 | — |
| 1996 | 13.31 | 0.12 | 0.40 | 0.16 | — |

Source: FAO, 1997.

would be at heightened risk of drought and would suffer potential increases in soil erosion. The Yunnan plateau, with generally abundant rainfall, would be subject to alternating drought and waterlogging, as well as to cold spells, and hence would also suffer yield losses (IPCC 1996, WG II, Section 13.6.3).

Studies for Japan indicate that the positive effects of $CO_2$ on rice yields generally would more than offset negative climatic effects in the central and northern areas. However, in southwest Japan, particularly in Kyushu, the effects on rice yields, on balance, would be negative (IPCC 1996, WG II, Section 13.6.3).

GISS-G1 and GFDL-A3 GCMs under 2x$CO_2$ equilibrium scenarios suggest that the production of spring wheat in Mongolia could be reduced significantly because of higher evapotranspiration. Adaptive measures—such as changing planting dates, using different varieties of spring wheat, or applying the ideal amount of nitrogen fertilizer at the optimum time—are potential responses that could modify these effects (Bayasgalan *et al.*, 1996). Climate change is projected to have favorable impacts on agriculture in the northern areas of Siberia and to cause a general northward shift of crop zones. Grain production in the steppes of southwestern Siberia is projected to fall by about 20% as the result of a more arid climate. These projections could change substantially if market reforms succeed in improving the efficiency and productivity of agriculture (IPCC 1996, WG II, Section 13.6.5).

For mountain regions in China, climatic warming would, in general, increase agricultural productivity—partly as a result of reduced periods of low temperature and partly because of the expansion of arable lands. Studies using simple empirical indices show that a warming of 1°C in mean temperature in the mountain regions of southwest China would cause a 170-m shift in the upper boundary for growing grain. There is uncertainty, however, related to the expansion of crop areas because soil is a very important limiting factor for grain production. In addition, the specific crop varieties that are used in mountain agriculture may need to be altered to keep up with global warming. In Japan, for example, although current varieties of rice respond positively only to small temperature increments, late-maturing varieties are likely to adapt well under much warmer conditions. For such crop varieties, under a temperature rise of 3°C, most of the land below 500 m on Hokkaido and up to 600 m on Tohoku could become viable for agriculture (Yoshino *et al.*, 1987).

Estimated net economic welfare impacts of climate change for three GCMs under 2x$CO_2$ equilibrium scenarios are shown in Table 10-8. According to the GFDL-A2 and UKMO-H3 GCMs, all countries/regions would experience negative impacts except the mainland of China. On the other hand, the GISS-G1 GCM estimated positive impacts for all countries except the Republic of Korea if no adaptation options were pursued. In China, climate change would occur against a steadily increasing demand for food, which is expected to continue through at least 2050. The increased annual cost of government investment only (excluding farmers' additional costs) in agriculture in response to climate change through 2050 was estimated at US$3.48 billion (17% of the cost of government

***Table 10-7:*** *Agricultural yield impact of selected climate change studies.*

| Country/Region | Yield Impact (%) | Direct $CO_2$ Effect | Author(s) | Reference[4] |
|---|---|---|---|---|
| **Rice** | | | | |
| China | -6 | Yes | Tao | 1 |
| China | -11 to -7 | Yes | Zhang | 1 |
| China | -78[1] to -6 | No | Jin *et al.* | 1 |
| China[2] | -37 to +15 | No | Jin *et al.* | 1 |
| China | -18 to -4 | Yes | Matthews *et al.* | 1 |
| China | -21 to 0 | No | Lin | 2 |
| Japan | +10 | No | Sugihara | 1 |
| Japan | -11 to +12 | No | Seino | 1 |
| Japan | -45 to +30 | Yes | Horie | 1 |
| Japan | -28 to +10 | Yes | Matthews *et al.* | 1 |
| Korea, Rep | -37 to +16 | Yes | Yun | 3 |
| Korea, Rep | -40 | Yes | Oh | 4 |
| Taiwan | +2 to +28 | Yes | Matthews *et al.* | 1 |
| **Wheat** | | | | |
| China | -8 | Yes | Tao | 1 |
| China | -21 to +55 | No | Lin | 2 |
| Japan | -41 to +8 | Yes | Seino | 1 |
| Mongolia | -67 to -19 | No | Lin | 2 |
| Russia[3] | -19 to +41 | Yes | Menzhulin and Koval | 5 |
| **Maize** | | | | |
| China | -4 to +1 | Yes | Tao | 1 |
| China | -19 to +5 | No | Lin | 2 |
| Japan | -31 to +51 | Yes | Seino | 1 |
| **Pasture** | | | | |
| Mongolia | -40 to +25 | No | Bolortsetseg *et al.* | 6 |

[1]This large negative percentage is a result of the modeled result at a single site in southwest China.
[2]For irrigated rice.
[3]Including European Russia.
[4]Sources: (1) IPCC 1996, WG II, Table 13-6; (2) Lin, 1996; (3) Yun, 1990; (4) Oh, 1995; (5) IPCC 1996, WG II, Table 13-8; and (6) Bolortsetseg and Tuvaansuren, 1996.

investment in agriculture in 1990) (IPCC 1996, WG II, Section 13.6.3).

### *10.2.3.2. Fisheries*

Even without major changes in atmospheric and oceanic circulation, local shifts in centers of production and mixes of species in marine and fresh waters are expected as ecosystems are displaced geographically and changed internally. The relocation of populations will depend on the presence of properties in changing environments to shelter all stages of the life cycle of a species. Under climatic warming, positive effects for saltwater fisheries—such as longer growing seasons, lower natural winter mortality, and faster growth rates in higher latitudes—may be offset by negative factors such as the alteration of established reproductive patterns, migration routes, and ecosystem relationships. Changes in abundance are likely to be more pronounced near major ecosystem boundaries. The rate of climate change may prove to be a major determinant of the abundance and distribution of new populations. Rapid change from physical forcing usually will favor production of smaller, low-priced, opportunistic species that discharge large numbers of eggs over long periods. Regionally, gains or losses in freshwater fisheries will depend on changes in the amount and timing of precipitation, on temperatures, and on species tolerances. For example, increased rainfall during a shorter period in winter could lead to reduced summer levels of river flows, lakes, and wetlands—and thus to reductions in freshwater fisheries. Marine stocks that reproduce in freshwater (e.g., salmon) or require reduced estuarine salinity will be similarly affected (IPCC 1996, WG II, Chapter 6 Executive Summary).

Aquaculture is particularly important to Temperate Asia. In 1994, the world's aquaculture production was about 18.6 million

***Table 10-8:*** *Estimated net economic welfare impacts of climate change projected by three GCMs under $2xCO_2$ equilibrium scenarios (millions of 1989 $US; negative numbers in parentheses).*

| Country or Region | GISS-G1 No Adaptation | GISS-G1 Adaptation | GFDL-A2 No Adaptation | GFDL-A2 Adaptation | UKMO-H3 No Adaptation | UKMO-H3 Adaptation |
|---|---|---|---|---|---|---|
| Mainland China | 1,039 | 2,535 | 80 | 2,199 | (275) | 3,183 |
| Other East Asia | 48 | 157 | (147) | (34) | (717) | (411) |
| Japan | 1,290 | 2,170 | (2,016) | (501) | (7,839) | (4,686) |
| South Korea | (176) | 174 | (658) | (166) | (2,185) | (1,169) |
| Former Soviet Union | 1,367 | 1,859 | (1,502) | (293) | (10,403) | (5,020) |
| Taiwan | 69 | 271 | (271) | (7) | (1,340) | (671) |

Source: Reilly *et al.*, 1993.
Note: Economic model simulations are based on yield shocks as estimated by Rosenzweig and Parry (1994). Economic impacts are measured as combined consumer and producer surplus and government budget changes.

tons (excluding seaweed), of which Temperate Asia accounted for 64.7% and China alone accounted for 57.2% (FAO, 1996a). With seaweeds included, the world's production in 1994 was 25.4 million tons, and the shares of Temperate Asia and China were 70.3% and 60.4%, respectively. During the period from 1984 to 1994, aquaculture in Temperate Asia demonstrated a yearly growth rate of 10% in volume and 11% in value (FAO, 1996b). Climate change impacts generally will be positive through faster growth, lower winter mortality rates, reduced ice cover, and reduced energy costs as a result of expanded regions of warmer water. Cultivation of warm-water species also may expand. Warming will require greater attention to possible oxygen depletion, fish diseases, and introduction of unwanted species (IPCC 1996, WG II, Section 16.2.3).

#### *10.2.3.3. Production Forestry*

Projections for 2010 indicate that the need for industrial roundwood could increase 38% (southern Temperate Asia) to 96% (eastern Temperate Asia). These requirements may result in serious shortages of boreal industrial roundwood, placing further stress on boreal forests (IPCC 1996, WG II, Section 15.4.3).

### ***10.2.4. Coastal Systems***

#### *10.2.4.1. Sea-Level Rise*

Current best estimates for sea-level rise—1.0–2.5 mm/yr—represent a rate two to five times higher than that experienced in the past 100 years. Even with assumed stabilization of global greenhouse gas emissions, sea level is projected to continue to rise beyond the year 2100 because of lags in climate response. Regional differences will exist as a result of wind and atmospheric pressure patterns, regional ocean density differences, and oceanic circulation. For Temperate Asia, sea-level rise is projected to be slightly below the global mean value around the Okhotsk Sea and along the coasts south of about 30°N and slightly above the mean elsewhere (IPCC 1996, WG II, Section 9.3.1).

Along most of the continental coast, relative sea level (i.e., sea level in relation to land) is an important factor for coastal environments. Deltaic coasts in China are facing severe problems of relative sea-level rise as a consequence of tectonically and anthropogenically induced land subsidence. Sea-level rise related to global warming will exacerbate these problems. In addition, saltwater intrusion problems will become more serious with sea-level rise in deltaic regions and coastal plains (IPCC 1996, WG II, Section 9.4.2).

**Box 10-2. Sea-Level Rise at Major Deltas of China (ESD-CAS, 1994)**

The Old Huanghe (Yellow River) delta, the Changjiang (Yangtze River) delta, and the Zhujiang (Pearl River) delta are major areas of economic activity in China. The metropolises of Tianjin, Shanghai, and Guangzhou (Canton) are located within these three deltas, respectively. The three deltas are located in regions with tectonic subsidence rates of about 2–3 mm/yr, 1–2 mm/yr, and 1–2 mm/yr, respectively (although hilly lands in the Zhujiang delta have a tectonic uplifting rate of 1 mm/yr). The Old Huanghe delta and the Changjiang delta have experienced severe land subsidence problems in the past as a result of groundwater extraction. Recent efforts to mitigate this problem have been successful in reducing the subsidence rate. It is estimated that these rates can be controlled within the range of 6–10 mm/yr for the Old Huanghe delta and 3–5 mm/yr for the Changjiang delta. In the Zhujiang delta, natural progradation of the coast and active land reclamation activities have resulted in a 0.5–1 mm/yr sea-level rise in the distributaries in the estuarine area—the same order of magnitude as the projected value due to climate change. This rate is expected to continue for some time. In the next 50 years, therefore, the expected eustatic sea-level rise due to climate change will not be a major factor in relative sea-level rise for the Old Huanghe and Changjiang deltas, although it may be for the Zhujiang delta.

#### 10.2.4.2. Coastal Flooding and Inundation

Tokyo, Osaka, and Nagoya are located in the coastal zone, and together they account for more than 50% of Japan's industrial production. In these metropolitan areas, about 860 $km^2$ of coastal land—an area supporting 2 million people and with physical assets worth $450 billion (1985 dollars)—already are below mean high-water level. With a 1-m rise in mean sea level, this area would expand by a factor of 2.7 to embrace 4.1 million people and assets worth $908 billion. The same sea-level rise would expand the flood-prone area from 6,270 $km^2$ to 8,900 $km^2$. The cost of adjusting existing protection measures has been estimated at about $80 billion (IPCC 1996, WG III, Section 6.5.2).

One of the potential threats that sea-level rise poses to the coastal environment is exacerbated beach erosion. Sandy beaches occupy 20–25% of the total length of the Japanese coast. About 120 $km^2$ of these beaches have been eroded over the past 70 years. According to Bruun's Rule, an additional 118 $km^2$ of beaches—57% of the remaining sandy beaches in Japan today—would disappear with a 30-cm sea-level rise. This percentage would increase to 82% or 90% if the sea level rose by 65 cm or 100 cm, respectively (Mimura, 1995).

Temperate Asia experiences tropical cyclones every year; these storms inundate and exacerbate flood situations in coastal areas, eroding and restructuring coastal formations. There is no conclusive evidence that the frequency or intensity of tropical cyclones would change as a result of climate change or that a systematic shift in their tracks would occur (IPCC 1996, WG II, Section 9.3.2). However, with the projected sea-level rise, even if the frequency and severity of storms remain unchanged in the future, storm surge could still present an increased hazard. It should be pointed out that China and other countries of the region have a long history of experience in the defense against sea encroachment, through the construction of dikes, seawalls, and other structures (ESD-CAS, 1994).

#### 10.2.4.3. Integrated Coastal Zone Management

Coastal oceans provide rich living and nonliving resources; serve as media for transportation and recreation; and receive, dilute, and transform massive quantities of wastes from human activities on land. Coastal oceans already are under stress as a result of a combination of factors (e.g., increased population pressure, habitat destruction, increased land-based pollutant loads, and increased nutrient inputs from rivers). The effects of climate change therefore will constitute additional—largely adverse—series of impacts on an already overstressed resource, with potential for synergistic relationships among stresses (IPCC 1996, WG II, Section 8.2.2).

The most effective adaptation strategy would embrace the tenets of integrated coastal zone management (ICZM). ICZM is an iterative, evolutionary, and continuous process. It involves comprehensive assessment, setting of objectives, and planning and management of coastal systems and resources; it takes into account traditional, cultural, and historical perspectives, as well as conflicting interests and uses (IPCC 1996, WG II, Box 9.5). At present, implementation of ICZM in Temperate Asia is in its exploratory phase; further studies are needed.

### 10.2.5. Human Settlements and Industry

#### 10.2.5.1. Human Settlements

If future climates resemble those projected by GCMs, wetter coasts, drier midcontinent areas, and sea-level rise may cause the gravest effects of climate change through sudden human migration, as millions of people are displaced by shoreline erosion, river and coastal flooding, or severe drought. For example, agricultural settlements in China are projected to be sensitive to drought conditions (IPCC 1996, WG II, Section 12.4.2).

Climate change may affect the intensity or the probability of the occurrence of extreme weather events. The impact of extreme events on human settlements can be affected by the grade of the extreme event, the level of economic and technological development, and the extent of countermeasures taken. In Japan, for example, although the number of deaths is large for the more severe typhoons, such deaths have been declining since World War II (see Figure 10-3). This decline is thought to have occurred because of increasingly effective mitigation measures taken during the past 40 years. These measures include conservation of mountain slopes, rivers, and coasts; harmonization of institutions and laws related to disaster mitigation; preparation of systems for disaster prevention, including better design; meteorological observation and information systems; promotion of people's consciousness with regard to preventing disasters; and development of communication systems for disaster occurrences. On the other hand, economic damages from typhoons have been rising in Japan during the same period because of the increasing values of properties along the coasts (IPCC 1996, WG II, Section 12.4.2).

In mountain lands and continental permafrost areas, cryospheric change could reduce slope stability and increase the incidence of natural hazards for people, buildings and other structures, pipelines, and communication links (IPCC 1996, WG II, Section 7.5).

During summer days, from the early morning to the time of maximum temperature, the temperature inside a big oasis in the arid regions of northwest China usually is 2–3°C lower than that outside the area; this gradient is called the "oasis effect" (Du and Maki, 1994; Yoshino and Liu, 1997). In the future, as the population increases in oases, so will the cultivated area and water consumption. This pattern could lower the moisture content in the oasis and hence weaken the oasis effect. Climate change thus could compound the hardship for human settlements in the oases.

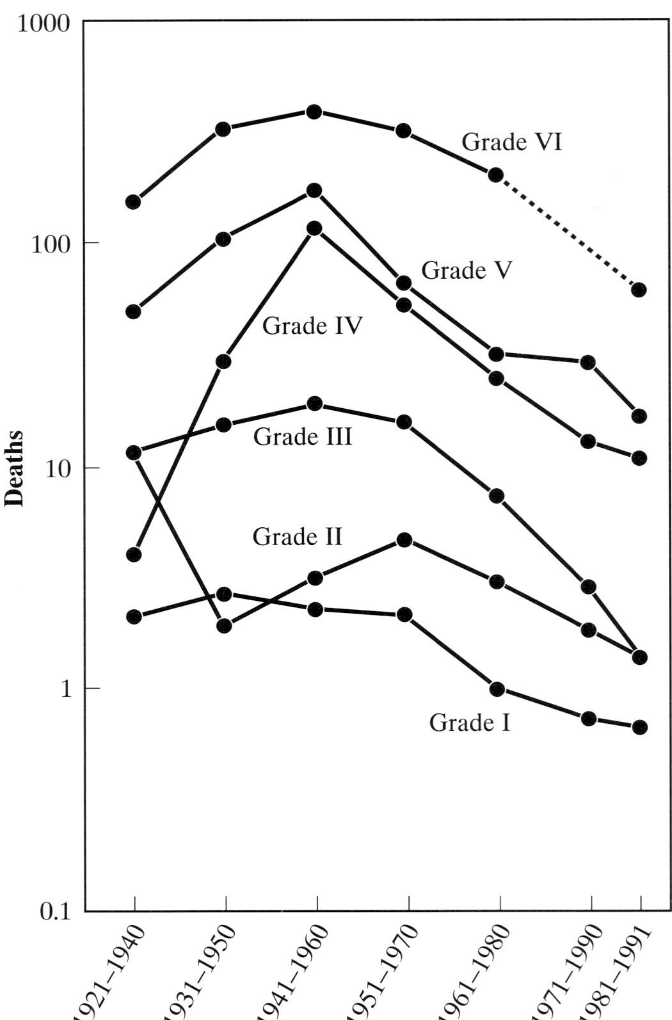


**Figure 10-3:** Change in numbers of deaths per typhoon in Japan during the past 70 years. Typhoon grade: I—weak; II—normal; III—strong; IV—stronger; V—violent; VI—super-violent (after IPCC 1996, WG II, Figure 12-2).

Ironically, some arid regions of northwest China also have been suffering from heavy rainfall and flood damages. In Xinjiang, during the period 1949–1990, about one-third of the total natural hazards and two-thirds of the total losses incurred by natural hazards were caused by floods (Yoshino and Liu, 1997). Under global warming, this tendency could be intensified because intense rainfall is likely to occur more frequently.

#### 10.2.5.2. *Industry*

Power-related industries could be affected both negatively and positively by climate change. As a result of climate change, demand for electricity for cooling purposes will rise in the summer. In Japan, this demand could rise by 5% by 2050, with a 10% increase in peak demand.

The rate of methane production in biogas generators increases at high temperatures and may be interrupted by low temperatures in the winter, when heating of the digesters is required. Therefore, in north China, for example, global warming may lengthen the periods of high yields and extend the areas in which biogas can be produced without heating the digester. Reduction of water resources, however, could jeopardize hydroelectric energy supplies, and existing petroleum production and distribution systems in the tundra zone could be disrupted because of thawing of permafrost (IPCC 1996, WG II, Chapter 11).

With climate warming, a longer shipping season—with significant cost savings in ice-breaking operations—is likely for rivers in northern Temperate Asia. The Arctic coast would have major periods during which coastal waters were open and navigable. On the other hand, thawing induced by climate warming could result in serious disruptions and increased maintenance costs for railways and highways through the permafrost zone from ground subsidence, side-slope slumpings, landslides, icings, and ice-mound growth. China, for example, has more than 3,000 km of railways and more than 13,000 km of highways in the permafrost zone that would be affected (IPCC 1996, WG II, Sections 7.5.1 and 11.5).

### *10.2.6. Human Health*

#### 10.2.6.1. *Heat Stress*

An increase in the frequency or severity of heat waves would cause an increase in (predominantly cardiorespiratory) illness and mortality. Although this increase in heat-related deaths would be partially offset by a reduction in cold-related deaths, there are insufficient data to quantify this trade-off. In addition, this balance would vary by location and according to adaptive responses (IPCC 1996, WG II, Chapter 18 Executive Summary).

Based on data collected for persons in Shanghai aged over 65 years for the period 1980–1989, the threshold temperature for heat-related mortality in summer is 34°C. Days with high afternoon temperatures, low wind speeds, and high humidity were associated with the greatest mortality increase. The transient models (GFDL-X2, UKMO-X6) estimate that, for the year 2050, heat-related deaths in Shanghai would increase to 3.6–7.1 times the present figure. Since the warming would be a gradual process, however, this increase could drop to 2.5–2.6 times the present figure if acclimatization of the population is taken into consideration. Climate change also may lead to more frequent occasions with both very hot weather and very high levels of air pollution. The interactive effects of both stresses on human health are not yet well understood (McMichael *et al.*, 1996).

#### 10.2.6.2. *Infectious Diseases*

Net climate change-related increases in the geographic distribution (elevation and latitude) of vector organisms of infectious diseases (e.g., malarial mosquitoes, schistosome-spreading snails) and changes in the life-cycle dynamics of vector and infective parasites would, in aggregate, increase the potential transmission of many vector-borne diseases in Temperate Asia. Increases in potential transmission areas for malaria, for example, are projected to occur in temperate regions. However, actual climate-related increases in malaria incidence would occur primarily in tropical, subtropical, and less well protected temperate-zone populations that currently are at the margins of endemically infected areas. Some localized decreases also may occur (IPCC 1996, WG II, Chapter 18 Executive Summary).

Increases in non-vector-borne infectious diseases—such as cholera, salmonellosis, and other food- and water-related infections—also could occur as a result of climatic impacts on water distribution, temperature, and microorganism proliferation (IPCC 1996, WG II, Chapter 18 Executive Summary). Recent evidence suggests that the "El Tor" biotype cholera may be classified as a water-based disease. It has spread as far northward as the Korean peninsula. Its mechanism of transmission is still subject to debate, however (McMichael *et al.*, 1996).

Disease surveillance could be strengthened and integrated with other environmental monitoring activities to develop early warning systems; early, environmentally sound public health interventions; and anticipatory social policies to reduce the risk of outbreaks and the subsequent spread of epidemics (IPCC 1996, WG II, Section 12.5.6).

#### *10.2.6.3. Pollution Hazards*

Experiments by Hatakeyama *et al.* (1991) showed that the higher temperatures are, the faster ground-level ozone forms and the longer high concentrations of ozone last. Therefore, global warming would accelerate the photochemical reaction rates among chemical pollutants in the atmosphere, increasing oxidants in many urban areas. High levels of photochemical oxidants are associated with eye irritation, severe respiratory irritation, increased frequency of asthmatic attacks of susceptible persons, and decreased pulmonary functions (Ando, 1993).

## 10.3. Integrated Analysis of Potential Vulnerabilities and Impacts

### *10.3.1. North China*

North China—including Beijing, Tianjin, the four provinces (Hebei, Henan, Shandong, Shanxi), parts of Anhui province, and parts of inner Mongolia—is an economic center of China. It also is a topographical and climatological entity. Its population is 371 million, and its area is 76.5 million ha, including 28.9 million ha of cultivated land. Because this region already is at risk from normal climate variability, it also is likely to be quite vulnerable to long-term secular shifts. The concept of regions at risk is used here to focus on four different managed ecosystems: water resources, agriculture, forests, and coastal zones.

#### *10.3.1.1. Water Resources*

Water resources in north China are sensitive and vulnerable to climate change because:

- Water resource availability in this area amounts to merely 500 $m^3$ per capita—half the critical value of the United Nations (UN) standard (1,000 $m^3$ per capita) for maintaining socioeconomic and environmental development.
- There already is a high level of water project development in this region. For example, in the Haihe basin, the ratio of water supply to total runoff has reached 0.87, but the ratio of water demand to total runoff already is 1.32. Because of the shortage of surface runoff for meeting water demand, depletion of groundwater resources is very serious.
- Depending on the different scenarios used for GCMs, changes in runoff could range between -16% and +7%; these shifts would occur mainly during the flood season. Such changes could unfavorably affect reservoir capability for storage and flood control. Although water shortages resulting from climate change are likely to be less important than those caused by population growth, economic development, and urbanization (for example, in the Hailuang River basin, water shortages resulting from climate change are projected to be about 6.9–9.5 billion $m^3$, whereas shortages from other factors are projected to be 25.2–47.8 billion $m^3$), the potential water deficiency due to climate change in moderately and extremely dry years may seriously exacerbate the current water shortage—and thus badly affect socioeconomic development in this region.

Trends of runoff for north China under four GCM scenarios are shown in Table 10-9. (At its lower reach, the Huanghe is

***Table 10-9:*** *Runoff trends under four GCM scenarios.*

| GCM | Runoff Trends |
|---|---|
| LLNL[1] | All decrease |
| UKMO-H3 | Huanghe and area to its south decrease<br>Others increase |
| OSU-B1 | Huaihe and area to its south increase<br>Others decrease slightly |
| GISS-G1 | Huanghe, Liaohe, and Songhuajiang increase<br>Others decrease |

[1]L. Gates, pers. comm.
Note: Under the GCM scenarios, subregions of northern China have different ranges of runoff change: Jing-Jin-Tang (-16 to +3%); Huaihe River Basin (-15 to +8%); middle reaches of Yellow River Basin (-12 to -5%).

elevated above its neighboring lands.) Both surface water and groundwater in north China seem to be quite sensitive to climatic variability, especially on the Huang-Hai plain, according to the model results. Under these scenarios, climate change will result in an additional shortfall of 0.15–1.4 billion $m^3$ of water in the Jing-Jin-Tang subregion, which would cause economic losses of US$50–800 million (constant 1990 values) in a normal year and US$230–2,270 million in a very dry year.

#### *10.3.1.2. Agriculture*

Three subregions in north China appear to be especially sensitive to climate change because of potential increases in the soil moisture deficit:

- *Areas along the Great Wall*—This area lies southeast of the transition belt between agriculture and animal husbandry. Warming and increased evapotranspiration, along with possible declines in precipitation, would make it difficult to maintain current crop patterns. The northern rangeland would gradually intrude into this area, creating a transition zone dominated by livestock. The northwestern part of this area would become an arid grassland.
- *The Huang-Hai plains*—Climatic warming may increase the moisture deficit (i.e., the difference between precipitation and evapotranspiration) by more than 70 mm and result in more frequent and severe spring droughts with hot, dry winds, damaging wheat production and limiting the present practice of double-cropping in succession. Dryland crops will suffer from drought.
- *The area north of the Huaihe River, including eastern Shandong*—This area lies along the southern edge of the temperate crop zone. Climate warming may cause the northward shift of subtropical crop areas. However, projected frequent waterlogging in the south and spring droughts in the north would inhibit the growth of subtropical crops.

The state and farmers can take steps to adapt agricultural production to the unfavorable impacts of possible climate change. Such strategies include allowing the sown acreage of grain to stabilize at a level of 0.8–0.9 ha per capita to attain the production target; strengthening irrigation capacity as one of the most beneficial means of maintaining agricultural production in the face of unfavorable climate change; and transforming medium- or low-yield farmland into high-yield farmland. To maintain the productivity of cultivated land, it is necessary to popularize a more optimal fertilizer mix and adopt the technique of subsoil application according to actual changes in soil conditions. It also is necessary to use and extend technology for agricultural adaptability—such as using superior species of crops, improving standardized cultural techniques under climatic variation, using dryland farming techniques, and developing feed crops instead of grain crops.

#### *10.3.1.3. Forests*

Forests in northern China have been seriously depleted over the past few centuries. Despite recent reforestation efforts, the forested area in the region is only 11.8% of the total land cover, which is lower than the mean value for China. Under the influence of projected climate change, the distribution pattern of many important tree species would be affected. For example, the present forests of *Pinus tabulaeformis*, a key temperate species widely distributed in northern China, will be reduced an additional 9.4% under a $2xCO_2$ equilibrium climate (Guo, 1995).

#### *10.3.1.4. Coastal Zones*

For the southern coast of Shangdong, estimated sea-level rise by the years 2030, 2050, and 2100 would be 1.1, 5.7, and 40.2 cm, respectively and, for the coast of Liaoning-Tianjin, 13.1, 22.5, and 69.0 cm, respectively (Du *et al.*, 1996). Construction of dikes and seawalls is likely to be the most common adaptation strategy in these areas; this practice has been used there throughout history to combat sea encroachment (ESD-CAS, 1994).

Integrated vulnerability to climate change for northern China, including the vulnerability of forests, is summarized in Table 10-10.

### ***10.3.2. Forests in Temperate Asia***

Forests, which are characterized by their composition of tree species, are among the important ecosystems of the environment. They are greatly affected by climate, soil, fire, industry, tourism, and pipeline/railway/road construction. Consequently, the expected impact of global warming on forests would vary from region to region.

In Temperate Asia, temperature seems to be the primary environmental parameter affecting the distribution of forests because sufficient precipitation exists for almost all forest types. Exceptions are in regions with heavy snow accumulation, windy mountain crests and slopes, or windy coastal zones.

In particular, the amount of snow accumulation seems to be an important parameter for forest growth. Accumulated snow protects trees from severe cold, but its weight can damage the branches and trunks. In Japan, large differences in snow accumulation between the Japan Sea side and the Pacific side are strongly reflected in the corresponding forest types.

The amount of snowfall may decrease as a result of global warming, resulting in spring and summer drought. Uncertainty remains, however, because changes in snowfall amounts depend on changes in the frequency of snow and freezing precipitation in the winter in the mountain regions; the wind speed and the difference between the air temperature and the sea-surface temperature over the Japan Sea in winter, which determine the amount of snow precipitation over mountain regions on the Japan Sea side; the ratio of snowfall to rainfall in winter; the dates of onset and termination of snowfall and the maximum depth of snow accumulation during winter; and the density of accumulated snow.

Almost all forest trees would grow better under a warmer climate, but increases in diseases, insect damages, and other meteorological hazards (e.g., severe storm damage) also may result in a shortened life for forest trees (Tsunekawa *et al.*, 1996a). When the original species in a forest decline, other species from neighboring forest zones more suitable for warmer conditions would grow. If there were an afforested region or broad cultivated area between natural forests, such changes would not occur. The impact of warming would occur noticeably on a time scale of decades to centuries, in areas where the distribution of plant communities is continuous.

Forests of the Hokuriku region on the Japan Sea side of central Japan are likely to respond to climatic change (Kojima, 1996). Of the four vegetation belts, the *Pinus pumila* zone—the highest belt in elevation—would be most seriously affected. It may remain only in small areas at higher elevations or even become fragmented by the upward advance of plant species adapted to a warmer climate. Uncertainties from the viewpoint of vegetation are the rate of geographical expansion of plant species, the time lag between climate change and soil change, and phenological (seasonal) and environmental adaptation of plant species.

Catastrophic scenarios include water deficit as a result of decreasing precipitation and/or increasing evapotranspiration due to warming; elevated minimum temperatures, resulting in the absence of low-temperature stimulation needed for germination, flowering, and fruition; and discord in the seasonal rhythm of air temperature and daytime length, which controls the phenology of plant growth for plants propagated to the north.

According to a study on potential vegetation distribution shifts in China (Tsunekawa *et al.*, 1996b), the northward shift rate of the deciduous genus *Queras* is projected to be 7 km/year—14–24 times faster than the migration rate recorded over the past 13,000 years in Europe. Difficulty in adaptation would be expected for such a rapid shift.

Forest and grassland fire is expected to occur more frequently and in broader areas in northern parts of Temperate Asia as a result of global warming and deforestation activities. In May 1987, one of the largest forest fires of the past 30 years occurred in China and Russia. In the Daxinganlin mountain range of northeast China, it was reported that the total area burned was 1,010 x $10^3$ ha, at 729 sites. The most serious wildfire in the forests and grasslands in Mongolia occurred in March 1996. The burned-out area was estimated to be about 8,000 x $10^3$ ha. Although the occurrence of forest fires varies greatly each year, such serious cases should be taken into consideration in global warming impact studies.

***Table 10-10:** Integrated vulnerability to climate change in northern China.*

| Sector | Scenarios | Method | Most Vulnerable Region | Summary of Results | Cross-Sector Impact |
|---|---|---|---|---|---|
| Water Resources (W) | LLNL[1]<br>UKMO-H3<br>OSU-B1<br>GISS-G1 | Climatic, hydrological, and socioeconomic indices | Hai-Luan River Basin, followed by the Huaihe River Basin | Runoff change of -16 to +17% | Decreased supply to (A) and reduction with (F) |
| Agriculture (A) | GFDL-A3<br>UKMO-H3<br>MPI-K1 | CERES and other crop models; moisture deficit and socioeconomic indices | Hebei, Shanxi, inner Mongolia, and along the Great Wall | Yield change of wheat (-6 to +42%), maize (-9 to +5%), rice (-21 to -7%), cotton (+21 to +53%) | Increased risk for (F) and increased demand for (W) |
| Forests (F) | LLNL[1]<br>UKMO-H3<br>OSU-B1<br>GISS-G1<br>GFDL-A3<br>MPI-K1 | Aridity and fuelwood supply indices | All areas | Productivity increase of +1 to 10%; area change of -57 to +12% (varying with species) | Increased risk from (A) and effect on (W) |
| Coast Zone (CZ) | Sea-level rise of 30–65 cm | IPCC 7-step method | Jing-Jin-Tang and Yellow River Delta | Likely and viable strategy of dikes and seawalls | Increased risk to (A) and (W) |

[1]L. Gates, pers. comm.
Source: Lin *et al.*, 1994.

## References


**Ando**, M., 1993: Health. In: *The Potential Effects of Climate Change in Japan* [Nishioka, S. and H. Harasawa (eds)]. Center for Global Environmental Research, Environmental Agency of Japan, pp. 87–93.

**Bayasgalan**, S.H., B. Bolortsetseg, D. Dagvadorj, and L. Natsagdorj, 1996: The impact of climate change on spring wheat yield in Mongolia and its adaptability. In: *Adapting Climate Change, Assessment and Issues* [Smith, J.B., N. Bhatti, G.V. Menzhulin, R. Benioff, M. Campos, B. Jallow, F. Rijsberman, M.I. Budyko, and R.K. Dixon (eds.)]. Springer-Verlag, New York, NY, 164–173.

**Bolortsetseg**, B. and G. Tuvaansuren, 1996: The potential impacts of climate change on pasture and cattle production in Mongolia. In: *Climate Change, Vulnerability and Adaptation in Asia and the Pacific* [Lin, E., W.C. Bolhofer, S. Huq, S. Lenhart, S.K. Mukherjee, J.B. Smith and J. Wisniewski (eds.)]. *Water, Air and Soil Pollution,* **91(1–2)**, 95–105.

**Davis**, S.D., V.H. Haywood, and A.C. Hamilton, 1995: *Centres of Plant Diversity. A Guide and Strategy for Their Conservation, Vol. 2: Asia, Australia and the Pacific.* World Wide Fund (WWF) for Nature and International Union for the Conservation of Nature and Natural Resources, Gland, Switzerland and Cambridge, U.K., 578 pp.

**DH-MWR**, 1987: *Water Resources Assessment for China.* Department of Hydrology, Ministry of Water Resources, China Water and Power Press, Beijing, 325 pp. (in Chinese and English)

**Ding**, Y. and X. Dai, 1994: Temperature in China during the last 100 years. *Monthly Meteorology,* **20**, 19–26.

**Dixon**, R.K., O.N. Krankina, and K.I. Kobak, 1996: Global climate change adaptation: examples from Russian boreal forests. In: *Adapting to Climate Change, Assessment and Issues* [Smith, J.B., N. Bhatti, G.V. Menzhulin, R. Benioff, M. Campos, B. Jallow, F. Rijsberman, M.I. Budyko, and R.K. Dixon (eds.)]. Springer-Verlag, New York, NY, pp. 359–373.

**Du**, M. and T. Maki, 1994: Climate differences between an oasis and its peral area in Turpan, Xianjiang, China. *JIRCAS Journal,* **1**, 47–55.

**Du**, M., M. Yoshino, Y. Fujita, S. Arizono, T. Maki, and J. Lei, 1996: Climate change and agricultural activities in the Taklamakan Desert, China, in recent years. *Jour. of Arid Land Studies,* **5**, 173–183.

**ESD-CAS**, 1994: *Impact of Sea-Level Rise on the Deltaic Regions of China and its Mitigation* [Earth Science Division, Chinese Academy of Sciences (eds.)]. Science Press, Beijing, China, 353 pp. (in Chinese)

**FAO**, 1995: *FAOSTAT-PC.* Food and Agriculture Organization, Rome, Italy, (on diskette).

**FAO**, 1996a: *FAO Fisheries Circular No. 815, Rev. 8.* Food and Agriculture Organization, Rome, Italy, 189 pp.

**FAO**, 1996b: East Asia. In: *The State of World Fisheries and Aquaculture*, Sofia, Bulgaria.

**FAO**, 1997: *FAOSTAT 1997.* Food and Agriculture Organization, Rome, Italy.

**Fukui**, E., 1977: *The Climate of Japan.* Kodansha Ltd., Tokyo, Japan and Elsevier Science Publishers, Amsterdam, Netherlands, 317 pp.

**Grishin**, S.Y., 1995: The boreal forests of north-eastern Eurasia. *Vegetatio,* **121**, 11–21.

**Guo**, Q., 1995: A study on the impacts of climate change on distribution of *Pinus Tabulaetormis* in China. *Forest Sciences*, **31(5)**, 393–403.

**Harazono**, Y., S.-G. Li, J.-Y. Shen, and Z.-Y. He, 1993: Seasonal micrometeorological changes over a grassland in Inner Mongolia. *Jour. Agricul. Met.,* **48(5)**, 711–714.

**Hatakeyama**, S., H. Akimoto, and N. Washida, 1991: Effects of temperature on the formation of photochemical ozone in a propene $N_{OX}$-air-irradiation system. *Environ. Sci. Technol.*, **25**, 1884–1890.

**Heywood**, V.H., 1995: *Global Biodiversity Assessment.* United Nations Environmental Programme, Cambridge University Press, Cambridge, United Kingdom, 1140 pp.

**IPCC**, 1996: *Climate Change 1995: The Science of Climate Change. Contribution of Working Group I to the Second Assessment Report of the Intergovernmental Panel on Climate Change* [Houghton, J.T., L.G. Meira Filho, B.A. Callander, N.Harris, A. Kattenberg, and K. Maskell (eds.)]. Cambridge University Press, Cambridge, United Kingdom, 572 pp.

- Kattenberg, A., F. Giorgi, H. Grassl, G.A. Meehl, J.F.B. Mitchell, R.J. Stouffer, T. Tokioka, A.J. Weaver, and T.M.L. Wigley, Chapter 6. *Climate Models – Projections of Future Climate,* pp. 289–284.

**IPCC**, 1996: *Climate Change 1995: Impacts, Adaptations and Mitigation of Climate Change: Scientific-Technical Analyses. Contribution of Working Group II to the Second Assessment Report of the Intergovernmental Panel on Climate Change* [Watson, R.T., M.C. Zinyowera, and R.H. Moss (eds.)]. Cambridge University Press, Cambridge, United Kingdom and New York, NY, USA, 880 pp.

- Kirschbaum, M.U.F. and A. Fischlin, Chapter 1. *Climate Change Impacts on Forests,* pp. 95–129.
- Allen-Diaz, B., Chapter 2. *Rangeland in a Changing Climate: Impacts, Adaptations, and Mitigation,* pp. 131–158.
- Beniston, M. and D.G. Fox, Chapter 5. *Impacts of Climate Change on Mountain Regions,* pp. 191–213.
- Oquist, M.G. and B.H. Svensson, Chapter 6. *Non-Tidal Wetlands,* pp. 215–239.
- Fitzharris, B.B., Chapter 7. *The Cryosphere: Changes and Their Impacts,* pp. 241–265.
- Ittekkot, V., Chapter 8. *Oceans,* pp. 267–288.
- Bijlsma, L., Chapter 9. *Coastal Zones and Small Islands,* pp. 289–324.
- Arnell, N., B. Bates, H. Lang, J.J. Magnuson, and P. Mulholland, Chapter 10. *Hydrology and Freshwater Ecology,* pp. 325–363.
- Moreno, R.A. and J. Skea, Chapter 11. *Industry, Energy, and Transportation: Impacts and Adaptation,* pp. 365–398.
- Scott, M.J., Chapter 12. *Human Settlements in a Changing Climate: Impacts and Adaptation,* pp. 399–426.
- Reilly, J., Chapter 13. *Agriculture in a Changing Climate: Impacts and Adaptation,* pp. 426–467.
- Kaczmarek, Z., Chapter 14. *Water Resources Management,* pp. 469–486.
- Solomon, A.M., Chapter 15. *Wood Production under Changing Climate and Land Use,* pp. 487–510.
- Everett, J.T., Chapter 16. *Fisheries,* pp. 511–537.
- McMichael, A.J., Chapter 18. *Human Population Health,* pp. 561–584.

**IPCC**, 1996: *Climate Change 1995: Economic and Social Dimensions of Climate Change. Contribution of Working Group III to the Second Assessment Report of the Intergovernmental Panel on Climate Change* [Bruce, J.P., H. Lee, and E.F. Haites (eds.)]. Cambridge University Press, Cambridge, United Kingdom, and New York, NY, USA, 448 pp.

- Pearce, D.W., W.R. Cline, A.N. Achanta, S. Fankhouser, R.K. Pachauri, R.S.L. Tol, and P. Vellinga, Chapter 6. *The Social Costs of Climate Change: Greenhouse Damage and the Benefits of Control,* pp. 179–224.

**Japan Meteorological Agency**, 1996: *Abnormal Weather Report, '94. Abnormal Weather and Climate Change of the World in the Recent Years, VI.* Printing Office of Ministry of Finance, Tokyo, 423 pp. (in Japanese)

**Kai**, K., M. Kainuma, and N. Murakoshi, 1996: Effects of global warming on the phenological observation in Japan. In: *Climate Change and Plants in East Asia* [Omasa, K., K. Kai, H. Toda, Z. Uchijima, and M. Yoshino (eds.)], Springer-Verlag, Tokyo, Japan, pp. 85–92.

**Kim**, Kwang-sik, Bong-nai Kwon, Sang-won Kim, Sung-sam Kim, Jung-knk Kim, Chin-nyun Kim, Chae-shif Rho, Chan-ho Park, Kwang-ho Lee, Doo-Hyeong Lee, Suk-woo Lee, Young-taek Lee, Chan Lee, Chang-hi Joung and Dong-wok Han, 1973: *The climate of Korea.* Iljisa Publishing House, Seoul, Korea, 446 pp. (in Korean)

**Kobak**, K.I., I.Y.E. Turchinovich, N. Yu. Kondrasheva, E.D. Schulze, W. Schulze, H. Koch, and N.N. Vygodskanya, 1996: Vulnerability and adaptation of the larch forest in eastern Siberia to climate change. *Water, Air, and Soil Pollution,* **92**, 119–127.

**Kojima**, S., 1992: *Forest Types and a Tentative Ecological Zonation of Kamtchatka Peninsula.* Proceedings of the Arctic Science Workshop, December 17, 1992, Arctic Environment Research Center, National Institute of Polar Research, Tokyo, Japan, pp. 92–96.

**Kojima**, S., 1996: Global climatic warming and possible responses of vegetation of Hokuriku region. *Japan. Jour. of Phytogeography and Taxonomy,* **44**, 9–18.

**Lin**, E., 1996: Agricultural vulnerability and adaptation to global warming in China. In: *Climate Change, Vulnerability and Adaptation in Asia and the Pacific* [Lin, E., W.C. Bolhofer, S. Huq, S. Lenhart, S.K. Mukherjee, J.B. Smith, and J. Wisniewski (eds.)]. *Water, Air and Soil Pollution,* **91(1–2)**, 63–73.

**Lin**, E., L. Chunzen, X. Deying, and D. Bilan, 1994: *Management of Vulnerable Natural Resources, in National Response Strategy for Global Climate Change: People's Republic of China.* Asia Development Bank Report TA 1690-PRC, Manila, Philippines, pp. 323–370.

**Liu**, C., 1997: The potential impact of climate change on hydrology and water resources in China. *Advances in Water Science,* **8(2)**, 21–32.

**McMichael**, A.J., A. Haines, R. Slooff, and S. Kovates (eds.), 1996: *Climate and Human Health.* World Health Organization, Geneva, Switzerland, 297 pp.

**Mijiddorj**, R., C.H. Yadamsren, and J. Baigalmaa, 1992: *Summer Rainfall Fluctuations in Gobi Region During Last Years.* Proceedings of the Symposium on Global Change and Gobi Desert, December 19, 1991, Hydrometeorological Research Institute, Centre of Projects on Development of Gobi, State Committee for Nature and Environmental Protection of Mongolia, Ulaanbaatar, Mongolia, pp. 64–68 (in Russian with English abstract).

**Mimura**, N., 1995: Responses of Coastal Landform to Sea-level Rise in Costal Zone in Japan. *Jour. Japan Society of Coastal Zone Studies,* **5**, 1–11 (in Japanese).

**Mooney**, H.A., J. Labchenco, R. Dirzo, and O.E. Sala, 1995: Biodiversity and ecosystem functioning: Basic principles. In: *Global Biodiversity Assessment* [Heywood, V.H., (ed.)]. Cambridge University Press, Cambridge, United Kingdom, pp. 279–325.

**Nakashizuka**, T., and S. Iida, 1995: Composition, dynamics and disturbance regime of temperate deciduous forests in Monsoon Asia. *Vegetatio,* **121**, 23–30.

**Nakatsugawa**, M., 1995: A study on effects of climate change on hydrologic processes in cold and snowy regions. *Civil Engineering Research Institute,* **106**, 5–22. (in Japanese)

**Nishimori**, M. and M. Yoshino, 1990: ENSO and its relation to generation, development and tracks of typhoons. *Geographical Review of Japan,* **63A(8)**, 530–540.

**Numata**, M., 1974: *The Flora and Vegetation of Japan.* Kodansha Ltd., Tokyo, Japan and Elsevier Science Publishers, Amsterdam, The Netherlands, 294 pp.

**Oh**, S., 1995: Study on impact of doubling atmospheric $CO_2$ climate change on lowland rice in Korea. *Jour. Korean Meteor. Soc.,* **31(3)**, 267–312.

**Omasa**, K., K. Kai, H. Toda, Z. Uchijima, and M. Yoshino, 1996: *Climate Change and Plants in East Asia.* Springer-Verlag, Tokyo, Japan, 215 pp.

**Pastor**, J., D.J. Mladenoff, Y. Haila, J. Bryant, and S. Payette, 1996: Biodiversity and ecosystem processes in boreal regions. In: *Functional Roles of Biodiversity: A Global Perspective* [Mooney, H.A,. (ed.)]. John Wiley and Sons, Ltd., Chichester, United Kingdom, 492 pp.

**Reilly**, R., N. Hohmann, and S. Kane, 1993: *Climate Change and Agriculture: Global and Regional Effects Using an Economic Model of International Trade.* MIT-CEEPR 93-012WP, August 1993, Massachusetts Institute of Technology Center for Energy and Environmental Policy Research, Cambridge, MA, USA, 78 pp.

**Rosenzweig**, C. and M. Parry, 1994: Potential impacts of climate change on world food supply. *Nature,* **367**, 133–138.

**Tokioka**, T., A. Noda, A. Kitoh, Y. Nikaidou, S. Nakagawa, T. Motoi, S. Yukimoto, and K. Takata, 1995: A transient $CO_2$ experiment with the MRI CGCM — quick report. *J. Meteor. Soc. Japan,* **73**, 817–826.

**Tsunekawa**, A., H. Ikeguhi, and K. Omasa, 1996a: Prediction of Japanese potential vegetation distribution in response to climatic change. In: *Climate Change and Plants in East Asia* [Omasa, K., K. Kai, H. Toda, Z. Uchijima, and H. Yoshino (eds.)], Springer-Verlag, Tokyo, Japan, pp. 57–65.

**Tsunekawa**, A., X. Zhang, G. Zhou, and K. Omasa, 1996b: Climatic change and its impacts on the vegetation distribution in China. In: *Climate Change and Plants in East Asia* [Omasa K, K. Kai, H. Toda, Z. Uchijima, and H. Yoshino (ed.)], Springer-Verlag, Tokyo, Japan, pp. 67–84.

**Ulziisaikhan**, V., 1996: *Impact Assessment of Climate Change on Forest Ecosystem in Mongolia.* Paper presented at the Regional Workshop on Climate Change Vulnerability and Adaptation in Asia and the Pacific, January 15–19, 1996, Manila, Philippines, pp. 1–10.

**United Nations Population Division**, 1993: *Annual Populations (1994 revision).* United Nations, New York, NY, USA (on diskettes).

**Xia**, X., 1993: *Desertification and Control of Blown Sand Disasters in Xianjiang.* Science Press, Beijing, China, 298 pp.

**Yan**, Z., and P. Marienicole, 1995: On the relationship between global thermal variations and wet/dry alterations in the Asia and African monsoon areas. *Acta Geographica Sinica,* **50(5)**, 471–479.

**Yatagai**, A. and T. Yasunari, 1994: Trends and decadal-scale fluctuations of surface air temperature and precipitation over China and Mongolia during the recent 40 year period (1951–1990). *Jour. Meteor. Society of Japan,* **72(6)**, 937–957.

**Yeh**, D.Z. and Y.X. Gao, 1979: *Meteorology of Qinghai-Xizhang (Tibet) Plateau.* Science Press, Beijing, China, 228 pp. (in Chinese with English abstract)

**Yoshino**, M., 1997: *Climate and Food Security.* Paper presented at the APN/SASCOM/IHDP/GCTE workshop on the Human Dimensions Programme, January 20–23, 1997, New Delhi, India, 15 pp.

**Yoshino**, M., T. Horie, H. Seino, H. Tsujii, T. Uchijima, and Z. Uchijima, 1987: The effect of climatic variations on agriculture in Japan. In: *The Impact of Climatic Variations on Agriculture, vol. 1: Assessments in Cool Temperature and Cold Regions* [Parry, M.L.,T.R. Carter, and N.T. Konijn (eds.)]. Reidl, Dordrecht, Netherlands, pp. 723–868.

**Yoshino**, M. and Y. Liu, 1997: Climate of the Turpan Basin: past and present. *Jour. of Arid Land Studies,* **6(2)**, 193–202.

**Yoshino**, M. and H.S.P. Ono, 1996: Variations in the plant phenology affected by global warming. In: *Climate Change and Plants in East Asia* [Omasa, K., K. Kai, H. Toda, Z. Uchijima, and H. Yoshino (eds.)]. Springer-Verlag, Tokyo, Japan, pp. 93–107.

**Yun**, J., 1990: Analysis of the climatic impact on Korean rice production under the carbon dioxide scenario. *J. Korean Meteor. Soc.,* **26(4)**, 263–274.

**Zhao**, Z., 1996: Climate change and sustainable development in China's semi arid regions. In: *Climate Variability, Climate Change and Social Vulnerability in the Semi-Arid Tropics.* Cambridge University Press, University Press, Cambridge, U.K., pp. 92–108.

# 11

# Tropical Asia

ROGER F. McLEAN (AUSTRALIA), S.K. SINHA (INDIA),
M.Q. MIRZA (BANGLADESH), AND MURARI LAL (INDIA)

Contributors:
*A. Achanta (India), S. Adhikary (Nepal), R.V. Cruz (Philippines), Lim Joo Tick (Malaysia), A.N. Purohit (India), A. Soegiarto (Indonesia), S. Soesanto (Indonesia), C. Tingsabadh (Thailand)*

# CONTENTS

# EXECUTIVE SUMMARY

The 16 countries of Tropical Asia range in size from about 61,000 ha (Singapore) to 300 million ha (India). The region is physiographically diverse and ecologically rich in natural and crop-related biodiversity. The present total population of the region is about 1.6 billion, and the population is projected to increase to 2.4 billion by 2025; although this population is principally rural, in 1995, the region included 6 of the 25 largest cities in the world. Exploitation of natural resources associated with rapid urbanization, industrialization, and economic development has led to increasing pollution, land degradation, and other environmental problems. Climate change represents a further stress. Over the long period of human occupation in the region, human use systems have developed some resilience to a range of environmental stresses. However, it is uncertain whether such resilience can continue in the face of rapid socioeconomic development, increasing population, and projected changes in climate.

### Climate Characteristics and Trends

Climate in Tropical Asia is characterized by seasonal weather patterns associated with the two monsoons and the occurrence of tropical cyclones in the two core areas of cyclogenesis (the northern Indian Ocean and the northwestern Pacific Ocean). Over the past 100 years, mean surface temperatures across the region have increased in the range of 0.3–0.8°C. No long-term trend in mean rainfall has been discernible over that period, although many countries have shown a decreasing trend in the past three decades. Similarly, no identifiable change in the number, frequency, or intensity of tropical cyclones has been observed in the region over the past 100 years; however, substantial decadal-scale variations have occurred.

### Ecological Systems

Substantial elevational shifts of ecosystems in the mountains and uplands of Tropical Asia are projected. At high elevations, weedy species can be expected to displace tree species, although the rates of vegetation change could be slow and constrained by increased erosion in the Greater Himalayas. Changes in the distribution and health of rainforest and drier monsoon forest will be complex. In Thailand, for instance, the area of tropical forest could increase from 45% to 80% of total forest cover; in Sri Lanka, a significant increase in dry forest and a decrease in wet forest could occur. Projected increases in evapotranspiration and rainfall variability are likely to have a negative impact on the viability of freshwater wetlands, resulting in shrinkage and dessication. Sea-level rise and increases in sea-surface temperature are the most probable major climate change-related stresses on coastal ecosystems. Coral reefs may be able to keep up with the rate of sea-level rise but may suffer bleaching from higher temperatures. Landward migration of mangroves and tidal wetlands is expected to be constrained by human infrastructure and human activities.

### Hydrology and Water Resources

The Himalayas play a critical role in the provision of water to continental monsoon Asia. Increased temperature and increased seasonal variability in precipitation are expected to result in accelerated recession of glaciers and increasing danger from glacial lake outburst floods. A reduction in flow of snow-fed rivers, accompanied by increases in peak flows and sediment yields, would have major impacts on hydropower generation, urban water supply, and agriculture. Availability of water from snow-fed rivers may increase in the short term but decrease in the long term. Runoff from rain-fed rivers may change in the future, although a reduction in snowmelt water would result in a decrease in dry-season flow of these rivers. Larger populations and increasing demands in the agricultural, industrial, and hydropower sectors will put additional stress on water resources. Pressure will be most acute on drier river basins and those subject to low seasonal flows. Hydrological changes in island and coastal drainage basins are expected to be small, apart from those associated with sea-level rise.

### Agriculture

The sensitivity of major cereal and tree crops to changes in temperature, moisture, and carbon dioxide ($CO_2$) concentration of the magnitudes projected for the region has been demonstrated in many studies. For instance, projected impacts on rice, wheat, and sorghum yields suggest that any increases in production associated with $CO_2$ fertilization will be more than offset by reductions in yield resulting from temperature and/or moisture changes. Although climate change impacts could result in significant changes in crop yields, production, storage, and distribution, the net effect of the changes regionwide is uncertain because of varietal differences and local differences in growing season, crop management, and so forth; noninclusion of possible diseases, pests, and microorganisms in crop model simulations; and the vulnerability of agricultural areas to episodic environmental hazards, including floods, droughts, and cyclones. Low-income rural populations that depend on traditional agricultural systems or on marginal lands are particularly vulnerable.

### Coastal Zones

Sea-level rise is the most obvious climate-related impact in coastal areas. Densely settled and intensively used low-lying coastal plains, islands, and deltas are especially vulnerable to coastal erosion and land loss, inundation and sea flooding, upstream movement of the saline/freshwater front, and seawater intrusion into freshwater lenses. Especially at risk are the large deltaic regions of Bangladesh, Myanmar, Viet Nam, and Thailand, and the low-lying areas of Indonesia, the Philippines, and Malaysia. Socioeconomic impacts could be felt in major cities, ports, and tourist resorts; artisinal and commercial fisheries; coastal agriculture; and infrastructure development. International studies have projected the displacement of several million people from the region's coastal zone in the event of a 1-m rise in sea level. The costs of response measures to reduce the impact of sea-level rise in the region could be immense.

### Human Health

The incidence and extent of some vector-borne diseases are expected to increase with global warming. Malaria, schistosomiasis, and dengue—which are significant causes of mortality and morbidity in Tropical Asia—are very sensitive to climate and are likely to spread into new regions on the margins of presently endemic areas as a consequence of climate change. Newly affected populations initially would experience higher case fatality rates. In presently vulnerable regions, increases in epidemic potential of 12–27% for malaria and 31–47% for dengue are anticipated, along with an 11–17% decrease for schistosomiasis. Waterborne and water-related infectious diseases, which already account for the majority of epidemic emergencies in the region, also are expected to increase when higher temperatures and higher humidity are superimposed on existing conditions and projected increases in population, urbanization rates, water quality declines, and other factors.

### Adaptation and Integration

Strategies for adapting to different climatic conditions will be quite diverse. For example, responses to impacts on agriculture will vary from region to region, depending on the local agro-climatic setting as well as the magnitude of climate change. New temperature- and pest-resistant crop varieties may be introduced, and new technologies may be developed to reduce crop yield losses. Countries in Tropical Asia could improve irrigation efficiency from current levels, to reduce total water requirements. Integrated approaches to river basin management, which already are used in a number of countries in the region, could be adapted regionwide. Such approaches could increase the effectiveness of adapting to the often-complex potential impacts of climate change that generally transcend political boundaries and encompass upstream and downstream areas. Similarly integrated approaches to coastal zone management can include current and longer-term issues, including climate change and sea-level rise.

## 11.1. Regional Characterization and Baseline Conditions

### 11.1.1. Introduction

This chapter is concerned with Tropical Asia, which extends over 80 degrees of longitude (from 70°E to 150°E) and 40 degrees of latitude (from 30°N to 10°S). The 16 countries that make up the region (see Box 11-1) range in area from about 61,000 ha (Singapore) to 300 million ha (India). Three states in the region (Nepal, Bhutan, and Laos) are landlocked. The region is physically diverse and ecologically rich in natural and crop-related biodiversity. The present total population is about 1.6 billion, and the population is projected to increase to 2.4 billion by 2025 (see Table 11-1); although the majority of this population is rural, in 1995, the region included 6 of the 25 largest cities in the world. Exploitation of natural resources associated with rapid urbanization, industrialization, and economic development has led to increasing pollution, declining water quality, land degradation, and other environmental problems.

Climate change represents an additional stress. Projected climate changes in the region include strengthening of monsoon circulation, increases in surface temperature, and increases in the magnitude and frequency of extreme rainfall events. Climate-related effects also will include sea-level rise. These changes could result in major impacts on the region's ecosystems and biodiversity; hydrology and water resources; agriculture, forestry, and fisheries; mountains and coastal lands; and human settlements and human health.

This chapter is based primarily on the Intergovernmental Panel on Climate Change (IPCC) Second Assessment Report (SAR) (IPCC, 1996), along with results from recent regional studies, such as the *Regional Workshop on Climate Change Vulnerability and Adaptation in Asia and the Pacific* held in Manila in January, 1996; national studies, such as the *Malaysian National Conference on Climate Change* held at the University of Agriculture in August, 1996; and other independent studies.

### 11.1.2. Physiography and Biogeography

Tropical Asia includes the major land masses of south and southeast Asia, as well as the long peninsulas that reach into the eastern Indian and western Pacific Oceans and the archipelagoes comprising the thousands of islands of Indonesia, Malaysia, India, and the Philippines (Figure 11-1). Physiographically, Tropical Asia is extremely complex and diverse. It contains the highest mountains on Earth, the deepest seas, and the largest number of islands—the latter making up what has been called the "maritime continent." The high mountains and the complex land-sea configuration have a strong influence on the weather and climate of the region.

Geologically, Tropical Asia straddles the boundary between the Indo-Australian and Pacific tectonic plates. The collision and subduction zones between these plates are characterized primarily by earthquakes, land movements, active volcanism, and volcanic eruptions, which are major environmental hazards. Earthquakes and land movements also are responsible for local differences in relative sea level. The massive eruptions of Krakatoa in Indonesia more than a century ago and of Mount Pinatubo in the Philippines in 1991 caused climatic cooling extending far beyond the region (Parker *et al.*, 1996).

Tropical Asia includes some large drainage basins, the sources of which extend into arid Asia to the north and west of the region. These basins include the Ganges, Brahmaputra, Meghna, Irrawaddy, and Salween Rivers, which drain into the Bay of Bengal and Andaman Sea; and the Nan, Mekong, and Red Rivers, which drain into the South China Sea. Varying impacts of climate change can be expected on the headwaters, broad valleys, and deltaic mouths of these catchments.

Biogeographically, the region includes the tropical Indo-Pacific borderlands, which have some of the greatest natural species diversity and productivity on Earth; coral reefs in the marine environment and tropical rainforests on land are prime examples. Ecological richness also is demonstrated by the region's crop and livestock diversity and by its large numbers of cultivars and varieties. The diversity of managed plants and animals is a result, in part, of climate and soil; it also reflects the length of human settlement in Tropical Asia, as well as its cultural and ethnic diversity. In addition, a number of animals, insects, and microbes have developed in parallel with the vegetation in this region.

Agriculture in Tropical Asia evolved on the basis of unique crop-related diversity, and only a few crops from other regions have become acceptable. Throughout the region, agriculture is critically important; it accounts for more than 30% of the gross domestic product (GDP) in at least seven countries. Climate-sensitive crops—such as rice, other grains and cereals, vegetables, and spices—are particularly important in the region. There is little doubt that agricultural systems in Tropical Asia have adapted to a range of environmental stresses over the region's long history of human settlement and land-use change. Whether such resilience can continue in the face of climate change and economic and population changes is uncertain, although it is expected that the processes

**Box 11-1. Countries in Tropical Asia**

| | |
|---|---|
| Bangladesh | Myanmar |
| Bhutan | Nepal |
| Brunei Darussalam | Papua New Guinea |
| Cambodia | Philippines |
| India | Singapore |
| Indonesia | Sri Lanka |
| Laos | Thailand |
| Malaysia | Viet Nam |

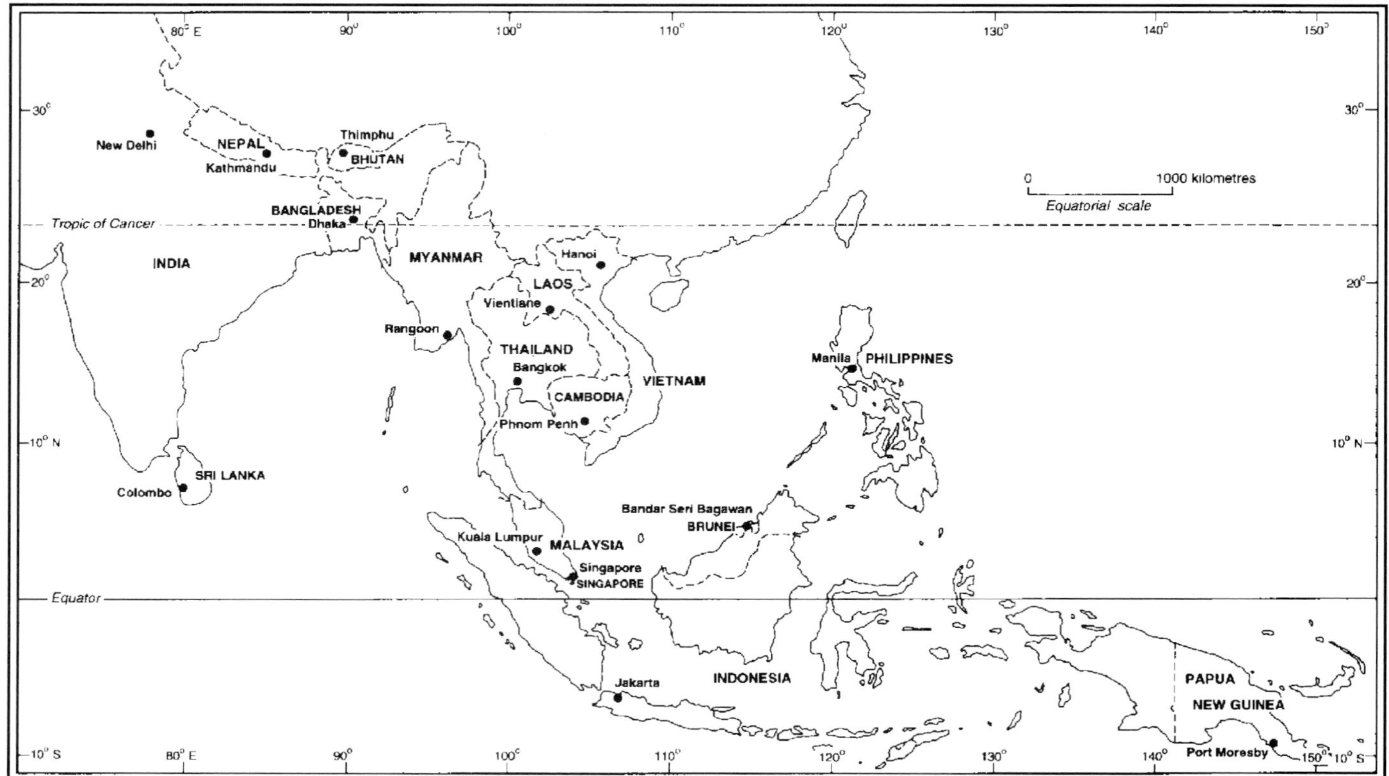


**Figure 11-1:** Location and countries covered in the Tropical Asia region.

of ongoing adaptation to changing environmental circumstances will continue.

### *11.1.3. Settlement, Population, and Economy*

Population estimates for the year 2025 for each country in Tropical Asia are given in Table 11-1, based on figures from WRI (1996); with the exception of Brunei, Singapore, Malaysia, and the Philippines, the populations of these countries are predominantly rural. By 2025, the region's total population is expected to grow from the present 1.6 billion to about 2.4 billion, with the largest increases expected in urban areas. In 1995, 6 of the world's 25 largest cities were located in Tropical Asia: Bombay (#5), Calcutta (#9), Jakarta (#11), Delhi (#17), Manila (#21), and Dhaka (#23); in terms of average annual growth rates during 1990–95, Dhaka was first (5.74%), Jakarta third (4.35%), and Bombay fifth (4.22%) among the 25 cities (see Table 11-2). In these and other cities of the region, rapid urban growth has been accompanied by a proliferation of slums and squatter settlements without access to basic infrastructure, clean water, and sanitation, with associated health risks (UNEP, 1997).

Rapid economic development and industrialization have characterized the economies of the region in recent years. In some countries (particularly in southeast Asia), there has been considerable expansion in the use of natural resources—including the exploitation of forests and fisheries, which has resulted in increasing environmental degradation.

## 11.2. Regional Climate

### *11.2.1. Present Climate Characteristics*

The climate of Tropical Asia is dominated by the two monsoons: The summer southwest monsoon influences the climate of the region from May to September, and the winter northeast monsoon controls the climate from November to February. The monsoons bring most of the region's precipitation and are the most critical climatic factor in the provision of drinking water and water for rain-fed and irrigated agriculture.

As a result of the seasonal shifts in weather, a large part of Tropical Asia is exposed to annual floods and droughts. The average annual flood covers vast areas throughout the region; in India and Bangladesh alone, floods cover 7.7 million ha and 3.1 million ha, respectively (GOI, 1992; Mirza and Ericksen, 1996). At least four types of floods are common: riverine flood, flash flood, glacial lake outburst flood, and breached landslide-dam flood (*bishayri*); the latter two are limited to mountainous regions of Nepal, Bhutan, Papua New Guinea, and Indonesia. Flash floods are common in the foothills, mountain borderlands, and steep coastal catchments; riverine floods occur along the courses of the major rivers, broad river valleys, and alluvial plains throughout the region.

Tropical cyclones also are an important feature of the weather and climate in parts of Tropical Asia. Two core areas of cyclogenesis exist in the region: one in the northwestern Pacific Ocean, which particularly affects the Philippines and Viet

**Table 11-1:** *Tropical Asia population: 1995 and 2025.*

| Country | Land Area (000 ha) | Population in 1995 (millions) | Estimated Population in 2025 (millions) | Urban Population in 1995 (%) | Estimated Urban Population in 2025 (%) |
|---|---|---|---|---|---|
| Bangladesh | 13,017 | 120.433 | 196.128 | 18 | 40 |
| Bhutan | 4,700 | 1.638 | 3.136 | 6 | 19 |
| Brunei | 527 | 0.285 | 0.389 | 58 | – |
| Cambodia | 17,652 | 10.251 | 19.686 | 21 | 44 |
| India | 297,319 | 935.744 | 1,392.086 | 27 | 45 |
| Indonesia | 181,157 | 197.588 | 275.598 | 35 | 61 |
| Laos | 23,080 | 4.882 | 9.688 | 22 | 45 |
| Malaysia | 32,855 | 20.140 | 31.577 | 54 | 73 |
| Myanmar | 65,755 | 46.527 | 75.564 | 26 | 47 |
| Nepal | 13,680 | 21.918 | 40.693 | 14 | 34 |
| Papua New Guinea | 45,286 | 4.302 | 7.532 | 16 | 32 |
| Philippines | 29,817 | 67.581 | 104.522 | 54 | 74 |
| Singapore | 61 | 2.848 | 3.355 | 100 | 100 |
| Sri Lanka | 6,463 | 18.354 | 25.031 | 22 | 43 |
| Thailand | 51,089 | 58.791 | 73.584 | 20 | 39 |
| Viet Nam | 32,549 | 74.545 | 118.151 | 21 | 39 |
| **Total** | 815,007 | 1,585.827 | 2,376.720 | | |

Source: WRI, 1996—Data Tables A.1 and 9.1.

Nam, and the other in the northern Indian Ocean, which particularly affects Bangladesh. Other extreme events include high-temperature winds, such as those that blow from the northwest into the Ganges valley during January.

In the megacities and large urban areas, high temperatures and heat waves also occur. These phenomena are exacerbated by the urban heat-island effect and air pollution.

Geographically much more extensive is the El Niño-Southern Oscillation (ENSO) phenomenon, which has an especially important influence on the weather and interannual variability of climate and sea level, especially in the western Pacific Ocean, South China Sea, Celebes Sea, and northern Indian Ocean. Indeed, the original historical record of El Niño events compiled by Quinn *et al.* (1978) considered the relationships among Indonesian droughts, the Southern Oscillation, and El Niño. For more recent analyses of historical ENSO teleconnections in the Eastern Hemipshere—based on teak tree-ring data from Java and a drought and famine chronology from India—see Whetton and Rutherford (1994, 1996).

The strength of such connections has been demonstrated in several other studies. Suppiah (1997) has found a strong correlation between the Southern Oscillation Index (SOI) and seasonal rainfall in the dry zone of Sri Lanka; Clarke and Liu (1994) relate recent variations in south Asian sea-level records to zonal ENSO wind stress in the equatorial Pacific. The influence of Indian Ocean sea-surface temperature on the large-scale Asian summer monsoon and hydrological cycle and the relationship between Eurasian snow cover and the Asian summer monsoon also have been substantiated (Sankar-Rao *et al.*,

**Table 11-2:** *Cities in Tropical Asia included in world's 25 largest cities, 1995.*

| City | Population (millions) | Rank in World Cities | Average Annual Growth Rate (1990–95) | Rank in World Cities |
|---|---|---|---|---|
| Bombay | 15.1 | 5 | 4.22 | 5 |
| Calcutta | 11.7 | 9 | 1.67 | 16 |
| Jakarta | 11.5 | 11 | 4.35 | 3 |
| Delhi | 9.9 | 17 | 3.80 | 6 |
| Manila | 9.3 | 21 | 3.05 | 8 |
| Dhaka | 7.8 | 23 | 5.74 | 1 |

Source: WRI, 1996—Data Table 1.1.

1996; Zhu and Houghton, 1996). Kripalani *et al.* (1996) studied rainfall variability over Bangladesh and Nepal and identified its connections with features over India.

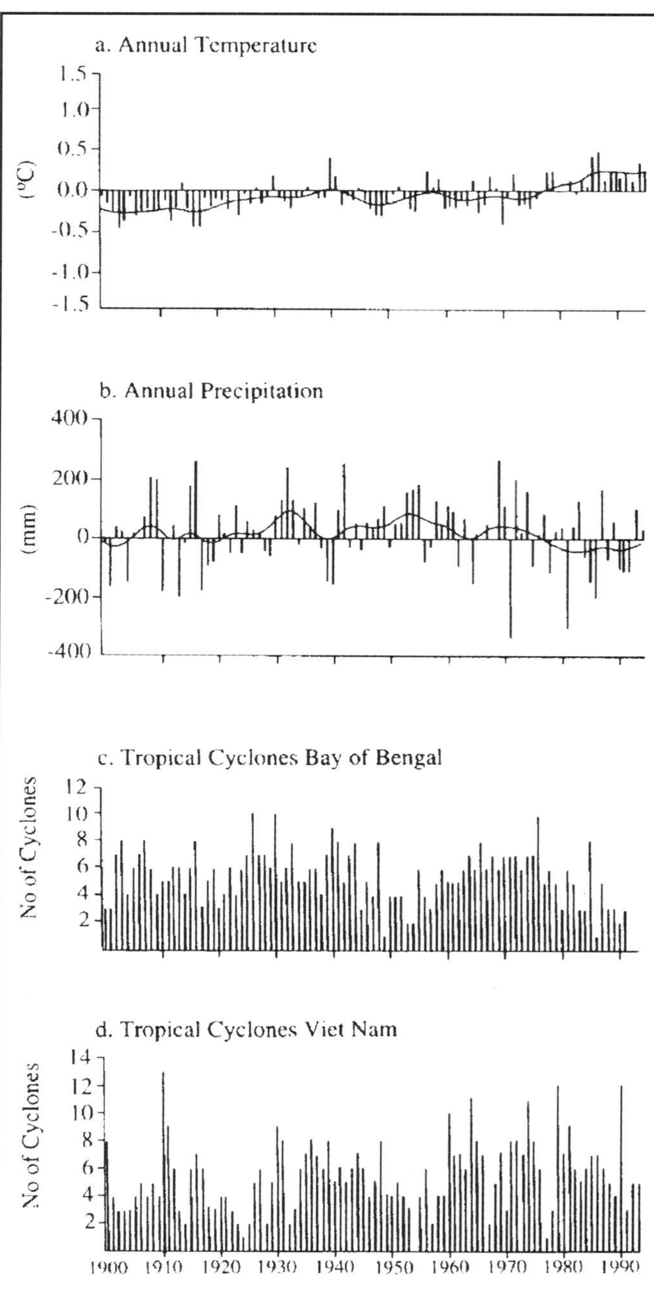


**Figure 11-2:** Time series of climate observations in Tropical Asia—a) observed annual temperature anomalies and b) observed precipitation anomalies, both relative to the 1961–90 means [longer term variations of the annual anomalies are emphasized by the smooth curve using a nine-point binomial filter (see Annex A)]; c) annual number of tropical cyclones over the Bay of Bengal, Arabian Sea, and land (Gadgill, 1996); and d) annual number of tropical cyclones that cross the coast of Viet Nam from the South China Sea (data supplied by Marine Hydrometeorological Center, Hydrometeorological Service, Hanoi, S.R. Viet Nam).

### *11.2.2. Recent Climate Trends and Variability*

#### *11.2.2.1. Temperature and Precipitation*

Over the past 100 years, mean surface temperatures have increased by 0.3–0.8°C across the region; the general increase is clearly evident in Figure 11-2a. Although there is no definite trend discernible in the long-term mean for precipitation for the region (Figure 11-2b) or in individual countries during this time period, many countries have shown a decreasing trend in rainfall in the past three decades. The southwest monsoon in India has shown definite changes in the period 1961–1990, in comparison with either 1901–1930 or 1931–1960, though no consistent longer-term trend is detectable. In Viet Nam, mean annual temperature has increased over the period 1895–1980, with net warming estimated at 0.27°C over the past two decades. Variations in rainfall also have been observed, with geographical variation. Since the 1960s, annual rainfall has been increasing in the north of Viet Nam and decreasing in the south (Granich *et al.*, 1993).

#### *11.2.2.2. Tropical Cyclones*

The frequencies of tropical cyclones in the core regions of cyclogenesis in Tropical Asia are quite different. Data for a 27-year period (1958–84) indicate that the average number of cyclones per year in the northern Indian Ocean is 5.5, whereas the northwestern Pacific experiences an average of 26.1 cyclones per year—representing 7% and 33%, respectively, of the total number of global cyclones (Climate Impact Group, 1992).

No identifiable changes in the number, frequency, or intensity of tropical cyclones or depressions have been observed in the northern Indian Ocean cyclone region (Bay of Bengal and Arabian Sea) over the past 100 years, although Gadgill (1995) has shown decadal-scale variations—with a rising trend during 1950–75 and a declining trend since that time (see Figure 11-2c). Similarly, there is evidence of substantial multidecadal variability in the northwestern Pacific, as shown by data from Viet Nam (see Figure 11-2d), though no clear evidence of long-term trends (Henderson-Sellars and Zhang, 1997). In Bangladesh, during the period 1948–1988, a total of 418 depressions and storms were formed, of which 79 were severe cyclonic storms (wind speed 69–117 km/hr). A time-series prepared by GOB (1989) indicates that, over the 40-year period, the highest number of depressions and storms formed in the 1960s, whereas the 1980s showed a decreasing trend. These findings are in broad agreement with regional trends.

#### *11.2.2.3. Trends in Climate-Related Phenomena: Floods and Droughts*

For the period 1871–1984, Parthasarathy *et al.* (1987) identified a range of 2–30 flood years (i.e., years when precipitation is at least 26% higher than normal) in the various meteorological subdivisions in India. In the same period, the range of severe flood years (i.e., precipitation more than 51% higher than normal) was between 1 and 14. Mirza *et al.* (1997) analyzed peak flood discharges recorded at various stations on the Ganges, Brahmaputra, and Meghna rivers in Nepal, India, and Bangladesh and found no conclusive increasing or decreasing trends. Similarly, no trends were detected in time-series data of flooded areas in various river basins of India and Bangladesh.

Droughts also can reach devastating proportions in Tropical Asia, although the incidence is variable in time and place. In India, Parthasarathy *et al.* (1987) identified a range of 1–12 severe drought years (i.e., precipitation 51% less than normal) for the various meteorological subdivisions of the subcontinent during the period 1871–1984. Chronically drought-affected areas cover the western parts of Rajasthan and the Kutch region of Gujrat (SAARC, 1992). In Bangladesh, about 2.7 million ha are vulnerable to annual drought; there is about a 10% probability that 41–50% of the country is experiencing drought in a given year (GOB, 1989). Drought or near-drought conditions also can occur in parts of Nepal (Sharma, 1979) and in Papua New Guinea and Indonesia, especially during El Niño events.

#### *11.2.2.4. Trends in Sea Level*

The IPCC Second Assessment Report (IPCC 1996, WG I, Section 7.2) indicates that:

- Global mean sea level has risen 10–25 cm over the last 100 years.
- There has been no detectable acceleration of sea-level rise during this century.

Warrick *et al.* (IPCC 1996, WG I, Section 7.4) suggest that the observed sea-level rise has been caused largely by increases in global temperatures and related factors, including thermal expansion of the ocean and melting of glaciers and ice caps. Changes in surface water and groundwater storage, along with tectonic movements and subsidence, also may have affected local sea levels.

Major difficulties in determining regional sea-level trends for Tropical Asia relate to the limited amount of historical tide-gauge data and the region's high decadal and interannual variability. For instance, only one station in the region (Bombay) is included in the list of stations with records exceeding 75 years used by Douglas (1992) to determine global acceleration in sea level; this study showed a -0.02 mm/yr acceleration for the 109-year record (1878–1987). With reference to Bangladesh, Warrick *et al.* (1996) indicate that existing tide-gauge data do not yet allow an unambiguous estimate of regional and local trends in relative sea-level change and their causes.

Studies of historical rates of relative sea-level rise in the South Asia Seas region, reported by Gable and Aubrey (1990), indicate an average annual relative sea-level rise of 0.67 mm/yr. In addition, during the past half-century or so, relative sea-level changes in the region have ranged from a fall (i.e., land emergence) of 1.33 mm/yr to a rise (i.e., land submergence) of 2.27 mm/yr.

In many parts of Tropical Asia, historical sea-level records appear to show not only a eustatic component but also local anthropogenic or meteorological effects. In Bangkok, for example, extraction of water from groundwater aquifers has resulted in accelerated land subsidence (i.e., a relative rise in sea level) of around 20 mm/yr since 1960, compared with an earlier trend of about 3 mm/yr (see IPCC 1996, WG II, Section 9.3.1); in contrast, high recent rates of sea-level rise in Manila have been blamed on coastal reclamation (Spencer and Woodworth, 1993). Analysis of time-series data for 1955–1990 indicates an average sea-level rise of 1.9 mm/yr at Hondau in North Viet Nam. This finding is in broad agreement with the observed rise in global mean sea level (Granich *et al.*, 1993). Meteorological effects generated by zonal interannual winds blowing along the Equator in the Pacific Ocean have been detected in the interannual sea-level signal that occurs along more than 8,000 km of Indian Ocean shoreline, extending from southern Java to Bombay (Clarke and Liu, 1994).

### *11.2.3. Projections of Future Climates: Regional Scenarios*

#### *11.2.3.1. Background Information*

The first detailed climate scenario for south and southeast Asia was developed by the Climate Impact Group (1992), as part of the Asian Development Bank's 1994 regional study on global environmental issues (including climate change); the methodology is outlined by Whetton (1994). Analyses were carried out using data obtained in experiments with four general circulation models (GCMs): the Canadian Climate Centre model (CCCJ1), the United Kingdom Meteorological Office model (UKMOH), the Geophysical Fluid Dynamics model (GFDLA), and the Australian CSIRO9 model. All of the experiments used an atmospheric model coupled to a simplified ocean model and were run to equilibrium conditions for present levels of greenhouse gases (GHGs) and for doubled $CO_2$ levels.

Comparisons of the control simulation of the GCMs with present-day climatology included results for the region's surface temperature, mean sea-level pressure, and precipitation. The broad-scale observed patterns are well simulated, including the summer and winter monsoons. Results described by Suppiah (1994), based on the aforementioned models, indicated strengthening of the monsoon circulation and an increase in wet-season rainfall under enhanced greenhouse conditions—that is, an increase in summer rainfall in the southwest monsoon region and an increase in winter rainfall in the northeast monsoon region.

#### *11.2.3.2. Temperature and Rainfall*

Temperature and rainfall scenarios are based on simulated changes averaged over two broad seasons: the southwest

***Table 11-3:*** *Temperature change scenarios for 2010 and 2070 (°C).*

| Region | Year 2010 | 2070 |
|---|---|---|
| Indonesia, Philippines, and coastal south and southeast Asia | 0.1–0.5 | 0.4–3.0 |
| Inland south and southeast Asia (not south Asia in June-July-August) | 0.3–0.7 | 1.1–4.5 |
| Inland south Asia in June-July-August | 0.1–0.3 | 0.4–2.0 |

Source: Whetton, 1994.

monsoon and the northeast monsoon. Temperature scenarios for Tropical Asia reported by Whetton (1994) and the Climate Impact Group (1992) suggest that temperature would increase throughout most of the region, although the amount of warming is projected to be less than the global average. Moreover, results presented in Table 11-3 indicate that there may be differences within the region, depending on proximity to the sea. Thus, warming is projected to be least in the islands and coastal areas throughout Indonesia, the Philippines, and coastal south Asia and Indo-China and greatest in inland continental areas of south Asia and Indo-China—except from June to August in south Asia, where reduced warming could occur.

In terms of rainfall, the models considered by Whetton (1994) suggested an April-to-September maximum over south Asia and the Indo-China peninsula and a minimum over Indonesia and areas near Australia. Projections of regionally averaged changes in rainfall for the years 2010 and 2070 are given in Table 11-4.

Other simulations of changes in rainfall indicated a tendency for an increase in wet-season rainfall in both monsoon regions, with changes ranging from -5% to +18% (Climate Impact Group, 1992). More consistent and much larger rainfall increases are projected for the south Asia subregion wet season, with values ranging from +17% to +59%. Changes in dry-season rainfall are less consistent and are estimated only as broad-scale regional rainfall changes; local-scale changes could be much greater.

Systematic increases in average rainfall intensity are a common feature in simulated daily rainfall experiments, along with associated increases in the projected frequency of heavy rainfall events. Whetton *et al.* (1994) conclude that there is reason for higher confidence in increasing rainfall intensity in south and southeast Asia under enhanced greenhouse conditions than in increases or decreases in total rainfall in particular regions.

### *11.2.3.3. Tropical Cyclones*

Current atmosphere-ocean general circulation model (AOGCM) simulations have limited ability to suggest likely changes in tropical cyclone activity; until high-resolution GCM simulations are available, it is not possible to say whether the frequency, area of occurrence, time of occurrence, mean intensity, or maximum intensity of tropical cyclones will change in Tropical Asia. However, more recent studies with models of higher resolution appear to be able to simulate tropical cyclone climatology (Bengtsson *et al*., 1995). Bengtsson *et al.* (1995) found a decrease in the number of tropical cyclones under enhanced greenhouse scenarios, especially in the Southern Hemisphere, although their geographical distribution remained unchanged.

A study by Holland (1997) indicated that tropical cyclones are unlikely to be more intense than the worst storms experienced under present-day climate conditions—although some potential exists for changes in cyclone intensity in tropical oceanic

***Table 11-4:*** *Rainfall scenarios for 2010 and 2070 (% change).*

| Region | 2010 Wet Season | 2010 Dry Season | 2070 Wet Season | 2070 Dry Season |
|---|---|---|---|---|
| *Southwest Monsoon Region* India, Pakistan, Bangladesh, Philippines (western part), and Viet Nam (except east coast) | 0 | 0 | 0 to 10 | -10 to +10 |
| *Northeast Monsoon Region* Indonesia, Philippines (east part), Viet Nam (east coast), Sri Lanka, and Malaysia | 0 to -5 | 0 | -5 to +15 | 0 to +10 |
| *South Asia Subregion* (15–30°N; 65–95°E) | 0 to +10 | -5 to +5 | +5 to +50 | -5 to +20 |

Source: Whetton, 1994.

regions, where sea-surface temperature (SST) is betwen 26°C and 29°C. According to Henderson-Sellars and Zhang (1997), recent studies indicate that the maximum potential intensities of cyclones will remain the same or undergo a modest increase of up to 10–20%; they add that these predicted changes are small compared with observed natural variations and fall within the uncertainty range in current studies. Lal *et al.* (1995a), in their ECHAM3-T106 experiment, found no significant change in the number and intensity of monsoon depressions in the Indian Ocean in a warmer climate. Similarly, likely changes in the ENSO phenomenon under enhanced greenhouse conditions and their possible impact on the interannual variability of the summer monsoon are not known.

The spatial patterns of increases in surface temperature during winter (DJF) and summer (JJA), simulated by the more recent coupled AOGCMs, tend to confirm, to a large extent, the projections of the Climate Impact Group (1992) for the monsoon Asian region. Area-averaged increases in temperature for doubled $CO_2$ conditions are, however, lower by 15–20% in the transient experiments (using AOGCMs and based on a 1% increase in $CO_2$) than in equilibrium experiments (using atmospheric GCMs and based on an instantaneous doubling of $CO_2$ and a mixed-layer ocean). The area-averaged increase in summer monsoon rainfall and its spatial distribution are highly variable among GCMs, although all models produce an increase in monsoon rainfall.

### *11.2.4. A Recent Scenario for the Indian Subcontinent*

The response of the monsoon climate to transient increases in GHGs and sulfate aerosols in the Earth's atmosphere has recently been examined by Lal *et al.* (1995b), using data generated by the MPI-ECHAM3 atmospheric model, coupled to a large-scale geostrophic ocean model (ECHAM3 + LSG). The authors identified the potential role of sulfate aerosols in obscuring GHG-induced warming over the Indian subcontinent during the past century and found that year-to-year variability in simulated monsoon rainfall for the past century is in fair agreement with observed climatology.

Lal *et al.* (1995b) then presented a scenario of climate change for the Indian subcontinent for the middle of the next century, taking into account projected emissions of GHGs and sulfate aerosols. They suggested an increase in annual mean maximum and minimum surface air temperatures of 0.7°C and 1.0°C over the land regions in the decade of the 2040s with respect to the 1980s. This warming would be less pronounced during the monsoon season than in the winter months. A significant decrease in the winter diurnal temperature range, with no appreciable change during the monsoon season, also is projected (Lal *et al.*, 1996). Moreover, projected warming of the land region of the Indian subcontinent is likely to be relatively lower in magnitude than that of the adjoining ocean, resulting in a decline in the land-sea thermal contrast—the primary factor responsible for the onset of summer monsoon circulation.

As a consequence—and contrary to simulations that consider only $CO_2$ forcing—Lal *et al.* (1995b) found a decline in mean summer monsoon rainfall of about 0.5 mm/day over the region, which is marginally above the range of interannual variability for the present-day atmosphere. This decline in summer monsoon rainfall resulting from combined GHG and aerosol forcing in the southeast Asian region also has been suggested in the UKMO GCM experiments (see Figure B-2).

Despite the great strides that have been made in objective modeling of climate and climate change through AOGCMs, there still are uncertainties associated with model projections, especially for climate change on regional scales. The problem is even more complicated when researchers attempt to project time-dependent regional climatic responses to future increases in radiative forcing from anthropogenic GHGs and aerosols. Because the anthropogenically induced sulfate aerosol burden has large spatial and temporal variations in the atmosphere, its regional-scale impacts could be in striking contrast to the impacts of GHGs—the concentrations of which are likely, in most cases, to change uniformly throughout the globe. Spatially localized radiative forcing resulting from anthropogenic aerosols is confined largely to the Northern Hemisphere and tends to yield a steepening of normalized meridional temperature gradient in that hemisphere. The effects of aerosols also yield distinct precipitation responses in the tropical region in general and over the Asian monsoon region in particular. The implications of enhanced aerosol loading on tropospheric clouds, which could strongly modulate the monsoon climate, still are not clear. Because anthropogenic sulfate aerosol loadings are projected to be substantial over the southeast Asian region, their impact on the Asian summer monsoon needs to be more carefully examined. Their precise magnitude, as well as the roles of other localized potential forcings (e.g., aerosols from biomass burning, increases in tropospheric ozone), also must be known before confident predictions of regional changes in the monsoon climate and its variability can be made.

## 11.3. Vulnerabilities and Potential Impacts

### *11.3.1. General Comments*

This section discusses impacts that could plausibly result from projected climate changes. In many studies relevant to Tropical Asia, however, specific scenarios are not identified; in several cases, assumptions also have been made about the magnitude and timing of potential climate changes. Four other points are of special relevance. First, climate change represents an important additional external stress on the numerous ecological and socioeconomic systems in Tropical Asia that already are adversely affected by air, water, or land pollution, as well as increasing resource demands, environmental degradation, and nonsustainable management practices. Second, most of Tropical Asia's ecological and socioeconomic systems are sensitive to the magnitude and the rate of climate change. They also have developed a resilience through their long history of adaptation to environmental and cultural changes; identifying resilient features is a particular challenge. Third, Tropical Asia's physical environment is extremely diverse, and traditional systems of land use are very closely adapted to these conditions. Such diversity has important implications for assessing the impacts of future climate change, which would vary greatly from area to area, depending not only on the climate change scenario but also on specific local conditions and changes in factors such as population and technology. Finally, the potential impacts of climate change in Tropical Asia rarely have been quantified. Thus, the results of impact studies usually are qualitative and often have been directed toward identifying the sensitivities of specific systems or sectors. In this regard, considerable progress has been made in recent years in assessing the sensitivity and vulnerability of some ecosystems, activities, resources, and environments in the region. This progress is illustrated by the following discussion. Figure 11-3 summarizes the potential impacts of some of the components of climate change in Tropical Asia.

### *11.3.2. Ecosystems and Biodiversity*

Tropical Asia is rich in natural and managed (crop-related) biodiversity. The richness in terrestrial and coastal ecosystems can be gauged from the following facts:

- The rainforests of southeast Asia contain about 10% of the world's floral diversity.

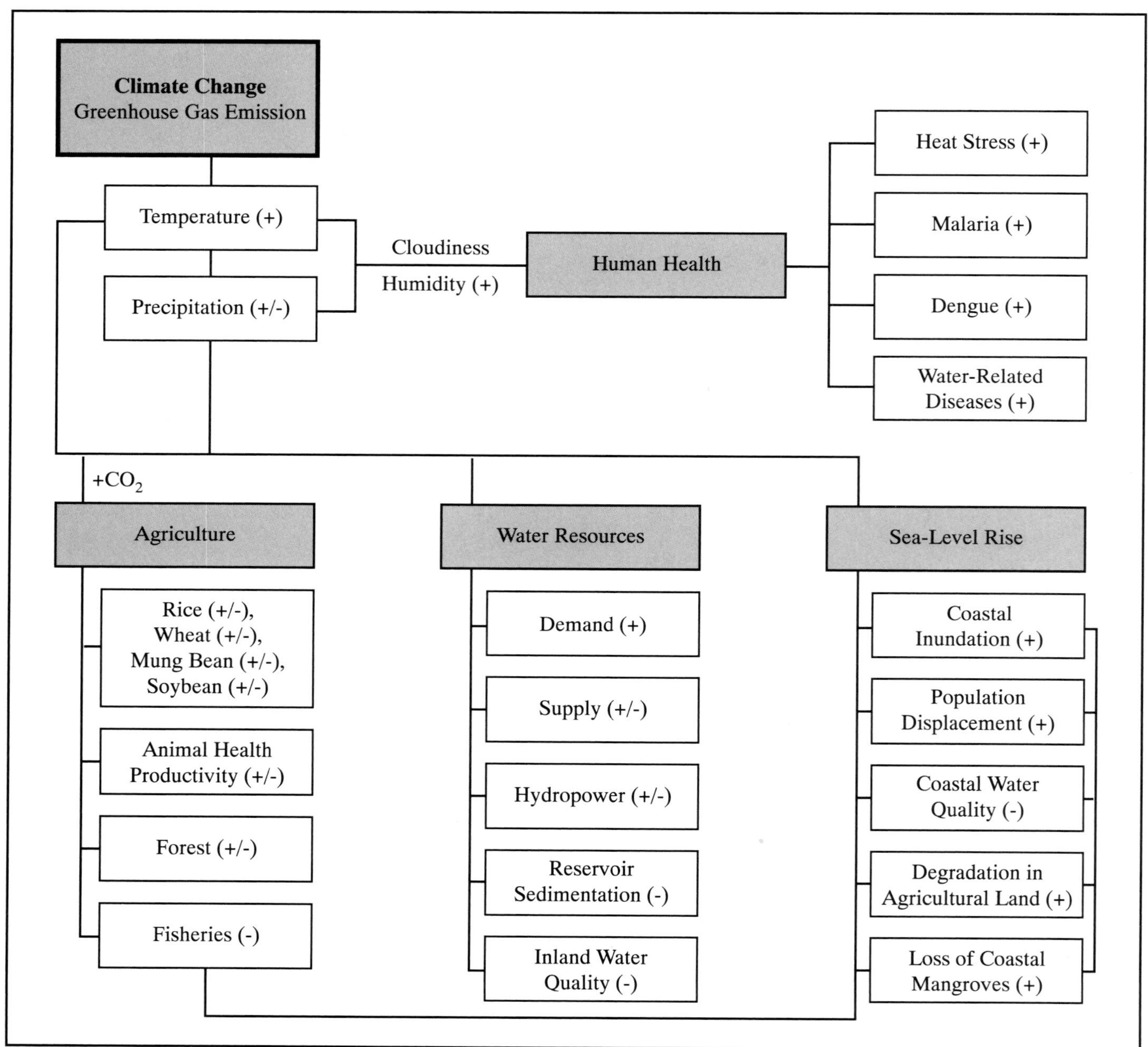


**Figure 11-3:** Possible climate change-related impacts in Tropical Asia (+ = projected increase, - = projected decrease).

- The Indo-Pacific borderlands are the center of global shallow-water diversity, containing the greatest number of coral, mangrove, sea grass, and coastal fish genera in the world.

In both cases, however, habitat loss and degradation have reached acute levels (UNEP, 1997).

Potential climate change-related impacts are highlighted in the following examples, along with other types of human-related impacts.

### *11.3.2.1. Terrestrial Ecosystems*

#### *11.3.2.1.1. Tropical forests*

Forests are extremely sensitive to climate change, as well as to other stresses. Section 1.4 of the Working Group II contribution to the SAR (IPCC, 1996) concludes that tropical forests are likely to be more affected by changes in land use than by climate change, as long as deforestation continues at its current high rate; that any degradation of these forests, whether it is caused by climate or land-use changes, will lead to an irreversible loss in biodiversity; and that tropical forests are likely to be more affected by changes in soil water availability (caused by the combined effects of changes in temperature and rainfall) than by changes in temperature per se.

Details on the potential impacts of climate change on the forests of Tropical Asia are provided in Section 11.3.4.2.

#### *11.3.2.1.2. Freshwater wetlands*

Freshwater wetlands or peatlands are accumulations of decomposed organic matter that have an ability to store water, reduce floodwater peaks, and provide water during dry periods. More than 10% (43 million ha) of global peatlands are located in the humid tropical zone of Asia; by far the largest area of peatlands is in Indonesia, especially the Kalimantan region, followed by Malaysia and Papua New Guinea (Rieley *et al*., 1995). Tropical peatlands require high rainfall input and are sensitive to drought and changes in land use; impacts on drainage, for example, compromise the sustained utilization of the peatland resource. Projected increases in evapotranspiration and increased rainfall variability are likely to have negative impacts on the viability of peatlands, resulting in dessication and shrinkage.

#### *11.3.2.1.3. Montane ecosystems*

Changes in high-elevation ecosystems can be expected as a consequence of global warming. For instance, the present distribution of species is projected to shift to higher elevations, although the rates of vegetation change are expected to be slow, and colonization success would be constrained by increased erosion and overland flows in the highly dissected and steep terrain of the Greater Himalayas. Weedy species with a wide ecological tolerance will have an advantage over others. High-elevation tree species—such as *Abies, Acer,* and *Betula*—prevail in cold climates because of their adaptations to chilling winters. Temperature increases would result in competition between such species and new arrivals. The sensitivity of plants to climatic factors and, in particular, water stress in the summit region of Mt. Kinabalu, Borneo—the highest mountain in southeast Asia—recently has been demonstrated (Kitayama, 1996).

#### *11.3.2.1.4. Crop biodiversity*

In addition to the biodiversity of the region's natural terrestrial ecosystems, Tropical Asia has a unique crop-related biodiversity. In particular, the Indochina-Indonesian and Hindustani regions have been identified as centers of diversity of a great many crops and other economically important plants that originate in these regions. Tropical Asia abounds in diverse cereal, tree, spice, and fruit species, which are endowed with unique genetic properties. Thousands of accessions of rice, cucurbits, legumes, mangoes, citrus, and other species and varieties have been collected and conserved. Maintenance of this diversity to allow further evolution to occur in response to climate change may be important for future generations, particularly when the effects of increasing human populations also are taken into account. Moreover, because of their interdependence and shared evolutionary history, the plants, animals, and microbes in the region strongly influence each other. For example, a change in insect population and diversity could influence the evolution of plant biodiversity, and vice-versa. These subjects—which are the basis of ecological sustenance, crop and animal improvement, and ultimately economic growth—remain unexplored.

### *11.3.2.2. Coastal Ecosystems*

The Indo-Pacific borderlands contain the greatest shallow-water and intertidal biodiversity on Earth. Coral reefs have the highest biodiversity of any marine ecosystem; coral reef macrobiota represent about 4–5% of the described global biota. Coral reef diversity and the majority of the region's coral genera are centered around the archipelagoes of the Philippines and Indonesia. Similar rich concentrations of species in the region occur in mangroves and tropical sea grasses (Woodroffe, 1990; Mukai, 1993).

#### *11.3.2.2.1. Coral reefs*

The effects of climate change on coral reefs have been reviewed by Wilkinson and Buddemeier (1994). Brown *et al.* (1994) have shown that coral reefs are particularly sensitive to prolonged increases in seawater temperature and increased irradiance. Oxidation stress during exposure to elevated temperatures also has been shown to cause coral bleaching (Lesser, 1997). In Tropical Asia, projected increases in surface air temperature could lead to increases in seawater temperature, posing a major threat to the health of coral reefs. In Indonesia, where severe bleaching took place as a result of seawater warming during an ENSO event in 1983, coral reefs have failed to show continued recovery beyond the initial recovery noted in 1988 (Brown and Suharsono, 1990).

#### *11.3.2.2.2. Mangroves*

Mangroves constitute a significant part of tropical coastal biodiversity. They are found throughout Tropical Asia, where they occupy more than 75,000 km$^2$, or 40% of the world's total area of mangroves. Indonesia alone, with 42,530 km$^2$, accounts for 23% of the world total; India (with 6,700 km$^2$), Malaysia (with 6,424 km$^2$), and Bangladesh (with 5,767 km$^2$) each have over 3% of the world total (Spalding, 1997). Over the past few decades, mangrove forests in Tropical Asia have declined considerably as a consequence of human activities. The Philippines and Thailand have experienced reductions of 60% and 55%, respectively, over 25 years. Between 1980 and 1990, the area of mangroves declined by more than 37% in Viet Nam and more than 12% in Malaysia. These figures suggest a loss of nearly 7,500 km$^2$ of mangrove forests in these four countries alone—representing more than 4% of the current global total.

Mangroves may be affected by climate change-related increases in temperature and sea-level rise. Although the temperature effect on growth and species diversity is not known, sea-level rise may pose a serious threat to these ecosystems. In Bangladesh, for instance, there is a threat to species in the three distinct ecological zones that make up the Sundarbans—the largest continuous mangrove area in the world. If the saline water front moves further inland, *Heritiera fomes* (the dominant species in the landward freshwater zone) could be threatened. Species in the other two ecological zones (*Excoecaria agallocha* in the moderately saltwater zone and *Ceriops decandra* in the saltwater zone) also could suffer. These changes could result in economic impacts: Direct employment supported by the Sundarbans is estimated to be in the range of 500,000–600,000 people for at least half of the year (ESCAP, 1987), and a large number of these people—who are directly employed in the industries that use raw materials from the Sundarbans (e.g., woodcutting; collection of thatching materials, honey, beeswax, and shells; fishing)—may lose their sources of income. Sea-level rise also may threaten a wide range of mammals, birds, amphibians, reptiles, and crustaceans living in the Sundarbans. A large part of the coastal area that now is protected by the Sunderbans will be vulnerable to cyclonic storms and surges. In Thailand, the potential impact of sea-level rise on the mangrove community is a serious issue (Aksorakaoe and Paphavasit, 1993).

### *11.3.3. Hydrology and Water Resources*

#### *11.3.3.1. Introduction*

The impact of climate change on the water resources of Tropical Asia may be significant. Whetton *et al.* (1994) suggest that increased evaporation (resulting from higher temperatures), combined with regional changes in precipitation characteristics (e.g., total amount, variability, and frequency of extremes), has the potential to affect mean runoff, frequency, and intensity of floods and droughts; soil moisture; and water supplies for irrigation and hydroelectric generation. Anglo *et*

***Table 11-5:*** *Water resources and use in Tropical Asia.*

| Country | Annual Internal Renewable (km$^3$) | Annual Withdrawal (km$^3$) | % of Water Resources | Sectoral Withdrawal (%) Domestic | Industry | Agriculture |
|---|---|---|---|---|---|---|
| Bangladesh | 2,357.0 | 22.50 | 1 | 3 | 1 | 96 |
| Bhutan | 95.5 | 0.02 | 0 | 36 | 10 | 54 |
| Cambodia | 496.1 | 0.52 | 0 | 5 | 1 | 94 |
| India | 2,085.0 | 380.00 | 18 | 3 | 4 | 93 |
| Indonesia | 2,530.0 | 16.59 | 1 | 13 | 11 | 76 |
| Laos | 270.0 | 0.99 | 0 | 8 | 10 | 82 |
| Malaysia | 456.0 | 9.42 | 2 | 23 | 30 | 47 |
| Myanmar | 1,082.0 | 3.96 | 0 | 7 | 3 | 90 |
| Nepal | 170.0 | 2.68 | 2 | 4 | 1 | 96 |
| Philippines | 323.0 | 29.50 | 9 | 18 | 21 | 61 |
| Singapore | 0.6 | 0.19 | 32 | 45 | 51 | 4 |
| Sri Lanka | 43.2 | 6.30 | 15 | 2 | 2 | 96 |
| Thailand | 179.0 | 31.90 | 18 | 4 | 6 | 90 |
| Viet Nam | 376.0 | 28.90 | 8 | 13 | 9 | 78 |

Source: WRI, 1996—Data Table 13.1.

*al.* (1996) confirm that water resources in the region are very sensitive not only to changes in temperature and precipitation but also to changes in tropical cyclones. For example, runoff in Nepal is affected by snow cover, the southwest monsoon, and cyclones—all of which may be affected by climate change.

Water resources in the region also are vulnerable to increasing demand resulting from population growth, urbanization, industrialization, and agriculture (Schreir and Shah, 1996). At present, agriculture is the predominant water use throughout the region (see Table 11-5).

### *11.3.3.2. Hydrological Systems*

The effects of climate change on hydrology in Tropical Asia would have many facets. In the Himalayas, the storage of precipitation in the form of snow and ice (in glaciers) over a long period provides a large water reservoir that regulates annual water distribution. The majority of rivers originating in the Himalayas have their upper catchments in snow-covered areas and flow through steep mountains. This factor and the perennial nature of the rivers provide excellent conditions for the development of hydropower resources, despite the temporal variability in the sources of runoff for Himalayan rivers. Studies of one large catchment in the western Himalayas (the Chenab, a tributary of the Indus) show that the average snowmelt and glacier-melt contribution to the annual flow is 49.1%; a significant proportion of runoff is derived from snow in the dry season, when water demand is highest (Singh *et al.*, 1997). Climate change-related increases in temperature also could increase the rate of snowmelt and reduce the amount of snowfall, if the winter is shortened.

If climate change does alter the rainfall pattern in the Himalayas, the impacts could be felt in the downstream countries—that is, India and Bangladesh. By and large, dry-season flow in the major Himalayan rivers in a given year results from the monsoon rainfall of the previous year. Catchments in Nepal supply about 70% of the dry-season flow of the Ganges River, and tributaries of the Brahmaputra River originating in Bhutan supply about 15% of the total annual flow of that river. If climate change disrupts these resources and alters mountain hydrological regimes, the effects will be felt not only in the montane core of Tropical Asia but also downstream, in countries that depend on this water resource.

Any change in the length of the monsoon also would be significant. For instance, if the southwest monsoon arrives later or withdraws earlier, soil moisture deficits in some areas may get worse. On the other hand, prolonged monsoons may contribute to more frequent flooding and increase the depth of inundation in many parts of the large river basins. If precipitation increases, dry-season river flow may increase because of increased recharge in the monsoon season. For instance, flooding is experienced on the east coast of Malaysia and in the coastal areas of Sarawak and Sabah almost every year during the rainy northeast monsoon season; these floods become severe and catastrophic with heavy rainstorms around the same time of the year (Sooryanarayana, 1995). Increases in rainfall during the northeast monsoon, as well as increases in the magnitude of extreme rainfall events—both of which have been projected with climate change—are expected to increase the frequency and intensity of flooding in the region.

Divya and Mehrotra (1995) examined regional effects of climate change on various components of the hydrological cycle—such as surface runoff, soil moisture, and evapotranspiration—by applying a conceptual model on a monthly time scale. The experiment was conducted using hypothetical scenarios in three drainage basins located in different agroclimatic zones of central India. The authors found that basin characteristics—such as soil type, moisture-holding capacity, and runoff coefficients—significantly influence basin runoff. More recently, Mirza (1997) modeled nine subbasins of the Ganges River using the GISS transient scenario (a global coupled ocean-atmosphere climate model with 4x5-degree resolution). The GCM experiment is based on 74 years, during which $CO_2$ increases by 1% (compounded) per year. Mirza (1997) found that changes in mean annual runoff in the range of 27–116% occurred in the subbasins at doubled $CO_2$ and that runoff was more sensitive to climate change in the drier subbasins than in the wetter subbasins.

### *11.3.3.3. Runoff and River Discharge*

Water resources in Tropical Asia could be significantly affected by climate change. Results from early greenhouse warming simulations in MPI-ECHAM3 model experiments reported in Lal (1994) suggest that mean annual surface runoff throughout the region could increase by about 15% by the end of the century, with the greatest increases in northeast India and Indonesia. However, a more recent study (Lal *et al.*, 1995b) that takes into account the combined effects of GHGs and aerosols suggests a future decline in surface runoff throughout the region.

In many parts of the region, the ratio of monsoon-to-dry-season runoff is very high (e.g., nearly 6:1 for the Ganges River). Because water requirements for agriculture and other water-use sectors are significantly higher during the dry season than during the monsoon, water supply in the dry season often cannot meet demand. Any change in the availability of water resources as a consequence of climate change may have a substantial effect on agriculture, navigation, fisheries, industrial and domestic water supply, reservoir storage and operation, and salinity control (Divya and Mehrotra, 1995; Mirza and Dixit, 1997).

Mirza (1997) applied empirical models using standardized precipitation scenarios from the CSIRO9, UKMO, GFDL, and LLNL GCMs to estimate changes in mean annual discharge of the Ganges and Brahmaputra rivers in Bangladesh. For the CSIRO9-E1 (Whetton *et al.*, 1993), GFDL-A2 (Whetherald and Manabe, 1986), and LLNL GCM experiments, the global warming value is equivalent to the climate sensitivity of the climate model (i.e., the equilibrium global mean temperature change for a doubling of $CO_2$-equivalent concentration). For the UKMO-X5 GCM experiment (Murphy and Mitchell, 1995), the changes are defined as the difference between the mean climate state of one decade in the perturbed integration minus the climate state of the equivalent decade in the control integration. In the study, a 2–6°C temperature increase was considered; under a 4°C global mean temperature change scenario, mean annual discharge of the Ganges River could increase by 27% (CSIRO9), 42% (UKMO), 15% (GFDL), or 2% (LLNL). Under the same temperature scenario, mean annual discharge of the Brahamaputra River may change by -0.1%, +13%, +9%, and +2% for the four GCMs, respectively. For higher temperature increases, the changes vary linearly.

The sediment load transported by these rivers is a nonlinear function of discharge. The implications of these increases in discharge on the sediment load of the Ganges and Brahamaputra—which already carry an extraordinarily heavy sediment load and have a high, though irregular, rate of downstream deposition—could be severe (Subramanian and Ramanathan, 1996).

Riebsame *et al.* (1995) conducted a study on the potential effect of climate change on water resources of the Mekong River basin. Equilibrium precipitation scenarios for doubled $CO_2$ from the GISS (Hansen *et al.*, 1983), GFDL (Mitchell *et al.*, 1990), and UKMO (Wilson and Mitchell, 1987) GCMs, as well as a transient scenario (for the year 2030) from the GISS GCM (Hansen *et al.*, 1988), were used in the assessment. Riebsame *et al.* (1995) found slight or no changes in annual river flows, but they did detect changes in seasonality. Reduced hydropower production and low-flow augmentation from a planned cascade of 13 dams also were projected.

### *11.3.4. Food and Fiber: Agriculture, Forestry, and Fisheries*

#### *11.3.4.1. Agriculture*

##### *11.3.4.1.1. Regional importance of agriculture*

Agriculture dominates land use and the economies of most countries in Tropical Asia. Arable land and permanent pasture occupy 15–35% of the land area in most countries; they occupy approximately 60% and 80% in India and Bangladesh, respectively (see Table 11-6). Agriculture is a key economic sector, employing more than half of the labor force and accounting for 10–63% of the GDP in 1993 in most countries of the region. Substantial foreign exchange earnings also are derived from exports of agricultural products.

Agriculture in Tropical Asia has evolved on the basis of regionally available crop biodiversity; only a few crops from other regions—such as wheat, potatoes, and tomatoes—have competed with endemic crops and become widely cultivated. Much of the land is intensively cropped, and there is a shortage of agricultural land in some countries (see Annex D). Most areas have the potential to produce a variety of crops through year-round cultivation; in several areas, more than one crop is grown annually on the same field. Many crop combinations are common, either individually or as intercrops. Because farmers in the region utilize a variety of cropping systems, studies of one specific crop in one area cannot be extrapolated to represent tropical agriculture in Asia or its impact on the agricultural economy in general. Some indication of the range of potential impacts is highlighted in the examples below.

***Table 11-6:*** *Land use in Tropical Asia, 1991–93.*

| Country | Land Area (000 ha) | Cropland (000 ha) | Permanent Pasture (000 ha) | Forest and Woodland (000 ha) | Other Land (000 ha) |
|---|---|---|---|---|---|
| Bangladesh | 13,017 | 9,700 | 800 | 1,896 | 818 |
| Bhutan | 4,700 | 133 | 272 | 3,100 | 1,194 |
| Brunei | 527 | 7 | 6 | 450 | 64 |
| Cambodia | 17,652 | 2,367 | 1,967 | 11,667 | 1,652 |
| India | 297,319 | 169,547 | 11,533 | 68,330 | 47,909 |
| Indonesia | 181,157 | 30,993 | 11,776 | 111,258 | 27,130 |
| Laos | 23,080 | 807 | 800 | 703 | 20,770 |
| Malaysia | 32,855 | 4,880 | 27 | 20,347 | 7,601 |
| Myanmar | 65,755 | 10,061 | 359 | 32,397 | 22,938 |
| Nepal | 13,680 | 2,354 | 2,000 | 5,750 | 3,576 |
| Papua New Guinea | 45,286 | 412 | 13,577 | 7,377 | 2,014 |
| Philippines | 29,817 | 9,177 | 1,277 | 13,600 | 5,764 |
| Singapore | 61 | 1 | | 3 | 57 |
| Sri Lanka | 6,463 | 1,903 | 439 | 2,126 | 1,995 |
| Thailand | 51,089 | 20,775 | 797 | 13,557 | 15,960 |
| Viet Nam | 32,549 | 6,607 | 328 | 9,639 | 15,975 |

Source: WRI, 1996—Data Table 9.1.

***Table 11-7:*** *Agricultural production in selected countries of Tropical Asia (average annual production, 1992–95).*

| **Country** | **Rice (000 MT)** | **Wheat (000 MT)** | **Maize (000 MT)** | **Pulses (000 MT)** | **Soybean (000 MT)** | **Meat (000 MT)** | **Milk (000 MT)** |
|---|---|---|---|---|---|---|---|
| Bangladesh | 26,120 | 1,144 | 3 | 525 | | 349 | 792 |
| Bhutan | 43 | 5 | 40 | 2 | | 8 | 32 |
| Cambodia | 2,161 | | 55 | 14 | 36 | 128 | 19 |
| India | 117,920 | 58,760 | 9,722 | 13,683 | 4,003 | 4,116 | 60,374 |
| Indonesia | 48,205 | | 7,387 | 374 | 1,686 | 1,759 | 413 |
| Laos | 1,454 | | 66 | 42 | 5 | 84 | 11 |
| Malaysia | 2,108 | | 39 | | | 220 | 42 |
| Myanmar | 17,472 | 132 | 243 | 965 | 35 | 169 | 536 |
| Nepal | 3,228 | 840 | 1,269 | 184 | 12 | 27 | 902 |
| Philippines | 10,025 | | 4,534 | 37 | 4 | 1,500 | 33 |
| Sri Lanka | 2,570 | | 30 | 40 | | 83 | 279 |
| Thailand | 20,217 | | 3,732 | 374 | 512 | 1,330 | 209 |
| Viet Nam | 22,989 | | 806 | 71 | 111 | 1,240 | 71 |
| **Total** | 274,512 | 60,881 | 27,926 | 16,311 | 6,404 | 11,013 | 63,713 |

Source: FAO Production Yearbook, 1996.

Rice is the most important cereal crop in Tropical Asia. The importance of rice relative to other crops, as well as to meat and milk, is indicated by recent production figures (see Table 11-7). In Bangladesh, Cambodia, Laos, Malaysia, Myanmar, Sri Lanka, and Viet Nam, rice constitutes 94–98% of total cereal production. Cereal production in Thailand and Indonesia consists of 80% and 84% rice, respectively, with the remainder predominantly maize. Rice production in India, Nepal, and the Philippines accounts for 50%, 52%, and 66% of the cereal produced; most of the remainder in India and Nepal is wheat, and maize is another important cereal in the Philippines. With regard to calorie intake in the monsoon region, 29–80% of the supply is contributed by rice; calorie intake from rice is highest in Bangladesh and lowest in Malaysia (Hossain and Fischer, 1995).

*11.3.4.1.2. Vulnerability of agriculture*

Anglo *et al.* (1996) note that changes in average climate conditions and climate variability will have a significant effect on agriculture in many parts of the region. Particularly vulnerable are low-income populations that depend on isolated agricultural systems. These communities include many areas that concentrate on the production of tropical crops, such as tea and coconuts, as well as regions with limited access to agricultural markets. Changes in climate variability would affect the reliability of agriculture and livestock production in the region. Increasing population also could place stress on agricultural production. During the past three decades, food production—largely rice and wheat—increased at a higher rate than did population growth, but these crops currently show signs of stagnating productivity; there also appears to be a decrease in the production of coarse grains in the region (Sinha, 1997).

Moreover, agricultural areas in Tropical Asia are vulnerable to many environmental hazards—including frequent floods, droughts, cyclones, and storm surges—that can damage life and property and severely reduce agricultural production. For example, on average during 1962–1988, Bangladesh annually lost about half of a million tons of rice—nearly 30% of the country's average annual food grain imports (BBS, 1989; Paul and Rashid, 1993)—as a result of floods. In India, the crop area affected by floods is nearly one-third of the average flood-prone area (GOI, 1992). More insidious changes in the groundwater level, as well as groundwater pollution, appear to be taking place in some high-productivity agricultural areas of the country (Sinha, 1997).

Two additional factors increase the vulnerability of the agricultural sector in the region. First, even though 35% of the region's land is irrigated (Annex D), a significant portion of agriculture is rain-fed. About 46 million ha, or 88.5%, of global rain-fed lowland rice is cultivated in south and southeast Asia (IRRI, 1993). In eastern India, approximately 80% of the 20 million ha of rice is grown in rain-fed lowlands (Zeigler and Puckridge, 1995). Yields for rice under rain-fed conditions in Asian tropical regions are low, compared with those of irrigated rice yields. A number of factors explain this pattern, including floods, droughts, temporary inundation from rainfalls, and tidal flows and coastal salinity (Hossain and Fischer, 1995).

Second, the region supports a large human population per hectare of cropland: 5.4 people per hectare in south Asia and 5.7 in southeast Asia. Although the rate of growth of rice yields is modest in many countries of the monsoon region, it has been estimated that per-hectare rice yields will need to be doubled by 2025 to meet demand. Even if yields in rain-fed lowlands can be doubled through the development of high-yielding varieties that are resistant to floods, droughts, and problem soils, the required rice yield from irrigated areas by 2025 will be about 8 tons/ha (compared with current yields of 4.9 tons/ha) (Hossain and Fischer, 1995).

*11.3.4.1.3. Climate change and crop yields*

The main climate variables that are important for crops, as for other plants, are air temperature and humidity, cloudiness, solar radiation, water, and atmospheric $CO_2$ concentration; the first and last of these figure explicitly in climate change projections.

Selected crop studies for south and southeast Asia are summarized in Section 13.6.2 of the Working Group II contribution to the SAR (IPCC, 1996). Yield impacts for rice for a number of countries and from four different studies are given in Table 11-8.

Estimated impacts of climate change on rice yields, as indicated in Table 11-8, are highly variable and depend on the rice/yield model, the choice of scenario, the region, the growing season, and other factors. Rice models, for instance, assume a constraint-free environment for water, fertilizers, pests, and other factors. Although there is an acknowledged need for further improvement in such models, these estimates clearly demonstrate that a single estimate of change in rice yield for the whole region would mean very little. Several local studies illustrate the fact that intraregional variability will be important in the future, as it is now.

In Bangladesh, the impact of climate change on high-yield rice varieties was studied by Karim *et al.* (1996), using the CERES-Rice model and several scenarios and sensitivity analyses. They found that:

- Increased $CO_2$ levels increased rice yields.
- Considerable spatial and temporal variations occurred.
- High temperatures reduced rice yields in all seasons and in most locations.
- The detrimental effect of temperature rise more than offset the positive effect of increased $CO_2$ levels.

Similarly, experiments in India reported by Sinha (1994) found that higher temperatures and reduced radiation associated with increased cloudiness caused spikelet sterility and reduced yields to such an extent that any increase in dry-matter production as a result of $CO_2$ fertilization proved to be no advantage in grain productivity. Similar studies conducted recently in Indonesia and the Philippines confirmed these results. Amien *et al.* (1996) found that rice yields in east Java could decline by 1% annually as a result of increases in temperature. In a study of northwest India, Lal *et al.* (1996) also found that reductions in yield resulting from a rise in surface air temperature offset the effects of elevated $CO_2$ levels; the projected net effect is a considerable reduction in rice yield.

Simulations of the impact of climate change on wheat yields for several locations in India using a dynamic crop growth model, WTGROWS, indicated that productivity depended on the magnitude of temperature change. In north India, a 1°C rise in mean temperature had no significant effect on potential yields, although an increase of 2°C reduced potential grain yields in most places (Aggarwal and Sinha, 1993). In a subsequent study, Rao and Sinha (1994) used the CERES-Wheat simulation model and scenarios from three equilibrium GCMs (GISS, GFDL, and UKMO) and the transient GISS model to assess the physiological effects of increased $CO_2$ levels. In all simulations, wheat yields were smaller than those in the current climate, even with the beneficial effects of $CO_2$ on crop yield; yield reductions were associated with a shortening of the wheat-growing season resulting from projected temperature increases. Karim *et al.* (1996) also have shown that wheat yields are vulnerable to climate change in Bangladesh. Studies of the productivity of sorghum showed adverse effects in rainfed areas of India (Rao *et al.*, 1995). Results were similar for corn yields in the Philippines (Buan *et al.*, 1996).

The likely impact of climate change on the tea industry of Sri Lanka was studied by Wijeratne (1996). He found that tea yield is sensitive to temperature, drought, and heavy rainfall. An increase in the frequency of droughts and extreme rainfall events could result in a decline in tea yield, which would be greatest in the low-country regions (<600 m). Other important crops of the region include rubber, oil palm, coconut, sugarcane, coffee, and spices, but almost no information is available on the impact of climate change on these crops.

*11.3.4.2. Forests and Forestry*

In Tropical Asia, two major types of closed-canopy forest have been distinguished (Collins *et al.*, 1991): rainforest and monsoon

***Table 11-8:*** *Yield impact of selected climate change studies for rice in Tropical Asia.*

| Geographic Scope | Yield Impact (%) | Reference |
|---|---|---|
| Bangladesh | -6 to +8 | 1 |
| Bangladesh | +10 | 2 |
| Bangladesh | -9 to +14 | 4 |
| Thailand | -17 to +6 | 1 |
| Thailand sites | -5 to +8 | 3 |
| Thailand | -12 to +9 | 4 |
| Philippines | -21 to +12 | 1 |
| Philippines | decrease | 2 |
| Philippines | -14 to +14 | 4 |
| Indonesia | -3 | 2 |
| Indonesia | approx. -4 | 3 |
| Indonesia | +6 to +23 | 4 |
| Malaysia | -22 to -12 | 3 |
| Malaysia | +2 to +27 | 4 |
| Myanmar | -14 to +22 | 4 |

Sources: IPCC 1996, WG II, Section 13.6.2; (1) Rosenzweig and Iglesias, 1994; (2) Qureshi and Hobbie, 1994; (3) Parry *et al.*, 1992; (4) Matthews *et al.*, 1994.

forest. Rainforests grow in ever-wet conditions where rainfall is heavy and spread throughout the year. These forests are evergreen or semi-evergreen and include lowland, montane, and swamp forests. Monsoon forests grow where rainfall is high but unevenly spread throughout the year. Most monsoon forest trees are deciduous; they shed their leaves in the dry season. These two types of forest cover about 2 million km$^2$ in the region, although many forests are degraded by logging, farming, and fire. In Malaysia, Myanmar, and Sumatra, forests are being modified by timber extraction. In Laos, Kalimantan, and Papua New Guinea, shifting cultivation is encroaching into the forests; in Bangladesh, India, the Philippines, Sri Lanka, Thailand, and Viet Nam, only scattered relicts of relatively undamaged forests remain (Collins *et al.*, 1991).

Studies of the potential regional impacts of climate change on the forests and forestry of Tropical Asia are limited, although a number of local studies have been carried out. Results of research from Thailand suggest that climate change would have a profound effect on the future distribution, productivity, and health of Thailand's forests. Using climate change scenarios generated by the UKMO and GISS GCMs, Boonpragob and Santisirisomboon (1996) estimated that the area of subtropical forests could decline from the current 50% to either 20% or 12% of Thailand's total forest cover (depending on the model used), whereas the area of tropical forests could increase from 45% to 80% of total forest cover. Somaratne and Dhanapala (1996) used the same model for Sri Lanka and estimated a decrease in tropical rainforest of 2–11% and an increase in tropical dry forest of 7–8%. A northward shift of tropical wet forests into areas currently occupied by tropical dry forests also is projected.

Teak is an important wood product throughout the region; in Java, it has been shown to be sensitive to variations in climate (Whetton and Rutherford, 1994). Using climate scenarios generated by ECHAM3, Achanta and Kanetkar (1996) have linked the precipitation effectiveness index (PEI) to net primary productivity of teak plantations in Kerala State, India. They estimate that a projected depletion of soil moisture would likely cause teak productivity to decline from 5.40 m$^3$/ha to 5.07 m$^3$/ha. The productivity of moist deciduous forests also could decline, from 1.8 m$^3$/ha to 1.5 m$^3$/ha.

In recent decades, deforestation has increased in Tropical Asia. The impacts have been investigated in a series of GCM simulations of the effects of deforestation on the local and regional climate (Zhang, 1994). In general, Zhang found that deforestation can modify land-surface characteristics, such as surface albedo and surface roughness, and thus redistribute the local surface energy budget. At the same time, modification of land-surface processes by deforestation may produce some disturbances of monsoon circulation over the southeast Asian region. Deforestation, along with the potential impacts of climate change, also may have a negative impact on sustainable-nutrition security in south Asia (Sinha and Swaminathan, 1991).

In semi-arid regions of Tropical Asia, tropical forests generally are sensitive to changes in temperature and rainfall, as well as changes in their seasonality. Fire—which also is influenced by these changes—significantly affects the structure, composition, and age diversity of forests in the region. Fire frequency is affected by intentional (e.g., slash and burn agriculture) and nonintentional changes in human land use, as well as by variations in climate (e.g., longer or shorter dry seasons).

#### *11.3.4.3. Fisheries*

Few studies have been done on the possible impact of climate change on fisheries in Tropical Asia, despite the fact that commercial and subsistence marine and freshwater fisheries and aquaculture are important for the food security and the economies of many countries in the region. In the Philippines, for instance, fisheries account for about 4% of gross national product (GNP). The fisheries sector employs nearly a million people—about 26% of whom are engaged in aquaculture operations, 6% in commercial fishing, and 68% in marine and freshwater municipal fishing (i.e., artisinal, small-scale, or traditional fisheries) (Lim *et al.*, 1995). Overexploitation of inshore and inland fisheries in most countries of south and southeast Asia threatens fishery resources as well as the livelihoods of the fishermen.

Section 16.2 of the Working Group II contribution to the SAR (IPCC, 1996) identified climate change impacts on several types of fisheries and considered health and infrastructural issues. For marine fisheries, slight changes in environmental variables—such as temperature, salinity, wind speed and direction, ocean currents, strength of upwelling, and predators—can sharply alter the abundance of fish populations (Glantz and Feingold, 1992). Rises in sea level also may cause saline water fronts to penetrate further inland, which could increase the habitat of brackish-water fisheries. Coastal inundation also could damage the aquaculture industry.

The effect on inland fisheries may be felt strongly because large sections of the region's population consume freshwater fish. Fisheries at higher elevations may be particularly affected by lower oxygen availability resulting from increases in temperature. In the plains, the timing and amount of precipitation may affect the migration of fish species from the river to the floodplains for spawning, dispersal, and growth. If the magnitude and extent of floods increase, more flood-control projects would further deplete floodplain fisheries. Current annual losses of openwater fish in Bangladesh from flood-control and drainage projects are estimated to be 65 kg/ha (Mirza and Ericksen, 1996).

### *11.3.5. Mountain and Upland Regions*

Scenarios of climate change in the mountainous regions of the world are highly uncertain; they are poorly resolved even in the highest-resolution GCMs (IPCC 1996, WG II, Section 5.2). A few impact studies have been carried out in the mountain regions of Tropical Asia, where topography and elevation also have an important influence on the regional climate.

In Nepal, Bhutan, and northern India, for instance, mountains provide food, fuel, and fresh water—which are needed for human survival and are fundamental resources for tourism and economic development. Tourism contributes about 24% of Nepal's foreign-exchange earnings (HMGN, 1992). Seasonal variations in the water resources of Nepal's Himalaya region are very high: The maximum-to-minimum discharge ratios for the Karnali, Sapt Gandaki, and Sapt Kosi rivers are 120:1, 180:1, and 105:1, respectively. This variation is due mainly to heavy and concentrated rainfall during a short time period; steep topography, which encourages higher surface runoff; and the high rate of deforestation in the watersheds (Uprety, 1988).

At high elevations in the Himalayas, an increase in temperature could result in faster recession of glaciers and an increase in the number and extent of glacial lakes—many of which, according to Watanabe *et al.* (1994), have formed in the past several decades. The rapid growth of such lakes could exacerbate the danger from glacial lake outburst floods (GLOFs), with potentially disastrous effects. GLOFs result from the failure of moraine dams when excessive hydrostatic pressure is exerted on dam walls as a result of increased melting and increased water depth in lakes. In a warmer world, GLOFs could be expected to increase in number and severity throughout the Himalayas of India, Nepal, and Bhutan. Recent examples, which may be suggestive of increasingly frequent occurrences in the future, include the catastrophic outburst of two glacial lakes in the Lunana area, northern Bhutan, in October 1994 (which resulted in the deaths of 21 persons and damage to villages, washed away bridges, and filled water with debris and large logs) and the 1985 outburst flood in Khumbu Himal, Nepal (Watanabe and Rothacher, 1996).

Devastating floods occur in the central Himalaya of Nepal during the monsoon months of June through September. Monsoon floods differ from GLOFs and floods caused by breaching of landslide dams: They are of longer duration and less sudden in onset. The incessant rainfall of August 1987 submerged parts of central and eastern Terai in waters of as much as 1 m (BNJST, 1989). Damage from floods takes several forms, including the destruction of footbridges that often provide the only link between remote mountain villages; demolition of irrigation diversions; mass-wasting by undercutting of steep, stream-adjacent slopes; and damage to floodplain agricultural land by erosion and sedimentation (Marston *et al.*, 1996).

The effects of climate change on soil erosion and sedimentation in mountain regions of Tropical Asia may be indirect but could be significant. An erosion rate in the range of 1–43 tons/ha—with an average of 18 tons/ha—was calculated in three small experimental plots in central Nepal. Over the 3-year period of the experiment, 40–96% of the annual losses occurred during two storms (Carver and Nakarmi, 1995). Uprety (1988) found a statistically significant correlation between monsoon precipitation in the Karnali and Spata Kosi watersheds in Nepal and sediment load. In a single storm in 1993, Sthapti (1995) reported a sediment load into the reservoir of the Kulekhani hydropower project in Nepal that was about 35 times higher than the design value. Increased severity and frequency of monsoonal storms and flooding in the Himalayas, which are expected outcomes of climate change, may significantly alter the area's erosion, river discharge, and sediment dynamics. Eventually, this may affect existing hydropower reservoirs, as well as those planned for construction in the Himalayas. Part of the generated sediment may be deposited on agricultural lands or in irrigation canals and streams, which will contribute to a deterioration in crop production and in the quality of agricultural lands.

### *11.3.6. Coastal Zones*

The impacts of climate change on coastal areas in Tropical Asia could be severe and in some areas catastrophic. Nicholls *et al.* (1995) estimate that (assuming no adaptation and no change in existing population) a 1-m rise in sea level could displace nearly 15 million, 7 million, and at least 2 million people from their homes in Bangladesh, India, and Indonesia, respectively; millions more could be threatened in Viet Nam, Myanmar, Thailand, and the Philippines. (Potential impacts on special tropical coastal ecosystems, including coral reefs and mangroves, are discussed in Section 11.3.2.)

Sea-level rise and heavy rainfall events are projected with some confidence to increase with climate change; these increases, superimposed on existing coastal problems, would have major impacts, regardless of whether there is any change in the frequency or intensity of tropical storms.

Projected impacts include:

- Land loss and population displacement
- Increased flooding of low-lying coastal areas
- Agricultural impacts (e.g., loss of yield and employment) resulting from inundation, salinization, and land loss
- Impacts on coastal aquaculture
- Impacts on coastal tourism, particularly the erosion of sandy beaches.

Nicholls *et al.* (1995) consider the first of these impacts to be the most serious coastal issue in the region. They estimate that a 1-m rise in sea level could lead to land loss in Bangladesh, India, Indonesia, and Malaysia of nearly 30,000; 6,000; 34,000; and 7,000 km$^2$, respectively. In Viet Nam, 5,000 km$^2$ of land could be inundated in the Red River delta, and 15,000–20,000 km$^2$ of land may be threatened in the Mekong delta.

Several studies have been undertaken on the vulnerability of Bangladesh to climate change, particularly to sea-level rise. A comprehensive summary appears in Warrick and Ahmad (1996). Recent geological studies suggest that the magnitude of tectonic subsidence in the Ganges-Brahmaputra delta of Bangladesh is greater than has been taken into account in some earlier projections; this subsidence is expected to result in higher estimates of relative sea-level rise. Around the city of

Dhaka, average subsidence is about 0.62 mm/yr; elsewhere, it can exceed 20 mm/yr (Alam, 1996).

The Ganges-Brahmaputra delta is one of the world's most densely populated areas, and the combined effects of subsidence and sea-level rise could cause serious drainage and sedimentation problems, in addition to coastal erosion and land loss. With higher sea level, more areas would be affected by cyclonic surges; inland freshwater lakes, ponds, and aquifers could be affected by saline and brackish-water intrusion. The present limit of tidal influence is expected to move further upstream, and increases in soil salinity, as well as surface-water and groundwater salinity, may cause serious water supply problems for drinking and irrigation over large areas (Alam, 1996). Reduced dry-season freshwater supply from upstream sources may further exacerbate salinity conditions in the coastal area of Bangladesh. These impacts clearly would have immense socioeconomic costs.

The other large deltaic areas in Tropical Asia—such as the Irrawaddy in Myanmar and the Mekong and Red in Viet Nam—as well as smaller deltaic regions and low-lying coastal plains in India, Thailand, Cambodia, Malaysia, Indonesia, and the Philippines will be affected in similar ways. Specific examples of vulnerability assessments and impact analyses for several south and southeast Asian countries are given in McLean and Mimura (1993), Chou (1994), Nicholls and Leatherman (1995), Erda *et al.* (1996), and Milliman and Haq (1996).

The magnitude of physical and socioeconomic impacts of sea-level rise alone on a small segment of the Malaysian coast in West Johor, which is occupied by 150,000 people, has been detailed by Teh and Voon (1992). In this example, a projected coastal embankment failure resulting from sea-level rise would result in a substantial reduction in the area under agriculture and extensive damage to houses, schools, and other social amenities. In addition, millions of dollars of revenue loss from the destruction of the three main crops (coconuts, oil palm, and rubber) ultimately could involve a population shift inland. This example is duplicated many times in the region, with the implication that greenhouse-induced sea-level rise may require a reassessment of existing and future coastal developments—as has been done in Malaysia (Midun and Lee, 1995).

In the previous example, the area was essentially rural and agricultural. The potential impact of sea-level rise on recreational and tourist sectors—for example, on the east coast of peninsular Malaysia (Wong, 1992) and at Phuket, Thailand (Wong, 1995)—and on infrastructure, harbors, and airports also would be severe. Indeed, Nicholls (1995) shows that present and future megacities of the region—including Jakarta, Bangkok, and Manila—are especially vulnerable. For Bangkok, the steady rise in sea level poses a threat for the investment, operation, and safety levels of the flood-control system, which could have an estimated annual pumping cost of up to US$20 million (Sabhasri and Suwarnarat, 1996).

### *11.3.7. Human Health*

Vector-borne diseases are significant causes of mortality and morbidity in Tropical Asia. Climate change may cause a change in the distribution of malaria, dengue, bilharzia, leishmaniasis, and schistosomiasis. The region also is particularly vulnerable to extreme weather events such as tropical cyclones, storms, and floods. The vulnerability of the population to these events is significantly increased by high population densities and poor sanitation; thus, waterborne and food-borne infectious diseases also are major problems. For example, waterborne and waterrelated diseases account for about 70% of the epidemic emergencies in India. The additional impacts of climate change, superimposed on existing conditions, could seriously exacerbate health problems.

#### *11.3.7.1. Malaria*

Malaria is the most serious and widespread tropical disease in the world today. In most areas of the Asia-Pacific region where malaria now occurs, its distribution was considerably reduced or eliminated during the 1960s and 1970s. In recent years, however, the situation has been worsening in frontier regions of economic development (Porter, 1994). The emergence and rapid spread of drug-resistant malaria have become major health concerns to all malaria-affected countries in the region (WHO, 1996a).

Climate has a direct influence on malaria mosquitoes because each species has a range of climate conditions suitable for its development (see Figure 11-4). Climate also has an indirect effect on malaria through its influence on suitable vegetation and vector breeding sites. Precipitation is important because mosquitoes require water to lay their eggs, as well as for the subsequent development of larvae.

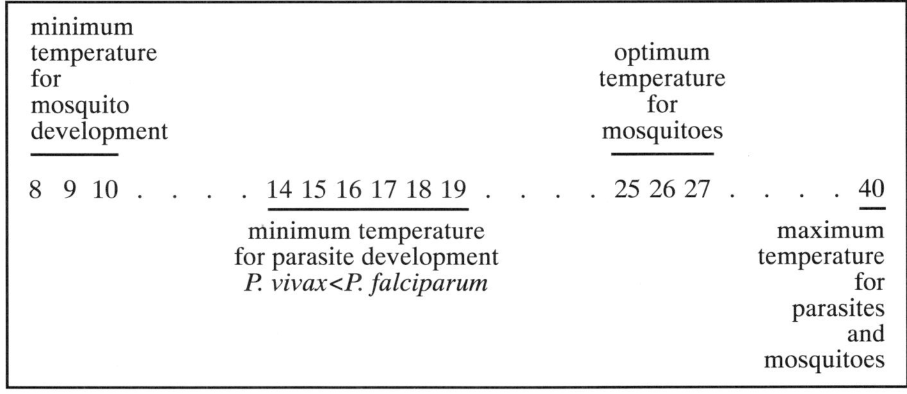


**Figure 11-4:** Critical temperatures in malaria epidemiology (°C) (adapted from WHO, 1996b).

Several estimates of the potential distribution of malaria have been produced, using climate change scenarios from a number of GCMs. These projections indicate that, in the future, malaria is most likely to extend its range into the fringes of established endemic areas (Martens *et al.*, 1995). In southeast Asia, this would include new areas of Indonesia—notably Sumatra and Irian Jaya—and Papua New Guinea (Porter, 1994). Martin and Lefebvre (1995) report that the intensity and extent of malaria's potential transmission would change significantly under scenarios generated by the five GCMs they used. In all cases, an increase in seasonal malaria at the expense of perennial malaria results; such an increase is most likely to lead to epidemics among unprepared or nonimmune populations. Similarly, Martens *et al.* (1995) conclude that newly affected populations would initially experience high case-fatality rates because of their lack of naturally acquired immunity.

Increased incidence of malaria associated with climate change may significantly strain the economies of many countries of the region. Picard and Mills (1992) estimated losses of five working days for infected persons, during the period from infection to recovery, in two districts in Nepal. Mills (1994) found that the economic consequences of malaria could be high in the areas without malaria-control programs. The high cost of alternative drugs has put enormous pressure on the scarce economic resources of the countries in Asia (WHO, 1996a). ISPAN (1992) estimated that a complete course of treatment in rural Bangladesh may cost one month's wage for an agricultural worker (about US$40), or roughly one-fifth of overall per capita income. Hammer (1993) discusses the economics of malaria treatment and cites examples from Nepal, Thailand, and Indonesia—where the economic return of malaria treatment was found to be substantial.

### *11.3.7.2. Dengue and Schistosomiasis*

Dengue and dengue hemorrhagic fever (DHF) are widespread in many countries of Tropical Asia. Although the transmission of DHF is said to have resulted from rapid urbanization, the disease vector *Aedes aegypti* exists in remote areas (WHO, 1996a). In Tropical Asia, children are particularly affected by dengue/DHF; these diseases cause many hospitalizations and deaths. According to WHO (1996a), dengue/DHF will continue to persist in Indonesia, Myanmar, and Thailand, where they are prevalent in endemic form. In recent years, sporadic cases and outbreaks also have been reported in Bangladesh, India, and Sri Lanka.

Studies suggest that climate change would likely affect the distribution, life cycle, and population dynamics of dengue (WHO, 1996b). According to Koopman *et al.* (1991), an increase of 3–4°C in average temperature may double the reproduction rate of the dengue virus. Results from a simple mathematical model developed in Indonesia suggest that, under the best-estimate climate change scenario, incidences of dengue may increase threefold in Indonesia (ADB, 1994b).

Schistosomiasis is a water-based infectious disease caused by five species of the trematode *Schistosoma.* The spread of this disease is largely attributed to the expansion of irrigation in tropical regions. According to Martens *et al.* (1995), climate change-related temperature increases would influence snail reproduction and growth, schistome mortality, infectivity and development in the snail, and human-water contact. Climate change impacts on the spread of schistosomiasis also may be indirect (WHO, 1996b); for example, expansion of irrigation to new areas may introduce schistosomiasis where endemic foci already exist.

A recent assessment of the sensitivity of malaria, dengue, and schistosomiasis (the world's most prevalent vector-borne diseases) to global warming suggests an increase in the extent of geographical areas susceptible to transmission of malarial plasmodium parasites, dengue flavivirus, and schistosoma worms. Martens *et al.* (1997) show that the transmission potential of the three associated vector-borne diseases is highly sensitive to climate changes on the periphery of the currently endemic areas and at higher elevations within such areas, including Tropical Asia. With reference to present endemic areas, their findings show that the potential increase in epidemic transmission of malaria and dengue may be estimated at 12–27% and 31–47%, respectively; in contrast, schistosomiasis transmission potential may be expected to exhibit a 11–17% decrease.

### *11.3.7.3. Other Diseases*

In recent years, incidence of visceral leishmaniasis has increased in some countries of Tropical Asia. During 1987–1990, visceral leishmaniasis reached epidemic form in the Indian state of Bihar and spread rapidly to surrounding areas. WHO (1996a) estimated that about 110 million people were at risk from visceral leishmaniasis. Major endemic foci are reported in border areas between India (states of Bihar and West Bengal), Bangladesh, and Nepal. In Bangladesh, visceral leishmaniasis already has reached epidemic form; the most vulnerable populations are poor and rural cattlekeepers. Reported cases appear to cluster close to flood-control embankments; there appears to be a significant risk that visceral leishmaniasis prevalence in some localities will increase as a result of flood-control and drainage projects (ISPAN, 1992). In a warmer climate, the incidence of visceral leishmaniasis also may increase (IPCC 1996, WG II, Section 18.3; WHO, 1996b).

Other waterborne and food-borne infectious diseases pose a great threat to public health in the tropical monsoon region. In 1995, in eight countries of the region (Bangladesh, Bhutan, India, Indonesia, Myanmar, Nepal, Sri Lanka, and Thailand), the total death toll from diarrhea was estimated to be 1.03 million; 71% and 12% of the deaths occurred in India and Bangladesh, respectively. Children under age 5 accounted for about 25% of these deaths (WHO, 1996a).

According to Colwell (1996), the major rivers of the Indian subcontinent that discharge into the Bay of Bengal carry huge amounts of agricultural and industrial waste, providing nutrients sufficient to convert coastal waters to eutrophic conditions. Brackish water extends some distance upriver for all rivers. Salinities between 5 ppm and 30 ppm, which were detected in inland coastal areas of Bangladesh, as well as in seawater (Huq *et al.*, 1984), are favorable for the growth of *V. cholerae*.

In the Bay of Bengal, evidence has been found—by synthesizing satellite remote-sensing, in-situ hydrographic and meteorological data sets, and cholera cases in Bangladesh—that cholera cases occur with a rise in ocean temperature (Colwell, 1996). Two peak periods for cholera outbreaks have been identified: from early April to mid-May and from early September to the end of November. During these periods, high sea-surface temperature, salinity, and concentrations of nutrients probably favor the growth of *V. cholerae*. Outbreaks during the monsoon have been substantially smaller. Colwell (1996) also indicated a possible link between outbreaks of cholera in Peru and neighboring countries and a warming ENSO event. Although the ENSO phenomenon has a substantial effect on the Indian summer monsoon, its linkage with cholera has not been investigated. Future increases in sea-surface temperature, as well as increased concentrations of pollutants in river flows under climate change scenarios, may create a more favorable environment for the growth of *V. cholerae* throughout the year in the coastal area of Bangladesh.

Essential steps in recognizing and mitigating the emergence of malaria, dengue, and bilharzia, as well as many other infectious diseases, include enhanced surveillance and response (Martens *et al.*, 1997). Dowlatabadi (1997) argues that because public health measures, case management, and land use play more significant roles in determining the prevalence of malaria than does climate, the most appropriate policy is to introduce or improve simple public health measures in the region; this approach would ensure that, even with climate change, developing countries would be far less susceptible to marginal increases in the potential prevalence of diseases such as malaria, schistosomiasis, dengue, and cholera.

## 11.4. Integrated Assessment of Potential Vulnerabilities and Impacts

Individual countries, regions, resources, sectors, and systems will be affected by climate change not in isolation but in interaction with one another. The direct effects of climate change to which this chapter alludes, such as changes in rice crop yields and inundation of coastal areas, will have further indirect effects.

Integration of impacts and adaptations to climate change can take many forms. Several different approaches can be considered for Tropical Asia, including the following:

1) *Country-Specific Studies*—These studies have attempted to summarize salient aspects of vulnerability, impacts, and adaptations, generally in a qualitative manner. Some simple examples for India, Indonesia, Malaysia, and Thailand are given in Section 6.5 of the Working Group III contribution to the SAR (IPCC, 1996) and in ADB (1994a, 1994b). In these and other national studies, the conclusions usually are strongly thematic or sector-based.
2) *Geographical Integration*—Linkages between different functional regions are difficult to investigate although, at a conceptual level, they are readily acknowledged. For example, drainage basins can provide functional links between climate change impacts in headwaters and the downstream areas they affect; at the same time, however, downstream areas also may be responding to locally specific climate change impacts. Some simple examples have been described in this chapter (e.g., Section 11.3.3.3). Functional adaptations—such as through total river catchment analyses or integrated coastal management—are extraordinarily difficult to implement because of jurisdictional sensitivities, even though conceptual and methodological hurdles may be relatively easy to overcome.
3) *Sectoral and Trade Integration*—There have been a number of attempts to integrate the various components of food supply (agriculture) in Tropical Asia, generally with simplified production change models in response to changes in environmental conditions (e.g,. temperature or precipitation and their derivatives, such as length of growing season and time of fruiting), some with yield adaptations and some without. These results are then integrated into an international trade-conditions model, and estimates of net economic welfare impacts are calculated. Reilly et al. (1993) have done this for Tropical Asia using three GCMs and $2xCO_2$ equilibrium scenarios.
4) *Comprehensive Monetary Estimates*—Another way of integrating the range of potential impacts of climate change is to derive a comprehensive monetary estimate, which adds all of a country's impacts, expressed in their dollar value. Expressing damage to marketed goods and services (e.g., land lost to sea-level rise, energy savings in winter) in monetary terms is more or less straightforward because the price is known. Expressing damage to non-marketed goods and services (e.g., wetland loss, mortality changes) in monetary terms can be done either through examining market transactions where such goods or services are implicitly traded (e.g., tourism, where landscape beauty is valued) or through interviewing people to determine their preferences (i.e., their willingness to pay to secure a benefit or their willingness to accept compensation for a loss). Such valuation techniques reflect economic circumstances and value systems in the Organisation for Economic Cooperation and Development (OECD). The estimates for non-OECD countries in Chapter 6 of the Working Group III contribution to the SAR (IPCC, 1996) are based on extrapolation from studies in the OECD. With these caveats in mind, Chapter 6 of the WG III contribution to the SAR (IPCC, 1996) reports best estimates for the annual impact resulting from a doubling of

atmospheric concentrations of $CO_2$ of about 2.1–8.6% of GDP in South and Southeast Asia. These figures compare with an estimated world impact of 1.4–1.9% of GDP. Many assumptions underlie these best guesses, however, and large uncertainties remain.

## References


**ADB**, 1994a: *Climate Change in Asia: Executive Summary*. Asian Development Bank, Manila, Philippines, 122 pp.

**ADB**, 1994b: *Climate Change in Asia: Indonesia Country Report*. Asian Development Bank, Manila, Philippines, 122 pp.

**Achanta**, A.N. and R. Kanetkar, 1996: *Impact of climate change on forest productivity: a case study of Kerala, India*. Paper presented at the Asian and Pacific Workshop on Climate Change Vulnerability and Adaptation Assessment, January 15–19, 1996, Manila, Philippines.

**Aggarwal**, P.K. and S.K. Sinha, 1993: Effect of probable increase in carbon dioxide and temperature on wheat yields in India. *Journal of Agricultural Meteorology,* **48(5)**, 811–814.

**Aksorakaoe**, S. and N. Paphavasit, 1993: Effect of sea level rise on the mangrove community in Thailand. *Malaysian Journal of Tropical Geography,* **24**, 29–34.

**Alam**, M., 1996: Subsidence of the Ganges-Brahmaputra delta of Bangladesh and associated drainage, sedimentation and salinity problems. In: *Sea-Level Rise and Coastal Subsidence: Causes, Consequences, and Strategies* [Milliman, J.D. and B.U. Haq (eds.)] Kluwer Academic Publishers, Dordrecht, The Netherlands, 169–192 pp.

**Amien**, I., P. Rejekiningrum, A. Pramudia, and E. Susanti, 1996: Effects of interannual climate variability and climate change on rice yield in Java, Indonesia, In: *Climate Change Variability and Adaptation in Asia and the Pacific* [Erda, L., W. Bolhofer. S. Huq, S. Lenhart, S.K. Mukherjee, J.B. Smith, and J. Wisniewski (eds.)]. Kluwer Academic Publishers, Dordrecht, Netherlands, pp. 29-39.

**Anglo**, E.G., W.C. Bolhofer, L. Erda, S. Huq, S. Lenhart, S.K. Mukherjee, J. Smith, and J. Wisniewski, 1996: *Regional Workshop on Climate Change Vulnerability and Adaptation in Asia and the Pacific: Workshop Summary.* January 15–19, 1996, Manila, Philippines, Kluwer Academic Publishers, Dordrecht, The Netherlands, 16 pp.

**BBS (Bangladesh Bureau of Statistics)**, 1989: *Statistical Year Book of Bangladesh*. Ministry of Planning, Dhaka, 729 pp.

**Bengsston**, L., M. Botzet, and M. Esch, 1995: Hurricane-type vortices in a general circulation model. *Tellus,* **47A**, 175–196.

**BNJST (Bangladesh-Nepal Joint Study Team)**, 1989: *Report on Flood Mitigation Measures and Multipurpose Use of Water Resources*. Government of the People's Republic of Bangladesh and His Majesty's Government of Nepal, 89 pp.

**Boonpragob**, K. and J. Santisirisomboon, 1996: Modeling potential impacts of climate changes of forest area in Thailand under climate change. In: *Climate Change Variability and Adaptation in Asia and the Pacific* [Erda, L., W. Bolhofer. S. Huq, S. Lenhart, S.K. Mukherjee, J.B. Smith, and J. Wisniewski] Kluwer Academic Publishers, Dordrecht, Netherlands, 107–117 pp.

**Brown**, B., R.P. Dunne, T.P. Scoffin, and M.D.A. le Tissier, 1994: Solar damage in intertidal corals. *Marine Ecological Progress Series,* **105**, 219–230.

**Brown**, B. and Suharsono, 1990: Damage and recovery of coral reefs affected by El Niño related to seawater warming in the Thousand Islands, Indonesia. *Coral Reefs,* **8**, 163–170.

**Buan**, R.D., A.R. Maglinao, P.P. Evangelista, and B.G. Pajuelas, 1996: Vulnerability of rice and corn to climate change in the Philippines. In: *Climate Change Variability and Adaptation in Asia and the Pacific* [Erda, L., W. Bolhofer. S. Huq, S. Lenhart, S.K. Mukherjee, J.B. Smith, and J. Wisniewski (eds.)]. Kluwer Academic Publishers, Dordrecht, The Netherlands, pp.41-51.

**Carver**, M. and G. Nakarmi, 1995: The effect of surface conditions on soil erosion and stream suspended sediments. In: *Challenges in Resource Dynamics in Nepal: Processes, Trends and Dynamics in Middle Mountain Watersheds* [Schreir, H., P.B. Shah, and S. Brown (eds.)]. Proceedings of a workshop by the International Centre for Integrated Mountain Development (ICIMOD) and International Development Research Centre (IDRC), Kathmandu, Nepal, 155–162 pp.

**Chou**, L.M. (ed.), 1994: *Implications of Expected Climate Changes in the East Asian Seas Region: an Overview*. RCU/EAS Technical Report Series No. 2, United Nations Environmental Programme, Bangkok, Thailand.

**Clarke**, A.J. and X. Liu, 1994: Interannual sea level in the northern and eastern Indian Ocean. *Journal of Physical Oceanography,* **24**, 1224–1235.

**Climate Impact Group**, 1992: *Climate Change Scenarios for South and Southeast Asia*. Commonwealth Scientific and Industrial Research Organisation (CSIRO), Division of Atmospheric Research, Aspendale, Mordialloc, Australia, 41 pp.

**Collins**, N.M., J.A. Sayer, and T.C. Whitmore (eds.), 1991: *Conservation Atlas of Tropical Forests, Asia and the Pacific*. BP, IUCN, and WCMC, Macmillan Press, London, United Kingdom.

**Colwell**, R.R., 1996: Global climate and infectious disease: the cholera paradigm. *Science,* **274**, 2025–2031.

**Divya**, and R. Mehrotra, 1995: Climate change and hydrology with emphasis on the Indian subcontinent. *Hydrological Sciences Journal,* **40**, 231–241.

**Douglas**, B., 1992: Global sea level acceleration. *Journal of Geophysical Research,* **97(12)**, 699–712, 706.

**Dowlatabadi**, H., 1997: Assessing the health impacts of climate change: an editorial essay. *Climate Change,* **35**, 137–144.

**Erda**, L., W. Bolhofer. S. Huq, S. Lenhart, S.K. Mukherjee, J.B. Smith, and J. Wisniewski (eds.), 1996: *Climate Change Variability and Adaptation in Asia and the Pacific*. Kluwer Academic Publishers, Dordrecht, The Netherlands, pp. 249.

**ESCAP**, 1987: *Coastal Environmental Management Plan for Bangladesh — Volume One: Summary*. Economic and Social Commission for Asia and the Pacific, Bangkok, Thailand, 32 pp.

**Gable**, F.J. and D.G. Aubrey, 1990: Potential coastal impacts of contemporary changing climate on South Asian Seas states. *Environmental Management,* **14(1)**, 33–46.

**Gadgill**, S., 1995: Climate change and agriculture — an Indian perspective. *Current Science,* **69(8)**, 649–659.

**Glantz**, M.H. and L.E. Feingold, 1992: Climate variability, climate change, and fisheries: a summary. In: *Climate Variability, Climate Change, and Fisheries* [Glantz, M.H. and L.E. Feingold (eds.)]. Cambridge University Press, London, United Kingdom, 417–438 pp.

**GOB (Government of Bangladesh)**, 1989: *Study on the Causes and Consequences of Natural Disasters and Protection and Preservation of the Environment*. Bangladesh Country Report, GOB, Dhaka, pp. 1–13.

**GOI (Government of India)**, 1992: *Eighth Five Year Plan (1992–1997)*. Planning Commission, New Delhi, India, 480 pp.

**Granich**, S., M. Kelly, and N.H. Ninh, 1993: *Global Warming and Vietnam*. Centre for Environment Research Education/University of East Anglia/International Institute for Environment and Development, London, United Kingdom.

**Hammer**, J.S., 1993: The economics of malaria. *The World Bank Research Observer,* **8(1)**, 1–22.

**Hansen**, J., I. Fung, A. Lacis, D. Rind, S. Lebedeff, R. Ruedy, and G. Russell, 1988: Global climate changes as forecast by the Goddard Institute of Space Studies three-dimensional model. *Journal of Geophysical Research,* **93**, 9341–9364.

**Hansen**, J., G. Russell, D. Rind, P. Stone, A. Lacis, S. Lebedeff, R. Ruedy, and L. Travis, 1983: Efficient three-dimensional global models for climate studies: models I and II. *Monthly Weather Review,* **3**, 609–662.

**Henderson-Sellers**, A., and H. Zhang, 1997: *Tropical Cyclones and Global Climate Change*. Report to the WMO/CAS/TMRP Committee on Climate Change Assessment (Project TC-2), World Meteorological Organization, Geneva, Switzerland, 44 pp. + 5 figures. http://www.bom.gov.au/bmrc/

**HMGN (His Majesty's Government of Nepal)**, 1992: *National Report on United Nations Conference on Environment and Development*. Kathmandu, 63 pp.

**Holland**, G.J., 1997: The maximum potential intensity of tropical cyclones. *Journal of Atmospheric Science* (submitted).

**Hossain**, M. and K.S. Fischer, 1995: Rice research for food security and sustainable agricultural development in Asia: achievements and future challenges. *Geojournal,* **35(3)**, 286–298.

**Huq**, A., P.A. West, E.B. Small, M.J. Huq, and R.R. Colwell, 1984: Influence of water temperature, salinity, and pH on survival and growth of toxigenic Vibriocholerae serovar O1 associated with live copepods in laboratory microcosm. *Applied Environmental Microbiology,* **48(2)**: 420-424

**IPCC**, 1996: *Climate Change 1995: The Science of Climate Change. Contribution of Working Group I to the Second Assessment Report of the Intergovernmental Panel on Climate Change* [Houghton, J.J., L.G. Meiro Filho, B.A. Callander, N. Harris, A. Kattenberg, and K. Maskell (eds.)]. Cambridge University Press, Cambridge, United Kingdom and New York, NY, USA, 572 pp.

- Warrick, R.A., C. Le Provost, M.F. Meier, J. Oerlemans, and P.L. Woodworth, Chapter 7. *Changes in Sea Level*, pp. 359–405.

**IPCC**, 1996. *Climate Change 1995: Impacts, Adaptations, and Mitigation of Climate Change: Scientific-Technical Analyses. Contribution of Working Group II to the Second Assessment Report of the Intergovernmental Panel on Climate Change* [Watson, R.T., M.C. Zinyowera, and R.H. Moss (eds.)]. Cambridge University Press, Cambridge, United Kingdom and New York, NY, USA, 880 pp.

- Kirschbaum, M.U.F. and A. Fischlin, Chapter 1. *Climate Change Impacts on Forests,* pp. 95–130.
- Beniston, M. and D.G. Fox, Chapter 5. *Impacts of Climate Change on Mountain Regions,* pp. 191–213.
- Bijlsma, L, Chapter 9. *Coastal Zones and Small Islands*, pp. 289–324.
- Reilly, J., Chapter 13. *Agriculture in a Changing Climate: Impacts and Adaptation*, pp. 427–467.
- Everett, J., Chapter 16. *Fisheries,* pp. 511–537.
- McMichael, A., Chapter 18. *Human Population Health,* pp. 561–584.

**IPCC**, 1996. *Climate Change 1995: Economic and Social Dimensions of Climate Change. Contribution of Working Group III to the Second Assessment Report of the Intergovernmental Panel on Climate Change* [Bruce, J.P., H. Lee, and E.F. Haites (eds.)]. Cambridge University Press, Cambridge, United Kingdom and New York, NY, USA, 448 pp.

- Pearce, D.W., W.R. Cline, A.N. Achanta, S. Fankhauser, R.K. Pachauri, R.S.J. Tol, and P. Vellinga, Chapter 6. *The Social Costs of Climate Change: Greenhouse Damage and Benefits of Control*, pp. 179–224.

**IRRI (International Rice Research Institute)**, 1993: *IRRI: 1993–1995: IRRI Rice Almanac*. IRRI, Manila, Philippines, 142 pp.

**ISPAN (Irrigation Support Project for Asia and the Near-East)**, 1992: *Impacts of Flood Control and Drainage on Vector-Borne Disease Incidence in Bangladesh*., Dhaka, Bangladesh, ISPAN, VA, USA, 34 pp.

**Karim**, Z., S.G., Hussain, and M. Ahmed, 1996: Asssessing impacts of climate variations on foodgrain production in Bangladesh. In: *Climate Change Variability and Adaptation in Asia and the Pacific* [Erda, L., W. Bolhofer. S. Huq, S. Lenhart, S.K. Mukherjee, J.B. Smith, and J. Wisniewski (eds.)]. Kluwer Academic Publishers, Dordrecht, Netherlands, pp. 53–62.

**Kitayama**, K., 1996: Climate of the summit region of Mount Kinabalu (Borneo) in 1992, an El Niño year. *Mountain Research and Development,* **16(1)**, 65–75.

**Koopman**, J.S., D.R. Prevots, M.A.V. Marin, H.G. Dantes, M.I.Z Aqino, I.M. Longimi, and J.S. Amor, 1991: Determinants and predictors of dengue infection in Mexico. *American Journal of Epidemiology,* **133**, 1168–1178.

**Kripalani**, R.H., S. Inamdar, and N.A. Sontakke, 1996: Rainfall variability over Bangladesh and Nepal: comparison and connections with features over India. *International Journal of Climatology,* **16**, 689–704.

**Lal**, M., 1994: Water resources of the South Asian region in a warmer atmosphere. *Advances in Atmospheric Sciences,* **11(2)**, 239–246.

**Lal**, M., L. Bengtsson, U. Cubasch, M. Esch, and U. Schlese, 1995a: Synoptic scale disturbances of Indian summer monsoon as simulated in a high resolution climate model. *Climate Research,* **5**, 243–258.

**Lal**, M., U. Cubasch, R. Voss, and J. Waszkewitz, 1995b: Effect of transient increase in greenhouse gases and sulphate aerosols on monsoon climate. *Current Science,* **69(9)**, 752–763.

**Lal**, M., K.K. Singh, L.S. Rathor, G. Srinivasan, and S.A. Saseendran, 1996: *Vulnerability of Rice and Wheat Yields in Northwest India to Future Changes in Climate.* Centre for Atmospheric Sciences, Indian Institute of Technology, Tech Report No. A/TR/1-96 (August 1996), New Delhi, India, 29 pp.

**Lal**, M., G. Srinivasan, and U. Cubasch, 1996: Implications of increasing greenhouse gases and aerosols on the diurnal temperature cycle of the Indian subcontinent. *Current Science,* **21(10)**, 746–752.

**Lim**, C.P., Y. Matsuda, and Y. Shigemi, 1995: Problems and constraints in Philippine municipal fisheries: the case of San Miguel Bay, Camarines Sur. *Environmental Management,* **19(6)**, 837–852.

**Lesser**, M.P., 1997: Oxidative stress causes coral bleaching during exposure to elevated temperatures. *Coral Reefs,* **16(3)**, 187–192.

**Marston**, R., J. Kleinman, and M. Miller, 1996: Geomorphic and forest cover controls on monsoon flooding, central Nepal Himalaya. *Mountain Research and Development,* **16**, 257–264.

**Martens**, W.J.M., T.H. Jetten, and D.A. Focks, 1997: Sensitivity of malaria, schistosomiasis and dengue to global warming. *Climate Change,* **35**, 145–156.

**Martens**, W.J.M., L.W. Niessen, J. Rotmans, T.H. Jetten, and A.J. McMichael, 1995: Potential impact of global climate change on malaria risk. *Environmental Health Perspectives,* **103(5)**, 458–464.

**Martin**, P. and M. Lefebvre, 1995: Malaria and climate: sensitivity of malaria potential transmission to climate. *Ambio,* **24 (4)**, 200–207.

**Matthews**, R.B., M.J. Kropff, D. Bachelet, and H.H. van Laar, 1994: *The Impact of Global Climate Change on Rice Production in Asia: a Simulation.* Report No. ERL-COR-821, U.S. Environmental Protection Agency, Environmental Research Laboratory, Corvallis, OR, USA.

**McLean**, R.F., and N. Mimura (eds.), 1993: *Vulnerability Assessment to Sea Level Rise and Coastal Zone Management.* Proceedings of the IPCC Eastern Hemisphere Workshop, August 3–6, 1993, Tsukuba, Japan, Department of Environment, Sport and Territories, Canberra, Australia, 429 pp.

**Midun**, Z. and S.C. Lee, 1995: Implications of a greenhouse-induced sea-level rise: a national assessment for Malaysia. *Journal of Coastal Research,* **S1(14)**, 96–115.

**Milliman**, J.D. and B.U. Haq, 1996: *Sea-level Rise and Coastal Subsidence: Causes, Consequences and Strategies.* Kluwer Academic Press, Dordrecht, Netherlands, 369 pp.

**Mills**, A., 1994: The economic consequences of malaria for households: a case-study in Nepal. *Health Policy,* **29**, 209–227.

**Mirza**, M.Q., 1997: The runoff sensitivity of the Ganges river basin to climate change and its implications. *Journal of Environmental Hydrology,* **5**, 1–13.

**Mirza**, M.Q. and A. Dixit, 1997: Climate change and water management in the GBM basins. *Water Nepal,* **5(1)**, 71–100.

**Mirza**, M.Q. and N.J. Ericksen, 1996: Impact of water control projects on fisheries resources in Bangladesh. *Environmental Management,* **20(4)**, 523–539.

**Mirza**, M.Q., N.J. Ericksen, R.A. Warrick, and G.J. Kenny, 1997: Are floods getting worse in the Ganges, Brahmaputra and Meghna basins? *Natural Hazards* (in press).

**Mitchell**, J.F.B., S. Manabe, T. Tokioka, and V. Meleshko, 1990: Equilibrium climate change. In: *Climate Change: The IPCC Scientific Assessment* [Houghton, J.T., G.J. Jenkins, and J.J. Ephraums (eds.)]. Cambridge University Press, Cambridge, United Kingdom, pp. 131–172.

**Mukai**, H., 1993: Biogeography of tropical seagrasses in the western Pacific. *Australian Journal of Marine and Freshwater Research,* **44**, 1–17.

**Murphy**, J.M. and J.F.B. Mitchell, 1995: Transient response of the Hadley Centre coupled ocean-atmosphere model to increasing carbon dioxide. Part II. Spatial and temporal structure of response. *Journal of Climate,* **8**, 57–80.

**Nicholls**, R.J. 1995: Coastal megacities and climate change. *Geojournal,* **37**, 369–379.

**Nicholls**, R.J. and S.P. Leatherman (eds.), 1995: Potential impacts of accelerated sea-level rise on developing countries. *Journal of Coastal Research,* **14**, pp. 324.

**Nicholls**, R.J., N. Mimura, and J.C. Topping, 1995: Climate change in south and south-east Asia: some implications for coastal areas. *Journal of Global Environmental Engineering,* **1**, 137–154.

**Parker**, D.E., H. Wilson, P.D. Jones, J.R, Christy, and C.K. Folland, 1996: The impact of Mount Pinatubo on world-wide temperatures. *International Journal of Climatology,* **16**, 487–497.

**Parry**, M.L., M. Baltran de Rozari, A.L. Chong, and S. Panich (eds.), 1992: *The Potential Socio-Economic Effects of Climate Change in South-East Asia.* United Nations Environment Programme, Nairobi, Kenya, pp. 126.

**Parthasarathy**, B., N.A. Sontake, A.A. Monot, and D.R. Kothawale, 1987: Drought-flood in the summer monsoon season over different meteorological subdivisions of India for the period 1871–1984. *Journal of Climatology,* **7**, 57–70.

**Paul**, B. and H. Rashid, 1993: Flood damage to rice crop in Bangladesh. *The Geographical Review,* **83(2)**, 151–159.

**Picard**, J. and A. Mills, 1992: The effect of malaria on work time: analysis of data from two Nepalese districts. *Journal of Tropical Medicine and Hygiene,* **95**, 382–389.

**Porter**, J., 1994: The impact of climate change on malaria distribution in the Asia-Pacific region. In: *Climate Impact Assessment Methods for Asia and the Pacific* [Jakeman, A.J. and A.B. Pittock (eds.)]. Proceedings of a regional symposium, 10–12 March 1993, organised by ANUTECH Pty. Ltd. on behalf of the Australian International Development Assistance Bureau (AIDAB), Canberra, Australia, 149–152 pp.

**Quinn**, W.H., D.O. Zopf, K.S. Short, and R.T. Kuo Yang, 1978: Historical trends and statistics of the Southern Oscillation, El Niño, and Indonesian droughts. *Fisheries Bulletin,* **76**, 663–678.

**Qureshi**, A. and D. Hobbie, 1994: *Climate Change in Asia: Thematic Overview.* Asian Development Bank, Manila, Philippines, 351 pp.

**Rao**, D.G., J.C. Katyal, S.K. Sinha, and K. Srinivas, 1995: Impacts of climate change on sorghum productivity in India: simulation study. In: *Climate Change and Agriculture: Analysis of Potential International Aspects,* American Society of Agronomy, Special Publication 59, Madison, WI, USA, 325–337 pp.

**Rao**, D.G. and S.K. Sinha, 1994: Impact of climate change on simulated wheat production in India, EPA Conference, New Delhi, India, pp. 1–8.

**Reilly**, J., N. Hohmann, and S. Kane, 1993: *Climate change and agriculture: The role of International Trade. American Economic Association Papers and Proceedings 83*, pp. 306–312.

**Riebsame**, W.E., K.M. Strzepek, L.L. Wescoat, Jr., R. Perritt, G.L. Gaile, J. Jacobs, R. Lieichenko, C. Magadza, H. Phien, B.J. Urbiztondo, P. Restrepo, W.R. Rose, M. Saleh, L.H. Ti, C. Tucci, and D. Yates, 1995: Complex river basins. In: *As Climate Changes: International Impacts and Implications* [Strzepek, K.M. and J.L. Smith (eds.)] Cambridge University Press, Melbourne, Australia, pp. 57–91.

**Rieley**, J.O., S. Page, and G. Sheffermann, 1995: Tropical peat swamp forests of southern Asia: ecology and environmental importance. *Malaysian Journal of Tropical Geography,* **26(2)**, 130–141.

**Rosenzweig**, C. and A. Iglesias, 1994: *Implications of Climate Change for International Agriculture: Crop Modeling Study.* EPA 230-B-94-003, U.S. Environmental Protection Agency, Washington, DC, USA, 312 pp.

**SAARC (South Asian Association for Regional Cooperation)**, 1992: *Regional Study on the Causes and Consequences of Natural Disasters and the Protection and Preservation of the Environment.* SAARC Secretariat, Kathmandu, Nepal, 147–149 pp.

**Sabhasri**, S. and K. Suwarnarat, 1996: Impact of sea level rise on flood control in Bangkok and vicinity. In: *Sea-Level Rise and Coastal Subsidence: Causes, Consequences and Strategies* [Milliman, J.D. and B.U. Haq (eds.)]. Kluwer Academic Publishers, Dordrecht, Netherlands, 343–355 pp.

**Sankar-Rao**, M., K.M. Lau, and S. Yang, 1996: On the relationship between Eurasian snow cover and the Asian summer monsoon. *International Journal of Climatology,* **16**, 605–616.

**Schreir**, H. and P.B. Shah, 1996: Water dynamics and population pressure in the Nepal Himalayas. *Geojournal,* **40(1–2)**, 45–51.

**Sharma**, C.K., 1979: Partial drought conditions in Nepal. *Hydrological Sciences Bulletin,* **24(3)**, 327–333.

**Singh**, P., S.K. Jain, and N. Kumar, 1997: Estimation of snow and glacier-melt contribution to the Chenab River, Western Himalaya. *Mountain Research and Development,* **17(1)**, 49–56.

**Sinha**, S.K., 1994: Response of tropical agrosystems to climate change. *Proceedings of the International Crop Science Congress,* **1**, 281–289.

**Sinha**, S.K, 1997: Global change scenario: current and future with reference to land cover change and sustainable agriculture — south and south-east Asian context. *Current Science,* **72(11)**, 846–854.

**Sinha**, S.K., and M.S. Swaminathan, 1991: Deforestation, climate change and sustainable nutrition security. *Climatic Change,* **19**, 201–209.

**Somaratne**, S. and A.H. Dhanapala, 1996: Potential impact of global climate change on forest in Sri Lanka. In: *Climate Change Variability and Adaptation in Asia and the Pacific* [Erda, L., W. Bolhofer. S. Huq, S. Lenhart, S.K. Mukherjee, J.B. Smith, and J. Wisniewski (eds.)] Kluwer Academic Publishers, Dordrecht, Netherlands, 129–135 pp.

**Sooryanarayana**, V., 1995: Floods in Malaysia: patterns and implications. *Malaysian Journal of Tropical Geography,* **26(1)**, 35–46.

**Spalding**, M., 1997: The global distribution and status of mangrove ecosystems. *Intercoast Network,* **1**, 20–21.

**Spencer**, N.E. and P.L. Woodworth, 1993: *Data Holdings of the Permanent Service for Mean Sea Level (November 1993).* Permanent Service for Mean Seal Level (PSMSL), Bidston Observatory, Birkenhead, United Kingdom, 81 pp.

**Sthapti**, K.M., 1995: Sedimentation of Lakes and Reservoirs with Special Reference to the Kulekhani Reservoir. In: *Challenges in Resource Dynamics in Nepal: Processes, Trends and Dynamics in Middle Mountain Watersheds* [Schreir, H., P.B. Shah, and S. Brown (eds.)]. Proceedings of a workshop, International Centre for Integrated Mountain Development (ICIMOD) and International Development Research Centre (IDRC), Kathmandu, Nepal, pp. 5–12.

**Subramanian**, V. and A.L. Ramanathan, 1996: Nature of sediment load in the Ganges-Brahmaputra River systems in India. In: *Sea-Level Rise and Coastal Subsidence.* [Milliman, J.D. and B.U. Haq (eds.)], Kluwer Academic Publishers, Dordecht, Netherlands, pp. 151–168.

**Suppiah**, R., 1994: The Asian monsoons: simulations from four GCMs and likely changes under enhanced greenhouse conditions. In: *Climate Impact Assessment Methods for Asia and the Pacific* [Jakeman, A.J. and B. Pittock (eds.)]. Proceedings of a regional symposium, organised by ANUTECH Pty. Ltd. on behalf of the Australian International Development Assistance Bureau 10–12 March 1993, Canberra, Australia, pp. 73–78.

**Suppiah**, R., 1997: Extremes of the southern oscillation phenomenon and the rainfall of Sri Lanka. *International Journal of Climatology,* **17**, 87–101.

**Teh**, T.S. and P.K. Voon, 1992: Impacts of sea level rise in West Johore, Malaysia. *Malaysian Journal of Tropical Geography,* **23(2)**, 93–102.

**UNEP**, 1997: *Global Environment Outlook.* United Nations Environment Programme and Oxford University Press, Nairobi, Kenya and Oxford, United Kingdom, 270 pp.

**Uprety**, B.K., 1988: *Impact of Riverbank Erosion, Flood Hazard and River Shifting in Nepal.* Paper presented at the International Symposium on the Impact of Riverbank Erosion, Flood Hazard and the Problems of Population Displacement. April 11–13, 1988, Dhaka, Bangladesh, 27 pp.

**Warrick**, R.A. and Q.K. Ahmad (eds.), 1996: *The Implications of Climate and Sea-Level Change for Bangladesh.* Kluwer Academic Publishers, Dordrecht, Netherlands, 415 pp.

**Watanabe**, T., J.D. Ives, and J.E. Hammond, 1994: Rapid growth of a glacial lake in Khumbu Himal, Nepal Himalaya: prospects for a catastrophic flood. *Mountain Research and Development,* **14**, 329–340.

**Watanabe**, T. and D. Rothacher, 1996: The 1994 Lugge Tsho glacial lake outburst flood, Bhutan Himalaya. *Mountain Research and Development,* **16**, 77–81.

**Whetherald**, R.T. and S. Manabe, 1986: An investigation of cloud cover change in response to thermal forcing. *Climatic Change,* **8**, 5–23.

**Whetton**, P., 1994: Constructing climate scenarios: the practice. In: *Climate Impact Assessment Methods for Asia and the Pacific* [Jakeman, A.J. and A.B. Pittock (eds.)]. Proceedings of a regional symposium, organised by ANUTECH Pty. Ltd. on behalf of the Australian International Development Assistance Bureau 10–12 March 1993, Canberra, Australia, pp. 21–27.

**Whetton**, P.H., A.M. Fowler, M.R. Haylock, and A.B. Pittock, 1993: Implications of climate change due to the enhanced greenhouse effect on floods and droughts in Australia. *Climatic Change,* **25**, 289–318.

**Whetton**, P., A.B. Pittock, and R. Suppiah, 1994: Implications of climate change for water resources in south and southeast Asia. In: *Climate Change in Asia: Thematic Overview.* Asian Development Bank, Manila, Philippines, pp. 57–103.

**Whetton**, P.H. and I. Rutherford, 1994: Historical ENSO teleconnections in the Eastern Hemisphere. *Climate Change,* **28**, 221–253.

**Whetton**, P.H. and I. Rutherford, 1996: Historical teleconnections in the Eastern Hemisphere: comparison with latest El Niño series of Quinn. *Climate Change,* **32**, 103–109.

**Wijeratne**, M.A., 1996: Vulnerability of Sri Lanka tea production to global climate change. In: *Climate Change Variability and Adaptation in Asia and the Pacific* [Erda, L., W. Bolhofer. S. Huq, S. Lenhart, S.K. Mukherjee, J.B. Smith, and J. Wisniewski (eds.)]. Kluwer Academic Publishers, Dordrecht, Netherlands, 249 pp.

**Wilkinson**, C.R. and R.W. Buddemeier, 1994: *Global Climate Change and Coral Reefs: Implications for People and Reefs.* Report of the UNEP-IOC-ASPEI-IUCN Global Task Team on the Implications of Climate Change on Coral Reefs, International Union for the Conservation of Nature and Natural Resources (IUCN), Gland, Switzerland, 124 pp.

**Wilson**, C.A. and J.F.B. Mitchell, 1987: A doubled $CO_2$ climate sensitivity experiment with a global climate model including a simple ocean. *Journal of Geophysical Research,* **92**, 13315–13343.

**Wong**, P.P., 1992: Geomorphology and tourism on the east coast of peninsular Malaysia. In: *The Coastal Zone of Peninsular Malaysia* [Tija, H.D. and S.M.S. Abdullah (eds.)]. Penerbit Universiti, Kebangsaan, Bangi, Malaysia, pp. 72–79.

**Wong**, P.P., 1995: Tourism-environment interaction in the western bays of Phuket Islands. *Malaysian Journal of Tropical Geography,* **26(1)**, 67–75.

**Woodroffe**, C.D., 1990: The impact of sea-level rise on mangrove shorelines. *Progress in Physical Geography*, **14**, 483–520.

**WHO (World Health Organization)**, 1996a: *Regional Health Report 1996*. WHO Regional Office for South-East Region, New Delhi, India, 80 pp.

**WHO (World Health Organization)**, 1996b: *Climate Change and Human Health* [McMichael, A.J., A. Haines, R. Slooff, and S. Kovats (eds.)]. WHO, Geneva, Switzerland, 279 pp.

**WRI (World Resources Institute)**, 1996: *World Resources: A Guide to the Global Environment, 1996–97*. World Resources Institute/United Nations Environment Programme/United Nations Development Programme/The World Bank. Oxford University Press, New York, 342 pp.

**Zeigler**, R.S., and D.W. Puckridge, 1995: Improving sustainable productivity in rice-based rainfed lowland systems of south and southeast Asia. *Geojournal,* **35(3)**, 307–324.

**Zhang**, H., 1994: Impacts of tropical deforestation in S.E. Asia. In: *Climate Impact Assessment Methods for Asia and the Pacific* [Jakeman, A.J. and B. Pittock (eds.)]. Proceedings of a regional symposium, organised by ANUTECH Pty. Ltd. on behalf of the Australian International Development Assistance Bureau 10–12 March 1993, Canberra, Australia, pp. 137–140.

**Zhu**, Y., and D.D. Houghton, 1996: The impact of Indian Ocean SST on the large-scale Asian summer monsoon and the hydrological cycle. *International Journal of Climatology,* **16**, 617–632.

# The Regional Impacts of Climate Change: Annexes

Prepared by IPCC Working Group II

# A

# Regional Trends and Variations of Temperature and Precipitation

THOMAS R. KARL (USA)

IPCC (1996) provided time-series plots and global maps depicting trends of precipitation and temperature. This Annex extends and updates these records for a broader number of contiguous regions.

Two data sets are used to represent near-surface air temperature change. Over land this includes a near-surface air temperature data set developed by Jones (1994), and over the oceans a sea-surface temperature data set developed by Folland and Parker (1995). Both of these data sets provide the basis for calculating global temperature change as reported in the IPCC Second Assessment Report (1996). Parker *et al.* (1994) describe the methodology used to aggregate land and ocean temperatures for grid cells that span both regions.

Precipitation changes are also calculated from two data sets. Land-surface precipitation data are derived and updated from Hulme (1991) and from Eischeid *et al.* (1995); the latter is referred to as the GHCN (version 1) data set in IPCC (1996). Both data sets were available with a resolution of 5°x5°. Since two data sets were used, a procedure was needed to integrate the data. A simple equal weighting scheme was used when both data sets had data available. For some grid cells, data were available from only one data set, and this provided additional coverage relative to the use of a single data set.

In Figures A-1 and A-2, the magnitude of the trends of precipitation and temperature for each 5°x5° grid cell is given by the area of circle centered in each cell; brown and blue circles reflect decreasing trends, and green and red circles increasing trends. Trends are given in %/century for precipitation and °C/century for temperature. Precipitation trends are expressed in percent relative to the 1961–90 average precipitation.

Time-series plots of the annual anomalies of precipitation and temperature (relative to 20th century averages) are depicted for each of the various regions delineated in Figure A-3. For those regions where a 5°x5° grid cell spanned two or more regions, the anomalies for that grid cell were used more than once—once for each region that intersected the grid cell. The time series depicted in Figures A-4 through A-13 show the anomalies from the 1961–90 means. Longer term variation of these annual anomalies are emphasized by a smooth curve using a nine-point binomial filter.

Time-series plots of precipitation are not provided for Antarctica, due to very poor spatial coverage and large data uncertainties. The trends of temperature for the small island states include both land and ocean sea-surface temperature trends for grid cells that include both land and ocean.

## References


**Eischeid**, J.K., C.B. Baker, T.R. Karl, and H.F. Diaz, 1995: The quality control of long-term climatological data using objective data analysis. *Journal of Applied Meteorology*, **34**, 2787–2795.

**Folland**, C.K. and D.E. Parker, 1995: Correction of instrumental biases in historical sea-surface temperature data. *Quarterly Journal of the Royal Meteorological Society*, **121**, 319–367.

**Hulme**, M., 1991: An intercomparison of model and observed global precipitation climatologies. *Geophysical Research Letters*, **18**, 1715–1718.

**IPCC**, 1996: *Climate Change 1995: The Science of Climate Change*. Contribution of Working Group I to the Second Assessment Report of the Intergovernmental Panel on Climate Change [Houghton, J.J., L.G. Meiro Filho, B.A. Callander, N. Harris, A. Kattenberg and K.Maskell (eds.)]. Cambridge University Press, Cambridge and New York, 572 pp.

**Jones**, P.D., 1994: Hemispheric surface air temperature variations: A reanalysis and an update to 1993. *Journal of Climate*, **3**, 1794–1802.

**Parker**, D.E., P.D. Jones, C.K. Folland, and A. Bevan, 1994: Interdecadal changes of surface temperature since the late 19th century. *Journal of Geophysical Research*, **99**, 14373–14399.


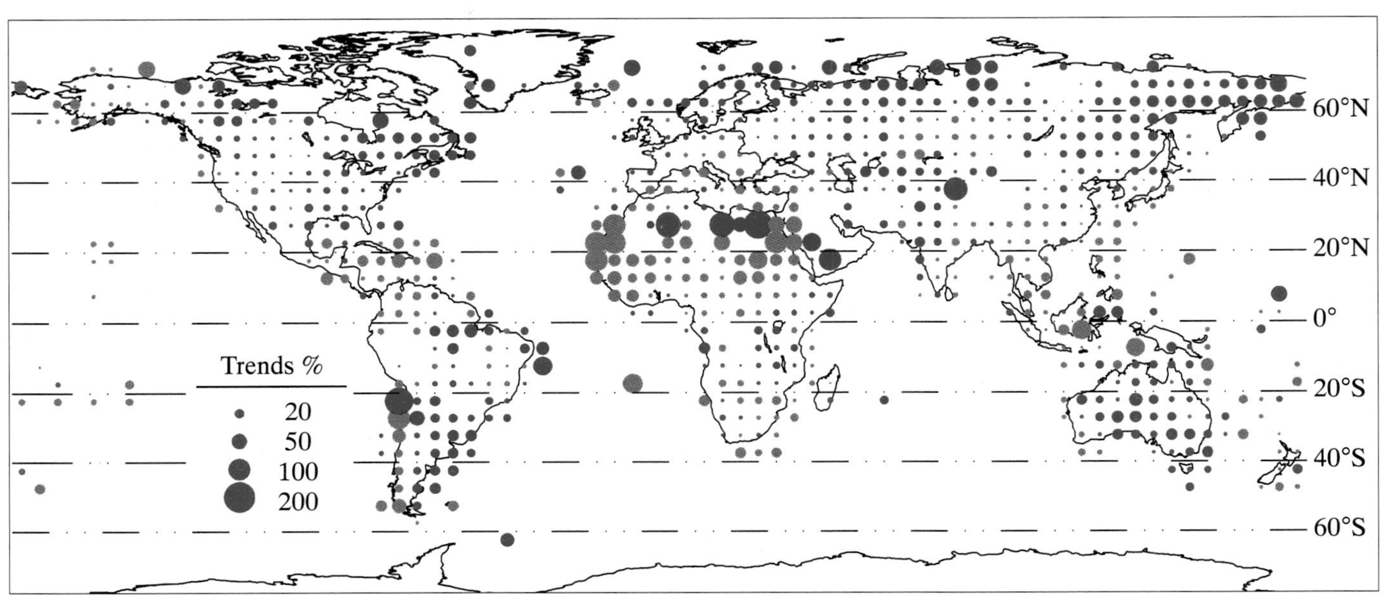


**Figure A-1**: Precipitation—Annual 1901–1995.

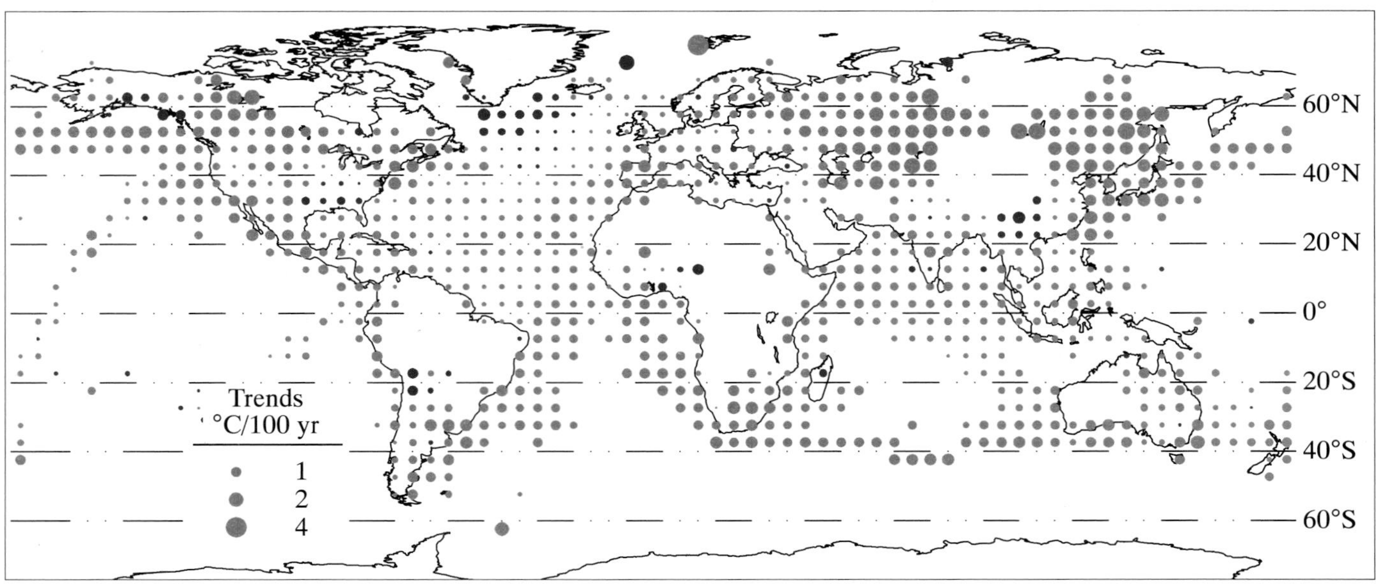


**Figure A-2**: Temperature—Annual 1901–1996.

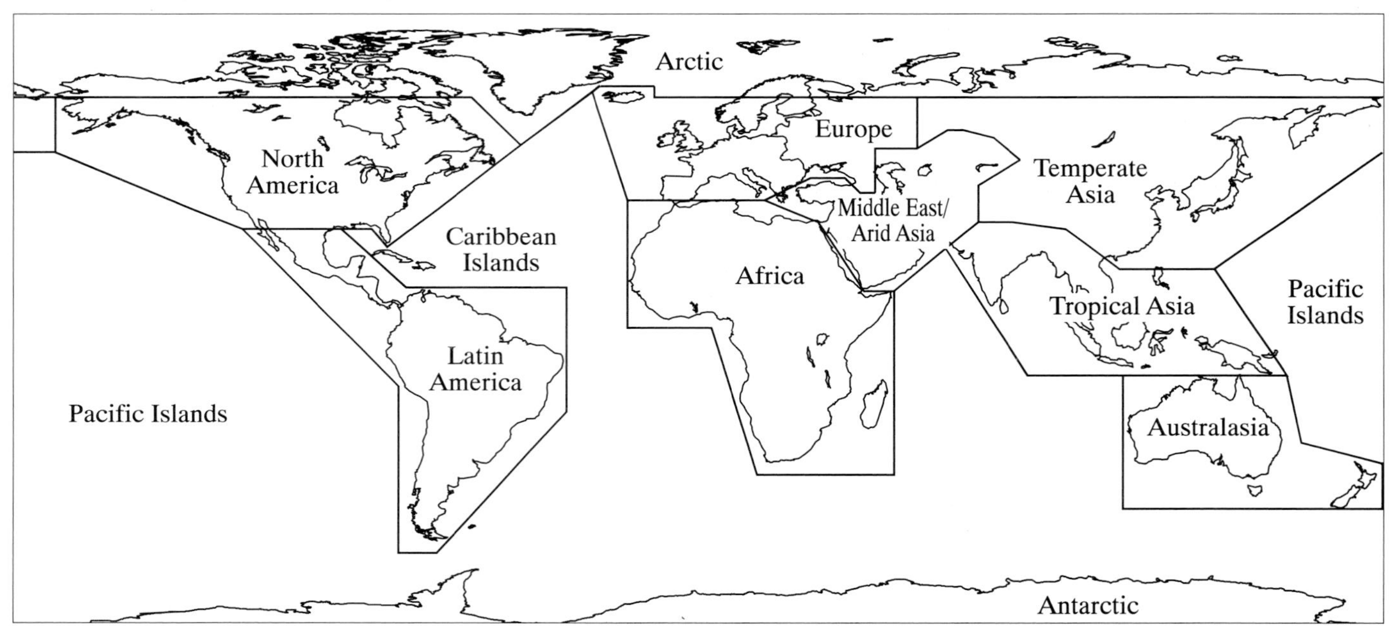


**Figure A-3**: GIS mask used to define the 10 regions covered in this Special Report.

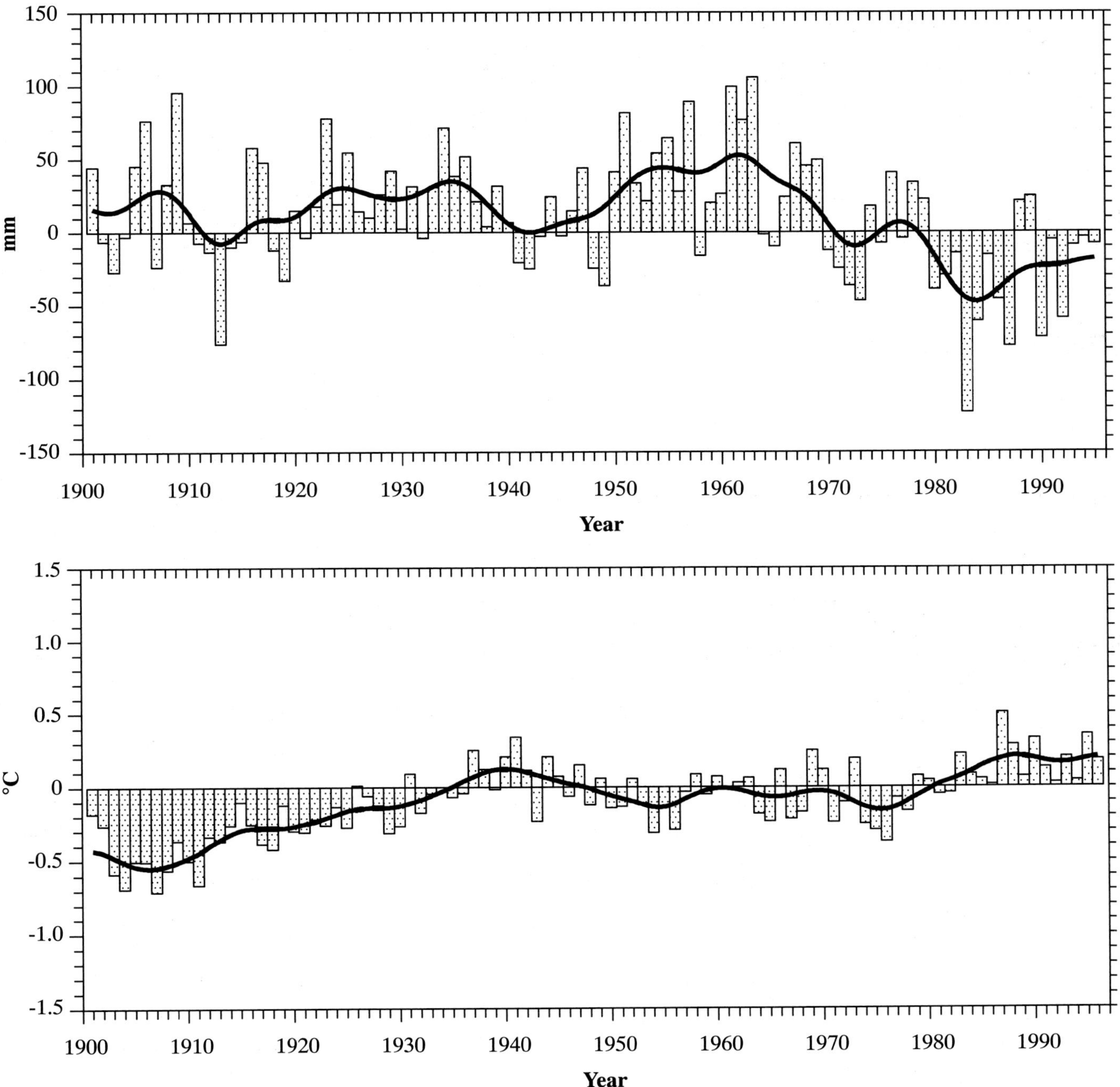


**Figure A-4**: Observed annual precipitation (top) and temperature (bottom) changes for the Africa region.

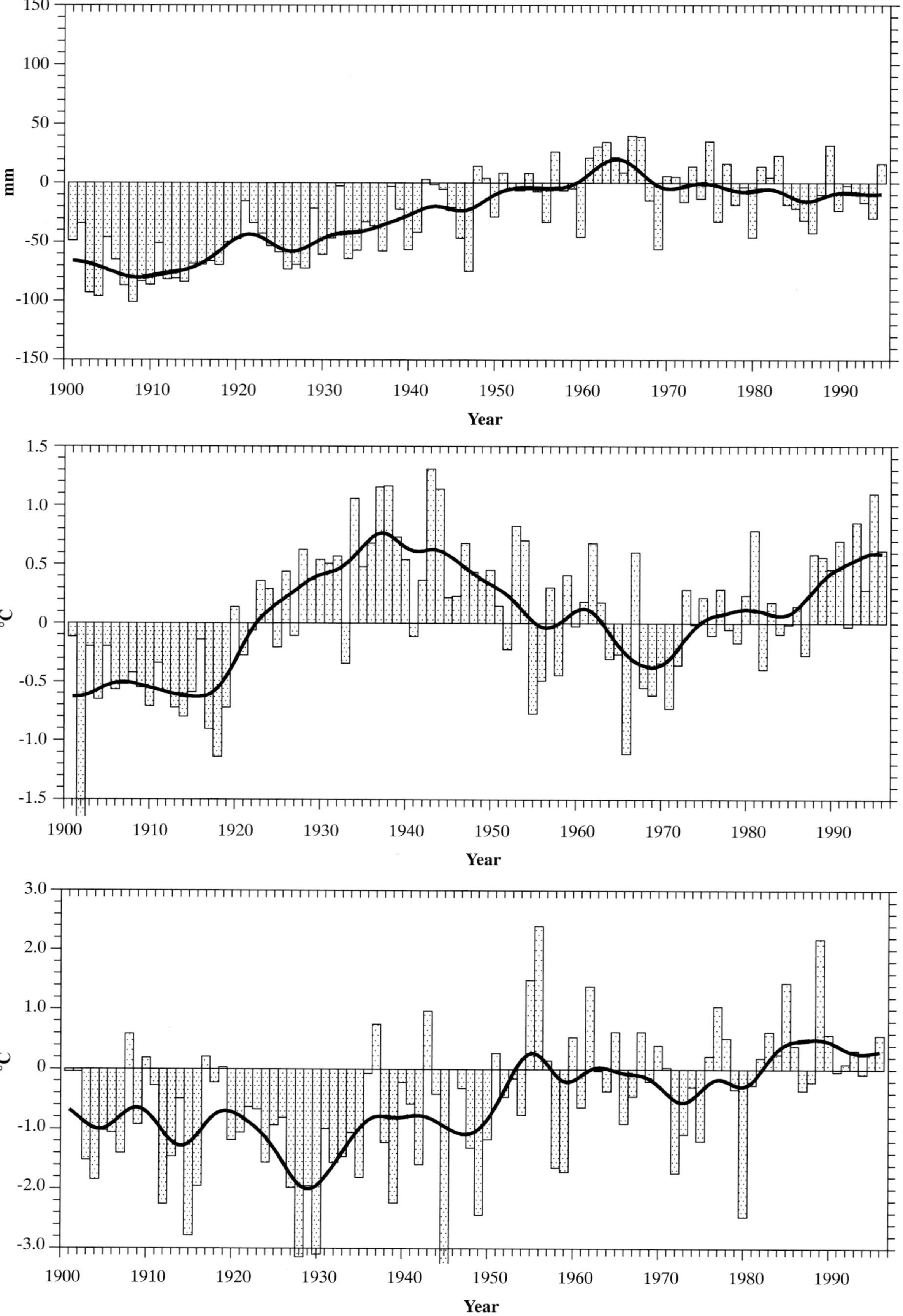


**Figure A-5**: Observed annual precipitation [top (Arctic)] and temperature [middle (Arctic) and bottom (Antarctica)] changes for the Arctic/Antarctica region.

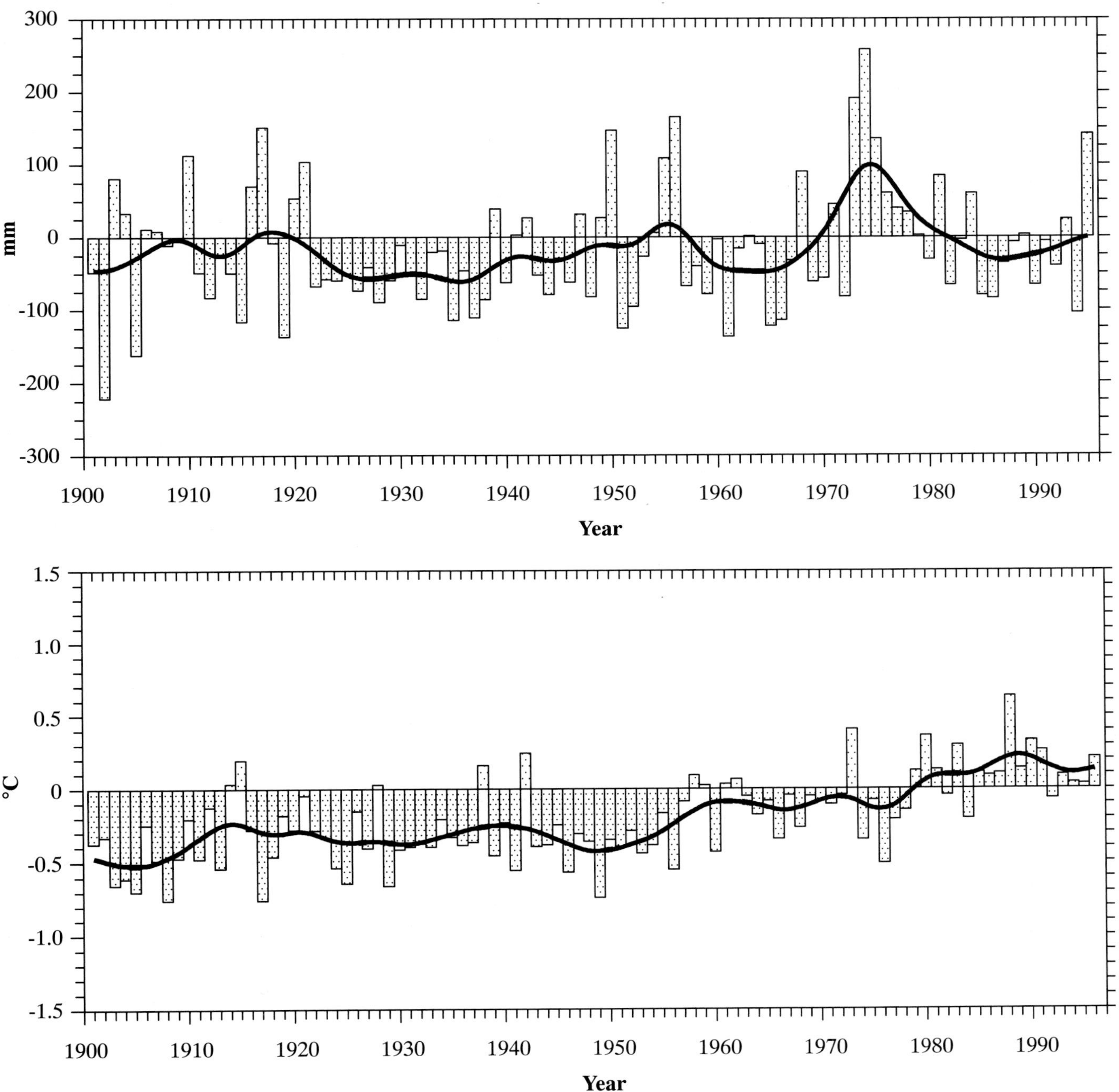


**Figure A-6**: Observed annual precipitation (top) and temperature (bottom) changes for the Australasia region.

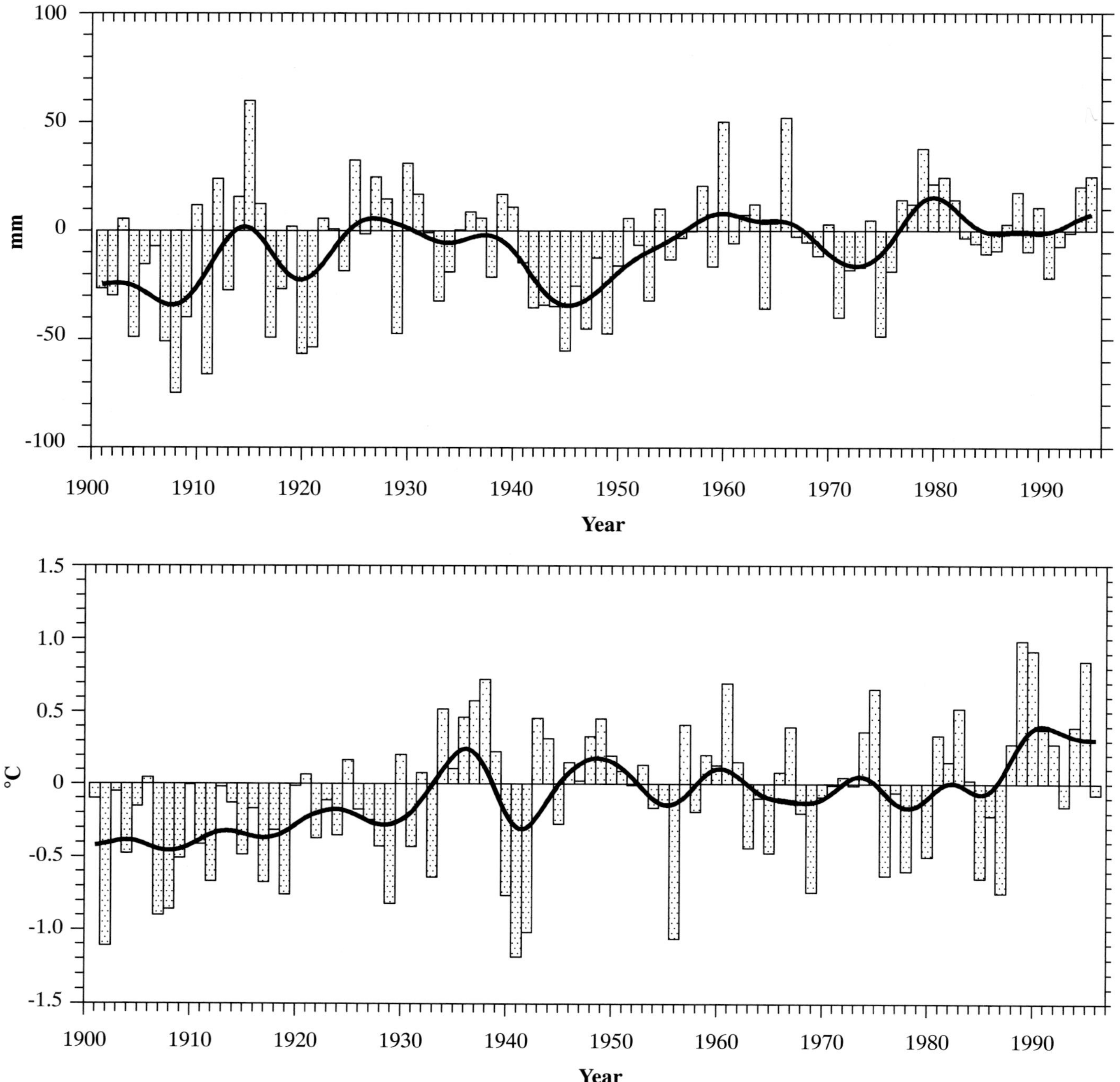


**Figure A-7**: Observed annual precipitation (top) and temperature (bottom) changes for the Europe region.

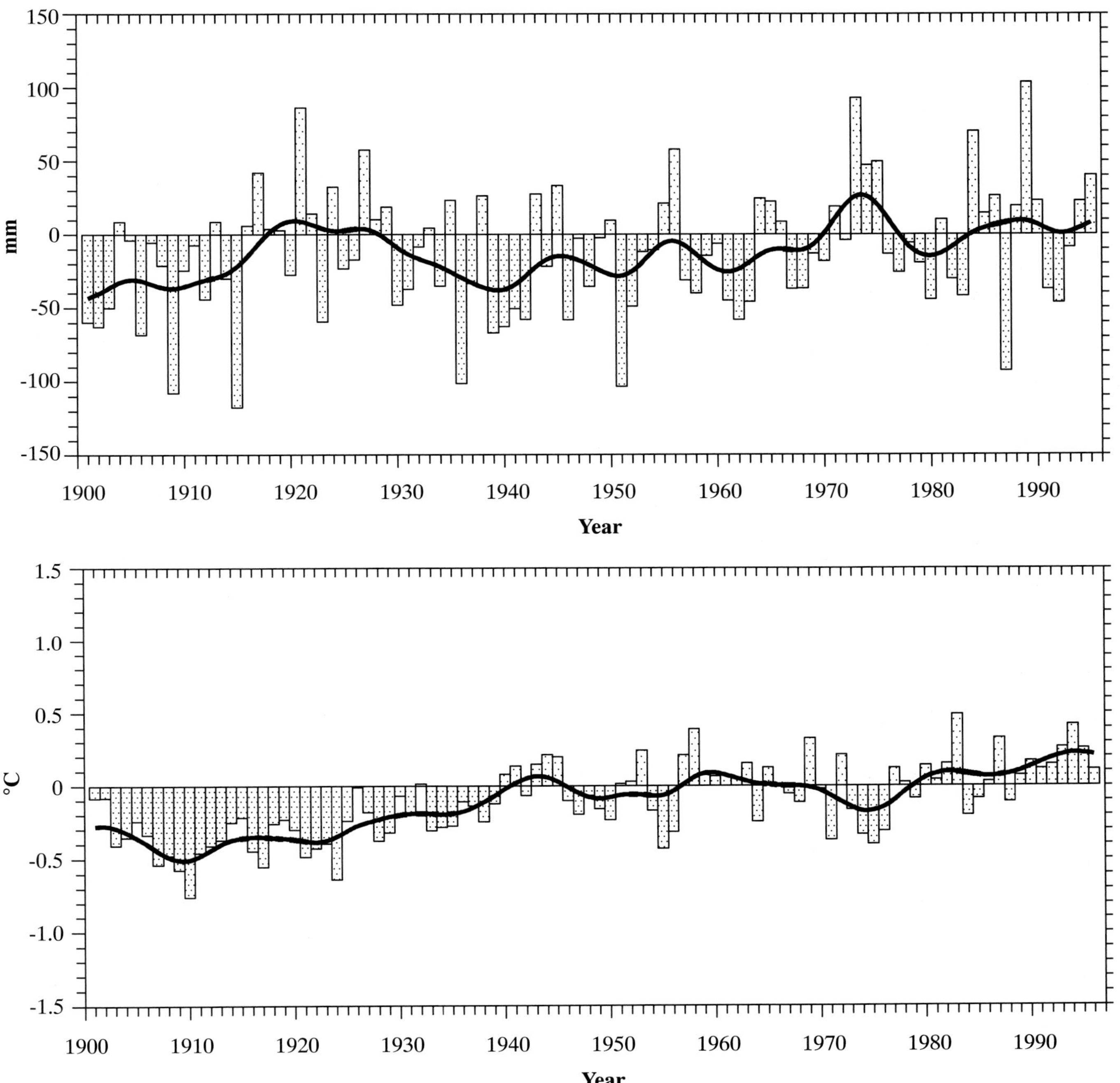


**Figure A-8**: Observed annual precipitation (top) and temperature (bottom) changes for the Latin America region.

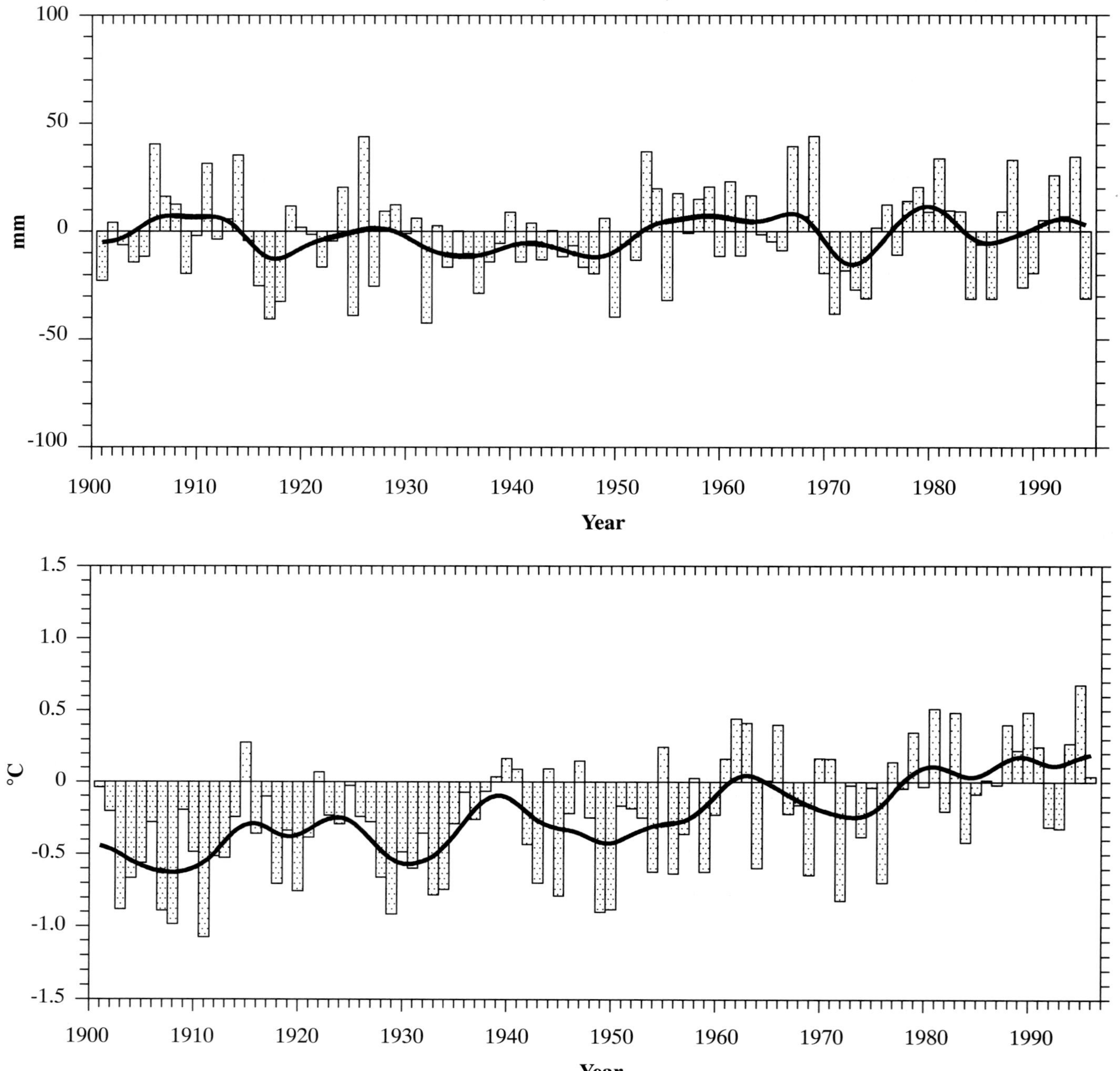


**Figure A-9**: Observed annual precipitation (top) and temperature (bottom) changes for the Middle East/Arid Asia region.

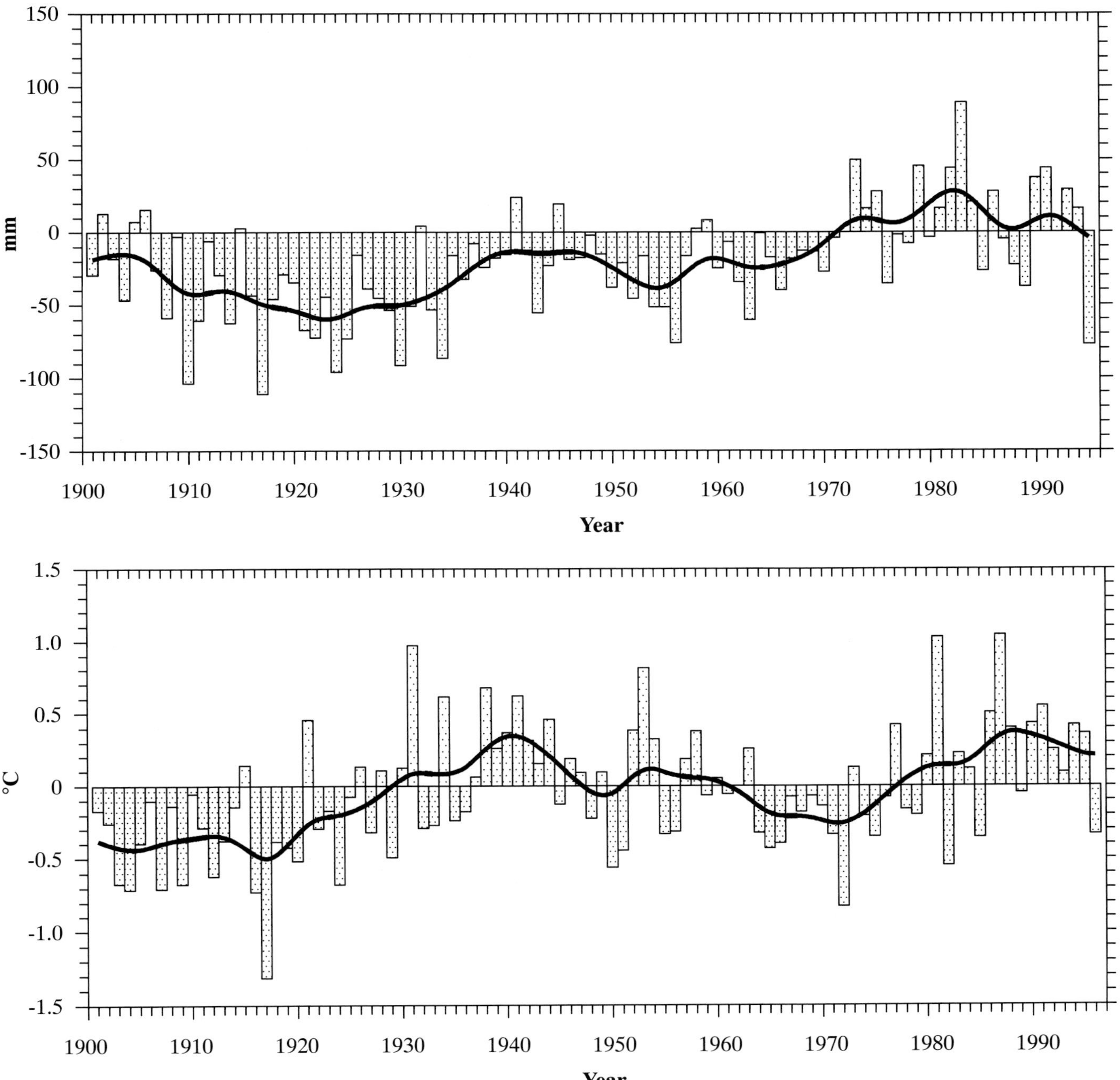


**Figure A-10**: Observed annual precipitation (top) and temperature (bottom) changes for the North America region.

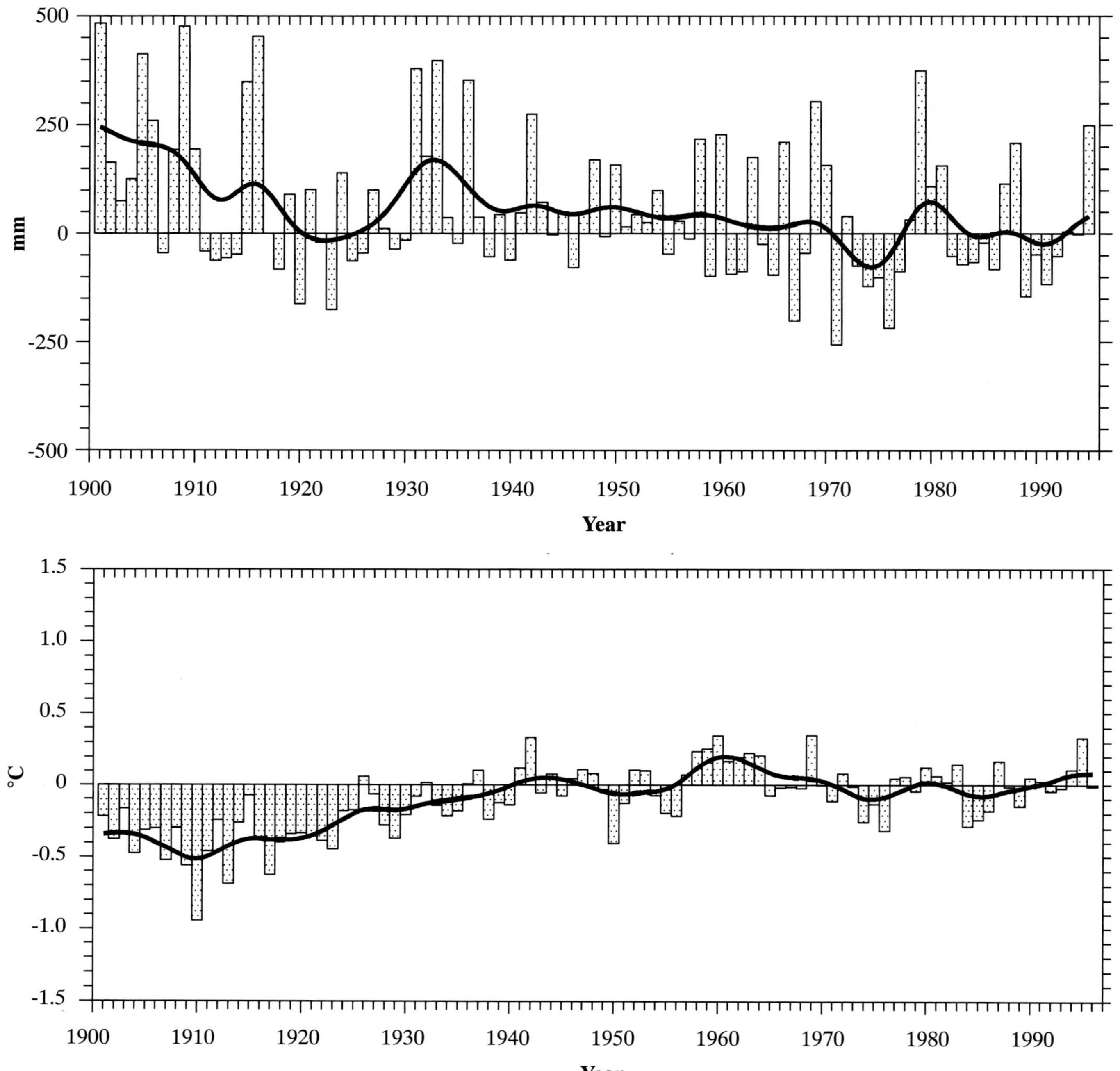


**Figure A-11**: Observed annual precipitation (top) and temperature (bottom) changes for the Caribbean island component of the Small Island States region.

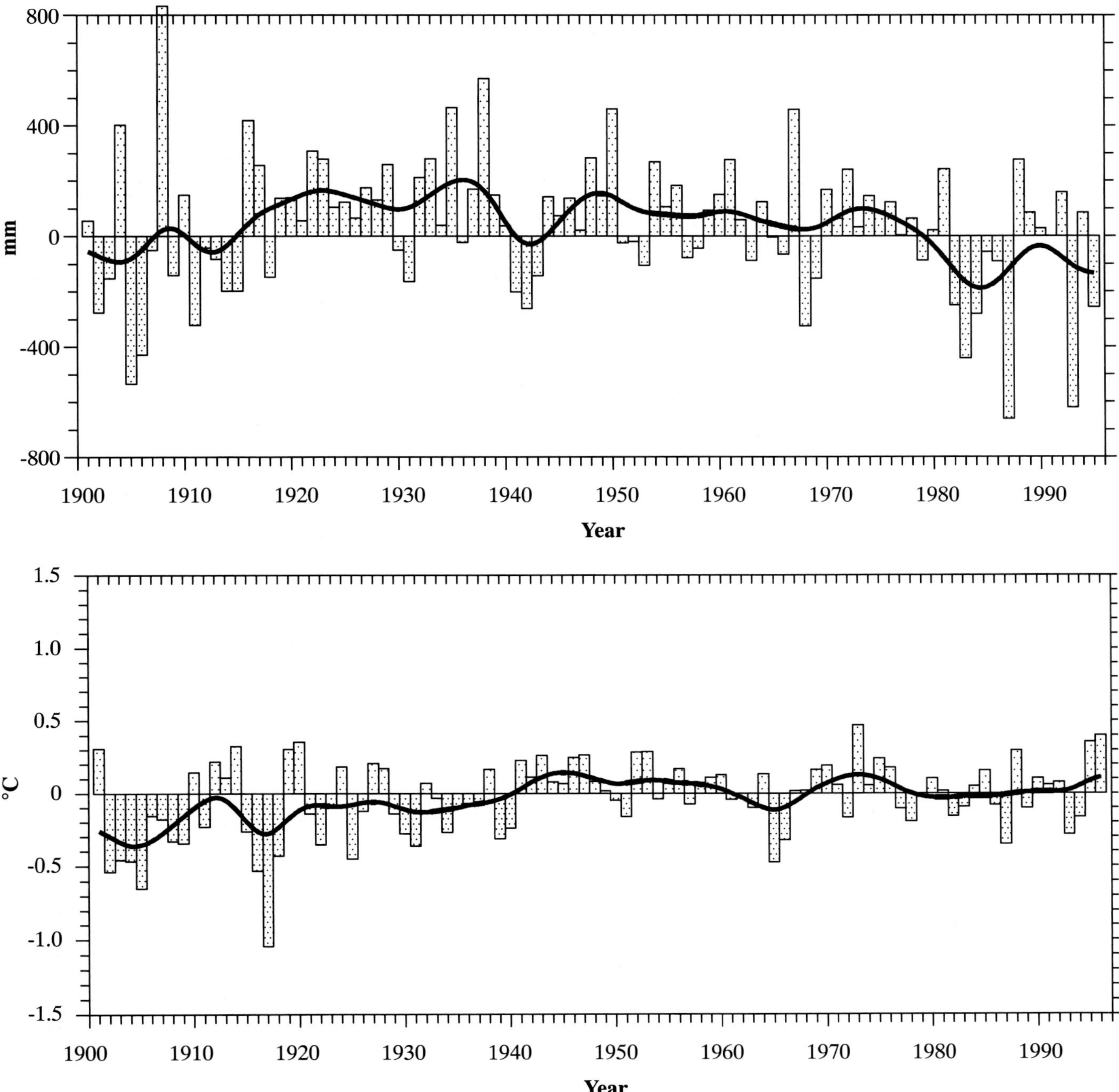


**Figure A-11 (continued)**: Observed annual precipitation (top) and temperature (bottom) changes for the Pacific island component of the Small Island States region.

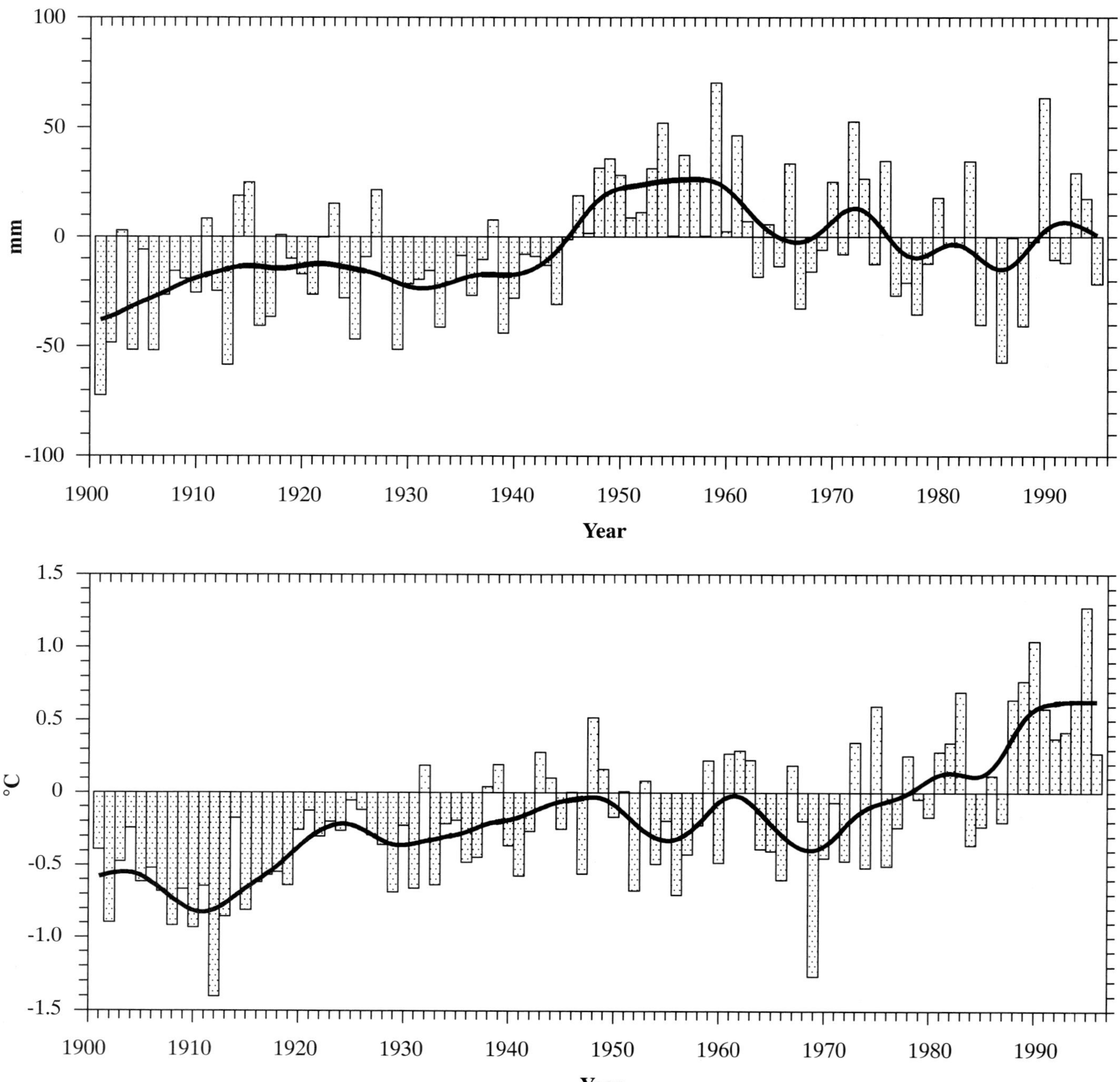


**Figure A-12**: Observed annual precipitation (top) and temperature (bottom) changes for the Temperate Asia region.

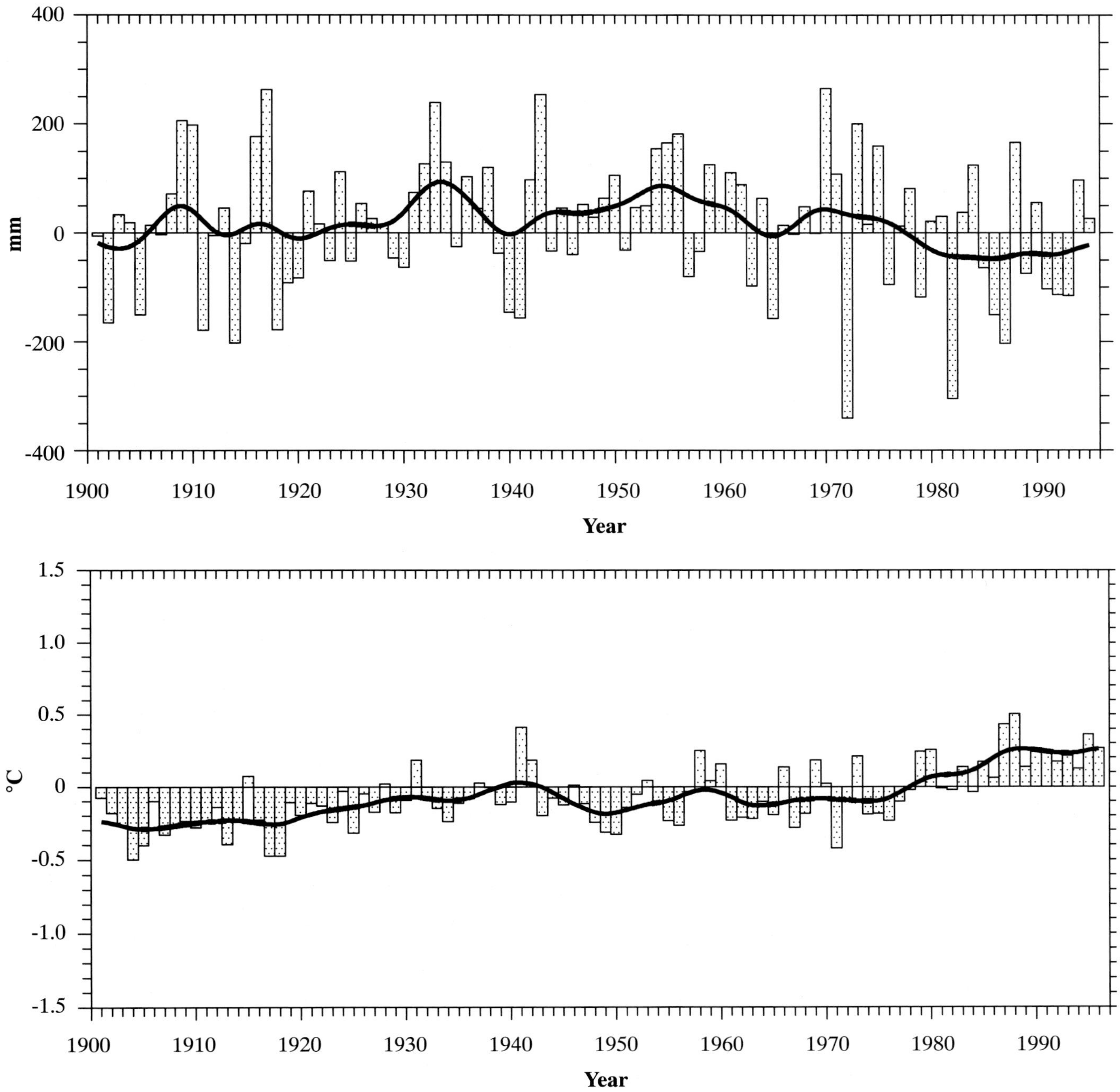


**Figure A-13**: Observed annual precipitation (top) and temperature (bottom) changes for the Tropical Asia region.

# B

# Simulation of Regional Climate Change with Global Coupled Climate Models and Regional Modeling Techniques

This Annex is based solely on Chapter 6 of the Working Group I contribution to the IPCC Second Assessment Report (1996).

F. GIORGI (USA), G.A. MEEHL (USA), A. KATTENBERG (THE NETHERLANDS), H. GRASSL (SWITZERLAND), J.F.B. MITCHELL (UK), R.J. STOUFFER (USA), T. TOKIOKA (JAPAN), A.J. WEAVER (CANADA), AND T.M.L. WIGLEY (USA)

# CONTENTS

In IPCC (1990) and IPCC (1992), very low confidence was placed on the climate change scenarios produced by general circulation model (GCM) equilibrium experiments on the sub-continental, or regional, scale (order of $10^5$–$10^7$ km$^2$). This was mainly attributed to coarse model resolution, limitations in model physics representations, errors in model simulation of present-day regional climate features, and wide inter-model range of simulated regional change scenarios. Since then, transient runs with Atmosphere-Ocean GCMs (AOGCMs) have become available that allow a similar regional analysis. In addition, different regionalization techniques have been developed and tested in recent years to improve the simulation of regional climate change. This section examines regional change scenarios produced by new coupled GCM runs. Following the 1990 and 1992 reports, emphasis is placed on the simulation of seasonally averaged surface air temperature and precipitation, although the importance of higher order statistics and other surface climate variables for impact assessment is recognized (Kittel *et al.*, 1995; Mearns *et al.*, 1995a,b).

## B.1. Regional Simulations by GCMs

In IPCC (1990), five regions were identified for analysis of regional climate change simulation: Central North America (CNA; 35-50°N, 85-105°W), South East Asia (SEA; 5-30°N, 70-105°E), Sahel (Africa) (SAH; 10-20°N, 20°W-40°E), Southern Europe (SEU; 35-50°N, 10 W- 45°E), and Australia (AUS; 12-45°S, 110-155°E). Output from different coupled model runs with dynamical oceans for these regions was analyzed by Cubash *et al.* (1994a), Whetton *et al.* (1996), and Kittel *et al.* (1997), while analysis over the Australian region from equilibrium simulations with mixed-layer ocean models was performed by Whetton *et al.* (1994). Results over two additional regions were analyzed by Raisanen (1995) for Northern Europe (NEU; land areas north of 50°N and west of 60°E) and Li *et al.* (1994) for East Asia (EAS; 15-60°N, 70-140°E). To summarize the findings of these works, Figure B-1 shows differences between region-average values at the time of $CO_2$ doubling and for the control run, and differences between control run averages and observations (hereafter referred to as bias), for winter and summer surface air temperature and precipitation. Note that these models contain increases of $CO_2$ only. Experiments including increased $CO_2$ and the effects of sulfate aerosols will be discussed later.

The biases are presented as a reference for the interpretation of the scenarios, because it can be generally expected that the better the match between control run and observed climate (i.e., the lower the biases), the higher the confidence in simulated

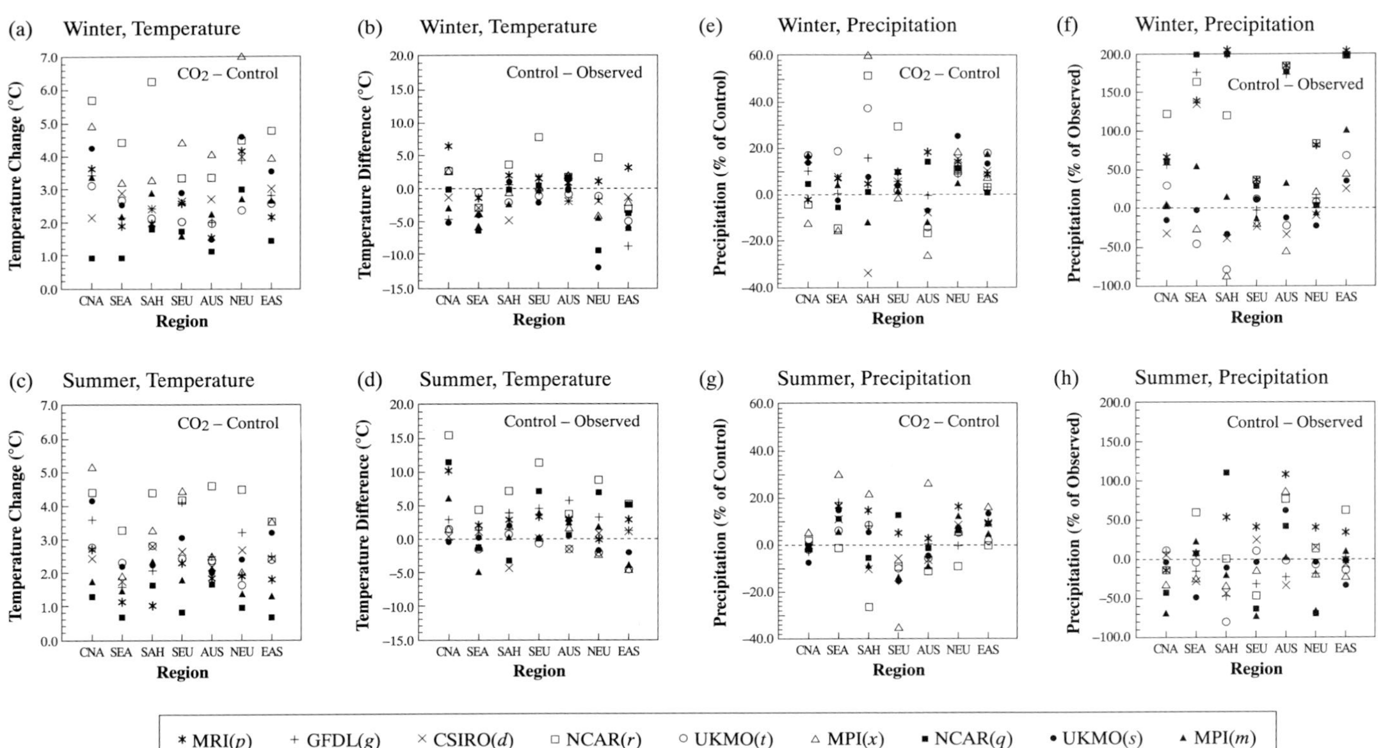


**Figure B-1:** Difference between averages at time of $CO_2$ doubling and control run averages ($CO_2$–Control) and difference between control run averages and observed averages (Control–Observed) as simulated by nine AOGCM runs over seven regions. Units are °C for temperature and percentage of control run, or observed, averages for precipitation. In (f) and (h), values in excess of 200% have been reported at the top end of the vertical scale. In (e), values in excess of 60% have been reported at the top end of the vertical scale. Winter averages are for December-January-February (DJF) for Northern Hemisphere regions, and June-July-August (JJA) for Australia; summer averages are for JJA for the Northern Hemisphere, and DJF for Australia. See Kittel *et al.* (1997) for further details.

change scenarios. The model runs are labeled *d*, *g*, *m*, *x*, *p*, *q*, *r*, *s*, and *t* as shown in the legend of the figure. Letter designations here refer to model descriptions in Table B-1. Note also in Table B-1 that the models employ different spatial resolutions and flux adjustments.

Scenarios produced by these transient experiments varied widely among models and from region to region, both for temperature and precipitation. Except for a few outliers, individual values of projected surface warming varied mostly in the range of ~1 to ~5°C [Figure B-1 (a,c)]—with the NCAR(*r*) and MPI(*q*) runs showing the least temperature sensitivity because, with a 1% linear increase in $CO_2$/year (Table B-1), $CO_2$ has increased only by a factor of 1.7 by the end of the 70-year simulation, compared to a doubling for the other model runs. For most regions, the inter-model range of simulated temperature increase was rather pronounced, about 3-5°C. With the exception of one or two outliers, the smallest inter-model range of simulated warming at the time of $CO_2$ doubling was over Australia in summer and the Sahel in winter, where the scenarios differed among models by no more than 1.3°C. It should be noted, however, that for a region such as Australia continental-scale agreement may come from canceling differences at the sub-continental scale.

The surface air temperature biases had positive and negative values both in winter and summer [Figure B-1 (b,d)]. However, biases were mostly negative in winter and positive in summer, an indication that the models tended to overestimate the seasonal temperature cycles. Most biases were in the range of -7 to 10°C, but values as large as ~15°C were found. The smallest biases were found over Australia and, with the exception of one or two models, South East Asia and Southern Europe. Over most regions, the inter-model range of temperature biases was of the order of 10°C (i.e., it was greater than the inter-model range of regional temperature increase). The surface temperature biases as well as the simulated regional warming scenarios were in the same range as those reported in IPCC (1990) for a number of equilibrium runs.

Regional precipitation biases spanned a wide range, with values as extreme as ~-90% or greater than 200% [Figure B-1 (f,h)]. The biases were generally larger in winter than in summer, and, overall, regions with the smallest biases were Southern Europe, Northern Europe, and Central North America. Regions receiving low winter precipitation (e.g., Sahel, South East Asia) tended to have large positive or negative biases, because small errors in control run values appear as large biases when reported in percentage terms.

Simulated precipitation sensitivity to doubled $CO_2$ was mostly in the range of -20 to 20% of the control value [Figure B-1 (e,g)]. The most salient features of simulated regional precipitation changes are summarized as follows:

- All models agreed in summer precipitation increases over East Asia and, except for one model, South East Asia—reflecting an enhancement of summer monsoonal flow (contrast this result to the experiments that include the effects of sulfate aerosols; see discussion below in relation to Figure B-2).
- All models agreed in winter precipitation increases over Northern Europe, East Asia, and, except for one model, Southern Europe. In the other cases, agreement was not found among models even on the sign of the simulated change.
- Regions with the smallest inter-model range of simulated precipitation change were Central North America, East Asia, and Northern Europe in summer and Southern Europe, Northern Europe, and East Asia in winter.
- Overall, the precipitation biases were greater than the simulated changes. A rigorous statistical analysis of the model results in Figure B-1 has not been carried out; however, it can be expected that, due to relatively high temporal and spatial variability in precipitation, temperature changes are more likely to be statistically significant than precipitation changes.

In summary, several instances occurred in which regional scenarios produced by all models agreed, at least in sign. In fact, regardless of whether flux correction was used, the range of model sensitivity was less than the range of biases (note that the scales in Figure B-1 are different for the sensitivities and the biases). However, the range of simulated scenarios of the model regional biases were still large, so that confidence in regional scenarios simulated by AOGCMs remains low. It should be pointed out that, while model agreement increases our confidence in the veracity of model responses, it does not necessarily guarantee their correctness because of possible systematic errors or deficiencies shared by all models. On the other hand, in spite of these errors, models are useful tools to study climate sensitivity (see IPCC 1996, WG I, Chapter 5). Even though models cannot exactly reproduce many details of today's climate, key processes that we know to exist in the real climate system are represented in these models (see IPCC 1996, WG I, Chapter 4). For example, the simulation of the seasonal cycle of winds, temperature, pressure, and humidity in both the horizontal and vertical provides us with a first-order qualification of the fidelity of the models' ability to capture these basic features of the Earth's climate. As another example, AOGCMs exhibit the ability to simulate essential responses of the climate system to various forcings (e.g., those involving El Niño sea surface temperature anomalies and aerosols from volcanic activity). This increases our confidence in the use of AOGCM sensitivity experiments to evaluate potential changes in important climate processes.

## B.2. Simulations with Greenhouse Gas and Aerosol Forcing

Two simulations (*x*,*w*) were forced with the historical increase in equivalent $CO_2$, then a 1%/yr increase in equivalent $CO_2$. The patterns of change are qualitatively similar to those in the experiments above. In Sections 6.2.1.2 and 8.4.2.3 of the

***Table B-1:** Summary of transient coupled AOGCM experiments used in this assessment. The scenario gives the rate of increase of $CO_2$ used; most experiments use 1%/yr, which gives a doubling of $CO_2$ after 70 years (IS92a gives a doubling of equivalent $CO_2$ after 95 years). The ratio of the transient response at the time of doubled $CO_2$ to the equilibrium (long-term) response to doubling $CO_2$ is given if known.*

| Center | Expt | Reference | Flux Adjusted? | Scenario | Warming at Doubling† | Equilibrium Warming | Ratio (%)† |
|---|---|---|---|---|---|---|---|
| BMRC | *a* | Power *et al.* (1993), Colman *et al.* (1995) | No | 1%/yr | 1.35 | 2.1 | 63 |
| CCC | *b* | G. Boer (pers. comm.) | Yes | 1%/yr | – | 3.5 | |
| COLA | *c* | E. Schneider (pers. comm.) | No | 1%/yr | 2.0 | – | |
| CSIRO | *d* | Gordon and O'Farrell (1997) | Yes | 1%/yr | 2.0 | 4.3 | 47 |
| GFDL | *e* | Stouffer (pers. comm.) | Yes | 0.25%/yr | *2.6* | 3.7 | |
| | *f* | Stouffer (pers. comm.) | Yes | 0.50%/yr | *2.4* | 3.7 | |
| | *g* | Manabe *et al.* (1991, 1992) | Yes | 1%/yr | 2.2 | 3.7 | 59 |
| | *h* | Stouffer (pers. comm.) | Yes | 2%/yr | *1.8* | 3.7 | |
| | *i* | Stouffer (pers. comm.) | Yes | 4%/yr | *1.5* | 3.7 | |
| | *j* | Stouffer (pers. comm.) | Yes | 1%/yr | – | – | |
| GISS | *k* | Russell *et al.* (1995), Miller and Russell (1995) | No | 1%/yr | 1.4 | – | |
| IAP | *l*[1] | Keming *et al.* (1994) | Yes | 1%/yr | 2.5 | – | |
| MPI | *m*[2] | Cubasch *et al.* (1992, 1994b), Hasselmann *et al.* (1993), Santer *et al.* (1994) | Yes | IPCC90A | *1.3* | 2.6 | *50* |
| | *n* | Cubasch *et al.* (1992), Hasselmann *et al.* (1993), Santer *et al.* (1994) | Yes | IPCC90D | na | 2.6 | |
| | *o* | | Yes | IPCC90A | *1.5* | – | |
| | *x*[3] | Hasselmann *et al.* (1995) | Yes | IPCC90A | na | 2.6 | |
| | *y*[4] | Hasselmann *et al.* (1995) | Yes | Aerosols | na | 2.6 | |
| MRI | *p* | Tokioka *et al.* (1995) | Yes | 1%/yr | 1.6 | – | |
| NCAR | *q* | Washington and Meehl (1989) | No | 1%/yr* | *2.3* | 4.0 | 58 |
| | *r*[5] | Washington and Meehl (1993, 1996), Meehl and Washington (1996) | No | 1%/yr | 3.8 | 4.6 | 83 |
| UKMO | *s* | Murphy (1995a,b), Murphy and Mitchell (1995) | Yes | 1%/yr | 1.7 | 2.7 | 64 |
| | *t*[6] | Johns *et al.* (1997), Keen (1995) | Yes | 1%/yr | 1.7 | 2.5 | 68 |
| | *w*[7] | Johns *et al.* (1997), Tett *et al.* (1997), Mitchell *et al.* (1995b), Mitchell and Johns (1997) | Yes | 1%/yr | na | 2.5 | |
| | *z*[8] | Johns *et al.* (1997), Tett *et al.* (1997), Mitchell *et al.* (1995b), Mitchell and Johns (1997) | Yes | Aerosols | na | 2.5 | |

na = not available
†Numbers in italics indicate simulations with other than a 1%/yr increase in $CO_2$.
*1%/yr of current $CO_2$ concentrations.
[1]Polar deep ocean quantities constrained.
[2]Three additional 50-year runs, each from different initial conditions.
[3]$CO_2$ from IPCC scenario 90A after greenhouse gas forcing from 1880 to 1990.
[4]As (1) with a representation of aerosol forcing, with increases after 1990 based on IS92a.
[5]Equilibrium model excluded sea ice dynamics. Coupled model has warmer than observed tropical sea surface temperatures and a vigorous ice albedo feedback (Washington and Meehl, 1996) contributing to the high sensitivity.
[6]Average of three experiments from different initial conditions.
[7]$CO_2$ increased by 1%/yr from 1990. Observed greenhouse gas forcing used from 1860 to 1990.
[8]As (7) with a representation of aerosol forcing, with increases of aerosol and greenhouse gases after 1990 based on IS92a.

Working Group I contribution to IPCC (1996), we saw that the inclusion of the direct sulfate aerosol forcing can improve the simulation of the patterns of temperature change over the last few decades. Here we consider the effect of combined greenhouse gas and direct sulfate aerosol forcing derived from IS92a on simulated patterns of temperature change to 2050 and beyond (*y*,*z*). Figure B-2 shows area averages of summer and winter surface temperature, precipitation, and soil moisture from two models—one set with $CO_2$ increase only, the other with $CO_2$ increase and the direct effects of sulfate aerosols. The areas considered include five from Figure B-1 (i.e., Central North America, South East Asia, Sahel, Southern Europe, and Australia).

Increasing $CO_2$ alone leads to positive radiative forcing everywhere, with the largest radiative heating in regions of clear skies and high temperatures (experiment *w* shown in Figure 6.7a of IPCC 1996, WG I). The surface temperature warms everywhere except in the northern North Atlantic (Figure 6.7b of IPCC 1996, WG I). In transient simulations to 2050, the inclusion of aerosols based on IS92a (*y*,*z*) reduces the global mean radiative forcing, and leads to negative radiative forcing over southern Asia. This leads to a muted warming or even small regions of cooling (*y*) in mid-latitudes. In (*z*), China continues to warm, albeit at a very reduced rate, even though the local net radiative forcing becomes increasingly negative (Figure 6.7c in IPCC 1996, WG I). The rate of warming over North America and western Europe, where the aerosol forcing weakens, remains below that in the simulation with greenhouse gases only (*w*). The cooling due to aerosols is amplified by sea ice feedbacks in the Arctic (Figure 6.7f in IPCC 1996, WG I).

In assessing these results, one should bear in mind the possible exaggeration of the sulfate aerosol concentrations under this scenario, the uncertainties in representing the radiative effects of sulfate aerosols, and neglect of other factors including the indirect effect of sulfates. Nevertheless, these experiments suggest that the direct effect of sulfate aerosols could have strong influence on future temperature changes, particularly in northern mid-latitudes.

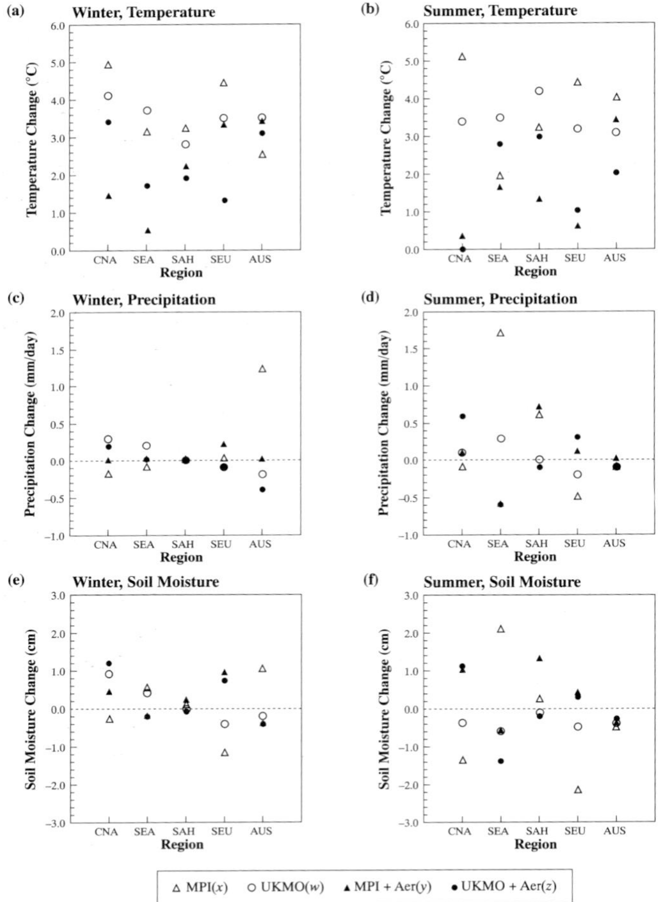


**Figure B-2:** Simulated regional changes from 1880–1889 to 2040–2049 (experiments *x*, *y*) or from pre-industrial to 2030–2050 (experiments *w*, *z*). Experiments *x* and *w* include greenhouse gas forcing only, whereas *y* and *z* also include direct sulfate aerosol effects (see Table B-1): (a) Temperature (December to February); (b) Temperature (June to August); (c) Precipitation (December to February); (d) Precipitation (June to August); (e) Soil Moisture (December to February); and (f) Soil Moisture (June to August).

## B.3. Seasonal Changes in Temperature, Precipitation, and Soil Moisture

IPCC (1990) reported some broad scale changes that were evident in most of the equilibrium $2xCO_2$ experiments that were then available. The detailed regional changes differed from model to model. In the transient experiments reported in IPCC (1992), it was found that the large-scale patterns of response at the time of doubling $CO_2$ were similar to the corresponding equilibrium experiments (IPCC, 1990), except that there was a smaller warming in the vicinity of the northern North Atlantic and the Southern Ocean in transient experiments. Here we summarize the main features in the seasonal (December to February and June to August) patterns of change in temperature, precipitation, and soil moisture in those experiments with a 1%/yr increase in $CO_2$ for which data were available. The changes are assessed at the time data of $CO_2$ doubling (after 70 years). In experiments *w*-*z*, we also contrast the continental-scale response under the IS92a scenario with and without aerosol forcing at around 2040.

### B.3.1. *Temperature*

With increases in $CO_2$, all models produce a maximum annual mean warming in high northern latitudes (Figures 6.6 and 6.7b of IPCC 1996, WG I). The warming is largest in late autumn and winter, largely due to sea ice forming later in the warmer climate. In summer, the warming is small; if the sea ice is removed with increased $CO_2$, then the thermal inertia of the mixed-layer prevents substantial warming during the short summer season, otherwise melting sea ice is present in both control and anomaly simulations, and there is no change in surface temperature (see Ingram *et al.*, 1989). The details of these changes are sensitive to parameterization of sea ice and, in particular, the specification of sea ice albedo (e.g., Meehl and Washington, 1995). In one simulation (*k*), there is a marked cooling over the northeastern Atlantic throughout the year, which leads to a cooling over part of northwest Europe in winter. There is little seasonal variation of the warming in low latitudes or over the southern circumpolar ocean.

When aerosol effects are included (*y,z* cf. *x,w* respectively), the maximum winter warming in high northern latitudes is less extensive. In mid-latitudes, there are some regions of cooling (e.g., over China), and the mean warming in the tropics is greater than in mid-latitudes. In northern summer, there are again regions of cooling in mid-latitudes and the greatest warming now occurs over Antarctica. Again, including the direct forcing by sulfate aerosols has a strong effect on simulated regional temperature changes, though the reader should bear in mind the limitations of these experiments as noted earlier.

### *B.3.2. Precipitation*

On increasing $CO_2$, all models produce an increase in global mean precipitation. Precipitation increases in high latitudes in winter (except in *k* around the Norwegian Sea where there is cooling and a reduction in precipitation), and in most cases the increases extend well into mid-latitudes (e.g., Figure 6.11a in IPCC 1996, WG I). The warming of the atmosphere leads to higher atmospheric water vapor content and enhanced poleward water vapor transport into the northern high latitudes—hence enhanced water vapor convergence and precipitation (e.g., Manabe and Wetherald, 1975). In the tropics, the patterns of change vary from model to model, with shifts or changes in intensity of the main rainfall maxima. However, many produce more rainfall over India and/or southeast Asia as seen in Figure B-1. This is consistent with an increase in atmospheric water vapor concentration leading to enhanced low level moisture convergence associated with the strong mass convergence into the monsoon surface pressure low. All models considered apart from *p* and *q* produce a general reduction in precipitation over southern Europe. In general, changes in the dry subtropics are small.

With the inclusion of aerosol forcing (*y,z*), there is only a small increase in global mean precipitation. The patterns of change in precipitation in northern winter are broadly similar to that in a parallel simulation with greenhouse gases only (*x,w* respectively), but less intense. In northern summer, there is a net reduction in precipitation over the Asian monsoon region (Figure B-2), because the aerosol cooling reduces the land-sea contrast and the strength of the monsoon flow. This is in contrast to the models run with $CO_2$ increase only that showed increases of monsoon precipitation (Figure B-1). Precipitation increases on average over southern Europe (it decreases when aerosol effects are omitted) and over North America, where changes were small with increases in greenhouse gases only.

There is now mounting evidence to suggest that a warmer climate will be one in which the hydrological cycle will in general be more intense (IPCC, 1992), leading to more heavy rain events (*ibid*, pp. 119). It should be noted, however, that as the GCM grid sizes are much larger than convective elements in the atmosphere, daily precipitation is poorly reproduced by GCMs.

### *B.3.3. Soil Moisture*

Soil moisture may be a more relevant quantity for assessing the impacts of changes in the hydrological cycle on vegetation than precipitation since it incorporates the integrated effects of changes in precipitation, evaporation, and runoff throughout the year. However, simulated changes in soil moisture should be viewed with caution because of the simplicity of the land-surface parameterization schemes in current models (e.g., experiments *a,e-i,m,n,p,q,* and *r* use an unmodified "bucket" formulation; see Section 5.3.2 of IPCC 1996, WG I).

Most models produce a general increase in soil moisture in the mean in high northern latitudes in winter, though in some (*a,k*) there are also substantial areas of reduction. The increases are due mainly to the increased precipitation discussed above, and the increased reaction of precipitation falling as rain in the warmer climate. At the low winter temperatures, the absolute change in potential evaporation is small, as expected from the Clausius-Clapyeron relation, so evaporation increases little even though temperature increases are a maximum in winter. Hence, the increase in soil moisture in high altitudes in winter is consistent with physical reasoning and the broad scale changes are unlikely to be model-dependent. However, it should be noted that in general the models considered here do not represent the effects of freezing on groundwater.

Most models produce a drier surface in summer in northern mid-latitudes. This occurs consistently over southern Europe (except *q*, which produces an excessively dry surface in winter in its control climate) and North America (except *d,k*, and *q*). The main factor in the drying is enhanced evaporation in summer (see Wetherald and Manabe, 1995): The absolute rate of increase in potential evaporation increases exponentially with temperature if other factors (wind, stability, and relative humidity) are unchanged.

As noted in the IPCC (1990), the following factors appear to contribute to summer drying:

- The soil in the control simulation is close to saturation in late winter or spring; this ensures that much of the extra precipitation in winter is not stored in the soil but lost as runoff.
- There is a substantial seasonal variation in soil moisture in the simulation of present climate; some of the simpler models may exaggerate the seasonal cycle of soil moisture (see Chapter 5 of IPCC 1996, WG I) leading to an exaggerated response in the warmer climate.

- In higher latitudes, earlier snowmelt leading to enhanced solar absorption and evaporation may contribute.
- Changes in soil moisture may be amplified by cloud feedbacks in regions where evaporation is being limited by low soil moisture values (e.g., Wetherald and Manabe, 1995).
- The drying is more pronounced in regions where precipitation is reduced in summer.

Given the varying response of different land-surface schemes to the same prescribed forcing (IPCC 1996, WG I, Chapter 5), the consistency from model to model of reductions over southern Europe in summer might be regarded as surprising. All models submitted (except *p,q*) produced a reduction in summer precipitation over southern Europe: Here changes in circulation and precipitation may be more important in determining soil moisture changes than the details of the land-surface scheme. Reductions over North America are less consistent, and there is a still wider model-to-model variation in the response over northern Europe and northern Asia.

With aerosol forcing included (*y,z*), the patterns of soil moisture change in northern winter are similar but weaker than with greenhouse gas forcing only (*x,w*). However, soil moisture increases over North America and southern Europe in summer when aerosol effects are included (*y,z*), presumably because of the reduced warming and its effect on evaporation, and because of increases in precipitation. The changes in the hydrological cycle are likely to be sensitive to the distribution of aerosol forcing and the coupled model used. However, it is clear that aerosol effects have a strong influence on simulated regional climate change.

## B.4. Simulations using Statistical Downscaling and Regional Climate Modeling Systems

Although computing power has substantially increased during the last few years, the horizontal resolution of present coupled GCMs is still too coarse to capture the effects of local and regional forcings in areas of complex surface physiography and to provide information suitable for many impact assessment studies. Since IPCC (1992), significant progress has been achieved in the development and testing of statistical downscaling and regional modeling techniques for the generation of high-resolution regional climate information from coarse-resolution GCM simulations.

### *B.4.1. Statistical Downscaling*

Statistical downscaling is a two-step process basically consisting of i) development of statistical relationships between local climate variables (e.g., surface air temperature and precipitation) and large-scale predictors, and ii) application of such relationships to the output of GCM experiments to simulate local climate characteristics. A range of statistical downscaling models have been developed (IPCC 1996, WG I), mostly for U.S., European, and Japanese locations where better data for model calibration are available. The main progress achieved in the last few years has been the extension of many downscaling models from monthly and seasonal to daily time scales, which allows the production of data more suitable for a broader set of impact assessment models (e.g., agriculture or hydrologic models).

When optimally calibrated, statistical downscaling models have been quite successful in reproducing different statistics of local surface climatology (IPCC 1996, WG I). Limited applications of statistical downscaling models to the generation of climate change scenarios has occurred showing that in complex physiographic settings local temperature and precipitation change scenarios generated using downscaling methods were significantly different from, and had a finer spatial scale structure than, those directly interpolated from the driving GCMs (IPCC 1996, WG I).

### *B.4.2. Regional Modeling*

The (one-way) nested modeling technique has been increasingly applied to climate change studies in the last few years. This technique consists of using output from GCM simulations to provide initial and driving lateral meteorological boundary conditions for high-resolution Regional Climate Model (RegCM) simulations, with no feedback from the RegCM to the driving GCM. Hence, a regional increase in resolution can be attained through the use of nested RegCMs to account for sub-GCM grid-scale forcings. The most relevant advance in nested regional climate modeling activities was the production of continuous RegCM multi-year climate simulations. Previous regional climate change scenarios were mostly produced using samples of month-long simulations (IPCC 1996, WG I). The primary improvement represented by continuous long-term simulations consists of equilibration of model climate with surface hydrology and simulation of the full seasonal cycle for use in impact models. In addition, the capability of producing long-term runs facilitates the coupling of RegCMs to other regional process models, such as lake models, dynamical sea ice models, and possibly regional ocean (or coastal) and ecosystem models.

Continuous month- or season-long to multi-year experiments for present-day conditions with RegCMs driven either by analyses of observations or by GCMs were generated for regions in North America, Asia, Europe, Australia, and Africa. Equilibrium regional climate change scenarios due to doubled $CO_2$ concentration were produced for the continental U.S., Tasmania, Eastern Asia, and Europe. None of these experiments included the effects of atmospheric aerosols.

In the experiments mentioned above, the model horizontal grid point spacing varied in the range of 15 to 125 km and the length of runs from 1 month to 10 years. The main results of the validation and present-day climate experiments with RegCMs can be summarized in the following points:

- When driven by analyses of observations, RegCMs simulated realistic structure and evolution of synoptic events. Averaged over regions on the order of $10^4$–$10^6$ $km^2$ in size, temperature biases were mostly in the range of a few tenths of °C to a few °C, and precipitation biases were mostly in the range of 10–40% of observed values. The biases generally increased as the size of the region decreased.

- The RegCM performance was critically affected by the quality of the driving large-scale fields, and tended to deteriorate when the models were driven by GCM output, mostly because of the poorer quality of the driving large-scale data compared to the analysis data (e.g., position and intensity of storm tracks).
- Compared to the driving GCMs, RegCMs generally produced more realistic regional detail of surface climate as forced by topography, large lake systems, or narrow land masses. However, the validation experiments also showed that RegCMs can both improve and degrade aspects of regional climate compared to the driving GCM runs, especially when regionally averaged.
- Overall, the models performed better at mid-latitudes than in tropical regions.
- The RegCM performance improved as the resolution of the driving GCM increased, mostly because the GCM simulation of large-scale circulation patterns improved with increasing resolution.
- Seasonal as well as diurnal temperature ranges were simulated reasonably well.
- An important problem in the validation of RegCMs has been the lack of adequately dense observational data, since RegCMs can capture fine structure of climate patterns. This problem is especially relevant in mountainous areas, where only a relatively small number of high-elevation stations are often available.

When applied to the production of climate change scenarios, nested model experiments showed the following (IPCC 1996, WG I):

- For temperature, the differences between RegCM- and GCM-simulated region-averaged change scenarios were in the range of 0.1 to 1.4°C. For precipitation, the differences between RegCM and GCM scenarios were more pronounced than for temperature, in some instances by one order of magnitude or even in sign. These differences between RegCM- and GCM-produced regional scenarios are due to the combined contributions of the different resolution of surface forcing (e.g., topography, lakes, coastlines) and atmospheric circulations, and in some instances the different behavior of model parameterizations designed for the fine- and coarse-resolution models. In summer, differences between RegCM and GCM results were generally more marked than in winter due to the greater importance of local processes.
- While the simulated temperature changes obtained with nested models were generally larger than the corresponding biases, the precipitation changes were generally of the same order of, or smaller than, the precipitation biases.

Finally, of relevance for the simulation of regional climate change is the development of a variable-resolution global model technique, whereby the model resolution gradually increases over the region of interest.

## B.5. Conclusions

Analysis of surface air temperature and precipitation results from regional climate change experiments carried out with AOGCMs indicates that the biases in present-day simulations of regional climate change and the inter-model variability in the simulated regional changes are still too large to yield a high level of confidence in simulated change scenarios. The limited number of experiments available with statistical downscaling techniques and nested regional models has shown that complex topographical features, large lake systems, and narrow land masses not resolved at the resolution of current GCMs significantly affect the simulated regional and local change scenarios, both for precipitation and (to a lesser extent) temperature (IPCC 1996, WG I). This adds a further degree of uncertainty in the use of GCM-produced scenarios for impact assessments. In addition, most climate change experiments have not accounted for human-induced landscape changes and only recently has the effect of aerosols been investigated. Both these factors can further affect projections of regional climate change.

Compared to the global-scale changes due to doubled $CO_2$ concentration, the changes at $10^4$–$10^6$ $km^2$ scale derived from transient AOGCM runs are greater. Considering all models, at the $10^4$–$10^6$ $km^2$ scale, temperature changes due to $CO_2$ doubling varied between 0.6 and 7°C and precipitation changes varied between -35 and 50% of control run values, with a marked inter-regional variability. Thus, the inherent predictability of climate diminishes with reduction in geographical scale. The greatest model agreement in the simulated precipitation change scenarios was found over the South East Asia (about -1 to 30%), Northern Europe (about -9 to 16%), Central North America (about -7 to 5%), and East Asia (about 0.1 to 16%) regions in summer, and Southern Europe (about -2 to 29%), Northern Europe (about 5 to 25%), and East Asia (about 0.5 to 18%) in winter. For temperature, the greatest model agreement in simulated warming occurred over Australia in summer (about 1.65 to 2.5°C, when excluding one outlier) and the Sahel in winter (about 1.8 to 3.15°C, when excluding one outlier). Regardless of whether flux correction was used, the range of model sensitivities was less than the range of biases, which suggests that models produce regional sensitivities that are more similar to each other than their biases.

The latest regional model experiments indicate that high-resolution information, on the order of a few 10s of km or less, may be necessary to achieve high accuracy in regional and local change scenarios in areas of complex physiography. In the last few years, substantial progress has been achieved in the development of tools for enhancing GCM information. Statistical methods were extended from the monthly/seasonal to the daily time scale, and nested model experiments were extended to the multi-year time scale. Also, variable- and high-resolution global models can be used to study possible feedbacks of mesoscale forcings on general circulation.

Regional modeling techniques, however, rely critically on the GCM performance in simulating large-scale circulation patterns at the regional scale, because they are a primary input to both empirical and physically based regional models. Although the regional performance of coarse-resolution GCMs is still somewhat poor, there are indications that features such as positioning of storm track and jet stream core are better simulated as the model resolution increases. The latest nested GCM/RegCM and variable-resolution model experiments, which employed relatively high-resolution GCMs and were run for long simulation times (up to 10 years) show an improved level of accuracy. Therefore, as a new generation of higher resolution GCM simulations become available, it is expected that the quality of simulations with regional and local downscaling models will also rapidly improve. In addition, the movement towards coupling regional atmospheric models with appropriately scaled ecological, hydrological, and mesoscale ocean models will not only improve the simulation of climatic sensitivity, but also provide assessments of the joint response of the land surface, atmosphere, and/or coastal systems to altered forcings.

## References


**Colman**, R.A., S.B. Power, B.J. MacAvaney and R.R. Dahni, 1995: A non-flux-corrected transient $CO_2$ experiment using the BMRC coupled atmosphere/ocean GCM. *Geophysical Research Letters*, **22**, 3047-3050.

**Cubasch**, U., K. Hasselmann, H. Höck, E. Maier Reimer, U. Mikolajewicz, B.D. Santer, and R. Sausen, 1992: Time-dependent greenhouse warming computations with a coupled ocean-atmosphere model. *Climate Dynamics*, **8**, 55-69.

**Cubasch**, U., G. Meehl, and Z.C. Zhao, 1994a: *IPCC WG I Initiative on Evaluation of Regional Climate Change*. Summary Report, 12 pp.

**Cubasch**, U., B.D. Santer, A. Hellach, G. Hegerl, H. Höck, E. Maier-Reimer, U. Mikolajewicz and A. Stössl, 1994b: Monte Carlo climate change forecasts with a global coupled ocean-atmosphere model. *Climate Dynamics*, **10**, 1-20.

**Gordon,** H.B. and S.P. O'Farrell, 1997: Transient climate change in the CSIRO coupled model with dynamical sea-ice. *Monthly Weather Review*, **125**, 875-907.

**Hasselmann**, K, R. Sausen, E. Maier-Reimer and R. Voss, 1993: On the cold start problem in transient simulations with coupled ocean-atmosphere models. *Climate Dynamics,* **9**, 53-61.

**Hasselmann** K., L. Bengtsson, U. Cubasch, G.C. Hegerl, H. Rodhe, E. Roeckner, H. von Storch, R. Voss, and J. Waszkewitz, 1995: Detection of anthropogenic climate change using a fingerprint method. In: *Proceedings of "Modern Dynamical Meteorology," Symposium in honor of Aksel Wiin-Nielsen, 1995* [P. Ditlevsen (ed.)]. ECMWF Press.

**Ingram**, W.J., C.A. Wilson, and J.F.B. Mitchell, 1989: Modeling climate change: An assessment of sea-ice and surface albedo feedbacks. *Journal of Geophysical Research*, **94**, 8609-8622.

**IPCC**, 1990: *Climate Change: The IPCC Scientific Assessment* [Houghton, J.T., G.J. Jenkins, and J.J. Ephraums (eds.)]. Cambridge University Press, Cambridge and New York, 365 pp.

**IPCC**, 1992. *Climate Change 1992: The Supplementary Report to the IPCC Scientific Assessment.* Report of the IPCC Scientific Assessment Working Group [Houghton, J.T., B.T. Callander, and S.K. Varney (eds.)]. Cambridge University Press, Cambridge and New York, 200 pp.

**IPCC**, 1996: *Climate Change 1995: The Science of Climate Change.* Contribution of Working Group I to the Second Assessment Report of the Intergovernmental Panel on Climate Change [Houghton, J.J., L.G. Meiro Filho, B.A. Callander, N. Harris, A. Kattenberg and K.Maskell (eds.)]. Cambridge University Press, Cambridge and New York, 572 pp.

**Johns**, T.C., R.E. Carnell, J.F. Crossley, J.M. Gregory, J.F.B. Mitchell, C.A. Senior, S.F.B. Tett, and R.A. Wood, 1997: The second Hadley Centre coupled ocean-atmosphere GCM: Model description, spinup and validation. *Climate Dynamics*, **13**, 103-134.

**Keen**, A.B., 1995: Investigating the effects of initial conditions on the response of the Hadley Centre coupled model. *Hadley Centre Internal Note no. 71*.

**Keming**, C., J. Xiangze, L. Wuyin, Y. Yongquiang, G. Yufu, and Z. Xuehong, 1994: Lecture in International Symposium on Global Change in Asia and the Pacific Region (GCAP). 8-10 Aug, 1994, Beijing.

**Kittel**, T.G.F., N.A. Rosenbloom, T.H. Painter, D.E. Schimel, and VEMAP modeling participants, 1995: The VEMAP integrated database for modeling United States ecosystem/vegetation sensitivity to climate change. *Journal of Biogeography*, **22**, 857-862.

**Kittel**, T.G.F., F. Giorgi and G.A. Meehl, 1997: Intercomparison of regional biases and doubled $CO_2$ sensitivity of coupled atmosphere-ocean general circulation model climate experiments. *Climate Dynamics* (in press).

**Li**, X., Z. Zongci, W. Shaowu, and D. Yohui, 1994: Evaluation of regional climate change simulation: A case study. *Proceedings of the IPCC Special Workshop on Article 2 of the United Nations Framework Convention on Climate Change*. Fortaleza, Brazil, 17-21 October 1994.

**Manabe**, S. and R.T. Wetherald, 1975: The effects of doubling the $CO_2$ concentration on the climate of a general circulation model. *Journal of Atmospheric Sciences*, **32**, 3-15.

**Manabe**, S., R.J. Stouffer, M.J. Spelman, and K. Bryan, 1991: Transient responses of a coupled-ocean atmosphere model to gradual changes of atmospheric $CO_2$. Part I: Annual mean response. *Journal of Climate,* **4**, 785-818.

**Manabe**, S., M.J. Spelman, and R.J. Stouffer, 1992: Transient responses of a coupled ocean-atmosphere model to gradual changes of atmospheric $CO_2$. Part II: Seasonal response. *Journal of Climate,* **5**, 105-126.

**Mearns**, L.O., F. Giorgi, L. McDaniel, and C. Shields, 1995a: Analysis of daily variability of precipitation in a nested regional climate model: Comparison with observations and doubled $CO_2$ results. *Global and Planetary Change*, **10**, 55-78.

**Mearns**, L.O., F. Giorgi, L. McDaniel, and C. Shields, 1995b: Analysis of variability and diurnal range of daily temperature in a nested regional climate model: Comparison with observations and doubled $CO_2$ results. *Climate Dynamics*, **11**, 193-209.

**Meehl**, G.A. and W.M. Washington, 1995: Cloud albedo feedback and the super greenhouse effect in a global coupled GCM. *Climate Dynamics*, **11**, 399-411.

**Meehl**, G.A. and W.M. Washington, 1996: El Niño-like climate change in a model with increased atmospheric $CO_2$ concentrations. *Nature*, **382**, 56-60.

**Miller**, J.R. and G.L. Russell,1995: Climate change and the Arctic hydrologic cycle as calculated by a global coupled atmosphere-ocean model. *Annual Glaciology*, **21**, 91-95.

**Mitchell**, J.F.B. and T.J. Johns, 1997: On modification of global warming by sulfate aerosols. *Journal of Climate*, **10**, 245-267.

**Mitchell**, J.F.B., R.A. Davis, W.J. Ingram, and C.A. Senior, 1995b: On surface temperature, greenhouse gases and aerosols: models and observations. *Journal of Climate,* **10**, 2364-2386.

**Murphy**, J.M., 1995a: Transient response of the Hadley Centre coupled model to increasing carbon dioxide. Part I – Control climate and flux adjustment. *Journal of Climate*, **8**, 36-56.

**Murphy**, J.M., 1995b: Transient response of the Hadley Centre coupled model to increasing carbon dioxide. Part III – Analysis of global-mean response using simple models. *Journal of Climate*, **8**, 496-514.

**Murphy**, J.M. and J.F.B. Mitchell, 1995: Transient response of the Hadley Centre coupled model to increasing carbon dioxide. Part II – Temporal and spatial evolution of patterns. *Journal of Climate*, **8**, 57-80.

**Power**, S., R. Colman, B. McAvaney, R. Dahni, A. Moore, and N. Smith, 1993: The BMRC coupled atmosphere/ocean/sea ice model. *The BMRC Research Report No 37.*

**Raisanen**, J., 1995: A comparison of the results of seven GCM experiments in Northern Europe. *Geophysica,* **30(1-2)**, 3-30.

**Russell**, G.L., J.R. Miller, and D. Rind, 1995: A coupled atmosphere-ocean model for transient climate change studies. *Atmosphere-Ocean*, **33**, 4.

**Santer**, B.D., W. Bruggemann, U. Cubasch, K. Hasselmann, E. Maier-Reimer, and U. Mikolajewicz, 1994: Signal-to-noise analysis of time-dependent greenhouse warming experiments. Part 1: Pattern analysis. *Climate Dynamics*, **9**, 267-285.

**Tett**, S.F.B., T.C. Johns, and J.F.B. Mitchell, 1997: Global and regional variability in a coupled AOGCM. *Climate Dynamics*, **13**, 303-323.

**Tokioka**, T., A. Noda, A. Kitoh, Y. Nikaidou, S. Nakagawa, T. Motoi, S. Yukimoto, and K. Takata, 1995: A transient $CO_2$ experiment with the MRI CGCM. *Journal of the Meteorological Society of Japan,* **74(4)**, 817-826.

**Washington**, W.M. and G.A. Meehl, 1989: Climate sensitivity due to increased $CO_2$: Experiments with a coupled atmosphere and ocean general circulation model. *Climate Dynamics*, **4**, 1-38.

**Washington,** W.M. and G.A. Meehl, 1993: Greenhouse sensitivity experiments with penetrative cumulus convection and tropical cirrus albedo effects. *Climate Dynamics*, **8**, 211-223.

**Washington**, W.M. and G.A. Meehl, 1996: High-latitude climate change in a global coupled ocean-atmosphere-sea ice model with increased atmospheric $CO_2$. *Journal of Geophysical Research*, **101**, 12795-12801.

**Wetherald**, R.T. and S. Manabe, 1995: The mechanisms of summer dryness induced by greenhouse warming. *Journal of Climate*, **8**, 3096-3108.

**Whetton**, P.H., P.J. Rayner, A.B. Pittock, and M.R. Haylock, 1994: An assessment of possible climate change in the Australian region based on an intercomparison of general circulation modeling results. *Journal of Climate*, **7**, 441-463.

**Whetton**, P., M. England, S. O'Farrell, I. Waterson, and B. Pittock, 1996: Global comparison of the regional rainfall results of enhanced greenhouse coupled and mixed layer ocean experiments: Implications for climate change scenario development. *Climatic Change*, **33**, 497-519.

# C

# Simulated Changes in Vegetation Distribution under Global Warming

RONALD P. NEILSON (USA)

Lead Authors:
*I.C. Prentice (UK), B. Smith (Australia)*

Contributors:
*T. Kittel (USA), D. Viner (UK)*

# CONTENTS

## C.1. Introduction

Global vegetation models (GVM) have in the past decade evolved from largely statistical-correlational to more process-based, rendering greater confidence in their abilities to address questions of global change. There are generally two classes of GVMs, biogeography models and biogeochemistry models. The biogeography models place emphasis on determination of what can live where, but either do not calculate or only partially calculate the cycling of carbon and nutrients within ecosystems. The biogeochemistry models simulate the carbon and nutrient cycles within ecosystems, but lack the ability to determine what kind of vegetation could live at a given location. BIOME3 has significantly blurred this model distinction (Haxeltine and Prentice, 1996). There are over 20 biogeochemistry models and about 5 biogeography models. Two of the biogeography models, MAPSS (Neilson, 1995) and BIOME3 were used to provide estimates of changes in vegetation distribution, density and hydrology for this IPCC special report. These are equilibrium models, which simulate the potential 'climax' vegetation that could live at any well-drained, upland site in the world under an 'average' seasonal climate. Equilibrium models provide useful 'snapshots' of what a terrestrial biosphere in equilibrium with its climate might look like, but can provide only inferential information about how the biosphere will make transitions from one condition to another. This is in contrast to other models, which simulate the timeseries of vegetation change at a point (Shugart and Smith, 1996), but which do not produce maps of vegetation distribution and function. Fully dynamic versions of the spatially-explicit GVMs are being developed and incorporate both biogeography and biogeochemistry processes, but the dynamic global vegetation models (DGVM) are not yet ready for assessment purposes (Neilson and Running, 1996). Several model intercomparison projects are underway and can serve to provide some context for the two models used here. One such intercomparison is the VEMAP process.

## C.2. VEMAP Model Intercomparison

The Vegetation/Ecosystem Modeling and Analysis Project (VEMAP) compared three biogeography models, MAPSS (Neilson, 1995), BIOME2 (Haxeltine *et al.*, 1996), and DOLY (Woodward and Smith, 1994; Woodward *et al.*, 1995) and three biogeochemistry models TEM (Raich *et al.*, 1991; McGuire *et al.*, 1992; Melillo *et al.*, 1993), CENTURY (Parton *et al.*, 1987; Parton *et al.*, 1988; Parton *et al.*, 1993), and BIOME-BGC (Hunt and Running, 1992; Running and Hunt, 1993). The two classes of global models were intercompared and loosely coupled for an assessment of both model capabilities and the potential impacts of global warming on U.S. ecosystems (VEMAP Members, 1995). The VEMAP process determined that all the models have roughly equal skill in simulating the current environment, but exhibit some divergences under alternative climates, in some cases producing vegetation responses of opposite sign.

Given the timeframe of this IPCC special report, only MAPSS and BIOME3 were able to provide global simulations. MAPSS and BIOME2 (a precursor to BIOME3) were found to produce generally similar results under the future climate scenarios of the VEMAP process. However, MAPSS is consistently more sensitive to water stress, producing a more xeric outcome under future climates and it also has a more sensitive response to elevated $CO_2$. That is, when incorporating a direct, physiological $CO_2$ effect, MAPSS produces a larger benefit to vegetation from increased water-use-efficiency (VEMAP Members, 1995).

## C.3. Biogeography Model Description

MAPSS (Mapped Atmosphere-Plant-Soil System; Neilson, 1995) and BIOME3 (Haxeltine and Prentice, 1996) are among a new generation of process-based, equilibrium biogeographic models (IPCC 1996, WG II, Section 1.3.4; VEMAP Members, 1995). The models simulate the distribution of potential global vegetation based on local vegetation and hydrologic processes and the physiological properties of plants. Both models simulate the mixture of vegetation lifeforms, such as trees, shrubs, and grasses, that can coexist at a site while in competition with each other for light and water. A set of physiologically-grounded 'rules' determines whether the woody vegetation will be broadleaved or needleleaved, or evergreen or deciduous, as well as other properties. The models also simulate the maximum carrying capacity, or vegetation density, in the form of leaf area that can be supported at the site, under the constraints of energy and water. Energy constraints, largely applicable to cold ecosystems, are prescribed in MAPSS, but are simulated by an explicit carbon flux model in BIOME3. The two models simulate a similar set of water balance processes, incorporating soil texture effects.

Thus, MAPSS and BIOME3 simulate the distribution of vegetation, such as forests, savannas, shrubland, grasslands and deserts over all non-wetland sites of the Earth, based on the relative densities or productivity of overstory and understory, vegetation leaf characteristics and thermal tolerances. The models simulate the distribution of generalized vegetation lifeforms (e.g. tree, shrub, grass; evergreen-deciduous; broadleaf-needleleaf), rather than species and assemble these into a vegetation type classification. There are currently 45 different vegetation types simulated by MAPSS and 18 by BIOME3. The vegetation types are hierarchical, representing biomes (e.g., boreal forest, temperate savanna, grassland, etc.) at the top and more detailed community-level descriptions at the lower end (e.g., subtropical, xeromorphic woodland). Only the top of the hierarchy is utilized in this analysis. Since both models simulate a full site water balance, they are also calibrated and tested hydrologic models, thereby, allowing estimates of impacts on water resources fully integrated with the simulated impacts on vegetation (Neilson and Marks, 1994; VEMAP Members, 1995). As equilibrium models, MAPSS and BIOME3 simulate vegetation distribution and hydrology under an average seasonal cycle of climate. They simulate an equilibrium land-surface biosphere under current or future climate, but not the transitional vegetation changes from one climate to another. Thus, the models show the long term potential consequences of climate change, but one can only infer immediate (1-10 year) effects.

MAPSS and BIOME3 contain algorithms that allow the incorporation of a direct physiological $CO_2$ effect. Elevated $CO_2$ concentrations can, among other effects, enhance productivity and increase the water-use-efficiency (WUE, carbon fixed per unit water transpired) of the vegetation thereby reducing the sensitivity of the vegetation to drought stress (IPCC 1996, WG II, Section A.2.3; Bazzaz *et al.*, 1996; Eamus, 1991). BIOME3 allows a direct $CO_2$ effect on both productivity and water-use-efficiency. MAPSS accomplishes the same effect by reducing stomatal conductance, which then results in increased leaf area, thus indirectly incorporating a productivity effect. Individual species exhibit variations in their expressions of direct $CO_2$ effects. For example, the productivity and WUE effects are not necessarily tightly coupled (Eamus, 1996a). However, since the models simulate functional types, rather than species, both models have generalized the direct $CO_2$ effects to all vegetation, with the exception in BIOME3 of differentiating $C_3$ and $C_4$ physiological types. In BIOME3 $C_4$ plants do not experience the elevated growth of $C_3$ plants, but do experience increased water-use-efficiency. The realized importance of the direct $CO_2$ processes in complex, mature ecosystems remains a matter of debate (Bazzaz et al., 1996).

A review of 58 studies indicated an average 32% increase in plant dry mass under a doubling of $CO_2$ concentration (Wullschleger, Post, and King, 1995). Norby (1996) documented an average 29% increase in annual growth per unit leaf area in seven broadleaf tree species under 2 x $CO_2$ over a wide range of conditions. Increased WUE, examined in another review, averaged about 30-40% as indexed by reductions in leaf conductance to water vapor (Eamus, 1991). If such responses were maintained in forests over many decades, they would imply a substantial potential for increased storage of atmospheric carbon, as well as conferring some increased tolerance to drought due to increased WUE. However, some species or ecosystems exhibit acclimation to elevated $CO_2$ by downregulating photosynthesis (Bazzaz, 1990; Grulke *et al.*, 1993; Grulke *et al.*, 1990); while others do not exhibit acclimation (Bazzaz, 1990; Teskey, 1997). Most of the early $CO_2$ research was done on juvenile trees in pots and growth chambers. New research is beginning to emerge which focuses on larger trees or intact forested ecosystems. Recent reviews of this newer literature (Eamus, 1996a; Curtis, 1996) indicate that acclimation may not be as prevalent when roots are unconstrained and also that leaf conductance may not be reduced and that both responses are dependent on the experimental conditions, the length of exposure and the degree of nutrient or water stress. These results imply that forests could produce more leaf area under elevated $CO_2$, but may not gain a benefit from increased WUE. In fact, with increased leaf area, transpiration should increase on a per tree basis and the stand would use more water. Elevated temperatures would increase transpiration even further, perhaps drying the soils and inducing a drought effect on the ecosystem (*ibid*). Prominent among the environmental influences that are thought to moderate long-term responses to elevated $CO_2$ is nitrogen supply (Kirschbaum *et al.*, 1994; McGuire *et al.*, 1995; Eamus, 1996b). Unless $CO_2$ stimulates an increase in N mineralization (Curtis *et al.*, 1995), productivity gains in high $CO_2$ are likely to be constrained by the system's N budget (Körner, 1995). Nitrogen limitations may constrain carbon gains to structural tissue, rather than leaves (Curtis *et al.*, 1995). Thus, in areas receiving large amounts of N deposition, a direct $CO_2$ response could result in large increases in leaf area, increasing transpiration and possibly increasing sensitivity to drought via rapid soil water depletion. Early growth increases may disappear as the system approaches carrying capacity as limited by water or nutrients (Körner, 1995). Shifts in species composition will likely result from different sensitivities to elevated $CO_2$ (Bazzaz *et al.*, 1996; Körner, 1995). Both MAPSS and BIOME3 have been operated with and without the direct $CO_2$ effects in this study in order to gauge the importance and sensitivity of the processes within the modeling framework. The direct effects of elevated $CO_2$ are imparted only in the ecological model processes and not in the GCMs. There are no feedbacks between the ecological and atmospheric models

The MAPSS model also contains a fire model that shifts some vegetation to a 'fire climax' state, such as in many grasslands or savannas. BIOME3 embeds these processes in the calibration. Neither model considers current or past land-use practices. Thus, some areas that the models indicate as grassland, for example, might actually be shrublands, due to either grazing or fire suppression. Although the two models do not simulate actual land-use, the 'potential' land-cover simulated by the models should provide an accurate estimate of the land-surface potential. That is, forests cannot be grown in deserts or shrublands without irrigation; and, agricultural productivity should be higher in a potential forest landscape than in a potential shrubland landscape (given similar soils). Changes in LAI can be interpreted as a change in the overall carrying capacity or standing crop of the site, regardless of whether it is in potential natural vegetation or under cultivation. Additions from irrigation or nitrogen could alter this conclusion, but it should hold for non-irrigated, upland systems. Thus, simulated changes in potential natural vegetation should be valuable indicators of general shifts in agricultural potential.

## C.4. Vegetation Classification

The vegetation classification from each model has been aggregated to ten broad classes for MAPSS and nine for BIOME3. The models are most accurate in differentiating the broad physiognomic divisions of Forest, Savanna, semi-arid lands (Shrublands and Grasslands) and Arid Lands. These differences are largely based on the relationship between leaf area and site water balance and the simulated changes in leaf area and site water balance should be generally reliable (especially with respect to the sign of the change). The aggregated vegetation classes used for this analysis are as follows.

1) *Tundra* is defined as the treeless vegetation which extends beyond treeline at high latitudes and altitudes,

regardless of whether it is dominated by dwarf shrubs or herbaceous plants.

2) *Taiga/Tundra* is the broad 'ecotonal' region of open woodland, which occurs at higher latitudes or elevations beyond the 'closed' Boreal forest. This type is not explicitly simulated by BIOME3, but rather is included in Boreal Conifer Forest.
3) *Boreal Conifer Forest* is the Taiga proper, i.e., relatively dense forest composed mainly of needle-leaved trees and occurring in cold-winter climates.
4) *Temperate Evergreen Forest* encompasses the wet temperate and subtropical conifer forests of the Northwest in North America, as well as subtropical evergreen broadleaf forests (e.g., in China) and the Nothofagus and Eucalyptus forests of the Southern Hemisphere.
5) *Temperate Mixed Forest* includes pure temperate broadleaf forests, such as oak-hickory, or beech-maple. It also includes mixtures of broadleaf and temperate evergreen types, such as the cool-mixed pine/fir and hardwood forests of the northeastern United States or the warm-mixed pine/hardwood forests of the southeastern U.S.
6) *Tropical Broadleaf Forest* includes both tropical evergreen forest and dense tropical drought-deciduous forests.
7) *Savanna/Woodlands* encompass all 'open' tree vegetation from high to low latitudes and elevations. The tropical dry savannas and drought deciduous forests are contained within this class. So too are the temperate pine savannas and 'pygmy' forests and the aspen woodlands adjacent to the boreal forest. Fire can play an important role in maintaining the open nature of these woodlands; while, grazing can increase the density of woody vegetation at the expense of grass.
8) *Shrub/Woodlands* are distinguished from the Savanna/Woodlands by their lower biomass and shorter stature. This is a drier vegetation type than the Savanna/Woodlands and encompasses most semi-arid vegetation types from Chaparral to mesquite woodlands to cold, semi-desert sage shrublands. The actual

***Table C-1:*** *Potential future biome area (percentage of current) simulated by the MAPSS and BIOME3 biogeography models under three older (IPCC 1990, WG I), equilibrium 2 x $CO_2$ GCM scenarios and under three newer (IPCC 1996, WG I), transient simulations from which 2 x $CO_2$ scenarios were extracted. The reported ranges include both ecological models under several GCM scenarios. The baseline areas estimates are outputs from each model. Since BIOME3 does not differentiate Taiga/Tundra from Boreal Forest, two different aggregations are presented. The Taiga/Tundra summaries are MAPSS data only; while the "Boreal + Taiga/Tundra" and "Total Forest + Taiga/Tundra" summaries are from both models. The ranges of percent change for Boreal Conifer are from both models (except FAR scenarios, which are MAPSS output). The Taiga/Tundra under the MAPSS simulations decreases in area in all scenarios; while, Boreal conifer increases in area. Were these two vegetation zones aggregated in MAPSS, they would exhibit either increases or decreases, as in the BIOME3 simulations. The decreases in Boreal Conifer, shown in the table, are BIOME3 simulations.*

| | **Baseline Area (Mha)** | | **With $CO_2$ Effect** | | **Without $CO_2$ Effect** |
|---|---|---|---|---|---|
| **Biome Type** | MAPSS | BIOME3 | FAR Scenarios | SAR Scenarios | SAR Scenarios |
| Tundra | 792 | 950 | 33-59% | 43-60% | 43-60% |
| Taiga/Tundra | 999 | | 35-62% | 56-64% | 56-64% |
| Boreal Conifer Forest | 1,024 | 1,992 | 109-133% | 64-116% | 68-111% |
| *Boreal + Taiga/Tundra* | 2,023 | 1,992 | 72-95% | 64-90% | 68-87% |
| Temperate Evergreen Forest | 1,142 | 816 | 104-121% | 104-137% | 84-109% |
| Temperate Mixed Forest | 744 | 1,192 | 125-161% | 139-199% | 104-162% |
| *Total Temperate Forest* | 1,886 | 2,008 | 116-125% | 137-158% | 107-131% |
| Tropical Broadleaf Forest | 1,406 | 1,582 | 71-151% | 120-138% | 70-108% |
| Savanna/Woodland | 2,698 | 2,942 | 90-130% | 78-89% | 100-115% |
| Shrub-Steppe | 994 | 1,954 | 61-70% | 70-136% | 81-123% |
| Grassland | 2,082 | 554 | 109-126% | 45-123% | 120-136% |
| *Total Shrub/Grassland* | 3,076 | 2,508 | 96-108% | 105-127% | 111-126% |
| Arid Lands | 1,470 | 1,351 | 71-72% | 59-78% | 83-120% |
| **Total Vegetation** | 13,351 | 13,333 | 100-101% | 100-101% | 100-101% |

Note: FAR = First Assessment Report (IPCC 1990, WG I); SAR = Second Assessment Report (IPCC 1996, WG I).

vegetation associated with this type is very susceptible to variation depending on soils, topography, fire, grazing and land-use history. Distinctions between shrub-steppe and grassland are sometimes difficult to quantify, given that each usually contains elements of both grass and woody vegetation. The relative abundance of the two functional types is considered in determining the classification, but there are no generally accepted rules to indicate how much woody vegetation is sufficient to label a region a shrubland, or conversely how much grass is required to label it a grassland.

9) *Grasslands* include both $C_3$ and $C_4$ grassland types in both temperate and tropical regions. Much of the grassland type is a 'fire climax' type that would be populated by shrubs either with the absence of fire, or with extensive grazing.
10) *Arid Lands* encompass all regions drier than grasslands, from hyper-arid to semi-arid, ranging from the "waterless" deserts such as the Namib to the 'semi-deserts' of central Asia and Patagonia. The regions could be more or less 'grassy' or 'shrubby' depending on disturbance and land-use history.

## C.5. Future Climate Scenarios

Estimation of the potential impacts of global warming should utilize several future climate scenarios, since the magnitude, timing and spatial details of global warming vary among climate models. Most published impacts studies were based on atmospheric General Circulation Model (GCM) doubled $CO_2$ radiative forcing equilibrium experiments with simple mixed-layer oceans. Doubled $CO_2$ radiative forcing (2 x $CO_2$) includes only about 50% actual $CO_2$ forcing with the balance arising from other greenhouse gases. More recent, transient experiments with coupled atmosphere-ocean GCMs have suggested a global average increase in temperature of about 1.0-3.5°C by the time of $CO_2$ doubling, estimated as 60-70 years from now (described in the IPCC Second Assessment Report, SAR; IPCC 1996, WG I, Section 6; Annex B). The most recent GCMs include sulfate aerosols in some experiments, which can cool the climate. The analysis presented here will rely both on the older 2 x $CO_2$ equilibrium GCM scenarios (described in the IPCC First Assessment Report, FAR; IPCC 1990, WG I, Section 3; Annex B), since most published analyses have relied on them, and on three new simulations, two from the Hadley Center (HADCM2GHG and HADCM2SUL; Johns *et al.*, submitted; Mitchell *et al.*, 1995; IPCC 1996, WG I, Sections 5, 6), and one from the Max Planck Institute for Meteorology (MPI-T106; Bengtsson, *et al.* 1995; Bengtsson, *et al.*, 1996; IPCC 1996, WG I, Section 6), which have been made using coupled atmosphere-ocean GCMs and considering sulfate aerosol forcing.

To allow direct comparison with the previously completed VEMAP simulations over the conterminous U.S. (VEMAP Members, 1995), the same three equilibrium GCM scenarios were utilized for the global simulations: UKMO (Mitchell and Warrilow, 1987); GFDL-R30 (IPCC 1990, WG I, Section 3; IPCC 1990, WG I, Section 5); and OSU (Schlesinger and Zhao, 1989). The coarse grid from each model was interpolated to a 0.5° x 0.5°, lat.-long. grid. Scenarios were constructed by applying ratios ((2 x $CO_2$)/(1 x $CO_2$)) of all climate variables (except temperature) back to a baseline longterm average monthly climate dataset (Leemans and Cramer, 1991). Ratios were used to avoid negative numbers (e.g., negative precipitation), but were not allowed to exceed 5, to prevent unrealistic changes in regions with normally low rainfall. Temperature scenarios were calculated as a difference ((2 x $CO_2$) - (1 x $CO_2$)) and applied to the baseline dataset.

The newer GCM scenarios are extracted from transient GCM simulations wherein trace gases were allowed to increase gradually over a long period of years, allowing the climate to adjust while incorporating inherent lags in the ocean-atmosphere systems. In order to run the equilibrium vegetation models under

***Table C-2:*** *Percentage area of current biomes which could undergo a loss of leaf area (i.e., biomass decrease) due to global warming under various older (FAR) and newer (SAR) GCM scenarios, and with or without a direct $CO_2$ effect (see Table C-1 for details), as simulated by the MAPSS and BIOME3 biogeography models (ranges include both models). The losses in leaf area generally indicate a less favorable water balance (drought).*

| | **With $CO_2$ Effect** | | **Without $CO_2$ Effect** |
|---|---|---|---|
| **Biome Type** | FAR Scenarios | SAR Scenarios | SAR Scenarios |
| Tundra | 1-3% | 0-1% | 0-2% |
| Taiga/Tundra | 1-5% | 1% | 2% |
| Boreal Conifer Forests | 39-67% | 0-20% | 3-69% |
| Temperate Evergreen Forests | 24-57% | 1-18% | 28-51% |
| Temperate Mixed Forests | 54-86% | 1-29% | 15-75% |
| Tropical Broadleaf Forests | 5-63% | 1-42% | 26-33% |
| Savanna/Woodlands | 10-21% | 7-17% | 38-75% |
| Shrub-Steppe | 26-45% | 1-24% | 20-59% |
| Grasslands | 33-37% | 5-46% | 43-75% |
| Arid Lands | 8-12% | 0-13% | 0-29% |

***Table C-3:** Percentage area of current biomes which could undergo a gain of leaf area (i.e., biomass increase) due to global warming under various older (FAR) and newer (SAR) GCM scenarios, and with or without a direct $CO_2$ effect (see Table C-1 for details), as simulated by the MAPSS and BIOME3 biogeography models (ranges include both models). The gains in leaf area generally indicate a more favorable water balance.*

| | **With $CO_2$ Effect** | | **Without $CO_2$ Effect** |
|---|---|---|---|
| **Biome Type** | FAR Scenarios | SAR Scenarios | SAR Scenarios |
| Tundra | 20-74% | 20-58% | 49-82% |
| Taiga/Tundra | 91-98% | 92-95% | 91-94% |
| Boreal Conifer Forests | 13-21% | 36-93% | 3-58% |
| Temperate Evergreen Forests | 20-41% | 46-67% | 7-18% |
| Temperate Mixed Forests | 4-26% | 50-91% | 9-21% |
| Tropical Broadleaf Forests | 7-40% | 16-87% | 0-7% |
| Savanna/Woodlands | 74-88% | 46-84% | 4-31% |
| Shrub-Steppe | 46-64% | 64-80% | 16-42% |
| Grasslands | 56-60% | 45-78% | 3-28% |
| Arid Lands | 51-57% | 53-80% | 23-66% |

the newer transient GCMs, a control climate is extracted as an average of either 30 years (Hadley Center) or 10 years (Max Planck Institute) of model output associated with present climate (e.g. 1961-1990). Likewise, a 30 or 10 year average is extracted from the time period approximating 2 x $CO_2$ forcing (e.g. 2070- 2099). These average climates are then used to drive the vegetation models. Note that because the vegetation models are equilibrium models, the results must be interpreted as indicating the potential vegetation, i.e., the climatically suitable vegetation. Time lags and transient responses of the vegetation to climate change are not considered here.

## C.6. Interpretation of Biogeographic Model Simulations

Each of the ten IPCC regions was supplied with a set of MAPSS and BIOME3 output. Included were figures of vegetation distribution under current and future climate, vegetation density change (indexed by leaf area change), and runoff change. Also included were summary tables of the areas of the different biomes within each region under current and future climate, a change matrix indicating the area shifts from current biome type to other types, the areas within each biome expected to undergo an increase or decrease in vegetation density (change in LAI) and the areas within each biome expected to undergo an increase or decrease in annual runoff. These results were supplied for each vegetation model and for each GCM scenario. MAPSS and BIOME3 were both run under the Hadley Center scenarios; BIOME3 alone was run under the Max Planck Institute scenario; and, MAPSS alone was run under the older OSU, GFDL-R30 and UKMO scenarios. The Hadley and MPI simulations were run both with and without a direct $CO_2$ effect (applied in the ecological models); while, the OSU, GFDL-R30 and UKMO scenarios were only run with the direct $CO_2$ effects incorporated, in keeping with the VEMAP analyses.

Since the regional maps are of a much smaller extent and include quantitative information, the detailed interpretation

***Table C-4:** Percentage area of current biomes which could undergo a loss of annual runoff due to global warming under various older (FAR) and newer (SAR) GCM scenarios, and with or without a direct $CO_2$ effect (see Table C-1 for details), as simulated by the MAPSS and BIOME3 biogeography models (ranges include both models).*

| | **With $CO_2$ Effect** | | **Without $CO_2$ Effect** |
|---|---|---|---|
| **Biome Type** | FAR Scenarios | SAR Scenarios | SAR Scenarios |
| Tundra | 19-32% | 16-45% | 28-46% |
| Taiga/Tundra | 79-90% | 71-79% | 76-82% |
| Boreal Conifer Forests | 1-25% | 3-53% | 33-81% |
| Temperate Evergreen Forests | 12-21% | 25-37% | 33-67% |
| Temperate Mixed Forests | 59-77% | 51-66% | 62-68% |
| Tropical Broadleaf Forests | 11-40% | 15-54% | 23-68% |
| Savanna/Woodlands | 14-19% | 37-60% | 31-46% |
| Shrub-Steppe | 43-61% | 23-44% | 18-42% |
| Grassland | 34-38% | 41-60% | 33-56% |
| Arid Lands | 24-26% | 1-20% | 2-20% |

***Table C-5:** Percentage area of current biomes which could undergo a gain of annual runoff due to global warming under various older (FAR) and newer (SAR) GCM scenarios, and with or without a direct $CO_2$ effect (see Table C-1 for details), as simulated by the MAPSS and BIOME3 biogeography models (ranges include both models).*

| | With $CO_2$ Effect | | Without $CO_2$ Effect |
|---|---|---|---|
| **Biome Type** | FAR Scenarios | SAR Scenarios | SAR Scenarios |
| Tundra | 67-80% | 36-82% | 32-70% |
| Taiga/Tundra | 10-20% | 20-28% | 18-23% |
| Boreal Conifer Forests | 74-98% | 41-95% | 14-63% |
| Temperate Evergreen Forests | 78-87% | 58-73% | 29-66% |
| Temperate Mixed Forests | 23-41% | 33-47% | 11-37% |
| Tropical Broadleaf Forests | 60-89% | 46-85% | 32-76% |
| Savanna/Woodlands | 80-84% | 31-60% | 51-59% |
| Shrub-Steppe | 23-44% | 15-45% | 23-48% |
| Grasslands | 38-41% | 19-32% | 17-40% |
| Arid Lands | 7-24% | 4-15% | 3-15% |

will be left to the regions and the following discussion will only address general features of the simulations, particularly the differences between the older and newer GCMs and the MAPSS and BIOME3 intercomparisons. Although each region received the full set of figures, only a subset will be presented here. The MAPSS and BIOME3 results are sufficiently similar that the ranges presented in Tables C-1 to C-5 encompass the output from both models to indicate the full range of uncertainties within the scope of these experiments and models.

### *C.6.1. Control Climate*

MAPSS and BIOME3 produce similar vegetation maps under current climate, but there are some differences. Some of the

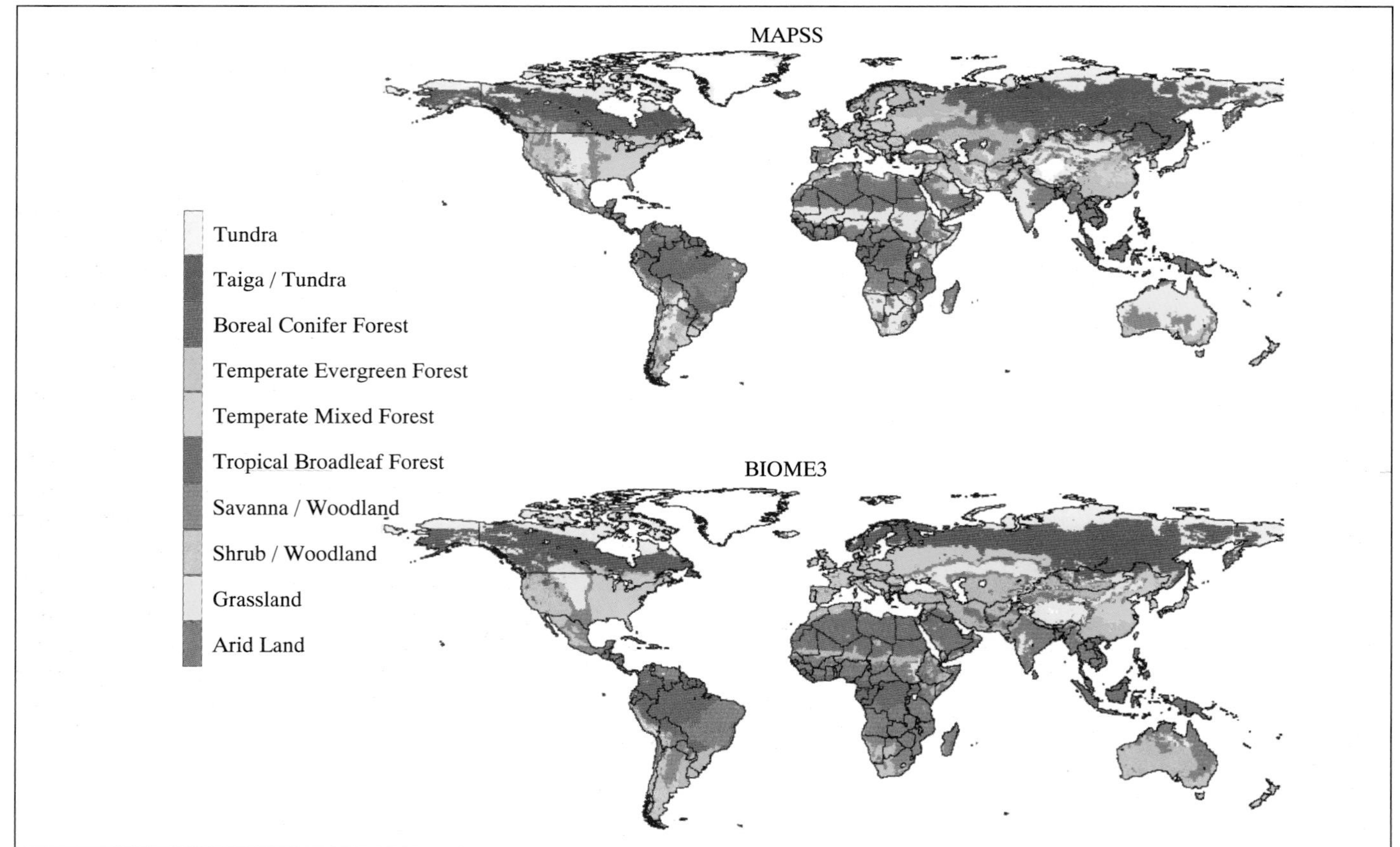


**Figure C-1:** The distribution of major biome types as simulated under current climate by the (a) MAPSS and (b) BIOME3 biogeography models.

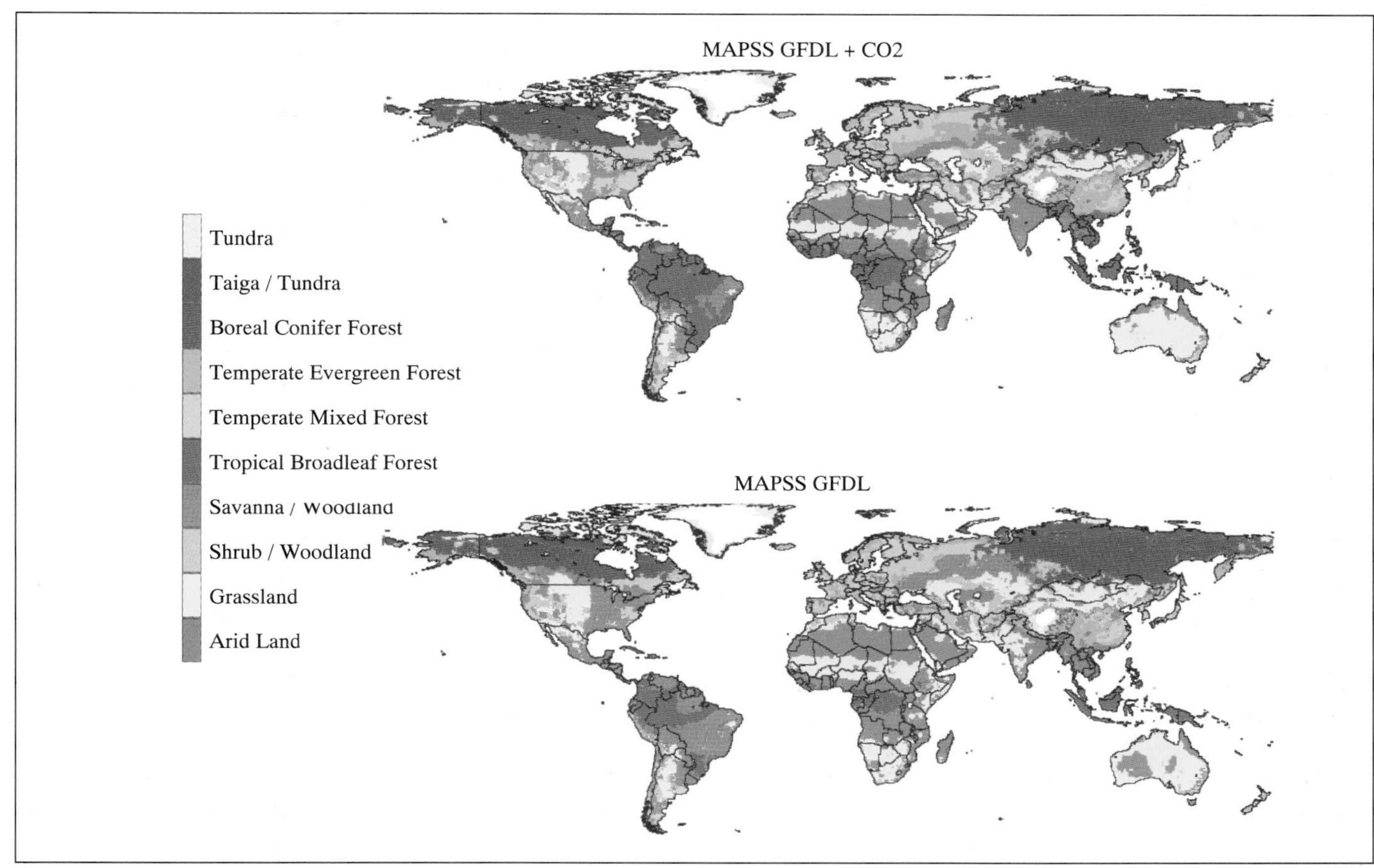


**Figure C-2:** The potential distribution of major biomes as simulated under the GFDL-R30 2 x $CO_2$ GCM experiment (Geophysical Fluid Dynamics Laboratory, slab ocean, no sulfate aerosols), by MAPSS, both (a) with and (b) without a direct, physiological $CO_2$ effect.

discrepancies between MAPSS and BIOME3 under current climate (Figure C-1) are due to questions of classification, especially in the drier types. For example, the Sahel region in Africa is labeled as 'shrub-steppe' in the original BIOME3 classification, but as various grassland types in MAPSS. The MAPSS grassland types do allow some shrubs, but the shrub density is usually reduced by the fire model, which assumes that there has been no reduction in fuel due to grazing. Were such land-use constraints included, the two models would be in better agreement on the classification. This classification difference between the models occurs over many of the drier parts of the world.

The models each appear to be better calibrated to their 'home' continents than either is to other continents. For example, MAPSS over-estimates the distribution of Temperate Evergreen Forests (conifers) in western Europe; while, BIOME3 overestimates the distribution of Temperate Mixed Forests (broadleaf) in western North America (Table C-1); yet, the two models are generally in agreement on the amount and location of temperate forests.

One area of significant departure from observed vegetation is the Pampas of southern South America. Both models simulate forests where grasslands are generally predominant. Various hypotheses for this discrepancy include unique soils, fire disturbance, rainfall seasonality and interannual variability of rainfall (VEMAP Members, 1995; Neilson, 1995; Neilson and Marks, 1994) and represent a focus for future research. Other local to regional errors in the MAPSS and BIOME3 classifications will be apparent to the knowledgeable reader. Reasons for these errors are many, but include 1) possible errors in the interpolated climate, 2) grazing, harvest, fire and other disturbances, and 3) missing or weak representation of some processes in the models. Globally, both models are reasonably accurate and are generally considered to be more accurate under altered climates than previous, empirical approaches (VEMAP Members, 1995). Empirical approaches cannot simultaneously simulate changes in vegetation distribution and changes in vegetation density and hydrology. Nor can they examine the sensitivity of the system to altered $CO_2$ concentrations. However, as the focus shifts to ever smaller regions or locales, the model uncertainty and the likelihood of error increases.

### *C.6.2. Future Vegetation Distribution*

Both MAPSS and BIOME3 produce large shifts of cold-limited vegetation boundaries into higher latitudes and elevations.

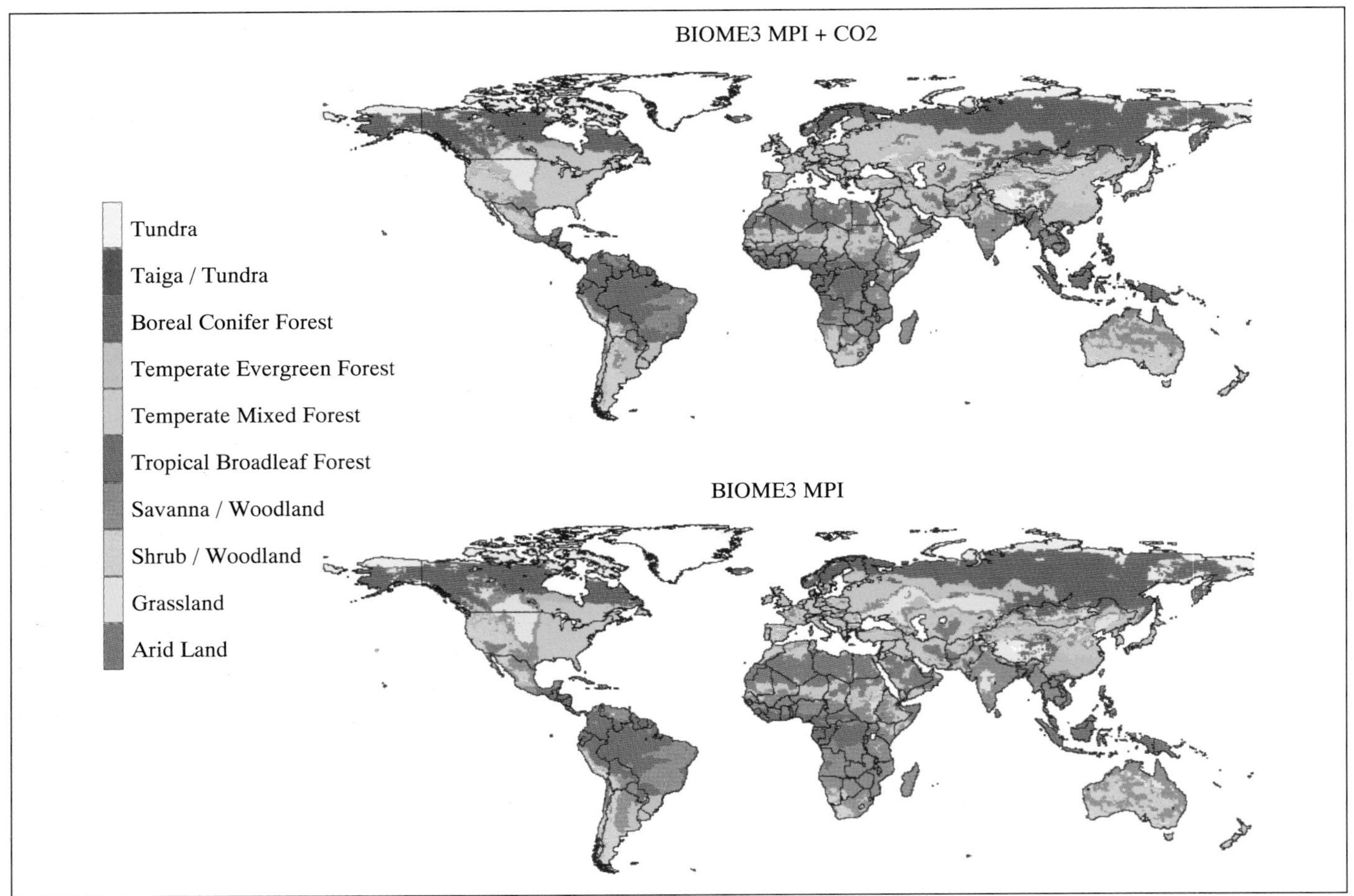


**Figure C-3:** The potential distribution of major biomes as simulated under the MPI-T106 GCM experiment (Max Planck Institute, 2 x $CO_2$ greenhouse gas radiative forcing, extracted from transient simulation, no sulfate aerosols), by BIOME3, both (a) with and (b) without a direct, physiological $CO_2$ effect.

However, the water-controlled boundaries may exhibit any direction of change, depending on the interaction of several variables including among others, the relative changes in temperature and precipitation, and whether or not the direct effects of $CO_2$ have been incorporated.

The older GCM scenarios tend to be hotter than the newer ones and produce a more dramatic change in vegetation distribution. The MAPSS results under GFDL-R30 serve to illustrate one of the older simulations. MAPSS and BIOME2 (Haxeltine *et al.*, 1996) were similar over the U.S. under this scenario (VEMAP Members, 1995), if both incorporated the direct $CO_2$ effects. MAPSS was far more xeric in response than BIOME2 without the direct $CO_2$ effects (*ibid*).

The Tundra decreases by as much as 1/3 to 2/3 of its present size, as does the Taiga/Tundra, under all scenarios and with both ecological models (Table C-1, Figures C-2 to C-5). The boreal forest expands in size under all scenarios ranging from 108% to 133% of its present size (MAPSS only). Since BIOME3 includes the Taiga/Tundra, which contracts under all warming scenarios (MAPSS simulations), with the Boreal Conifer Forest, which expands under all scenarios (MAPSS simulations), the net change simulated by BIOME3 usually indicates a loss of Boreal Forest. However, the aggregation of the two types in BIOME3 hides the observation that the two vegetation types (as defined above) tend to change in opposite sign with respect to area, i.e., Taiga/Tundra decreasing, Boreal Conifer increasing. The two models are quite consistent in the simulated response of the combined biomes (Table C-1). Temperate forests (inclusive of both types) increase in area (107% to 158%). Tropical forests could either expand or contract, largely dependent on the inclusion of the direct $CO_2$ effect, but also dependent on the severity of the scenario. Savanna/woodlands expand or contract, depending on whether or not they are encroached upon by neighboring forests or semi-arid lands, again reflecting whether or not direct $CO_2$ effects are considered and on the scenario. BIOME3 shows a competitive displacement of tropical savannas by neighboring forests, due to the superior competitive ability in the model of the $C_3$ trees over the $C_4$ grasses under elevated $CO_2$. The total area of grasslands and shrublands in these simulations remains largely unchanged or expands by as much as 27%, depending on the $CO_2$ effect and the scenario. If the direct effects of $CO_2$ are included, arid lands tend to contract in all scenarios, shifting to less arid types (Table C-1, Figures C-2 to C-5). Without the direct $CO_2$, arid lands either expand or contract in area, depending on the climate scenario and the ecological model.

### *C.6.3. Change in Vegetation Density (LAI)*

Although temperature-controlled vegetation boundaries shift predictably in all cases, water-controlled boundaries could shift any direction, reflecting either more or less beneficial water status. Likewise, vegetation change does not simply consist of shifts in the boundaries between homogeneous blocks of vegetation. Indeed, changes in vegetation density (via leaf area index, LAI) may often be more informative, since changes in LAI in water-limited areas generally indicate a change in the site water status and carrying capacity (Tables C-2, C-3; Figures C-6 to C-9). The change in LAI could also be taken as an indication of what could happen in the near term, since changes in LAI can occur in a matter of a few years while adjustments of vegetation structure and composition take much longer.

MAPSS and BIOME3 produce generally similar maps of change in LAI when forced by the same scenario, except that MAPSS produces a consistently stronger drought effect (compare Figures C-6 and C-7 and Figures C-8 and C-9). For example, within the U.S., when not including a direct $CO_2$ effect (Figure C-9), both models indicate increases in LAI in the Southwest and either an increase or no change in LAI over most of the eastern U.S. Both models simulate a decline or no change (BIOME3) over much of the western U.S. (excluding the SW). Both models produce increases in LAI over parts of the Sahara/Sahel, either with or without the direct $CO_2$ effect under all scenarios, both old and new. Likewise, both models under all scenarios indicate some increases in LAI over much of the arid interior of Australia. In general, there appear to be increases in LAI in already low LAI regions, either arid or cold. The increases in cold regions are due to expansion of forests into non-forested areas. The increases in arid areas are due to increased rainfall, a consequence of a generally more vigorous hydrologic cycle. There are many other consistencies between the two biogeography models with respect to the relative regional or subregional simulated LAI changes. A more complete discussion of simulated LAI patterns from the VEMAP models over the conterminous U.S. is in preparation.

If the direct effects of $CO_2$ are minimal and the future scenarios are relatively warm, decreases in LAI could occur over very large forested areas, ranging up to nearly 2/3 or more of the areas of boreal, temperate and tropical forests (Table C-2). By contrast, if the direct effects of $CO_2$ are strong and scenarios

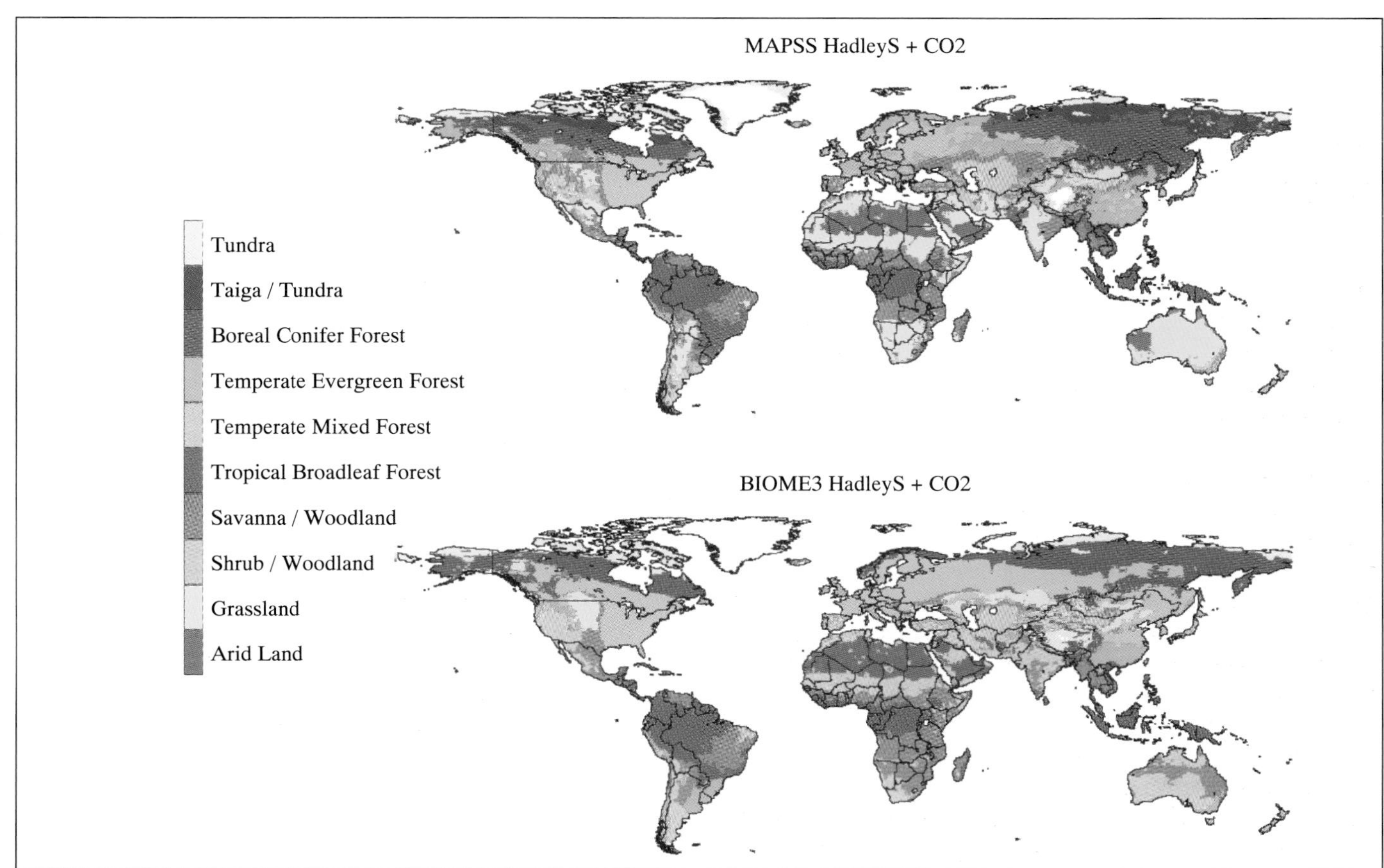


**Figure C-4:** The potential distribution of major biomes as simulated under the HADCM2SUL GCM experiment (Hadley Center, 2 x $CO_2$ greenhouse gas radiative forcing, extracted from transient simulation, plus sulfate aerosols), by (a) MAPSS and (b) BIOME3. Both models have incorporated a direct, physiological $CO_2$ effect.

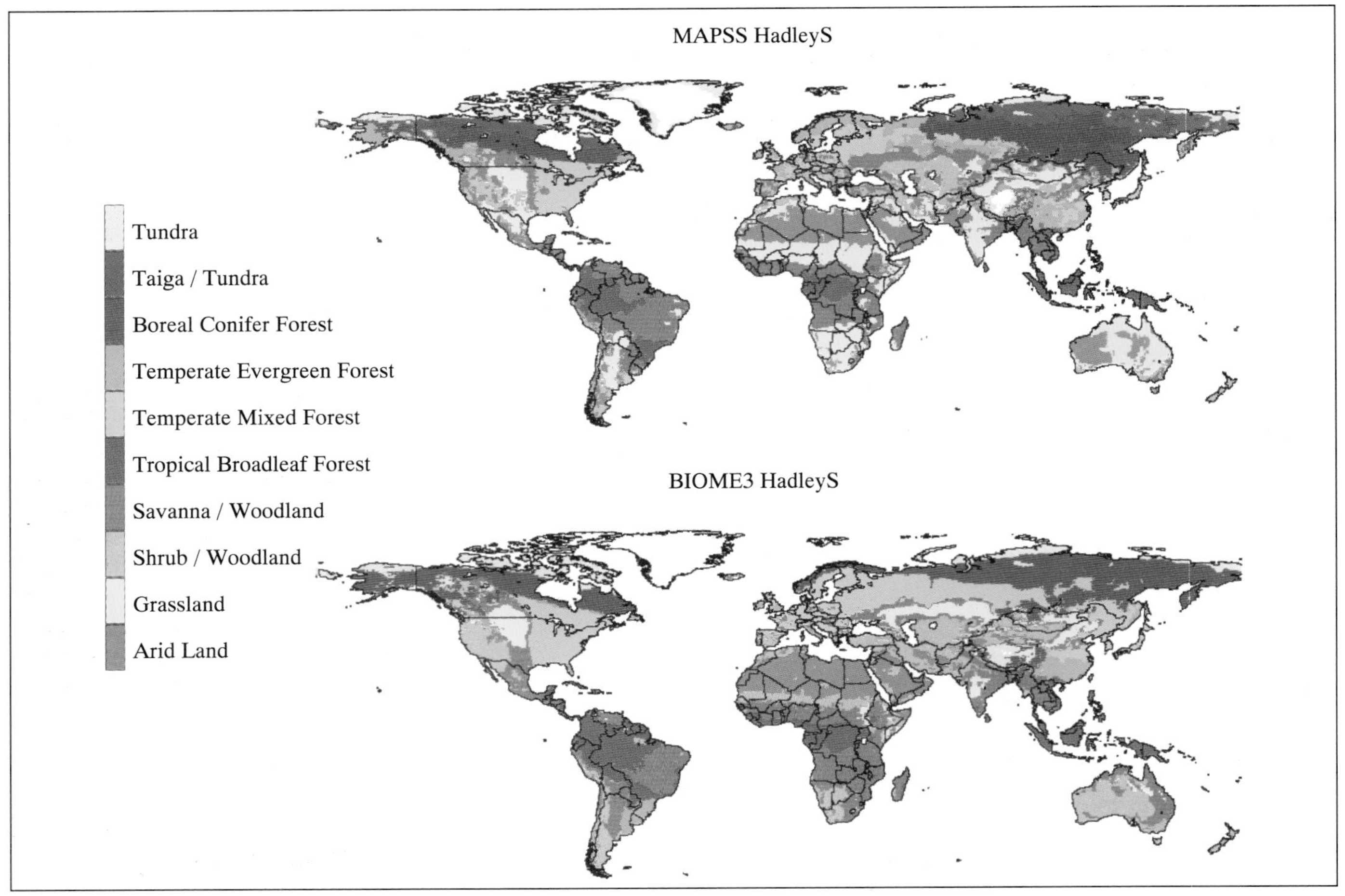


**Figure C-5:** The potential distribution of major biomes as simulated under the HADCM2SUL GCM experiment (Hadley Center, 2 x $CO_2$ greenhouse gas radiative forcing, extracted from transient simulation, plus sulfate aerosols), by (a) MAPSS and (b) BIOME3. A direct, physiological $CO_2$ effect is not incorporated in either model.

are not too warm, then all forest vegetation zones could experience increased biomass over as much as 2/3 or more of their areas (Table C-3). More likely, the responses will be intermediate with large regional contrasts, decreases in vegetation density in some areas, increases in others. Even though these are equilibrium simulations, a simulated decline in LAI generally implies a less favorable water balance and a loss of vegetation density. These losses imply a process of loss over some time period. We can only draw inferences about how rapidly such losses would occur, based on the simulated amount of loss. The regions that could experience declining LAI (Figures C-6 to C-9), would exhibit spatial gradients in response from mild decline grading into potentially catastrophic dieback. All reaches along the decline gradients would experience drought stress, which could trigger other responses, such as pest infestations and fire. Following disturbance by drought, infestation or pests, new vegetation, either of the same or of a different type would grow, but to a lower density.

Including both equilibrium and 'transient' scenarios, MAPSSwas run under four different scenarios (not counting the sulfate scenario, HADSUL). These range in global temperature increase (delta T) at the time of 2 x $CO_2$ from 1.7 (HADGHG) to 5.2°C (UKMO) (Annex B). In general, the areas of forest decline within individual biomes (incorporating a direct $CO_2$ effect) increase linearly with increasing delta T in the temperate and boreal forests; while, the areas of increased forest density decrease with increasing delta T. Tropical forests exhibit a similar pattern across the three FAR scenarios, but under the cooler HADGHG scenario show a large decline as simulated by MAPSS. By contrast, BIOME3, under the HADGHG scenario, shows almost no change in tropical forest density. Interestingly, adjacent tropical savannas increase in density in both ecological models under the HADGHG scenario.

### C.6.4. *Equilibrium vs. "Transient" Scenarios and the Importance of Elevated $CO_2$*

The newer climate scenarios (IPCC 1996, WG I, Section 6), extracted from transient GCM simulations, are as a group quite different from the older, equilibrium scenarios (IPCC 1990, WG I, Section 3), in terms of the simulated ecological responses that these scenarios produce. All of the older scenarios produce large regions showing LAI declines (especially in temperate to high latitudes), as well as gains, even when the direct effects of $CO_2$ are included (MAPSS simulations, Figure C-6, OSU and UKMO scenarios not shown). By contrast, under the newer scenarios, if a direct $CO_2$ effect is assumed, then there are very few regions with declines in LAI, as simulated by both MAPSS and BIOME3 (Figures C-7, C-8); rather, most of the world is simulated with an increased LAI. Actual increases in LAI could be limited by nitrogen availability in some areas, although elevated soil temperatures could increase decomposition, releasing more nitrogen (McGuire *et al.*, 1995; VEMAP Members, 1995). The first-order differences between the older and newer scenarios are likely due to the smaller global temperature increases in the newer climate scenarios, which came from GCMs that had not attained their full equilibrium temperature changes.

### C.6.5. *Sulfate Aerosols*

The incorporation of sulfate aerosols produced a cooling effect in the HADCM2SUL run compared to the HADCM2GHG run, which lacked the sulfate forcing (GHG runs are not shown). The vegetation response to the sulfate forcing is observable in the model output from both MAPSS and BIOME3, but is relatively small compared to the differences between the newer and older climate scenarios. The newer scenarios produce widespread enhanced vegetation growth, even without the sulfate effect, if direct $CO_2$ effects are included and widespread decline if the $CO_2$ effects are excluded. The presence of the sulfate-induced cooling produces a much smaller amplitude effect on the vegetation than does the presence or absence of the direct effects of elevated $CO_2$ on water-use-efficiency.

### C.6.6. *Change in Annual Runoff*

Changes in annual runoff (Figure C-10) were mapped for all scenarios from both MAPSS and BIOME3. The changes in runoff are more stable among the different climate scenarios than are the simulated changes in LAI. The relative stability of simulated runoff change may reflect that runoff is a passive drainage process; whereas, evapotranspiration is a biological process and a function of the product of LAI and stomatal conductance. If stomatal conductance is reduced, e.g., via a direct $CO_2$ effect, LAI will compensate by increasing and runoff will show little change (Neilson and Marks, 1994). Some of the

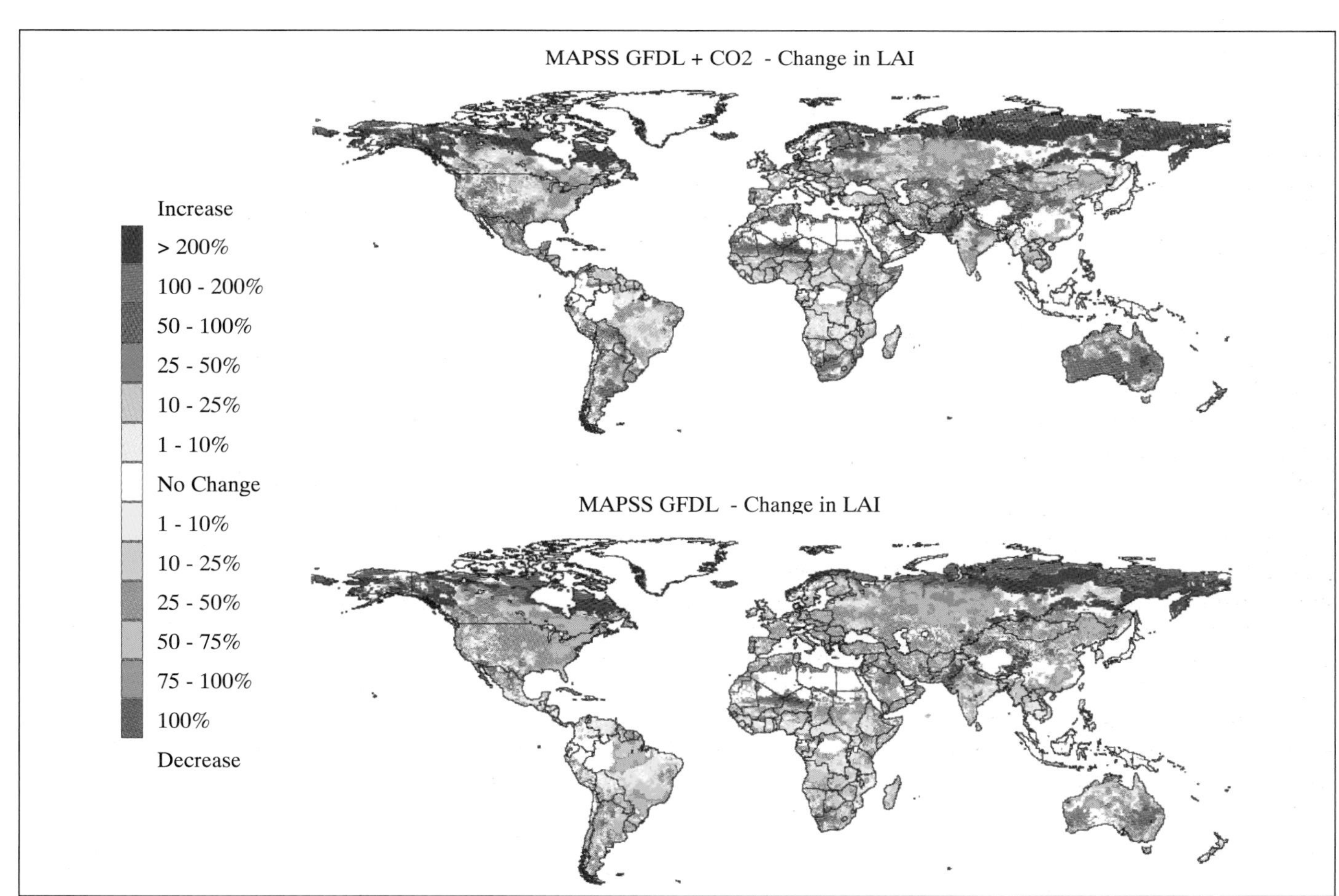


**Figure C-6:** The potential change in vegetation leaf area index (LAI), which can be considered as an index of vegetation density or biomass, as simulated under the GFDL-R30 2 x $CO_2$ GCM experiment (Geophysical Fluid Dynamics Laboratory, slab ocean, no sulfate aerosols), by MAPSS, both (a) with and (b) without a direct, physiological $CO_2$ effect.

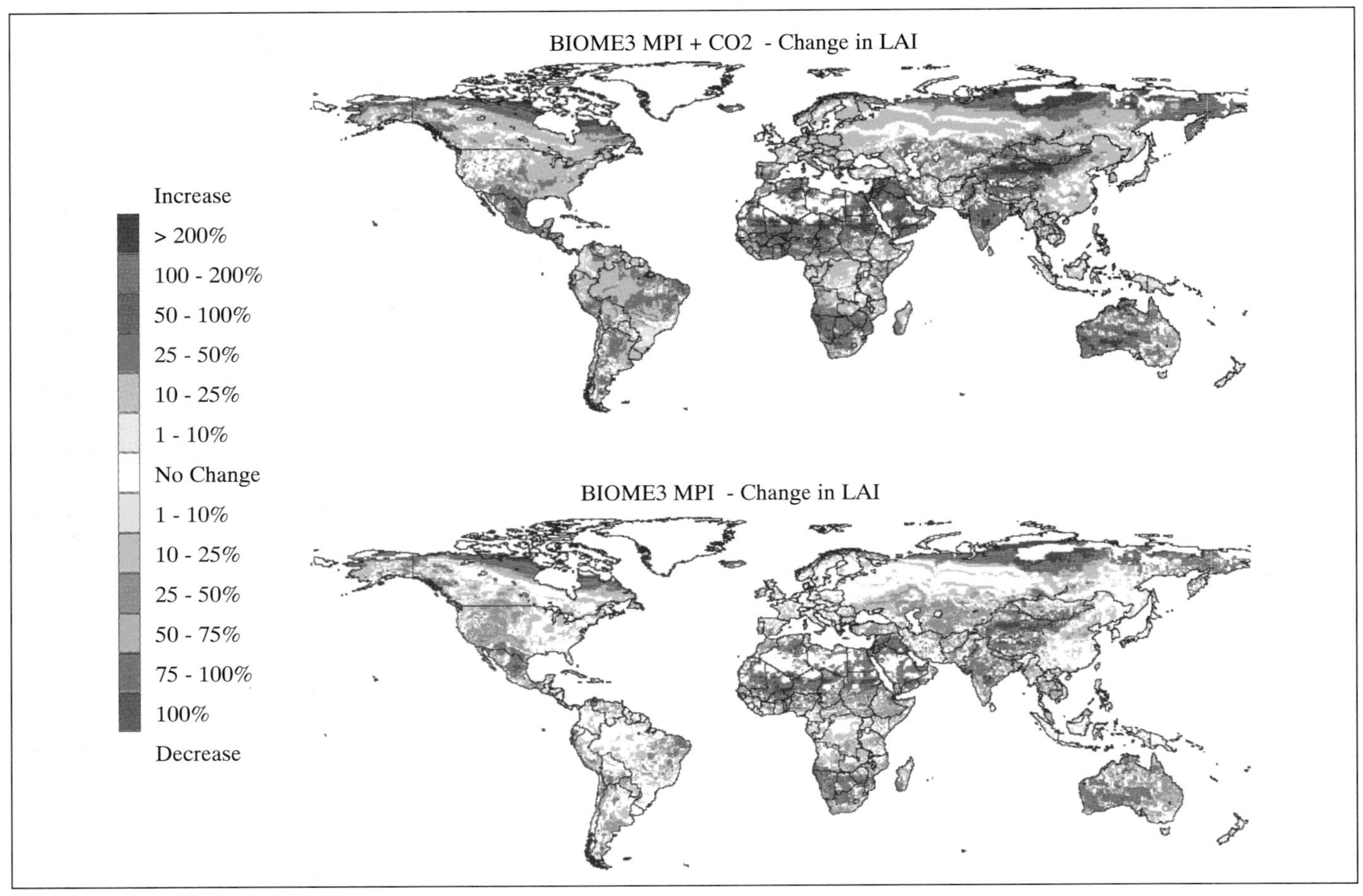


**Figure C-7:** The potential change in vegetation leaf area index (LAI), which can be considered as an index of vegetation density or biomass, as simulated under the MPI-T106 GCM experiment (Max Planck Institute, 2 x $CO_2$ greenhouse gas radiative forcing, extracted from transient simulation, no sulfate aerosols), by BIOME3, both (a) with and (b) without a direct, physiological $CO_2$ effect.

obvious differences between MAPSS and BIOME3 can be attributed to structural differences in the models. BIOME3 calculates water balance daily, even though all inputs are monthly; whereas, MAPSS calculates water balance monthly. This difference alone could be causing the more extreme responsiveness of MAPSS, which shows both larger runoff increases and larger losses in different regions. On the other hand, MAPSS uses a 3-layer soil with roots only in the top two layers; while BIOME3 uses a 2-layer soil with roots in both layers. The third layer in MAPSS provides a consistent base flow and might explain why MAPSS produces runoff in some drier regions, such as the western U.S., while BIOME3 does not. The hydrology models in both MAPSS and BIOME3, although process-based, are considered prototypes for eventual replacement by more elaborate models (see for example, the PILPS model intercomparison study; Love and Henderson-Sellers, 1994).

In general, MAPSS and BIOME3 produce similar regional patterns in the estimated changes in runoff. Although the magnitude of the changes are different, there are broad similarities in the sign of the change (but, clearly not in all regions). The largest area of regional difference between the two models is in interior Eurasia (Figure C-10).

Runoff generally increases in the Tundra, due to higher temperatures, more precipitation and more melting (Tables C-4, C-5). It decreases in the Taiga/Tundra due to encroachment of high-density boreal forest into low density vegetation (hence, higher transpiration). Runoff results are varied in the temperate forests, but Temperate Mixed forests tend to present a higher likelihood of reduced runoff over large areas (range 51% to 88% of the area under all scenarios) than of increased runoff (range 11% to 47% of the area under all scenarios, Tables C-4, C-5). Even the most benign scenarios indicate a minimum of 51% of the area of the world's temperate evergreen forests could experience a runoff decline; whereas, a maximum of 47% of the area would experience increased runoff. Temperate Evergreen Forests exhibit a greater likelihood of increased runoff over large areas (range 29% to 87% of the area under all scenarios) than decreased runoff (range 11% to 68%), but the overlap in these increase and decrease ranges indicates the degree of uncertainty in the simulations. However, much of the increased runoff in the Temperate

Evergreen forested areas is due to increased winter runoff, which is not necessarily available for use by ecosystems, irrigation or domestic purposes. Runoff from tropical forest areas could either increase or decrease over large areas, depending largely on the importance of the direct $CO_2$ effects. Runoff from drier vegetation types is regionally variable and exhibits both increases and decreases, depending on the direct $CO_2$ effects and regional rainfall patterns.

## C.7. Conclusions

MAPSS and BIOME3 produce qualitatively similar results under alternative future climate scenarios, with or without including a direct, physiological $CO_2$ effect. However, MAPSS produces consistently stronger drought effects with increasing temperature than does BIOME3. When under common scenarios, the two models produce similar subregional sensitivities. That is, if two adjacent subregions show opposite sign responses under a future climate in MAPSS, they will tend to exhibit the same relationship in BIOME3, but the overall sensitivity will be lower in BIOME3. The newer scenarios, constructed from transient GCM experiments, are consistently less xeric, as measured by simulated changes in leaf area index (LAI), than the older, equilibrium GCM scenarios. Under the newer scenarios, both ecological models indicate an overall increase in LAI (although nutrient constraints could limit or delay the increase), when a direct physiological $CO_2$ effect is included. However, if the direct $CO_2$ effects are not included, both models indicate a general reduction in global vegetation density.

The changes in vegetation leaf area (LAI) simulated by both MAPSS and BIOME3 are analogous to the changes in soil water content reported by earlier GCM experiments (IPCC 1996, WG I, Section 6). Those earlier experiments maintained a fixed vegetated land surface. That is, the vegetation type and density were not allowed to respond to changes in either climate or elevated $CO_2$ concentration. Therefore, as evaporative demand went up in those simulations, soil water content decreased, or was totally depleted. MAPSS and BIOME3, however, absorb those processes directly in the vegetation response. Both models simulate water-limited LAI by maximizing the LAI that can be supported and just barely transpire available soil water. Thus, in the equilibrium solution to LAI, soil water is fully utilized and can't change much under altered

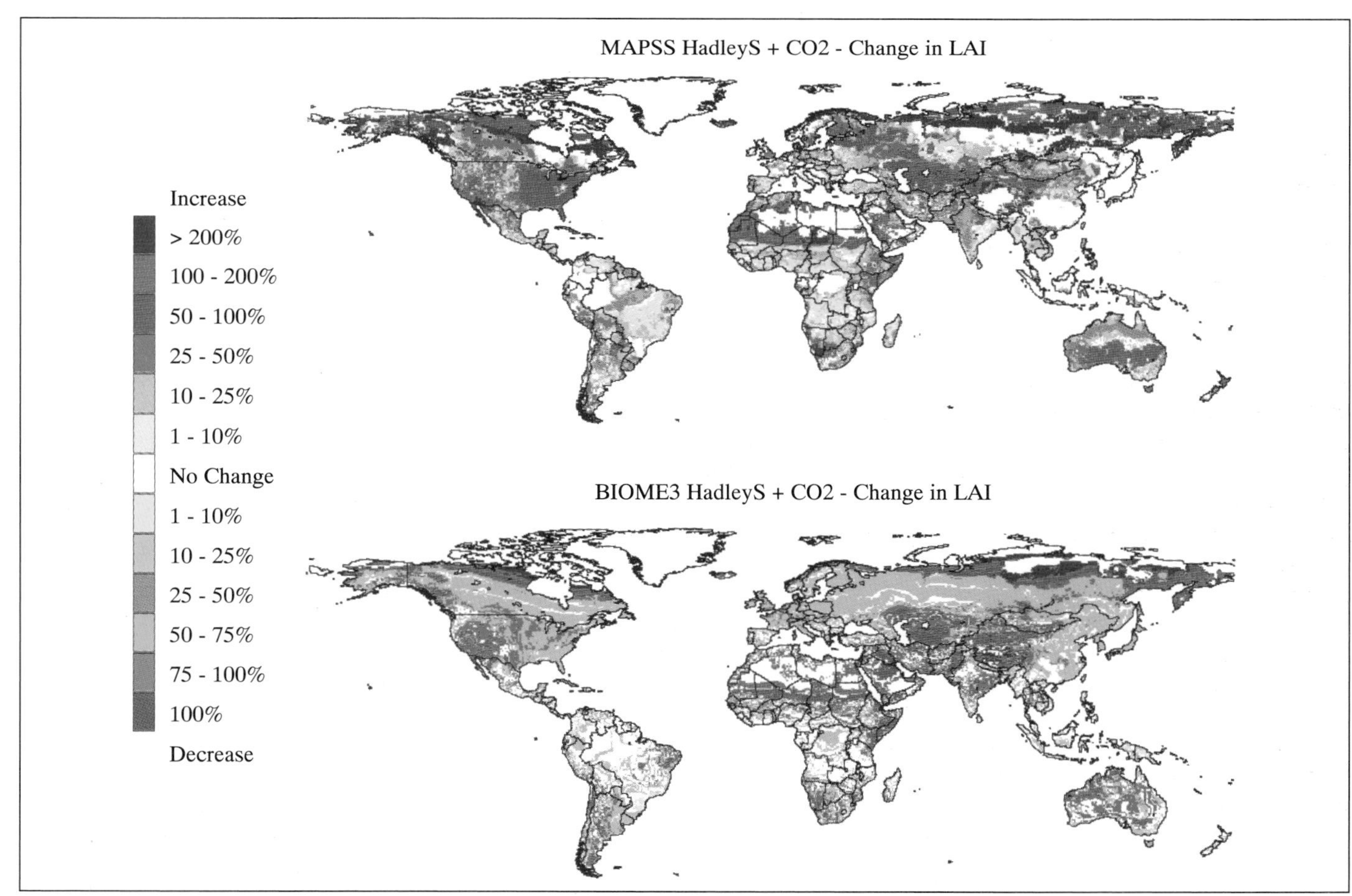


**Figure C-8:** The potential change in vegetation leaf area index (LAI), which can be considered as an index of vegetation density or biomass, as simulated under the HADCM2SUL GCM experiment (Hadley Center, 2 x $CO_2$ greenhouse gas radiative forcing, extracted from transient simulation, plus sulfate aerosols), by (a) MAPSS and (b) BIOME3. Both models have incorporated a direct, physiological $CO_2$ effect. This figure is a companion to Figure C-4.

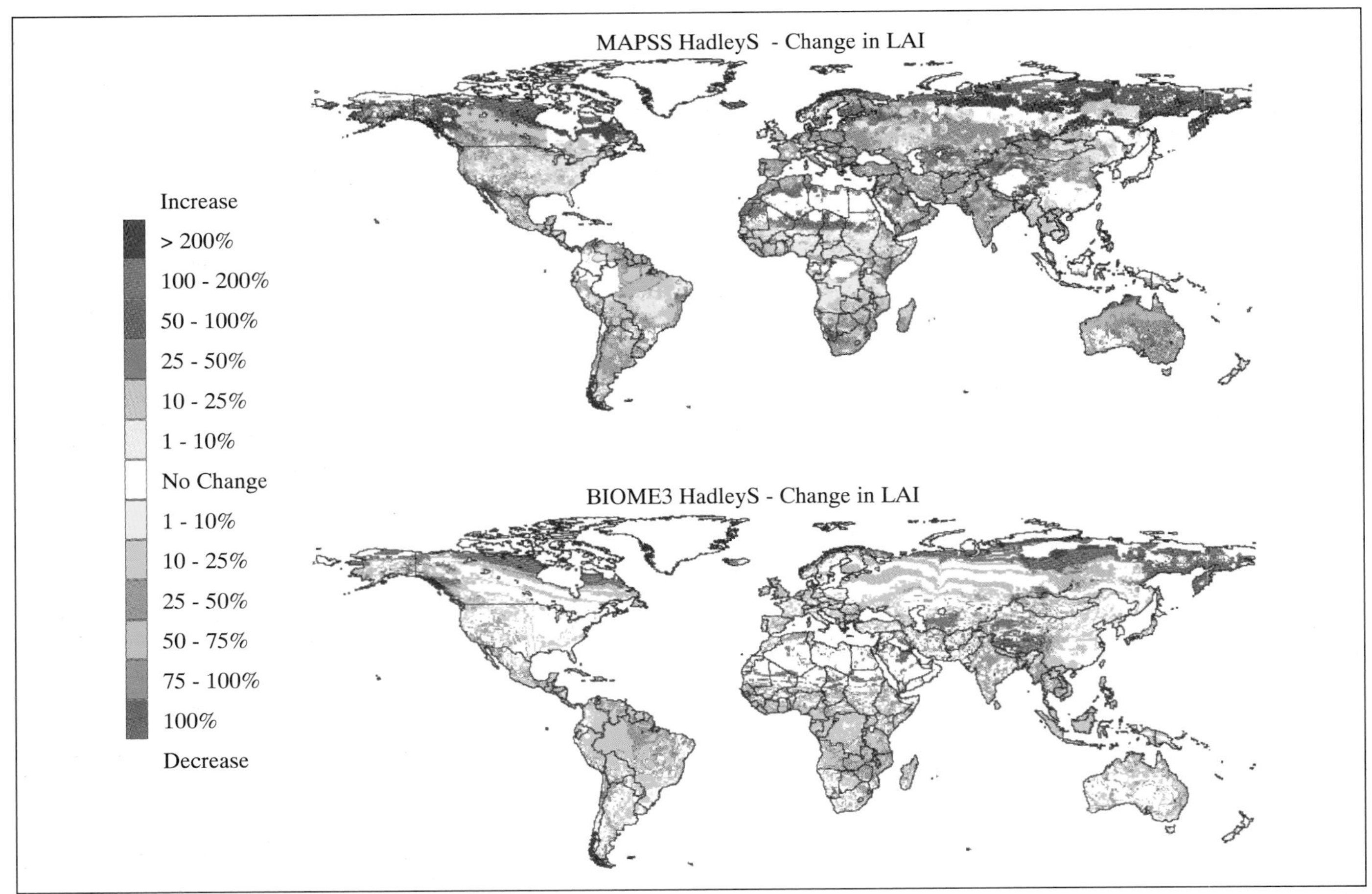


**Figure C-9:** The potential change in vegetation leaf area index (LAI), which can be considered as an index of vegetation density or biomass, as simulated under the HADCM2SUL GCM experiment (Hadley Center, 2 x $CO_2$ greenhouse gas radiative forcing, extracted from transient simulation, plus sulfate aerosols), by (a) MAPSS and (b) BIOME3. A direct, physiological $CO_2$ effect is not incorporated in either model. This figure is a companion to Figure C-5.

climates. Changes in the site water balance are, therefore, indicated by changes in LAI.

The newer climate scenarios used in this analysis are relatively cool in comparison to other possible new scenarios (IPCC 1996, WG I, Section 6). Therefore, the analyses presented here must be considered as a relatively conservative subset of the possible future ecological responses.

Although many of the simulations from both MAPSS and BIOME3 indicate potentially large expansions of tropical and in some cases temperate forests, actual expansions would be limited by urban and agricultural land-use constraints, unsuitable soils in some areas and slow dispersal rates, among other factors. Even so, if a forest is anticipated to expand into a region formerly indicated as shrubland, any agriculture in the region might expect an increase in potential productivity, and vice versa. Such changes between forest and shrubland are usually underlain by a change in LAI, which reflects the site water balance. An increase or decrease in LAI indicates a change in the water or energy balance and the potential biomass density or carrying capacity that could be supported on the site, regardless of whether the biomass is 'natural' or agricultural (dryland agriculture only).

The results presented here are for steady-state, or equilibrium conditions and do not directly indicate how the systems would behave in their transient responses toward a new equilibrium. For example, in areas where LAI is indicated to decline, it may be that equilibrium runoff is indicated to increase. However, one hypothesis is that during the processes of LAI decline, increased evaporative demand could cause reductions in runoff, before the vegetation becomes sufficiently drought-stressed for the LAI to be reduced. After further time, if the vegetation is sufficiently drought stressed, a rapid dieback could occur and might be facilitated by pests and fire. Were vegetation to undergo such a large dieback, then transpiration demand would be temporarily reduced and runoff could increase substantially. Thus, before a new equilibrium is attained with new vegetation growth, streams could go through a dry to wet oscillation. These possible hydrologic responses to vegetation change are, however, of a different timeframe (years) than possible short term floods and droughts that could occur simply due to increased variation of extreme weather events (IPCC 1996, WG I, Section 6).

At least two contrasting, transient trajectories of vegetation change are possible. If a large, direct $CO_2$ benefit were to occur, vegetation could increase in growth and biomass under relatively cool, early warming conditions, only to experience drought-stress and decline or dieback under the hotter, later stages of warming. Alternatively, if direct $CO_2$ benefits are more muted, vegetation could become drought-stressed and experience decline or dieback within the next few decades even under mild warming. Expansion of forests into cooler zones would likely lag behind decline and dieback in warmer zones, producing a transient reduction in forest area, possible increases in pests and fire, and possibly large releases of $CO_2$ to the atmosphere (King and Neilson, 1992; Smith and Shugart, 1993).

## References


**Bazzaz,** F.A. 1990: The Response of Natural Ecosystems to the Rising Global $CO_2$ Levels. *Annual Review of Ecology and Systematics*, **21**, 167-196.

**Bazzaz,** F.A., S.L. Bassow, G.M. Berntson, and S.C. Thomas, 1996: Elevated $CO_2$ and terrestrial vegetation: implications for and beyond the global carbon budget. In: *Global Change and Terrestrial Ecosystems* [Walker, B. and W. Steffen (eds.)]. Cambridge University Press, Cambridge, pp. 43-76.

**Bengtsson,** L., M. Botzet, and M. Esch, 1995: Hurricane-type vortices in a general circulation model. *Tellus*, **47A**, 1751-1796.

**Bengtsson,** L., M. Botzet, and M. Esch, 1996: Will greenhouse gas-induced warming over the next 50 years lead to higher frequency and greater intensity of hurricanes? *Tellus*, **48A**, 57-73.

**Curtis,** P.S. 1996: A meta-analysis of leaf gas exchange and nitrogen in trees grown under elevated carbon dioxide. *Plant, Cell and Environment*, **19**, 127-137.

**Curtis,** P.S., C.S. Vogel, K.S. Pregitzer, D.R. Zak, and J.A. Teeri, 1995: Interacting effects of soil fertility and atmospheric $CO_2$ on leaf area growth and carbon gain physiology in *Populus* X *euramericana* (Dode) Guinier. *New Phytologist*, **129**, 253-263.

**Eamus**, D., 1991: The interaction of rising $CO_2$ and temperatures with water use efficiency. *Plant, Cell and Environment*, **14**, 843-852.

**Eamus,** D., 1996a: Responses of field grown trees to $CO_2$ enrichment. *Commonwealth Forestry Review*, **75**, 39-47.

**Eamus,** D. 1996b: Tree responses to $CO_2$ enrichment: $CO_2$ and temperature interactions, biomass allocation and stand-scale modeling. *Tree Physiology*, **16**, 43-47.

**Grulke,** N.E., J.L. Hom, and S.W. Roberts, 1993: Physiological adjustment of two full-sib families of ponderosa pine to elevated $CO_2$. *Tree Physiology*, **12**, 391-401.

**Grulke,** N.E., G.H. Riechers, W.C. Oechel, U. Hjelm, and C. Jaeger. 1990: Carbon balance in tussock tundra under ambient and elevated atmospheric $CO_2$. *Oecologia*, **83**, 485-494.

**Haxeltine,** A. and I.C. Prentice, 1996: BIOME3: An equilibrium terrestrial biosphere model based on ecophysiological constraints, resource availability and competition among plant functional types. *Global Biogeochemical Cycles*, **10(4)**, 693-710.


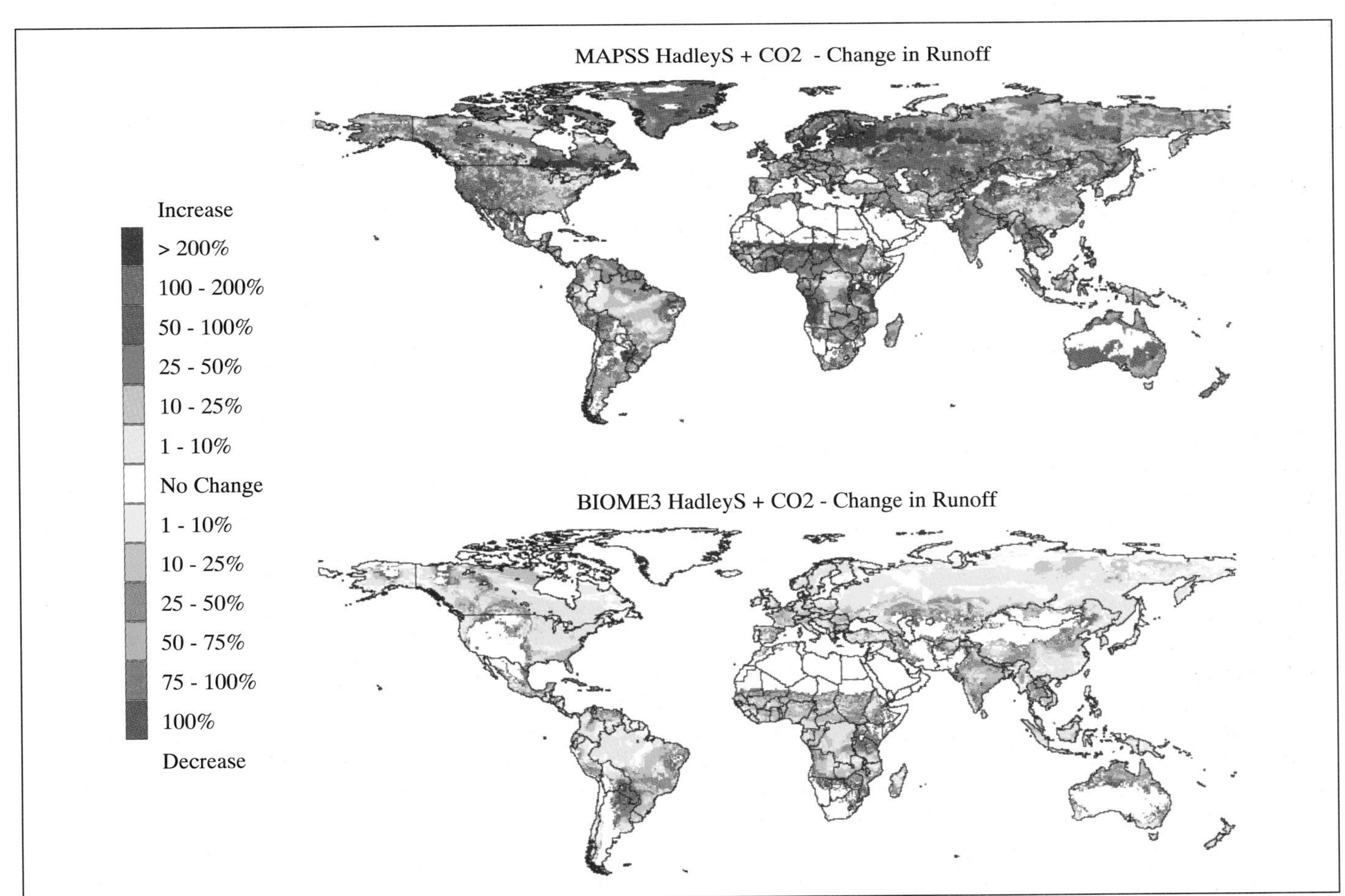


**Figure C-10:** The potential change in annual runoff, as simulated under the HADCM2SUL GCM experiment (Hadley Center, 2 x $CO_2$ greenhouse gas radiative forcing, extracted from transient simulation, plus sulfate aerosols), by (a) MAPSS and (b) BIOME3. Both models have incorporated a direct, physiological $CO_2$ effect. This figure is a companion to Figures C-4 and C-8.

**Haxeltine,** A., I.C. Prentice, and I.D. Creswell, 1996: A coupled carbon and water flux model to predict vegetation structure. *Journal of Vegetation Science*, **7(5)**, 651-666.

**Hunt,** E.R. and S.W. Running, 1992: Simulated dry matter yields for aspen and spruce stands in the North American boreal forest. *Canadian Journal of Remote Sensing*, **18**, 126-133.

**IPCC**, 1990: *Climate Change: The IPCC Scientific Assessment* [Houghton, J.T., G.J. Jenkins, and J.J. Ephraums (eds.)]. Cambridge University Press, Cambridge and New York, 365 pp.
- Cubasch, U. and R.D. Cess, Chapter 3. *Processes and Modelling,* pp. 69-91.
- Mitchell, J.F.B., S. Manabe, V. Meleshko, and T. Tokioka, Chapter 5. *Equilibrium Climate Change — and its Implications for the Future*, pp. 131-172.

**IPCC**, 1996: *Climate Change 1995: The Science of Climate Change. Contribution of Working Group I to the Second Assessment Report of the Intergovernmental Panel on Climate Change* [Houghton, J.T., L.G. Meira Filho, B.A. Callander, N. Harris, A. Kattenberg, and K. Maskell (eds.)]. Cambridge University Press, Cambridge and New York, 572 pp.
- Gates, W.L., A. Henderson-Sellers, G.J. Boer, C.K. Folland, A.Kitoh, B.J. McAvaney, F. Semazzi, N. Smith, A.J. Weaver, and Q.-C. Zeng, Chapter 5. *Climate Models —Evaluation*, pp. 235-284.
- Kattenberg, A., F. Giorgi, H. Grassl, G.A. Meehl, J.F.B. Mitchell, R.J. Stouffer, T. Tokioka, A. J. Weaver, and T.M.L. Wigley, Chapter 6. *Climate Models —Projections of Future Climate*, pp. 289-357.

**IPCC**, 1996. *Climate Change 1995: Impacts, Adaptations and Mitigation of Climate Change: Scientific-Technical Analyses. Contribution of Working Group II of the Second Assessment Report of the Intergovernmental Panel on Climate Change* [Watson, R.T., M.C. Zinyowera, and R.H. Moss (eds.)]. Cambridge University Press, Cambridge and New York, 880 pp.
- Kirschbaum, M.U.F., Chapter A. *Ecophysiological, Ecological, and Soil Processes in Terrestrial Ecosystems: A Primer on General Concepts and Relationships*, pp. 57-74.
- Kirschbaum, M.U.F. and A. Fischlin, Chapter 1. *Climate Change Impacts on Forests*, pp. 95-129.

**Johns,** T.C., R.E. Carnell, J.F. Crossley, J.M. Gregory, J.F.B. Mitchell, C.A. Senior, S.F.B. Tett, and R.A. Wood, 1997: The second Hadley Centre coupled ocean-atmosphere GCM: Model description, spinup and validation. *Climate Dynamics* (submitted).

**Kirschbaum,** M.U.F., D.A. King, H.N. Comins, R.E. McMurtrie, B.E. Medlyn, S. Pongracic, D. Murty, H. Keith, R.J. Raison, P.K. Khanna, and D.W. Sheriff. 1994: Modelling forest response to increasing $CO_2$ concentration under nutrient-limited conditions. *Plant, Cell and Environment*, **17**, 1081-1099.

**King,** G.A. and R.P. Neilson, 1992: The transient response of vegetation to climate change: A potential source of $CO_2$ to the atmosphere. *Water, Air, Soil Pollution*, **64**, 365-383.

**Körner,** C. 1995. Towards a better experimental basis for upscaling plant responses to elevated $CO_2$ and climate warming. *Plant, Cell and Environment*, **18**, 1101-1110.

**Leemans,** R. and W.P. Cramer, 1991: *The IIASA database for mean monthly values of temperature, precipitation, and cloudiness on the global terrestrial grid.* International Institute for Applied Systems Analysis RR-91-18, pp. 1-62.

**Love,** P.K. and A. Henderson-Sellers, 1994: Land-surface climatologies of AMIP-PILPS models and identification of regions for future investigation (PILPS Phase 3a). IGPO Publication Series 13, pp. 1-48.

**McGuire,** A.D., J.M. Melillo, and L.A. Joyce, 1995: The role of nitrogen in the response of forest net primary production to elevated atmospheric carbon dioxide. *Annual Review of Ecology and Systematics*, **26**, 473-503.

**McGuire,** A.D., J.M. Melillo, L.A. Joyce, D.W. Kicklighter, A.L. Grace, B. Moore, and C.J. Vorosmarty, 1992: Interactions between carbon and nitrogen dynamics in estimating net primary productivity for potential vegetation in North America. *Global Biogeochemical Cycles*, **6**, 101-124.

**Melillo,** J.M., A.D. McGuire, D.W. Kicklighter, B. Moore, C.J. Vorosmarty, and A.L. Schloss, 1993: Global climate change and terrestrial net primary production. *Nature*, **363**, 234-240.

**Mitchell,** J.F.B., T.C. Johns, J.M. Gregory, and S. Tett, 1995: Climate response to increasing levels of greenhouse gases and sulphate aerosols. *Nature*, **376**, 501-504.

**Mitchell,** J.F.B. and D. A. Warrilow, 1987: Summer dryness in northern mid latitudes due to increased $CO_2$. *Nature,* **330**, 238-240.

**Neilson,** R.P., 1995: A model for predicting continental-scale vegetation distribution and water balance. *Ecological Applications*, **5**, 362-385.

**Neilson,** R.P. and D. Marks, 1994: A global perspective of regional vegetation and hydrologic sensitivities from climatic change. *Journal of Vegetation Science*, **5**, 715-730.

**Neilson,** R.P. and S.W. Running, 1996: Global dynamic vegetation modelling: coupling biogeochemistry and biogeography models. In: *Global Change and Terrestrial Ecosystems* [Walker, B. and W. Steffen (eds.)]. Cambridge University Press, Cambridge, pp. 451-465.

**Norby,** R.J. 1996: Forest canopy productivity index. *Nature*, **381**, 564.

**Parton,** W.J., D.S. Schimel, C.V. Cole, and D. Ojima, 1987: Analysis of factors controlling soil organic levels of grasslands in the Great Plains. *Soil Science Society of America*, **51**, 1173-1179.

**Parton,** W.J., J.M.O. Scurlock, D.S. Ojima, T.G. Gilmanov, R.J. Scholes, D.S. Schimel, T. Kirchner, J.-C. Menaut, T. Seastedt, E.G. Moya, A. Kamnalrut, and J.I. Kinyamario, 1993: Observations and modeling of biomass and soil organic matter dynamics for the grassland biome worldwide. *Global Biogeochemical Cycles,* **7**, 785-809.

**Parton,** W.J., J.W.B. Stewart, and C.V. Cole, 1988: Dynamics of C, N, P and S in grassland soils: a model. *Biogeochemistry*, **5**, 109-131.

**Prentice,** I.C., W. Cramer, S.P. Harrison, R. Leemans, R.A. Monserud, and A.M. Solomon, 1992: A global biome model based on plant physiology and dominance, soil properties and climate. *Journal of Biogeography,* **19**, 117-134.

**Raich,** J.W., E.B. Rastetter, J.M. Melillo, D.W. Kicklighter, P.A. Steudler, B.J. Peterson, A.L. Grace, B. Moore, and C.J. Vorosmarty, 1991: Potential net primary productivity in South America: Application of a global model. *Ecological Applications*, **1**, 399-429.

**Running,** S.W. and E.R. Hunt, 1993: Generalization of a forest ecosystem process model for other biomes, BIOME-BGC, and an application for global-scale models. In: *Scaling Physiological Processes: Leaf to Globe* [Ehleringer, J.R. and C.B. Field (eds.)]. Academic Press, Inc., San Diego, CA, p. 141-158.

**Schlesinger,** M.E. and Z.C. Zhao, 1989: Seasonal climatic change introduced by double $CO_2$ as simulated by the OSU atmospheric GCM/mixed-layer ocean model. *Journal of Climate,* **2**, 429-495.

**Shugart,** H.H. and T.M. Smith, 1996: A review of forest patch models and their application to global change research. *Climatic Change*, **34(2)**, 131-153.

**Smith,** T.M. and H.H. Shugart, 1993: The transient response of terrestrial carbon storage to a perturbed climate. *Nature*, **361**, 523-526.

**Teskey,** R.O. 1997: Combined effects of elevated $CO_2$ and air temperature on carbon assimilation of Pinus taeda trees. *Plant, Cell and Environment*, **20**, 373-380.

**VEMAP Members**, 1995: Vegetation/ecosystem modeling and analysis project: Comparing biogeography and biogeochemistry models in a continental-scale study of terrestrial ecosystem responses to climate change and $CO_2$ doubling. *Global Biogeochemical Cycles*, **9**, 407-437.

**Woodward,** F.I. and T.M. Smith, 1994: Predictions and measurements of the maximum photosynthetic rate at the global scale. In: *Ecophysiology of Photosynthesis* [Schulze, E.-D. and M.M. Caldwell (eds.)]. Ecological Studies, vol. 100. Springer-Verlag, New York, pp. 491-509.

**Woodward,** F.I., T.M. Smith and W.R. Emanuel, 1995: A global land primary productivity and phytogeography model. *Global Biogeochemical Cycles*, **9**, 471-490.

**Wullschleger,** S.D., W.M. Post, and A.W. King. 1995: On the Potential for a $CO_2$ Fertilization Effect in Forest Trees: An Assessment of 58 Controlled-Exposure Studies and Estimates of the Biotic Growth Factor. In: Biotic Feedbacks in the Global Climate System [Woodwell, G.M. and F.T. Mackenzie (eds.)]. Oxford University Press, New York, pp. 85-107.

# D

# Socioeconomic Baseline Data

## Sources (see page 479)

| | |
|---|---|
| Total Population | UN Population Division, Annual Populations (1994 Revision), Median Estimate [HD16101] |
| Current Population Density | UN Population Division, Annual Populations (1994 Revision), Median Estimate [HD16104] |
| Projected Population Density | UN Population Division, Annual Populations (1994 Revision), Median Estimate [HD16104] |
| Total Urban Population | United Nations, Urban and Rural Areas 1950-2025 (1995 Revision) [HD16405] |
| Urban Population in Coastal Cities | UN Statistical Office, UN Office for Ocean Affairs and the Law of the Sea, and Offshore Magazine [HD16419] |
| GDP per Capita in Constant PPP | Penn World Tables [EI5124] |
| GDP from Agriculture | World Bank World Tables 1994 Update [EI15105] |
| GDP from Industry | World Bank World Tables 1994 Update [EI15106] |
| GDP from Services | World Bank World Tables 1994 Update [EI15107] |
| GDP Annual Growth Rate | World Bank World Tables 1994 Update [EI15122] |
| Total Land Area | FAO 1995 [LC17101] |
| Arable and Permanent Cropland | FAO 1995 [LC17102] |
| Permanent Pasture | FAO FAOSTAT 95 [LC17104] |
| Forest and Woodland | FAO FAOSTAT 95 [LC17105] |
| Other Land | FAO FAOSTAT 95 [LC17106] |
| Water Resources per Capita | Compiled by World Resources Institute [WA22102] |
| Domestic Annual Water Withdrawals | Compiled by World Resources Institute [WA22109] |
| Industry Annual Water Withdrawals | Compiled by World Resources Institute [WA22109] |
| Agriculture Annual Water Withdrawals | Compiled by World Resources Institute [WA22109] |
| Irrigated Land | FAO FAOSTAT 95 [FA18203 |
| Agricultural Labor Force | FAO FAOSTAT 95 [FA18509] |
| Total Labor Force | FAO FAOSTAT 95 [FA18508] |
| Cattle Stocks | FAO FAOSTAT 95 [FA18301] |
| Sheep Stocks | FAO FAOSTAT 95 [FA18302] |
| Goat Stocks | FAO FAOSTAT 95 [FA18303] |
| Pig Stocks | FAO FAOSTAT 95 [FA18304] |
| Equines (Horses, Mules, Asses) | FAO FAOSTAT 95 [FA18305] |
| Buffalo Stocks | FAO FAOSTAT 95 [FA18306] |
| Camel Stocks | FAO FAOSTAT 95 [FA18307] |
| Total Commercial Energy Consumption | UN Energy Statistics Yearbook 1993 [EM21201] |
| Traditional Fuel Consumption | UN FAO [EM21209] |
| Commercial Hydroelectric Production | UN Energy Statistics Yearbook 1993 [EM21106] |
| Known and Endemic Mammal Species | World Conservation Monitoring Centre [BI20401] |
| Known and Endemic Bird Species | World Conservation Monitoring Centre [BI20402] |
| Known and Endemic Plant Species | World Conservation Monitoring Centre [BI20501] |

***Table D-1:** Africa region.*

| | Population and Human Development | | | | | Economic Conditions | | | | |
|---|---|---|---|---|---|---|---|---|---|---|
| **Country** | Total Pop. (1000s) 1995 | Current Pop. Density (persons/ km²) 1995 | Projected Pop. Density (persons/ km²) 2025 | Total Urban Pop. (1000s) 1995 | Urban Pop. in Coastal Cities (1000s) 1980 | GDP per Capita in Constant PPP ('85 IN$) 1992 | GDP from Agri-culture (%) 1993 | GDP from Industry (%) 1993 | GDP from Services (%) 1993 | GDP Annual Growth Rate (%) 1993 |
| Algeria | 27939 | 12 | 19 | 15591 | 3493 | 2719 | 13.47 | 43.30 | 43.23 | -2.20 |
| Angola | 11072 | 9 | 21 | 3569 | 1132 | X | X | X | X | X |
| Benin | 5409 | 48 | 109 | 1691 | 585 | X | 35.78 | 12.86 | 51.36 | 3.30 |
| Botswana | 1487 | 3 | 5 | 418 | X | X | 5.67 | 47.28 | 47.04 | 3.00 |
| Burkina Faso | 10319 | 38 | 79 | 2809 | X | 514 | X | X | X | 0.40 |
| Burundi | 6393 | 230 | 485 | 480 | X | 569 | 51.88 | 21.11 | 27.01 | -1.20 |
| Cameroon | 13233 | 28 | 61 | 5938 | 854 | 1029 | 28.61 | 24.85 | 46.54 | -4.90 |
| Central African Republic | 3315 | 5 | 10 | 1301 | X | 514 | 49.88 | 14.14 | 35.99 | -2.50 |
| Chad | 6361 | 5 | 10 | 1362 | X | 408 | 43.61 | 21.57 | 34.82 | -2.90 |
| Congo | 2590 | 8 | 17 | 1523 | 217 | 2240 | 11.43 | 35.18 | 53.39 | -2.10 |
| Cote d'Ivoire | 14253 | 44 | 114 | 6211 | 1495 | 1104 | 37.42 | 24.06 | 38.52 | -1.10 |
| Dem. Republic of Congo | 43901 | 19 | 45 | 12766 | 102 | X | X | X | X | X |
| Djibouti | 577 | 25 | 45 | 478 | 211 | X | 2.85 | 21.20 | 75.95 | X |
| Egypt | 62931 | 63 | 97 | 28170 | 4246 | 1869 | 17.87 | 22.38 | 59.74 | 0.50 |
| Equatorial Guinea | 400 | 14 | 28 | 169 | 181 | X | 47.20 | 25.84 | 26.96 | 7.30 |
| Eritrea | 3531 | 30 | 60 | 607 | X | X | 13.15 | 20.67 | 66.18 | X |
| Ethiopia | 55053 | 50 | 115 | 7371 | 760 | X | 60.45 | 10.30 | 29.25 | 7.70 |
| Gabon | 1320 | 5 | 10 | 660 | 155 | 3622 | 8.24 | 44.93 | 46.83 | 2.50 |
| Gambia | 1118 | 99 | 186 | 286 | 109 | X | 27.53 | 14.63 | 57.84 | 1.50 |
| Ghana | 17453 | 73 | 159 | 6333 | 1336 | 956 | 47.55 | 16.08 | 36.37 | 4.80 |
| Guinea | 6700 | 27 | 61 | 1981 | 696 | 740 | 23.92 | 31.13 | 44.94 | 4.50 |
| Guinea-Bissau | 1073 | 30 | 55 | 238 | 174 | 634 | 44.79 | 18.98 | 36.22 | 2.90 |
| Kenya | 28261 | 49 | 109 | 7817 | 489 | 914 | 28.92 | 17.56 | 53.52 | 1.00 |
| Lesotho | 2050 | 68 | 137 | 473 | X | 952 | 10.08 | 47.19 | 42.74 | 6.20 |
| Liberia | 3039 | 27 | 65 | 1366 | 465 | X | X | X | X | X |
| Libya | 5407 | 3 | 7 | 4649 | 1496 | X | X | X | X | X |
| Madagascar | 14763 | 25 | 59 | 4003 | 570 | 608 | 33.95 | 13.78 | 52.26 | 1.90 |
| Malawi | 11129 | 94 | 189 | 1505 | X | 496 | 39.20 | 18.12 | 42.68 | 8.80 |
| Mali | 10795 | 9 | 20 | 2909 | X | X | 42.39 | 15.36 | 42.25 | 7.70 |
| Mauritania | 2274 | 2 | 4 | 1224 | 238 | 837 | 27.65 | 30.16 | 42.19 | 5.00 |
| Morocco | 27028 | 61 | 91 | 13071 | 5543 | 2173 | 14.30 | 32.36 | 53.34 | 0.20 |
| Mozambique | 16004 | 20 | 44 | 5481 | 1109 | 711 | 33.15 | 12.01 | 54.83 | 19.10 |
| Namibia | 1540 | 2 | 4 | 576 | 76 | 2774 | 9.84 | 26.90 | 63.27 | 0.30 |
| Niger | 9151 | 7 | 18 | 1558 | X | X | 38.52 | 17.89 | 43.58 | 1.40 |
| Nigeria | 111721 | 121 | 258 | 43884 | 4383 | 978 | 33.52 | 42.63 | 23.86 | 2.90 |
| Reunion | 653 | 260 | 359 | 442 | 279 | X | X | X | X | X |
| Rwanda | 7952 | 302 | 600 | 483 | X | 762 | 40.54 | 21.47 | 38.00 | 3.20 |
| Senegal | 8312 | 42 | 86 | 3512 | 1378 | X | 19.51 | 19.13 | 61.36 | -2.00 |
| Sierra Leone | 4509 | 63 | 121 | 1632 | 453 | 734 | X | X | X | 0.70 |
| Somalia | 9250 | 15 | 33 | 2382 | 1186 | X | X | X | X | X |
| South Africa | 41465 | 34 | 58 | 21073 | 4272 | 3068 | 4.56 | 39.45 | 55.99 | 1.10 |
| Sudan | 28098 | 11 | 23 | 6915 | 356 | X | X | X | X | X |
| Swaziland | 855 | 49 | 95 | 267 | X | X | 11.76 | 38.70 | 49.54 | 2.00 |
| Tanzania | 29685 | 31 | 67 | 7230 | 1750 | X | 55.98 | 14.45 | 29.57 | X |
| Togo | 4138 | 73 | 165 | 1276 | 324 | 530 | 48.56 | 17.98 | 33.47 | -12.70 |
| Tunisia | 8896 | 54 | 81 | 5093 | 2476 | 3075 | 17.89 | 31.34 | 50.77 | 2.70 |
| Uganda | 21297 | 90 | 204 | 2670 | X | 547 | 52.66 | 12.35 | 34.99 | 6.40 |
| Zambia | 9456 | 13 | 25 | 4071 | X | X | 33.69 | 36.33 | 29.98 | 6.80 |
| Zimbabwe | 11261 | 29 | 50 | 3619 | X | X | 15.18 | 36.39 | 48.43 | 2.00 |

0 = zero or less than half the unit of measure; X = not available.

***Table D-1:*** *Africa region (continued).*

| Country | Land Cover/Use: Total Land Area (1000 ha) 1993 | Arable and Permanent Cropland (1000 ha) 1993 | Permanent Pasture (1000 ha) 1993 | Forest and Woodland (1000 ha) 1993 | Other Land (1000 ha) 1993 | Water: Water Resources per Capita (m³) 1995 | Domestic Annual Withdraws (%) 1995 | Industry Annual Withdraws (%) 1995 | Agriculture Annual Withdraws (%) 1995 |
|---|---|---|---|---|---|---|---|---|---|
| Algeria | 238174 | 7850 | 30700 | 4000 | 195624 | 528 | 25 | 15 | 60 |
| Angola | 124670 | 3500 | 29000 | 51900 | 40270 | 16618 | 14 | 10 | 76 |
| Benin | 11062 | 1880 | 442 | 3400 | 5340 | 4770 | 23 | 10 | 67 |
| Botswana | 56673 | 420 | 25600 | 26500 | 4153 | 9886 | 32 | 20 | 48 |
| Burkina Faso | 27360 | 3565 | 6000 | 13800 | 3995 | 2713 | 19 | 0 | 81 |
| Burundi | 2568 | 1360 | 915 | 85 | 208 | 563 | 36 | 0 | 64 |
| Cameroon | 46540 | 7040 | 2000 | 35900 | 1600 | 20252 | 46 | 19 | 35 |
| Central African Rep | 62298 | 2020 | 3000 | 46700 | 10578 | 42534 | 21 | 5 | 74 |
| Chad | 125920 | 3256 | 45000 | 32400 | 45264 | 6760 | 16 | 2 | 82 |
| Congo | 34150 | 170 | 10000 | 21100 | 2880 | 321236 | 62 | 27 | 11 |
| Cote d'Ivoire | 31800 | 3710 | 13000 | 7080 | 8010 | 5451 | 22 | 11 | 67 |
| Dem. Republic of Congo | 226705 | 7900 | 15000 | 173800 | 30005 | 23211 | 61 | 16 | 23 |
| Djibouti | 2318 | X | X | 6 | X | X | X | X | X |
| Egypt | 99545 | 2800 | 4950 | 31 | 91764 | 923 | 6 | 9 | 85 |
| Equatorial Guinea | 2805 | 230 | 104 | 1300 | 1171 | 75000 | 81 | 13 | 6 |
| Eritrea | 10100 | 1280 | 4800 | 2000 | 2020 | 2492 | X | X | X |
| Ethiopia | 110100 | 13930 | 44750 | 25000 | 26420 | 1998 | 11 | 3 | 86 |
| Gabon | 25767 | 460 | 4700 | 19900 | 707 | 124242 | 72 | 22 | 6 |
| Gambia | 1000 | 180 | 90 | 280 | 450 | 7156 | 7 | 2 | 91 |
| Ghana | 22754 | 4320 | 5000 | 7900 | 5534 | 3048 | 35 | 13 | 52 |
| Guinea | 24572 | 730 | 5500 | 14460 | 3882 | 33731 | 10 | 3 | 87 |
| Guinea-Bissau | 2812 | 340 | 1080 | 1070 | 322 | 25163 | 60 | 4 | 36 |
| Kenya | 56914 | 4520 | 21300 | 16800 | 14294 | 1069 | 20 | 4 | 76 |
| Lesotho | 3035 | 320 | 2000 | X | 715 | 2551 | 22 | 22 | 56 |
| Liberia | 9675 | 375 | 5700 | 1700 | 1900 | 76341 | 27 | 13 | 60 |
| Libya | 175954 | 2170 | 13300 | 840 | 159644 | 111 | 11 | 2 | 87 |
| Madagascar | 58154 | 3105 | 24000 | 23200 | 7849 | 22827 | 1 | 0 | 99 |
| Malawi | 9408 | 1700 | 1840 | 3700 | 2168 | 1678 | 10 | 3 | 86 |
| Mali | 122019 | 2503 | 30000 | 6900 | 82616 | 6207 | 2 | 1 | 97 |
| Mauritania | 102522 | 208 | 39250 | 4410 | 58654 | 5013 | 6 | 2 | 92 |
| Morocco | 44630 | 9920 | 20900 | 8970 | 4840 | 1110 | 5 | 3 | 92 |
| Mozambique | 78409 | 3180 | 44000 | 14000 | 17229 | 12997 | 9 | 2 | 89 |
| Namibia | 82329 | 662 | 38000 | 18000 | 25667 | 29545 | 29 | 3 | 68 |
| Niger | 126670 | 3605 | 8915 | 2500 | 111650 | 3552 | 16 | 2 | 82 |
| Nigeria | 91077 | 32385 | 40000 | 11300 | 7392 | 2506 | 31 | 15 | 54 |
| Reunion | 250 | 48 | 12 | 88 | 102 | X | X | X | X |
| Rwanda | 2467 | 1170 | 450 | 550 | 297 | 792 | 5 | 2 | 94 |
| Senegal | 19253 | 2350 | 3100 | 10450 | 3353 | 4740 | 5 | 3 | 92 |
| Sierra Leone | 7162 | 540 | 2200 | 2040 | 2382 | 35485 | 7 | 4 | 89 |
| Somalia | 62734 | 1020 | 43000 | 16000 | 2714 | 1459 | 3 | 0 | 97 |
| South Africa | 122104 | 13179 | 81378 | 8200 | 19347 | 1206 | 17 | 11 | 72 |
| Sudan | 237600 | 12975 | 110000 | 44240 | 70385 | 5481 | 4 | 1 | 94 |
| Swaziland | 1720 | 191 | 1070 | 120 | 339 | 5275 | 2 | 2 | 96 |
| Tanzania | 88359 | 3500 | 35000 | 33500 | 16359 | 2998 | 9 | 2 | 89 |
| Togo | 5439 | 2430 | 200 | 900 | 1909 | 2900 | 62 | 13 | 25 |
| Tunisia | 15536 | 4952 | 3103 | 676 | 6805 | 443 | 9 | 3 | 89 |
| Uganda | 19965 | 6770 | 1800 | 5500 | 5895 | 3099 | 32 | 8 | 60 |
| Zambia | 74339 | 5273 | 30000 | 28700 | 10366 | 12267 | 16 | 7 | 77 |
| Zimbabwe | 38685 | 2876 | 4856 | 8800 | 22153 | 1776 | 14 | 7 | 79 |

0 = zero or less than half the unit of measure; X = not available; total water withdrawals may exceed 100% due to groundwater drawdowns or river inflows.

***Table D-1:*** *Africa region (continued).*

| | Agriculture/Food | | | | | | | | | |
|---|---|---|---|---|---|---|---|---|---|---|
| **Country** | Irrigated Land (1000 ha) 1993 | Agricultural Labor Force (1000s) 1993 | Total Labor Force (1000s) 1993 | Cattle Stocks (1000s) 1994 | Sheep Stocks (1000s) 1994 | Goat Stocks (1000s) 1994 | Pig Stocks (1000s) 1994 | Equines (horses/ mules/asses) (1000s) 1994 | Buffalo Stocks (1000s) 1994 | Camel Stocks (1000s) 1994 |
| Algeria | 555 | 1468 | 6441 | 1370 | 17850 | 2820 | 6 | 508 | X | 125 |
| Angola | 75 | 2687 | 3918 | 3280 | 255 | 1570 | 820 | 6 | X | X |
| Benin | 10 | 1360 | 2321 | 1223 | 940 | 1198 | 555 | 7 | X | X |
| Botswana | 2 | 287 | 473 | 2800 | 344 | 2475 | 17 | 193 | X | X |
| Burkina Faso | 20 | 4199 | 5020 | 4261 | 5686 | 7400 | 551 | 468 | X | 12 |
| Burundi | 14 | 2752 | 3032 | 380 | 350 | 850 | 80 | 0 | X | X |
| Cameroon | 21 | 2754 | 4717 | 4867 | 3770 | 3767 | 1380 | 51 | X | X |
| Central African Rep | X | 877 | 1475 | 2800 | 152 | 1340 | 480 | 0 | X | X |
| Chad | 14 | 1458 | 2041 | 4621 | 2152 | 3178 | 17 | 467 | X | 593 |
| Congo | 1 | 533 | 908 | 68 | 111 | 305 | 56 | 0 | X | X |
| Cote d'Ivoire | 68 | 2616 | 4964 | 1232 | 1251 | 976 | 404 | 0 | X | X |
| Dem. Republic of Congo | 10 | 9312 | 14510 | 1696 | 1012 | 4317 | 1185 | 0 | X | X |
| Djibouti | X | X | X | 190 | 470 | 507 | X | 8 | X | 62 |
| Egypt | 3246 | 6119 | 15674 | 3070 | 3382 | 3210 | 27 | 1661 | 3200 | 133 |
| Equatorial Guinea | X | 79 | 152 | 5 | 36 | 8 | 5 | 0 | X | X |
| Eritrea | 28 | X | X | 1550 | 1510 | 1400 | X | 0 | X | 69 |
| Ethiopia | 0 | 16250 | 22320 | 29450 | 5000 | 16700 | 20 | 630 | X | 1000 |
| Gabon | 4 | 355 | 546 | 38 | 170 | 83 | 165 | 0 | X | X |
| Gambia | 15 | 332 | 414 | 414 | 121 | 150 | 11 | 46 | X | X |
| Ghana | 6 | 2865 | 5961 | 1680 | 3288 | 3337 | 595 | 15 | X | X |
| Guinea | 93 | 1903 | 2650 | 1658 | 435 | 460 | 33 | 3 | X | X |
| Guinea-Bissau | 17 | 361 | 467 | 494 | 263 | 276 | 312 | 7 | X | X |
| Kenya | 66 | 8252 | 10908 | 11000 | 5500 | 7438 | 107 | 2 | X | 815 |
| Lesotho | 3 | 682 | 884 | 663 | 1691 | 1010 | 78 | 292 | X | X |
| Liberia | 2 | 702 | 1025 | 36 | 210 | 220 | 120 | 0 | X | X |
| Libya | 470 | 162 | 1252 | 50 | 3500 | 600 | X | 70 | X | 120 |
| Madagascar | 1087 | 4169 | 5556 | 10288 | 740 | 1300 | 1558 | 0 | X | X |
| Malawi | 28 | 3084 | 4267 | 980 | 196 | 890 | 245 | 2 | X | X |
| Mali | 1 | 2524 | 3187 | 5542 | 5173 | 7380 | 63 | 712 | X | 260 |
| Mauritania | 49 | 445 | 705 | 1011 | 4800 | 3100 | X | 173 | X | 1000 |
| Morocco | 1258 | 2868 | 8441 | 2431 | 15594 | 4431 | 10 | 1555 | X | 36 |
| Mozambique | 118 | 6225 | 7716 | 1250 | 119 | 389 | 174 | 20 | X | X |
| Namibia | 6 | 154 | 472 | 2036 | 2620 | 1639 | 18 | 137 | X | X |
| Niger | 66 | 3645 | 4241 | 1986 | 3700 | 5900 | 39 | 532 | X | 370 |
| Nigeria | 957 | 28219 | 44279 | 16717 | 14455 | 25497 | 6926 | 1237 | X | 18 |
| Reunion | 6 | 25 | 248 | 25 | 2 | 32 | 86 | 0 | X | X |
| Rwanda | 4 | 3348 | 3690 | 610 | 400 | 1100 | 130 | 0 | X | X |
| Senegal | 71 | 2613 | 3362 | 2800 | 4600 | 3200 | 329 | 879 | X | 16 |
| Sierra Leone | 29 | 910 | 1518 | 362 | 302 | 168 | 51 | 0 | X | X |
| Somalia | 180 | 2539 | 3714 | 5000 | 13000 | 12000 | 9 | 46 | X | 6000 |
| South Africa | 1270 | 1886 | 14716 | 12584 | 29134 | 6402 | 1511 | 454 | X | X |
| Sudan | 1946 | 5066 | 8943 | 21751 | 22870 | 16449 | X | 699 | X | 2856 |
| Swaziland | 67 | 217 | 340 | 620 | 27 | 434 | 32 | 13 | X | X |
| Tanzania | 150 | 10688 | 13495 | 13376 | 3955 | 9682 | 335 | 178 | X | X |
| Togo | 7 | 1054 | 1538 | 250 | 1250 | 2048 | 934 | 6 | X | X |
| Tunisia | 385 | 625 | 2917 | 660 | 7100 | 1420 | 6 | 367 | X | 231 |
| Uganda | 9 | 6611 | 8361 | 5100 | 1980 | 3350 | 880 | 17 | X | X |
| Zambia | 46 | 1972 | 2918 | 3300 | 69 | 620 | 295 | 2 | X | X |
| Zimbabwe | 193 | 2818 | 4228 | 4500 | 550 | 2530 | 280 | 129 | X | X |

0 = zero or less than half the unit of measure; X = not available.

***Table D-1:** Africa region (continued).*

| Country | Energy: Total Commercial Energy Consumption (PJ) 1993 | Energy: Traditional Fuel Consumption (TJ) 1991 | Energy: Commercial Hydroelectric Production (PJ) 1993 | Biodiversity: Known Mammal Species (#) 1990s | Biodiversity: Endemic Mammal Species (#) 1990s | Biodiversity: Known Bird Species (#) 1990s | Biodiversity: Endemic Bird Species (#) 1990s | Biodiversity: Known Plant Species (#) 1990s | Biodiversity: Endemic Plant Species (#) 1990s |
|---|---|---|---|---|---|---|---|---|---|
| Algeria | 1183 | 19076 | 1.27 | 92 | 2 | 375 | 1 | 3100 | 250 |
| Angola | 26 | 56054 | 4.95 | 276 | 7 | 909 | 13 | 5000 | 1260 |
| Benin | 7 | 48223 | 0.00 | 188 | 0 | 423 | 0 | 2000 | X |
| Botswana | X | 13197 | 0.00 | 164 | 0 | 550 | 0 | X | 17 |
| Burkina Faso | 8 | 84547 | 0.00 | 147 | 0 | 453 | 0 | 1100 | X |
| Burundi | 3 | 43522 | 0.41 | 107 | 0 | 596 | 0 | 2500 | X |
| Cameroon | 36 | 114320 | 9.53 | 297 | 13 | 874 | 8 | 8000 | 156 |
| Central African Rep | 3 | 33751 | 0.28 | 209 | 2 | 662 | 0 | 3600 | 100 |
| Chad | 1 | 35462 | 0.00 | 134 | 1 | 532 | 0 | 1600 | X |
| Congo | 24 | 21852 | 1.54 | 200 | 2 | 569 | 0 | 4350 | 1200 |
| Cote d'Ivoire | 109 | 102843 | 3.95 | 230 | 1 | 694 | 0 | 3517 | 62 |
| Dem. Republic of Congo | 73 | 365076 | 21.68 | 415 | 28 | 1096 | 22 | 11000 | 1100 |
| Djibouti | 18 | 0 | 0.00 | X | X | X | X | X | X |
| Egypt | 1226 | 45365 | 30.67 | 98 | 7 | 439 | 0 | 2066 | 70 |
| Equatorial Guinea | 2 | 4367 | 0.01 | 184 | 3 | 322 | 3 | 3000 | 66 |
| Eritrea | X | 0 | 0.00 | 112 | 0 | 537 | 0 | X | X |
| Ethiopia | 45 | 414055 | 4.09 | 255 | 31 | 813 | 28 | 6500 | 1000 |
| Gabon | 32 | 26415 | 2.56 | 190 | 2 | 629 | 0 | 6500 | X |
| Gambia | 3 | 8939 | 0.00 | 108 | 0 | 504 | 0 | 966 | X |
| Ghana | 67 | 151523 | 22.01 | 222 | 1 | 725 | 1 | 3600 | 43 |
| Guinea | 15 | 35275 | 0.68 | 190 | 1 | 552 | 0 | 3000 | 88 |
| Guinea-Bissau | 3 | 4124 | 0.00 | 108 | 0 | 319 | 0 | 1000 | 12 |
| Kenya | 90 | 344388 | 10.77 | 359 | 21 | 1068 | 6 | 6000 | 265 |
| Lesotho | X | 6486 | 0.00 | 33 | 0 | 281 | 0 | 1576 | 2 |
| Liberia | 5 | 48492 | 0.63 | 193 | 0 | 581 | 1 | 2200 | 103 |
| Libya | 457 | 5234 | 0.00 | 76 | 5 | 323 | 0 | 1800 | 134 |
| Madagascar | 15 | 75959 | 1.25 | 105 | 77 | 253 | 103 | 9000 | 6500 |
| Malawi | 11 | 132511 | 2.80 | 195 | 0 | 645 | 0 | 3600 | 49 |
| Mali | 7 | 53504 | 0.76 | 137 | 0 | 622 | 0 | 1741 | 11 |
| Mauritania | 39 | 79 | 0.09 | 61 | 1 | 541 | 0 | 1100 | X |
| Morocco | 297 | 13725 | 1.59 | 105 | 4 | 416 | 0 | 3600 | 625 |
| Mozambique | 14 | 147370 | 0.18 | 179 | 1 | 678 | 0 | 5500 | 219 |
| Namibia | X | 0 | 0.00 | 154 | 3 | 609 | 1 | 3128 | X |
| Niger | 15 | 46886 | 0.00 | 131 | 0 | 482 | 0 | 1170 | X |
| Nigeria | 705 | 1009506 | 11.52 | 274 | 6 | 862 | 2 | 4614 | 205 |
| Reunion | 23 | 5167 | 1.79 | X | X | X | X | X | X |
| Rwanda | 7 | 52771 | 0.83 | 151 | 0 | 666 | 0 | 2288 | 26 |
| Senegal | 38 | 49439 | 0.00 | 155 | 0 | 610 | 0 | 2062 | 26 |
| Sierra Leone | 6 | 29662 | 0.00 | 147 | 0 | 622 | 0 | 2090 | 74 |
| Somalia | X | 71411 | 0.00 | 171 | 11 | 649 | 10 | 3000 | 500 |
| South Africa | 3578 | 131424 | 2.97 | 247 | 27 | 790 | 7 | 23000 | X |
| Sudan | 48 | 220068 | 3.38 | 267 | 11 | 937 | 0 | 3132 | 50 |
| Swaziland | X | 18487 | 0.00 | 47 | 0 | 485 | 0 | 2636 | 4 |
| Tanzania | 30 | 329752 | 2.25 | 322 | 14 | 1005 | 19 | 10000 | 1122 |
| Togo | 9 | 10352 | 0.02 | 196 | 1 | 558 | 0 | 2000 | X |
| Tunisia | 218 | 30791 | 0.23 | 78 | 1 | 356 | 0 | 2150 | X |
| Uganda | 16 | 136987 | 2.81 | 338 | 6 | 992 | 3 | 5000 | X |
| Zambia | 51 | 129891 | 27.88 | 229 | 3 | 736 | 1 | 4600 | 211 |
| Zimbabwe | 208 | 69940 | 6.09 | 270 | 1 | 648 | 0 | 4200 | 95 |

0 = zero or less than half the unit of measure; X = not available; all plant species includes flowering plants only.

***Table D-2:** Australasia region.*

| Country | Population and Human Development: Total Pop. (1000s) 1995 | Current Pop. Density (persons/km²) 1995 | Projected Pop. Density (persons/km²) 2025 | Total Urban Pop. (1000s) 1995 | Urban Pop. in Coastal Cities (1000s) 1980 | Economic Conditions: GDP per Capita in Constant PPP ('85 IN$) 1992 | GDP from Agri-culture (%) 1993 | GDP from Industry (%) 1993 | GDP from Services (%) 1993 | GDP Annual Growth Rate (%) 1993 |
|---|---|---|---|---|---|---|---|---|---|---|
| Australia | 18088 | 2 | 3 | 15318 | 10568 | 14458 | X | X | X | X |
| New Zealand | 3575 | 13 | 16 | 3077 | 2279 | 11363 | X | X | X | X |

| Country | Land Cover/Use: Total Land Area (1000 ha) 1993 | Arable and Permanent Cropland (1000 ha) 1993 | Permanent Pasture (1000 ha) 1993 | Forest and Woodland (1000 ha) 1993 | Other Land (1000 ha) 1993 | Water: Water Resources per Capita ($m^3$) 1995 | Domestic Annual Withdraws (%) 1995 | Industry Annual Withdraws (%) 1995 | Agriculture Annual Withdraws (%) 1995 |
|---|---|---|---|---|---|---|---|---|---|
| Australia | 764444 | 46486 | 413800 | 145000 | 159158 | 18963 | 65 | 2 | 33 |
| New Zealand | 26799 | 3800 | 13500 | 7380 | 2119 | 91469 | 46 | 10 | 44 |

| Country | Agriculture/Food: Irrigated Land (1000 ha) 1993 | Agricultural Labor Force (1000s) 1993 | Total Labor Force (1000s) 1993 | Cattle Stocks (1000s) 1994 | Sheep Stocks (1000s) 1994 | Goat Stocks (1000s) 1994 | Pig Stocks (1000s) 1994 | Equines (horses/mules/asses) (1000s) 1994 | Buffalo Stocks (1000s) 1994 | Camel Stocks (1000s) 1994 |
|---|---|---|---|---|---|---|---|---|---|---|
| Australia | 2107 | 387 | 8584 | 24732 | 132609 | 241 | 2740 | 274 | X | X |
| New Zealand | 285 | 132 | 1546 | 8550 | 50135 | 484 | 430 | 80 | X | X |

| Country | Energy: Total Commercial Energy Consumption (PJ) 1993 | Traditional Fuel Consumption (TJ) 1991 | Commercial Hydroelectric Production (PJ) 1993 | Biodiversity: Known Mammal Species (#) 1990s | Endemic Mammal Species (#) 1990s | Known Bird Species (#) 1990s | Endemic Bird Species (#) 1990s | Known Plant Species (#) 1990s | Endemic Plant Species (#) 1990s |
|---|---|---|---|---|---|---|---|---|---|
| Australia | 3917 | 109059 | 136.87 | 252 | 198 | 751 | 353 | 15000 | 14074 |
| New Zealand | 565 | 489 | 76.58 | 10 | 4 | 287 | 76 | 2160 | 1942 |

0 = zero or less than half the unit of measure; X = not available; all plant species includes flowering plants only.

***Table D-3:** Europe region.*

| Country | Total Pop. (1000s) 1995 | Current Pop. Density (persons/ km²) 1995 | Projected Pop. Density (persons/ km²) 2025 | Total Urban Pop. (1000s) 1995 | Urban Pop. in Coastal Cities (1000s) 1980 | GDP per Capita in Constant PPP ('85 IN$) 1992 | GDP from Agri- culture (%) 1993 | GDP from Industry (%) 1993 | GDP from Services (%) 1993 | GDP Annual Growth Rate (%) 1993 |
|---|---|---|---|---|---|---|---|---|---|---|
| | **Population and Human Development** | | | | | **Economic Conditions** | | | | |
| Albania | 3441 | 120 | 162 | 1285 | 622 | X | X | X | X | X |
| Andorra | X | X | X | X | X | X | X | X | X | X |
| Armenia | 3599 | 121 | 159 | 2473 | X | X | 48.00 | 29.84 | 22.16 | -14.80 |
| Austria | 7968 | 95 | 99 | 4424 | X | 12955 | X | X | X | X |
| Azerbaijan | 7558 | 87 | 117 | 4216 | X | X | X | X | X | -13.00 |
| Belarus | 10141 | 49 | 48 | 7215 | X | X | 16.86 | 53.73 | 29.41 | -9.50 |
| Belgium | 10113 | 331 | 341 | 9809 | 1968 | 13484 | X | X | X | X |
| Belgium/Luxembourg | X | X | X | X | X | X | X | X | X | X |
| Bosnia and Herzegovina | 3459 | 68 | 88 | 1695 | X | X | X | X | X | X |
| Bulgaria | 8769 | 79 | 70 | 6201 | 857 | 5208 | 12.98 | 38.16 | 48.86 | -4.70 |
| Croatia | 4495 | 80 | 75 | 2896 | X | X | 11.50 | 30.02 | 58.48 | X |
| Czech Republic | 10296 | 131 | 135 | 6736 | X | X | 6.17 | 39.95 | 53.87 | X |
| Denmark | 5181 | 120 | 118 | 4414 | 3980 | 14091 | X | X | X | X |
| Estonia | 1530 | 34 | 32 | 1118 | X | X | 8.08 | 29.17 | 62.75 | X |
| Finland | 5107 | 15 | 16 | 3225 | 1539 | 12000 | X | X | X | X |
| France | 57981 | 105 | 111 | 42203 | 9380 | 13918 | X | X | X | X |
| Georgia | 5457 | 78 | 88 | 3190 | X | X | 58.03 | 21.79 | 20.18 | -40.00 |
| Germany | 81591 | 229 | 214 | 70616 | X | 14709 | X | X | X | X |
| Greece | 10451 | 79 | 75 | 6817 | 5252 | X | X | X | X | -0.10 |
| Hungary | 10115 | 109 | 101 | 6541 | X | 4645 | 5.60 | 28.40 | 65.99 | -3.30 |
| Iceland | 269 | 3 | 3 | 246 | 186 | 12618 | X | X | X | X |
| Ireland | 3553 | 51 | 55 | 2043 | 1766 | 9637 | X | X | X | X |
| Italy | 57187 | 190 | 174 | 38101 | 21232 | 12721 | X | X | X | X |
| Latvia | 2557 | 40 | 36 | 1863 | X | X | 14.90 | 31.61 | 53.49 | -9.50 |
| Liechtenstein | X | X | X | X | X | X | X | X | X | X |
| Lithuania | 3700 | 57 | 59 | 2667 | X | X | 20.54 | 41.47 | 38.00 | -16.20 |
| Luxembourg | 406 | 157 | 170 | 361 | X | 16798 | X | X | X | X |
| Macedonia | 2163 | 84 | 100 | 1294 | X | X | X | X | X | X |
| Moldova | 4432 | 132 | 152 | 2293 | X | X | 34.59 | 47.69 | 17.72 | -14.00 |
| Monaco | X | X | X | X | X | X | X | X | X | X |
| Netherlands | 15503 | 380 | 398 | 13801 | 7764 | 13281 | X | X | X | X |
| Norway | 4337 | 13 | 15 | 2667 | 2324 | 15518 | X | X | X | X |
| Poland | 38388 | 119 | 129 | 24853 | 1842 | 3826 | 6.33 | 38.59 | 55.08 | 4.00 |
| Portugal | 9823 | 106 | 105 | 3496 | 2352 | X | X | X | X | -0.50 |
| Romania | 22835 | 96 | 92 | 12650 | 573 | X | 20.51 | 39.88 | 39.61 | 1.20 |
| Russian Federation | 147000 | 9 | 8 | 111736 | 18372 | X | X | X | X | -12.00 |
| San Marino | X | X | X | X | X | X | X | X | X | X |
| Slovak Republic | 5353 | 109 | 123 | 3146 | X | X | 6.69 | 43.90 | 49.41 | -4.40 |
| Slovenia | 1946 | 96 | 90 | 1236 | X | X | 5.64 | 36.04 | 58.32 | X |
| Spain | 39621 | 78 | 74 | 30292 | 13903 | 9802 | X | X | X | X |
| Sweden | 8780 | 20 | 22 | 7296 | 4018 | 13986 | X | X | X | X |
| Switzerland | 7202 | 174 | 189 | 4379 | X | 15887 | X | X | X | X |
| Ukraine | 51380 | 85 | 81 | 36099 | X | X | 34.72 | 47.15 | 18.13 | -18.50 |
| United Kingdom | 58258 | 239 | 252 | 52119 | 26765 | 12724 | X | X | X | X |
| Yugoslavia | 10849 | 106 | 112 | 6134 | 1236 | X | X | X | X | X |

0 = zero or less than half the unit of measure; X = not available; shaded cells contain data originally attributed to former country identity (e.g., former Yugoslavia) or data for more than one country (e.g., Belgium/Luxembourg).

***Table D-3:** Europe region (continued).*

| | **Land Cover/Use** | | | | | **Water** | | | |
|---|---|---|---|---|---|---|---|---|---|
| **Country** | Total Land Area (1000 ha) 1993 | Arable and Permanent Cropland (1000 ha) 1993 | Permanent Pasture (1000 ha) 1993 | Forest and Woodland (1000 ha) 1993 | Other Land (1000 ha) 1993 | Water Resources per Capita ($m^3$) 1995 | Domestic Annual Withdraws (%) 1995 | Industry Annual Withdraws (%) 1995 | Agriculture Annual Withdraws (%) 1995 |
| Albania | 2740 | 702 | 424 | 1048 | 566 | 6190 | 6 | 18 | 76 |
| Andorra | X | X | X | X | X | X | X | X | X |
| Armenia | 2840 | X | X | X | X | 3687 | 13 | 15 | 72 |
| Austria | 8273 | 1498 | 1954 | 3240 | 1581 | 11333 | 33 | 58 | 9 |
| Azerbaijan | 8610 | 2000 | 2200 | 950 | X | 4364 | 4 | 22 | 74 |
| Belarus | 20760 | 6248 | 3106 | 7000 | X | 7277 | 32 | 49 | 19 |
| Belgium | 3282 | X | X | X | X | 1236 | 11 | 85 | 4 |
| Belgium/Luxembourg | 3282 | 794 | 688 | 700 | 1100 | X | X | X | X |
| Bosnia and Herzegovina | 5100 | 940 | 1000 | 2000 | 1160 | X | X | X | X |
| Bulgaria | 11055 | 4310 | 1811 | 3877 | 1057 | 23378 | 3 | 76 | 22 |
| Croatia | 5592 | 1313 | 1093 | 2100 | 1086 | 13660 | X | X | X |
| Czech Republic | 7728 | 3293 | 873 | 2629 | 933 | 5653 | 41 | 57 | 2 |
| Denmark | 4243 | 2542 | 197 | 445 | 1059 | 2509 | 30 | 27 | 43 |
| Estonia | 4227 | 1143 | 313 | 2022 | 749 | 11490 | 5 | 92 | 3 |
| Finland | 30461 | 2580 | 106 | 23186 | 4589 | 22126 | 12 | 85 | 3 |
| France | 55010 | 19439 | 10764 | 14931 | 9876 | 3415 | 16 | 69 | 15 |
| Georgia | 6970 | 1000 | 2000 | 2700 | X | 11942 | 21 | 37 | 42 |
| Germany | 34927 | 12116 | 5251 | 10700 | 17560 | 2096 | 11 | 70 | 20 |
| Greece | 12890 | 3494 | 5250 | 2620 | 1526 | 5612 | 8 | 29 | 63 |
| Hungary | 9234 | 4973 | 1157 | 1764 | 1340 | 11864 | 9 | 55 | 36 |
| Iceland | 10025 | 6 | 2274 | 120 | 7625 | 624535 | 31 | 63 | 6 |
| Ireland | 6889 | 923 | 4690 | 320 | 956 | 14073 | 16 | 74 | 10 |
| Italy | 29406 | 11860 | 4300 | 6770 | 6476 | 2920 | 14 | 27 | 59 |
| Latvia | 6205 | 1711 | 803 | 2839 | 852 | 13297 | 42 | 44 | 14 |
| Liechtenstein | X | X | X | X | X | X | X | X | X |
| Lithuania | 4551 | 3008 | 460 | 2000 | X | 6541 | 7 | 90 | 3 |
| Luxembourg | X | X | X | X | X | X | X | X | X |
| Macedonia | 2543 | 663 | 634 | 1000 | 246 | X | X | X | X |
| Moldova | 3297 | 2193 | 421 | 421 | 262 | 3093 | 7 | 70 | 23 |
| Monaco | X | X | X | X | X | X | X | X | X |
| Netherlands | 3392 | 934 | 1051 | 350 | 1057 | 5805 | 5 | 61 | 34 |
| Norway | 30683 | 890 | 123 | 8330 | 21340 | 90385 | 20 | 72 | 8 |
| Poland | 30442 | 14668 | 4047 | 8785 | 2942 | 1464 | 13 | 76 | 11 |
| Portugal | 9195 | 3160 | 840 | 3300 | 1895 | 7085 | 15 | 37 | 48 |
| Romania | 23034 | 9941 | 4852 | 6682 | 1559 | 9109 | 8 | 33 | 59 |
| Russian Federation | 1699580 | 133900 | 76200 | 778500 | X | 30599 | 17 | 60 | 23 |
| San Marino | X | X | X | X | X | X | X | X | X |
| Slovak Republic | 4808 | 1613 | 835 | 1991 | 369 | 5753 | X | X | X |
| Slovenia | 2012 | 301 | 558 | 1020 | 133 | X | X | X | X |
| Spain | 49944 | 19656 | 10300 | 16137 | 3851 | 2809 | 12 | 26 | 62 |
| Sweden | 41162 | 2780 | 576 | 28000 | 9806 | 20501 | 36 | 55 | 9 |
| Switzerland | 3955 | 467 | 1114 | 1252 | 1122 | 6943 | 23 | 73 | 4 |
| Ukraine | 57935 | 34417 | 7473 | 10331 | 5714 | 4496 | 16 | 54 | 30 |
| United Kingdom | 24160 | 6127 | 11048 | 2438 | 4547 | 1219 | 20 | 77 | 3 |
| Yugoslavia | X | X | X | 2700 | X | X | X | X | X |

0 = zero or less than half the unit of measure; X = not available; shaded cells contain data originally attributed to former country identity (e.g., former Yugoslavia) or data for more than one country (e.g., Belgium/Luxembourg); total water withdrawals may exceed 100% due to groundwater drawdowns or river inflows.

***Table D-3:** Europe region (continued).*

| | Agriculture/Food | | | | | | | | | |
|---|---|---|---|---|---|---|---|---|---|---|
| **Country** | Irrigated Land (1000 ha) 1993 | Agricultural Labor Force (1000s) 1993 | Total Labor Force (1000s) 1993 | Cattle Stocks (1000s) 1994 | Sheep Stocks (1000s) 1994 | Goat Stocks (1000s) 1994 | Pig Stocks (1000s) 1994 | Equines (horses/ mules/asses) (1000s) 1994 | Buffalo Stocks (1000s) 1994 | Camel Stocks (1000s) 1994 |
| Albania | 341 | 756 | 1631 | 630 | 1900 | 1280 | 86 | 196 | 2 | X |
| Andorra | X | X | X | X | X | X | X | X | X | X |
| Armenia | X | X | X | X | 720 | 16 | 80 | 12 | X | X |
| Austria | 4 | 190 | 3777 | 2430 | 324 | 40 | 3800 | 65 | X | X |
| Azerbaijan | 1000 | X | X | 1621 | 4339 | 200 | 115 | 35 | 10 | 30 |
| Belarus | 100 | X | X | 5851 | 289 | 30 | 4175 | 223 | 20 | X |
| Belgium | X | X | X | X | X | X | X | X | X | X |
| Belgium/Luxembourg | 1 | 68 | 4314 | 3289 | 160 | 9 | 6948 | 21 | X | X |
| Bosnia and Herzegovina | 2 | X | X | 390 | 600 | X | 223 | 50 | 1 | X |
| Bulgaria | 1237 | 485 | 4397 | 750 | 3763 | 676 | 2071 | 433 | 17 | X |
| Croatia | 3 | X | X | 519 | 444 | 108 | 1347 | 32 | X | X |
| Czech Republic | 24 | 709 | 8468 | 2113 | 196 | 45 | 4071 | 18 | X | X |
| Denmark | 435 | 119 | 2887 | 2082 | 82 | X | 10864 | 17 | X | X |
| Estonia | X | X | X | 463 | 83 | 0 | 424 | 5 | X | X |
| Finland | 64 | 185 | 2562 | 1230 | 79 | 6 | 1300 | 49 | X | X |
| France | 1485 | 1182 | 26026 | 20112 | 10452 | 1055 | 13383 | 370 | X | X |
| Georgia | 400 | X | X | 1050 | 1300 | 85 | 650 | 20 | 20 | X |
| Germany | 475 | 1662 | 40268 | 15891 | 2360 | 88 | 26044 | 530 | X | X |
| Greece | 1314 | 904 | 4008 | 608 | 9604 | 5557 | 1143 | 200 | 1 | X |
| Hungary | 206 | 518 | 5186 | 1002 | 1280 | 36 | 5002 | 76 | X | X |
| Iceland | X | 9 | 141 | 77 | 500 | X | 22 | 82 | X | X |
| Ireland | X | 166 | 1347 | 6308 | 5991 | 9 | 1487 | 68 | X | X |
| Italy | 2710 | 1429 | 23748 | 7683 | 10370 | 1346 | 8200 | 388 | 92 | X |
| Latvia | X | X | X | 995 | 133 | 5 | 737 | 28 | X | X |
| Liechtenstein | X | X | X | X | X | X | X | X | X | X |
| Lithuania | X | X | X | 1650 | 48 | 9 | 1200 | 78 | X | X |
| Luxembourg | X | X | X | X | X | X | X | 0 | X | X |
| Macedonia | 83 | X | X | 276 | 2444 | X | 181 | 62 | 1 | X |
| Moldova | 311 | X | X | 916 | 1373 | 72 | 1165 | 57 | X | X |
| Monaco | X | X | X | X | X | X | X | X | X | X |
| Netherlands | 560 | 204 | 6314 | 4629 | 2198 | 13 | 13991 | 66 | X | X |
| Norway | 97 | 99 | 2190 | 1003 | 2316 | 89 | 745 | 22 | X | X |
| Poland | 100 | 3710 | 19674 | 7696 | 870 | X | 19466 | 721 | X | X |
| Portugal | 630 | 644 | 4540 | 1322 | 5991 | 836 | 1487 | 275 | X | X |
| Romania | 3102 | 2118 | 11848 | 3597 | 11499 | 776 | 9262 | 785 | 0 | X |
| Russian Federation | 4000 | 16389 | 144909 | 48900 | 41078 | 2622 | 28600 | 2520 | 140 | 10 |
| San Marino | X | X | X | X | X | X | X | X | X | X |
| Slovak Republic | 80 | X | X | 916 | 397 | 12 | 2179 | 0 | X | X |
| Slovenia | 2 | X | X | 504 | 21 | 8 | 620 | 0 | X | X |
| Spain | 3453 | 1380 | 14821 | 5000 | 23838 | 2739 | 18188 | 412 | X | X |
| Sweden | 115 | 153 | 4442 | 1830 | 483 | X | 2168 | 86 | X | X |
| Switzerland | 25 | 122 | 3458 | 1700 | 425 | 52 | 1680 | 57 | X | X |
| Ukraine | 2605 | X | X | 21607 | 6118 | 745 | 15298 | 731 | X | X |
| United Kingdom | 108 | 523 | 28567 | 11735 | 29300 | X | 7910 | 184 | X | X |
| Yugoslavia | 0 | 2104 | 10955 | 1809 | 2752 | X | 4000 | 82 | 19 | X |

0 = zero or less than half the unit of measure; X = not available; shaded cells contain data originally attributed to former country identity (e.g., former Yugoslavia) or data for more than one country (e.g., Belgium/Luxembourg).

***Table D-3:** Europe region (continued).*

| | Energy | | | Biodiversity | | | | | |
|---|---|---|---|---|---|---|---|---|---|
| Country | Total Commercial Energy Consumption (PJ) 1993 | Traditional Fuel Consumption (TJ) 1991 | Commercial Hydroelectric Production (PJ) 1993 | Known Mammal Species (#) 1990s | Endemic Mammal Species (#) 1990s | Known Bird Species (#) 1990s | Endemic Bird Species (#) 1990s | Known Plant Species (#) 1990s | Endemic Plant Species (#) 1990s |
| Albania | 43 | 15199 | 11.99 | 68 | 0 | 306 | 0 | 2965 | 24 |
| Andorra | X | X | X | X | X | X | X | X | X |
| Armenia | 49 | 0 | 10.80 | X | 3 | X | 0 | X | X |
| Austria | 966 | 29615 | 137.00 | 83 | 0 | 414 | 0 | 2950 | 35 |
| Azerbaijan | 546 | 0 | 8.73 | X | 0 | X | 0 | X | X |
| Belarus | 1249 | 0 | 0.07 | X | 0 | X | 0 | X | X |
| Belgium | 1976 | 5595 | 3.67 | 58 | 0 | 429 | 0 | 1400 | 1 |
| Belgium/Luxembourg | X | 0 | 0.00 | X | X | X | X | X | X |
| Bosnia and Herzegovina | 29 | 0 | 14.40 | X | 0 | X | 0 | X | X |
| Bulgaria | 965 | 12846 | 6.99 | 81 | 0 | 374 | 0 | 3505 | 320 |
| Croatia | 263 | 0 | 15.64 | X | 0 | X | 0 | X | X |
| Czech Republic | 1659 | 0 | 5.75 | X | 0 | X | 0 | X | X |
| Denmark | 762 | 4871 | 0.10 | 43 | 0 | 439 | 0 | 1200 | 1 |
| Estonia | 214 | 0 | 0.00 | 65 | 0 | 330 | 0 | 1630 | X |
| Finland | 1014 | 29803 | 48.96 | 60 | 0 | 425 | 0 | 1040 | X |
| France | 9153 | 101044 | 244.42 | 93 | 0 | 506 | 9 | 4500 | 133 |
| Georgia | 159 | 0 | 23.40 | X | 2 | X | 0 | X | X |
| Germany | 13724 | 0 | 77.27 | 76 | 0 | 503 | 0 | 2600 | 6 |
| Greece | 989 | 13218 | 9.15 | 95 | 2 | 398 | 0 | 4900 | 742 |
| Hungary | 990 | 23681 | 0.60 | 72 | 0 | 363 | 0 | 2148 | 38 |
| Iceland | 54 | 0 | 16.07 | 11 | 0 | 316 | 0 | 340 | 1 |
| Ireland | 428 | 489 | 3.64 | 25 | 0 | 417 | 0 | 892 | X |
| Italy | 6749 | 48428 | 160.14 | 90 | 3 | 490 | 0 | 5463 | 712 |
| Latvia | 187 | 0 | 10.35 | 83 | 0 | 325 | 0 | 1153 | X |
| Liechtenstein | X | X | X | X | X | X | X | X | X |
| Lithuania | 368 | 0 | 1.41 | 68 | 0 | 305 | 0 | 1200 | X |
| Luxembourg | 160 | 196 | 1.67 | X | X | X | X | X | X |
| Macedonia | 139 | 0 | 3.24 | X | 0 | X | 0 | X | X |
| Moldova | 234 | 0 | 1.35 | 68 | 0 | 270 | 0 | X | X |
| Monaco | X | X | X | X | X | X | X | X | X |
| Netherlands | 3306 | 2298 | 0.33 | 55 | 0 | 456 | 0 | 1170 | X |
| Norway | 904 | 9449 | 430.24 | 54 | 0 | 453 | 0 | 1650 | 1 |
| Poland | 4056 | 27819 | 12.87 | 79 | 0 | 421 | 0 | 2300 | 3 |
| Portugal | 603 | 5639 | 31.45 | 63 | 1 | 441 | 2 | 2500 | 150 |
| Romania | 1762 | 19364 | 45.96 | 84 | 0 | 368 | 0 | 3175 | 41 |
| Russian Federation | 30042 | 0 | 630.63 | X | X | X | X | X | X |
| San Marino | X | X | X | X | X | X | X | X | X |
| Slovak Republic | 672 | 0 | 10.89 | X | X | X | X | X | X |
| Slovenia | 194 | 0 | 10.88 | 69 | 0 | 361 | 0 | X | X |
| Spain | 3359 | 18420 | 92.80 | 82 | 4 | 506 | 5 | X | X |
| Sweden | 1660 | 122254 | 271.37 | 60 | 0 | 463 | 0 | 4916 | 941 |
| Switzerland | 985 | 14481 | 131.82 | 75 | 0 | 400 | 0 | 1650 | 1 |
| Ukraine | 8058 | 0 | 40.45 | X | 1 | X | 0 | 2927 | 1 |
| United Kingdom | 1039 | 4150 | 20.47 | 50 | 0 | 590 | 1 | 1550 | 16 |
| Yugoslavia | 381 | 29554 | 36.05 | X | X | X | X | X | X |

0 = zero or less than half the unit of measure; X = not available; shaded cells contain data originally attributed to former country identity (e.g., former Yugoslavia) or data for more than one country (e.g., Belgium/Luxembourg); all plant species includes flowering plants only.

***Table D-4:*** *Latin America region.*

| | Population and Human Development | | | | | Economic Conditions | | | | |
|---|---|---|---|---|---|---|---|---|---|---|
| **Country** | Total Pop. (1000s) 1995 | Current Pop. Density (persons/km²) 1995 | Projected Pop. Density (persons/km²) 2025 | Total Urban Pop. (1000s) 1995 | Urban Pop. in Coastal Cities (1000s) 1980 | GDP per Capita in Constant PPP ('85 IN$) 1992 | GDP from Agri-culture (%) 1993 | GDP from Industry (%) 1993 | GDP from Services (%) 1993 | GDP Annual Growth Rate (%) 1993 |
| Argentina | 34587 | 13 | 17 | 30463 | 12273 | X | 5.99 | 30.68 | 63.33 | 6.00 |
| Belize | 215 | 9 | 17 | 101 | X | 4253 | 19.37 | 27.80 | 52.82 | 4.20 |
| Bolivia | 7414 | 7 | 12 | 4505 | X | 1721 | X | X | X | 4.00 |
| Brazil | 161790 | 19 | 27 | 126599 | 25616 | 3882 | X | X | X | 5.00 |
| Chile | 14262 | 19 | 26 | 11966 | 3212 | 4890 | X | X | X | 5.80 |
| Colombia | 35101 | 31 | 43 | 25526 | 2926 | 3380 | X | X | X | 5.10 |
| Costa Rica | 3424 | 67 | 110 | 1702 | 1050 | 3569 | 15.29 | 25.78 | 58.94 | 6.10 |
| Ecuador | 11460 | 40 | 63 | 6698 | 1529 | 2830 | 12.11 | 37.62 | 50.27 | 2.10 |
| El Salvador | 5768 | 274 | 463 | 2599 | 1680 | 1876 | 8.57 | 24.97 | 66.45 | 5.10 |
| French Guiana | 147 | X | X | 112 | X | X | X | X | X | X |
| Guatemala | 10621 | 98 | 199 | 4404 | 780 | 2247 | 25.16 | 19.45 | 55.40 | 4.00 |
| Guyana | 835 | 4 | 5 | 302 | 213 | X | X | X | X | 7.90 |
| Honduras | 5654 | 50 | 95 | 2482 | 583 | 1385 | 19.73 | 30.35 | 49.91 | 3.90 |
| Mexico | 93674 | 48 | 70 | 70535 | 6529 | 6253 | 8.45 | 28.38 | 63.17 | 0.40 |
| Nicaragua | 4433 | 34 | 70 | 2787 | 1166 | X | 30.31 | 20.19 | 49.50 | -1.10 |
| Panama | 2631 | 35 | 50 | 1401 | 989 | 3332 | 10.16 | 18.28 | 71.56 | 5.80 |
| Paraguay | 4960 | 12 | 22 | 2613 | X | 2178 | 26.40 | 20.90 | 52.70 | 3.70 |
| Peru | 23780 | 19 | 29 | 17175 | 6975 | 2092 | 11.00 | 43.18 | 45.82 | 6.50 |
| Suriname | 423 | 3 | 4 | 213 | 140 | X | 22.05 | 24.24 | 53.71 | 0.00 |
| Uruguay | 3186 | 18 | 21 | 2877 | 1511 | 5185 | 9.03 | 27.27 | 63.71 | 1.10 |
| Venezuela | 21844 | 24 | 38 | 20281 | 5158 | 7082 | 5.04 | 41.91 | 53.05 | -1.00 |

| | Land Cover/Use | | | | | Water | | | |
|---|---|---|---|---|---|---|---|---|---|
| **Country** | Total Land Area (1000 ha) 1993 | Arable and Permanent Cropland (1000 ha) 1993 | Permanent Pasture (1000 ha) 1993 | Forest and Woodland (1000 ha) 1993 | Other Land (1000 ha) 1993 | Water Resources per Capita (m³) 1995 | Domestic Annual Withdraws (%) 1995 | Industry Annual Withdraws (%) 1995 | Agriculture Annual Withdraws (%) 1995 |
| Argentina | 273669 | 27200 | 142000 | 50900 | 53569 | 28739 | 9 | 18 | 73 |
| Belize | 2280 | 57 | 48 | 2100 | 75 | 74419 | 10 | 0 | 90 |
| Bolivia | 108438 | 2380 | 26500 | 58000 | 21558 | 40464 | 10 | 5 | 85 |
| Brazil | 845651 | 48955 | 185000 | 488000 | 123696 | 42957 | 22 | 19 | 59 |
| Chile | 74880 | 4257 | 13600 | 16500 | 40523 | 32814 | 6 | 5 | 89 |
| Colombia | 103870 | 5460 | 40600 | 50000 | 7810 | 30483 | 41 | 16 | 43 |
| Costa Rica | 5106 | 530 | 2340 | 1570 | 666 | 27745 | 4 | 7 | 89 |
| Ecuador | 27684 | 3020 | 2090 | 15600 | 6974 | 27400 | 7 | 3 | 90 |
| El Salvador | 2072 | 730 | 610 | 104 | 628 | 3285 | 7 | 4 | 89 |
| French Guiana | 8815 | 12 | 9 | 7300 | 1494 | X | X | X | X |
| Guatemala | 10843 | 1880 | 2602 | 5813 | 548 | 10922 | 9 | 17 | 74 |
| Guyana | 19685 | 496 | 1230 | 16500 | 1459 | 288623 | 1 | 0 | 99 |
| Honduras | 11189 | 2015 | 1533 | 6000 | 1641 | 11216 | 4 | 5 | 91 |
| Mexico | 190869 | 24730 | 74499 | 48700 | 42940 | 3815 | 6 | 8 | 86 |
| Nicaragua | 11875 | 1270 | 5500 | 3200 | 1905 | 39477 | 25 | 21 | 54 |
| Panama | 7443 | 660 | 1490 | 3260 | 2033 | 54732 | 12 | 11 | 77 |
| Paraguay | 39730 | 2270 | 21700 | 12850 | 2910 | 63306 | 15 | 7 | 78 |
| Peru | 128000 | 3430 | 27120 | 84800 | 12650 | 1682 | 19 | 9 | 72 |
| Suriname | 15600 | 68 | 21 | 15000 | 511 | 472813 | 6 | 5 | 89 |
| Uruguay | 17481 | 1304 | 13520 | 930 | 1727 | 38920 | 6 | 3 | 91 |
| Venezuela | 88205 | 3915 | 17800 | 30000 | 36490 | 60291 | 43 | 11 | 46 |

0 = zero or less than half the unit of measure; X = not available; total water withdrawals may exceed 100% due to groundwater drawdowns or river inflows.

***Table D-4:** Latin America region (continued).*

| Country | Irrigated Land (1000 ha) 1993 | Agricultural Labor Force (1000s) 1993 | Total Labor Force (1000s) 1993 | Cattle Stocks (1000s) 1994 | Sheep Stocks (1000s) 1994 | Goat Stocks (1000s) 1994 | Pig Stocks (1000s) 1994 | Equines (horses/ mules/asses) (1000s) 1994 | Buffalo Stocks (1000s) 1994 | Camel Stocks (1000s) 1994 |
|---|---|---|---|---|---|---|---|---|---|---|
| | **Agriculture/Food** | | | | | | | | | |
| Argentina | 1700 | 1166 | 12067 | 50000 | 20000 | 3408 | 2200 | 3634 | X | X |
| Belize | 2 | X | X | 59 | 4 | 1 | 26 | 9 | X | X |
| Bolivia | 175 | 1015 | 2537 | 6012 | 7789 | 1517 | 2331 | 1041 | X | X |
| Brazil | 2800 | 13189 | 58463 | 151600 | 20500 | 12200 | 30450 | 9260 | 1435 | X |
| Chile | 1265 | 572 | 4942 | 3692 | 4649 | 600 | 1407 | 488 | X | X |
| Colombia | 530 | 2872 | 11327 | 25700 | 2540 | 960 | 2635 | 3332 | X | X |
| Costa Rica | 120 | 251 | 1140 | 1694 | 3 | 2 | 252 | 126 | X | X |
| Ecuador | 556 | 1006 | 3585 | 4963 | 1728 | 345 | 2540 | 967 | X | X |
| El Salvador | 120 | 628 | 1794 | 1256 | 5 | 15 | 325 | 123 | X | X |
| French Guiana | 2 | X | X | 10 | 4 | 1 | 10 | 0 | X | X |
| Guatemala | 125 | 1435 | 2901 | 2210 | 440 | 78 | 720 | 165 | X | X |
| Guyana | 130 | 66 | 311 | 190 | 131 | 79 | 50 | 3 | X | X |
| Honduras | 74 | 952 | 1788 | 2286 | 14 | 28 | 603 | 269 | X | X |
| Mexico | 6100 | 8962 | 31885 | 30702 | 5905 | 10450 | 18000 | 12611 | X | X |
| Nicaragua | 88 | 418 | 1206 | 1650 | 4 | 6 | 535 | 301 | X | X |
| Panama | 32 | 219 | 945 | 1437 | 0 | 5 | 295 | 161 | X | X |
| Paraguay | 67 | 719 | 1581 | 8000 | 386 | 122 | 3300 | 415 | X | X |
| Peru | 1280 | 2523 | 7612 | 4000 | 11600 | 1713 | 2405 | 1414 | X | X |
| Suriname | 60 | 24 | 157 | 98 | 9 | 9 | 37 | 0 | 1 | X |
| Uruguay | 140 | 160 | 1230 | 10316 | 23441 | 15 | 230 | 484 | X | X |
| Venezuela | 190 | 732 | 7440 | 15071 | 550 | 1850 | 2250 | 1007 | X | X |

| Country | Total Commercial Energy Consumption (PJ) 1993 | Traditional Fuel Consumption (TJ) 1991 | Commercial Hydroelectric Production (PJ) 1993 | Known Mammal Species (#) 1990s | Endemic Mammal Species (#) 1990s | Known Bird Species (#) 1990s | Endemic Bird Species (#) 1990s | Known Plant Species (#) 1990s | Endemic Plant Species (#) 1990s |
|---|---|---|---|---|---|---|---|---|---|
| | **Energy** | | | **Biodiversity** | | | | | |
| Argentina | 2019 | 115572 | 86.93 | 320 | 47 | 976 | 19 | 9000 | 1100 |
| Belize | 4 | 3833 | 0 | 125 | 0 | 533 | 0 | 2750 | 150 |
| Bolivia | 86 | 19232 | 4.95 | 316 | 20 | 1274 | 16 | 16500 | 4000 |
| Brazil | 3800 | 2020612 | 845.13 | 394 | 96 | 1635 | 177 | 55000 | X |
| Chile | 539 | 83618 | 63.45 | 91 | 16 | 448 | 15 | 5125 | 2698 |
| Colombia | 829 | 235405 | 100.70 | 359 | 28 | 1695 | 62 | 50000 | 1500 |
| Costa Rica | 63 | 35263 | 14.26 | 205 | 6 | 850 | 7 | 11000 | 950 |
| Ecuador | 245 | 74195 | 21.14 | 302 | 23 | 1559 | 37 | 18250 | 4000 |
| El Salvador | 72 | 38894 | 6.48 | 135 | 0 | 420 | 0 | 2500 | 17 |
| French Guiana | 12 | 645 | 0 | X | X | X | X | X | X |
| Guatemala | 72 | 103649 | 6.92 | 250 | 3 | 669 | 1 | 8000 | 1171 |
| Guyana | 15 | 4370 | 0.02 | 193 | 1 | 737 | 0 | 6000 | X |
| Honduras | 43 | 58138 | 8.15 | 173 | 1 | 684 | 1 | 5000 | 148 |
| Mexico | 4941 | 248036 | 93.65 | 450 | 140 | 1026 | 89 | 25000 | 12500 |
| Nicaragua | 52 | 38877 | 1.11 | 200 | 2 | 750 | 0 | 7000 | 40 |
| Panama | 61 | 16157 | 8.26 | 218 | 14 | 929 | 8 | 9000 | 1222 |
| Paraguay | 51 | 54999 | 113.08 | 305 | 2 | 600 | 0 | 7500 | X |
| Peru | 314 | 87615 | 41.09 | 344 | 45 | 1678 | 109 | 17121 | 5356 |
| Suriname | 24 | 1225 | 4.06 | 180 | 0 | 673 | 0 | 4700 | X |
| Uruguay | 77 | 28176 | 26.27 | 81 | 1 | 365 | 0 | 2184 | 40 |
| Venezuela | 2083 | 21878 | 170.90 | 305 | 16 | 1296 | 42 | 20000 | 8000 |

0 = zero or less than half the unit of measure; X = not available; all plant species includes flowering plants only.

***Table D-5:*** *Middle East/Arid Asia region.*

| Country | Total Pop. (1000s) 1995 | Current Pop. Density (persons/km²) 1995 | Projected Pop. Density (persons/km²) 2025 | Total Urban Pop. (1000s) 1995 | Urban Pop. in Coastal Cities (1000s) 1980 | GDP per Capita in Constant PPP ('85 IN$) 1992 | GDP from Agri-culture (%) 1993 | GDP from Industry (%) 1993 | GDP from Services (%) 1993 | GDP Annual Growth Rate (%) 1993 |
|---|---|---|---|---|---|---|---|---|---|---|
| | **Population and Human Development** | | | | | **Economic Conditions** | | | | |
| Afghanistan | 20141 | 31 | 69 | 4026 | X | X | X | X | X | X |
| Bahrain | 564 | 831 | 1360 | 509 | 279 | X | 0.94 | 41.70 | 57.36 | X |
| Iran | 67283 | 41 | 75 | 39716 | 872 | 3685 | 20.75 | 36.36 | 42.88 | 2.80 |
| Iraq | 20449 | 47 | 97 | 15258 | 0 | X | X | X | X | X |
| Israel | 5629 | 267 | 371 | 5098 | 2826 | 9843 | X | X | X | X |
| Jordan | 5439 | 56 | 123 | 3887 | 70 | X | 7.96 | 26.24 | 65.80 | 5.80 |
| Kazakstan | 17111 | 6 | 8 | 10218 | X | X | X | X | X | -12.90 |
| Kuwait | 1547 | 87 | 157 | 1501 | 1190 | X | 0.49 | 54.82 | 44.69 | X |
| Kyrgyz | 4745 | 24 | 36 | 1847 | X | X | X | X | X | -16.40 |
| Lebanon | 3009 | 289 | 425 | 2622 | 2016 | X | X | X | X | X |
| Oman | 2163 | 10 | 29 | 285 | 62 | X | X | X | X | 7.00 |
| Pakistan | 140497 | 176 | 358 | 48742 | 5215 | 1432 | 24.81 | 25.29 | 49.90 | 2.60 |
| Qatar | 551 | 50 | 73 | 503 | 197 | X | X | X | X | X |
| Saudi Arabia | 17880 | 8 | 20 | 14339 | 1954 | X | X | X | X | X |
| Syria | 14661 | 79 | 181 | 7676 | 266 | X | X | X | X | X |
| Tajikistan | 6101 | 43 | 82 | 1964 | X | X | X | X | X | X |
| Turkey | 61945 | 79 | 117 | 42598 | 9928 | 3807 | 15.09 | 30.39 | 54.51 | X |
| Turkmenistan | 4099 | 8 | 14 | 1839 | X | X | X | X | X | X |
| United Arab Emirates | 1904 | 23 | 35 | 1600 | 824 | X | 2.21 | 57.50 | 40.29 | X |
| Uzbekistan | 22843 | 51 | 84 | 9430 | X | X | 22.98 | 35.99 | 41.03 | -2.40 |
| Yemen | 14501 | 27 | 64 | 4877 | 927 | X | 21.00 | 24.02 | 54.98 | X |

| Country | Total Land Area (1000ha) 1993 | Arable and Permanent Cropland (1000ha) 1993 | Permanent Pasture (1000ha) 1993 | Forest and Woodland (1000ha) 1993 | Other Land (1000ha) 1993 | Water Resources per Capita (m³) 1995 | Domestic Annual Withdraws (%) 1995 | Industry Annual Withdraws (%) 1995 | Agriculture Annual Withdraws (%) 1995 |
|---|---|---|---|---|---|---|---|---|---|
| | **Land Cover/Use** | | | | | **Water** | | | |
| Afghanistan | 65209 | 8054 | 30000 | 1900 | 25255 | 2482 | 1 | 0 | 99 |
| Bahrain | 68 | 2 | 4 | X | 62 | X | X | X | X |
| Iran | 163600 | 18150 | 44000 | 11400 | 90050 | 1746 | 4 | 9 | 87 |
| Iraq | 43737 | 5450 | 4000 | 192 | 34095 | 5340 | 3 | 5 | 92 |
| Israel | 2062 | 435 | 145 | 126 | 1356 | 382 | 16 | 5 | 79 |
| Jordan | 8893 | 405 | 791 | 70 | 7627 | 314 | 29 | 6 | 65 |
| Kazakstan | 266980 | 34800 | 186562 | 9600 | X | 9900 | 4 | 17 | 79 |
| Kuwait | 1782 | 5 | 137 | 2 | 1638 | 103 | 64 | 32 | 4 |
| Kyrgyz | 19130 | 1420 | 8700 | 700 | X | 13003 | 3 | 7 | 90 |
| Lebanon | 1023 | 306 | 10 | 80 | 627 | 1854 | 11 | 4 | 85 |
| Oman | 21246 | 63 | 1000 | X | 20183 | 892 | 3 | 3 | 94 |
| Pakistan | 77088 | 21250 | 5000 | 3480 | 47358 | 3331 | 1 | 1 | 98 |
| Qatar | 1100 | 7 | 50 | X | 1043 | X | X | X | X |
| Saudi Arabia | 214969 | 3740 | 120000 | 1800 | 89429 | 254 | 45 | 8 | 47 |
| Syria | 18378 | 5775 | 8060 | 650 | 3893 | 3662 | 7 | 10 | 83 |
| Tajikistan | 14060 | 849 | 3545 | 537 | 9129 | 16604 | 5 | 7 | 88 |
| Turkey | 76963 | 27535 | 12378 | 20199 | 16851 | 3117 | 24 | 19 | 57 |
| Turkmenistan | 48810 | 1480 | 30800 | 4000 | X | 17573 | 1 | 8 | 91 |
| United Arab Emirates | 8360 | 39 | 200 | 3 | 8118 | 1047 | 11 | 9 | 80 |
| Uzbekistan | 42540 | 4500 | 20800 | 1300 | X | 5674 | 4 | 12 | 84 |
| Yemen | 52797 | 1481 | 16065 | 2000 | 33251 | 359 | 5 | 2 | 93 |

0 = zero or less than half the unit of measure; X = not available; total water withdrawals may exceed 100% due to groundwater drawdowns or river inflows.

***Table D-5:*** *Middle East/Arid Asia region (continued).*

| Country | Agriculture/Food | | | | | | | | | |
|---|---|---|---|---|---|---|---|---|---|---|
| | Irrigated Land (1000 ha) 1993 | Agricultural Labor Force (1000s) 1993 | Total Labor Force (1000s) 1993 | Cattle Stocks (1000s) 1994 | Sheep Stocks (1000s) 1994 | Goat Stocks (1000s) 1994 | Pig Stocks (1000s) 1994 | Equines (horses/ mules/asses) (1000s) 1994 | Buffalo Stocks (1000s) 1994 | Camel Stocks (1000s) 1994 |
| Afghanistan | 3000 | 3234 | 6133 | 1500 | 14200 | 2150 | X | 1483 | X | 265 |
| Bahrain | 1 | 3 | 207 | 16 | 29 | 17 | X | 0 | X | 1 |
| Iran | 9400 | 4325 | 17029 | 7100 | 45400 | 23500 | 0 | 2288 | 300 | 140 |
| Iraq | 2550 | 1041 | 5672 | 1100 | 6320 | 1050 | X | 196 | 100 | 15 |
| Israel | 180 | 82 | 2134 | 362 | 330 | 100 | 100 | 11 | X | 10 |
| Jordan | 63 | 58 | 1190 | 42 | 2100 | 555 | X | 26 | X | 18 |
| Kazakstan | 2200 | X | X | 9347 | 33524 | 684 | 2445 | 1440 | 105 | 55 |
| Kuwait | 2 | X | X | 12 | 150 | 15 | X | 0 | X | 1 |
| Kyrgyz | 900 | X | X | 1061 | 7077 | 219 | 165 | 310 | X | 50 |
| Lebanon | 86 | 69 | 929 | 80 | 258 | 456 | 41 | 46 | X | 1 |
| Oman | X | 169 | 454 | 144 | 149 | 739 | X | 26 | X | 96 |
| Pakistan | 17110 | 18413 | 38226 | 18146 | 28975 | 41340 | X | 4335 | 18887 | 1121 |
| Qatar | 8 | X | X | 12 | 170 | 150 | X | 1 | X | 43 |
| Saudi Arabia | 435 | 1877 | 5126 | 203 | 7257 | 4150 | X | 101 | X | 415 |
| Syria | 906 | 768 | 3414 | 770 | 12000 | 1200 | 1 | 225 | 1 | 3 |
| Tajikistan | 639 | X | X | 1250 | 2000 | 845 | 40 | 85 | X | 50 |
| Turkey | 3674 | 11676 | 25598 | 11910 | 37541 | 10133 | 9 | 1463 | 316 | 2 |
| Turkmenistan | 1300 | X | X | 1104 | 6000 | 314 | 159 | 45 | X | 40 |
| United Arab Emirates | 5 | 18 | 864 | 65 | 333 | 861 | X | 0 | X | 148 |
| Uzbekistan | 4000 | X | X | 5291 | 8600 | 968 | 391 | 270 | 85 | 20 |
| Yemen | 360 | 1679 | 3124 | X | 6947 | 3232 | X | 503 | X | 173 |

| Country | Energy | | | Biodiversity | | | | | |
|---|---|---|---|---|---|---|---|---|---|
| | Total Commercial Energy Consumption (PJ) 1993 | Traditional Fuel Consumption (TJ) 1991 | Commercial Hydroelectric Production (PJ) 1993 | Known Mammal Species (#) 1990s | Endemic Mammal Species (#) 1990s | Known Bird Species (#) 1990s | Endemic Bird Species (#) 1990s | Known Plant Species (#) 1990s | Endemic Plant Species (#) 1990s |
| Afghanistan | 22 | 50649 | 1.71 | 123 | 1 | 460 | 0 | 3500 | 800 |
| Bahrain | 276 | 0 | 0.00 | X | X | X | X | X | X |
| Iran | 3264 | 28613 | 39.60 | 140 | 5 | 502 | 1 | X | X |
| Iraq | 933 | 1026 | 2.16 | 81 | 1 | 381 | 1 | X | X |
| Israel | 505 | 126 | 0.11 | 92 | 3 | 500 | 0 | X | X |
| Jordan | 147 | 79 | 0.08 | 71 | 0 | 361 | 0 | 2200 | X |
| Kazakstan | 3381 | 0 | 27.46 | X | 4 | X | 0 | X | X |
| Kuwait | 471 | 0 | 0.00 | 21 | 0 | 321 | 0 | 234 | X |
| Kyrgyz | 150 | 0 | 32.05 | X | X | X | X | X | X |
| Lebanon | 121 | 4639 | 1.30 | 54 | 0 | 329 | 0 | X | X |
| Oman | 162 | 0 | 0.00 | 56 | 2 | 430 | 0 | 1018 | 73 |
| Pakistan | 1135 | 296244 | 77.44 | 151 | 3 | 671 | 0 | 4929 | 372 |
| Qatar | 558 | 182 | 0.00 | X | X | X | X | X | X |
| Saudi Arabia | 2933 | 0 | 0.00 | 77 | 0 | 413 | 0 | 1729 | X |
| Syria | 565 | 117 | 24.15 | X | X | X | X | X | X |
| Tajikistan | 258 | 0 | 61.62 | X | 2 | X | 0 | X | X |
| Turkey | 1979 | 95689 | 122.22 | 116 | 1 | 418 | 0 | 8472 | 2675 |
| Turkmenistan | 555 | 0 | 0.10 | X | 0 | X | 0 | X | X |
| United Arab Emirates | 1039 | 0 | 0.00 | 25 | 0 | 360 | 0 | X | X |
| Uzbekistan | 1903 | 0 | 26.49 | X | 0 | X | 0 | X | X |
| Yemen | 123 | 3165 | 0.00 | 66 | 2 | 366 | 8 | X | 135 |

X = not available; shaded cells contain data originally attributed to Peoples' Republic of Yemen or Republic of Yemen; all plant species includes flowering plants only.

***Table D-6:*** *North America region.*

| Country | Population and Human Development: Total Pop. (1000s) 1995 | Current Pop. Density (persons/km²) 1995 | Projected Pop. Density (persons/km²) 2025 | Total Urban Pop. (1000s) 1995 | Urban Pop. in Coastal Cities (1000s) 1980 | Economic Conditions: GDP per Capita in Constant PPP ('85 IN$) 1992 | GDP from Agri-culture (%) 1993 | GDP from Industry (%) 1993 | GDP from Services (%) 1993 | GDP Annual Growth Rate (%) 1993 |
|---|---|---|---|---|---|---|---|---|---|---|
| Canada | 29463 | 3 | 4 | 22593 | 3066 | 16362 | X | X | X | X |
| United States | 263250 | 28 | 35 | 200695 | 60324 | 17945 | X | X | X | 3 |

| Country | Land Cover/Use: Total Land Area (1000 ha) 1993 | Arable and Permanent Cropland (1000 ha) 1993 | Permanent Pasture (1000 ha) 1993 | Forest and Woodland (1000 ha) 1993 | Other Land (1000 ha) 1993 | Water: Water Resources per Capita (m³) 1995 | Domestic Annual Withdraws (%) 1995 | Industry Annual Withdraws (%) 1995 | Agriculture Annual Withdraws (%) 1995 |
|---|---|---|---|---|---|---|---|---|---|
| Canada | 922097 | 45500 | 27900 | 494000 | 354697 | 98462 | 18 | 70 | 12 |
| United States | 957311 | 187776 | 239172 | 286200 | 244163 | 9413 | 13 | 45 | 42 |

| Country | Agriculture/Food: Irrigated Land (1000 ha) 1993 | Agricultural Labor Force (1000s) 1993 | Total Labor Force (1000s) 1993 | Cattle Stocks (1000s) 1994 | Sheep Stocks (1000s) 1994 | Goat Stocks (1000s) 1994 | Pig Stocks (1000s) 1994 | Equines (horses/mules/asses) (1000s) 1994 | Buffalo Stocks (1000s) 1994 | Camel Stocks (1000s) 1994 |
|---|---|---|---|---|---|---|---|---|---|---|
| Canada | 710 | 387 | 13756 | 12306 | 691 | 28 | 11200 | 429 | X | X |
| United States | 20700 | 2600 | 126205 | 100988 | 9600 | 2009 | 57904 | 3944 | X | X |

| Country | Energy: Total Commercial Energy Consumption (PJ) 1993 | Traditional Fuel Consumption (TJ) 1991 | Commercial Hydroelectric Production (PJ) 1993 | Biodiversity: Known Mammal Species (#) 1990s | Endemic Mammal Species (#) 1990s | Known Bird Species (#) 1990s | Endemic Bird Species (#) 1990s | Known Plant Species (#) 1990s | Endemic Plant Species (#) 1990s |
|---|---|---|---|---|---|---|---|---|---|
| Canada | 9198 | 67018 | 1165.28 | 193 | 7 | 578 | 3 | 2920 | 147 |
| United States | 81751 | 916422 | 995.27 | 428 | 101 | 768 | 70 | 16302 | 4036 |

0 = zero or less than half the unit of measure; X = not available; all plant species includes flowering plants only.

***Table D-7:** Small Island States region.*

| | Population and Human Development | | | | | Economic Conditions | | | | |
|---|---|---|---|---|---|---|---|---|---|---|
| Country | Total Pop. (1000s) 1995 | Current Pop. Density (persons/ km²) 1995 | Projected Pop. Density (persons/ km²) 2025 | Total Urban Pop. (1000s) 1995 | Urban Pop. in Coastal Cities (1000s) 1980 | GDP per Capita in Constant PPP ('85 IN$) 1992 | GDP from Agri-culture (%) 1993 | GDP from Industry (%) 1993 | GDP from Services (%) 1993 | GDP Annual Growth Rate (%) 1993 |
| Antigua and Barbuda | 66 | X | X | 24 | X | X | X | X | X | 3.50 |
| Bahamas | 276 | 20 | 27 | 239 | X | X | X | X | X | X |
| Barbados | 262 | 609 | 719 | 124 | 100 | X | X | X | X | 1.50 |
| Cape Verde | 392 | 97 | 182 | 213 | 125 | 1085 | 12.85 | 15.26 | 71.89 | 4.00 |
| Comoros | 653 | 292 | 736 | 201 | 89 | 527 | 39.45 | 12.25 | 48.30 | 1.20 |
| Cook Islands | X | X | X | X | X | X | X | X | X | X |
| Cuba | 11041 | 100 | 114 | 8389 | 6628 | X | X | X | X | X |
| Cyprus | 742 | 80 | 100 | 401 | 291 | 9203 | 5.69 | 24.85 | 69.46 | X |
| Dominica | 71 | X | X | X | X | X | X | X | X | X |
| Dominican Republic | 7823 | 161 | 229 | 5051 | 2787 | 2250 | 15.49 | 22.96 | 61.55 | 2.90 |
| Fed. States of Micronesia | X | X | X | X | X | X | X | X | X | X |
| Fiji | 784 | 43 | 64 | 319 | 244 | X | X | X | X | 1.70 |
| Grenada | 92 | X | X | X | X | X | 13.72 | 19.35 | 66.92 | X |
| Haiti | 7180 | 259 | 473 | 2266 | 1216 | X | 38.62 | 15.52 | 45.86 | X |
| Jamaica | 2447 | 223 | 300 | 1314 | 1016 | X | 8.40 | 40.89 | 50.72 | 1.20 |
| Kiribati | 79 | X | X | 28 | X | X | X | X | X | X |
| Maldives | 254 | 854 | 1876 | 68 | X | X | X | X | X | X |
| Malta | 366 | 1159 | 1336 | 327 | 303 | X | 3.17 | 34.86 | 61.96 | X |
| Marshall Islands | X | X | X | X | X | X | X | X | X | X |
| Mauritius | 1117 | 547 | 726 | 453 | 410 | 6167 | 9.86 | 33.42 | 56.73 | 5.60 |
| Nauru | 11 | X | X | 11 | X | X | X | X | X | X |
| Palau | X | X | X | X | X | X | X | X | X | X |
| Samoa | 171 | 61 | 108 | 36 | X | X | X | X | X | X |
| St. Kitts and Nevis | 41 | X | X | 18 | X | 4799 | X | X | X | X |
| St. Lucia | X | X | X | 69 | X | X | 10.80 | 20.88 | 68.32 | 3.60 |
| St. Vincent/The Grenadines | 112 | X | X | 52 | X | X | X | X | X | X |
| Sao Tome and Principe | 133 | X | X | 62 | X | X | X | X | X | X |
| Seychelles | 73 | X | X | 40 | X | X | 4.14 | 18.95 | 76.91 | 3.90 |
| Solomon Islands | 378 | 13 | 29 | 65 | X | X | X | X | X | X |
| Tonga | 98 | X | X | 40 | X | X | X | X | X | X |
| Trinidad and Tobago | 1306 | 255 | 353 | 938 | 623 | X | 2.53 | 42.92 | 54.55 | -2.40 |
| Tuvalu | 10 | X | X | 4 | X | 547 | X | X | X | X |
| Vanuatu | 169 | 14 | 27 | 33 | X | X | X | X | X | X |

0 = zero or less than half the unit of measure; X = not available.

***Table D-7:*** *Small Island States region (continued).*

| | Land Cover/Use | | | | | Water | | | |
|---|---|---|---|---|---|---|---|---|---|
| Country | Total Land Area (1000 ha) 1993 | Arable and Permanent Cropland (1000 ha) 1993 | Permanent Pasture (1000 ha) 1993 | Forest and Woodland (1000 ha) 1993 | Other Land (1000 ha) 1993 | Water Resources per Capita ($m^3$) 1995 | Domestic Annual Withdraws (%) 1995 | Industry Annual Withdraws (%) 1995 | Agriculture Annual Withdraws (%) 1995 |
| Antigua and Barbuda | 44 | 8 | 4 | X | 32 | X | X | X | X |
| Bahamas | 1001 | 10 | 2 | 324 | 665 | X | X | X | X |
| Barbados | 43 | 16 | 2 | 5 | 20 | X | X | X | X |
| Cape Verde | 403 | 45 | 25 | 1 | 332 | X | X | X | X |
| Comoros | 223 | 100 | 15 | 40 | 68 | X | X | X | X |
| Cook Islands | X | X | X | X | X | X | X | X | X |
| Cuba | 10982 | 3340 | 2970 | 2608 | 2064 | 3125 | 9 | 2 | 89 |
| Cyprus | 924 | 158 | 4 | 123 | 639 | X | X | X | X |
| Dominica | 75 | 17 | 200 | 50 | -192 | X | X | X | X |
| Dominican Republic | 4838 | 1450 | 2 | 600 | 2786 | 2557 | 5 | 6 | 89 |
| Fed. States of Micronesia | X | X | X | X | X | X | X | X | X |
| Fiji | 1827 | 260 | 175 | 1185 | 207 | 36416 | 20 | 20 | 60 |
| Grenada | 34 | 11 | 1 | 3 | 19 | X | X | X | X |
| Haiti | 2756 | 910 | 495 | 140 | 1211 | 1532 | 24 | 8 | 68 |
| Jamaica | 1083 | 219 | 257 | 185 | 422 | 3392 | 7 | 7 | 86 |
| Kiribati | 73 | 37 | X | 2 | X | X | X | X | X |
| Maldives | 30 | 3 | 1 | 1 | 25 | X | X | X | X |
| Malta | 32 | 13 | X | X | X | X | X | X | X |
| Marshall Islands | X | X | X | X | X | X | X | X | X |
| Mauritius | 203 | 106 | 7 | 44 | 46 | 1979 | 16 | 7 | 77 |
| Nauru | 2 | X | X | X | X | X | X | X | X |
| Palau | X | X | X | X | X | X | X | X | X |
| Samoa | X | X | X | 134 | X | X | X | X | X |
| St. Kitts and Nevis | 36 | 14 | 1 | 6 | 15 | X | X | X | X |
| St. Lucia | 61 | 18 | 3 | 8 | 32 | X | X | X | X |
| St. Vincent/TheGrenadines | 39 | 11 | 2 | 14 | 12 | X | X | X | X |
| Sao Tome and Principe | 100 | X | X | X | X | X | X | X | X |
| Seychelles | 45 | 7 | X | 5 | X | X | X | X | X |
| Solomon Islands | 2799 | 57 | 39 | 2450 | 253 | 118254 | 40 | 20 | 40 |
| Tonga | 72 | 48 | 4 | 8 | 12 | X | X | X | X |
| Trinidad and Tobago | 513 | 122 | 11 | 235 | 145 | 3905 | 27 | 38 | 35 |
| Tuvalu | 3 | X | X | X | X | X | X | X | X |
| Vanuatu | 1219 | 144 | 25 | 914 | 136 | X | X | X | X |

0 = zero or less than half the unit of measure; X = not available; total water withdrawals may exceed 100% due to groundwater drawdowns or river inflows.

***Table D-7:*** *Small Island States region (continued).*

| | Agriculture/Food | | | | | | | | | |
|---|---|---|---|---|---|---|---|---|---|---|
| **Country** | Irrigated Land (1000 ha) 1993 | Agricultural Labor Force (1000s) 1993 | Total Labor Force (1000s) 1993 | Cattle Stocks (1000s) 1994 | Sheep Stocks (1000s) 1994 | Goat Stocks (1000s) 1994 | Pig Stocks (1000s) 1994 | Equines (horses/ mules/asses) (1000s) 1994 | Buffalo Stocks (1000s) 1994 | Camel Stocks (1000s) 1994 |
| Antigua and Barbuda | X | X | X | 16 | 1 | 12 | 4 | 3 | X | X |
| Bahamas | X | X | X | 6 | 40 | 19 | 15 | 0 | X | X |
| Barbados | X | 8 | 138 | 25 | 41 | 5 | 30 | 5 | X | X |
| Cape Verde | 3 | 57 | 140 | 18 | 7 | 137 | 111 | 15 | X | X |
| Comoros | X | 199 | 257 | 50 | 15 | 128 | X | 5 | X | X |
| Cook Islands | X | X | X | | X | 8 | X | X | X | X |
| Cuba | 910 | 850 | 4727 | 4500 | 310 | 95 | 1503 | 617 | X | X |
| Cyprus | 39 | 64 | 334 | 61 | 285 | 200 | 370 | 8 | X | X |
| Dominica | X | X | X | 9 | 8 | 10 | 5 | 0 | X | X |
| Dominican Republic | 230 | 817 | 2478 | 2450 | 134 | 587 | 900 | 616 | X | X |
| Fed. States of Micronesia | X | X | X | X | X | X | X | X | X | X |
| Fiji | 1 | 95 | 254 | 334 | 6 | 205 | 115 | 44 | X | X |
| Grenada | X | X | X | 4 | 12 | 11 | 3 | 1 | X | X |
| Haiti | 75 | 1848 | 2988 | 800 | 85 | 910 | 200 | 690 | X | X |
| Jamaica | 35 | 328 | 1274 | 335 | 2 | 442 | 180 | 37 | X | X |
| Kiribati | X | 3 | 27 | X | 0 | X | 9 | 0 | X | X |
| Maldives | 78 | X | X | X | 0 | X | X | 0 | X | X |
| Malta | X | 5 | 138 | 20 | 6 | 5 | 111 | 2 | X | X |
| Marshall Islands | X | X | X | X | X | X | X | X | X | X |
| Mauritius | 17 | 94 | 440 | 34 | 7 | 95 | 17 | 0 | X | X |
| Nauru | X | X | X | X | 0 | X | 3 | 0 | X | X |
| Palau | X | X | X | X | X | X | X | X | X | X |
| Samoa | X | X | X | X | 0 | X | X | 0 | X | X |
| St. Kitts and Nevis | X | X | X | 5 | 4 | 10 | 2 | 0 | X | X |
| St. Lucia | 1 | X | X | 12 | 16 | 12 | 13 | 3 | X | X |
| St. Vincent/TheGrenadines | 1 | X | X | 6 | 6 | 6 | 9 | 1 | X | X |
| Sao Tome and Principe | X | X | X | 4 | 0 | 5 | 2 | 0 | X | X |
| Seychelles | X | X | X | 2 | 0 | 5 | 18 | 0 | X | X |
| Solomon Islands | X | X | X | 13 | 0 | X | 55 | 0 | X | X |
| Tonga | X | X | X | 10 | 0 | 16 | 94 | 11 | X | X |
| Trinidad and Tobago | 22 | 33 | 476 | 55 | 14 | 52 | 48 | 5 | 9 | X |
| Tuvalu | X | X | X | X | 0 | X | 13 | 0 | X | X |
| Vanuatu | X | X | X | 132 | 0 | 11 | 59 | 3 | X | X |

0 = zero or less than half the unit of measure; X = not available.

***Table D-7:** Small Island States region (continued).*

| | Energy | | | Biodiversity | | | | | |
|---|---|---|---|---|---|---|---|---|---|
| **Country** | Total Commercial Energy Consumption (PJ) 1993 | Traditional Fuel Consumption (TJ) 1991 | Commercial Hydroelectric Production (PJ) 1993 | Known Mammal Species (#) 1990s | Endemic Mammal Species (#) 1990s | Known Bird Species (#) 1990s | Endemic Bird Species (#) 1990s | Known Plant Species (#) 1990s | Endemic Plant Species (#) 1990s |
| Antigua and Barbuda | 4 | 0 | 0.00 | X | X | X | X | X | X |
| Bahamas | 24 | 0 | 0.00 | X | X | X | X | X | X |
| Barbados | 14 | 1676 | 0.00 | X | X | X | X | X | X |
| Cape Verde | 2 | 0 | 0.00 | X | X | X | X | X | X |
| Comoros | 1 | 0 | 0.01 | X | X | X | X | X | X |
| Cook Islands | X | X | X | X | X | X | X | X | X |
| Cuba | 369 | 204977 | 0.38 | 31 | 12 | 342 | 22 | 6004 | 3229 |
| Cyprus | 63 | 240 | 0.00 | X | X | X | X | X | X |
| Dominica | 1 | 0 | 0.06 | X | X | X | X | X | X |
| Dominican Republic | 148 | 25348 | 6.35 | 20 | 0 | 254 | 0 | 5000 | 1800 |
| Fed. States of Micronesia | X | X | X | X | X | X | X | X | X |
| Fiji | 11 | 11829 | 1.39 | 4 | 1 | 109 | 26 | 1307 | 760 |
| Grenada | 2 | 0 | 0.00 | X | X | X | X | X | X |
| Haiti | 9 | 56611 | 0.59 | 3 | 0 | 220 | 0 | 4685 | 1623 |
| Jamaica | 104 | 6011 | 0.32 | 24 | 3 | 262 | 25 | 2746 | 923 |
| Kiribati | X | 0 | 0.00 | X | X | X | X | X | X |
| Maldives | 1 | 0 | 0.00 | X | X | X | X | X | X |
| Malta | 24 | 0 | 0.00 | X | X | X | X | X | X |
| Marshall Islands | X | X | X | X | X | X | X | X | X |
| Mauritius | 21 | 16793 | 0.37 | 4 | 2 | 81 | 9 | 700 | 325 |
| Nauru | 2 | 0 | 0.00 | X | X | X | X | X | X |
| Palau | X | X | X | 214 | 57 | 708 | 80 | 10000 | X |
| Samoa | X | 736 | 0.00 | X | X | X | X | X | X |
| St.Kitts and Nevis | 1 | 495 | 0.00 | X | X | X | X | X | X |
| St. Lucia | 2 | 0 | 0.00 | X | X | X | X | X | X |
| St. Vincent/TheGrenadines | 1 | 0 | 0.14 | X | X | X | X | X | X |
| Sao Tome and Principe | 1 | 0 | 0.03 | X | X | X | X | X | X |
| Seychelles | 2 | 0 | 0.00 | X | X | X | X | X | X |
| Solomon Islands | 2 | 3224 | 0.00 | 53 | 19 | 223 | 44 | 2780 | 30 |
| Tonga | 1 | 0 | 0.00 | X | X | X | X | X | X |
| Trinidad and Tobago | 267 | 2825 | 0.00 | 100 | 1 | 433 | 1 | 1982 | 236 |
| Tuvalu | 16 | 0 | 0.00 | X | X | X | X | X | X |
| Vanuatu | 1 | 235 | 0.00 | X | X | X | X | X | X |

0 = zero or less than half the unit of measure; X = not available; all plant species includes flowering plants only.

***Table D-8:*** *Temperate Asia region.*

| Country | Total Pop. (1000s) 1995 | Current Pop. Density (persons/km²) 1995 | Projected Pop. Density (persons/km²) 2025 | Total Urban Pop. (1000s) 1995 | Urban Pop. in Coastal Cities (1000s) 1980 | GDP per Capita in Constant PPP ('85 IN$) 1992 | GDP from Agriculture (%) 1993 | GDP from Industry (%) 1993 | GDP from Services (%) 1993 | GDP Annual Growth Rate (%) 1993 |
|---|---|---|---|---|---|---|---|---|---|---|
| | **Population and Human Development** | | | | | **Economic Conditions** | | | | |
| China | 1221462 | 127 | 159 | 369736 | 38936 | 1493 | 19.48 | 47.59 | 32.93 | 12.20 |
| Hong Kong | 5865 | 5612 | 5681 | 5574 | 4614 | 16471 | X | X | X | X |
| Japan | 125095 | 331 | 322 | 97120 | 78349 | 15105 | X | X | X | 0.10 |
| Mongolia | 2410 | 2 | 2 | 1468 | X | X | 20.83 | 46.34 | 32.83 | -1.40 |
| North Korea | 44995 | 454 | 550 | 36572 | 16911 | X | 7.07 | 43.38 | 49.55 | 5.50 |
| South Korea | 23917 | 198 | 277 | 14650 | 5973 | X | X | X | X | X |

| Country | Total Land Area (1000ha) 1993 | Arable and Permanent Cropland (1000ha) 1993 | Permanent Pasture (1000ha) 1993 | Forest and Woodland (1000ha) 1993 | Other Land (1000ha) 1993 | Water Resources per Capita (m³) 1995 | Domestic Annual Withdraws (%) 1995 | Industry Annual Withdraws (%) 1995 | Agriculture Annual Withdraws (%) 1995 |
|---|---|---|---|---|---|---|---|---|---|
| | **LandCover/Use** | | | | | **Water** | | | |
| China | 932641 | 95975 | 400000 | 130496 | 306169 | 2292 | 6 | 7 | 87 |
| Hong Kong | 99 | 7 | 1 | 22 | 69 | X | X | X | X |
| Japan | 37652 | 4463 | 661 | 25100 | 7428 | 4373 | 17 | 33 | 50 |
| Mongolia | 156650 | 1401 | 125000 | 13750 | 16499 | 10207 | 11 | 27 | 62 |
| North Korea | 9873 | 2055 | 90 | 6460 | 1268 | 1469 | 19 | 35 | 46 |
| South Korea | 12041 | 2000 | 50 | 7370 | 2621 | 2801 | 11 | 16 | 73 |

| Country | Irrigated Land (1000ha) 1993 | Agricultural Labor Force (1000s) 1993 | Total Labor Force (1000s) 1993 | Cattle Stocks (1000s) 1994 | Sheep Stocks (1000s) 1994 | Goat Stocks (1000s) 1994 | Pig Stocks (1000s) 1994 | Equines (horses/mules/asses) (1000s) 1994 | Buffalo Stocks (1000s) 1994 | Camel Stocks (1000s) 1994 |
|---|---|---|---|---|---|---|---|---|---|---|
| | **Agriculture/Food** | | | | | | | | | |
| China | 49872 | 463121 | 710441 | 90906 | 111649 | 105990 | 402846 | 26344 | 22416 | 373 |
| Hong Kong | 2 | 32 | 3066 | 2 | 0 | 0 | 97 | 1 | 0 | X |
| Japan | 2782 | 3508 | 63817 | 4989 | 25 | 31 | 10621 | 28 | X | X |
| Mongolia | 80 | 310 | 1115 | 2779 | 14392 | 6469 | 49 | 2100 | X | 415 |
| North Korea | 1335 | 4387 | 20228 | 3200 | 4 | 520 | 6300 | 6 | X | X |
| South Korea | 1460 | 3782 | 12223 | 1330 | 396 | 305 | 3368 | 52 | X | X |

| Country | Total Commercial Energy Consumption (PJ) 1993 | Traditional Fuel Consumption (TJ) 1991 | Commercial Hydroelectric Production (PJ) 1993 | Known Mammal Species (#) 1990s | Endemic Mammal Species (#) 1990s | Known Bird Species (#) 1990s | Endemic Bird Species (#) 1990s | Known Plant Species (#) 1990s | Endemic Plant Species (#) 1990s |
|---|---|---|---|---|---|---|---|---|---|
| | **Energy** | | | **Biodiversity** | | | | | |
| China | 29679 | 2017764 | 546.48 | 394 | 77 | 1244 | 67 | 30000 | 18000 |
| Hong Kong | 386 | 2374 | 0.00 | X | X | X | X | X | X |
| Japan | 17505 | 9663 | 379.69 | 132 | 38 | 583 | 21 | 4700 | 2000 |
| Mongolia | 105 | 13188 | 0.00 | 134 | 0 | 390 | 0 | 2272 | 229 |
| North Korea | 4504 | 25788 | 21.62 | 49 | 0 | 372 | 0 | 2898 | 224 |
| South Korea | 2925 | 40412 | 86.40 | X | 0 | 390 | 0 | 2898 | 107 |

0 = zero or less than half the unit of measure; X = not available; total water withdrawals may exceed 100% due to groundwater drawdowns or river inflows; all plant species includes flowering plants only.

***Table D-9:*** *Tropical Asia region.*

| Country | Population and Human Development: Total Pop. (1000s) 1995 | Current Pop. Density (persons/ km²) 1995 | Projected Pop. Density (persons/ km²) 2025 | Total Urban Pop. (1000s) 1995 | Urban Pop. in Coastal Cities (1000s) 1980 | Economic Conditions: GDP per Capita in Constant PPP ('85 IN$) 1992 | GDP from Agri-culture (%) 1993 | GDP from Industry (%) 1993 | GDP from Services (%) 1993 | GDP Annual Growth Rate (%) 1993 |
|---|---|---|---|---|---|---|---|---|---|---|
| Bangladesh | 120433 | 836 | 1362 | 22034 | 1809 | 1510 | 30.47 | 17.50 | 52.03 | 4.40 |
| Bhutan | 1638 | 35 | 67 | 105 | X | X | 40.57 | 29.49 | 29.94 | X |
| Brunei | 285 | 49 | 74 | 165 | X | X | X | X | X | X |
| Cambodia | 10251 | 57 | 109 | 2123 | 50 | X | 47.29 | 14.46 | 38.25 | X |
| India | 935744 | 285 | 423 | 250681 | 37317 | 1282 | 31.36 | 27.33 | 41.31 | 3.10 |
| Indonesia | 197588 | 104 | 145 | 69992 | 29166 | 2102 | 18.79 | 39.42 | 41.79 | 6.60 |
| Laos | 4882 | 21 | 41 | 1060 | X | X | 51.34 | 18.15 | 30.51 | 5.90 |
| Malaysia | 20140 | 61 | 96 | 10814 | 3997 | 5746 | X | X | X | 8.50 |
| Myanmar | 46527 | 69 | 112 | 12188 | 3923 | X | 63.04 | 8.55 | 28.41 | X |
| Nepal | 21918 | 156 | 289 | 2996 | X | X | 43.14 | 21.29 | 35.57 | 2.90 |
| Papua New Guinea | 4302 | 9 | 16 | 690 | 322 | 1606 | 25.96 | 42.77 | 31.28 | 15.40 |
| Philippines | 67581 | 225 | 348 | 36614 | 17736 | 1689 | 21.68 | 32.92 | 45.40 | 1.70 |
| Singapore | 2848 | 4608 | 5430 | 2848 | 2414 | 12653 | 0.19 | 36.76 | 63.06 | X |
| Sri Lanka | 18354 | 280 | 382 | 4108 | 2433 | 2215 | 24.64 | 25.60 | 49.75 | 6.90 |
| Thailand | 58791 | 115 | 143 | 11787 | 5698 | 3942 | 9.96 | 39.24 | 50.80 | 7.80 |
| Viet Nam | 74545 | 225 | 356 | 15479 | 5585 | X | 29.29 | 28.41 | 42.31 | X |

| Country | LandCover/Use: Total Land Area (1000ha) 1993 | Arable and Permanent Cropland (1000ha) 1993 | Permanent Pasture (1000ha) 1993 | Forest and Woodland (1000ha) 1993 | Other Land (1000ha) 1993 | Water: Water Resources per Capita (m³) 1995 | Domestic Annual Withdraws (%) 1995 | Industry Annual Withdraws (%) 1995 | Agriculture Annual Withdraws (%) 1995 |
|---|---|---|---|---|---|---|---|---|---|
| Bangladesh | 13017 | 9694 | 600 | 1900 | 823 | 19571 | 3 | 1 | 96 |
| Bhutan | 4700 | 134 | 273 | 3100 | 1193 | 57998 | 36 | 10 | 54 |
| Brunei | 527 | 7 | 6 | 450 | 64 | X | X | X | X |
| Cambodia | 17652 | 2400 | 2000 | 11600 | 1652 | 48590 | 5 | 1 | 94 |
| India | 297319 | 169650 | 11400 | 68500 | 47769 | 2228 | 3 | 4 | 93 |
| Indonesia | 181157 | 30987 | 11800 | 111774 | 26596 | 12804 | 13 | 11 | 76 |
| Laos | 23080 | 805 | 800 | 12500 | 8975 | 55305 | 8 | 10 | 82 |
| Malaysia | 32855 | 4880 | 27 | 22304 | 5644 | 22642 | 23 | 30 | 47 |
| Myanmar | 65755 | 10087 | 359 | 32408 | 22901 | 23255 | 7 | 3 | 90 |
| Nepal | 13680 | 2354 | 2000 | 5750 | 3576 | 7756 | 4 | 1 | 95 |
| Papua New Guinea | 45286 | 415 | 80 | 42000 | 2791 | 186192 | 29 | 22 | 49 |
| Philippines | 29817 | 9190 | 1280 | 13600 | 5747 | 4779 | 18 | 21 | 61 |
| Singapore | 61 | 1 | X | 3 | X | 211 | 45 | 51 | 4 |
| Sri Lanka | 6463 | 1900 | 440 | 2100 | 2023 | 2354 | 2 | 2 | 96 |
| Thailand | 51089 | 20800 | 800 | 13500 | 15989 | 3045 | 4 | 6 | 90 |
| Viet Nam | 32549 | 6700 | 330 | 9650 | 15869 | 5044 | 13 | 9 | 78 |

0 = zero or less than half the unit of measure; X = not available; total water withdrawals may exceed 100% due to groundwater drawdowns or river inflows.

*Table D-9: Tropical Asia region (continued).*

| Country | Irrigated Land (1000 ha) 1993 | Agricultural Labor Force (1000s) 1993 | Total Labor Force (1000s) 1993 | Cattle Stocks (1000s) 1994 | Sheep Stocks (1000s) 1994 | Goat Stocks (1000s) 1994 | Pig Stocks (1000s) 1994 | Equines (horses/ mules/asses) (1000s) 1994 | Buffalo Stocks (1000s) 1994 | Camel Stocks (1000s) 1994 |
|---|---|---|---|---|---|---|---|---|---|---|
| | **Agriculture/Food** | | | | | | | | | |
| Bangladesh | 3100 | 25041 | 37645 | 24130 | 1070 | 28050 | X | 0 | 874 | X |
| Bhutan | 34 | 635 | 704 | 435 | 59 | 42 | 75 | 58 | 4 | X |
| Brunei | 1 | X | X | 1 | 0 | 0 | 14 | 0 | 10 | X |
| Cambodia | 92 | 2532 | 3673 | 2589 | X | X | 2154 | 21 | 829 | X |
| India | 48000 | 223898 | 341944 | 192980 | 44809 | 118347 | 11780 | 2734 | 78825 | 1520 |
| Indonesia | 4597 | 35655 | 77797 | 11595 | 6411 | 12281 | 8720 | 714 | 3512 | X |
| Laos | 125 | 1460 | 2077 | 1137 | X | 153 | 1605 | 29 | 1308 | X |
| Malaysia | 340 | 2240 | 7592 | 686 | 336 | 356 | 3098 | 5 | 186 | X |
| Myanmar | 1068 | 8566 | 18971 | 9691 | 304 | 1113 | 2589 | 130 | 2130 | X |
| Nepal | 850 | 7748 | 8488 | 6546 | 914 | 5525 | 612 | 0 | 3176 | X |
| Papua New Guinea | X | 1241 | 1941 | 105 | 4 | 2 | 1033 | 2 | X | X |
| Philippines | 1580 | 11074 | 24415 | 1825 | 30 | 2800 | 8227 | 210 | 2630 | X |
| Singapore | X | 12 | 1322 | 0 | 0 | 1 | 150 | 0 | 0 | X |
| SriLanka | 550 | 3400 | 6642 | 1600 | 19 | 500 | 90 | 2 | 870 | X |
| Thailand | 4400 | 19057 | 30600 | 7593 | 98 | 136 | 4931 | 162 | 4257 | X |
| Viet Nam | 1860 | 20392 | 34887 | 3438 | X | 300 | 15043 | 134 | 3009 | X |

| Country | Total Commercial Energy Consumption (PJ) 1993 | Traditional Fuel Consumption (TJ) 1991 | Commercial Hydroelectric Production (PJ) 1993 | Known Mammal Species (#) 1990s | Endemic Mammal Species (#) 1990s | Known Bird Species (#) 1990s | Endemic Bird Species (#) 1990s | Known Plant Species (#) 1990s | Endemic Plant Species (#) 1990s |
|---|---|---|---|---|---|---|---|---|---|
| | **Energy** | | | **Biodiversity** | | | | | |
| Bangladesh | 313 | 276590 | 2.79 | 109 | 0 | 684 | 0 | 5000 | X |
| Bhutan | 2 | 11723 | 5.83 | 99 | 0 | 543 | 0 | 5446 | 75 |
| Brunei | 121 | 771 | 0.00 | X | X | X | X | X | X |
| Cambodia | 7 | 53841 | 0.25 | 123 | 0 | 429 | 0 | X | X |
| India | 9338 | 2823738 | 254.40 | 316 | 44 | 1219 | 55 | 15000 | 5000 |
| Indonesia | 2658 | 1464776 | 43.51 | 436 | 198 | 1531 | 393 | 27500 | 17500 |
| Laos | 5 | 38525 | 3.09 | 172 | 0 | 651 | 1 | X | X |
| Malaysia | 996 | 90197 | 17.73 | 286 | 27 | 736 | 9 | 15000 | 3600 |
| Myanmar | 71 | 192832 | 5.22 | 251 | 6 | 999 | 4 | 7000 | 1071 |
| Nepal | 19 | 205666 | 3.15 | 167 | 1 | 824 | 2 | 6500 | 315 |
| Papua New Guinea | 33 | 59799 | 1.66 | 214 | 57 | 708 | 80 | 10000 | X |
| Philippines | 787 | 381814 | 15.35 | 153 | 97 | 556 | 183 | 8000 | 3500 |
| Singapore | 745 | 0 | 0.00 | 45 | 1 | 295 | 0 | 2000 | 2 |
| Sri Lanka | 78 | 89415 | 13.67 | 88 | 13 | 428 | 23 | 3000 | 890 |
| Thailand | 1628 | 526385 | 13.32 | 265 | 7 | 915 | 3 | 11000 | X |
| Viet Nam | 316 | 250800 | 29.91 | 213 | 7 | 761 | 10 | >7000 | 1260 |

0 = zero or less than half the unit of measure; X = not available; all plant species includes flowering plants only.

## References


HD16101 and HD16104

**United Nations Population Division**, 1993: *Annual Populations* (The 1994 Revision), United Nations, New York, on diskette.

HD16405

**United Nations Population Division**, 1995: *Urban and Rural Areas, 1950-2025* (The 1994 Revision), U.N., New York, on diskette.

HD16419

**UN Statistical Office, UN Office for Ocean Affairs and the Law of the Sea, Offshore Magazine**, and other sources.

EI15124

**Summers**, Robert *et al.*, 1994: *The Penn World Table*, Mark 5.6, University of Pennsylvania, Philadelphia.

EI15105, EI15106, EI15107, EI15122

**World Bank**, 1994: *World Tables 1994*, World Bank, Washington, D.C.

LC17101, LC17102, LC17104, LC17105, LC17106

**United Nations Food and Agriculture Organization**, 1995: *FAOSTAT-PC*, FAO, Rome, on diskette.

FA18203, FA18509, FA18508, FA18301, FA18302, FA18303, FA18304, FA18305, FA18306, FA18307

**United Nations Food and Agriculture Organization**, 1995: *FAOSTAT-PC*, FAO, Rome, on diskette.

WA22102, WA22109

Compiled by the World Resources Institute from the following sources:

**Belyaev,** A.V., 1990: Institute of Geography, U.S.S.R. National Academy of Sciences, Moscow, personal communication.

**Economic Commission for Europe**, 1992: *The Environment in Europe and North America*, United Nations, New York, pp. 15–23.

**European Communities Commission**, 1990: *Environment Statistics 1989,* Office des Publications Officielles des Communautes Europeennes, Luxembourg, p. 130.

**Forkasiewicz**, J and J. Margat, 1980, *Tableau Mondial de Donnees Nationales d Economie de l Eau, Ressources et Utilisation*, Departement Hydrogeologie, Orleans, France.

**Gleick,** P., 1995: Pacific Institute, Oakland, California, personal communication.

**Margat**, J., 1988: Bureau de Recherches Geologiques et Minieres, Orleans, France, personal communicatio).

**Organisation for Economic Cooperation and Development (OECD)**, 1995: *OECDEnvironmental Data Compendium*, OECD, Paris.

**Solley**, W.B., R.R. Pierce, and H.A. Perlman, 1993: *Estimated Use of Water in the United States in 1990*, U.S. Geological Survey Circular, No. 1081, U.S. Geological Survey, Reston, Virginia.

**United Nations Economic Commission for Europe (ECE)**, 1995: *ECE Environmental Statistical Database*, on diskette, Statistical Division, UN/ECE.

**United Nations Food and Agriculture Organization**, 1995: *Water Resources of African Countries, A Review*, FAO, Rome, pp. 14-15.

EM21201, EM21106

**United Nations Statistical Division (UNSTAT)**, 1995: *1993 Energy Statistics Yearbook*, UNSTAT, New York.

EM21209

**United Nations Food and Agriculture Organization**, unpublished. Estimates were prepared by the Food and Agriculture Organization of the United Nations after an assessment of the available consumption data. Data were supplied by the answers to questionnaires or come from official publications.

BI20401, BI20402, BI20501

**World Conservation Monitoring Centre**, 1995: unpublished data, Cambridge, U.K.

**World Conservation Monitoring Centre**, 1994: *Biodiversity Data Sourcebook*, World Conservation Press, Cambridge, U.K.

**World Conservation Monitoring Centre**, 1992: *Global Biodiversity Status of the Earth's Living Resources*, Chapman and Hall, London.

# E

# Color Plates

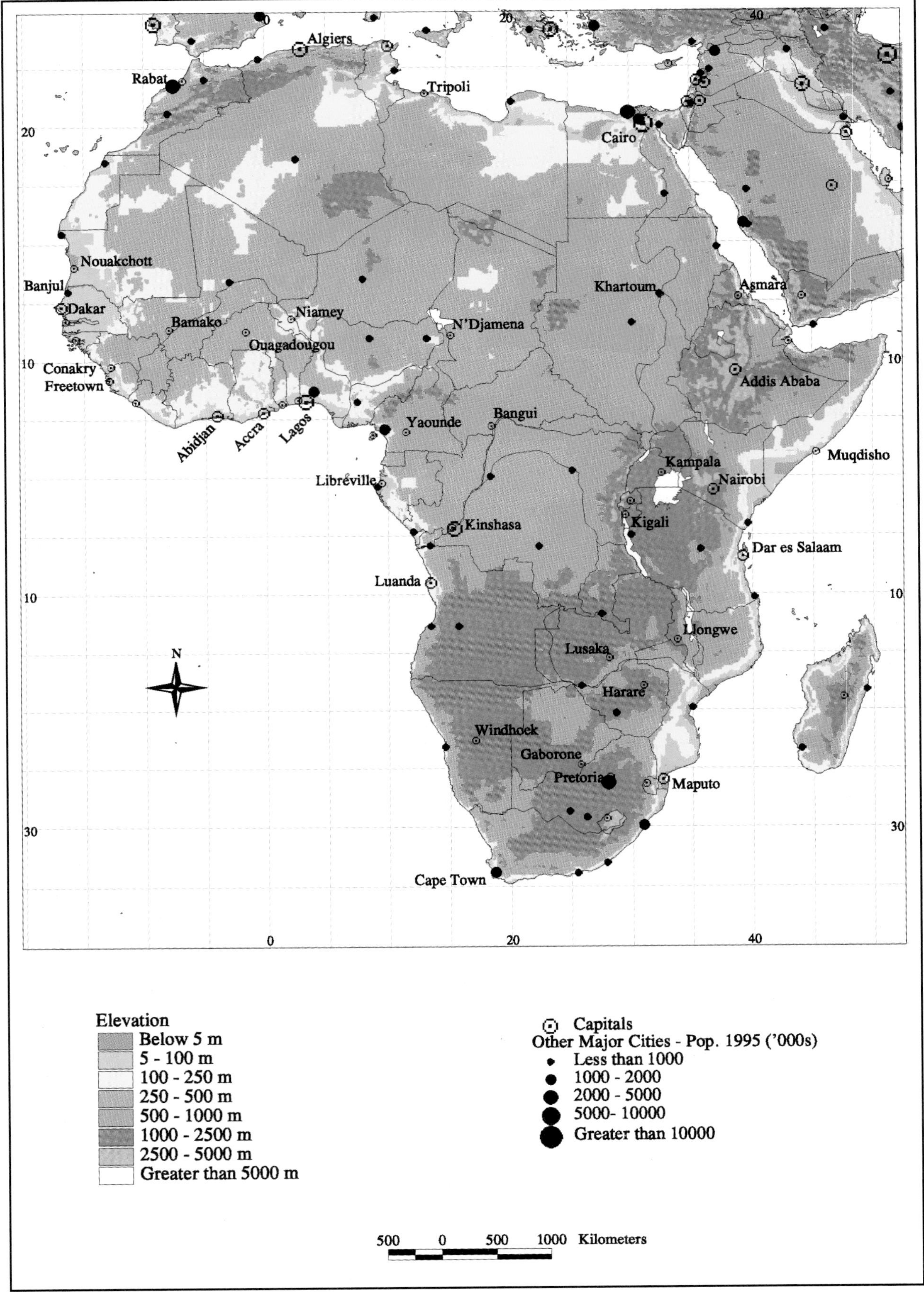


**Figure 2-1:** The Africa region (courtesy of David Gray/The World Bank).

**Box 2-10. The Potential Impact of Long-Term Climate Change on Vector-Borne Diseases: The Case of Malaria**

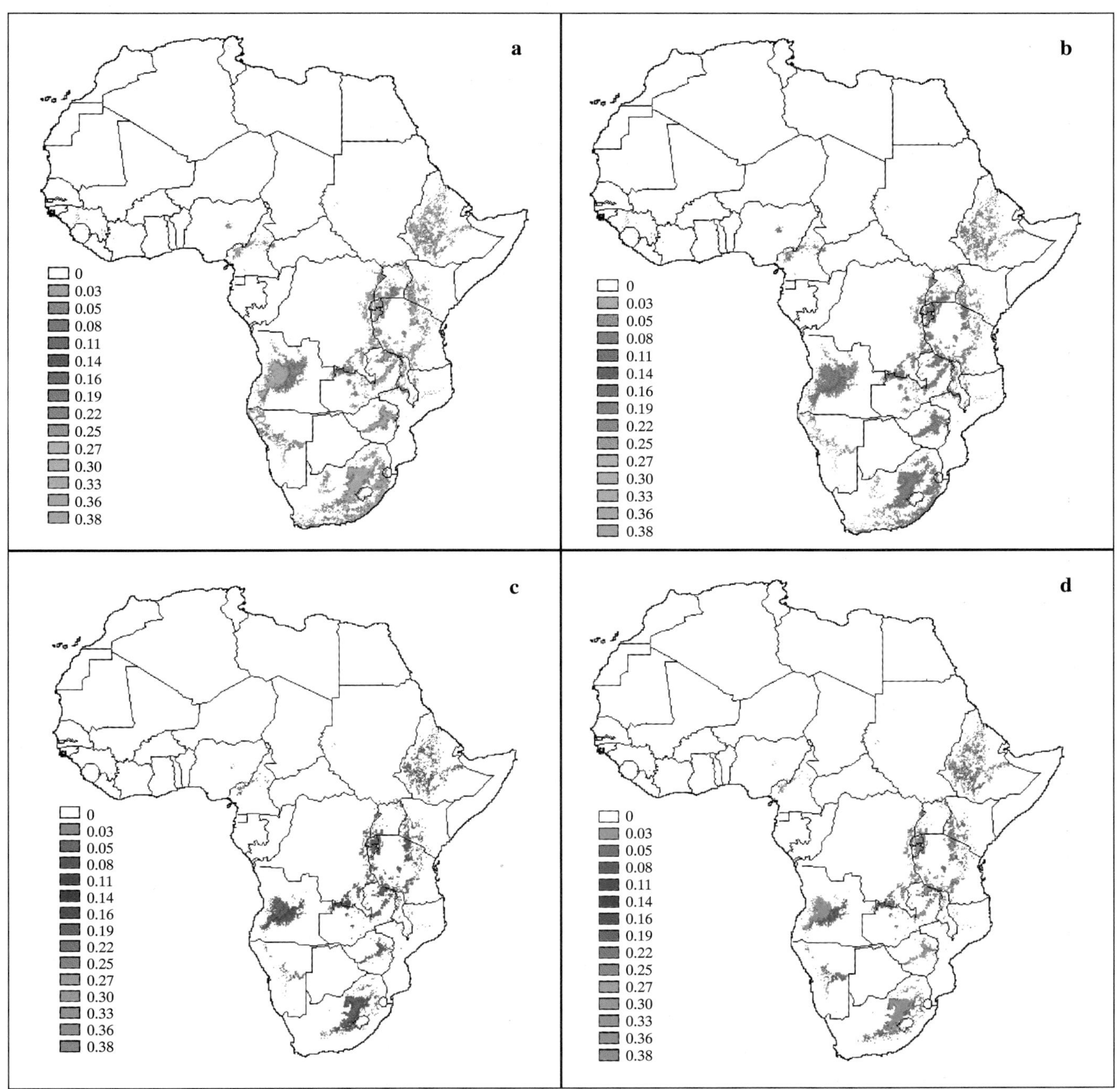


*Degree of Change: a) 5-month model with 1°C increase, b) 5-month model with 0.5°C increase, c) 5-month model with 0.5°C increase at elevation >1,300 m, and d) 5-month model with 1°C increase at elevation >1,300 m.*

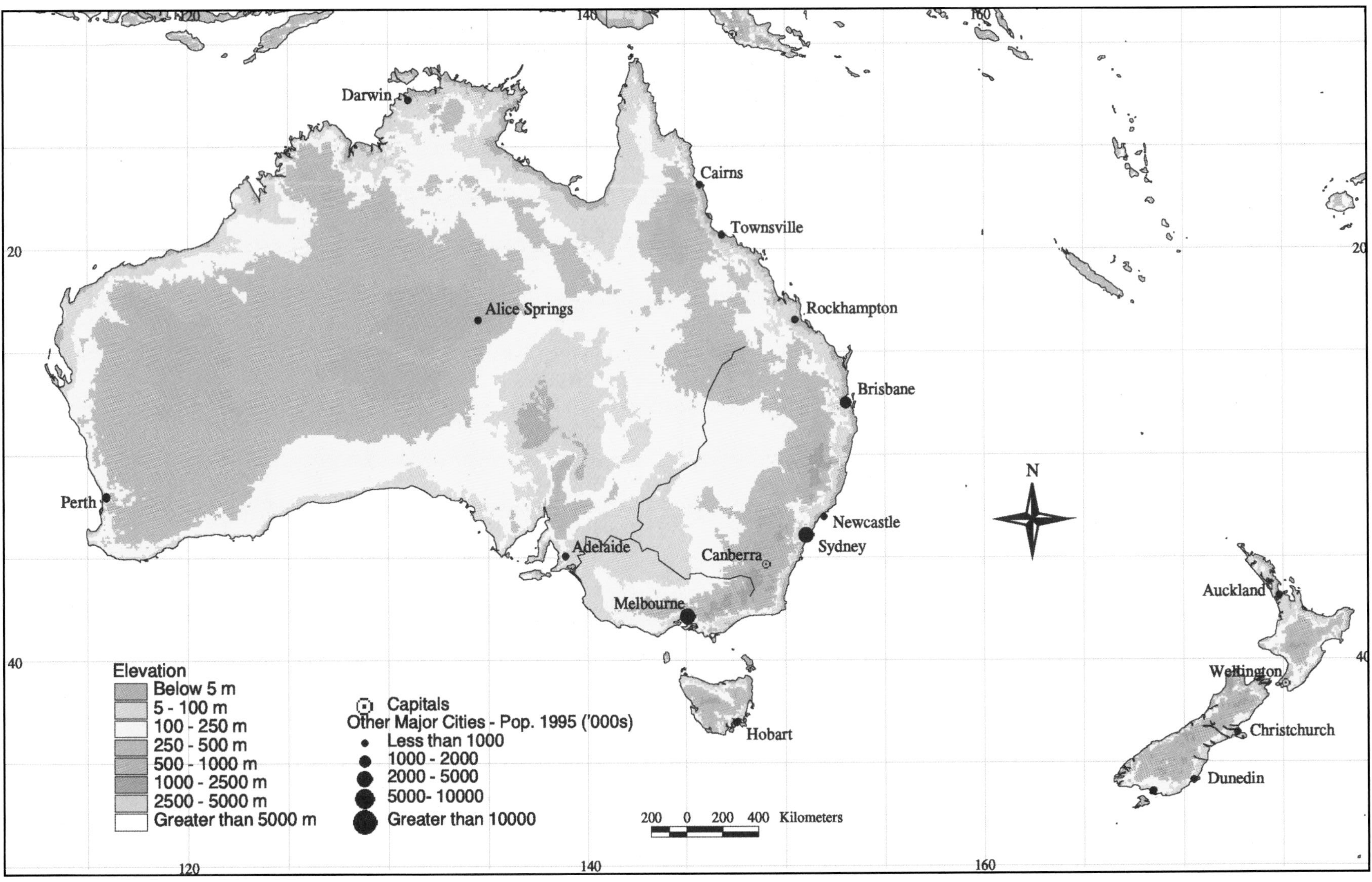


**Figure 4-1:** The Australasia region (courtesy of David Gray/The World Bank).

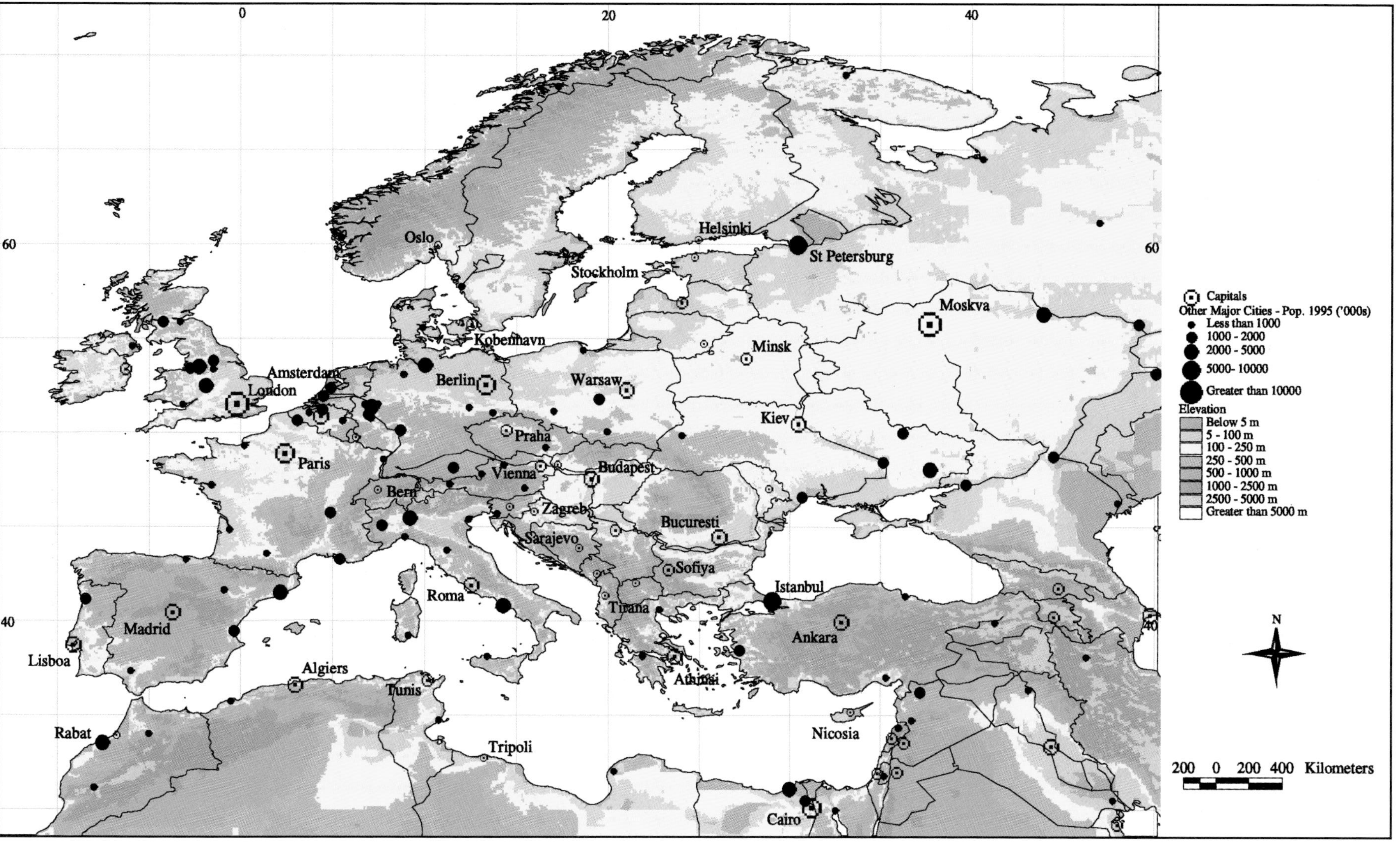


**Figure 5-1:** The Europe region (courtesy of David Gray/The World Bank).

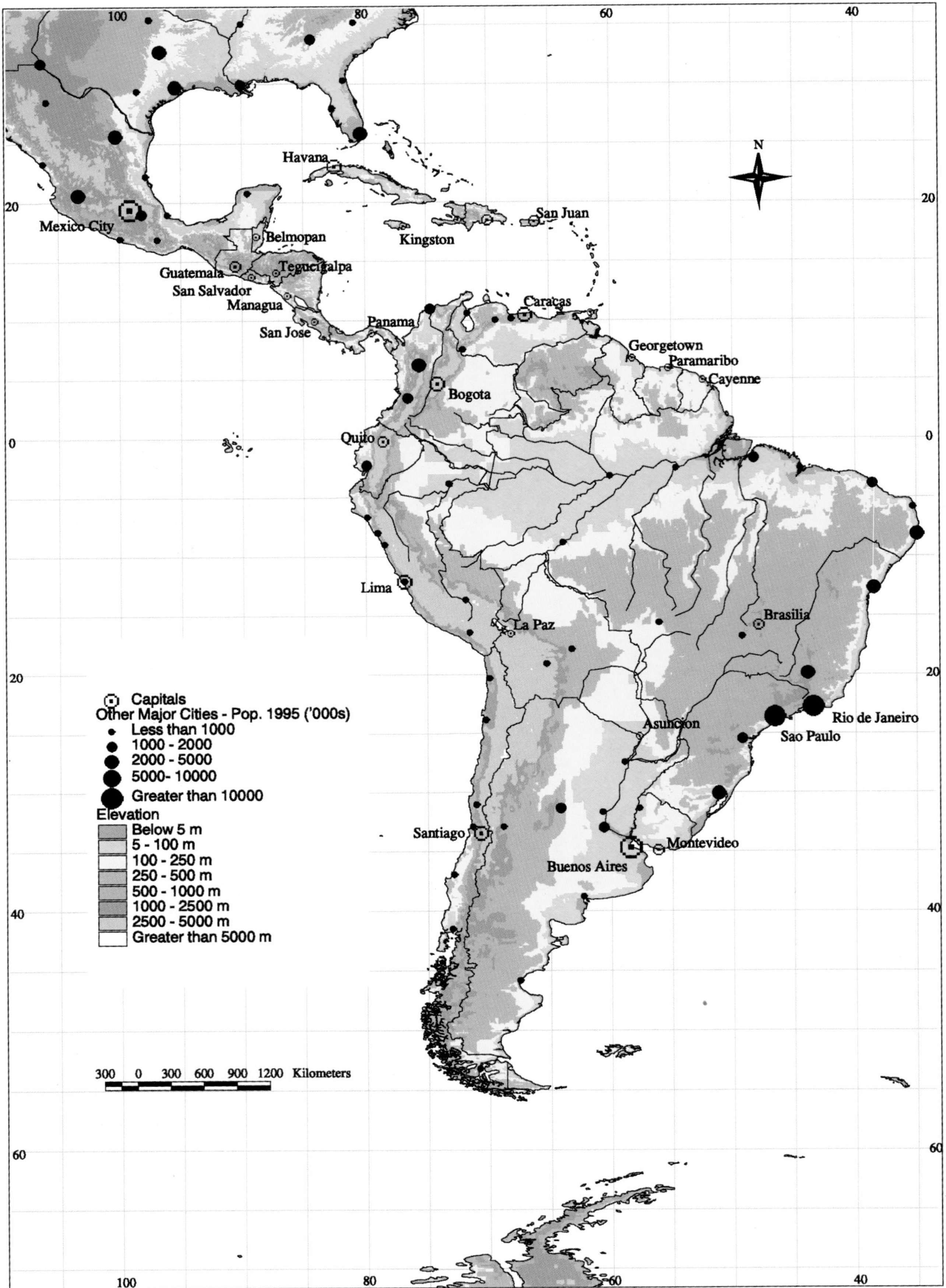


**Figure 6-1:** The Latin America region (courtesy of David Gray/The World Bank).

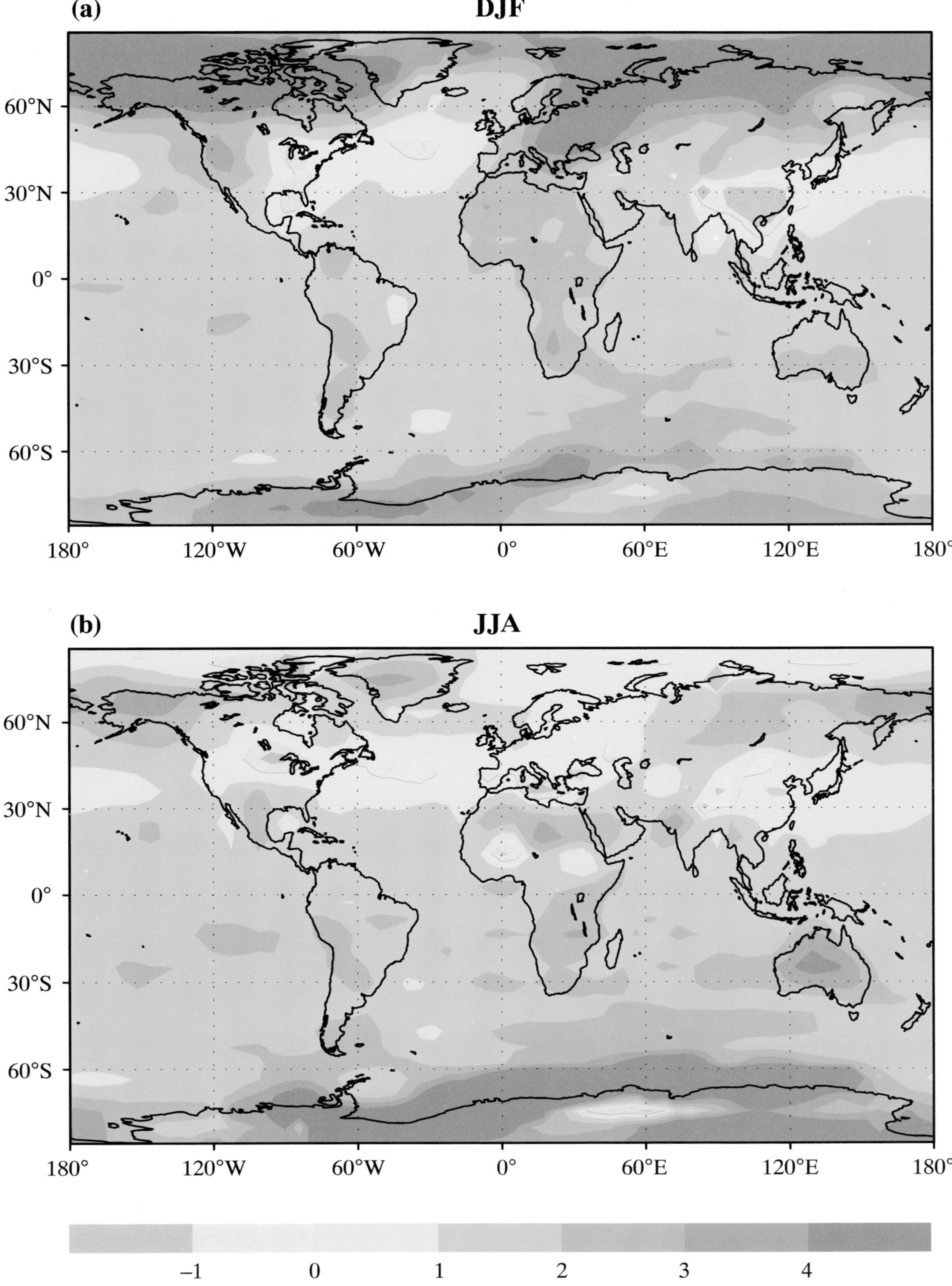


**Figure 6-2:** Seasonal change in surface temperature from 1880–1889 to 2040–2049 in simulations with aerosol effects included. Contours are at every 1°C (IPCC 1996, WG I, Figure 6.10).

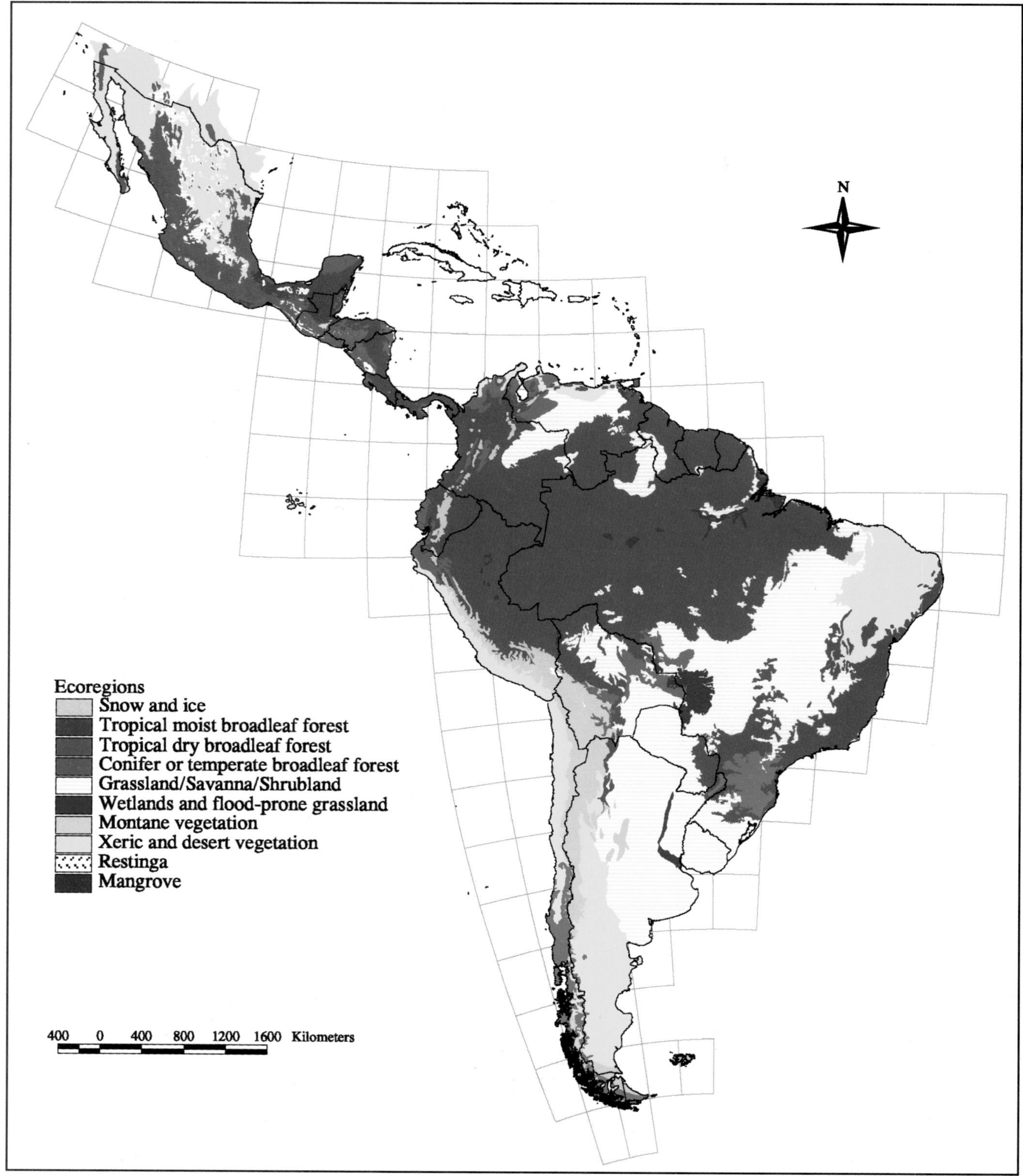


**Figure 6-4:** Major biomes in Latin America (courtesy of David Gray/The World Bank).

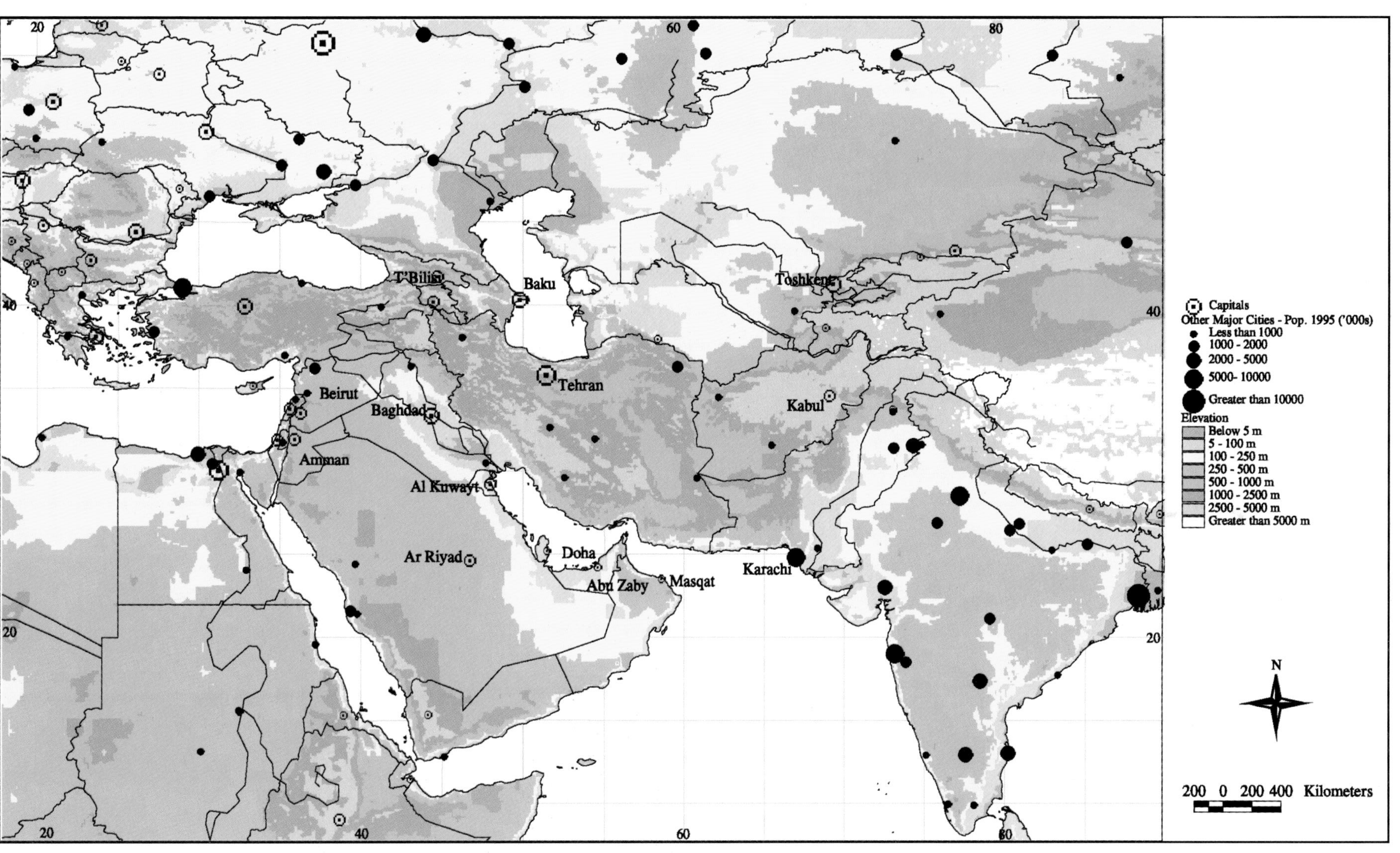


**Figure 7-1:** The Middle East and Arid Asia region (courtesy of David Gray/The World Bank).

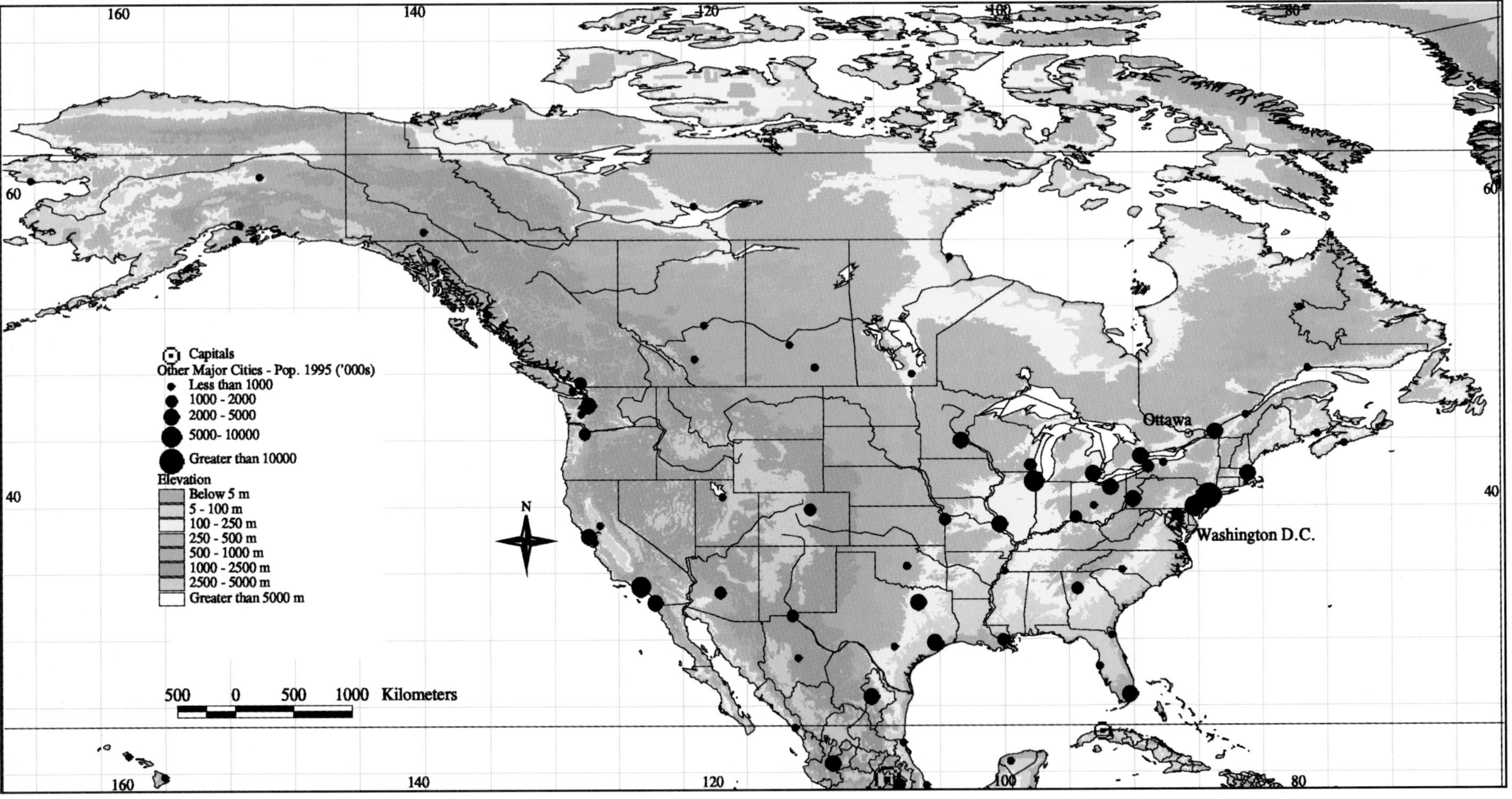


**Figure 8-1:** The North America region (courtesy of David Gray/The World Bank).

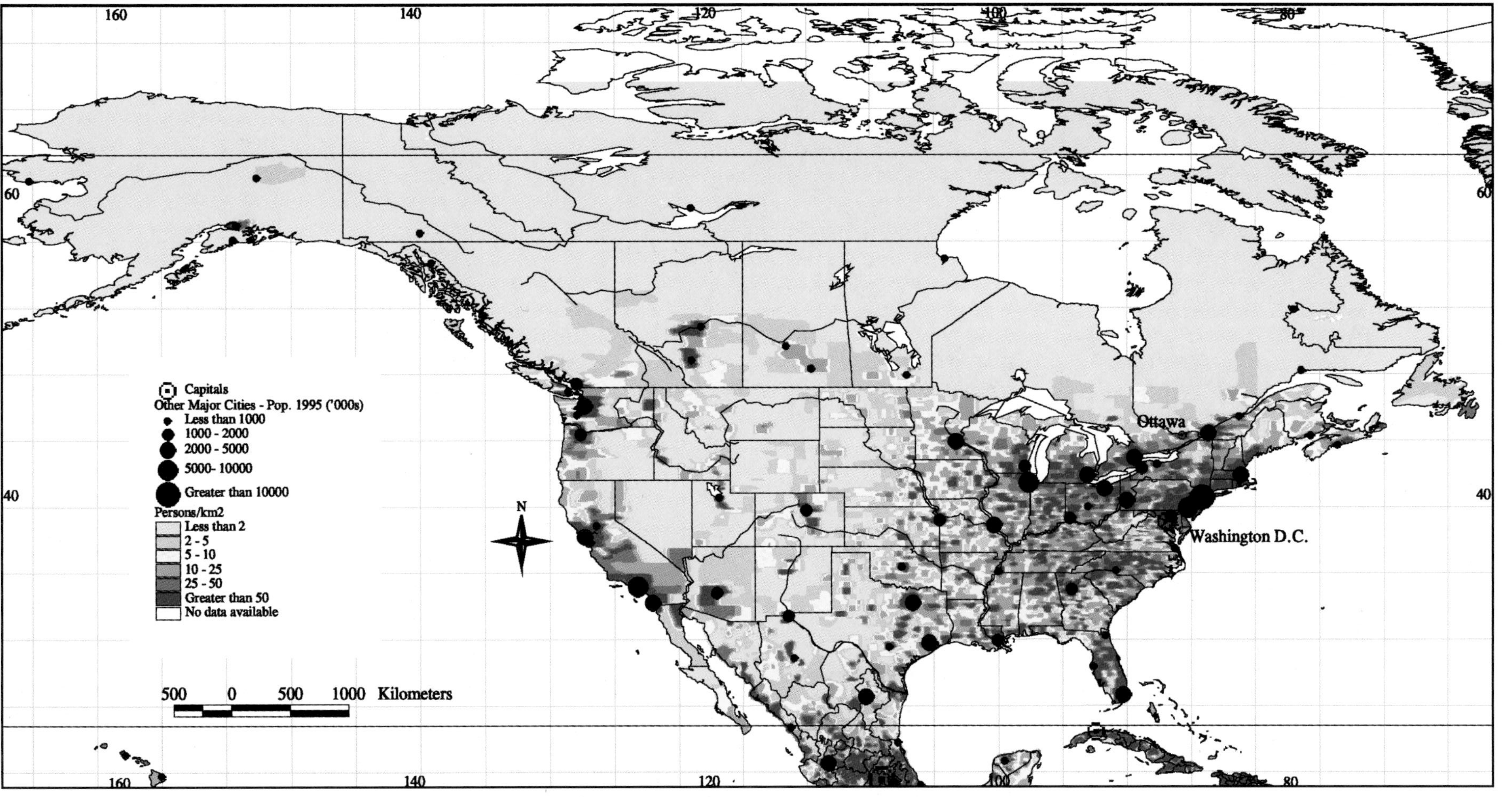


**Figure 8-2:** North American population density (courtesy of David Gray/The World Bank).

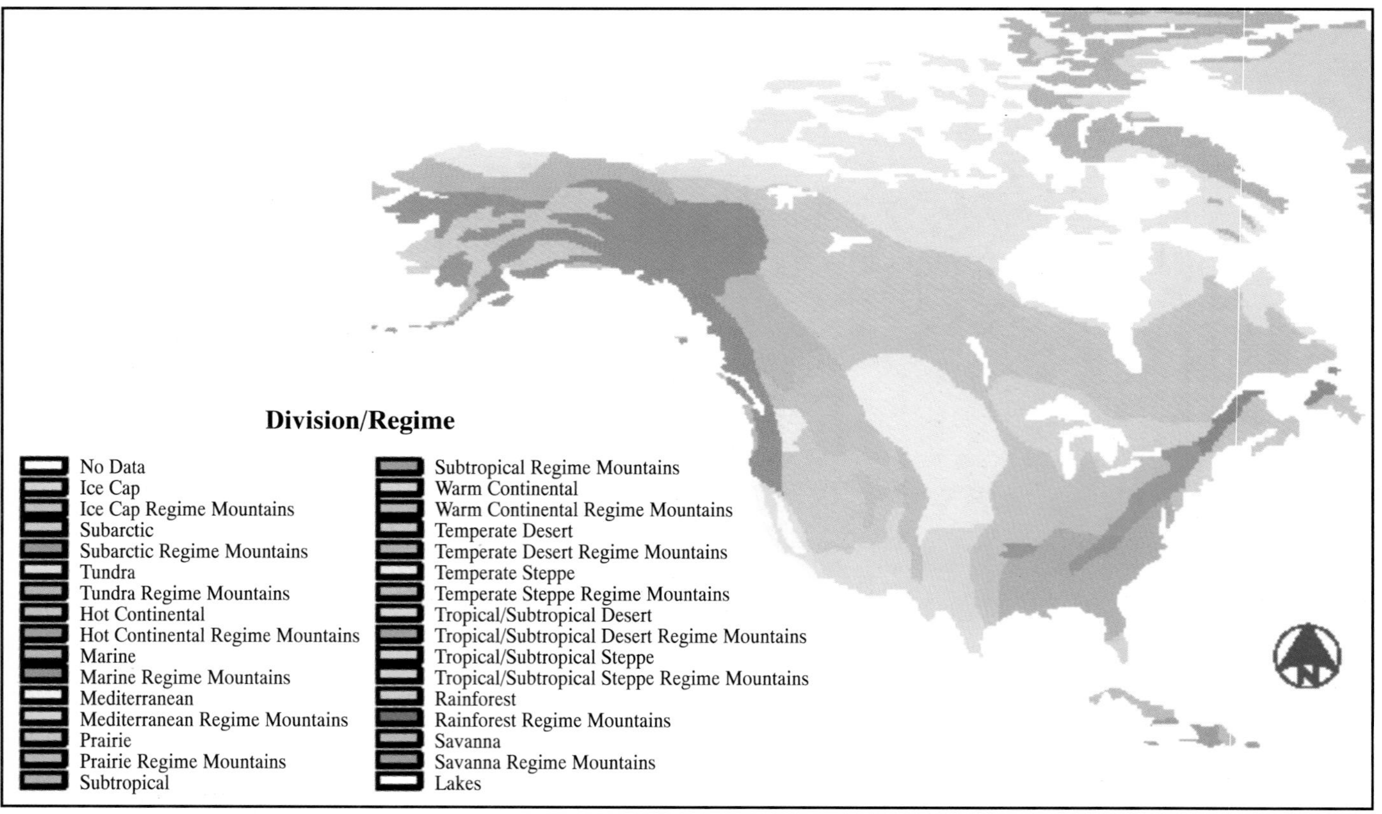


**Figure 8-3:** Ecoregions for North America (courtesy of R.A. Washington-Allen/ORNL), based on data provided by the National Geophysical Data Center.

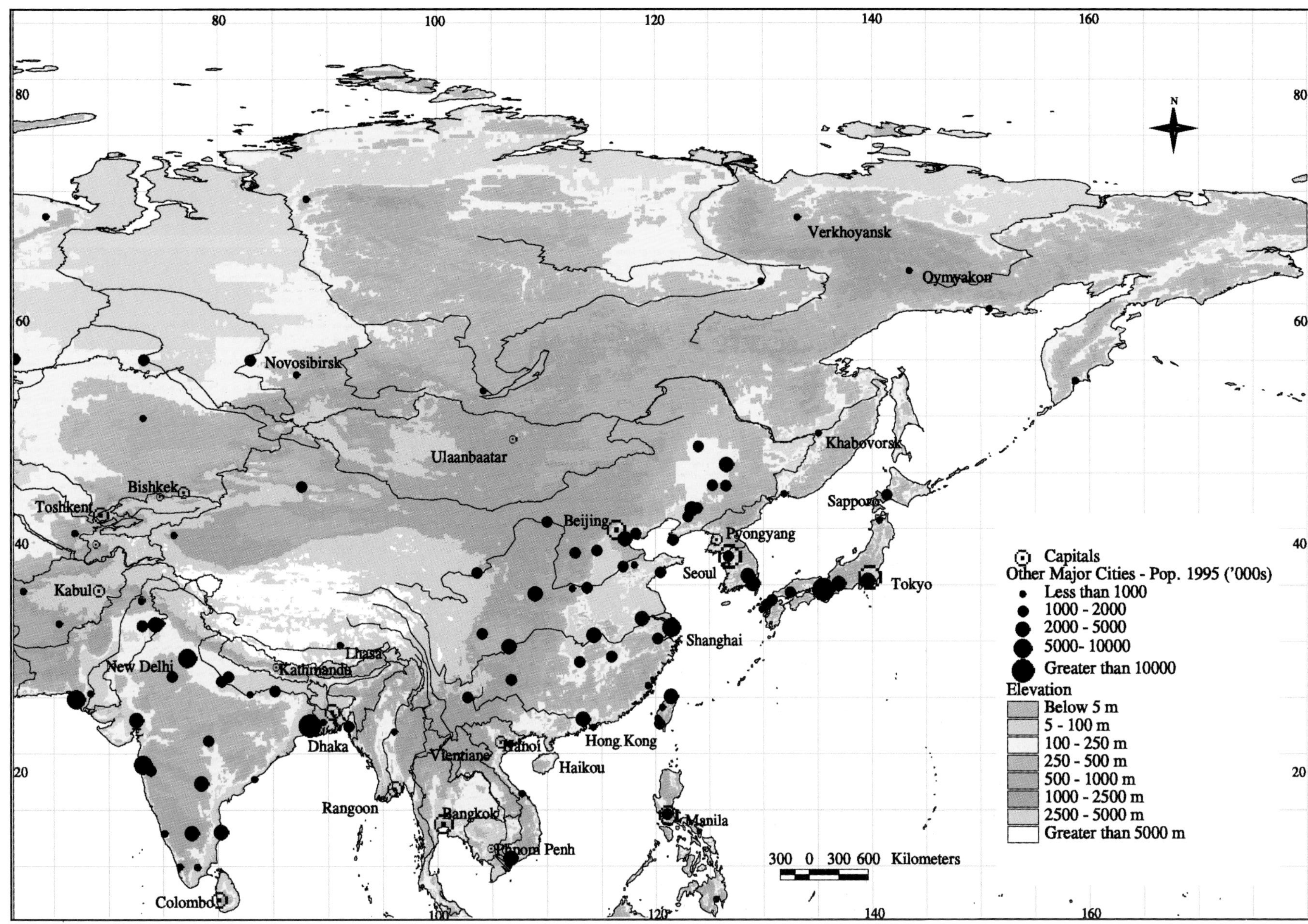


**Figure 10-1:** The Temperate Asia region (courtesy of David Gray/The World Bank).

# F

# Glossary of Terms

**acclimatization**
physiological adaptation to climatic variations

**active layer**
the top layer of soil in permafrost that is subjected to seasonal freezing and thawing

**adaptability**
the degree to which adjustments are possible in practices, processes, or structures of systems to projected or actual changes of climate; adaptation can be spontaneous or planned, and can be carried out in response to or in anticipation of changes in conditions

**afforestation**
forest stands established artificially on lands that previously have not supported forests for more than 50 years

**agroclimatic**
climatic conditions as they relate to agricultural production; discrete set of zones each of which identifies areas capable of like types and levels of agricultural production

**albino**
the surface reflectivity of the globe

**alevin**
a young fish; especially, a newly hatched salmon when still attached to the yolk sac

**algal blooms**
a reproductive explosion of algae in a lake, river, or ocean

**alpine**
the biogeographic zone made up of slopes above timberline and characterized by the presence of rosette-forming herbaceous plants and low shrubby slow-growing woody plants

**anadromous species**
species of fish, such as salmon, that spawn in fresh water and then migrate into the ocean to grow to maturity

**anaerobic**
living, active, or occurring in the absence of free oxygen

**annual plants**
terrestrial plants that complete their life cycle in one growing season; plants that die off each year during periods of temperature and moisture stress but leave behind seeds to germinate during the next favorable climatic season

**anoxia**
a deficiency of oxygen, especially of such severity as to result in permanent damage

**anoxic**
greatly deficient in oxygen

**anthropogenic**
caused or produced by humans

**anticyclone system**
a system of winds that rotates about a center of high atmospheric pressure (clockwise in the Northern Hemisphere and counterclockwise in the Southern Hemisphere)

**aquifer**
permeable water-bearing formation capable of yielding exploitable quantities of water

**arbovirus**
any of various viruses transmitted by arthropods and including the causative agents of dengue fever, yellow fever, and some encephalitis

**arid lands**
ecosystems with <250 mm precipitation per year

**autochthonous**
indigenous; formed or originating in the place where found

**autonomous adaptation**
adaptation that occurs without specific human intervention

**baseline scenario**
the set of predicted levels of economic growth, energy production and consumption, and greenhouse gas emissions assumed as the starting point for an analysis of mitigation options

**basin**
drainage area of a stream, river, or lake

**billabong**
cut-off meander

**biodiversity**
the number of different species or functional groups of flora and fauna found in an area or ecosystem

**biofuels**
fuels obtained as a product of biomass conversion (e.g., alcohol or gasohol)

**biogas**
a gas composed principally of a mixture of methane and carbon dioxide produced by anaerobic digestion of biomass

**biogeography**
the study of the geographical distribution of living organisms

**biomass**
the total quantity of living matter in a particular habitat; plant and organic waste materials used as fuel and feedstock in place of fossil fuels

**biome**
a grouping of similar plant and animal communities into broad landscape units that occur under similar environmental conditions

**bog**
a poorly drained (usually acid) area rich in accumulated plant material, frequently surrounding a body of open water, and having a characteristic flora (as of sedges, heaths, and sphagnum)

**bottom-up modeling**
a modeling approach that arrives at economic conclusions from an analysis of the effect of changes in specific parameters on narrow parts of the total system

**building stock**
the residential and/or commercial structures extant in a society or a geographic area

**$C_3$ plants**
plants that produce a three-carbon compound during photosynthesis, including most trees and agricultural crops such as rice, wheat, soybeans, potatoes, and vegetables

**$C_4$ plants**
plants that produce a four-carbon compound during photosynthesis; mainly of tropical origin, including grasses and the agriculturally important crops maize, sugar cane, millet, and sorghum

**calorie (food)**
1000 (technical) calories

**calving**
the breaking away of a mass of ice from a floating glacier, ice front, or iceberg

**CAM**
variant of the $C_4$ photosynthetic pathway in which most gas exchange occurs at night; occurs primarily in succulents (e.g., cacti)

**carbon intensity**
$CO_2$ emissions per unit of energy or economic output

**carbon sequestration**
the biochemical process through which carbon in the atmosphere is absorbed by biomass such as trees, soils, and crops

**carbon sinks**
chemical processes that absorb carbon dioxide

**carbon stocks**
the amount of carbon that is stored in carbon sinks

**carbon tax**
a levy exacted by a government on the use of carbon-containing fuels for the purpose of influencing human behavior (specifically economic behavior) to use less fossil fuels

**carrying capacity**
the number of individuals in a population that the resources of a habitat can support

**catchment**
area having a common outlet for its surface runoff

**Chagas' disease**
a parasitic disease caused by the Trypanosoma cruzi and transmitted by triatomine bugs in the Americas, with two clinical periods: Acute (fever, swelling of the spleen, edemas) and chronic (heart disorder that may produce high fatality, or digestive syndrome)

**$CO_2$ fertilization**
the enhancement of plant growth as a result of elevated atmospheric $CO_2$ concentration

**communicable disease**
infectious disease caused by transmission of an infective biological agent (virus, bacterium, protozoan, or multicellular macroparasite)

**cryosphere**
all global snow, ice, and permafrost

**Dengue fever**
an infectious viral disease spread by mosquitoes, the first infection of which is often called breakbone fever and is characterized by severe pain in joints and back, fever, and rash; a subsequent infection is usually characterized by fever, bleeding from bodily orifices, and sometimes death

**desert**
an ecosystem with $<100$ mm precipitation per year

**determinate crops**
crops characterized by sequential flowering from the central or uppermost bud to the lateral or basal buds; also, crops characterized by growth in which the main stem ends in an inflorescence and stops growing with only branches from the main stem having further and similarly restricted growth

**diapause**
period of suspended growth or development and reduced metabolism in the life cycle of many insects, when organism is more resistant to unfavorable environmental conditions than in other periods

**dissolved load**
the amount of particles in a stream or other water source that arises as a result of erosion diurnal climate a climate with uniform amplitudes of temperature throughout the year

**econometric**
an approach to studying a problem through use of mathematical and statistical methods in the field of economics to develop and verify theories

**economies in transition**
national economies that are moving from a period of heavy government control toward lessened intervention, increased privatization, and greater use of competition

**ecotax**
a levy exacted by a government for the purpose of influencing human behavior (specifically economic behavior) to follow an ecologically benign path

**ecotone**
transition area between adjacent ecological communities (e.g., between forests and grasslands), usually involving competition between organisms common to both

**ecotopic**
tendency or involving adjustment to specific habitat conditions

**edaphic**
of or relating to the soil; factors inherent in the soil

**El Niño**
an irregular variation of ocean current that, from January to February, flows off the west coast of South America, carrying warm, low-salinity, nutrient-poor water to the south; does not usually extend farther than a few degrees south of the Equator, but occasionally it does penetrate beyond 12°S, displacing the relatively cold Peruvian current; usually short-lived effects, but sometimes last more than a year, raising sea-surface temperatures along the coast of Peru and in the equatorial eastern Pacific Ocean, having disastrous effects on marine life and fishing

**endemic**
restricted or peculiar to a locality or region

**endemic infection**
a sustained, relatively stable, pattern of infection within a specified population

**energy efficiency**
ratio of energy output of a conversion process or of a system to its energy input; also known as first-law efficiency

**energy intensity**
ratio between the consumption of energy to a given quantity; usually refers to the amount of primary or final energy consumed per unit of gross domestic or national product

**epidemic**
appearance of an abnormally high number of cases of infection in a given population; can also refer to noninfectious diseases (e.g., heart disease) or to acute events such as chemical toxicity

**epilimnion**
the water layer overlying the thermocline of a lake

**euphotic zone**
the upper layers of a body of water into which sufficient light penetrates to permit photosynthesis

**eustatic sea-level rise**
worldwide rise in sea level

**eutrophication**
the process by which a body of water (often shallow) becomes (either naturally or by pollution) rich in dissolved nutrients with a seasonal deficiency in dissolved oxygen

**evapotranspiration**
loss of water from the soil both by evaporation from the surface and transpiration from the plants growing thereon

**exergy**
the maximum amount of energy that under given (ambient) thermodynamic conditions can be converted into any other form or energy; also known as availability or work potential

**exergy efficiency**
the ratio of (theoretical) minimum exergy input to actual input of a process or a system; also known as second-law efficiency

**extant**
currently or actually existing

**extinction**
complete disappearance of an entire species

**extirpation**
disappearance of a species from part of its range; local extinction

**fallow**
land left unseeded after plowing; uncultivated

**fast ice**
sea or lake ice that remains tied to the coast (usually >2 m above sea level)

**feedback**
when one variable in a system triggers changes in a second variable that in turn ultimately affects the original; a positive feedback intensifies the effect, and a negative reduces the effect

**fen**
low land covered wholly or partly with water unless artificially drained

**forest**
an ecosystem in which the dominant plants are trees; woodlands are distinguished from forests by their lower density of trees

**forestation**
generic term for establishing forest stands by reforestation and afforestation

**forest decline**
premature, progressive loss of tree and stand vigor and health

**frazil ice**
fine spicules or plates of ice in suspension in water

**friable soils**
soils that are easily crumbled or pulverized

**geomorphic**
of or related to the form of the Earth or its surfaces

**greenhouse gas**
any gas that absorbs infrared radiation in the atmosphere

**gross primary production**
the amount of carbon fixed in photosynthesis by plants

**ground ice**
ice present within rock, sediments, or soil

**groundwater recharge**
process by which external water is added to the zone of saturation of an aquifer, either directly into a formation or indirectly by way of another formation

**halocarbons**
chemicals containing carbon and members of the halogen family

**halophyte species**
a plant (as saltbush or sea lavender) that grows in salty soil and usually has a physiological resemblance to a true xerophyte

**heath**
any of the various low-growing shrubby plants of open wastelands, usually growing on acidic, poorly drained soils

**heat island**
an area within an urban area characterized by ambient temperatures higher than those of the surrounding area because of the absorption of solar energy by materials like asphalt

**herbaceous**
flowering, non-woody plants

**herbivore**
an animal that feeds on plants

**hydroperiod**
the depth, frequency, duration, and season of wetland flooding

**hypolimnion**
the part of a lake below the thermocline made up of water that is stagnant and of essentially uniform temperature except during the period of overturn

**ice cap**
a dome-shaped glacier covering a highland area (considerably smaller in extent than ice sheets)

**ice jam**
an accumulation of broken river or sea ice caught in a narrow channel

**ice sheet**
a mass of snow and ice of considerable thickness and large area greater than 50,000 $km^2$

**ice shelf**
a floating ice sheet of considerable thickness attached to a coast (usually of great horizontal extent with a level or gently undulating surface); often a seaward extension of ice sheets

**icing**
a sheet-like mass of layered ice formed by the freezing of water as it emerges from the ground or through fractures in river or lake ice

**immunosuppression**
reduced functioning of an individual's immune system

**incidence**
the number of cases of a disease commencing, or of persons falling ill, during a given period of time within a specified population

**industrial ecology**
the set of relationships of a particular industry with its environment; often refers to the conscious planning of industrial processes so as to minimize their negative interference with the surrounding environment

**industrialization**
the conversion of a society from one based on manual labor to one based on the application of mechanical devices

**infiltration**
flow of water through the soil surface into a porous medium

**infrastructure**
the basic installations and facilities upon which the operation and growth of a community depend, such as roads; schools; electric, gas, and water utilities; transportation and communications systems; and so on

**inoculation**
the introduction of a pathogen or antigen into a living organism to stimulate the production of antibodies

**inselberg**
an isolated mountain or granite outcropping

**isohyet**
a line on a map or chart indicating equal rainfall

**international dollars**
values obtained using special conversion factors that equalize the purchasing powers of different currencies (i.e., the number of units of a country's currency required to buy the same amounts of goods and services in the domestic market as $1 would buy in the "average" country), thus equalizing dollar prices in every country so that cross-country comparisons of GDP reflect differences in quantities of goods and services free of price-level differentials

**isotherms**
geographic bands of similar temperatures

**keystone species**
a species that has a central servicing role affecting many other organisms and whose demise is likely to result in the loss of a number of species and lead to major changes in ecosystem function

**land use**
the purpose an area of the Earth is put to (e.g., agriculture, forestry, urban dwellings, or transportation corridors) or its character (e.g., swamp, grassland, or desert)

**lapse rate**
the rate of temperature decrease with increase in altitude

**leaching**
the removal of soil elements or applied chemicals through percolation

**legume**
plants that through a symbiotic relationship with soil bacteria are able to fix nitrogen from the air (e.g., peas, beans, alfalfa, clovers)

**lichen**
symbiotic organisms consisting of an alga and fungus important to the weathering and breakdown of rocks

**life-cycle cost**
the cost of a good or service over its entire lifetime

**littoral zone**
a coastal region; the shore zone between high and low watermarks

**low emissivity**
a property of materials that hinders or blocks the transmission of a particular band of radiation (e.g., that in the infrared)

**macroeconomic**
pertaining to a study of economics in terms of whole systems, especially with reference to general levels of output and income and to the interrelations among sectors of the economy

**malaria**
endemic or epidemic parasitic disease caused by species of the genus Plasmodium (protozoa) and transmitted by mosquitoes of the genus Auopheles; produces high fever attacks and systemic disorders, and kills ~2 million people every year

**market equilibrium**
the point at which demand for goods and services equals the supply; often described in terms of the level of prices, determined in a competitive market, that "clears" the market

**market penetration**
the percentage of all its potential purchasers to which a good or service is sold per unit time

**miombo**
deciduous tropical woodland and dry forest ecosystems dominated by trees in the genera *Brachystegia*, *Julbernardia*, and *Isoberlinia* of the family *Fabaceae*, subfamily *Caesalpinioideae*

**mitigation**
an anthropogenic intervention to reduce the emissions or enhance the sinks of greenhouse gases

**monsoon**
wind in the genereal atmospheric circulation typified by a seasonal persistent wind direction and by a pronounced change in direction from one season to the next

**montane**
the biogeographic zone made up of relatively moist, cool upland slopes below timberline and characterized by the presence of large evergreen trees as a dominant life form

**mopane**
woodland of *Colophospermum mopane*—a multi-purpose hardwood tree species used for fodder, house building, and fuelwood

**moraine**
an accumulation of Earth and stones carried and finally deposited by a glacier

**morbidity**
the rate of occurrence of disease or other health disorder within a population, taking account of the age-specific morbidity rates; health outcomes include, for example, chronic disease incidence or prevalence, rates of hospitalization, primary care consultations, disability-days (e.g., of lost work), and prevalence of symptoms

**morphology**
the form and structure of an organism or any of its parts

**morphometry**
measurement of external form

**mortality**
the rate of occurrence of death within a population within a specified time period; calculation of mortality takes account of age-specific death rates, and can thus yield measures of life expectancy and the extent of premature death

**mycosis**
infection with or disease caused by a fungus

**net ecosystem production**
the net gain or loss of carbon from an ecosystem or region

**net primary production**
the increase in plant biomass or carbon of a unit of a landscape; gross primary production (all carbon fixed through photosynthesis) minus plant respiration equals net primary production

**nitrification**
the oxidation of ammonium salts to nitrites and the further oxidation of nitrites to nitrates

**$NO_x$**
any of several oxides of nitrogen

**non-tidal wetlands**
areas of land not subject to tidal influences where the water table is at or near the surface for some defined period of time, leading to unique physiochemical and biological processes and conditions characteristic of water-logged systems

**northern wetlands**
wetlands in the boreal, subarctic, and arctic regions of the northern hemisphere

**obligate species**
species restricted to one particularly characteristic mode of life

**orography**
the branch of physical geography that deals with mountains and mountain systems

**pack ice**
any area of sea, river, or lake ice other than fast ice

**paleoecology**
the branch of ecology concerned with identifying and interpreting the relationships of ancient plants and animals to their environment

**paludism**
malaria

**pancake ice**
new ice about 0.3 to 3 m in diameter, with raised rims about the circumference from striking other pieces

**peat**
unconsolidated soil material consisting largely of partially decomposed organic matter accumulated under conditions of excess moisture or other conditions that decrease decomposition rates

**pelagic**
of, relating to, or living or occurring in the open sea

**perennial plants**
plants that persist for several years, usually with new herbaceous growth from a perennating part

**permafrost**
perennially frozen ground that occurs wherever the temperature remains below 0°C for several years

**phenology**
the study of natural phenomena that recur periodically (e.g., blooming, migrating) and their relation to climate and seasonal changes

**photochemical smog**
a mix of photochemical oxidant air pollutants produced by the reaction of sunlight with primary air pollutants

**photoperiodic response**
response to the lengths of alternating periods of light and dark as they affect the timing of development

**physiographic**
of, relating to, or employing a description of nature or natural phenomena

**phytophagous insects**
insects that feed on plants

**potential evapotranspiration**
maximum quantity of water capable of being evaporated in a given climate from a continuous stretch of vegetation (i.e., includes evaporation from the soil and transpiration from the vegetation of a specified region in a given time interval, expressed as depth)

**potential production**
estimated production of a crop under conditions when nutrients and water are available at optimum levels for plant growth and development; other conditions such as daylength, temperature, soil characteristics, etc., determined by site characteristics

**prevalence**
the proportion of persons within a population who are currently affected by a particular disease

**primary energy**
the energy that is embodied in resources as they exist in nature (e.g., coal, crude oil, natural gas, uranium, or sunlight); the energy that has not undergone any sort of conversion

**purchasing power parity (PPP)**
GDP estimates based on the purchasing power of currencies rather than on current exchange rates; such estimates are a blend of extrapolated and regression-based numbers, using the results of the International Comparison Program (ICP); PPP estimates tend to lower per capita GDPs in industrialized countries and raise per capita GDPs in developing countries

**radiative forcing**
a change in average net radiation at the top of the troposphere resulting from a change in either solar or infrared radiation due to a change in atmospheric greenhouse gases concentrations; perturbance in the balance between incoming solar radiation and outgoing infrared radiation

**rangeland**
unimproved grasslands, shrublands, savannas, and tundra

**reference scenario**
the set of predicted levels of economic growth, energy production and consumption, and greenhouse gas emissions (and underlying assumptions) with which other scenarios examining various policy options are compared

**reforestation**
forest stands established artificially on lands that have supported forests within the last 50 years

**reserves**
those occurrences of energy sources or minerals that are identified and measured as economically and technically recoverable with current technologies and prices

**resources**
those occurrences of energy sources or minerals with less certain geological and/or economic/technical recoverability characteristics, but that are considered to become potentially recoverable with foreseeable technological and economic development

**respiration**
the metabolic process by which organisms meet their internal energy needs and release $CO_2$

**riparian**
relating to or living or located on the bank of a natural watercourse (as a river) or sometimes of a lake or a tidewater

**runoff**
water (from precipitation or irrigation) that does not evaporate or seep into the soil but flows into rivers, streams, or lakes, and may carry sediment

**ruderal**
pertaining to or inhabiting highly disturbed sites; weedy

**salinization**
the accumulation of salts in soils

**saltation**
the transportation of particles by currents of water or wind in such a manner that they move along in a series of short intermittent leaps

**seasonal climate**
a climate characterized by both warm and cold periods through the year

**semi-arid lands**
ecosystems that have >250 mm precipitation per year, but are not highly productive; usually classified as rangelands

**senescence**
the growth phase in a plant or plant part (as a leaf) from full maturity to death

**sensitivity**
the degree to which a system will respond to a change in climatic conditions (e.g., the extent of change in ecosystem composition, structure and functioning, including net primary productivity, resulting from a given change in temperature or precipitation)

**sequestration**
to separate, isolate or withdraw; usually refers to removal of $CO_2$ from atmosphere by plants or by technological measures

**set-aside program**
a generic term covering a variety of government programs—primarily in the U.S., Canada, and Europe that require farmers to remove a portion of their acreage from production for purposes of controlling yield, soil conservation, etc.

**shelterbelt**
a natural or artificial forest maintained for protection against wind or snow

**silt**
unconsolidated or loose sedimentary material whose constituent rock particles are finer than grains of sand and larger than clay particles

**slake**
to cause to heat and crumble by treatment with water

**slip faces**
the lee side of a dune where the slope approximates the angle of rest of loose sand (usually ~33°)

**smog**
see photochemical

**smolt**
a young salmon or sea trout about two years old that is at the stage of development when it assumes the silvery color of the adult and is ready to migrate to the sea

**snowpacks**
a seasonal accumulation of slow-melting snow

**soil erosion**
the process of removal and transport of the soil by water and/or wind

**southern oscillation**
a large-scale atmospheric and hydrospheric fluctuation centered in the equatorial Pacific Ocean; exhibits a nearly annual pressure anomaly, alternatively high over the Indian Ocean and high over the South Pacific; its period is slightly variable, averaging 2.33 years; the variation in pressure is accompanied by variations in wind strengths, ocean currents, sea- surface temperatures, and precipitation in the surrounding areas

**sphagnum moss**
a genus of moss that covers large areas of wetlands in the northern hemisphere; sphagnum debris is usually a major constituent of the peat in these areas

**stakeholders**
person or entity holding grants, concessions, or any other type of value which would be affected by a particular action or policy

**stochastic events**
events involving a random variable, chance, or probability

**stomata**
the minute openings in the epidermis of leaves through which gases interchange between the atmosphere and the intercellular spaces within leaves

**succession**
transition in the composition of plant communities following disturbance

**susceptibility**
probability for an individual or population of being affected by an external factor

**sustainable**
a term used to characterize human action that can be undertaken in such a manner as to not adversely affect environmental conditions (e.g., soil, water quality, climate) that are necessary to support those same activities in the future

**symbionts**
organisms that live together to mutual benefit [e.g., nitrogen-fixing bacteria that live with a plant (legume)]

**synoptic**
relating to or displaying atmospheric and weather conditions as they exist simultaneously over a broad area

**taiga**
coniferous forests of northern North America and Eurasia

**talik**
a layer of unfrozen ground occurring between permafrost and the active layer

**technical calorie**
the amount of heat needed to raise the temperature of 1 g of water 1°C at 15°C

**thermohaline circulation**
circulation driven by density gradients, which are controlled by temperature and salinity

**thermokarst**
irregular, hummocky topography in frozen ground caused by melting of ice

**thermophilic species**
species growing at high temperatures

**timberline**
the upper limit of tree growth in mountains or high latitudes

**transpiration**
the emission of water vapor from the surfaces of leaves or other plant parts

**tsunami**
a large tidal wave produced by a submarine earthquake, landslide, or volcanic eruption

**ungulate**
a hoofed typically herbivorous quadruped mammal (as a ruminant, swine, camel, hippopotamus, horse, tapir, rhinoceros, elephant, or hyrax)

**upwelling**
transport of deeper water to the surface, usually caused by horizontal movements of surface water

**urbanization**
the conversion of land from a natural state or managed natural state (such as agriculture) to cities

**vector**
an organism, such as an insect, that transmits a pathogen from one host to another

**vernalization**
the act or process of hastening the flowering and fruiting of plants by treating seeds, bulbs, or seedlings so as to induce a shortening of the vegetative period

**vulnerability**
the extent to which climate change may damage or harm a system; it depends not only on a system's sensitivity, but also on its ability to adapt to new climatic conditions

**wadi**
a water course that is dry except during the rainy season; the stream or flush that runs through it

**water-use efficiency**
carbon gain in photosynthesis per unit water lost in evapotranspiration; can be expressed on a short-term basis as the ratio of photosynthetic carbon gain per unit transpirational water loss, or on a seasonal basis as the ratio of net primary production or agricultural yield to the amount of available water

**winter dormancy**
period without biochemical activity in plant tissues

**xeric**
requiring only a small amount of moisture

**xerophyte**
a plant structurally adapted for life and growth with a limited water supply, especially by means of mechanisms that limit transpiration or that provide for the storage of water

# G

# Acronyms, Chemical Symbols, and Units

## ACRONYMS

| | |
|---|---|
| 3-D | 3-dimensional |
| ACCS | African Climate Change Scenario |
| ADB | Asian Development Bank |
| aET | actual evapotranspiration |
| AIM | Asian-Pacific Integrated Model |
| AOGCM | atmosphere-ocean general circulation model |
| AUS | Australia |
| BBS | Bangladesh Bureau of Statistics |
| BCTE | Bureau of Transport and Communications Economics (Australia) |
| BMRC | Bureau of Meteorology Research Centre (Australia) |
| BNJST | Bangladesh-Nepal Joint Study Team |
| BP | before present |
| BRS | Bureau of Resource Sciences |
| CAP | Common Agricultural Policy |
| CAST | Council for Agricultural Science and Technology |
| CCC | Canadian Climate Centre |
| CCIRG | Climate Change Impacts Review Group |
| CDC | Centers for Disease Control and Prevention |
| CDEA | Commission on Development and Environment for Amazonia |
| CERES | Clouds and Earth's Radiant Energy System |
| CNA | Central North America |
| COP | Conference of the Parties |
| CPUE | catch-per-unit-effort |
| CSIRO | Commonwealth Scientific and Industrial Research Organization |
| CZMP | coastal zone management plan |
| DENR | Department of Environment and Natural Resources |
| DEST | Department of Environment, Sport and Territories (Australia) |
| DGVM | dynamic global vegetation model |
| DHF | dengue hemorrhagic fever |
| DH-MWR | Department of Hydrology, Ministry of Water Resources (China) |
| DJF | winter (December-January-February)—Northern Hemisphere |
| DNA | deoxyribonucleic acid |
| DOC | dissolved organic carbon |
| DTR | diurnal temperature range |
| EAS | East Asia |
| ECHAM | European Centre/Hamburg Model |
| ECSN | European Climate Support Network |
| EEZ | exclusive economic zone |
| EN | El Niño |
| ENSO | El Niño-Southern Oscillation |
| ESCAP | Economic and Social Commission for Asia and the Pacific |
| ESD-CAS | Earth Science Division, Chinese Academy of Sciences |
| EU | European Union |
| EWS | early-warning system |
| FAO | Food and Agriculture Organization |
| FAR | First Assessment Report |
| FASOM | Forest and Agricultural Sector Optimization Model |
| FCCC | Framework Convention on Climate Change |
| FEMA | Federal Emergency Management Agency (United States) |
| FSU | former Soviet Union |
| GCM | general circulation model |
| GDP | gross domestic product |
| GEF | Global Environment Facility |
| GEO | Global Environment Outlook |
| GFDL | Geophysical Fluid Dynamics Laboratory |
| GHG | greenhouse gas |
| GIS | Geographic Information Systems |
| GISS | Goddard Institute for Space Studies |
| GLASOD | Global Assessment of Soil Degradation |
| GLOF | glacial lake outburst flood |
| GLSLB | Great Lakes-St. Lawrence Basin Project |
| GNP | gross national product |
| GOB | Government of Bangladesh |
| GOI | Government of India |
| GTZ | Gesellschaft für Technische Zusammenarbeit (Germany) |
| GVA | Global Vulnerability Analysis |
| GVM | global vegetation model |
| GYC | general yield class |
| HEWS | health early warning systems |
| HIV | human immunodeficiency virus |
| HMGN | His Majesty's Government of Nepal |
| HSSW | high-salinity shelf water |
| ICIMOD | International Centre for Integrated Mountain Development |
| ICLARM | International Center for Living Aquatic Resources Management |
| ICST | Ivorian Country Study Team |
| ICZM | integrated coastal zone management |
| IDRC | International Development Research Centre |
| IIASA | International Institute for Applied Systems Analysis |
| IMAGE | Integrated Model to Assess the Greenhouse Effect |
| IPCC | Intergovernmental Panel on Climate Change |
| IQQM | Integrated Quantity and Quality Model |
| IRC | Insurance Research Council |
| IRRI | International Rice Research Institute |
| ISPAN | Irrigation Support Project for Asia and the Near-East |
| ITCZ | Inter-Tropical Convergence Zone |
| IUCN | International Union for the Conservation of Nature and Natural Resources |

JJA summer (June-July-August)—Northern Hemisphere
LAI leaf area index
LLNL Lawrence Livermore National Laboratory
LPG liquefied petroleum gas
MAPSS Mapped Atmosphere-Plant-Soil System
MARA/ARMA Mapping Malaria Risk in Africa
MBIS Mackenzie Basin Impact Study
MERCOSUR Mercado Común del Sur/Common Market of the South
MOPT Ministerio de Obras Publicas y Transportes (Spain)
MPI Max-Planck Institut für Meteorologie (Germany)
MRI Meteorological Research Institute (Japan)
NAAQS National Ambient Air Quality Standard (United States)
NCAR National Center for Atmospheric Research (United States)
NEU Nothern Europe
NGO non-governmental organization
NRC National Research Council (United States)
NSW New South Wales
NWR National Wildlife Refuge
NWT Northwest Territories
OECD Organisation for Economic Cooperation and Development
OIES Office for Interdisciplinary Earth Studies
ORNL Oak Ridge National Laboratory (United States)
OSU Oregon State University
OTA Office of Technology Assessment (United States)
P precipitation
PAA Port Autonome d'Abidjan
PAHO Pan American Health Organization
PAN peroxyacetal nitrate
PCCC Proyecto Centroamericano de Cambios Climáticos
PCSD The President's Council on Sustainable Development
PE potential evaporation
PEI precipitation effectiveness index
pET potential evapotranspiration
P:PET precipitation:potential evapotranspiration
PPP purchasing power parity
Pr precipitation
PV photovoltaic
PWWS Philadelphia Hot Weather-Health Watch/Warning System
R runoff
RCM regional climate model
RMA Resource Management Act (New Zealand)
RMSF Rocky Mountain spotted fever
SAARC South Asian Association for Regional Cooperation
SACZ South Atlantic Convergence Zone
SADC Southern African Development Community
SAH Sahel
SAR Second Assessment Report
SARDC Southern African Research and Documentation Center
SBSTA Subsidiary Body for Scientific and Technological Advice
SEA South East Asia
SEU Southern Europe
SO Southern Oscillation
SOI Southern Oscillation Index
SPCZ South Pacific Convergence Zone
SPREP South Pacific Regional Environment Programme
SSA sub-Saharan Africa
SST sea-surface temperature
T temperature
TAR Third Assessment Report
$T_{max}$ daily maximum temperature
TMF tropical moist forest
$T_{min}$ daily minimum temperature
UKMO United Kingdom Meteorological Office
UN United Nations
UNEP UN Environment Programme
UNESCO UN Educational, Scientific and Cultural Organization
UNFCCC UN Framework Convention on Climate Change
USCSP U.S. Country Studies Program
USDA U.S. Department of Agriculture
USEPA U.S. Environmental Protection Agency
UV-B ultraviolet-B
VBD vector-borne disease
VEMAP Vegetation/Ecosystem Modeling and Analysis Project
WFC UN World Food Council
WG Working Group
WHO World Health Organization
WMO World Meteorological Organization
WRI World Resources Institute
WSC Weddell Scotia Confluence
WUE water-use efficiency

## CHEMICAL SYMBOLS

C carbon
$CH_4$ methane
$CO_2$ carbon dioxide
$H_2$ molecular hydrogen
N nitrogen
$NO_x$ nitrogen oxide
$N_2O$ nitrous oxide

## UNITS

### *SI (Systéme Internationale) Units*

| Physical Quantity | Name of Unit | Symbol |
|---|---|---|
| length | meter | m |
| mass | kilogram | kg |
| time | second | s |
| thermodynamic temperature | kelvin | K |
| amount of substance | mole | mol |

| Fraction | Prefix | Symbol | Multiple | Prefix | Symbol |
|---|---|---|---|---|---|
| $10^{-1}$ | deci | d | 10 | deca | da |
| $10^{-2}$ | centi | c | $10^{2}$ | hecto | h |
| $10^{-3}$ | milli | m | $10^{3}$ | kilo | k |
| $10^{-6}$ | micro | $\mu$ | $10^{6}$ | mega | M |
| $10^{-9}$ | nano | n | $10^{9}$ | giga | G |
| $10^{-12}$ | pico | p | $10^{12}$ | tera | T |
| $10^{-15}$ | femto | f | $10^{15}$ | peta | P |
| $10^{-18}$ | atto | a | $10^{18}$ | exa | E |

### *Special Names and Symbols for Certain SI-Derived Units*

| Physical Quantity | Name of SI Unit | Symbol for SI Unit | Definition of Unit |
|---|---|---|---|
| force | newton | N | $kg\ m\ s^{-2}$ |
| pressure | pascal | Pa | $kg\ m^{-1}\ s^{-2}$ (= $Nm^{-2}$) |
| energy | joule | J | $kg\ m^{2}\ s^{-2}$ |
| power | watt | W | $kg\ m^{2}\ s^{-3}$ (= $Js^{-1}$) |
| frequency | hertz | Hz | $s^{-1}$ (cycle per second) |

### *Decimal Fractions and Multiples of SI Units Having Special Names*

| Physical Quantity | Name of Unit | Symbol for Unit | Definition of Unit |
|---|---|---|---|
| length | ångstrom | Å | $10^{-10}$ m = $10^{-8}$cm |
| length | micrometer | $\mu$m | $10^{-6}$m = $\mu$m |
| area | hectare | ha | $10^{4}\ m^{2}$ |
| force | dyne | dyn | $10^{-5}$ N |
| pressure | bar | bar | $10^{5}\ N\ m^{-2}$ |
| pressure | millibar | mb | 1hPa |
| weight | ton | t | $10^{3}$ kg |

### *Non-SI Units*

| | |
|---|---|
| °C | degrees Celsius (0°C = ~273K); Temperature differences are also given in °C rather than the more correct form of "Celsius degrees" |
| Btu | British Thermal Unit |
| kWh | kilowatt-hour |
| $MW_e$ | megawatts of electricity |
| ppmv | parts per million ($10^{6}$) by volume |
| ppbv | parts per billion ($10^{9}$) by volume |
| pptv | parts per trillion ($10^{12}$) by volume |
| tce | tons of coal equivalent |
| toe | tons of oil equivalent |
| tWh | terawatt-hour |

# H

# Authors, Contributors, and Expert Reviewers of the Regional Impacts Special Report

**Argentina**

| | |
|---|---|
| Vicente Barros | University of Buenos Aires, Department of Atmospheric Sciences |
| Marcelo Cabido | Universidad Nacional de Córdoba (CONICET), Instituto Multidisciplinario de Biología Vegetal |
| Osvaldo F. Canziani | Instituto de Estudios e Investigaciones (IEIMA) |
| Rodolfo Carcavallo | Department of Entomology |
| Sandra Diaz | Universidad Nacional de Córdoba (CONICET), Instituto Multidisciplinario de Biología Vegetal |
| Gillermo Funes | Universidad Nacional de Córdoba (CONICET), Cátedra de Biogeografía |
| Juan Carlos Labraga | Universidad Nacional de Córdoba (CONICET) |
| Carlos A. Rinaldi | Instiuto Antarctico Argentino |
| Walter M. Vargas | University of Buenos Aires (IEIMA) |
| Ernesto F. Viglizzo | Universidad Nacional de Córdoba (CONICET) |

**Austria**

| | |
|---|---|
| Klaus Radunsky | Austrian Federal Environmental Agency |
| Shokri Ghanem | Organization of the Petroleum Exporting Countries |

**Australia**

| | |
|---|---|
| N. Abel | CSIRO Wildlife and Ecology |
| M.P. Austin | CSIRO Wildlife and Ecology |
| S. Barlow | Bureau of Resource Sciences |
| M. Barson | Bureau of Resource Sciences |
| Bryson Bates | CSIRO |
| P. Beggs | Macquarie University |
| D. Bennett | Hassall and Associates |
| D. Black | NSW Dept. of Land and Water Conservation |
| R. Braaf | Macquarie University |
| R. Braithwaite | CSIRO Tourism Research Program |
| R. Buxton | CSIRO Wildlife and Ecology |
| J. Carter | Queensland Department of Natural Resources |
| J. Conroy | University of Western Sydney |
| Elizabeth Curran | Bureau of Meteorology |
| S. Davies | University of Adelaide |
| L. Dobes | Bureau of Transport and Communication Economics |
| Terry Done | Australian Institute of Marine Science |
| M. Finlayson | Environmental Research Institute |
| Roger M. Gifford | CSIRO |
| Angela Gillman | Environment Australia |
| Habiba Gitay | Australian National University |
| Dean Graetz | CSIRO Office of Space Science and Applications |
| N. Hall | Australian Bureau of Agricultural and Resource Economics |
| Kevin Hennessy | CSIRO Atmospheric Research |
| N. Holbrook | Macquarie University |
| David Hopley | James Cook University |
| Mark Howden | Bureau of Resource Sciences |
| B. Hunt | CSIRO Atmospheric Research |
| W. Kininmonth | Bureau of Meteorology National Climate Centre |
| Miko Kirschbaum | CSIRO Forestry and Forest Products |
| T. Koslow | CSIRO Division of Fisheries |
| S. Lake | Monash University |
| S. Li | Macquarie University |
| J. Lutze | CSIRO Plant Industry |
| N. Marshman | Technical Services Australia (RTZ/CRA) |
| Gregory M. McKeon | Queensland Department of Natural Resources |
| Roger McLean | Australian Defence Force Academy |
| Heather McMaster | Macquarie University |
| R.E. McMurtrie | University of NSW |
| A. Moore | CSIRO Plant Industry |
| Neville Nicholls | Bureau of Meteorology Research Centre |
| Ian Noble | Australian National University |
| A. Norton | Charles Sturt University |
| W. Osborne | University of Canberra |
| Barrie Pittock | CSIRO Climate Impact Group |
| Neil Plummer | National Climate Centre |
| P. Reyenga | Bureau of Resource Sciences |
| H. Ross | Australian National University |
| H. Schaap | Electricity Supply Association |
| D. Smith | Australian National University |
| M. Stafford Smith | CSIRO Wildlife and Ecology |
| W. Steffen | GCTE Core Project Office |
| R. Suppiah | CSIRO Atmospheric Research |
| Robert Sutherst | CSIRO Division of Entomology |
| R. Taplin | Macquarie University |
| Brian H. Walker | CSIRO Wildlife and Ecology |
| George R. Walker | Alexander Howden Reinsurance |
| Y-P. Wang | CSIRO Atmospheric Research |
| P. Waterman | Environmental Management Services |
| Peter Whetton | CSIRO Division of Atmosphere Research |
| D. White | Agro-Ecosystems Consulting |
| Jann E. Williams | Charles Stuart University |
| T. Yonow | CRC for Tropical Pest Management |

**Bangladesh**

| | |
|---|---|
| M.Q. Mirza | University of Waikato |

**Barbados**

| | |
|---|---|
| Leonard Nurse | Coastal Conservation Unit |

**Belgium**

| | |
|---|---|
| Renate Christ | European Commission |

**Bolivia**

| | |
|---|---|
| Martín de Zuviría | |

**Brazil**

| | |
|---|---|
| Carlos C. Cerri | Universidade de Sao Paolo |
| Cleber Galvao | |

**Canada**

| | |
|---|---|
| F. Andrey | University of Waterloo |
| Mike Apps | Canadian Forestry Service |

M. Brklacich — Carleton University
Ross Brown — Environment Canada
Denis D'Amours — Department of Fisheries and Oceans
F. Diamond — Environment Canada
Kirsty Duncan — University of Windsor
Larry Dyke — Natural Resources Canada
Anne Gunn — Government Northwest Territories
Bob Jefferies — University of Toronto
Grace Koshida — Atmospheric Environment Service
John Legg — NRCan
Abdel R. Maarouf — Environment Canada
Jay Malcolm — University of Toronto
Dave Martell — University of Toronto
Barrie Maxwell — Atmospheric Environment Service
Linda Mortsch — Environment Canada
Terry Root — Environment Canada
Brian Stocks — Canadian Forestry Service
Roger B. Street — Atmospheric Environment Service
Hague Vaughan — Canada Centre for Inland Waters
Andrew J. Weaver — University of Victoria

**Chile**
Patricio Aceituno — University of Chile
Humberto Fuenzalida-Ponce — Universidad de Chile
Terence Lee — Economic Commission for Latin America and the Caribbean

**China**
Ding Yihui — China Meteorological Administration
Liu Chunzhen — Hydrological Forecasting and Water Control Center
Lin Erda — Chinese Academy of Agricultural Sciences
Su Jilan — Second Institute of Oceanography

**Costa Rica**
Marcos Campos — Comision Centroamericana de Ambiente y Desarollo

**Cote d'Ivoire**
Sekou Touré — ENSTP

**Cuba**
Lino Naranjo Diaz — Ministry of Science, Technology, and Environment
Ada L. Perez — Institute of Physical Planning
Antonio J. Lopez Almiral — Ministry of Science, Technology, and Environment
Luis R. Paz — Ministry of Science, Technology, and Environment
Elias Ramirez Cruz — Ministry of Science, Technology, and Environment
Leda Menendez Carreras — Institute of Ecology and Systematics
Avelino G. Suarez — Institute of Ecology and Systematics

**Czech Republic**
Ivana Nemesova — Institute of Atmospheric Physics

**Egypt**
Mohamed El-Raey — Institute of Graduate Studies and Research

**Finland**
Timothy Carter — Finnish Meteorological Institute
Kaija Hakala — Agricultural Research Center
Timo Karjalainen — European Forest Institute
Peter Kuhry — Arctic Center

**France**
Richard Delécolle — Unite de Bioclimatologie

**Germany**
Wolfgang Cramer — Potsdam Institute for Climate Impact Research
Georg Hörmann — Kiel University
Venugopalan Ittekkot — University of Hamburg
Ferenc Toth — Potsdam Institute for Climate Impact Research

**India**
Amrita Achanta — Climate Change Secretariat (UNFCCC)
Murari Lal — Centre for Atmospheric Sciences
S.K. Sinha — Indian Agricultural Research Institute

**Indonesia**
Aprilani Soegiarto — Indonesian Institute of Sciences
Sri Soewasti Soesanto — Health Ecology Research and Development

**Iran**
Bohloul Alijani

**Israel**
Uriel N. Safriel — The Blaustein Institute for Desert Research

**Italy**
R. Welcomme — FAO

**Japan**
Hideo Harasawa — Center for Global Environmental Research
Nobuo Mimura — Ibaraki University
Tatsushi Tokioka — Japan Meteorological Agency
Masatoshi Yoshino — Aichi University

**Kazakhstan**
Olga Pilifosova — Climate Study Laboratory

**Kenya**
K.A. Edwards — UN Environment Programme
S.H. Mwandoto — Kenya Meteorological Department
Joseph Kagia Njihia — Kenya Meteorological Department
Laban J. Ogallo — University of Nairobi
H.W.O. Okoth-Ogendo — The National Council for Population and Development
Peter Usher — UN Environment Programme

**Korea**
Suam Kim — Korea Ocean Research and Development Institute
Byong-Lyol Lee — National Institute of Agricultural Sciences and Technology

**Malaysia**
Lim Joo Tick — Malaysian Meteorological Service

**Maldives**
Mohamed Ali — Ministry of Planning, Natural Resources, and Environment

**Mexico**
Carlos Gay-Garcia — Environment Department
P. Grace — International Maize and Wheat Improvement Center
Ana Rosa Moreno — Pan American Health Organization

**Nepal**
Sharad P. Adhikary — Water and Energy Commission Secretariat

**New Zealand**
Reid E. Basher — National Institute of Water and Atmospheric Research
Bruce D. Campbell — AgResearch Grasslands Research Centre
Blair Fitzharris — University of Otago
J. Gibb — Coastal Management Consultants
S. Hales — Wellington School of Medicine
John Hay — The University of Auckland
C. Hickey — National Institute of Water and Atmospheric Research
R. Ibbit. — National Institute of Water and Atmospheric Research
Gavin J. Kenny — University of Waikato
R.Kirk — University of Canterbury
Dick Martin — Crop and Food Research Institute
Piers McLaren — Forest Research Institute
N. Mitchell — University of Auckland
T. Murray — National Institute of Water and Atmospheric Research
L. Paul — National Institute of Water and Atmospheric Research
J. Renwick — National Institute of Water and Atmospheric Research
Richard Warrick — University of Waikato
H. Weinstein — Wellington School of Medicine
David Whitehead — Landcare Research
D. Wilson — Crop and Food Research Institute
D. Wratt — National Institute of Water and Atmospheric Research
Alastair Woodward — Wellington School of Medicine

**Nigeria**
Larry F. Awosika — Nigerian Institute for Oceanography and Marine Research

**Norway**
Egil Sakshaug — University of Trondheim

**Papua New Guinea**
Graham Sem — University of Papua New Guinea

**Peru**
Eduardo Calvo — Concejo Nacional del Ambiente

**Philippines**
Rex Victor Cruz — University of Phillipines

**Russia**
Kira I. Kobak — State Hydrological Institute

**Scotland**
Andrew D Dlugolecki — General Accident, Fire, and Life Assurance Corp.

**Senegal**
Amadou Bachirou Diop — Ministère de l'Equipement

**Sierra Leone**
Ogunlade Davidson — University of Sierra Leone

**Singapore**
Poo Poo Wong — National University of Singapore

**Slovenia**
Andrej Kranjc — Hydrometeorological Institute of Slovenia
Zoran Stojic — IBE Consulting Engineers

**South Africa**
Timm Hoffman — National Botanical Institute
Alec Joubert — University of the Witwaterstrand
A.C. Kruger — South African Weather Bureau
Robert Scholes — CSIRO
David le Sueur — University of Natal
Colleen Voegel — University of the Witwatersrand

**Spain**
Ana Iglesias — Ciudad Universitaria

**Sweden**
Ulf Molau — University of Goteberg
Mats Oquist — University of Agricultural Sciences
I. Colin Prentice — University of Lund
Benjamin Smith — University of Lund

**Switzerland**
Martin Beniston — University of Fribourg
Hartmut Grassl — World Meteorological Organization
John Innes — Swiss Federal Institute for Forest, Snow and Landscape Research

**Thailand**
Charit Tingsabadh — Chulalongkorn University

**The Gambia**
Bubu P. Jallow — Department of Water Resources

**The Netherlands**
Richard Klein — Vrije Universiteit, Institute for Environmental Studies
Rik Leemans — National Institute of Public Health and Environmental Protection
Pim Martens — Maastricht University
Richard S.J. Tol — Vrije Universiteit, Institute for Environmental Studies
Pier Vellinga — Vrije Universiteit, Institute for Environmental Studies

| | |
|---|---|
| **Uganda** | |
| Eric L. Edroma | Uganda Institute of Ecology |
| **United Kingdom** | |
| Nigel Arnell | University of Southampton |
| Thomas E. Downing | University of Oxford |
| David Drewry | NERC |
| Michael Hulme | University of East Anglia |
| Sari Kovats | London School of Hygiene and Tropical Medicine |
| Anthony McMichael | London School of Hygiene and Tropical Medicine |
| John Mitchell | Hadley Center |
| Robert Nicholls | Middlesex University |
| Jim Skea | University of Sussex |
| David Viner | University of East Anglia |
| Eric Wolff | British Antarctic Survey |
| **United States** | |
| Richard Adams | Oregon State University |
| Richard Ball | Department of Energy |
| Susan Bassow | Office of Science and Technology Policy |
| Heather Benway | National Oceanic and Atmospheric Administration |
| Paul Berkman | Ohio State University |
| Julio Betancourt | U.S. Geological Survey |
| Suzanne Bolton | National Marine Fisheries Service |
| Jerry Brown | International Permafrost Association |
| David J. Campbell | Michigan State University |
| Lee De Cola | U.S. Geological Survey |
| Doug Demaster | National Marine Fisheries Service |
| Barbara Allen-Diaz | University of California, Berkeley |
| Paul Desanker | Michigan Technological University |
| Robert Dixon | U.S. Country Studies |
| David Jon Dokken | IPCC Working Group II TSU |
| Linda Duguay | National Science Foundation |
| Jerry Elwood | Department of Energy |
| Paul R. Epstein | Harvard Medical School |
| John Everett | National Oceanic and Atmospheric Administration |
| Paul Filmer | National Science Foundation |
| Penny Firth | National Science Foundation |
| Ann Fisher | Pennsylvania State University |
| Mary Gant | HHS/NIEHS |
| Filippo Giorgi | National Center for Atmospheric Research |
| Miquel A. Gonzalez-Meler | Duke University |
| Christy Goodale | University of New Hampshire |
| James A. Graham | USAID/CARPE |
| David Goodrich | National Oceanic and Atmospheric Administration |
| Anne Grambsch | Environmental Protection Agency |
| John F. Griffiths | Texas A&M University |
| Duane Gubler | Centers for Disease Control |
| Ronald Hellman | City University of New York |
| Roger Hewitt | National Marine Fisheries Service |
| Harry J Hillaker, Jr. | Iowa State Climatologist |
| William Hunt | Colorado State University |
| Robert B. Jackson | University of Texas at Austin |
| Anthony Janetos | NASA |
| Jennifer Jenkins | University of New Hampshire |
| Linda Joyce | Rocky Mountain Forest and Range Experiment Station |
| Sally Kane | National Oceanic and Atmospheric Administration |
| Thomas Karl | U.S. Department of Commerce |
| Barry D. Keim | University of New Hampshire |
| John Kelmelis | Department of Interior |
| Mary Kidwell | University of Arizona |
| Timothy Kittel | National Center for Atmospheric Research |
| Kalee Kreider | Greenpeace USA Climate Campaign |
| Neil Leary | U.S. Environmental Protection Agency |
| Alice Welsh Leeds | National Science Foundation |
| Frances Li | National Science Foundation |
| Michael C. MacCracken | U.S.Global Change Research Program |
| Norman MacDonald | Certified Consulting Meteorologist |
| George Maul | Florida Institute of Technology |
| Herman Mayeux | U.S. Department of Agriculture |
| Mack McFarland | IPCC Working Group II TSU |
| Laura VanWie McGrory | IPCC Working Group II TSU |
| David McGuire | University of Alaska, Fairbanks |
| Steven McNulty | U.S. Department of Agriculture |
| Gerald Meehl | National Center for Atmospheric Research |
| Robert Mendelsohn | Yale University |
| Richard Moss | IPCC Working Group II TSU |
| Patrick Mulholland | Oak Ridge National Laboratory |
| Ron Neilson | U.S.Department of Agriculture Forest Service |
| Richard J. Norby | Oak Ridge National Laboratory |
| Dennis Ojima | Natural Resource Ecology Labaratory (NREL) |
| Scott Ollinger | University of New Hampshire |
| Michael C. Oppenheimer | Environmental Defense Fund |
| Florence Ormond | IPCC Working Group II TSU |
| Jonathan A. Patz | Johns Hopkins School of Public Health |
| Charles J. Peckham | LMR, Inc. |
| Wayne Polley | U.S. Department of Agriculture |
| Wilfred Mac Post | Oak Ridge National Laboratory |
| Bradley Reed | EROS Data Center |
| John Reilly | U.S. Department of Agriculture |
| Alan Robock | University of Maryland |
| Michael J. Sale | Oak Ridge National Laboratory |
| Joel D. Scheraga | U.S. Environmental Protection Agency |
| Miranda Schreurs | University of Maryland |
| Michael Scott | Battelle Pacific Northwest Laboratory |
| Clive Shiff | John Hopkins University |
| David Smith | Universtiy Museum |
| Joel B. Smith | RCG/Hagler, Bailly, Inc |
| David Shriner | Oak Ridge National Laboratory |
| Allen Solomon | U.S. Environmental Protection Agency |
| Christopher C. Spaur | U.S. Army Corps of Engineers |
| Eugene Stakhiv | U.S. Army Corps of Engineers |
| Ron Stouffer | U.S. Department of Commerce |
| Kenneth Strzepek | University of Colorado |
| Regina Tannon | University Corporation Atmospheric Research |
| Melissa Taylor | U.S.Global Change Research Program |

David Theobald — Colorado State University
James Titus — U.S. Environmental Protection Agency
Cynthia Tynan — National Marine Fisheries Service
Anandu D. Vernekar — University of Maryland
Wei-Chyung Wang — State University of New York, Albany
Elizabeth C. Weatherhead — National Oceanic and Atmospheric Administration
Thompson Webb III — Brown University
Gunter Weller — University of Alaska/Fairbanks
Wayne M. Wendland — Illinois State Water Survey
Tom Wigley — University Corporation for Atmospheric Research
Gunter Weller — University of Alaska
Mark Weltz — U.S. Department of Agriculture
David Yates — University Corporation for Atmospheric Research

**Uruguay**

Andrés Saizar — Ministry of Housing, Territorial Ordering, and Environment

**Venezuela**

Rigoberto Andressen — University of Merida
Luis J. Mata — UNFCCC Secretariat
Martha Perdomo — Ministerio del Ambiente

**Western Samoa**

James Aston — South Pacific Regional Environment Program

**Zimbabwe**

Chris H.D. Magadza — Universitry of Zimbabwe
Wish Marume — Department of Meteorological Services
Shakespeare Maya — Southern Centre for Energy and Environment
L.S. Unganai — Department of Meteorological Services
Marufu C. Zinyowera — Department of Meteorological Services

# I

# List of Major IPCC Reports

**Climate Change—The IPCC Scientific Assessment**
The 1990 Report of the IPCC Scientific Assessment Working Group (also in Chinese, French, Russian, and Spanish)

**Climate Change—The IPCC Impacts Assessment**
The 1990 Report of the IPCC Impacts Assessment Working Group (also in Chinese, French, Russian, and Spanish)

**Climate Change—The IPCC Response Strategies**
The 1990 Report of the IPCC Response Strategies Working Group (also in Chinese, French, Russian, and Spanish)

**Emissions Scenarios**
Prepared for the IPCC Response Strategies Working Group, 1990

**Assessment of the Vulnerability of Coastal Areas to Sea Level Rise–A Common Methodology**
1991 (also in Arabic and French)

**Climate Change 1992—The Supplementary Report to the IPCC Scientific Assessment**
The 1992 Report of the IPCC Scientific Assessment Working Group

**Climate Change 1992—The Supplementary Report to the IPCC Impacts Assessment**
The 1992 Report of the IPCC Impacts Assessment Working Group

**Climate Change: The IPCC 1990 and 1992 Assessments**
IPCC First Assessment Report Overview and Policymaker Summaries, and 1992 IPCC Supplement

**Global Climate Change and the Rising Challenge of the Sea**
Coastal Zone Management Subgroup of the IPCC Response Strategies Working Group, 1992

**Report of the IPCC Country Studies Workshop**
1992

**Preliminary Guidelines for Assessing Impacts of Climate Change**
1992

**IPCC Guidelines for National Greenhouse Gas Inventories**
Three volumes, 1994 (also in French, Russian, and Spanish)

**IPCC Technical Guidelines for Assessing Climate Change Impacts and Adaptations**
1995 (also in Arabic, Chinese, French, Russian, and Spanish)

**Climate Change 1994—Radiative Forcing of Climate Change and an Evaluation of the IPCC IS92 Emission Scenarios**
1995

**Climate Change 1995—The Science of Climate Change – Contribution of Working Group I to the Second Assessment Report**
1996

**Climate Change 1995—Impacts, Adaptations, and Mitigation of Climate Change: Scientific-Technical Analyses – Contribution of Working Group II to the Second Assessment Report**
1996

**Climate Change 1995—Economic and Social Dimensions of Climate Change – Contribution of Working Group III to the Second Assessment Report**
1996

**Climate Change 1995—IPCC Second Assessment Synthesis of Scientific-Technical Information Relevant to Interpreting Article 2 of the UN Framework Convention on Climate Change**
1996 (also in Arabic, Chinese, French, Russian, and Spanish)

**Technologies, Policies, and Measures for Mitigating Climate Change – IPCC Technical Paper I**
1996 (also in French and Spanish)

**An Introduction to Simple Climate Models used in the IPCC Second Assessment Report – IPCC Technical Paper II**
1997 (also in French and Spanish)

**Stabilization of Atmospheric Greenhouse Gases: Physical, Biological and Socio-economic Implications – IPCC Technical Paper III**
1997 (also in French and Spanish)

**Implications of Proposed $CO_2$ Emissions Limitations – IPCC Technical Paper IV**
1997 (also in French and Spanish)

ENQUIRIES: IPCC Secretariat, c/o World Meteorological Organization, P.O. Box 2300, CH1211, Geneva 2, Switzerland.